Birkhäuser

Contemporary Mathematicians

Joseph P.S. Kung

Editor

For further volumes:
http://www.springer.com/series/4817

Patrizia Pucci • Vicentiu D. Radulescu
Hans Weinberger
Editors

James Serrin. Selected Papers

Volume 1

Birkhäuser

Editors
Patrizia Pucci
Università di Perugia
Perugia, Italy

Vicentiu D. Radulescu
King Abdulaziz University
Jeddah, Saudi Arabia

Hans Weinberger
University of Minnesota
Minneapolis, USA

ISBN 978-3-0348-0684-8 ISBN 978-3-0348-0845-3 (eBook)
DOI 10.1007/978-3-0348-0845-3
Springer Basel Heidelberg New York Dordrecht London

Library of Congress Control Number: 2014939300

Mathematics Subject Classification (2010): 26, 34, 35, 76

Printed on acid-free paper

Springer Basel is part of Springer Science+Business Media (www.birkhauser-science.com)

Table of Contents

Volume 1

Volume 2

Introduction

This volume is part of a two-volumes set of the Selected Works of James Serrin. The purpose of this set is to make available to a wider audience some of the main mathematical contributions of James Serrin. We feel that this is useful, because some of his papers are not as well known as they should be, while others are not easily available since they were published in volumes with only limited circulation.

The first volume is composed of the basic contributions to mathematics made by Professor Serrin, while the second volume is devoted to works which are more directly related to applications in the physical sciences. Each volume is grouped into a large variety of fields, and into topics within the fields, as shown in the table of contents. A special feature of the volumes is that, in most cases, the papers in a particular topic are preceded by a commentary by an expert on the subject. These commentaries are intended to indicate how the papers are related to other work, including future work, on this topic. We wish to express our gratitude and thanks to all the authors of these commentaries.

In order to give an overview of a few very long papers without overburdening the volumes, Professor Serrin chose to omit some details of the proofs in the reprinted versions. Such omissions are clearly marked in the articles.

The table of contents assigns capital Roman numerals to the fields, and, if necessary, Arabic numerals to the topics within a field. The field and topic numbers appear before the commentary, if any, and the reprinted articles on this topic. The papers within a topic are designated by lower case Roman numerals. A commentary on a topic uses the lower case numerals to refer to the papers in this topic, which appear just below the commentary.

We also draw the reader's attention to the final Section VIII in Volume 2, containing various essays of James Serrin. These include assessments of the work of *Jean Leray*, *Eberhard Hopf* and *Clifford Truesdell*, and some further comments on special topics, in particular an essay on the position of applied mathematics in scientific thought. In addition, the two papers in Section I.3 constitute a discussion of elliptic partial differential equations in the second half of the previous century, while the paper I.2 (ii) contains a historical essay on the origin of the minimal surface equation (see pages 491–493 of the article), and the paper I.6 (ii) includes an essay on the history of *the Liouville theorem* for entire solutions of elliptic partial differential equations (see pages 139–141 of the article).

A complete list of the publications of James Serrin appears at the end of each volume. Articles in this list which are reprinted in these volumes are indicated by boldface item numbers.

The articles in these volumes are reprinted with the kind permission of the copyright holders, who are indicated at the tops of the articles.

P. Pucci, V. Rădulescu, H. Weinberger

James Serrin 1990, Minneapolis, MN

A Brief Biography of James B. Serrin

Patrizia Pucci

I am particularly honored to have the opportunity to write a brief biography of one of my distinguished teachers, with whom I have also had the good fortune to collaborate for so many years, Professor James Serrin.

The outstanding contributions which Professor Serrin has made until the last days of his life to the field of mathematical analysis and its applications are well known. These volumes are dedicated to his work, including commentaries on his mathematical discoveries, many of which are now in the form of basic principles, whose importance should remain undiminished far into the future. His name is attached to several theorems and methods, and many of his early results have become classical and are included in modern textbooks on partial differential equations, the calculus of variations, real analysis, fluid mechanics and thermodynamics. Among these are: the Gilbarg–Serrin theorem on isolated singularities, the Serrin overdetermined symmetry theorem, the Meyers–Serrin theorem, the work on quasilinear elliptic equations, the Alexandrov–Serrin moving plane technique, the Serrin integral of the calculus of variations, the Peletier–Serrin uniqueness theorem, the Serrin version of the second law of thermodynamics, the Pucci–Serrin variational identity, as well as other more recent contributions.

As is evident from these volumes, James Serrin worked in a variety of fields of mathematics. His elegant contributions arose from the desire to understand the essence of deep problems which arise from the study of nature, rather than to create artificial problems to which known techniques can be adapted. From a list of his most famous results one can also see a further quality of his thought: the willingness to collaborate with his colleagues and to generously share his ideas with them. This aspect of his personality was made particularly clear to me at the conference organized in Minneapolis in May 1991 to celebrate his sixty-fifth birthday, at which a distinguished group of colleagues spoke about his mathematical career.

Another aspect of his temperament was the elegance of his taste. This is evident in much of his work, in which each of his theorems has the beauty of an essential form. This is a further reason why his work has influenced others so much, and why many of his theorems have opened new directions in mathematical research.

Professor Serrin was born in Chicago on November 1, 1926. He lived in Evanston, where he studied at Evanston Township High School. He attended Northwestern University for two years, majoring in Electrical Engineering. After receiving the Bachelor of Arts degree from Western Michigan College, he went to Indiana University, where in 1951 he received his doctorate with a thesis on the hydrodynamical theory of cavitation, written under the supervision of Professor *David Gilbarg.* Before the submission of his thesis he had already published the paper *Free boundaries and jets in the theory of cavitation* written in collaboration with David Gilbarg [1]. The title of the thesis was *The Existence and Uniqueness of Flows Solving Four Free Boundary Problems*, but it was published under the title *Existence theorems for some hydrodynamical free boundary problems* [4] in the first issue of the Journal of Rational Mechanics and Analysis in 1952. This journal was later renamed the Archive for Rational Mechanics and Analysis.

Following his initial appointment as Fine Instructor of Mathematics at Princeton University in 1951, he was a C.L.E. Moore Instructor of Mathematics at the Massachusetts Institute of Technology in the period 1952–1954.

In Boston he met and married Barbara West in September 1952, who gave him the warmth of a lovely family and three beautiful daughters: Martha, Elizabeth and Janet. She shared with him all the anxieties of life, as well as the recognition which his work received.

In 1954 he joined the faculty of the University of Minnesota and remained there until his death. At Minnesota he was Head of the School of Mathematics from 1964 to 1965, and was appointed Regents' Professor of Mathematics in 1969. Even after his retirement in 1995, he kept his famous office on the first floor of *Vincent Hall*, which he used until the last days of his life, and where many visitors and colleagues stopped to talk to him, not only about mathematics, and were fascinated by his personality and his mind, which remained lucid until the end of his days. In fact, James Serrin continued to do research and to write excellent papers from the beginning of his career until 2012, as documented by his presence at recent international conferences. Indeed, from the database of MathSciNet it appears that he has more than 4420 citations from more than 2400 different authors and has an h-index of 37.

Professor Serrin received many honors for his work. He was awarded the degree *laurea honoris causa* by the University of Sussex in 1972, by the Universities of Ferrara and Padua in 1992 and by the University *François–Rabelais* at Tour in 2005. From the American Mathematical Society he received the *George David Birkhoff Award in Applied Mathematics* in 1973, and from Indiana University the *Distinguished Alumni Award* in 1979. He was elected as a member of the *National Academy of Sciences* in 1980 and of the *American Academy of Arts and Sciences* in 1984, and as a fellow of the *American Academy for the Advancement of Science* in 1980. To commemorate his sixtieth birthday, a series of papers dedicated to him in the *Archive for Rational Mechanics and Analysis* were collected in the volume *Analysis and Continuum Mechanics* (Springer, 1989), whose dedicatory preface was written by Professor *Clifford Truesdell.* Two conferences were held to celebrate his sixty-fifth birthday: one at the University of Minnesota in 1991, and one at the University of Ferrara in 1992. A conference for his seventy-fifth birthday was held at the University of Perugia in 2002, and there was a conference for his eightieth birthday at the University of Perugia in 2008.

He was president of the *Society for Natural Philosophy* in 1969–1970 and served on the council of the *American Mathematical Society* in 1972–1974. From 1969 until 1986 he was a co-editor of the *Archive for Rational Mechanics and Analysis*, where his dedicated editorial efforts further enhanced its reputation among the great journals of mathematical analysis. He also served on the editorial boards of the *Journal of Differential Equations* from 1976 to 2012; *Communications in Partial Differential Equations* from 1978 to 2000; *Rendiconti del Circolo Matematico di Palermo* from 1980 to 2012; *Atti del Seminario Matematico e Fisico dell'Università di Modena* from 1985 to 2012; *Asymptotic Analysis* from 1988 to 2012; *Differential and Integral Equations* and *Advances in Differential*

Equations from 1990 to 2012; and *Bollettino dell'Unione Matematica Italiana* from 2008 to 2012. Indeed, he was a member of the *Unione Matematica Italiana* since 1990.

Professor Serrin was a visiting professor at Stanford University, the University of Chicago, The Johns Hopkins University, the University of Sussex, the Mittag–Leffler Institute, the Universities of Naples, Perugia and Modena, Oxford University, and the National Science Council in Taiwan. His list of invited addresses, symposia and departmental colloquia is almost endless. I particularly recall here his superb lectures at Cortona and Milan in 1981, and at Varenna in 1982, which I attended as a student. As further recognition of his achievements in mathematical research, Professor Serrin delivered invited addresses at the *International Congresses of Mathematicians* in Nice, 1970, on *Boundary curvatures and the solvability of Dirichlet's Problem* [71] and in Warsaw, 1983, on *The structure and laws of thermodynamics* [100]. He lectured at numerous major scientific conferences until his last days.

During his teaching career, James Serrin directed the research of 14 mathematicians who obtained their doctoral degrees in pure and applied mathematics. He was more than cordial with aspiring young people and his office door was always open to all who desire to discuss problems with him. His generosity went even further, since Barbara and James shared their marvelous house with many colleagues: young or old, unknown or famous. I recall what *Laurent Véron* wrote: *What exquisite hosts Barbara and Jim Serrin are! In their house, with antiques and memories, paintings and books, I had the privilege to have lunch on beautiful Wedgwood china, before attending a concert in their company.*

Professor Serrin's publications include more than 200 papers, whose range can be seen from the publication list which appears in this volume. He wrote extensive and important papers in partial and ordinary differential equations, in the calculus of variations, in real analysis, in fluid mechanics, in the constitutive theory of materials, and in the modern foundations of phenomenological thermodynamics. Many of these papers are gathered in these volumes. His fundamental contributions in these fields were the basis for the award of the *laurea honoris causa* in Padua on the four hundredth anniversary celebration of the association of Galileo at the University of Padua in 1992.

I must add here the fundamental article [18] on fluid mechanics *Mathematical principles of classical fluid mechanics* (Monograph), Handbuch der Physik, **VIII.1**, Springer (1959), pp. 125–263 (Russian translation: Foreign Literature Publ. House, Moscow 1963, 265 pages), one of the major post-war treatises on this field, which has served as a handbook for countless students of the subject.

In 2007 James Serrin and I published the monograph *The Maximum Principle*, Birkhäuser, providing a clear explanation of the various maximum principles available for second order elliptic equations, from their beginnings in linear theory to recent work on nonlinear equations, operators and inequalities. While simple in essence, these results lend themselves to a quite remarkable number of subtle uses when combined appropriately with other notions. The book, besides careful discussion of standard maximum principles, contains a number of new results.

A more comprehensive essay on the life and work of James Serrin written by myself appeared in [*An appreciation of James Serrin*, Nonlinear analysis and continuum mechanics (Ferrara, 1992), 1–13, Springer, New York, 1998, x+148]. The volume *Nonlinear analysis and continuum mechanics* contains papers from the *Conference in Honor of James Serrin on his 65th birthday* held at the University of Ferrara, Ferrara, October 26–30, 1992 and was edited by Giuseppe Buttazzo, Giovanni Paolo Galdi, Ermanno Lanconelli and myself.

On August 23, 2012 surrounded by the love of his wife Barbara and daughters Martha, Elisabeth, and Janet, James Serrin passed away at his home in Minneapolis. Regrettably, this happened before these volumes could be published. James was an enthusiastic participant in the preparation of these volumes and was looking forward to their publication.

James Serrin was a man of great culture, who loved to talk, and with whom it was very pleasant to have a conversation. My long association with him has been a wonderful experience for me, not only for the great pleasure of viewing the elegance and beauty of his way of thinking about mathematics, but also of sharing with him and Barbara the quiet moments in which his highly literate life appeared. I just mention his interest in classical music and his sophisticated and ample collection of rare records; his sincere love for art and architecture and his splendid paintings (many by Professor Serrin's brother Richard) and the many fine objects collected together with his wife Barbara over the years; his large library with impressive and almost unique books on art; and the collection of photographs taken during his scientific travels. The love for the beauty of my country grew from the close relationship he had to his brother, Richard W. Serrin, an artist who has lived in Firenze with his family since the 1960s. The goodness of a life never trivial or mediocre is evident from all these interests, which he shared with his devoted wife Barbara on a daily basis.

I close by saying that the great master James Serrin was for me above all a close friend since the dearly beloved *Lamberto Cesari* introduced him to me at the University of Michigan in August 1980. We worked together from then on. James was a great mathematician and a great man: the international mathematical community lost an irreplaceable member and his loss is inconsolable for anyone who, like me, had the good fortune of getting to know him. I hope that these volumes will serve to celebrate and preserve his memory.

Acknowledgement. I wish to express my deepest thanks to Patrizia Pucci, Hans F. Weinberger and Vicentiu Radulescu for their immeasurable help in bringing these volumes to fruition.

Further, I want to convey to the contributors my sincere thanks for their excellent commentaries, which add so much to the value of the volumes.

James Serrin

Minneapolis, Minnesota
March, 2012

James Serrin and Patrizia Pucci, Perugia, Italy, 1996.

James Serrin and Laurent Veron, Technion University, Haifa, Israel, 2010.

Marie-Francoise Veron, University of Tours, 2012.

I. Stationary Partial Differential Equations

I.1. Local Behavior of Solutions

Marie-Françoise Bidaut-Véron and Laurent Véron

Mathematics Subject Classification (2010). 35J15, 35J25, 35B45, 35B50, 35B53, 35J62, 35J75, 35J92, 35J93.

Keywords. Elliptic equations; a priori estimates; Liouville theorems; Harnack inequalities; removable singularities; singular solutions.

One of the main aspects of Serrin's contribution to the theory of partial differential equations lies in the fact that the methods introduced allow one to study a wide range of nonlinear problems even when the methods were originally applied to linear equations. Typical in this matter are his works on Harnack inequalities and isolated singularities. This work includes the four papers

[i] *On the Harnack inequality for linear elliptic equations*

[ii] *Local Behavior of solutions of quasi-linear equations*

[iii] *Isolated singularities of solutions of quasi-linear equations*

[iv] *The Harnack inequality for quasilinear elliptic equations in* R^2,

which are reprinted below.

Paper [i] deals with the Harnack inequality for solutions of second order linear elliptic equations

$$Lu = \sum_{i,j=1}^{n} a_{ij} \frac{\partial^2 u}{\partial x_i \partial x_j} + \sum_i b_i \frac{\partial u}{\partial x_i} + cu = 0 \qquad (c \leq 0). \tag{1}$$

It states that if $u > 0$ satisfies (1) in a ball B_2 of $\mathbb{R}^n$, then for any x, y in the ball $\overline{B}_1$ with the same center as B_2,

$$\frac{1}{K} u(x) \leq u(y) \leq K u(x), \tag{2}$$

and the constant K depends only on n and coefficients of L. This Harnack inequality, straightforward to obtain for positive harmonic functions, is known for a long time since the early works of Lichtenstein and Feller. It is proved here with weaker smoothness assumptions on the coefficients of L: uniform boundedness if $n = 2$ and Dini continuity if $n > 2$. The elegance of the proofs are characteristic of Serrin's relationship with mathematics. If $n = 2$, he uses a geometric construction and the Jordan curve theorem. As a corollary he extends the historical result of Liouville and Bernstein that any entire solution of the uniformly elliptic equation

$$A \frac{\partial^2 u}{\partial x^2} + 2B \frac{\partial^2 u}{\partial x \partial y} + C \frac{\partial^2 u}{\partial y^2} = 0 \tag{3}$$

P. Pucci et al. (eds.), *James Serrin. Selected Papers: Volume 1*, Contemporary Mathematicians,
DOI 10.1007/978-3-0348-0845-3_1, © Springer Basel 2014

in $\mathbb{R}^2$ which is bounded from below is a constant. If $n > 2$, the proof is based upon the construction of a quasi-representation of L-harmonic functions (i.e., solutions of $Lu = 0$) by mean of boundary integrals. This is done by freezing the coefficients of L and by a delicate use of the Dini continuity.

Paper [ii] below is widely considered as one of the main breakthroughs in the theory of quasilinear equations. It deals with equations in divergence form

$$-\operatorname{div} A(x,u,Du) + B(x,u,Du) = 0 \tag{4}$$

in a domain $\Omega \subset \mathbb{R}^n$. One of its remarkable aspects is the minimality of the assumptions on the operators A and B: Caratheodory functions satisfying for all $(x,r,q) \in \Omega \times \mathbb{R} \times \mathbb{R}^n$

$$\begin{aligned} &\text{(i)} && |A(x,r,q)| + |B(x,r,q)| \le c_1|q|^{p-1} + c_2|r|^{p-1} + c_3 \\ &\text{(ii)} && A(x,r,q).p \ge c_1^{-1}|q|^p - c_4|r|^p - c_5 \end{aligned} \tag{5}$$

for some positive constants c_j and $p > 1$. The prototype of such an operator is the p-Laplacian $\Delta_p u = \operatorname{div}(|Du|^{p-2}Du)$, which has, however, a monotonicty property which is not required for A. By adapting Moser's iteration method to the nonlinear case, it is shown that all positive weak solutions of (4) satisfy an Harnack type inequality (up to an additive constant); with this estimate, they turn out to be Hölder continuous. In the same paper the question of removability of singularities is considered. This question had already been widely studied for linear equations, in particular by Serrin himself, but very little was known in the nonlinear case except for the two-dimensional minimal surface equation, for which Bers discovered the removability of isolated singularities by mean of the Weiertsrass representation [2]. The following statement is proved in [ii].

Theorem 1. *Assume $1 < p \le s \le n$, $\delta > 0$, Q is a compact subset of Ω with $C^{1,s}$-Bessel capacity zero and $u \in L^{(1+\delta)\theta}_{\mathrm{loc}}(\Omega)$ is a continuous weak solution of (4) in $\Omega \setminus Q$ where $\theta = \frac{s(p-1)}{s-p}$. Then u can be defined on the set Q so that the resulting function is a continuous solution of (4) in the entire domain Ω.*

This result extends the classical removability theorem for bounded harmonic functions which asserts that a set with zero $C^{1,2}$ (or Newtonian) capacity is removable for a bounded harmonic function. Another type of removability result exploits the dimension m of the singular set Q.

Theorem 2. *Assume that Q is a compact m-dimensional submanifold of Ω, $p - m \le n$, $\delta > 0$ and u is a continuous weak solution of (4) in $\Omega \setminus Q$. If*

$$\begin{aligned} &\text{(i)} && u \in L^{(1+\delta)\theta}_{\mathrm{loc}}(\Omega) && \text{if } m < n - \alpha \\ &\text{(ii)} && u(x) = O(|\ln(\operatorname{dist}(x,Q))|)^{1-\delta} && \text{if } m = n - \alpha \end{aligned} \tag{6}$$

and $\theta = \frac{(n-m)(p-1)}{n-m-p}$, the conclusion of Theorem 1 holds.

A long series of extensions to semi-linear equations such as

$$-\Delta u + |u|^{q-1}u = 0 \qquad (q > 1) \tag{7}$$

due to Baras, Brezis, Pierre, Véron and others followed these fundamental results; moreover this subject is still very active in various situations such as boundary singularities, nonlinear parabolic equations, fully nonlinear equations, Hamilton–Jacobi equations and convection equations. The last part of this seminal article is devoted to the description of singular solutions of

$$-\operatorname{div} A(x,u,Du) = 0 \tag{8}$$

in $\Omega \setminus \{0\}$ where Ω is an open neighborhood of 0. Although no explicit fundamental solution exists, it is shown that any solution u of (8) which is bounded from below satisfies

$$c^{-1} \leq \frac{u(x)}{\mu_p(x)} \leq c \tag{9}$$

near 0, where $\mu_p(x) = |x|^{\frac{p-N}{p-1}}$ if $1 < p < N$ and $\mu_N(x) = -\ln|x|$. This result has been a crucial step to obtain the precise behaviour of positive p-harmonic functions in a punctured domain [6]. Another beautiful application of Serrin's Harnack inequality is a geometric Liouville property due to Reschetnyak [7]: *Any bounded map $f : \mathbb{R}^N \mapsto \mathbb{R}^N$ with bounded distortion is a constant.* Actually, with the help of Serrin's work, Reschetnyak proves that any entire bounded p-harmonic function is constant.

Paper [iii] is a delicate extension of the characterization (in the sense of (9)) of isolated singularities of positive solutions of (4). In the introduction Serrin wrote

> "*The final chapter of* [ii] *treated a variety of problems concerning the behavior of solutions at an isolated singularity. In that chapter I found it necessary to impose the condition $B = 0$, a condition which had not been required in the preceding parts of* [i]. *The main purpose of the present paper is to show that this additional condition can be removed, and thus to complete, in an important way, the theory of the earlier paper.*"

One of the most important *applications* of [iii] is the Gidas and Spruck paper [5], which deals with the equation

$$-\Delta u = u^q. \tag{10}$$

Furthermore, this study is followed by a remarkable extension due to Serrin and Zou in [8] to

$$-\operatorname{div}\left(|\nabla u|^{p-2}\nabla u\right) u = u^q. \tag{11}$$

In each of these two articles, Serrin's idea is that the control of $|u|^{q+1-p}$ in some L^r-space implies the Harnack inequality, so that the search for *a priori* estimates in PDE reduces to a priori estimates in differential inequalities. This control is achieved by differentiating the equation and multiplying it by suitable test functions. It should be noted that the other approach for obtaining a priori estimates for positive solutions of (10), based on the asymptotic symmetry of solutions, finds its origin in the reflection of solutions through hyperplanes, an idea introduced by Alexandrov in a geometric setting and Serrin in the PDE framework.

The most recent paper [iv] presents a nonlinear Harnack inequality for a wide class of quasilinear inequalities in the plane of the form

$$-B(\nabla u) - \leq \operatorname{div}(A(|\nabla u|)\nabla u) \leq B(\nabla u) + f(u), \tag{12}$$

where A and B satisfy growth conditions. Define $\Omega(t) = tA(t)$ and suppose that $\Omega'(t) > 0$ for $t > 0$, $0 < B(p) \leq \mu\Omega(|p|)$ for any $p \in \mathbb{R}^2$ with $|p| \leq 1$, and f is continuous, non-decreasing on $[0, \infty)$, with $f(0) = 0$. Assume, in addition, that either $f \equiv 0$ near 0 or, if $f > 0$ on $(0, \infty)$, that

$$\int_0^1 \frac{ds}{H^{-1}(F(s))} = \infty \tag{13}$$

where

$$F(s) = \int_0^s f(\tau)d\tau \text{ and } H(s) = \int_0^s \tau\Omega'(\tau)d\tau.$$

Then a solution of (12) *in* $B_R(0)$ *satisfies*

$$u(x, y) \geq \Phi_R(u(0, 0)) \tag{14}$$

in $B_{R/2}(0)$, *where* Φ_R *is a strictly increasing function defined on* $[0, \infty)$ *which vanishes at* 0. The surprising aspect of the result is that its proof is based upon maximum principle only and uses integral structural assumptions instead of the usual pointwise estimates on A and f. A companion inequality

$$u(x, y) \leq \Psi_R(u(0, 0)), \tag{15}$$

is also proved. Although the results are restricted to 2 dimensions, they allow the treatment of the difficult case of the mean curvature equation, which does not satisfy the conditions of [iii].

References

[1] P. Baras and M. Pierre, *Singularités éliminables pour des équations semi-linéaires*, Ann. Inst. Fourier (Grenoble) 34, 185–206 (1984).

[2] L. Bers, *Isolated singularities of minimal surfaces*, Ann. of Math. (2) 53, 364–386 (1951).

[3] H. Brezis and L. Véron, *Removable singularities of some nonlinear elliptic equations*, Arch. Rat. Mech. Anal. 75, 1–6 (1980).

[4] L. Carleson, *Selected problems on exceptional sets*, Uppsala, Mimeographed, (1961).

[5] B. Gidas and J. Spruck, *Global and local behavior of positive solutions of nonlinear elliptic equations*, Comm. Pure Appl. Math. 34, 525–598 (1981).

[6] S. Kichenassamy and L. Véron *Singular solution of the p-Laplace equation*, Math. Annalen 275, 599–615 (1986).

[7] Y.G. Reshetnyak, *Mappings with bounded deformation as extremals of Dirichlet type integrals*, Sibirsk. Mat. Zh. 9, 652–666 (1966).

[8] J. Serrin and H. Zou, *Cauchy-Liouville and universal boundedness theorems for quasilinear elliptic equations and inequalities*, Acta Math. 189, 79–142 (2002).

Marie-Françoise Bidaut-Véron and Laurent Véron
Laboratoire de Mathématiques et Physique Théorique
CNRS UMR 6083
Faculté des Sciences
Université François Rabelais
F-37200 Tours, France
e-mail: veronmf@univ-tours.fr
veronl@univ-tours.fr

J. Anal. Math. 4 (1956), 292–308. **Reprint I.1.[i]**

ON THE HARNACK INEQUALITY FOR LINEAR ELLIPTIC EQUATIONS

By

James Serrin

in Minneapolis, Minn., U.S.A.

If $u(P)$, $P=(x_1, \dots, x_n)$ is a positive harmonic function defined in a sphere of radius a and center O, then we have the classical inequality of Harnack

$$\frac{a^{n-2}(a-\rho)}{(a+\rho)^{n-1}} u(O) \leqq u(P) \leqq \frac{a^{n-2}(a+\rho)}{(a-\rho)^{n-1}} u(O), \tag{1}$$

where $\rho = |O-P|$ is the distance from O to P. This result is an immediate consequence of the Poisson integral formula

$$u(P) = \int K(P, Q)\, u(Q)\, dS_Q,$$

where the integral is over the sphere $|O-Q|=a$, and

$$K(P, Q) = \frac{1}{a\omega_n} \frac{a^2-\rho^2}{|P-Q|^n}.$$

Inequality (1) implies the following important result (Kellogg [6], p. 263): Let $u(P)$ be a positive harmonic function in a domain T, and let S be a closed subregion of T. Then there exists a positive constant K, depending only on S and T, such that the inequality

$$u(P) \leqq K u(Q) \tag{2}$$

holds for any two points P and Q in S. The point of (2) is that a positive harmonic function which is small at one point in its domain of definition T must be uniformly small in any closed subregion of T.

Lichtenstein [7] and Feller [3] have demonstrated inequalities similar to (1) and (2) for positive solutions of elliptic equations of the type

$$Lu = a_{ij}(P)\, u_{ij} + b_i(P)\, u_i + c(P)\, u = 0 \qquad (c \leqq 0), \tag{3}$$

where $P=(x_1, \dots, x_n)$,

$$u_i = \frac{\partial u}{\partial x_i}, \quad a_{ij} u_{ij} = \sum_{i,j=1}^{n} a_{ij} \frac{\partial^2 u}{\partial x_i\, \partial x_j}, \quad \text{etc.}$$

This work was supported by the Office of Naval Research. It was completed in part while ihe author was at Stanford University ih the summer of 1955.

Their method, which is basically the same as that indicated above for harmonic functions, makes use of the Green's function belonging to the adjoint of (3), and consequently applies only to equations whose coefficients are sufficiently often differentiable.

In this paper we shall improve the results of Lichtenstein and Feller by materially lightening the smoothness properties required of the coefficients. For the case $n=2$ we shall assume only that the coefficients are bounded and that L is uniformly elliptic [1]; when $n>2$ it will be necessary to assume additionally that the coefficients a_{ij} are continuous. The results we obtain are thus applicable in approximately the same generality as modern existence theories, and in particular may be useful in the study of nonlinear equations. All the proofs are elementary in character, the chief mathematical tool being the maximum principle [5]. In another paper [4] some applications are made to the study of isolated singularities of solutions of elliptic differential equations. (It should be mentioned that Lichtenstein and Feller give no indication of the dependence of the Harnack constants on the moduli of the coefficients (3). On the other hand, we are always able to state this dependence precisely; this is important in the applications to be made in [4]).

Theorems 1 and 2 treat the case of $n=2$ independent variables. Theorem 3 is a Liouville theorem for elliptic equations, closely related to a well-known result of S. Bernstein on surfaces of non-positive Gauss curvature. Finally in Theorems 4 and 5 we give two inequalities for the case $n>2$, each valid under different hypotheses. The inhomogeneous equation $Lu=f(P)$ is discussed in § 4.

In what follows the operator L defined in (3) is assumed to be *uniformly elliptic*; that is, we suppose there exists a constant $\lambda>0$ such that, for all points P and all real n-tuples $\xi_1, \ldots, \xi_n$,

$$\lambda^{-1}(\Sigma\xi_i^2) \leq a_{ij}(P)\,\xi_i\,\xi_j \leq \lambda(\Sigma\xi_i^2)\,. \tag{4}$$

The coefficients b_i and c are assumed bounded as follows:

$$(\Sigma b_i^2)^{1/2} \leq M\,, \qquad -M \leq c \leq 0\,. \tag{5}$$

Also from (4) we obtain the bound

$$|a_{ij}| \leq \lambda \qquad i, j = 1, \ldots, n\,. \tag{6}$$

We shall always assume that the solutions $u(x)$ of (3) are in class C^2.

1. Bers and Nirenberg [2] have obtained a Harnack inequality under similar hypotheses. The proof we give is simpler and considerably more elementary than theirs.

1. The Harnack inequality for $n=2$ variables.

In this section we shall use coordinates $P=(x, y)$ and write (3) in the form

(7) $$Lu = Au_{xx} + 2Bu_{xy} + Cu_{yy} + Du_x + Eu_y + Fu = 0 \quad (F \leqq 0).^{(2)}$$

We begin with a simple but fundamental construction.

Let $\mathcal{E}$ denote the closed region bounded by an arbitrary ellipse in the xy-plane. For the purposes of the present paragraph it is convenient to introduce coordinates so that $\mathcal{E}$ is given by

$$\sigma^2 = \frac{x^2}{a^2} + \frac{y^2}{b^2} \leqq 1\,, \qquad a \geqq b > 0\,.$$

The subregion

$$\frac{1}{2} \leqq \sigma \leqq 1\,, \quad x \geqq \frac{1}{2}a\,,$$

of $\mathcal{E}$ will be called a m i n o r a n t r e g i o n. We shall associate with each minorant region R a function

$$v = v(P) = \frac{e^{-\alpha\sigma^2} - e^{-\alpha}}{e^{-\alpha/4} - e^{-\alpha}}\,,$$

where α is a positive constant to be chosen so that

(8) $$Lv \geqq 0 \text{ in } R\,.$$

It is always possible to satisfy (8), for

$$(e^{-\alpha/4} - e^{-\alpha})\, Lv = e^{-\alpha\sigma^2} \left\{4\alpha^2 \left(A\frac{x^2}{a^4} + 2B\frac{xy}{a^2b^2} + C\frac{y^2}{b^4}\right)\right.$$
$$\left. - 2\alpha\left(\frac{A}{a^2} + \frac{C}{b^2} + \frac{Dx}{a^2} + \frac{Ey}{b^2}\right) + F\right\} - Fe^{-\alpha}\,,$$
$$\geqq e^{-\alpha\sigma^2}\left\{\alpha^2\frac{1}{\lambda a^2} - 2\alpha\left(\frac{2\lambda}{b^2} + \frac{M}{b}\right) - M\right\},^{(3)}$$

and the last expression is positive for sufficiently large α. For definiteness

2. We use this form of equation (3) for reasons of typographical clarity; conditions (4)—(6) on the coefficients are of course to be carried over in the natural way, (e.g. in place of (5) we have $(D^2+E^2)^{1/2} \leqq M$, $-M \leqq F \leqq 0$, etc.) Several of the letters $A, \dots, F$ will be used later for other purposes, but this should not cause confusion.

3. The following inequalities were used:

$$A\frac{x^2}{a^4} + 2B\frac{xy}{a^2b^2} + C\frac{y^2}{b^4} \geqq \frac{1}{a^2}\cdot\lambda^{-1}\left(\frac{x^2}{a^2} + \frac{y^2}{b^2}\right) \geqq \frac{1}{4\lambda a^2}\,,$$
$$\left|\frac{Dx}{a^2} + \frac{Ey}{b^2}\right| \leqq \frac{1}{b}(D^2+E^2)^{1/2}\left(\frac{x^2}{a^2} + \frac{y^2}{b^2}\right)^{1/2} \leqq \frac{M}{b}\,.$$

we choose α to be the specific (positive) value for which the expression in braces vanishes. The function v thus obtained will be called the minorant function for R.

On the curved boundary of R we have $v=0$, on the straight boundary $v \leq 1$, and at interior points $0<v<1$. For a given region R the minorant function depends only on the moduli λ and M belonging to L.

Theorem 1. *Let $u(P)$ be a positive solution of $Lu=0$ in a circle of radius $a \leq 1$ and center O. Then there exists a constant $K>0$, depending only on λ and M, such that $u(P) \geq Ku(O)$ in the circle $0 \leq \rho = |O-P| \leq \frac{1}{3}a$.*

Proof. It is sufficient to consider the case $a=1$, for the case $a<1$ may be reduced to this one by a change of variables $(x, y) \to (ax, ay)$. (In the new variables (7) becomes

$$Au_{xx} + 2Bu_{xy} + Cu_{yy} + aDu_x + aEu_y + a^2Fu = 0$$

for which conditions (4) and (5) remain true.)

In the circle $\rho \leq a = 1$ consider the point set $u > \frac{1}{2}u(O) \equiv \varkappa$, (defining $\varkappa$), and let G denote the component containing O. By the maximum principle it is clear that G must contain some point Q on $\rho = 1$. Hence there exists a smooth non-intersecting curve Γ which extends from O to Q and which has the property

$$u > \varkappa \qquad \text{on } \Gamma .$$

For convenience we take the coordinates so that O is the origin and Q is (0,1). Now let two minorant regions R_1 and R_2, with parameters $a=3$, $b=\frac{1}{2}$, be placed as shown in Figure 1 with their major axes horizontal and their vertices at $(\frac{1}{2}, \frac{1}{4}\sqrt{3})$ and $(-\frac{1}{2}, \frac{1}{4}\sqrt{3})$. The curved boundaries of R_1 and R_2 will be denoted by $\dot{R}_1$ and $\dot{R}_2$ respectively; the corresponding minorant functions will be called v_1 and v_2.

Consider any point P in the intersection of R_1 and R_2. There exists at least one region D in $\rho<1$ such that D contains P and either

(i) D is contained in R_1, and the boundary of D consits of arcs of $\dot{R}_1$ and arcs ot Γ, or

(ii) D is contained in R_2, and the boundary of D consists of arcs of $\dot{R}_2$ and arcs of Γ. [4]

4. To see this, let S denote the region $(R_1+R_1)\cap(\rho<1)$. It is clear that there

(In Figure 1 the region D for a representative point P is shaded). Suppose now that case (i) holds (as in Figure 1). Then the function $u - \varkappa v_1$ is defined in D. Because

$$u - \varkappa v_1 > \varkappa(1 - v_1) > 0 \qquad \text{on } \Gamma,$$
$$u - \varkappa v_1 = u > 0 \qquad \text{on } \dot{R}_1,$$

we have $u - \varkappa v_1 > 0$ on the boundary of D; moreover $L(u - \varkappa v_1) \leqq 0$ in D. Hence by the maximum principle,

$$u > \varkappa v_1 \quad \text{in } D,$$

and in particular $u(P) > \varkappa v_1(P)$. If case (ii) holds, the same argument gives $u(P) > \varkappa v_2(P)$. Therefore, wherever P is in the intersection of R_1 and R_2,

(9) $$u(P) > \varkappa \operatorname{Min}[v_1(P), v_2(P)].$$

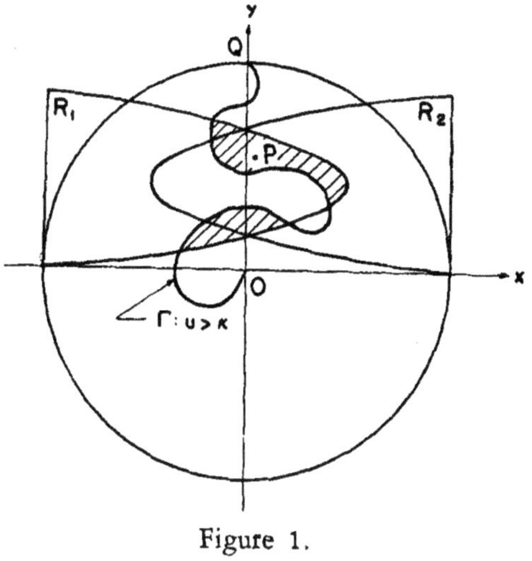

Figure 1.

Now let a minorant region R_3, with parametets $a = 2 + \frac{1}{2}\sqrt{3}$, $b = \frac{1}{2}$, be oriented as shown in Figure 2 with its major axis vertical and its vertex at $(0, -1)$. It is evident that the straight boundary of R_3 lies interior to the intersection of R_1 and R_2, as shown in Figure 2. On the straight boundary of R_3 we have from (9),

exists at least one arc $\widehat{pq}$ of Γ which lies entirely in S and which joins the "lower" boundary of S to the "upper" boundary of S. Obviously $\widehat{pq}$ divides S into exactly two parts S_1 and S_2 (let S_1 be the "left hand" region, S_2 the "right hand" region).

If P lies in S_1, we take $D = R_2 \cap S_1$; if P lies in S_2 (as in Figure 1) we take $D = R_1 \cap S_2$. It is easily seen that D has the desired properties.

(10) $$u > \varkappa K_1$$

where $K_1 > 0$ is the minimum of v_1 (or of v_2) on this boundary.[5] Applying the maximum principle to $u - \varkappa K_1 v_3$ in R_3, we obtain, with the help of (10),

(11) $$u > \varkappa K_1 v_3 \qquad \text{in } R_3;$$

(v_3 is the minorant function for R_3). Now one easily checks that R_3 contains the circle $\rho \leqq \frac{1}{3}$ in its interior, hence

$$K_2 = \min_{\rho \leqq 1/3} v_3(P) > 0 .$$

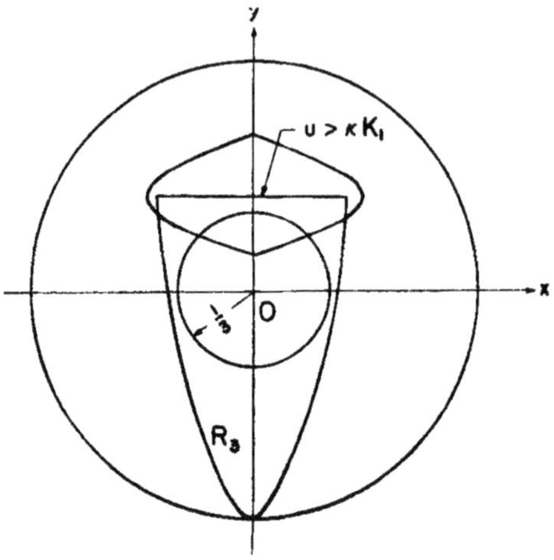

Figure 2.

Evidently $K_2 = K_2(\lambda, M)$. Setting $K = \frac{1}{2} K_1 K_2$ and recalling that $\varkappa = \frac{1}{2} u(O)$, it follows from (11) that

$$u(P) > 2\varkappa K = K u(O) \quad \text{for} \quad 0 \leqq \rho \leqq \tfrac{1}{3} .$$

This completes the proof.

Theorem 1 can be generalized so as to apply to positive solutions of the inhomogeneous equation $Lu = f(P)$. The details are presented in § 4. We turn now to various consequences of Theorem 1.

Corollary. Under the hypotheses of Theorem 1 we have

$$u(P) \leqq K^{-1} u(O)$$

for $0 \leqq \rho \leqq \frac{1}{4} a$, where K is the constant of Theorem 1.

5. $K_1 = v_1(\frac{1}{4}\sqrt{3}, \frac{1}{4}\sqrt{3}) = K_1(\lambda, M)$.

Proof. Let P be a point in the circle $0 \leqq \rho \leqq \frac{1}{4}a$, and let C be a circle of radius $\frac{3}{4}a$ about P. Then since $u>0$ in C we have by Theorem 1,

$$u(P') \geqq Ku(P), \tag{12}$$

if $|P-P'| \leqq \frac{1}{3}\cdot\frac{3}{4}a = \frac{1}{4}a$. Taking $P'=O$ in (12) we get $u(O) \geqq Ku(P)$, which is the desired inequality.

Theorem 2. *Let $u(P)$ be a positive solution of $Lu=0$ in a domain T, and let S be a closed region in T. Then there exists a positive constant K, depending only on λ, M, S, and T, such that the inequality*

$$u(P) \leqq Ku(Q) \tag{2'}$$

holds for any two points P and Q in S.

Theorem 2 is obtained from the preceding corollary by a standard procedure of potential theory ([6], p. 263). We shall omit the details.

Theorem 3. *If $u(P)$ is a solution over the whole plane of the uniformly elliptic equation*

$$Au_{xx} + 2Bu_{xy} + Cu_{yy} = 0, \tag{13}$$

and if u is bounded from below, then $u \equiv$ Const.

Remarks. This result generalizes the well-known Liouville theorem of potential theory. An example of S. Bernstein shows that one cannot relax the hypothesis of *uniform* ellipticity to mere *pointwise* ellipticity $(AC-B^2>0)$. Namely, the function e^{x-y^2} satisfies the equation

$$(2+4y^2)u_{xx} + 4yu_{xy} + u_{yy} = 0$$

and is at the same time positive and non-constant. On the other hand, Bernstein has proved that pointwise ellipticity does imply that a *bounded* solution over the whole plane is constant [1]. The Liouville theorem above thus forms a complement to Bernstein's theorem.

In the following paper [4] we shall give a more general result than Theorem 3, but if only the Liouville theorem is desired the present route is shortest.

Proof of Theorem 3. Let u_0 = greatest lower bound of u. Let us assume for contradiction that $u \not\equiv u_0$. Then by the maximum principle $w = u - u_0 > 0$. Since g.l.b. $w=0$ there exists a point P such that

$$w(P) < Kw(O),$$

where K is the constant of Theorem 1. Now Theorem 1 applies to equation (13) without the restriction $a \leqq 1$, and in particular it applies when $a = 3|O-P|$; thus

$$w(P) \geqq Kw(O).$$

This contradiction establishes Theorem 3.

Note. From Theorem 1 we can also obtain the following result, analogous to (1): *Let $u(P)$ be a positive solution of $Lu=0$ in a circle of radius $a \leqq 1$ and center O. Then there exists a continuous monotonically decreasing function $f(t)$; $0 \leqq t \leqq 1$, $f(0)=1$, $f(1)=0$, and a continuous monotonically increasing function $g(t)$, $0 \leqq t \leqq 1$, $g(0)=1$, both functions depending only on λ and M, such that*

$$f\left(\frac{\rho}{a}\right) u(O) \leqq u(P) \leqq g\left(\frac{\rho}{a}\right) u(O). \tag{1'}$$

2. The Harnack inequality for $n>2$ variables.

It is apparent that the preceding proof cannot be generalized to more than two variables. If certain mild smoothness conditions are required of the coefficients, however, a Harnack inequality may nevertheless be obtained.

Theorem 4. *Let $u(P)$ be a positive solution of $Lu=0$ in a sphere of radius $a \leqq 1$ and center O. Let the coefficients a_{ij} of L possess a modulus of continuity $\varphi(s)$ such that*

$$|a_{ij}(P) - a_{ij}(Q)| \leqq \varphi(|P-Q|) \tag{14}$$

for any two points P and Q satisfying $|O-P| \leqq a$, $|O-Q| = a$; suppose also that

$$\int_{+0} \varphi(s) s^{-1} ds < \infty \quad , \quad d\varphi(s) \geqq 0 .^{(6)} \tag{15}$$

Then there exists a constant $K>0$, depending only on λ, M, and Δ, where

---

6. This type of condition is called a Dini condition; the most important case is $\varphi(s)=As^{\gamma}$, $0<\gamma\leqq 1$.

$$\Delta = \int_0^{2a} \varphi(s)\, s^{-1}\, ds\,,$$

such that the inequality

$$K^{-1} u(O) \leqq u(P) \leqq K u(O)$$

holds for $0 \leqq \rho \leqq \frac{1}{3}a$.

Remark. It should be noted that the coefficients a_{ij} are required to be continuous only at points of the circumference $\rho = a$. This seems quite unnatural since the conclusion asserts a property of the solution for $\rho \leqq \frac{1}{3}a$; nevertheless we have been unable to obtain a proof dispensing with the continuity condition.

Proof. The method to be used is an adaptation of the one for harmonic functions outlined in the introduction. We cannot use a Poisson formula, of course, but two other integrals are developed to serve the same purpose. As in the proof of Theorem 1, it is sufficient to consider the case $a = 1$.

We shall construct two functions, $K_+(P\,,Q)$ and $K_-(P\,,Q)$,

$$\rho = |O - P| \leqq 1\,, \quad |O - Q| = 1\,, \quad P \neq Q\,,$$

having the following properties:

(i) Considered as functions of P for fixed Q,

$$LK_+ \geqq 0 \quad \text{and} \quad LK_- \leqq 0\,.$$

(ii) For any point Q_0, $|O - Q_0| = 1$, and any continuous function $g(Q)$,

$$\lim_{P \to Q_0} \int K_{\pm}(P\,,Q)\, g(Q)\, dS_Q = g(Q_0)\,.$$

(iii) There exist positive constants α and β, depending only on λ, M, and Δ, such that if $\rho \leqq \frac{1}{3}$,

$$K_+(P\,,Q) \geqq \alpha \quad \text{and} \quad K_-(P\,,Q) \leqq \beta\,.$$

Once these functions are obtained, the proof is easy. We define

$$u_{\pm}(P) = \int K_{\pm}(P\,,Q)\, u(Q)\, dS_Q\,.$$

Then by property (i),

$$Lu_+ \geqq 0 \quad \text{and} \quad Lu_- \leqq 0 \qquad\qquad \text{for } \rho < 1,$$

and by property (ii), $u_{\pm} = u$ on $\rho = 1$. Therefore by the maximum principle

$$u_+ \leqq u \leqq u_-\,. \tag{16}$$

Now $u(Q) > 0$, so by property (iii),

(17) $$u_+(P) \geqq \alpha \int u(Q)\,dS_Q \qquad \text{and} \qquad u_-(P) \leqq \beta \int u(Q)\,dS_Q$$

for $0 \leqq \rho \leqq \frac{1}{3}$. From (16) and (17) we get, for $\rho \leqq \frac{1}{3}$,

$$u(P) \leqq u_-(P) \leqq \beta \int u(Q)\,dS_Q \leqq \beta\alpha^{-1} u(O),$$

$$u(P) \geqq u_+(P) \geqq \alpha \int u(Q)\,dS_Q \geqq \alpha\beta^{-1} u(O).$$

This proves Theorem 4, with $K = \beta\alpha^{-1}$.

It remains therefore only to determine the functions $K_\pm(P, Q)$. We may clearly assume without loss of generality that

(18) $$\det a_{ij} = 1\,.$$

Let the *elliptic distance* R from $Q = (z_1, \ldots, z_n)$ to $P = (x_1, \ldots, x_n)$ be defined by

$$R = [A_{ij}(Q)(x_i - z_i)(x_j - z_j)]^{1/2},$$

where A_{ij} is the inverse matrix to a_{ij}, $(a_{ij} A_{jk} = \delta_{ik})$.

Since the characteristic values of A_{ij} are inverse to those of a_{ij} we have

(19) $$\lambda^{-1} r^2 \leqq R^2 \leqq \lambda r^2 \qquad r = |P - Q|,$$

and also

(20) $$|A_{ij}| \leqq \lambda\,, \qquad i, j = 1, \ldots, n\,.$$

Now define

(21) $$k = k(P, Q) = \frac{1 - \rho^2}{R^n}\,.$$

The analogy with the kernel in Poisson's integral should be noted; for this reason $k(P, Q)$ might be called a *kernel parametrix*. It seems reasonable to choose $K_\pm$ in the form

$$K_\pm = f_\pm(k)\,.$$

To see that this is possible we need some preliminary estimates. These occupy the following three paragraphs.

For simplicity in making the estimates, it is convenient to think of Q as a fixed parameter and to introduce a specific coordinate system. We take Q as origin and orient the axes so that O is on the positive x_n-axis. Then $r^2 = |P - Q|^2 = \Sigma x_i^2$, $R^2 = \Sigma A_{ij} x_i x_j$, and

$$k = \frac{2x_n}{R^n} - \frac{r^2}{R^n} = U_1 - U_2,$$

the last equality defining U_1 and U_2 in the obvious way. One may easily verify that

(22) $$a_{ij}(Q)\,\Gamma_{ij} = 0, \qquad \text{where } \Gamma = R^{2-n}.$$

Now, if numbers γ_i, $i = 1, \ldots, n$, are defined by the system of linear equations

$$A_{ij}(Q)\,\gamma_i = \delta_{jn}, \qquad j = 1, \ldots, n,$$

(this is possible since the matrix $A_{ij}(Q)$ is non-singular), then

$$\Sigma \gamma_i\, \partial\Gamma / \partial x_i = (2-n) R^{-n} \gamma_i A_{ij}(Q) x_j = (2-n) R^{-n} x_n;$$

thus U_1 is expressed as a linear combination of the derivatives of Γ. Therefore because of (22),

(23) $$a_{ij}(Q)\, U_{1,ij} = 0.$$

The following bounds are apparent, if we observe that any differentiation of U increases the degree of the denominator over the numerator by one, and then apply (19) and (20),

(24) $$\begin{aligned} &0 \leqq k \leqq C_1 r^{1-n}, && |k_i| \leqq C_1 r^{-n}, \\ &|U_{1,ij}| \leqq C_1 r^{-1-n}, && |U_{2,ij}| \leqq C_1 r^{-n}; \end{aligned}$$

here $C_1 = C_1(\lambda, n)$. We also have

(25) $$\Sigma k_i^2 = |\operatorname{grad} k|^2 \geqq C_2 r^{-2n}, \qquad C_2 = C_2(\lambda, n) > 0.$$

To prove this, let P be an arbitrary point and differentiate k with respect to r along the ray QP. If ϑ is the angle between QP and the plane $x_n = 0$, we get

$$\frac{dk}{dr}(P) = -\frac{2\sin\vartheta}{R^n} - (n-2)\frac{k}{r};$$

(the differentiation is simplified by noting that R/r is constant on the ray). Since $k \geqq 0$, it follows that

(26) $$|\operatorname{grad} k| \geqq \left|\frac{dk}{dr}\right| \geqq \frac{2\sin\vartheta}{R^n}.$$

Similarly, let us differentiate k with respect to ρ along the ray OP. We get

$$\frac{dk}{d\rho}(P) = -\frac{2\rho}{R^n} - n\frac{k}{R}\frac{dR}{d\rho}.$$

Now by (19),

$$\left|\frac{dR}{d\rho}\right| \leqq |\operatorname{grad} R| \leqq \lambda^{1/2}$$

$$\frac{k}{R} \leqq \frac{2\lambda^{1/2}\sin\vartheta}{R^n}.$$

Hence, using the fact that $\rho \geqq \cos\vartheta$,

$$|\operatorname{grad} k| \geqq \left|\frac{dk}{d\rho}\right| \geqq \frac{2\cos\vartheta - 2n\lambda\sin\vartheta}{R^n}. \tag{27}$$

From inequalities (26) and (27) it follows that, for any P,

$$|\operatorname{grad} k| \geqq \frac{2}{R^n\sqrt{2+2n\lambda+n^2\lambda^2}}.$$

(25) follows at once.

Finally it is necessary to obtain a certain majorant for the function

$$A = A(P) = (a_{ij}k_ik_j)^{-1}Lk.$$

From (4) and (25),

$$a_{ij}k_ik_j \geqq \lambda^{-1}(\Sigma k_i^2) \geqq \lambda^{-1}C_2 r^{-2n}. \tag{28}$$

Also, in virtue of (23),

$$Lk = \{a_{ij}(P) - a_{ij}(Q)\}\, U_{1,ij} - a_{ij}(P)\, U_{2,ij} + b_ik_i + ck.$$

Hence using (4), (6), (15), and (24),

$$|Lk| \leqq C_1[n^2\varphi(r)r^{-1-n} + n^2\lambda r^{-n} + nMr^{-n} + Mr^{1-n}]. \tag{29}$$

From (28) and (29) we have

$$|A| \leqq C_3[\varphi(r)r^{n-1} + r^n], \qquad C_3 = C_3(\lambda, M, n). \tag{30}$$

For purposes which will be apparent in a moment, (30) must be given in altered form. Let us set

$$s = s(P) = \left(\frac{k}{C_1}\right)^{\frac{1}{1-n}},$$

where C_1 is the constant of (24). Then by (24), $s \geqq r$, hence

$$|A| \leqq C_3[\varphi(s)s^{n-1} + s^n].$$

On the other hand we always have $r \leqq 2$, hence also

$$|A| \leqq C_3[\varphi(2)2^{n-1} + 2^n].$$

The desired majorant is now

$$\Omega(s) = C_3\begin{cases} \varphi(2)2^{n-1} + 2^n & \text{for } s = s(P) \geqq 2, \\ \varphi(s)s^{n-1} + s^n & \text{for } s = s(P) \leqq 2. \end{cases}$$

This completes the preliminary estimates.

Under the assumption $K_{\pm} = f_{\pm}(k)$ we have (dropping the subscripts temporarily),

$$LK = f'' a_{ij} k_i k_j + f' LK + c(f - kf')$$
$$= (a_{ij} k_i k_j)(f'' + Af') - c \int_0^k kf'' \, dk + cf(0) .$$

Now choose f as a solution of the ordinary[7] differential equation

(31) $$f'' = \pm \Omega(s) f' \quad \text{with} \quad f(0) = 0,\ f' > 0.$$

Since $|A| \leqq \Omega$ it is clear that property (i) is fulfilled. A particular solution of (31) for which (ii) holds is

$$f_{\pm} = \frac{1}{\omega_n} \int_0^k \exp\left[\mp \int_x^\infty \Omega(s)\, dk\right] dx .$$

To prove this, we begin by showing that the inner integral is convergent. Since $k = C_1 s^{1-n}$ we have

$$\int_0^\infty \Omega(s)\, dk = (n-1) C_1 \int_0^\infty \Omega(s) s^{-n} \, ds$$
$$= C_1 C_3 [(n-1)\Delta + \varphi(2) + 2n]$$
$$\leqq C_4(\lambda, M, \Delta, n), \text{[8]}$$

and the integral converges. It follows that $f_{\pm}$ is well defined and that

(32) $$\lim_{k \to \infty} \frac{f_{\pm}(k)}{k} = \frac{1}{\omega_n},$$

that is, $f_{\pm}$ is asymptotic to k/ω_n at the singularity Q. Since

$$1 \leqq \exp \int_k^\infty \Omega(s)\, dk \leqq \exp C_4,$$

it follows that $K_{\pm} = f_{\pm}(k)$ is bounded by positive multiples of k,

(33) $$C_5 k \leqq K_{\pm} \leqq C_6 k ,$$

where C_5 and C_6 depend on λ, M, Δ, and n. Using (18), (21), (32), and (33) one can show that property (ii) holds; the procedure involves standard

7. Recall that f and s are both functions of k.

8. To obtain the inequality we have assumed that $\varphi(s) \leqq 2\max|a_{ij}| \leqq 2\lambda$ as is always possible.

methods of potential theory ([6], p. 242), but would be somewhat lengthy and tedious. We shall therefore omit the details.

Property (iii) follows at once from (33), (21), and (19), and the proof is complete.

Theorem 4′. *Let $u(P)$ be a positive solution of $Lu=0$ in a domain T, and let S be a closed region in T. Let the coefficients a_{ij} of L satisfy the condition,*

$$|a_{ij}(P)-a_{ij}(Q)| \leqq \varphi(|P-Q|),$$

uniformly for all P, Q in T, where φ satisfies (15). Then there exists a constant $K>0$, depending only on λ, M, φ, S and T, such that the inequality

$$u(P) \leqq Ku(Q) \tag{2''}$$

holds for any two points P and Q in S.

The proof is standard.

3. The Harnack inequality for $n>2$ variables (continued).

Condition (14) of Theorem 4 is somewhat unnatural, as we have already remarked. In an attempt to avoid this hypothesis, I obtained the following result. It is not a true Harnack inequality, but in some cases it serves the same purpose.

Theorem 5. *Let $u(P)$ be a positive solution of $Lu=0$ in a sphere of radius $a \leqq 1$ and center O. Suppose that*

$$|\operatorname{grad} u| \leqq N. \tag{34}$$

Then there exists a continuous monotonically increasing function $\Phi(t)$, $t \geqq 0$, $\Phi(0)=0$, depending only on λ, M, and N, such that

$$u(P) \leqq \Phi[u(O)]$$

throughout $0 \leqq \rho \leqq a$.

Proof. As usual it is sufficient to consider the case $a=1$.

Let $h=\max u(P)=u(P_0)$, (P_0 is on $\rho=1$). By (34) we have

$$u(P) \geqq \tfrac{1}{2}h \qquad \text{in } C,$$

where C denotes the sphere of radius $h/2N$ about P_0. If $h \geqq 2N$ this implies

$$u(O) \geqq \tfrac{1}{2}h \qquad (h \geqq 2N). \tag{35}$$

For $h < 2N$ we shall obtain a different lower bound for $u(0)$. By the methods of § 1 one may easily construct minorant functions and minorant regions in n-space (in appropriately oriented coordinates a minorant region is defined by

$$\frac{1}{2} \leqq \frac{x_1^2}{a^2} + \frac{x_2^2 + \ldots + x_n^2}{b^2} \leqq 1, \qquad x_1 \geqq \frac{1}{2} a, \qquad a \geqq b > 0).$$

We place a minorant region R, with parameters $a = 4$, $b = h/2N$, so that the vertex is tangent to the circumference $\rho = 1$, and the major axis contains the ray OP_0. Let R' denote the part of R outside C.

Now one checks easily that R' is contained in $\rho \leqq 1$, that R' contains O, and that the boundary of R' consists of part of the boundary of C and part of the curved boundary of R. In virtue of this we find

$$w = u - \tfrac{1}{2} hv > 0$$

on the boundary of R', where v is the minorant function for R. Since $Lw \leqq 0$ we obtain from the maximum principle, $w > 0$ in R', and in particular

$$u(O) > \tfrac{1}{2} hv(O) \qquad (h < 2N). \tag{36}$$

From (35) and (36) it follows that

$$u(O) \geqq F(h),$$

where F is a continuous monotonically increasing function, depending only on λ, M, and N, such that $F(0) = 0$. The function Φ inverse to F has the same properties, and also

$$\Phi[u(O)] \geqq h \geqq u(P),$$

proving the theorem.

Theorem 5′. *Let $u(P)$ be a positive solution of $Lu = 0$ in a domain T, and let S be a closed region in R. Suppose that*

$$|\operatorname{grad} u| \leqq N. \tag{34′}$$

Then there exists a continuous monotonically increasing function $\Phi(t)$, $t \geqq 0$, $\Phi(0) = 0$, depending only on λ, M, N, S, and T, such that the inequality

$$u(P) \leqq \Phi[u(Q)] \tag{2‴}$$

holds for any two points P and Q in S.

Remarks. 1) The proofs of Theorems 5, 5′ depend on the existence of a modulus of continuity for the solution. This is provided for by the hypothesis $|\operatorname{grad} u| \leqq N$, although less stringent conditions could be allowed.

2) If the coefficients a_{ij} are continuous and u is bounded in some region containing T in its interior, condition (34′) is a consequence of a theorem of Morrey ([8], p. 123). (Another proof of this theorem will appear in a forthcoming work of L. Nirenberg). Since in the application of Morrey's theorem the bound N in (34′) will depend on the modulus of continuity of a_{ij} and on the bound on u, as well as on the other constants, we have in this case that Φ depends only on λ, M, S, T, the modulus of continuity of a_{ij}, and the bound on the solution.

4. The inhomogeneous equation $Lu = f(P)$.

In this section we shall prove the following generalization of Theorem 1:

Let $u(P)$ be a positive solution of the inhomogeneous equation $Lu = f$ in the circle of radius $a \leqq 1$ and center O. Let f be bounded, $|f| \leqq B'$. Then there exist three positive constants, K, K' and K'', depending only on λ and M, such that

$$u(P) \geqq Ku(O) - K'B',$$

for $0 \leqq \rho \leqq aK''$.

Proof. We need consider only the case $a = 1$. Now one easily checks that

$$\begin{aligned} L(\rho^2) &= 2(A + C) + 2Dx + 2Ey + F\rho^2 \\ &\geqq 4\lambda^{-1} - M(2\rho + \rho^2) \geqq \lambda^{-1}, \end{aligned}$$

if $0 \leqq \rho \leqq \rho_0 = \operatorname{Min}[1, \sqrt{1 + 3/\lambda M} - 1]$. It follows that

$$L(u + \lambda B'\rho^2) \geqq 0, \qquad 0 \leqq \rho \leqq \rho_0.$$

Therefore, as in Theorem 1, there exists a smooth curve Γ which extends from O to a point Q on the circle $\rho = \rho_0$ and which has the property

$$u + \lambda B'\rho^2 > \tfrac{1}{2} u(O) \equiv \varkappa \qquad \text{on } \Gamma.$$

Now define

$$w = u + \lambda B'(2\rho_0^2 - \rho^2). \tag{37}$$

Then $Lw \leqq 0$ in $0 \leqq \rho \leqq \rho_0$ (recall that $F \leqq 0$), and $w > \varkappa$ on Γ.

With these preliminaries out of the way we may continue exactly as in the proof of Theorem 1, obtaining

$$w > 2\varkappa K \quad \text{for} \quad 0 \leqq \rho \leqq \tfrac{1}{3}\rho_0, \tag{38}$$

where K is the constant of Theorem 1. From (37) and (38) it follows that

$$u > Ku(O) - K'B' \qquad \text{for } 0 \leqq \rho \leqq K'',$$

where $K' = 2\lambda\rho_0^2$, $K'' = \frac{1}{3}\rho_0$. This completes the proof.

Corresponding to Theorem 2 we have the following result, for which proof is omitted.

Let $u(P)$ be a positive solution of $Lu = f$ in a domain T, and let S be a closed region in T. Suppose $|f| \leqq B'$. Then there exists a positive constant K, depending only on λ, M, S, and T, such that the inequality

$$u(P) \leqq K[u(Q) + B']$$

holds for any two points P and Q in S.

Similar generalizations of the results of §§ 2 and 3 can be obtained if so desired. In particular Theorems 4′ and 5′ may be extended to positive solutions of the inhomogeneous equation

$$Lu = f(P), \qquad |f| \leqq B',$$

merely by replacing $u(Q)$ in (2″) or (2‴) by $u(Q) + B'$.

REFERENCES

1. S. Bernstein, Ueber ein geometrisches Theorem und seine Anwendung auf die partiellen Differentialgleichungen vom elliptischen Typus, *Math. Zeit.*, 26 (1927) pp. 551—558.

2. L. Bers and L. Nirenberg, On linear and non-linear elliptic boundary value problems in the plane, *Convegno Internazionale sulle Equazioni Derivate Parziali*, 1954, pp. 141—167.

3. W. Feller, Ueber die Lösungen der linearen partiellen Differentialgleichungen zweiter Ordnung vom elliptischen Typus. *Math. Ann.* 102 (1930), pp. 633—649.

4. D. Gilbarg and J. Serrin, On isolated singularities of solutions of second order elliptic differential equations, *J. d'Analyse Math.* (following in this issue).

5. E. Hopf. Elementare Betrachtungen ueber die Lösungen partieller Differentialgleichungen zweiter Ordnung vom elliptischen Typus, *Sitzungsberichte Preuss. Akad. Wiss.* 19 (1927) pp. 147—152.

6. O. D. Kellogg. Foundations of Potential Theory, Springer, Berlin, 1929.

7. L. Lichtenstein, Beiträge zur Theorie der linearen partiellen Differentialgleichungen zweiter Ordnung vom elliptischen Typus. Unendliche Folgen positiver Lösungen, *Rend. Circ. Math. Palermo*, 33 (1912), pp. 201—211.

8. C. B. Morrey, Second order elliptic systems of differential equations, in Contributions to the theory of partial differential equations, *Ann. of Math. Studies* No. 33, Princeton, 1954, pp. 101—159.

University of Minnesota
Minneapolis, Min. U.S.A.

(Received January 6, 1956)

Acta Math. 111 (1964), 247–302.

Reprint I.1.[ii]

LOCAL BEHAVIOR OF SOLUTIONS OF QUASI-LINEAR EQUATIONS

BY

JAMES SERRIN

University of Minnesota, U.S.A.

This paper deals with the local behavior of solutions of quasi-linear partial differential equations of second order in $n \geqslant 2$ independent variables. We shall be concerned specifically with the *a priori* majorization of solutions, the nature of removable singularities, and the behavior of a positive solution in the neighborhood of an isolated singularity. Corresponding results are for the most part well known for the case of the Laplace equation; roughly speaking, our work constitutes an extension of these results to a wide class of non-linear equations.

Throughout the paper we are concerned with real quasi-linear equations of the general form

$$\operatorname{div} \mathcal{A}(x, u, u_x) = \mathcal{B}(x, u, u_x). \tag{1}$$

Here $\mathcal{A}$ is a given vector function of the variables x, u, u_x, $\mathcal{B}$ is a given scalar function of the same variables, and $u_x = (\partial u/\partial x_1, \ldots, \partial u/\partial x_n)$ denotes the gradient of the dependent variable $u = u(x)$, where $x = (x_1, \ldots, x_n)$. The structure of (1) is determined by the functions $\mathcal{A}(x, u, p)$ and $\mathcal{B}(x, u, p)$. We assume that they are defined for all points x in some connected open set (domain) Ω of the Euclidean number space E^n, and for all values of u and p. Furthermore, they are to satisfy inequalities of the form

$$\left.\begin{aligned} |\mathcal{A}| &\leqslant a|p|^{\alpha-1} + b|u|^{\alpha-1} + e, \\ |\mathcal{B}| &\leqslant c|p|^{\alpha-1} + d|u|^{\alpha-1} + f, \\ p\cdot\mathcal{A} &\geqslant a^{-1}|p|^{\alpha} - d|u|^{\alpha} - g. \end{aligned}\right\} \tag{2}$$

Here $\alpha > 1$ is a fixed exponent, a is a positive constant, and the coefficients b through g are measurable functions of x, contained in certain definite Lebesgue classes over Ω (see Chapter I).

The generality of these assumptions naturally requires that equation (1) be interpreted in a weak sense. Let D be a subdomain of Ω, and let u be a function having strong derivatives u_x which are locally of class L_α over D. Then u will be called a *weak solution* (or simply a *solution*) of (1) in D if

$$\int (\phi_x \cdot \mathcal{A} + \phi \mathcal{B})\, dx = 0 \tag{3}$$

for any continuously differentiable function $\phi = \phi(x)$ with compact support in D. Obviously any function which satisfies (1) in the classical sense would be a solution in the sense just defined, though of course not conversely.

Before turning to a description of results, it is worth noting that the above structure includes linear elliptic equations, where $\alpha = 2$, and also the Euler equations of a wide class of regular as well as non-regular variational problems. The reader may consult reference [22], where this observation is made more explicit.

The main body of the paper is divided into three chapters. In Chapter I we consider various *a priori* estimates concerning the majorization of solutions. To begin with, Theorem 1 states that a solution u in D is essentially bounded on any compact subset D' of D, the bound depending only on the structure of (1), on the L_α norm of u over D, and on the geometry of D and D'. If u is continuous in the closure of D, one can further estimate the maximum of u in terms of its L_α norm together with the maximum of its boundary values (Theorem 3). The most important result of the first chapter is an inequality of Harnack type, Theorem 5, which generalizes to non-linear equations a recent result of Moser for linear equations. This theorem is basic for much of the following work. Finally, in Theorem 8 we show that solutions of (1) are necessarily Hölder continuous, possibly after redefinition on a set of measure zero. A more detailed outline of the contents of this chapter is impossible here; the reader is referred directly to the paper for more specific conclusions.

The proofs are based on the iteration technique introduced by Moser in references [16] and [17], and at the same time make strong use of the general Sobolev inequalities. We must also remark on the papers of Stampacchia, Morrey, and Ladyzhenskaya and Uraltseva, whose spirit is much the same as that of the first chapter here. In particular, Ladyzhenskaya and Uraltseva have proved by quite different methods the Hölder continuity of bounded solutions of (1), under conditions rather similar to (2).

The second chapter deals with the general problem of removable singularities. If we consider Laplace's equation, it is known that a set of capacity zero constitutes a removable singularity for a bounded harmonic function, while, on the other hand, a single point is removable provided only that the solution is $o(\log r)$ or $o(r^{2-n})$ in the neighborhood. For sets of intermediate size a corresponding removable singularity theorem was recently discovered by Carleson [1, p. 78], the idea there being to relate the Hausdorff dimension of the singular set to the Lebesgue class of the solution. The following result extends Carleson's theorem to all equations of the general form (1), (2).

Let Q be a compact set of s-capacity zero,(1) *where $\alpha \leqslant s \leqslant n$, and let D be a domain in Ω. Suppose that u is a continuous solution of* (1) *in the set $D-Q$, and that*

$$u \in L_{\theta(1+\delta)},$$

where $\theta = s(\alpha-1)/(s-\alpha)$ and δ is some positive number. Then u can be defined on the set Q so that the resulting function is a continuous solution of (1) *in the entire domain D.*

In order to see this result more clearly, let us consider its implications for the Laplace equation (where $\alpha = 2$). When $n = 2$ nothing new is obtained, for then $s = 2$ and we simply regain the result that a set of ordinary capacity zero is removable for a bounded (L_∞) solution. When $n > 2$, however, s can vary from 2, where we obtain the usual result, to n, where we get the result that a set of n-capacity zero is removable provided that $u \in L_{n/(n-2)+\delta}$. Since a single point has n-capacity zero, this case of the theorem is seen to be associated with the usual growth condition at an isolated singularity. In other words, when $n > 2$ the above result constitutes an interpolation theorem of the desired sort. (Actually our result is not quite as sharp for Laplace's equation as the one obtained by Carleson, though, of course, it does apply to a larger class of equations. We add that our work was done independently of Carleson's, and that the overlap was discovered only after the manuscript had been submitted for publication. Other work of a similar nature is due to Picone [19].)

In addition to the above result we shall also prove the following removable singularity theorem, which has the advantage of a considerably less abstract hypothesis.

Let Q be a smooth manifold of dimension $m \leqslant n-\alpha$, and let D be a domain in Ω. Suppose that u is a continuous solution of (1) *in the set $D-Q$, and that*

$$u \in L_{\theta(1+\delta)} \qquad \text{if} \quad m < n-\alpha,$$

$$u = O(|\log \xi|^{1-\delta}) \quad \text{if} \quad m = n-\alpha,$$

where $\theta = (n-m)(\alpha-1)/(n-m-\alpha)$, δ is some positive number, and ξ is the distance from Q. Then u can be defined on Q so that the resulting function is a continuous solution of (1) *in all of D.*

---

(1) Cf. Section 7. We note that any non-empty set has positive $(n+\varepsilon)$-capacity, so that only values $s \leqslant n$ need be considered.

The preceding theorems raise obvious questions concerning the existence of solutions of (1) having non-removable singularities. In Chapter III we investigate this problem in considerably more detail for the special equation

$$\operatorname{div} \mathcal{A}(x, u, u_x) = 0 \quad (\alpha \leqslant n). \tag{4}$$

It is shown that at an isolated singularity a positive solution of (4) has precisely the order of growth $r^{(\alpha-n)/(\alpha-1)}$ if $\alpha < n$ and $\log 1/r$ if $\alpha = n$.(1) For linear equations of the special form $(a_{ij}(x)\, u,_i),_j = 0$ a corresponding result is due to Royden [20]; cf also [10, 17]. A weaker version for the non-linear case was given earlier by the author [21]. Our present proof is basically the same as in [20], though the reader will see that the idea there serves only as a theme upon which numerous variations have been played. We may also mention that in the case of linear equations of the general form $a_{ij} u,_{ij} + b_i u,_i = 0$ there are corresponding isolated singularity theorems, requiring however, some continuity of the coefficients, and not necessarily providing an explicit order of growth at the singularity [3, 4, 6].

In Section 13 we show under suitable conditions that there exist solutions of (4) with precisely the behavior indicated above. A corresponding result for linear equations is due to Littman, Stampacchia, and Weinberger. The final section of the paper contains some further results for linear equations. Although these theorems are quite special in comparison with the rest of the work in the paper, they have an interest in their own right, and indicate to some extent the underlying differences between linear and non-linear equations.

I wish finally to thank Professor Hans Weinberger for his interest in this work, and for several quite helpful suggestions.

I. Majorization of solutions

In this chapter we shall consider various *a priori* estimates for solutions of the equation

$$\operatorname{div} \mathcal{A}(x, u, u_x) = \mathcal{B}(x, u, u_x). \tag{5}$$

It will be assumed that the functions $\mathcal{A}(x, u, p)$ and $\mathcal{B}(x, u, p)$ are defined for all values of u and p, and for all points x in some fixed domain Ω. Moreover, we suppose that they satisfy inequalities of the form

(1) An explicit example is provided by the equation

$$\operatorname{div} (u_x |u_x|^{\alpha-2}) = 0,$$

which has the solutions $Ar^{(\alpha-n)/(\alpha-1)} + B$ and $A \log r + B$ when $\alpha \neq n$ and $\alpha = n$ respectively.

$$\left.\begin{aligned} |\mathcal{A}| &\leqslant a|p|^{\alpha-1}+b|u|^{\alpha-1}+e, \\ |\mathcal{B}| &\leqslant c|p|^{\alpha-1}+d|u|^{\alpha-1}+f, \\ p\cdot\mathcal{A} &\geqslant |p|^{\alpha}-d|u|^{\alpha}-g, \end{aligned}\right\} \tag{6}$$

for $x \in \Omega$ and all values of u and p. Here $\alpha > 1$ is a fixed exponent, a is a positive constant, and b through g are measurable functions on Ω (the slightly simpler form of (6), as compared with (2), can be achieved by a simple normalization). If $1 < \alpha < n$ we assume that b through g are in the respective Lebesgue classes

$$b, e \in L_{n/(\alpha-1)}; \quad c \in L_{n/(1-\varepsilon)}; \quad d, f, g \in L_{n/(\alpha-\varepsilon)}, \tag{7}$$

ε being some positive number less than or equal to one. If $\alpha = n$ we suppose that b through g satisfy

$$b, e \in L_{n/(n-1-\varepsilon)}; \quad c \in L_{n/(1-\varepsilon)}; \quad d, f, g \in L_{n/(n-\varepsilon)}, \tag{8}$$

again with ε some positive number less than or equal to one. Since the remaining case $\alpha > n$ is somewhat anomalous, we shall discuss it separately in Section 5. In any event, the discrepancy between the conditions required for the various cases $1 < \alpha < n$, $\alpha = n$, and $\alpha > n$ seems to be an essential part of the situation. Finally, certain alternative hypotheses will be considered in the concluding section of the chapter.

Now let D be a subdomain of Ω, and let $u = u(x)$ be a function having strong derivatives which are locally of class L_α over D. As already explained in the introduction, u will be called a *solution* of (5) in D if

$$\int (\phi_x \cdot \mathcal{A} + \phi\mathcal{B})\, dx = 0 \tag{9}$$

for any continuously differentiable function $\phi = \phi(x)$ with compact support in D. In writing equation (9) it is tacitly assumed that the functions $\mathcal{A} = \mathcal{A}(x, u, u_x)$ and $\mathcal{B} = \mathcal{B}(x, u, u_x)$ are measurable, as will certainly be the case in all reasonable situations. Since u_x is locally in L_α, it follows that also

$$u \in \begin{cases} L_{\alpha^*} & \text{if} \quad \alpha < n, \\ L_\sigma & \text{if} \quad \alpha = n \end{cases}$$

locally in D, where $\alpha^* = \alpha n/(n-\alpha)$ is the Sobolev conjugate of α, and σ is any positive real number. Thus, using assumptions (7) and (8), it is a straightforward consequence of Hölder's inequality that $\mathcal{A}(x, u, u_x)$ is locally in $L_{\alpha/(\alpha-1)}$ while $\mathcal{B}(x, u, u_x)$ is locally in $L_{(\alpha^*)'}$ if $\alpha < n$, or locally in $L_{1+\delta}$ if $\alpha = n$, where $(\alpha^*)'$ is the Hölder conjugate of α^*, and $\delta = \varepsilon/2(n-\varepsilon)$. It follows therefore that if u is a solution of (5), then (9) holds not only for continuously differentiable functions ϕ, but in fact for any ϕ with strong derivatives in L_α and with compact support in D. This remark will be of considerable importance later on.

0. Preliminary lemmas

The theorems of the following sections require some preparatory results which we group together here.

LEMMA 1. *Let a_i, $i=1, \ldots, N$, be non-negative real numbers, and let α be a positive exponent. Then*

$$\lambda \sum a_i^\alpha \leqslant \left(\sum a_i\right)^\alpha \leqslant \Lambda \sum a_i^\alpha,$$

where $\lambda = \operatorname{Min}(1, N^{\alpha-1})$ and $\Lambda = \operatorname{Max}(1, N^{\alpha-1})$.

Proof. We may assume without loss of generality that $\alpha \neq 1$ and $\sum a_i = 1$. The maximum of the quantity $\sum a_i^\alpha$ is then easily found to occur either when all the a_i are equal, or (endpoint maximum) when all but one a_i are zero. This proves the left-hand inequality. The right-hand result is obtained the same way.

LEMMA 2. *Let α be a positive exponent, and let a_i, β_i, $i=1, \ldots, N$, be two sets of N real numbers such that* $0 < a_i < \infty$ and $0 \leqslant \beta_i < \alpha$. *Suppose that z is a positive number satisfying the inquality*

$$z^\alpha \leqslant \sum a_i z^{\beta_i}.$$

Then

$$z \leqslant C \sum (a_i)^{\gamma_i},$$

where C depends only on N, α, and β_i, and where $\gamma_i = (\alpha - \beta_i)^{-1}$.

Proof. We make use of Young's inequality

$$ab \leqslant \frac{1}{p} a^p + \frac{1}{q} a^q,$$

where $1/p + 1/q = 1$. Then if $0 < \beta < \alpha$ and $a > 0$ we have

$$az^\beta = \left(\frac{a}{\varepsilon}\right) \cdot (\varepsilon z^\beta) \leqslant \frac{\beta}{\alpha} (\varepsilon z^\beta)^{\alpha/\beta} + \frac{\alpha - \beta}{\alpha} \left(\frac{a}{\varepsilon}\right)^{\alpha/(\alpha-\beta)}$$

Putting $\varepsilon = (\alpha/2N\beta)^{\beta/\alpha}$ yields therefore

$$az^{\beta} \leqslant (1/2N)\, z^{\alpha} + (2Na)^{\alpha/(\alpha-\beta)}. \tag{10}$$

Note also that (10) holds trivially when $\beta = 0$. Now applying (10) to each term of the sum $\sum a_i z^{\beta_i}$, we obtain at once

$$z^{\alpha} \leqslant \sum \{1/2N)\, z^{\alpha} + (2Na_i)^{\alpha/(\alpha-\beta_i)}\},$$

whence by transposition there arises

$$z^{\alpha} \leqslant 2 \sum (2Na_i)^{\alpha\gamma_i}.$$

The required conclusion follows immediately from Lemma 1.

The next four lemmas are general calculus inequalities, due basically to Sobolev and Morrey. A simple proof of Lemmas 3 and 4 is given in reference [18], and Lemma 5 can be obtained by quite similar methods. Although Lemma 6 is essentially well-known, we include a proof for completeness.

Now let $\psi = \psi(x)$ be a measurable function on an open set D, and let p be a real number, $1 \leqslant p \leqslant \infty$. The L_p norm of ψ is defined by

$$\|\psi\|_{p,D} = \left(\int_D |\psi|^p\, dx\right)^{1/p}, \quad (p < \infty); \qquad \|\psi\|_{\infty,D} = \operatorname{ess\,sup}_D |\psi|.$$

For simplicity we shall write $\|\psi\|_p$ rather than $\|\psi\|_{p,D}$ when the domain D is apparent from the context.

LEMMA 3 (Morrey). *Let ψ be a strongly differentiable function on the unit ball $|x| < 1$, and suppose that $\|\psi_x\|_\alpha$ is finite for some $\alpha > n$. Then ψ is (essentially) Hölder continuous, with*

$$|\psi(x) - \psi(y)| \leqslant \text{Const.}\, |x-y|^{1-n/\alpha}\, \|\psi_x\|_\alpha,$$

the constant depending only on α and n.

LEMMA 4 (Sobolev). *Let ψ be a strongly differentiable function with compact support in E^n, and suppose that $\|\psi_x\|_\alpha$ is finite for some $\alpha < n$. Set $\alpha^* = \alpha n/(n-\alpha)$. Then ψ is in L_{α^*} and*

$$\|\psi\|_{\alpha^*} \leqslant \text{Const.}\, \|\psi_x\|_\alpha,$$

where the constant depends only on α and n.

In the next lemma, we use ϱ to denote the distance from a point x to the hyperplane $x_1 = \ldots = x_s = 0$, $1 \leqslant s \leqslant n$, while σ is some fixed positive number.

LEMMA 5. *Let ψ be a strongly differentiable function defined in the domain $\varrho > \sigma$ and having compact support in E^n. Suppose that $\|\psi_x\|_\alpha$ is finite for some $\alpha < n$. Then ψ is in L_{α^*} and*

$$\|\psi\|_{\alpha^*} \leqslant \text{Const.}\, \|\psi_x\|_\alpha$$

the constant depending only on α and n.

LEMMA 6. *Let ψ be a strongly differentiable function defined in an open ball S of radius h. Then if $\psi_S \equiv |S|^{-1} \int_S \psi\, dx$, we have*

$$\|\psi - \psi_S\|_1 \leqslant \text{Const.}\, h\, \|\psi_x\|_1,$$

the constant depending only on n.

Proof. Let x and y be two points in S, and set $\varrho = |x-y|$. Then, assuming that $\psi \in C^1$, we have

$$\psi(y) - \psi(x) = \int_0^\varrho (\partial\psi/\partial\varrho')\, d\varrho' \leqslant \int_0^{2h} |\psi_x|\, d\varrho'$$

(setting $|\psi_x| \equiv 0$ at points outside S). Multiplying both sides by $dy = \varrho^{n-1} d\varrho\, d\omega$ and integrating over S yields

$$|S| \cdot |\psi_S - \psi(x)| \leqslant \frac{(2h)^n}{n} \iint |\psi_x|\, d\varrho'\, d\omega = \frac{(2h)^n}{n} \int_S |\psi_x|\, \varrho^{-n+1}\, dy.$$

Then by the well-known convolution inequality of Young,

$$|S| \cdot \|\psi_S - \psi\|_1 \leqslant \frac{(2h)^n}{n} \|\psi_x\|_1 \cdot \|\varrho^{-n+1}\|_1 = \text{Const.}\, h^{n+1} \|\psi_x\|_1.$$

This completes the proof when ψ is continuously differentiable. The general case follows by standard approximation arguments.

The final lemma is due to John and Nirenberg. Its use in connection with the *a priori* estimation of solutions of differential equations already occurs in the paper of Moser which we have discussed in the introduction.

LEMMA 7. *Let ψ be a summable function on the unit ball S_0, and suppose that*

$$\int_S |\psi - \psi_S|\, dx \leqslant |S|$$

for every open ball $S \subset S_0$. Then there exist constants λ and μ, depending only on n, such that

$$\int_{S_0} e^{\lambda\psi} dx \cdot \int_{S_0} e^{-\lambda\psi} dx \leqslant \mu^2. \tag{11}$$

The conclusion of the theorem of John and Nirenberg is actually that

$$\int_{S_0} e^{\lambda|\psi-\psi_0|}\,dx \leqslant \mu,$$

but (11) is an immediate consequence of this inequality.

We close the section by noting a useful derivation rule in the theory of strong differentiation. Let G be a piecewise smooth function of the real variable u, $-\infty < u < +\infty$, with corners at the points $l_1, \ldots, l_N$. Suppose also that G satisfies a uniform Lipschitz condition. Then if $u(x)$ is strongly differentiable in a domain D, the composite function $G(u(x))$ is also strongly differentiable in D, and

$$G(u)_x = \begin{cases} G'(u)\,u_x, & \text{where} \quad u \neq l_1, \ldots, l_N, \\ 0, & \text{where} \quad u = l_1, \ldots, l_N. \end{cases} \tag{12}$$

Formula (12) of course defines only a particular representative of $G(u)_x$, as is only natural since strong derivatives are, technically, equivalence classes of measurable functions. In particular, changing u and u_x on a set of measure zero alters the representative but not the equivalence class.

An interesting consequence of (12) is that the strong derivative of a function u is zero almost everywhere on any set where u is constant. Indeed set

$$G(u) = \mathrm{Max}\,(0, u), \qquad H(u) = \mathrm{Min}\,(0, u),$$

and note that
$$u = G(u) + H(u), \qquad u_x = G(u)_x + H(u)_x.$$

Evaluating the second relation on the set where $u = 0$ yields the desired conclusion for the case when the constant is zero. The general result follows at once.

1. Majorization of $|u|$

We shall show in this section that weak solutions of (5) are locally bounded. This is of course obvious in case $\alpha > n$ by virtue of Lemma 3. The remaining cases are considerably more difficult, and will be treated separately.

Let P be a fixed point in the basic domain Ω, and let $S(R)$ denote the open ball of radius R centered at P. For simplicity we use the symbol $\|\cdot\|_{p,R}$ to denote the L_p norm of a function over $S(R)$.

Theorem 1. *Let u be a weak solution of equation* (5) *defined in some ball $S(2R) \subset \Omega$. Suppose that $\alpha < n$ and that conditions* (6) *and* (7) *hold. Then*

$$\|u\|_{\infty,R} \leqslant CR^{-n/\alpha}(\|u\|_{\alpha,2R} + kR^{n/\alpha})$$

and

$$\|u_x\|_{\alpha,R} \leqslant CR^{-1}(\|u\|_{\alpha,2R} + kR^{n/\alpha}),$$

where C and k are constants depending only on the structure of (5). *In particular*

$$C = C(\alpha, n, \varepsilon;\ a, \|b\|,\ R^{\varepsilon}\|c\|,\ R^{\varepsilon}\|d\|),$$

$$k = (\|e\| + R^{\varepsilon}\|f\|)^{1/(\alpha-1)} + (R^{\varepsilon}\|g\|)^{1/\alpha},$$

the norms of the coefficients b through g being taken in the respective Lebesgue spaces (7).

Proof. We consider first the case when $R=1$, with the solution correspondingly defined in the open ball $S(2) \subset \Omega$. Set

$$\bar{u} = |u| + k, \quad x \in S(2),$$

where $k = (\|e\| + \|f\|)^{1/(\alpha-1)} + \|g\|^{1/\alpha}$. Then obviously

$$\left.\begin{aligned} |\mathcal{A}| &\leqslant a|p|^{\alpha-1} + \bar{b}|\bar{u}|^{\alpha-1}, \\ |\mathcal{B}| &\leqslant c|p|^{\alpha-1} + \bar{d}|\bar{u}|^{\alpha-1}, \\ p\cdot\mathcal{A} &\geqslant |p|^{\alpha} - \bar{d}|\bar{u}|^{\alpha}, \end{aligned}\right\} \tag{13}$$

with $\bar{b} = b + k^{1-\alpha}e$, $\bar{d} = d + k^{1-\alpha}f + k^{-\alpha}g$. Moreover, the norms of $\bar{b}$ and $\bar{d}$ are bounded,

$$\|\bar{b}\| \leqslant \|b\| + 1, \quad \|\bar{d}\| \leqslant \|d\| + 2.$$

The proof now rests on an appropriate choice of the test function in relation (9).

For fixed numbers $q \geqslant 1$ and $l > k$ we define the functions

$$F(\bar{u}) = \begin{cases} \bar{u}^q & \text{if } k \leqslant \bar{u} \leqslant l, \\ ql^{q-1}\bar{u} - (q-1)l^q & \text{if } l \leqslant \bar{u}, \end{cases}$$

and

$$G(u) = \operatorname{sign} u \cdot \{F(\bar{u})\,F'(\bar{u})^{\alpha-1} - q^{\alpha-1}k^{\beta}\}, \quad -\infty < u < \infty,$$

where q and β are related by $\alpha q = \alpha + \beta - 1$. Evidently F is a continuously differentiable function of the variable $\bar{u}$, which is linear beyond the "cutoff" value l. Similarly G is a piecewise smooth function of u, with corners at $u = \pm(l-k)$. Now let $\eta = \eta(x)$ be a non-negative smooth function with compact support in $S(2)$, and put

$$\phi(x) = \eta^{\alpha}G(u), \quad (u = u(x)).$$

Since G is linear in u for $|u| > l-k$, it follows from the preceding section that ϕ has strong derivatives in L_α. Thus ϕ is admissible in (9).

Now in the set where $|u| \neq l-k$ we have obviously

$$\phi_x = \alpha\eta^{\alpha-1}\eta_x G + \eta^\alpha G' u_x,$$

with

$$G' = \begin{cases} q^{-1}\beta(F')^\alpha & \text{if } |u| < l-k, \\ (F')^\alpha & \text{if } |u| > l-k. \end{cases}$$

Therefore, using (13) and the fact that $|G| \leqslant F(F')^{\alpha-1}$ one has

$$\begin{aligned} \phi_x \cdot \mathcal{A} + \phi\mathcal{B} &\geqslant \eta^\alpha G' \{|u_x|^\alpha - \bar{d}|\bar{u}|^\alpha\} - \alpha\eta^{\alpha-1}|\eta_x G|\{a|u_x|^{\alpha-1} + \bar{b}|\bar{u}|^{\alpha-1}\} \\ &\quad - \eta^\alpha |G| \{c|u_x|^{\alpha-1} + \bar{d}|\bar{u}|^{\alpha-1}\} \\ &\geqslant |\eta F' \bar{u}_x|^\alpha - \alpha a|\eta_x F| \cdot |\eta F' \bar{u}_x|^{\alpha-1} - \alpha\bar{b}|\eta_x F| \cdot |\eta F' \bar{u}|^{\alpha-1} - c\eta F|\eta F' \bar{u}_x|^{\alpha-1} \\ &\quad - \bar{d}\{q^{-1}\beta|\eta F' \bar{u}|^\alpha + \eta F|\eta F' \bar{u}|^{\alpha-1}\}, \end{aligned} \tag{14}$$

valid wherever $|u| \neq l-k$.(1) The last inequality may be further simplified by setting $v = v(x) = F(\bar{u})$. Since $\bar{u}F' \leqslant qF$ we get

$$\begin{aligned} \phi_x \cdot \mathcal{A} + \phi\mathcal{B} &\geqslant |\eta v_x|^\alpha - \alpha a|\eta_x v| \cdot |\eta v_x|^{\alpha-1} - \alpha q^{\alpha-1}\bar{b}|\eta_x v| \cdot (\eta v)^{\alpha-1} \\ &\quad - c\eta v|\eta v_x|^{\alpha-1} - (1+\beta)\, q^{\alpha-1}\bar{d}(\eta v)^\alpha. \end{aligned} \tag{15}$$

In the set where $|u| = l-k$ we have $\phi_x = \alpha\eta^{\alpha-1}\eta_x G$ and $u_x = \bar{u}_x = 0$ (a.e.), so that (15) holds also on this set. We may therefore integrate (15) over $S(2)$, with the result

$$\begin{aligned} \|\eta v_x\|_\alpha^\alpha \leqslant \alpha a \int |\eta_x v| \cdot |\eta v_x|^{\alpha-1} dx + \alpha q^{\alpha-1} \int \bar{b}|\eta_x v| \cdot (\eta v)^{\alpha-1} dx \\ + \int c\eta v\, |\eta v_x|^{\alpha-1} dx + (1+\beta)\, q^{\alpha-1} \int \bar{d}(\eta v)^\alpha dx. \end{aligned} \tag{16}$$

Here for simplicity we have written $\|\cdot\|_\alpha$ rather than $\|\cdot\|_{\alpha,2}$.

The terms on the right-hand side of the preceding inequality can be estimated by using Hölder's inequality together with Lemma 4. Thus

$$\int |\eta_x v| \cdot |\eta v_x|^{\alpha-1} dx \leqslant \|\eta_x v\|_\alpha \|\eta v_x\|_\alpha^{\alpha-1},$$

$$\begin{aligned} \int \bar{b}|\eta_x v| \cdot (\eta v)^{\alpha-1} dx &\leqslant \|\bar{b}\|_{n/(\alpha-1)} \|\eta_x v\|_\alpha \|\eta v\|_{\alpha^*}^{\alpha-1} \\ &\leqslant C\|\eta_x v\|_\alpha \|(\eta v)_x\|_\alpha^{\alpha-1} \\ &\leqslant C\{\|\eta_x v\|_\alpha^\alpha + \|\eta_x v\|_\alpha \|\eta v_x\|_\alpha^{\alpha-1}\}. \end{aligned}$$

(1) The substitution $|u_x| = |\bar{u}_x|$ is obviously valid in the set where $|u| \neq 0$. It also holds when $u = 0$, since $u_x = 0$ almost everywhere on this set.

The letter C here denotes a constant (which usually changes from one line to the next) *depending only on the quantities listed in the statement of the theorem*. Similarly

$$\begin{aligned}\int c\eta v\,|\eta v_x|^{\alpha-1}\,dx &= \int c\,(\eta v)^{\varepsilon}(\eta v)^{1-\varepsilon}\,|\eta v_x|^{\alpha-1}\,dx \\ &\leqslant \|c\|_{n/(1-\varepsilon)}\,\|\eta v\|_{\alpha}^{\varepsilon}\,\|\eta v\|_{\alpha*}^{1-\varepsilon}\,\|\eta v_x\|_{\alpha}^{\alpha-1} \\ &\leqslant C\,\|\eta v\|_{\alpha}^{\varepsilon}\{\|\eta_x v\|_{\alpha}^{1-\varepsilon}\,\|\eta v_x\|_{\alpha}^{\alpha-1}+\|\eta v_x\|_{\alpha}^{\alpha-\varepsilon}\},\end{aligned}$$

and

$$\begin{aligned}\int \bar d(\eta v)^{\alpha}\,dx &= \int \bar d(\eta v)^{\varepsilon}(\eta v)^{\alpha-\varepsilon}\,dx \\ &\leqslant \|\bar d\|_{n/(\alpha-\varepsilon)}\,\|\eta v\|_{\alpha}^{\varepsilon}\,\|\eta v\|_{\alpha*}^{\alpha-\varepsilon} \\ &\leqslant C\,\|\eta v\|_{\alpha}^{\varepsilon}\{\|\eta_x v\|_{\alpha}^{\alpha-\varepsilon}+\|\eta v_x\|_{\alpha}^{\alpha-\varepsilon}\}.\end{aligned}$$

If the four previous estimates are inserted into the right-hand side of (16), and we set

$$z=\|\eta v_x\|_{\alpha}/\|\eta_x v\|_{\alpha}, \qquad \zeta=\|\eta v\|_{\alpha}/\|\eta_x v\|_{\alpha},$$

the result may be written

$$z^{\alpha}\leqslant C[z^{\alpha-1}+q^{\alpha-1}(1+z^{\alpha-1})+\zeta^{\varepsilon}(z^{\alpha-1}+z^{\alpha-\varepsilon})+(1+\beta)\,q^{\alpha-1}\,\zeta^{\varepsilon}(1+z^{\alpha-\varepsilon})]. \tag{17}$$

Applying Lemma 2 and simplifying the result, one obtains (since $1+\beta\leqslant(\alpha+1)\,q$)

$$z\leqslant Cq^{\alpha/\varepsilon}(1+\zeta),$$

or in terms of the original quantities

$$\|\eta v_x\|_{\alpha}\leqslant Cq^{\alpha/\varepsilon}(\|\eta v\|_{\alpha}+\|\eta_x v\|_{\alpha}). \tag{18}$$

Another use of Lemma 4 yields finally

$$\|\eta v\|_{\alpha*}\leqslant Cq^{\alpha/\varepsilon}(\|\eta v\|_{\alpha}+\|\eta_x v\|_{\alpha}). \tag{19}$$

The preceding inequalities (18) and (19) are the basic estimates of the paper; they will appear in one form or another in almost all of the following results.

To proceed with the proof of Theorem 1, let h and h' be real numbers satisfying $h'<h\leqslant 2$. Let the function η be chosen so that $\eta=1$ in $S(h')$, $0\leqslant\eta\leqslant 1$ in $S(h)$, and identically zero outside $S(h)$. We can do this, moreover, in such a way that $\operatorname{Max}|\eta_x|=2(h-h')^{-1}$. Setting this function into (18) and (19) yields immediately

$$\|v_x\|_{\alpha,h'}\leqslant Cq^{\alpha/\varepsilon}(h-h')^{-1}\|v\|_{\alpha,h} \tag{20}$$

and

$$\|v\|_{\alpha*,h'}\leqslant Cq^{\alpha/\varepsilon}(h-h')^{-1}\|v\|_{\alpha,h}. \tag{21}$$

We may let $l \to \infty$ in inequality (21). Since $v \to \bar{u}^q$, one obtains by virtue of Lebesgue's monotone convergence theorem,

$$\|\bar{u}^q\|_{\alpha^*,h'} \leqslant Cq^{\alpha/e}(h-h')^{-1}\|\bar{u}^q\|_{\alpha,h}. \tag{22}$$

Note that this is valid irrespective of the finiteness of either norm. Finally, (22) may be simplified by setting

$$p = \alpha q = \alpha + \beta - 1, \quad \varkappa = \alpha^*/\alpha = n/(n-\alpha),$$

whence it becomes

$$\|\bar{u}\|_{\varkappa p,h'} \leqslant [C(p/\alpha)^{\alpha/e}(h-h')^{-1}]^{\alpha/p}\|\bar{u}\|_{p,h}. \tag{23}$$

The required conclusion follows by iteration of this inequality. We set, for $\nu = 0, 1, 2, \ldots$,

$$p_\nu = \varkappa^\nu \alpha,$$

and $h_\nu = 1 + 2^{-\nu}$, $h'_\nu = h_{\nu+1}$, whence (23) becomes

$$\|\bar{u}\|_{p_{\nu+1}, h_{\nu+1}} \leqslant C^{1/\varkappa^\nu} K^{\nu/\varkappa^\nu}\|\bar{u}\|_{p_\nu, h_\nu},$$

where $K = 2\varkappa^{\alpha/e}$. Iteration yields

$$\|\bar{u}\|_{p_{\nu+1}, h_{\nu+1}} \leqslant C^{\Sigma 1/\varkappa^\nu} K^{\Sigma \nu/\varkappa^\nu}\|\bar{u}\|_{\alpha,2} \leqslant C\|\bar{u}\|_{\alpha,2},$$

since both series are convergent. Letting $\nu \to \infty$, and observing that $\|\bar{u}\|_{\infty,1} \leqslant \lim \|\bar{u}\|_{p_\nu, h_\nu}$, there results

$$\|\bar{u}\|_{\infty,1} \leqslant C\|\bar{u}\|_{\alpha,2}.$$

Recalling that $\bar{u} = |u| + k$ then yields the conclusion

$$\|u\|_{\infty,1} \leqslant C\{\|u\|_{\alpha,2} + k\}.$$

This proves the first part of the theorem for the case $R = 1$. The second part follows immediately by setting $q = 1$, $h' = 1$, $h = 2$ in (20).

To complete the proof it remains only to show that the general case $R \neq 1$ can be obtained from the special case $R = 1$. This involves only a simple change of linear dimension, however, which the reader may carry out for himself. Theorem 1 is therefore completely proved.

THEOREM 2. *Let u be a weak solution of equation* (5) *defined in some open ball $S(2R) \subset \Omega$. Suppose that $\alpha = n$ and that conditions* (6) *and* (8) *hold. Then*

$$\|u\|_{\infty,R} \leqslant CR^{-1}(\|u\|_{n,2R} + kR)$$

and $$\|u_x\|_{n,R} \leqslant CR^{-1}(\|u\|_{n,2R} + kR),$$

where C and k are constants depending only on the structure of (5). *In particular*

$$C = C(n, \varepsilon;\ a, R^\varepsilon\|b\|,\ R^\varepsilon\|c\|,\ R^\varepsilon\|d\|)$$

and $$k = (R^\varepsilon\|e\| + R^\varepsilon\|f\|)^{1/(n-1)} + (R^\varepsilon\|g\|)^{1/n}.$$

Proof. Although this follows the same pattern as the proof of Theorem 1, certain alterations are necessary because the Sobolev inequalities must be applied in a slightly different way. Up to and including inequality (16) there are, of course, no changes. The various integrals in (16) are then estimated as follows, with $\tilde{\alpha} = n(1 + \varepsilon/2n)^{-1}$ and $\varepsilon' = \varepsilon(n+1)/2n$,

$$\begin{aligned}\int \bar{b}\,|\eta_x v|\cdot(\eta v)^{n-1}\,dx &\leqslant \|b\|_{n/(n-1-\varepsilon)}\,\|\eta_x v\|_{n/(1+\varepsilon')}\,\|\eta v\|_{\tilde{\alpha}^*}^{n-1}\\ &\leqslant C\,\|\eta_x v\|_{n/(1+\varepsilon')}\,\|(\eta v)_x\|_{\tilde{\alpha}}^{n-1}\\ &\leqslant C\,\|\eta_x v\|_n\,\|(\eta v)_x\|_n^{n-1}\\ &\leqslant C\{\|\eta_x v\|_n^n + \|\eta_x v\|_n\,\|\eta v_x\|_n^{n-1}\},\end{aligned}$$

$$\begin{aligned}\int c\eta v\,|\eta v_x|^{n-1}\,dx &= \int c(\eta v)^{\varepsilon/2}(\eta v)^{1-\varepsilon/2}\,|\eta v_x|^{n-1}\,dx\\ &\leqslant \|c\|_{n/(1-\varepsilon)}\,\|\eta v\|_{\tilde{\alpha}}^{\varepsilon/2}\,\|\eta v\|_{\tilde{\alpha}^*}^{1-\varepsilon/2}\,\|\eta v_x\|_{\tilde{\alpha}}^{n-1}\\ &\leqslant C\,\|\eta v\|_n^{\varepsilon/2}\{\|\eta_x v\|_n^{1-\varepsilon/2}\,\|\eta v_x\|_n^{n-1} + \|\eta v_x\|_n^{n-\varepsilon/2}\},\end{aligned}$$

and $$\begin{aligned}\int \bar{d}(\eta v)^n\,dx &= \int \bar{d}(\eta v)^{\varepsilon/2}(\eta v)^{n-\varepsilon/2}\,dx\\ &\leqslant \|\bar{d}\|_{n/(n-\varepsilon)}\,\|\eta v\|_{\tilde{\alpha}}^{\varepsilon/2}\,\|\eta v\|_{\tilde{\alpha}^*}^{n-\varepsilon/2}\\ &\leqslant C\,\|\eta v\|_n^{\varepsilon/2}\{\|\eta_x v\|_n^{n-\varepsilon/2} + \|\eta v_x\|_n^{n-\varepsilon/2}\}.\end{aligned}$$

This leads exactly as in Theorem 1 to the inequality

$$\|\eta v_x\|_n \leqslant Cq^{2n/\varepsilon}(\|\eta v\|_n + \|\eta_x v\|_n).$$

Application of the Sobolev inequality with exponent $\tilde{\alpha}$ gives next, as in Theorem 1,

$$\|v\|_{\tilde{\alpha}^*, h'} \leqslant Cq^{2n/\varepsilon}(h - h')^{-1}\,\|v\|_{n,h}.$$

The rest of the proof is exactly the same as Theorem 1, except that now $\varkappa = \tilde{\alpha}^*/n = 2n/\varepsilon$.

Remark. Both Theorems 1 and 2 remain true for the equation

$$\operatorname{div} \mathcal{A}(x, u, u_x) = \mathcal{B}(x, u, u_x) + \mathsf{C}(x, u),$$

where the functions $\mathcal{A}$ and $\mathcal{B}$ are the same as before, while C is subject only to the simple restriction $u\mathsf{C}(x, u) \geqslant 0$. The only change which this requires in the proof is to note that the integral of the left side of (15) is now equal to $-\int \phi \mathsf{C}\, dx$, and is therefore non-positive. Then (16) holds as before, and the rest of the proof is identical.

2. Uniform estimates. A maximum principle

The estimate of Theorem 1 can be made uniform over the whole domain of definition of the solution, provided that u is continuous in the neighborhood of the boundary. More precisely we have the following result, whose statement will be restricted for simplicity to the case $\alpha < n$.

THEOREM 3. *Let u be a weak solution of* (5) *in a domain $D \subset \Omega$. Suppose $u \leqslant M$ on the boundary of D, in the sense that for every $\varepsilon' > 0$ there exists a neighborhood of the boundary in which $u \leqslant M + \varepsilon'$. Assume also that $\alpha < n$ and that conditions* (6) *and* (7) *hold.*(1) *Then*

$$\max u \leqslant C(|D|^{-1/\alpha} \|\tilde{u}\|_{\alpha, D} + k) + M, \quad (\tilde{u} = \operatorname{Max}(0, u - M)), \tag{24}$$

where C and k depend only on the structure of (5). *In particular*

$$C = C(\alpha, n, \varepsilon;\ |D|^{\varepsilon/n} \|c\|,\ |D|^{\varepsilon/n} \|d\|)$$

and

$$k = (|D|^{\varepsilon/n} \|f\|)^{1/(\alpha-1)} + (|D|^{\varepsilon/n} \|g\|)^{1/\alpha},$$

where the norms of the coefficients c through g are taken in the respective Lebesgue spaces (7).

Remark. The symbol *max* in (24) of course stands for *essential supremum.* This agreement will be followed also in later theorems.

We should also mention that a result similar to Theorem 3 was obtained by Stampacchia [23, 24] and Mazya [11] n the case of linear divergence structure equations.

(1) The inequality $|\mathcal{A}| \leqslant a|p|^{\alpha-1} + b|u|^{\alpha-1} + e$ is in fact unnecessary, except that we must of course require $\mathcal{A}(x, u, u_x) \in L_{\alpha/(\alpha-1)}$.

Proof. We assume first that $|D|=1$, and define for $x \in D$,

$$\bar{u} = \operatorname{Max}(M+\varepsilon', u) + k - M - \varepsilon'.$$

Then obviously we have

$$|\mathcal{B}| \leqslant c|p|^{\alpha-1} + \bar{d}|\bar{u}|^{\alpha-1}, \quad p\cdot\mathcal{A} \geqslant |p|^{\alpha} - \bar{d}|\bar{u}|^{\alpha}, \tag{25}$$

with $\|\bar{d}\|$ bounded as in the proof of Theorem 1. The proof now rests on an appropriate choice of the test function ϕ. We consider two real functions $F(\bar{u})$ and $G(u)$, the function $F(\bar{u})$ being the same as in Theorem 1, while

$$G(u) = F(\bar{u})\, F'(\bar{u})^{\alpha-1} - q^{\alpha-1}k^{\beta}, \qquad -\infty < u < \infty.$$

Evidently G is a piecewise smooth function of u, with corners at $u = M+\varepsilon'$ and $M+\varepsilon'+(l-k)$. Moreover, $G\equiv 0$ for $u \leqslant M+\varepsilon'$ and G is linear for $u > M+\varepsilon'+(l-k)$. Thus it is clear that the function

$$\phi(x) = G(u), \quad (u = u(x)),$$

is admissible in (9).

For $u > M+\varepsilon'$ and different from $M+\varepsilon'+(l-k)$ we have $\phi_x = G'(u)\,u_x$, with

$$G' = \begin{cases} q^{-1}\beta(F')^{\alpha} & \text{if } \; u < M+\varepsilon'+(l-k), \\ (F')^{\alpha} & \text{if } \; u > M+\varepsilon'+(l-k). \end{cases}$$

Hence, setting $v = v(x) = F(\bar{u})$, we obtain as in the proof of Theorem 1

$$\phi_x\cdot\mathcal{A} + \phi\mathcal{B} \geqslant |v_x|^{\alpha} - cv|v_x|^{\alpha-1} - (1+\beta)\,q^{\alpha-1}\bar{d}v^{\alpha}, \tag{26}$$

valid in the set indicated above. Outside this range we have $\phi_x = 0$ and $\bar{u}_x = 0$ almost everywhere, so that (26) in fact holds almost everywhere in D. Therefore, integrating over D and using (9), we have

$$\|v_x\|_{\alpha}^{\alpha} \leqslant \int cv|v_x|^{\alpha-1}\,dx + (1+\beta)\,q^{\alpha-1}\int \bar{d}v^{\alpha}\,dx. \tag{27}$$

These terms on the right side can be estimated by using Hölder's inequality together with Lemma 4.(1) Thus

(1) The function v has not compact support, of course, but $v-k^q$ does. Thus from Lemma 4

$$\|v-k^q\|_{\alpha^*} \leqslant \text{Const.}\,\|(v-k^q)_x\|_{\alpha},$$

and hence $$\|v\|_{\alpha^*} \leqslant \text{Const.}\,\|v_x\|_{\alpha} + k^q|D|^{1/\alpha^*} \leqslant \text{Const.}\,\|v_x\|_{\alpha} + |D|^{-1/n}\,\|v\|_{\alpha}.$$

$$\begin{aligned}\int cv\,|v_x|^{\alpha-1}\,dx &= \int cv^{\varepsilon}v^{1-\varepsilon}|v_x|^{\alpha-1}\,dx\\ &\leqslant \|c\|_{n/(1-\varepsilon)}\|v\|_{\alpha}^{\varepsilon}\|v\|_{\alpha*}^{1-\varepsilon}\|v_x\|_{\alpha}^{\alpha-1}\\ &\leqslant C\{\|v\|_{\alpha}^{\varepsilon}\|v_x\|_{\alpha}^{\alpha-\varepsilon}+\|v\|_{\alpha}\cdot\|v_x\|_{\alpha}^{\alpha-1}\},\end{aligned}$$

$$\begin{aligned}\int \bar{d}v^{\alpha}\,dx &= \int \bar{d}v^{\varepsilon}v^{\alpha-\varepsilon}\,dx\\ &\leqslant \|\bar{d}\|_{n/(\alpha-\varepsilon)}\|v\|_{\alpha}^{\varepsilon}\|v\|_{\alpha*}^{\alpha-\varepsilon}\\ &\leqslant C\{\|v\|_{\alpha}^{\varepsilon}\|v_x\|_{\alpha}^{\alpha-\varepsilon}+\|v\|_{\alpha}^{\alpha}\}.\end{aligned}$$

Inserting these estimates into (27), and setting $z=\|v_x\|_{\alpha}/\|v\|_{\alpha}$, yields

$$z^{\alpha}\leqslant C[z^{\alpha-\varepsilon}+z^{\alpha-1}+q^{\alpha}(z^{\alpha-\varepsilon}+1)],$$

whence by Lemma 2, $$z\leqslant Cq^{\alpha/\varepsilon},$$

that is $$\|v_x\|_{\alpha}\leqslant Cq^{\alpha/\varepsilon}\|v\|_{\alpha}. \tag{28}$$

Applying Lemma 4 once more (see the preceding footnote) gives finally

$$\|v\|_{\alpha*}\leqslant Cq^{\alpha/\varepsilon}\|v\|_{\alpha}. \tag{29}$$

We may now carry out the iteration process of Theorem 1, with the important exception that there is no necessity to reduce the radius at each step. The final result is clearly

$$\|\bar{u}\|_{\infty}\leqslant C\,\|\bar{u}\|_{\alpha}. \tag{30}$$

Recalling the definition of $\bar{u}$ this is easily seen to imply

$$\max u\leqslant M+\varepsilon'+C(\|\bar{u}\|_{\alpha}+k).$$

To obtain the required conclusion we let $\varepsilon'\to 0$, and then normalize back to the given value of $|D|$.

THEOREM 4. *Let u satisfy the hypotheses of Theorem 3. Then there exists a constant D_0, depending only on the structure of equation* (5), *such that if $|D|\leqslant D_0$ then* $\max u\leqslant M+Ck$.

Proof. It is convenient to make use of the calculations already carried out during the proof of Theorem 3. Retaining the assumption that $|D|=1$, we obtain respectively from Hölder's inequality, Lemma 4 (footnote), and inequality (28) written for $q=1$,

$$\|\bar{u}\|_\alpha \leqslant \|\bar{u}\|_{\alpha*} \leqslant \text{Const.}\,\|\bar{u}_x\|_\alpha + k \leqslant C\|\bar{u}\|_\alpha + k. \tag{31}$$

Now suppose that in the original definition of $\bar{u}$ the constant k is replaced by $\theta^{-1}k$ where θ is a (small) positive number to be determined later. Then in place of (31) we obtain

$$\|\bar{u}\|_\alpha \leqslant C\,\|\bar{u}\|_\alpha + \theta^{-1}k. \tag{31'}$$

It will be important to assess the dependence of C on the parameters $\|c\|$, $\|d\|$, and θ. Assuming that $\|c\|$, $\|d\|$, and θ are all $\leqslant 1$, say, one easily finds that

$$C = \text{Const.}\ (\|c\| + \|d\|^{1/\alpha} + \theta^{(\alpha-1)/\alpha}),$$

where the Const. depends only on α, n, and ε. Clearly if $\theta\,(\leqslant 1)$ is chosen suitably small, and if $\|c\| + \|d\| \leqslant \theta$, then we may take $C \leqslant \frac{1}{2}$. Consequently we have from (31')

$$\|\bar{u}\|_\alpha \leqslant 2\,\theta^{-1}k.$$

Substituting this into (30) yields finally $\|\bar{u}\|_\infty \leqslant Ck$, that is to say, $\max u \leqslant M + Ck$. The proof is completed by observing that the values of $\|c\|$, $\|d\|$ which actually arise through normalization to unit volume are $|D|^{\varepsilon/n}\|c\|$ and $|D|^{\varepsilon/n}\|d\|$, whence we may take $D_0 = \theta^{n/\varepsilon}(\|c\| + \|d\|)^{-n/\varepsilon}$.

Remarks. If $c = d = 0$ in (6) then $D_0 = \infty$ and the maximum principle holds for arbitrary bounded regions D.

Finally, we observe that these results also hold if the hypothesis $u \leqslant M$ on the boundary is replaced by the assumption that u is the strong limit in $W^1_\alpha(D)$ of continuous functions which are $\leqslant M$ on the boundary of D. We leave the details to the reader.

3. The Harnack inequality

The following theorem of Harnack type will be used to estimate the Hölder continuity of solutions of (5), and will also be of importance in our later discussion of the behavior of a solution near an isolated singularity. The main idea of the proof is the same as in Moser's celebrated paper, though the details are considerably more delicate and involved.

THEOREM 5. *Let u be a non-negative weak solution of equation* (5) *in some open ball $S(3R) \subset \Omega$. Assume that $\alpha < n$ and that conditions* (6) *and* (7) *hold. Then*

$$\max u \leqslant C(\min u + k) \quad \text{in } S(R),$$

where C and k are constants depending only on the structure of (5), *as in Theorem 1.*

Proof. It is enough to prove the result for the case $R=1$, with the solution correspondingly defined in an open ball $S(3)$. By Theorem 1 the solution is bounded on any compact subset of $S(3)$. Thus if η is a non-negative smooth function with compact support in $S(3)$, then

$$\phi(x)=\eta^{\alpha}\bar{u}^{\beta}, \quad \bar{u}=u+k+\varepsilon'$$

is admissible in (9) for any real value of β and any $\varepsilon'>0$. It is our purpose to insert this function into (9), and in this way to derive an estimate similar to (18). The resulting calculations are slightly different in the ranges

$$\beta<1-\alpha, \quad 1-\alpha<\beta<0, \quad \beta>0,$$

while the values $\beta=1-\alpha$ and $\beta=0$ are singular. We take up these various cases in order.

I. $\beta>0$. Here the calculation is essentially the same as that of Theorem 1, with $v=\bar{u}^q$ and q and β related by $\alpha q=\alpha+\beta-1$. In carrying out the details it is helpful to note that $\bar{u}^q$ and $\bar{u}^{\beta}$ here correspond to $F(\bar{u})$ and Const. $G(u)$ in Theorem 1. The main difference is that there is no longer a cutoff value l, so that we must always use the formulas of Theorem 1 corresponding to the range $|u|<l-k$. This means, in particular, that the factor $q^{-1}\beta$ should be inserted in front of the first term on the right sides of (14) and (15), and in front of the term on the left side of (16) and (17). Thus we have finally

$$\beta z^{\alpha}\leqslant C[qz^{\alpha-1}+q^{\alpha}(1+z^{\alpha-1})+q\zeta^{\varepsilon}(z^{\alpha-1}+z^{\alpha-\varepsilon})+(1+\beta)\,q^{\alpha}\zeta^{\varepsilon}(1+z^{\alpha-\varepsilon})], \tag{32}$$

the notation being the same as in Theorem 1. Applying Lemma 2 then leads to

$$z\leqslant Cq^{\alpha/\varepsilon}(1+\beta^{-1})^{1/\varepsilon}(1+\zeta),$$

where we have used the fact that $q>(\alpha-1)/\alpha$. Thus

$$\|\eta v_x\|_{\alpha}\leqslant Cq^{\alpha/\varepsilon}(1+\beta^{-1})^{1/\varepsilon}(\|\eta v\|_{\alpha}+\|\eta_x v\|_{\alpha}). \tag{33}$$

Choosing η as in Theorem 1 there results, in conclusion,

$$\|v\|_{\alpha^*,h'}\leqslant Cq^{\alpha/\varepsilon}(h-h')^{-1}(1+\beta^{-1})^{1/\varepsilon}\|v\|_{\alpha,h}, \quad (v=\bar{u}^q). \tag{34}$$

II. $1-\alpha<\beta<0$. This goes as Case I, except that now $q^{-1}\beta$ is negative and inequalities (14) and (15) must run the opposite way. Otherwise there are no essential changes, and we obtain finally

$$\|v\|_{\alpha^*,h'} \leqslant C(h-h')^{-1}(1-\beta^{-1})^{1/\varepsilon}\|v\|_{\alpha,h}, \quad (v=\bar{u}^q), \tag{35}$$

(the term $q^{\alpha/\varepsilon}$ can be omitted since $0<q<1$).

III. $\beta<1-\alpha$. This goes exactly as Case I, except that q is negative and absolute values are necessary. The result is

$$\|v\|_{\alpha^*,h'} \leqslant C(h-h')^{-1}(1+|q|)^{\alpha/\varepsilon}\|v\|_{\alpha,h}, \quad (v=\bar{u}^q)$$

(here one uses the fact that $|\beta|>\alpha-1$). Since the case $\beta=0$ is trivial there remains only case

IV. $\beta=1-\alpha$. Substituting the functions

$$\phi=\eta^\alpha\bar{u}^{1-\alpha}, \quad \phi_x=\alpha(\eta/\bar{u})^{\alpha-1}\eta_x+(1-\alpha)(\eta/\bar{u})^\alpha u_x$$

directly into relation (9), we obtain with the help of (13)

$$\begin{aligned}\phi_x\cdot\mathcal{A}+\phi\mathcal{B} &\leqslant (1-\alpha)(\eta/\bar{u})^\alpha\{|u_x|^\alpha-\bar{d}|\bar{u}|^\alpha\}\\ &\quad+\alpha|\eta_x|\cdot(\eta/\bar{u})^{\alpha-1}\{a|u_x|^{\alpha-1}+\bar{b}|\bar{u}|^{\alpha-1}\}+\eta^\alpha\bar{u}^{1-\alpha}\{c|u_x|^{\alpha-1}+\bar{d}|\bar{u}|^{\alpha-1}\}\\ &=(1-\alpha)|\eta v_x|^\alpha+\alpha a|\eta_x|\cdot|\eta v_x|^{\alpha-1}+\alpha\bar{b}\eta^{\alpha-1}|\eta_x|+c\eta|\eta v_x|^{\alpha-1}+\alpha\bar{d}\eta^\alpha,\end{aligned}$$

where $v=v(x)=\log\bar{u}$. Hence integrating over $S(3)$ there results

$$(\alpha-1)\|\eta v_x\|_\alpha^\alpha \leqslant \alpha a\int|\eta_x|\cdot|\eta v_x|^{\alpha-1}dx+\alpha\int\bar{b}\eta^{\alpha-1}|\eta_x|\,dx+\int c\eta|\eta v_x|^{\alpha-1}dx+\alpha\int\bar{d}\eta^\alpha dx. \tag{36}$$

Before estimating the integrals on the right-hand side, it is convenient to specialize the function η. Let S be an arbitrary open ball of radius h contained in $S(2)$. We choose η so that $\eta=1$ in S and $0\leqslant\eta\leqslant1$ in $S(3)-S$. This may be done, moreover, in such a way that the support of η is contained in a concentric sphere about S of radius $(3/2)h$, and so that $\operatorname{Max}|\eta_x|=3/h$. Then by Hölder's inequality we have the estimates

$$\int|\eta_x|\cdot|\eta v_x|^{\alpha-1}dx \leqslant Ch^{(n-\alpha)/\alpha}\|\eta v_x\|_\alpha^{\alpha-1},$$

$$\int\bar{b}\eta^{\alpha-1}|\eta_x|\,dx \leqslant Ch^{n-\alpha},$$

$$\int c\eta|\eta v_x|^{\alpha-1}dx \leqslant Ch^{(n-\alpha)/\alpha}\|\eta v_x\|_\alpha^{\alpha-1},$$

$$\int\bar{d}\eta^\alpha dx \leqslant Ch^{n-\alpha},$$

where conditions (7) have been used at each stage. Inserting the preceding inequalities into (36) there follows

$$\|\eta v_x\|_\alpha^\alpha \leqslant C[h^{(n-\alpha)/\alpha}\|\eta v_x\|_\alpha^{\alpha-1} + h^{n-\alpha}].$$

Thus by Lemma 2 we have $\|v_x\|_{\alpha,S} \leqslant Ch^{(n-\alpha)/\alpha}$, since $\eta \equiv 1$ in S. Finally by Lemma 6 and a simple use of Hölder's inequality we obtain

$$\|v - v_S\|_1 \leqslant Ch\|v_x\|_1 \leqslant Ch^{1+n(\alpha-1)/\alpha}\|v_x\|_\alpha \leqslant Ch^n,$$

that is

$$\int_S |v - v_S|\, dx \leqslant C|S|, \quad (v = \log \bar{u}). \tag{37}$$

This estimate, valid for any ball S contained in $S(2)$, completes the discussion of case IV.

This being accomplished, the required conclusion will now be obtained by an iteration process somewhat similar to that of Theorem 1. For any real number $p \neq 0$ we define

$$\Phi(p, h) = \left(\int_{S(h)} |\bar{u}|^p\, dx\right)^{1/p},$$

(thus for $p \geqslant 1$ we have $\Phi(p, h) = \|\bar{u}\|_{p,h}$). Now from (37) and Lemma 7 it follows that

$$\int_{S(2)} e^{p_0 v}\, dx \cdot \int_{S(2)} e^{-p_0 v}\, dx \leqslant 2^{2n}\mu^2,$$

where $p_0 = \lambda/C$; the constant C is here the same as in (37), while λ and μ depend only on the dimension. Since $v = \log \bar{u}$ this inequality may be rewritten simply

$$\Phi(p_0, 2) \leqslant C\Phi(-p_0, 2). \tag{38}$$

Next, putting $p = \alpha q = \alpha + \beta - 1$ in (34) and (35), taking the qth root of each side, and combining the results in a single inequality, we obtain

$$\Phi(\varkappa p, h') \leqslant [C(h - h')^{-1}(1 + |\beta|^{-1})^{1/\varepsilon}(1 + p)^{\alpha/\varepsilon}]^{\alpha/p}\, \Phi(p, h), \tag{39}$$

where $\varkappa = \alpha^*/\alpha$ and p is any positive number other than $\alpha - 1$. We wish to iterate this inequality, beginning with $\Phi(p_0, 2)$ and setting generally

$$p_\nu = \varkappa^\nu p_0, \quad \nu = 0, 1, 2, \ldots,$$

and $h_\nu = 1 + 2^{-\nu}$, $h'_\nu = h_{\nu+1}$. In order that (39) be applicable at each stage, the sucessive iterates p_ν must avoid the point $p = \alpha - 1$. To accomplish this in a definite way we shall in fact choose a new initial value $p_0' \leqslant p_0$ so that the point $p = \alpha - 1$ lies midway between some two consecutive iterates of p_0'. The value p_0' being thus fixed, we observe that at all stages of the iteration process one has

$$|\beta| = |p - (\alpha - 1)| \geqslant \frac{\alpha(\alpha - 1)}{2n - \alpha}.$$

The term $(1 + |\beta|^{-1})^{1/\varepsilon}$ in (39) can thus be absorbed into the general constant C. Application of the iteration process just described then leads, as in the proof of Theorem 1, to the inequality

$$\max_{S(1)} \bar{u} \leqslant C\Phi(p_0', 2). \tag{40}$$

It remains to examine the inequality of Case III. Again setting $p = \alpha q$ and taking qth roots, this inequality becomes

$$\Phi(\varkappa p, h') \geqslant [C(h - h')(1 + |p|)^{\alpha/\varepsilon}]^{\alpha/p}\,\Phi(p, h),$$

where the sign is reversed since p and q are now negative. Iterating as before, with $p_\nu = -\varkappa^\nu p_0$ and $h_\nu = 1 + 2^{-\nu}$, $h_\nu' = h_{\nu+1}$, there results without difficulty

$$\Phi(p_{\nu+1}, h_{\nu+1}) \geqslant C^{-1}\Phi(-p_0. 2).$$

Letting $\nu \to \infty$ then establishes the inequality

$$\min_{S(1)} \bar{u} \geqslant C^{-1}\Phi(-p_0, 2). \tag{41}$$

The proof is now easily completed. From (40), (38), and (41), and a simple application of Hölder's inequality, (¹) we have

$$\max_{S(1)} \bar{u} \leqslant C\Phi(p_0', 2) \leqslant C\Phi(p_0, 2) \leqslant C\Phi(-p_0, 2) \leqslant C \min_{S(1)} \bar{u}.$$

Since $\bar{u} = u + k + \varepsilon'$ this implies in turn $\max u \leqslant C(\min u + k + \varepsilon')$. Letting $\varepsilon' \to 0$ concludes the demonstration.

THEOREM 6. *Let u be a non-negative weak solution of (5) in some open ball $S(3R) \subset \Omega$. Assume that $\alpha = n$ and that conditions (6) and (8) hold. Then*

$$\max u \leqslant C(\min u + k) \quad in \quad S(R),$$

where C and k are constants depending only on the structure of (5), as in Theorem 2.

(¹) To obtain $\Phi(p_0', 2) \leqslant C\Phi(p_0, 2)$. In particular we find

$$\Phi(p_0', 2) \leqslant |S(2)|^{1/p_0' - 1/p_0}\,\Phi(p_0, 2),$$

where the exponent can be assumed less than $(\varkappa - 1)/p_0$ since surely one can choose $p_0' > \varkappa^{-1} p_0$.

The proof may be omitted, since it is simply a modification of the preceding argument, in exactly the same way as Theorem 2 was a modification of Theorem 1.

A Harnack inequality for the case $\alpha > n$ will be proved in Section 5.

4. Applications. Hölder continuity of solutions

The Harnack inequalities of the foregoing section can be extended without difficulty so as to apply to arbitrary regions D. In particular, the following result holds.

THEOREM 7. *If u is a non-negative weak solution of (5) in a domain D, and if D' is any compact subset of D, then*

$$\max u \leqslant C'(\min u + k') \quad in \quad D',$$

where C' depends only on the structure of equation (5) and on the domains D and D', and $k' = (\|e\| + \|f\|)^{1/(\alpha-1)} + \|g\|^{1/\alpha}$.

The proof is standard, depending only on covering D' with spheres and on a simple chaining argument.

We turn now to a fundamental estimate of the Hölder continuity of solutions of (5). It is enough to consider the case $\alpha \leqslant n$, for when $\alpha > n$ the result is immediate from Morrey's lemma. Also, if $\alpha < n$, it will be supposed that e is in a more restrictive Lebesgue class than originally supposed, namely

$$e \in L_{n/(\alpha-1-\varepsilon)},$$

(ε should now be less than both 1 and $\alpha - 1$). In this case the parameter k in Theorems 5 and 6 has the form

$$k = (R^{\varepsilon}\|e\| + R^{\varepsilon}\|f\|)^{1/(\alpha-1)} + (R^{\varepsilon}\|g\|)^{1/\alpha}.$$

We may now prove

THEOREM 8. *Let u be a weak solution of (5) in a domain $D \subset \Omega$, Then u is (essentially) Hölder continuous in D. Moreover, if $|u| \leqslant L$ in D and if D' is any compact subset of D, then*

$$|u(x) - u(y)| \leqslant H(k' + L)\,|x - y|^{\lambda}, \quad x, y \in D',$$

where $\lambda = \lambda(\alpha, n, \varepsilon; a, \|b\|, \|c\|, \|d\|)$, and H depends only on the structure of (5) and on the geometry of D and D'.

Proof. It will first be shown that u is Hölder continuous at any fixed point P in D. The general result will then follow easily.

Thus let P be a point of D, and let R be chosen so that the ball $S(R)$ has compact closure in D. By a suitable change of variables we may suppose $R=1$. The expressions

$$M(r)=\max_{S(r)} u, \quad \mu(r)=\min_{S(r)} u$$

are then well defined for $0<r\leqslant 1$, and it follows that both functions

$$\bar{u}=M-u \quad \text{and} \quad \bar{\bar{u}}=u-\mu$$

are non-negative in $S(r)$. Obviously $\bar{u}$ satisfies the differential equation

$$\operatorname{div} \bar{\mathcal{A}}(x,\bar{u},\bar{u}_x)=\bar{\mathcal{B}}(x,\bar{u},\bar{u}_x)$$

with $\bar{\mathcal{A}}(x,\bar{u},\bar{p})=\mathcal{A}(x,M-\bar{u},-\bar{p})$ and a similar formula for $\bar{\mathcal{B}}$. These quantities clearly obey the inequalities

$$|\bar{\mathcal{A}}|\leqslant a|\bar{p}|^{\alpha-1}+\bar{b}|\bar{u}|^{\alpha-1}+\bar{e}, \quad \text{etc.,}$$

where $\bar{b}=2^\alpha b$, $\bar{e}=e+2^\alpha bL^{\alpha-1}$, etc. Thus we may apply Theorem 5 (or Theorem 6) to $\bar{u}$ in the open ball $S(r)$, with the result

$$M-\mu'=\max_{S(r/3)}\bar{u}\leqslant C(\min_{S(r/3)}\bar{u}+\bar{k})=C(M-M'+\bar{k}), \quad 0<r\leqslant 1, \tag{42}$$

where $M'=M'(r)=M(r/3)$, $\mu'=\mu'(r)=\mu(r/3)$. In this inequality the constant C depends on the variables listed in the theorem, while

$$\bar{k}=(r^\varepsilon\|\bar{e}\|+r^\varepsilon\|\bar{f}\|)^{1/(\alpha-1)}+(r^\varepsilon\|\bar{g}\|)^{1/\alpha}.$$

In the same way, we have

$$M'-\mu=\max_{S(r/3)}\bar{\bar{u}}\leqslant C(\min_{S(r/3)}\bar{\bar{u}}+\bar{k})=C(\mu'-\mu+\bar{k}). \tag{43}$$

Adding (42) and (43) and transposing terms then gives

$$M'-\mu'\leqslant\frac{C-1}{C+1}(M-\mu)+\frac{2C\bar{k}}{C+1}. \tag{44}$$

Now $\bar{k}\leqslant k_0 r^{\varepsilon/\alpha}$ where $k_0=(\|\bar{e}\|+\|\bar{f}\|)^{1/\alpha-1}+(\|\bar{g}\|)^{1/\alpha}$. Thus setting

$$\vartheta=\frac{C-1}{C+1}, \quad \tau=\frac{2Ck_0}{C+1},$$

and denoting the oscillation $M-\mu$ of u in $S(r)$ by $\omega(r)$, we may write (44) in the form

$$\omega(r/3)\leqslant\vartheta\{\omega(r)+\tau r^{\varepsilon/\alpha}\},\quad 0<r\leqslant 1.$$

Since $\omega(r)$ is an increasing function, it is clear that for any number $s\geqslant 3$ one has also

$$\omega(r/s)\leqslant\vartheta\{\omega(r)+\tau r^{\varepsilon/\alpha}\},\quad 0<r\leqslant 1.$$

Iteration of this relation from $r=1$ to successively smaller radii yields

$$\omega(s^{-\nu})\leqslant\vartheta^{\nu}\{\omega(1)+\tau[1+(\vartheta s^{\varepsilon/\alpha})^{-1}+\ldots+(\vartheta s^{\varepsilon/\alpha})^{-\nu+1}]\},\tag{45}$$

valid for $\nu=1,2,3,\ldots$. Now choose s according to the relation

$$\vartheta s^{\varepsilon/\alpha}=2,\quad (s\geqslant 4).$$

whence from (45) follows

$$\omega(s^{-\nu})\leqslant\vartheta^{\nu}\{\omega(1)+2\tau\}.\tag{46}$$

For any fixed ϱ, $0<\varrho\leqslant s^{-1}$, let ν be chosen such that $s^{-\nu-1}<\varrho\leqslant s^{-\nu}$. Then by virtue of (46),

$$\omega(\varrho)\leqslant\omega(s^{-\nu})\leqslant\vartheta^{\nu}\{\omega(1)+2\tau\}.\tag{47}$$

Now one easily checks that $\omega(1)+2\tau\leqslant C(L+k)$. In addition, if γ is defined by the relation $2^{-\gamma}=\vartheta$, then we have $\vartheta=s^{-\lambda}$ where $\lambda=\varepsilon\gamma/\alpha(\gamma+1)>0$. Therefore (47) implies

$$\omega(\varrho)\leqslant C(L+k)\,\varrho^{\lambda},\quad \varrho\leqslant 2^{-(\gamma+1)\alpha/\varepsilon},$$

or when $R\neq 1$,

$$\omega(\varrho)\leqslant C(L+k)\,(\varrho/R)^{\lambda},\quad \varrho\leqslant 2^{-(\gamma+1)\alpha/\varepsilon}R.\tag{48}$$

This proves that u is essentially Hölder continuous at P.

It is a trivial, though not entirely simple, task to show that we can redefine u on a set of measure zero so that the resulting function is Hölder continuous at *every* point of D. Leaving aside this demonstration, the first part of the theorem is therefore proved. The second part follows easily from the estimate (48), using a simple chaining argument (choose $R=\operatorname{Min}(1,\xi)$, where ξ is the distance from D' to the boundary of D).

5. The Harnack inequality for $\alpha>n$

In the foregoing sections we have considered the majorization and continuity of solutions when $\alpha\leqslant n$. The situation for $\alpha>n$ is somewhat different, since by Morrey's lemma solutions are necessarily bounded and Hölder continuous on compact subsets of their domain of definition. For this reason we may confine our discussion solely to the Harnack inequality.

We retain the basic structure (6), and assume further that

$$b, e \in L_{\alpha/(\alpha-1)}; \quad c \in L_\alpha; \quad d, f, g \in L_1. \tag{49}$$

Then the following result holds.

THEOREM 9. *Let u be a non-negative weak solution of* (5) *in some open ball $S(2R) \subset \Omega$. Assume that $\alpha > n$ and that conditions* (6) *and* (49) *hold. Then for any two points x, y in $S(R)$ we have*

$$u(x) \leqslant \{u(y) + k\} \exp\{C(|x-y|/R)^{1-n/\alpha}\}, \tag{50}$$

where $$C = C(\alpha, n; a, R^{(\alpha-n)/\alpha'}\|b\|, R^{(\alpha-n)/\alpha}\|c\|, R^{\alpha-n}\|d\|)$$

and $$k = (R^{(\alpha-n)/\alpha'}\|e\| + R^{\alpha-n}\|f\|)^{1/(\alpha-1)} + (R^{\alpha-n}\|g\|)^{1/\alpha},$$

the norms of the coefficients being taken in the respective Lebesgue spaces (49).

Proof. Suppose first that $R = 1$. Let η be a smooth function with compact support in $S(2)$, such that $\eta = 1$ in $S(1)$, and $0 \leqslant \eta \leqslant 1$ in $S(2) - S(1)$. Now set

$$\phi(x) = \eta^\alpha \bar{u}^{1-\alpha}$$

in (9), where $\bar{u} = u + k + \varepsilon'$. Then, as in Case IV of the proof of Theorem 5, we find

$$(\alpha - 1)\|\eta v_x\|_\alpha^\alpha \leqslant \alpha \int a|\eta_x| \cdot |\eta v_x|^{\alpha-1} dx + \alpha \int \bar{b}\eta^{\alpha-1}|\eta_x| dx + \int c\eta|\eta v_x|^{\alpha-1} dx + \alpha \int \bar{d}\eta^\alpha dx.$$

Here $v = \log \bar{u}$, and $\bar{b}$ and $\bar{d}$ are defined as in Theorem 1. We may assume that $\operatorname{Max}|\eta_x| = 2$, hence by Hölder's inequality

$$\|\eta v_x\|_\alpha^\alpha \leqslant C[1 + \|\eta v_x\|_\alpha^{\alpha-1}].$$

Application of Lemma 2 then yields (since $\eta \equiv 1$ in $S(1)$)

$$\|v_x\|_\alpha \leqslant C, \tag{51}$$

the integral being taken over the unit ball $S(1)$.

Since $\|v_x\|_\alpha$ is finite we conclude from Morrey's lemma that v is Hölder continuous in $S(1)$,

$$|v(x) - v(y)| \leqslant C|x-y|^{1-n/\alpha}.$$

But $v=\log \bar{u}$, so that this implies

$$\bar{u}(x) \leqslant \bar{u}(y)\, e^{C|x-y|^{1-n/\alpha}}, \quad (x, y \in S(1)).$$

Recalling the definition of $\bar{u}$, the required inequality now follows immediately. The case $R \neq 1$ offers no additional difficulty, and the theorem is proved.

As an immediate consequence of Theorem 9 we have the result that

$$\max u \leqslant C(\min u + k);$$

this is, of course, considerably weaker than the actual inequality (50).

6. Generalizations

We begin by noting that certain equations which nominally do not fall into the categories above can in fact be considered as special cases. In particular, in case $\alpha < n$ let us suppose that (6) is replaced by

$$|\mathcal{A}| \leqslant a(|p|^{\alpha-1} + |u|^{(\alpha-1)/\varsigma} + e),$$
$$|\mathcal{B}| \leqslant a(|p|^{\alpha-\varsigma} + |u|^{(\alpha-\varsigma)/\varsigma} + f),$$
$$p \cdot \mathcal{A} \geqslant |p|^{\alpha} - a(|u|^{\alpha/\varsigma} + g),$$

where $\varsigma = (n - \alpha + \alpha\varepsilon)/n$. If we now consider a *fixed solution* u, we can set

$$b(x) = a|u|^{(\alpha-1)(1-\varsigma)/\varsigma}, \quad c(x) = a|u_x|^{1-\varsigma}, \quad d(x) = a|u|^{\alpha(1-\varsigma)/\varsigma}$$

and it is easily verified (assuming that $u_x \in L_\alpha$, $u \in L_{\alpha*}$) that conditions (7) are in fact satisfied. Thus the conclusions of Theorems 1, 3, 4, 7, and 8 remain valid, with the exception that the coefficient C now must depend on the W^1_α norm of the solution. A similar result holds when $\alpha \geqslant n$, but it is not necessary to carry this out in detail.

A slightly different situation arises if we know *a priori* that u is *bounded*. Then under genuinely weaker conditions it is possible to carry through the arguments leading to the Harnack inequality and the Hölder continuity of solutions. Since a result of this sort will not be required in the later sections of the paper, we shall not pause to consider the situation in any detail. It may be remarked, however, that the arguments of references [8], [9] are in general based on such an *a priori* estimate for the magnitude of u, which partially explains the difference between their hypotheses and ours.

In conclusion, there has been some interest in the situation when the "inhomogeneous" terms e, f, g do not lie in the respective Lebesgue spaces (7). For this case we have the following results, analogous to Theorems 1, 3, and 4.

THEOREM 1'. *Let the hypotheses of Theorem 1 be satisfied, with the exception that* $e \in L_{r*}$, $f \in L_r$, $g \in L_t$, *where*

$$\frac{1}{\alpha-1}\left(\frac{1}{r}-\frac{\alpha}{n}\right)=\frac{1}{\alpha}\left(\frac{1}{t}-\frac{\alpha}{n}\right)=\frac{1}{\sigma}$$

and σ *is some number in the range* $\alpha^* \leqslant \sigma < \infty$. *Then*

$$\|u\|_{\sigma,R} \leqslant CR^{n/\sigma - n/\alpha}(\|u\|_{\alpha,2R} + kR^{n/\alpha}),$$

where C *and* k *depend only on the structure of equation* (5).

Proof. This proceeds in the same way as Theorem 1, except that we take the initial substitution in the form $\bar{u}=|u|$. It is clear that certain terms involving e, f, g must then be carried along in relations (14) through (23). In particular, to the right side of (16) one must add the expression

$$\alpha q^{\alpha-1}\int e\,|\eta_x v|\cdot(\eta v)^{\beta/q-1}\,dx + q^{\alpha-1}\int f(\eta v)^{\beta/q}\,dx + \beta q^{\alpha-1}\int g(\eta v)^{\alpha(q-1)/q}\,dx.$$

These integrals can be estimated only if $q \leqslant \sigma/\alpha^*$. For example, with this restriction on q one has

$$\begin{aligned}\int e\,|\eta_x v|\cdot(\eta v)^{\beta/q-1}\,dx &\leqslant \|e\|_{r*}\,\|\eta_x v\|_\alpha\,\|\eta v\|_{\alpha*}^{\beta/q-1}\\ &\leqslant C(\|e\|_{r*}^{\alpha q/(\alpha-1)} + \|\eta_x v\|_\alpha^\alpha) + \text{Const.}\,\|\eta v_x\|_\alpha^\alpha,\end{aligned}$$

where the constant in front of $\|\eta v_x\|_\alpha^\alpha$ can be taken as small as we like. The remaining two integrals likewise have similar estimates. Thus, by following the steps of Theorem 1, one obtains in place of (23)

$$\|u\|_{\varkappa p,h'} \leqslant C(h-h')^{-\alpha/p}\{\|u\|_{p,h} + \|e\|_{r*}^{1/(\alpha-1)} + \|f\|_r^{1/(\alpha-1)} + \|g\|_t^{1/\alpha}\},$$

valid for $p=\alpha q \leqslant \sigma/\varkappa$. The required conclusion now follows by a *finite* iteration of this inequality.

THEOREM 3'. *Let the hypotheses of Theorem 3 hold, with the exception that* e, f, g *are assumed to be in the preceding Lebesgue classes. Then*

$$\|\tilde{u}\|_{\sigma,D} \leqslant C|D|^{1/\sigma-1/\alpha}(\|\tilde{u}\|_{\alpha,D} + |D|^{1/\alpha}k),$$

where $\tilde{u}=\text{Max}(0,u-M)$ *and* C *and* k *are constants depending only on the structure of* (5).

This is proved by the argument of Theorem 3, beginning with the substitution $\bar{u} = \mathrm{Max}\,(0, u - M - \varepsilon')$. The details can be omitted, as also in the case of

THEOREM 4'. *Let u satisfy the hypotheses of Theorem 3'. Then there exists a con-constant D_0, depending only on the structure of equation* (5), *such that if* $|D| \leqslant D_0$ *then*

$$\|\tilde{u}\|_{\sigma, D} \leqslant C |D|^{1/\sigma} k,$$

where $\tilde{u} = \mathrm{Max}\,(0, u - M)$.

There are also similar results for the case $\sigma < \alpha^*$, though the proofs are quite a bit more delicate. Since this case is somewhat artificial (recall that u is always locally in class L_{α^*}), the discussion may be omitted.

II. Removable singularities

The methods of Chapter I will here be used to prove two basic theorems concerning the removable singularities of solutions of equation (5). We shall retain the structure outlined in the opening discussion of Chapter I, with the exception that e will be assumed in $L_{n/(\alpha-1-\varepsilon)}$ rather than $L_{n/(\alpha-1)}$.

We may further confine the discussion to the case $\alpha \leqslant n$, for the alternate case $\alpha > n$ can admit no removable singularity theorems of the ordinary sort. Indeed, consider the equation

$$\mathrm{div}\,(u_x |u_x|^{\alpha-2}) = 0$$

with $\alpha > n$. It is easily checked that $u = r^{(\alpha-n)/(\alpha-1)}$ is a bounded solution in the domain $0 < r < 1$. On the other hand, the origin is definitely a non-removable singularity for this solution, since if ϕ is a continuously differentiable function with compact support in the unit open ball, then

$$\int \phi_x \cdot u_x |u_x|^{\alpha-2} dx = -\left(\frac{\alpha - n}{\alpha - 1}\right)^{\alpha-1} \omega_n \phi(0),$$

and this need not vanish. That is, *when $\alpha > n$ there can exist bounded solutions with non-removable isolated singularities*. The above example indicates in addition the type of behavior one can expect at an isolated singularity when $\alpha < n$, and will therefore be useful in testing the sharpness of some of our later results.

7. Statement of results

In the following it is convenient to restrict consideration to continuous solutions of (5), that is, weak solutions which have furthermore the property of being continuous functions. This involves no real loss of generality, of course, since by Theorem 8 any weak solution can be made continuous by appropriately redefining it on a set of measure zero.

We shall require also the notion of capacity. Let Q be a bounded set in E^n. The s-capacity of Q, $1 \leqslant s < \infty$, is defined to be

$$\inf \int |\psi_x|^s \, dx,$$

where the infimum is taken over all continuously differentiable functions ψ which are $\geqslant 1$ on Q and have compact support in E^n (if $s \geqslant n$, we require compact support in some fixed sphere $|x| < R_0$). When $s=2$ this definition reduces to the familiar one of potential theory. We may now state our fundamental result.

THEOREM 10. *Let Q be a compact set of s-capacity zero, where $\alpha \leqslant s \leqslant n$, and let D be a domain in Ω. Suppose that u is a continuous solution of* (5) *in the set $D-Q$, and that for some $\delta > 0$ we have*

$$u \in L_{\theta(1+\delta)}, \quad \theta = s(\alpha-1)/(s-\alpha). \tag{52}$$

Then u can be defined on the set Q so that the resulting function is a continuous solution of (5) *in all of D. The exponent θ in* (52) *is best possible.*

As a special case, if the α-capacity of Q is zero, then Q is a removable singularity for any bounded solution of (5).

This theorem has several corollaries which further clarify its meaning, and relate the size of the exceptional set to the classical concepts of *potential theoretic capacity* and *Hausdorff dimension*. (Definitions of these concepts appear in references [1] and [26], and in other places; they may be omitted here since the corollaries will not be used in the sequel.)

COROLLARY 1. *Let Q be a compact manifold of Hausdorff dimension m, and let D be a domain in Ω. Suppose that $m < n-\alpha$. Let u be a continuous solution of* (5) *in $D-Q$, such that for some $\delta > 0$*

$$u \in L_{\theta(1+\delta)}, \quad \theta = (n-m)(\alpha-1)/(n-m-\alpha). \tag{53}$$

Then u can be defined on the set Q so that the resulting function is a continuous solution of (5) *in all of D.*

Proof. Let Λ_p denote the Hausdorff p-measure of the manifold Q. Then clearly $\Lambda_{m+\varepsilon}=0$ for any $\varepsilon>0$. This implies (cf. [26]) that the s-capacity of Q is zero, where $s=n-m-\varepsilon$. On the other hand, by (53)

$$u \in L_{\varkappa}, \quad \varkappa=\frac{s(\alpha-1)}{(s-\alpha)}(1+\delta'),$$

provided ε is suitably small. It now follows from Theorem 10 that Q is removable.

COROLLARY 2. *Let Q be a compact manifold of potential theoretic λ-capacity zero, and let D be a domain in Ω. Suppose also that $0\leqslant\lambda<n-\alpha$ or that $\lambda=n-\alpha$ and $\alpha\geqslant 2$. Let u be a continuous solution of* (5) *in $D-Q$, such that for some $\delta>0$*

$$u \in L_{\theta(1+\delta)}, \quad \theta=(n-\lambda)(\alpha-1)/(n-\lambda-\alpha).$$

Then u can be defined on the set Q so that the resulting function is a continuous solution of (5) *in all of D.*

Proof. If $0\leqslant\lambda<n-\alpha$, then by [26], Theorem A, the s-capacity of Q is zero, where $s=n-\lambda-\varepsilon$. Thus Q is removable, as in Corollary 1. If $\lambda=n-\alpha$ and $\alpha\geqslant 2$, then $\lambda\leqslant n-2$ and by [26], Theorem A, the α-capacity of Q is zero. Since $u\in L_\infty$ in the present case, we conclude once more from Theorem 10 that Q is removable.

Another slightly different version of Theorem 10 can be given if we replace the Lebesgue class hypothesis on u by a strict growth property. For this purpose we understand that a smooth manifold of co-dimension s is a compact set in E^n with the property that there exists a C^1 diffeomorphism $x\rightarrow y$ of E^n onto itself such that the image of the manifold lies in the hyperplane $y_1=\ldots=y_s=0$.

THEOREM 11. *Let Q be a smooth manifold of co-dimension s, where $\alpha\leqslant s\leqslant n$, and let D be a domain in Ω. Suppose that u is a continuous solution of* (5) *in $D-Q$, and that for some $\delta>0$ we have*

$$u=\begin{cases} O(\xi^{(\alpha-s)/(\alpha-1)+\delta}) & if \quad s>\alpha,\\ O(|\log\xi|^{1-\delta}) & if \quad s=\alpha,\end{cases}$$

where ξ is the distance from Q. Then u can be defined on the manifold Q so that the resulting function is a solution of (5) *in the whole domain D.* (1)

(1) If Q is a single point, and r denotes the distance to this point, the hypothesis of Theorem 11 becomes $u=O(r^{(\alpha-n)/(\alpha-1)+\delta})$ or $O(|\log r|^{1-\delta})$. When $\alpha=2$ this further reduces to $u=O(r^{2-n+\delta})$ or $O(|\log r|^{1-\delta})$, as might be expected. Earlier results of this type appear in [6] and [21].

When $s > \alpha$ this theorem is a special case of Corollary 1 (if desired, the reader may show directly that Q has s-capacity zero and so reduce the result directly to Theorem 10). The case $s = \alpha$, on the other hand, requires its own separate proof.

The proof of Theorem 10 will be given in the next section. Following this we prove the logarithmic case of Theorem 11. Finally, in Section 10 we prove a lemma on capacity which was crucial at one stage in the demonstration of Theorem 10.

There is some interest in the existence of solutions with non-removable singularities. In the following chapter we shall take up this question in some detail for the special equation $\operatorname{div} A(x, u_x) = 0$. Here we shall only remark that *if a positive solution of* (5) *has a non-removable isolated singularity at a point* P, *then the solution necessarily tends to infinity at* P. Indeed by Theorem 10 or 11 there must exist a sequence of points x_ν, $\nu = 1, 2, \ldots$, tending to P, such that $u(x_\nu) \to \infty$. We now wish to apply the Harnack inequality (Theorem 7) in annular domains about P. For simplicity of notation suppose P is the origin. Then by normalizing each sphere $|x| = |x_\nu|$ to radius 1 and applying Theorem 7 we obtain

$$\mu_\nu = \min_{|x| = |x_\nu|} u(x) \to \infty.$$

Finally by Theorem 4 one sees that (for sufficiently large ν)

$$u(x) \geqslant \operatorname{Min}(\mu_\nu, \mu_{\nu+1}) - Ck, \quad |x_{\nu+1}| \leqslant |x| \leqslant |x_\nu|.$$

The required conclusion follows at once. It should be noted that this argument is essentially the same as one in reference [6].

8. Proof of Theorem 10

It is enough to show that u can be made a continuous solution in the neighborhood of any point of D. Thus let P be in D and let R be such that the ball $S(2R)$ is also in D. By a suitable change of variables it may be supposed that $R = 1$. As in the proof of Theorem 1 we define

$$\bar{u} = |u| + k, \quad x \in S(2) - Q,$$

where $k = 1 + (\|e\| + \|f\|)^{1/(\alpha-1)} + \|g\|^{1/\alpha}$. We now introduce an appropriate test function $\phi(x)$. Let η and $\bar{\eta}$ be non-negative smooth functions, η having compact support in $S(2)$, and $\bar{\eta}$ vanishing in some negborhood of Q. Then since u is locally bounded and has strong derivatives which are locally in L_α, it is clear that the function

$$\phi(x) = (\eta\bar{\eta})^\alpha \operatorname{sign} u \cdot \{\bar{u}^\beta - k^\beta\}$$

is admissible in (9) for any $\beta > 0$. The resulting calculation [1] is essentially the same as in Theorems 1 and 5, and gives the results

$$\|\eta\bar{\eta}v_x\|_\alpha \leq Cq^{\alpha/\varepsilon}(1+\beta^{-1})^{1/\varepsilon}(\|\eta\bar{\eta}v\|_\alpha + \|(\eta\bar{\eta})_x v\|_\alpha), \tag{54}$$

$$\|\eta\bar{\eta}v\|_{\alpha*} \leq Cq^{\alpha/\varepsilon}(1+\beta^{-1})^{1/\varepsilon}(\|\eta\bar{\eta}v\|_\alpha + \|(\eta\bar{\eta})_x v\|_\alpha), \tag{55}$$

where $v = \bar{u}^q$ and $\alpha q = \alpha + \beta - 1$.

We now require a lemma, whose proof will be deferred until Section 10. In order to formulate the lemma more simply we use the notation $U(Q)$ to denote the class of smooth functions $\bar{\eta}(x)$ which satisfy $0 \leq \bar{\eta} \leq 1$ and vanish in some neighborhood of Q.

Lemma 8. *Let Q be a compact set of s-capacity zero, $1 \leq s \leq n$. Then there exists a sequence of functions $\bar{\eta}^{(\nu)}$ contained in $U(Q)$ such that $\bar{\eta}^{(\nu)} \to 1$ almost everywhere, while $\|\bar{\eta}_x^{(\nu)}\|_s \to 0$.*

Now consider the term $\|\bar{\eta}_x v\|_\alpha$ which appears in both (54) and (55). We have, if $s > \alpha$,

$$\|\bar{\eta}_x v\|_\alpha^\alpha = \int |\bar{\eta}_x|^\alpha \bar{u}^{\alpha q}\, dx \leq \|\bar{\eta}_x\|_s^\alpha \left(\int \bar{u}^{\alpha q s/(s-\alpha)}\, dx\right)^{1-\alpha/s}.$$

Choosing $\beta = \beta_0 = \delta(\alpha - 1)$ one finds that $\alpha q s/(s-\alpha) = (\alpha+\beta-1)s/(s-\alpha) = \theta(1+\delta)$, whence using the hypothesis (52),

$$\|\bar{\eta}_x v\|_\alpha \leq \text{Const.}\ \|\bar{\eta}_x\|_s. \tag{56}$$

On the other hand, if $s = \alpha$ then $\theta = \infty$ and by hypothesis $u \in L_\infty$ and $\|\bar{\eta}_x v\|_\alpha \leq$ Const. $\|\bar{\eta}_x\|_\alpha$. This being shown, let us replace $\bar{\eta}$ in (54) and (55) by the elements $\bar{\eta}^{(\nu)}$ given by Lemma 8. Then letting $\nu \to \infty$, we obtain from the dominated convergence theorem

$$\|\eta v_x\|_\alpha \leq Cq^{\alpha/\varepsilon}(\|\eta v\|_\alpha + \|\eta_x v\|_\alpha), \tag{57}$$

$$\|\eta v\|_{\alpha*} \leq Cq^{\alpha/\varepsilon}(\|\eta v\|_\alpha + \|\eta_x v\|_\alpha), \tag{58}$$

where the norms are taken over $S(2) - Q$, and the constant C now depends also on the value of δ.

If η is now chosen as in Theorem 1, inequality (58) may be written in the form

$$\Phi(\varkappa p_0, h') \leq [C(p_0/\alpha)^{\alpha/\varepsilon}(h-h')^{-1}]^{\alpha/p_0}\Phi(p_0, h), \tag{59}$$

where $p_0 = \alpha + \beta_0 - 1 = (\alpha - 1)(1+\delta)$, and $\Phi(p, h) = (\int_{S(h)-Q} \bar{u}^p\, dx)^{1/p}$.

---

[1] For simplicity we suppose that $\alpha < n$. The case $\alpha = n$ requires only trivial changes (cf. Theorem 2).

We wish to proceed as in Theorem 1, obtaining a corresponding inequality for each $p>p_0$. This requires a more delicate choice of the test function than previously. For $q \geqslant q_0 = p_0/\alpha$ and $l>k$, we define

$$F(\bar u)=\begin{cases}\bar u^q & \text{if } k\leqslant \bar u\leqslant l,\\ q_0^{-1}[ql^{q-q_0}u^{q_0}+(q_0-q)\,l^q] & \text{if } l\leqslant \bar u,\end{cases}$$

and
$$G(u)=\operatorname{sign} u\cdot\{F(\bar u)\,F'(\bar u)^{\alpha-1}-q^{\alpha-1}k^{\beta}\},\qquad -\infty<u<\infty,$$

where q and β are related by $\alpha q=\alpha+\beta-1$. Evidently F is a continuously differentiable function of $\bar u$, and G is a piecewise smooth function of u, with corners at $u=\pm(l-k)$. Moreover, these functions have the properties

$$F\leqslant (q/q_0)\,l^{q-q_0}\bar u^{q_0},\qquad \bar u F'\leqslant qF,$$

and
$$G'\geqslant\begin{cases}q^{-1}\beta(F')^{\alpha} & \text{if } |u|<l-k,\\ q_0^{-1}\beta_0(F')^{\alpha} & \text{if } |u|>l-k.\end{cases}$$

The first two of these inequalities are quite simple to establish. The third arises from the calculations (1)

$$\begin{aligned}G'(u)&=(F')^{\alpha}+(\alpha-1)\,FF''(F')^{\alpha-2}\\&=(F')^{\alpha}+(\alpha-1)\left(\frac{q-1}{q}\right)(F')^{\alpha}=q^{-1}\beta(F')^{\alpha}\end{aligned}$$

if $\bar u<l$, and
$$\begin{aligned}G'(u)&=(F')^{\alpha}+(\alpha-1)\,FF''(F')^{\alpha-2}\\&\geqslant(F')^{\alpha}+(\alpha-1)\left(\frac{q_0-1}{q_0}\right)(F')^{\alpha}=q_0^{-1}\beta_0(F')^{\alpha}\end{aligned}$$

when $\bar u>l$. We may now substitute

$$\phi(x)=(\eta\bar\eta)^{\alpha}G(u),\qquad (u=u(x))$$

into (9). Since $|G|\leqslant F(F')^{\alpha-1}$, one finds exactly as in Theorem 1 that

$$\|\eta\bar\eta v\|_{\alpha*}\leqslant Cq^{\alpha/\varepsilon}(\|\eta\bar\eta v\|_{\alpha}+\|(\eta\bar\eta)_x v\|_{\alpha}),$$

where $v=v(x)=F(\bar u)$.

Since $v\leqslant \text{Const.}\,\bar u^{q_0}$, it is clear as in the earlier part of the proof that $\|\bar\eta_x v\|_{\alpha}\leqslant \text{Const.}\,\|\bar\eta_x\|_s$. Thus replacing $\bar\eta$ by $\bar\eta^{(\nu)}$ and letting $\nu\to\infty$, we obtain

$$\|\eta v\|_{\alpha*}\leqslant Cq^{\alpha/\varepsilon}(\|\eta v\|_{\alpha}+\|\eta_x v\|_{\alpha}).$$

(1) For the range $\bar u>l$ we use the relation $q_0\leqslant 1$, which is in turn equivalent to $\beta_0\leqslant 1$ and hence to $\delta\leqslant(\alpha-1)^{-1}$. But this last inequality can always be assumed without loss of generality.

Here we may let $l\to\infty$. Since $v\to\bar{u}^q$ there results by Lebesgue's theorem

$$\|\eta\bar{u}^q\|_{\alpha*}\leqslant Cq^{\alpha/\varepsilon}(\|\eta\bar{u}^q\|_\alpha+\|\eta_x\bar{u}^q\|_\alpha).$$

Finally, choosing η as in Theorem 1 (and as in the earlier part of the proof) yields

$$\Phi(\varkappa p,h')\leqslant[C(h-h')^{-1}(p/\alpha)^{\alpha/\varepsilon}]^{\alpha/p}\,\Phi(p,h), \tag{60}$$

where $p=\alpha q\geqslant p_0$. We emphasize that (60) is valid whether or not the integrals involved are finite.

Now iterate the inequalities (59), (60) starting with $p=p_0$. This clearly yields the conclusion

$$\|\bar{u}\|_\infty\leqslant C\|\bar{u}\|_{p_0}, \tag{61}$$

where the left hand norm is over the set $S(1)-Q$ and the right-hand norm is over $S(2)-Q$. But $\|\bar{u}\|_{p_0}\leqslant\|\bar{u}\|_{\theta(1+\delta)}$, so that the right side of (61) is finite. Thus we have shown that u is uniformly bounded on the set $S(1)-Q$.

It remains to establish that u can be made a continuous solution of (5) in all of $S(1)$. Choosing η so that $\eta\equiv1$ in $S(1)$, we have from (57)

$$\|v_x\|_{\alpha,1}\leqslant\text{Const.}\,\|v\|_{\alpha,2},$$

that is

$$\int\bar{u}^{\beta_0-1}|u_x|^\alpha\,dx\leqslant\text{Const.}\int\bar{u}^{\alpha q_0}\,dx\leqslant\text{Const.}$$

Since $\beta_0\leqslant1$ and $\bar{u}$ is bounded in $S(1)-Q$, this proves that u_x is in L_α on the set $S(1)-Q$.

A set of capacity zero is also a set of measure zero (see the corollary in section 10). We shall show that if u is arbitrarily set equal to zero on Q the resulting function is strongly differentiable in $S(1)$. Indeed, for any smooth function ϕ with compact support in $S(1)-Q$, we have

$$\int u\phi_x\,dx=-\int\phi u_x\,dx.$$

Setting $\phi=\eta\bar{\eta}$, where η has compact support in $S(1)$, then yields

$$\int u(\eta\bar{\eta}_x+\bar{\eta}\eta_x)\,dx=-\int\eta\bar{\eta}u_x\,dx.$$

Thus replacing $\bar{\eta}$ by $\bar{\eta}^{(\nu)}$ and letting $\nu\to\infty$, we obtain from the dominated convergence theorem

$$\int u\eta_x\,dx = -\int \eta u_x\,dx, \tag{62}$$

the integrals being evaluated over the set $S(1)-Q$. If one arbitrarily sets $u_x=0$ on Q, relation (62) becomes valid over all of $S(1)$. But this is just the condition that u be strongly differentiable in $S(1)$, and the assertion is proved.

Finally, the functions $\mathcal{A}(x,u,u_x)$ and $\mathcal{B}(x,u,u_x)$ are clearly in $L_{\alpha/(\alpha-1)}$ and $L_{(\alpha^*)'}$ over $S(1)$, and

$$\int (\phi_x\cdot\mathcal{A}+\phi\mathcal{B})\,dx=0$$

when ϕ has compact support in $S(1)-Q$. Again setting $\phi=\eta\bar{\eta}$, we easily obtain in the limit as $\nu\to\infty$,

$$\int (\eta_x\cdot\mathcal{A}+\eta\mathcal{B})\,dx=0,$$

valid whenever η has compact support in $S(1)$. It follows that u, as defined over Q, is a weak solution of (5) in the ball $S(1)$. Finally by Theorem 8 we can redefine u on a set of measure zero so that it is Hölder continuous in $S(1)$. Since this redefinition cannot effect the values of u on $S(1)-Q$, where it is already continuous, we see that by suitably assigning values to u on the set Q, it becomes a continuous solution of (5) in $S(1)$, that is, in a non-empty neighborhood of the point P. This completes the main part of the proof. It remains only to show that the exponent $\theta=s(\alpha-1)/(s-\alpha)$ is best possible, and to prove Lemma 8.

In order to show that θ is the best possible exponent, consider the equation

$$\operatorname{div}(u_x|u_x|^{\alpha-2})=0,$$

and let Q be the hyperplane $x_1=\ldots=x_s=0$. Clearly Q has s-capacity zero. Now one checks easily that the function

$$u=\begin{cases}\varrho^{(\alpha-s)/(\alpha-1)}, & \alpha<s\\ \log\varrho, & \alpha=s,\end{cases} \tag{63}$$

is a solution, where ϱ denotes the distance to Q. Moreover, if $\alpha<s$ then

$$u\in L_{\theta-\varepsilon'}$$

for any $\varepsilon'>0$, while if $\alpha=s$ then

$$u\in L_\sigma$$

for any $\sigma<\infty$. Thus an exponent smaller than θ in (52) would allow the solution (63) into competition, and the conclusion of the theorem would become impossible.

9. The logarithmic case of Theorem 11

By a smooth change of variables, leaving the form of inequalities (6) invariant, it may be assumed that Q is contained in the set

$$x_1 = \ldots = x_s = 0, \quad |x| \leqslant 3, \tag{64}$$

and that D in turn includes the ball $|x| < 2$. In fact, we may even suppose that Q is precisely (64), for this at most enlarges the singular set.

Now, retaining the notation of the preceding section, let us make the same substitution

$$\phi(x) = (\eta\bar{\eta})^\alpha \operatorname{sign} u \cdot \{\bar{u}^\beta - k^\beta\}, \quad (\beta > 0),$$

in relation (9). Then we find, by separating out the terms involving $\bar{\eta}_x$,

$$\begin{aligned} q^{-1}\beta \| \eta\bar{\eta}v_x \|_\alpha^\alpha \leqslant \alpha a \int |\eta_x v| \cdot |\eta\bar{\eta}v_x|^{\alpha-1} dx + \alpha q^{\alpha-1} \int \bar{b} |\eta_x v| \cdot (\eta\bar{\eta}v)^{\alpha-1} dx \\ + \int c\eta v |\eta\bar{\eta}v_x|^{\alpha-1} dx + (1+\beta) q^{\alpha-1} \int \bar{d} (\eta\bar{\eta}v)^\alpha dx \\ + \alpha \int |\bar{\eta}_x v| \cdot (a |\eta v_x|^{\alpha-1} + q^{\alpha-1} \bar{b} (\eta v)^{\alpha-1}) dx, \end{aligned} \tag{65}$$

where it is assumed that both η and $\bar{\eta}$ are less than or equal to one. Next, let ϱ denote the distance to Q, and for fixed $\sigma > 0$ choose

$$\bar{\eta}(x) = \begin{cases} 0 & \text{if} \quad \varrho \leqslant \sigma \\ 1 & \text{if} \quad \varrho \geqslant \sigma + h, \end{cases}$$

and linear in the interval $\sigma \leqslant \varrho \leqslant \sigma + h$. Now consider the effect of letting $h \to 0$ while keeping σ fixed. Since $|\bar{\eta}_x| = h^{-1}$ it is clear from Lebesgue's differentiation theorem that, for almost all values of σ,

$$\int |\bar{\eta}_x v| \cdot (a |\eta v_x|^{\alpha-1} + q^{\alpha-1} \bar{b} |\eta v|^{\alpha-1}) dx \to \oint_{\varrho=\sigma} v (a |\eta v_x|^{\alpha-1} + q^{\alpha-1} \bar{b} |\eta v|^{\alpha-1}) ds.$$

The limit process on the remaining terms of (65) is trivial, hence we have for almost all values of σ,

$$\begin{aligned} q^{-1}\beta \| \eta v_x \|_\alpha^\alpha \leqslant \alpha a \int |\eta_x v| \cdot |\eta v_x|^{\alpha-1} dx + \alpha q^{\alpha-1} \int \bar{b} |\eta_x v| \cdot (\eta v)^{\alpha-1} dx \\ + \int c\eta v |\eta v_x|^{\alpha-1} dx + (1+\beta) q^{\alpha-1} \int \bar{d} (\eta v)^\alpha dx \\ + \alpha \oint_{\varrho=\sigma} v (a |\eta v_x|^{\alpha-1} + q^{\alpha-1} \bar{b} |\eta v|^{\alpha-1}) ds, \end{aligned}$$

the volume integrals being taken over the set where $\varrho > \sigma$. The first four terms on the right may be estimated as in Theorem 1 (we assume $\alpha < n$ for simplicity), with the exception that Lemma 4 must be replaced by the slightly different Lemma 5. Thus we obtain with the help of Lemma 2,

$$\|\eta v_x\|_\alpha \leqslant C(1+\beta^{-1})^{1/\varepsilon} q^{\alpha/\varepsilon} (\|\eta v\|_\alpha + \|\eta_x v\|_\alpha) + C\beta^{-1/\alpha} q I(\sigma)^{1/\alpha}, \tag{66}$$

where
$$I(\sigma) = \oint_{\varrho=\sigma} v(|\eta v_x|^{\alpha-1} + \mathfrak{b}|\eta v|^{\alpha-1})\,ds.$$

Now by Hölder's inequality and Lemma 1

$$I(\sigma) \leqslant \left(2\oint v^\alpha ds\right)^{1/\alpha} \left(\oint (|\eta v_x|^\alpha + \mathfrak{b}^{\alpha/(\alpha-1)}|\eta v|^\alpha)\,ds\right)^{(\alpha-1)/\alpha},$$

while by hypothesis

$$\oint v^\alpha ds = \oint \bar{u}^{\alpha q} ds \leqslant \bar{C}\sigma^{s-1}\,|\log\sigma|^{(\alpha+\beta-1)(1-\delta)},$$

where $\bar{C}$ is some constant. Since $s = \alpha$ in the present case, one has by setting $\beta = \beta_0 = \delta(\alpha-1)$ and $\varepsilon' = \delta^2$,

$$\oint v^\alpha ds \leqslant \bar{C}(\sigma\,|\log\sigma|^{1-\varepsilon'})^{\alpha-1}.$$

Furthermore, since $s > 1$ it is easy to see that the terms $\|\eta v\|_\alpha$ and $\|\eta_x v\|_\alpha$ in (66) are uniformly bounded as $\sigma \to 0$. Let us set

$$J = J(\sigma) = \int_{\varrho>\sigma} |\eta v_x|^\alpha dx, \quad \Psi = \Psi(x) = \mathfrak{b}^{\alpha/(\alpha-1)}|\eta v|^\alpha.$$

Then noting that $J'(\sigma) = -\oint |\eta v_x|^\alpha ds$ and using the preceding estimates, a straightforward calculation establishes that (66) may be written in the form

$$(J-K)^{\alpha/(\alpha-1)} \leqslant \bar{C}\sigma\,|\log\sigma|^{1-\varepsilon'}\left(-J' + \oint \Psi\,ds\right), \tag{67}$$

where
$$K = C^\alpha q^{\alpha^2/\varepsilon}(\|\eta v\|_\alpha + \|\eta_x v\|_\alpha)^\alpha,$$

the norms in the preceding line being taken over the set where $\varrho > 0$, that is, over the complement of Q.

We assert that $J \leqslant K$ for all $\sigma > 0$. Indeed, suppose to the contrary that at some value $\sigma = \sigma_0$ we have $J(\sigma_0) > K$. Since J increases as σ decreases we have also $J > K$ in the interval $0 < \sigma \leqslant \sigma_0$. Thus setting $H(\sigma) = J(\sigma) - K$ in (67) it may be rewritten

$$\frac{1}{\sigma |\log \sigma|^{1-\varepsilon'}} \leqslant -\frac{\bar{C}H'}{H^{\alpha/(\alpha-1)}} + \frac{\bar{C}}{H(\sigma_0)^{\alpha/(\alpha-1)}} \oint \Psi\, ds,$$

valid for almost all values of σ in the interval $0 < \sigma \leqslant \sigma_0$. Integration of both sides from σ to σ_0 yields

$$\frac{1}{\varepsilon'}\left(|\log \sigma|^{\varepsilon'} - |\log \sigma_0|^{\varepsilon'}\right) \leqslant (\alpha-1)\bar{C}\left[\frac{1}{H(\sigma_0)^{1/(\alpha-1)}} - \frac{1}{H(\sigma)^{1/(\alpha-1)}}\right] + \frac{\bar{C}}{H(\sigma_0)^{\alpha/(\alpha-1)}} \int_{\sigma<\varrho<\sigma_0} \Psi\, dx. \tag{68}$$

But
$$\int \Psi\, dx \leqslant \left(\int \bar{b}^{n/(\alpha-1)}\, dx\right)^{\alpha/n} \left(\int |\eta v|^{\alpha n/(n-\alpha)}\, dx\right)^{1-\alpha/n},$$

and the right side is uniformly bounded as $\sigma \to 0$, (recall that $u = O(|\log \varrho|^{1-\delta})$). The right hand side of (68) therefore remains bounded as $\sigma \to 0$, while the left-hand side becomes infinite. This contradiction proves the assertion $J \leqslant K$. Interpreting this in terms of the original functions yields

$$\|\eta v_x\|_\alpha \leqslant C q^{\alpha/\varepsilon}(\|\eta v\|_\alpha + \|\eta_x v\|_\alpha),$$

where the norms are taken over the complement of Q, and the constant C depends only on δ and on the structure of equation (5). Finally, applying the Sobolev inequality (Lemma 5) we obtain

$$\|\eta v\|_{\alpha*} \leqslant C q^{\alpha/\varepsilon}(\|\eta v\|_\alpha + \|\eta_x v\|_\alpha).$$

The preceding two inequalities are the analogues in the present case of (57) and (58) in the foregoing section.

The remainder of the proof is exactly the same as before, except that the limit process $\nu \to \infty$ must be replaced by the present technique of differential equations. Theorem 11 is thus completely proved.

10. Capacity

Here we shall prove Lemma 8, and also present some simple results about s-capacity which will be useful in the sequel.

Proof of Lemma 8. For simplicity we shall consider only the the case $s < n$, the remaining case $s = n$ being treated by similar methods. Now according to hypothesis, there exists a sequence of continuously differentiable functions $\psi^{(\nu)}$ with compact support in $E^n, \geqslant 1$ on Q, and such that

$$\int |\psi_x^{(\nu)}|^s \, dx \leqslant 1/\nu. \tag{69}$$

The function $2\psi^{(\nu)}$ is > 1 in some neighborhood of Q. Consider then the truncated functions

$$\bar{\psi}^{(\nu)} = \begin{cases} 0, & \text{where } \psi^{(\nu)} < 0, \\ 2\psi^{(\nu)}, & \text{where } 0 \leqslant 2\psi^{(\nu)} \leqslant 1, \\ 1, & \text{where } 2\psi^{(\nu)} > 1. \end{cases}$$

Evidently $\bar{\psi}^{(\nu)}$ is strongly differentiable, equals one in some neighborhood of Q, and satisfies

$$\int |\bar{\psi}_x^{(\nu)}|^s \, dx \leqslant 2^s \int |\psi_x^{(\nu)}|^s \, dx \leqslant 2^s/\nu. \tag{70}$$

Moreover, $\bar{\psi}^{(\nu)}$ has compact support in E^n. Therefore by Lemma 4 and (70),

$$\| \bar{\psi}^{(\nu)} \|_{s*} \leqslant \text{Const.}/\nu.$$

Consequently $\bar{\psi}^{(\nu)}$ converges to zero in measure, whence for some subsequence (still denoted by $\bar{\psi}^{(\nu)}$) we have

$$\bar{\psi}^{(\nu)} \to 0 \quad \text{almost everywhere.} \tag{71}$$

This being shown, it is easy to see that the sequence $\bar{\eta}^{(\nu)} = 1 - \bar{\psi}^{(\nu)}$ fulfills the conditions of the lemma, with the single exception that each $\bar{\eta}^{(\nu)}$ is only Lipschitz continuous and not continuously differentiable. This defect can be removed by a suitable mollification, or alternately one can easily justify the direct substitution of $\bar{\eta}^{(\nu)}$ for $\bar{\eta}$. This completes the proof.

Corollary. *If Q is a compact set of s-capacity zero, $1 \leqslant s \leqslant n$, then $|Q| = 0$.*

Proof. The function $\bar{\psi}^{(\nu)}$ above is equal to one on Q for each value of ν. On the other hand, $\bar{\psi}^{(\nu)} \to 0$ almost everywhere. Hence obviously $|Q| = 0$.

Lemma 9. *If ψ is any continuous strongly differentiable function with compact support in E^n, and if $\psi \geqslant 1$ on Q, then*

$$\int |\psi_x|^s \, dx \geqslant \text{Cap}_s \, Q.$$

Proof. If the result were not true we could construct a continuously differentiable function $\bar{\psi}$ with compact support in E^n, such that $\bar{\psi} \geqslant 1$ on Q and $\| \bar{\psi}_x \|_s^s < \text{Cap}_s \, Q$, which would be impossible. The construction involves first multiplying ψ by $(1+\varepsilon')$, so that the resulting function is $\geqslant(1+\varepsilon')$ on Q, and then forming a smooth integral average. The details may be omitted.

LEMMA 10. *For a ball $S(R)$ of radius R we have*

$$\operatorname{Cap}_\alpha S(R)=\begin{cases}\omega_n\left(\dfrac{n-\alpha}{\alpha-1}\right)^{\alpha-1}R^{n-\alpha} & (1<\alpha<n),\\[2mm] \omega_n\log\,(R_0/R)^{1-n} & (\alpha=n).\end{cases}$$

Proof. Suppose first that $\alpha<n$, and set $\tau=(\alpha-n)/(\alpha-1)$. The function

$$h=h(r)=\frac{r^\tau-R_0^\tau}{R^\tau-R_0^\tau},$$

where $r=|x|$ and $R_0>R$, is then a solution of the differential equation

$$\operatorname{div}\,(u_x|u_x|^{\alpha-2})=0, \tag{72}$$

that is, the Euler equation of the variational problem $\int|u_x|^\alpha dx=\text{Minimum}$. Evidently $h=0$ when $r=R_0$ and $h=1$ when $r=R$. By standard comparison arguments in the calculus of variations if follows that

$$\|\psi_x\|_\alpha\geqslant\|h_x\|_\alpha\quad(\text{over } R<r<R_0),$$

where ψ is any continuously differentiable function $\geqslant 1$ on $S(R)$ and with compact support in $r<R_0$. Therefore it is clear that

$$\operatorname{Cap}_\alpha S(R)\geqslant\lim_{R_0\to\infty}\|h_x\|_\alpha^\alpha.$$

Similarly by setting $\psi(x)=0$ for $r\geqslant R_0$, $\psi(x)=h(r)$ for $R<r<R_0$, and $\psi(x)=1$ for $r\leqslant R$, and using Lemma 9 we have

$$\operatorname{Cap}_\alpha S(R)\leqslant\lim_{R_0\to\infty}\|h_x\|_\alpha^\alpha.$$

The required conclusion follows immediately from the calculation

$$\|h_x\|_\alpha^\alpha=\omega_n|\tau/(R^\tau-R_0^\tau)|^{\alpha-1}.$$

The case $\alpha=n$ is handled in the same way, except that $h(r)=\log\,(r/R_0)/\log\,(R/R_0)$.

This result indicates the close connection between α-capacity and equation (72), a connection well-known if $\alpha=2$.

19 – 642946 *Acta mathematica*. 111. Imprimé le 9 juin 1964

11. A remark concerning the case $\alpha=1$

The result of Theorem 10 can be generalized to apply to equation (5) even for the case $\alpha=1$, that is, when the functions $\mathcal{A}$ and $\mathcal{B}$ satisfy

$$|\mathcal{A}| \leqslant \text{Const.}, \quad |\mathcal{B}| \leqslant f,$$

and

$$p \cdot \mathcal{A} \geqslant |p| - d\,|u| - g$$

with $d, f, g \in L_{n/(1-\varepsilon)}$. Indeed we have the following

THEOREM 10′. *Let Q be a compact set of 1-capacity zero, and let D be a domain in Ω. Suppose that u is a solution of (5) in the set $D-Q$, and that the functions $\mathcal{A}$ and $\mathcal{B}$ satisfy the conditions above. Then, if*

$$u \in L_\delta$$

for some $\delta>0$, we can define u on the set Q so that the resulting function is a solution of (5) in all of D.

Proof. Let $S(2)$ be an open ball in D, and define

$$\bar{u} = |u| + k, \quad \bar{u}_l = \text{Min}\,(l, \bar{u}),$$

where $k = 1 + \|g\|$, and l is some large constant. We introduce the test function

$$\phi = (\eta\bar{\eta})\,\text{sign}\,u \cdot \{\bar{u}_l^\beta - k^\beta\}, \quad x \in S(2) - Q,$$

as in the proof of Theorem 10. In the set where $\bar{u} < l$ we have

$$\phi_x \cdot \mathcal{A} + \varphi \mathcal{B} \geqslant (\eta\bar{\eta})\,|v_x| - \text{Const.}\ (\eta\bar{\eta})_x v - (\beta\bar{d} + f)\,\eta\bar{\eta} v,$$

where we have put $v = \bar{u}_l^\beta$. Similarly, when $\bar{u} \geqslant l$

$$\begin{aligned}\phi_x \cdot \mathcal{A} + \varphi \mathcal{B} &\geqslant -\text{Const.}\ (\eta\bar{\eta})_x l^\beta - f\,\eta\bar{\eta}\, l^\beta \\ &\geqslant (\eta\bar{\eta})\,|v_x| - \text{Const.}\ (\eta\bar{\eta})_x v - (\beta\bar{d} + f)\,\eta\bar{\eta} v\end{aligned}$$

since $v = l^\beta$ in this set. Hence as in the proof of Theorem 1 (or Theorem 10), we may derive the relations

$$\|\eta\bar{\eta} v_x\|_1 \leqslant C(1+\beta)^{1/\varepsilon}\,(\|\eta\bar{\eta} v\|_1 + \|(\eta\bar{\eta})_x v\|_1)$$

and

$$\|\eta\bar{\eta} v\|_{1*} \leqslant C(1+\beta)^{1/\varepsilon}\,(\|\eta\bar{\eta} v\|_1 + \|(\eta\bar{\eta})_x v\|_1).$$

Now the term $\|\bar{\eta}_x v\|$ which appears in both preceding lines satisfies the inequality $\|\bar{\eta}_x v\|_1 \leqslant l^\beta \|\bar{\eta}_x\|_1$. Hence replacing $\bar{\eta}$ by $\bar{\eta}^{(\nu)}$ and letting $\nu \to \infty$, we obtain

$$\|\eta v_x\|_1 \leqslant C(1+\beta)^{1/\varepsilon}(\|\eta v\|_1 + \|\eta_x v\|_1) \tag{73}$$

and

$$\|\eta v\|_{1*} \leqslant C(1+\beta)^{1/\varepsilon}(\|\eta v\|_1 + \|\eta_x v\|_1). \tag{74}$$

Here we may let $l \to \infty$. Since $v \to \bar{u}^\beta$ and $v_x \to (\bar{u}^\beta)_x$, it follows from the monotone convergence theorem that (73) and (74) hold with v replaced by $\bar{u}^\beta$.

Iteration of (74) beginning with $\beta = \delta$ establishes that $\bar{u}$ is bounded in $S(1) - Q$. The rest of the proof is then the same as in Theorem 10. It should be remarked, however, that since there is no analogue of Theorem 8 for the case $\alpha = 1$ the *continuity* of the solution remains an open problem; in special cases (e.g. the minimal surface equation in two dimensions) one may, of course, be able to settle this question by an independent investigation.

In conclusion, we note that Theorem 10′ is quite similar in both hypothesis and conclusion to a well known removable singularity theorem due to Finn.

III. Isolated singularities

A detailed description of the behavior of solutions at an isolated singularity seems to require some specialization of equation (5). We shall consider the equation

$$\operatorname{div} \mathcal{A}(x, u, u_x) = 0, \quad (\alpha \leqslant n), \tag{75}$$

which is, of course, sufficiently general to include the Euler equations of variational problems with integrand independent of u. It will be assumed that the function $\mathcal{A}(x, u, p)$ satisfies the conditions

$$|\mathcal{A}| \leqslant a|p|^{\alpha-1} + b|u|^{\alpha-1} + e, \quad p \cdot \mathcal{A} \geqslant |p|^\alpha - d|u|^\alpha - g, \tag{76}$$

where b, d, e and g are measurable functions defined in the basic domain Ω and contained in the respective Lebesgue classes

$$b \in L_{n/(\alpha-1)}; \quad e \in L_{n/(\alpha-1-\varepsilon)}; \quad d, g \in L_{n/(\alpha-\varepsilon)}; \tag{77}$$

(when $\alpha = n$ we require further that $b \in L_{n/(n-1-\varepsilon)}$).

In Section 12 we shall consider the general behavior of a solution at an isolated non-removable singularity. The final sections of the paper establish the existence of solutions with positive singularities under suitable additional conditions.

12. Behavior of solutions at an isolated singularity

Under the assumptions (76, 77) noted above we have the following basic result.

THEOREM 12. *Let u be a continuous solution of (75) in the set $D-\{0\}$, where D is a domain in Ω. Suppose that $u \geqslant L$, for some constant L. Then either u has a removable singularity at 0, or else*

$$u \approx \begin{cases} r^{(\alpha-1)/(\alpha-1)}, & \alpha < n, \\ \log 1/r, & \alpha = n, \end{cases} \tag{78}$$

in the neighborhood of the origin. (Here $f \approx g$ means that $C' \leqslant f/g \leqslant C''$ where C' and C'' are positive constants.)

Proof. Let Θ be a strongly differentiable function with compact support in D, which is identically 1 in some neighborhood of the origin. We assert that

$$\int \Theta_x \cdot \mathcal{A}\, dx = \text{Const.} = K, \tag{79}$$

where the constant is independent of the particular choice of Θ.

In order to see this, let Θ and $\overline{\Theta}$ be two functions satisfying the above condition (that is, with compact support in D and identically 1 in a neighborhood of 0). Then $\phi = \overline{\Theta} - \Theta$ has compact support in $D-\{0\}$. Therefore

$$\int (\overline{\Theta} - \Theta)_x \cdot \mathcal{A}\, dx = 0,$$

or in other words,

$$\int \overline{\Theta}_x \cdot \mathcal{A}\, dx = \int \Theta_x \cdot \mathcal{A}\, dx.$$

This proves the assertion.

Now assume that the singularity at 0 is not removable. We must then prove that u has the asymptotic behavior (78) at 0. Let R be chosen so that D contains the ball $r = |x| \leqslant R$. We may assume without loss of generality, moreover, that $u < 0$ on the circumference $|x| = R$. By the remark at the end of Section 7 it is clear that $u \to \infty$ as $x \to 0$. Hence there exists some constant $\sigma_0 > 0$ such that $u > 0$ for $|x| \leqslant \sigma_0$. Let $M = M(\sigma)$ and $\mu = \mu(\sigma)$ be respectively the maximum and minimum of u on a given circumference $|x| = \sigma < \sigma_0$. Furthermore, for $\sigma < |x| < R$ let us define the function

$$v = v(x, \sigma) = \begin{cases} 0 & \text{if} \quad u(x) \leqslant 0, \\ u(x) & \text{if} \quad 0 < u(x) < \mu, \\ \mu & \text{if} \quad u(x) \geqslant \mu. \end{cases}$$

We may suppose the definition of v is extended to all of E^n by setting $v=0$ for $|x| \geqslant R$ and $v=\mu$ for $|x| \leqslant \sigma$. Then v is strongly differentiable in all space, has compact support in D, and is identically equal to μ in some neighborhood of the origin.

Now for fixed $\sigma < \sigma_0$, and $\mu = \mu(\sigma)$, $v = v(x, \sigma)$, we have by (79)

$$\mu K = \int v_x \cdot \mathcal{A}(x, u, u_x)\, dx = \int v_x \cdot \mathcal{A}(x, v, v_x)\, dx, \tag{80}$$

since $v=u$ and $v_x = u_x$ almost everywhere in the set where $v_x \neq 0$ (see the remarks at the conclusion of Section 0). By inequality (76), moreover,

$$\int v_x \cdot \mathcal{A}(x, v, v_x)\, dx \geqslant \int (|v_x|^\alpha - d\,|v|^\alpha - g)\, dx. \tag{81}$$

The second term on the right may be estimated by Hölder's inequality and Lemma 4, thus (¹)

$$\int d\,|v|^\alpha dx \leqslant \|d\|_{n/\alpha} \|v\|_{\alpha*}^\alpha \leqslant \text{Const. } \|d\|_{n/\alpha} \|v_x\|_\alpha^\alpha. \tag{82}$$

Now the radius R introduced at the beginning of the proof may be taken as small as we please, and therefore it is clear that we may suppose Const. $\|d\|_{n/\alpha} \leqslant \frac{1}{2}$. Hence by (81) and (82)

$$\int v_x \cdot \mathcal{A}(x, v, v_x)\, dx \geqslant \tfrac{1}{2} \int |v_x|^\alpha dx - C. \tag{83}$$

Moreover, by Lemma 9 and Lemma 10, since $v \equiv \mu$ for $|x| \leqslant \sigma$,

$$\int |v_x|^\alpha dx \geqslant \omega_n \left(\frac{n-\alpha}{\alpha - 1}\right)^{\alpha - 1} \sigma^{n-\alpha} \mu^\alpha \tag{84}$$

(assuming $\alpha < n$). Thus from (80), (83), and (84) follows

$$\mu^\alpha \leqslant \text{Const. } (\mu K + C) \begin{cases} \sigma^{\alpha - n}, & \alpha < n, \\ |\log \sigma|^{n-1}, & \alpha = n, \end{cases} \tag{85}$$

valid for $\sigma < \sigma_0$.

(¹) The calculation is given only for the case $\alpha < n$. If $\alpha = n$ only slight changes are necessary.

We assert that $K > 0$. Indeed in the contrary case we obtain from (85)

$$\mu \leqslant \text{Const.}\begin{cases} \sigma^{(\alpha-n)/\alpha}, & \alpha < n, \\ |\log \sigma|^{1-1/n}, & \alpha = n. \end{cases}$$

But then by the Harnack inequality, (Theorem 7), the value $M = M(\sigma)$ obeys the same relation. Thus $u = O(r^{(\alpha-n)/\alpha})$ or $O(|\log r|^{1-1/n})$ depending on whether $\alpha < n$ or $\alpha = n$. Therefore by Theorem 11 the singularity at 0 is removable, which contradicts our initial assumption. The assertion thus being established, it now follows from (85) and Lemma 2 that

$$\mu \leqslant \text{Const.}\begin{cases} \sigma^{(\alpha-n)/(\alpha-1)}, & \alpha < n, \\ |\log \sigma|, & \alpha = n. \end{cases} \tag{86}$$

The next step in the proof is to obtain a reverse inequality (cf. (92) below) for the value M. To this end we introduce a new comparison function $V = V(x, \sigma)$ according to the formula

$$V = \begin{cases} 0 & \text{when } |x| \geqslant R \\ \text{Max}\,(0, u) & \text{when } \sigma_0 < |x| < R \\ u & \text{when } \sigma \leqslant |x| \leqslant \sigma_0 \\ \text{Min}\,(M, u) & \text{when } 0 < |x| < \sigma \\ M & \text{when } |x| = 0. \end{cases}$$

Evidently V is continuous and strongly differentiable, has compact support in D, and $V \equiv M$ in some neighborhood of the origin. Moreover, for fixed $\sigma < \sigma_0$, and $\tau = (\alpha - n)/(\alpha - 1)$, we set (¹)

$$H = H(r, \sigma) = \begin{cases} 0 & \text{when } |x| \geqslant \sigma_0 \\ M\dfrac{r^\tau - \sigma_0^\tau}{\sigma^\tau - \sigma_0^\tau} & \text{when } \sigma < |x| < \sigma_0 \\ M & \text{when } |x| \leqslant \sigma. \end{cases}$$

Again H is strongly differentiable, has compact support in D, and $H \equiv M$ in a neighborhood of the origin. Thus letting Δ denote the annulus $\sigma < r < \sigma_0$, we have by (79),

$$MK = \int H_x \cdot \mathcal{A}\, dx \leqslant \int_\Delta \left\{ \frac{1}{\alpha}\left|\frac{H_x}{\varepsilon}\right|^\alpha + \frac{\alpha - 1}{\alpha}\, |\varepsilon \mathcal{A}|^{\alpha/(\alpha-1)} \right\} dx, \tag{87}$$

using Young's inequality. Now by (76)

(¹) Cf. the preceding footnote.

$$\int_{\Delta} |\mathcal{A}|^{\alpha/(\alpha-1)} dx \leq C \int_{\Delta} \{|u_x|^{\alpha} + b^{\alpha/(\alpha-1)} |u|^{\alpha} + e^{\alpha/(\alpha-1)}\} dx$$
$$= C \int_{\Delta} \{|u_x|^{\alpha} - 2d|u|^{\alpha} - 2g + (2d + b^{\alpha/(\alpha-1)}) |u|^{\alpha} + (2g + e^{\alpha/(\alpha-1)})\} dx. \tag{88}$$

The integral on the right side of (88) is only increased if it is extended over the set of points in $0 < |x| < R$ where $V = u$. Denoting this set by Δ' we have

$$\int_{\Delta'} (2d + b^{\alpha/(\alpha-1)}) |u|^{\alpha} dx \leq \int (2d + b^{\alpha/(\alpha-1)}) |V|^{\alpha} dx$$
$$\leq \text{Const. } (\|d\|_{n/\alpha} + \|b\|_{n/(\alpha-1)}^{\alpha/(\alpha-1)}) \cdot \|V_x\|_{\alpha}^{\alpha}$$
$$= \text{Const. } (\|d\|_{n/\alpha} + \|b\|_{n/(\alpha-1)}^{\alpha/(\alpha-1)}) \cdot \int_{\Delta'} |u_x|_{\alpha} dx, \tag{89}$$

since $V_x = u_x$ almost everywhere in Δ' and $V_x = 0$ almost everywhere in the complement of Δ'. Again supposing that R is suitably small, (89) implies

$$\int_{\Delta'} (2d + b^{\alpha/(\alpha-1)}) |u|^{\alpha} dx \leq \int_{\Delta'} |u_x|^{\alpha} dx.$$

Substituting this into (88), and then using (76), yields

$$\int_{\Delta} |\mathcal{A}|^{\alpha/(\alpha-1)} dx \leq C\left(1 + \int_{\Delta'} u_x \cdot \mathcal{A}\, dx\right) = C\left(1 + \int V_x \cdot \mathcal{A}\, dx\right) = C(1 + MK). \tag{90}$$

Combining (87) and (90) there results

$$MK \leq (1/\alpha\varepsilon^{\alpha}) \|H_x\|_{\alpha}^{\alpha} + C\varepsilon^{\alpha/(\alpha-1)} (1 + MK).$$

Thus choosing ε so that $C\varepsilon^{\alpha/(\alpha-1)} = \frac{1}{2}$, we obtain

$$MK \leq \text{Const. } \|H_x\|_{\alpha}^{\alpha} + 1.$$

On the other hand, by the calculation of Lemma 10,

$$\|H_x\|_{\alpha}^{\alpha} = \omega_n \left(\frac{n-\alpha}{\alpha-1}\right)^{\alpha-1} (\sigma^{\tau} - \sigma_0^{\tau})^{1-\alpha} M^{\alpha}.$$

Therefore
$$CM^{\alpha} \geq (MK - 1) \begin{cases} (\sigma^{\tau} - \sigma_0^{\tau})^{\alpha-1}, & \alpha < n, \\ |\log(\sigma/\sigma_0)|^{n-1}, & \alpha = n, \end{cases} \tag{91}$$

valid for $\sigma < \sigma_0$. It has already been noted that $u \to \infty$ as $x \to 0$. Hence there exists some number $\sigma_1, 0 < \sigma_1 < \sigma_0$, such that for all $\sigma \leq \sigma_1$,

$$MK \geqslant 2, \quad (\sigma/\sigma_0)^\tau \geqslant 2.$$

It follows from (81), therefore, that

$$M \geqslant \text{Const.} \begin{cases} \sigma^{(\alpha-n)/(\alpha-1)}, & \alpha < n, \\ |\log \sigma|, & \alpha = n, \end{cases} \tag{92}$$

valid for all $\sigma \leqslant \sigma_1$.

What has been shown, then, is (by (86)) that the minimum of u grows at most at the rate $r^{(\alpha-n)/(\alpha-1)}$, while the maximum grows at least this fast. The required asymptotic estimate then follows from the Harnack inequality. Indeed, by (86) and Theorem 7 we have

$$M \leqslant C'(\mu + k') \leqslant \text{Const. } \sigma^{(\alpha-n)/(\alpha-1)}, \quad (\sigma \leqslant \sigma_1),$$

while by (92) and Theorem 7

$$\mu \geqslant M/C' - k' \geqslant \text{Const. } \sigma^{(\alpha-n)/(\alpha-1)}, \quad (\sigma \leqslant \sigma_1)$$

(in applying Theorem 7 one first normalizes each sphere $|x| = \sigma$ to unit radius). Thus (78) holds in a neighborhood of the origin, when $\alpha < n$. The result for $\alpha = n$ is obtained in the same way, and Theorem 12 is completed.

13. Existence of solutions with isolated singularities

The very light hypotheses required for the proof of Theorem 12 do not seem strong enough to prove the general *existence* of solutions having positive isolated singularities. We shall therefore restrict consideration in this section to the equation

$$\operatorname{div} \mathcal{A}(x, u_x) = 0, \quad (\alpha \leqslant n), \tag{93}$$

where the function $\mathcal{A}$ is independent of u. In addition to the usual conditions on $\mathcal{A}$, we shall suppose that (93) also has the following four properties.

P1. For all $x \in \Omega$ and all values of p and q

$$(p - q) \cdot \{\mathcal{A}(x, p) - \mathcal{A}(x, q)\} \geqslant 0,$$

with equality holding if and only if $p = q$.

P2. For smooth boundaries and smooth (continuous) boundary data there exist smooth (continuous) solutions of (93) taking on the given boundary data.

P3. The uniform limit of solutions of (93) is also a solution of (93), and

P4. Let Γ denote a spherical annulus in Ω with center at the origin. By *P2* there exists a solution v of (93) in Γ, taking the constant values m_0 on the outer circumference and m on the inner. Also by virtue of assumption *P1* a weak maximum principle holds for the difference of any two solutions. Consequently at any point P in Γ the values $v(P)$ increase monotonically with m. *As the final assumption, we suppose that $v(P)$ tends to infinity as m does.*

We observe that *P1* is easily verified for linear elliptic equations, and also for variational problems whose integrands are strictly convex in the variable p. More generally if $\mathcal{A}$ is of class C^1 in p, then *P1* follows from the condition that $\mathcal{A}_{i,k}\xi_i\xi_k$ be positive definite.

Assumption *P2* is satisfied for a wide class of equations of the form (93), as shown by the work of Morrey, Ladyzhenskaya and Uraltseva, Gilbarg, and Stampacchia. It would be superfluous to elaborate on this, however, for on the one hand new classes of equations for which *P2* holds will certainly be discovered, while on the other it seems that only in pathological circumstances will *P2* generally fail.

Turning to assumption *P3*, it is not hard to see that it holds for linear equations, and for equations admitting an *a priori* estimate of the continuity modulus of the first derivative of solutions. *P3* may also be established in case $\mathcal{A}(x,p)$ is continuous in all its variables, though this is more delicate and is stated merely for the record.

Assumption *P4* is of rather a different sort, and we shall therefore note here a class of equations for which it is valid. In particular, suppose that $\mathcal{A}$ is of class C^1 in x and p, at least for all suitably large values of p, and let

$$\lambda = \lambda(x,p), \quad \Lambda = \Lambda(x,p)$$

be respectively the minimum and maximum eigenvalues of the quadratic form $\mathcal{A}_{i,k}\xi_i\xi_k$ (by condition *P1* it is clear that $\lambda \geqslant 0$). We now assume that the quantities

$$\Lambda/\lambda \quad \text{and} \quad |\sum \partial \mathcal{A}_i/\partial x_k|/\lambda|p| \tag{94}$$

are uniformly bounded for suitably large values of p.(1) Then *P4* holds.

To see this, suppose for definiteness that the annulus Γ has outer radius 1 and inner radius σ, and consider the function

(1) This condition is satisfied by many of the equations studied in the recent literature (cf. [5], [8], [9], [15], [25], etc.).

$$k = k(r) = (m - m_0)\,\frac{r^\gamma - 1}{\sigma^\gamma - 1} + m_0, \quad (\gamma = \text{Const.} < 0).$$

Then
$$\operatorname{div} \mathcal{A}(x, k_x) = (\mathcal{A}_{i,k}\,\xi_i\,\xi_k)\,k'' + (\mathcal{A}_{i,i} - \mathcal{A}_{i,k}\,\xi_i\,\xi_k)\,k'/r + \sum \frac{\partial \mathcal{A}_i}{\partial x_i}$$
$$\geqslant \lambda k'' + (n-1)\Lambda k'/r - \left|\sum \frac{\partial \mathcal{A}_i}{\partial x_k}\right|,$$

where $\xi = x/r$ and the primes denote differentiation with respect to r. Now let B be a bound for the quantities (94) when $p \geqslant N$, say. We choose γ according to the relation

$$1 - \gamma = nB.$$

Therefore $|k_x| = -k' = -\gamma(m - m_0)(\sigma^\gamma - 1)^{-1} r^{\gamma-1}$ is greater than N for all m sufficiently large. Consequently for large m,

$$\operatorname{div} \mathcal{A}(x, k_x) \geqslant -\lambda\gamma(m - m_0)(\sigma^\gamma - 1)^{-1} r^{\gamma-2} \cdot \{(1-\gamma) - nB\} = 0.$$

Thus by virtue of the maximum principle one has $v \geqslant k$, where v is the solution postulated in assumption *P4*. But $k(P) \to \infty$ as $m \to \infty$, hence so does $v(P)$, completing the demonstration.

This completes our discussion of assumptions *P1* through *P4*. We emphasize that the restrictions placed on (93) by these conditions are not particularly heavy, so that the following result holds in considerable generality.

THEOREM 13. *Suppose that equation* (93) *admits the properties P1 through P4 listed above. Let smooth* (*continuous*) *boundary data* $\psi = \psi(x)$ *be assigned on the sphere* $|x| = 1$, *it being assumed that* $|x| < 1$ *is contained in* Ω.

Then there exists a one parameter, linearly ordered, family of solutions $G = G(x)$ *in the domain* $0 < |x| < 1$, *taking on the given boundary values and satisfying*

$$G \approx \begin{cases} r^{(\alpha-n)/(\alpha-1)}, & \alpha < n, \\ \log 1/r, & \alpha = n, \end{cases} \tag{95}$$

in the neighborhood of the origin. The value of G may be assigned arbitrarily at a given point P, subject only to the restriction $G(P) > w(P)$, *where w denotes the solution of* (93) *in the ball* $|x| < 1$ *which takes on the assigned boundary values.*

Remark. By *linearly ordered*, we mean that if G_1 and G_2 are two different members of the family, then either $G_1 \leqslant G_2$ or $G_1 \geqslant G_2$ in $0 < |x| < 1$.

Proof of Theorem 13. We shall first construct a solution $G = G(x)$ in $0<|x|<1$ taking on the assigned boundary data, and with a prescribed value $\chi > w(P)$ at some fixed point P. It will then be shown that this solution satisfies (95), and finally that it can be assumed to increase as χ increases. By making the change of dependent variables $\bar{u} = u + \text{const.}$, with the constant suitably chosen, it may be supposed without loss of generality that $w \geqslant 0$. We assume this done, and then drop the bar on u.

Now let Γ_σ denote the annulus $\sigma<|x|<1$, and let $v_{\sigma,m}$ be the particular solution of (93) in Γ_σ which obeys the boundary conditions

$$v=\psi \text{ on } |x|=1, \quad v=w+m \text{ on } |x|=\sigma.$$

Obviously such a solution exists by assumption *P2*. We assert that m can be chosen in such a way that

$$v_{\sigma,m}(P)=\chi \tag{96}$$

(it is assumed that σ is small enough for P to be contained in Γ_σ). Indeed, if $m=0$ then obviously $v_{\sigma,m}=w$, while by the maximum principle the value $v_{\sigma,m}(P)$ continuously increases as m increases. By assumption *P4* it is evident that $v_{\sigma,m}(P)$ tends to infinity as m does. Hence there exists a first value $m=m(\sigma)$ such that (96) holds. We shall henceforth write v_σ for the function $v_{\sigma,m(\sigma)}$.

Now each function v_σ is a positive solution of (93) in the corresponding annulus Γ_σ, and satisfies $v_\sigma(P)=\chi$. If we write $\theta=|x_P|$, then by the Harnack inequality (Theorem 7) we have for fixed C' and k'

$$v_\sigma(x) \leqslant C'(\chi+k') \quad \text{on} \quad |x|=\theta,$$

so long as $\sigma \leqslant \theta/2$. Let W be the solution of (93) in the annulus Γ_θ taking on the boundary values

$$W=\psi \quad \text{on} \quad |x|=1, \qquad W=C'(\chi+k') \quad \text{on} \quad |x|=\theta.$$

Then according to the maximum principle

$$w \leqslant v_\sigma \leqslant W \quad \text{in} \quad \Gamma_\theta,$$

valid for all $\sigma \leqslant \theta/2$. We assert also that in each fixed annulus $h<|x|<1-h$ the functions v_σ are uniformly bounded and equicontinuous, provided that $\sigma \leqslant h/4$. Indeed by the Harnack inequality it is not hard to see that v_σ must be uniformly bounded in $h/2<|x|<1$, when $\sigma \leqslant h/4$. But then by Theorem 8 these functions are equicontinuous in $h<|x|<1-h$, as asserted.

Now consider the family $\{v_\sigma\}$ as $\sigma \to 0$. It is clear from what has already been shown that there is a sequence of values $\sigma_1, \sigma_2, \ldots$ such that

$$v_{\sigma_\nu}(x) \to \text{limit}, \quad 0 < |x| < 1,$$

the convergence being uniform in any fixed annulus $h < |x| < 1 - h$, (Ascoli's theorem). We denote the limit by $G(x)$. Clearly $G(x)$ is a continuous function satisfying $G(P) = \chi$. Moreover,

$$w \leqslant G \leqslant W \quad \text{in} \quad \Gamma_\theta.$$

It thus follows that G is *continuous* in $0 < |x| \leqslant 1$ and takes the assigned boundary values on $|x| = 1$, as required.

That G is a solution of (93) in $0 < |x| < 1$ follows immediately from property $P3$. It still must be shown that G has the asymptotic behavior (95). By Theorem 12, however, any solution which is bounded below must either satisfy (95) or have a removable singularity at 0. It is therefore enough to show that G cannot have a removable singularity. Thus suppose for contradiction that G could be defined at 0 so as to be a solution in the entire ball $|x| < 1$. Then since $G = \psi$ on $|x| = 1$ we would necessarily have $G \equiv w$ by the maximum principle. This is impossible, however, since $G(P) = \chi > w(P)$. Thus (95) is established.

It remains to show that the above construction leads to a one parameter linearly ordered family of solutions. Let a dense denumerable set of values χ be chosen in the interval $(w(P), \infty)$. Then it is clear that, by a diagonal process, a *fixed* sequence $\sigma_1, \sigma_2, \ldots$ can be used to define the functions G corresponding to every χ in the set. But in this case the functions v_{σ_ν} corresponding to a particular value of χ are less than or equal to those belonging to a larger value of χ (recall the unique definition of each function v_σ). It follows that if χ_1 and χ_2 are two values in the set, with $\chi_1 < \chi_2$, then also $G_1 \leqslant G_2$. Having thus obtained a monotone family of solutions G for the values of χ in a dense set, it is now a simple matter to construct solutions for the omitted values of χ, by taking limits of those already constructed. We may omit the details of this process, which depend of course on Theorem 8 and property $P3$. This completes the proof of Theorem 13.

14. Linear equations

The results of the preceding sections can be sharpened somewhat in case (93) is linear, that is, of the form

$$\frac{\partial}{\partial x_i}\left(a_{ij}(x)\frac{\partial u}{\partial x_j}\right)=0, \tag{97}$$

where $\lambda\xi^2 \leqslant a_{ij}\xi_i\xi_j \leqslant \Lambda\xi^2$ and λ and Λ are positive constants. We note that assumptions *P1* through *P4* are satisfied in the present case—*P1* quite obviously, *P2* on the basis of the work of Ladyzhenskaya and Uraltseva, *P3* in view of Theorem 1 and the weak compactness of L_2, and finally *P4* by virtue of linearity. This being established, it follows that both Theorems 12 and 13 hold for equation (97), with $\alpha=2$.(1) In addition, we have the following supplementary results.

THEOREM 14. *Let G be a particular solution of* (97) *in the set $D-\{0\}$, such that $G\approx r^{2-n}$ or $G\approx \log 1/r$ depending on whether $n>2$ or $n=2$. Then every non-negative solution of* (97) *in $D-\{0\}$ has the form*

$$u=\text{Const. } G+w,$$

where w is a solution of (97) *in the entire domain D.*

Remark. It follows that in the case of equation (97) the one parameter family of solutions given by Theorem 13 is unique.

Theorem 14 is an exact analogue of Theorem 5 of reference [6], and is demonstrated by the same method. Note that the proof in [6] is restricted to two dimensions because the Harnack inequality used there was proved only for two variables; since in the present case we have a Harnack inequality irrespective of dimension, it is clear that the argument carries over intact.

THEOREM 15. *Let u be a continuous solution of* (97) *in the set $D-\{0\}$, where D is a domain in Ω. Suppose that $u=O(r^{2-n})$ or $u=O(\log 1/r)$, depending on whether $n>2$ or $n=2$. Then either u has a removable singularity at* 0, *or else (possibly after multiplication by* -1)

$$u\approx\begin{cases} r^{2-n}, & n>2,\\ \log 1/r, & n=2,\end{cases} \tag{98}$$

in the neighborhood of the origin.

Proof. For a suitably large constant A we have

$$u+AG>0$$

in the neighborhood of the origin, where G is some particular solution satisfying

(1) As we have remarked in the introduction, these special results are due originally to Royden and to Littman, Stampacchia, and Weinberger.

$G \approx r^{2-n}$ or $G \approx \log 1/r$ (such a function exists by Theorem 13). The function $u + AG$ is therefore a positive solution of (97) in some neighborhood of the origin. Consequently by Theorem 12 we have $u + AG \approx r^{2-n}$ or $\log 1/r$, and the required conclusion follows at once.

The above argument is essentially due to Gevrey, who proved a similar result for linear equations of the form $a_{ij}u,_{ij} + b_i u,_i + cu = 0$ with certain smoothness conditions placed on the coefficients a_{ij}.

Remarks. For the Laplace equation the result of Theorem 15 can be considerably improved. In fact in this case the hypothesis can be weakened to read

$$u = o(r^{1-n}),$$

without affecting the conclusion. For the general class of equations under consideration here, however, there is no immediately analogous result—that is, *the order of growth $O(r^{2-n})$ or $O(\log 1/r)$ is best possible in Theorem 15.*

Indeed, consider equation (97) with

$$a_{ij} = \delta_{ij} + (a-1)\, x_i x_j / r^2, \tag{99}$$

where a is a constant greater than one (cf. [6], p. 336). One easily verifies that

$$\xi^2 \leqslant a_{ij}\, \xi_i\, \xi_j \leqslant a\, \xi^2,$$

so that (99) is an allowable set of coefficients. We consider solutions of the form

$$u = H(x)\, f(r), \tag{100}$$

where H is a harmonic polynomial of degree m, and $r = |x|$. After a straightforward calculation we find that (100) satisfies (97), (99) provided $f = r^{2-m-n-\varepsilon}$ and a and ε are related by

$$a = \frac{m(m+n-2)}{\varepsilon(\varepsilon+n-2)}, \qquad (0 < \varepsilon < m).$$

The solution (100) is therefore $O(r^{2-n-\varepsilon})$, that is, *there exist solutions of linear equations of the form* (97) *which are $O(r^{2-n-\varepsilon})$ for any preassigned $\varepsilon > 0$ and yet are not $\approx r^{2-n}$.*

Note. This work was partially supported by the United States Air Force Office of Scientific Research under Grant No $AF - AFOSR - 62 - 101$.

References

[1]. Carleson, L., *Selected problems on exceptional sets*. Uppsala, 1961. (Mimeographed.)

[2]. Finn, R., Isolated singularities of solutions of non-linear partial differential equations. *Trans. Amer. Math. Soc.* 75 (1953), 385–404.

[3]. Gevrey, M., Sur une généralisation du principe des singularités positives de M. Picard. *C. R. Acad. Sci. Paris*, 211 (1940), 581–584.

[4]. Gilbarg, D., Some local properties of elliptic equations. *Proceedings of Symposia on Pure Mathematics, Vol. 4, Partial Differential Equations*, pp. 127–142. Amer. Math. Soc., Providence, R.I., 1961.

[5]. —— Boundary value problems for non-linear elliptic equations. *Nonlinear Problems*, pp. 151–160. Univ. of Wisconsin Press, Madison, 1963.

[6]. Gilbarg, D. & Serrin, J., On isolated singularities of solutions of second order elliptic differential equations. *J. d'Analyse Math.*, 4 (1955–1956), 309–340.

[7]. John, F. & Nirenberg, L., On functions of bounded mean oscillation. *Comm. Pure Appl. Math.*, 14 (1961), 415–426.

[8]. Ladyzhenskaya, O. A. & Uraltseva, N. N., Quasi-linear elliptic equations and variational problems with many independent variables. *Uspehi Mat. Nauk*, 16 (1961), 19–92; translated in *Russian Math. Surveys*, 16 (1961), 17–91.

[9]. —— On the smoothness of weak solutions of quasi-linear equations in several variables and of variational problems. *Comm. Pure Appl. Math.*, 14 (1961), 481–495.

[10]. Littman, W., Stampacchia, G. & Weinberger, H., Regular points for elliptic equations with discontinuous coefficients. *Ann. Scuola Norm. Sup. Pisa*, Ser. III, 17 (1963), 45–79.

[11]. Mazya, V. G., Some estimates for the solutions of second order elliptic equations. *Dokl. Akad. Nauk SSSR*, 137 (1961), 1057–1059; translated in *Soviet Math. Dokl.*, 2 (1961), 413–415.

[12]. —— p-conductance and theorems of imbedding certain function spaces into the space C. *Dokl. Akad. Nauk SSSR*, 140 (1961), 299–302; translated in *Soviet Math. Dokl.*, 2 (1961), 1200–1203.

[13]. Morrey, C. B., Jr., Second order elliptic equations and Hölder continuity. *Math. Z.*, 72 (1959), 146–164.

[14]. —— Multiple integral problems in the calculus of variations and related topics. *Ann. Scuola Norm. Sup. Pisa*, 14 (1960), 1–62.

[15]. —— Existence and differentiability theorems for variational problems for multiple integrals. *Partial Differential Equations and Continuum Mechanics*, pp. 241–270. Univ. of Wisconsin Press, Madison, 1961.

[16]. Moser, J., A new proof of De Giorgi's theorem concerning the regularity problem for elliptic differential equations. *Comm. Pure Appl. Math.*, 13 (1960), 457–468.

[17]. —— On Harnack's theorem for elliptic differential equations. *Comm. Pure Appl. Math.*, 14 (1961), 577–591.

[18]. Nirenberg, L., On elliptic partial differential equations. *Ann. Scuola Norm. Sup. Pisa*, 13 (1959), 1–48.

[19]. Picone, M., Sur la théorie d'un équation aux derivées partielles classique de la physique mathématique. *C. R. Acad. Sci. Paris*, 266 (1948), 1945–1947. Cf. also, Sulle singularitá delle soluzioni di una classica equazione a derivate parziali delle fisica matematica. *Atti 3° Congresso U.M.I.*, pp. 69–71. Rome, 1951.

[20]. Royden, H., The growth of a fundamental solution of an elliptic divergence structure equation. *Studies in Mathematical Analysis and Related Topics: Essays in honor of G. Polya*, pp. 333–340. Stanford Univ. Press, Stanford, 1962.

[21]. SERRIN, J., Dirichlet's principle in the calculus of variations. *Proceedings of Symposia on Pure Mathematics, Vol. 4, Partial Differential Equations*, pp. 17–22. Amer. Math. Soc., Providence, R.I., 1961.

[22]. —— A Harnack inequality for nonlinear equations. *Bull. Amer. Math. Soc.*, 69 (1963), 481–486.

[23]. STAMPACCHIA, G., Régularisation des solutions de problèmes aux limites elliptiques à données discontinues. *Intern. Symposium on Linear Spaces*, pp. 399–408. Jerusalem, 1960.

[24]. —— Problemi al contorno ellittici, con dati discontinui dotati di soluzioni hölderiane. *Ann. Pura Appl.*, 51 (1960), 1–38.

[25]. —— On some multiple integral problems in the calculus of variations. *Comm. Pure Appl. Math.*, 16 (1963), 382–422.

[26]. WALLIN, H., A connection between α-capacity and L^p-classes of differentiable functions. *Ark. Mat.*, 5 (1964), 331–341.

Received October 17, 1963

Acta Math. 113 (1965), 219–240

Reprint I.1.[iii]

ISOLATED SINGULARITIES OF SOLUTIONS OF QUASI-LINEAR EQUATIONS

BY

JAMES SERRIN

University of Minnesota, U.S.A.

In a previous paper in this journal [1] we have studied the local behavior of solutions of non-linear second order partial differential equations having the form

$$\operatorname{div} \mathcal{A}(x, u, u_x) = \mathcal{B}(x, u, u_x). \tag{1}$$

Here $\mathcal{A}$ is a given vector function of the variables x, u, u_x, $\mathcal{B}$ is a given scalar function of the same variables, and $u_x = (\partial u/\partial x_1, \ldots, \partial u/\partial x_n)$ denotes the gradient of the dependent variable $u = u(x)$, where $x = (x_1, \ldots, x_n)$. The final chapter of [1] treated a variety of problems concerning the behavior of solutions at an isolated singularity. In that chapter I found it necessary to impose the condition $\mathcal{B} \equiv 0$, a condition which had not been required in the preceding parts of [1]. The main purpose of the present paper is to show that this additional condition can be removed, and thus to complete, in an important way, the theory of the earlier paper.

We assume as always that the functions $\mathcal{A}$ and $\mathcal{B}$ are defined for all points x in some connected open set (domain) Ω of the Euclidean number space E^n, and for all values of u and u_x. Furthermore, they are to satisfy inequalities of the general form

$$\begin{aligned} |\mathcal{A}(x,u,p)| &\leqslant a|p|^{\alpha-1} + b|u|^{\alpha-1} + e \\ |\mathcal{B}(x,u,p)| &\leqslant c|p|^{\alpha-1} + d|u|^{\alpha-1} + f \\ p\cdot\mathcal{A}(x,u,p) &\geqslant |p|^{\alpha} - d|u|^{\alpha} - g, \end{aligned} \tag{2}$$

where the exponent α is a fixed number in the range $1 < \alpha \leqslant n$, the coefficient a is a positive real number, and the coefficients b through g are measurable functions of x contained in the respective Lebesgue classes

$$b, e \in L_{n/(\alpha-1-\varepsilon)}; \quad c \in L_{n/(1-\varepsilon)}; \quad d, f, g \in L_{n/(\alpha-\varepsilon)},$$

where $\varepsilon > 0$, (if $\alpha < n$, the Lebesgue class of b can be weakened to $L_{n/(\alpha-1)}$). We shall then prove the following result, generalizing Theorem 12 of [1] to the case of a non-vanishing right hand side $\mathcal{B}(x, u, u_x)$.

THEOREM 1. *Let u be a continuous solution of* (1) *in the set $D-\{0\}$, where D is a domain in Ω. Suppose that $u \geqslant L$ for some constant L. Then either u has a removable singularity at* 0, *or else*

$$u \approx \begin{cases} r^{(\alpha-n)/(\alpha-1)}, & \alpha < n, \\ \log 1/r, & \alpha = n, \end{cases} \tag{3}$$

in the neighborhood of the origin. (*Here $f \approx g$ means that $C' \leqslant f/g \leqslant C''$, where C' and C'' are positive constants.*)

This result applies in particular to solutions of the variational problem

$$\delta \int F(x, u, u_x)\, dx = 0,$$

provided the integrand obeys certain natural conditions (cf. [2]). In the case of quadratic variational problems, for example, Theorem 1 implies that the order of growth at a positive isolated singularity is r^{2-n} or $\log 1/r$, depending on whether $n > 2$ or $n = 2$. Earlier results of a similar nature are noted in the introductory paragraphs of [1].

Theorem 1 is proved in the following two sections. In contrast with the simple statement of the theorem, the intricacy of the proof comes as something of a surprise, lending its own interest to the result. The method of proof furthermore shows that a solution with a positive isolated singularity has certain of the attributes of a *fundamental solution* (cf. Theorem 3). In Section 4 we show under suitable conditions that there exists a solution of (1) with precisely the asymptotic behavior (3). This result is a generalization of Theorem 13 of [1], both in the equation treated and in the weaker structure required. For linear equations

$$\frac{\partial}{\partial x_i}\left(a_{ij}\frac{\partial u}{\partial x_j} + b_i u + e_i\right) = c_i \frac{\partial u}{\partial x_i} + du + f \tag{4}$$

it is possible to obtain somewhat more detailed conclusions. In Section 5 we consider two results generalizing Theorems 14 and 15 of reference [1]. Although we shall not present the details here, it is also worth noting that the results of the paper may be used to construct a Green's function $G(x, y)$ for (4), having the usual properties of positivity and symmetry, and yelding a representation formula for solutions of the Dirichlet problem with zero boundary data.

The final section of the paper is more or less unrelated to the earlier part. Here we take the opportunity to correct an error in [1] occurring in the statement and proof of the maximum principle. We also point out several places in [1] where the results can be slightly improved.

It is assumed throughout that the reader is familiar with Chapters I and II of [1]. Moreover, the notation and terminology of that paper will be used as needed. In particular, we recall that a *continuous solution of* (1) *in a domain* D is a function u which is continuous and has strong derivatives which are locally of class L_α over D, and is such that

$$\int(\phi_x\cdot \mathcal{A}+\phi \mathcal{B})\,dx=0$$

for any continously differentiable function $\phi=\phi(x)$ with compact support in D.

1. Proof of Theorem 1. First Part

We shall assume that the singularity at 0 is *not* removable. To prove the theorem it must therefore be shown that u has the asymptotic behavior (3) in the neighborhood of the origin. We shall restrict our discussion, in fact, to the ball $S=\{|x|\leqslant R\}$ in D, where R is chosen so that the Lebesgue norms $\|b\|$, $\|c\|$, and $\|d\|$ over S are suitably small (how small will be determined in the course of the proof, but in any case will depend only on α, n, and ε). It may be assumed without loss of generality that $u<0$ on the circumference $|x|=R$; indeed, if this is not already the case, it can be accomplished simply by the subtraction of a suitable constant from u, a device which affects the structure of (1) only by increasing the relatively unimportant coefficients e, f, and g.

Lemma 1. *Let* Θ *be a stongly* (L_α) *differentiable function with compact support in* $|x|<R$, *which is identically* 1 *in some neighborhood of the origin. Then*

$$\int\{\Theta_x\cdot \mathcal{A}+(\Theta-1)\mathcal{B}\}\,dx=\text{Const.}=\overline{K},$$

where the constant is independent of the particular choice of Θ.

Proof. Let Θ and $\overline{\Theta}$ be two functions satisfying the conditions of the lemma. Then $\phi=\overline{\Theta}-\Theta$ has compact support in $0<|x|<R$, and

$$\int\{(\bar{\Theta}-\Theta)_x\cdot\mathcal{A}+(\bar{\Theta}-\Theta)\mathcal{B}\}\,dx=0.$$

Since $\Theta\equiv 1$ near 0 this can be rewritten

$$\int\{\bar{\Theta}_x\cdot\mathcal{A}+(\bar{\Theta}-1)\mathcal{B}\}\,dx=\int\{\Theta_x\cdot\mathcal{A}+(\Theta-1)\mathcal{B}\}\,dx,$$

proving the lemma.

Since we have assumed that the singularity at 0 is *not* removable, it is clear from the remark on page 278 of [1] that $u\to\infty$ as $x\to 0$. Hence there exists a constant σ_0 such that $u\geqslant 1$ for $0<|x|\leqslant\sigma_0$. Let $m=m(\sigma)$ denote the minimum of u on a given sphere $|x|=\sigma$, $(\sigma\leqslant\sigma_0)$. For $\sigma\leqslant|x|\leqslant R$ let us define

$$v=v(x,\sigma)=\begin{cases}0 & \text{if } u(x)\leqslant 0\\ u & \text{if } 0<u(x)<m\\ m & \text{if } u(x)\geqslant m.\end{cases}$$

We suppose the definition of v to be extended to the entire ball S by setting $v\equiv m$ when $|x|\leqslant\sigma$. Then v is a strongly differentiable function with compact support in $|x|<R$, and is identically equal to m in a neighborhood of the origin.

Thus for fixed $\sigma\leqslant\sigma_0$, and $m=m(\sigma)$, $v=v(x,\sigma)$, we have by Lemma 1,

$$\begin{aligned}m\bar{K}&=\int\{v_x\cdot\mathcal{A}(x,u,u_x)+(v-m)\mathcal{B}(x,u,u_x)\}\,dx\\&=\int\{v_x\cdot\mathcal{A}(x,v,v_x)+(v-m)\mathcal{B}(x,u,u_x)\}\,dx,\end{aligned}\tag{5}$$

since $v=u$ and $v_x=u_x$ almost everywhere in the set where $v_x\neq 0$. Now using inequality (2),

$$\int v_x\cdot\mathcal{A}(x,v,v_x)\,dx\geqslant\int(|v_x|^\alpha-dv^\alpha-g)\,dx.\tag{6}$$

The second term on the right may be estimated by the Hölder and Sobolev inequalities, thus [1]

$$\int dv^\alpha\,dx\leqslant\|d\|_{n/\alpha}\|v\|^\alpha_{\alpha^*}\leqslant\text{Const.}\;\|d\|_{n/\alpha}\|v_x\|^\alpha_\alpha.\tag{7}$$

The radius R introduced at the beginning of the proof can be chosen so small that the coefficient of $\|v_x\|^\alpha_\alpha$ is $\leqslant\frac{1}{3}$. Hence by (6) and (7)

[1] This calculation is given for the case $\alpha<n$. If $\alpha=n$ only slight changes are necessary.

$$\int v_x \cdot \mathcal{A}(x, v, v_x)\, dx \geqslant \tfrac{2}{3} \int |v_x|^\alpha dx - C, \tag{8}$$

where C is an appropriate constant. Moreover, by Lemmas 9 and 10 of [1], since $v \equiv m$ for $|x| \leqslant \sigma$.

$$\int |v_x|^\alpha dx \geqslant \omega_n m^\alpha \begin{cases} \left(\dfrac{n-\alpha}{\alpha-1}\right)^{\alpha-1} \sigma^{n-\alpha}, & \alpha < n, \\ (\log R/\sigma)^{1-n}, & \alpha = n. \end{cases} \tag{9}$$

Inequalities (8) and (9) serve to estimate the first term on the right hand side of (5) from below. The remaining term can be estimated by setting

$$B(\sigma) \equiv \int_\sigma^R |\mathcal{B}(x, u, u_x)|\, dx \text{ (1)}$$

and noting that $|v - m| \leqslant m$. Thus we have

$$\int (v - m)\, \mathcal{B}(x, u, u_x)\, dx \geqslant -m B(\sigma). \tag{10}$$

It follows from (5), therefore, that

$$m^\alpha \leqslant \text{Const. } [(B(\sigma) + \bar{K})m + C] \begin{cases} \sigma^{\alpha-n}, & \alpha < n, \\ (\log R/\sigma)^{n-1}, & \alpha = n. \end{cases} \tag{11}$$

Since $m \geqslant 1$ we obtain finally

$$m^{\alpha-1} \leqslant \text{Const. } (B(\sigma) + 1) \begin{cases} \sigma^{\alpha-n}, & \alpha < n, \\ (\log R/\sigma)^{n-1}, & \alpha = n, \end{cases} \tag{12}$$

valid for $\sigma \leqslant \sigma_0$. (Many of the multiplicative factors which appear in the proof are denoted simply by Const. These factors are usually different from line to line, but they have this in common: they can be computed explicitly and they are independent of σ.)

LEMMA 2. *The function* $\mathcal{B}(x, u, u_x)$ *is integrable over* $0 < |x| \leqslant R$, *and*

$$\int_{r<\sigma} |\mathcal{B}(x, u, u_x)|\, dx \leqslant \text{Const.} \begin{cases} \sigma^\varepsilon, & \alpha < n, \\ \sigma^{\varepsilon/2}, & \alpha = n. \end{cases}$$

Proof. By the Harnack principle (cf. Theorem 7 of [1], and the corresponding remark in Section 6 of this paper) it is clear that for any fixed σ less than $\sigma_0/4$ one has

(1) We shall frequently write $\int_a^b$ instead of $\int_{a<r<b}$.

$$(1-\beta\sigma^{\varepsilon})^{-1}\leqslant 2, \qquad (1-\beta\sigma^{\varepsilon})^{-1}\leqslant \exp(2\beta\sigma^{\varepsilon}),$$

and consequently

$$B(\sigma)\leqslant 2\beta\sigma^{\varepsilon}+\exp(2\beta\sigma^{\varepsilon})\cdot B(2\sigma), \quad (\sigma\leqslant\hat{\sigma}).$$

By successive iteration

$$\begin{aligned} B(2^{-k}\hat{\sigma}) &\leqslant 2^{-k\varepsilon}+\exp(2^{-k\varepsilon})\cdot B(2^{1-k}\hat{\sigma}) \\ &\leqslant 2^{-k\varepsilon}+\exp(2^{-k\varepsilon})\cdot 2^{(1-k)\varepsilon}+\exp(2^{-k\varepsilon}+2^{(1-k)\varepsilon})\cdot B(2^{(2-k)}\hat{\sigma}) \\ &\leqslant \exp\left(\sum_1^k 2^{-j\varepsilon}\right)\cdot\left\{\sum_1^k 2^{-j\varepsilon}+B(\hat{\sigma})\right\} \\ &\leqslant \text{Const.}\,(B(\hat{\sigma})+1), \end{aligned}$$

since $\sum 2^{-j\varepsilon}<\infty$. This estimate holds for any $k=1,2,\dots$, hence $B(\sigma)$ is uniformly bounded as $\sigma\to 0$. That is, $\mathcal{B}(x,u,u_x)$ is integrable over $0<|x|\leqslant R$.

To complete the proof for the case $\alpha<n$, we observe that

$$\int_0^{\sigma}|\mathcal{B}(x,u,u_x)|\,dx=\sum_1^{\infty}\{B(2^{-j}\sigma)-B(2^{1-j}\sigma)\}\leqslant\beta\sigma^{\varepsilon}(B(0)+1)\sum_1^{\infty}2^{-j\varepsilon}=\text{Const.}\ \sigma^{\varepsilon},$$

using (16) and the fact that $B(\sigma)$ is uniformly bounded. If $\alpha=n$, we obtain in place of (16)

$$B(\sigma)-B(2\sigma)\leqslant\beta\sigma^{\varepsilon}(B(\sigma)+1)(\log R/\sigma)^{n-1}\leqslant\text{Const.}\ \sigma^{\varepsilon/2}(B(\sigma)+1)$$

for some appropriate constant, and the required conclusion follows exactly as before. This proves the lemma.

We may now return to inequality (12). Since $B(\sigma)$ is uniformly bounded, (12) implies

$$m\leqslant\text{Const.}\begin{cases}\sigma^{(\alpha-n)/(\alpha-1)}, & \alpha<n,\\ \log R/\sigma, & \alpha=n,\end{cases} \tag{17}$$

valid for any $\sigma\leqslant\sigma_0$. In the next section we shall complete the proof of Theorem 1 by showing that a certain reverse inequality is also valid.

2. Completion of the proof

We begin with a result analogous to Lemma 1 of the preceding section.

LEMMA 3. *Let Θ be a strongly (L_α) differentiable function with compact support in $|x|<R$, which is identically 1 in some neighborhood of the origin. Then*

$$\int(\Theta_x\cdot\mathcal{A}+\Theta\mathcal{B})\,dx=\text{Const.}=K,$$

where the constant is independent of the particular choice of Θ.

Proof. The integral $\int \Theta \mathcal{B}\,dx$ is well defined in view of Lemma 2. The result then follows by a repetition of the proof of Lemma 1.

LEMMA 4. $K>0$.

Proof. Suppose for contradiction that $K \leqslant 0$. Then for fixed $\sigma \leqslant \sigma_0$, and $m=m(\sigma)$, $v=v(x,\sigma)$, we have from Lemma 3

$$0 \geqslant \int \{v_x \cdot \mathcal{A}(x,u,u_x) + v\,\mathcal{B}(x,u,u_x)\}\,dx$$
$$= \int v_x \cdot \mathcal{A}(x,v,v_x)\,dx + \int_{S_1} v\,\mathcal{B}(x,v,v_x)\,dx + \int_{S_2} m\,\mathcal{B}(x,u,u_x)\,dx, \tag{18}$$

where $S_1=\{\sigma \leqslant |x| \leqslant R\} \cap \{u<m\}$ and $S_2=\{0<|x|<\sigma\} \cup \{u \geqslant m\}$, (this calculation should be compared with the corresponding one in Section 1).

As in Section 1, [1]

$$\int v_x \cdot \mathcal{A}(x,v,v_x)\,dx \geqslant \tfrac{2}{3}\int |v_x|^\alpha\,dx - C. \tag{19}$$

Similarly
$$\int_{S_1} v\,|\mathcal{B}(x,v,v_x)\,dx \leqslant \int (cv\,|v_x|^{\alpha-1} + dv^\alpha + vf)\,dx,$$

and if R is suitably small

$$\int cv\,|v_x|^{\alpha-1}\,dx \leqslant \|c\|_n\,\|v\|_{\alpha*}\,\|v_x\|_\alpha^{\alpha-1} \leqslant \text{Const.}\ \|c\|_n\,\|v_x\|_\alpha^\alpha \leqslant \tfrac{1}{12}\|v_x\|_\alpha^\alpha,$$

and
$$\int dv^\alpha\,dx \leqslant \tfrac{1}{3}\|v_x\|_\alpha^\alpha, \quad \text{(cf. (7))}.$$

Finally,
$$\int vf\,dx \leqslant \|f\|_{\alpha n/(\alpha n+\alpha-n)}\,\|v\|_{\alpha*} \leqslant \text{Const.}\ \|v_x\|_\alpha \leqslant \tfrac{1}{12}\|v_x\|_\alpha^\alpha + \text{Const.}$$

using Young's inequality (with the trick). Combining the last four inequalities yields

$$\int_{S_1} v\,|\mathcal{B}(x,v,v_x)|\,dx \leqslant \tfrac{1}{2}\|v_x\|_\alpha^\alpha + C. \tag{20}$$

Hence from (18) (19), (20), and the capacity inequality (9), there results easily

$$m^\alpha \leqslant \text{Const.}\ [m\int_{S_2} |\mathcal{B}(x,u,u_x)|\,dx + C]\begin{cases}\sigma^{\alpha-n}, & \alpha<n,\\ (\log R/\sigma)^{n-1}, & \alpha=n,\end{cases} \tag{21}$$

valid for $\sigma \leqslant \sigma_0$. This should be compared with (11) in Section 1.

---

(1) Cf. (8). We assume $\alpha<n$ for simplicity here.

On the basis of (21) we assert that for some appropriate constant $A \geqslant 1$,

$$m(\sigma) \leqslant A \begin{cases} \sigma^{\tau(1-\delta)}, & \alpha < n, \\ (\log R/\sigma)^{1-1/n}, & \alpha = n, \end{cases} \tag{22}$$

where $\tau = (\alpha - n)/(\alpha - 1)$ and $\delta = \varepsilon/(n - \alpha + \varepsilon)$. (¹)

Consider first the case $\alpha < n$. Since (22) holds trivially for any σ such that $m \leqslant \sigma^{\tau(1-\delta)}$, we need consider only those values of σ for which $m > \sigma^{\tau(1-\delta)}$. Let θ be a fixed number, to be determined in a moment, and consider the point set

$$\theta\, \sigma^{1-\delta} \leqslant |x| \leqslant \sigma_0/2.$$

By application of the Harnack principle, we have for any point x in this set,

$$\begin{aligned} u(x) &\leqslant \text{Const. } m(|x|) \leqslant \text{Const. } |x|^{\tau} \\ &\leqslant \text{Const. } \theta^{\tau} \sigma^{\tau(1-\delta)} < \text{Const. } \theta^{\tau} m(\sigma) = m(\sigma) \end{aligned}$$

provided θ is chosen appropriately large (note that (17) was used at the second step in this chain of inequalities). It follows that the set S_2 is contained in the union of the sets

$$\{0 < |x| < \sigma\}, \quad \{0 < |x| < \theta\, \sigma^{1-\delta}\}, \quad \text{and} \quad \{\sigma_0/2 < |x| < R\}.$$

Since $m \to \infty$ as $\sigma \to 0$, there will be no points of S_2 in the last of these sets when σ is small enough, say $\sigma \leqslant \sigma_1$. By making σ_1 even smaller, if necessary, the first set is contained in the second. In summary, then, if σ_1 is chosen appropriately small, the set S_2 will be contained in $\{0 < |x| < \theta\, \sigma^{1-\delta}\}$ whenever $\sigma \leqslant \sigma_1$. Thus for $\sigma \leqslant \sigma_1$, making use of Lemma 2,

$$\int_{S_2} |\mathcal{B}(x, u, u_x)|\, dx \leqslant \int_{r < \theta\sigma^{1-\delta}} |\mathcal{B}(x, u, u_x)|\, dx \leqslant \text{Const. } \sigma^{\varepsilon(1-\delta)}.$$

Substitution into (21) yields

$$m^{\alpha} \leqslant \text{Const. } (m\, \sigma^{\alpha - n + \varepsilon(1-\delta)} + \sigma^{\alpha - n}).$$

Hence after a short calculation (cf. Lemma 2 of [1]),

$$m \leqslant \text{Const. } (\sigma^{\tau(1-\delta)} + \sigma^{(\alpha - n)/\alpha}).$$

Since $(\alpha - n)/\alpha \geqslant \tau(1-\delta)$, (cf. the footnote below), the assertion follows for $\sigma \leqslant \sigma_1$. For $\sigma > \sigma_1$ it is clear that (22) holds for *some* constant A, whence in all cases (22) is verified.

(¹) In case $\alpha < n$ the assertion requires $\varepsilon \leqslant (n-\alpha)/(\alpha - 1)$. This restriction, however, clearly involves no loss of generality.

Before turning to the case $\alpha = n$ we observe that (22), together with Theorem 11 of [1], implies that the singularity at 0 is removable. This contradiction of our basic assumption establishes Lemma 4 in the case $\alpha < n$.

We now consider (22) in the case $\alpha = n$. Since it holds trivially for any σ such that $m \leqslant (\log R/\sigma)^{1-1/n}$, we may suppose at the outset that $m > (\log R/\sigma)^{1-1/n}$. Put $y = (\log R/\sigma)^{1-1/n}$. Let θ be a fixed number, to be determined in a moment, and consider the point set

$$Re^{-2\theta y} \leqslant |x| \leqslant \sigma_0/2.$$

By application of the Harnack principle, we have for any point x in this set,

$$\begin{aligned} u(x) &\leqslant \text{Const. } m(|x|) \leqslant \text{Const. } \log R/|x| \\ &\leqslant \text{Const. } \theta y < \text{Const. } \theta m = m \end{aligned}$$

provided θ is chosen appropriately small. Thus the set S_2 is contained in the union of the sets

$$\{0 < |x| < \sigma\}, \quad \{0 < |x| < Re^{-2\theta y}\}, \quad \text{and} \quad \{\sigma_0/2 < |x| < R\}.$$

As before, if σ_1 is chosen appropriately small the set S_2 will be contained in

$$\{0 < |x| < Re^{-2\theta y}\}$$

whenever $\sigma \leqslant \sigma_1$. Thus by Lemma 2,

$$\int_{S_2} |\mathcal{B}(x, u, u_x)| \, dx \leqslant \int_{r < Re^{-2\theta y}} |\mathcal{B}(x, u, u_x)| \, dx \leqslant \text{Const. } e^{-\varepsilon\theta y}.$$

Substitution into (21) yields

$$m^n \leqslant \text{Const. } (\log R/\sigma)^{n-1} (me^{-\varepsilon\theta y} + 1),$$

whence after a simple calculation

$$m \leqslant \text{Const. } (\log R/\sigma)^{1-1/n} \{y^{1/(n-1)} e^{-\varepsilon\theta y/(n-1)} + 1\}.$$

Since the expression in braces is uniformly bounded for $0 \leqslant y < \infty$ the assertion follows for $\sigma \leqslant \sigma_1$. For $\sigma > \sigma_1$ it is clear that (22) holds for *some* constant A, whence (22) is verified in all cases.

This being shown, it now follows from Theorem 11 of [1] that 0 is again a removable singularity. But this contradicts our basic assumption, and Lemma 4 is proved.

Now let $M = M(\sigma)$ denote the maximum value of u on a given circumference $|x| = \sigma$, $(\sigma \leqslant \sigma_0)$. Our goal is to obtain the important inequality (29), reverse to (17). To this end we introduce a second auxiliary function $V = V(x, \sigma)$ according to the formula

$$V = \begin{cases} \operatorname{Max}(0,u) & \text{for } \sigma \leqslant |x| \leqslant R \\ \operatorname{Min}(M,u) & \text{for } 0 < |x| < \sigma \\ M & \text{at } x = 0. \end{cases}$$

Evidently V is strongly differentiable, has compact support in $|x|<R$, and $V \equiv M$ in some neighborhood of the origin.

Let σ_2, $(\sigma_2 \leqslant \sigma_0)$, be chosen so that $\int_0^{\sigma_2} |\mathcal{B}(x,u,u_x)|\,dx \leqslant \frac{1}{2}K$; this choice is allowable by virtue of Lemmas 2 and 4. For fixed $\sigma < \sigma_2$, we define

$$\Theta = \Theta(x,\sigma) = \begin{cases} 0 & \text{for } \sigma_2 < |x| \leqslant R \\ \dfrac{r^\tau - \sigma_2^\tau}{\sigma^\tau - \sigma_2^\tau} & \text{for } \sigma \leqslant |x| \leqslant \sigma_2 \\ 1 & \text{for } |x| < \sigma, \end{cases}$$

where $\tau = (\alpha - n)/(\alpha - 1)$, (we restrict the discussion to the case $\alpha < n$, for simplicity). Clearly Θ is strongly differentiable, has compact support in $|x|<R$, and $\Theta \equiv 1$ in some neighborhood of the origin. Thus by Lemma 3, Hölder's inequality, and the choice of σ_2,

$$K = \int (\Theta_x \cdot \mathcal{A} + \Theta\,\mathcal{B})\,dx \leqslant \|\Theta_x\|_\alpha \|\mathcal{A}\|_{\alpha/(\alpha-1)} + \tfrac{1}{2}K.$$

Since $\|\Theta_x\|_\alpha^\alpha = \omega_n |\tau|^{\alpha-1} (\sigma^\tau - \sigma_2^\tau)^{1-\alpha}$, this yields at once

$$K'(\sigma^\tau - \sigma_2^\tau) \leqslant \int |\mathcal{A}|^{\alpha/(\alpha-1)}\,dx, \tag{23}$$

where K' is a positive constant and the integral is to be taken over the set $\sigma \leqslant |x| \leqslant \sigma_2$.

The next problem is to estimate this integral. Since $\sigma_2 \leqslant \sigma_0$, it is clear that $V \equiv u$ in the domain of integration. Thus the integral will only be increased if it is extended over the entire set where $V = u$. Denoting this set by Δ, and using inequality (2), we have

$$\begin{aligned} \int |\mathcal{A}|^{\alpha/(\alpha-1)}\,dx &\leqslant \text{Const.} \int_\Delta \{|u_x|^\alpha + b^{\alpha/(\alpha-1)} u^\alpha + e^{\alpha/(\alpha-1)}\}\,dx \\ &= \text{Const.} \int_\Delta \{(|u_x|^\alpha - 2du^\alpha - 2g) + (2d + b^{\alpha/(\alpha-1)})u^\alpha + (2g + e^{\alpha/(\alpha-1)})\}\,dx. \end{aligned} \tag{24}$$

Now by the Hölder and Sobolev inequalities,

$$\begin{aligned} \int_\Delta (2d + b^{\alpha/(\alpha-1)})\,u^\alpha\,dx &\leqslant \int (2d + b^{\alpha/(\alpha-1)})\,V^\alpha\,dx \\ &\leqslant \text{Const.}\,(\|d\|_{n/\alpha} + \|b\|_{n/(\alpha-1)}^{\alpha/(\alpha-1)})\,\|V_x\|_\alpha^\alpha \\ &= \text{Const.}\,(\|d\|_{n/\alpha} + \|b\|_{n/(\alpha-1)}^{\alpha/(\alpha-1)}) \int_\Delta |u_x|^\alpha\,dx, \end{aligned} \tag{25}$$

since $V_x = u_x$ almost everywhere in Δ and $V_x = 0$ almost everywhere in the complement of Δ. Supposing that R is suitably small, (25) implies

$$\int_\Delta (2d + b^{\alpha/(\alpha-1)})\, u^\alpha dx \leqslant \int_\Delta |u_x|^\alpha dx.$$

Substituting this into (24), and then using (2), yields

$$\int |\mathcal{A}|^{\alpha/(\alpha-1)} dx \leqslant \text{Const.}\left(1 + \int_\Delta u_x \cdot \mathcal{A}\, dx\right) = \text{Const.}\left(1 + \int V_x \cdot \mathcal{A}\, dx\right). \tag{26}$$

Now by Lemma 3

$$\int V_x \cdot \mathcal{A}\, dx = MK - \int V\mathcal{B}\, dx \leqslant MK + B(0) \max V, \tag{27}$$

where $B(0) = \int_0^R |\mathcal{B}(x, u, u_x)|\, dx$. We assert that

$$\max V \leqslant \text{Const. } M. \tag{28}$$

Indeed this is obvious from the definition of V, except in the set $\sigma \leqslant |x| \leqslant R$. But by applying the maximum principle (cf. Theorem 8 below) to u in this set, we find

$$\max_{\sigma \leqslant r \leqslant R} u \leqslant \text{Const. } M;$$

(application of Theorem 8 requires that the measure of the set $\sigma \leqslant |x| \leqslant R$ be suitably small, which in turn can be accomplished by choosing R suitably small at the beginning of the proof; note also that $M \geqslant 1$, which allows us to absorb the additive constant appearing in Theorem 8). Inequality (28) now follows from the relation $V = \text{Max}\,(0, u)$ in $\sigma \leqslant |x| \leqslant R$. The required estimate for $\int |\mathcal{A}|^{\alpha/(\alpha-1)} dx$ finally results by combining (26), (27), and (28), thus

$$\int |\mathcal{A}|^{\alpha/(\alpha-1)} dx \leqslant \text{Const. } M.$$

Inserting this into (23) yields the estimate

$$M \geqslant 2K''(\sigma^\tau - \sigma_2^\tau), \quad (\sigma > \sigma_2),$$

where K'' is a *positive* constant. For suitably small σ this implies

$$M \geqslant K'' \sigma^{(\alpha-n)/(\alpha-1)} \quad \text{if} \quad \alpha < n. \tag{29}$$

By a similar calculation, which may be omitted here, we find also

$$M \geqslant K'' \log R/\sigma \quad \text{if} \quad \alpha = n. \tag{29'}$$

The required asymptotic behavior of u is a consequence of inequalities (17) and (29). Indeed by (17) and the Harnack principle we have

$$M \leqslant \text{Const. } m \leqslant \text{Const. } \sigma^{(\alpha-n)/(\alpha-1)}$$

valid for $\sigma \leqslant \sigma_0$, while by (29) and the Harnack principle,

$$m \geqslant \text{Const. } M \geqslant \text{Const. } \sigma^{(\alpha-n)/(\alpha-1)}$$

valid for all suitably small σ. Thus (3) holds in a neighborhood of the origin, when $\alpha < n$. The result for $\alpha = n$ is obtained in the same way, and Theorem 1 is completed.

3. Further results

The conclusion of Theorem 1 may be augmented by several further results concerning the behavior of a solution in the neighborhood of a positive isolated singularity.

THEOREM 2. *Under the hypotheses of Theorem* 1, *if the singularity at* 0 *is* not *removable, then for all sufficiently small values of* σ.

$$\int_\sigma^{2\sigma} |u_x|^\alpha dx \leqslant \text{Const.} \begin{cases} \sigma^{(\alpha-n)/(\alpha-1)}, & \alpha < n, \\ (\log 1/\sigma)^n, & \alpha = n. \end{cases}$$

Proof. According to the argument preceding inequality (15),

$$\int_\sigma^{2\sigma} |u_x|^\alpha dx \leqslant \text{Const. } [\sigma^{-1} \|u\|_\alpha + \sigma^{(n-\alpha)/\alpha}]^\alpha \leqslant \text{Const. } \sigma^{n-\alpha} m(\sigma)^\alpha.$$

The required conclusion then follows from (17).

COROLLARY 1. *We have*

$$u_x \in L_{\theta-\delta}(S), \quad (S = \{|x| \leqslant R\}), \tag{30}$$

where $\theta = n(\alpha-1)/(n-1)$ *and* δ *is any positive number. Moreover, if* $\theta > 1$ (*i.e. if* $\alpha > 2 - 1/n$) *then* $u \in W^1_{\theta-\delta}(S)$.

Proof. By Hölder's inequality and Theorem 2

$$\int_\sigma^{2\sigma} |u_x|^{\theta-\delta} dx \leqslant \text{Const.} \begin{cases} \sigma^{\delta(n-1)/(\alpha-1)}, & \alpha < n, \\ \sigma^\delta (\log 1/\sigma)^{n-\delta}, & \alpha = n. \end{cases}$$

Thus by the argument at the close of Lemma 2, $|u_x|^{\theta-\delta}$ is integrable over S, and (30) is proved.

Now let ϕ be a continuously differentiable function with compact support, and let S_σ denote the ball of radius σ about 0. Then obviously

$$\int_{D-S_\sigma} (u\phi_x + \phi u_x)\,dx = -\oint_{r=\sigma} u\phi\mathbf{n}ds, \tag{31}$$

where $\mathbf{n}$ denotes the unit normal to the sphere $r=\sigma$. By Theorem 1

$$\oint_{r=\sigma} u\phi\mathbf{n}ds = O(\sigma^{(\alpha-n)/(\alpha-1)+n-1}),$$

and this tends to zero with σ when $\theta>1$. Consequently, if $\theta>1$ we may let $\sigma\to 0$ in (31) to obtain

$$\int (u\phi_x + \phi u_x)\,dx = 0.$$

Thus $u \in W^1_{\theta-\delta}(S)$. (In fact, we have shown that u is in $W^1_{\theta-\delta}$ over any proper subdomain of D.)

The following result should be compared with Lemma 2.

COROLLARY 2. *The function $\mathcal{A}(x, u, u_x)$ is integrable over $0<|x|\leqslant R$, and*

$$\int_{r<\sigma} |\mathcal{A}(x, u, u_x)|\,dx \leqslant \text{Const.} \begin{cases} \sigma & , \quad \alpha<n, \\ \sigma(\log 2R/\sigma)^{n-1}, & \alpha=n. \end{cases}$$

Proof. According to inequality (2), we have

$$|\mathcal{A}| \leqslant a\,|u_x|^{\alpha-1} + b\,|u|^{\alpha-1} + e.$$

Now by Hölder's inequality and Theorem 2

$$\int_\sigma^{2\sigma} |u_x|^{\alpha-1} dx \leqslant \text{Const.} \begin{cases} \sigma & , \quad \alpha<n, \\ \sigma(\log 1/\sigma)^{n-1}, & \alpha=n, \end{cases}$$

valid for all sufficiently small σ. Moreover, using Theorem 1 and the Lebesgue class conditions on b and e it is easy to see that

$$\int_\sigma^{2\sigma} (bu^{\alpha-1} + e)\,dx \leqslant \text{Const. } \sigma.$$

The required conclusion follows from the summation argument at the close of Lemma 2.

THEOREM 3. *Under the hypotheses of Theorem 1, if the singularity at* 0 *is* not *removable then*

$$\operatorname{div} \mathcal{A}(x, u, u_x) - \mathcal{B}(x, u, u_x) = -K\delta(0)$$

in the sense of distributions; that is, for any continuously differentiable function $\phi = \phi(x)$ *with compact support in* D,

$$\int (\phi_x \cdot \mathcal{A} + \phi \mathcal{B})\, dx = K\phi(0). \tag{32}$$

Proof. Let η be a non-negative smooth function which vanishes outside the ball S_σ, is identically 1 in $S_{\sigma/2}$, and elsewhere satisfies $0 \leqslant \eta \leqslant 1$. We may obviously suppose that $|\eta_x| \leqslant 3/\sigma$. The function

$$\Theta = (1-\eta)\phi + \eta\phi(0)$$

has compact support in D and is identically equal to $\phi(0)$ in a neighborhood of the origin. By Lemma 3,

$$\int (\Theta_x \cdot \mathcal{A} + \Theta \mathcal{B})\, dx = K\phi(0),$$

that is

$$\int \{(1-\eta)(\phi_x \cdot \mathcal{A} + \phi\mathcal{B}) + (\phi(0) - \phi)\eta_x \cdot \mathcal{A} + \phi(0)\eta\mathcal{B}\}\, dx = K\phi(0). \tag{33}$$

Now both $\mathcal{A}$ and $\mathcal{B}$ are locally summable in D. Hence if $\sigma \to 0$,

$$\int (1-\eta)(\phi_x \cdot \mathcal{A} + \phi\mathcal{B})\, dx \to \int (\phi_x \cdot \mathcal{A} + \phi\mathcal{B})\, dx.$$

Likewise $\int \eta \mathcal{B} dx \to 0$, while by Corollary 2 one easily obtains

$$\left| \int (\phi(0) - \phi)\eta_x \cdot \mathcal{A} dx \right| \leqslant \text{Const.}\ \sigma^{1-\varepsilon} \to 0.$$

Thus letting $\sigma \to 0$ in (33) we obtain (32), and the proof is complete.

Remark. Theorem 3 shows that every solution of (1) with a positive isolated singularity has the attributes of a fundamental solution, as in the case of linear equations.

4. Existence of solutions with isolated singularities

The very light hypotheses required for the proof of Theorem 1 do not seem strong enough to imply the general existence of solutions with isolated singularities. Accordingly, we shall suppose here that (1) is subject to certain further conditions. In particular, *it will be assumed that for any domain D of sufficiently small diameter, the following two properties hold:*

P1. The Dirichlet problem with continuous boundary data is uniquely solvable for any smoothly bounded domain in D, and the solutions are continuous functions of the data, in the uniform topology.

P2. Let Γ denote a smoothly bounded annular domain in D. By *P1* there exists a solution v in Γ assuming the continuous boundary data $\psi_0(x)$ on the outer boundary and $\psi_1(x)+m$ on the inner boundary. *We suppose that at each point P in Γ, the value $v(P)$ tends to infinity with m.*

These conditions should be compared with the corresponding, but stronger, conditions imposed in [1].

A general discussion of these properties is beyond the scope of the paper. We may remark, however, that *P1* is satisfied for a wide variety of equations (including linear equations), provided that the domains in questions are suitably small. We believe, moreover, that *P2* is a consequence of the general structure of (1). Since a proof of this apparently involves an effort at least comparable to that of the preceding sections of the paper, we shall rest content here with imposing *P2* as an additional assumption on (1).

THEOREM 4. *Let continuous data $\psi(x)$ be assigned on the boundary of a smooth domain D, it being assumed that D is small enough so that P1 and P2 are valid, and also so that the maximum principle (Theorem 8) holds.*

Let 0 be a point of D. Then there exists a family of solutions $G=G(x)$ in $D-\{0\}$, taking on the given boundary values and satisfying

$$G \approx \begin{cases} r^{(\alpha-n)/(\alpha-1)}, & \alpha < n, \\ \log 1/r \quad , & \alpha = n, \end{cases}$$

in the neighborhood of 0. The values of G may be assigned arbitrarily at any point $P \neq 0$ in D, subject only to the restriction $G(P) > w(P)$, where w denotes the unique solution of (1) in D which takes on the assigned boundary values.

Proof. This is for the most part a duplication of the proof of Theorem 13 of [1]. Applications of the (simple) maximum principle must be replaced with continuous dependence arguments, which may safely be left to the reader. The main difference lies in guaranteeing that each solution v_σ is non-negative. Let L denote the minimum of w over D. Then each v_σ has boundary values which are $\geqslant L$. Thus by the maximum principle we have $v_\sigma \geqslant \text{Const.}\, L - \text{Const.}$, where the coefficient of L may even be negative. In any case, the functions v_σ are uniformly bounded below. By making the change of variables $\bar{u} = u + \text{Const.}$, with the constant suitably chosen, it may therefore be supposed without loss of generality that $v_\sigma \geqslant 0$.

5. Linear equations

The results of the preceding sections can be sharpened somewhat in case (1) is linear, that is, of the form

$$\frac{\partial}{\partial x_i}\left(a_{ij}(x)\frac{\partial u}{\partial x_j} + b_i(x)\,u + e_i(x)\right) = c_i(x)\frac{\partial u}{\partial x_i} + d(x)\,u + f(x). \tag{34}$$

Here it is assumed that the coefficients $a_{ij}(x)$ are bounded measurable functions satisfying the ellipticity condition

$$a_{ij}\xi_i\xi_j \geqslant \lambda\xi^2, \quad \lambda = \text{Const.} > 0,$$

and that the coefficients b_i through f fall into the Lebesgue classes

$$b_i,\, c_i,\, e_i \in L_{n/(1-\varepsilon)}; \quad d,\, f \in L_{n/(2-\varepsilon)}.$$

The various conditions (2) are easily seen to be met with $\alpha = 2$, and it follows that Theorems 1 through 3 hold for (34), again with $\alpha = 2$.

Now let $\mathfrak{M}$ denote the class of smoothly bounded domains D such $|D| \leqslant D_0$, D_0 being the constant in Theorem 8. We assert that properties *P1* and *P2* hold for any domain D of class $\mathfrak{M}$.

Indeed, the uniqueness and continuous dependence of solutions of the Dirichlet problem is a direct consequence of Theorem 8 applied to the difference of two solutions. The existence of solutions is naturally a more difficult matter. For smooth data, the result can be obtained (in outline) by first smoothing the coefficients, then showing that the solution (known to exist by the classical Schauder theory) is uniformly continuous in the closure of the domain in question (this involves the usual straightening of the boundary), and finally tending to a limit. When the data is only continuous, existence can be obtained by a standard approximation process based on the existence theorem for smooth data. An alternate discussion based on functional analysis is given in reference [3]. Property *P2* is obvious, by the superposition property of linear equations.

This being the case, it follows that for any domain D in the class $\mathfrak{M}$ there exists a family of solutions of (34) which satisfies the conclusion of Theorem 4. In addition, we have the following supplementary results, in which $(34)_0$ refers to equation (34) with $e_i \equiv f \equiv 0$.

THEOREM 5. *Let G be a particular solution of $(34)_0$ in the set $D-\{0\}$, such that $G \approx r^{2-n}$ or $G \approx \log 1/r$ near 0, depending on whether $n>2$ or $n=2$. Then every solution of (34) in $D-\{0\}$ which in the neighborhood of 0 is bounded below by a multiple of r^{2-n}, or $\log 1/r$, has the form*

$$u = \text{Const. } G + w,$$

where w is a solution of (34) in the entire domain D.

Proof. This result is an analogue of Theorem 5 of [4]. It is clearly enough to carry out the proof when D is in the class $\mathfrak{M}$. This being the case, we may suppose by virtue of property *P1* that both u and G are zero on the boundary of D, and that $e_i \equiv f \equiv 0$. Moreover, by adding a suitable multiple of G to u it can be assumed without loss of generality that $u \geqslant 0$ near 0.

Under these assumptions it follows from Theorem 8 that both u and G are non-negative in D. The proof of Theorem 5, reference [4], can now be taken over almost word for word (and indeed even simplifies a bit).

Remark. In view of Theorem 5, the family of solutions given by Theorem 4 is unique and depends continuously on a single multiplicative parameter.

The following result is a slight generalization of Theorem 5 in that the existence of a solution G is not required.

THEOREM 6. *Let u be a continuous solution of (34) in the set $D-\{0\}$, which in the neighborhood of 0 is bounded below by a multiple of r^{2-n}, or $\log 1/r$. Then either u has a removable singularity at 0, or else (possibly after multiplication by -1)*

$$u \approx \begin{cases} r^{2-n}, & n>2, \\ \log 1/r, & n=2, \end{cases}$$

in the neighborhood of the origin.

Proof. Let D' be a suitably small neighborhood of the origin. Then there exists a solution G of $(34)_0$ in the set $D'-\{0\}$, such that $G \approx r^{2-n}$ or $G \approx \log 1/r$. Consequently, by Theorem 5,

$$u \equiv \text{Const. } G + w$$

in the set $D'-\{0\}$, and the conclusion follows at once.

6. Corrections and additions to reference [1]

1. As they are stated, Theorems 3, 4, 3′, and 4′ of [1] are valid only in case $M=0$. The correct versions of Theorems 3 and 4 are as follows:

THEOREM 7. *Let u be a weak solution of* (1) *in a domain $D \subset \Omega$. Suppose that $u \leqslant M$ on the boundary of D, and that conditions* (2) *hold. Then*

$$\max u \leqslant M + C'|M| + C\{|D|^{-1/\alpha} \|\tilde{u}\|_{\alpha,\,D} + k\}, \quad (\tilde{u} = \text{Max}\,(0, u-M)),$$

where C, C', and k depend only on the structure of (1). *In particular*

$$k = \{|D|^{\varepsilon/n} \|f\|\}^{1/(\alpha-1)} + \{|D|^{\varepsilon/n} \|g\|\}^{1/\alpha},$$

while C' tends to zero as $\|d\|$ tends to zero.

THEOREM 8 (Maximum principle). *Let u satisfy the hypotheses of Theorem 7. Then there exists a constant D_0, depending only on the structure of equation* (1), *such that if $|D| \leqslant D_0$ then*

$$\max u \leqslant M + C'|M| + Ck.$$

Proof of Theorems 7 and 8. The error in the original proof came in asserting the general validity of (25) on page 262. When $M=0$, $k>0$, however, (25) holds in the set $u \geqslant \varepsilon'$, and the proof as given is correct. The case $M=0$, $k=0$ may then be obtained by a trivial approximation argument. Thus, if $M=0$, we have shown that

$$\max u \leqslant C(|D|^{-1/\alpha} \|\tilde{u}\|_{\alpha,\,D} + k), \quad (\tilde{u} = \text{Max}\,(0, u)), \tag{35}$$

or in the case of Theorem 8,

$$\max u \leqslant Ck. \tag{36}$$

The general result can be obtained by applying (35) and (36) to the new dependent variable $u^* = u - M$. This change affects the structure of (1) by replacing f and g respectively with

$$f + 2^{\alpha-1} M^{\alpha-1} d, \quad g + 2^{\alpha-1} M^{\alpha} d.$$

Hence k must be replaced in (35) and (36) by

$$\{|D|^{\varepsilon/n} \|f + 2^{\alpha-1} M^{\alpha-1} d\|\}^{1/\alpha-1} + \{|D|^{\varepsilon/n} \|g + 2^{\alpha-1} M^{\alpha} d\|\}^{1/\alpha}.$$

The latter expression is easily seen to be less than

$$\text{Const.}\; k + C'|M|,$$

where C' tends to zero as $|D|^{\varepsilon/n} \|d\|$ tends to zero. This completes the proof.

Theorems 3′ and 4′ are corrected in a similar way. We may omit the details. Theorem 3 was not used in [1], and Theorem 4 was applied only at one point on page 278. Here the inequality

$$u(x) \geqslant \text{Min } (\mu_\nu, \mu_{\nu+1}) - Ck$$

must be replaced by

$$u(x) \geqslant (1 - C') \text{ Min } (\mu_\nu, \mu_{\nu+1}) - Ck,$$

according to Theorem 8. For sufficiently large ν we have $C' \leqslant \frac{1}{2}$, and the conclusion $u \to \infty$ as $x \to 0$ follows as before.

We note that the maximum principles above can be given wider validity in several directions. First, the results clearly apply not only to the equation div $\mathcal{A} = \mathcal{B}$, but also to the differential inequality div $\mathcal{A} \geqslant \mathcal{B}$. More generally, if

$$\text{div } \mathcal{A}(x, u, u_x) \geqslant \mathcal{B}(x, u, u_x)$$

whenever $u > M_0$, then Theorems 7 and 8 hold for all $M \geqslant M_0$. This result has particular application to the linear equation (34) when the coefficient d is non-negative and $M \geqslant 0$. Indeed, in this case it is clear that we may omit the term $C'|M|$ in the conclusions of Theorems 7 and 8, and moreover neglect $\|d\|$ in the determination of the constants C and D_0.

Also, the maximum principle holds for all exponents α in the range $1 \leqslant \alpha < \infty$, assuming for $\alpha > n$ that $b, e \in L_{\alpha/(\alpha-1)}$, $c \in L_\alpha$, and $d, f, g \in L_1$. This extension requires no essentially new ideas in the proofs already given.

2. The phrasing of Theorem 7 of [1] does not indicate the specific dependence of the coefficient C' on the domains D and D'. It is therefore worth pointing out that this dependence is completely expressed by a single number N, *the number of spheres (of radius* $\leqslant 1$) *required for the chaining argument.* In particular, for pairs of domains D, D' and $\bar{D}$, $\bar{D}'$ which are geometrically similar and both contained in a bounded subset of Ω, it may be assumed that the corresponding constants C' and $\bar{C}'$ are the same. This fact plays an important part in our various applications of the Harnack principle over annular regions (cf., for example, the proof of Lemma 2).

3. In the same way, Theorem 8 of [1] is not stated in as sharp a form as could be desired. The following version is preferable.

THEOREM 9. *Let u be a weak solution of* (1) *in a domain $D \subset \Omega$. Then u is* (*essentially*) *Hölder continuous in D. Moreover, if $|u| \leqslant L$ then for x, $y \in D$*

$$|u(x)-u(y)| \leqslant H(L+k')\left|\frac{x-y}{R}\right|^{\lambda},$$

where H and λ depend only on the structure of (1), *and R is the maximum distance of x or y to the boundary of D, (or $R=1$ if this is smaller).*

To see this, it is enough to replace the final sentence of the original proof by the following argument:

Let x be the point further from the boundary of D. If $|x-y| \leqslant 2^{-(\gamma+1)\alpha/\varepsilon}R$, then by inequality (48)

$$|u(x)-u(y)| \leqslant C(L+k')\left|\frac{x-y}{R}\right|^{\lambda}.$$

On the other hand, if $|x-y| \geqslant 2^{-(\gamma+1)\alpha/\varepsilon}R$, then since $|u| \leqslant L$ we have

$$|u(x)-u(y)| \leqslant 2L \leqslant CL\left|\frac{x-y}{R}\right|^{\lambda}.$$

This completes the proof.

4. We observe (without proof) that assumption (8) in Chapter I of [1] can be replaced by the weaker condition (7). Condition (8), is required, however in Chapters II and III. Finally, though we shall not carry out the details, one can show that the hypothesis of Theorem 1 of the present paper can we weakened slightly to read

$$b, e \in L_{n/(\alpha-1)}; \quad c \in L_{n/(1-\varepsilon)}; \quad d, f, g \in L_{n/(\alpha-\varepsilon)},$$

except that for $\alpha = n$ we require $b \in L_{n/(n-1-\varepsilon)}$.

5. One further reference should be included in the bibliography of [1], namely

GEVREY, M., Sur certaines propriétés des fonctions harmoniques et leur extension aux équations aux dérivées partielles *C. R. Acad. Sci. Paris*, 183 (1926), 546–548.

In this paper there occurs for the first time a removable singularity theorem of interpolatory type for solutions of elliptic equations.

Note: This work was partially supported by the United States Air Force Office of Scientific Research under Grant No. AF-AFOSR-63-373.

References

[1]. SERRIN, J., Local behavior of solutions of quasi-linear equations. *Acta Math.*, 111 (1964), 247–302.

[2]. SERRIN, J., A Harnack inequality for nonlinear equations. *Bull. Amer. Math. Soc.*, 69 (1963), 481–486.

[3]. MORREY, C. B. Jr., Second order elliptic equations and Hölder continuity. *Math. Z.*, 72 (1959), 146–164.

[4]. GILBARG, D. & SERRIN, J., On isolated singularities of solutions of second order elliptic differential equations. *J. d'Analyse Math.*, 4 (1955–56), 309–340.

Received August 25, 1964

J. Anal. Math. 85 (2001), 307–321. **Reprint I.1.[iv]**

THE HARNACK INEQUALITY IN $\mathbb{R}^2$ FOR QUASILINEAR ELLIPTIC EQUATIONS

By

P. Pucci and J. Serrin

Abstract. We use maximum principle techniques to obtain a Harnack inequality for two-dimensional elliptic operators.

1 Introduction

We use maximum principle techniques to obtain a Harnack inequality. Although the final result is restricted to plane domains, the class of equations to which it applies is in many respects more extensive than in previous work. We consider the model quasilinear elliptic differential equation

$$\operatorname{div}\{A(|Du|)Du\} - f(u) = 0, \qquad u > 0, \tag{1.1}$$

in $\mathbb{R}^2$ and, more generally, the two-sided differential inequality

$$-B(Du) \le \operatorname{div}\{A(|Du|)Du\} \le B(Du) + f(u), \qquad u > 0. \tag{1.2}$$

Throughout the paper, let $\Omega(t) = tA(t)$ for $t > 0$ and $\Omega(0) = 0$, and assume the following natural conditions.

(Ω1) $\Omega \in C[0,\infty) \cap C^{1,1}(0,\infty)$.

(Ω2) $\Omega'(t) > 0$ for $t > 0$ (ellipticity).

(B) B is continuous in $\mathbb{R}^2$ and continuously differentiable in $\mathbb{R}^2\backslash\{0\}$. Moreover,

$$0 \le B(p) \le \mu\Omega(|p|) \quad (\mu = \text{pos. const.}) \qquad \text{for all } p \in \mathbb{R}^2 \text{ with } |p| < 1.$$

(F) f is continuous and non-decreasing on $[0,\infty)$, with $f(0) = 0$.

In addition to these assumptions, several further technical conditions will be placed on the function Ω to allow the derivations to proceed. It is convenient to defer their statements for the moment (see (Ω3)–(Ω4) at the end of the introduction). We note particularly, however, that (1.1) is a special case of (1.2), and also that the prime examples of the *m-Laplace operator* $A(t) = t^{m-2}$, $m > 1$, and the *mean curvature operator* $A(t) = 1/\sqrt{1+t^2}$ are covered by our hypotheses; see below.

Now define

$$F(u) = \int_0^u f(v)dv, \qquad u \geq 0,$$

and

$$G(t) = \int_0^t \Omega(s)ds, \qquad H(t) = tG'(t) - G(t) = \int_0^t s\Omega'(s)ds, \qquad t \geq 0.$$

Clearly F is non-negative, while G is strictly convex and H is strictly increasing and positive by (Ω2). Let us denote by B_R the disk in $\mathbb{R}^2$ centered at $(0,0)$ with radius $R > 0$. Then the following main result holds.

Theorem 1. *Let the function* Ω *satisfy* (Ω1)–(Ω4), *let* B *satisfy condition* (B) *and assume* f *obeys* (F). *Also, suppose either* $f \equiv 0$ *on* $[0, \delta]$, $\delta > 0$, *or that* $f > 0$ *on* $\mathbb{R}^+$ *and*

$$\int_0^1 \frac{ds}{H^{-1}(F(s))} = \infty. \tag{1.3}$$

Then for every $R > 0$, *there is a strictly increasing continuous function* Φ_R : $[0, \infty) \to [0, \infty)$, *with* $\Phi_R(0) = 0$, *such that any* C^1 *distribution solution* u *of the differential inequality* (1.2) *in the disk* B_R *satisfies the Harnack condition*

$$u(x,y) \geq \Phi_R(u(0,0)) \qquad \textit{for all } (x,y) \in B_{R/3}. \tag{1.4}$$

Our method of proof does not make obvious the dependence of the function Φ_R on R or on the parameters of (1.2). For a given function A, however, it is worth noting that the dependence of Φ_R on f arises from the divergence rate as $t \to 0$ of the integral

$$\int_t^1 \frac{ds}{H^{-1}(F(s))}.$$

The main previous results concerning the Harnack inequality for general quasi-linear elliptic operators are due to the second author of this paper [S2] and to Lieberman [L], both of whom studied the general divergence structure equation

$$\operatorname{div} A(x,u,Du) + B(x,u,Du) = 0 \tag{1.5}$$

in n-dimensional space. In comparison with that work, the restriction here to $\mathbb{R}^2$ is a serious drawback. On the other hand, for the inequality (1.2) itself, conditions

(Ω1)–(Ω4) are weaker than those earlier required. For example, in [S2] the operator $A(t)$ is closely related to the degenerate Laplace case $\Omega(t) = t^{m-1}$, while on the other hand Lieberman asks that (Ω1)–(Ω3) and [L, (1.1)′] hold. In our notation, (1.1)′ takes the form

$$(1.6) \qquad t\Omega'(t) \geq \delta\Omega(t), \qquad t \geq 0, \qquad \delta > 0,$$

and in particular (Ω2)–(Ω3) and (1.6) imply (Ω4) with $c_1(\theta) = c\theta^{c-1}/\delta$.

It should be noted as well that the mean curvature operator does not obey the conditions either of [S2] or of [L]; see [L, (1.1)′]. Also, more importantly, in [L] the Harnack principle depends on a given bound $0 \leq u \leq M$ for the solution, a strong condition which is not required here. Finally, our basic requirement on the nonlinearity f, namely the integral condition (1.3), is more general than the assumptions of either [S2] or [L]. Suppose, e.g., that f satisfies Lieberman's conditions (1.1)′ (that is, (1.6) above) and (1.3c)″. Then, by (1.3c)″, we have

$$0 \leq f(u) \leq \gamma\Omega(u), \qquad u \geq 0;$$

so, with the help of (1.6), one easily derives

$$H(t) \geq \frac{\delta}{\delta+1} t\Omega(t), \quad t \geq 0; \qquad F(u) \leq \gamma u\Omega(u), \quad u \geq 0.$$

Hence $F(u) \leq H(du)$ for $d = \max\{1, (\delta+1)\gamma/\delta\}$. Thus $H^{-1}(F(u)) \leq du$, and so (1.3) holds.

For the special case of the mean curvature and other closely related operators, Trudinger [T] has also given a Harnack inequality in n dimensions, but again only for bounded solutions, with the constant in the Harnack principle depending on the bound. Moreover, in the context of inequality (1.2) and the mean curvature operator $A(t) = 1/\sqrt{1+t^2}$, Trudinger for example requires the conditions

$$0 \leq B(p) \leq \text{Const.}\{|p|^2/(1+|p|) + |p|/(1+|p|)^{1/2}\}$$

and

$$0 \leq f(u) \leq \text{Const.}\, u$$

(cf. [T, (3.2)]), which are stronger than conditions (B) and (1.3), respectively. Finally, we note reference [G], which considers viscosity solutions of the minimal surface and other related equations, always when $f \equiv 0$.

Theorem 1 has the following important companion result.

Theorem 2. *Under the assumptions of Theorem* 1, *for every* $R > 0$ *there exists a strictly increasing continuous function* $\Psi_R : [0, M_R) \to [0, \infty)$, *with* $\Psi_R(0) = 0$, *such that any* C^1 *distribution solution* u *of* (1.2) *in the disk* B_R, *with* $u(0,0) < M_R$, *satisfies*

$$u(x,y) \le \Psi_R(u(0,0)) \qquad \textit{for all } (x,y) \in B_{R/4}. \tag{1.7}$$

In Proposition 3 and Theorem 3 (Sections 2 and 3), we give a criterion for the value M_R to be infinite, in which case the restriction $u(0,0) < M_R$ in Theorem 2 can be omitted.

The importance of condition (1.3) for the Harnack inequality is emphasized by its appearence also in the Keller–Osserman theorem [K], [O] (case of the Laplace operator) and in the strong maximum principle for quasilinear equations [V], [D], [PSZ]. It should be noted as well that the standard extension of the Harnack inequality to arbitrary domains is also true in the present circumstances.

The further conditions for the function Ω, whose statement has been deferred, can now be given.

(Ω3) There is a positive constant c such that

$$t\Omega'(t) \le c\Omega(t) \quad \text{for all } t > 0.$$

(Ω4) For all $\theta \ge 1$, there is a number $c_1 = c_1(\theta) \ge 1$ such that

$$\Omega'(\lambda t) \le c_1\Omega'(t) \quad \text{for all } \lambda \in [1, \theta] \text{ and } t > 0.$$

Conditions (Ω1)–(Ω4) are notably weak. For example, they hold for the following examples:

1. The generalized mean curvature operator $A(t) = (1+t^2)^{-s/2}$, $s \le 1$. Here one can take $c = \max\{1,\ 1-s\}$; similarly $c_1 = 1$ if $s = 1$ and $c_1 = \theta^q$, $q = \max\{2,\ -s\}$ if $s < 1$. Note also that $\Omega(\infty) = 1$ for $s = 1$, while $\Omega(\infty) = \infty$ for $s < 1$.

2. The m-Laplace operator $A(t) = t^{m-2}$, $m > 1$. Here one takes $c = m - 1$, $c_1 = \theta^{m-2}$.

3. $A(t) = t^{m-2} + t^{m_1-2}$, $1 < m \le m_1$. Now $c = m_1 - 1$ and $c_1 = \theta^{m_1-2}$. See [BFP] for applications in quantum physics.

4. $A(t) = \tan^{-1} t/t$. Here $c = 1$ and $c_1 = 1$, while $\Omega(\infty) = \pi/2$.

5. $A(t) = (1+t^2)^{-s/2}t^{m-2}$, $s \le m-1$, $m > 1$. Then $c = \max\{m-1,\ m-s-1\}$ and $c_1 = \theta^q$, $q = \max\{m,\ m-s-2\}$. Also, $\Omega(\infty) = 1$ for $s = m-1$, otherwise $\Omega(\infty) = \infty$.

In the final section of the paper, we apply our results to the Euler equation for the variational problem

$$\delta \int a(x,y,u)[G(|Du|) + F(u)]\,dxdy = 0,$$

where the integrands $G(t)$ and $F(u)$ are as given above and $a = a(x,y,u)$ is a positive C^1 function of its variables.

2 Construction of comparison functions

The proof of Theorem 1 uses the method given in [S1]. Let E denote the closed region bounded by an arbitrary ellipse in $\mathbb{R}^2$. It is convenient to introduce *local Cartesian coordinates* (ξ,η) on E, so that E is given by

$$\sigma^2 = \frac{\xi^2}{a^2} + \frac{\eta^2}{b^2} \le 1, \qquad a \ge b > 0$$

(note that the coordinates (ξ,η) on E are related by translation and rotation to the global coordinates (x,y) on $\mathbb{R}^2$). The subset

$$\tfrac{1}{2} \le \sigma \le 1, \qquad \xi \ge \tfrac{1}{2}a$$

of E will be called a *minorant region* and denoted by Σ. To each minorant region Σ, we associate a C^2 function $v = v(\sigma) : \Sigma \to \mathbb{R}^+$ (where $\sigma \in [\frac{1}{2}, 1]$ is the local coordinate on Σ defined above), called a *minorant function* and constructed in such a way that

$$\text{(2.1)} \qquad \begin{aligned} &\operatorname{div}\{A(|Dv|Dv)\} - B(Dv) - f(v) \ge 0, \quad |Dv| < 1 \qquad \text{in } \Sigma, \\ &v(1) = 0; \qquad v' < 0, \quad v'' \ge 0 \quad \text{in } [\tfrac{1}{2}, 1]; \end{aligned}$$

here $'$ denotes differentiation with respect to the variable σ.[1]

To carry out the construction, we need the following lemmas; it is convenient in stating and proving these to replace the variables (ξ,η) by (x,y)—the corresponding results for the variables (ξ,η) then follow from the translational and rotational invariance of (2.1).

Lemma 1. *Let $\sigma^2 = x^2/a^2 + y^2/b^2$, $s^2 = x^2/a^4 + y^2/b^4$. Then if $v = v(\sigma)$ and $v'(\sigma) < 0$, we have*

(i) $\quad Dv = \left(\frac{x}{a^2}, \frac{y}{b^2}\right)\frac{v'}{\sigma}, \qquad |Dv| = \frac{s}{\sigma}|v'|,$

[1] By standard abuse of notation, the letter v denotes either the mapping $v : \Sigma \to \mathbb{R}^+$ or the ordinary real function $v : [\frac{1}{2}, 1] \to \mathbb{R}^+$. The distinction will always be clear from the context.

(ii) $A(|Dv|)Dv = -\frac{1}{s}\left(\frac{x}{a^2}, \frac{y}{b^2}\right)\Omega(|Dv|)$,

(iii) $\operatorname{div}\left\{\frac{1}{s}\left(\frac{x}{a^2}, \frac{y}{b^2}\right)\right\} = \frac{1}{a^2b^2}\frac{\sigma^2}{s^3}$,

(iv) $\frac{1}{s}\left(\frac{x}{a^2}, \frac{y}{b^2}\right)\cdot D\{\Omega(|Dv|)\} = \left(\frac{a^2-b^2}{a^3b^3}\right)^2\frac{x^2y^2}{\sigma^3 s^2}\,\Omega'(|Dv|)|v'|$
$\qquad - \frac{s^2}{\sigma^2}\,\Omega'(|Dv|)v''$.

Proof. The formulas in (i) are obvious; and (ii) then follows at once, since $v' < 0$ by assumption. Identity (iii) is proved by direct calculation. Again, a straightforward calculation and the use of (i) gives

$$\frac{\sigma}{s}\left(\frac{x}{a^2}, \frac{y}{b^2}\right)\cdot D\{\Omega(|Dv|)\} = -\left\{\left[\frac{1}{s^2}\left(\frac{x^2}{a^6}+\frac{y^2}{b^6}\right) - \frac{s^2}{\sigma^2}\right]v' + \frac{s^2}{\sigma}v''\right\}\Omega'(|Dv|).$$

But

$$\sigma^2\left(\frac{x^2}{a^6}+\frac{y^2}{b^6}\right) - s^4 = \left(\frac{a^2-b^2}{a^3b^3}\right)^2 x^2y^2,$$

which immediately gives (iv), since $v' < 0$. □

Lemma 2. *We have*

(i) $bs \le \sigma \le as$, (ii) $\left(\frac{xy}{s\sigma}\right)^2 \le \frac{a^3b^3}{4}$.

Proof. Condition (i) follows at once from the definition of s and σ and the assumption $a \ge b > 0$. To prove (ii), note that

$$\frac{1}{a^3b^3}\left(\frac{xy}{s\sigma}\right)^2 = \frac{1}{\sigma^2}\left(\frac{x}{a}\frac{y}{b}\right)\cdot\frac{1}{s^2}\left(\frac{x}{a^2}\frac{y}{b^2}\right) \le \frac{1}{4},$$

which follows by applying the Cauchy–Schwarz inequality to the two products in parentheses. □

Lemma 3. *Let* $(\Omega 3)$ *and* $(\Omega 4)$ *hold. Then*

(i) $\Omega'(|Dv|)|v'| \le ac\Omega(|v'|/b)$;

(ii) $\frac{s^2}{\sigma^2}\Omega'(|Dv|) \ge c_2\Omega'(|v'|/b)$, *where* $c_2^{-1} = a^2c_1(a/b)$.

Proof. To prove (i), we observe by Lemma 1 (i) that

$$\begin{aligned}\Omega'(|Dv|)|v'| &= \Omega'((s/\sigma)|v'|)|v'| = \frac{\sigma}{s}\Omega'(t)t \qquad (t = (s/\sigma)|v'|)\\ &\le ac\Omega(t) = ac\Omega((s/\sigma)|v'|) \le ac\Omega(|v'|/b)\end{aligned}$$

by Lemma 2 (i) and assumptions $(\Omega 2)$, $(\Omega 3)$, proving (i). To obtain (ii) note that

$$\frac{s^2}{\sigma^2}\Omega'(|Dv|) = \frac{s^2}{\sigma^2}\Omega'((s/\sigma)|v'|) \ge \frac{1}{a^2}\Omega'((s/\sigma)|v'|) \ge c_2\Omega'(|v'|/b),$$

where at the second step we have used Lemma 2 (i) and at the final step (Ω4) with $t = (s/\sigma)|v'|$ and $\lambda = \sigma/bs \in [1, a/b]$. □

Lemma 4. *Let* $w = v/b$. *Then*

$$-\mathrm{div}\{A(|Dv|)Dv\} \leq \frac{1}{\ell}\{\Omega(|w'|)\}' + \frac{\kappa}{\ell\sigma}\Omega(|w'|),$$

where

$$\kappa = c_1(\theta)\left[\theta^3 + \tfrac{1}{4}\left(\theta^2 - 1\right)^2 c\right], \qquad \ell = a\theta c_1(\theta), \qquad \theta = a/b.$$

Proof. By direct calculation, using Lemma 1 (ii)–(iv),

$$\begin{aligned}-\mathrm{div}\{A(|Dv|)Dv\} =& \Omega(|Dv|)\mathrm{div}\left\{\frac{1}{s}\left(\frac{x}{a^2}, \frac{y}{b^2}\right)\right\} + \frac{1}{s}\left(\frac{x}{a^2}, \frac{y}{b^2}\right)\cdot D\{\Omega(|Dv|)\}\\ =& \frac{1}{a^2b^2}\frac{\sigma^2}{s^3}\Omega(|Dv|) + \left(\frac{a^2-b^2}{a^3b^3}\right)^2\frac{x^2y^2}{\sigma^3 s^2}\Omega'(|Dv|)|v'|\\ &- \frac{s^2}{\sigma^2}\Omega'(|Dv|)v''\\ \leq& \frac{1}{a^2b^2}\frac{a^3}{\sigma}\Omega(|Dv|) + \frac{1}{a^3b^3}\left(\frac{a^2-b^2}{2}\right)^2\frac{1}{\sigma}\Omega'(|Dv|)|v'|\\ &- \frac{s^2}{\sigma^2}\Omega'(|Dv|)v''\end{aligned}$$

by Lemma 2 (i), (ii). In turn, from Lemma 2 (i) again and Lemma 3 (i), (ii), we get

$$\begin{aligned}-\mathrm{div}\{A(|Dv|)Dv\} &\leq \frac{1}{a\theta\sigma}\left[\theta^3 + \tfrac{1}{4}(\theta^2-1)^2 c\right]\Omega(|v'|/b) - c_2\Omega'(|v'|/b)v''\\ &= \frac{\kappa}{\ell\sigma}\Omega(|v'|/b) + \frac{1}{\ell}\{\Omega(|v'|/b)\}',\end{aligned}$$

where κ, ℓ and θ are as given in the statement of Lemma 4; here we have used the facts that $v' < 0$, $v'' \geq 0$ and $\Omega'(t) > 0$ for $t > 0$. Lemma 4 follows since $w = v/b$. □

We can now complete the construction of the minorant function $v = bw$ by choosing w as a solution of the following initial value problem (see (2.1) and Lemma 4):

$$\{\Omega(|w'|)\}' + \frac{k}{\sigma}\Omega(|w'|) + \ell f(bw) = 0, \qquad k = \kappa + \mu\ell, \tag{2.2}$$

$$w(1) = 0, \qquad w'(1) = -\alpha < 0. \tag{2.3}$$

A short calculation shows that if w is a solution of the problem (2.2)–(2.3) over the *full interval* $[\frac{1}{2}, 1]$, and if $w' > -1$, then $v = bw : \Sigma \to \mathbb{R}^+$ is a minorant function on Σ. (Here one uses condition (B), the relation $|Dv| = (s/\sigma)|v'| = (bs/\sigma)|w'| \leq |w'| < 1$ and $\Omega(|Dv|) \leq \Omega(|w'|)$, since Ω is increasing.)

The content of the next result is that such a solution w exists.

Proposition 1. *Suppose either that $f \equiv 0$ on $[0, \delta]$, $\delta > 0$, or that $f > 0$ on $\mathbb{R}^+$ and* (1.3) *holds.*

Then there exists a number $\alpha_0 > 0$ such that the initial value problem (2.2)–(2.3) has a unique convex C^2 solution $w = w_\alpha = w_\alpha(\sigma)$ on the entire interval $[\frac{1}{2}, 1]$, with $w' > -1$, provided that $\alpha \in (0, \alpha_0)$.

Moreover, w_α depends continuously on α;

$$w'_\alpha < 0, \quad w''_\alpha \geq 0 \quad \text{in } [\tfrac{1}{2}, 1];$$

$$w_\alpha > w_\beta \quad \text{if } \alpha > \beta > 0;$$

$$w_\alpha(\sigma) \geq \alpha(1-\sigma), \qquad \lim_{\alpha \to 0} w_\alpha(\tfrac{1}{2}) = 0;$$

and the limit $m_0 = \lim_{\alpha \to \alpha_0} w_\alpha(\frac{1}{2})$ exists and is finite.

Proof. 1. *Existence.* By Lemma 2 of [PS], there exists a convex piecewise C^2 solution $z = z(\sigma)$, $\sigma \in [\frac{1}{2}, 1]$, of the differential inequality

$$\{\Omega(|z'|)\}' + \frac{k}{\sigma}\Omega(|z'|) + \ell f(bz) \leq 0; \qquad z(1) = 0, \quad z'(1) = -\beta, \tag{2.4}$$

provided that β is sufficiently small, say $\beta < \beta_0$, for some appropriate positive number β_0. From [PS] it also follows that z, $z' \to 0$ uniformly as $\beta \to 0$.

We can now construct the required solution w of (2.2)–(2.3) by a standard continuation procedure. Fix $\alpha > 0$. By the Schauder fixed point argument (see [FLS, Proposition A1]), there exists a C^2 solution w of (2.2), (2.3) on some non-trivial interval $(\sigma_0, 1]$. Observe that $w'' > 0$ by (2.2). Consequently, $w > z_0$ on $(\sigma_0, 1)$, where $z_0 = z_0(\sigma) = \alpha(1-\sigma)$.

Now let α be restricted by $0 < \alpha < \beta_0$ and choose β such that $\alpha < \beta < \beta_0$. Let z be the solution of (2.4) which was constructed above, and which of course is defined on the entire interval $[\frac{1}{2}, 1]$. Then $w < z$ and $|w'| < |z'|$ on $(\sigma_0, 1)$ by standard comparison arguments and the fact that Ω is increasing and f is non-decreasing. By the same reasoning, w can be continued backward, with $z_0 < w < z$, $|w'| < |z'|$, $w'' \geq 0$, until one reaches $\sigma = 1/2$.

Finally, if β_0 is made even smaller if necessary, then $|w'| < |z'| < 1$. This choice of β_0 we now call α_0.

To complete the proof, we note some important properties of the solution $w = w_\alpha$.

2. *Uniqueness and continuous dependence on* α. This follows from standard theory of ordinary differential equations applied to the inverse function $\sigma = \sigma(w)$, which satisfies the corresponding equation

$$\Omega'\left(\frac{1}{|\sigma_w|}\right)\sigma_{ww} + \frac{k}{\sigma}\Omega\left(\frac{1}{|\sigma_w|}\right)\sigma_w^3 + \ell f(bw)\sigma_w^3 = 0, \qquad w \geq 0,$$

with $\sigma(0) = 1$, $\sigma_w(0) = -1/\alpha$.

3. *Monotonicity of* $\alpha \mapsto w_\alpha$. As earlier, this is a consequence of standard comparison arguments and the fact that f is non-decreasing. That $w_\alpha(\frac{1}{2})$ has a (finite) limit as $\alpha \to \alpha_0$ follows immediately.

Remark. The condition $|w'| < |z'| < 1$ forces $\alpha_0 < 1$ and $m_0 < 1/2$. In some cases, however, the principal inequality of (B) holds without the restriction $|p| < 1$. In this case, the condition $|Dv| < 1$ is not required in (2.1); and, in turn, the proof of Proposition 1 can be carried out without the condition $|z'| < 1$. It can then happen that $\beta_0 = \infty$, hence also $\alpha_0 = \infty$ and $m_0 = \infty$. We take up this situation in more detail in Proposition 3 below.

Now define $m(\alpha) = w_\alpha(\frac{1}{2})$, $\alpha \in J = (0, \alpha_0)$. The properties of the function m are given by the following.

Proposition 2. *Under the assumptions of Proposition 1, we have*
(i) $m(\alpha) \to 0$ *as* $\alpha \to 0$,
(ii) m *is continuous and strictly increasing on* J,
(iii) $m(\alpha) > \alpha/2$,
(iv) $m_0 = \lim_{\alpha\to\alpha_0} m(\alpha)$.

This follows directly from Proposition 1.

It is convenient in the sequel to have $m_0 = \infty$ in all cases. Indeed, this can always be maintained by replacing $m(\alpha)$ by a new function $\hat{m}(\alpha)$, also defined on J and obeying (i)–(iv), but additionally such that $\hat{m}(\alpha) \geq m(\alpha)$ and $\lim_{\alpha\to\alpha_0} \hat{m}(\alpha) = \infty$. Obviously, then, for all $\alpha \in J$ we have

(2.5) $$\hat{m}(\alpha) \geq w_\alpha(\tfrac{1}{2}), \qquad \alpha \in J.$$

In the main proofs of Section 3, it is only Proposition 2 which is needed, together with (2.5). Thus, in the sequel, we can drop the"hat" from $\hat{m}$ without confusion.

Proposition 3. *Let* $\Omega(\infty) = \infty$. *Under the assumptions of Proposition* 1, *we have* $\alpha_0 = \infty$ *if the principal inequality of* (B) *holds for all* $p \in \mathbb{R}^2$ *and if, moreover, either* $f \equiv 0$ *for* $u \geq 0$ *or*

(2.6) $$\int_1^\infty \frac{ds}{H^{-1}(F(s))} = \infty.$$

Proof. First note that $H(\infty) = \infty$, so that the integral in (2.6) is well-defined.

Because of the strengthened form of (B), the condition $|Dv| < 1$ in (2.1) is not needed in the construction of a minorant function v. It is therefore enough to show that problem (2.4) has a solution over the entire interval $[\frac{1}{2}, 1]$ for *all* $\beta > 0$. Suppose not. Then there are $\beta > 0$ and $\bar{\sigma} \in (\frac{1}{2}, 1)$ such that the corresponding local solution z is piecewise C^2 in $(\bar{\sigma}, 1]$ and $z(\sigma) \to \infty$ as $\sigma \to \bar{\sigma}$; see [PS, proof of Lemma 2]. In particular, by [PS, (2.16)] and the fact that $\Omega(\infty) = \infty$, the case $z(\bar{\sigma}) < \infty$, $z'(\bar{\sigma}) = -\infty$ cannot occur at the blow-up point $\bar{\sigma}$.

Also, as in [PS, (2.24)], we have

$$|z'(\sigma)| \le H^{-1}(\nu F(bz(\sigma))) \qquad \text{for all } \sigma \in (\bar{\sigma}, \rho], \tag{2.7}$$

where $\nu > 1$ and $\rho \in (\bar{\sigma}, 1)$ are appropriate constants depending only on β, b, θ, μ, f and Ω. Integration of (2.7) on $(\bar{\sigma}, \rho]$ gives

$$\int_{\gamma}^{\infty} \frac{ds}{H^{-1}(F(s))} \le \frac{\nu b}{2},$$

where $\gamma = \nu b z(\rho)$. This contradicts (2.6) and completes the proof. □

3 Proof of Theorem 1

We begin with two lemmas.

Lemma 5. *Assume Ω satisfies conditions $(\Omega 1)$–$(\Omega 2)$. Let u_1 and u_2 be respective C^1 distribution solutions of*

$$\operatorname{div}\{A(|Du_1|)Du_1)\} - \hat{B}(u_1, Du_1) \le 0, \qquad u_1 \ge 0, \tag{3.1}$$

and of

$$\operatorname{div}\{A(|Du_2|)Du_2)\} - \hat{B}(u_2, Du_2) \ge 0, \qquad u_2 \ge 0, \tag{3.2}$$

in a bounded domain D, where the function $\hat{B}(u,p)$ is non-decreasing in the variable u and continuously differentiable in the variable p for $p \in \mathbb{R}^2\setminus\{0\}$. Suppose also that u_1 and u_2 are continuous on $\overline{D}$ and satisfy $|Du_1| + |Du_2| > 0$ there.

Then $u_1 \ge u_2$ on ∂D implies $u_1 \ge u_2$ in D.

This is the weak comparison principle given in [PSZ, Lemma 5] for $D \subset \mathbb{R}^n$ and for $\hat{A}(p) = pA(|p|)$.

Lemma 6. *Assume that Ω satisfies conditions $(\Omega 1)$–$(\Omega 2)$ and that the function $B(p)$ obeys (B). Let u be a solution of $(1.2)_{left}$ in a bounded domain D and let $d > 0$ be such that the set $U_d = \{x \in D \,:\, u(x) > d\}$ is non-empty. Then $\partial K \cap \partial D$ is non-empty for any component K of U_d.*

Proof. Suppose $\partial K \cap \partial D = \emptyset$. Since ∂K and ∂D are compact, $\text{dist}(\partial K, \partial D) > 0$ and $\overline{K} \subset D$. It then follows from the definition of U_d and K that $u = d$ on ∂K.

We now wish to apply Lemma 5. Here it can be assumed without loss of generality that $D \subset B_R$ for some $R > 0$. Let $\epsilon > 0$ and define

$$u_1 = u_1(y) = d + \int_R^y \Omega^{-1}(\epsilon e^{-\mu s})\, ds \quad \text{for } y \in [-R, R],$$

where μ is the constant in condition (B). One easily checks that

$$\text{div}\{A(|Du_1|)Du_1)\} = \{\Omega(|Du_1|)\}' = -\mu\Omega(|Du_1|) \quad \text{in } D.$$

Let ϵ be taken so small that $|Du_1| < 1$ in D. Then by (B), u_1 obeys (3.1) with $\hat{B}(u,p) = -B(p)$, while also $|Du_1| > 0$ in B_R. Next put $u_2 = u$, so by $(1.2)_{\text{left}}$ the inequality (3.2) holds, again with $\hat{B}(u,p) = -B(p)$. Obviously, $u_1 > u_2$ on ∂K. Hence by Lemma 5, with $D = K$, we obtain $u = u_2 \le u_1 = u_1(y)$. Letting $\epsilon \to 0$ then yields $u \le d$ in K, an immediate contradiction. □

It almost goes without saying that Lemmas 5 and 6 also hold in arbitrary dimensions $n \ge 2$.

For the remainder of the section, we follow [S1]; see also [PW, pages 111–117] and [GT, pages 41–44].

Let u be a positive solution of (1.2) in B_R. From Lemma 6, with $D = B_R$, $d = u(0,0)/2$ and K the component of U_d containing $(0,0)$, it follows at once that there is a curve Γ in U_d joining $(0,0)$ and some point $Q \in \partial B_R$. By rotation of coordinates, it can be assumed without loss of generality that $Q = (0, R)$; see [S1, Figure 1 on page 296] or [PW, Figure 17 on page 112]. As in [S1] (see also [PW, pages 112–113]), we introduce two ellipses E_1 and E_2, with horizontal major axes and $a = 3R$, $b = R/2$ (so $\theta = a/b = 6$), and with respective centers at $(-5R/2, R/2)$ and $(5R/2, R/2)$. Let Σ_1 and Σ_2 be corresponding minorant regions, chosen so that $E_1 \cap B_R \subset \Sigma_1$ and $E_2 \cap B_R \subset \Sigma_2$. Clearly, $T \equiv \Sigma_1 \cap \Sigma_2 \subset B_R$; and, in fact, T contains the line segment τ joining the vertices $(-R/2, R/2)$ and $(R/2, R/2)$ of E_1 and E_2.

We complete the construction by introducing a third ellipse E_3, with vertical major axis and $a = 3R$ and $b = R/2$, and with center at $(0, 2R)$. Let Σ_3 be the associated minorant region which is contained in B_R; in particular, Σ_3 is bounded above by a proper subsegment $\hat{\tau}$ of τ.

A simple calculation shows that the closed disk $\overline{B_{R/3}}$ is contained in the interior of Σ_3; see [S1, Figure 2 on page 297]. In fact, the local coordinate σ in Σ_3 satisfies $\sigma \le 0.9542 (\approx 2\sqrt{2}/3)$ at all points in $B_{R/3}$. Also, for future use, one finds easily that the local coordinate σ in Σ_1, or in Σ_2, is such that $\sigma \ge (10 + \sqrt{3})/12$ on $\hat{\tau}$.

Let $\hat{\alpha}$ be defined by $bm(\hat{\alpha}) = \frac{1}{2}u(0,0)$, and put $\hat{w} = w_{\hat{\alpha}}$, the solution of (2.2)–(2.3) with $\alpha = \hat{\alpha}$. Also let $\sigma_1 : \Sigma_1 \to [\frac{1}{2}, 1]$, $\sigma_2 : \Sigma_2 \to [\frac{1}{2}, 1]$ be, respectively, the local coordinates σ on Σ_1 and Σ_2. Then

$$v_1 = b\hat{w}(\sigma_1) : \Sigma_1 \to \mathbb{R}^+ \quad \text{and} \quad v_2 = b\hat{w}(\sigma_2) : \Sigma_2 \to \mathbb{R}^+$$

are minorant functions associated to Σ_1 and Σ_2.

In particular, for points on the straight boundary segment $\xi = a/2$ of Σ_1, it is evident that $v_1 \le b\hat{w}(\frac{1}{2}) \le bm(\hat{\alpha}) = \frac{1}{2}u(0,0)$; similarly, on the boundary segment $\xi = a/2$ of Σ_2, we have $v_2 \le \frac{1}{2}u(0,0)$. In turn,

$$v_1 \le \tfrac{1}{2}u(0,0) \quad \text{in } \Sigma_1, \qquad v_2 \le \tfrac{1}{2}u(0,0) \quad \text{in } \Sigma_2.$$

Then, following [S1] and using the comparison Lemma 5 with

$$\hat{B}(u,p) = B(p) + f(u),$$

and with $u_1 = u$ (so that $(1.2)_{\text{right}}$ implies that u_1 satisfies (3.1)) and u_2 equal to either v_1 or v_2 (so that by (2.1) u_2 satisfies (3.2)), we get

$$u(x,y) \ge \min\{v_1, v_2\} = b\min\{\hat{w}(\sigma_1), \hat{w}(\sigma_2)\} \qquad \text{on } \tau. \tag{3.3}$$

Note here that $|Dv_1| > 0$, $|Dv_2| > 0$, so Lemma 5 remains applicable. Moreover, since $\hat{w}'' \ge 0$ and $\hat{w}(1) = 0$, $\hat{w}'(1) = -\hat{\alpha}$, we find that

$$\hat{w}(\sigma_1),\ \hat{w}(\sigma_2) \ge \left(1 - \frac{10+\sqrt{3}}{12}\right)\hat{\alpha} \qquad \text{on } \hat{\tau}. \tag{3.4}$$

In the same way, let $\hat{\gamma}$ be such that $m(\hat{\gamma}) = \frac{1}{12}(2-\sqrt{3})\hat{\alpha}$, and denote by $\hat{\phi}$ the solution of (2.2)–(2.3) with $\alpha = \hat{\gamma}$. Finally, let σ_3 be the local coordinate σ on Σ_3, and consider the minorant function $v_3 = b\hat{\phi}(\sigma_3) : \Sigma_3 \to \mathbb{R}^+$ associated with Σ_3. Then from (3.3), (3.4), and from Lemma 5 with $u_1 = u$ and $u_2 = v_3$ (and again for $\hat{B}(u,p) = B(p) + f(u)$), we obtain

$$u(x,y) \ge v_3(x,y) \quad \text{in } \Sigma_3. \tag{3.5}$$

From the main construction above, one sees that

$$v_3(x,y) \ge \frac{3-2\sqrt{2}}{3}b\hat{\gamma} = \frac{3-2\sqrt{2}}{6}R\hat{\gamma} \quad \text{on } B_{R/3} \tag{3.6}$$

since $b = R/2$. Combining the inequalities (3.5) and (3.6) yields, for $(x,y) \in B_{R/3}$,

$$u(x,y) \ge \frac{3-2\sqrt{2}}{6}Rm^{-1}\left(\frac{2-\sqrt{3}}{12}m^{-1}\left(\frac{u(0,0)}{R}\right)\right) \equiv \Phi_R(u(0,0)),$$

where

$$\Phi_R(t) = \frac{3-2\sqrt{2}}{6} Rm^{-1}\left(\frac{2-\sqrt{3}}{12} m^{-1}\left(\frac{t}{R}\right)\right).$$

The existence of the inverse function m^{-1} follows from Proposition 2. By the agreement $m_0 = \infty$, it is evident that m^{-1} maps $[0, \infty)$ strictly monotonically onto $[0, \alpha_0)$. Hence Φ_R is strictly increasing on $[0, \infty)$, and the proof is complete. □

Proof of Theorem 2. Define

$$M_R = \begin{cases} \dfrac{3-2\sqrt{2}}{6} Rm^{-1}\left(\dfrac{2-\sqrt{3}}{12}\alpha_0\right) & \text{if } \alpha_0 < \infty, \\ \infty & \text{if } \alpha_0 = \infty, \end{cases} \tag{3.7}$$

so in particular Φ_R is strictly increasing from $[0, \infty)$ onto $[0, M_R)$, and admits an inverse, say $\Psi_R = \Phi_R^{-1} : [0, M_R) \to [0, t_0)$. The conclusion now follows exactly as in the proof of Corollary 1 in [S1]. □

Theorem 3. *Let the hypotheses of Proposition 3 hold. Then $M_R = \infty$.*

Proof. By Proposition 3, we have $\alpha_0 = \infty$, so the conclusion is an immediate consequence of the definition (3.7) of M_R. □

Remark. Conditions (1.3) and (2.6) place non-trivial restrictions on the behavior of the function f for u near 0 and u near ∞. For example, if $A(t) = t^{m-2}$ (m-Laplace operator), then $H(t) = mt^m/(m-1)$. Thus a function f with polynomial behavior $f \approx u^p$ near $u = 0$ and $f \approx u^q$ near $u = \infty$ satisfies (1.3), (2.6) only when $p \geq m-1$, $q \leq m-1$.

4 A variational example

Here we consider the Euler equation for the variational problem

$$\delta \int a(x, y, u)\left[G(|Du|) + F(u)\right] dxdy = 0,$$

where the integrands $G(t)$ and $F(u)$ are given at the beginning of the introduction, and $a = a(x, y, u)$ is a positive C^1 function satisfying

$$0 \leq a_u/a \leq \text{Const.} \tag{4.1}$$

One checks easily that the Euler equation takes the form

$$\begin{aligned} \operatorname{div}\{A(|Du|)Du\} = &- \frac{Da \cdot Du + a_u|Du|^2}{a} A(|Du|) \\ &+ f(u) + \frac{a_u}{a}\left[G(|Du|) + F(u)\right]. \end{aligned} \tag{4.2}$$

We show that this can be put into the form (1.2). First, since we always deal with bounded domains (e.g., B_R), there exists $\hat{\mu} > 0$ such that

$$0 \le a_u/a \le \hat{\mu}, \qquad |Da|/a \le \hat{\mu};$$

moreover, since f and Ω are non-decreasing,

$$F(u) \le uf(u), \qquad G(t) \le t\Omega(t).$$

Hence, noting particularly that $a_u \ge 0$, we see that, as required,

$$-B(|Du|) \le \operatorname{div}\{A(|Du|)Du\} \le \hat{\mu}\Omega(|Du|) + \hat{f}(u), \tag{4.3}$$

where $B(p) = \hat{\mu}(1 + |p|)\Omega(|p|)$ and $\hat{f}(u) = (1 + \hat{\mu} u)f(u)$. Here $B(p) \le 2\hat{\mu}\Omega(|p|)$ when $|p| < 1$, so (B) holds with $\mu = 2\hat{\mu}$.

Since $\hat{f}$ obviously satisfies (F) whenever f does, and since equally clearly

$$\int_0^1 \frac{ds}{H^{-1}(\hat{F}(s))} = \infty$$

if (1.3) holds, we have the following result.

Theorem 4. *Let the function Ω satisfy* (Ω1)–(Ω4), *and assume that f obeys* (F). *Suppose also that either $f \equiv 0$ on $[0, \delta]$, $\delta > 0$, or $f > 0$ on $\mathbb{R}^+$ and*

$$\int_0^1 \frac{ds}{H^{-1}(F(s))} = \infty. \tag{1.3}$$

Then, for every $R > 0$, there is a strictly increasing continuous function Φ_R : $[0, \infty) \to [0, \infty)$, with $\Phi_R(0) = 0$, such that any C^1 distribution solution u of the Euler equation (4.2) *in the disk B_R satisfies the Harnack condition*

$$u(x, y) \ge \Phi_R(u(0, 0)) \qquad \textit{for all } (x, y) \in B_{R/3}.$$

It is obvious that Theorem 2 also holds for equation (4.2).

In conclusion, it is not hard to see that in order to have $M_R = \infty$, one needs the condition (4.1) to hold in the stronger form

$$(1 + u)a_u/a \le \text{Const.}$$

(so that one can take $\hat{f}(u) = \text{Const.}\, f(u)$), and also that either $f \equiv 0$ or (2.6) is satisfied. Here one uses the explicit form of the right hand side of (4.3), from which it follows that the restriction $|Dv| < 1$ is *not* required for the construction of the minorant function v; see the Remark before Proposition 2.

Acknowledgement. This work is partially supported by CNR, Progetto Strategico *Modelli e Metodi per la Matematica e l'Ingegneria*, by the Italian *Ministero della Università e della Ricerca Scientifica e Tecnologica* under the auspices of *Gruppo Nazionale di Analisi Funzionale e sue Applicazioni* of the CNR and by the Project *Metodi Variazionali ed Equazioni Differenziali Non Lineari*.

REFERENCES

[BFP] V. Benci, D. Fortunato and L. Pisani, *Soliton like solutions of a Lorentz invariant equation in dimension* 3, Rev. Math. Phys. **10** (1998), 315–344.

[D] J. I. Diaz, *Nonlinear Partial Differential Equations and Free Boundaries*, Research Notes in Math. **106**, 1985.

[FLS] B. Franchi, E. Lanconelli and J. Serrin, *Existence and uniqueness of ground states of quasilinear elliptic equations*, Adv. Math. **118** (1996), 174–243.

[GT] D. Gilbarg and N. Trudinger, *Elliptic Partial Differential Equations of Second Order*, 2nd edition, Springer-Verlag, Berlin, 1983.

[G] G. Gregori, *Generalized solutions for a class of non-uniformly elliptic equations in divergence form*, Comm. Partial Differential Equations **22** (1997), 581–617.

[K] J. Keller, *On solutions of* $\Delta u = f(u)$, Comm. Pure Appl. Math. **10** (1957), 503–510.

[L] G. M. Lieberman, *The natural generalization of the natural conditions of Ladyzhenskaya and Ural'tseva for elliptic equations*, Comm. Partial Differential Equations **16** (1991), 311–361.

[O] R. Osserman, *On the inequality* $\Delta u \geq f(u)$, Pacific J. Math. **7** (1957), 1641–1647.

[PS] P. Pucci and J. Serrin, *A note on the strong maximum principle for elliptic differential inequalities*, J. Math. Pures Appl. **79** (2000), 57–71.

[PSZ] P. Pucci, J. Serrin and H. Zou, *A strong maximum principle and a compact support principle for singular elliptic inequalities*, J. Math. Pures Appl. **78** (1999), 769–789.

[PW] M. H. Protter and H. F. Weinberger, *Maximum Principles in Differential Equations*, Prentice-Hall, Englewood Cliffs, NJ, 1967.

[S1] J. Serrin, *On the Harnack inequality for linear elliptic equations*, J. Analyse Math. **4** (1954–56), 292–308.

[S2] J. Serrin, *Local behavior of solutions of quasi-linear equations*, Acta Math. **111** (1964), 247–302.

[T] N. S. Trudinger, *Harnack inequalities for nonuniformly elliptic divergence structure equations*, Invent. Math. **64** (1981), 517–531.

[V] J. L. Vazquez, *A strong maximum principle for some quasilinear elliptic equations*, Appl. Math. Optim. **12** (1984), 191–202.

Patrizia Pucci
DIPARTIMENTO DI MATEMATICA
UNIVERSITÁ DI PERUGIA
PERUGIA, ITALY
email: pucci@dipmat.unipg.it

James Serrin
DEPARTMENT OF MATHEMATICS
UNIVERSITY OF MINNESOTA
MINNEAPOLIS, MINNESOTA, USA
email: serrin@math.umn.ed

(Received September 24, 2000 and in revised form March 18, 2001)

I.2. The Dirichlet Problem – Theory

Yehuda Pinchover

Mathematics Subject Classification (2010). Primary 35J25; Secondary 35J15, 35B50, 35J62, 35J92, 35J93.

Keywords. Dirichlet problem, barrier, minimal surface equation, p-Laplacian, maximum principle, quasilinear elliptic operator.

James Serrin's research on the theory of the Dirichlet problem for *non-uniformly* elliptic equations in non-divergence form, which is the topic under discussion, was fundamental, and generated a broad and lasting impact on the study of second-order elliptic and parabolic equations. It is our aim to provide a short historical setting for his works [i–iii] below, and to summarize some of the significant related research that followed.

Commentary on [i] and [ii] below: The pioneering work of S. Bernstein on the solvability of the Dirichlet problem for quasilinear elliptic equations with two independent variables was developed and extended by many authors including Leray, Schauder, Morrey, Bers, and Nirenberg, and in higher dimensions by Gilbarg, Stampacchia, Trudinger and Oskolkov. In [10], Ladyzhenskaya and Uraltseva proved interior estimates of solutions and gradient estimates up to the boundary for the quasilinear elliptic Dirichlet problem in higher dimensions. Using [10], James Serrin developed in his remarkable papers [i] and [ii] a far-reaching general theory for quasilinear *non-uniformly* elliptic equations in non-divergence form, encompassing many classical examples like the minimal surface equation and the equation of specified mean curvature. In particular, Serrin obtained necessary conditions and sufficient conditions (curvature restrictions) for the solvability, in the classical sense, of the Dirichlet problem in a given domain for arbitrarily given smooth boundary data.

The results of [i] and [ii] are presented in detail in the classical book of Gilbarg and Trudinger [8]. We note that the smoothness assumptions on the boundary of the domain and on the Dirichlet data in [ii] can be slightly reduced. See, for example, [13, 20]. For a variational approach to the prescribed mean curvature Dirichlet problem in the class of BV-functions see for example [7].

Serrin's result concerning the existence and uniqueness of graphs of prescribed mean curvature in the Euclidean space was extended to the (fully nonlinear) prescribed m mean curvature equation by Trudinger [23], and to a broad class of Riemannian manifolds by many authors. See for example [1, 21], and the references therein.

The author acknowledges the support of the Israel Science Foundation (grant No. 587/07) founded by the Israel Academy of Sciences and Humanities.

Serrin's proof requires the analysis of the distance function and the curvature properties of the boundary. In recent studies of partial differential equations and the associated coercivity inequalities there has been an increasing amount of attention devoted to properties of the distance function and the effects of curvature of the boundary, see for example [11, 12, 16], and the references therein.

Finally, parabolic analogs of the results in [ii] were obtained by many authors. See the notes in [14, p. 248].

Commentary on [iii] below: Meyers–Serrin's work on the exterior Dirichlet problem [iii] led naturally to an extensive investigation of the asymptotic behavior of solutions of linear and also quasilinear elliptic equations near the isolated singular point ∞, see for example [2, 5, 22] and the references therein. The exterior Dirichlet problem was studied from a probabilistic point of view in [4], while the *selfadjoint* case was studied in [9]. In [6] Friedman used the techniques of [iii] to show the existence and uniqueness of an *entire* solution which vanishes at infinity of the equation $Lu = f$, under the assumptions that $L\mathbf{1} = 0$ and that $f = O(|x|^{-2-\varepsilon})$ at infinity.

The effective dimension (which is also called Meyers–Serrin dimension) introduced in [iii] plays an important role in analyzing the validity of the Liouville theorem [18], the existence of positive super-solutions to semi-linear non-divergence type elliptic equations in exterior domains [15, 19], and the stabilization of solutions of the Cauchy problem and the initial-boundary value problem for second-order linear parabolic equations with time dependent coefficients [3].

The results in [iii] are closely related to criticality and the Martin boundary theory of second-order linear elliptic operators in a general domain D in $\mathbb{R}^d$ or a noncompact manifold. For instance, suppose that L is defined on D, and assume that $L\mathbf{1} = 0$. Then the existence of an anti-barrier (respectively, the violation of the extended maximum principle) implies the criticality/recurrency (respectively, subcriticality/transiency) of the operator L in D [17, Corollary 3.10]. In particular, if L is critical/recurrent in D, then Problem II (with $f = 0$) in an exterior subdomain of D is well set.

This Commentary is dedicated to James Serrin on the occasion of his 85th birthday.

References

[1] P. Bayard, *Dirichlet problem for space-like hypersurfaces with prescribed scalar curvature in* $\mathbb{R}^{n,1}$. Calc. Var. Partial Differential Equations **18** (2003), 1–30.

[2] H. Berestycki, and L. Nirenberg, *Asymptotic behaviour via the Harnack inequality.* In: Nonlinear Analysis, 135–144, Sc. Norm. Super. di Pisa Quaderni, Scuola Norm. Sup., Pisa, 1991.

[3] V.N. Denisov, On the behavior of solutions of parabolic equations for large time values. Russian Math. Surveys **60** (2005), 721–790.

[4] M.I. Freidlin, *The exterior Dirichlet problem for the class of bounded functions.* Teor. Verojatnost. i Primenen **11** 1966, 463–471.

[5] M. Fraas and Y. Pinchover, *Positive Liouville theorems and asymptotic behavior for p-Laplacian type elliptic equations with a Fuchsian potential*, Confluentes Mathematici **3**, 291–323.

[6] A. Friedman, *Bounded entire solutions of elliptic equations.* Pacific J. Math. **44** (1973), 497–507.

[7] M. Giaquinta, *On the Dirichlet problem for surfaces of prescribed mean curvature.* Manuscripta Math. **12** (1974), 73–86.

[8] D. Gilbarg, and N.S. Trudinger, *Elliptic Partial Differential Equations of Second Order.* Reprint of the 1998 edition. Classics in Mathematics. Springer-Verlag, Berlin, 2001.

[9] C.E. Kenig, and W.-M. Ni, *An exterior Dirichlet problem with applications to some nonlinear equations arising in geometry.* Amer. J. Math. **106** (1984), 689–702.

[10] O.A. Ladyzhenskaya, and N.N. Uraltseva, *Linear and Quasilinear Elliptic Equations.* Nauka, Moscow, 1964 [Russian]; English transl., Academic Press, San Diego, 1968.

[11] R.T. Lewis, J. Li, and Y.Y. Li, *A geometric characterization of a sharp Hardy inequality*, J. Funct. Anal. **262** (2012), 3159–3185.

[12] Y.Y. Li, and L. Nirenberg, *The distance function to the boundary, Finsler Geometry and the singular set of viscosity solutions of some Hamilton–Jacobi equations.* Comm. Pure Appl. Math. **58** (2005), 85–145.

[13] G.M. Lieberman, *The Dirichlet problem for quasilinear elliptic equations with continuously differentiable boundary data.* Comm. Partial Differential Equations **11** (1986), 167–229.

[14] G.M. Lieberman, *Second Order Parabolic Differential Equations.* World Scientific Publishing Co., Inc., River Edge, NJ, 1996.

[15] V. Kondratiev, V. Liskevich, and Z. Sobol, *Positive super-solutions to semi-linear second-order non-divergence type elliptic equations in exterior domains. Trans.* Amer. Math. Soc. **361** (2009), 697–713.

[16] M. Marcus, V.J. Mizel, and Y. Pinchover, *On the best constant for Hardy's inequality in* $\mathbb{R}^n$, Trans. Amer. Math. Soc. **350** (1998), 3237–3255.

[17] Y. Pinchover, *On positivity, criticality, and the spectral radius of the shuttle operator for elliptic operators.* Duke Math. J. **85** (1996), 431–445.

[18] R. Redheffer, *A classification of solutions of certain nonlinear differential inequalities with application to theorems of Liouville type.* Math. Z. **192** (1986), 453–465.

[19] R. Redheffer, *A class of quasilinear differential inequalities whose solutions are ultimately constant.* Arch. Rational Mech. Anal. **99** (1987), 165–187.

[20] J. Serrin, *The Dirichlet problem for surfaces of constant mean curvature.* Proc. London Math. Soc. (3) **21** (1970), 361–384.

[21] J. Spruck, *Interior gradient estimates and existence theorems for constant mean curvature graphs in* $M^n \times \mathbb{R}$. Pure Appl. Math. Q. 3 (2007), 785–800.

[22] J. Serrin, and H.F. Weinberger, *Isolated singularities of solutions of linear elliptic equations.* Amer. J. Math. *88* (1966), 258–272.

[23] N.S. Trudinger, *Apriori bounds and necessary conditions for solvability of prescribed curvature equations.* Manuscripta Math. **67** (1990), 99–112.

Yehuda Pinchover
Department of Mathematics
Technion – Israel Institute of Technology
Haifa 32000, Israel
e-mail: `pincho@tx.technion.ac.il`

Proc. Natl. Acad. Sci. USA 58 (1967), 1829–1835. **Reprint I.2.[i]**

THE DIRICHLET PROBLEM FOR QUASILINEAR EQUATIONS WITH MANY INDEPENDENT VARIABLES

BY JAMES SERRIN

DEPARTMENT OF MATHEMATICS, UNIVERSITY OF MINNESOTA, MINNEAPOLIS

Communicated by Charles B. Morrey, Jr., July 5, 1967

The purpose of this note is to outline some results of a study of the Dirichlet problem for elliptic partial differential equations of the form

$$\mathcal{A}(x,u,Du)D^2u = \mathcal{B}(x,u,Du), \qquad x = (x_1,\ldots,x_n). \tag{1}$$

Here Du and D^2u denote the set of first and second derivatives of the dependent variable u, and $\mathcal{A}(x,u,Du)$ and $\mathcal{B}(x,u,Du)$ are, respectively, a given matrix and a given scalar function. We suppose throughout that $\mathcal{A}$ and $\mathcal{B}$ are continuously differentiable for all real values of u and Du, and for all points x in any domain under consideration. The ellipticity of (1) requires further that the coefficient matrix satisfy

$$\xi\mathcal{A}\xi > 0, \tag{2}$$

where ξ is an arbitrary nonvanishing real vector.

In what follows, in order to present a simple and unified body of results, we shall from the very beginning endow $\mathcal{A}$ with a certain degree of additional structure (§1). Nevertheless, we shall maintain suitable latitude to include the minimal surface equation, the equation for surfaces of constant mean curvature, and a number of other nonuniformly elliptic equations of classical interest.

Roughly speaking, the conclusions are in the form of necessary and sufficient conditions for the Dirichlet problem to be solvable for arbitrarily assigned smooth boundary data. The work thus constitutes a development and extension of the theory initiated by Bernstein in a series of papers[1–4] in 1910–1912 and applied by him to a number of cases of great generality. Bernstein's theory was subsequently sharpened and enlarged in a fundamental paper of Leray[5] which appeared in 1939. In comparing this work with ours, we note first of all that their considerations were restricted to the case of two independent variables, and involve certain regularity assumptions on the behavior of $\mathcal{A}$ and $\mathcal{B}$ for large values of Du which we avoid here. Bernstein had also concluded for certain equations (including that governing surfaces of constant mean curvature) that the Dirichlet problem was not generally solvable for arbitrary boundary data, even when the domain in question was convex and arbitrarily small. While strictly true, this statement nevertheless avoids the problem of determining precisely those domains for which the Dirichlet problem is in fact solvable. It is one of the goals of this paper (as it was of Leray's) to consider questions of just this sort. We add finally that some of the recent results of Finn[6] and of Jenkins and Serrin[7] on the solvability of the Dirichlet problem for the minimal surface equation appear as special cases in the present theory.

Some examples illustrating our conclusions are listed at the end of this note.

1. Concerning the structure of the coefficient matrix, it will be assumed (in some analogy with the definition of uniform ellipticity) that for all suitably large values of the real vector $p = (p_1,\ldots,p_n)$

$$\lambda|p|^m \le \frac{p\mathfrak{A}(x,u,p)p}{\text{Trace } \mathfrak{A}(x,u,p)} \le \Lambda|p|^m, \tag{3}$$

where m is a real constant, and λ, Λ are positive bounded quantities for x,u in any compact set. From (2) and (3) it is evident that $m \le 2$. Following Bernstein, the difference $g = 2 - m$ is called the *genre* of equation (1).

We shall require of the function $\mathfrak{B}(x,u,p)$ very little in the way of structure, and what we do require is appropriately left to the statement of individual theorems. It is worth remarking, however, that the Dirichlet problem will generally not be well-set when the degree of $\mathfrak{B}/\text{Trace } \mathfrak{A}$ in p is greater than the maximum of the numbers 1 and m, while conversely, if the degree is at most this large, then the problem is well-set, at least in appropriately restricted domains (see Theorems 1 to 5). In section 6 we indicate various results which hold without condition (3).

2. In order to establish the existence of solutions of the Dirichlet problem in a given (bounded) domain Ω, we use the classical technique of *a priori estimation* of the C^1 norm of eventual solutions. The fact that such estimates lead to the existence of solutions was already known to Bernstein in the case when $n = 2$, and has recently been established in higher dimensions through the work of Ladyzhenskaya and Uraltseva.[8, 9] The estimation of the norm itself proceeds in several steps.

(A) The maximum of the solution is estimated in a neighborhood of the boundary of Ω, in terms of the boundary data and the geometry of the domain.

(B) Assuming the preceding result, one determines an estimate for the gradient of the solution at the boundary of Ω.

(C) An estimate is then obtained for the gradient of the solution in the entire domain, depending on the results derived in the previous steps.

Although a uniform bound for the solution may already be known from step (A), such a result follows in any event from (C) and the boundedness of Ω. Hence, steps (A) through (C) in fact supply a program for the estimation of the C^1 norm of a solution. (Actually, one must estimate the C^1 norm of the solutions of a *family* of homotopically related problems, but the program remains the same.)

As we shall see, step (B) occupies a central position in this procedure, in that conditions which are sufficient for its validity are at the same time necessary for the existence of solutions of the Dirichlet problem for arbitrary data. To some extent, the same goes for step (A) when the *genre* is greater than one. The matter is rather different for step (C), and for step (A) when $g \le 1$. For on the one hand (see the remarks below), it appears that necessary and sufficient conditions for the validity of these steps must involve quite delicate considerations, while, on the other hand, there do exist a number of well-known methods for verifying them. For example, if $\mathfrak{B}(x,M,0) \ge 0$ and $\mathfrak{B}(x,-M,0) \le 0$ for some positive constant M, and if either $\partial\mathfrak{B}/\partial u > 0$ or $\partial\mathfrak{B}/\partial u \ge 0$ and $\partial\mathfrak{A}/\partial u = 0$, then ($A$) follows from a simple application of the maximum principle. Similarly, if $\mathfrak{A}$ depends only on p, and $\mathfrak{B}$ depends only on u and p, then (C) can be obtained from the maximum principle and a translation argument; and there are still further techniques available.

To illustrate the difficulty in determining a necessary and sufficient condition for (A), consider the simple equation in two independent variables

$$r + t = (x^2 + y^2 - 2)(p^2 + q^2) + \alpha, \tag{4}$$

where α is a constant. Here steps (*B*) and (*C*) can be verified for the domain $x^2 + y^2 < 1$ and for arbitrary boundary data of class C^2. On the other hand, step (*A*) succeeds if and only if $\alpha > -4$.[10] Since the basic structure of (4) appears to be essentially the same regardless of the value of α, it is clear that an over-all theory for (*A*) will not be an easy matter.

In view of the above remarks, we shall concentrate heavily on step (*B*) in what follows.

3. Consider then a bounded domain Ω in E^n whose boundary Γ is of class C^2. We shall say that the Dirichlet problem is *well-set* in Ω if it is solvable for arbitrarily given C^2 boundary values f on Γ.

THEOREM 1. *Suppose $g \leq 1$ and that steps (A) and (C) above can be verified. Then the Dirichlet problem in Ω is well-set whenever $\mathcal{B}/\mathcal{E} = O(\log|p|)$ as $p \to \infty$, where $\mathcal{E} = p\mathcal{A}(x,u,p)p$.*

(In Theorem 1, as well as later, order terms are assumed to be uniform for x,u in any compact set.)

We have already observed that the Dirichlet problem for $g > 1$ is generally not well-set unless $\mathcal{B}/|p|$ Trace $\mathcal{A}$ is bounded for large p. Thus, when $g > 1$, we may assume without loss of generality that $\mathcal{B}$ has the asymptotic behavior

$$\frac{|\mathcal{B}(x,u,p)|}{|p| \text{ Trace } \mathcal{A}(x,u,p)} \leq B(x,u) + o(1) \tag{5}$$

as $p \to \infty$, where $B(x,u)$ is a continuous function of its arguments. (More specific behavior, analogous to the following relation (6), can also be treated.) Now let $\bar{x}$ be a point on the boundary Γ. We define the value of the function $B(x,u)$ at $\bar{x}$ to be the number $B(\bar{x},f(\bar{x}))$, and we let $\varkappa$ be the minimum normal curvature of Γ at $\bar{x}$. The following result then holds.

THEOREM 2. *Suppose $g > 1$, and assume that step (C) can be verified. Then the Dirichlet problem in Ω is well-set whenever $\varkappa > B$ at each point of the boundary.*

Theorem 2 can be considerably strengthened if we impose some additional structure on the coefficient matrix. In particular, let us suppose that

$$\frac{\mathcal{A}(x,u,p)}{\text{Trace } \mathcal{A}(x,u,p)} = A(x,u,\nu) + o(1), \qquad \nu = p/|p| \tag{6}$$

as $p \to \infty$, where $A(x,u,\nu)$ is a continuous function of its arguments (if $n = 2$ and $g > 0$, then necessarily $A_{ij} = \delta_{ij} - \nu_i\nu_j$). In most situations of importance, including the examples in section 5 as well as the case $n = 2$, the matrix A is centrally symmetric, that is, satisfies the condition $A(x,u,\nu) = A(x,u,-\nu)$. We shall treat only this case in what follows.

The function A can be used to introduce a modified mean curvature of the boundary of Ω. At a point $\bar{x}$ on Γ let $\bar{\nu}$ be the unit inward normal to Γ. Moreover, let κ_i be the principal curvatures of Γ, and let λ_i be unit vectors in the corresponding principal directions, $i = 1,\ldots,n-1$. We introduce $(n-1)$-tuples $\mathbf{a},\boldsymbol{\varkappa}$ according to the relations

$$\mathbf{a} = (\lambda_i \bar{A} \lambda_i), \qquad \boldsymbol{\varkappa} = (\varkappa_i),$$

where $\bar{A} = A(\bar{x},f(\bar{x}),\bar{\nu})$. Then the *mean curvature* $(\mathcal{A},f)$ of Γ at $\bar{x}$ is defined by

$$H = \mathbf{a}\cdot\boldsymbol{\varkappa}. \tag{7}$$

In connection with this definition it is worth noting that the expression $\mathbf{a}\cdot\boldsymbol{\kappa}$ is invariant under orthogonal transformations of E^n. In fact, if we consider A as an auxiliary metric tensor, the expression $\sqrt{\lambda_i A \lambda_i}$ is simply the length of the vector λ_i in this metric. With this interpretation, whenever $\bar{A}$ is proportional on Γ to the underlying Euclidean metric, H becomes the ordinary mean curvature. We observe finally that H agrees with the ordinary curvature of the boundary when $n = 2$ (assuming $g > 0$).

The following results can then be established under conditions (5) and (6).

THEOREM 3. *Suppose $g > 1$ and that step (C) above can be verified. Then the Dirichlet problem is well-set in Ω whenever $H > B$ at each point of Γ.*

THEOREM 4. *Suppose $1 < g \leq 2$ and that steps (A) and (C) above can be verified. Suppose also in conditions (5) and (6) that*

(a) *A is independent of u and B is nondecreasing in u,*
(b) *A and B are Lipschitz continuous,*
(c) *the error terms are $o(|p|^{1-g})$.*[11]

Then the Dirichlet problem is well-set in Ω whenever $H \geq B$ at each point of Γ.

For equations with two independent variables, Bernstein established restricted versions of Theorem 1 and the case $B = 0$ of Theorem 2, and Leray obtained essentially the result of Theorem 4. It should be emphasized, in any case, that Theorems 1–4 do not in themselves prove the existence of solutions of the Dirichlet problem, since the validity of step (C) or of steps (A) and (C) has been taken for granted.

The proof of the theorems involves determination of a family of universal barrier functions in boundary neighborhoods N of Γ. Assume that these neighborhoods have the form $0 \leq d < a$, where d is the normal distance from Γ into Ω. For Theorems 1 and 4 the barrier $v(x)$ must satisfy $\mathfrak{A}(x,v,Dv)D^2v < \mathfrak{B}(x,v,Dv)$ in N and

$$v = f \text{ when } d = 0, \qquad v = M \text{ when } d = a.$$

Given step (A) and the existence of a barrier function for all sufficiently small values of a, one can easily obtain an upper bound for $\partial u/\partial \bar{\nu}$ on Γ. Combining this with a similar lower bound establishes the estimate required for step (B). For Theorems 2 and 3 the barrier function is required to satisfy the slightly different conditions $\mathfrak{A}(x,v,Dv)D^2v < \mathfrak{B}(x,v,Dv)$ in N and

$$v = f \text{ when } d = 0, \qquad \partial v/\partial \bar{\nu} = \infty \text{ when } d = a.$$

In this case the gradient bound of step (B) is obtained directly without the intervention of step (A).

4. The conditions on $\mathfrak{B}$ and H given in the preceding theorems are in a sense the best possible, as the following result shows.

THEOREM 5. *Consider the Dirichlet problem in Ω for equation (1), where $\mathfrak{A}$ and $\mathfrak{B}$ are independent of u. Let $\gamma \geqslant 0$, $\delta > 0$ be arbitrary constants, and set $\mathfrak{C} = \mathfrak{B}/|p|$ Trace $\mathfrak{A}$. Then there exists arbitrarily smooth boundary data for which no solution is possible whenever*

I. *$g \leq 1$ and $\mathfrak{B}/\mathcal{E} \geq (\log|p|)^{1+\delta}$ for large p, or*
II. *$g > 1$ and one of the following holds:*
$\mathfrak{C} \geq \gamma - o(1)$ for large p, and $H < \gamma$ at some point on Γ,

$\mathcal{C} \geq \gamma + |p|^{1-g}(\log|p|)^{1+\delta}$ *for large* p, *and* $H \leq \gamma$ *on some open set on* Γ,

$\mathcal{C}$ *tends to infinity as* p *tends to infinity, uniformly for* x *in* Ω.

(In the second to last case, we require the error term in (6) to be $O(|p|^{1-g})$ and the function $\mathcal{A}$ to be Lipschitz continuous; also in this case, if $g > 2$, the expression $|p|^{1-g}$ should be replaced by $|p|^{-1}$.)

The proof of Theorem 5 is based on a form of the maximum principle due to Bernstein, Leray, and Finn, and again involves the construction of a suitable comparison function. It is possible to state more delicate nonexistence theorems than Theorem 5, in which $\mathcal{A}$ and $\mathcal{B}$ depend also on the variable u. Although such results are worthwhile in themselves and useful in the discussion of examples, they are omitted here for simplicity.

5. In the papers noted earlier, Bernstein presented a great many examples of equations which partially or completely fall under his theory. It is of interest to reconsider these in the light of the present results, and to state multidimensional generalizations where possible. In so doing, we remark that the verification of steps (*A*) and (*C*) occasionally requires methods which have not been indicated here.

It is appropriate to begin with the classical equation for surfaces of constant mean curvature $\mathcal{H}$. Here

$$\mathcal{A}_{ij} = (1 + |p|^2)\delta_{ij} - p_i p_j, \qquad \mathcal{B} = n\mathcal{H}(1 + |p|^2)^{3/2}; \tag{8}$$

the full equation has the following form when written out for two independent variables,

$$(1 + q^2)t - 2pqs + (1 + p^2)r = 2\mathcal{H}(1 + p^2 + q^2)^{3/2}.$$

Evidently, $p\mathcal{A}p = |p|^2$, Trace $\mathcal{A} = n + (n-1)|p|^2$, $m = 0$, $g = 2$, and deg $\mathcal{C} = 0$, where $\mathcal{C} = \mathcal{B}/|p|\text{Trace }\mathcal{A}$. Thus, in (5), (6), and (7)

$$A_{ij} = \frac{1}{n-1}(\delta_{ij} - \nu_i\nu_j), \qquad B = \frac{n}{n-1}|\mathcal{H}|, \qquad H = \frac{1}{n-1}\Sigma\kappa_i,$$

where the error terms in (5) and (6) are of degree -2. Using Theorems 4 and 5 we obtain the following conclusion.[12]

Let Ω *be a bounded domain in* E^n, *whose boundary* Γ *is of class* C^2. *Then the Dirichlet problem for surfaces of constant mean curvature is well-set for arbitrary* C^2 *boundary data if and only if the mean curvature* H *of* Γ *is* $\geq \frac{n}{n-1}|\mathcal{H}|$ *at each point of* Γ.

The preceding result can be generalized to variable mean curvatures $\mathcal{H} = \mathcal{H}(x)$ provided we require in addition that $|\text{grad }\mathcal{H}| < \frac{n}{n-1}\mathcal{H}^2$ in Ω.

If $\mathcal{H} = cu$ in (8), where c is a positive constant, the resulting equation is that satisfied by the free surface of a stationary fluid subject to gravity and surface tension. *The corresponding Dirichlet problem is well-set provided that* $H \geq c|f(x)|/(n-1)$ *at each point of the boundary of* Γ. *Moreover, the factor* $c/(n-1)$ *is best possible.*

A second example, due to Bernstein, arises from the variational problem $\delta \int \sqrt{1 + u^2 + |p|^2}dx = 0$, namely

$$\mathcal{A}_{ij} = (1 + u^2 + |p|^2)\delta_{ij} - p_i p_j, \qquad \mathcal{B} = u(1 + u^2 + 2|p|^2).$$

Here $m = 0$, $g = 2$, and deg $\mathcal{C} = -1$. Using Theorem 4 and a refinement of Theorem 5 we find that *the Dirichlet problem is well-set if and only if* $H \geq 0$ *at each point of* Γ. We note that without the term u^2 in the integrand this becomes the standard minimal surface equation, which is also solvable for arbitrary boundary data when $H \geq 0$ at each point of the boundary.

Related to the preceding equations is the example

$$\mathcal{A}_{ij} = (1 + |p|^2)\delta_{ij} - p_i p_j, \qquad \mathcal{B} = cu(1 + |p|^2)^{\theta/2} \tag{9}$$

where θ *is constant. When* $\theta \leq 2$, *the Dirichlet problem is well-set if and only if* $H \geq 0$ *on* Γ; *and when* $2 < \theta < 3$, *it is well-set whenever* $H > 0$ *on* Γ. *The same result can be demonstrated when the factor cu is omitted.* (The case $n = 2$ is discussed by Bernstein.[1])

Although the formulation (3) makes it quite apparent that the individual eigenvalues of the matrix $\mathcal{A}$ are not in themselves relevant to the solution of Dirichlet's problem, this is even more strongly emphasized by the example

$$\mathcal{A}_{ij} = \delta_{ij} + p_i p_j, \qquad \mathcal{B} = cu(1 + |p|^2)^{\theta/2}. \tag{10}$$

Here the eigenvalues (disregarding multiplicities) are exactly the same as in (9), but $m = 2$ and $g = 0$. *Thus the Dirichlet problem is well-set for arbitrary domains of class* C^2 *if and only if* $\theta \leq 4$. If the factor cu is omitted from $\mathcal{B}$ in (10), the results are somewhat different. *In this case, the Dirichlet problem is well-set for arbitrary domains of class* C^2 *if and only if* $\theta \leq 3$. *If* $3 < \theta \leq 4$, *there exists a function* $R(\theta,n)$ *such that the Dirichlet problem is well-set in domains which are contained in a closed ball of radius* $R(\theta,n)$; *on the other hand, the equation allows no solutions whatsoever in any domain containing such a closed ball.* As θ increases from 3 to 4, R decreases from infinity to a limit value $> \pi/2$.

A final example, in which the *genre* is different from 2, arises from the variational problem $\delta \int (1 + |p|^g)^{1/g} dx = 0$ where $g > 1$. The Euler-Lagrange equation has coefficients

$$\mathcal{A}_{ij} = (1 + |p|^g)|p|^2\delta_{ij} + (g - 2 - |p|^g)p_i p_j.$$

It is easily checked that $m = 2 - g$ and that g is the *genre*. *When* $1 < g \leq 2$, *a necessary and sufficient condition for the Dirichlet problem to be well-set is that the mean curvature of the boundary be everywhere nonnegative. Moreover, when* $g > 2$ *this condition remains necessary, while a sufficient condition is that the mean curvature be everywhere positive.*

6. To a surprising degree, the preceding theorems can be obtained under individually weaker hypotheses on the coefficient matrix $\mathcal{A}$. To begin with, Theorems 1, 4, and the first part of Theorem 5 continue to hold if, in place of (3), we suppose merely that

$$\frac{(p - p_0)\mathcal{A}(x,u,p)(p - p_0)}{\text{Trace } \mathcal{A}(x,u,p)} \geq \lambda|p|^m, \tag{3'}$$

where λ is positive for x,u,p_0 in any bounded set. Similarly, Theorems 2, 3, and the second part of Theorem 5 remain valid if only the right-hand inequality of (3) is assumed. (One could also generalize the concept of *genre* by recasting its definition in terms of the convergence or divergence of certain integrals, as suggested by Nitsche.[13])

Even when all conditions of *genre* are dropped, the following result holds.

THEOREM 6. *Suppose that* $\mathcal{B}$ *satisfies condition* (5) *with* B *nondecreasing in* u, *and assume that steps* (A) *and* (C) *can be verified. Then the Dirichlet problem for* (1) *is well-set in smoothly bounded domains* Ω *whenever* $\varkappa > B$ *at each point of the boundary. If, in addition,* $\mathcal{A}$ *obeys the asymptotic estimate* (6) *with* A *independent of* u, *then the Dirichlet problem is well-set whenever* $H > B$ *at each point of the boundary.*

As an example, consider the "false" minimal surface equation

$$(1 + q^2)r + 2pqs + (1 + p^2)t = 0. \tag{11}$$

Here the *genre* depends on the direction (p,q), so that (3) does not hold. However, on the basis of the results above we can establish that the Dirichlet problem for (11) is well-set in a domain Ω whenever $\varkappa \geq 0$ in the neighborhood of boundary points whose normal vector is directed along a coordinate axis. It is interesting to contrast this result with the corresponding one for the minimal surface equation.

[1] Bernstein, S., "Conditions necessaires et suffisantes pour la possibilité du problème de Dirichlet," *Compt. Rend.*, **150**, 514–515 (1910).

[2] Bernstein, S., "Sur la géneralisation du problème de Dirichlet. II," *Math. Ann.*, **69**, 82–136 (1910).

[3] Bernstein, S., "Sur les surfaces définies au moyen de leur courbure moyenne ou totale," *Ann. Sci. École Norm. Sup.*, **27**, 233–256 (1910).

[4] Bernstein, S., "Sur les équations du calcul des variations," *Ann. Sci. École Norm. Sup.*, **29**, 431–485 (1912).

[5] Leray, J., "Discussion d'un problème de Dirichlet," *J. Math. Pures Appl.*, **18**, 249–284 (1939).

[6] Finn, R., "Remarks relevant to minimal surfaces and to surfaces of constant mean curvature," *J. Analyse Math.*, **14**, 139–160 (1965).

[7] Jenkins, H., and J. Serrin, "The Dirichlet problem for the minimal surface equation in higher dimensions," to appear in *J. Reine Angew. Math.*

[8] Ladyzhenskaya, O. A., and N. N. Uraltseva, "On the Hölder continuity of the solutions and the derivatives of linear and quasi-linear equations of elliptic and parabolic types," *Trudy Mat. Inst. Steklov.*, **73**, 172–220 (1964).

[9] Trudinger, N. S., thesis: "Quasilinear elliptic partial differential equations in n variables," Stanford University (1966); cf. also, Trudinger, N. S., *Bull. Am. Math. Soc.*, **73**, 410–413 (1967).

[10] Indeed, if $\alpha > -4$, it is possible to construct a superfunction in $x^2 + y^2 \leq 1$, while if $\alpha \leq -4$, it can be shown that (4) has no bounded solutions whatsoever in this region. *Thus the Dirichlet problem for* (4) *can be solved in* $x^2 + y^2 < 1$ *if and only if* $\alpha \geq -4$.

[11] If $g = 2$, the error term in (6) should be $o(|p|^{-1})$, or, more generally, of the form $|p|^{-1}\{A'(x, u, \nu) + o(1)\}$, where $\xi A'\xi \leq 0$ if $n > 2$ or $\lambda + \det A' > 0$ if $n = 2$.

[12] Serrin, J., "The Dirichlet problem for surfaces having prescribed mean curvature," *Notices Am. Math. Soc.*, **14**, 135 (1967).

[13] Nitsche, J. C. C., "On differential equations of mode 2," *Proc. Am. Math. Soc.*, **16**, 902–908 (1965), especially p. 907.

Philos. Trans. R. Soc. Lond., Ser. A, 264 (1969), 413–496. **Reprint I.2.[ii]**

Pages 447–475 have been removed for the sake of brevity.

THE PROBLEM OF DIRICHLET FOR QUASILINEAR ELLIPTIC DIFFERENTIAL EQUATIONS WITH MANY INDEPENDENT VARIABLES

By J. SERRIN
University of Minnesota

(*Communicated by W. K. Hayman, F.R.S.—Received* 14 *May* 1968)

CONTENTS

This paper is concerned with the existence of solutions of the Dirichlet problem for quasilinear elliptic partial differential equations of second order, the conclusions being in the form of necessary conditions and sufficient conditions for this problem to be solvable in a given domain with arbitrarily assigned smooth boundary data. A central position in the discussion is played by the concept of global barrier functions and by certain fundamental invariants of the equation. With the help of these invariants we are able to distinguish an important class of 'regularly elliptic' equations which, as far as the Dirichlet problem is concerned, behave comparably to uniformly elliptic equations. For equations which are not regularly elliptic it is necessary to impose significant restrictions on the curvatures of the boundaries of the underlying domains in order for the Dirichlet problem to be generally solvable; the determination of the precise form of these restrictions constitutes a second primary aim of the paper. By maintaining a high level of generality throughout, we are able to treat as special examples the minimal surface equation, the equation for surfaces having prescribed mean curvature, and a number of other non-uniformly elliptic equations of classical interest.

Vol. 264. A. 1153. (Price £2 4s.; U.S. $5.70) 54 [Published 8 May 1969

Introduction

Within recent years the study of the Dirichlet problem for second-order quasilinear elliptic equations has reached a certain degree of completeness, and, in particular, methods have been developed which reduce the existence question to one of obtaining *a priori* estimates for the solution and its gradient. I propose here to treat this latter problem, presenting sufficient detail to show the pre-eminent role which certain fundamental invariants of the equation play in the decision as to whether the Dirichlet problem is or is not generally solvable for a given equation and a given domain.

We consider, specifically, quasilinear elliptic partial differential equations of the form

$$\mathscr{A}(x, u, Du)\, D^2u = \mathscr{B}(x, u, Du). \tag{1}$$

Here $x = (x_1, \ldots, x_n)$ denotes points in the underlying n-dimensional Euclidean space, Du and D^2u denote respectively the gradient vector and Hessian matrix of the dependent variable $u = u(x)$, and $\mathscr{A}$ and $\mathscr{B}$ are respectively a given matrix and a given scalar function of the variables indicated. The multiplication convention implicit in (1) is the natural one which makes a scalar of the left-hand side; that is, using component notation,

$$\mathscr{A}\, D^2u = \mathscr{A}_{ij}\, \partial^2 u / \partial x_i\, \partial x_j$$

where summation from 1 to n over repeated indices is understood.

The structure of (1) is determined by the functions $\mathscr{A}(x, u, p)$ and $\mathscr{B}(x, u, p)$. We suppose throughout that they are defined and continuously differentiable for all real values of u and p, and for all points x in the closure of any particular domain under consideration. Ellipticity of (1) requires that the coefficient matrix satisfy the condition

$$\xi \mathscr{A} \xi > 0 \tag{2}$$

for arbitrary non-vanishing real vectors $\xi = (\xi_1, \ldots, \xi_n)$. Moreover, we can obviously assume without loss of generality that $\mathscr{A}$ is symmetric. Beyond this, any additional structure which will be required of the coefficient matrix and inhomogeneous term is best left to the statement of individual theorems. In any event, we shall maintain a fairly high level of generality throughout the paper, and accordingly will be able to treat as special examples the minimal surface equation, the equation for surfaces of prescribed mean curvature, and a number of other uniformly and non-uniformly elliptic equations of classical interest.

Roughly speaking, the conclusions are in the form of necessary conditions and sufficient conditions for the problem of Dirichlet in a given domain to be solvable for arbitrarily assigned smooth boundary data. The work thus constitutes a development and extension of the theory initiated by Bernstein in a series of papers from 1907 to 1912 and applied by him to a number of cases of great generality. Bernstein's work was subsequently sharpened and enlarged in an important and fundamental paper of Leray which appeared in 1939. In comparing this work with ours, we note first of all that their considerations were restricted to the case of two independent variables and involve various regularity assumptions on the behaviour of $\mathscr{A}$ and $\mathscr{B}$ for large values of p which we avoid here. Bernstein had also concluded for certain equations (including that governing surfaces of constant mean curvature) that the Dirichlet problem was not generally solvable for arbitrary smooth data, even when the domain in question is convex and arbitrarily small. While strictly true, this statement nevertheless sidesteps the problem of determining precisely those domain for which the Dirichlet problem is in fact generally solvable. It is one of the goals of this paper (as it was of Leray's) to consider questions of just this sort.

In higher dimensions, the Dirichlet problem for non-uniformly elliptic equations has been comparatively less studied. We mention here specifically the work of Gilbarg and of Stampacchia, and recent papers of Trudinger and Oskolkov, and also a treatment by Jenkins & Serrin on the minimal surface equation. Some of the relations between this work and ours will be discussed below (see in particular §§ 9, 10 and 13).

The formal work of the paper is divided into four chapters. In the first one we give a precise statement of the problem under consideration and outline the general procedures to be followed in its study. The fundamental existence theorem upon which the further results of the paper are based is essentially due to Ladyzhenskaya & Uraltseva, though the form in which it has been stated does not seem to have appeared previously in the literature.

The second chapter is in many ways the central one of the paper; in it we treat in detail the various *a priori* estimates required by the existence theory. The results themselves are strongly influenced by the structure of the equation, and in particular by the behaviour of certain fundamental invariants. Following a discussion of maximum principles and global barrier functions, we introduce the concept of regular ellipticity. As far as the Dirichlet problem is concerned, the class of regularity elliptic equations behaves comparably to the class of uniformly elliptic equations. On the other hand, for equations which are not regularly elliptic it is frequently necessary to impose significant and important restrictions on the curvatures of the boundaries of the underlying domains in order to be able to solve the Dirichlet problem for arbitrarily given smooth data. An intimation of this state of affairs occurs already in the work of Bernstein and Leray, and in the paper by Jenkins & Serrin noted above. Bernstein had in fact seen the importance of convexity in the solution of Dirichlet's problem in two variables, while Leray had found a further relation between the second derivative of the boundary curve and the inhomogeneous term $\mathscr{B}$. For the minimal surface equation in higher dimensions, on the other hand, the mean curvature of the boundary of the underlying domain is the significant invariant. These various conditions are here seen as special cases of the criteria developed in §§ 9 and 10.

In §§ 12 and 13 we consider a number of interior estimates for the gradient vector, assuming that a boundary estimate is already available. This work is a clarification and elaboration of fundamental ideas due to Bernstein; we are in addition able to extend the results to a large class of equations in more than two variables, an extension which had not originally been thought possible in view of the inadequate number of critical point relations available when $n > 2$. Chapter II closes with a summary of various existence theorems which are direct consequences of the earlier *a priori* estimates.

Chapter III deals with certain cases of the non-solvability of Dirichlet's problem for arbitrary smooth data. It is shown that the results of chapter II are in many ways best possible, in that when their hypotheses fail to hold one may construct boundary data for which the Dirichlet problem cannot be solved. Similar results for two independent variables have been given by Bernstein, Leray, and Finn, and indeed our proof method is based on an interesting form of the maximum principle used by these authors.

The paper is concluded with an extended discussion of various examples and special cases. These examples possess a considerable amount of independent interest, besides serving as concrete illustrations of the general theory. It should also be pointed out that, more often than not, the results of this chapter go beyond mere specialization of some earlier theorem of the paper. The following proposition may be singled out as indicative of the conclusions obtained (see also Serrin 1967):

Let Ω be a bounded domain in n-dimensional Euclidean space, whose boundary is of class C^2. Then the Dirichlet problem in Ω for the equation of constant mean curvature

$$[(1+|Du|^2)I-DuDu]D^2u = n\Lambda(1+|Du|^2)^{\frac{3}{2}}$$

has a unique solution for arbitrary smooth boundary data, if and only if the mean curvature of the boundary surface is everywhere greater than or equal to $[n/(n-1)]\,|\Lambda|$.

In the above example the notation *DuDu* stands for the dyadic matrix generated by the gradient vector, while I as usual denotes the identity matrix.

Particular importance attaches to this proposition in view of the geometric interest of the equation, the paucity of known exact solutions, and the number of conjectures which have been made concerning the Dirichlet problem in this case. Because of these circumstances I have included in the final part of the paper a fairly complete discussion of the background of this theorem, and have also stated generalizations for the case of prescribed (non-constant) mean curvatures and for surfaces with a single valued spherical projection. Similar theorems are also given for the closely related differential equation which describes the free surface of a stationary fluid under the combined action of gravity and surface tension.

Although the paper deals with both uniformly and non-uniformly elliptic equation on an equal footing, the former case has in fact received such careful and far-reaching study at the hands of other writers that our conclusions must be regarded for the most part as a contribution to the theory of non-uniformly elliptic equations. In this regard, it may be worthwhile to point out specifically the main theorems of the paper, as opposed to the various methods and techniques of *a priori* estimation used to attain them. In particular, we mention the five existence theorems of § 14, the non-existence theorems of §§ 16 and 18, and the four theorems of § 19 concerning the equation

$$\mathscr{A}(x,u,Du)\,D^2u = 0. \tag{1'}$$

As already mentioned, the examples discussed in chapter IV all have some special interest, either for their geometric significance or because they illustrate some particular point of the theory.

Theorems are numbered consecutively in each separate section. To avoid ambiguity in later references to a theorem, the section number is added as a distinguishing mark. Thus theorem 1 of § 8 is referred to in later sections as theorem 8·1.

Chapter I

In this chapter is given a precise statement of the Dirichlet problem and the general procedures to be followed in its study are outlined. The last section contains an important identity which is basic to a number of constructions later on in the paper.

1. The fundamental existence theorem

Let Ω be a bounded domain (connected open set) in n-dimensional Euclidean space. We shall use the notation $\partial\Omega$ to denote the boundary of Ω and $\overline{\Omega}$ to denote the closure of Ω. The Dirichlet problem for equation (1) then consists in determining a function $u = u(x)$ such that:

(i) u is twice continuously differentiable in Ω and continuous in $\overline{\Omega}$,

(ii) u takes on given continuous boundary values f on $\partial\Omega$, and

(iii) $\mathscr{L}u \equiv 0$ in Ω, where the operator $\mathscr{L}$ is defined by

$$\mathscr{L}u \equiv \mathscr{A}(x, u, Du)\, D^2u - \mathscr{B}(x, u, Du).$$

In order to establish the existence of solutions of the Dirichlet problem one has available the classical technique of *a priori* estimation of the norms of eventual solutions. In particular, the following important proposition holds.

Let the boundary of Ω and the assigned boundary values f be of class C^3. Also let τ be an arbitrary real number in the closed interval $[0, 1]$. Suppose there exists a constant M, independent of τ, such that the conditions

(*i*) $v \in C^2(\overline{\Omega})$,

(*ii*) $v = \tau f$ *on* $\partial\Omega$,

(*iii*) $\mathscr{A}(x, v, Dv)\, D^2v = \tau\mathscr{B}(x, v, Dv)$ *in* Ω

imply $\sup_\Omega (|v| + |Dv|) \leqslant M$. *Then the Dirichlet problem for equation* (1) *in* Ω *has at least one solution for the given boundary values f. Moreover, this solution is twice continuously differentiable in the closure of* Ω.

The proof of this theorem involves two different sets of ideas—first, an *a priori* estimate concerning the equicontinuity of the gradient vector Dv, and second, a topological argument showing that solutions of (*i*), (*ii*), (*iii*) exist for each value of τ, $0 \leqslant \tau \leqslant 1$. For two dimensions, such a programme was initiated and carried through by Bernstein under the assumption that the solutions v are unique. Leray & Schauder (1934) then provided a general topological framework for the proof, avoiding Bernstein's uniqueness condition and at the same time giving a wide scope to the allowable homotopy. Moreover, the derivation of the requisite *a priori* estimates was greatly simplified by the work of Morrey and Nirenberg; cf. Nirenberg (1952). Thus the theorem for the case of two dimensions may be considered classical.

The topological apparature of Leray & Schauder applies unchanged in higher dimensions, but the *a priori* estimates concerning the equicontinuity of Dv are now a much deeper matter. For the case of divergence structure equations these estimates were found by a number of authors on the basis of work of De Giorgi, Nash, and Moser, and corresponding existence theorems were stated by Ladyzhenskaya & Uraltseva (1961), Gilbarg (1963), and others. The first derivation of the requisite *a priori* estimates for the general case was given by Ladyzhenskaya & Uraltseva (1964). The above theorem follows as a consequence of their work and the Leray–Schauder theorem; for completeness we shall include the proof, using a method due to Gilbarg.

Let w be an element of the Banach space $C^{1+\gamma}(\overline{\Omega})$, where γ is a real number in the open interval $(0, 1)$ which will be fixed later. Consider the linear elliptic equation

$$a(x)\, D^2 W = b(x) \tag{3}$$

where $$a(x) = \mathscr{A}(x, w(x), Dw(x)), \quad b(x) = \mathscr{B}(x, w(x), Dw(x)).$$

Since the coefficients $a(x)$ and $b(x)$ are Hölder continuous in the closure of Ω, Schauder's theorem implies the existence of a unique solution $W \in C^{2+\gamma}(\overline{\Omega})$ which takes on the given boundary values f. This process clearly defines a mapping T of $C^{1+\gamma}(\overline{\Omega})$ into itself, namely

$$w \to W = Tw.$$

Obviously T maps bounded sets in $C^{1+\gamma}(\overline{\Omega})$ into bounded sets in $C^{2+\gamma}(\overline{\Omega})$, whence by Arzela's theorem and an easy argument (see the appendix at the end of this section) it follows that T is completely continuous.

The proposition will now be proved if we can show that T has a fixed point. To this end we apply the Leray–Schauder fixed point theorem in a simplified version due to Schaeffer. According to that result, T will have a fixed point if there exists a constant M', independent of τ, such that

$$v = \tau T v \tag{4}$$

implies $\|v\| \leqslant M'$. Here τ denotes an arbitrary real number in $[0, 1]$, while the norm of course is that in the Banach space $C^{1+\gamma}(\overline{\Omega})$.

Let v then be a solution of (4). By the definition of the mapping T it is evident that $v = \tau f$ on $\partial\Omega$ and

$$\mathscr{A}(x, v, Dv)\, D^2 v = \tau \mathscr{B}(x, v, Dv) \quad \text{in } \Omega. \tag{5}$$

Consequently, by the principal hypothesis of the proposition, we have

$$\sup_\Omega (|v| + |Dv|) \leqslant M. \tag{6}$$

To complete the proof it must further be shown that the quantity

$$\frac{|Dv(x) - Dv(y)|}{|x-y|^\gamma} \tag{7}$$

is uniformly bounded for all x, y in Ω and all τ in $[0, 1]$.

In view of (6), the arguments x, v, Dv in the coefficients of (5) are uniformly bounded. Consequently, as far as the solution v is concerned, (5) may be considered uniformly elliptic and the functions $\mathscr{A}$ and $\tau\mathscr{B}$ uniformly of class C^1 in their arguments. Thus by an important theorem of Ladyzhenskaya & Uraltseva (1964), there exists a constant γ (depending on M, on the structure of (1), an on norms for the boundary and the boundary data, but independent of τ) such that the quantity (7) is uniformly bounded. Making this choice of γ from the very beginning thereby completes the proof.

It is of interest to examine how the smoothness of the boundary and of the boundary data is used in the proof. To begin with, it is required in order to apply Schauder's theorem and thus to determine the function W; and, secondly, it is needed for certain boundary estimates in the theorem of Ladyzhenskaya & Uraltseva. Actually, Hölder continuity of the second derivatives of the boundary representations and the boundary data would suffice equally here, but this generalization would be pedantic. In fact, Trudinger (1967) has shown that a slightly modified version of the main proposition continues to hold without any conditions on the boundary data beyond continuity. Since Trudinger's modification requires further topological apparatus, and since the result as stated is both simple and adequate to our purposes, it will be unnecessary to consider these generalizations here.

Remark. The above result remains true if we replace conditions (*ii*) and (*iii*) by the more general conditions

(*ii'*) $v = k(\tau) f$ *on* $\partial\Omega$,

(*iii'*) $\mathscr{A}(x, v, Dv; \tau)\, D^2 v = \mathscr{B}(x, v, Dv; \tau)$ *in* Ω,

where the functions $\mathscr{A}(x, u, p; \tau)$, $\mathscr{B}(x, u, p; \tau)$, and $k(\tau)$ are continuously differentiable in their arguments, satisfy $\xi\mathscr{A}\xi > 0$ for $\xi \neq 0$, and finally are such that

$$\mathscr{A}(x, u, p; 1) = \mathscr{A}(x, u, p), \quad \mathscr{B}(x, u, p; 1) = \mathscr{B}(x, u, p), \quad k(1) = 1$$

and

$$\mathscr{B}(x, u, p; 0) = 0 \quad \text{for} \quad |p| \leqslant 1, \qquad k(0) = 0.$$

In order to prove this version of the result we consider the Dirichlet problem

$$\mathscr{A}(x, w, Dw; \tau)\, D^2 W = \mathscr{B}(x, w, Dw; \tau) + \mathscr{B}(x, w, Dw; 0)\left(\frac{Dw \cdot DW}{|Dw|^2} - 1\right)$$

with $W = k(\tau) f$ on the boundary. This defines a mapping T from $C^{1+\gamma}(\overline{\Omega}) \times [0, 1]$ into $C^{1+\gamma}(\overline{\Omega})$, namely

$$(w, \tau) \to W = T(w, \tau),$$

and it is required to show that $T(\cdot, 1)$ has a fixed point. To this end we observe that T is continuous and maps bounded sets into sequentially compact sets. Moreover, if $v = T(v, \tau)$ then v satisfies (*i*), (*ii'*), and (*iii'*). Hence as in the proof already given, there exists a constant M' independent of τ such that $\|v\| \leqslant M'$. Finally, $T = 0$ when $\tau = 0$. It follows therefore from the Leray–Schauder theorem (see Browder 1966) that T has a fixed point when $\tau = 1$, completing the proof.

Appendix

We shall show here that the mapping Tw defined above is completely continuous. To begin with, we recall that T maps bounded sets in $C^{1+\gamma}(\overline{\Omega})$ into bounded sets in $C^{2+\gamma}(\overline{\Omega})$. Hence by Arzela's theorem it is evident that T maps bounded sets in $C^{1+\gamma}(\overline{\Omega})$ into sequentially compact sets in $C^{1+\gamma}(\overline{\Omega})$. It thus remains only to prove that T is continuous.

Let $w \to w_0$ in $C^{1+\gamma}(\overline{\Omega})$. We must then show that $Tw \to Tw_0$ in the same space. Suppose for contradiction that this were not true. Since the elements Tw are sequentially compact in $C^2(\overline{\Omega})$ as well as in $C^{1+\gamma}(\overline{\Omega})$, it follows that there is a sequence $\{w_m\}$ such that

$$w_m \to w_0 \quad \text{in} \quad C^{1+\gamma}(\overline{\Omega}), \quad W_m = Tw_m \to \hat{W} \quad \text{in} \quad C^2(\overline{\Omega})$$

where $\hat{W} \neq Tw_0$. On the other hand, by (3) we have

$$\mathscr{A}(x, w_m, Dw_m)\, D^2 W_m = \mathscr{B}(x, w_m, Dw_m)$$

whence letting m tend to infinity there results

$$\mathscr{A}(x, w_0, Dw_0)\, D^2 \hat{W} = \mathscr{B}(x, w_0, Dw_0).$$

Consequently $\hat{W} = Tw_0$, and this contradiction completes the proof.

2. The existence programme

The proposition which was established in the previous section requires, for its application, the estimation of the C^1 norms of solutions of a family of Dirichlet problems. We propose to carry out this estimation in a series of steps which are described below and which are more or less unrelated to each other. If a particular equation possesses an appropriate structure so that all the various steps in this estimation process can be carried out, then the corresponding Dirichlet problem will have at least one solution.

The existence programme, then, will be based on the following succession of steps:

(A) The maximum absolute value of the solution is estimated in Ω.

(B) Assuming the preceding step, one determines an estimate for the gradient of the solution at the boundary of Ω.

(C) An estimate is then obtained for the gradient of the solution in the entire domain, depending on the results derived in the preceding steps.

Of course, it almost goes without saying that in this programme we are obliged to consider a *family* of Dirichlet problems and that the corresponding estimates must be independent of the homotopy parameter τ. On the other hand, in presenting *a priori* estimates for the various steps of the programme it will not be necessary to retain the parameter τ in evidence, so long as the dependence of the estimates on the structure of the equation and on the given domain and data is set out in each case.

While there is some tendency for the existence programme as described to appear fragmentary, I think this need not be taken seriously: the subject itself is broad and simply does not in the nature of things possess a high degree of unity. In fact, looked at from another point of view, the requisite classification of non-linear structures in order to obtain meaningful results has its own particular interest.

The following chapter treats in detail the *a priori* estimates required by steps (A), (B), and (C) of the existence programme. Before turning to this, however, we shall derive an important identity which will be basic to a number of later constructions. This identity is contained in § 4; the next section contains certain preliminary material for the derivation.

3. The distance function

Consider an $(n-1)$-dimensional surface S of class C^3 in n-dimensional Euclidean space, having no self-intersections and no boundary. We shall be interested in the function $d = d(x)$, defined as the distance from x to the surface S. The following lemma holds.

Lemma 1. *Let the normal curvatures of S be bounded in absolute value by K. Then d is of class C^2 at all points whose distance from S is less than $1/K$.*

Proof. For points x whose distance from S is less than $1/K$, we define $y = y(x)$ to be the (unique) point on S nearest to x.

Now let $\bar{x}$ be a fixed point whose distance from S is less than $1/K$. We consider a specific coordinate frame in which the x_n axis is oriented along the normal to S at $\bar{y} = y(\bar{x})$. In the neighbourhood of the point $\bar{y}$ the surface S can then be represented in the form

$$x_n = \phi(x_1, \ldots, x_{n-1})$$

where ϕ is three times continuously differentiable. Moreover, for all points x near $\bar{x}$ the following relations hold,

$$x_i = y_i - \frac{\phi_i(y)}{\sqrt{\{1+|D\phi(y)|^2\}}} d, \quad i = 1, \ldots, n-1; \quad x_n = \phi(y) + \frac{d}{\sqrt{\{1+|D\phi(y)|^2\}}}, \tag{8}$$

where $\phi_i = \partial\phi/\partial x_i$, $D\phi$ is the gradient vector of ϕ, and $y = y(x)$. (There is a small point of notation here—the arguments of ϕ, ϕ_i, and $D\phi$ are, to be absolutely precise, $y_1, \ldots, y_{n-1}$.)

Relations (8) have the general form

$$x = f(y_1, \ldots, y_{n-1}, d) \tag{9}$$

where f is a function of class C^2 in its arguments. The function $d(x)$, which can be considered as arising by inversion of (9), will then be of class C^2 at $\bar{x}$ provided the Jacobian of the transformation is non-zero at this point.

In order to compute this Jacobian it is clearly allowable to suppose that the coordinates $x_1, \ldots, x_{n-1}$ lie along the principal directions of S at the point $\bar{y}$. In these coordinates the gradient $D\phi$ vanishes at $\bar{y}$ and the Hessian matrix $D^2\phi$ is diagonal there; consequently, since the diagonal elements are precisely the principal curvatures of S at $\bar{y}$, an easy calculation yields the expression

$$\prod_1^{n-1} (1-k_i d)$$

for the Jacobian at $\bar{x}$, where the numbers k_i are the principal curvatures in question. This completes the proof of the lemma.

In what follows we shall require formulae for the first and second derivatives of d at the point x. It will be sufficient to evaluate these in the special coordinate system used in the proof of lemma 1—that is, with the x_n axis taken along the normal to S at $\bar{y}$ and directed towards $\bar{x}$, and the remaining axes lying along the principal directions of S at $\bar{y}$.

LEMMA 2. *In the special coordinate system introduced above, we have at* $\bar{x}$

$$Dd = (0, \ldots, 0, 1)$$

and

$$D^2 d = -\left(\frac{k_1}{1-k_1 d}, \ldots, \frac{k_{n-1}}{1-k_{n-1} d}, 0\right)_{\text{diagonal}}$$

where $k_1, \ldots, k_{n-1}$ are the principal curvatures of S at $\bar{y}$.

Proof. Put $\xi = (y, \ldots, y_{n-1}, d)$ and write (8) in component form

$$x_i = f_i(\xi). \tag{10}$$

Differentiating (10) with respect to x_j yields

$$\delta_{ij} = f_{i,l}\xi_{l,j} \tag{11}$$

where the Jacobian matrix $(f_{i,l})$ is diagonal at $\bar{x}$ with elements

$$(1-k_1 d, \ldots, 1-k_{n-1} d, 1).$$

Evaluating (11) at $\bar{x}$, it follows that $(\xi_{l,j})$ is also diagonal there, with elements

$$\left(\frac{1}{1-k_1 d}, \ldots, \frac{1}{1-k_{n-1} d}, 1\right).$$

The final row of this matrix is just Dd, proving the first part of the lemma.

If we now differentiate (11) with respect to x_m there results

$$0 = f_{i,l}\,\xi_{l,jm} + f_{i,lk}\,\xi_{l,j}\,\xi_{k,m}$$

where $i, j, m = 1, \ldots, n$. Putting $i = n$ and evaluating this identity at $\bar{x}$ yields

$$0 = d_{,jm} + f_{n,jm}(1-k_j d)^{-1}(1-k_m d)^{-1} \tag{12}$$

(if either j or m equals n, the corresponding factor $1-k_j d$ or $1-k_m d$ should be suppressed). Now the matrix $(f_{n,jm})$ is easily found by direct calculation to be diagonal at $\bar{x}$ with elements

$$(k_1(1-k_1 d), \ldots, k_{n-1}(1-k_{n-1} d), 0).$$

Inserting this into (12), the lemma follows at once.

It is perhaps worth emphasizing that the principal curvatures $k_1, \ldots, k_{n-1}$ in lemma 2 are those arising when the unit normal to S is directed *towards* the point $\bar{x}$.

4. A basic identity

Consider a bounded domain Ω whose boundary is of class C^3. Let $d = d(x)$ denote the distance from points x in $\bar{\Omega}$ to the boundary surface. It follows from the results of the previous section that d is twice continuously differentiable for $0 \leqslant d < d_0$, where $d_0 = 1/K$ and K is an absolute bound for the normal curvatures of $\partial\Omega$.

For points x in the boundary strip $0 \leqslant d \leqslant a$, we define

$$v(x) = g(x) + h(d)$$

where g is twice continuously differentiable in $\bar{\Omega}$, and h is continuous for $0 \leqslant d \leqslant a$ and twice continuously differentiable with respect to d for $0 < d < a$. We suppose that

$$h' > 0 \quad \text{and} \quad a < d_0,$$

the latter condition guaranteeing that v is of class C^2 in $0 < d < a$. Our purpose in what follows is to evaluate the expression

$$\mathscr{L}v = \mathscr{A}(x, v, Dv)\,D^2 v - \mathscr{B}(x, v, Dv).$$

The result is conveniently expressed formally as

Lemma 1. *For x in the boundary strip $0 < d < a$ we have*

$$\mathscr{L}v = \mathscr{F}h''/h'^2 - \mathscr{J}h' + \mathscr{A}D^2 g - \mathscr{B}.$$

The notation here has the following meaning: first,

$$\mathscr{A} = \mathscr{A}(x, v, p), \quad \mathscr{B} = \mathscr{B}(x, v, p)$$

where $$p = p_0 + \nu h', \quad p_0 = Dg,$$

and ν is the inner unit normal vector at the (unique) point $y(x)$ on $\partial\Omega$ nearest to x; and second,

$$\mathscr{F} = (p-p_0)\mathscr{A}(p-p_0), \quad \mathscr{J} = \sum_1^{n-1} \frac{\lambda_i \mathscr{A} \lambda_i}{1-k_i d} k_i,$$

where $k_1, \ldots, k_{n-1}$ and $\lambda_1, \ldots, \lambda_{n-1}$ are respectively the principal curvatures and principal directions of $\partial\Omega$ at $y(x)$.

Proof. In order to evaluate $\mathscr{L}v$ at a particular point $\bar{x}$ in the boundary strip $0 < d < a$ it is convenient to introduce new coordinates (denoted with tildes) so that the $\tilde{x}_n$ axis coincides with the normal direction into Ω at $\bar{y} = y(\bar{x})$. We may further assume that the axes $\tilde{x}_1, \ldots, \tilde{x}_{n-1}$ are alined along the principal directions of the boundary surface at $\bar{y}$.

According to well known orthogonal transformation formulae, we have

$$\mathscr{L}v = \tilde{\mathscr{A}}(\tilde{x}, v, \tilde{D}v)\, \tilde{D}^2 v - \tilde{\mathscr{B}}(\tilde{x}, v, \tilde{D}v)$$

where

$$\tilde{\mathscr{A}} = \mathscr{T}\mathscr{A}\mathscr{T}^{-1}, \quad \tilde{\mathscr{B}} = \mathscr{B},$$

and $\mathscr{T}$ is the orthogonal transformation relating the original and new coordinate systems, that is

$$\mathscr{T} = \begin{pmatrix} \lambda_1 \\ \vdots \\ \lambda_{n-1} \\ \nu \end{pmatrix}.$$

Using the definition of v together with lemma 2 of the previous section, we find easily that

$$\tilde{D}^2 v = \tilde{D}^2 g - \left(\frac{k_1}{1-k_1 d} h', \ldots, \frac{k_{n-1}}{1-k_{n-1} d} h', -h''\right)_{\text{diagonal}} \quad \text{at } \bar{x}.$$

Consequently at $\bar{x}$,

$$\mathscr{L}v = \tilde{\mathscr{A}}\tilde{D}^2 g - \tilde{\mathscr{B}} - \sum_1^{n-1} \frac{\tilde{\mathscr{A}}_{ii} k_i}{1-k_i d} h' + \tilde{\mathscr{A}}_{nn} h''. \tag{13}$$

Reverting to the original coordinates, one has

$$\tilde{\mathscr{A}}\tilde{D}^2 g = \mathscr{A} D^2 g, \quad \tilde{\mathscr{B}} = \mathscr{B},$$

and

$$\tilde{\mathscr{A}}_{ii} = \lambda_i \mathscr{A} \lambda_i \quad (i = 1, \ldots, n-1), \quad \tilde{\mathscr{A}}_{nn} = \nu \mathscr{A} \nu.$$

Moreover, at $\bar{x}$ the arguments of $\mathscr{A}$ and $\mathscr{B}$ obviously are $\bar{x}$, v, and $Dv = Dg + h'Dd = p_0 + \nu h'$. By inserting these relations into (13), and noting that $\nu = (p-p_0)/h'$, the proof is completed.

A slightly different version of lemma 1 will also be useful. We formulate this as

LEMMA 2. *For x in the boundary strip $0 < d < a$ we have also*

$$\mathscr{L}v = \mathscr{F}\frac{h'' + Hh'}{h'^2} - \mathscr{K}h' + \mathscr{A}D^2 g - \mathscr{B}$$

where H denotes the mean curvature of $\partial\Omega$ at $y(x)$,

$$\mathscr{K} = \sum_1^{n-1} \frac{\lambda_i \mathscr{A} \lambda_i}{1-k_i d} k_i + (\nu \mathscr{A} \nu)\, H,$$

and the remaining notation is the same as in lemma 1.

To obtain this result we simply add the quantity $\tilde{\mathscr{A}}_{nn} Hh'$ to the fourth term on the right-hand side of (13), and correspondingly subtract this quantity from the third term.

55-2

The quantity $\mathscr{K}$ is a linear combination of the principal curvatures of the boundary surface. We note particularly the inequality

$$\mathscr{K} \geqslant k \operatorname{trace} \mathscr{A}$$

where k stands for the minimum (signed) normal curvature of $\partial\Omega$ at the point $y(x)$. To prove this result, first observe that

$$\frac{k_i}{1-k_i d} \geqslant k_i \geqslant k, \quad H \geqslant k,$$

whence
$$\mathscr{K} \geqslant \left(\sum_1^{n-1} \lambda_i \mathscr{A} \lambda_i + \nu \mathscr{A} \nu\right) k.$$

The quantity in parentheses is simply the trace of $\mathscr{A}$ in the special coordinate frame $\lambda_1, \ldots, \lambda_{n-1}, \nu$. Since trace is an orthogonal invariant, the required inequality is proved.

By the same argument we find
$$\mathscr{I} \geqslant k \sum_1^{n-1} \lambda_i \mathscr{A} \lambda_i.$$

Since $k \geqslant -K$, where K is an absolute bound for the curvatures of $\partial\Omega$, it follows easily that $\mathscr{I} \geqslant -K \operatorname{trace} \mathscr{A}$.

Chapter II

This chapter treats in detail the various *a priori* estimates required by steps (A), (B), and (C) of the existence programme described in § 2.

5. Basic definitions and agreements

We consider throughout this chapter a bounded domain Ω in n-dimensional Euclidean space, whose boundary $\partial\Omega$ is of class C^3. It will be assumed further that the boundary data under consideration is also of class C^3.

Under these assumptions, it is clear that there exist C^3 extensions of the boundary data into the closure of Ω, while conversely any C^3 function defined in the closure of Ω generates C^3 boundary data by restriction. In consequence, *we may suppose without loss of generality that the boundary data in question is determined by a given function f in $C^3(\overline{\Omega})$.*

We let

$$c_0 = \sup_\Omega |f|, \quad c_1 = \sup_\Omega |Df|, \quad c_2 = \sup_\Omega \|D^2 f\|,$$

where the norm $\|\mathscr{A}\|$ of a matrix is defined to be the root square of the sum of the squares of the components of $\mathscr{A}$. We also let K be an upper bound for the unsigned normal curvatures of the boundary surface. Note that then the distance function $d = d(x)$ is of class C^2 in the boundary strip $0 \leqslant d < d_0$, where $d_0 = 1/K$.

For simplicity in what follows it will be convenient to normalize equation (1) by requiring

$$\operatorname{trace} \mathscr{A} = 1. \tag{14}$$

This condition will be tacitly understood in the formulation of all further results. As an immediate consequence of (2) and (14), the eigenvalues of $\mathscr{A}$ must lie between zero and one. Consequently we have

$$0 < \xi \mathscr{A} \xi < 1 \quad \text{for} \quad |\xi| = 1. \tag{15}$$

We shall be concerned throughout the paper with certain invariant functions, and it is convenient to place their definition at a central location. In particular, we set

$$\mathscr{C} = \mathscr{C}(x, u, p) = \mathscr{B}(x, u, p)/|p| \qquad (p \neq 0),$$

$$\mathscr{E} = \mathscr{E}(x, u, p) = p \mathscr{A}(x, u, p) p,$$

$$\mathscr{F} = \mathscr{F}(x, u, p, p_0) = (p - p_0) \mathscr{A}(x, u, p)(p - p_0).$$

It is clear that these expressions remain unchanged under orthogonal transformations of the underlying Euclidean space.

The invariant $\mathscr{E}$ was introduced in 1912 by Bernstein in his fundamental paper on the calculus of variations. In particular, Bernstein considered equations in two independent variables satisfying the condition

$$\mu_1 |p|^l \leqslant \mathscr{E} \leqslant \mu_2 |p|^l \quad (|p| \geqslant 1), \tag{16}$$

where μ_1 and μ_2 are positive constants and l is a real number (actually Bernstein considered only integral values of l, but this restriction is unnecessary). It is evident that $l \leqslant 2$; the difference $g = 2 - l$ was called by Bernstein the *genre* of the equation.

We note, for example, that uniformly elliptic equations have genre zero, while the minimal surface equation on the other hand has genre two.† In the following work we shall not, however, restrict considerations solely to equations which have a well-defined genre.

Step (A) of the existence programme is discussed in § 6.

In §§ 7 through 11 we present several estimates for step (B), which apply under differing hypotheses concerning the structure of equation (1). Finally, in §§ 12 and 13 we treat step (C) of the existence programme. The last section of the chapter includes various existence theorems which are direct consequences of the earlier *a priori* estimates.

In what follows, a solution of (1) in Ω will always be understood to be a function u in $C^2(\Omega)$ such that $\mathscr{L}u \equiv 0$ in Ω. For convenience, we introduce one further notational convention: if any one of the variables x, u, or p in some relation is not specifically quantified we assume that its range is, respectively, the closure of Ω, the real numbers, or the real n-tuples.

6. Elementary maximum principles

We present in this section a number of maximum principles, all but one ultimately based on the well-known sign contradiction method at a maximum value of the dependent variable. While no claim is made for completeness, we have, nevertheless, attempted to give a useful selection of results, with the requirements of non-linear equations particularly in mind. Theorems 3, 4 and 6 seem to be new.

Theorem 1. *Let $\omega(x)$ be a function in $C^2(\Omega)$ such that*

$$\mathscr{L}(\omega+b) \leqslant 0 \quad in \quad \Omega$$

for all constants $b > 0$. Suppose also that u is a solution of (1) *in Ω such that* $\limsup\,(u-\omega) \leqslant 0$ *as one approaches any point of the boundary. Then $u \leqslant \omega$ in Ω.*

Proof. Let $v = u-\omega$ and $\kappa = \sup_\Omega v$. Suppose for contradiction that κ is greater than zero. Then according to the boundary conditions there must exist a point P in Ω where $v = \kappa$. If there is more than one such point we fix P specifically so that any neighbourhood of P contains points where $v < \kappa$; this is surely possible since $v < \kappa$ near the boundary.

Clearly there exists a neighbourhood N of P where

$$0 < v \leqslant \kappa, \quad v \not\equiv \kappa.$$

Now by hypothesis $\qquad \mathscr{A}(x, \omega+b, D\omega)\, D^2\omega - \mathscr{B}(x, \omega+b, D\omega) \leqslant 0,$

this inequality being valid for all x in Ω and all positive real numbers b. Consequently it will hold for any positive function $b(x)$. Choosing $b = v$ in particular, and restricting consideration to the neighbourhood N (hence $b > 0$), we obtain

$$\mathscr{A}(x, u, D\omega)\, D^2\omega - \mathscr{B}(x, u, D\omega) \leqslant 0 \quad \text{in} \quad N. \tag{17}$$

Moreover, since u is a solution of (1),

$$\mathscr{A}(x, u, Du)\, D^2u - \mathscr{B}(x, u, Du) = 0 \quad \text{in} \quad N. \tag{18}$$

† It is easy to construct equations having any given non-negative genre g, for example,

$$[I + (1+|Du|^2)^{\frac{1}{2}g-1}(|Du|^2 I - DuDu)]D^2u = 0.$$

Subtracting (17) from (18) and applying the mean value theorem, it is easy to establish that

$$\mathscr{A}(x, u, Du)\, D^2 v + \text{linear function of } Dv \geqslant 0 \quad \text{in } N.$$

But then Hopf's maximum principle implies $v \equiv \kappa$ in N, which is a contradiction. This completes the proof of the theorem.

The next four theorems in this section are consequences of specific choices of the function $\omega(x)$.

THEOREM 2. *Suppose that*

$$u\mathscr{B}(x, u, 0) \geqslant 0 \quad \text{for} \quad |u| > M.$$

Let u be a solution of (1) *in* Ω, *such that* $\limsup |u| \leqslant m$ *as one approaches any point of the boundary. Then* $|u| \leqslant \max(m, M)$ *in* Ω.

Proof. Let $\omega(x) = \max(m, M)$. Then

$$\mathscr{L}(\omega + b) = -\mathscr{B}(x, \max(m, M) + b, 0) \leqslant 0.$$

Consequently, by theorem 1, $u \leqslant \omega$ in Ω. Replacing u with $-u$ we obtain a similar result, and the theorem is thereby proved.

THEOREM 3. *Let Ω be contained in a ball of radius R, and assume that*

$$\operatorname{sign} u \,.\, \mathscr{C}(x, u, p) \geqslant -1/R$$

for $|u| \geqslant M$ *and* $|p| \geqslant l(>0)$. *Let u be a solution of* (1) *in* Ω, *such that* $\limsup |u| \leqslant m$ *as one approaches points of the boundary. Then*

$$|u| < \max(m, M) + 2lR \quad in \quad \Omega.$$

Proof. Let O be the centre of the ball containing Ω, and let r denote distance from O. Consider the function

$$\omega(x) = \max(m, M) + lR(\mathrm{e} - \mathrm{e}^{r/R}), \quad 0 \leqslant r \leqslant R.$$

By direct calculation, $|D\omega| = l\,\mathrm{e}^{r/R}$ and

$$\mathscr{L}(\omega + b) = l\,\mathrm{e}^{r/R}\left\{\left(\frac{1}{r} - \frac{1}{R}\right)\frac{\vec{r}\mathscr{A}\vec{r}}{r^2} - \frac{1}{r} - \mathscr{C}\right\}$$

$$\leqslant l\,\mathrm{e}^{r/R}\left\{-\frac{1}{R} - \mathscr{C}\right\} \leqslant 0, \quad 0 < r \leqslant R,$$

where we have written $\vec{r}$ for the radius vector from O and have used the main hypothesis at the last step (note that $\omega + b > M$ and $|D\omega| \geqslant l$).

We assert that $u \leqslant \omega$ in Ω. Suppose for contradiction that this were not true. Then by theorem 1 and the fact that $u \leqslant \omega$ on the boundary of Ω, the maximum of $u - \omega$ must occur at $r = 0$. This is impossible, however, in view of the functional form of ω at $r = 0$. Thus the assertion holds, and

$$u \leqslant \omega < \max(m, M) + 2lR \quad \text{in} \quad \Omega.$$

Replacing u with $-u$ we obtain a similar result, and the theorem is proved.

COROLLARY. *Suppose that* $\mathscr{B} = o(|p|)$ *as* $|u|, |p| \to \infty$. *Let u be a solution of* (1) *in* Ω, *such that* $\limsup |u| \leqslant m$ *as one approaches points of the boundary. Then one can give an* a priori *estimate for* $|u|$ *in* Ω.

The actual estimate depends on bounds for the order term. Since we shall not need this result, the details can be omitted.

THEOREM 4. *Suppose that for some fixed direction* ν *and for all* $\rho > 0$ *we have*

$$\frac{|\mathscr{B}(x,u,p)|}{\mathscr{E}(x,u,p)} \leqslant \phi(\rho), \quad p = \rho\nu,$$

where $\phi(\rho)$ *is a positive continuous function such that*

$$\int_0^\infty \frac{\mathrm{d}\rho}{\rho^2\phi(\rho)} = \infty.$$

Let u *be a solution of* (1) *in* Ω, *such that* $\limsup |u| \leqslant m$ *as one approaches any point of the boundary. Then* $|u| \leqslant L$ *in* Ω, *where* L *depends only on* m, ϕ *and the diameter of* Ω.

Proof. Let L denote the supporting plane of Ω which has normal direction ν, and which is such that ν is directed toward the side of L containing Ω. Also let d denote distance from L, and let $\bar{\delta}$ be the diameter of Ω.

For any function $\omega(d)$ which is twice continuously differentiable and satisfies $\omega' > 0$ in the interval $0 \leqslant d \leqslant \bar{\delta}$, we have by an easy calculation

$$\mathscr{L}(\omega+b) = \mathscr{E}(x,\omega+b,p)\,\omega''/\omega'^2 - \mathscr{B}(x,\omega+b,p), \quad p = \omega'\nu.$$

Now let α and β be positive numbers such that

$$\bar{\delta} = \int_\alpha^\beta \frac{\mathrm{d}\rho}{\rho^2\phi(\rho)}$$

and let ω and d be parametrically related by the formulae

$$\omega = \int_\rho^\beta \frac{\mathrm{d}\rho}{\rho\phi(\rho)} + m, \quad d = \int_\rho^\beta \frac{\mathrm{d}\rho}{\rho^2\phi(\rho)}$$

where $\alpha \leqslant \rho \leqslant \beta$. We find easily that $\omega' = \rho$ and $\omega'' = -\omega'^2\phi(\omega')$. Consequently, using the main hypothesis of the theorem, $\mathscr{L}(\omega+b) \leqslant 0$. Therefore by theorem 1

$$u \leqslant \omega < m + \int_\alpha^\beta \frac{\mathrm{d}\rho}{\rho\phi(\rho)} \quad \text{in} \quad \Omega.$$

Since a similar result holds for $-u$, the proof is complete.

Remarks. A rather obvious generalization of theorem 4 can be given in which the main hypothesis is stipulated only for sufficiently large values of $|u|$, and also by using sign $u.\mathscr{B}(x,u,p)$ rather than $|\mathscr{B}(x,u,p)|$. We also note that a related result was proved by Ladyzhenskaya & Uraltseva (1961, page 79).

THEOREM 5. *Suppose that* $\mathscr{A}$ *is independent of* u *and that* $\partial\mathscr{B}/\partial u \geqslant 0$. *Let* $\omega(x)$ *be a function in* $C^2(\Omega)$ *satisfying the condition*

$$\mathscr{L}\omega \leqslant 0 \quad in \quad \Omega.$$

Suppose also that u *is a solution of* (1) *in* Ω *such that* $\limsup (u-\omega) \leqslant 0$ *as one approaches points of the boundary. Then* $u \leqslant \omega$ *in* Ω.

This well known result is a direct consequence of theorem 1, for in view of the hypotheses on $\mathscr{A}$ and $\mathscr{B}$ we have

$$\mathscr{L}(\omega+b) \leqslant \mathscr{L}\omega \leqslant 0$$

for any positive constant b.

There is one further maximum principle rather similar to theorem 5, but applying when equation (1) has divergence structure. While this result will not be of immediate use, we shall nevertheless include the proof because of its special interest.

We suppose particularly that (1) can be written in the form

$$\partial\{A_i(x, u, Du)\}/\partial x_i = B(x, u, Du)$$

where A_i and B are respectively of class C^2 and C^1 in their arguments. We suppose further that the $(n+1)$ by $(n+1)$ matrix

$$\begin{pmatrix} \partial A_1/\partial p_1 & \dots & \partial A_1/\partial p_n & \partial A_1/\partial u \\ \vdots & & \vdots & \vdots \\ \partial A_n/\partial p_1 & \dots & \partial A_n/\partial p_n & \partial A_n/\partial u \\ \partial B/\partial p_1 & \dots & \partial B/\partial p_n & \partial B/\partial u \end{pmatrix} \tag{19}$$

is non-negative definite (here $(p_1, \dots, p_n)$ is the position variable for Du, as usual).

THEOREM 6. *Suppose that* (1) *has divergence structure and that the matrix* (19) *is non-negative definite.*

Assume that $u \in C^2(\overline{\Omega})$ *and* $v \in C^2(\overline{\Omega})$ *are solutions of* (1) *in* Ω, *such that* $u - v \leqslant 0$ *on the boundary. Then* $u \leqslant v$ *in* Ω.

Proof. Let $\kappa = \sup(u-v)$ in Ω, and suppose for contradiction that $\kappa > 0$. For $\epsilon < \kappa$ put

$$w = \max(u - v - \epsilon, 0).$$

Clearly w is Lipschitz continuous in Ω and vanishes in some neighbourhood of the boundary. Consequently a straightforward process of integration by parts yields

$$\int \{Dw \cdot A(x, u, Du) + wB(x, u, Du)\}\,\mathrm{d}x = 0$$

and

$$\int \{Dw \cdot A(x, v, Dv) + wB(x, v, Dv)\}\,\mathrm{d}x = 0,$$

and integrations taking place over the set where $w > 0$. If we now subtract these equations, there results

$$\int \{(Du - Dv) \cdot (A(x, u, Du) - A(x, v, Dv)) + (u-v)\,(B(x, u, Du) - B(x, v, Dv))\}\,\mathrm{d}x$$
$$-\epsilon \int \{B(x, u, Du) - B(x, v, Dv)\}\,\mathrm{d}x = 0.$$

Letting ϵ tend to zero then yields

$$\int \{(Du - Dv) \cdot (A(x, u, Du) - A(x, v, Dv)) + (u-v)\,(B(x, u, Du) - B(x, v, Dv))\}\,\mathrm{d}x = 0,$$

the integration now extending over the set where $u > v$.

The integrand is non-negative by the definiteness of the matrix (19), and consequently vanishes on the set where $u > v$. Now by a well-known application of the mean value theorem, the integrand can be written in the form

$$(Du - Dv)\frac{\widetilde{\partial A}}{\partial p}(Du - Dv) + (Du - Dv) \cdot \left(\frac{\widetilde{\partial A}}{\partial u} + \frac{\widetilde{\partial B}}{\partial p}\right)(u - v) + \frac{\widetilde{\partial B}}{\partial u}(u-v)^2,$$

where the notation has an obvious meaning and the tilde denotes evaluation at some intermediate value of the variables. The matrix $\partial A/\partial p$ is positive definite due to ellipticity, and moreover the intermediate value in question lies in a compact set of the u, p space when x is in $\bar{\Omega}$. Thus there exists a positive constant λ such that

$$(Du-Dv)\,(\widetilde{\partial A}/\partial p)\,(Du-Dv) \geqslant \lambda|Du-Dv|^2$$

for x in $\bar{\Omega}$. Furthermore, by the Cauchy–Schwarz inequality

$$\begin{aligned}(Du-Dv)\cdot\left(\frac{\widetilde{\partial A}}{\partial u}+\frac{\widetilde{\partial B}}{\partial p}\right)(u-v) &\geqslant -\text{const.}|Du-Dv|\,.\,|u-v| \\ &\geqslant -\tfrac{1}{2}\lambda\,|Du-Dv|^2-\text{const.}(u-v)^2.\end{aligned}$$

Consequently we find $$|Du-Dv| \leqslant \text{const.}(u-v)$$

throughout the set where $u > v$.

This, however, is impossible. For let P by any point where $u > v$ and let Q be some nearest point to P where $u = v$. (Such a point exists since $u \leqslant v$ at the boundary of Ω.) By integrating the preceding differential inequality along the straight line from Q to P we then get

$$0 < u(P)-v(P) \leqslant [u(Q)-v(Q)]\exp\,(\text{const.}|P-Q|) = 0,$$

and the theorem is thereby proved.

Perhaps the most interesting application of theorem 6 is to first-order multiple integral variational problems in which the integrand $F(x, u, Du)$ is jointly convex in u and Du for each fixed value of x. For other maximum principles for elliptic equations, see Redheffer (1962).

7. Global barriers

Let a, M be positive constants, and let N denote the boundary neighbourhood $0 \leqslant d < a$. A *global barrier* is a function v in $C^2(N)$ which satisfies the condition

$$\mathscr{L}(v+b) \leqslant 0 \quad \text{in} \quad N$$

for all positive constants b, and has the form

$$v(x) = f(x) + h(x),$$

where h is continuous in the closure of N and satisfies

$$h = 0 \quad \text{when} \quad d = 0, \qquad h = M \quad \text{when} \quad d = a.$$

A set of global barrier functions, defined for all $M > 0$, will be called a *global barrier family.*

The following lemma shows that when step (A) of the existence programme can be carried out, step (B) can be reduced to the problem of determining an appropriate global barrier.

LEMMA. *Let $u \in C^2(\bar{\Omega})$ be a solution of the Dirichlet problem in Ω. Suppose that $u \leqslant m$ and that there exists a global barrier function corresponding to $M = m + c_0$. Then*

$$\partial u/\partial n \leqslant L \quad \text{on} \quad \partial\Omega$$

for every direction $\vec{n}$ into Ω. Here the bound L depends only on the global barrier function in question.

Proof. Let $v = f + h$ be the global barrier function. Clearly $u = v$ on $\partial\Omega$. Also, recalling the definition of c_0 as the supremum of $|f|$ in Ω, we have

$$u \leqslant v \quad \text{when} \quad d = a.$$

Consequently by theorem 6·1 $\qquad u \leqslant v \quad \text{in} \quad N.$

It now follows (since both u and v are differentiable on $\partial\Omega$) that

$$\frac{\partial u}{\partial n} \leqslant \frac{\partial v}{\partial n} \quad \text{on} \quad \partial\Omega$$

for every direction $\vec{n}$ into Ω. Setting $L = \max_{\partial\Omega} |Dv|$ completes the proof.

8. Regularly elliptic equations

In this section we shall impose an appropriate degree of structure on the coefficient matrix $\mathscr{A}$ in order to carry out step (B) for an arbitrary domain Ω of class C^3. The definitions and agreements of § 5 will be maintained throughout.

We say that equation (1) is *m-regularly elliptic* provided that

$$\frac{1+|\mathscr{C}|}{\mathscr{E}} \leqslant \Phi(|p|) \quad \text{for} \quad |u| \leqslant m, \quad p \neq 0, \tag{20}$$

where $\Phi(\rho)$, $0 < \rho < \infty$, is a decreasing continuous function satisfying the condition

$$\int^{\infty} \frac{d\rho}{\rho^2 \Phi(\rho)} = \infty. \tag{21}$$

Before proceeding further, it is worthwhile considering this definition in relation to several particular examples.

1. Suppose that $\mathscr{E} \geqslant \mu |p|$ for $|p| \geqslant 1$ and that for any $m > 0$

$$|\mathscr{B}|/\mathscr{E} \leqslant \gamma(m) \log(1+|p|) \quad \text{when} \quad |u| \leqslant m, \quad |p| \geqslant 1. \tag{22}$$

Then we can take $\qquad \Phi(\rho) = \dfrac{\mu^{-1} + \gamma \log(1+\rho)}{\rho}$

and (1) is regularly elliptic for each m.

If in particular equation (1) has a genre $g \leqslant 1$, then it is regularly elliptic provided that (22) holds. An even more special case occurs for uniformly elliptic equations, these being regularly elliptic provided that

$$|\mathscr{B}| \leqslant \gamma(m) \, |p|^2 \log(1+|p|)$$

for $|u| \leqslant m$ and $|p| \geqslant 1$.

2. It may be worth presenting a specific example of a non-uniformly elliptic equation which is regularly elliptic, namely

$$(1+u_x^2) \, u_{xx} + 2u_x u_y u_{xy} + (1+u_y^2) \, u_{yy} = 0$$

where x and y are independent variables. (The regular ellipticity is obvious since the genre is zero. On the other hand, the equation is not uniformly elliptic since its coefficient matrix has the same eigenvalues as the minimal surface equation.)

3. If an equation has genre $g > 1$ then it cannot be regularly elliptic, as one easily checks. In particular, neither the minimal surface equation nor the equation for surfaces of constant

mean curvature are regularly elliptic. It is this feature which leads to the convexity and curvature restrictions which have proved necessary for the solution of the Dirichlet problem for these equations.

We may now turn to the main result of the section, after first proving a simple preliminary

LEMMA. *Suppose that* $|p_0| \leqslant c$. *Then*

$$\tfrac{1}{2}\mathscr{E} - 2c^2 \leqslant \mathscr{F} \leqslant 2\mathscr{E} + 2c^2.$$

Proof. Since $\mathscr{A}$ is positive definite and has unit trace, it is clear that for arbitrary real vectors ξ and ζ

$$|\xi\mathscr{A}\zeta| \leqslant \tfrac{1}{2}(\xi\mathscr{A}\xi + \zeta\mathscr{A}\zeta) \leqslant \tfrac{1}{2}(|\xi^2| + \zeta\mathscr{A}\zeta).$$

Setting $\xi = \mathscr{A}p$ and $\zeta = p$ then yields $|\mathscr{A}p|^2 \leqslant p\mathscr{A}p$. Now

$$\mathscr{F} = (p-p_0)\,\mathscr{A}(p-p_0) = p\mathscr{A}p - 2p_0\mathscr{A}p + p_0\mathscr{A}p_0.$$

For $|p_0| \leqslant c$ we have obviously $|p_0\mathscr{A}p| \leqslant c|\mathscr{A}p| \leqslant c\sqrt{(p\mathscr{A}p)}$. Hence, recalling the definition of $\mathscr{E}$, it follows that

$$\mathscr{E} - 2c\sqrt{\mathscr{E}} \leqslant \mathscr{F} \leqslant \mathscr{E} + 2c\sqrt{\mathscr{E}} + c^2.$$

The required conclusion is now an immediate consequence of Cauchy's inequality.

THEOREM 1. *Let* $u \in C^2(\bar{\Omega})$ *be a solution of the Dirichlet problem for equation* (1) *in* Ω. *Assume that* $|u| \leqslant m$ *in* Ω. *Then if* (1) *is m-regularly elliptic we have*

$$|Du| \leqslant L \quad on \quad \partial\Omega$$

where L depends only on c_0, c_1, c_2, m, K, *and the function* Φ. (K, c_0, c_1, c_2 *are defined in* §5.)

Proof. We shall assume to begin with that (20) holds without the added restriction $|u| \leqslant m$; that is, we assume

$$\frac{1+|\mathscr{C}|}{\mathscr{E}} \leqslant \Phi(|p|) \tag{23}$$

for all x in the closure of Ω, all real u, and all $p \neq 0$. This assumption will be removed at the end of the proof.

Our goal is to construct an appropriate global barrier family and then to apply the lemma of the previous section. Let

$$v(x) = f(x) + h(d), \quad 0 \leqslant d \leqslant a,$$

where $a < d_0$. We assume that h is twice continuously differentiable with respect to d and that

$$h(0) = 0, \quad h(a) = M, \quad h'(d) \geqslant \alpha, \tag{24}$$

α being a positive constant whose value will be specified later. In order to evaluate $\mathscr{L}(v+b)$ we may apply lemma 4·1. Since $\mathscr{J} \geqslant -K$, this yields

$$\mathscr{L}(v+b) \leqslant \mathscr{F}h''/h'^2 + Kh' + \mathscr{A}D^2f - \mathscr{B}, \tag{25}$$

the arguments of $\mathscr{A}$, $\mathscr{B}$, and $\mathscr{F}$ being x, $v+b$, and $p = p_0 + \nu h'$ where $p_0 = Df$.

Now $\mathscr{E} \geqslant 1/\Phi$, and Φ necessarily tends to zero as $|p|$ tends to infinity (since the integral (21) diverges). Consequently there exists a constant α_1, depending only on c_1 and Φ, such that

$$\mathscr{E} \geqslant 8c_1^2 \quad \text{for} \quad |p| \geqslant \alpha_1.$$

Now choose

$$\alpha = \max(c_1 + \alpha_1, MK, 1).$$

Since $p = p_0 + \nu h'$ and $|p_0| \leqslant c_1$ it is easy to check that

$$\alpha_1 \leqslant h' - c_1 \leqslant |p| \leqslant 2h'.$$

We are now in position to estimate the coefficients $\mathscr{A}$, $\mathscr{B}$, and $\mathscr{F}$ in the inequality for $\mathscr{L}(v+b)$.

To begin with, since the eigenvalues of $\mathscr{A}$ all lie between 0 and 1, we have

$$|\mathscr{A} D^2 f| \leqslant \|D^2 f\| \leqslant c_2.$$

Next

$$|\mathscr{B}| \leqslant |p|\,|\mathscr{C}| \leqslant 2h'\,|\mathscr{C}|,$$

and finally, using the preliminary lemma and the fact that $\mathscr{E} \geqslant 8c_1^2$,

$$\mathscr{F} \geqslant \tfrac{1}{2}\mathscr{E} - 2c_1^2 \geqslant \tfrac{1}{4}\mathscr{E}.$$

Inserting these estimates into (25) yields

$$\begin{aligned} \mathscr{L}(v+b) &\leqslant \mathscr{F} h' \left\{ \frac{h''}{h'^3} + 4\frac{c_2+K+2|\mathscr{C}|}{\mathscr{E}} \right\} \\ &\leqslant \mathscr{F} h' \left\{ \frac{h''}{h'^3} + c\Phi(|p|) \right\} \quad (c = 4c_2+4K+8) \\ &\leqslant \mathscr{F} h' \left\{ \frac{h''}{h'^3} + c\Phi(h'-c_1) \right\}, \end{aligned}$$

the last step holding since Φ is a decreasing function of $|p|$.

We now choose h' so that the expression in braces vanishes and so that conditions (24) hold. In order to do this, define β by means of the relation

$$cM = \int_\alpha^\beta \frac{d\rho}{\rho^2 \Phi(\rho - c_1)}$$

(this is possible since the integral diverges). Next, let h and d be parametrically related by

$$ch = \int_\rho^\beta \frac{dp}{\rho^2\Phi(\rho-c_1)}, \quad cd = \int_\rho^\beta \frac{dp}{\rho^3\Phi(\rho-c_1)}$$

where $\alpha \leqslant \rho \leqslant \beta$. One easily checks that $h' = \rho \geqslant \alpha$ and that the quantity in braces vanishes. Moreover, it is evident that

$$h(0) = 0, \quad h(a) = M,$$

where

$$a = \frac{1}{c}\int_\alpha^\beta \frac{d\rho}{\rho^3\Phi(\rho-c_1)},$$

(note that $a < M/\alpha \leqslant d_0$).

To summarize, we have constructed for each $M > 0$ a global barrier function with $a < d_0$. By the lemma of the previous section, therefore,

$$\partial u/\partial n \leqslant L \quad \text{on} \quad \partial\Omega$$

where L depends only on the global barrier function corresponding to $M = m + c_0$. The latter depends, moreover, only on the quantities listed in the statement of the theorem. Replacing u with $-u$ in the equation leaves the construction unaltered. Hence we also have

$$\partial u/\partial n \geqslant -L \quad \text{on} \quad \partial\Omega,$$

completing the proof of the theorem subject to the initial assumption (23).

If (23) does not hold, we consider a new equation with coefficient matrix $\hat{\mathscr{A}}$ defined by

$$\hat{\mathscr{A}}(x,u,p) = \begin{cases} \mathscr{A}(x,-m,p) & \text{if } u<-m, \\ \mathscr{A}(x,u,p) & \text{if } -m\leqslant u\leqslant m, \\ \mathscr{A}(x,m,p) & \text{if } u>m, \end{cases}$$

and with a similarly defined inhomogeneous term $\hat{\mathscr{B}}$. Clearly u is also a solution of the equation $\hat{\mathscr{A}}D^2u = \hat{\mathscr{B}}$, and this fact together with the evident relation

$$\frac{1+|\hat{\mathscr{C}}|}{\hat{\mathscr{E}}} \leqslant \Phi(|p|), \quad p \neq 0,$$

allows us to repeat the previous proof word for word, completing the demonstration of the theorem.

9. Boundary estimates depending on curvature

In this section we consider step (B) for equations which are not regularly elliptic. In such cases it turns out that the curvatures of the boundary surface must be restricted in order to obtain boundary estimates.

We shall consider the following conditions on the asymptotic behaviour of the invariants $\mathscr{A}$ and $\mathscr{C}$ for large values of p, namely

$$\mathscr{A}(x,u,p) = \mathscr{A}_0(x,\sigma)+o(1), \qquad \sigma = p/|p|, \tag{26}$$

and

$$\mathscr{C}(x,u,p) = \mathscr{C}_0(x,u,\sigma)+o(1), \quad \sigma = p/|p|, \tag{27}$$

as $p\to\infty$, where $\mathscr{A}_0(x,\sigma)$ and $\mathscr{C}_0(x,u,\sigma)$ are continuously differentiable functions of their arguments and where

$$\partial\mathscr{C}_0/\partial u \geqslant 0. \tag{28}$$

In these relations, as well as later in the paper, *order terms are assumed to be uniform for* (x,u) *in any compact set, unless otherwise stated.* We emphasize that condition (28) will be tacitly assumed whenever we deal with the asymptotic relation (27).

The matrix $\mathscr{A}_0$ can be used to introduce a generalized mean curvature of the boundary surface. Let y be a point on the boundary and let ν be the unit inner normal at y. Also let $k_1,\dots,k_{n-1}$ and $\lambda_1,\dots,\lambda_{n-1}$ be respectively the principal curvatures and principal directions of $\partial\Omega$ at y. We then put

$$A = A(y) = \mathscr{A}_0(y,\nu)$$

and define

$$\mathscr{H} = \mathscr{H}(y) = \sum_1^{n-1} (\lambda_i A\lambda_i)k_i + (\nu A\nu)H,$$

where H is the ordinary mean curvature of $\partial\Omega$ at y. The function which arises upon replacing ν by $-\nu$ in this construction will be denoted by $\mathscr{I}(y)$.

It is worth emphasizing that $\mathscr{H}$ and $\mathscr{I}$ are *averages* of the principal curvatures† and that both quantities are orthogonal invariants, exactly as is the case with the mean curvature.

† That is, $\mathscr{H} = \sum_1^{n-1} a_i k_i$ where

$$a_i \geqslant 0, \quad \Sigma a_i = 1, \tag{*}$$

with a similar relation holding for $\mathscr{I}$. Conditions (*) are a direct consequence of two facts, (i) the numbers $(\lambda_i A\lambda_i)$ and $\nu A\nu$ are the diagonal elements of the matrix A in the coordinate frame $\lambda_1,\dots,\lambda_{n-1},\nu$, and (ii) the matrix $\mathscr{A}_0$ necessarily has unit trace and is non-negative definite.

Moreover, if the matrix A is proportional on $\partial\Omega$ to the Euclidean metric tensor, then $\mathscr{H} = \mathscr{I} = H$. This condition automatically holds in two dimensions, so that $\mathscr{H} = \mathscr{J} = \kappa$ in this case, where κ is the curvature of the boundary curve.

To illustrate the constructions of $\mathscr{A}_0$, $\mathscr{C}_0$, $\mathscr{H}$ and $\mathscr{I}$, consider in particular the equation for surfaces of constant mean curvature in n dimensions, namely

$$(1+|Du|^2)\,\Delta u - DuDu\,D^2u = n\Lambda(1+|Du|^2)^{\frac{3}{2}}.$$

After normalizing to unit trace, we find

$$\mathscr{A}_0 = \frac{I-\sigma\sigma}{n-1}, \quad \mathscr{C}_0 = \frac{n}{n-1}\Lambda;$$

moreover, the error terms in (26) and (27) are of degree -2. Evidently $\lambda_i A\lambda_i = (n-1)^{-1}$ and $\nu A\nu = 0$, and we have therefore

$$\mathscr{H} = \mathscr{I} = H$$

(this also follows from the fact that A is proportional to the Euclidean metric on $\partial\Omega$).

Theorem 1. *Let $u \in C^2(\bar{\Omega})$ be a solution of the Dirichlet problem for equation* (1) *in Ω. Assume that $|u| \leqslant m$ in Ω. If* (26) *and* (27) *hold, and if both conditions*

$$\mathscr{H} > -\mathscr{C}_0(y,f,\nu), \quad \mathscr{I} > \mathscr{C}_0(y,f,-\nu) \qquad (29)$$

are satisfied at each point y of the boundary surface, then

$$|Du| \leqslant L \quad on \quad \partial\Omega$$

where L depends only on c_0, c_1, c_2, m, K, a lower bound for the differences in (20), *bounds for the error terms in* (26) *and* (27), *and C^1 norms of the functions $\mathscr{A}_0$ and $\mathscr{C}_0$.*

Proof. According to the argument used in the final step of the proof of theorem 8·1, we can assume without loss of generality that the error terms in (26) and (27), as well as the norms of $\mathscr{A}_0$ and $\mathscr{C}_0$, are uniformly bounded with respect to u. As before, we shall carry out the proof by constructing an appropriate global barrier family and applying the lemma of § 7.

We put

$$v(x) = f(x) + h(d), \quad 0 \leqslant d \leqslant a$$

where $a < d_0$. It is further assumed that h is twice continuously differentiable with respect to d and that

$$h(0) = 0, \quad h(a) = M, \quad h'(d) \geqslant \alpha, \qquad (30)$$

α being a positive constant which will be determined later. By lemma 4·2

$$\mathscr{L}(v+b) = \mathscr{F}\frac{h''+Hh'}{h'^2} - \mathscr{K}h' + \mathscr{A}D^2f - \mathscr{B}$$

the arguments of $\mathscr{A}$, $\mathscr{B}$, and $\mathscr{F}$ being x, $v+b$, and $p = p_0 + \nu h'$.

Let $\theta > 0$ be chosen so that

$$\mathscr{H} \geqslant -\mathscr{C}_0(y,f,\nu) + 5\theta$$

at each point of the boundary surface (this can surely be done since the terms in (29) are continuous functions of y).

An easy estimate yields $|p| \geqslant \alpha - c_1$ and

$$\|\mathscr{A}(x,v+b,p) - \mathscr{A}_0(x,\nu)\| \leqslant \|\mathscr{A}(x,v+b,p) - \mathscr{A}_0(x,\sigma)\| + \frac{2c_1}{|p|}\sup\left\|\frac{\partial\mathscr{A}_0}{\partial\sigma}\right\|.$$

Moreover, if $y = y(x)$ denotes the point on the boundary nearest x, we have

$$\|\mathscr{A}_0(x,\nu) - \mathscr{A}_0(y,\nu)\| \leqslant d \sup \left\| \frac{\partial \mathscr{A}_0}{\partial x} \right\|.$$

Consequently, when α is suitably large and d is sufficiently small, the difference

$$\mathscr{A}(x, v+b, p) - A(y)$$

becomes arbitrarily small. In particular, for $\alpha \geqslant \alpha_1$ and $d \leqslant a_1$,

$$\mathscr{K} \geqslant \sum_1^{n-1} (\lambda_i \mathscr{A} \lambda_i) k_i + (\nu \mathscr{A} \nu) H \geqslant \mathscr{H} - \theta. \tag{31}$$

Here a_1 and α_1 depend only on θ, c_1, K, bounds for the error term $\|\mathscr{A} - \mathscr{A}_0\|$, and a bound for the C^1 norm of $\mathscr{A}_0$.

Similarly, for α suitably large we have

$$\mathscr{C}(x, v+b, p) \geqslant \mathscr{C}_0(x, v+b, \nu) - \theta \geqslant \mathscr{C}_0(x, f(x), \nu) - \theta$$

since $\mathscr{C}_0$ is increasing in its second argument. Since $|Df| \leqslant c_1$ we may assume further that

$$\mathscr{C}_0(x, f(x), \nu) \geqslant \mathscr{C}_0(y, f(y), \nu) - \theta \quad \text{for} \quad d \leqslant a_2.$$

This leads to the required estimate

$$\begin{aligned} \mathscr{B}(x, v+b, p) &= |p|\,\mathscr{C}(x, v+b, p) \\ &\geqslant |p|\,\{\mathscr{C}_0(y, f, \nu) - 2\theta\} \geqslant h'\{\mathscr{C}_0(y, f, \nu) - 3\theta\} \end{aligned} \tag{32}$$

provided $\alpha \geqslant \alpha_2$ and $d \leqslant a_2$.

Now choose

$$\alpha = \max(\alpha_1, \alpha_2, M/a_1, M/a_2, MK, c_2/\theta).$$

Then according to (31) and (32) we have (provided $a < \min(a_1, a_2, d_0)$)

$$\mathscr{L}(v+b) \leqslant \mathscr{F} \frac{h'' + Hh'}{h'^2} + h'\{-\mathscr{H} - \mathscr{C}_0(y, f, \nu) + 5\theta\} \leqslant \mathscr{F} \frac{h'' + Kh'}{h'^2},$$

where at the last line we have used the definition of θ and the fact that $H \leqslant K$.

We now define

$$h(d) = M \frac{1 - e^{-Kd}}{1 - e^{-Ka}}, \quad \text{where} \quad a = \frac{1}{K} \log\left(1 + \frac{MK}{\alpha}\right).$$

It is easy to check that $h'' + Kh' = 0$ and that conditions (30) hold. Moreover, $a < \min(a_1, a_2, d_0)$, so that finally

$$\mathscr{L}(v+b) \leqslant 0 \quad \text{for} \quad 0 < d < a.$$

The rest of the proof is almost exactly the same as that of theorem 8·1. The only significant change is that when u is replaced by $-u$ we must simultaneously replace $\mathscr{A}(x, u, p)$ with $\mathscr{A}(x, -u, -p)$ and $\mathscr{B}(x, u, p)$ with $-\mathscr{B}(x, -u, -p)$. This accounts for the second condition in (29), in which $\mathscr{H}$ has been replaced by $\mathscr{I}$ and $\mathscr{C}_0(y, f, \nu)$ by $-\mathscr{C}_0(y, f, -\nu)$.

Remark. It is clear from the above proof that a result similar to theorem 1 can be obtained under a milder structure condition than (27), namely

$$\mathscr{C}_0^-(x, u, \sigma) - o(1) \leqslant \mathscr{C}(x, u, p) \leqslant \mathscr{C}_0^+(x, u, \sigma) + o(1).$$

In this case (29) should be replaced by the pair of inequalities

$$\mathscr{H} > -\mathscr{C}_0^+(y,f,\nu) \quad \text{and} \quad \mathscr{I} > \mathscr{C}_0^-(y,f,-\nu).$$

The rather more complicated nature of these conditions is the main reason for emphasizing the structure (27) and the condition (29).

Even with no structure imposed on the coefficient matrix, it is still possible to obtain a result of interest.

THEOREM 2. *Let $u \in C^2(\bar{\Omega})$ be a solution of the Dirichlet problem for equation* (1) *in* Ω. *Assume that $|u| \leqslant m$ in Ω. Then if* (27) *holds, and if the minimum normal curvature k of the boundary surface satisfies the condition*

$$k > \mp\mathscr{C}_0(y,f,\pm\nu) \tag{33}$$

at each point of y of the boundary, we have

$$|Du| \leqslant L \quad on \quad \partial\Omega$$

where L depends only on c_0, c_1, c_2, m, K, a lower bound for the differences in (33), *a bound for the error term in* (27), *and the C^1 norm of the function $\mathscr{C}_0$.*

The proof is exactly the same as before, except that we use the relation $\mathscr{K} \geqslant k$ in place of the more difficult inequality (31).

Comment. Theorem 2 includes as a special case the boundary estimates used by Gilbarg and Stampacchia in their study of non-uniformly elliptic equations of the form

$$\partial\{A_i(Du)\}/\partial x_i = 0 \quad A_i \in C^2.$$

In fact, when this equation is written out we have $\mathscr{B} \equiv 0$, so that (33) becomes precisely the convexity condition imposed by these authors.

10. BOUNDEDLY NON-LINEAR EQUATIONS

With a very slight strengthening of the asymptotic structure of $\mathscr{A}$ and $\mathscr{C}$, the equality sign can be allowed in conditions (29) and (33). Although at first glance there in no great gain in this, the fact that a number of important examples can be included here makes the stronger result well worth stating.

We say that equation (1) is *(c, m)-boundedly non-linear* provided that (26) and (27) hold and

$$\frac{\|\mathscr{A}-\mathscr{A}_0\| + |\mathscr{C}-\mathscr{C}_0| + |p|^{-1}}{\mathscr{F}} \leqslant \Phi(|p|) \tag{34}$$

for $|u| \leqslant m$, $|p_0| \leqslant c$, $|p| > c$, where $\Phi(\rho)$ is a decreasing continuous function satisfying the condition

$$\int^{\infty} \frac{d\rho}{1+\rho^2\Phi(\rho)} = \infty.$$

Let us consider some particular cases.

1. Suppose that $\mathscr{F} \geqslant \mu(m,c) > 0$ and

$$\frac{\|\mathscr{A}-\mathscr{A}_0\| + |\mathscr{C}-\mathscr{C}_0|}{\mathscr{F}} \leqslant \gamma(m,c)\frac{\log(1+|p|)}{|p|}$$

when $|p| \geqslant 1+c$. Then we can take

$$\Phi(\rho) = \frac{\mu^{-1}+\gamma\log(1+\rho)}{\rho}, \quad \rho \geqslant 1+c$$

and (1) is boundedly non-linear for each m and c.

If equation (1) has a well-defined genre g, $1 < g < 2$, then it is boundedly non-linear provided

$$\|\mathscr{A}-\mathscr{A}_0\|+|\mathscr{C}-\mathscr{C}_0| \leqslant \gamma(m)\,|p|^{1-g} \quad \text{for} \quad |u| \leqslant m,\ |p| \geqslant 1$$

(the lemma of § 8 implies $\mathscr{F} \approx |p|^{2-g}$ for large p). Conversely, if $g > 2$ then (1) cannot be boundedly non-linear.

2. The case where (1) has genre two is particularly interesting in view of the large number of examples of this type, including the minimal surface equation and the equation for surfaces having constant mean curvature. The following lemma states a sufficient condition for an equation of genre two to be boundedly non-linear.

Lemma. *Assume that equation* (1) *has genre two and that*

$$\mathscr{A}-\mathscr{A}_0 = o(|p|^{-1}), \qquad \mathscr{C}-\mathscr{C}_0 = O(|p|^{-1})$$

as p tends to infinity. Then (1) *is boundedly non-linear. If for large p the least eigenvalue of $\mathscr{A}$ is bounded below by a positive multiple of $|p|^{-2}$, then we can replace $o(|p|^{-1})$ by $O(|p|^{-1})$ in the order condition for $\mathscr{A}-\mathscr{A}_0$.*

Proof. In view of the result of paragraph 1, it is enough to show that $\mathscr{F} \geqslant \mu$ for $|p| \geqslant 1+c$. Consider first the case where $\mathscr{A}-\mathscr{A}_0 = o(|p|^{-1})$. From (26) we have obviously

$$\mathscr{E} = [\sigma\mathscr{A}_0\sigma+o(1)]\,|p|^2.$$

Since $\mathscr{E} = O(1)$ by the genre assumption, it follows that $\sigma\mathscr{A}_0\sigma = 0$. Moreover, $\mathscr{A}_0$ must surely be non-negative definite. Consequently we have

$$\begin{aligned}\mathscr{F} = (p-p_0)\mathscr{A}(p-p_0) &\geqslant (p-p_0)(\mathscr{A}-\mathscr{A}_0)(p-p_0)\\ &= p(\mathscr{A}-\mathscr{A}_0)p+o(1)\\ &= p\mathscr{A}p+o(1) = \mathscr{E}+o(1).\end{aligned}$$

Now $\mathscr{E} \geqslant \mu_1$ for $|p| \geqslant 1$. Hence $\mathscr{F} \geqslant \frac{1}{2}\mu_1$ for all sufficiently large p, and the required conclusion follows at once from the fact that $\mathscr{F}$ is positive and continuous for $c < |p| < \infty$.

When the least eigenvalue of $\mathscr{A}$ is of the order $|p|^{-2}$, we have obviously

$$\mathscr{F} = (p-p_0)\mathscr{A}(p-p_0) \geqslant \text{const.}\frac{|p-p_0|^2}{|p|^2}$$

and the final part of the lemma is an immediate consequence of this relation.

3. We have remarked in the previous section that the error terms $\mathscr{A}-\mathscr{A}_0$ and $\mathscr{C}-\mathscr{C}_0$ for the minimal surface equation and the equation for surfaces of constant mean curvature are of degree -2. By the above lemma, therefore, these equations are boundedly non-linear.

We now turn to the main results of the section, which slightly extend the conclusions of theorems 9·1 and 9·2. Note that the expressions $\mathscr{H}$ and $\mathscr{I}$ were defined in § 9.

Theorem 1. *Let $u \in C^2(\Omega)$ be a solution of the Dirichlet problem for equation* (1) *in Ω. Assume that $|u| \leqslant m$ in Ω, and that* (1) *is (c_1, m)-boundedly non-linear. Then if both conditions*

$$\mathscr{H} \geqslant -\mathscr{C}_0(y,f,\nu), \quad \mathscr{I} \geqslant \mathscr{C}_0(y,f,-\nu) \tag{35}$$

hold at each point y of the boundary surface, we have

$$|Du| \leqslant L \quad on \quad \partial\Omega$$

where L depends only on c_0, c_1, c_2, m, K, the function $\Phi(p)$, and C^1 norms of the functions $\mathscr{A}_0$ and $\mathscr{C}_0$.

Proof. We may suppose without loss of generality that

$$\frac{\|\mathscr{A}-\mathscr{A}_0\|+|\mathscr{C}-\mathscr{C}_0|+|p|^{-1}}{\mathscr{F}} \leqslant \Phi(|p|) \quad (|p_0|\leqslant c_1,\ |p|>c_1).$$

We carry out the proof, as usual, by constructing an appropriate global barrier family and applying the lemma of § 7.

Choosing v in the form $\quad v(x)=f(x)+h(d), \quad 0\leqslant d\leqslant a,$

where $a<d_0$ and where h satisfies the conditions

$$h(0)=0, \quad h(a)=M, \quad h'(d)\geqslant \alpha, \tag{36}$$

we find by lemma 4·2

$$\mathscr{L}(v+b)=\mathscr{F}\frac{h''+Hh'}{h'^2}-\mathscr{K}h'+\mathscr{A}D^2f-\mathscr{B}$$

where the arguments of $\mathscr{A}$, $\mathscr{B}$, and $\mathscr{F}$ are x, $v+b$, and $p=p_0+\nu h'$.

As in the proof of theorem 9·1,

$$\|\mathscr{A}(x,v+b,p)-A(y)\|\leqslant\|\mathscr{A}-\mathscr{A}_0\|+\frac{2c_1}{|p|}\sup\left\|\frac{\partial\mathscr{A}_0}{\partial\sigma}\right\|+d\sup\left\|\frac{\partial\mathscr{A}_0}{\partial x}\right\|.$$

It follows (see (31)) that

$$\mathscr{K}\geqslant\mathscr{H}-\sqrt{n}K\|\mathscr{A}-\mathscr{A}_0\|-\text{const.}(|p|^{-1}+d).$$

Similarly (see (32))

$$\begin{aligned}\mathscr{B}(x,v+b,p)&\geqslant|p|\{\mathscr{C}_0(y,f,\nu)-|\mathscr{C}-\mathscr{C}_0|-\text{const.}(|p|^{-1}+d)\}\\&\geqslant h'\{\mathscr{C}_0(y,f,\nu)-2|\mathscr{C}-\mathscr{C}_0|-\text{const.}(|p|^{-1}+d)\}\end{aligned}$$

provided $\alpha\geqslant c_1$. (The constants in the last three lines are of course not the same, though all depend only on c_1, K, and the C^1 norms of the functions $\mathscr{A}_0$ and $\mathscr{C}_0$.)

Before inserting the preceding estimates into the identity for $\mathscr{L}(v+b)$, it is convenient to obtain a bound for d in terms of $|p|$. To this end we assume $h''<0$, a condition which will be satisfied by the eventual choice of h. Then, for $\alpha\geqslant c_1$,

$$d\leqslant 2h'|p|^{-1}d\leqslant 2M|p|^{-1},$$

since by the convexity of h we have $h'(d)d\leqslant h(d)-h(0)\leqslant M$.

With the help of the above estimates, it follows that

$$\begin{aligned}\mathscr{L}(v+b)\leqslant\mathscr{F}\frac{h''+Kh'}{h'^2}&+h'\{-\mathscr{H}-\mathscr{C}_0(y,f,\nu)+2c_2|p|^{-1}\\&+\sqrt{n}K\|\mathscr{A}-\mathscr{A}_0\|+2|\mathscr{C}-\mathscr{C}_0|+\text{const.}(1+M)|p|^{-1}\}\end{aligned}$$

provided $\alpha\geqslant c_1$. Using the principal hypothesis (35) and setting

$$c=2+2c_2+\sqrt{n}K+\text{const.}(1+M)$$

yields finally

$$\begin{aligned}\mathscr{L}(v+b)&\leqslant h'\mathscr{F}\left\{\frac{h''}{h'^3}+c\left(\frac{1}{h'^2}+\Phi(|p|)\right)\right\},\\&\leqslant h'\mathscr{F}\left\{\frac{h''}{h'^3}+c\left(\frac{1}{h'^2}+\Phi(h'-c_1)\right)\right\}\end{aligned}$$

since Φ is a decreasing function. (It should be noted here that the constant c depends only on c_1, c_2, K, M, and C^1 norms of the functions $\mathscr{A}_0$ and $\mathscr{C}_0$.)

57-2

Now choose
$$\alpha = \max(c_1, MK)$$
and define β by means of the relation
$$cM = \int_\alpha^\beta \frac{d\rho}{1+\rho^2\Phi(\rho-c_1)}$$
(this is always possible since the integral is divergent). The function $h(d)$ is finally determined parametrically by the formulae
$$ch = \int_\rho^\beta \frac{d\rho}{1+\rho^2\Phi(\rho-c_1)}, \quad cd = \int_\rho^\beta \frac{d\rho}{\rho[1+\rho^2\Phi(\rho-c_1)]}$$
where $\alpha \leqslant \rho \leqslant \beta$. One easily checks that $h' = \rho \geqslant \alpha$ and that (36) holds with
$$a = \frac{1}{c}\int_\alpha^\beta \frac{d\rho}{\rho[1+\rho^2\Phi(\rho-c_1)]}$$
(note that $a < M/\alpha \leqslant d_0$). Moreover
$$h'' = -ch'[1+h'^2\Phi(h'-c)]$$
so that $h'' < 0$ and $\mathscr{L}(v+b) \leqslant 0$.

The remaining part of the proof is by now standard, and consequently may be omitted.

THEOREM 2. *Let $u \in C^2(\overline{\Omega})$ be a solution of the Dirichlet problem for equation* (1) *in Ω. Assume that $|u| \leqslant m$ in Ω, and that* (34) *holds with the term $\|\mathscr{A}-\mathscr{A}_0\|$ omitted. Then if the minimum normal curvature k of the boundary surface satisfies the condition*
$$k \geqslant \mp\mathscr{C}_0(y, f, \pm\nu) \tag{37}$$
at each point y of the boundary surface, we have
$$|Du| \leqslant L \quad \text{on} \quad \partial\Omega$$
where L depends only on c_0, c_1, c_2, m, K, the function $\Phi(\rho)$, and the C^1 norm of the function $\mathscr{C}_0$.

The proof is exactly the same as before, except that we use the relation $\mathscr{K} \geqslant k$ in place of the more difficult estimate for $\mathscr{K}$ in terms of $\mathscr{H}$.

The minimal surface equation is boundedly nonlinear, as we remarked earlier; moreover, in this case $\mathscr{H} = \mathscr{I} = H$, where H is the mean curvature of the boundary surface. It is therefore evident from theorem 1 that step (B) in the existence programme can be carried out if
$$H \geqslant 0$$
at each point of the boundary surface. Now step (A) is trivial for the minimal surface equation, and step (C) follows from the fact that the derivatives of a solution satisfy the maximum principle. *Hence the Dirichlet problem for the minimal surface equation in a C^3 domain Ω is solvable for arbitrarily given C^3 boundary data provided that $H \geqslant 0$ at each boundary point.* This result was first proved by Howard Jenkins and the present author, and in fact holds equally for C^2 domains and continuous boundary data (see § 19).

11. MIXED EQUATIONS

It may happen that (1) is regularly elliptic for certain directions of the vector p and boundedly non-linear for other directions. As an example, consider the 'false' minimal surface equation
$$(1+u_y^2)\,u_{xx} + 2u_x u_y u_{xy} + (1+u_x^2)\,u_{yy} = 0 \tag{38}$$

where x and y are independent variables. Putting $p = (p_1, p_2)$ and recalling the normalization to unit trace, we find easily that

$$\mathscr{E} = \frac{|p|^2 + 4p_1^2 p_2^2}{|p|^2+2}.$$

If the vector $\sigma = p/|p|$ is bounded away from both coordinate directions, and $|p| \geqslant 1$, then

$$\mu_1 |p|^2 \leqslant \mathscr{E} \leqslant \mu_2 |p|^2$$

for appropriate constants μ_1 and μ_2. Thus (38) is regularly elliptic for these directions.

When σ lies in a coordinate direction, on the other hand, then $\mathscr{E} \leqslant 1$ and the condition of regular ellipticity fails. The condition of bounded non-linearity holds, however, since (by an easy calculation)

$$\mathscr{F} \geqslant \frac{1}{2+(1+c)^2} \quad \text{for} \quad |p| \geqslant 1+c.$$

In the following theorem we present a fairly simple result indicating the type of conclusions which can be drawn in such mixed cases.

Theorem. *Let $u \in C^2(\bar{\Omega})$ be a solution of the Dirichlet problem for equation* (1) *in* Ω.

Assume that $|u| \leqslant m$ in Ω, that (1) *is m-regularly elliptic for all values of p whose directions $\sigma = p/|p|$ lie in some open set R of the unit sphere, and finally that* (1) *is (c_1, m)-boundedly non-linear.*† *Then if*

$$\mathscr{H} \geqslant -\mathscr{C}_0(y, f, \nu), \quad \mathscr{I} \geqslant \mathscr{C}_0(y, f, -\nu)$$

at each point y of the boundary surface where $\nu \notin R$, we have

$$|Du| \leqslant L \quad \text{on} \quad \partial\Omega$$

where L depends only on c_0, c_1, c_2, m, K, the function $\Phi(\rho)$, and C^1 norms of the functions $\mathscr{A}_0$ and $\mathscr{C}_0$.

The proof is a direct combination of the demonstrations of theorems 8·1 and 10·1. The assumption that $\Phi(\rho)$ is the same in both conditions (20) and (34) is used in order to construct a single function $h(d)$ satisfying the requirements of each proof separately.

Returning to equation (38), let R be any fixed set $\{|\sigma_1| > \epsilon, |\sigma_2| > \epsilon\}$, where $\epsilon > 0$. Then

$$\frac{1}{\mathscr{E}} \leqslant \frac{\text{const.}}{|p|^2} \quad \text{for} \quad |p| \geqslant 1, \quad \sigma \in R$$

while clearly

$$\frac{\|\mathscr{A} - \mathscr{A}_0\| + |p|^{-1}}{\mathscr{F}} \leqslant \frac{\text{const.}}{|p|} \quad \text{for} \quad |p| \geqslant 1+c.$$

Choosing $\Phi(\rho) = \text{const.}/\rho$, $\rho \geqslant 1+c$, it is apparent that the hypotheses of the theorem are satisfied. Since steps (A) and (C) are trivial for the present case (for example, use theorem 6·2 and the fact that the first derivatives of a solution of (38) satisfy a maximum principle), we have therefore proved the following result:

The Dirichlet problem for equation (38) *in a C^3 domain Ω is solvable for arbitrarily given C^3 boundary data provided that the curvature κ of the boundary curve is non-negative in the neighbourhood of each boundary point whose tangent lies along one of the coordinate directions.*

The condition that the boundary and the data be of class C^3 can be lightened to class C^2 by an approximation argument; we shall, however, omit the details.

† We suppose the function $\Phi(\rho)$ to be the same in both conditions (20) and (34); doubtless this assumption could be removed by a suitable modification of the proof.

It is interesting to contrast this result with the corresponding theorem for the minimal surface equation. The rather trivial reversal of the sign of the middle term in these equations, though leaving the eigenvalues of the coefficient matrix unaltered and apparently affecting the structure only slightly, in fact has a profound effect on the type of Dirichlet problem which can be solved.

12. Interior estimates for $|Du|$

In contrast to the work of preceding sections, the derivation of interior estimates for the gradient of a solution naturally divides into two cases, depending on whether $n = 2$ or $n > 2$. For $n = 2$ the results go back to Bernstein; we shall follow his ideas in outline, though not in detail, when treating that case.

We first introduce some notation. If $\mathscr{W}$ is a given scalar function of the variables x, u, p we shall write

$$\mathscr{W}_x = (\partial\mathscr{W}/\partial x_1, \ldots, \partial\mathscr{W}/\partial x_n), \quad \mathscr{W}_u = \partial\mathscr{W}/\partial u, \quad \mathscr{W}_p = (\partial\mathscr{W}/\partial p_1, \ldots, \partial\mathscr{W}/\partial p_n).$$

Also for $p \neq 0$ we put
$$\dot{\mathscr{W}} = \dot{\mathscr{W}}(x, u, p) = \sigma \cdot \mathscr{W}_x + |p|\mathscr{W}_u \tag{39}$$

where $\sigma = p/|p|$. When $\mathscr{W}$ is a vector or matrix function it is convenient to retain the same notation, with an obvious interpretation.

We shall require two further invariants of (1), namely

$$\mathscr{E}^* = \frac{\mathscr{E}}{|p|^2} \quad \text{and} \quad \mathscr{D} = \frac{\mathscr{C}}{1-\mathscr{E}^*}, \quad p \neq 0.$$

Note that $\mathscr{E}^* < 1$ since trace $\mathscr{A} = 1$. For the remainder of the section it will be supposed that $n = 2$. The following result then holds.

Theorem 1. *Suppose that* $\dot{\mathscr{D}} + \mathscr{D}^2 \geqslant 0$ *for* $|p| \geqslant L'$.

Let $u \in C^1(\overline{\Omega}) \cap C^3(\Omega)$ *be a solution of* (1) *in* Ω, *such that* $|Du| \leqslant L$ *on* $\partial\Omega$. *Then* $|Du| \leqslant \max(L, L')$ *in* Ω.

Proof. Put $w = |Du|^2$ and let Ω' denote the open subset of Ω where $w \neq 0$. We claim that

$$\mathscr{A}D^2 w = 2(1-\mathscr{E}^*)(\dot{\mathscr{D}} + \mathscr{D}^2)w + \mathscr{M} \cdot Dw, \qquad x \in \Omega'$$

where the arguments of $\mathscr{A}$, $\mathscr{D}$, and $\mathscr{E}^*$ are x, u, Du and where $\mathscr{M} = \mathscr{M}(x)$ is a continuous function on Ω'.

Before proving this identity, we note that the result of the theorem follows from it as an immediate consequence of the strong maximum principle.

In deriving the identity, it is convenient to use standard component notation for partial derivatives. Writing (1) in the form
$$\mathscr{A}_{ij} u_{ij} = \mathscr{B}$$

and applying the operator $u_k(\partial/\partial x_k)$ to both sides, there results

$$\mathscr{A}_{ij} u_k u_{ijk} + |Du|\, \dot{\mathscr{A}}_{ij} u_{ij} + \frac{\partial\mathscr{A}_{ij}}{\partial p_l} u_k u_{kl} u_{ij} = |Du|\, \dot{\mathscr{B}} + \frac{\partial\mathscr{B}}{\partial p_l} u_k u_{kl}.$$

Now $w = u_k u_k$, whence
$$w_i = 2u_k u_{ki}, \quad w_{ij} = 2u_k u_{kij} + 2u_{ki} u_{kj}. \tag{40}$$

Inserting these expressions into the preceding line gives

$$\tfrac{1}{2}\mathscr{A}_{ij}w_{ij} = \mathscr{A}_{ij}u_{ik}u_{jk} + |Du|\,(\dot{\mathscr{B}} - \dot{\mathscr{A}}_{ij}u_{ij}) + \tfrac{1}{2}w_l\left(\frac{\partial\mathscr{B}}{\partial p_l} - \frac{\partial\mathscr{A}_{ij}}{\partial p_l}u_{ij}\right). \tag{41}$$

Let us further temporarily introduce the classical (p, q, r, s, t) notation for the partial derivatives of functions of two independent variables; that is, we put

$$p = u_1, \quad q = u_2, \quad r = u_{11}, \quad s = u_{12}, \quad t = u_{22}.$$

Then clearly

$$\left.\begin{aligned} pr+qs &= \tfrac{1}{2}w_1,\\ ps+qt &= \tfrac{1}{2}w_2,\\ Ar+2Bs+Ct &= D, \end{aligned}\right\} \tag{42}$$

the last line simply being equation (1) in the classical notation. Setting

$$E = Aq^2 - 2Bpq + Cp^2$$

and solving the linear system (42) for r, s, t, we find that

$$\begin{aligned} r &= q^2D/E + \text{linear function of } Dw,\\ s &= -pqD/E + \text{linear function of } Dw,\\ t &= p^2D/E + \text{linear function of } Dw, \end{aligned}$$

valid for $p^2+q^2 \neq 0$. From these results, it follows that

$$\dot{\mathscr{A}}_{ij}u_{ij} = D\dot{E}/E + \text{linear function of } Dw$$

and

$$\begin{aligned} \mathscr{A}_{ij}u_{ki}u_{kj} &= A(r^2+s^2) + 2B(rs+st) + C(s^2+t^2)\\ &= (r+t)\,(Ar+2Bs+Ct) + (A+C)\,(s^2-rt)\\ &= wD^2/E + \text{linear function of } Dw. \end{aligned}$$

Inserting the preceding expressions into (41) and rearranging terms proves the required identity (note here that $D = \mathscr{B}$ and $E = (p^2+q^2)\,(1-\mathscr{E}^*)$).

For the special case $\mathscr{A} = \mathscr{A}(p)$ the result of theorem 1 was obtained by Bernstein (1910, page 129; see also Leray 1939, page 255).

Theorem 1 gives an important method for dominating the gradient of a solution in terms of the gradient at the boundary. Bernstein noted (1912, pages 459–461) that the result could be further refined by a change of dependent variables, though in order to carry out the construction in this case it is also necessary to have an *a priori* estimate for the modulus of the solution itself. Because Bernstein's proof contains a serious inaccuracy (i.e. formula (7) on page 459 applies only if the coefficients of the second derivatives are independent of (x, y, z), while on the other hand in the application of this formula on page 460 the coefficients of the transformed equation definitely involve the variable z), it is worthwhile presenting a complete discussion here. There are two essentially different situations to consider, roughly corresponding to the case when (1) is boundedly non-linear and when it is regularly elliptic. The results may be compared with those of Leray (1939, page 263).

THEOREM 2. *Suppose that $n = 2$ and that for some constant $\mu > 0$ the invariant function $\mathscr{E}$ satisfies the conditions*

$$p\cdot\mathscr{E}_p \leqslant (1-\mu)\,\mathscr{E}, \quad \mathscr{E} \geqslant \mu \tag{43}$$

at least for all sufficiently large values of p, say $|p| \geqslant l$. Assume that as p tends to infinity

$$\mathscr{E}_x, \mathscr{E}_u, \mathscr{E}_p = O(\mathscr{E}) \tag{44}$$

and that (27) *holds with the remainder $\mathscr{R} = \mathscr{C} - \mathscr{C}_0$ satisfying*

$$\mathscr{R}, \mathscr{R}_x, \mathscr{R}_u, |p|\,\mathscr{R}_p = O(\mathscr{E}/|p|). \tag{45}$$

Let $u \in C^1(\overline{\Omega}) \cap C^3(\Omega)$ be a solution of (1) *satisfying $|u| \leqslant m$ in Ω and $|Du| \leqslant L$ on $\partial\Omega$. Then*

$$|Du| \leqslant M \quad in \quad \Omega$$

where M depends only on μ, m, l, L, bounds for the order terms listed above, and on the C^1 norm of the function $\mathscr{C}_0$.

Remarks. Conditions of the type (43), (44), (45) are consistent with the assumption that equation (1) is boundedly non-linear, and indeed one of the main applications of theorem 2 will be to the equation for surfaces having prescribed mean curvature.

Although the hypotheses appear somewhat complicated at first sight, the reader will notice that they amount roughly to the statement that differentiation of the invariants $\mathscr{E}$ and $\mathscr{R}$ with respect to x and u does not alter their order, while differentiation with respect to p operates as if the functions were rational. Thus conditions (44) and (45) are in fact 'natural' assumptions in the sense of Ladyzhenskaya & Uraltseva. This being the case, it is apparent that the crucial additional requirements for theorem 2 are the existence of an interior estimate for $|u|$ and the condition $p \cdot \mathscr{E}_p \leqslant (1-\mu)\,\mathscr{E}$.

Proof of theorem 2. Since $|u| \leqslant m$, we may assume without loss of generality that the order terms in (44) and (45), as well as the norm of the function $\mathscr{C}_0$, are bounded independently of u.

Now introduce a new dependent variable $\bar{u}$ by means of the transformation

$$u = \varphi(\bar{u}) \quad (\varphi'(\bar{u}) > 0).$$

One checks without difficulty that $\bar{u}$ satisfies the equation

$$\overline{\mathscr{A}}\, D^2 \bar{u} = \overline{\mathscr{B}} \tag{46}$$

with

$$\overline{\mathscr{A}} = \mathscr{A} \quad \text{and} \quad \overline{\mathscr{B}} = \varphi'^{-1}(\mathscr{B} + \omega \mathscr{E}),$$

where the arguments of $\mathscr{A}$, $\mathscr{B}$, and $\mathscr{E}$ are x, u, $p = \varphi'(\bar{u})\bar{p}$ and where

$$\omega = -\varphi''/\varphi'^2.$$

In order to apply the result of theorem 1 to the new equation, we note that

$$\overline{\mathscr{C}} = \mathscr{C} + \omega \mathscr{E}/|p|, \quad \overline{\mathscr{E}}^* = \mathscr{E}^*.$$

Moreover, recalling that p depends on $\bar{u}$, an easy calculation yields

$$\bar{\sigma} \cdot \frac{\partial}{\partial x} + |\bar{p}| \frac{\partial}{\partial \bar{u}} = \sigma \cdot \frac{\partial}{\partial x} + |p| \left(\frac{\partial}{\partial u} - \omega p \cdot \frac{\partial}{\partial p} \right).$$

Thus we find, with regard to equation (46),

$$\dot{\overline{\mathscr{D}}} + \overline{\mathscr{D}}^2 = \delta \left(\frac{\mathscr{C} + \omega \mathscr{E}/|p|}{1 - \mathscr{E}^*} \right) + \left(\frac{\mathscr{C} + \omega \mathscr{E}/|p|}{1 - \mathscr{E}^*} \right)^2,$$

where δ is an abbreviation for the differential operator written out in full on the preceding line. The expression $\dot{\overline{\mathscr{D}}}+\overline{\mathscr{D}}^2$ will be non-negative provided that

$$I \equiv (1-\mathscr{E}^*)\,\delta\left(\mathscr{C}+\frac{\omega}{|p|}\mathscr{E}\right)+\left(\mathscr{C}+\frac{\omega}{|p|}\mathscr{E}\right)\delta\mathscr{E}^* \geqslant 0. \tag{47}$$

We shall show that (47) holds for all sufficiently large values of p when

$$\varphi(\bar{u}) = (1/a)\log\bar{u}$$

and a is chosen to be a suitably large constant. For this choice of φ it is easy to see that $\omega = a =$ constant.

Now $\delta\mathscr{E} = \sigma\cdot\mathscr{E}_x+|p|\,\mathscr{E}_u-a|p|p\cdot\mathscr{E}_p$, whence (44) and the first inequality of (43) imply

$$-(C+(1-\mu)\,a)\,|p|\,\mathscr{E} \leqslant \delta\mathscr{E} \leqslant C(1+a)\,|p|^2\mathscr{E} \tag{48}$$

for all p sufficiently large, say $|p| \geqslant \alpha_1$; here we use the letter C as a generic bound depending only on the order terms in (44) and (45) and the norm of the function $\mathscr{C}_0$. Similarly, from (45) and the second inequality of (43),

$$\delta\mathscr{C} = \delta\mathscr{C}_0+\delta\mathscr{R} \geqslant \sigma\cdot\frac{\partial\mathscr{C}_0}{\partial x}+\delta\mathscr{R} \geqslant -C(\mu^{-1}+1+a)\,\mathscr{E} \tag{49}$$

provided $|p| \geqslant \alpha_2$; here we have used the facts that $\partial\mathscr{C}_0/\partial u \geqslant 0$ and $p\cdot\partial\mathscr{C}_0/\partial p = 0$. From (48) and (49) we obtain

$$\delta(\mathscr{C}+a\mathscr{E}/|p|) \geqslant (\mu a^2-(1+2a+\mu^{-1})\,C)\,\mathscr{E}$$

and also, since $\delta\mathscr{E}^* = |p|^{-2}\delta\mathscr{E}+2a\mathscr{E}/|p|$,

$$((1+\mu)\,a-C)\frac{\mathscr{E}}{|p|} \leqslant \delta\mathscr{E}^* \leqslant \left(\frac{2a}{|p|}+(1+a)\,C\right)\mathscr{E}.$$

We now write $\mathscr{C}+a\mathscr{E}/|p| = \mathscr{C}_0+(\mathscr{R}+a\mathscr{E}/|p|)$ and estimate the two parts separately: thus

$$\mathscr{C}_0 \geqslant -C, \quad \mathscr{R}+\frac{a\mathscr{E}}{|p|} \geqslant (a-C)\frac{\mathscr{E}}{|p|}$$

for $|p| \geqslant \alpha_3$.

Let us assume a final choice of a satisfying $a > C$. Using the estimates of the previous paragraph, the function I defined by (47) then obeys

$$I \geqslant \mathscr{E}[(1-\mathscr{E}^*)\,(\mu a^2-(1+2a+\mu^{-1})\,C)+\mathscr{E}^*(a-C)\,((1+\mu)\,a-C)-(2aC+(1+a)\,C^2)]$$

provided $|p| \geqslant \max(\alpha_1,\alpha_2,\alpha_3,1) \equiv L'$. Assuming $\mu < 1$, as is certainly allowable, it is easy to see that

$$I \geqslant \mathscr{E}[\mu a^2-(4C+C^2)\,a-(1+\mu^{-1})\,C-C^2].$$

Now choose a (necessarily greater than C) so that the quantity in square brackets vanishes.

We have thus shown that $\dot{\overline{\mathscr{D}}}+\overline{\mathscr{D}}^2 \geqslant 0$ provided $|\bar{p}| \geqslant L'/\min\varphi'$. Therefore by theorem 1

$$|D\bar{u}| \leqslant \max(L,L')/\min\varphi'$$

and hence
$$|Du| \leqslant \frac{\max\varphi'}{\min\varphi'}\max(L,L') \leqslant \max(L,L')\exp(2am),$$

completing the proof of theorem 2.

Remark. The hypothesis $\mathscr{E} \geqslant \mu$ was used during the proof only to estimate the term $\partial\mathscr{C}_0/\partial x$ in inequality (49). Consequently, this condition can be dropped from (43) if $\mathscr{C}_0$ is independent of x.

The following result is the analogue of theorem 2 for the case of regular ellipticity. Again the crucial assumption is an *a priori* estimate for $|u|$.

THEOREM 3. *Suppose that* $n = 2$ *and that for some constant* $\mu > 0$ *the invariant function* $\mathscr{E}^*$ *satisfies the condition*

$$\mathscr{E}^* \leqslant 1-\mu. \tag{50}$$

Assume further that

$$\mathscr{B}, \mathscr{B}_x, \mathscr{E}_x = O(\mathscr{E}) \tag{51}$$

and

$$\mathscr{B}_u, \mathscr{B}_p, \mathscr{E}_u, \mathscr{E}_p = O(\mathscr{E}/|p|). \tag{52}$$

Let $u \in C^1(\overline{\Omega}) \cap C^3(\Omega)$ *be a solution of* (1) *satisfying* $|u| \leqslant m$ *in* Ω *and* $|Du| \leqslant L$ *on* $\partial\Omega$. *Then*

$$|Du| \leqslant M \quad in \quad \Omega,$$

where M *depends only on* μ, m, L, *and bounds for the order terms listed above.*

Proof. As in the preceding proof we can assume without loss of generality that the order terms are bounded independently of u. Moreover, as in that proof, it will be enough to show that (47) holds for an appropriate choice of the function $\varphi(\bar{u})$ and for suitably large values of p.

Now assume $\omega > 0$. Then

$$\delta\mathscr{C} \geqslant -C(|p|^{-1}+\omega)\,\mathscr{E}, \quad |\delta\mathscr{E}| \leqslant C(1+\omega|p|)\,\mathscr{E}$$

for $|p| \geqslant \alpha_1$, where C is a generic bound depending only on the order terms in (51) and (52). Consequently, since

$$\delta\omega = |p|\frac{\omega'}{\varphi'} \quad \text{and} \quad \delta|p|^N = -N\omega|p|^{N+1},$$

we have

$$\delta\left(\mathscr{C}+\frac{\omega}{|p|}\mathscr{E}\right) \geqslant \left(\frac{\omega'}{\varphi'}+\omega^2-(1+\omega)\,(|p|^{-1}+\omega)\,C\right)\mathscr{E}$$

and

$$|\delta\mathscr{E}^*| \leqslant (2\omega+(|p|^{-1}+\omega)\,C)\frac{\mathscr{E}}{|p|}.$$

Since also

$$\left|\mathscr{C}+\frac{\omega}{|p|}\mathscr{E}\right| \leqslant (C+\omega)\frac{\mathscr{E}}{|p|} \quad (|p| \geqslant \alpha_2)$$

we obtain the following estimate for I:

$$\begin{aligned} I &\geqslant \mathscr{E}[(1-\mathscr{E}^*)\,(\omega'/\varphi'+\omega^2-(1+\omega)\,(|p|^{-1}+\omega)\,C)-\mathscr{E}^*(C+\omega)\,(2\omega+(|p|^{-1}+\omega)\,C)] \\ &\geqslant \mu\mathscr{E}[\omega'/\varphi'-A(\omega^2+\omega+|p|^{-1})] \end{aligned}$$

where in the last step we have used (50) and put

$$A = (2+C)^2/\mu;$$

here $|p| \geqslant \max(\alpha_1, \alpha_2, 1)$.

We now choose $\varphi(\bar{u})$ according to the inverse relation

$$\bar{u} = \int_{-2m-1}^{\varphi-m-1} \frac{\mathrm{d}t}{(1-\mathrm{e}^{2At})^{1/2A}}. \tag{53}$$

Since u varies in the range $(-m, m)$ the integral is clearly well defined and $\varphi'(\bar{u}) > 0$.

Chapter IV

In this chapter we shall give an extended discussion of various examples and special cases in order to provide concrete illustrations of the general theory developed earlier. Sections 22 through 24 are devoted to the case of surfaces having prescribed mean curvature, where the results have special geometric significance beyond their simple interest as examples.

20. The minimal surface operator

Here we shall consider various equations arising when the left-hand side of (1) has the form

$$[(1+|Du|^2)\,I-DuDu]\,D^2u$$

as in the case of the minimal surface equation. We note that the coefficient matrix $\mathscr{A}$, after normalization to unit trace, is

$$\frac{(1+|p|^2)\,I-pp}{n+(n-1)\,|p|^2}$$

and that the corresponding invariants $\mathscr{H}$ and $\mathscr{I}$ are given by the simple relation (cf. §9)

$$\mathscr{H}=\mathscr{I}=H$$

where H devotes the ordinary mean curvature of the boundary surface in question.

Perhaps the most important case concerns surfaces of constant mean curvature. As is well known, a smooth surface in $(n+1)$-dimensional Euclidean space with the non-parametric representation $z=u(x)$ has mean curvature Λ provided that the function $u=u(x)$ satisfies

$$[(1+|Du|^2)\,I-DuDu]\,D^2u=n\Lambda(1+|Du|^2)^{\frac{3}{2}}.$$

The corresponding Dirichlet problem in a given domain Ω has the geometric content that the surface $z=u(x)$ is defined over Ω and takes on given ordinate values on the boundary set $\partial\Omega$. In particular, when $n=2$ the surface $z=u(x,y)$ spans a given simple closed curve in the three dimensional (x, y, z)-space.

As already remarked in the introduction of the paper, the following result holds:

Let Ω be a bounded domain in n-dimensional Euclidean space, whose boundary is of class C^2. Then the Dirichlet problem in Ω for surfaces of constant mean curvature is solvable for arbitrarily given boundary data of class C^2, if and only if the mean curvature H of the boundary surface $\partial\Omega$ satisfies the condition

$$H\geqslant\frac{n}{n-1}|\Lambda| \tag{93}$$

at each point of the boundary. The solution is unique if it exists.

Proof. We first show that (93) is necessary. Suppose in fact that (93) fails at a single point. If $\Lambda\geqslant 0$, then theorem 18·1 shows that there exists boundary data having arbitrarily small absolute value for which no solution of the Dirichlet problem is possible. If $\Lambda<0$ the same proof applies, if we merely replace u with $-u$.

The proof of sufficiency is more difficult. We shall give here an argument applicable only in two dimensions. The proof for higher dimensions will be deferred until §22, where in fact we shall derive an even stronger result.

Suppose then that $n = 2$. In this case (93) implies that Ω is contained in a ball of radius $1/2|\Lambda|$. The hypotheses of theorem 6·3 are then satisfied, and consequently the solvability of the Dirichlet problem is guaranteed by theorem 14·3. The uniqueness of solutions follows from theorem 6·5, whatever the dimension. This completes the demonstration of the theorem, subject of course to the deferred existence proof in higher dimensions.

As we have already remarked, when $n = 2$ condition (93) implies that Ω is convex and contained in a ball of radius $1/2|\Lambda|$. In higher dimensions no such simple conclusions are available, and (93) can in fact be satisfied even by domains which are not simply connected. For example, the torus generated by revolving the disk $(x-a)^2+y^2 = b^2$ about the y axis satisfies the condition whenever $b < 1/3|\Lambda|$ and $a \geqslant 2b+3|\Lambda|b^2/(1-3|\Lambda|b)$.

The minimal surface operator also appears in the equation of Laplace and Gauss governing the free surface of a stationary fluid under the combined action of gravity and surface tension. The following result holds:

Let Ω be a bounded domain in n-dimensional Euclidean space, whose boundary is of class C^2. Then the Dirichlet problem in Ω for the Gauss–Laplace equation

$$[(1+|Du|^2)I-DuDu]\,D^2u = cu(1+|Du|^2)^{\frac{3}{2}}, \quad c>0 \tag{94}$$

is uniquely solvable for given C^2 boundary values f provided that

$$|f| \leqslant (n-1)\,H/c \tag{95}$$

at each point of the boundary. Moreover, condition (95) is best possible.

Proof. Existence follows directly from theorem 14·3 (with theorem 6·2 used at the final step.) By saying that (95) is best possible we mean that, given any domain Ω and any positive number ϵ, then the following statements hold:

(A) If $H < 0$ at a single point of the boundary there exists smooth boundary data g such that $|g| \leqslant \epsilon$ and yet for which no solution of the Dirichlet problem for (94) exists.

(B) If $H \geqslant 0$ at each point of the boundary there exists smooth boundary data g such that $|g| \leqslant (n-1)\,H/c+\epsilon$ and yet for which no solution of the Dirichlet problem exists.

The first assertion follows by setting $f = 0$ and using theorem 18·2 (together with the fact that its conclusion remains true even if (85) holds only at a single point P of the boundary where f takes its greatest value).

In case (B), let P be an arbitrary point on the boundary and let smooth boundary values f be assigned such that $\quad f = (n-1)\,H/c+\frac{1}{2}\epsilon \quad \text{at} \quad P$

and $\quad f \leqslant \min\{f(P), (n-1)\,H/c+\frac{1}{2}\epsilon\}$ elsewhere.

The conclusion then follows exactly as in case (A), completing the proof of the theorem.

Corollary. *Let Ω be a ball of radius R. Then the Dirichlet problem above is solvable for arbitrarily given C^2 boundary values f satisfying the condition $|f| \leqslant (n-1)/cR$. Conversely, if $|f| > (n-1)/cR$ at a single point of the boundary, then the Dirichlet problem is not well set.*

It would be interesting to know similarly whether the failure of condition (94) at a single point implies the existence of nearby data for which the Dirichlet problem cannot be solved.

Because of their general interest we shall consider two further equations,

$$[(1+|Du|^2)\,I-DuDu]\,D^2u = n\Lambda(1+|Du|^2)^{\frac{1}{2}\theta} \tag{96}$$

and

$$[(1+|Du|^2)\,I-DuDu]\,D^2u = cu(1+|Du|^2)^{\frac{1}{2}\sigma} \tag{97}$$

where Λ, c, and θ are constants, $c > 0$. The case $\theta = 3$ has already been discussed in the foregoing work. We may therefore restrict the following treatment to values $\theta \neq 3$.

When $\theta > 3$ it is evident that both equations are irregularly elliptic; accordingly the Dirichlet problem is not well posed in that case, whatever the domain.

When $\theta < 3$ a maximum principle is available for both equations (theorems 6·3 and 6·2), and $\mathscr{C}_0 = 0$. Also when $\theta \leqslant 2$ both equations are boundedly non-linear. Finally theorems 12·1 and 13·1 are applicable, with

$$\dot{\mathscr{D}} = 0, \qquad \dot{\mathscr{S}} = 0$$

and

$$\dot{\mathscr{D}} = c\,\frac{(1+|p|^2)^{\frac{1}{2}\theta-1}}{n-1} > 0, \quad \dot{\mathscr{S}} = c\,\frac{(1+|p|^2)^{\frac{1}{2}\theta-1}}{n} > 0,$$

respectively. The argument of theorem 14·3 therefore may be applied to prove the existence of solutions of the Dirichlet problem for (96) and (97) when $\theta < 3$.

On the other hand, according to theorem 18·1 the Dirichlet problem for $\theta < 3$ cannot generally be solved if $H < 0$ at some point of the boundary. A slight modification of the proof of the antecedent lemma 17·3 in fact allows this result to be strengthened: when $2 < \theta < 3$ the Dirichlet problem cannot always be solved if $H = 0$ at some point on the boundary (for this modification alone, we make the tacit assumption that the boundary of Ω is of class C^3 in the neighbourhood of points where $H = 0$). Combining the preceding results, we obtain the following theorem.

Let Ω be a bounded domain in n dimensional Euclidean space, whose boundary is of class C^2. Then the Dirichlet problem for equations (96) *and* (97) *in Ω is solvable for arbitrarily given C^2 boundary values under the following circumstances:*

(*i*) *when $\theta \leqslant 2$, if and only if $H \geqslant 0$ at each point of the boundary, and*

(*ii*) *when $2 < \theta < 3$, if and only if $H > 0$ at each point of the boundary.*

When $\theta > 3$ the Dirichlet problem is not generally solvable, whatever the domain.

In a short paper published in 1910 Bernstein listed both equations (96) and (97) as examples (though with n equal to 2 and θ restricted to *integer* values), and showed that the corresponding Dirichlet problems are solvable for arbitrary (analytic) boundary data in an arbitrary strictly convex (analytic) domain if and only if $\theta \leqslant 2$.

In the same paper Bernstein introduced a further pair of equations, namely

$$(1+u_x^2)\,u_{xx} + 2u_x u_y u_{xy} + (1+u_y^2)\,u_{yy} = (1+u_x^2+u_y^2)^{\frac{1}{2}\theta}$$

and

$$(1+u_x^2)\,u_{xx} + 2u_x u_y u_{xy} + (1+u_y^2)\,u_{yy} = u(1+u_x^2+u_y^2)^{\frac{1}{2}\theta},$$

with the parameter θ once more being restricted to integer values. A recent note of mine (1967, page 1834) called attention to the n-dimensional generalizations of these equations, and gave precise conditions for the solvability of the corresponding Dirichlet problems. The interested reader should be able to supply the necessary justification of those results without undue difficulty.

21. Variational equations

Both the equation for surfaces having constant mean curvature and the equation for fluid interfaces under the action of gravity and surface tension can be obtained from an appropriate variational problem. It is the purpose of this section to present several further variational equations of interest, including higher dimensional generalizations of the problems introduced by Bernstein in his memoir on the calculus of variations (1912).

We begin with the example

$$\delta \int \sqrt{(1+|x|^2+|Du|^2)}\, dx = 0,$$

for which the corresponding Euler–Lagrange equation has the form

$$[(1+|x|^2+|Du|^2)\, I - DuDu]\, D^2u = x \cdot Du.$$

By theorems 14·3 and 18·1 the Dirichlet problem in a bounded domain Ω with C^2 boundary is solvable for arbitrary boundary values of class C^2 if and only if $H \geqslant 0$ at each point of the boundary. By theorem 6·5 the solution is unique if it exists.

Consider next the problem

$$\delta \int \sqrt{(1+u^2+|Du|^2)}\, dx = 0,$$

whose Euler–Lagrange equation has the form

$$[(1+u^2+|Du|^2)\, I - DuDu]\, D^2u = u(1+u^2+2|Du|^2).$$

Again by theorems 14·3 and 18·1 the Dirichlet problem is solvable for arbitrarily given boundary values if and only if $H \geqslant 0$ at each point of the boundary (note in particular that an *a priori* bound for $|u|$ is given by theorem 6·2). Uniqueness of the solution follows from theorem 6·6.

In both preceding cases the slight modification of the integrand from the area expression $\sqrt{(1+|Du|^2)}$ has no effect on the type of Dirichlet problem which can be solved. A rather different situation arises for the variational problem

$$\delta \int \phi(x) \sqrt{(1+|Du|^2)}\, dx = 0$$

where $\phi(x)$ is a given positive function of class C^2. The corresponding Euler–Lagrange equation is

$$[(1+|Du|^2)\, I - DuDu]\, D^2u = -\frac{D\phi \cdot Du}{\phi}(1+|Du|^2);$$

thus while the left-hand side involves only the minimal surface operator, the right-hand side yields the relation

$$\mathscr{C}_0 = -\frac{1}{n-1}\frac{\sigma \cdot D\phi}{\phi}.$$

Hence the geometric conditions required for the solvability of Dirichlet's problem necessarily differ from those for the minimal surface equation. In particular, by theorem 14·3 the Dirichlet problem in a bounded domain Ω with C^2 boundary is solvable for arbitrary boundary values of class C^2 provided that

$$H \geqslant \frac{1}{n-1}\frac{\nu \cdot D\phi}{\phi} \tag{98}$$

at each point of the boundary. It can be shown that this condition is also necessary for the solvability of the Dirichlet problem for arbitrary data, though we shall not do so here.

The special case $\phi = \sqrt{(1+|x|^2)}$ was remarked by Bernstein (1912, page 479). If for definiteness we take Ω to be convex and to contain the origin, *then* (98) *is always satisfied.*

62-2

When Ω is a ball of radius R whose centre lies at a distance S from the origin, (98) becomes

$$\frac{1}{R} \geqslant \frac{1}{n-1} \frac{S^2-R^2-|x|^2}{2R(1+|x|^2)},$$

that is
$$2(n-1)+(2n-1)|x|^2 \geqslant S^2-R^2.$$

The minimum value of $|x|$ encountered on the surface of the ball is $|R-S|$. Thus we obtain the following solvability condition for the Dirichlet problem in this special case,

$$(S-R)\left(\frac{n}{n-1}R-S\right) \leqslant 1;$$

in particular the problem is solvable whenever

(1) Ω contains the origin, or
(2) the radius of Ω is less than $2(n-1)$, or
(3) the centre of Ω lies in a ball of radius $2\sqrt{\{n(n-1)\}}$ about the origin.

This result serves to emphasize the delicacy of the situation when any but the simplest equations are treated. We observe specifically that even if the Dirichlet problem is solvable for a given ball Σ, it need not be solvable for all balls contained in Σ, nor even for all smaller concentric balls!

22. Surfaces having prescribed mean curvature

In this section we shall generalize the result of § 20 to include cases where the surface $z = u(x)$ has a prescribed mean curvature $\Lambda(x)$ of class C^1.

An example will make clear that beyond the simple geometric condition (93), some additional requirements on the behaviour of $\Lambda(x)$ are necessary in order for the Dirichlet problem to be solvable. Consider in particular a smooth function $\Lambda(x)$ which is identically 2 when $|x| \leqslant \frac{1}{2}$ and vanishes identically when $|x| \geqslant 1$. Then (93) is satisfied if Ω is the unit ball, but according to a result of Bernstein (1910, page 243) no solution of the Dirichlet problem can possibly exist, no matter what boundary data is assigned.

In the following result we shall give a family of sufficient conditions, depending on a parameter a, such that the Dirichlet problem is solvable for a given domain Ω whenever $\Lambda(x)$ satisfies any one of the conditions in the family.

For points y on the boundary of Ω, let $k_1(y), \ldots, k_{n-1}(y)$ be the principal curvatures of the boundary surface, listed in decreasing order of size. Let $d = d(x)$ denote as usual the distance to the boundary of Ω, and let Ω_1 be the (open) subset of Ω consisting of points x such that

(1) the closed ball of radius $d(x)$ centred at x touches the boundary of Ω at a single point $y = y(x)$, and

(2) $d(x) < 1/k_1(y)$.

For x in $\bar{\Omega}$ we can now define

$$H(x) = \begin{cases} \dfrac{1}{n-1}\sum \dfrac{k_i(y)}{1-k_i(y)\,d(x)}, & x \in \Omega_1 \cup \partial\Omega, \\ \infty, & x \notin \Omega_i \cup \partial\Omega. \end{cases}$$

Clearly $H(x)$ agrees with the ordinary mean curvature of the boundary surface for points x on $\partial\Omega$. Moreover, although the fact is unnecessary for our present purposes, we remark that when the boundary is of class C^3 the function $H(x)$ can be interpreted as the mean curvature of the level surface $d =$ constant passing through the point x.

The maximum value of $d(x)$ in Ω will be called the internal radius of Ω, and will be denoted by d_1. We can now state the main result of the section:

Let Ω be a bounded domain in n-dimensional Euclidean space, whose boundary is of class C^2. Let $\Lambda(x)$ be of class C^1 in the closure of Ω, and suppose that there is some constant a, $0 \leqslant a \leqslant 1/d_1$, such that

$$\frac{n}{n-1}|\Lambda(x)| \leqslant (1-ad)\,H(x)+\frac{a}{n-1} \quad \textit{in} \quad \Omega. \tag{99}$$

Then the Dirichlet problem in Ω for the equation

$$[(1+|Du|^2)\,I-DuDu]\,D^2u = n\Lambda(x)\,(1+|Du|^2)^{\frac{3}{2}}$$

is solvable for arbitrarily given C^2 boundary values if and only if

$$H \geqslant \frac{n}{n-1}|\Lambda(y)|$$

at each point y of the boundary. The solution is unique if it exists.

The proof is somewhat different in the three cases $0 < a < 1/d_1$, $a = 1/d_1$, and $a = 0$, and accordingly we shall consider these situations separately.

Case 1. $0 < a < 1/d_1$

Supposing that the boundary of Ω is of class C^3, we assert that

$$|u| \leqslant \sup|f|+1/a \quad \text{in} \quad \Omega. \tag{100}$$

Assuming this for the moment, the result then follows exactly as in the proof of theorem 14·3, with (100) taking the place of the final hypothesis on the coefficient $\mathscr{B}$.

It is therefore sufficient to prove (100). Let

$$h(d) = \surd\{2ad-(ad)^2\}/a, \quad 0 \leqslant d < 1/a.$$

Since $0 \leqslant h < 1/a$ the assertion will follow provided we can establish the relation

$$|u(x)| \leqslant \sup|f|+h(d) \quad d = d(x), \tag{101}$$

for all x in Ω.

Suppose for contradiction that (101) fails at some point where $u > 0$. Since (101) necessarily holds on the boundary of Ω, it follows that there exists a positive constant κ and a point $\bar{x}$ in Ω such that

$$u(x) \leqslant \sup|f|+h(d)+\kappa \quad \text{in} \quad \Omega \tag{102}$$

with equality holding at $\bar{x}$. There are now two cases to consider, depending on whether $\bar{x}$ is in the set Ω_1 or in its complement.

I. $\bar{x} \in \Omega_1$. According to the work of § 3, one sees easily that relations (8) hold and $d(x)$ is of class C^2 in Ω_1. Hence, letting $v(x)$ denote the function on the right-hand side of (102), we find by direct computaion (as in the proof of lemma 4·1) that

$$\begin{aligned}\frac{[(1+|Dv|^2)\,I-DvDv]\,D^2v}{(1+|Dv|^2)^{\frac{3}{2}}} &= \frac{h''}{(1+h'^2)^{\frac{3}{2}}}-\frac{h'}{(1+h'^2)^{\frac{1}{2}}}\sum\frac{k_i(y)}{1-k_i(y)\,d(x)}\\ &= -a-(1-ad)\,(n-1)\,H(x) \leqslant n\Lambda(x)\end{aligned}$$

for x in Ω_1. With the help of a standard maximum principle argument this implies $u \equiv v$ in Ω_1, which is impossible since $u < v$ on $\partial\Omega$.

II. $\bar{x} \notin \Omega_1$. Let $\bar{y}$ be any point where the closed ball of radius $d(\bar{x})$ centred at $\bar{x}$ touches the boundary of Ω. Then for any point x on the closed line segment joining $\bar{y}$ to $\bar{x}$ we have

$$d(x) = |x-\bar{y}|.$$

Since $h' > 0$, it follows from (102) that $Du \neq 0$ at $\bar{x}$. The level set $u = \text{constant} = \bar{u}$ passing through $\bar{x}$ is consequently a surface of class C^2 in the neighbourhood of $\bar{x}$. This being the case, there exists a closed ball Σ tangent to this surface at $\bar{x}$, such that $u \geqslant \bar{u}$ in Σ. But then by (102) it is clear that $d \geqslant d(\bar{x})$ in Σ, and *a fortiori* $d \geqslant d(\bar{x})$ on $\partial\Sigma$.

If the centre of Σ is x_0 and its radius σ_0, then according to the geometric meaning of $d(x)$ the ball of radius $d(\bar{x})+\sigma_0$ centred at x_0 must be contained in $\bar{\Omega}$. This implies first of all that x_0 is on the extension of the line segment joining $\bar{y}$ to $\bar{x}$, and in turn that $\bar{x}$ is in Ω_1. This is contrary to assumption, and accordingly we have proved that (101) must hold whenever $u > 0$. Since a similar argument applies to points where $u < 0$, this completes the proof of (101).

Case 2. $a = 1/d_1$

Let $\bar{a}$ be an arbitrary constant, $1/2d_1 < \bar{a} < 1/d_1$. If Λ satisfies (99) for $a = \bar{a}$, then by what has already been shown in case 1,

$$|u| \leqslant \sup|f| + 2d_1 \quad \text{in} \quad \Omega$$

(provided as always that the boundary of Ω is of class C^3). The solvability of the Dirichlet problem when $a = 1/d_1$ can now be established as in the proof of theorem 14·3, since each approximating problem obviously can be chosen so that (99) holds for some $a < 1/d_1$.

Case 3. $a = 0$

As in the proof of case 1, it is sufficient to establish an *a priori* estimate for $|u|$ when the boundary of Ω is of class C^3. To this end, we put

$$v(x) = \sup|f| + h(d), \quad 0 \leqslant d \leqslant d_1,$$

where

$$h(d) = \frac{e^{cd_1}}{c}(1-e^{-cd}), \quad d = d(x),$$

and $c = n \sup|\Lambda(x)|+1$. Now for $x \in \Omega_1$

$$\begin{aligned}\frac{[(1+|Dv|^2)I - DvDv]D^2v}{(1+|Dv|^2)^{\frac{3}{2}}} &= \frac{h''}{(1+h'^2)^{\frac{3}{2}}} - \frac{h'}{(1+h'^2)^{\frac{1}{2}}}(n-1)H(x) \\ &\leqslant \frac{h''}{(1+h'^2)^{\frac{3}{2}}} + \frac{h'}{(1+h'^2)^{\frac{1}{2}}} n\Lambda(x) \\ &\leqslant n\Lambda(x) + \frac{h''}{(1+h'^2)^{\frac{3}{2}}} + c\left\{1 - \frac{h'}{(1+h'^2)^{\frac{1}{2}}}\right\} \\ &\leqslant n\Lambda(x) + \frac{h''+ch'}{(1+h'^2)^{\frac{3}{2}}} = n\Lambda(x),\end{aligned}$$

the second to last step holding by virtue of the fact that $h' \geqslant 1$. This being shown, the proof of case 1 carries over with only minor changes to the present situation. Thus we have

$$|u(x)| \leqslant \sup|f| + h(d) \leqslant \sup|f| + (e^{cd_1}-1)/c$$

in Ω. This completes the proof of case 3, and consequently also of the main result.

The necessity of the condition $H \geqslant (n/n-1)|\Lambda|$ follows directly from theorem 18·1 in case Λ does not change sign. If Λ takes on both positive and negative values the argument is somewhat more complicated but follows the same basic lines (we shall omit the details). Finally, the uniqueness of the solution is an immediate consequence of theorem 6·5.

The result of case 3 is worth stating as a separate corollary because of its special interest.

Let Ω be a bounded domain in n-dimensional Euclidean space, whose boundary is of class C^2. Then the Dirichlet problem in Ω for surfaces having prescribed mean curvature $\Lambda(x)$ is solvable for arbitrarily given C^2 boundary values provided that

$$|\Lambda(x)| \leqslant \frac{n-1}{n} H(x)$$

at each point x in the closure of Ω. The solution is unique if it exists.

Since $H(x)$ obviously takes its minimum value on the boundary of Ω, this corollary clearly includes as a special case the theorem stated in § 20. The results of § 20 are thus completely proved.

The case $a = 1/d_1$ of the main result is also of individual interest, and yields the sufficient condition:

$$H \geqslant \frac{n}{n-1}|\Lambda(y)|, \quad y \text{ on } \partial\Omega$$

and

$$|\Lambda(x)| \leqslant \frac{1}{nd_1} + \left(1-\frac{1}{n}\right)\left(1-\frac{d}{d_1}\right) H(x), \quad x \text{ in } \Omega.$$

By dropping the final term on the right-hand side we get the simple, though not especially strong, sufficient condition $|\Lambda(x)| \leqslant 1/nd_1$.

It is unfortunately not particularly easy to give an explicit formula for $H(x)$ for any but the simplest domains. On the other hand, for a given domain Ω it is always possible to employ in (99) the function $\tilde{H}(x)$ corresponding to any domain $\tilde{\Omega}$ containing Ω. A case of particular importance occurs when $\tilde{\Omega}$ is a ball of radius R. If r denotes distance from the centre of $\tilde{\Omega}$, then

$$d(x) = R - r, \quad \tilde{H}(x) = 1/r$$

and

$$(1-ad)\,\tilde{H}(x) + \frac{a}{n-1} = \frac{1-aR}{r} + \frac{an}{n-1}.$$

We have proved the following result:

Let Ω be a domain of class C^2 in n-dimensional Euclidean space, contained in a ball of radius R. Let $\Lambda(x)$ be of class C^1 in the closure of Ω, and suppose that there is some constant a, $0 \leqslant a \leqslant 1/R$, such that

$$|\Lambda(x)| \leqslant a + \frac{n-1}{n}\,\frac{1-aR}{r} \quad \text{in } \Omega. \tag{99'}$$

Then the Dirichlet problem in Ω for surfaces having prescribed mean curvature $\Lambda(x)$ is solvable for arbitrarily given boundary values of class C^2 if and only if

$$H \geqslant \frac{n}{n-1}|\Lambda(y)|$$

at each point y of the boundary. The solution is unique if it exists.

It is interesting to observe that *condition* (99′) *is best possible*, in the sense that when $\Lambda(x)$ is subject only to the restrictions

$$|\Lambda(x)| \leq a + \frac{n-1}{n}\frac{1-aR}{r} + \epsilon \quad \text{in} \quad \Omega, \quad |\Lambda(x)| \leq \frac{n-1}{n} H \quad \text{on} \quad \partial\Omega$$

it may not be possible to solve the corresponding Dirichlet problem. To prove this, suppose that Ω is itself a ball of radius R, and consider a smooth non-negative function $\Lambda(x)$ such that

$$\Lambda(x) \leq \frac{n-1}{n}\frac{1}{R} \quad \text{when} \quad r = R$$

and

$$\Lambda(x) \equiv a + \frac{n-1}{n}\frac{1-aR}{r} + \epsilon \quad \text{when} \quad \delta < r < R - \delta$$

where δ will be determined later. Using the identity

$$\frac{[(1+|Du|^2)I - DuDu]D^2u}{(1+|Du|^2)^{\frac{3}{2}}} = \operatorname{div}\left(\frac{Du}{(1+|Du|^2)^{\frac{1}{2}}}\right)$$

we obtain, for any eventual solution u of the Dirichlet problem in Ω,

$$\int_\Omega n\Lambda(x)\,\mathrm{d}x \leq \oint_{\partial\Omega} \frac{|Du|}{(1+|Du|^2)^{\frac{1}{2}}}\,\mathrm{d}s \leq \omega_n R^{n-1},$$

where ω_n denotes the area of the unit sphere in n dimensions. Also

$$\int_\Omega n\Lambda(x)\,\mathrm{d}x > \int_{\delta<r<R-\delta} (n-1)\left(\frac{1-aR}{r} + \frac{an}{n-1} + \epsilon\right)\mathrm{d}x$$
$$> \omega_n\left[\left(\frac{1}{R} + \epsilon\frac{n-1}{n}\right)(R-\delta)^n - \left(\frac{1}{\delta} + \epsilon\frac{n-1}{n}\right)\delta^n\right].$$

Combining the preceding two inequalities and letting $\delta \to 0$ yields

$$\omega_n R^{n-1} \geq \omega_n\left(1 + \epsilon R\frac{n-1}{n}\right)R^{n-1}$$

which is impossible. Hence a contradiction occurs for all sufficiently small values of δ, proving our result.

The value of the preceding result lies in the fact that it does not require a specific calculation of the function $H(x)$ for the given domain Ω. We conclude the section with another theorem having the same property.

Let Ω be a bounded domain in n-dimensional Euclidean space, whose boundary is of class C^2. Let $\Lambda(x)$ be of class C^1 in the closure of Ω, and suppose that

$$|D\Lambda| \leq \frac{n}{n-1}\Lambda^2 \quad \text{in} \quad \Omega. \tag{103}$$

Then the Dirichlet problem in Ω for surfaces having prescribed mean curvature $\Lambda(x)$ is solvable for arbitrarily given C^2 boundary values if and only if

$$H \geq \frac{n}{n-1}|\Lambda(y)| \tag{104}$$

at each point y of the boundary. The solution is unique if it exists.

Proof. We shall show that (103) and (104) imply

$$|\Lambda(x)| \leqslant \frac{n-1}{n} H(x), \quad x \in \bar{\Omega}, \tag{105}$$

which will obviously complete the proof (see case 3 of the main result).

The rate of change of $H(x)$ with respect to d in a direction normal to the boundary is clearly given by

$$\frac{\mathrm{d}H(x)}{\mathrm{d}d} = \frac{1}{n-1} \Sigma \left(\frac{k_i(y)}{1-k_i(y)\, d(x)}\right)^2, \quad x \in \Omega_1.$$

Consequently, letting a prime denote summation over all values of i for which $k_i(y)$ is positive, we have

$$\frac{\mathrm{d}H(x)}{\mathrm{d}d} \geqslant \frac{1}{n-1} \Sigma' \left(\frac{k_i(y)}{1-k_i(y)\, d(x)}\right)^2$$
$$\geqslant \frac{1}{(n-1)^2} \left(\Sigma' \frac{k_i(y)}{1-k_i(y)\, d(x)}\right)^2 \geqslant H(x)^2$$

since $H(x) \geqslant 0$ in the present case because of (104). Thus for x in Ω_1

$$\frac{\mathrm{d}}{\mathrm{d}d}\left\{\frac{n-1}{n} H(x)\right\} \geqslant \frac{n}{n-1}\left\{\frac{n-1}{n} H(x)\right\}^2. \tag{106}$$

If we now integrate both (103) and (106) along normals to the boundary, and make use of (104) as an initial condition, it follows that

$$|\Lambda(x)| \leqslant \frac{n-1}{n} H(x), \quad x \in \Omega_1 \cup \partial\Omega.$$

Since (105) obviously holds at the remaining points of Ω, this completes the proof.

23. Surfaces of prescribed mean curvature in central projection

Let Σ be a smooth surface possessing a one-to-one central projection onto a domain Ω of the unit sphere in $(n+1)$ dimensions. If the radial values of Σ are assigned on the boundary of Ω, and if the mean curvature of Σ is prescribed as a function over Ω, we say that Σ is a solution of the Dirichlet problem for surfaces of prescribed mean curvature (in central projection). We shall adopt the sign convention that the mean curvature of Σ is based on the normal to the surface on the same side as the extended radius vector. The following result then holds.

Let Ω be a domain on the unit sphere in $(n+1)$ dimensions, whose closure is contained in an open hemisphere and whose boundary is of class C^2. Let $\Lambda(x)$ be a non-negative function of class C^1 defined over the closure of Ω. Then the Dirichlet problem in Ω for surfaces having prescribed mean curvature $\Lambda(x)$ (in central projection) is solvable for arbitrarily given radial values f of class C^2 assigned on the boundary of Ω, provided that

$$H_g \geqslant \frac{n}{n-1} \Lambda(y) f(y)$$

at each point y of the boundary, where H_g denotes the geodesic mean curvature of the boundary surface (with respect to the underlying unit sphere). The solution is unique if it exists.

The assigned radial values f are of course necessarily positive, so that (just as in the parallel projection problem studied in the previous section) the mean curvature of the boundary must be non-negative. In particular, when $n = 2$ the restriction of Ω to a hemisphere is already a consequence of the solvability condition.

If $\Lambda(x)$ is identically zero the solution is a minimal surface in central projection. For this case the solvability condition becomes simply

$$H_g \geqslant 0 \tag{107}$$

at each point of the boundary. The case $n = 2$ of this result is due to Radó (1932), his proof resting on the general solution of Plateau's problem for Jordan boundary curves. *We note that since the assigned boundary values do not appear in* (107), *this condition is also necessary.*

For non-constant curvatures it is remarkable that no restrictions need be placed on $\Lambda(x)$ beyond the simple condition that it does not take negative values. In this respect the Dirichlet problem in central projection is much simpler than the Dirichlet problem in parallel projection.

In order to prove the result we must first introduce an appropriate analytical structure for the problem. Let Σ be a surface with one-to-one central projection onto the domain Ω. If P is a point of Σ whose projection on the unit sphere is the point x, we write $v = v(x)$ for the radial distance of P (that is, the distance from P to the centre of the sphere). We next introduce a system of local coordinates over Ω. To this end it can be supposed without loss of generality that Ω lies in the upper hemisphere of the unit sphere, and it is convenient to use stereographic coordinates based on projection from the South pole onto the tangent space at the North pole.

With these coordinates, we may in what follows consider x to be a Cartesian n-tuple $(x_1, \ldots, x_n)$ in the tangent space, and we may similarly interpret Ω as a domain in this space.

LEMMA. *The surface Σ has mean curvature $\Lambda(x)$ provided that $v = v(x)$ satisfies the equation*

$$\begin{aligned}[(a^2v^2+|Dv|^2)I-DvDv]\,D^2v &= an\Lambda v(a^2v^2+|Dv|^2)^{\frac{3}{2}} \\ &\quad +\tfrac{1}{2}ax\cdot Dv\{(n-2)\,a^2v^2+(n-1)\,|Dv|^2\}+a^2v\{na^2v^2+(n+1)\,|Dv|^2\}\end{aligned} \tag{108}$$

for x in Ω, where $a = 4/(4+|x|^2)$. Moreover, we have

$$H_g = \frac{H}{a}+\frac{x\cdot\nu}{2}$$

where H denotes the ordinary mean curvature of the boundary of Ω considered as a surface in the tangent space endowed with its Euclidean metric.

This lemma is proved at the end of the section. Granting the result, in order to prove the first part of the theorem we must show that the Dirichlet problem for equation (108) in Ω has a positive solution v for arbitrarily given positive boundary data f of class C^2, provided that

$$H \geqslant a\left(\frac{n}{n-1}\Lambda(y)f-\tfrac{1}{2}y\cdot\nu\right) \tag{109}$$

at each point y of the boundary.

The invariant $\mathscr{C}_0$ for (108) is given by

$$\mathscr{C}_0(x, v, \sigma) = a\left(\frac{n}{n-1}\Lambda(x)\, v + \tfrac{1}{2}\sigma\cdot x\right)$$

and $\partial\mathscr{C}_0/\partial v \geqslant 0$ since $\Lambda(x) \geqslant 0$. Using (109), it is therefore clear that the solvability conditions (62) in theorem 14·3 are satisfied. Consequently, were it not for the singularity in equation (108) when $v = 0$, theorem 14·3 could now be applied to obtain the existence of a solution of the given Dirichlet problem.

This being the case, it is natural to look for some artifice which avoids the fact that (108) is singular when $v = 0$. To this end, let κ be a small positive constant, and consider the modified coefficient matrix

$$(a^2 g(v) + |D^2 v|)\, I - DvDv,$$

where

$$g(v) = \begin{cases} v^2 & \text{for } |v| \geqslant \kappa, \\ \tfrac{1}{2}\kappa^2 & \text{for } |v| \leqslant \tfrac{1}{2}\kappa \end{cases}$$

and $v^2 \leqslant g(v) \leqslant \kappa^2$ otherwise. Let (108) be written in the form $\mathscr{L}v = 0$, and let $\bar{\mathscr{L}}v = 0$ be the corresponding equation with modified coefficient matrix. By theorem 14·3 the Dirichlet problem for $\bar{\mathscr{L}}v = 0$ has a solution v corresponding to the given boundary values f. We assert that v is positive, and, moreover, that if κ is chosen suitably small, then v satisfies the given equation $\mathscr{L}v = 0$ as required.

To prove this, consider a family of comparison functions of the form

$$w(x) = h(r^2), \quad r = |x|,$$

where $h, h', h'' > 0$. Clearly (since $x\cdot Dw \geqslant 0$ and $g(w) \geqslant w^2$)

$$\frac{\bar{\mathscr{L}}w}{a^2 g(w) + |Dw|^2} \geqslant \Delta w - \frac{DwDwD^2w}{a^2 g(w) + |Dw|^2} - an\Lambda w(a^2w^2 + |Dw|^2)^{\frac{1}{2}} - \tfrac{1}{2}(n-1)\, ax\cdot Dw - (n+1)a^2 w$$

$$\geqslant (n-1)\,(2 - ar^2)\, h' - an\Lambda h(a^2h^2 + 4r^2h'^2)^{\frac{1}{2}} - (n+1)\, a^2 h.$$

By assumption the closure of Ω lies in the open upper hemisphere of the unit sphere, and correspondingly we have $r < 2$ there. Thus there exists a constant $\epsilon > 0$ such that

$$ar^2 \leqslant 2 - \epsilon \quad \text{in} \quad \Omega.$$

Hence $\bar{\mathscr{L}}w \geqslant 0$ in Ω if $\quad \epsilon(n-1)\, h' \geqslant (n+1)\, h + n\Lambda h(h^2 + 16h'^2)^{\frac{1}{2}}.$

This inequality, as well as the conditions $h, h', h'' > 0$, is satisfied by the family of functions

$$h = \alpha \exp\left\{\frac{n+2}{\epsilon(n-1)}(r^2 - 4)\right\}, \quad 0 < \alpha \leqslant \kappa_1$$

where κ_1 is a suitably small constant depending only on ϵ, n, and the supremum of Λ in Ω.

We can now prove the assertion. First, we have $v \geqslant 0$ according to theorem 6·2. Then, using the family of comparison functions above, we find by an easy application of the maximum principle of Eberhard Hopf (1927, page 152) that

$$v \geqslant \min(\kappa_1, \kappa_2) \exp\left\{-4\frac{n+2}{\epsilon(n-1)}\right\} \quad \text{in} \quad \Omega$$

where κ_2 (> 0) is the infimum of the given boundary values. Choosing κ to be the constant on the right-hand side of the last inequality completes the proof of the assertion. Hence we have established the existence of a positive solution of (108) taking on any given positive twice continuously differentiable boundary values f.

Uniqueness of the solution is essentially a consequence of a theorem of Alexandrov (1956). In fact, a stronger result holds: if Σ and $\bar{\Sigma}$ both possess one-to-one central projections onto a domain Ω, if the radial values f of Σ and $\bar{f}$ of $\bar{\Sigma}$ on the boundary of Ω satisfy $f \leqslant \bar{f}$, and if the mean curvatures $\Lambda(x)$ and $\bar{\Lambda}(x)$ finally satisfy the condition $\Lambda(x) \geqslant \bar{\Lambda}(x)$, then the radial values of Σ never exceed the radial values of $\bar{\Sigma}$. For completeness we include a proof of this result which is independent of the work of Alexandrov (though it uses the same basic idea).

Let $v(x)$ and $\bar{v}(x)$ be the radial values in question, and suppose for contradiction that $v(x)$ exceeds $\bar{v}(x)$ at some point x in Ω. Then there exists a constant $\bar{\alpha} > 1$ and a point $\bar{x}$ in Ω such that

$$v(x) \leqslant \bar{\alpha}\bar{v}(x) \quad \text{in} \quad \Omega, \qquad v(\bar{x}) = \bar{\alpha}\bar{v}(\bar{x}).$$

Let $u(x) = \bar{\alpha}\bar{v}(x)$. Using the fact that $\bar{v}$ is a solution of (108), we find that

$$[(a^2u^2+|Du|^2)I-DuDu]D^2u = an(\bar{\Lambda}/\bar{\alpha})\,u\,(a^2u^2+|Du|^2)^{\frac{3}{2}}$$
$$+\tfrac{1}{2}ax\cdot Du\{(n-2)\,a^2u^2+(n-1)\,|Du|^2\}+a^2u\{na^2u^2+(n+1)\,|Du|^2\}.$$

Since $\bar{\Lambda}/\bar{\alpha} \leqslant \Lambda$ we conclude from the maximum principle of Eberhard Hopf that $u \equiv v$ in Ω, which contradicts the given boundary conditions. This completes the proof of the theorem, subject of course to the deferred lemma.

Proof of the lemma

It is well known (and easily verified using rectangular Cartesian coordinates) that the local mean curvature Λ of a surface satisfies the variational equation

$$\delta(\mathrm{d}A)+n\Lambda\,\delta(\mathrm{d}V) = 0,$$

variations in the volume V being considered positive when they occur in the direction of the normal defining Λ.

For a surface Σ defined by the function $v(x)$ as in the lemma, we have

$$\mathrm{d}V = \frac{1}{n+1}v^{n+1}\,\mathrm{d}\omega(x) = \frac{a^n}{n+1}v^{n+1}\,\mathrm{d}x$$

where $\mathrm{d}\omega$ is the element of area on the unit sphere, and $a = 4/(4+|x|^2)$; see figure 2. In the same way, using the fact that stereographic projection is conformal, it is not hard to obtain

$$\mathrm{d}A = v^n\sqrt{\{1+|Dv|^2/a^2v^2\}}\,\mathrm{d}\omega(x) = (av)^{n-1}\sqrt{\{a^2v^2+|Dv|^2\}}\,\mathrm{d}x.$$

If we put

$$F(x,v,p) = (av)^{n-1}\sqrt{\{a^2v^2+|p|^2\}}, \quad G(x,v) = \frac{na^n}{n+1}v^{n+1}$$

it follows from variational calculus that

$$\operatorname{div}\{F_p(x,v,Dv)\} = F_v+\Lambda G_v$$

which upon expansion yields equation (108).

To prove the second part of the lemma, note that since stereographic projection is conformal the relation between H and H_g must necessarily be linear,

$$H_g = \alpha H+\beta\cdot\nu.$$

To determine the coefficients we can thus resort to special cases where the boundary of Ω is a plane or a sphere.

Considering first the boundaries $x \cdot \nu = 0$ we find that β must be directed along a ray from the origin, that is $\beta = \gamma x$ where γ is a scalar. Taking next the case $x \cdot \nu =$ constant and evaluating at the point $x = r\nu$, we get

$$H = 0, \quad H_g = \frac{\sin \frac{1}{2}\theta}{\cos \frac{1}{2}\theta} = \tfrac{1}{2}r, \quad \beta \cdot \nu = \gamma r$$

whence $\gamma = \frac{1}{2}$. Finally, putting $|x| = r = R$, we find at any point on this sphere

$$H = 1/r, \quad H_g = \frac{\cos\theta}{\sin\theta} = \frac{4-r^2}{4r}, \quad \beta \cdot \nu = -\tfrac{1}{2}r$$

so that $\alpha = \frac{1}{4}(4+r^2)$, completing the proof of the lemma.

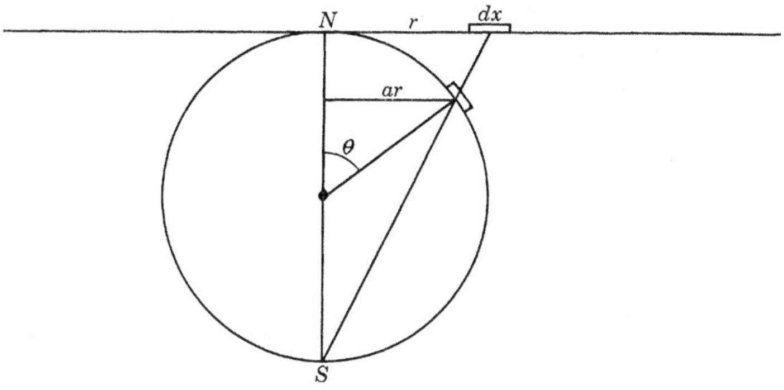

FIGURE 2

Remark. The above results, as well as those of the preceding section, can be generalized to the case where the mean curvature Λ is allowed to depend on v and Dv in addition to x. We shall consider these matters in more detail in another paper to appear later.

24. The differential equation of constant mean curvature

In 1760 Lagrange published a new method—essentially the variational technique still used—for determining the differential equation to be satisfied by functions which minimize an integral expression under appropriate boundary conditions and side conditions (Euler's earlier method had been based on geometrical considerations and was moreover not particularly suited to problems involving functions of more than one independent variable). As an example illustrating the advantages of his general procedure Lagrange considered the problem of minimizing the area integral

$$\iint \sqrt{\{1+u_x^2+u_y^2\}}\, dx\, dy$$

63-3

of a surface $z = u(x, y)$ while keeping the volume integral

$$\iint u \,dx\,dy$$

fixed. He was correspondingly led to the differential equation

$$\frac{\partial}{\partial x}\left(\frac{u_x}{\sqrt{\{1+u_x^2+u_y^2\}}}\right)+\frac{\partial}{\partial y}\left(\frac{u_y}{\sqrt{\{1+u_x^2+u_y^2\}}}\right) = \text{constant} = 2\Lambda \tag{110}$$

which upon expansion yields the well-known equation for surfaces of constant mean curvature. This is the first partial differential equation to have been obtained by the methods of the calculus of variations, and is justly famous on this account as well as for its obvious geometric interest.

In the final part of his memoir, Lagrange noted that spheres (or more precisely their upper and lower hemispheres) generate a four-parameter family of exact solutions of equation (110), but beyond this he did not pursue the problem from the point of view either of the calculus of variations or of the differential equation itself. Further analysis of Lagrange's equation awaited the later interpretation of Λ as the mean curvature of the surface $z = u(x, y)$. This interpretation in particular established that the upper and lower parts of right circular cylinders form a new family of exact solutions.

While the minimal surface equation ($\Lambda = 0$) shortly became the object of serious investigation, owing to the transformation formulae initiated by Monge and Legendre, no corresponding analytical techniques were found for dealing with the case of non-zero curvature. Accordingly it was not until 1841 that a further family of exact solutions was discovered by Delaunay. Delaunay proposed to determine the surface of revolution having constant mean curvature, and found that the generating curve must be such as would be traced out by the focus of a conic section, if the conic section itself rolls without slipping along the axis in question.† The particular case of a rolling parabola accounts for the catenary generating the minimal surface of revolution. Although no further exact solutions have been discovered since Delaunay's surfaces of revolution, a result of Bonnet (1853) should be noted for its related interest. If we consider a surface S of constant mean curvature Λ, where Λ may be taken positive without loss of generality, then the two parallel surfaces at distances $1/2\Lambda$ and $1/\Lambda$ in the direction of the preferred normal vector to S have, respectively, constant Gaussian curvature $4\Lambda^2$ and constant negative mean curvature $-\Lambda$. A proof can be given on the basis of the formulae in § 4. In fact, one easily establishes that the mean and Gaussian curvatures $\bar{\Lambda}$ and $\bar{K}$ of the parallel surfaces at distance d from S are given by

$$\bar{\Lambda} = \frac{\Lambda - Kd}{1-2\Lambda d+Kd^2}, \quad \bar{K} = \frac{K}{1-2\Lambda d+Kd^2},$$

where K is the Gaussian curvature of S, and the result follows at once.‡

The fact that there are few known exact solutions is of course neither unduly surprising nor seriously disturbing, though it naturally indicates the importance of general theoretical investigations. From the beginning of the eighteenth century mathematicians were aware

† Euler had found the differential equations corresponding to Delaunay's solutions as early as 1732, though he did not attempt their integration. The generating curves are illustrated by Minkowski (1906, page 582).

‡ See also W. Blaschke, *Differentialgeometrie*, Bd. I, p. 119. Berlin: Springer (1930).

of the necessity to formulate and solve problems in the calculus of variations corresponding to definite boundary conditions. In fact, already in 1785 Meusnier had clearly stated the Plateau problem: among all surfaces passing through a given closed curve not lying in a plane, to find that one whose area is least. Similarly, an early Dirichlet type problem for equation (110) was proposed by Gergonne (1816): Find the surface of minimum area among all those which are bounded by the sides of a square, and which include between themselves and the square a given volume. In view of the difficulty of such problems it is not to be wondered at that few attempted their study until well into this century.

In a series of remarkable papers appearing between the years 1907 and 1912 Bernstein reopened the Dirichlet problem for elliptic partial differential equations of the type (1) in two independent variables. While these papers were concerned for the most part with a fairly general situation, the example of surfaces of constant mean curvature still figures prominently and was undoubtedly an important stimulus to Bernstein's thought. At the beginning of his work Bernstein evidently considered equation (110) as unexceptional, for he states (1907, page 1027) that the corresponding Dirichlet problem in a circular domain is solvable for arbitrarily given continuous boundary data. By 1910, however, he had realized the various difficulties in the problem—in particular, that the inhomogeneous term on the right-hand side grows too rapidly in $|Du|$ for such a conclusion to be true.

At this time (1910, page 243) he obtained the remarkable theorem that if a domain Ω properly contains a circle of radius $1/|\Lambda|$ then no solution of (110) can possibly exist over Ω, irrespective of the boundary data which might be assigned.† On page 244 Bernstein suggested a procedure for determining when a solution exists, and applied this to the particular case of an elliptical domain with zero boundary data. He writes furthermore: 'Il serait, sans doute, intéressant d'indiquer la marche systématique à suivre pour poivoir décider dans chaque cas particulier, par un nombre fini d'essais, si le problème est possible, ou non. Je n'ais pas l'intention de m'occupier ici de cette question.' Apart from the result discussed in the following paragraph, in fact, Bernstein never returned to this question.

Equation (110) appears again in a note in the *Comptes Rendus, Paris* (1910, page 515) where it is stated (among other things) that the Dirichlet problem is not generally solvable for convex domains. This result is of course a consequence of the theorem noted in the previous paragraph; Bernstein's general argument, however, is directed to the case when the domain may be arbitrarily small as well as convex, as is apparent from the later published proof of the result (1912, pages 465–469). This work is of considerable interest for the non-solvability of Dirichlet's problem, even though it is not sufficiently delicate to make a distinction between those convex domains for which a solution of Dirichlet's problem for equation (110) is always possible and those for which it is not (cf. § 20). Bernstein did not in fact realize that such a distinction might exist, for in a footnote to the theorem on page 465 he remarks: 'Il n'est pas douteux, d'ailleurs, qu'en complétent convenablement la démonstration on trouveriat que tous les contours jouissent de la méme propriété.'

A similar situation occurs in his treatment (1912, page 479) of the variational problem

$$\delta\int\sqrt{(1+|x|^2)}\sqrt{(1+|Du|^2)}\,dx = 0, \quad n = 2.$$

† Other proofs and generalizations of this fact have been given by Heinz (1955), Chern (1965), and Finn (1965).

He states there that this extremum problem does not, in general, admit a solution even for convex domains (compare § 21 of the present work). The adjoining remark is prophetic: 'Ce probléme, qui diffère si peu du problème de Plateau, rend compte, il me semble, des difficultés qu'on a du recontrer en cherchant des démonstrations directes de l'existence d'une surface minima pasant par un contour donné'. The example which Bernstein notes at the foot of page 483 provides still another instance where a more delicate treatment is required (see § 20).

To conclude the discussion of Bernstein's work, attention should be drawn to the final paragraphs of the memoir of 1912. That the result stated there, 'Si le problème de Dirichlet est toujours possible pour toute fonction analytique sur un contour convex, il est également possible si la fonction est supposée seulement continue sur le contour', cannot be correct follows from an example of Finn (1954, page 399). The problem is of course that of obtaining interior estimates for the gradient of a solution in terms of a maximum bound, but independent of the particular boundary data. While such estimates are available for certain equations in two independent variables (Finn 1963; Serrin 1963) and for most uniformly elliptic equations (cf. Ladyzhenskaya & Uraltseva 1964), there are scarcely any non-uniformly elliptic equations in higher dimensions for which such conclusions are known (in this regard, E. De Giorgi has recently shown the existence of such an interior estimate for the minimal surface equation in higher dimensions; see § 19).

During the period between Bernstein's memoir of 1912 and Leray's fundamental paper of 1939, I am not aware of any contributions to the problem of Dirichlet for surfaces of constant mean curvature. We have already commented in the introduction on the relation between Leray's work and Bernstein's. It does not seem to have been noticed before, however, that Leray's results can be applied to the Dirichlet problem for equation (110), provided a certain degree of care is taken. In particular, the following result can be obtained.

Let Ω be a bounded domain in two-dimensional Euclidean space, whose boundary is of class C^5. Then the Dirichlet problem for equation (110) *in Ω is solvable for arbitrarily given boundary values of class C^5 provided that $\kappa \geqslant 2|\Lambda|$ at each point of the boundary, where κ denotes the curvature of the boundary curve.*

To see this, one may use corollary II on page 282 of Leray's paper (a straightforward but somewhat tedious calculation verifies the hypotheses) together with the argument given in § 20 of the present paper.

In a review article which appeared in 1961, Akhiezer & Petrovsky attribute to Bernstein the result that the Dirichlet problem for surfaces of constant mean curvature is solvable in any region which is entirely inside a circle of radius $1/|\Lambda|$. This statement seems to have resulted from reading more into some of Bernstein's remarks than was actually intended.

The question of uniqueness of surfaces having prescribed mean curvature is included in the important work of Alexandrov (1956–58) on surfaces in the large defined by functions of their principal curvatures. In addition to considering various Dirichlet-type problems in parallel and central projection, Alexandrov also obtained general results concerning the uniqueness of compact surfaces. These latter results imply in particular that a compact non-self-intersecting surface of constant mean curvature is a sphere, generalizing earlier work of Liebmann (1900) who required that the surface be convex. While this result applies equally in any number of dimensions, a complementary theorem of Hopf (1951) states that a simply connected compact surface of constant mean curvature in three dimensions is a sphere. The literature on these and related questions is fairly extensive, and the interested reader is referred to Alexandrov's paper for further details.

In another recent paper, Finn (1965) has obtained the following result: *Let* $u = u(x, y)$ *be a solution of* (110) *defined over the circular domain* $x^2+y^2 < 1/\Lambda^2$. *Then*

$$u \equiv \text{const.} + \sqrt{\{(1/\Lambda^2) - x^2 - y^2\}}.$$

This is the natural analogue for equation (110) of Bernstein's celebrated theorem that a minimal surface $z = u(x, y)$ defined over the entire plane must itself be a plane. There are finally some conclusions of a differential geometric character concerning surfaces of constant mean curvature (e.g. the lines of curvature on such a surface form an isothermal net, etc.) for which we refer the reader to standard treaties on classical differential geometry or the theory of surfaces.

The discussion would not be complete without mention of the important series of papers of Heinz (1954) and Werner (1957, 1960) concerning the existence of surfaces of constant mean curvature Λ in three-dimensional Euclidean space with arbitrary Jordan curves as assigned boundary. The surfaces in question are defined parametrically, and consequently (as in the problem of Plateau for minimal surfaces) may have self-intersections, multiple coverings, and branch points at which the local representation is singular. These matters aside, the elegant sufficient condition for solvability due to Werner is that the Jordan curves in question should be contained in a ball of radius R where $1/R > 2|\Lambda|$. It is remarkable that this condition almost exactly parallels the solvability condition $\kappa \geqslant 2|\Lambda|$ for the Dirichlet problem for arbitrary assigned data.

This work has been in progress over a number of years and accordingly I have discussed it with a great many of my friends. To all of these, and particularly to D. E. Edmunds and J. C. C. Nitsche, I express my heartiest thanks and gratitude. The manuscript was written at the University of Sussex during a sabbatical leave of absence from the University of Minnesota. I am most grateful to both these institutions for the opportunity which they provided.

References

Alexandrov, A. D. 1956 Uniqueness theorems for surfaces in the large. *Vestnik Lenin. Univ.* **11**, 5–17.

Bernstein, S. 1907 Méthode générale pour la résolution du problème de Dirichlet. *C. R. Acad. Sci. Paris* **144**, 1025–1027.

Bernstein, S. 1910 Conditions nécessaires et suffisantes pour la possibilité du problème de Dirichlet. *C. R. Acad. Sci. Paris* **150**, 514–515.

Bernstein, S. 1910 Sur la géneralisation du problème de Dirichlet, II. *Math. Ann.* **69**, 82–136.

Bernstein, S. 1910 Sur les surfaces définies au moyen de leur courbure moyenne et totale. *Ann. Sci. École Norm. Sup*, **27**, 233–256.

Bernstein, S. 1912 Sur les équations du calcul des variations. *Ann. Sci. École Norm. Sup.* **29**, 431–485.

Bombieri, E., De Giorgi, E. & Miranda, M. 1969 Una maggiorazioni a priori relativa alle ipersuperfici minimali non parametriche. *Arch. Rat. Mech. Anal.* **32**.

Browder, F. 1966 *Problèmes Non-linéaires* (*Seminar de Mathématiques Supérieures*). University of Montreal Press.

Finn, R. 1965 Remarks relevant to minimal surfaces and to surfaces of constant mean curvature. *J. d'Anal. Math.* **14**, 139–160.

Gilbarg, D. 1963 Boundary value problems for nonlinear elliptic equations in *n* variables, pp. 151–159. *Nonlinear problems.* Madison: University of Wisconsin Press.

Hartman, P. & Stampacchia, G. 1966 On some nonlinear elliptic differential functional equations. *Acta Math.* **115**, 271–310.

Hopf, E. 1927 Elementare Bemerkungen über die Lösungen partieller Differentialgleichungen zweiter Ordnung von Elliptischen Typen. *Sber. Preuss. Akad. Wiss.* **19**, 147–152.

Jenkins, H. & Serrin, J. 1968 The Dirichlet problem for the minimal surface equation in higher dimensions. *J. Reine Ang. Math.* **229**, 170–187.

Ladyzhenskaya, O. A. & Uraltseva, N. N. 1961 Quasilinear elliptic equations and variational problems with many independent variables. *Uspekhi Mat. Nauk.* **16**, 19–92. Translated in *Russ. Math. Surveys* **16**, 17–91.

Ladyzhenskaya, O. A. & Uraltseva, N. N. 1964 On Hölder continuity of solutions and their derivatives of linear and quasilinear elliptic and parabolic equations. *Trudy Steklov Inst. Leningrad* **73**, 172–220.

Ladyzhenskaya, O. A. & Uraltseva, N. N. 1964 *Linear and quasilinear elliptic equations.* Moscow.

Leray, J. 1939 Discussion d'un problème de Dirichlet. *J. Math. Pures Appl.* **18**, 249–284.

Leray, J. & Schauder, J. 1934 Topologie et équations fonctionelles. *Ann. Sci. École Norm. Sup.* **51**, 45–78.

Nirenberg, L. 1952 On nonlinear elliptic partial differential eqations and Hölder continuity. *Comm. Pure Appl. Math.* **6**, 103–156.

Nitsche, J. C. C. 1965 On differential equations of mode 2. *Proc. Am. Math. Soc.* **16**, 902–908.

Oskolkov, A. P. 1967 *A priori* estimates for the first derivatives of solutions of the Dirichlet problem for non-uniformly elliptic quasilinear equations. *Trudy Mat. Inst. Steklov* (*Akad. Nauk. SSSR*) **102**, 105–127.

Radó, T. 1932 Contributions to the theory of minimal surfaces. *Acta Szeged* **6**, 1–20. Cf. also *The problem of plateau.* pp. 35–37. Berlin: Springer (1932).

Redheffer, R. 1962 An extension of certain maximum principles. *Monatsh. Math.* **66**, 32–42. Cf. also *J. Reine Ang. Math.* **211**, 70–77.

Schaeffer, H. 1955 Über die Methode der a priori Schranken. *Math. Ann.* **129**, 415–416.

Schauder, J. 1934 Über lineare elliptische Differentialgleichungen zweiter Ordnung. *Math. Z.* **38**, 257–282. Cf. also Courant, R. & Hilbert, D. *Methods of mathematical physics*, vol. 2, pp. 331–346. New York: Interscience (1962).

Serrin, J. 1967 The Dirichlet problem for quasilinear equations with many independent variables. *Proc. Nat. Acad. Sci.* **58**, 1829–1835.

Serrin, J. (To appear). The Dirichlet–Alexandrov problem for surfaces having constant mean curvature.

Stampacchia, G. 1963 On some regular multiple integral problems in the calculus of variations. *Comm. Pure Appl. Math.* **16**, 383–421.

Trudinger, N. 1966 Quasilinear elliptic partial differential equations in *n* variables. Dissertation, Stanford University. Cf. also *Bull. Am. Math. Soc.* **73**, 410–413.

Trudinger, N. 1967 On the Dirichlet problem for quasilinear uniformly elliptic equations in *n* variables. *Arch. Rat. Mech. Anal.* **27**, 108–119.

Valentine, F. A. 1964 *Convex sets.* New York: McGraw-Hill.

Supplementary bibliography for section 24

Akhiezer, N. I. & Petrovski, I. G. 1961 S. N. Bernstein's contribution to the theory of partial differential equations. *Uspekhi Mat. Nauk* **16**, 5–20.

Alexandrov, A. D. 1956–58 Uniqueness theorems for surfaces in the large. *Vestnik Lenin. Univ.* **11** (no. 19), 5–17; **12** (no. 7), 15–44; **13** (no. 7), 14–26; **13** (no. 19), 5–8. Translated in *Am. Math. Soc. Translations* (Series 2) **21**, 341–403, 412–416.

Bernstein, S. 1907 Méthode générale pour la résultion du problème de Dirichlet. *C. R. Acad. Sci. Paris* **144**, 1025–1027.

Bernstein, S. 1910 Conditions nécessaires et sufficisantes pour la possibilité du problème de Dirichlet. *C. R. Acad. Sci. Paris* **150**, 514–515.

Bernstein, S. 1910 Sur les surfaces définies au moyen de leur courbure moyenne et totale. *Ann. Sci. École Norm. Sup.* **27**, 233–256.

Bernstein, S. 1912 Sur les équations du calcul des variations. *Ann. Sci. École Norm. Sup.* **29**, 431–485.

Bonnet, O. 1853 Sur une propriété de maximum relative à la sphére. *Nouv. Ann. Math.* **12**, 433–438.

Chern, S. 1965 On the curvatures of a piece of hypersurface in Euclidean space. *Abh. Math. Sem. Univ. Hamburg* **29**, 77–91.

Delaunay, Ch. 1841 Sur la surface de révolution dont la courbure moyenne est constante. *J. Math. Pures Appl.* **6**, 309–315. Cf. also Sturm, J. *ibid.*, 315–320.

Euler, L. 1738 Problematis isoperimetrici in latissimo sensu accepti solutio generalis. *Comm. acad. sci. Petrop.* **6** (1732/3), 123–155. *Opera omnia* **25**, 13–40 (especially p. 30).

Finn, R. 1954 On equations of minimal surface type. *Ann. Math.* **60**, 397–416.

Finn, R. 1963 New estimates for equations of minimal surface type. *Arch. Rat. Mech. Anal.* **14**, 337–375.

Finn, R. 1965 Remarks relevant to minimal surfaces and to surfaces of constant mean curvature. *J. d'Anal. Math.* **14**, 139–160.

Gergonne, J. D. 1816 Problèmes de géométrie, dépendant de la méthode des variations. *Ann. Math. Pures Appl.* **7**, 156.

Heinz, E. 1954 Über die Existenz einer Fläche konstanter mittlerer Krümmung bei vorgegebener Berandung. *Math. Ann.* **127**, 258–287.

Heinz, E. 1955 Über Flächen mit eineindeutiger Projection auf eine Ebene, deren Krümmung durch Ungleichungen eingeschränkt sind. *Math. Ann.* **129**, 451–454.

Hopf, H. 1951 Über Flächen mit einer Relation zwischen den Hauptkrümmungen. *Math. Nachr.* **4**, 232–249.

Lagrange, J. L. 1760 Essai d'une nouvelle méthode pour déterminer les maxima et les minima des formulae integrales indéfinies. *Misc. Taur.* **2**, 173–195. *Oeuvres de Lagrange*, vol. 1, pp. 333–362.

Leray, J. 1939 Discussion d'un problème de Dirichlet. *J. Math. Pures Appl.* **18**, 249–284.

Liebmann, H. 1900 Über die Verbiegung der geschlossenen Flächen positiver Krümmung. *Math. Ann.* **53**, 81–112.

Meusnier, M. 1785 Mémoire sur la courbure des surfaces. *Mém. Divers Savans* (*Etrangers*) **10**, 477–510.
Minkowski, H. 1906 Kapillarität. *Ency. Math. Wiss.* **V 1**, 559–613.
Serrin, J. 1963 *A priori* estimates for solutions of the minimal surface equation. *Arch. Rat. Mech. Anal.* **14**, 376—383; **28**, 149–154.
Werner, H. 1957 Das Problem von Douglas für Flächen konstanter mittlerer Krümmung. *Math. Ann.* **133**, 303–319.
Werner, H. 1960 The existence of surfaces of constant mean curvature with arbitrary assigned Jordan curves as assigned boundary. *Proc. Am. Math. Soc.* **11**, 63–70.

This paper is dedicated
to the memory of
Sergei Natanovich Bernstein
1880–1968

J. Math. Mech. 9 (1960), 513–538. Reprint I.2.[iii]

The Exterior Dirichlet Problem for Second Order Elliptic Partial Differential Equations

NORMAN MEYERS & JAMES SERRIN

In this paper we shall treat the exterior Dirichlet problem for linear elliptic equations of second order in $n \geqq 2$ independent variables. Our main concern will be with the existence and uniqueness of solutions, and with the striking effect which the asymptotic behavior of the coefficients has on the results.

It will be convenient to preface our considerations with some remarks concerning the special case of the Laplace equation. In treating the exterior Dirichlet problem for that equation it is clear that some restriction on the behavior of the solution at distant points of space is necessary in order to insure unicity. A primary condition is that the solution be bounded. Indeed, for the case $n = 2$ this condition alone suffices for uniqueness, as is well known. On the other hand, in three or more dimensions there are an infinity of bounded solutions corresponding to any given boundary data, so that some additional restriction must be imposed. Since every bounded harmonic function defined in an exterior region tends to a limit at infinity, it is natural to assign this limit value as the specifying condition. The exterior problem then becomes uniquely solvable, and, moreover, as the limit value ranges over the real numbers, every bounded solution satisfying the given data is obtained.

Consider now the general elliptic equation of second order, which we write in the form

$$Lu = a_{ik}(x)u_{ik} + b_i(x)u_i + c(x)u = f(x), \tag{1}$$

where $x = (x_1, \cdots, x_n)$ and

$$u_i = \frac{\partial u}{\partial x_i}, \qquad a_{ik}u_{ik} = \sum_1^n a_{ik} \frac{\partial^2 u}{\partial x_i \, \partial x_k}, \quad etc.$$

By the exterior Dirichlet problem we mean the following: Let $\mathcal{E}$ be a given open region in the Euclidean number space E^n, containing a neighborhood of infinity and having a smooth internal boundary $F\mathcal{E}$. A continuous function $\phi(x)$ is prescribed on $F\mathcal{E}$. We wish then to find a function $u = u(x)$ which is of

Journal of Mathematics and Mechanics, Vol. 9, No. 4 (1960).

class C^2 and satisfies $Lu = f$ in $\mathcal{E}$, and which continuously takes on the values $\phi(x)$ on the boundary $F\mathcal{E}$. The problem is further specified by certain conditions on the behavior of u at infinity. Corresponding to the preceding discussion of the Laplace equation, we shall consider two different cases:

I. *The function u shall tend to an assigned limit l as x tends to infinity.*

II. *The function u shall be bounded in $\mathcal{E}$.*

The problems arising from these two conditions will be referred to respectively as the exterior Dirichlet problem I and the exterior Dirichlet problem II. A problem will be said to be *well-set* for a given equation if for any assigned data there exists exactly one corresponding solution.

It will be observed that problem I is well-set for the Laplace equation in $n \geqq 3$ independent variables, while problem II is well-set only for $n = 2$ variables. Such a simple distinction does not hold for the general equation (1), nor even for the special equation

$$\Lambda u = a_{ik}u_{ik} = 0. \tag{2}$$

Indeed, there are simple equations of the form (2) in two independent variables for which problem I is well-set, while conversely there are equations in three variables for which problem II is well-set (*cf.* section 9). These examples are all the more remarkable because the operator Λ can in each case be uniformly elliptic. One of the main purposes of this paper is to clarify the rather anomolous behavior indicated by these examples, and to give criteria by which one can predict ahead of time whether problem I or problem II will be well-set for a particular equation. As a byproduct, this investigation has yielded new results even for the equation $\Delta u + cu = f$.

In connection with our work, we must mention certain previous investigations of Giraud and Krzyzański. The former author was the first to make a general study of the exterior problem for elliptic equations. His work, although a highly difficult exercise in the integral equation technique, is quite restricted both as to the nature of the operator L and the type of results obtained. Krzyzański's results are more similar to ours in method and outlook, but the specific problems which he treats are not immediately related to those discussed here. Altogether, the exterior Dirichlet problem (important though it may be) seems to be only superficially touched in the literature.

Description of results. Part I of the paper deals specifically with equations of the form (2), these being of considerable interest in their own right. In section 1 the concept of a ϕ-sequence is used to prove the existence of solutions of the exterior Dirichlet problem (certain technical continuity requirements on the coefficients in Λ may be taken for granted at this time). It is then shown that problem I is well-set if and only if there exists a barrier at infinity. The idea here is essentially the same as that used to describe a regular boundary point in potential theory, but its application to the exterior problem seems not to have been explicitly noticed before. In Theorems 2 and 3 we state several sufficient conditions for the existence of a barrier. Theorem 3, in particular, is so simple that we repeat it here: *If* $a_{ik} \to a^0_{ik}$ *as* x *tends to infinity, and if the matrix* (a^0_{ik}) *has at least three positive eigenvalues, then there exists a barrier and problem* I *is well-set.* The Laplace equation with $n \geqq 3$ occurs as a special case under this criterion. On the other hand, from this result it also follows that problem I can be well-set for certain non-uniformly elliptic equations.

The results for problem II have a comparable degree of completeness. In particular, this problem is well-set for an equation of the form (2) if and only if the equation enjoys an extended maximum principle. With the help of this result and the concept of an anti-barrier we then obtain certain conditions for problem II to be well-set (Theorems 5–7). An unexpected feature which arises here is the duality between problems I and II, which may be seen by comparing Theorems 1, 2, and 3, respectively, with Theorems 5, 6, and 7.

In the preceding two paragraphs we have discussed conditions under which one or the other of the two exterior problems will be well-set. The question naturally arises, do there exist equations for which neither problem is well-set? In section 3 a partial answer to this question is given: *For a uniformly elliptic equation of the form* (2), *whose coefficients are continuous at infinity, one or the other of the two problems is well-set. Moreover, among the solutions for varying data must occur every bounded solution. The requirement that the coefficients tend to limits at infinity can be weakened somewhat, and even entirely dropped when* $n = 2$.

Part II of the paper deals with the full elliptic equation (1). For the most part it will be supposed that $c \leqq 0$, this requirement being essentially necessary in the problem (*cf.*, however, section 8). The existence and uniqueness theorems obtained for equation (1) are of the same generality as those already discussed in connection with equation (2). The reader is referred particularly to Theorems 10 and 12, which are believed to be new even for the special equation $\Delta u + cu = f$. It should be observed that the added difficulties posed by equation (1), in comparison to equation (2), are due primarily to the strong effect which the function $f(x)$ exerts on the equation.

The simple and elegant Theorem 13 generalizes an interesting result of Yosida and Atkinson concerning the asymptotic behavior of solutions of the equation $\Delta u + cu = 0, c \leqq -m^2$. Similarly, Theorem 15 has certain aspects in common with a recent theorem of Picone.

In Part III of the paper we present several examples illustrating the various possibilities which can arise in the solution of an exterior Dirichlet problem. These examples form an integral part of the paper, though they also have considerable independent interest. The result of Section 10 concerning the existence of bounded solutions of the Poisson equation may be cited in particular.

The final part of the paper deals with the exterior problem corresponding to Neumann and other non-Dirichlet type data on $F\mathcal{E}$. These problems can be treated in much the same way as the exterior Dirichlet problem, and this is our justification for including the material here. Theorem 16 is concerned with the exterior problem when a fixed limit zero is assigned at infinity (problem I_0), and Theorem 17 when only a boundedness condition is imposed (problem II). With certain reservations on the kind of data assigned (for which it is best to refer directly to Part IV) the results may be roughly summarized by the statement that the exterior problems for non-Dirichlet data are well-set if and only if they are well-set for Dirichlet data.

Preliminary hypotheses and results. The following assumptions concerning the operators L and Λ will be made throughout the paper:

1°. The coefficients a_{ik}, b_i, and c, and the function f shall be Hölder continuous on bounded subsets of $\mathcal{E} + F\mathcal{E}$.

2°. At each point of $\mathcal{E} + F\mathcal{E}$, and for all real n-tuples $(\xi_1, \cdots, \xi_n) \neq 0$,

$$a_{ik}(x)\xi_i\xi_k > 0.$$

By virtue of these hypotheses, equations (1) and (2) will be uniformly elliptic in any bounded part of $\mathcal{E}$. On the other hand, the coefficients *need not* be continuous at infinity and equations (1) and (2) are not necessarily uniformly elliptic over $\mathcal{E}$. One of the purposes of the paper is to avoid these specializing assumptions.

In our existence theorems the required solution will be found by taking limits of solutions defined only on bounded subsets of the given exterior domain $\mathcal{E}$. This procedure is facilitated by the concept of a ϕ-sequence. Let $\{\mathcal{S}_n\}$, $n = 1, 2, 3, \cdots$, be a sequence of open balls centered at the origin, with radii r_n tending monotonically to infinity and with $F\mathcal{E} \subset \mathcal{S}_1$. The function sequence $\{u_n\}$ will be called a *ϕ-sequence for equation* (1) if each function $u_n = u_n(x)$ satisfies (1) in the set $\mathcal{E} \cap \mathcal{S}_n$ and continuously takes on the values $\phi(x)$ on $F\mathcal{E}$.

Lemma 1. *Every locally bounded ϕ-sequence has a subsequence which converges to a solution in $\mathcal{E}$ taking on the boundary values $\phi(x)$.*

Proof. Hypotheses 1° and 2° above allow the application of SCHAUDER's interior estimates to the individual functions u_n. It follows that the sequence $\{u_n\}$ is locally equicontinuous as well as locally bounded, whence one can extract a subsequence which converges uniformly on compact subsets of $\mathcal{E}$ to a solution u. That this solution takes on the boundary values $\phi(x)$ may be proved with the aid of SCHAUDER's boundary estimates, or it may be demonstrated directly by the technique of barriers. We shall omit these rather standard arguments and only observe that, whichever method is employed, some degree of smoothness must be demanded of the boundary $F\mathcal{E}$. When the barrier method is adopted, in particular, it is enough that $F\mathcal{E}$ have bounded curvature. This latter condition will be assumed throughout the paper.

The smoothness assumption on $F\mathcal{E}$ also guarantees the *existence* of ϕ-sequences, according to standard theorems of elliptic partial differential equations. Indeed, one can even preassign the values of u_n on the sphere $r = r_n$, a convenience which we shall occasionally utilize.

For later use, we make the following definition: A function $\delta(t)$, $t_0 \leqq t < \infty$, will be called a *Dini function* if and only if the (improper) integral

$$\int^{\infty} \delta(t) \frac{dt}{t}$$

exists and is finite. The functions $t^{-\alpha}$, $(\log t)^{-1-\alpha}$, and $\sin t$, where $\alpha > 0$, are examples of Dini functions.

Part I

1. The exterior problem I. In this and the following two sections we shall deal with the special equation

$$\Lambda u = a_{ik} u_{ik} = 0. \tag{2}$$

This is done mainly for simplicity. As we shall see later, many of the results carry over to the general equation (1). We begin with the following simple observation:

There cannot be more than one solution of problem I.

This is an immediate consequence of the maximum principle, since the difference of two solutions would have zero data on $F\mathcal{E}$ and would tend to zero as $r \to \infty$. Here r denotes, as it will throughout the paper, the distance Ox, that is, $r^2 = \sum_{i=1}^{n} x_i^2$.

We now introduce the concept of a barrier. A function $v = v(x)$ will be called a *barrier at infinity for the equation* $\Lambda u = 0$ if (i) v is defined and positive in some neighborhood of infinity, (ii) v tends to zero as $r \to \infty$, and (iii) v is twice continuously differentiable and $\Lambda v \leqq 0$.

Property (iii) can be replaced by the following weaker condition: (iii*) v is continuous; if u satisfies $\Lambda u = 0$ in some region $\mathcal{R}$ and if $u \leqq v$ at every point of $F\mathcal{R}$, then $u \leqq v$ throughout $\mathcal{R}$.

Theorem 1. *Problem* I *is well-set for the equation* $\Lambda u = 0$ *if and only if this equation has a barrier at infinity.*

Proof. In view of the preceding observation on uniqueness, it is enough to show that Problem I is solvable for arbitrary assigned data if and only if there exists a barrier.

But if Problem I is always solvable, let v be that solution with boundary values equal to one and limit at infinity equal to zero. Then v is clearly a barrier.

On the other hand, suppose that v is a barrier. We first extend v so that it is defined over all of $\mathcal{E}$. To this end, let R be a number so large that v is defined for $r \geqq R$. Let $v_0 = \min_{r=R} v$ and define

$$v^* = \begin{cases} \min(v_0, v) & \text{for } r \geqq R \\ v_0 & \text{for } r < R. \end{cases} \tag{3}$$

It can easily be shown that v^* has the property (iii*).

Now we are in a position to construct a solution u such that $u = \varphi$ on $F\mathcal{E}$ and $\lim u = l$ as $r \to \infty$. First define

$$\bar{v} = cv^* + l \quad \text{and} \quad \underline{v} = -cv^* + l,$$

where c is a constant chosen so large that $\bar{v} \geqq \varphi$ and $\underline{v} \leqq \varphi$ on $F\mathcal{E}$. Let $\{u_n\}$ be a φ-sequence with $u_n = l$ on $r = r_n$. Because v^* satisfies (iii*), it follows that

$$\bar{v} \geqq u_n \geqq \underline{v}. \tag{4}$$

By Lemma 1, the sequence $\{u_n\}$ determines a solution u in $\mathcal{E}$ with boundary values φ. Finally, since (4) holds for each function u_n, it must also hold for the limit function u, which implies that u tends to l at infinity. This completes the proof.

We turn now to several conditions which will guarantee existence of a barrier. It will be convenient to introduce the auxiliary function

$$\mathfrak{a} = \mathfrak{a}(x) = \frac{\sum a_{ii}x_k x_k}{\sum a_{ik}x_i x_k} = \frac{\sum a_{ii}}{Q},$$

where $Q \equiv (1/r^2) \sum a_{ik}x_i x_k$. Since $\sum a_{ii}$ is the sum of the eigenvalues of the matrix (a_{ik}), while Q is a value of the quadratic form associated with (a_{ik}), the function $\mathfrak{a}$ is evidently an orthogonal (but not an affine) invariant under coordinate changes in E^n. In general we have $1 < \mathfrak{a} < \infty$, while for uniformly elliptic equations $\mathfrak{a}$ is bounded away from both 1 and ∞. In the special case of Laplace's equation, $\mathfrak{a} \equiv n$; in view of later theorems the reader may find it convenient to look upon $\mathfrak{a}$ as the "effective dimension" of the equation $\Lambda u = 0$.

Theorem 2. *If $\mathfrak{a} \geqq 2 + \epsilon$ in some neighborhood of infinity, where $\epsilon = \epsilon(r)$ is such that*

$$\Delta(t) \equiv \exp\left\{-\int^t \epsilon(s)\,\frac{ds}{s}\right\} \tag{5}$$

is a Dini function, then problem I *is well-set for the equation $\Lambda u = 0$.*

Proof. We shall find a barrier which is a function of r alone. To this end, consider a function $v = v(r)$ such that $v' < 0$. Then

$$\Lambda v \equiv Q\left[v'' + \frac{v'}{r}(\mathfrak{a} - 1)\right] \leqq Q\left[v'' + \frac{v'}{r}(1 + \epsilon)\right]. \tag{6}$$

The particular function

$$v = \int_r^\infty \exp\left\{-\int^t \epsilon(s)\,\frac{ds}{s}\right\}\frac{dt}{t}$$

makes the final expression in (6) equal to zero, and obviously satisfies $v' < 0$. It follows immediately that v is a barrier.

Remarks. Theorem 2 is sharp; for given any $\epsilon(r) > -1$ such that

$\exp\{-\int^t (\epsilon(s)/s)ds\}$ is *not* a Dini function, there exists an equation with $\mathcal{A} = 2 + \epsilon$ for which problem I is *not* well-set (section 9). However, the real importance of condition (5) does not lie in this result, but rather in the fact that it can be satisfied even in two dimensions by uniformly elliptic equations (section 9). Hence the scope of problem I is by no means restricted to higher dimensions. Furthermore, in all dimensions the condition allows certain types of non-uniform ellipticity which can become more pronounced as the dimension increases. The following geometric discussion, together with Theorem 3, justifies this statement.

Let us replace the condition $\mathcal{A} \geqq 2 + \epsilon$ by the stronger, though geometrically more illuminating, condition

$$\frac{\sum a_{ii}}{\lambda} \geqq 2 + \epsilon, \tag{7}$$

where $\lambda = \lambda(x) =$ largest eigenvalue of $(a_{ik}(x))$. In the discussion we shall also assume that the coefficients are normalized so that $\sum a_{ii} \equiv 1$, whence (7) takes the simple form

$$\lambda \leqq \frac{1}{2 + \epsilon}. \tag{8}$$

Now let H denote the $(n - 1)$-dimensional set in R^n which is the convex hull of those points having one coordinate equal to one and all others equal to zero. Moreover, let $g(x)$ be the mapping from $\mathcal{E}$ to H which takes x into those points of H whose coordinates are the eigenvalues of $(a_{ik}(x))$, in any order. Since Λ is an elliptic operator, it is clear that $g(x)$ cannot lie on any face of H. Moreover, Λ is uniformly elliptic if and only if the closure of $g(\mathcal{E})$ has no points in common with a face of H.

Let H' denote the convex hull of those points in R^n which have two coordinates equal to $\frac{1}{2}$ and all others zero. If $g(r)$ denotes the image of points for which $\sum x_i^2 = r^2$, condition (8) will be fulfilled if $g(r)$ is contained in the interior of H' and does not approach too rapidly those $(n - 2)$-dimensional faces where one coordinate equals $\frac{1}{2}$. The situation is easily visualized in the case $n = 3$ (see Figure 1). The case $n = 4$ can also be illustrated if we use barycentric coordinates in the 3-dimensional set H. Since the vertices of H are equidistant in R^4, it follows that H can be represented as a tetrahedron in R^3, and that H' is an octahedron inscribed in H (Figure 2). These remarks can be summarized by the statement that if $g(E)$ is interior to H then (2) is elliptic, while if $g(\infty)$ is in H' (and does not approach certain faces of H' too rapidly) then problem I is well-set.

Since the quantity $\mathcal{A}$ is not an affine invariant, it is clear that the decision as to whether Theorem 2 applies to a given equation is related to the particular coordinates used to represent the operator Λ. The following theorem, although somewhat weaker in its result, is free from this defect.

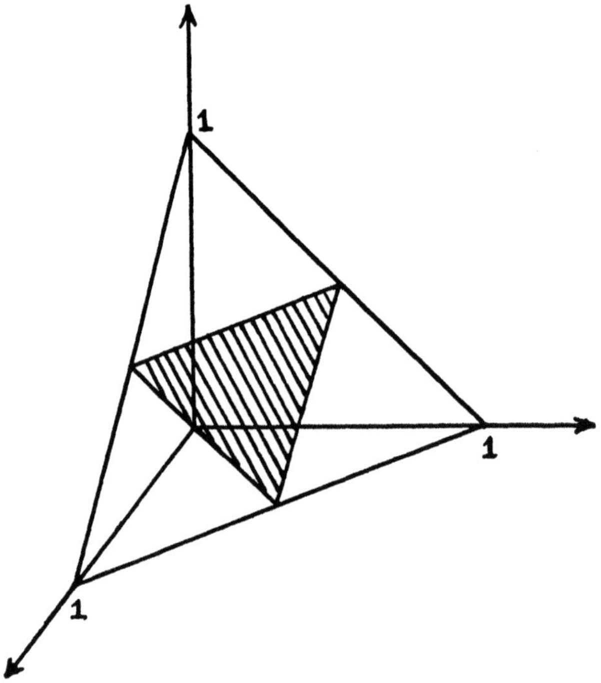

Figure 1. The sets H and H' as subsets of R^3, (H' shaded).

Theorem 3 *(Three eigenvalue criterion). If* $(a_{ik}) \to (a_{ik}^0)$, $r \to \infty$, *and if* (a_{ik}^0) *has at least three positive eigenvalues (equivalently, if Rank* $(a_{ik}^0) \geqq 3$*), then problem* I *is well-set for the equation* $\Lambda u = 0$.

Proof. Let ν be the number of positive eigenvalues of (a_{ik}^0), $3 \leqq \nu \leqq n$. By means of an affine coordinate transformation the matrix (a_{ik}^0) can be put in diagonal form with ones and zeros on the diagonal. In this new system, lim $\sum a_{ii} = \nu$ and lim sup $Q \leqq 1$ as $r \to \infty$. Thus lim inf $A \geqq \nu$, and we can apply Theorem 2 with ϵ any constant satisfying $0 < \epsilon < \nu - 2$. The corresponding barrier has the form $v = \text{Const.}\ r^{-\epsilon}$, as is evident from formula (7).

Remark. An estimate for the asymptotic behavior of a solution of the exterior Dirichlet problem I can be obtained from the relation

$$u - l = O(v)$$

where v is an arbitrary barrier function (see the proof of Theorem 1). Thus, *under the hypotheses of Theorem* 3, *a solution of the exterior problem* I *has the asymptotic behavior*

$$u - l = O(r^{2-\nu+\delta}),$$

in which ν *is the rank of* (a_{ik}^0) *and* δ *is an arbitrary positive constant* This result is a generalization of Theorem 3′ of [2], insofar as the hypotheses on the coefficients a_{ik} are concerned.

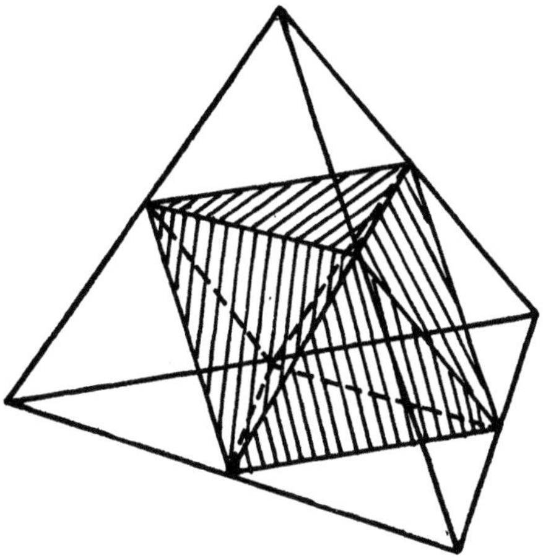

Figure 2. The relation between the sets H and H' in the case $n = 4$, (H' shaded).

2. The exterior problem II. As in the preceding section we confine our attention to the special equation $\Lambda u = 0$. To begin with, we have the simple result: *For an equation of the form* $\Lambda u = 0$, *problem* II *is always solvable.*

This fact, which is dual to the observation made at the beginning of section 1, may be proved as follows. Let $\{u_n\}$ be a φ-sequence such that $u_n = 0$ on $r = r_n$. Obviously $|u_n|$ is bounded by max $|\varphi|$, and the result follows from Lemma 1.

Theorem 4. *Problem* II *is well-set for the equation* $\Lambda u = 0$ *if an only if this equation enjoys the following extended maximum principle: For every bounded solution u defined on a set* $r \geqq R$,

$$\limsup_{r\to\infty} |u| \leqq \max_{r=R} |u|. \tag{9}$$

Proof. Suppose first that the equation $\Lambda u = 0$ satisfies the extended maximum principle. Then there can be at most one solution of problem II, as one sees from a simple application of (9) and the ordinary maximum principle. Consequently problem II is well-set.

On the other hand, suppose there exists a bounded solution w which violates the extended maximum principle, *i.e.*,

$$\limsup_{r\to\infty} |w| > \max_{r=R} |w|.$$

From the ordinary maximum principle, it follows that R can be chosen so large that the region $r > R$ is contained in E. We will now show that problem II has more than one solution for zero data, from which the theorem immediately follows.

Define w^* by means of a φ-sequence based on the region $r > R$ and the boundary values $\varphi = w$ on $r = R$; we may further suppose that $w_n^* = 0$ on $r = r_n$. Evidently

$$\limsup_{r\to\infty} |w^*| \leqq \max_{r=R} |w|,$$

whence $v = w - w^*$ is a bounded solution of our equation such that

(10) $$v = 0 \quad \text{on} \quad r = R \quad \text{and} \quad \limsup_{r\to\infty} |v| > 0.$$

Now consider a φ-sequence in $\mathcal{E}$ with $\varphi = 0$ on $F\mathcal{E}$ and with $u_n = \sup |v|$ on $r = r_n$. (10) and the maximum principle imply that $u_n \geqq |v|$. Therefore $u \geqq |v|$. Thus $u = 0$ on $F\mathcal{E}$ and u is not the zero solution, thereby completing the proof.

The next problem is to formulate conditions under which the equation $\Lambda u = 0$ will enjoy an extended maximum principle. To this end, let us say that a function $v = v(x)$ is an *anti-barrier*[1] *at infinity for the equation* $\Lambda u = 0$ if (i) v tends to $+\infty$ as $r \to \infty$ and (ii) v is twice continuously differentiable and $\Lambda v \leqq 0$.

Theorem 5. *Problem* II *is well-set for the equation* $\Lambda u = 0$ *if this equation has an anti-barrier at infinity.*

It suffices to show that $\Lambda u = 0$ satisfies the extended maximum principle. The proof is exactly the same as that of Theorem 1 of reference [2] and will therefore be omitted. Just as Theorems 4 and 5 are dual to Theorem 1, so is the following result dual to Theorem 2.

Theorem 6. *If* $\mathcal{A} \leqq 2 + \epsilon$ *in some neighborhood of infinity, where* $\epsilon = \epsilon(r)$ *is such that*

$$\Delta(t) \equiv \exp\left\{-\int^t \epsilon(s)\,\frac{ds}{s}\right\}$$

is not a Dini function, then problem II *is well-set for the equation* $\Lambda u = 0$.

Proof. We shall determine an anti-barrier which is a function of r alone. To this end, consider a function $v = v(r)$ satisfying $v' > 0$. Then

(11) $$\Lambda v = Q\left[v'' + \frac{v'}{r}(\mathcal{A} - 1)\right] \leqq Q\left[v'' + \frac{v'}{r}(1 + \epsilon)\right].$$

The particular function

(12) $$v = \int^r \exp\left\{-\int^t \epsilon(s)\,\frac{ds}{s}\right\}\frac{dt}{t}$$

makes the final expression in (11) equal to zero, and obviously has the property $v' > 0$. Thus $\Lambda v \leqq 0$. Moreover, by hypothesis the integral (12) tends to infinity as $r \to \infty$. It follows, then, that v is an anti-barrier, and the proof is complete.

Theorem 7. *Consider the equation* $\Lambda u = 0$ *in* $n = 2$ *variables. If* $a_{ik} \to \delta_{ik}$ *as* $r \to \infty$ *according to a Dini modulus, then problem* II *is well-set.*

---

[1] The concept of an anti-barrier appears also in references [4] and [7].

Proof. From the hypothesis it is clear that $\mathfrak{A} \leqq 2 + \epsilon$ where ϵ is a Dini function. But if ϵ is a Dini function then Δ is not. Thus Theorem 6 applies, and problem II is well-set. The restriction on the rate at which $a_{ik} \to \delta_{ik}$ is essential, as shown by an example in section 9.

Remark. Although not immediately related to the subject of the paper, it may be of interest to point out another property of operators Λ which possess an anti-barrier at infinity, namely the following Liouville theorem: *Let $u = u(x, y)$ be defined and non-negative in E^n, and suppose that $\Lambda u \leqq 0$, where Λ has an anti-barrier at infinity. Then $u \equiv$ constant.*

The proof rests on the observation that u must satisfy an extended maximum principle,

$$\liminf_{r\to\infty} u \geqq \min_{r=R} u. \tag{9'}$$

This inequality is proved in the same way as Theorem 5. Now if (9′) holds and u is defined throughout E^n, then certainly u takes on its minimum at some *finite* point. The ordinary maximum principle then implies that u constant, and the assertion is proved.

For equations $\Lambda u = 0$ which do not possess an anti-barrier a Liouville theorem still may be valid, but the result lies much deeper (*cf.* reference [2]).

3. Uniformly elliptic equations. Under the assumption that the operator Λ is uniformly elliptic, certain further conclusions can be drawn. The main additional tool which is necessary is the following limit theorem for uniformly elliptic equations ([2], Theorem 3):

Let Λ be a uniformly elliptic operator, and, if $n \geqq 3$, assume that the coefficients a_{ik} tend to limits at infinity. Let $u = u(x)$ be a bounded solution of $\Lambda u = 0$ in $r \geqq R$. Then u tends to a limit at infinity.

It is an immediate corollary of this theorem that if Λ satisfies the above hypotheses, then all possible bounded solutions of $\Lambda u = 0$ will be obtained by varying the assigned data in either exterior problem I or exterior problem II. Moreover, under the above hypotheses one or the other of these problems must be well-set. For $n \geqq 3$ this fact is nothing more than a special case of Theorem 3; the result for $n = 2$, however, is of separate interest:

Theorem 8. *Suppose that $n = 2$, and that Λ is uniformly elliptic. Then either problem* I *or problem* II *is well-set for the equation $\Lambda u = 0$.*

Proof. If problem II is *not* well-set, then for some set of boundary values ϕ there are two distinct solutions u and v. Clearly u and v tend to distinct limits u_0 and v_0 at infinity. Letting u_0 be the larger limit, we see that the function $(u_0 - v_0) - (u - v)$ is a barrier at infinity. Problem I must therefore be well-set.

The results of Theorems 1, 4, 5, and 8 may be summarized in the following diagram, where the letter "U" placed over an arrow denotes that the proof makes use of the limit theorem for uniformly elliptic equations.

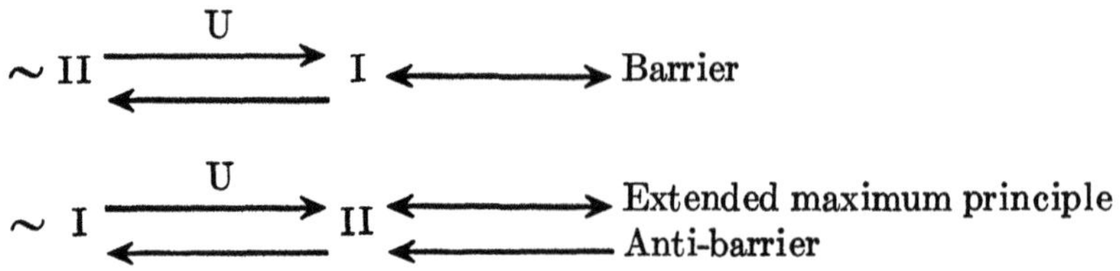

Part II

4. The general homogeneous equation. We consider briefly in this section the equation

$$Lu = 0 \qquad (c \leqq 0). \tag{13}$$

The results are not radically different from those already stated, except that, insofar as exterior problem I is concerned, *we require that the limit l be zero.* Non-zero limiting values are more naturally treated in connection with the general inhomogeneous equation.

With the proviso noted above, one sees easily that Theorems 1, 4, and 5 remain true for equation (13). The sufficient conditions for problems I and II, which are stated in Theorems 2 and 6, respectively, are likewise valid if the function $\mathfrak{A}$ is replaced by $\mathfrak{A}^*$, where

$$\mathfrak{A}^* = \frac{\sum (a_{ii} + b_i x_i)}{Q}. \tag{14}$$

Analogues of Theorems 3 and 7 can also be stated but these require additional restrictions on the growth of the coefficients at infinity.

5. The inhomogeneous equation. In this and the following sections we shall consider the full equation

$$Lu = f \qquad (c \leqq 0), \tag{15}$$

and we shall suppose that (15) is subject to the normalization

$$Q = \frac{1}{r^2} \sum a_{ik} x_i x_k = 1. \tag{16}$$

This normalization can be accomplished without loss of generality provided the origin is not in $\mathcal{E} + F\mathcal{E}$; the latter condition will be assumed henceforth. The restriction noted for c is more or less inherent in the problem, though in certain cases it can be lightened somewhat. This aspect of the problem will be put off to a later section. So far as problem I is concerned, it is sufficient to study the special case where the limit value l is zero, for the dependent variable $w = u - l$ satisfies an equation of the form (15) and tends to zero at infinity. Exterior problem I with the assigned limit $l = 0$ will be referred to as problem I_0. We shall require the following simple Sturmian-type lemma.

Lemma 2. *Consider the ordinary differential equation*

$$\psi'' + B(t)\psi' + C(t)\psi = F(t), \qquad a \leqq t < b,$$

where B, C, F *are continuous functions satisfying* $C \leqq 0, F \geqq 0$. *If* $\psi(a) = \psi'(a) = 0$, *then* $\psi, \psi' \geqq 0$ *for* $a \leqq t < b$.

Proof. Consider the equation

$$\psi_n'' + B(t)\psi_n' + C(t)\psi_n = F(t) + \frac{1}{n} \tag{17}$$

with the initial conditions $\psi_n(a) = \psi_n'(a) = 0$. From (17) we have $\psi_n''(a) \geqq 1/n$, implying that ψ_n is positive in a neighborhood of a. Now let $a < t \leqq c$ be any interval in which $\psi_n > 0$. Then since

$$\psi_n'' + B\psi_n' = F + \frac{1}{n} - C\psi_n = G > 0,$$

we have

$$\psi_n' = \frac{1}{H(t)} \int_a^t G(s)H(s)\, ds, \qquad a \leqq t \leqq c,$$

where $H(t) = \exp \{\int B(t)dt\}$. Evidently $\psi_n' > 0$ in $a < t \leqq c$, whence, by an obvious argument ψ_n, $\psi_n' > 0$ in the whole interval $a < t < b$. Letting $n \to \infty$ we get $\psi_n \to \psi$, $\psi_n' \to \psi'$, from which the conclusion of the lemma follows.

Lemma 3. *Suppose that* f *vanishes in a neighborhood of infinity. Then if problem* I_0 *is well-set for* $Lu = 0$ $(c \leqq 0)$, *it is well-set for* $Lu = f$.

Proof. Since problem I_0 is well-set for the homogeneous equation, it is sufficient to consider only the case where $\varphi \equiv 0$ on $F\mathcal{E}$.

Let R be so large that $f = 0$ for $r \geqq R$. Also choose functions $B = B(r)$ and $C = C(r)$ such that

$$B \geqq \frac{1}{r}(\mathcal{A}^* - 1), \qquad C \leqq c,$$

where $\mathcal{A}^*$ is defined by (14). Finally, let $F = F(r)$ be a continuous function satisfying $F \geqq |f|$ and $F = 0$ for $r \geqq R$. An easy computation shows that if $\psi = \psi(r) \geqq 0$, $\psi' \leqq 0$, and

$$\psi'' + B\psi' + C\psi = F, \tag{18}$$

then $L(\psi) \geqq |f|$. From here on let ψ denote the particular solution of (18) such that $\psi(R) = \psi'(R) = 0$. Then $\psi \equiv 0$ for $r \geqq R$, while

$$\psi \geqq 0, \qquad \psi' \leqq 0 \quad \text{for} \quad r \leqq R,$$

the inequalities following from Lemma 2.

Now let $\{u_n\}$ be a φ-sequence for $Lu = f$ with $\varphi = 0$ on $F\mathcal{E}$. It may be assumed that $r_n \geqq R$ for all n, and also that $u_n = 0$ on $r = r_n$. Since $L\psi \geqq |Lu_n|$,

$$\psi \pm u_n \leqq \max_{F\mathcal{E}} \psi \quad \text{in} \quad \mathcal{E} \cap (r \leqq r_n).$$

Furthermore, because $\psi \geqq 0$ it follows that $|u_n| \leqq \max_{F\mathcal{E}} \psi$. In other words, the sequence $\{u_n\}$ is uniformly bounded. By Lemma 1, a subsequence $\{u_m\}$ convergences to a solution u of $Lu = f$ with $u = \varphi = 0$ on $F\mathcal{E}$. It remains only to show that $u \to 0$ as $r \to \infty$.

To do this, choose a barrier v for the homogeneous equation such that

$$v \geqq 1 + |u| \quad \text{on} \quad r = R.$$

This can certainly be done, since problem I_0 is well-set for $Lu = 0$. For all but a finite number of indices m, we infer that

$$|u_m| \leqq v, \qquad R \leqq r \leqq r_m .$$

Letting $m \to \infty$ yields $|u| \leqq v$ for $r \geqq R$, and hence u tends to zero as $r \to \infty$.

The concept of a barrier is again fundamental in treating problem I_0, though it assumes a slightly different form. We shall say that a *pair* of functions $\bar{v}$ and $\underline{v}$ constitutes a *barrier at infinity for equation* (15) if and only if (i) $\bar{v} > \underline{v}$ in some neighborhood of infinity, (ii) $\bar{v}$, $\underline{v}$ tend to zero as $r \to \infty$, and (iii) $\bar{v}$, $\underline{v}$ are twice continuously differentiable and $L\bar{v} \leqq f$, $L\underline{v} \geqq f$.

Theorem 9. *Problem* I_0 *is well-set for equation* (15) *if and only if this equation has a barrier at infinity.*

Remark. In proving Theorem 9 the principle difficulty to be overcome is that, unlike the homogeneous case, there appears to be no simple method for extending the barrier to all of $\mathcal{E}$. But for this, the proof of Theorem 1 would carry over to Theorem 9 more or less intact. Lemma 3 is designed to circumvent this difficulty.

Proof of Theorem 9. If problem I_0 is well-set, then we can easily construct a barrier by the method of Theorem 1.

On the other hand, problem I_0 obviously cannot have more than one solution, so that it remains only to show that if equation (15) has a barrier, then problem I_0 is solvable. Suppose then that the given barrier functions are defined for $r \geqq \frac{1}{2} R$. By the methods of Lemma 1 we can easily construct a function w such that

$$\underline{v} \leqq w \leqq \bar{v}, \qquad Lw = f \qquad (r > \tfrac{1}{2}R).$$

We leave the details to the reader.

Now let w^* be a C^2 function defined over $\mathcal{E} + F\mathcal{E}$ such that $w^* \equiv w$ for $r \geqq R$; and introduce into the equation (15) the new dependent variable $u^* = u - w^*$. Evidently $Lu^* = g$, where $g \equiv 0$ for $r \geqq R$. Problem I_0 is well-set for the equation $Lu^* = 0$, since $v = \bar{v} - \underline{v}$ constitutes a barrier (see section 4). Therefore, by Lemma 3, problem I_0 is solvable for $Lu^* = g$. Thus it must also be solvable for equation (15), completing the proof.

The following theorem is a direct analogue of Theorem 2.

Theorem 10. *If in some neighborhood of infinity*

$$\mathcal{A}^* \geqq 2 + \epsilon \quad \text{and} \quad r^2|f| \leqq \delta(r),$$

where ϵ *is a positive constant and* $\delta(r)$ *is a Dini function, then problem* I_0 *is well-set for equation* (15).

Proof. Suppose $v = v(r)$ and $v > 0$, $v' \leqq 0$. Then by (14) and (16)

$$Lv \equiv v'' + \frac{v'}{r}(\mathcal{A}^* - 1) + cv \leqq v'' + \frac{v'}{r}(1 + \epsilon),$$

since $c \leqq 0$. Now consider the particular function

$$v = \int_r^\infty t^{-1-\epsilon} \int_1^t s^{-1+\epsilon}\, \delta(s)\, ds\, dt.$$

The fact that the integral converges follows by interchanging orders of integration; thus

$$v = \frac{1}{\epsilon}\left[r^{-\epsilon} \int_1^r s^\epsilon\, \delta(s) \frac{ds}{s} + \int_r^\infty \delta(s) \frac{ds}{s}\right].$$

In these formulas we have tacitly assumed δ to be defined over the interval $1 \leqq r < \infty$, and we shall also suppose $\delta > 0$. Both these conditions can be imposed without loss of generality. It is now easily verified that

$$v > 0, \qquad v' < 0, \qquad v'' + \frac{v'}{r}(1 + \epsilon) = -\frac{\delta}{r^2},$$

while also $v \to 0$ as $r \to \infty$. The function pair $\bar{v} = v(r)$, $\underline{v} = -v(r)$ thus constitutes a barrier at infinity for equation (15).

Corollary. *If in some neighborhood of infinity,*

$$\mathcal{A}^* \geqq 2 + \epsilon, \qquad r^2(|c| + |f|) \leqq \delta(r),$$

where ϵ *is a positive constant and* $\delta(r)$ *is a Dini function, then problem* I *is well-set for equation* (15).

That the condition imposed on f in the above theorem and corollary is sharp is shown by the example of section 10.

6. The inhomogeneous equation. Problem II. Uniqueness for the inhomogeneous equation is equivalent to uniqueness for the homogeneous equation; the relevant facts are found in section 4. The existence of solutions of Problem II is, however, less trivial.

Theorem 11. *Problem* II *is solvable for equation* (15) *if and only if there exist functions* $\bar{w}$ *and* $\underline{w}$ *which in some neighborhood of infinity are bounded, of class* C^2, *and satisfy* $L\underline{w} \geqq f \geqq L\bar{w}$.

Proof. The argument is almost a duplication of the proof of Theorem 9. To begin with, by the method of Lemma 3 we find that problem II is solvable whenever f vanishes in some neighborhood of infinity. The general case is then reduced to this one by the method of Theorem 9, with the functions $\bar{w}$ and $\underline{w}$ playing the roles of $\bar{v}$ and $\underline{v}$. The "barrier" functions $\bar{w}$, $\underline{w}$ which enter explicitly into the solution of problem II for inhomogeneous equations are implicitly present as constants in the homogeneous case.

The next result is a direct analogue of Theorem 6.

Theorem 12. *If*

$$\mathcal{a}^* \leqq 2 + \epsilon(r), \qquad r^2|f| \leqq \delta(r),$$

where $\epsilon(r)$ *and* $\delta(r) \log r$ *are Dini functions, then problem* II *is well-set for equation* (15).

Proof. In order to show that problem II has a solution we must construct functions $\bar{w}$ and $\underline{w}$ satisfying the conditions of the preceding theorem.

Consider the function

$$w = \int_1^r \Delta(t)\left(\int_t^\infty \frac{\delta(s)}{\Delta(s)}\frac{ds}{s}\right)\frac{dt}{t}, \qquad r \geqq 1,$$

where $\Delta(t)$ is defined by (5). From the hypothesis on ϵ, it is clear that there exist positive constants a and b such that $a \leqq \Delta(t) \leqq b$, $t \geqq 1$. The inner integral is therefore convergent, and w is well defined. Moreover, w is a bounded function,

$$w \leqq \frac{b}{a}\int_1^\infty \left(\int_t^\infty \delta(s)\frac{ds}{s}\right)\frac{dt}{t} = \frac{b}{a}\int_1^\infty \delta(s)\log s\,\frac{ds}{s}.$$

Finally, since $w \geqq 0$, $w' \geqq 0$ we have

$$Lw = w'' + \frac{w'}{r}(\mathcal{a}^* - 1) + cw \leqq w'' + \frac{w'}{r}(1 + \epsilon) = -\delta(r)r^{-2} \leqq -|f|.$$

Assuming, as we may, that $\delta > 0$, it follows that $\bar{w} = w$, $\underline{w} = -w$ satisfy the conditions of Theorem 11. Thus problem II is solvable.

On the other hand, problem II has at most one solution since the homogeneous equation $Lu = 0$ admits only one solution whenever $\Delta(t)$ is *not* a Dini function (see section 4). But if $\epsilon(r)$ *is* a Dini function, then certainly $\Delta(t)$ is *not*.

Corollary. *Consider equation* (15) *in* $n = 2$ *variables. Suppose*

$$a_{ik} \to \delta_{ik}, \qquad rb_i \to 0, \qquad r^2 f \to 0$$

as $r \to \infty$, *the convergence in each case being governed by a Hölder modulus. Then the exterior Dirichlet problem* II *is well-set.*

Even for the equation $\Delta u = f$ this corollary seems to be new. The conditions of Theorem 12 are best possible, as shown by the last example of section 10.

7. The special case $c \leqq -m^2 < 0$. In the quite special case that $c(x)$ is bounded away from 0, certain more precise conclusions can be drawn (for previous work on this case, see references [1], [8]). We begin with the following uniqueness theorem, of interest even outside the context of the paper.

Theorem 13. *Let u be a solution of the homogeneous equation* (15) *in $\mathcal{E}$, with $u = 0$ on $F\mathcal{E}$. Suppose that in some neighborhood of infinity $M \leqq \mathfrak{A}^* \leqq N$, $c \leqq -m^2$, where M, N, and m are constants ($m > 0$). Under these conditions, if*

$$u = o(r^{\frac{1}{2}(1-N)}e^{mr}) \quad as \quad r \to \infty,$$

then $u \equiv 0$ in $\mathcal{E}$.

Proof. A simple computation establishes that

$$\begin{aligned} L(r^p e^{mr}) &= e^{mr}[(m^2 + c)r^p + m(2p + \mathfrak{A}^* - 1)r^{p-1} + p(p + \mathfrak{A}^* - 2)r^{p-2}] \\ &\leqq e^{mr}[m(2p + N - 1)r^{p-1} + O(r^{p-2})]. \end{aligned}$$

Consequently

$$L[r^{\frac{1}{2}(1-N)}e^{mr}(1 + r^{-\frac{1}{2}})] \leqq -mr^{-\frac{1}{2}(2+N)}e^{mr}[1 + O(r^{-\frac{1}{2}})],$$

and the latter expression is negative for r suitably large. The function

$$v = r^{\frac{1}{2}(1-N)}e^{mr}(1 + r^{-\frac{1}{2}})$$

is therefore an anti-barrier. Hence any function which is $o(v)$ satisfies the extended maximum principle (see reference [2], Theorem 1). Since $u = o(v)$ and has zero boundary data, it follows that $u \equiv 0$ in $\mathcal{E}$.

Theorem 13 allows us to establish conditions under which problem II is well-set. In particular, we have

Theorem 14. *If in some neighborhood of infinity $\mathfrak{A}^*$ and f are bounded and $c \leqq -m^2 < 0$, then problem* II *is well-set for equation* (15).

Proof. From Theorem 13, there cannot be more than one solution. On the other hand, taking $\bar{w}$ and $\underline{w}$ to be constants, by Theorem 11 there exists a solution of problem II.

8. Relaxing the condition $c \leqq 0$. As we have remarked earlier, the condition $c \leqq 0$ is more or less essential in deriving the conclusions of this paper. It may happen, however, that by a transformation of the dependent variable our equation can be changed from a form in which c assumes positive values into a form in which $c \leqq 0$. In such a case, the preceding work can be used to investigate the exterior Dirichlet problem. Theorem 15, which is an extension of Theorem 10, bears upon this problem.

Theorem 15. *Consider equation* (15), *except that the condition $c \leqq 0$ is now dropped. Suppose that*

$$\mathcal{A}^* \geqq 2 + \epsilon, \qquad r^2 c \leqq \frac{\alpha(\epsilon - \alpha)}{1 + \beta r^\alpha}, \qquad r^2|f| \leqq \delta(r)$$

where ϵ, α, β *are positive constants,* $\alpha \leqq \epsilon$, *and* $\delta(r)$ *is a Dini function. Then problem* I_0 *is well-set.*

Proof. (By the same method it is possible to extend Theorem 12, but we have omitted this result because of its awkward statement.) Introduce the new dependent variable u' by means of the substitution

$$u = (1 + Ar^{-\alpha})u', \qquad A = \frac{1}{\beta}.$$

Then $L'u' = f'$, where $a'_{ik} = a_{ik}$,

$$b'_i = b_i - 2\alpha r^{-2} a_{ik} x_k (1 + \beta r^\alpha)^{-1}, \qquad f' = (1 + Ar^{-\alpha})^{-1} f,$$

and

$$r^2 c' = r^2 c - \alpha(\mathcal{A}^* - 2 - \alpha)(1 + \beta r^\alpha)^{-1} \leqq 0.$$

In addition, for r sufficiently large,

$$\mathcal{A}^{*\prime} = \mathcal{A}^* - 2\alpha(1 + \beta r^\alpha)^{-1} > 2 + \tfrac{1}{2}\epsilon,$$

while $|f'| \leqq$ const. $\delta(r) r^{-2}$. Theorem 10 therefore applies to $L'u' = f'$, and consequently problem I_0 is well-set for $Lu = f$.

Corollary. If the condition on c *in Theorem* 15 *is replaced by the hypothesis that* $\mathcal{E}$ *is contained in the exterior of a sphere of radius* R *and if*

$$r^2 c < \alpha(\epsilon - \alpha)\left(\frac{R}{r}\right)^\alpha \quad \text{in} \quad \mathcal{E} + F\mathcal{E},$$

then the conclusion of the theorem remains valid.

Proof. In the set $\mathcal{E} \cap (R < r \leqq 2R)$ there exists a constant k, $0 < k < 1$, such that

$$r^2 c < \alpha(\epsilon - \alpha)\left[\left(\frac{R}{r}\right)^\alpha - k\right].$$

Thus in $\mathcal{E}$ we have

$$r^2 c < \frac{\alpha(\epsilon - \alpha)}{1 + \beta r^\alpha}, \qquad \beta = (2R)^{-\alpha} \operatorname{Min}\left(\frac{k}{1 - k}, 2^\alpha - 1\right),$$

and Theorem 15 applies.

Part III

9. Examples. In this and in the following section we present some equations which illustrate the essential sharpness of our results.

Consider the particular equation

$$\left(\delta_{ij} + \frac{g(r)}{r^2} x_i x_j\right) u_{ij} = 0, \qquad g > -1, \tag{19}$$

which was first introduced in reference [2]. The eigenvalues of the coefficient matrix are 1 and $1 + g$, the former having multiplicity $n - 1$. Thus (19) is elliptic, and even uniformly elliptic if g is bounded away from -1 and $+\infty$. Furthermore,

$$\mathfrak{a} = \frac{n + g}{1 + g} = 2 + \epsilon,$$

where

$$\epsilon = \epsilon(r) = \frac{n - 2 - g}{1 + g}, \qquad -1 < \epsilon < \infty.$$

It is easily verified that (19) possesses a radially symmetric solution

$$u = u(r) = \int^r \exp\left\{-\int^t \epsilon(s)\,\frac{ds}{s}\right\}\frac{dt}{t}. \tag{20}$$

In further considerations the asymptotic behavior of u will be essential. Since $u > 0$, $u' > 0$, it is clear that either

I. $u \to$ finite limit l, or

II. $u \to +\infty$ as $r \to \infty$.

In the first case the function $l - u$ is a barrier, hence the exterior Dirichlet problem I is well-set. In the second case, u itself is an anti-barrier and problem II is well-set. Since these cases are mutually exhaustive, it follows that *for equation* (19), *either problem* I *or problem* II *is well-set depending on whether the function*

$$\Delta(t) = \exp\left\{-\int^t \epsilon(s)\,\frac{ds}{s}\right\}$$

is or is not a Dini function. This shows that Theorems 2 and 6 are best possible, as already stated.

In addition, since the exterior problem is shown to be governed only by the asymptotic behavior of $\epsilon(r)$, it is clear that neither the dimension of the underlying space, nor the hypothesis of uniform ellipticity, is of any aid (in itself) in determining which of the problems I or II is well-set.

Finally, we shall show that Theorem 7 is sharp, in that the conditions $a_{ik} \to \delta_{ik}$, $n = 2$, do not in themselves guarantee that problem II is well-set. Indeed, in equation (19) set $n = 2$ and

$$g = -\frac{2}{2 + \log r}.$$

Then as $r \to \infty$ we have $g \to 0$ and $a_{ik} \to \delta_{ik}$. However, (20) yields

$$u(r) = l + k\,(\log r)^{-1},$$

which tends to a finite limit. Hence problem I is well-set. It is only when the limit δ_{ik} is attained with sufficient rapidity (say according to a Dini modulus) that problem II will necessarily be well-set. This example may give pause to those who, on the sole basis of the fact $a_{ik} \to \delta_{ik}$ as $r \to \infty$, presume that their equation has all the usual qualitative properties of the Laplace equation.

10. The Poisson equation. By means of the Poisson equation we shall now prove the sharpness of the asymptotic conditions imposed on the right hand side, $f(x)$, in Theorems 10 and 12. Consider, then, the equation

$$\Delta u = f(x) \qquad (n \geqq 3). \tag{21}$$

We know from Theorem 10 that if $r^2|f|$ is bounded by a Dini function, then problem I is well-set. *Suppose, however, that*

$$r^2 f \geqq \psi(r) \geqq 0, \tag{22}$$

where $\psi(r)$ *is* not *a Dini function. Then* (21) *can* not *have a solution bounded from above, which is defined in a neighborhood of infinity.*[2]

Proof. Let $u = u(x)$ be a solution of (21) defined for $r \geqq R$, and suppose $u \leqq K < \infty$. Also let

$$G(x, h) = \frac{1}{n-2}\left[\frac{1}{d^{n-2}} - \left(\frac{R}{rd'}\right)^{n-2}\right]$$

be the Green's function for the exterior region $r > R$, where $d = |x - y|$ and $d' = |x' - y|$. We observe that for a fixed x, $r > R$,

$$G(x, y) \geqq (\text{pos. const.})\, |y|^{2-n} \quad \text{as} \quad |y| \to \infty. \tag{23}$$

Denote by $\{f_\nu\}$ a sequence of non-negative Hölder continuous functions converging upward to f and vanishing outside a compact set. Finally, set

$$u_\nu(x) = K - \frac{1}{\omega}\int_{|y|>R} G(x, y) f_\nu(y)\, dy$$

(ω = area of unit sphere). Now $\Delta u_\nu = f_\nu \leqq f$, and $u_\nu = K$ when $r = R$, $r = \infty$. It follows from the maximum principle that $u_\nu \geqq u$ everywhere.

From the Lebesgue monotone convergence theorem, however,

$$\lim_{\nu\to\infty} \int G(x, y) f_\nu(y)\, dy = \int G(x, y) f(y)\, dy.$$

[2] This can be phrased alternatively as a result about differential inequalities: If

$$\Delta u \geqq r^{-2}\psi(r) \geqq 0,$$

where $\psi(r)$ *is not a Dini function, then* $\limsup_{r\to\infty} u = +\infty$.

The latter integral is divergent by virtue of (22) and (23); hence, for each fixed x, $u_\nu(x) \to -\infty$. But this contradicts the condition $u_\nu \geqq u$, thus completing the proof.

The corresponding situation when $n = 2$ is a little more delicate. Consider the equation

$$\Delta u = f \qquad (n = 2). \tag{24}$$

From Theorem 12, if $r^2|f|$ is bounded by $\delta(r)$ and if $\delta \log r$ is a Dini function, then problem II is well-set. *However, if*

$$r^2 f \geqq \psi(r) \geqq 0,$$

where $\psi \log r$ *is* not *a Dini function, then* (24) *can have no bounded solutions. If* ψ *itself is* not *a Dini function then* (24) *can have no solutions bounded from above.*

Proof. As before, let u be a solution of (24) defined for $r \geqq R$ and such that $u \leqq K < \infty$. Let

$$G(x, y) = \log \frac{1}{d} - \log \frac{R}{r\,d'}$$

be the Green's function for $r > R$, and observe that for a fixed x

$$G(x, y) \to \log \frac{r}{R} \quad \text{as} \quad |y| \to \infty. \tag{25}$$

With f_ν defined as in the preceding proof, set

$$u_\nu(x) = K - \frac{1}{2\pi} \int_{|y|>R} G(x, y) f_\nu(y)\, dy.$$

Then $\Delta u_\nu = f_\nu \leqq f$ and $u_\nu = K$ on $r = R$. We now assert that $u_\nu \geqq u$ everywhere (this is no longer an immediate consequence of the maximum principle, as it was before). To prove the assertion set $w = u - u_\nu$, and observe that $\Delta w \geqq 0$ and $w = 0$ on $r = R$. Also w is bounded above, since u_ν has some finite lower bound (see equation (26)). It follows then from the maximum principle that

$$w \leqq \epsilon \log \frac{r}{R} \quad \text{for any} \quad \epsilon > 0,$$

and this in turn shows that $w \leqq 0$, as asserted.

By the Lebesgue theorem,

$$\lim_{\nu \to \infty} \int G(x, y) f_\nu(y)\, dy = \int G(x, y) f(y)\, dy.$$

Since for fixed x and large y we have $G \geqq$ (pos. const.), the latter integral is divergent whenever ψ is *not* a Dini function. This proves the second part of the theorem.

To prove the first part, we have from (25) and the symmetry of the Green's function

(26) $$\lim_{r\to\infty} u_\nu = K - \frac{1}{2\pi}\int \log\frac{|y|}{R}\, f_\nu(y)\, dy.$$

Since

$$\lim_{\nu\to\infty}\int \log\frac{|y|}{R}\, f_\nu(y)\, dy = \int \log\frac{|y|}{R}\, f(y)\, dy$$

and $u \leqq u_\nu$, it follows from (26) that

(27) $$\limsup_{r\to\infty} u \leqq K - \frac{1}{2\pi}\int \log\frac{|y|}{R}\, f(y)\, dy = -\infty,$$

which implies that u is not bounded.

Notice that we have proved more than was asserted, for (27) implies that u has a limit at infinity. Therefore if $\psi \log r$ is *not* a Dini function and u is bounded from above, $\lim_{r\to\infty} u = -\infty$.

Part IV

11. Non-Dirichlet type data. Preliminary considerations. The final part of the paper is devoted to the exterior problem corresponding to non-Dirichlet type data assigned on the boundary $F\mathcal{E}$. In order to formulate the boundary conditions which we shall consider, let $\mathbf{v}$ be a field of unit vectors defined on $F\mathcal{E}$ and directed into $\mathcal{E}$. Also let

$$\frac{\partial u}{\partial \nu} = \mathbf{v}\cdot\nabla u \qquad x \in F\mathcal{E},$$

be the inward directional derivative of u along $\mathbf{v}$, and let $k(x)$ be a fixed continuous function on $F\mathcal{E}$ with $k(x) \leqq 0$. The applied boundary condition is then

(28) $$\frac{\partial u}{\partial \nu} + ku = \varphi$$

where $\varphi = \varphi(x)$ is an arbitrary prescribed function on $F\mathcal{E}$. When $k \equiv 0$ and $\mathbf{v}$ is normal to $\mathcal{E}$, this is the Neumann problem. Other cases are sometimes referred to as the third boundary value problem; we shall not, however, insist on this terminology. As in the case of the exterior Dirichlet problem, it suffices to consider here the following two conditions on the behavior of a solution at infinity:

I_0. *The solution u shall tend to the limit zero as x tends to infinity.*

II. *The solution u shall be bounded in $\mathcal{E}$.*

These problems are conveniently designated as the exterior problem I_0 and the exterior problem II; no misunderstanding should result from the same nomenclature having been adopted already for the Dirichlet problem.

Now let $\{\mathcal{S}_n\}$ be a sequence of open balls centered at the origin, with radii r_n tending monotonically to infinity and with $F\mathcal{E} \subset \mathcal{S}_1$. The function sequence u_n will be called a *φ-sequence* if each function $u_n = u_n(x)$ satisfies (15) in $\mathcal{E} \cap \mathcal{S}_n$ and (28) on $F\mathcal{E}$. Under suitable technical continuity conditions which will be given shortly, we have the following result, analogous to Lemma 1.

Lemma 4. *Every locally bounded φ-sequence has a subsequence which converges to a solution in $\mathcal{E}$ satisfying the boundary condition* (28).

Proof. The fact that a subsequence $\{u_n\}$ exists which converges uniformly on compact subsets of $\mathcal{E}$ to a solution u follows exactly as in Lemma 1. The only difficulty lies in showing that u satisfies (28), and this necessarily requires some degree of smoothness of $F\mathcal{E}$, as well as of the vector field $\mathbf{v}$, the function k, and the data φ. If, firstly, the local representations of $F\mathcal{E}$ are of class $C^{2+\lambda}$, and $\mathbf{v}$, k, and φ are in $C^{1+\lambda}$ on $F\mathcal{E}$, then standard Schauder-type boundary estimates imply that the second derivatives of u are equicontinuous in $(\mathcal{E} + F\mathcal{E}) \cap \mathcal{S}_1$. From this it is immediate that u satisfies (28). The result can, however, be obtained from less strict assumptions, as we shall now show.

Suppose in fact that the frontier $F\mathcal{E}$ has bounded curvature,[3] so that E. Hopf's boundary point maximum principle will hold. In addition, denoting by $\mathcal{S}$ an arbitrary open ball containing $F\mathcal{E}$, let us assume that the mixed problem

$$(29)\qquad \begin{cases} Lu = f \quad \text{in} \quad \mathcal{E} \cap \mathcal{S} \\ \dfrac{\partial u}{\partial \nu} + ku = \varphi \quad \text{on} \quad F\mathcal{E}, \qquad u = \psi \quad \text{on} \quad F\mathcal{S} \end{cases}$$

is solvable for arbitrary functions ψ of class C^2 on $F\mathcal{S}$. Sufficient conditions for this last assumption to hold are that $\mathbf{v}$, k, and φ should be in C^λ on $F\mathcal{E}$ (see [5], §23). To show now that u satisfies (28), let w be the solution of (29) such that $\psi \equiv u$ on $F\mathcal{S}$. Setting $v = u_n - w$ we obtain $Lv = 0$ in $\mathcal{E} \cap \mathcal{S}$, and

$$\frac{\partial v}{\partial \nu} + kv = 0 \quad \text{on} \quad F\mathcal{E}, \qquad v = u_n - u \quad \text{on} \quad F\mathcal{S}.$$

It follows then from the boundary point principle that $|v| \leqq \mathrm{Max}_{F\mathcal{S}} |u_n - u|$, whence $\lim |u_n - w| = 0$. Thus $u \equiv w$, and condition (28) is certainly fulfilled.

The smoothness assumptions on $F\mathcal{E}$ and on $\mathbf{v}$, k, and φ mentioned in the foregoing proof clearly guarantee the *existence* of φ-sequences, and, as in the case of the Dirichlet problem, one can even preassign the values of u_n on the sphere $r = r_n$. It will be tacitly assumed in the remainder of the paper that the above conditions hold.

12. The exterior problem I_0. The theorem of this section gives a necessary and sufficient condition for the exterior problem I_0 to be well-set for the in-

[3] This is stronger than necessary (as it was also in the case of the Dirichlet problem), but it is in the spirit of our approach not to place stress on the technical details of best possible boundary hypotheses.

homogenous equation (15). Theorems 2, 3, and 10 can then be applied if one wishes specific criteria for a particular equation.

Theorem 16. *The exterior problem* I_0 *is well-set for the equation* $Lu = f$ $(c \leqq 0)$ *and the boundary condition*

$$Mu = \frac{\partial u}{\partial \nu} + ku = \varphi$$

if and only if the exterior Dirichlet problem I_0 *is well-set.*

Proof. The uniqueness of the solution is an immediate consequence of the boundary point principle and the ordinary maximum principle. It suffices then to show that problem I_0 is solvable if and only if there exists a barrier.

To begin with, suppose problem I_0 is solvable, and let $\bar{v}$ and $\underline{v}$ be solutions corresponding to $\varphi = 1$ and $\varphi = -1$, respectively. From the boundary point lemma it follows that $\bar{v} > \underline{v}$, and therefore the functions $\bar{v}$ and $\underline{v}$ constitute a barrier (see the definition stated prior to Theorem 9).

On the other hand, suppose that a barrier at infinity exists. Then by Theorem 9 the exterior Dirichlet problem I_0 is solvable. Let w_1 be the solution such that

$$Lw_1 = f \quad \text{in} \quad \mathcal{E}, \qquad w_1 = 0 \quad \text{on} \quad F\mathcal{E},$$

and let w_2 be the solution defined by

$$Lw_2 = 0 \quad \text{in} \quad \mathcal{E}, \qquad w_2 = 1 \quad \text{on} \quad F\mathcal{E}.$$

Evidently $0 < w_2 < 1$ in $\mathcal{E}$. Moreover $Mw_2 < 0$, so that there exists a positive constant C with the property

$$C \cdot Mw_2 \leqq -|\varphi - Mw_1|.$$

Now let $\{u_n\}$ be a φ-sequence for the equation $Lu = f$, with $u_n = w_1$ on $r = r_n$. Setting $w = Cw_2 \pm (u_n - w_1)$ we obtain easily $Lw = 0$ in $\mathcal{E} \cap \mathcal{S}_n$, and

$$Mw \leqq 0 \quad \text{on} \quad F\mathcal{E}, \qquad w \geqq 0 \quad \text{on} \quad F\mathcal{S}_n.$$

It follows then from the boundary point lemma that $w \geqq 0$; thus we have proved

$$|u_n| \leqq Cw_2 + |w_1|. \tag{30}$$

This shows that the sequence $\{u_n\}$ is uniformly bounded. Therefore by Lemma 4 there exists a solution u of $Lu = f$ which satisfies $Mu = \varphi$ on $F\mathcal{E}$. It remains only to show that u tends to zero as x tends to infinity. This, however, follows directly from (30), and the proof is completed.

13. The exterior problem II. Our object is to determine conditions under which this problem will be well-set. To begin with, the uniqueness of solutions of problem II under the boundary condition (28) requires that two conditions be satisfied: *First, the equation* $Lu = 0$ *must enjoy an extended maximum principle, and, second, at least one of the functions* c *and* k *must take on non-zero values.* Indeed, if these conditions are satisfied, then the difference of two solutions must (in the first place) be constant, and then by the second condition the constant must be zero. On the other hand, the failure of either condition allows one to construct more than one solution of a given boundary problem. To see this, observe that if $c \equiv k \equiv 0$, then two solutions can differ by a constant; while if the extended maximum principle fails, then (as in the proof of Theorem 4) we can construct a non-trivial bounded solution of $Lu = 0$ satisfying $Mu = 0$ on $F\mathcal{E}$. This discussion shows, by the way, that the exterior Neumann problem II is not well-set if $c \equiv 0$.

Having disposed of the uniqueness question, we turn now to the more difficult problem of existence. The following result is an analogue of Theorem 11.

Theorem 17. *Consider the exterior problem* II *for the equation* $Lu = f$ $(c \leqq 0)$ *and the boundary condition*

$$Mu = \frac{\partial u}{\partial \nu} + ku = \varphi,$$

and suppose that at least one of the functions c and k takes on non-zero values. Then this problem is solvable if and only if the exterior Dirichlet problem II *is solvable.*

Remarks. The condition on c and k is essential. Indeed, if $c \equiv k \equiv 0$ there may be no bounded solutions; *e.g.* Laplace's equation in two dimensions has no bounded solution in $\mathcal{E}$ unless $\int_{F\mathcal{E}} (\partial u/\partial n)\, dS = 0$.

In order to apply this theorem to a specific equation one can use the criteria for the existence of a solution of the Dirichlet problem contained in Theorems 11, 12, and 14. Moreover, if the equation is homogeneous then the exterior Dirichlet problem is always solvable.

Proof of Theorem 17. Suppose first that problem II is solvable. If w is *any* solution, we can set $\bar{w} = \underline{w} = w$ and thus verify the condition of Theorem 11.

On the other hand, let us assume that the exterior Dirichlet problem II is solvable. Then without loss of generality we can suppose that our problem is reduced to the homogeneous case $f \equiv 0$ (let w be *any* solution of the exterior Dirichlet problem II, and consider the new unknown $u - w$). With this understanding, let a boundary function φ^* be defined as follows:

(1) If $k \equiv 0$, set $\varphi^* \equiv 1$.

(2) If $k \equiv 0$, choose on $F\mathcal{E}$ two non-empty, disjoint, closed sets Γ_1 and Γ_2 with the property that $k < 0$ on $F\mathcal{E} - \Gamma_1$.

We then take for φ^* a smooth function satisfying

$$\varphi^* = \begin{cases} 1 & \text{on} \quad \Gamma_1 \\ 0 & \text{on} \quad \Gamma_2 . \end{cases}$$

Now let w be a solution of the exterior Dirichlet problem obtained from a φ^*-sequence with $w_n^* = 0$ on $r = r_n$. From the maximum principle we have

$$0 < w^* < 1 \quad \text{in} \quad \mathcal{E};$$

for case (2) this is obvious, while for case (1) it is a consequence of the fact the $c \not\equiv 0$.

In case (1), from the boundary point principle we have $Mw^* < 0$. In case (2), for the same reason, we have $Mw^* < 0$ on Γ_1; therefore, since $k < 0$ on the complement of Γ_1,

$$M(w^* + C_1) < 0 \quad \text{on} \quad F\mathcal{E}$$

provided C_1 is a suitably large constant. Finally, in either case, there exists a positive constant C_2 such that

$$C_2 \cdot M(w^* + C_1) \leqq -|\varphi| \quad \text{on} \quad F\mathcal{E}.$$

Now let $\{u_n\}$ be a φ-sequence with $u_n = 0$ on $r = r_n$. Exactly as in the proof of Theorem 16 we derive

$$|u_n| \leqq C_2(w^* + C_1) \leqq C_2(1 + C_1).$$

Thus $\{u_n\}$ is bounded, and Lemma 4 implies that there exists a bounded solution of the problem in question. This completes the proof of Theorem 17.

The above research was supported in part by the United States Air Force Office of Scientific Research under Contract No. AF 49(638)-262.

Bibliography

[1] F. V. Atkinson, *On a theorem of K. Yosida*, Proc. Japan Academy, **28** (1952), pp. 327–329.

[2] D. Gilbarg & J. Serrin, *On isolated singularities of solutions of second order elliptic differential equations*, J. d'Analyse Mathématique, **4** (1954–6), pp. 309–340.

[3] G. Giraud, *Nouvelle généralisation des problèmes relatif aux opérations du type elliptique*, Ann. Societé Polonaise de Mathématique, **14** (1935), pp. 74–115.

[4] M. Krzyzański, *Sur le probleme de Dirichlet pour l'equation lineare du type elliptique dans un domaine non borne*, Rend. Accademia Nazionale dei Lincei (8), **4** (1948), pp. 408–416.

[5] C. Miranda, Equazioni alle derivate parziali di tipo ellittico. Springer, Berlin, 1955 [especially Section 49].

[6] M. Picone, *Intorno alla teoria di una classica equazioni a derivate parziali della Fisica matematica*, Ann. Societé Polonaise de Mathématique, **21** (1948). pp. 161–169.

[7] S. Simoda & M. Nagumo, *Sur la solution bornée de l'equation aux derivees partielles du type elliptique*, Proc. Japan Academy, **27** (1951), pp. 334–339.

[8] K. Yosida, *A theorem of Liouville's type for meson equation*, Proc. Japan Academy, **27** (1951), pp. 214–215.

University of Minnesota
Minneapolis, Minnesota

I.3. The Dirichlet Problem – Surveys

Ireneo Peral

Abstract. We comment on a couple of surveys by James Serrin about the solvability of the Dirichlet problem for elliptic equations, a subject in which Serrin's contributions have been determinant.

Mathematics Subject Classification (2010). Primary 00B10, 35J25; Secondary 3500, 35J93.

Keywords. Dirichlet problem, equation invariants, Hilbert's problems.

Introduction

The XXth Century was an epoch in which History has been accelerated by the scientific achievements and its influence in our lifestyle. Moreover, the first half of the Century was a convulse and turbulent time, dominated by different totalitarian ideologies and dramatically struck by the two World Wars and all their consequences.

However, from the point of view of Science, the XXth Century has been a Golden Age. All the modern theories in Physics, Chemistry and Biology and their applications in Technology have deeply transformed human life. From the progress in the medical health care to the new and revolutionary systems of communications, every daily aspect that one can think has been transformed in an unbelievable way. Almost all this progress depends strongly on the mathematical results obtained during the XXth Century. The role of the Mathematics in this huge progress of Science has been decisive.

The XXth Century, among other events, started for Mathematics with David Hilbert's seminal conference in the 1900 edition of the International Congress of Mathematics (ICM) in Paris. More precisely, the developments in the theory of Partial Differential Equations during the last hundred years are impressive and they mean a huge leap forward in this particular area. We consider the ICM events as argumental line precisely because [1] is Serrin's talk in the 1970 edition of the ICM at Nice and [2] tells us about the 20th problem proposed by Hilbert in the 1900 edition of the ICM at Paris.

A couple of masterpieces

By reading the surveys [1] and [2] one really enjoys both the beauty of the mathematical contents and also the precise, elegant and didactic style. The two surveys by James Serrin are a couple of masterpieces that only a great mathematician could write. Both of them are wonderful samples of the progress of the partial differential equations area during the XXth Century, and the decisive contributions by Serrin to this theory.

P. Pucci et al. (eds.), *James Serrin. Selected Papers: Volume 1*, Contemporary Mathematicians, DOI 10.1007/978-3-0348-0845-3_3, © Springer Basel 2014

I. In [1], Serrin starts pointing out how *certain invariants of the equation play a central role in the question of existence of solutions.* The equation under consideration is the following,

$$\mathfrak{A}(x,u,Du)D^2u = \mathfrak{B}(x,u,Du),$$

where $\mathfrak{A},\mathfrak{B}$ are smooth enough and $\mathfrak{A}$ is an elliptic matrix. We assume a domain $\Omega \subset \mathbb{R}^n$ with smooth boundary and we prescribe Dirichlet's boundary data f given by a sufficiently regular function. He takes his pioneering result about the solvability of the *prescribed constant mean curvature* problem as a motivation to present in a short and clear way the necessity of some conditions on the curvature of the boundary.

Previously, Serrin describes in a magistral way the general continuity procedure that allows to solve the Dirichlet problem for a large class of second-order quasilinear elliptic equations. For the basic existence theorem for the Dirichlet problem, by using some results by O.A. Ladyzhenskaya and N.N. Uraltseva, the celebrated method by J. Leray and J. Schauder and his own version in the reference [19] of the survey. Recall that in the continuity method we assume our original problem immersed in a continuous one-parametric family of problems,

$$\mathfrak{A}(x,v,Dv)D^2v = \mathfrak{B}(x,v,Dv;\tau),\ x\in\Omega, \quad v=\tau f \text{ on } \partial\Omega,$$

where $\mathfrak{B}(x,v,Dv;0) = 0$ and $\mathfrak{B}(x,v,Dv;1) = \mathfrak{B}(x,v,Dv)$. The result is that if we are able to find a bound $M > 0$ such that for all solution $v \in \mathcal{C}^2(\overline{\Omega})$, $||v||_\infty + ||Dv||_\infty \le M$, for all $\tau \in [0,1]$, then the original problem has a solution.

Serrin pointed out how to obtain this result in three steps: i) L^∞ estimates for v; ii) Assuming L^∞-estimates, find a bound for the gradient on the boundary; iii) Assuming i), ii), estimates for $||Dv||_\infty$ in Ω.

Step i) is based on some Maximum Principle. For ii) and iii), Serrin gives precise conditions in Theorem 1 and Theorem 2, respectively, in [1]. This dissection allows to understand deeply the role played by *certain invariants of the equation.*

To finish this rather incomplete description of the contents we point out that many examples related to problems in geometry and physics are considered in the last part of the survey, most of them Serrin's original contributions.

II. The survey [2] begins with a nice historical reference to the Hilbert twentieth problem in the 1900 list of problems at ICM in Paris and its influence in the mathematics of the XX century. Reading the list of references one can realize that the best mathematical minds took care of the solution of Hilbert's proposal. It is worth pointing out that Serrin's contributions quoted in [2] as Serrin [1], [2], [3] and [4], are milestones of the theory of quasilinear elliptic equations. Serrin next, point out how the works by Schauder introducing topological arguments are important to solve the Dirichlet and other boundary problems.

The relation of the Dirichlet problem for quasilinear elliptic equations with the Calculus of Variations in the case of Lagrangians of the form $F(x,u,p)$, where $F \in \mathcal{C}^2$ and $\left(\dfrac{\partial^2 F}{\partial p_i \partial p_j}\right) > 0$ is given by the corresponding Euler–Lagrange equation, $\mathrm{div} D_p F(x,u,p) = F_u(x,u,p)$. (Remember the result by Serrin in Acta Math. 102 (1959), pp. 1–32, to obtain the weak lower semi-continuity of the functional associated to F.) This method started by Hilbert and Serrin describes some of the main contributions and contributors. The key point is that in this way we first find a *weak solution* and then study some stronger regularity. Notice that then we can say that the Hilbert's *twentieth problem closes the nineteenth.* In this sense the results by De Giorgi, Nash, and Moser in the linear case and by Serrin in the nonlinear case are a historical reference.

The survey continues going back to the continuity method described in [1]. Serrin closes the paper with very interesting reflections about the state of the art by the time and the presentation of more general problems, for instance fully nonlinear equations, more general boundary conditions and other extensions that show that in fact Hilbert's proposal is not a single problem but a real program.

Epilogue

Given the limitations of space we cannot give more details and we limit ourselves to recommend the reading of the surveys and enjoy the great mathematical ideas contained in both articles.

But I also recommend to read the papers as an historical contribution by a great mathematician, protagonist and witness of this Golden Age that was the XXth Century.

References

[1] James Serrin, *Boundary Curvatures and the Solvability of Dirichlet's Problem.* Actes, Congrès Intern. Math., 1970, Tome 2, pp. 867–875.

[2] James Serrin, *The Solvability of Boundary Value Problem*, Proccedings of Symposia in Pure Mathematics. Volume 28, 1976, pp. 507–524.

Ireneo Peral
Departamento de Matemáticas
Universidad Autónoma de Madrid
E-28049 Madrid, Spain
e-mail: `ireneo.peral@uam.es`

Actes du Congrès International des Mathématiciens (Nice, 1970),
Tome 2, pp. 867–875, Gauthier-Villars, Paris, 1971.

Reprint I.3.[i]

BOUNDARY CURVATURES AND THE SOLVABILITY OF DIRICHLET'S PROBLEM

by James SERRIN

In recent years the theory of the Dirichlet problem for second order quasilinear elliptic differential equations has been brought to a significant level of completeness, and in particular it has been found that certain invariants of the equation play a central role in the question of existence of solutions. Moreover, in certain cases, governed by the behavior of these invariants, it turns out to be necessary to impose curvature restrictions on the boundaries of the underlying domains in order for the Dirichlet problem to be generally solvable for given smooth boundary data. The purpose of this note is to give a very brief outline of this theory, and to present a number of examples illustrating its application to particular situations.

We consider throughout quasilinear elliptic equations of the form

$$\mathfrak{A}(x\,,u\,,Du)\,D^2u = \mathfrak{B}(x\,,u\,,Du) \tag{1}$$

where $x = (x_1, \dots, x_n)$ denotes points of E^n, Du and D^2u denote respectively the gradient vector and Hessian matrix of the dependent variable $u = u(x)$, and $\mathfrak{A}$ and $\mathfrak{B}$ are respectively a given symmetric matrix and a given scalar function of the variables indicated. The multiplication convention on the left hand side is the natural contraction $\mathfrak{A}_{ij} D^2_{ij}\, u$.

The structure of (1) is determined by the functions $\mathfrak{A}(x\,,u\,,p)$ and $\mathfrak{B}(x\,,u\,,p)$, which are assumed to be defined and continuously differentiable for all real u and p, and for all points x in the closure of any domain under consideration. Ellipticity is then expressed by the condition $\xi\,\mathfrak{A}\,\xi > 0$ for non-vanishing real vectors $\xi = (\xi_1, \dots, \xi_n)$. The Dirichlet problem for (1) in a domain Ω consists in the determination of a solution u of class $C^2(\Omega)$, which continuously takes on assigned values f on the boundary of Ω.

As a typical example, consider the equation

$$[(1 + |Du|^2)I - DuDu]\,D^2u = n\Lambda(1 + |Du|^2)^{3/2} \tag{2}$$

which governs surfaces of constant mean curvature Λ. Here the following result holds [17], [19], [20], [9].

Let Ω be a bounded domain in n-dimensional Euclidean space, whose boundary is of class C^2. Then the Dirichlet problem for equation (2) *in Ω is solvable for arbitrarily given continuous boundary data if and only if*

$$H \geqslant \frac{n}{n-1}\,|\Lambda| \tag{3}$$

at each point of the boundary of Ω, where H denotes the mean curvature of the boundary surface.

When $n = 2$ and $\Lambda = 0$, (2) reduces to the minimal surface equation ; the curvature condition (3) then implies that Ω should be convex, a well known result of Bernstein and Finn. In connection with this example, it is also of interest to consider the equation

$$(4) \qquad [I + Du\,Du]\,D^2u = c(1 + |Du|^2)^{3/2}$$

where c is a constant. The coefficient matrix has exactly the same eigenvalues as before (though with different multiplicities when $n > 2$), but now the Dirichlet problem is solvable for all C^2 domains and C^2 data ; this contrasting result emphasizes the delicate relation between the structure of (1) and the need for boundary curvature restrictions.

1.

The basic existence theorem for the Dirichlet problem can be stated as follows ([12], [15] ; for the particular form given, see [19], Section 1).

Assume that the boundary of Ω and the boundary data f are of class C^3. Let τ be any real number in $[0, 1]$. Suppose there exists a number M, independent of τ, such that if v is a member of $C^2(\bar{\Omega})$ with

$$\mathcal{A}(x, v, Dv)\,D^2v = \mathcal{B}(x, v, Dv\,;\tau) \quad \text{in } \Omega, \quad v = \tau f \quad \text{on } \partial\Omega$$

then $\sup_\Omega(|v| + |Dv|) \leqslant M$. Here $\mathcal{B}(x, u, p\,;\tau)$ is assumed to be a continuously differentiable function of all its arguments satisfying the conditions

$$\mathcal{B}(x, u, p\,;0) = 0 \quad when \quad |p| \leqslant 1, \qquad \mathcal{B}(x, u, p\,;1) = \mathcal{B}(x, u, p).$$

Then the Dirichlet problem for equation (1) in Ω has at least one solution $u \in C^2(\bar{\Omega})$ corresponding to the given boundary data f.

This proposition reduces the question of existence to that of obtaining appropriate *a priori* estimates ; accordingly, the existence of solutions of the Dirichlet problem for smooth domains and smooth data can be established by the following succession of steps :

(A) Obtain an estimate for the maximum absolute value of the solution v in Ω.

(B) Assuming the result of (A), obtain an estimate for the gradient of the solution at the boundary of Ω.

(C) Assuming the results of (A) and (B), obtain an estimate for the gradient of the solution in all of Ω.

In this program we are of course obliged to consider a *family* of Dirichlet problems corresponding to all values of the homotopy parameter τ, and the estimates must be independent of τ. Despite this it will suffice to discuss the various steps in terms of the original Dirichlet problem, since the details stand out more clearly there and can easily be carried over to the case when the homogeneous parameter is present.

There are a variety of methods available for treating Part (A) of the existence program, based on the maximum principle and other devices. Perhaps the simplest result in this regard is the following : *Suppose that* $u\mathcal{B}(x, u, 0) \geqslant 0$ *for large* $|u|$. *Then Part (A) of the existence program can be carried out.* While there is no guarantee that this or any other of these methods will apply to a particular problem under consideration, we are at least fortunate that many applications are presently covered, including all those to be discussed in Section 2.

Part (B) is the source of the boundary curvature restrictions discussed earlier. We introduce the orthogonal invariants

$$\mathcal{C}(x, u, p) = |p|^{-1}\, \mathcal{B}(x, u, p), \qquad \mathcal{E}(x, u, p) = p\, \mathcal{A}(x, u, p) p,$$

where it is assumed for the moment all that the equation is normalized so that Trace $\mathcal{A} = 1$. The following result is proved in [19].

THEOREM 1. – *Suppose there exists an increasing continuous function* $\Phi(\rho)$ *such that*

$$\frac{\mathcal{E}}{1 + |\mathcal{C}|} \geqslant \Phi(|p|), \qquad \int^{\infty} \Phi(\rho) \frac{d\rho}{\rho^2} = \infty .$$

Then Part (B) of the existence program can be carried out (1).

Suppose on the contrary that there exists a continuous function $\Psi(\rho)$ *such that*

$$\frac{\mathcal{E}}{1 + |\mathcal{C}|} \leqslant \Psi(|p|), \qquad \int^{\infty} \Psi(\rho) \frac{d\rho}{\rho^2} < \infty .$$

Then the following conclusions hold : I. If $|\mathcal{C}| \to \infty$ *as* $p \to \infty$ *then for any smooth domain there exists smooth boundary data for which the Dirichlet problem is not solvable. II. If* $\mathcal{C}$ *is bounded, then there exist smooth domains and smooth boundary data for which the Dirichlet problem has no solution.*

When the hypotheses of the first part of the theorem hold we say that (1) is *regularly elliptic.* When the hypotheses of the second part hold, the equation is called *irregularly elliptic* in case I, and *singularly elliptic* in case II. It is easy to show that this classification is invariant under arbitrary C^3 diffeomorphisms of E^n. If (1) is irregularly elliptic we cannot expect the Dirichlet problem to be well-set for any domain ; conversely, if the equation is singularly elliptic, Part (B) of the existence program can be carried out for certain domains, provided the boundary satisfies appropriate curvature restrictions. Because of space limitations we must refer the reader to [19] for a discussion of these conditions.

In any case, it should be noted that equation (2) is singularly elliptic and correspondingly requires a boundary curvature condition for the solvability of the Dirichlet problem (that is, inequality (3)), while equation (4) is regularly elliptic and is consequently solvable for all smooth domains and data. It may also be observed that any uniformly elliptic equation for which $\mathcal{B}(x, u, p) = O(|p|^2)$ is regularly elliptic. (Order relations here and later in the paper are assumed to apply as $p \to \infty$, uniformly for (x, u) in any compact set).

---

(1) For this conclusion it is actually enough to have such a function $\Phi(\rho)$ corresponding to each compact set of values (x, u).

automatically satisfied. We observe moreover that case (1) covers variational problems $\delta \int F(Du)\,dx = 0$, cases (2) – (4) apply to variational problems $\delta \int F(x, u, |Du|)\,dx = 0$, and case (2) can be used for uniformly elliptic equations if we set $\mathfrak{A}' = \mathfrak{A}$, and $\mathfrak{C} = 0$. (For related work, see [13], [16], [21]).

The existence program outlined above applies to smooth boundary data. By compactness arguments it is possible to reduce the degree of smoothness required of the data, provided that appropriate interior estimates are available for the gradient of solutions. Estimates of this type are discussed in the paper of Uraltseva included in this volume ; in particular, these estimates can be used to lighten the smoothness requirements on the data in some of the examples given in the following section.

2.

In this section we consider a number of examples which can be treated by the theory outlined above. When specific references are not given, the result is new.

1. Let M denote the minimal surface operator (i.e. the differential operator on the left side of (2)), and consider the equations

$$Mu = n\Lambda(1 + |Du|^2)^{\theta/2}, \qquad Mu = cu(1 + |Du|^2)^{\theta/2}$$

where $\Lambda \neq 0$ and $c > 0$ are constants (1). Then the corresponding Dirichlet problem in a domain Ω with C^2 boundary is solvable for arbitrarily given boundary values of class C^2 in the following circumstances : when $\theta \leqslant 2$, if and only if $H \geqslant 0$ at each boundary point ; when $2 < \theta < 3$, if $H > 0$ at each boundary point ; when $\theta = 3$, if and only if $H \geqslant [n/(n-1)]\,|\Lambda|$ and $H \geqslant c|f|/(n-1)$, respectively. When $\theta > 3$ both equations are irregularly elliptic and the Dirichlet problem is not generally solvable for any domain [3], [18], [19].

2. Let $M'u = [I + DuDu]\,D^2u$ and consider the equations

$$M'u = c(1 + |Du|^2)^{\theta/2}, \qquad M'u = cu(1 + |Du|^2)^{3/2}.$$

Then the corresponding Dirichlet problem in a domain with C^2 boundary is solvable for arbitrarily given C^2 boundary values in the following circumstances : for the first equation, if and only if $\theta \leqslant 3$; for the second equation, if and only if $\theta \leqslant 4$. (Both equations are regularly elliptic for $\theta \leqslant 4$ and irregularly elliptic for $\theta > 4$; for the first equation, however, Part (A) of the existence program can be treated for arbitrary domains only when $\theta \leqslant 3$: if $3 < \theta \leqslant 4$ the domain must have a suitably small diameter [18]).

3. Suppose (1) is uniformly elliptic (with Trace $\mathfrak{A} = 1$) and

$$\mathfrak{A}_x\,, p \cdot \mathfrak{A}_p = O(1)\,, \qquad \mathfrak{A}_u = o(1)$$

$$\mathfrak{B}\,, \mathfrak{B}_x = O(|p|^2)\,, \qquad p \cdot \mathfrak{B}_p \leqslant O(|p|^2)\,, \qquad \mathfrak{B}_u \geqslant -o(|p|^2).$$

---

(1) Included in this example are soap films under the influence of gravity (first equation with $\theta = 2$) and the upper surface of a fluid under the combined action of gravity and surface tension (second equation with $\theta = 3$).

In connection with Theorem 1 it would be valuable to prove that the Dirichlet problem is not well-posed in the sense of Hadamard for any irregularly elliptic equation, or for any singularly elliptic equation unless the boundary curvature condition is satisfied.

For Part (C) of the existence program to be carried out it seems necessary to assume a special form for the matrix $\mathfrak{A}$. Whether this can be avoided remains an open problem, though, as we shall see, a number of important situations are covered by the theory as it now stands. In any event, we shall specifically assume here that $\mathfrak{A}$ has the form

$$\mathfrak{A}'(x, u, p) + p\mathfrak{C}(x, u, p) + \mathfrak{C}(x, u, p)p$$

where $\mathfrak{A}'$ is a positive definite matrix and $p\mathfrak{C}$ and $\mathfrak{C}p$ are dyadics. Here $\mathfrak{A}'$ and $\mathfrak{C}$ are assumed to be continously differentiable, and $\mathfrak{A}'$ should additionally obey at least one of the following conditions :

(I) $\mathfrak{A}'$ depends only on p,

(II) $\mathfrak{A}'_x = O(\sqrt{(\lambda\mathcal{E})})$, $\quad \mathfrak{A}'_u, p \cdot \mathfrak{A}'_p = O(\sqrt{(\lambda\mathcal{E})}/|p|)$

where λ is the smallest eigenvalue of $\mathfrak{A}'$ and $p \cdot \mathfrak{A}'_p$ denotes the contraction $p_i \partial \mathfrak{A}'/\partial p_i$.

THEOREM 2. – *Part (C) of the existence program can be carried out in the following cases :*

(1) *Condition (I) holds and* $|p|^{-1} p \cdot \mathfrak{C}_x + |p|\,\mathfrak{C}_u + \mathfrak{C}^2/\text{Trace}\,\mathfrak{A}' \geqslant 0$ *for large*

(2) *Condition (II) holds (with O replaced by o for* $\mathfrak{A}'_x$ *and* $\mathfrak{A}'_u$*),*

$$p \,.\, \mathcal{E}_p \leqslant O(\mathcal{E})$$

and

$$\mathcal{B}, \mathcal{B}_x, \mathcal{E}_x, \mathcal{E}_u = O(\mathcal{E}), \quad -\mathcal{B}_u \leqslant o(\mathcal{E}), \quad p \cdot \mathcal{B}_p \leqslant O(\mathcal{E}).$$

(3) *Condition (II) holds and (with O replaced by o for* $p \,.\, \mathfrak{A}'_p$*), and*

$$\mathfrak{A}' = o(\sqrt{(\lambda\mathcal{E})}/|p|),$$

$$p \,.\, \mathcal{E}_p \leqslant O(\mathcal{E}),$$

while also

$$\mathcal{B}, \mathcal{B}_x, \mathcal{E}_x, \mathcal{E}_u = O(\mathcal{E}), \qquad -\mathcal{B}_u, p\ \mathcal{B}_p \leqslant O(\mathcal{E}).$$

(4) *Condition (II) holds (with O replaced by o for* $p \cdot \mathfrak{A}'_p$*), and*

$$p \cdot \mathcal{E}_p \leqslant \mathcal{E} + o(\mathcal{E}),$$

while also

$$\mathfrak{C}_x, \mathcal{E}_x, \mathcal{E}_u = O(\mathcal{E}), \qquad -\mathfrak{C}_u, p \,.\, \mathfrak{C}_p \leqslant O(\mathcal{E}/|p|).$$

Case (1) is proved in [19], Section 13 ; the remaining parts will appear in a forthcoming paper. Note that when $n = 2$ any positive definite matrix can be written in the form $\mathcal{G}(I + p\mathfrak{C} + \mathfrak{C}p)$ so that conditions (I) and (II) are

Moreover, assume that $u\mathcal{B}(x, u, 0) \geqslant 0$ for all sufficiently large $|u|$. Then the Dirichlet problem for (1) is solvable for all smooth domains and data. (A weaker version of this result appears in [12]).

4. Consider the Euler-Lagrange equation associated with the variational problem

$$\delta \int \phi(u) \sqrt{a^2 + |Du|^2}\, dx = 0$$

where $\phi(u)$ and $a = a(x, u)$ are positive functions of class C^2, with $u(a\,\phi)_u \geqslant 0$ for all sufficiently large $|u|$. Then the corresponding Dirichlet problem in a domain with C^2 boundary is solvable for arbitrarily given C^2 boundary values if and only if $H \geqslant 0$ at each point of the boundary.

5. Consider the variational problem $\delta \int F(|Du|)\, dx = 0$, where $F(t)$ is of class C^3 and $F'' > 0$ (it is assumed that $F'(0) = F'''(0) = 0$ for consistency). Let $f = tF''/F'$. Suppose first $\int^{\infty} f\, dt = \infty$. If $f \geqslant \Phi(t)$, where

$$\Phi \searrow 0\ , \quad \int^{\infty} \Phi\, dt = \infty\,,$$

then the Euler-Lagrange equation is regularly elliptic and the Dirichlet problem is solvable for arbitrarily given C^2 boundary data on any domain of class C^2. (It should be noted in passing that $\lim t^{-1}F = \infty$ implies $\int^{\infty} f\, dt = \infty$).

Suppose next that $\int^{\infty} f\, dt < \infty$. If $tf \geqslant \Phi(t)$, where Φ satisfies the same condition as before, then the Dirichlet problem is solvable for arbitrarily given C^2 boundary data if and only if $H \geqslant 0$ at each point of the boundary. Finally, whatever may be the asymptotic behavior of f, the Dirichlet problem is solvable for smooth domains and data if $H > 0$ at each point of the boundary.

6. Let S be an oriented n-dimensional surface embedded in E^{n+1}, with unit normal vector $\vec{n}$, and consider the regular parametric variational problem

$$\delta \int_S \mathfrak{F}(\vec{n})\, dA = 0\,,$$

where $\mathfrak{F}$ is of class C^3 on the unit sphere. The corresponding Euler-Lagrange equation for a non-parametric extremal is singularly elliptic, and boundary curvature restrictions are required for the associated Dirichlet problem. It is interesting to note that the required conditions depend *only* on the form of the reduced function $\mathfrak{F}(\vec{n}_0)$, where $\vec{n}_0$ is restricted to be orthogonal to the particular projection under consideration [11].

7. Consider equation (2), but with the mean curvature $\Lambda = \Lambda(x, u\,; \vec{n})$ a prescribed continuously differentiable function of the position (x, u) and the unit normal $\vec{n}$ on the surface $z = u(x)$. We assume that Ω is contained in a ball of radius R and that its boundary is of class C^2. Suppose finally

(5) $\Lambda_u \geqslant 0$, $\quad 1 + R\Lambda \operatorname{sign} u \geqslant 0 \quad$ for $|u|$ suitably large.

Then the Dirichlet problem in Ω is solvable for given C^2 boundary values f provided

$$H \geqslant \pm \frac{n}{n-1} \Lambda(y, f; \pm \vec{\nu}) \tag{6}$$

at each point y of the boundary, where $\vec{\nu}$ denotes the inner normal vector to the boundary. The solution is unique if it exists.

If Λ is independent of u, then condition (6) is necessary for the Dirichlet problem to be solvable for arbitrarily given smooth boundary values. On the other hand, if Λ depends only on u, the second condition of (5) can be dropped ; if $\Lambda = \Lambda(u; \vec{n})$ this condition can again be omitted provided the equality sign does not hold in (6).

8. A similar conclusion can be obtained for surfaces of prescribed mean curvature in central projection, that is, when the surfaces can be represented in the form $r = \nu(x)$, where x denotes points on the unit sphere S^n and r is the distance from the origin. In fact, let Ω be a smoothly bounded domain on the unit sphere whose closure is contained in the upper hemisphere $x_{n+1} > 0$. Let the mean curvature $\Lambda = \Lambda(x, r; \vec{n})$ be a prescribed continuously differentiable function on $E^{n+1} \times S^n$, such that $(r\Lambda)_r \geqslant 0$. Then the corresponding Dirichlet problem over Ω is solvable for (positive) radial values $\nu = f$ of class C^2 assigned on the boundary of Ω, provided that

$$H_g \geqslant \frac{n}{n-1} f(y) \Lambda(y, f; \vec{\nu})$$

at each point y of the boundary, where H_g is the geodesic mean curvature of the boundary and $\vec{\nu}$ is the unit inner normal (as a vector on S^n).

9. In connection with the preceding example, Alexandrov [1] has proposed relations of the type

$$\Lambda(x, r; \vec{n}) = \frac{\gamma(x, \vec{n})}{r} \qquad \text{and} \qquad \Lambda(x, r; \vec{n}) = \frac{\epsilon(x, \vec{n})}{s}$$

where γ and ϵ are continuously differentiable functions and $s = rx \cdot \vec{n}$ is the distance from the tangent plane at (x, ν) to the origin. (Note that Alexandrov's functions are invariant under similarities about the origin). The above formulae do not satisfy the smoothness requirements of the preceding case and so require a separate treatment. In this regard we shall in fact consider the more general relations

$$\Lambda(x, r; \vec{n}) = \frac{\gamma(x, r; \vec{n})}{r} \quad \text{and} \quad \Lambda(x, r; \vec{n}) = \frac{\epsilon(x, r; \vec{n})}{s}.$$

For the first relation, assume that γ is continuously differentiable and satisfies the conditions

$$\frac{\partial\gamma}{\partial r} \geqslant 0\,, \qquad \frac{n}{n-1}\;|\gamma(x\,,0\,;\vec{n})| < \frac{x_{n+1}}{\sqrt{1-x_{n+1}^2}} \tag{7}$$

Then the associated Dirichlet problem over Ω is solvable for radial values f of class C^2 assigned on the boundary of Ω provided that

$$H_g \geqslant \pm \frac{n}{n-1}\;\gamma(y\,,f\,;\pm\vec{\nu}) \tag{8}$$

at each point y of the boundary. If γ is independent of r, then condition (8) is necessary if the Dirichlet problem is to be solvable for arbitrarily given smooth boundary values. If γ is independent of x, the second condition of (7) can be omitted provided the equality sign does not hold in (8).

When $\Lambda = \epsilon/s$ the relevant differential equation is irregularly elliptic (assuming $\epsilon \neq 0$). Consequently, for any domain there exist smooth boundary values for which the associated Dirichlet problem is not solvable.

10. The geometric problems discussed in examples 7, 8, and 9, lead one to propose similar boundary value problems when the Gauss curvature, or other combinations of the principal curvatures of a surface, are prescribed functions of the space variables and the surface normal. The theory of such problems should form an interesting area of research.

In conclusion, note that the list of references contains several relevant papers not mentioned earlier due to limitation of space.

REFERENCES

[1] Alexandrov A.D. — Uniqueness theorems for surfaces in the large. *Vestnik Lenin Univ.*, 11, 1956, p. 5-17.

[2] Bakkelman I.J. — Mean curvature and quasilinear elliptic equations. *Siberskii Mat. Z.*, 9, 1968, p. 1014-1040.

[3] Bernstein S. — Conditions nécessaires et suffisantes pour la possibilité du problème de Dirichlet. *C.R. Acad. Sci. Paris*, 150, 1910, p. 514-515.

[4] Bernstein S. — Sur les surfaces définies au moyen de leur courbure moyenne et totale. *Ann. Sci. Ecole Norm. Sup.*, 27, 1910, p. 233-256.

[5] Bernstein S. — Sur les équations du calcul des variations. *Ann. Sci. Ecole Norm. Sup.*, 29, 1912, p. 431-485.

[6] Finn R. — Remarks relevant to minimal surfaces and to surfaces of constant mean curvature. *J. d'Anal. Math.*, 14, 1965, p. 139-160.

[7] Gilbarg D. — *Boundary value problems for nonlinear elliptic equations in n variables. Nonlinear problems.* Madison : Univ. of Wisc. Press, 1963, p. 151-159.

[8] Hartman P. & Stampacchia G. — On some nonlinear elliptic differential functional equations. *Acta Math.*, 115, 1966, p. 271-310.

[9] Heinz E. — To appear.

[10] JENKINS H. & SERRIN J. — The Dirichlet problem for the minimal surface equation in higher dimensions. *J. Reine Ang. Math.*, 229, 1968, p. 170-187.
[11] JENKINS H. & SERRIN J. — *The Dirichlet problem for regular parametric variational problems.* To appear.
[12] LADYZHENSKAYA O.A. & URALTSEVA N.N. — *Linear and quasilinear elliptic equations.* Moscow, 1964.
[13] LADYZHENSKAYA O.A. & URALTSEVA N.N. — Global estimates of the gradient of solutions of quasilinear elliptic and parabolic equations. *Sem. Mat. Inst. Steklov*, 14, 1969, p. 127-151.
[14] LERAY J. — Discussion d'un problème de Dirichlet. *J. Math. Pures Appl.*, 18, 1939, p. 249-284.
[15] LERAY J. & SCHAUDER J. — Topologie et équations fonctionelles. *Ann. Sci. Ecole Norm. Sup.*, 51, 1934, p. 45-78.
[16] OSKOLKOV A.P. — A priori estimates for the first derivatives of solutions of the Dirichlet problem for non-uniformly elliptic quasilinear equations. *Trudy Mat. Inst. Steklov*, 102, 1967, p. 105-127.
[17] SERRIN J. — The Dirichlet problem for surfaces having prescribed mean curvature. *Notices Amer. Math. Soc.*, 14, 1967, p. 135.
[18] SERRIN J. — The Dirichlet problem for quasilinear equations with many independent variables. *Proc. Nat. Acad. Sci.*, 58, 1967, p. 1829-1835.
[19] SERRIN J. — The problem of Dirichlet for quasilinear elliptic differential equations with many independent variables. *Phil. Trans. Roy. Soc. London*, A 264 1969, p. 413-496.
[20] SERRIN J. — The Dirichlet problem for surfaces of constant mean curvature, *Proc. London Math. Soc.*, 21, 1970, p. 361-384.
[21] TRUDINGER N.S. — Some existence theorems for quasilinear non-uniformly elliptic equations in divergence form. *J. Math. Mech.*, 17, 1968.

University of Minnesota
Dept. of Mathematics,
Minneapolis
Minnesota 55 455 (USA)

In: F. E. Browder (ed.), *Mathematical Developments Arising from Hilbert Problems* (Proc. Symposia in Pure Math., vol. 28) (1976), pp. 507–524. **Reprint I.3.[ii]**

THE SOLVABILITY OF BOUNDARY VALUE PROBLEMS

James Serrin

Among the prophetic problems in Hilbert's famous list one must surely include the twentieth, the general problem of boundary values for elliptic partial differential equations. This subject, only a seedling in the year 1900, has burst into flower during our century, has developed in directions surely Hilbert never imagined, and today encompasses a vast area of work which to a mathematician of seventy-five years ago would seem little short of astonishing. A list of the mathematicians who have significantly developed the theory is itself impressive: beginning with the work which Hilbert himself contributed in 1900 in the form of a legitimization of Riemann's version of the Dirichlet principle, other important advances through the first half-century were supplied by Serge Bernstein, Jacques Hadamard, Henri Lebesgue, Eberhard Hopf, Richard Courant, O. Perron, Norbert Wiener, Jules Schauder, Jean Leray, K.O. Friedrichs, G. Giraud, C.B. Morrey, Jr., Lars Garding, and others. An even longer list would be required for the 25 years since 1950, while a complete bibliography would be monumental.

The twentieth problem, like so many of the others in Hilbert's list, consisted as much in a program as in a specific problem requiring some definitive answer, and in just this fact we can see one facet of Hilbert's genius and breadth. The first sentences of the twentieth problem show clearly this programmatic aspect. I quote, with some paraphrasing: "An important problem ... is the question concerning existence of solutions of partial differential equations when the values on the boundary of the region are prescribed. [The methods now available for treating this problem], however, seem to be generally not capable of direct extension to the case where [differential relations are prescribed along the boundary], nor can they be extended immediately to the case where the inquiry is not for potential surfaces but, say, for surfaces of least area, or for [other more complicated cases of boundary value problems]."

From the context of the twentieth problem, it is apparent that Hilbert had in mind particularly equations of the form

$$\sum_{i,j=1}^{n} a_{ij}(x)\frac{\partial^2 u}{\partial x_i \partial x_j} + \sum_{i=1}^{n} b_i(x)\frac{\partial u}{\partial x_i} + c(x)u = f(x), \quad x \text{ in } \Omega,$$

$$\sum_{i,j=1}^{n} a_{ij}(x,u,p)\frac{\partial^2 u}{\partial x_i \partial x_j} + b(x,u,p) = 0, \quad x \text{ in } \Omega,$$

where Ω is a bounded open subset of R^n, $u = u(x) : \Omega \to K$,

$$p = p(x) = (\partial u/\partial x_1, \ldots, \partial u/\partial x_n),$$

and the leading coefficient matrix a_{ij} is positive definite:

$$(a_{ij}(x)) > 0, \quad x \in \bar{\Omega},$$

$$(a_{ij}(x, u, p)) > 0, \quad x \in \bar{\Omega}, \quad u \in R^1, \ p \in R^n.$$

The first equation is linear of second order, the second quasilinear; note that the second case includes the first. It is of course reasonable and proper to extend the domain of Hilbert's program to include equations of higher order, and even general systems of equations. In this form, however, the scope of the twentieth problem becomes little less than encyclopedic, and a far from easy subject for a general lecture. For this reason I shall confine myself for the most part to the equations noted, particularly since they are also immediately relevant to the direct query with which Hilbert closes the twentieth problem. This question asks, very specifically,

"Has not every regular variational problem a solution, provided certain assumptions regarding given boundary conditions are satisfied (say that the functions concerned in these boundary conditions are continuous and have ... one or more derivatives), and provided also if need be that the notion of a solution shall be suitably extended?"

By a regular variational problem, we shall mean what Hilbert meant, a problem of the following sort: find a function $u : \bar{\Omega} \to R$ which is of class $C^1(\Omega) \cap C^0(\bar{\Omega})$ and is such that, among all functions of this class,

$$I[u] = \int_\Omega F(x, u(x), p(x))\, dx = \text{Minimum};$$

we shall assume, moreover, that the admissible functions also satisfy the following Dirichlet-type boundary condition:

$$u(x) = \varphi(x) \quad \text{for } x \in \partial\Omega$$

where $\varphi(x)$ is an assigned continuous function on $\partial\Omega$. Here the function F is given, and $F(x, u, p)$ satisfies the regularity (or convexity) condition

$$F \in C^2, \quad \left(\frac{\partial^2 F}{\partial p_i \partial p_j}\right) > 0 \qquad \text{for } x \in \bar{\Omega}, \ u \in R, \ p \in R^n.$$

(Hilbert even assumes that F is analytic in its arguments and restricts to $n = 2$, though we shall not be quite this special.)

This problem is linked to partial differential equations by means of its Euler-Lagrange equation. That is, if u minimizes the expression $I[u]$ and if it is sufficiently smooth, then u satisfies the partial differential equation

$$\sum_{i=1}^{n} \frac{\partial}{\partial x_i}\{F_{p_i}(x, u, p)\} = F_u(x, u, p),$$

where the subscripts on F denote partial differentiation with respect to the indicated variable. If we carry out the differentiations we get

$$\sum_{i,j=1}^{n} F_{p_i p_j}(x,u,p)\frac{\partial^2 u}{\partial x_i \partial x_j} + \sum_{i=1}^{n}(F_{p_i u} p_i + F_{p_i x_i}) = F_u .$$

Evidently, every smooth minimizing function is a solution of this equation, while conversely any solution obeying the given boundary condition is a prime candidate for the minimizing function. Note that the quadratic form is positive definite by virtue of the regularity condition; thus this equation is a special example of the nonlinear equation noted before.

Several main ideas have become dominant in studying the existence of solutions of elliptic partial differential equations satisfying given boundary conditions, namely:

I. Continuation of solutions along a parameter, for which the problem varies from a known situation to a desired one.

II. The a priori estimation of the magnitude of the partial derivatives, depending only on the structure of the equation, the boundary data, and the domain.

III. A functional-analytic approach, guaranteeing the existence of a weak or generalized solution of the given problem.

Since the first two techniques are of crucial importance to the study of quasilinear equations, and since the third is surely required in the discussion of Hilbert's concluding question, let me discuss for a moment the background of these techniques.

In a remarkable series of papers in the years 1910–1912 Serge Bernstein, who was at the time just 30 years of age, had been able to obtain the existence of solutions of analytic quasilinear second order equations in $n = 2$ variables by a combination of analytic continuation and successive approximations. The difficult and laborious nature of this reasoning obscured for 20 years the brilliant underlying idea: that in order to prove the existence of solutions it is enough to show that for any eventual solution the corresponding partial derivatives appearing in the equation can be estimated and bounded (in an appropriate norm), the bounds depending only on the structure of the equation, on the boundary data, and on the smoothness of the domain. For, with these a priori bounds in hand, one could then obtain the existence of solutions by a homotopy argument involving continuous deformation of the given problem to a simpler one, the solution of the latter either being known or else available by inspection. It is remarkable, nevertheless, that Bernstein could carry out such an ambitious program at the time, and, even more remarkable, that he could already so much foreshadow present results as to obtain special conclusions of the following sort: the Dirichlet boundary value problem for the minimal surface equation

$$\left(1+\left(\frac{\partial u}{\partial y}\right)^2\right)\frac{\partial^2 u}{\partial x^2} - 2\frac{\partial u}{\partial x}\frac{\partial u}{\partial y}\frac{\partial^2 u}{\partial x \partial y} + \left(1+\left(\frac{\partial u}{\partial x}\right)^2\right)\frac{\partial^2 u}{\partial y^2} = 0$$

is solvable for all suitably smooth boundary data when the domain Ω has a smooth strictly convex boundary.

In 1934 Jules Schauder clarified the reasoning involved in Bernstein's approach. Moreover, adopting and refining the two methods of continuation and a priori estimation, he was able to treat the Dirichlet boundary value problem ($u = \varphi$ given on $\partial\Omega$) for linear second order equations (i.e. the first displayed equation in this article). His result is impeccable, bringing this aspect of Hilbert's program to the following positive and definite conclusion:

Provided $c \leqq 0$ and provided the coefficients, the right hand side f, the given data φ, and the domain are all suitably smooth, there exists one and only one solution of the problem. For general c, if one replaces c by $c+\lambda$ then a Fredholm alternative exists.

Coupled with the work of Giraud (occurring at roughly the same period) on other boundary conditions, including that of Neumann type

$$\nu_i \frac{\partial u}{\partial x_i} = \psi(x) \quad \text{for } x \in \partial\Omega \quad (\psi(x) \text{ given on } \partial\Omega)$$

where $\nu = (\nu_1, \ldots, \nu_n)$ denotes the normal vector to $\partial\Omega$, Schauder's conclusions constitute an unquestioned high point in Hilbert's program. It should be added that Giraud's methods today are supplemented by more modern techniques developed by Fiorenza and by Agmon, Douglas and Nirenberg.

The third thread of modern work was directly initiated by Hilbert. Following the sentences I quoted at the beginning of the talk, Hilbert had gone on to say: "It is my conviction that it will be possible to prove existence theorems by means of a general principle whose nature is indicated by Dirichlet's principle." Hilbert then refers to a paper of his which had just appeared in the Jahresberichte der Deutschen Mathematische-Vereinigung; from this we may gather his meaning: namely, given a regular variational problem, then from a sequence of functions u_n such that $I[u_n] \to$ Minimum one should construct another sequence convergent to a solution of the corresponding Euler-Lagrange partial differential equation. This is the first statement of the so-called direct method of the calculus of variations, later developed by Courant, Tonelli, Morrey and others. This approach is of course patently inapplicable for obtaining the existence of solutions of partial differential equations which do not arise (as the Euler-Lagrange equation does) from some variational problem. Nevertheless, when reinterpreted, and when filled out with appropriate results of functional analysis (at the simplest level, that any continuous linear functional on a Hilbert space can be represented as an inner product), it provided a tool of tremendous power not only for variational problems, but also for the resolution of linear boundary value problems. The procedure indeed is independent of the order of the equation, is extensible to the case of systems of equations, and even allows one to treat boundary conditions other than those of simple Dirichlet type. The flaw in the method, if such there be, is that the solutions which are obtained have derivatives only in a generalized sense and satisfy the equation only in a correspondingly weak form. Whether the solution has derivatives in a classical sense, and a fortiori whether they satisfy the equation in a normal sense is left aside: never mind, the daring gamble pays off!

Mathematicians have turned to the problem of proving that such "generalized solutions" are "regular," namely possess further properties of differentiation and, ultimately, enough smoothness so as to satisfy the differential equation in a classical sense (Nirenberg, Agmon, Douglas, Browder, Schechter). Of course, such "regularity" proofs are the counterpart for the functional analytic approach of the "a priori estimates" of the continuation approach. In this respect, Hilbert's twentieth problem closes with the nineteenth, for the problem of existence of classical solutions becomes precisely the problem of regularity of generalized solutions. While Hilbert apparently did not expect such a nicely packaged outcome, his brilliant insight nevertheless accurately pin-pointed two of the most crucial issues in the modern theory of elliptic partial differential equations.

Let us turn to the question of existence of solutions of equations of the form

$$\text{(Q)}\qquad \sum_{i,j=1}^{n} a_{ij}(x,u,p)\frac{\partial^2 u}{\partial x_i \partial x_j} = b(x,u,p), \quad x \text{ in } \Omega,$$

with Dirichlet boundary conditions

$$u = \varphi(x) \quad \text{for } x \text{ on } \partial\Omega,$$

and with

$$b,\ a_{ij} \in C^1, \quad a_{ij}\xi_i\xi_j > 0 \quad \text{for all } x,u,p.$$

As I have remarked, this situation includes partial differential equations arising from regular variational problems. We have the following theorem, essentially due to Leray and Schauder:

Let the boundary $\partial\Omega$ and the assigned boundary data be of class C^3. Also let τ be an arbitrary real number in the closed interval $[0,1]$. Suppose there exists a constant M, independent of τ, such that the conditions

(i) $v \in C^2(\bar{\Omega})$,

(ii) $v = \tau\varphi$ on $\partial\Omega$,

(iii) $a_{ij}(x,v,q)\,\partial^2 v/\partial x_i\partial x_j = \tau b(x,v,q)$ in Ω, $q = (\partial v/\partial x_1, \ldots, \partial v/\partial x_n)$,

are incompatible with $\|v\|_{C^2(\bar{\Omega})} > M$. Then the Dirichlet problem is solvable in the class $C^2(\bar{\Omega})$.

The proof involves Schauder's continuation of a solution along the parameter τ, from $\tau = 0$ where the problem (i), (ii), (iii) has the unique solution $v = 0$, to $\tau = 1$, which is essentially the given problem. The apparatus guaranteeing that the continuation exists throughout the full homotopy range $[0,1]$ consists of a topological and an analytical part. The topological portion is much deeper than that required by Bernstein[1] and in fact is a special case of the Leray-Schauder fixed point theory. The latter was in fact developed primarily in order to prove this theorem! Let me add that Browder has given an elegant and simple proof of this special case, independent of the Brouwer index. The analytical part of the continuation procedure, needless to say, relies on a priori estimates, in fact precisely those estimates developed by Schauder for the linear theory above.

The theorem as it stands is useful only if we can justify its main hypothesis, namely prove the further a priori estimate $\|v\|_{C^2(\bar{\Omega})} \leq M$ for solutions of (i), (ii), (iii). Except for the case $n = 2$, treated by Leray and Schauder and by others, the problem of estimating $\|v\|_{C^2(\bar{\Omega})}$ remained open for 30 years. It was finally solved by Ladyzhenskaya and Uraltseva, who showed that one can replace the condition $\|v\|_{C^2(\bar{\Omega})} > M$ in the hypothesis by $\|v\|_{C^1(\bar{\Omega})} > M$. As we shall see in a moment, this improvement allows a far-reaching existence theorem to be established for quasilinear equations. The theorem of Ladyzhenskaya and Uraltseva is consequently one of the key results in the structure of Hilbert's program. Although the original proof was fairly difficult, there now exists a simplified version due to Trudinger (on the basis of earlier papers of De Giorgi, Moser, and Serrin). This theorem (and its comparatively elementary proof) stands now as a deservedly high point of mathematical analysis.

With its help, the existence of solutions of the Dirichlet problem for equation (Q) now splits into three steps:

(A) Estimates for $\sup_\Omega |v|$.

(B) Estimates for $\sup_{\partial\Omega} |q|$, assuming (A).

(C) Estimates for $\sup_\Omega |q|$, assuming (A) and (B).

None of these steps need be possible for a given equation, as can be shown by counterexamples. On the other hand, step (A) is frequently quite simple, involving in many situations only the application (sometimes in a rather sophisticated form) of the maximum principle of E. Hopf; moreover step (C) is known to hold under very wide conditions on the structure of (Q), as has been shown by Ladyzhenskaya and Uraltseva, Serrin, Simon and others. Step (B) is the focus of main interest.

We introduce the orthogonal invariants

$$\mathcal{E}(x,u,p) = p_i a_{ij}(x,u,p) p_j, \qquad \mathcal{J}(x,u,p) = \frac{\mathcal{E}(x,u,p)}{|b| + |p| \text{ trace } (a_{ij})}.$$

Then the following Main Result holds (Serrin, 1969).

Suppose there exists an increasing continuous function $\Phi(\rho)$ such that

$$\mathcal{J} \geq \Phi(|p|), \qquad \int^\infty \Phi(\rho) \frac{d\rho}{\rho} = \infty.$$

[1] Or by Schauder in his earlier work on linear equations, where the functional analytic apparatus requires only a contraction mapping.

Then step (B) can be carried out.

Suppose on the contrary that there exists a continuous function $\Psi(\rho)$ such that

$$\mathscr{I} \leqq \Psi(|p|), \qquad \int^{\infty} \Psi(\rho) \frac{d\rho}{\rho} < \infty.$$

Then the following results hold:

(a) If $|b|/|p|$ trace $(a_{ij}) \to \infty$ as $|p| \to \infty$, then for any smooth domain there exists smooth data for which the Dirichlet problem has no solution;

(b) If $|b|/|p|$ trace (a_{ij}) is bounded, then there exist smooth domains and smooth data for which the Dirichlet problem has no solution.

In the latter case, in fact, it turns out to be necessary to impose certain curvature conditions on the boundary $\partial\Omega$ in order for the Dirichlet problem to be solvable for arbitrarily given smooth data. These conditions have the general form

$$\sum_{i=1}^{n-1} a_i \varkappa_i \geqq d, \qquad x \in \partial\Omega,$$

where the coefficients a_i and d depend only on the structure of the equation, the boundary values, and the principal directions on the boundary, and where the $\varkappa_i$ are the principal curvatures.

Here is one example where the results are nearly definitive. We consider the variational problem [2]

$$\int F(|p|)\, dx = \text{Minimum}.$$

Regulairty of the integrand forces $F''(t) > 0$ for $t \geqq 0$ and $F'(t) > 0$ for $t > 0$. Now put

$$G(t) = t\,\frac{F''(t)}{F'(t)}.$$

Then we have the following conclusions (Serrin, 1971):

(a) If $G \geqq \Phi(t)$, where Φ is decreasing and $\int^{\infty} \Phi\, dt = \infty$, then the Dirichlet problem is solvable.

Examples: $F = |p|^2$, $G = 1$, $\Phi = 1$ (Laplace equation),
$F \approx |p|^m$, $G \approx (m-1)$, $\Phi = m - 1$ $(m > 1)$.

(b) If $\int G\, dt < \infty$ and $tG \geqq \Phi(t)$, where Φ is decreasing and $\int^{\infty} \Phi\, dt = \infty$, then the Dirichlet problem is solvable if and only if the mean curvature H of the boundary is nonnegative at each boundary point. (For $n = 2$ this is due to Bernstein and to Finn.)

Examples: $F = \sqrt{1 + |p|^2}$ (minimal surface equation),
$F = (1 + |p|^{\alpha})^{1/\alpha}$, $G = (\alpha - 1)/(1 + |t|^{\alpha})$ $(\alpha > 1)$.

We may now turn to Hilbert's question concerning the existence of solutions of regular variational problems. I will propose that the appropriate answer to Hilbert's

[2] This problem is the most general first order variational problem in which the integrand is independent of u, and invariant under translations and rotations of the x-space.

problem is both YES and NO; indeed with some reformation the answer also remains OPEN!

NO. Consider the integral

$$F(x,u,p) = \sqrt{1+|p|^2} + 2\Lambda u, \quad \Lambda = \text{const} > 0.$$

Then all the conditions of Hilbert are satisfied; the corresponding partial differential equation

$$(1+|p|^2)\Delta u - p_i p_j \frac{\partial^2 u}{\partial x_i \partial x_j} = 2\Lambda (1+|p|^2)^{3/2}$$

represents the classical case of a surface of constant mean curvature Λ, and indeed Hilbert mentions just this situation as an example of a regular variational problem.

Now take

$$\Omega = \{|x| \leqq 2\Lambda\} \text{ and } \varphi = 0 \text{ on } \partial\Omega.$$

We shall show that this problem has no solution, that in fact there is no finite valued function for which the minimum is attained. To see this, it is enough to show that the minimum in question is $-\infty$, that is, that there exist functions for which $I[u]$ is arbitrarily large negatively. Indeed, consider the radial symmetric function u_M whose graph is indicated in the figure.

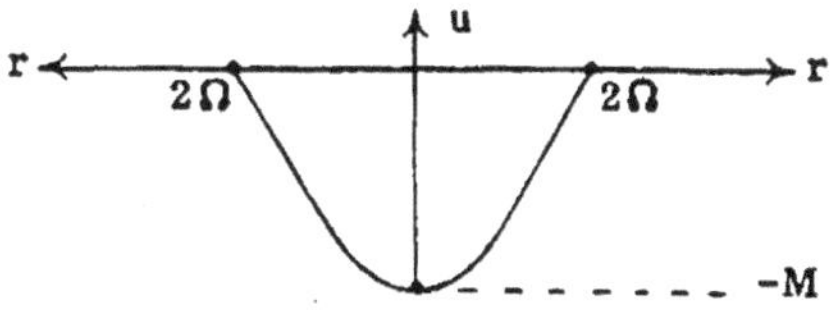

Then $I[u_M] \to -\infty$ as $M \to \infty$, as one easily shows.

It may be objected that the difficulty here arises because $I[u]$ is not <u>bounded below</u> on the class of admissible functions u. Let us therefore add this condition, even in the strong form that $F \geqq 0$. I will make the case that the answer is still

NO. Consider the integral

$$F(x,u,p) = \sqrt{1+|p|^2},$$

again one discussed by Hilbert, corresponding to surfaces of least area spanning the graph of the given boundary values. Take as domain Ω a <u>non-convex</u> smoothly bounded set of the form

and data for which the minimal surface equation has no (non-parametric) solution. Then the variational problem can have no solution either, as can be proved by an easy additional argument.

The difficulty here arises because of the non-convexity of the boundary, which, in view of the earlier discussion, prevents solutions from taking on arbitrary boundary values.

In spite of the negative answers so far given, there are still two ways available to restore an affirmative answer to Hilbert's query. Since almost all problems need redefinition at some stage, we may ask that the integrals in Hilbert's problem should be restricted to those for which the first part of the main result above applies. Since this includes a great variety of situations, including the "natural" integrands of Ladyzhenskaya and Uraltseva, we thus obtain a large and important class of regular variational problems for which the answer to Hilbert's question is YES.

Moreover, even when the first part of the theorem does not apply but case (b) of the second part holds, there is still a large class of domains for which the answer is YES, these domains being distinguished by curvature restrictions of the form noted earlier.

There is still another way to obtain a positive answer; for in fact we have not yet heeded Hilbert's final (and clairvoyant) admonition "provided also if need be that the notion of a solution be sufficiently extended." Let me give an example: we have seen that the variational problem with $F = \sqrt{1 + |p|^2}$ need not always have a solution if the domain does not satisfy appropriate curvature conditions. We therefore modify the notion of solution as follows: consider functions

$$u \in C^2(\Omega) \cap W^{1,1}(\Omega) \quad \text{(hence } u \in L^1(\partial\Omega))$$

and say that u is a solution of the variational problem provided that

$$I[u] = \int_\Omega \sqrt{1 + |Du|^2}\, dx + \int_{\partial\Omega} |u - \varphi| ds = \text{Minimum}.$$

Then there is always a solution; if it takes on the right boundary values it is even the solution of the classical problem, and if there exists a classical solution that will be the solution here (M. Miranda).

Whether generalized solutions can be defined and will exist for all regular variational problems, or even all those for which $F \geqq 0$, is however OPEN. [3]

I should like to conclude with a small selection of open problems, directly in the spirit of Hilbert's statement, but more specifically oriented.

1. Consider a quasilinear equation, a smooth domain, and smooth boundary data. As we have seen, a classical solution need not exist unless the boundary of Ω satisfies a curvature condition. Nevertheless one may still introduce "generalized solutions." For variational type problems this can be attempted by direct methods, as I indicated in discussing Miranda's theorem. For non-variational type problems is there an alternate method? Perron's process comes to mind here, and is seen to work whenever differences of solutions satisfy a maximum principle. Does the Perron upper solution then equal the Perron lower solution, as it does for potential theory? For a regular variational problem does the Perron solution agree with the solution attained by direct methods? (In partial answer to this question R. Temam has informed me (private communication, 1974) that one necessarily has the relation: lower Perron

[3] Some recent work in this direction is reported by Temam.

solution $\leq$ generalized solution of the variational problem $\leq$ upper Perron solution.]
At a boundary point where the "generalized solution does not take on the "correct" boundary value, is the solution nevertheless continuous in $\bar{\Omega}$ and does the graph of the solution have a tangent plane orthogonal to the x-space?

2. The general boundary condition

$$\varphi(x, u, Du) = 0 \quad \text{for } x \in \partial\Omega$$

is important; special cases include the Neumann condition, the so-called third boundary condition, and the "natural" condition

$$a_{ij}(x, u, Du) \cdot \vec{n} = g(x, u), \quad g(x, u) \text{ given on } \partial\Omega.$$

For uniformly elliptic quasilinear equations, Ladyzhenskaya and Uraltseva have proved the existence of classical solutions corresponding to this boundary condition, provided suitable smoothness is assumed for the domain and data. Recently Spruck has treated the capillary surface problem (an equation which Gauss showed could be obtained from a variational problem), a much more difficult situation since the Euler-Lagrange equation falls in case (b) of our main result, but nevertheless one where the boundary condition is again of "natural" type. Remarkably, Spruck found that no boundary curvature conditions were required. Is this the general situation, or is it due to the fact that he treated the "natural" condition?

3. The Dirichlet problem in unbounded domains. While a reasonably complete theory for such problems in the linear case has been given by Meyers and Serrin, the quasilinear situation is relatively untouched.

4. Fully nonlinear equations

$$F(x, u, Du, D^2u) = 0$$

have been studied only in special circumstances, and few results are available. The partial differential equations for surfaces of constant positive Gaussian curvature, and, in general, for Weingarten surfaces are important special cases. For these cases, in particular, is it possible to distinguish domains and data for which the Dirichlet boundary value problem is solvable? What about other boundary conditions?

5. When the integral $F(x, u, p)$ satisfies the conditions

$$F(x, u, p) \geq \mu|p|^m - \nu, \qquad \mu, \nu > 0, \ m > 1,$$

one can define generalized solutions of the variational problem even without the convexity condition. Is there any way to characterize these solutions? In particular, is it true that at almost every point in Ω the generalized solution satisfies the condition $(f_{p_i p_j}(x, u, Du)) \geq 0$ so that the generalized solution is an elliptic-parabolic solution of the Euler-Lagrange equation?

These problems remind us of Hilbert's words "The supply of problems in mathematics is inexhaustible, for as soon as one is solved numerous others come forth in its place." With an abundance of problems it is reasonable to predict that the final quarter of this century will find elliptic equations to be of as great interest as they have been during the seventy-five years since Hilbert's talk. Moreover, I have mentioned only problems which are definite and special, and even in this area I have restricted consideration to second order equations, echoing Hilbert's premise that "it is just such definite and special problems which attract us the most and from which the most lasting influence is exerted." Nevertheless, Hilbert stressed broader implications as well, significantly choosing variational problems as his example. I think it worth emphasizing here, in the same way, that interest in elliptic equations need not rest solely on boundary value problems. There are other directions of research of equal significance. I would stress the discovery of qualitative features of solutions, such as the behavior of solution at singularities, uniqueness theory and monotonicity theorems, and the structure of singular sets for solutions of higher order nonlinear equations. Bifurcation theory and the theory of variational inequalities are only in their beginning stages, and offer further significant directions of research.

Hilbert closed his lecture with these words, which seem to me applicable not only to the present problem but to the entire meeting. "The problems here are merely samples, yet they show how rich, how manifold and extensive the mathematical science of today is. The question is therefore urged on us whether mathematics is doomed to the fate of those other sciences that have split into separate branches, and whose representatives scarcely understand one another and whose connection becomes ever more loose. ... And, with the extension of mathematical knowledge will it not finally become impossible for the single investigator to embarce all departments of this knowledge?"

I am not sure how to answer those questions today, but I am sure that with the bewildering variety of available problems mathematics will remain a vital science and individual mathematicians will continue to be beckoned, some by general theories, others by the richness of detail to be found in special problems. It seems to me that the next twenty-five years will provide us with "gifted masters and many zealous and enthusiastic disciples" who will make the coming years as vital as those since Hilbert's talk.

Note. This paper is essentially the same as the lecture given at De Kalb on May 24. Its style is thus that of a talk, rather than an expository research paper, and accordingly it states few detailed conclusions. So far as the results mentioned in the paper are concerned, these can of course be found in the references listed at the close of the paper. At the same time, many topics were necessarily omitted. These include, most importantly, an outline of the elegant proof of the Leray-Schauder-Ladyzhenskaya-Uraltseva theorem, a discussion of the remarkable work of Ladyzhenskaya, Uraltseva, and Simon on interior estimates for the gradient of solutions of equation (Q) in cases where the equation is severely nonlinear, a review of much of the most recent work (1973-1974) on the boundary behavior of solutions, a discussion of various boundary value problems alternative to the Dirichlet problem, and, finally, treatment of higher order linear and nonlinear problems. The author had intended to discuss all of these issues except the last, but the necessary time to survey the large amount of literature involved turned out not to be available. To consider higher order problems a full treatise would have been needed, so that from the beginning the author intended that these should not be included. Indeed, were Hilbert's twentieth problem to be given a full and definitive treatment in all its varied ramifications, the resulting paper must needs have grown to elephantine size (the tract of Carlo Miranda, outlining the theory of elliptic equations up to 1965 already fills a large volume), and would in fact have been beyond the powers of any one, or even two or three, men. For all these reasons, then, it may in fact be just as valuable for the purposes at hand - namely, to introduce mathematicians to the general ideas involved in Hilbert's problems - to reproduce the lecture just as given, with all its flaws and also with all its enthusiasm for a beautiful subject.

Bibliography

The following list refers to specific matters discussed in the paper as well as to topics mentioned only in the note above. While not exhaustive, it nevertheless contains the major papers which have contributed significantly to the elucidation of Hilbert's twentieth problem, together with (more than) enough material for orientation to modern work. For convenience, books and monographs have been grouped separately at the end of the main list.

Agmon, S.
[1] The L^p approach to the Dirichlet problem. Ann. Scuola Norm. Sup. Pisa 13(1959), 405-448.

Agmon, S., Douglis, A., and Nirenberg L.
[1] Estimates near the boundary for solutions of elliptic partial differential equations satisfying general boundary conditions. I, II. Comm. Pure Appl. Math. 12(1959), 623-727; 17(1964), 35-92.

Beirao Da Veiga, H.
[1] On the $W^{2,p}$ regularity for solutions of mixed problems. J. Math. Pures Appl. 53(1974), 279-290.

Berstein, S.
[1] Sur la généralisation du problème de Dirichlet. Matn. Ann. 69(1910), 82-136.

[2] Sur les surfaces définies au moyen de leur courbure moyenne et totale. Ann. Aci. Ecole Norm. Sup. 27(1910), 233-256.

[3] Sur les équations du calcul des variations. Ann. Sci. Ecole Norm. Sup. 29 (1912), 431-485.

Bombieri, E. De Giorgi, E., and Miranda, M.
[1] Una maggiorazioni a priori relativa alle ipersuperfici minimali non parametriche. Arch. Rational Mech. Analysis 32(1969), 255-267.

Browder, F.E.
[1] Functional analysis and partial differential equations. Math. Ann. 138(1959), 55-79; 141(1962), 81-226.

[2] A priori estimates for elliptic and parabolic equations. Proc. Symp. Pure Math. IV. Amer. Math. Soc.: Providence (1961), 73-81.

[3] Nonlinear elliptic boundary value problems. Bull. Amer. Math. Soc. 69 (1963), 862-874; Trans. Amer. Math. Soc. 117(1965), 530-550.

[4] Remarks on the direct method of the calculus of variations. Arch. Rational Mech. Analysis 20(1965), 251-258.

[5] Existence theorem for nonlinear partial differential equations. Summer Institute in Global Analysis. Amer. Math. Soc. 1969.

Caccioppoli, R.
[1] Limitazioni integrale pur le soluzioni di un'equazione lineare ellitico a derivate parziali. Giorn. Mat. Battaglini 80(1950-51), 186-212.

Concus, P. and Finn, R.
[1] On the class of capillary surfaces. Proc. Nat. Acad. Sci. 63(1969), 292-299.

Cordes, H.O.
[1] Vereinfachter Beweis der Existenz einer Apriori-Hölderkonstante. Math. Ann. 138(1959), 155-178.

De Giorgi, E.
[1] Sulla differenziabilità e l'analiticità delle estremali degli integrali multipli regolari. Mem. Acc. Sci. Torino 3(1957), 1-19.

Douglis, A., and Nirenberg, L.
[1] Interior estimates for elliptic systems of partial differential equations. Comm. Pure Appl. Math. 8(1955), 503-538.

Edmunds, D.E.
[1] Quasilinear second order elliptic and parabolic equations. Bull. London Math. Soc. 2(1970), 5-28.

Fichera, G.
[1] Sul problema misto per le equazione lineari delle derivate parziali del secondo ordine di tipo ellitico. Rev. Math. Pures Appl. 9(1964), 3-10.

Finn, R.
[1] On equations of minimal surface type. Ann. Math. 60(1954), 397-416.

[2] New estimates for equations of minimal surface type. Arch. Rational Mech. Analysis 14(1963), 337-375.

[3] Remarks relevant to minimal surfaces and to surfaces of constant mean curvature. J. d'Anal. Math. 14(1965), 139-160.

Finn, R. and Gilbarg, D.
[1] Three-dimensional subonic flows and asymptotic estimates for elliptic partial differential equations. Acta Math. 98(1957), 265-296.

Florenza, R.
[1] Sui problemi di derivata obliqua per le equazioni ellittiche. Ricerche Mat. Napoli 8(1959), 83-110.

[2] Sulla hölderianità della soluzioni dei problemi di derivata obliqua regolare del secondo ordine. Ricerche Mat. Napoli 14(1965), 102-123.

[3] Sui problemi di derivata obliqua per le equazioni ellittiche quasi lineari. Ricerche Mat. Napoli 15(1966), 74-186.

Friedrichs, K.O.
[1] The identity of weak and strong extensions of differential operators. Trans. Amer. Math. Soc. 55(1944), 132-155.

Garding, L.
[1] Le problème de Dirichlet pour les équations aux dérivées partielles elliptiques homogènes à coefficients constants. C.R. Acad. Sci. Paris 230(1950), 1030-1032.

[2] Dirichlet's problem for linear elliptic partial differential equations. Math. Scand. 1(1953), 55-72.

[3] Some trends and problems in linear partial differential equations. Proc. Int. Congr. Math., Edinbourgh (1958), 87-102.

Gerhardt, C.

[1] Existence, regularity, and boundary behavior of generalized surfaces of prescribed mean curvature. Math. Zeit. 139(1974), 173-198.

[2] On the capillarity problem with constant volume. Preprint, 1974.

Giaquinta, M.

[1] On the Dirichlet problem for surfaces of prescribed mean curvature. Manuscripta Math. 12(1974), 73-86.

Gilbarg, D.

[1] Boundary value problems for nonlinear elliptic equations in n variables. Symp. Nonlinear Problems, Univ. Wisconsin Press (1962), 151-159.

Giusti, E.

[1] Sulla regolarità parziale delle soluzioni di sistemi ellittici quasi-linear di ordine arbitrario. Ann. Scuola Norm. Sup. Pisa 23(1969), 115-141.

[2] Superfici cartesiane di area minima. Rend. Sem. Mat. Milano 40(1970), 1-21.

Haar, A.

[1] Über das Plateausche Problem. Math. Ann. 97(1927), 124-258.

Heinz, E.

[1] Interior gradient estimates for surfaces $z = f(x,y)$ with prescribed mean curvature. J. Diff. Geometry 5(1971), 149-157.

Hilbert, D.

[1] Über das Dirichletsche Prinzip. Jahresber. Deutsch. Math. Vereinig. 8(1900), 184-188; Math. Ann. 59(1904), 161-186.

Hopf, E.

[1] Elementare Bemerkung über die Lösungen partieller Differentialgleichungen zweiter Ordung von Elliptischen Typen. Sber. Preuss. Akad. Wiss. 19(1927), 147-152.

Hörmander, L.

[1] On the regularity of the solutions of boundary problems. Acta Math. 99(1958), 225-264.

Ivanov, A.V.

[1] On interior estimates of the first derivatives of solutions of quasilinear non-uniformly elliptic and parabolic equations of general form. Sem. Steklov Inst. 14 (1969), 29-47; 19(1970), 79-94.

Jenkins, H. and Serrin J.

[1] Variational problems of minimal surface type, I, II, III. Arch. Rational Mech. Analysis 12(1963), 185-212; 21(1966), 321-342; 29(1968), 304-322.

[2] The Dirichlet problem for the minimal surface equation in higher dimensions. J. Reine Angew. Math. 229(1968), 170-187.

Koselev, A.I.

[1] A priori estimates in L^p and generalized solutions of elliptic equations and systems. Uspehki. Mat. Nauk. 13(1958), 29-88; English transl; Amer. Math. Soc. 20(1962), 105-171.

Ladyzhenskaya, O.A. and Uraltseva, N.N.

[1] On Hölder continuity of solutions and their derivatives of linear and quasi-linear elliptic and parabolic equations. Trudy Steklov Math. Inst. 73(1964), 172-220.

[2] Local estimates for gradients of solutions of non-uniformly elliptic and parabolic equations. Comm. Pure Appl. Math. 23(1970), 677-703.

Lebesgue, H.

[1] Sur le problème de Dirichlet. Rend. Circ. Math. Palermo 24(1907), 371-402.

Leray, J.
[1] Majoration des dérivées secondes des solutions d'un problème de Dirichlet. J. Math. Pures Appl. 17(1938), 89–104.

[2] Discussion d'un problème de Dirichlet. J. Math. Pures Appl. 18(1939), 249–284.

Leray, J. and Lions, J. L.
[1] Quelques résultats de Visik sur les problémes elliptiques non linéaires par les méthodes de Minty-Browder. Bull. Soc. Math. France 93(1965), 97–107.

Leray, J. and Schauder, J.
[1] Topologie et équations functionelles. Ann. Sci. Ecole Norm. Sup. 16(1934), 45–78.

Lions, J. L.
[1] Quelques résultats d'existence dans les équations aux dérivées partielles non linéaires. Bull. Soc. Math. de France 87(1959), 245–273.

Littman, W., Stampacchia, G., and Weinberger, H. F.
[1] Regular points for elliptic equations with discontinuous coefficients. Ann. Scuola Norm. Sup. Pisa 17(1963), 43–78.

Magenes, E.
[1] Sur les problèmes aux limites pour les équations linéaires elliptiques. Colloques internationaux sur les équations aux dérivées partielles. Paris (1962).

[2] Sul problema di Dirichlet per le equazioni lineari ellittiche in due variabili. Ann. Mat. Pura Appl. 48(1959), 257–279.

Meyers, N. and Serrin, J.
[1] The exterior Dirichlet problem for second order elliptic partial differential equations. J. Math. Mech. 9(1960), 513–538.

Miranda, M.
[1] Un teorema di esistenza ed unicità per il problema dell' area minima in n variabili. Ann. Scuola Norm. Sup. Pisa (1965), 233–249.

[2] Un principe di massimo forte per le frontiere minimali e una applicazione alla risoluzione del problema al contorno per l'equazione delle superfici di area minima. Rend. Sem. Mat. Univ. Padova 45(1971), 344–366.

Morrey, C. B., Jr.
[1] Second order elliptic equations in several variables and Hölder contimuity. Math. Zeit. 72(1959), 146–164.

[2] Existence and differentiability theorems for variational problems for multiple integrals. Univ. of Wisconsin Press: Madison (1961), 241–270.

Moser, J.
[1] A new proof of De Giorgi's theorem concerning the regularity problem for elliptic differential equations. Comm. Pure Appl. Math. 13(1960), 457–468.

[2] On Harnack's theorem for elliptic differential equations. Comm. Pure Appl. Math. 14(1961), 577–591.

Nirenberg, L.
[1] On nonlinear partial differential equations and Hölder continuity. Comm. Pure Appl. Math. 6(1953), 103–156.

[2] Remarks on strongly elliptic differential equations. Comm. Pure Appl. Math. 8(1955), 649–675.

[3] Estimates and existence of solutions of elliptic equations. Comm. Pure Appl. Math 9(1956), 506–529.

[4] On elliptic partial differential equations. Ann Scuola Norm. Sup. Pisa 13(1959), 115–162.

Oskolkov, A.P.
[1] A priori estimates for the first derivatives of solutions of the Dirichlet problem for non-uniformly elliptic quasilinear equations. Trudy Steklov Math. Inst. 102(1967), 105-127.

Peetre, J.
[1] Another approach to elliptic boundary problems. Comm. Pure Appl. Math. 14(1961), 711-731.

Perron, O.
[1] Eine neue Behandlung der Randwertaufgabe fur $\Delta u = 0$. Math. Zeit. 18(1923), 42-54.

Protter, M.H. and Weinberger, H.F.
[1] A maximum principle and gradient bound for linear elliptic equations. Indiana Univ. Math. J. 23(1973), 239-249.

Radó, T.
[1] Uber zweidimensionale reguläre Variationsprobleme Math. Ann. 101(1929), 620-632.

Schauder, J.
[1] Uber den Zusammenhang zwischen der Eindeutigkeit und Lösbarkeit partieller Differentialgleichungen zweiter Ordnung von elliptischen Typus. Math. Ann. 106 (1932), 661-721.

[2] Uber das Dirichletsche Problem im Grossen für nichtlineare elliptische Differentialgleichungen. Math. Zeit. 37(1933), 623-634 and 768.

[3] Uber lineare elliptische Differentialgleichungen zweiter Ordnung. Math. Zeit. 38(1934), 257-282.

[4] Équations du type elliptique, problèmes linéaires. Enseign. Math. 35(1936), 126-139.

Schechter, M.
[1] General boundary-value problems for elliptic partial differential equations. Comm. Pure Appl. Math. 12(1959), 457-486.

[2] Remarks on elliptic boundary value problems. Comm. Pure Appl. Math. 12 (1959), 561-578.

[3] Mixed boundary problems for general elliptic equations. Comm. Pure Appl. Math 13(1960), 183-201.

[4] Various types of boundary conditions for elliptic equations. Comm. Pure Appl. Math. 13(1960), 407-425.

[5] Some unusual boundary value problems. Proc. Symp. Pure Math. 4, Amer. Math. Soc. (1961), 109-113.

[6] On L^p estimates and regularity, I, II, III. Amer. J. Math. 85(1963), 1-13; Math. Scand. 13(1963), 47-69; Ricerche Mat. Napoli 13(1964), 192-206.

Serrin, J.
[1] On the definition and properties of certain variational integrals. Trans. Amer. Math. Soc. 101(1961), 139-167.

[2] Local behavior of solutions of quasilinear equations. Acta Math. 111(1964), 247-302.

[3] The problem of Dirichlet for quasilinear elliptic differential equations with many independent variables. Phil. Trans. Royal Soc. London 264(1969), 413-496.

[4] The Dirichlet problem for surfaces of constant mean curvature. Proc. London Math. Soc. 21(1970), 361-384.

[5] Gradient estimates for solutions of nonlinear elliptic and parabolic equations. Symposium on Nonlinear Functional Analysis, Univ. of Wisconsin: Madison, 1971.

[6] Boundary curvatures and the solvability of Dirichlet's problem. Proc. Int. Congr. Math., Nice, 1970. Tome 2, 867-875. Gauthier-Villars: Paris, 1971.

Simon, L.
[1] Interior gradient estimates for non-uniformly elliptic partial differential equations of divergence form. Thesis, Univ. of Adelaide, 1971.

Simon, L. and Spruck, J.
[1] Existence and regularity of a capillary surface with prescribed contact angle. Preprint, 1975.

Spruck, J.
[1] Infinite boundary value problems for surfaces of constant mean curvature. Arch. Rational Mech. Analysis 49(1972), 1-31.

[2] On the existence of a capillary surface with prescribed contact angle. Preprint, 1974.

Stampacchia, G.
[1] Problemi al contorno ellittici, con data discontinui, dotati di soluzioni hölderiane. Ann. Mat. Pura Appl. 51(1960), 1-37.

[2] On some regular multiple integral problems in the calculus of variations. Comm. Pure Appl. Math. 16(1963), 383-421.

[3] Le problème de Dirichlet pour les équations elliptiques du second ordre à coefficients discontinues. Ann. Inst. Four. 15(1965), 189-258.

Temam, R.
[1] Solutions généralisées de certaines équations du type hypersurfaces minima. Arch. Rational Mech. Analysis 44(1971), 121-156.

Tonelli, L.
[1] Sur la semi-continuité des integrales double de calcul des variations. Acta Math. 53(1929), 325-346.

Trudinger, N.S.
[1] On Harnack type inequalities and their application to quasilinear elliptic equations. Comm. Pure Appl. Math. 20(1967), 721-747.

[2] Some existence theorems for quasilinear, non-uniformly elliptic equations in divergence form. J. Math. Mech. 18(1969), 909-919.

[3] A new proof of the interior gradient bound for the minimal surface equation in n dimensions. Proc. Nat. Acad. Sci. 69(1972), 821-823.

[4] Gradient estimates and mean curvature. Math. Zeit. 131(1973), 165-175.

Wiener, N.
[1] The Dirichlet Problem. J. Math. and Physics 3(1924), 127-146.

Books and Monographs

Agmon, S. Lectures on Elliptic Boundary Value Problems. Van Nostrand Math. Studies 2: New York, 1965.

Bers, L., John F. and Schechter, M. Partial Differential Equations. Interscience: New York 1964.

Browder, F. Problèmes Nonlinéaires. Lecture Notes, Univ. of Montreal, 1966.

Courant, R. and Hilbert, D. Methods of Mathematical Physics, vol. II. Interscience: New York, 1962.

Courant, R. Dirichlet's Principle, Conformal Mapping and Minimal Surfaces. Interscience: New York, 1950.

Ekeland, I. and Temam, R. Analyse Convexe et Problémes Variationnels. Dunod-Gauthier-Villars: Paris, 1974.

Fichera, G. Linear Elliptic Differential Systems and Eigenvalue Problems. Lecture Notes in Mathematics No. 8. Springer-Verlag: Berlin, 1965.

Friedman, A. Partial Differential Equations of Parabolic Type. Prentice-Hall: Englewood Cliffs, 1964.

Friedman, A. Partial Differential Equations. Holt, Rinehart and Winston: New York, 1969.

Hadamard, J. Leçons sur le Calcul des Variations. A. Hermann: Paris, 1910.

Hörmander, L. Linear Partial Differential Operators. Springer-Verlag: Berlin, 1963.

Ladyzhenskaya, O.A., Uraltseva N.N. Quasilinear Elliptic Equations and Variational Problems with many Independent Variables. Uspekhi Mat. Nauk. 16(1961), 19-92; Translated in Russ. Math. Surveys 16, 17-91.

Ladyzhenskaya, O.A., Uraltseva N.N. Linear and Quasilinear Elliptic Equations. Moscow, 1964. Academic Press: New York, 1968. Revised edition, Moscow, 1973.

Lions, J.L. Equations Différentielles Opérationelles et Problèmes aux Limites. Springer-Verlag: Berlin, 1961.

Lions, J.L. Lectures on Elliptic Partial Differential Equations. Tata Inst.: Bombay, 1967.

Lions, J.L. Quelques Méthodes de Résolution des Problèmes aux Limites Non Linéaires, Dunod-Gauthier-Villares: Paris, 1969.

Miranda, C. Partial Differential Equations of Elliptic Type. Springer-Verlag: Berlin, 1970.

Morrey, C.B., Jr. Multiple Integral Problems in the Calculus of Variations and Related Topics. Univ. California Publ. Math. 1(1943), 1-130.

Morrey, C.B., Jr. Multiple Integrals in the Calculus of Variations. Springer-Verlag: Berlin, 1966.

Necas, J. Les Méthodes Directes en Théorie des Equations Élliptiques. Masson: Paris, 1967.

Protter, M.H. and Weinberger, H.F. Maximum Principles in Differential Equations. Prentice-Hall: Englewood Cliffs, 1967.

Rosenbloom, P.C. Linear Partial Differential Equations. Surveys in Appl. Math. V. Wiley: New York, 1958.

Simader, C. On Dirichlet's Boundary Value Problem. Lecture Notes in Mathematics No. 268. Springer-Verlag: Berlin, 1972.

Stampacchia, G. Équations Elliptiques du Second Ordre á Coefficients Discontinues. Lecture Notes, Univ. of Montreal, 1966.

Tonelli, L. Fondamenti di Calcolo delle Variazioni, Vols. I and II, Zanichelli: Bologna, 1921-1923.

Department of Mathematics
University of Minnesota

Johannes C. C. Nitsche, Louis Nirenberg, James Serrin, John Ball and Hans Weinberger in the second row, Minneapolis, MN, 1999.

Gregory I. Barenblatt, Nina N. Ural'tseva and Juan Luis Vázquez.

I.4. Minimal Surface Equation

Louis Nirenberg

Mathematics Subject Classification (2010). 35J25, 53A10.

Keywords. Minimal surfaces, variational problems in infinite-dimensional spaces, boundary value problems for second-order elliptic equations.

Papers [i] and [ii] below are concerned with the Dirichlet problem for the minimal surface and similar equations.

In [i] the authors give a beautiful answer to a longstanding problem: the Dirichlet problem for the minimal surface equation in a bounded domain D in $\mathbb{R}^n$. The problem is to find a solution u of the minimal surface equation

$$(1 + |\nabla u|^2)\Delta u - u_i u_j u_{ij} = 0 \quad \text{in } \Omega$$

with prescribed values f on $\partial\Omega$.

When $n = 2$ it was long known that a solution exists for any given continuous boundary values when D is convex, and need not exist for nonconvex D. People believed that convexity was also needed in higher dimensions. In this paper the correct condition on D is discovered:

Theorem 1. *If D is a bounded C^2 domain, the Dirichlet problem is well posed for any C^2 boundary data if and only if the mean curvature of $\partial\Omega$ (computed using the interior normal) is everywhere non-negative.*

The paper also gives an improvement of a 1911 result of Korn. The authors' result is

Theorem 2. *Let D be a bounded C^2 domain. There exists a constant C depending only on the geometry of D and on the C^2 norm of the boundary data such that the Dirichlet problem in D is well posed for C^2 boundary data if and only if the oscillation of f on $\partial D < C$. If the mean curvature of ∂D is everywhere non-negative, then $C = \infty$.*

They also prove some regularity of the solution in $\bar{\Omega}$ under conditions on f.

The constant C is computed explicitly.

They also show that if the mean curvature of $\partial\Omega$ is somewhere negative, then there exist smooth boundary data, with arbitrarily small absolute value, for which no solution exists. This result is generalized to nonsmooth boundaries in Section 4.

If D is convex, they also obtain existence of solutions for general second-order quasi-linear equations.

Basic tools in the proof of existence are the maximum principle and the construction of suitable supersolutions for the problem – and estimates of derivatives of the solution.

DOI 10.1007/978-3-0348-0845-3_4, © Springer Basel 2014

The results in the paper are remarkable and have had much influence on later work for general nonlinear equations.

Paper [ii] below extends results of an earlier one [1] that considered the Dirichlet problem for the minimal surface equation in a two-dimensional bounded domain, where the boundary data may be $\pm\infty$ on some arcs on the boundary.

Paper [ii] extends the results to variational problems of the form

$$\delta \int F(p,q)dx\,dy = O, \quad p = u_x, \quad q = u_y,$$

where $F(p,q)$ is a C^2 function defined in the (p,q) plane, satisfying for some constants A and E,

$$G := F_{pp}F_{qq} - F_{pq}^2 > 0, \quad F_{pp} > 0,$$
$$F - pF_p - qF_q > A,$$
$$(1+p^2)F_{pp} + 2pqFpq + (I+q^2)F_{qq} < 2E\sqrt{G}\sqrt{1+p^2+q^2}.$$

A solution is then a C^2 function satisfying the Euler–Lagrange equation

$$F_{pp}u_{xx} + 2F_{pq}u_{xy} + F_{qq}u_{yy} = 0, \qquad p = u_x,\ q = u_y.$$

Further conditions on the second derivative of F are assumed.

The authors consider the Dirichlet problem in a bounded domain in $\mathbb{R}^2$ whose boundary consists of a finite number of open convex curves and their endpoints. To some of these arcs $C_1, C_2, \ldots$, they assign continuous boundary values. To arcs $A_1, A_2, \ldots$, they assign boundary values $+\infty$, and to arcs $B_1, B_2, \ldots$, they assign boundary values $-\infty$.

They show that if the problem is solvable, then necessarily the A_i and B_i are straight segments.

They consider a domain P (polygon) with a finite number of arcs A_i, B_i, C_i. If the perimeter of D is given a counter-clockwise orientation then the formal sum of its oriented boundary segments forms a one-chain Γ. Γ can be uniquely expressed in the form

$$\Gamma = A - A' = B - B'.$$

Here A is composed of those arcs of Γ in the family $\{A_i\}$ and A' is the complement of A in Γ.

The authors define the functional $I(\gamma)$ on arcs γ to be the length of γ in the nonsymmetric Finsler metric induced by the function $f(p,q)$. For the minimal surface equation, $F = \sqrt{1+p^2+q^2}$, and $I(\gamma)$ reduces to the Euclidean length of γ. The Finsler metric is defined by $H(\Gamma)$.

The main result of the paper is

Theorem. *If the family of arcs $\{C\}$ is not empty, the Dirichlet problem stated above is solvable if and only if*

$$I(A) < I(A'), \quad I(-B) < I(-B') \tag{$*$}$$

for each finite polygon P as above. If the family $\{C\}$ is empty, the result is the same except that $(*)$ should be replaced by $I(A) = I(-B)$. In the first case the solution is unique, in the second, it is unique up to an additive constant.

The maximum principle plays a fundamental role. Also they introduce a conjugate function of u. The proofs include ingenious arguments.

References

[1] H. Jenkins and J. Serrin. Variational problems of minimal surface type. II. Boundary value problems for the minimal surface equation. *Arch. Rational Mech. Anal.* **21** (1966), 321–342.

Louis Nirenberg
Courant Institute of Mathematical Sciences
New York University
251 Mercer Street
New York, NY 10012, USA
e-mail: nirenl@cims.nyu.edu

J. Reine Angew. Math. 223 (1968), 170–187.   **Reprint I.4.[i]**

The Dirichlet problem for the minimal surface equation in higher dimensions

By *Howard Jenkins* and *James Serrin* at Minneapolis

Let D be a bounded domain in the Euclidean space E^n, with boundary Γ. The Dirichlet problem for the minimal surface equation in D consists in determining a function $u = u(x)$, which in the domain S satisfies the minimal surface equation

$$(1) \qquad Lu = (1 + |\nabla u|^2)\Delta u - u_i u_j u_{ij} = 0 \qquad (|\nabla u|^2 = u_i u_i),$$

and which continuously takes on given boundary values f on Γ. Here $x = (x_1, \ldots, x_n)$ denotes points of E^n, the subscript notation for partial differentiation has been employed, and repeated indices are to be summed from 1 to n. If $\mathcal{F}$ is a class of functions defined on Γ, the Dirichlet problem is said to be well posed for boundary data in the class $\mathcal{F}$ if it is uniquely solvable for any given function f in $\mathcal{F}$.

When $n = 2$ it is well known that a solution of the Dirichlet problem exists for arbitrary continuous boundary data, provided the domain D is convex. Proofs of this fact were given in successively increasing generality by Bernstein (1910) and Haar (1927), and the full result was obtained by Radó (1930). In a recent paper the present authors gave another proof, in many ways simpler than that of Radó. That the Dirichlet problem is not necessarily solvable when D is non-convex was pointed out by Bernstein, and in fact Finn (1965) has definitively shown that for every non-convex domain there exists continuous boundary data for which Dirichlet's problem is not solvable. From the results just noted, it follows that when $n = 2$ the Dirichlet problem in D is well posed for arbitrary continuous boundary data if and only if D is convex (this formulation was suggested by R. Osserman).

In higher dimensions it has generally been assumed that convexity would continue to be a necessary condition for the Dirichlet problem to be well posed for arbitrary data (see for example the remarks in [6], page 152). This assumption in fact plays an important role in the work of Gilbarg and of Stampacchia, where the existence of solutions for $n > 2$ is established for smooth boundary data and strictly convex smoothly bounded domains. We shall show here, however, that *convexity of the domain is not the appropriate generalization of the two dimensional situation, but rather it is the mean curvature of the boundary which is crucial.*

More specifically, let D be a domain in E^n whose boundary Γ is of class C^2. The sign of the mean curvàture of the boundary is then fixed by choosing an orientation for the normal vector on Γ; for definiteness we take this to be directed into D. (Thus, for example, a sphere will have constant *positive* mean curvature.) The following result then holds.

Theorem 1. *Let D be a bounded domain in E^n whose boundary is of class C^2. Then the Dirichlet problem for the minimal surface equation in D is well posed for C^2 boundary data if and only if the mean curvature of Γ is everywhere non-negative.*

When the dimension n is greater than two, the condition of non-negative mean curvature can be satisfied not only by non-convex domains, but even by domains which are multiply connected. For example, the torus generated by revolving the disk $(x-a)^2 + y^2 = b^2$ about the y axis has non-negative mean curvature whenever $a \leqq 2b$. It is, in fact, only in the case of two dimensions that the condition of non-negative mean curvature implies that the domain D is convex.

When the boundary data is subject to certain restrictions, the Dirichlet problem may be solvable even with no conditions on the curvature of Γ. In fact, as long ago as 1832 this journal published a short note by S. D. Poisson in which he announced an existence proof for the problem of least area, provided the spatial curve (or curves) bounding the area lie suitably near a fixed plane. Poisson's formulation of this result was somewhat vague, and in any case he never published the proof. In 1909 the problem was taken up again by Korn in a classical paper. Korn assumed that the boundary curve Γ and the boundary data f are of class $C^{2,\lambda}$. Then, supposing that the $C^{2,\lambda}$ norm of f is sufficiently small, he demonstrated the solvability of Dirichlet's problem by using the method of successive approximations (cf. also C. Muntz, J. reine angew. Math. **139** (1911), and [4], pp. 368—369).

The spatial boundary curve in Korn's formulation must lie nearly in some *horizontal* plane L. As far as Poisson's problem is concerned this involves no loss of generality, since the minimal surface equation is invariant under rotations of the ambient space. On the other hand, Korn's result does require that the boundary curve be near L in a particularly strong topology. Here we shall improve Korn's theorem by lightening the hypotheses on the boundary and the boundary data, and by requiring nearness roughly speaking only in the uniform topology (presumably more in line with Poisson's original intentions). The precise result, which applies equally in any number of dimensions, is as follows.

Theorem 2. *Let D be a bounded domain in E^n whose boundary is of class C^2. Then there exists a positive quantity $\mathcal{B}$, depending only on the geometry of D and on uniform norms $\|f\|_1$ and $\|f\|_2$ of the first and second derivatives of the boundary data, such that the Dirichlet problem in D is well posed for C^2 boundary values f whenever*

$$h < \mathcal{B}(\|f\|_1, \|f\|_2, D) \tag{2}$$

where $2h$ denotes the oscillation of f on Γ. If the mean curvature of Γ is everywhere non-negative then $\mathcal{B} = \infty$.

The function $\mathcal{B}(\|f\|_1, \|f\|_2, D)$ is given explicitly in Section 3. In terms of the mean curvature of Γ it tends to zero as the minimum mean curvature tends to minus infinity, increases to infinity as the minimum mean curvature increases to zero, and, as noted in the statement of the theorem, is infinite when the mean curvature is everywhere non-negative. Thus Theorem 2 extends the validity of Theorem 1 by making explicit the relationship between the mean curvature of the boundary of D and the solvability of Dirichlet's problem.

While we are unable to establish the strict necessity of a condition of the form (2) in order to solve the Dirichlet problem, we are nevertheless able to obtain a closely related conclusion.

Theorem 3. *Let D be a bounded domain in E^n whose boundary is of class C^2. If the mean curvature of Γ is anywhere non-negative, then there exists smooth boundary data having arbitrarily small absolute value for which no solution of the Dirichlet problem is possible.*

This result shows in particular that (2) cannot be replaced by any condition of the form $h < \mathscr{B}(D)$ without destroying the validity of the theorem. In more geometric terms, the spatial boundary in Poisson's problem not only must lie near a fixed plane, but the requisite nearness depends on the first and second derivatives of the data (see also Theorem 2′, Section 3).

Theorem 3 can be generalized so as to apply without assumptions on the smoothness of the boundary. We take this up in Section 4. Moreover, under the assumptions of Theorem 2 we remark that the solution of the Dirichlet problem is of class C^1 in the closure of D, and if the boundary and the data are of class C^3 then the solution is even of class C^2 in the closure of D.

Since Theorem 1 is an obvious corollary of Theorem 2 and Theorem 3, it is enough to prove the latter results. This will be done in Sections 3 and 4. Theorem 2, in particular, is a consequence of the fundamental Lemma 3, which establishes an *a priori* estimate for the gradient of solutions of the Dirichlet problem in terms of the boundary data and the geometry of D. This lemma in turn depends upon a classical barrier argument and the construction of a certain interesting supersolution of the minimal surface equation. It is perhaps worth emphasizing that in our proof of Theorem 2 the particular structure of the minimal surface equation is used only in the construction of this supersolution. *Thus similar results for a wide class of quasilinear equations can be obtained immediately, once corresponding supersolutions for these equations are constructed.*

In Section 5 we shall consider the situation for two dimensions. In this case several further conclusions can be drawn in view of the availability of interior estimates for the gradient of a solution.

The last section of the paper applies the preceding results (especially Theorem 3) to show that convexity of the domain, even in higher dimensions, is closely connected with the solvability of Dirichlet's problem for general second order quasi-linear equations. Thus, if one maintains the degree of generality considered by Gilbarg and Stampacchia in their work on the Dirichlet problem, their condition of convexity is essentially best possible.

Finally, the authors particularly wish to thank Professor J. C. C. Nitsche for many helpful conversations during the course of this work.

1. Preliminary results

We assume throughout, unless otherwise stated to the contrary, that D is a bounded domain (connected open set) in E^n whose boundary is of class C^2. Our existence proof will depend heavily upon an *a priori* estimate for the gradient of solutions of the minimal surface equation which continuously take on assigned data on the boundary of D. The following elementary but useful result shows that in order to obtain an estimate for the gradient at boundary points of D it is sufficient to determine an appropriate supersolution at each boundary point.

Proposition I. *Let u be a solution of the minimal surface equation in D which is continuously differentiable in the closure of D. Let P be a point on the boundary of D.*

If Ω is a neighbourhood of P, and if there exists a supersolution v defined in $\Omega \cap \overline{D}$ such that $v(P) = u(P)$ and $u \leqq v$ at every boundary point of $\Omega \cap D$, then

$$\frac{\partial u}{\partial \nu}(P) \leqq | \nabla v(P) |$$

for every direction ν into the domain D at P.

Although Proposition I is basically well known, we include a proof for completeness. By hypothesis $Lv \leqq 0$. The function $w = u - v$ consequently satisfies a linear elliptic inequality in $G = \Omega_\cap D$ of the form $a_{ik} w_{ik} + b_i w_i \geqq 0$, (see [4], pages 323—24). Since $w = u - v \leqq 0$ on the boundary of G, it follows from the maximum principle that $w \leqq 0$ throughout G.

Thus $u \leqq v$ in G. Since both u and v are continuously differentiable in $\Omega_\cap \bar{D}$ and $u(P) = v(P)$, the required conclusion follows at once.

In this and the following section we shall use Proposition I to obtain an estimate for the boundary gradient of a solution of the Dirichlet problem, and will then derive an *a priori* estimate for the gradient of a solution in the entire domain. We shall need some preliminary computations before introducing the requisite supersolution v.

Let $\varphi(x_1, \ldots, x_{n-1})$ be a function of class C^2 defined in the ball $r < a$, where $r = (x_1^2 + \cdots + x_{n-1}^2)^{\frac{1}{2}}$. Assume that $|\nabla\varphi| < 1$, and consider the surface

$$z = \varphi(x_1, \ldots, x_{n-1}), \quad r < a, \tag{3}$$

in E^n, where we have put $x_n = z$ in order to emphasize the special role of this coordinate. Let the principal curvatures of the surface (3) be bounded in absolute value by K. Then we have the following

Lemma. *Under the circumstances above*

$$|\Delta\varphi| \leqq \sqrt{2}\, nK \quad \text{and} \quad \sum_{ij} \varphi_{ij}^2 \leqq 2(n+2)K^2, \tag{4}$$

where $\varphi_i = \partial\varphi/\partial x_i$ *and* $\varphi_{ij} = \partial^2\varphi/\partial x_i \partial x_j$.

As is evident from the context, the indices i, j here run from 1 to $n-1$ rather than from 1 to n. A similar remark will be tacitly understood in later formulae of this section.

Proof. Let Q be an arbitrary point in $r < a$. We may suppose without loss of generality that the matrix $((\varphi_{ij}))$ is diagonal at Q, since (4) is invariant under coordinate rotations in E^{n-1}. Now consider the normal curvatures $\varkappa_i$ of the x_i-parametric curves on the surface (3), $i = 1, \ldots, n-1$. An easy calculation shows that

$$\varkappa_i = \frac{\varphi_{ii}}{(1+\varphi_i^2)(1+|\nabla\varphi|^2)^{\frac{1}{2}}}.$$

Then since $|\nabla\varphi| < 1$ we have $|\varphi_{ii}| \leqq \sqrt{2}\,(1+\varphi_i^2)K$. Consequently

$$|\Delta\varphi| \leqq \sqrt{2}\, K \sum_i (1+\varphi_i^2) \leqq \sqrt{2}\, nK,$$

while at Q there holds $\sum_{ij} \varphi_{ij}^2 = \sum_i \varphi_{ii}^2 \leqq 2(n+2)K^2$, and the lemma is proved.

Now let α, β, and c be positive constants such that

$$\alpha < \beta, \quad c < \frac{\pi}{2\beta}, \quad \frac{\pi}{2\alpha} \leqq a, \tag{5}$$

and (as above) let $\varphi(x_1, \ldots, x_{n-1})$ be a function of class C^2 defined for $r < a$ and such that $|\nabla\varphi| < 1$. Consider the domain Ω in E^n determined by the inequalities

$$\varphi + c - \frac{\pi}{2\beta} < z < \varphi + c - \frac{\alpha}{\beta} r, \quad r < \frac{\pi}{2\alpha} \tag{6}$$

(this domain has the general form shown in Figure 1). In Ω we define the functions

$$s = \sec \alpha r, \quad t = \tan \alpha r$$

and

$$S = \sec \beta(\varphi + c - z), \quad T = \tan \beta(\varphi + c - z).$$

With these definitions (and recalling that x_n is to be identified with z) the following result holds.

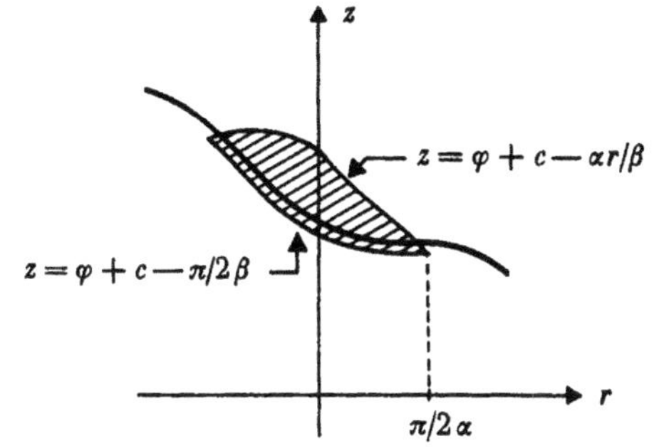

Figure 1. The domain Ω. The surface $z = \varphi$ is shown by a heavy line.

Lemma 1. *Let*

$$w(x_1, \ldots, x_n) = \frac{1}{\alpha} \log s - \frac{1}{\beta} \log S, \quad x \in \Omega.$$

Then

$$Lw \leqq -s^2 S^2 \{\beta - 2(n + \sqrt{n+2}) K - (3n - 5)\alpha\} - (n-1)(1 + |\nabla\varphi|^2)^{\frac{3}{2}} H T^3,$$

where $H = H(x_1, \ldots, x_{n-1})$ *is the mean curvature of the surface* $z = \varphi(x_1, \ldots, x_{n-1})$.

Proof. Put $\nu_i = x_i/r$ and $\vec{\nu} = (\nu_1, \ldots, \nu_{n-1})$. Then we find easily

$$w_i = t\nu_i - T\varphi_i$$

$$w_{ij} = \alpha s^2 \nu_i \nu_j + \frac{t}{r}(\delta_{ij} - \nu_i \nu_j) - \beta S^2 \varphi_i \varphi_j - T\varphi_{ij}$$

$$w_{in} \equiv w_{iz} = \beta S^2 \varphi_i$$

for $i, j = 1, \ldots, n-1$. Moreover, $w_n \equiv w_z = T$ and $w_{nn} \equiv w_{zz} = -\beta S^2$. Thus

$$1 + |\nabla w|^2 = s^2 + T^2(1 + |\nabla\varphi|^2) - 2tT\vec{\nu}\cdot\nabla\varphi$$

(7)
$$\Delta w = \alpha s^2 + (n-2)\frac{t}{r} - \beta S^2(1 + |\nabla\varphi|^2) - T\Delta\varphi.$$

Proceding in this fashion, one obtains after some calculation

(8)
$$\begin{aligned} Lw &= (1 + |\nabla w|^2)\Delta w - w_i w_j w_{ij} \qquad (i, j = 1, \ldots, n) \\ &= -(n-1)(1 + |\nabla\varphi|^2)^{\frac{3}{2}} H T^3 + (\alpha - \beta) s^2 S^2 \\ &\quad + T\Delta\varphi\{2tT(\vec{\nu}\cdot\nabla\varphi) - s^2\} \\ &\quad + tT\vec{\nu}\cdot\Phi\cdot\{t\vec{\nu} - 2T\nabla\varphi\} \\ &\quad + (n-2)\frac{t}{r}\{1 + |\nabla w|^2\} \\ &\quad - \left\{(\beta S^2 - \alpha T^2)s^2 + \frac{t}{r}T^2\right\}\{|\nabla\varphi|^2 - (\vec{\nu}\cdot\nabla\varphi)^2\} - \beta S^2(\vec{\nu}\cdot\nabla\varphi)^2 \end{aligned}$$

where Φ denotes the Hessian matrix $((\varphi_{ij}))$, and where we have used the formula

$$(1 + |\nabla\varphi|^2)\Delta\varphi - \varphi_i\varphi_j\varphi_{ij} = (n-1)(1 + |\nabla\varphi|^2)^{\frac{3}{2}} H.$$

To estimate the second line on the right hand side of (8), note that

$$|\nabla\varphi| < 1 \text{ and } |\Delta\varphi| \leqq \sqrt{2}\, nK.$$

Hence

$$T\Delta\varphi\{2tT(\vec{\nu}\cdot\nabla\varphi) - s^2\} \leqq T|\Delta\varphi|(2tT + s^2) \leqq \left(1 + \frac{1}{\sqrt{2}}\right) ns^2S^2K\ ^{1)}.$$

Similarly

$$tT\vec{\nu}\cdot\Phi\cdot\{t\vec{\nu} - 2T\nabla\varphi\} \leqq (t^2T + 2tT^2)\left(\sum_{ij}\varphi_{ij}^2\right)^{\frac{1}{2}} \leqq \left(1 + \frac{1}{\sqrt{2}}\right)\sqrt{n+2}\, s^2S^2K.$$

In order to estimate the fourth line, note that $t/r \leqq \alpha s$ (that is, $\sin\theta \leqq \theta$). Then from (7)

$$\frac{t}{r}(1 + |\nabla w|^2) \leqq \alpha s\{s^2 + 2T(t+T)\} \leqq \alpha\{s^3 + 2s^2S^2\}.$$

Now $\alpha r < \beta(\varphi + c - z)$ in Ω, so that $s \leqq S$. Consequently the fourth line is bounded by $3(n-2)\alpha s^2S^2$.

Since $\beta > \alpha$ the fifth line is clearly non-positive and may be omitted. Inserting the preceding estimates into (8) completes the proof of Lemma 1.

2. The gradient bound

We now turn to the construction of a supersolution v and the determination of a gradient bound for solutions of the Dirichlet problem.

By assumption the boundary of D is compact and of class C^2. Thus there exists a constant $\varrho > 0$ such that if P is a point of Γ and Σ denotes the ball of radius ϱ about P, then $\Gamma \cap \Sigma$ consists of a single component and can be represented in the form $\bar{x}_n = \varphi(\bar{x}_1, \ldots, \bar{x}_{n-1})$, where $(\bar{x}_1, \ldots, \bar{x}_n)$ are appropriately chosen coordinates in E^n. We may furthermore assume that the $\bar{x}_n$ axis is directed along the inner normal to Γ at P, and that $|\nabla\varphi| < 1$. This being the case, it is evident that the domain of φ contains a ball of radius $\varrho' = \varrho/\sqrt{2}$ about the origin. Also, using the fact that Γ is compact and of class C^2, its principal curvatures must be uniformly bounded. We let K denote an absolute bound for these curvatures.

Let f be a function of class C^2 on Γ. Then in $\Gamma \cap \Sigma$ we may write $f = \psi(\bar{x}_1, \ldots, \bar{x}_{n-1})$ where ψ is of class C^2. With this understanding, we define in an obvious notation

$$f_1(P) = \{\sum_i \psi_i(P)^2\}^{\frac{1}{2}}, \quad f_2(P) = 2\{\sum_{ij}\psi_{ij}(P)^2\}^{\frac{1}{2}}$$

and set $\|f\|_1 = \max_\Gamma |f_1(P)|$, $\|f\|_2 = \max_\Gamma |f_2(P)|$. Finally, let $2h$ denote the oscillation of f on Γ.

Lemma 2. *Suppose that u is a solution of the minimal surface equation in D which is continuously differentiable in the closure of D and takes on the boundary values f on Γ. Suppose also that the mean curvature of Γ everywhere satisfies an inequality of the form*

$$H \geqq -\alpha/\mu e^\mu$$

---

[1]) We have used the simple trigonometrical inequality

$$\frac{2tT^2 + s^2T}{s^2S^2} \leqq \frac{T + T^2}{S^2} \leqq \frac{1+\sqrt{2}}{2}.$$

where

$$\alpha \geqq \mathrm{Max}\left(\frac{\pi}{\varrho'}, K, \| f \|_2\right),$$

and

$$\mu = 8n(2\alpha h + \log\sqrt{1 + \| f \|_1^2}).$$

Then at each boundary point of D we have $\sqrt{1 + |\nabla u|^2} < 4e^\mu$.

Proof. Let P be an arbitrary point of Γ. For the purposes of estimating ∇u at P we can assume $f(P) = 0$. The proof then depends on the construction of an appropriate supersolution and the application of Proposition I.

Introduce coordinates $(\bar{x}_1, \ldots, \bar{x}_n)$ in the neighborhood of P, as in the preceding discussion. Then as previously noted, f is given by a function $\psi(\bar{x}_1, \ldots, \bar{x}_{n-1})$. Let r_0 be the unique number between 0 and $\pi/2\alpha$ such that $\tan \alpha r_0 = |\nabla \psi(P)|$. We change variables according to the translation

$$\text{(9)} \qquad \begin{cases} x_i = \bar{x}_i + r_0 \dfrac{\psi_i(P)}{|\nabla \psi(P)|}, & i = 1, \ldots, n-1 \\ x_n = \bar{x}_n. \end{cases}$$

In the new variables we write $f = \psi(x_1, \ldots, x_{n-1})$ for the boundary data and $x_n = \varphi(x_1, \ldots, x_{n-1})$ for the surface neighborhood $\Gamma_\cap \Sigma$. This should not be a source of confusion. Since the domain of φ in the original variables contains a ball of radius ϱ' about the origin, and since $r_0 < \pi/2\alpha \leqq \varrho'/2$, it is evident that the domain of φ in the new variables contains the ball $r \leqq \varrho'/2$ $(= a)$.

Let β and c be constants satisfying (5), whose precise values will be fixed later. We consider the function

$$v(x_1, \ldots, x_n) = 2h + w(x_1, \ldots, x_n), \quad x \in \Omega$$

where w is defined in Lemma 1 and Ω is determined by the inequalities (6), (recall here that $x_n = z$). In order to apply Proposition I it is necessary to establish a number of facts:

(i) Ω is a neighborhood of P

(ii) $v(P) = 0$

(iii) $u \leqq v$ at every boundary point of $\Omega_\cap D$

(iv) $Lv \leqq 0$ in $\Omega_\cap \bar{D}$.

According to (ii) we must have

$$0 = 2h + \frac{1}{\alpha} \log \sec \alpha r_0 - \frac{1}{\beta} \log \sec \beta c,$$

that is

$$\text{(10)} \qquad c = \frac{1}{\beta} \sec^{-1}(e^{\beta\lambda}), \quad \lambda = 2h + \frac{1}{\alpha} \log \sec \alpha r_0$$

where the inverse secant is taken in the first quadrant so that the second inequality of (5) holds.

In what follows we can assume $h > 0$, for otherwise the lemma is obviously valid. Then since $\beta > \alpha$ it is clear that $\beta c > \alpha r_0$. This being the case, and since $z = \varphi$, $r = r_0$

at P, one easily checks that P is in the domain Ω. It remains therefore to prove properties (iii) and (iv).

Property (iii). Using the results already established together with the condition $|\nabla\varphi| < 1$, it is not difficult to show that Ω is contained in the ball Σ of radius ϱ about P.[2]) Now by construction the intersection of D with Σ consists of a single component where $z > \varphi$; consequently

$$\Omega_\cap D = \left\{\varphi < z < \varphi + c - \frac{\alpha}{\beta} r\right\}, \text{ see Figure 2.}$$

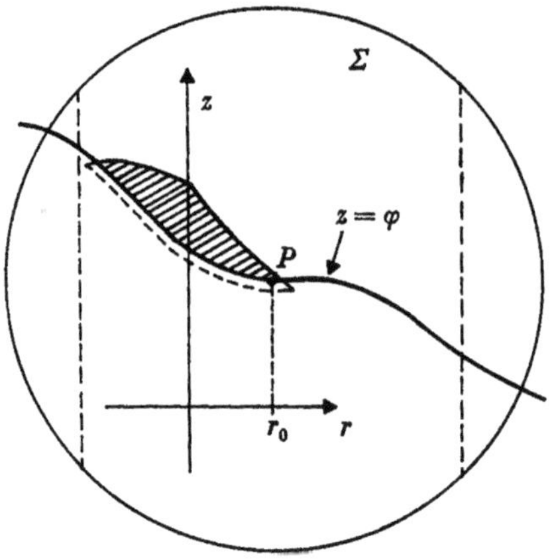

Figure 2. The domain $\Omega_\cap D$. The boundary of Ω is shown partly by a dotted line. The vertical dotted lines on either side describe the domain of φ in E^{n-1}, a ball of radius $\varrho' = \varrho/\sqrt{2}$ about P.

The boundary of $\Omega_\cap D$ consists of two parts. On the part where $z = \varphi + c - \alpha r/\beta$ we have

$$v = 2h + \left(\frac{1}{\alpha} - \frac{1}{\beta}\right)\log\sec\alpha r \geqq 2h.$$

On the other hand, the set $z = \varphi + c - \alpha r/\beta$ is contained in D. Since $f(P) = 0$ and the oscillation of f is $2h$, an application of the maximum principle yields $|u| \leqq 2h$ in D. Thus $u \leqq v$ on this portion of the boundary.

On the remaining portion of the boundary the value of v is given in local coordinates $(x_1, \ldots, x_{n-1})$ by the function

$$\omega(x_1, \ldots, x_{n-1}) = \frac{1}{\alpha}\log\left(\frac{\sec\alpha r}{\sec\alpha r_0}\right).$$

In view of (9) we have evidently $\omega(P) = \psi(P) = 0$ and $\nabla\omega(P) = \nabla\psi(P)$.

Let $\vec{\tau} = (\tau_1, \ldots, \tau_{n-1})$ be a unit vector in E^{n-1}. The second derivative of ω in the direction $\vec{\tau}$ is given by

$$\omega_{ij}\tau_i\tau_j = \left(\alpha s^2 - \frac{t}{r}\right)(\vec{\nu}\cdot\vec{\tau})^2 + \frac{t}{r} \geqq \alpha$$

[2]) Consider the cases where the function φ adopts its greatest and its least values. Then we find in fact that Ω is contained in the set $r < \pi/2\alpha$, $|z - z_P| < \pi/\alpha$.

since $\alpha \leqq t/r \leqq \alpha s$, (see the proof of Lemma 1). On the other hand, using the condition $|\nabla \varphi| < 1$ it is not hard to show that

$$|\psi_{ij}\tau_i\tau_j| \leqq \|f\|_2.$$

Thus in the ball $r \leqq \pi/2\alpha$ the second directional derivatives of ω are everywhere greater than those of ψ. Combining this with the conclusions of the previous paragraph we have $\omega \geqq \psi$ in $r \leqq \pi/2\alpha$, that is $v \geqq f = u$ on the remaining part $z = \varphi$ of the boundary of $\Omega_\cap D$.

Property (*iv*). It must be shown that $Lv \leqq 0$ in $\Omega_\cap \bar{D}$. For this purpose the constant β is still at our disposal. Since $K \leqq \alpha$, we have by Lemma 1 and a short calculation

$$(11) \qquad Lv \leqq -s^2S^2\{\beta - (8n-1)\alpha\} - (n-1)(1 + |\nabla\varphi|^2)^{\frac{3}{2}} HT^3$$

(this application of Lemma 1 uses the fact the principal curvatures of Γ and the form of the operator L are orthogonal invariants in E^n). By hypothesis

$$|\nabla\varphi| < 1, \quad H \geqq -\alpha/\mu e^\mu.$$

Moreover, in $\Omega_\cap \bar{D}$ we have $T \leqq \tan\beta c \leqq e^{\beta\lambda}$ where λ is defined by (10). Finally $\mu \geqq 8n\alpha\lambda$. Thus in $\Omega_\cap \bar{D}$

$$-(n-1)(1 + |\nabla\varphi|^2)^{\frac{3}{2}} HT \leqq \frac{2^{\frac{3}{2}}(n-1)\alpha e^{\beta\lambda}}{\mu e^\mu} \leqq \frac{1}{\lambda} e^{(\beta - 8n\alpha)\lambda - 1}.$$

Inserting this estimate into (11) yields

$$Lv \leqq -s^2S^2\left\{\beta - (8n-1)\alpha - \frac{1}{\lambda} e^{\lambda[\beta - 8n\alpha - 1/\lambda]}\right\}.$$

Choosing $\beta = (8n-1)\alpha + 1/\lambda$ we find $Lv \leqq 0$ in $\Omega_\cap \bar{D}$, as required.

With the properties (i)—(iv) now verified, Proposition I implies that

$$\frac{\partial u}{\partial \nu}(P) \leqq |\nabla v(P)|.$$

We require then an estimate for $|\nabla v(P)|$. By the proof of Lemma 1

$$1 + |\nabla v|^2 = s^2 + T^2(1 + |\nabla\varphi|^2) - 2tT\vec{\nu}\cdot\nabla\varphi$$
$$\leqq s^2 + 2T(t + T) \leqq 2s^2S^2.$$

At P, however,

$$S^2 = \sec^2\beta c = e^{2\beta\lambda} \leqq s^{-2}e^{2(\mu+1)}.$$

Consequently $1 + |\nabla v(P)|^2 \leqq 2e^{2(\mu+1)}$. In the same way, we can establish that

$$\frac{\partial \mu}{\partial \nu}(P) \geqq -|\nabla v'(P)|$$

where v' is a similarly constructed function, with $1 + |\nabla v'(P)|^2 \leqq 2e^{2(\mu+1)}$.

Combining the estimates of the preceding paragraph, we obtain

$$1 + |\nabla u(P)|^2 \leqq 2e^{2(\mu+1)},$$

and the proof of Lemma 2 is completed.

Lemma 3. *Under the hypotheses of Lemma* 2, *we have*

$$\sqrt{1 + |\nabla u|^2} < 4e^\mu$$

everywhere in the closure of D.

Proof. By differentiating the minimal surface equation with respect to x_i we find that each component of the gradient vector ∇u satisfies a linear elliptic equation. Consequently $|\nabla u|$ must take its maximum value on the boundary of D, and Lemma 3 follows from Lemma 2.

3. Proof of Theorem 2

In the following work we shall require an important result originally due to Gilbarg, which is stated below as Proposition II. For the proof see [6], Sections 2—5, where in fact a considerably stronger result appears, and for related theorems see [10], Section 7.

Proposition II. *Let D be a bounded domain in E^n whose boundary is of class C^3. Consider the Dirichlet problem for the minimal surface equation in D with boundary values σf on Γ, where σ is a constant and f is a function of class C^3.*

Suppose that for $0 \leqq \sigma \leqq 1$ one can establish an a priori *bound*

$$|\nabla u| < M \quad \text{for } x \in D$$

for the gradient of any solution which is continuously differentiable in the closure of D, where M is independent of σ. Then there exists a solution of the minimal surface equation in D which takes the boundary values f on Γ and is of class C^2 in the closure of D.

A similar result was obtained by Bernstein for the case of two independent variables (see [1], pages 233—236).

From Proposition II and Lemma 3 it follows that the boundary value problem for the minimal surface equation is solvable provided Γ and f are of class C^3 and the condition $H \geqq -\alpha/\mu e^\mu$ is satisfied (we can then take $M = 4e^\mu$). Lemma 4 below gives a more useful formulation of this result, and indeed establishes Theorem 2 for the case where the boundary and the data are of class C^3.

We require some preliminary definitions. Let

$$\mathscr{A} = \operatorname{Max}(\pi/\varrho', K, \|f\|_2), \quad \mathscr{C} = \mathscr{A}/(1 + \|f\|_1^2)^{8n},$$

the right hand sides depending only on the domain D, the assigned data, and the dimension (see Section 2). Also put

$$\mathscr{H} = \max_\Gamma(-H)$$

where H is the mean curvature of Γ. We now define

$$\mathscr{B}(\|f\|_1, \|f\|_2, D) = \frac{1}{32n\mathscr{A}} \begin{cases} \mathscr{C}/\mathscr{H} & \text{if } \mathscr{H} \geqq \mathscr{C} \\ 1 + \log(\mathscr{C}/\mathscr{H}) & \text{if } 0 < \mathscr{H} < \mathscr{C} \\ \infty & \text{if } \mathscr{H} \leqq 0. \end{cases}$$

As a function of the mean curvature of Γ, the quantity $\mathscr{B}$ is continuous in the extended reals, tends to zero as the minimum mean curvature tends to minus infinity, and is infinite when the mean curvature is everywhere non-negative.

Lemma 4. *Let D be a bounded domain in E^n whose boundary is of class C^3. Then the Dirichlet problem for the minimal surface equation in D is well posed for C^3 boundary values f whenever*

$$h \leqq \mathscr{B}(\|f\|_1, \|f\|_2, D) \tag{12}$$

where $2h$ denotes the oscillation of f on Γ. The solution is of class C^2 in the closure of D.

Proof. It is enough to show that condition (12) implies the existence of a constant $\alpha \geqq \mathscr{A}$ such that $H \geqq -\alpha/\mu e^\mu$ at each point of Γ, where

$$\mu = 8n(2\alpha h + \log\sqrt{1 + \|f\|_1^2}).$$

We shall take

$$\alpha = \mathscr{A}\ \mathrm{Max}\,(1, \mathscr{H}/\mathscr{C}).$$

If H is everywhere non-negative there is nothing to prove. Assume then that the mean curvature is negative at some places, that is, $\mathscr{H} > 0$. Since $h \leqq \mathscr{B}$ we have then after a short calculation

$$\mu \leqq \frac{1}{2}\left(1 + \log\frac{\alpha}{\mathscr{H}}\right).$$

Since $y \leqq e^{y-1}$ for any real y, the preceding inequality yields

$$\mu e^{\mu} \leqq e^{2\mu-1} \leqq \alpha/\mathscr{H}.$$

Consequently at each point of Γ we have $H \geqq -\mathscr{H} \geqq -\alpha/\mu e^{\mu}$, and the proof of Lemma 4 is completed.

The solution given by Lemma 4 has the uniform gradient bound $|\nabla u| < M = 4e^{\mu}$, that is

$$(13) \qquad |\nabla u| < 4\{(1 + \|f\|_1^2)e^{4\alpha h}\}^{4n} \text{ [3]}.$$

This estimate can be supplemented by a uniform modulus of continuity for ∇u. Indeed, according to a result of Ladyzhenskaya and Uraltseva ([10], Theorem 11) the gradient of the resulting solution is uniformly Hölder continuous in D with Hölder moduli depending only on n, ϱ, K, $\|f\|_1$, $\|f\|_2$, and M.[4] Since in the present case M depends in turn on n, ϱ, K, h, $\|f\|_1$, and $\|f\|_2$, it follows that ∇u has uniform Hölder moduli in D depending only on the quantities

$$n,\ \varrho,\ K,\ h,\ \|f\|_1, \text{ and } \|f\|_2.$$

We may now prove Theorem 2.

Rather than carry out a series of standard constructions, it will be sufficient to outline the general procedure. As a first step, let the boundary values f (which are by hypothesis of class C^2 on Γ) be extended to a function, still denoted by f, of class C^2 in E^n. Next approximate f by a sequence of C^3 functions f_m which converge to f in the C^2 topology in some domain containing the closure of D. Also let D be approximated by a sequence of domains D_m whose boundaries are of class C^3, in such a way that the associated quantities ϱ_m, K_m, and H_m converge to ϱ, K, and H as D_m tends to D. According to the hypothesis of Theorem 2, then, it is apparent that there exists an integer m_0 such that

$$h_m \leqq \mathscr{B}(\|f_m\|_1, \|f_m\|_2, D_m)$$

for all $m \geqq m_0$.

This being the case, Lemma 4 implies the existence of a solution u_m of the minimal surface equation in D_m which takes the boundary values f_m on Γ_m. Moreover, for all sufficiently large m we have $|\nabla u_m| < M + 1$, where M denotes the quantity on the right hand side of (13).

By standard techniques in the theory of elliptic equations, the sequence $\{u_m\}$ converges to a unique solution of the minimal surface equation in D which takes the boundary values f on Γ. Since the constructions outlined above can be assumed known, and since solutions are obviously unique, this completes the proof of Theorem 2.

[3]) The quantity $\alpha = \mathscr{A}\mathrm{Max}\,(1, \mathscr{H}/\mathscr{C})$ can be rewritten in the more concrete form

$$\mathrm{Max}\left(\frac{\pi}{\varrho'}, K, \|f\|_2, (1 + \|f\|_1^2)^{8n}\mathscr{H}\right).$$

[4]) This result appears also in Gilbarg's proof of Proposition II.

In the above proof the functions ∇u_m are uniformly Hölder continuous, as follows from the construction and the remarks after the proof of Lemma 4. Hence ∇u is uniformly Hölder continuous in D, and *a fortiori* u is of class C^1 in the closure of D.

Inspection of the argument used in proving Theorem 2 shows that actually the following result holds.

Theorem 2'. *Let D be a bounded domain in E^n whose boundary is of class C^2. Suppose that U is a solution of the minimal surface equation in some domain containing the closure of D. Then there exists a positive quantity $\mathscr{B}'(\|f\|_1, \|f\|_2, D, \max_D |\nabla U|)$ such that the Dirichlet problem in D is well posed for C^2 boundary values f whenever*

$$h < \mathscr{B}'(\|f\|_1, \|f\|_2, D, \max_D |\nabla U|)$$

where $2h$ denotes the oscillation of $f - U$ on the boundary of D.

Deferring the proof of Theorem 2' for a moment, we note that the result essentially generalizes Poisson's problem by replacing the fixed plane with a piece $\mathscr{S}$ of a minimal surface having a single valued projection in some fixed direction. That is, if the spatial boundary lies "suitably near" $\mathscr{S}$ then it can be spanned by a minimal surface lying near $\mathscr{S}$ and also having a single valued projection. So far as we know, there have been no previous results of this type concerning the perturbation of a minimal surface in space.

From another point of view, Theorem 2' can be considered as providing a solution of the Plateau problem for a spatial boundary curve lying near a piece of a minimal surface. The existence of a spanning minimal surface for this case of course also follows from the classical work of Douglis and Radó (cf. [3]). In their work, however, branch points and self-intersections may occur, while here the spanning surface not only has no self-intersections but can in fact be expressed in non-parametric form. Unfortunately, we are at present unable to determine whether our solution is unique, or whether it provides the least area, among all (parametric) surfaces spanning the same boundary curve. We conjecture that this is the case, though for the moment this is certain only when the competing surfaces are restricted to those which can be expressed in non-parametric form.

Proof of Theorem 2'. This depends basically on a modification of Lemma 2 to take into account the fact that $2h$ is now the oscillation of $f - U$ rather than of f on Γ. In the proof of Lemma 2 the quantity $2h$ serves as an upper bound for $|u|$ on the set $\Omega_\cap D$. In the present case this bound is inadequate and must be modified.

First of all, without loss of generality U can be normalized so that $|f - U| \leqq h$ on Γ. Then by the maximum principle applied to the difference $u - U$ we find

$$|u - U| \leqq h \text{ in } D. \tag{14}$$

Now by assumption $f(P) = 0$, so that $|U(P)| \leqq h$. Moreover, by the remark in footnote 2, the domain Ω is contained in a ball of radius $\sqrt{2}\,\pi/\alpha$ about P. Hence

$$|U| \leqq h + \frac{\sqrt{2}\,\pi}{\alpha} \max |\nabla U| \text{ in } \Omega_\cap D$$

and consequently by (14)

$$|u| \leqq 2h + \frac{\sqrt{2}\,\pi}{\alpha} \max |\nabla U| \text{ in } \Omega_\cap D.$$

This being shown, if we now replace $2h$ by $2h + \frac{\sqrt{2}\,\pi}{\alpha} \max |\nabla U|$ in the statement of Lemma 2, that result will hold equally in the present situation.

The effect of this change is to make

$$\mu = 8n(2\alpha h + \sqrt{2}\,\pi \max |\nabla U| + \log \sqrt{1 + \|f\|_1^2}),$$

an effect which may alternately be accomplished by replacing $1 + \|f\|_1^2$ with $(1 + \|f\|_1^2) \exp(2\sqrt{2}\,\pi \max |\nabla U|)$. Carrying out this change in Lemmas 3 and 4 and the proof of Theorem 2, we see that Theorem 2′ holds with

$$\mathscr{B}'(\|f\|_1, \|f\|_2, D, \max |\nabla U|) = \mathscr{B}(\sqrt{1 + \|f\|_1^2}\, e^{5 \operatorname{Max}|\nabla U|}, \|f\|_2, D).$$

As before, the solution is of class C^1 in the closure of D.

Remark. The norms $\|f\|_1$, $\|f\|_2$, and the oscillation h of f on Γ are necessarily related by an inequality of the form

$$\|f\|_1 \leqq \text{Const.}\,(h + \|f\|_2) \tag{15}$$

where the constant depends on the geometry of the boundary set, as follows from ordinary calculus inequalities. In consequence (as pointed out by Norman Meyers) it would be possible to eliminate $\|f\|_1$ from the condition $h < \mathscr{B}$ of Theorem 2. A similar remark applies equally to Theorem 2′.

Inequality (15) also implies that for a given domain D and a given bound on the norm $\|f\|_2$ of the data, the norm $\|f\|_1$ must itself be bounded in order to have the condition $h < \mathscr{B}$ satisfied. On the other hand, for any given domain it is easy to see that this condition can be fulfilled (and consequently the Dirichlet problem solved) for data having arbitrarily large norms $\|f\|_1$ and $\|f\|_2$.

4. Proof of Theorem 3

This depends on an interesting comparison theorem, which we state as our third proposition.

Proposition III. *Let D be a bounded domain in E^n whose boundary is the union of two closed sets γ and γ', with γ' being of class C^1. Let u be a solution of the minimal surface equation in D which is continuous on γ and continuously differentiable on γ'. Also let v be a supersolution in D which is continuous in the closure of D and such that $\partial v/\partial n = -\infty$ at every interior point of γ'.*

Under these conditions, if $u \leqq v$ on γ then $u \leqq v$ throughout D.

Proof. If $u \leqq v$ on γ' the result follows from the maximum principle, as in the proof of Proposition I. Thus suppose for contradiction that

$$d = \max_{\gamma'} (u - v) > 0.$$

Then by the maximum principle, $u \leqq v + d$ thoughout D. By the definition of d there is an interior point P on γ' where $u = v + d$. In combination with the previous inequality, this yields

$$\frac{\partial u}{\partial n} \leqq \frac{\partial v}{\partial n} \text{ at } P$$

which contradicts the assumption that $\partial v/\partial n = -\infty$ on γ'.

Proposition III in its present form is due to Finn [5], although the underlying idea was known and used by both Bernstein and Leray ([1], page 243; [2], pages 462—469; and [11], page 260).

It is convenient at this stage to introduce some simple terminology. Let D be a domain in E^n with boundary Γ. If P is a point of Γ and N is a neighborhood of P in E^n, the set $\Gamma \cap N$ is called a *boundary neighborhood of* P. A subset of Γ is called *open* if it contains a boundary neighborhood of each of its points. Theorem 3 is now an immediate consequence of the following result.

Lemma 5. *Let D be a bounded domain in E^n whose boundary is of class C^2, and let P be a point of Γ where $H < 0$. Then for any $\varepsilon > 0$ there exists a boundary neighborhood Γ' of P such that if u is a solution of the minimal surface equation in D which continuously takes on the boundary values f, then*

$$|u| < \max_{\Gamma - \Gamma'} |f| + \varepsilon \quad \text{for } x \in D.$$

Assuming Lemma 5 for the moment, in order to prove Theorem 3 it suffices to choose smooth boundary values f such that $f \equiv 0$ on $\Gamma - \Gamma'$ and $f = 2\varepsilon$ at P.

Proof of Lemma 5. We introduce coordinates $(\bar{x}_1, \ldots, \bar{x}_n)$ in the neighborhood of P, as in Section 2. The $\bar{x}_n$ axis is directed along the inner normal to Γ at P, the boundary set $\Gamma \cap \Sigma$ consists of a single component with the representation

$$\bar{x}_n = \varphi(\bar{x}_1, \ldots, \bar{x}_{n-1}),$$

and $|\nabla\varphi| < 1$. We may suppose without loss of generality that $\bar{x}_n = 0$ at P, i. e. the origin of coordinates is at P. With this choice, the bars on the coordinate symbols will now be omitted for convenience in printing.

Write $R^2 = x_1^2 + \cdots + x_n^2$ for the polar coordinate in E^n, and let a be a constant less than $\varrho/\sqrt{2}$ whose value will be fixed later (here ϱ is the radius of Σ). After a straightforward calculation we find

$$L\left[a \cosh^{-1} \frac{R}{a}\right] = \frac{(n-2)aR}{(R^2 - a^2)^{\frac{3}{2}}}, \quad R > a.$$

Putting $D' = D \cap \{R > a\}$, the boundary of D' is then the union of two parts $\gamma = \Gamma \cap \{R \geqq a\}$ and $\gamma' = \bar{D} \cap \{R = a\}$. We may apply Proposition III to the functions u and v in D', where

$$v = b - a \cosh^{-1}(R/a).$$

Indeed $\partial v/\partial n = -\infty$ at each interior point of γ', while $u \leqq v$ on γ provided that

$$b = \operatorname{Max}_\gamma |f| + a \cosh^{-1}(d/a),$$

d being the diameter of D. Thus $u \leqq v$ in D', and *a fortiori*

$$u \leqq b \text{ in } D \cap \{R \geqq a\}. \tag{16}$$

Consider next the domain E formed by intersecting the ball $R < a$ with the set $z > \varphi + \varepsilon'$, where ε' is a small positive constant. After some calculation we find that in E

$$L[c\sqrt{z - \varphi - \varepsilon'}] = -\frac{c}{8(z - \varphi - \varepsilon')^{\frac{3}{2}}}\{(n-1)(1 + |\nabla\varphi|^2)^{\frac{3}{2}} c^2 H + 2(1 + |\nabla\varphi|^2) + 4(z - \varphi - \varepsilon')\Delta\varphi\};$$

here c is a constant, $z = x_n$, and

$$H = \frac{(1 + |\nabla\varphi|^2)\Delta\varphi - \varphi_i\varphi_j\varphi_{ij}}{(n-1)(1 + |\nabla\varphi|^2)^{\frac{3}{2}}}$$

is the mean curvature of Γ in the local coordinates $(x_1, \ldots, x_{n-1})$. Let a_0 be so small that both $H < \frac{1}{2}H(P)$ and $\Delta\varphi < 0$ in $r < a_0$. The latter inequality is possible since $\Delta\varphi(P) = (n-1)H(P)$. We now suppose further that $a \leqq a_0$. Then putting

$$c = 2/\sqrt{(n-1)\,|H(P)|},$$

it follows that the function

$$w = b + c(\sqrt{2a} - \sqrt{z - \varphi - \varepsilon'})$$

is a supersolution in E.

The boundary of E consists of two parts. Evidently $\partial w/\partial z = -\infty$ on the part where $z = \varphi + \varepsilon'$. Also $u \leqq w$ on the part $R = a$, in view of (16). A final application of Proposition III therefore yields $u \leqq w$ in E.

Letting ε' tend to zero, it follows easily that $u \leqq b + c\sqrt{2a}$ in $D_\cap\{R < a\}$. By virtue of (16), the same inequality holds in all of D. Since this construction applies equally to the solution $-u$, we obtain (using the explicit expressions for the constants b and c)

$$|u| \leqq \max_\gamma |f| + a\cosh^{-1}(d/a) + \sqrt{8a/(n-1)\,|H(P)|} \quad \text{for } x \in D.$$

Here a is any constant less than both $\varrho/\sqrt{2}$ and a_0. Since the last two terms tend to zero as a tends to zero, the lemma follows by choosing a sufficiently small and letting $\Gamma' = \Gamma - \gamma$.

Remark. Theorem 3 can be generalized so as to apply without assumptions on the smoothness of the boundary of D. Indeed, it is enough to suppose that at some point P on Γ there is an internally touching surface element σ of class C^2 whose mean curvature is negative. (The sign of the mean curvature is here based on the convention that the normal vector to σ is directed toward the side of σ which near P contains only points interior to D, cf. Figure 3. This is clearly consistent with our earlier agreement concerning the sign of the mean curvature of Γ.)

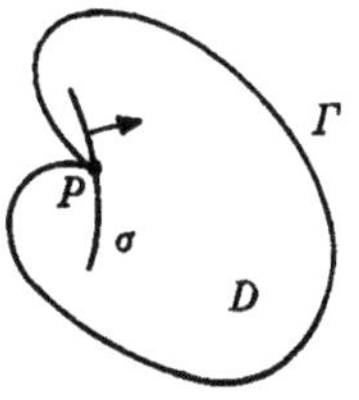

Figure 3. Configuration for an internally touching surface element σ at P.

If the surface σ is represented by $z = \varphi(x_1, \ldots, x_{n-1})$ as in the proof of Lemma 5, then by following the steps of that proof one finds

$$|u| \leqq \max_\gamma |f| + a\cosh^{-1}(d/a) + \sqrt{8a/(n-1)\,|H(P)|},$$

valid in the set $\{R < a\}_\cap \{z > \varphi\}$ for all sufficiently small a. Consequently, by choosing a suitably small, we see that $|f(P)|$ must be less than $\max_{\Gamma-\Gamma'} |f| + \varepsilon$ in order for the Dirichlet problem in D to be solvable. Thus again there obviously exists boundary data having arbitrarily small absolute value, for which no solution of the Dirichlet problem is possible.

When $n = 2$ this result can be sharpened further. In fact, in this case it is enough to suppose simply that D is non-convex, for then (by a result of Léja and Wilkosz, cf. [18], Theorem 4. 8) there must be some boundary point P where there is an internally touching arc of a circle with negative curvature. We thus obtain the following extension of the result of Finn which was stated in the introduction.

Let D be a bounded nonconvex domain in E^2. Then there exists smooth boundary data having arbitrarily small absolute value, such that no solution of the Dirichlet problem for the minimal surface equation is possible.

5. The Dirichlet problem in two dimensions

In Theorems 2 and 2′ the smoothness of the boundary of D has been used in order to obtain an overall estimate for the gradient of a solution. In two dimensions this condition can be lightened, though generally not omitted altogether.

We say that the boundary of D is locally convex at a point P if there exists a ball N with center at P such that $N \cap P$ is convex. A boundary set is called locally convex if it is locally convex at each point. Then so far as the Dirichlet problem in two dimensions is concerned, we need not require smoothness of either the boundary or the data on locally convex portions of the boundary. On the other hand, where the boundary is not locally convex some conditions of smoothness are absolutely essential. More precisely, we have the following result.

Theorem 4. *Let D be a bounded domain in E^2 whose boundary is the union of two open arcs Γ_0 and Γ_2, where Γ_0 is locally convex and Γ_2 is of class C^2. Consider boundary values f which are continuous on Γ_0 and of class C^2 on Γ_2. Then there exists a positive quantity $\mathscr{B}''$, depending only on the geometry of D and the norms $\| f \|_1$ and $\| f_2 \|$ restricted to Γ_2, such that the Dirichlet problem in D is well posed whenever*

$$h < \mathscr{B}''(\| f \|_1, \| f \|_2, D)$$

where $2h$ denotes the oscillation of f on the boundary of D.

Note that the norms $\| f \|_1$ and $\| f \|_2$ can always be assumed finite, if necessary by slightly shortening Γ_2.

The statement of the Theorem 4 has been restricted to the case of single arcs Γ_0 and Γ_2 only for simplicity in the discussion; alternating locally convex and C^2 arcs on each connected component of the boundary of D can in fact be handled by essentially the same methods. Finally, one can easily extend Theorem 4 to cover the case where the spatial boundary curve lies near a solution U of the minimal surface equation, as in Theorem 2′.

Proof of Theorem 4. By virtue of the Perron method, which is applicable here because bounded families of solutions of the minimal surface equation in E^2 are compact, it is enough to show that a barrier can be constructed at each point of the boundary of D[5]).

Now by the results of [8], Section 1, a barrier exists at each point of Γ_0. Consider then the boundary arc $\Gamma - \Gamma_0$ which by hypothesis is a subset of Γ_2. As in Section 2, there exists a constant $\varrho > 0$ such that if P is a point of $\Gamma - \Gamma_0$ and Σ denotes the ball of radius ϱ about P, then $\Gamma \cap \Sigma$ consists of a single component with the C^2 representation $\bar{x}_2 = \varphi(\bar{x}_1)$. Under these circumstances we constructed in the proof of Lemma 2 a supersolution $v(x)$, $x \in \Omega \cap D$, having the following properties:

(i) Ω is a neighborhood of P

(ii) $v(P) = 0$

(iii) $v \geqq \begin{cases} f - f(P) & \text{on } \Omega \cap \Gamma \\ 2h & \text{on the remaining boundary of } \Omega \cap D, \end{cases}$

[5]) The applicability of Perron's process to the minimal surface equation was remarked by J. C. C. Nitsche ([12], p. 201); see also [8], pp. 339—40, where the essential details are carried out.

providing, however, that the hypothesis of Lemma 2 holds. According to Lemma 4, this hypothesis is implied by the condition $h \leqq \mathscr{B}(\| f \|_1, \| f \|_2, D)$, the norms $\| f \|_1, \| f \|_2$ of course being restricted to Γ_2. Taking $\mathscr{B}'' = \mathscr{B}$, the condition of Theorem 1 thus suffices for the existence of v. Since v has the attributes of a barrier at P, the proof of Theorem 4 is completed.

Concerning the necessity of smoothness hypotheses where the boundary is not locally convex, we note that by Theorem 3 there exists data having arbitrary small absolute value for which no solution of the Dirichlet problem is possible. Thus conditions involving only the magnitude of the data cannot possibly suffice for the general solvability of Dirichlet's problem. If the boundary itself is not smooth the situation can be even worse. By slightly "tilting" the tetrahedron example considered by Nitsche [13], one can obtain data whose derivatives *up to any order* are arbitrarily small and yet for which no solution of the Dirichlet problem is possible.

6. A general convexity condition

It was shown by Gilbarg and by Stampacchia that the Dirichlet problem for an elliptic equation of the form

$$\frac{\partial}{\partial x_k} A_k(u_1, \ldots, u_n) = 0, \quad \left(u_i = \frac{\partial u}{\partial x_i}, \quad A_k \in C^2\right) \tag{17}$$

can always be solved in a strictly convex domain, provided that the boundary and the boundary data are sufficiently smooth. (A domain is called strictly convex if the principal curvatures of the boundary are all positive.)

Since the minimal surface equation can be written in the form (17) with $A_k = u_k / \sqrt{1 + |\nabla u|^2}$, our results would seem to indicate that the requirement of convexity is unnecessarily restrictive for any *particular* equation under consideration. Nevertheless, there remains a close connection between convexity and the solvability of the Dirichlet problem for the *class* of equations (17). The following theorem shows that in order for the Dirichlet problem in a fixed domain D to be solvable for *all* equations of the form (17) and *all* smooth data, then it is necessary for the domain to be convex.

Theorem 5. *Let D be a non-convex domain whose boundary is of class C^2. Then there exists an elliptic equation of the form* (17) *and smooth boundary data having arbitrarily small absolute value, for which no solution of the Dirichlet problem is possible.*

Proof. We assert that the principal curvatures of Γ cannot be everywhere non-negative. Indeed, if this were the case then obviously D would admit a local supporting plane at each point of Γ, and according to a theorem of Tietze, D would be convex.

Thus there must exist a point P on Γ where one of the principal curvatures of Γ is negative. As in Section 2, we choose coordinates $(x_1, \ldots, x_n)$ in E^n with the x_n axis directed along the inner normal to Γ at P. In the neighborhood of P the boundary then has the representation

$$x_n = \varphi(x_1, \ldots, x_{n-1})$$

with $|\nabla\varphi| = 0$ at P. We may further assume that the principal directions of Γ at P are parallel to the x_1 through x_{n-1} coordinates axes. The principal curvatures of Γ at P are then identical with the quantities $\varphi_{ii}(P)$, $i = 1, \ldots, n-1$, and without loss of generality we may suppose that $\varphi_{11}(P) < 0$.

Now consider the affine transformation T defined by

$$\bar{x}_1 = \frac{x_1}{a}, \quad \bar{x}_j = x_j, \quad j = 2, \ldots, n.$$

Let $\bar{D}$, $\bar{\Gamma}$, and $\bar{P}$ denote respectively the images of D, Γ, and P under T. The mean curvature of $\bar{\Gamma}$ at $\bar{P}$ is evidently given by

$$\bar{H} = \frac{1}{n-1}\{a^2\varphi_{11}(P) + \sum_2^{n-1} \varphi_{jj}(P)\}.$$

We choose a so large that $\bar{H} < 0$. Then according to Theorem 3 there exists smooth boundary data $\bar{f}$ on $\bar{\Gamma}$ such that the Dirichlet problem for the minimal surface equation in $\bar{D}$ is not solvable.

Transforming the data $\bar{f}$ on $\bar{\Gamma}$ into corresponding data f on Γ by means of the inverse relation T^{-1}, and similarly transforming the minimal surface equation in $\bar{D}$ into a corresponding equation in D, completes the proof of the theorem. Indeed, the data f can be assumed to have arbitrarily small absolute value, the resulting equation in D has the form (17) with

$$A_1 = \frac{a^2 u_1}{(1 + a^2u_1^2 + u_2^2 + \cdots + u_n^2)^{\frac{1}{2}}}, \quad A_j = \frac{u_j}{(1 + a^2u_1^2 + u_2^2 + \cdots + u_n^2)^{\frac{1}{2}}},$$

and the Dirichlet problem clearly can have no solution.

Note. This research was partially supported by the Air Force Office of Scientific Research under Grant No. Af-AFOSR-883-65.

References

[1] *S. Bernstein*, Sur les surfaces définies au moyen de leur courbure moyenne ou totale, Ann. Sci. École Norm. Sup. **27** (1910), 233—256.

[2] *S. Bernstein*, Sur les équations du calcul des variations, Ann. Sci. École Norm. Sup. **29** (1912), 431—485.

[3] *R. Courant*, Dirichlet's principle, conformal mapping, and minimal surfaces, Interscience, New York 1950.

[4] *R. Courant* and *D. Hilbert*, Methods of mathematical physics, Vol. II, Interscience, New York 1962.

[5] *R. Finn*, Remarks relevant to minimal surfaces and to surfaces of constant mean curvature, J. d'Analyse Mathématique **14** (1965), 139—160.

[6] *D. Gilbarg*, Boundary value problems for nonlinear elliptic equations in *n* variables, Nonlinear Problems, University of Wisconsin Press, Madison (1963), 151—160.

[7] *A. Haar*, Über das Plateausche Problem, Math. Ann. **97** (1927), 124—258.

[8] *H. Jenkins* and *J. Serrin*, Variational problems of minimal surface type II. Boundary value problems for the minimal surface equation, Arch. Rational Mech. Analysis **21** (1966), 321—342.

[9] *A. Korn*, Über Minimalflächen, deren Randkurven wenig von ebenen Kurven abweichen, Abh. König. Preuß. Akad. Wiss. Berlin Phys.-Math. Cl. II (1909), 1—37.

[10] *O. A. Ladyzhenskaya* and *N. N. Uraltseva*, Quasi-linear elliptic equations and variational problems with many independent variables, Usp. Mat. Nauk **16** (1961), 17—93. English translation, Russian Mathematical Surveys **16** (1961).

[11] *J. Leray*, Discussion d'un problème de Dirichlet, J. Math. Pures Appl. **18** (1939), 249—284.

[12] *J. C. C. Nitsche*, On new results in the theory of minimal surfaces, Bull. Amer. Math. Soc. **71** (1965), 195—270.

[13] *J. C. C. Nitsche*, On the non-solvability of Dirichlet's problem for the minimal surface equation, J. Math. Mech. **14** (1965), 779—788.

[14] *S. D. Poisson*, Note sur le surface dont l'aire est un minimum entre des limites données, J. reine angew. Math. 8 (1832), 361—362.

[15] *T. Radó*, The problem of the least area and the problem of Plateau, Math. Z. **32** (1930), 763—796.

[16] *G. Stampacchia*, On some multiple integral problems in the calculus of variations, Comm. Pure Appl. Math. **16** (1963), 382—422.

[17] *H. Tietze*, Bemerkungen über konvexe und nicht-konvexe Figuren, J. reine angew. Math. **160** (1929), 67—69.

[18] *F. A. Valentine*, Convex Sets, McGraw-Hill, New York 1964.

Eingegangen 10. Oktober 1966

Arch. Ration. Mech. Anal. 29 (1968), 304–322. **Reprint I.4.[ii]**

Variational Problems of Minimal Surface Type

III. The Dirichlet Problem with Infinite Data

Howard Jenkins & James Serrin

This is the third in a series of papers on variational problems of minimal surface type. In the first of these papers, hereafter referred to as I, we have investigated the local behaviour of solutions with particular regard to the existence of an interior estimate for the gradient, and we have shown that a Harnack principle applies. The second paper (II) contains an existence and uniqueness theory for the Dirichlet problem (with infinite boundary data) for the minimal surface equation. It is our purpose here to generalize the results of II to the full class of variational problems of minimal surface type.

We consider in particular non-parametric variational problems of the form

$$\delta \int F(p,q)\,dx\,dy=0, \qquad p=u_x, \qquad q=u_y, \tag{1}$$

where $F(p, q)$ is a function of class C^2 throughout the p, q plane, satisfying the conditions

$$F_{pp}F_{pq}-F_{pq}^2>0, \qquad F_{pp}>0, \tag{2}$$

$$F-pF_p-qF_q>A, \tag{3}$$

$$\frac{(1+p^2)F_{pp}+2pq\,F_{pq}+(1+q^2)F_{qq}}{\sqrt{1+p^2+q^2}\sqrt{F_{pp}F_{qq}-F_{pq}^2}}<2E \tag{4}$$

for some constants A and E. A solution of the variational problem (1) is then a twice continuously differentiable function $u=u(x, y)$ satisfying the Euler-Lagrange equation

$$F_{pp}u_{xx}+2F_{pq}u_{xy}+F_{qq}u_{yy}=0. \tag{5}$$

In addition to conditions (2), (3), (4) we shall assume throughout that the second derivatives of F are uniformly Hölder continuous on compact subsets of the p, q plane.

In I it was shown that variational problems of minimal surface type are closely related to regular parametric variational problems of the form

$$\delta \int \mathscr{F}(\boldsymbol{r})\,ds\,dt=0 \tag{6}$$

where $\boldsymbol{r}=(\sigma, \tau, \omega)$ and σ, τ, ω are the Jacobian matrices

$$\frac{\partial(y,z)}{\partial(s,t)}, \quad \frac{\partial(z,x)}{\partial(s,t)}, \quad \frac{\partial(x,y)}{\partial(s,t)}.$$

Indeed, a solution of (6) which can be represented in the form $z=u(x,y)$ is also a solution of a non-parametric variational problem whose integrand F necessarily satisfies conditions (2), (3), and (4). On the other hand, any variational problem (1) can be written in the form (6) where the integrand $\mathscr{F}$ is defined in the open half space $\omega<0$ by the equation

$$(7)\qquad \mathscr{F}(\sigma,\tau,\omega)=-\omega F(-\sigma/\omega,\,-\tau/\omega).$$

We shall show later that when F satisfies (2), (3), and (4), and is furthermore appropriately normalized, then the function $\mathscr{F}$ defined by (7) can be extended by continuity to the closed half space $\omega\leqq 0$. Moreover the extended function satisfies the condition $\mathscr{F}(\lambda r)=\lambda\mathscr{F}(r)$ for all $\lambda>0$.

This extension allows us to associate with (1) a corresponding one-dimensional parametric variational problem

$$(8)\qquad \delta\int \mathscr{F}_0(\vec{r})\,dt=0$$

where $\vec{r}=(\lambda,\mu)=(dx/dt,\,dy/dt)$, and

$$(9)\qquad \mathscr{F}_0(\lambda,\mu)=\mathscr{F}(\mu,\,-\lambda,\,0).$$

If γ is an oriented piecewise smooth arc in the plane we may correspondingly define

$$(10)\qquad I(\gamma)=\int_\gamma \mathscr{F}_0(\vec{r})\,dt;$$

this integral is intimately associated with our later study. For the moment, we simply notice that for the minimal surface equation $F=\sqrt{1+p^2+q^2}$ and $\mathscr{F}_0=|\vec{r}|$. Therefore in this case $I(\gamma)$ reduces to the Euclidean length of γ (the same result holds when the integrand F is radially symmetric; see the remark in Section 2). In more general cases $I(\gamma)$ can be interpreted as the length of γ with respect to the non-symmetric Finsler metric defined by the function $\mathscr{F}_0(\vec{r})$.

In what follows it will be useful to use the language of chains. In particular, by a one-chain we shall mean a formal sum of open oriented segments. If Γ, Δ are one-chains, then the chains $-\Gamma$ and $\Delta+\Gamma$ are defined in the obvious way, and $\Delta-\Gamma=\Delta+(-\Gamma)$. In addition, we define $I(\Gamma)$ as the sum of the integrals (10) over the individual arcs comprising Γ.

We turn now to a specific description of the results of the paper. Let D be a bounded domain in E^2 whose boundary consists of a finite number of open convex arcs* together with their endpoints. To each of these arcs we assign either continuous data or the value plus infinity or the value minus infinity — *the problem of Dirichlet is then to determine a solution of* (1) *in D which takes on this given data at the boundary.*

The arcs on which the values plus infinity and minus infinity are assigned will be denoted respectively by $A_1, A_2, \ldots$ and $B_1, B_2, \ldots$, and the remaining arcs will be named $C_1, C_2, \ldots$. We observe that each A_i and each B_i must be

* A boundary arc is called convex if there is a neighbourhood of each of its points whose intersection with D is a convex domain. We observe specifically that a straight segment is a special case of a convex boundary arc.

straight in order for the problem to be solvable at all. Indeed, according to the straight line lemma (Section 5) if one of these arcs were not straight the solution would necessarily be infinite in a neighbourhood of this arc, an obvious absurdity. In what follows, then, it will be assumed that all the A_i and B_i are straight.

Now let $\mathscr{P}$ be a polygonal domain lying in D, whose vertices are drawn from the set of endpoints of the arcs A_i, B_i and C_i. If the perimeter of $\mathscr{P}$ is given a counterclockwise orientation, then the formal sum of its oriented boundary segments forms a one-chain Γ. Moreover Γ can be uniquely expressed in the forms

$$\Gamma = A - A' = B - B'$$

where A and B are composed of those arcs of Γ which are in the families $\{A_i\}$ and $\{B_i\}$ respectively. The following main result then holds.

Theorem. *If the family of arcs $\{C_i\}$ is not empty, the Dirichlet problem stated above is solvable if and only if*

$$(*) \qquad I(A) < I(A'), \quad I(-B) < I(-B')$$

for each polygon $\mathscr{P}$ chosen as above. If the family $\{C_i\}$ is empty the result is the same, except that if $\mathscr{P}$ coincides with D condition $(*)$ should be replaced by $I(A) = I(-B)$.

In the first case the solution is unique, in the second, unique up to an additive constant.

For the special case of the minimal surface equation this result reduces to Theorem 4 of II. As in that case it is possible to single out particular situations where the results are especially simple. For example, if D is a quadrilateral with sides A_1, C_1, A_2, C_2, in that order, then the necessary and sufficient condition for a solution of the corresponding Dirichlet problem to exist becomes

$$I(A_1) + I(A_2) < I(-C_1) + I(-C_2),$$

that is, the sum of the generalized lengths of the sides on which the data plus infinity is prescribed should be less than the sum of the generalized lengths of the sides (orientation reversed) on which continuous data is prescribed. If the sides of D are A_1, B_1, A_2, B_2, in that order, then the condition is slightly different but equally simple, namely

$$I(A_1) + I(A_2) = I(-B_1) + I(-B_2).$$

Should both families $\{A_i\}$ and $\{B_i\}$ each contain at most one segment, then condition $(*)$ is automatically fulfilled in view of the triangle inequality (Section 5). As a special case of this, the Dirichlet problem admits a unique solution for arbitrary continuous data assigned on each arc of the boundary of D. (A similar result, but for convex domains D, was established by FINN for equations of minimal surface type.)

We remark finally that the formulation of the problem allows D to be multiply connected, and even to have re-entrant boundary segments. Naturally re-entrant segments are to be considered as having two sides, on each of which data is assigned.

The first sections of the paper develop and assemble preliminary material, much of which has importance and significance in its own right. Once this material is available, the proof of the main result can be given by exactly the same method as in the minimal surface case. Indeed, the matter becomes essentially axiomatic, and it suffices simply to outline the procedure. This we do in Section 6. In the earlier parts of the paper we shall make free use of the results of I and II in order to shorten the exposition.

1. A Preliminary Existence Theorem

Let $u=u(x, y)$ be a solution of (5) in a domain D. In reference I the authors showed that $|\nabla u|$ and $|\nabla^2 u|$ can be estimated at a point P in terms of a bound on $|u|$ and the distance from P to the boundary of D. Since we are assuming that the coefficients of the Euler equations (5) are Hölder continuous on compact subsets of the p, q plane, Schauder type estimates may then be used to prove the existence of a bound on the Hölder moduli of the second derivatives of u. Thus by standard procedures we obtain the following

Compactness Principle. *Let $\{u_n\}$ be a uniformly bounded sequence of solutions of* (5) *in a domain D. Then there exists a subsequence which converges to a solution in D, the convergence being uniform on every compact subset of D.*

According to a result in the Appendix the vector $\nabla F=(F_p, F_q)$ has bounded length. Consequently the following result also holds ([7], pp. 205 and 209):

General Maximum Principle. *Let D be a bounded domain and let E be a finite set of points on the boundary ∂D. Suppose that u_1 and u_2 are two solutions of* (5) *in D such that* $\liminf(u_1-u_2)\geqq 0$ *as one approaches any point of $\partial D-E$. Then $u_1\geqq u_2$ in D.*

It is well known (see for example [8], Theorem VI) that the Dirichlet problem for (5) can be solved in strictly convex domains for sufficiently smooth boundary data. With the aid of the compactness principle and the maximum principle this result can easily be generalized to the case where the assigned data on the boundary is continuous. Finally, by using the Perron method (II, pp. 339–340) we may infer the existence of solutions in bounded domains which assume prescribed boundary values at each boundary point at which a barrier exists.

We now show that a barrier exists at each point where the boundary is convex. Let Q be such a point. It is sufficient to find, for each pair of numbers M and δ, a neighbourhood N of Q and a non-negative function $w=w(x, y)$ defined in $\bar{N}\cap\bar{D}$ such that

i) N is contained in a disc of radius δ about Q,

ii) $w\geqq M$ on $\partial N\cap D$,

iii) $w=0$ at Q,

iv) w is a super-solution and $-w$ is a sub-solution of (5).

According to reference [5], in order for w to be a super-solution and $-w$ to be a sub-solution it is sufficient that $z=w(x, y)$ defines a surface whose mean curvature H and Gaussian curvature K are negative and satisfy the inequality $H^2/|K| \geqq E^2-1$. (For completeness we shall give an analytical proof of this result in the Appendix, independent of the argument in reference [5].)

For the particular function

$$v(x,y)=\frac{\delta}{2}\left(\frac{1}{\lambda}\log\cos\frac{2\lambda x}{\delta}-\log\cos\frac{2y}{\delta}\right), \qquad \lambda>1,\ \delta>0$$

a calculation yields $H<0$, $K<0$ in the rectangle $|x|<\delta\pi/4\lambda$, $|y|<\delta\pi/4$, and

$$4\frac{H^2}{|K|}=\frac{(\sqrt{\lambda}-1/\sqrt{\lambda})^2}{1-\sin^2\frac{2\lambda x}{\delta}\sin^2\frac{2y}{\delta}}\geqq\left(\sqrt{\lambda}-\frac{1}{\sqrt{\lambda}}\right)^2.$$

If λ is now chosen greater than $(E+\sqrt{E^2-1})^2$, the above criterion will be satisfied and v will be a super-solution in the rectangle for any positive number δ. A barrier at Q may now be constructed by a rotation and translation of the function $w=M+v$ restricted to the domain $|y|<\lambda x<\delta\pi/4$, cf. II, p. 324. We have thus proved the following

Preliminary Existence Theorem. *Let D be a bounded domain whose boundary consists of a finite number of open convex arcs C_i together with their endpoints. Then there exists a solution of* (1) *in D which takes on assigned bounded continuous data on the arcs C_i.*

This result remains true even without the assumption that the data is bounded, as we have remarked at the end of the introductory section.

2. Basic Lemmas

Here we shall establish some results which are more specific to our purposes than those of the preceding section.

To begin with, we observe that the quantities $\mu=F-pF_p-qF_q$ and $|\nabla F|$ are bounded functions,

$$(11) \qquad |\mu|\leqq \text{Constant}, \quad |\nabla F|\leqq\text{Constant}, \quad \text{for all } p, q.$$

The first of these inequalities is a simple consequence of the fact that $\mu(hp, hq)$ is a decreasing function of h for $h>0$, whence $A\leqq\mu\leqq F(0)$. The second result is proved in the Appendix.

In view of (11), the integrand $F(p,q)$ can be normalized by the addition of a linear function of p, q and multiplication by a constant so that

$$(12) \qquad \nabla F(0)=0, \quad \operatorname{Sup}|\nabla F|=1, \quad 1<\mu\leqq b,$$

where b is a suitable constant (one can in fact take $b=F(0)$).

This being done, we assert that there is a positive constant c such that

$$(13) \qquad 1+c\rho\leqq F\leqq\rho+b, \quad \rho=\sqrt{p^2+q^2}.$$

Indeed, from the third relation of (12) we have $1 < F(0) \leqq b$, and the right hand side follows at once. The left hand side is then an easy consequence of the convexity of the surface $Z = F(p, q)$ and the fact that $\nabla F(0) = 0$.

We can now carry out the extension of the function $\mathscr{F}$ to the full half space $\omega \leqq 0$. Let $\boldsymbol{r} = (\sigma, \tau, 0)$ be a vector in the plane $\omega = 0$, and let $\vec{r} = (\sigma, \tau)$ be the associated 2-vector. Putting $\boldsymbol{k} = (0, 0, 1)$, we then define

$$\mathscr{F}(\boldsymbol{r}) = \lim_{h \to \infty} \mathscr{F}(\boldsymbol{r} - h^{-1}\boldsymbol{k}) = \lim_{h \to \infty} h^{-1} F(h\vec{r}) = \lim_{h \to \infty} \vec{r} \cdot \nabla F(h\vec{r}). \tag{14}$$

The latter limit exists since $\vec{r} \cdot \nabla F(h\vec{r})$ is a bounded function of h with a non-negative derivative.

It must be shown that the extended function $\mathscr{F}$ is continuous and positively homogeneous of degree one. With regard to the first of these properties, it is clearly sufficient to consider only points of the plane $\omega = 0$. Letting $\boldsymbol{r}$ and $\boldsymbol{r}_1$ be vectors in this plane, and letting $\vec{r}$ and $\vec{r}_1$ be the associated 2-vectors, we have by (7), for $\varepsilon > 0$,

$$\left|\mathscr{F}(\boldsymbol{r} - \varepsilon\boldsymbol{k}) - \mathscr{F}\left(\boldsymbol{r}_1 - \frac{1}{h}\boldsymbol{k}\right)\right| \leqq \frac{1}{h}|F(h\vec{r}) - F(h\vec{r}_1)| + \left|\mathscr{F}(\boldsymbol{r} - \varepsilon\boldsymbol{k}) - \mathscr{F}\left(\boldsymbol{r} - \frac{1}{h}\boldsymbol{k}\right)\right|.$$

Using the fact that $|\nabla F| \leqq 1$ and letting $h \to 0$ then yields

$$|\mathscr{F}(\boldsymbol{r} - \varepsilon\boldsymbol{k}) - \mathscr{F}(\boldsymbol{r}_1)| \leqq |\boldsymbol{r} - \boldsymbol{r}_1| + |\mathscr{F}(\boldsymbol{r} - \varepsilon\boldsymbol{k}) - \mathscr{F}(\boldsymbol{r})|.$$

Since $\mathscr{F}(\boldsymbol{r} - \varepsilon\boldsymbol{k}) \to \mathscr{F}(\boldsymbol{r})$ as $\varepsilon \to 0$, the continuity of $\mathscr{F}$ follows at once. Next, it is evident from (7) that $\mathscr{F}$ is positively homogeneous at least for $\omega < 0$, and the same result for $\omega \leqq 0$ follows by continuity.

Remark. If F is radially symmetric, that is, depends on p, q only through the combination $p^2 + q^2$, then (13) holds in the stronger form

$$1 + \rho \leqq F \leqq \rho + b. \tag{13'}$$

Moreover, using (12) and (14), it is clear that in this case $\mathscr{F}(\boldsymbol{r}) = |\vec{r}|$ when $\boldsymbol{r} = (\sigma, \tau, 0)$. Hence $\mathscr{F}_0(\vec{q}) = |\vec{q}|$ and $I(\gamma)$ is the ordinary Euclidean length of γ.

Now consider the function $G(\sigma, \tau)$ which is obtained from $F(p, q)$ by the projective transformation

$$\sigma = \frac{p}{F}, \quad \tau = \frac{q}{F}, \quad G = -\frac{1}{F}.$$

The domain of G is a bounded convex set, and the second derivatives of G in this domain satisfy the inequalities

$$G_{\sigma\sigma} G_{\tau\tau} - G_{\sigma\tau}^2 > 0, \quad G_{\sigma\sigma} > 0,$$

and

$$\frac{G_{\sigma\sigma} + G_{\tau\tau}}{\sqrt{G_{\sigma\sigma} G_{\tau\tau} - G_{\sigma\tau}^2}} \leqq 2\mathscr{E} \tag{15}$$

where $\mathscr{E}$ is a constant depending only on the normalized integrand F (see I, p. 191, or part 2 of the Appendix).

Our interest in G stems from the fact that it is the integrand of a variational problem dual to (1). Indeed, as in I put

$$d\xi=(F-qF_q)\,dx+qF_p\,dy, \qquad d\eta=pF_q\,dx+(F-pF_p)\,dy.$$

It is well known that corresponding to any given solution $u(x,y)$ of (1) the differentials $d\xi$ and $d\eta$ are exact, and thereby define functions $\xi(x,y)$ and $\eta(x,y)$ by integration. Moreover, since

$$\xi_x\eta_y-\xi_y\eta_x=F(F-pF_p-qF_q)>1$$

the mapping $(x,y)\to(\xi,\eta)$ is locally one-to-one. Finally, the transformed solution $u(\xi,\eta)$ satisfies the equation

(16) $$G_{\sigma\sigma}u_{\xi\eta}+2G_{\sigma\tau}u_{\xi\eta}+G_{\tau\tau}u_{\eta\eta}=0, \qquad \sigma=u_\xi, \quad \tau=u_\eta.$$

In view of (15) this equation is uniformly elliptic.

The functions ξ,η also define a mapping of any solution surface $z=u(x,y)$ into the ξ,η plane. Let $d\zeta^2=(1+p^2)dx+2pq\,dx\,dy+(1+q^2)\,dy^2$ be the Euclidean metric on such a surface. We assert that the ratio $(d\xi^2+d\eta^2)/d\zeta^2$ is bounded from zero. Indeed, putting $a=1+p^2$, $b=pq$, $c=1+q^2$, $e=(F-qF_q)^2+(pF_q)^2$, $f=qF_p(F-qF_q)+pF_q(F-pF_p)$, $g=(F-pF_p)^2+(qF_p)^2$, we have

$$\frac{d\xi^2+d\eta^2}{d\zeta^2}=\frac{e\,dx^2+2f\,dx\,dy+g\,dy^2}{a\,dx^2+2b\,dx\,dy+c\,dy^2}.$$

Let α_1,α_2 denote respectively the minimum and maximum of the above ratio, and put $W=\sqrt{1+p^2+q^2}$, $\mu=F-pF_p-qF_q$ and $\theta=pF_q-qF_p$. After a calculation one finds

$$\alpha_1+\alpha_2=\frac{ag-2bf+ec}{ac-b^2}=\frac{F^2+W^2\mu^2+\theta^2}{W^2}, \qquad \alpha_1\alpha_2=\frac{eg-f^2}{ac-b^2}=\frac{F^2\mu^2}{W^2}.$$

Since α_1,α_2 are positive we have then

$$\frac{1}{\alpha_1}<\frac{\alpha_1+\alpha_2}{\alpha_1\alpha_2}=\frac{W^2}{F^2}+\frac{1}{\mu^2}+\frac{\theta^2}{F^2\mu^2}.$$

By (12) and (13) the right hand side is bounded by $3/c^2$, so that finally

(17) $$\frac{d\xi^2+d\eta^2}{d\zeta^2}>\frac{c^2}{3}.$$

We conclude this section with an analogue of Lemma 1 of II. We need first the following result from the theory of uniformly elliptic equations.

Let $v(x,y)$ be a positive solution of the equation

(18) $$a(x,y)\,v_{xx}+2b(x,y)\,v_{xy}+c(x,y)\,v_{yy}=0$$

in the domain $x^2+y^2<r^2$. Suppose moreover that (18) *is uniformly elliptic, that is,*

$$\frac{a+c}{\sqrt{ac-b^2}}\leqq 2e$$

for some constant e. Then there exists a constant k, depending only upon e, such that

$$|\nabla v(0,0)| \leqq k \frac{v(0,0)}{r}.$$

Proof. If $|v| \leqq m$ then according to a lemma of ROSENBLOOM ([6], p. 371) we have $|\nabla v(0,0)| < Km/r$, where K depends only upon e. In our case v may be unbounded; nevertheless, since it is positive we may infer from the Harnack inequality ([9], Theorem 1) that $|v| \leqq K' v(0,0)$ for $x^2+y^2 \leqq (r/2)^2$, the constant K' depending only upon e. Applying ROSENBLOOM's lemma in this disc completes the proof.

Now let $u=u(x,y)$ be a solution of (5) in a convex domain D, and let S denote the surface $z=u(x,y)$. For $P \in D$, let d be the distance from P to the boundary of D, let P' be any boundary point of D whose distance from P is d, and let $\vec{v}$ be the unit vector from P to P'. Also let P^* denote the point on S directly over P and let r be the geodesic distance on S from P^* to the boundary of S. The following estimate then holds.

Lemma 1. *For any $\varepsilon>0$ there exists a positive constant δ such that if $d/r<\delta$ then one of the following pairs of inequalities,*

$$|\vec{v}\cdot\nabla F - \mathscr{F}(\mathbf{v})| \leqq \varepsilon \quad \textit{and} \quad \vec{v}\cdot\nabla u>0$$

or

$$|\vec{v}\cdot\nabla F + \mathscr{F}(-\mathbf{v})| \leqq \varepsilon \quad \textit{and} \quad \vec{v}\cdot\nabla u<0,$$

*holds at P.**

Proof. Let the functions ξ, η be as defined above, and assume that the constants of integration have been chosen so that $\xi(P)=\eta(P)=0$. Since the mapping $(x,y)\to(\xi,\eta)$ is locally one-to-one, it is clear that the functions ξ, η define a locally one-to-one mapping of S such that the point P^* maps onto the origin. Let Δ be the largest circle about the origin in the ξ, η plane in which the inverse mapping is one-to-one. We assert that the radius of Δ is $\geqq cr/2$. To prove this, note that if the radius of Δ is less than $cr/2$ then (17) implies that the image on S of each radius of Δ is a curve from P^* to an interior point of S. The mapping is therefore one-to-one in a neighbourhood of every boundary point of Δ, contradicting the definition of Δ.

To prove the lemma, we may assume without loss of generality that $\vec{v}$ points in the positive x-direction. Now consider the function $x(\xi,\eta)$ in Δ. We find easily that

$$(19) \qquad x_\xi = \frac{F-pF_p}{F\mu} = \tau G_\tau - G, \qquad x_\eta = \frac{-qF_p}{F\mu} = -\tau G_\sigma.$$

Consequently, by direct calculation and the use of (16),

$$(20) \qquad G_{\sigma\sigma}x_{\xi\xi} + 2G_{\sigma\tau}x_{\xi\eta} + G_{\tau\tau}x_{\eta\eta} = 0, \qquad \sigma=u_\xi, \quad \tau=u_\eta.$$

Thus $x(\xi,\eta)$ is a solution of a uniformly elliptic equation in Δ. From our construction and from the convexity of D,

$$x - x_P \leqq d, \qquad (x,y)\in D,$$

* Here $\mathbf{v}$ is the three-vector $(v_1, v_2, 0)$ associated with $\vec{v}=(v_1,v_2)$.

so that the function $d+x_P-x(\xi,\eta)$ is a non-negative solution of (20) in Δ. Applying the result preceding Lemma 1, we find that

$$|\nabla x(0,0)|\leqq\frac{2k}{c}\cdot\frac{d}{r}, \tag{21}$$

where k depends only on $\mathscr{E}$ (see (15)), that is, only on the form of the integrand F.

Now using (19) and putting $\lambda=2kd/cr$, the preceding inequality may be rewritten

$$\rho^2 F_p^2-2pFF_p+F^2(1-\lambda^2\mu^2)\leqq 0 \qquad \text{at } P.$$

If we regard the left hand expression as a quadratic polynomial in F_p, then its discriminant must be non-negative. This immediately yields the inequality

$$|q|/\rho\leqq\lambda\mu \qquad \text{at } P. \tag{22}$$

Next, making use of (19), (21), (22), (12), and (13) it follows that

$$\frac{1}{F}=x_\xi-\frac{qF_q}{F\mu}\leqq|x_\xi|+\frac{\rho|F_q|}{F}\cdot\frac{|q|}{\rho\mu}\leqq(1+1/c)\lambda.$$

But since $\rho+b\geqq F$ this implies

$$\rho\geqq\frac{c}{(c+1)\lambda}-b \qquad \text{at } P.$$

To complete the proof, we put $\vec{p}=(p,q)$, $\boldsymbol{p}=(p,q,0)$ and observe that

$$\vec{p}\cdot\nabla F=F(\vec{p})-\mu=\mathscr{F}(\boldsymbol{p}-\boldsymbol{k})-\mu.$$

Division by ρ then yields the identity

$$\frac{\vec{p}}{\rho}\cdot\nabla F=\mathscr{F}\left(\frac{\boldsymbol{p}}{\rho}-\frac{\boldsymbol{k}}{\rho}\right)-\frac{\mu}{\rho}.$$

Now by (22), $\vec{p}/\rho$ approximates either $\vec{v}$ or $-\vec{v}$ when λ is small; at the same time ρ becomes arbitrarily large. Since $\mathscr{F}$ is continuous, $|\nabla F|\leqq 1$, and $|\mu|\leqq b$, this proves the lemma.

3. The Conjugate Function

If u is a solution of (5) in a domain D, the function ψ which arises by integrating the exact differential

$$d\psi=F_p\,dy-F_q\,dx$$

is called the conjugate function of u. This function plays a basic role in the solution of the infinite data Dirichlet problem. Clearly $|\nabla\psi|=|\nabla F|<1$, so that if D is a domain having a piecewise smooth boundary, then ψ is Lipschitz continuous in $\bar{D}$ (boundary slits and reentrant boundary segments are taken to be two sided).

Now let γ be an oriented piecewise smooth arc lying in D, and let s denote arc length along γ. Let $\vec{v}$ be the right hand unit normal on γ. We assert that

$$\frac{d\psi}{ds}=\vec{v}\cdot\nabla F<\mathscr{F}(\boldsymbol{v}), \qquad \boldsymbol{v}=(v_1,v_2,0).$$

For using (14) and the fact that $|\nabla F|$ is bounded, we have

$$\mathscr{F}(v)=\lim_{h\to\infty} h^{-1}F(h\vec{v})=\lim_{h\to\infty} h^{-1}F(\vec{p}+h\vec{v})=\lim_{h\to\infty}\vec{v}\cdot\nabla F(\vec{p}+h\vec{v})$$

where $\vec{p}=(p,q)$. The assertion then follows since $\vec{v}\cdot\nabla F(\vec{p}+h\vec{v})$ is a bounded increasing function of h. Similarly, letting $h\to-\infty$ we obtain $\vec{v}\cdot\nabla F>-\mathscr{F}(-v)$. If $\vec{t}$ denotes the unit tangent on γ, then by (9) we have $\mathscr{F}(v)=\mathscr{F}_0(\vec{t})$. Consequently

$$-\mathscr{F}_0(-\vec{t})<\frac{d\psi}{ds}<\mathscr{F}_0(\vec{t})$$

and

$$-I(-\gamma)<\int_\gamma d\psi<I(\gamma). \tag{23}$$

Suppose next that γ lies in the closure of D and that the boundary of D consists of a finite number of convex arcs (as in the introduction). Then we can approximate γ by piecewise smooth arcs in D, from which it follows at once that

$$-I(-\gamma)\leqq\int_\gamma d\psi\leqq I(\gamma). \tag{24}$$

Moreover, if γ is a piecewise smooth curve which bounds a subregion of D, then

$$\int_\Gamma d\psi=0. \tag{25}$$

The next lemma shows that inequality (23) can persist even for boundary arcs of D.

Lemma 2. *Let u be a solution of* (5) *in a domain D. Suppose in addition that u is continuous on $D\cup\gamma$ where γ is an oriented convex boundary arc of D. Then*

$$-I(-\gamma)<\int_\gamma d\psi<I(\gamma).$$

Proof. By the mean value theorem

$$(\vec{p}_1-\vec{p})\cdot\{\nabla F(\vec{p}_1)-\nabla F(\vec{p})\}>0 \tag{26}$$

when $\vec{p}\neq\vec{p}_1$. The result now follows exactly as in the proof of Lemma 3 of II.

Lemma 3. *Let D be a domain bounded in part by a straight segment T, oriented so that the right hand normal on T is the outer normal to D. Let u be a solution of* (5) *in D which takes the boundary value plus infinity on T. Then*

$$\int_T d\psi=I(T).$$

If u is a solution in D which takes the boundary value minus infinity on T, then

$$\int_T d\psi=-I(-T).$$

The proof follows by using Lemma 1 to estimate $\int_{T_\delta} d\psi$, where T_δ is a segment in D, parallel to T and at a distance δ from it. The reader is referred to the proof of Lemma 4 of II for the details (note that when $u=\infty$ on T it is the first inequality of Lemma 1 which applies, while if $u=-\infty$ on T then the second inequality must be used).

The same method of proof yields the following generalization of Lemma 3.

Lemma 4. *Let D be a domain bounded in part by a straight segment T, oriented as in Lemma 3. Let $\{u_n\}$ be a sequence of solutions of (5) in D, each u_n being continuous in $D\cup T$. If the sequence $\{u_n\}$ diverges uniformly to infinity on T while remaining uniformly bounded on compact subsets of D, or if the sequence diverges uniformly to minus infinity on compact subsets of D while remaining bounded on T, then*

$$\lim_{n\to\infty} \int_T d\psi_n = I(T).$$

On the other hand, if the sequence diverges uniformly to minus infinity on T while remaining uniformly bounded on compact subsets of D, or if the sequence diverges uniformly to plus infinity on compact subsets of D while remaining bounded on T, then

$$\lim_{n\to\infty} \int_T d\psi_n = -I(-T).$$

4. The Straight Line Lemma and the Triangle Inequality

The existence proof of the following section depends heavily on the limit behaviour of a monotone increasing family of solutions. The main result required in the discussion of this limit behaviour is the fundamental

Straight Line Lemma. *Let D be a bounded domain lying entirely to one side of a straight line L. Let the boundary of D consist of a part C disjoint from L and a remaining part C' contained in L. Suppose that u is a solution of (5) in D such that $m\leqq u\leqq M$ on C. Then for any compact subset S of $D\cup C$ there is a number N, depending only upon the diameter of D and the distance from S to L, such that*

$$m-N\leqq u\leqq M+N \qquad \textit{in } S.$$

This result is an easy consequence of the maximum principle together with the existence of a super-function ϕ^+ and a sub-function ϕ^- which are defined in the interior of a triangle, take respectively non-negative and non-positive values on two sides of the triangle, and become respectively positively and negatively infinite at interior points of the third side. The required functions ϕ^+ and ϕ^- are here supplied by the function $v(x, y)$ constructed in Section 1. The proof is carried out in detail in I, Lemma 6, and also in [5], Theorem 1.

The remaining technical tool for the study of monotone increasing families of solutions is a triangle inequality for the integral $I(\gamma)$. If P, Q are points in the (x, y) plane, we denote by $\overrightarrow{PQ}$ the open oriented segment from P to Q. The following conclusion then holds.

Triangle Inequality. *Let P, Q, R be non-collinear points in the (x, y) plane. Then $I(\overrightarrow{PR})<I(\overrightarrow{PQ})+I(\overrightarrow{QR})$.*

The proof hinges on the auxiliary assertion that the set in the (λ, μ) plane defined by $\mathscr{F}_0(\lambda, \mu)=1$ is a strictly convex closed curve surrounding the origin.

Before proving this assertion, we show how it implies the required inequality. Put

$$a=\mathscr{F}_0(\overrightarrow{PQ}), \quad b=\mathscr{F}_0(\overrightarrow{QR}).$$

By construction (see (9) and (14)) it is evident that a and b are greater than zero. Hence if $\vec{r}_1$ and $\vec{r}_2$ are non-collinear vectors satisfying $\mathscr{F}_0=1$ we have by convexity

$$\mathscr{F}_0\left(\frac{a}{a+b}\vec{r}_1+\frac{b}{a+b}\vec{r}_2\right)<1.$$

Setting $\vec{r}_1=\overrightarrow{PQ}/a$, $\vec{r}_2=\overrightarrow{QR}/b$ and using the positive homogeneity of $\mathscr{F}_0$ now yields easily

$$\mathscr{F}_0(\overrightarrow{PR})<\mathscr{F}_0(\overrightarrow{PQ})+\mathscr{F}_0(\overrightarrow{QR}),$$

which is equivalent to the required conclusion.

To prove the auxiliary assertion, consider the surface $\mathscr{F}(\boldsymbol{r})=1$ in the half space $\omega\leqq 0$. As shown in I, p. 191, this surface is convex about the origin and can be represented in the form $\omega=G(\sigma,\tau)$ where the domain of G is a bounded convex set $\mathscr{D}$ with $G=0$ on its boundary. Therefore, using (9), the strict convexity of the set $\mathscr{F}_0=1$ is equivalent to the strict convexity of $\mathscr{D}$.*

It is therefore enough to prove that $\mathscr{D}$ is strictly convex. Let us suppose for contradiction that this is not the case; that is, we suppose that the boundary of $\mathscr{D}$ contains a straight segment l. Without loss of generality it can be assumed that l has the form

$$\sigma=\sigma_0, \quad \alpha<\tau<\beta, \quad (\sigma_0>0).$$

We assert that $\nabla G=(G_\sigma, G_\tau)$ tends to a constant value as points of l are approached from $\mathscr{D}$.

Consider first G_τ. Let (σ_0,τ_0) be an arbitrary fixed point of l, and let τ_1 be a number between τ_0 and β. Using the convexity of G we find that

$$G(\sigma_0+h,\tau_1)-G(\sigma_0+h,\tau_0)\geqq(\tau_1-\tau_0)\,G_\tau(\sigma_0+h,\tau_0), \qquad h_0<h<0.$$

Now by direct calculation $\nabla G=\nabla F/\mu$, so that $|\nabla G|<1$. Therefore, since $G=0$ on l, we have

$$G_\tau(\sigma_0+h,\tau_0)\leqq 2|h|/(\tau_1-\tau_0).$$

Similarly, letting τ_2 be a number between α and τ_0, we get

$$G_\tau(\sigma_0+h,\tau_0)\geqq -2|h|/(\tau_0-\tau_2).$$

Consequently G_τ tends uniformly to zero as one approaches compact subsets of l.

Before considering the remaining component G_σ of ∇G, we observe that (15) implies

$$4\mathscr{E}^2G_{\sigma\tau}^2\leqq 2(2\mathscr{E}^2-1)\,G_{\sigma\sigma}G_{\tau\tau}-G_{\sigma\sigma}^2-G_{\tau\tau}^2.$$

---

* For particular integrands F it may be quite simple to verify directly that the domain $\mathscr{D}$ is strictly convex; in these cases the remaining argument can be avoided. A particularly important situation where this occurs is when F arises from a regular parametric variational problem, for then the surface $\mathscr{F}(\boldsymbol{r})=1$ has everywhere positive Gaussian curvature.

The minimal surface equation provides a specific example, the domain $\mathscr{D}$ in this case being circular.

By Cauchy's inequality the first term on the right hand side cannot be larger than $(2\mathscr{E}^2-1)^2 G_{\tau\tau}^2+G_{\sigma\sigma}^2$, whence

$$|G_{\sigma\tau}|\leqq\sqrt{\mathscr{E}^2-1}\,G_{\tau\tau}.$$

Consequently, for $\alpha<\tau<\beta$ and for fixed τ_0,

$$|G_\sigma(\sigma_0+h,\tau)-G_\sigma(\sigma_0+h,\tau_0)|\leqq\sqrt{\mathscr{E}^2-1}\,|G_\tau(\sigma_0+h,\tau)-G_\tau(\sigma_0+h,\tau_0)|.$$

Now $G_\sigma(\sigma_0+h,\tau)$ is a bounded monotone function of h. Hence we can allow h to tend to zero in the above inequality, with the result that

$$\lim G_\sigma(\sigma_0+h,\tau)=\lim G_\sigma(\sigma_0+h,\tau_0)$$

uniformly in τ. Thus G_σ tends to a constant upon approach to l.

Next we observe that the function pair (G_σ, G_τ) generates a quasi-conformal mapping of the domain $\mathscr{D}$. To prove this, note that by (15)

$$G_{\sigma\sigma}^2+2G_{\sigma\tau}^2+G_{\tau\tau}^2\leqq(G_{\sigma\sigma}+G_{\tau\tau})^2\leqq4\mathscr{E}^2(G_{\sigma\sigma}G_{\tau\tau}-G_{\sigma\tau}^2),$$

as required (the mapping $(x,y)\to(u,v)$ is quasi-conformal provided that the condition $u_x^2+u_y^2+v_x^2+v_y^2\leqq2K|u_x v_y-u_y v_x|$ is satisfied). Furthermore (G_σ, G_τ) is one-to-one since G is convex.

A contradiction now results from the fact that a one-to-one quasi-conformal mapping of a Jordan domain $\mathscr{D}$ cannot attain constant values on a non-degenerate boundary arc. (This fact can be proved by reducing it to the same assertion about conformal mappings. To be precise, let Φ be a quasi-conformal homeomorphism of a Jordan domain $\mathscr{D}$, and let g be a conformal map of $\Phi(\mathscr{D})$ onto the unit disk. Then $g\circ\Phi$ is a quasi-conformal homeomorphism of $\mathscr{D}$ onto the unit disk. By a well-known result [1, page 47] the mapping $g\circ\Phi$ can be extended to a homeomorphism of the *closure* of $\mathscr{D}$. Consequently, if Φ is constant on a non-degenerate boundary arc l, then g^{-1} must be constant on the circular arc $g\circ\Phi(l)$. But this is clearly impossible, and the required assertion is proved.)

We have thus proved that $\mathscr{D}$ is strictly convex, and consequently that the equation $\mathscr{F}_0=1$ defines a strictly convex closed curve. Since the domain $\mathscr{D}$ contains the origin, it is also clear that the curve $\mathscr{F}_0=1$ surrounds the origin. This completes the proof of the triangle inequality.

Lemma 5. *If condition* (*) *of the main theorem is satisfied, then no two segments* A_i *and no two segments* B_i *can meet at a convex corner of* D.

Proof. Suppose in fact that two segments A_i did meet at a convex corner of D. Let α be the boundary arc of D consisting of A_1, A_2 and their endpoints, and let γ be the curve of shortest length joining the outer endpoints of A_1 and A_2 and homotopic to α in $\bar{D}$.

Clearly γ is a polygonal arc different from α. Moreover, the vertices of γ are drawn from the endpoints of the families $\{A_i\}$, $\{B_i\}$, and $\{C_i\}$; and the corners of γ (if there are any) are convex in the same sense as the corner of α. Now let $\mathscr{P}$ be the polygonal domain bounded by α and γ. Assuming that A_1 and A_2 are positively oriented, the associated boundary chain of $\mathscr{P}$ has the form

$$\Gamma=A_1+A_2-L_1-\cdots-L_n$$

where $L_1, \ldots, L_n$ are oriented segments of γ. Evidently

(27) $$I(L_1)+\cdots+I(L_n)<I(A_1)+I(A_2)$$

(to see this, divide $\mathscr{P}$ into n triangles by extending each of the segments L_i in the positive direction until it reaches α, and then apply the triangle inequality).

Since (27) stands in contradiction to (*), and since a similar argument applies if two segments B_i meet at a convex corner of D, the lemma is proved.

5. The Maximum Principle for Infinite Data

In this section we shall give a generalization of the maximum principle to the case where the solutions involved take on infinite boundary values. This result immediately yields the uniqueness of solutions of the Dirichlet problem with infinite data.

Maximum Principle. *Let D be a bounded domain whose boundary ∂D includes two families of open segments $\{A_i\}$ and $\{B_i\}$.* Let u_1 and u_2 be two solutions of* (5) *in D, both of which take the boundary value plus infinity on each A_i and minus infinity on each B_i.*

Let E be a finite point set on ∂D, and let C denote the set of points of $\partial D-E$ which are not in the closure of the families $\{A_i\}$ and $\{B_i\}$. If C is non-empty, assume that $\liminf(u_2-u_1)\geqq 0$ *as one approaches any point of C. If C is null, assume that $u_2\geqq u_1$ at some point of D.*

Then in either case, $u_2\geqq u_1$ throughout D.

Proof. It can be assumed for simplicity that the set E is empty, since only minor and obvious changes are required in the proof otherwise.

We consider first the case when C is non-empty. Let ε, M be fixed positive numbers with ε small and M large, and define

$$\phi(x,y)=\begin{cases} M-\varepsilon & \text{when} \quad u_2-u_1\geqq M \\ u_2-u_1-\varepsilon & \text{when} \quad \varepsilon<u_2-u_1<M \\ 0 & \text{when} \quad u_2-u_1\leqq\varepsilon. \end{cases}$$

Clearly ϕ is a strongly differentiable continuous function in D, vanishing in the neighbourhood of any point of C.

Also let $D_{\varepsilon\delta}$ denote the domain consisting of those points of D at distance $>\varepsilon$ from the endpoints of the segments A_i and B_i and $>\delta$ from the boundary of D, where $\delta<\varepsilon$. Since ε is small, the boundary of $D_{\varepsilon\delta}$ consists of straight segments $L_1, L_2, \ldots$ near the segments A_i and B_i, circular arcs $E_1, E_2, \ldots$ near their endpoints, and a remaining set N near C. By taking δ suitably near zero, we may assume in addition that

(i) $\phi\equiv 0$ in a neighbourhood of N, and

(ii) $|d\psi_1/ds-d\psi_2/ds|\leqq 2\varepsilon$ on the segments $L_1, L_2, \ldots$.

That (ii) can in fact be attained follows from Lemma 1 and the relation

$$\frac{d\psi}{ds}=\vec{v}\cdot\nabla F$$

on these segments.

* It is tacitly assumed that the families $\{A_i\}$ and $\{B_i\}$ are finite.

With $\delta=\delta(\varepsilon)$ thus chosen, let

(28) $$J=\int_{D_{\varepsilon\delta}} \nabla\phi\cdot\{\nabla F(\nabla u_2)-\nabla F(\nabla u_1)\}\,dx\,dy.$$

By partial integration, using (5) and condition (i) above, one finds

$$J=\sum_i\Big(\int_{L_i}\phi\,d\tilde\psi+\int_{E_i}\phi\,d\tilde\psi\Big),\qquad \tilde\psi=\psi_2-\psi_1.$$

From (ii) and the conditions $|\phi|<M$, $|d\psi/ds|<1$, we obtain further

(29) $$J<2\varepsilon M\Big\{\sum_i(|A_i|+2\pi)+\sum_i(|B_i|+2\pi)\Big\}.$$

Now $\nabla\phi=0$ almost everywhere outside the set where $\varepsilon<u_2-u_1<M$, while in that set

$$\nabla\phi=\nabla u_2-\nabla u_1.$$

Accordingly the integrand on the right side of (28) is non-negative (see (26)). Consider a sequence of values of ε decreasing to zero such that the corresponding values $\delta(\varepsilon)$ also decrease to zero. The set $D_{\varepsilon\delta}$ is therefore increasing, and the monotone convergence theorem may be applied to J. Thus from (29) we obtain

$$\int\{\nabla u_2-\nabla u_1\}\cdot\{\nabla F(u_2)-\nabla F(u_1)\}\,dx\,dy=0$$

where the integration takes place over the set $0<u_2-u_1<M$. It follows that the integrand is zero in this set. But since M can be taken arbitrarily large, this in turn implies

$$\nabla u_2\equiv\nabla u_1\qquad\text{whenever } u_2>u_1.$$

Hence $u_2\equiv u_1+\text{constant}$ in any component of D where $u_2>u_1$. But then, assuming the existence of such a component, $u_2\equiv u_1+\kappa$ throughout D, where $\kappa>0$. This contradicts the given boundary condition on C, whence we must have $u_2\leqq u_1$ throughout D.

The same proof applies when C is empty, except that we must of course delete reference to C and N.

6. Proof of the Main Theorem

In the preceding sections we have assembled the necessary preliminary material to carry out the proof of the theorem stated in the introduction. The proof itself is exactly the same as the proof of Theorem 4 in II, and consequently it will be sufficient here simply to point out the corresponding steps.

First, the uniqueness assertion follows directly from the infinite data maximum principle of Section 5. The preliminary existence theorem which is required next (see II, p. 339) is given in Section 1. The remaining proof of Theorem 4 in II requires the following results: the maximum principle, the compactness principle, the straight line lemma, the monotone convergence theorem, results analogous to those in Section 3 here, Lemmas 6, 7, 8 of II, and the fact that no two segments A_i and no two segments B_i can meet at a convex corner of D.

The only ones of these results which have not been proved above are the montone convergence theorem and the analogues of Lemmas 6, 7, 8.

The monotone convergence theorem for our case is a consequence of the straight line lemma, the Harnack principle (I, Theorem 3), and the compactness principle. The reader is referred to I, p. 210, for a proof. The refinements of the monotone convergence theorem stated on p. 330 of I (including Lemma 6) depend only on the straight line lemma, the triangle inequality and the results of Section 3.

Lemmas 7 and 8, finally, are consequences of the maximum principle, the preliminary existence theorem, the straight line lemma, and a barrier type argument. A suitable function for the latter has already been given in Section 1. This completes the proof of our main result.

Remarks. It may appear paradoxical that we are able to prove existence of solutions of the Dirichlet problem in non-convex domains, when this problem is in general not well posed for continuous data. The explanation, however, lies in the observation that the solution is not required to fulfill any conditions at the endpoints of the arcs A_i, B_i, and C_i.

In this connection, it is worth noting that the solutions which we obtain have many of the qualitative features of solutions of the Dirichlet problem for convex domains. In particular, the following results hold:

(i) If u_1 and u_2 are two solutions and $u_2 \geqq u_1$ on the arcs C_i, then $u_2 \geqq u_1$ throughout D.

(ii) Solutions depend continuously upon the data assigned on the arcs C_i.

(iii) If the assigned boundary data has a removable discontinuity at at convex boundary point, then the solution also has a removable discontinuity there.

Appendix

1. We are to show that $|\nabla F| \leqq$ Constant, using conditions (2) and (3). Consider the surface $Z = F(p, q)$ in the three dimensional space (p, q, Z). By (2) this surface is convex, and by (3) the tangent plane to the surface at $(p, q, F(p, q))$ intersects the Z-axis at the point with ordinate $\mu = F - pF_p - qF_q \geqq A$. Hence it is geometrically evident that for $\rho = \sqrt{p^2 + q^2} \geqq 1$ the surface lies below the cone

$$Z = a\rho + A, \qquad a = \operatorname*{Max}_{\rho = 1} \{F(p, q) - A\}.$$

Thus there exists a constant a' such that

$$F \leqq a\rho + a' \qquad \text{for all } p, q. \tag{30}$$

For the rest of the argument it is convenient to resort to vector notation. By the convexity of F we have for any two vectors $\vec{p}$ and $\vec{p}_1$,

$$F(\vec{p}_1) - F(\vec{p}) \geqq (\vec{p}_1 - \vec{p}) \cdot \nabla F(\vec{p}).$$

Setting $\vec{p}_1 = \vec{p} + h\, \nabla F(\vec{p})$ and using (30) then yields

$$h\, |\nabla F(\vec{p})|^2 \leqq a(|\vec{p}| + h\, |\nabla F(\vec{p})|) + a' - F(\vec{p}).$$

If we now divide both sides by h and let h tend to infinity there results $|\nabla F| \leqq a$, and the proof is complete.

2. Here we shall give a short demonstration of (15). First of all, it is apparent that we can assume $q=0$ without loss of generality. Now

$$G_\sigma = \frac{F_p}{\mu}, \quad G_\tau = \frac{F_q}{\mu}, \quad (\mu = F - pF_p - qF_q),$$

whence by direct calculation and evaluation at $q=0$

$$\frac{G_{\sigma\sigma}+G_{\tau\tau}}{\sqrt{G_{\sigma\sigma}G_{\tau\tau}-G_{\sigma\tau}^2}} = \frac{(F^2+p^2F_q^2)F_{pp}+2\mu pF_qF_{pq}+\mu^2F_{qq}}{F\mu\sqrt{F_{pp}F_{qq}-F_{pq}^2}}. \tag{31}$$

In view of (2), (12), and (13), the right hand side is less than

$$\frac{4b}{c}\cdot\frac{(1+p^2)F_{pp}+F_{qq}}{\sqrt{1+p^2}\sqrt{F_{pp}F_{qq}-F_{pq}^2}}.$$

Therefore, using (4), we find that (15) holds with $\mathscr{E}=4bE/c$.

This argument also allows us to give a direct proof that (4) holds when F arises from a regular parametric variational problem, independent of the rather complicated technique used in I. We first observe that if F arises from a regular variational problem then the function $G(\sigma,\tau)$ which represents the (normalized) surface $\mathscr{F}(\boldsymbol{r})=1$ in non-parametric form can be extended to a domain $\mathscr{D}'$ properly including $\mathscr{D}$, and $G_{\sigma\sigma}G_{\tau\tau}-G_{\sigma\tau}^2>0$ in $\mathscr{D}'$. Thus there must exist some constant $\mathscr{E}$ for which (15) holds.

On the other hand, by (2)

$$|2\mu pF_qF_{pq}| \leqq \frac{2+c^2}{2}p^2F_q^2F_{pp} + \frac{2}{2+c^2}\mu^2F_{qq}.$$

Hence using (12) and (13) the right hand side of (31) is greater than

$$\frac{c^2}{\sqrt{2}\,b^2}\operatorname{Min}\left(\frac{1}{2},\frac{b}{2+c^2}\right)\cdot\frac{(1+p^2)F_{pp}+F_{qq}}{\sqrt{1+p^2}\sqrt{F_{pp}F_{qq}-F_{pq}^2}}.$$

Consequently (4) holds with $E=5b^2\mathscr{E}/c^2$ provided F denotes the normalized integrand. But this completes the proof, since the left hand side of (4) is invariant under normalization.

3. Let

$$Lu = Au_{xx}+2Bu_{xy}+Cu_{yy}$$

be a quasilinear elliptic operator satisfying the minimal surface type condition

$$\frac{(1+p^2)A+2pqB+(1+q^2)C}{\sqrt{1+p^2+q^2}\sqrt{AC-B^2}} \leqq 2E, \quad (E \geqq 1).$$

Also let $u=u(x,y)$ be a function such that the surface $z=u(x,y)$ has negative mean and Gaussian curvatures satisfying

$$\frac{H^2}{|K|} \geqq E^2-1.$$

Then $Lu \leqq 0$, that is, u is a super-solution.

Proof. Let P be a point at which Lu is to be evaluated. By obvious invariance properties, we may assume without loss of generality that $q=u_y=0$ at P. Introducing the classical r, s, t notation for the second derivatives of u, the assertion then reduces to showing

$$A r+2B s+C t \leqq 0$$

provided that $r t<s^2$ and

$$\frac{(1+p^2)A+C}{\sqrt{1+p^2}\sqrt{AC-B^2}} \leqq 2E, \qquad \frac{(1+p^2)r+t}{\sqrt{1+p^2}\sqrt{s^2-rt}} \leqq -2\sqrt{E^2-1}.$$

If we now write $A'=\lambda A$, $C'=\lambda^{-1}C$, $r'=\lambda r$, $t'=\lambda^{-1}t$ where $\lambda=\sqrt{1+p^2}$, the assertion reduces further (after dropping the primes) to proving that

$$A r+2B s+C t \leqq 0$$

when $A+C \leqq 2E\sqrt{AC-B^2}$ and $r+t \leqq -2\sqrt{E^2-1}\sqrt{s^2-rt}$.

In showing this, invariance properties allow an even further reduction to the case where $s=0$, $r>0$, $t<0$. Hence we must finally establish the inequality

$$A r+C t \leqq 0 \tag{32}$$

when $A+C \leqq 2E\sqrt{AC-B^2}$ and $r+t \leqq -2\sqrt{E^2-1}\sqrt{r|t|}$. The last conditions imply

$$\sqrt{\frac{A}{C}}+\sqrt{\frac{C}{A}} \leqq 2E, \qquad \sqrt{\frac{|t|}{r}}-\sqrt{\frac{r}{|t|}} \geqq 2\sqrt{E^2-1}.$$

Therefore

$$\sqrt{\frac{A}{C}} \leqq E+\sqrt{E^2-1}, \qquad \sqrt{\frac{|t|}{r}} \geqq E+\sqrt{E^2-1}$$

and (32) follows at once.

The same proof establishes that $Lu \geqq 0$ when $H>0$, $K<0$, and $H^2/|K| \geqq E^2-1$. Thus for the original function u we have $L(-u) \geqq 0$ and $-u$ is a sub-solution.

Acknowledgement: This work was partially supported by the United States Air Force Office of Scientific Research under Grants AF-AFOSR-883-67 and AF-AFOSR-1301-67.

References

1. Ahlfors, L., Lectures on Quasiconformal Mappings. Van Nostrand, New York 1966.
2. Finn, R., New estimates for equations of minimal surface type. Arch. Rational Mech. Anal. **14**, 337–375 (1963).
3. Jenkins, H., & J. Serrin, Variational problems of minimal surface type, I. Arch. Rational Mech. Anal. **12**, 185–212 (1963).

4. JENKINS, H., & J. SERRIN, Variational problems of minimal surface type, II. Boundary value problems for the minimal surface equation. Arch. Rational Mech. Anal. **21**, 321–342 (1966).
5. JENKINS, H., Super-solutions for quasi-linear elliptic equations. Arch. Rational Mech. Anal. **16**, 402–410 (1964).
6. MEYERS, N., On a class of non-uniformly elliptic quasi-linear equations in the plane. Arch. Rational Mech. Anal. **12**, 367–391 (1963).
7. NITSCHE, J. C. C., Über ein verallgemeinertes Dirichletsches Problem für die Minimalflächengleichung und hebbare Unstetigkeiten ihrer Lösungen. Math. Ann. **158**, 203–214 (1965).
8. NIRENBERG, L., On non-linear elliptic partial differential equations and Hölder continuity. Comm. Pure Appl. Math. **6**, 103–156 (1953).
9. SERRIN, J., On the Harnack inequality for linear elliptic equations. J. d'Analyse Mathématique **4**, 292–308 (1956).

Department of Mathematics
University of Minnesota
Minneapolis

(Received December 14, 1967)

Errata. Condition (3) with $A > 0$ is more restrictive than necessary, e.g., when $F = \sqrt{1+p^2+q^2}$ we have $F - pF_p - qF_q = 1/F \to 0$ as $p^2+q^2 \to \infty$, so necessarily $A \le 0$.

In particular, the condition $A > 0$ should be replaced by $A \in \mathbf{R}$. We also add the further assumption $F_p(0) = F_q(0) = 0$. Then, as in Section 2, one shows that there are positive constants a, b such that the new function $\bar{F} = 1 + F/a$ satisfies (2), (4), and (12). The arguments of the rest of the paper now carry over with F replaced by $\bar{F}$.

[*Example.* When $F = \sqrt{1+p^2+q^2}$ we take $A = 0$ (best possible) and $E = 1$. Then one gets $a = \sqrt{2}$, see Part 1 of the Appendix, while $b = 1 + 1/a = 1 + 1/\sqrt{2}$.]

It may be noted finally that Part 3 of the Appendix is irrelevant to the paper, and can be deleted.

I.5. Singularities of Solutions

Moshe Marcus

Mathematics Subject Classification (2010). 35J15, 35J60.

Keywords. Second-order elliptic equations, nonlinear elliptic equations, weak solutions.

In this review we concentrate on several of Serrin's publications in which he studied the behavior of solutions of second-order elliptic equations with isolated singularities and related questions. The contribution of Serrin in this area has long ago acquired classical status. Its influence is immense and it continues to increase as the research in this area spreads into new directions.

The papers under consideration, [i] to [iv] below, contain just a small (but important) part of this body of work. The main topics studied in these four papers are:

1. Behavior of positive solutions of linear and quasilinear equations at an isolated singularity.
2. Removable singularities, including the case of singular sets.
3. Behavior of signed solutions of a class of linear equations at an isolated singularity or at infinity.
4. Liouville theorems.
5. Non-existence of positive singular solutions of a class of semilinear equations.

In paper [i] Gilbarg and Serrin studied mainly topics 1. and 4. with respect to uniformly elliptic equations in $\mathbb{R}^n$ of the form

$$\mathcal{L}_1 u := \sum_{i,k=1}^{n} a_{i,k} u_{i,k} + \sum_{i=1}^{n} b_i u_i + cu = 0 \tag{1}$$

and

$$\mathcal{L}_2 u := \sum_{i,k=1}^{n} (a_{i,k} u_i)_k + \sum_{i=1}^{n} b_i u_i = 0. \tag{2}$$

For the latter equation, some results are obtained assuming only pointwise ellipticity, not necessarily uniform.

This paper contains many important results, some of which are proved to be optimal. They served as an essential link between earlier classical results and later developments. We note that the paper appeared in 1956, before the work of De Giorgi and Nash on regularity of solutions and the work of Moser on Harnack inequalities for equations in divergence form. We describe below a few of the results contained in this paper. We also

DOI 10.1007/978-3-0348-0845-3_5,

mention the most relevant classical results to which they are related and briefly discuss some of the later developments.

For $n = 2$ Bôcher [B03] proved that any positive solution of (1) (with smooth coefficients) in a punctured disk $B_1'(0) = B_1(0) \setminus \{0\}$ has the form

$$u = cG + w, \tag{3}$$

where $c \geq 0$ is a constant, G is a solution asymptotic to $\ln \frac{1}{r}$ and w is a solution in $B_1(0)$.

In [i] the authors prove that, if $n = 2$, a general form of Bôcher's theorem holds, assuming only *boundedness of the coefficients and uniform ellipticity.* Specifically, under these assumptions, they prove:

Theorem 1. *Suppose that* (1) *possesses a fundamental solution* G *at the origin.* (*This means that* G *is a solution in* $B_1'(0)$ *such that* $G \to \infty$ *as* $r \to 0$.) *Then every positive solution of* (1) *is of the form* (3) *where* w *is a* C^1 *solution in the disk.*

They also show that, under the assumptions of the theorem, (1) may have fundamental solutions with asymptotic behavior other than $\ln \frac{1}{r}$. However, if the coefficients are Hölder continuous, equation (1) possesses a Green's function obeying the standard estimates and (3) holds for $n \geq 2$ (see Gevrey [G40]).

For equations in divergence form the authors prove the following remarkable result:

Theorem 2. *Assume that* $\mathcal{L}_2$ *is elliptic but not necessarily uniformly elliptic. Further assume that the coefficients* $a_{i,k}$ *are bounded, that* $a_{i,k} = a_{k,i}$ *and that* $b_i \equiv 0$. *Let* u *be a non-constant function such that* $\mathcal{L}_2 u \geq 0$ *in* $B_1'(0)$ *and put* $M = \max_{|x|=1} u(x)$. *If, for some* $\delta > 0$,

$$u(x) = \begin{cases} O(|\ln r|^{1-\delta}) & \text{if } n = 2 \\ O(r^{2-n-\delta}) & \text{if } n > 2 \end{cases} \tag{4}$$

as $r \to 0$ *then* $\limsup_{r\to 0} u < M$.

The authors provide an example showing that the divergence structure is essential for the validity of this result. An analogue of Theorem 1 for equation (2) is also established.

Other results deal with the limiting behavior at infinity and lead to important extensions of Liouville's theorem. We mention one of these:

Theorem 3. *Let* u *be a solution of* (1) *in the exterior of the ball* $B_1(0)$. *Assume that* $b_i = O(\frac{1}{r})$ *as* $r \to \infty$ *and that* $c \equiv 0$. *In addition, if* $n = 2$ *suppose that* u *is bounded below; if* $n > 2$ *suppose that* u *is bounded and that the coefficients tend to a limit at infinity. Then* u *tends to a limit at infinity.*

Corollary 1. *Under these conditions, if* u *is a solution in the entire space then* u *is a constant.*

Among the numerous later developments in the study of the above topics we mention only a couple of results. Following the work of Moser [M61] on Harnack inequalities, Royden [R62] and Littman, Stampacchia and Weinberger [LSW63] proved (in the case $n > 2$): *Any solution* u *of*

$$\sum_{i,k=1}^{n} (a_{i,k} u_i)_k = 0, \quad a_{i,k} = a_{k,i} \tag{5}$$

(measurable, bounded coefficients) in $B_1'(0)$ such that $u \to +\infty$ at the origin is bounded above and below by positive multiples of $|x|^{2-n}$. Furthermore a fundamental solution exhibiting this behavior exists.

In [Acta64] Serrin studied a large class of non-linear equations which includes, in particular (5). Among the many results of this seminal paper one finds the following: *Any solution u of (5) in $B_1'(0)$ such that $u(x) = O(|x|^{2-n})$ at the origin must be of the form (3) where G is a particular solution behaving like $|x|^{2-n}$ and w is a regular solution in $B_1(0)$.*

This study was continued in the paper of Serrin and Weinberger [ii] in which the authors considered signed solutions. We state just one of the beautiful results established in this paper. The result is formulated for $n > 2$; an analogous result holds for $n = 2$.

Theorem 4. *There exists a number $\delta > 0$, depending only on the ellipticity constant of (5) such that the following holds. If u is a solution of (5) in $B_1'(0)$ and*

$$\text{either } \limsup_{r\to 0} |x|^{n-2+\delta} u \leq 0 \quad \text{or } \liminf_{r\to 0} |x|^{n-2+\delta} u \geq 0$$

then u satisfies (3) where w is a regular solution in $B_1(0)$ and G is a fundamental solution (which is known to behave like $|x|^{2-n}$).

Thus, any singular solution other than a fundamental solution must approach both $+\infty$ and $-\infty$, as $r \to 0$, at a rate definitely faster than $\pm|x|^{2-n}$.

In [iii] Serrin studies the question of removable singular sets for elliptic equations of minimal surface type and equations arising from a bounded conservation law. Among other results, he obtains an extension to this class of equations of a theorem of De Giorgi and Stampacchia [DS66] for the minimal surface equation.

In [iv] Ni and Serrin study non-existence results for singular solutions of equations of the form

$$\Delta u + f(u) = 0 \tag{6}$$

where $f \in C([0, \infty))$, $f(0) = 0$ and f is superlinear at $+\infty$. At the time it was well known that the Dirichlet problem for (6) in a ball, with $f(t) = t^q$, has no positive solution if $q \geq (n+2)/(n-2)$. One of the results obtained in [iv] is that, in this case, there is no positive solution even in the punctured ball. The argument is based on an extension of Pohozaev's identity and subtle a priori estimates for the growth of solutions at an isolated singularity. Similar results are proved for a quasilinear analogue.

References

[B03] M. Bôcher, *Singular points of functions which satisfy partial differential equations of elliptic type*, Bull. Am. Math. Soc. 9 (1903), pp. 455–465.

[G40] M. Gevrey, *Sur une généralisation du principe des singularités positives de M. Picard*, C.R. Acad. Sci. Paris, 211 (1940) pp. 581–584.

[M61] J. Moser, *On Harnack's theorem for elliptic differential equations*, Comm. Pure App. Math., 14 (1961) pp. 577–591.

[R62] H. Royden, *The growth of a fundamental solution of an elliptic divergence structure equation*, Studies in Mathematical Analysis and Related Topics; Essays in honor of G. Polya, (1962) pp. 333–340.

[LSW63] W. Littman, G. Stampacchia and H. Weinberger, *Regular points for elliptic equations with discontinuous coefficients*, Ann. Sc. Norm. Sup. Pisa, Ser. III, 17 (1963) pp. 45–79.

[Acta64] J.Serrin, *Local behavior of solutions of quasilinear equations, Acta Math.*, 111 (1964), pp. 247–302.

[DS66] E. De Giorgi and G. Stampacchia, *Sulle singolarità eliminabili delle ipersuperficie minimali,* (Italian) Atti del Convegno sulle Equazioni alle Derivate Parziali (Nervi, 1965) Edizioni Cremonese, Rome. (1966) pp. 55–58.

Moshe Marcus
Department of Mathematics
Technion
Haifa 32000, Israel
e-mail: marcusm@math.technion.ac.il

J. Anal. Math. 4 (1956), 309–340. **Reprint I.5.[i]**

ON ISOLATED SINGULARITIES OF SOLUTIONS OF SECOND ORDER ELLIPTIC DIFFERENTIAL EQUATIONS

By

D. Gilbarg and James Serrin

in Bloomington, Indiana   in Minneapolis, Minn.

1. Introduction and Summary.

In this paper we investigate the local behavior of solutions of linear elliptic partial differential equations in $n \geqq 2$ independent variables. The topics include an extended form of the maximum principle for solutions having a singular point, the asymptotic and limit behavior of solutions, Liouville's theorem, and the characterization of fundamental solutions. Generally speaking, our results hold under weak continuity assumptions on the coefficients, these assumptions being the same as, or weaker than, those made in recent existence theories (such as [3], [13]). We also emphasize an interesting qualitative connection between the continuity properties of the coefficients and the behavior of the corresponding solutions.

In § 2—5 we are concerned with uniformly elliptic equations of the form,

$$(1)_n \qquad Lu = a_{ik}(x)\, u_{ik} + b_i(x)\, u_i + c(x)\, u = f(x),$$

where $x = (x_1, \ldots, x_n)$,

$$u_i = \frac{\partial u}{\partial x_i}, \quad a_{ik}\, u_{ik} = \sum_{i,k=1}^{n} a_{ik} \frac{\partial^2 u}{\partial x_i \partial x_k}, \quad \text{etc.}$$

We follow these notational conventions throughout the paper, and, unless otherwise stated, the solutions are assumed in class C^2.

It is well-known that solutions of (1) enjoy a maximum principle, namely, if $c \equiv 0$, $f \geqq 0$, then a non-constant solution cannot take on a maximum in the interior of its domain of definition. § 2 discusses an extended form of this maximum principle which applies to solutions defined in the punctured sphere

$$(2) \qquad S_0: \quad 0 < r = (\Sigma x_i^2)^{1/2} \leqq r_0 .$$

We shall say that a solution $u(x)$ defined in S_0 satisfies the maximum

This work was supported by the Office of Naval Research. It was completed in part while the authors were at Stanford University in the summer 1955.

principle if, on setting $u(0) = \limsup_{x \to 0} u(x)$, the resulting function does not take on a maximum in $0 \leqq r < r_0$ unless it is constant. Theorem 1, which is due essentially to Gevrey [7], shows that if the coefficients a_{ik} are Dini continuous at $r = 0$, if the b_i are bounded, and if $c \equiv 0$, $f \geqq 0$, then a solution in S_0 may be allowed an order of growth $o(\log r)$ if $n = 2$ and $o(r^{2-n})$ if $n > 2$, and still enjoy the maximum principle. In Theorem 1′ we prove that if the a_{ik} are continuous at $r = 0$, but with no special modulus of continuity, then the same conclusion follows for $n > 2$ provided the solution is $O(r^{2-n+\delta})$ for some $\delta > 0$, no statement being made for $n = 2$. The fundamental significance of the modulus of continuity in the hypothesis is brought out clearly by Example 1 of § 7 which shows that Theorem 1′ is essentially sharp. That is, if the coeficients a_{ik} are only continuous at the singular point, then for $n = 2$ the solution, even if bounded, need not satisfy the maximum principle, and for $n > 2$ there may be unbounded solutions with order of growth $o(r^{2-n})$ at $r = 0$. Furthermore, for $n > 2$ if the coefficients are discontinuous, there may be bounded solutions which do not satisfy the maximum principle.

The same example shows that the Riemann theorem on removable singularities does not hold for $n = 2$ if the coefficients are only continuous [1]; indeed as this example shows, a bounded solution might not even be Dini continuous at an isolated singularity. Since it appears that some minimum continuity condition, such as Dini or Hölder continuity, is required of the coefficients in L to insure removability of singularities, the question naturally arises as to what smoothness, if any, is guaranteed a solution at the singular point by weaker types of continuity or by uniform ellipticity alone. For $n = 2$ the answer is relatively complete. In this case Theorem 2 shows, without any continuity assumptions on the coefficients, that a solution bounded on one side must have a limit, possibly infinite, at the singular point. As we have already remarked, Example 1 of § 7 demonstrates that the limit need not be approached even Dini continuously. For $n > 2$ there is reason to believe the situation is slightly better: we conjecture that if the coefficients a_{ik} are continuous then a bounded solution must be of class C^1 at an isolated singularity, although almost certainly it will not

1. The theorem is true, however, for Hölder continuous coefficients; cf. Gevrey [7].

(necessarily) be of class C^2 there. We have been able to prove part of this conjecture: namely, for the case $n > 2$, Theorem 2 asserts that if the coefficients are continuous at the singular point, then a bounded solution, or one having $O(r^{2-n+\delta})$ order of growth, has a finite limit at the point. The conclusions of Theorem 2 can be stated at infinity as well as at finite points (Theorem 3); it is furthermore possible to assert (for $n > 2$) that the limit is approached at least with the order $r^{2-n+\delta}$ as $r \to \infty$ if the coefficients are just continuous at infinity, and as r^{2-n} if they are Dini continuous (Theorem 3'). As a by-product of these limit theorems we obtain an extension of Liouville's theorem to uniformly elliptic equations in n-variables. Such a theorem is known to be untrue for non-uniformly elliptic equations in $n > 2$ variables; (cf. Hopf [11]).

In § 5 we seek to characterize the solutions of (1) that are positive in the neighborhood of an isolated singular point. Bôcher [4] showed in 1903 that any positive solution of the equation

$$\Delta u + a(x, y) u_x + b(x, y) u_y + c(x, y) u = 0 \tag{3}$$

in S_0 necessarily has the form

$$\text{Const.}\, G + w, \qquad \text{Const.} \geqq 0,$$

where w is a solution of (3) in the closed disc $0 \leqq r \leqq r_0$, and G is a solution asymptotic to $\log \frac{1}{r}$. The coefficients a, b, c were assumed analytic, but this condition is not essential. Bôcher's method depends on the use of the Green's function belonging to the adjoint of (3) and also on the existence of a solution G asymptotic to $\log \frac{1}{r}$. His method would therefore apply, as he pointed out, to much more general equations than (3) as soon as certain existence theorems not available in his day were obtained; thus the method is at present applicable to any equation $Lu = 0$ whose coefficients are sufficiently often differentiable. (If $n > 2$, G should of course be asymptotic to r^{2-n} rather than $\log \frac{1}{r}$). In Theorem 5 we prove a general form of Bôcher's theorem for $n = 2$ variables, assuming only boundedness of the coefficients and uniform ellipticity. This theorem has the important consequence that *the asymptotic character of a fundamental solution is an inherent property of the particular equation* (cf. the remarks in § 5). Theorem 6 discusses the case $n \geqq 2$ when the coefficients of L are Hölder continuous.[2]

The proofs of Theorems 1—6 all rest to some extent on comparison arguments using Hopf's maximum principle [10]. The chief innovation in method is the application in §§ 4 and 5 of a Harnack-type inequality due to Serrin [15]. A significant feature of the method is that it allows infinity to be treated in the same way as a finite point. The greater generality of the results for $n = 2$ over those for $n > 2$ (when $n = 2$ no continuity assumptions are necessary, in contrast with $n > 2$) reflects a corresponding difference in the Harnack inequality for these two cases. Should the Harnack inequality be proved for $n > 2$ in the same generality as for $n = 2$, then the results of §§ 4 and 5 would also have the same generality for all n.

In § 6 we study elliptic equations with divergence structure

$$(4)_n \qquad Du = (a_{ik} u_i)_k = 0 \,, \qquad (i, k = 1, 2, \ldots, n)$$

and, more generally,

$$(5)_n \qquad Du + b_i u_i = 0 \,,$$

the domain of definition of coefficients and solution again being the punctured sphere S_0: $0 < r \leqq r_0$. The effect of the special structure of equation (4) is manifested in the strengthened form taken by the maximum principle in S_0. Specifically, an analogue of Theorem 1′ is proved for (4), but assuming only boundedness of the coefficients at $r = 0$ and pointwise rather than uniform ellipticity (Theorem 7); moreover, it should be noted that the case $n = 2$ is no longer exceptional, as it is in Theorem 1′. The proof is based on the method of differential inequalities. The feature that makes possible treatment of non-uniformly elliptic equations by this method is the use, initiated by Finn [5], of integrals based on the actual quadratic form defined by the a_{ik} in (4), rather than Dirichlet-type integrals, as is customary when dealing with uniformly elliptic equations.

If the equations are uniformly elliptic, as we assume in the remainder

2. The result of Theorem 6 was stated by Gevrey [7] in 1926 and proved by him [8] (with the omission of some details) in 1940. Our proof is different from Gevrey's, and, it seems to us, both simpler and more elementary. Mention should also be made of the work of Hartman and Wintner [9], who proved Bôcher's theorem for equations of the type

$$\triangle u + b_i(x) u_i + c(x) u = 0 \,,$$

assuming only continuity of b_i and c.

of the discussion, then the maximum principle in Theorem 7 can be sharpened to prove the existence of a limit at the singular point. This is shown by Theorem 8 for $n=2$ and by the work of Nirenberg[3] for $n>2$. Theorem 8, which is the analogue for the case $n=2$ of Theorem 2, proves that a positive solution of $(5)_2$ must have a limit at $r=0$ (or $r=\infty$); no continuity assumptions are made concerning the coefficients at the singular point. Nirenberg complements this result by proving that a bounded solution of $(4)_n$, or one having a suitable order of growth, has a finite limit at $r=0$ (or $r=\infty$), provided the coefficients are continuous at the point in question. In all these statements the limits, when finite, are approached Hölder continuously. That these limit theorems cannot be obtained for non-uniformly elliptic equations is shown by a counterexample in § 7.

Finally, Theorem 9 characterizes the positive solutions of $(5)_2$ in S_0, showing that within a bounded solution the fundamental solution is uniquely determined by its property of approaching $+\infty$ at $r=0$. This result is the exact analogue of Theorem 5, but whether there are equations (5) with fundamental solutions having other than $\log r$ behavior at $r=0$, as is possible for $(1)_2$, remains an unsettled problem.

Preliminary hypotheses and definitions.

In § 2—5 of the following we shall be working with the elliptic operator L defined by (1). We take as the domain of definition of the coefficients either the punctured sphere,

$$(2) \qquad S_0: \ 0 < r = |x| = \left(\sum_{i=1}^{n} x_i^2\right)^{1/2} \leqq r_0 ,$$

or the exterior domain $r \geqq r_0$. The functions operated on by L are assumed of class C^2 in S_0 (or $r \geqq r_0$), but may be singular at $r=0$ (or $r=\infty$). The continuity properties of the coefficients will be stated explicitly in the hypotheses of the various theorems, and unless so stated, no continuity will be assumed. We shall assume always in § 2—5 that L is *uniformly elliptic*; that is, we suppose there exists a constant $\lambda>0$ such that, for all points x and all real n-tuples $\xi_1, \dots, \xi_n$,

$$(6) \qquad \lambda^{-1}\left(\sum_{i=1}^{n} \xi_i^2\right) \leqq a_{ik}(x)\,\xi_i\xi_k \leqq \lambda\left(\sum_{i=1}^{n} \xi_i^2\right).$$

3. To appear in a subsequent publication.

Finally, the coefficients b_i and c are assumed bounded:

$$|b_i| \leqq B, \quad |c| \leqq B. \tag{7}$$

The inhomogeneous term f in (1) is also assumed bounded, although an explicit bound need not be stated.

We recall for later use certain definitions· A function $f(x)$ satisfies a Dini condition at x_0, or is Dini continuous at x_0, if

$$|f(x)-f(x_0)| \leqq \varphi(|x-x_0|),$$

where $\varphi(r)$, the Dini modulus of continuity, is an increasing function such that

$$\int_{+0} \frac{\varphi(r)}{r}\, dr < \infty .$$

A special case of the Dini condition is the Hölder condition,

$$\varphi(r) = Kr^{\alpha}, \quad 0<\alpha<1 .$$

We shall say that f is uniformly Hölder continuous in a region if the same constants K, α can be used for all points of the region. If x_0 is at infinity, and f_0 is the limit value of f at infinity, then f is Dini continuous at infinity provided

$$|f(x)-f_0| \leqq \varphi(|x|),$$

where $\varphi(r)$ is a monotone decreasing function such that

$$\int^{+\infty} \frac{\varphi(r)}{r}\, dr < \infty ;$$

f is Hölder continuous at infinity if $\varphi(r) = Kr^{-\alpha}$, $0<\alpha<1$.

2. An extended form of the maximum principle.

We have already specified the general form of the elliptic operator L with which we shall deal. For such an operator it is well-known that a maximum principle holds, namely, if $c \equiv 0$, then a non-constant function $u(x)$ satisfying the inequality $Lu \geqq 0$ cannot take on a maximum value in the interior of its domain of definition [4]. It is the purpose of this section to extend this result by showing that in certain general circumstances a function satisfying $Lu \geqq 0$ cannot approach its maximum at an isolated

4. If $c \leqq 0$, then a non-constant function $u(x)$ satisfying $Lu \geqq 0$ cannot take on a positive (or zero) maximum in the interior of its domain of definition.

boundary point of its domain of definition, or in other words, at one of its isolated singularities.

Theorem 1. *Assume that the coefficients a_{ik} of L satisfy a Dini condition at $r=0$. Let $u(x)$ be a non-constant function satisfying the inequality $Lu \geqq 0$ in the punctured sphere S_0, and let $c \equiv 0$. We set $M = \max_{r=r_0} u$. Then if*

$$u = \begin{cases} o(\log r), & n=2 \\ o(r^{2-n}), & n>2, \quad \text{as } r \to 0, \end{cases} \tag{8}$$

it follows that $u < M$ in $0<r<r_0$, and furthermore $\limsup_{x\to 0} u(x) < M$. If $c \leqq 0$, the same conclusion follows with $M = \max_{r=r_0} (0, u)$.[5]

In case $n>2$ and $u = O(r^{2-n+\delta})$, $\delta>0$, the Dini condition may be dropped from the hypotheses of Theorem 1. We formulate this as

Theorem 1'. *Suppose $n>2$, and assume that the coefficients a_{ik} of L are continuous at $r=0$. Let $u(x)$ be a non-constant function obeying $Lu \geqq 0$ in S_0, with either $c \equiv 0$ or $c \leqq 0$, and suppose*

$$u = O(r^{2-n+\delta}), \quad \text{as } r \to 0, \tag{8'}$$

for some $\delta > 0$. Then u satisfies the conclusion of Theorem 1.

In § 7 we shall show by counterexample that Theorem 1' cannot be essentially sharpened. We first prove Theorem 1' and then Theorem 1.

Proof of Theorem 1'. It clearly involves no loss of generality to assume $a_{ik}(0) = \delta_{ik}$, for any other case may be reduced to this one by means of an appropriate linear coordinate transformation. (It is here that we use the hypothesis of uniform ellipticity). The cases $c \equiv 0$ and $c \leqq 0$ are treated concurrently.

Define the functions $g = r^{2-n}$ and $h = r^{2-n-\delta'} = g^{1-\gamma}$, where $0<\delta'<\delta$ and $\gamma = (n-2)^{-1}\delta'$. Then setting $L^* = L - c$, we have

---

5. This result is not essentially different from one obtained by Gevrey [7], [8], as part of a theorem on removable singularities (see footnote 1). Theorem 1 is stated and proved here for completeness and to emphasize its independent interest. We recall that the hypotheses on the operator L were stated at the close of §1.

(9) $$L^*h = (1-\gamma)g^{-1-\gamma}(gL^*g - \gamma\, a_{ik}\, g_i\, g_k)\,.$$

Since g is harmonic for $r>0$, and since $a_{ik} \to \delta_{ik}$ as $r\to 0$, we easily derive

(10) $$L^*g = (a_{ik}-\delta_{ik})\,g_{ik} + b_i\, g_i \leqq \varepsilon_1(r)\, r^{-n},$$

where $\varepsilon_1(r)\to 0$ as $r\to 0$. Moreover,

(11) $$a_{ik}\, g_i\, g_k \geqq (n-2)^2\,[1-\varepsilon_2(r)]\, r^{2-2n},$$

where $\varepsilon_2(r)\to 0$ as $r\to 0$. Let $r_1<r_0$ be chosen so small that

(12) $$\varepsilon_1 - \gamma(n-2)^2(1-\varepsilon_2) \leqq 0 \quad \text{in} \quad S_1:\ 0<r\leqq r_1\,.$$

The preceding inequalities then imply $Lh = L^*h + ch \leqq 0$ in S_1. Now if $c\equiv 0$ set $M_1 = \max\limits_{r=r_1} u$, otherwise $M_1 = \max\limits_{r=r_1}(0\,,u)$, and define

$$v = u - \varepsilon h - M_1\,, \qquad \varepsilon>0\,.$$

Then obviously $Lv \geqq 0$ in S_1 and $v\leqq 0$ on $r=r_1$. Moreover, by (8′), $v\leqq 0$ for all sufficiently small r (we may of course assume $2-n+\delta'<0$). It follows from the maximum principle, therefore, that $v<0$ in S_1, and hence (ε being arbitrary) that $u\leqq M_1$ in S_1. Applying the maximum principle in the annular domain $\frac{1}{2}r_1<r<r_0$ we infer

(13) $$\max_{r=r_1} u < \max(M\,,M_1)\,.$$

If $c\equiv 0$, this implies $M_1<M$, hence $u<M$ in $0<r<r_0$ and

$$\limsup_{x\to 0} u \leqq M_1 < M\,,$$

as desired.

If $c\not\equiv 0$, then (13) is possible only if $M_1<M$ or $M=M_1=0$. In the former case the desired conclusion is obtained exactly as before. In the latter case it is clear that $u<0$ in $0<r<r_0$, but it is not at once evident that

(14) $$\limsup_{x\to 0} u<0\,.$$

To prove (14) we set $u=(1+Ar^2)\,U$, where $A=\text{const.}>0$. Then U satisfies an inequality of the form

(15) $$a_{ik}\, U_{ik} + b'_i\, U_i \geqq 0$$

in some region $0<r<r_2$. Now the term cU is missing from the left hand side of (15), and $U<0$ on $r=r_2$. From the portion of the proof already completed we may therefore conclude

$$\limsup_{x\to 0} U < 0\,.$$

This inequality clearly implies (14), and the proof is complete.

Proof of Theorem 1. Let us define

$$h = \int_0^g \exp K\left(r + \int_0^r \varphi(s)\, s^{-1}\, ds\right) dg$$

where $g = r^{2-n}$ or $g = \log(r_0/r)$, depending on whether $n > 2$ or $n = 2$, and K is a positive constant. If the coefficients a_{ik} are Hölder continuous it suffices to choose h in the simpler form $h = g(1 + Kr^{\alpha})$. As in the preceding proof we have

$$Lh \leqq 0 \quad \text{in} \quad S_0,$$

provided K is chosen suitably large. Because of the convergence of the integral $\int_0^r \varphi(s)\, s^{-1}\, ds$ it is clear that $h \sim g$ as $r \to 0$. Hence, as in the preceding proof, $v = u - \varepsilon h - M < 0$ in S_0. Since ε is arbitrary this implies $u \leqq M$. The strong form of the maximum principle shows finally that equality cannot occur in S_0, and the theorem follows readily.

Remarks. The full strength of the condition $a_{ik} \to a_{ik}(0)$ was not used in the proof of Theorem 1′. In fact, the whole argument remains valid as long as $\limsup_{x \to 0} a_{ik}$ and $\liminf_{x \to 0} a_{ik}$ do not differ by too much, the permissible variation depending on the modulus of ellipticity λ, the bound B and δ. On the other hand, a too great variation in a_{ik} will destroy the validity of the maximum principle even for bounded solutions; see § 7. In other words, uniform ellipticity in itself is not sufficient to assure the maximum principle. We can also allow in both Theorems 1 and 1′ a certain degree of singularity in the coefficients b_i, specifically $b_i = O(r^{-1})$ as $r \to 0$, but we shall not pursue this matter further.

3. Auxiliary lemmas.

It is convenient to collect here several lemmas which will be useful in later sections. These lemmas depend on the following Harnack-type inequalities, derived in [15] (see in particular § 4).

Let $u(x)$ be a positive solution of equation (1) with $c \leqq 0$ and $|f| \leqq B'$. Let S be a closed subregion of the domain T in which u is defined. Then the following inequalities hold:

Harnack Inequality 1. Suppose $n = 2$. Then there exists a constant $K > 0$, depending only on S, T, and the constants λ and B, such that the inequality

$$u(x) \leqq K[u(y)+B']$$

holds for any two points x, y in S.

Harnack Inequality 2. Suppose $n>2$, and let the coefficients a_{ik} of L satisfy a uniform Dini condition in T. Then there exists a constant $K>0$, depending only on S, T, λ, B, and the modulus of continuity of a_{ik}, such that the inequality

$$u(x) \leqq K[u(y)+B']$$

holds for any two points x, y in S.

Harnack Inequality 3. Suppose $n>2$, and let $D=\max|\nabla u|$ in T. Then there exists a continuous monotonically increasing function $\Phi(t)$, $t \geqq 0$, $\Phi(0)=0$, depending only on S, T, and the constants λ, B, and D, such that the inequality

$$u(x) \leqq \Phi[u(y)+B']$$

holds for any two points x, y in S.

The necessary lemmas will now be given in their order of application in the paper.

Lemma 1a. Suppose $n=2$, and let $u(x)$ be a positive solution of

(16) $$Lu=f \qquad (|f| \leqq B',\ c \leqq 0)$$

in a region $0<r<r'$. Then there exists a constant $K>0$, depending only on λ and B, such that if $|x|=|x'| \leqq \operatorname{Min}[1, \frac{1}{2}r']$, then

(17) $$u(x) \leqq K[u(x')+|x'|^2 B'] .$$

Proof: Let x, x' be two points on a circle of radius $R \leqq \operatorname{Min}[1, \frac{1}{2}r']$ about the origin. We introduce new variables $\xi=(\xi_1, \xi_2)$ in the equation $Lu=0$ by means of the transformation $x=R\xi$. Then $w(\xi)=u(x)$ satisfies (for $n=2$)

(16′) $$a_{ik} w_{ik}+R b_i w_i+R^2 c w=R^2 f ,$$

where a_{ik}, b_i, and c are of course considered now as functions of ξ. The conditions (6), (7) are satisfied by the coefficients of (16′), and $w>0$ in the domain T: $0<|\xi|<2$.

Let ξ, ξ' be the images of x, x'. Then $|\xi|=|\xi'|=1$, and ξ, ξ' are contained in the subregion $\frac{1}{2} \leq \xi \leq \frac{3}{2}$ of T. Applying Harnack Inequality 1 to the positive solution w in T we get

$$w(\xi) \leq K[w(\xi') + R^2 B']$$

where K depends only on λ and B. Returning to the original variables we obtain (17).

Lemma 1b. *Suppose $n > 2$, and let $u(x)$, $0 < u \leq M$ be a bounded positive solution of*

$$(16)_n \qquad Lu = f, \qquad (|f| \leq B', \; c \leq 0)$$

in a region $0 < r < r'$. Suppose the coefficiens a_{ik} of L tend to limits $a_{ik}(0)$ as $r \to 0$. Then there exists a constant $\rho > 0$ and a continuous monotonically increasing function $\Phi(t)$, $t \geq 0$, $\Phi(0) = 0$, depending only on λ, B, B', M, and the modulus of continuity of a_{ik} at the origin, such that if $|x| = |x'| \leq \rho$, then

$$(18) \qquad u(x) \leq \Phi[u(x') + |x'|^2 B'].$$

Proof. As in the preceding lemma we introduce new variables $\xi = (\xi_1, \ldots, \xi_n)$ by $x = R\xi$, so that $w(\xi) = u(x)$ satisfies $(16')_n$. In the domain T', $0 < |\xi| < 2$, we have $0 < w \leq M$. Moreover, if R is sufficiently small, the coefficients a_{ik} differ from $a_{ik}(0)$ in T' by as little as desired. In particular, there is a constant $\rho > 0$, such that, for any $R \leq \rho$ the coefficients a_{ik} satisfy in T' the hypotheses of a theorem of Morrey ([13], p. 123)(6). This theorem asserts (in the form needed here) that if v is a solution of (1) in a domain Γ, under the assumptions (6), (7), $|f| < B'$, and $|v| \leq M$, and if the oscillation ω of a_{ik} is sufficiently small in Γ

$$(\omega = \text{l. u. b.}\, |a_{ik}(x) - a_{ik}(y)|, \; x, y \in \Gamma),$$

then in any fixed closed subregion of Γ the first derivatives of v are uniformly bounded by a quantity depending only on λ, B, B', M, and ω. Applying this result to $(16')_n$ we conclude that if $R \leq \rho$ and if T is the subregion $\frac{1}{2} \leq |\xi| \leq \frac{3}{2}$ of T', then

$$\max_T |\nabla w| \leq N,$$

6. In the proof of Morrey's Theorem it is assumed that a_{ik}, b_i, c, f are measurable, and this additional hypothesis will be tacitly assumed in all applications of Lemma 1b. A simpler proof of this theorem will appear in a forthcoming work of L. Nirenberg.

where N depends only on λ, B, B', M, and the oscillation ω in T'. If ξ, ξ' are the images of x, x' as in Lemma 1a, and S is the region $\frac{3}{4} \leqq |\xi| \leqq \frac{5}{4}$, we obtain from Harnack Inequality 3,

$$w(\xi) \leqq \Phi[w(\xi') + R^2 B'],$$

where $\Phi(t)$ now depends on λ, B, B', M, and ω. Returning to the original variables we get the desired inequality (18).

We remark that existence of the limit $a_{ik}(0)$ is not necessary as such in the proof of Lemma 1b; rather, it suffices that $|\limsup_{x\to 0} a_{ik} - \liminf_{x\to 0} a_{ik}|$ be suitably small, so that Morrey's theorem becomes applicable.

The following Lemmas 1c and 1d are similar in content and proof to the preceding. We shall therefore state them without proof.

Lemma 1c. *Suppose $n>2$, and let $u(x)$ be a positive solution of $(16)_n$ in the region $0<r<r'$. Let the coefficients a_{ik} of L satisfy a uniform Dini condition. Then there exists a constant $K>0$, depending only on λ, B, and the modulus of continuity of a_{ik}, such that if*

$$|x| = |x'| \leqq \mathrm{Min}[1, \tfrac{1}{2}r'],$$

then

$$u(x) \leqq K[u(x') + |x'|^2 B'].$$

Lemma 1d. *Suppose $n>2$, and let $u(x)$ be a positive solution of $(16)_n$ in the region $0<r<r'$. Suppose that*

$$|\nabla u| \leqq N/r.$$

Then there exists a continuous monotonically increasing function $\Phi(t)$, $t \geqq 0$, $\Phi(0)=0$, depending only on λ, B, and N, such that if $|x| = |x'| \leqq \mathrm{Min}[1, \frac{1}{2}r']$, then

$$u(x) \leqq \Phi[u(x') + |x'|^2 B'].$$

Remark 1. It should be noted that condition (7) on the coefficients b_i and c can be relaxed in the preceding lemmas to

$$(7') \qquad b_i = O\left(\frac{1}{r}\right), \quad c = O\left(\frac{1}{r^2}\right), \quad \text{as } r \to 0.$$

Remark 2. The singular point of u, which until now has been at the origin, could just as well be the point at infinity and the proofs would be basically unchanged. In this case condition (7) must be replaced by

$$(7'') \qquad b_i = O\left(\frac{1}{r}\right), \quad c = O\left(\frac{1}{r^2}\right) \quad \text{as } r \to \infty .$$

4. Limit theorems.

In this section we shall supplement the maximum principles of §2 with results on the limit behavior of solutions at an isolated singularity. We shall assume of the coefficients merely boundedness ($n = 2$) or continuity ($n > 2$). As already remarked in the Introduction, one cannot expect the Riemann theorem on removable singularities to hold in this generality. Nevertheless, it is still possible to obtain significant results in the form of limit theorems. As a case in point, we shall prove that *if $u = u(x, y)$ is a solution in S_0 (or in $r \geqq r_0$) of the uniformly elliptic equation*

$$(19) \qquad Au_{xx} + 2Bu_{xy} + Cu_{yy} = 0 ,$$

and if u is bounded on one side, then u tends to a limit (possibly infinity) as $r \to 0$ (or ∞). We emphasize that the coefficients in this theorem are subject to no conditions other than uniform ellipticity.

The following Liouville-type theorem is a corollary of the result just stated. *If $u(x, y)$ is a solution of (19) over the entire plane, and if u is bounded on one side, then $u \equiv$ Const.*[7] For $n > 2$ we shall prove a limit theorem similar to the one stated above, and it will have as a consequence another Liouville theorem: *If $u(x)$ is a bounded solution over all space of the uniformly elliptic equation*

$$a_{ik}(x)\, u_{ik} = 0 ,$$

and if the coefficients a_{ik} are continuous at infinity, then $u \equiv$ const. This seems to be the first Liouville theorem proved for a general elliptic equation in $n > 2$ variables. In the above Liouville theorems and in the limit theorems leading to them one cannot relax the hypothesis of uniform ellipticity to mere pointwise ellipticity. This is shown by an example of E. Hopf [11]. We turn now to the formal statement and proof of these results.

Theorem 2. *Let $u(x)$ be a solution of (1) in the punctured*

7. Bers and Nirenberg [3] have recently given an independent proof of this theorem; cf. also Theorem 3 of Serrin [15].

sphere S_0. If $n=2$ suppose that $u \geqq -M$; if $n>2$ suppose that the coefficients a_{ik} are continuous at $r=0$ and that $|u| \leqq M$.[8] Then u tends to a limit as $r \to 0$.

Proof. We show first that the general case of this theorem can be deduced from the special case $c \leqq 0$. Suppose therefore that the theorem is true when $c \leqq 0$. Define a new dependent variable U by setting

$$u = (1 - Ar^2)\, U \,, \qquad A = \text{const.} > 0 \,.$$

Then U satisfies the equation

$$(1 - Ar^2)\, L^* U - 4A\, a_{ij}\, x_i\, U_j + c' U = f \,,$$

where $L^* = L - c$ and

$$c' = [-2A \Sigma_i a_{ii} - 2A \Sigma_i x_i b_i + c\,(1 - Ar^2)] \,.$$

If we take $A = B\lambda/n$, then $c' \leqq 0$ for r sufficiently small. It follows that U tends to a limit as $r \to 0$, and consequently so does u.

In the sequel we may thus assume $c \leqq 0$. The plan of the proof is the same whether $n=2$ or $n>2$. We present these cases separately, however, since the idea stands out more clearly for $n=2$.

Case I. $n=2$. Let $u_0 = \liminf\limits_{x \to 0} u(x)$; obviously $u_0 \geqq -M$. If $u_0 = +\infty$, then $\lim u = +\infty$, and the theorem is proved. We may thus suppose u_0 finite. There then exists for any $\varepsilon > 0$ a number r' such that

$$v = u - u_0 + \varepsilon > 0 \qquad \text{for } 0 < r < r'.$$

v satisfies the equation $Lv = f + c\,(\varepsilon - u_0) = f'$, and f' is bounded, say $|f'| < B'$. Since $\liminf v = \varepsilon$, let x_m, $m = 1, 2, \ldots$, be a sequence of points such that

$$v\,(x_m) < 2\varepsilon \qquad \text{as } x_m \to 0\,;$$

we may suppose $|x_m| < \operatorname{Min}(1\,, \tfrac{1}{2} r')$. Then by Lemma 1a there exists a constant $K > 0$ with the property that

$$v\,(x) \leqq K\,[v\,(x_m) + |x_m|^2 B'] < K\,[2\varepsilon + |x_m|^2 B'] \tag{20}$$

$$\text{for } |x| = |x_m| \text{ and } m = 1, 2, \ldots \,.$$

Now for sufficiently large A and sufficiently small r'' we have

$$L\,(v + Ar^2) \geqq 0 \quad \text{in} \quad 0 < r < r''.$$

Let m_0 be so large that $|x_m| \leqq r''$ for $m \geqq m_0$. We may then apply the maximum principle to $v + Ar^2$ in the annuli $|x_{m+1}| \leqq r \leqq |x_m|$, $m \geqq m_0$.

8. This condition can be relaxed to $u = O\,(r^{2-n+\delta})$ by use of Theorem 1'.

Making use of (20), we obtain the inequality

$$v + Ar^2 \leqq K(2\varepsilon + |x_m|^2 B') + A\,|x_m|^2\,,$$

valid in the annuli $|x_{m+1}| \leqq r \leqq |x_m|$, $m \geqq m_0$. It follows that

$$v(x) \leqq 2K\varepsilon + (A + KB')\,|x_m|^2 \qquad \text{if } |x| \leqq |x_m|,\ m \geqq m_0\,. \tag{21}$$

Letting x, and hence x_m, tend to zero in (21), we obtain

$$\limsup v(x) \leqq 2K\varepsilon\,,$$

or equivalently,

$$\limsup u \leqq u_0 + (2K - 1)\varepsilon\,. \tag{22}$$

Now let $\varepsilon \to 0$, and obtain

$$\limsup u \leqq u_0 = \liminf u\,.$$

This implies the existence of $\lim\limits_{x \to 0} u(x)$, proving the theorem for $n = 2$.

Case II. $n > 2$. The preceding argument carries over to the present case, with only minor changes arising from the use of Lemma 1b instead of Lemma 1a. Specifically, inequality (20) is replaced by

$$v(x) \leqq \Phi\,[v(x_m) + |x_m|^2 B'] \quad \text{for } |x| = |x_m| \leqq \rho$$

where Φ and ρ are defined in Lemma 1b. Proceeding as before, we obtain in place of (22) the inequality

$$\limsup u \leqq u_0 + \Phi(2\varepsilon) - \varepsilon\,;$$

hence $\limsup u \leqq u_0$, and the desired limit exists. This completes the proof of Theorem 2.

By virtue of the remark following the proof of Lemma 1b we infer that for $n > 2$ Theorem 2 remains valid when the coefficients a_{ik} are discontinuous, provided the oscillation of a_{ik} at $r = 0$ is sufficiently small. An analogous statement can be made about the following Theorem 3.

For the case $c \equiv f \equiv 0$ the proof of Theorem 2 applies as well to a solution of (1) defined in the exterior of some sphere (or circle), with the point at infinity considered as a singularity. The sequence $|x_m|$ should now tend to infinity, of course, and application of Lemma 1a (or Lemma 1b) now requires the additional hypothesis $b_i = O\left(\frac{1}{r}\right)$; (see the remarks concluding § 3). The only other changes necessary in the proof are simplifications arising from the fact that now $c \equiv f \equiv f' \equiv 0$. We may therefore state

Theorem 3. *Let $u(x)$ be a solution of*

$$Lu = 0 \qquad (c \equiv 0) \tag{23}$$

in the exterior region $r \geqq r_0$, and let $b_i = O(\frac{1}{r})$ as $r \to \infty$. If $n=2$ suppose that $u \geqq -M$; if $n>2$ suppose that the coefficients a_{ik} tend to a limit at infinity and that $|u| \leqq M$. Then u tends to a limit at infinity.

The Liouville theorems discussed at the beginning of this section are consequences of Theorem 3. We remark that the condition $|u| \leqq M$ in Theorems 2 and 3, which makes possible the application of Lemma 1b, can be replaced by $u \geqq -M$ if the coefficients a_{ik} are Dini continuous (in the above proofs one then uses Lemma 1c rather than Lemma 1b).

For $n>2$ the limit behavior in Theorem 3 is made precise by the following

Theorem 3′. Let $u(x)$ be a bounded solution of

$$Lu = 0 \qquad (c \equiv 0) \tag{23}$$

in the exterior region $r \geqq r_0$. Suppose $n>2$, and assume that the coefficients a_{ik} of L tend to limits as $r \to \infty$, and that $b_i = O(1/r)$ as $r \to \infty$. Then $u(x)$ tends to a limit u_0 as $r \to \infty$, and furthermore

$$u - u_0 = O(r^{2-n+\delta}) \qquad \text{for every } \delta > 0 .$$

If in addition the coefficients a_{ik} are Dini continuous at infinity then

$$u - u_0 = O(r^{2-n}) .$$

Proof. We first construct a suitable comparison function, following the same procedure as in Theorems 1 and 1′. It may be supposed that a linear transformation has already taken (23) into an equation with coefficients a_{ik} approaching δ_{ik} at infinity. As in Theorem 1′ let us define the functions,

$$g = r^{2-n} \quad \text{and} \quad h = r^{2-n+\delta'} = g^{1-\gamma},$$

where $0 < \delta' < \delta$, and $\gamma = (n-2)^{-1}\delta' > 0$. We observe that the relations (9), (10), and (11) are again true, where now $r \to \infty$, and the inequality (12) holds in some region $G : r \geqq r_1$. It follows that $Lh \leqq 0$ in G.

For convenience let us assume that the limit u_0 (known to exist from Theorem 3) is zero, so that $u \to 0$ at ∞. Let $\alpha > 0$ be a constant such that $u - \alpha h \leqq 0$ on $r = r_1$. For any $\varepsilon > 0$ the function $u - \alpha h - \varepsilon$ has the property that it is non-positive on $r = r_1$ and also for all sufficiently large r. From this fact and the maximum principle ($L(u - \alpha h - \varepsilon) \geqq 0$ in G) we infer that $u - \alpha h - \varepsilon \leqq 0$ in G. Since ε is arbitrary we conclude that $u \leqq \alpha h$ in G. By taking $-u$ in place of u and repeating the argument, we obviously obtain $u \geqq -\beta h$ in G for some $\beta > 0$, and hence we can assert, as required,

$$u = O(r^{2-n+\delta'}) = o(r^{2-n+\delta}) .$$

The proof follows the same lines when the coefficients are Dini continuous at infinity. In this case the construction of the comparison function proceeds as in Theorem 1 rather than Theorem 1′.

We remark that, for a fixed value of δ the preceding proof remains valid if the coefficients a_{ik} are discontinuous, provided the oscillation of a_{ik} at infinity is sufficiently small. However, if the oscillation is too great, it is possible, as an example in § 7 shows, that the solution is not even Dini continuous at infinity. Whether or not a limit always exists in this case is still an open problem.

It is possible to state another type of limit theorem when the boundedness conditions on the solution u are replaced with conditions on the gradient of u. Part of the interest in this result derives from the fact that the coefficients are free from continuity conditions.

Theorem 4. *Let $u(x)$ be a solution of (1) in the punctured sphere S_0. If $|\nabla u| = O(r^{-1})$ as $r \to 0$ then u tends to a limit as $r \to 0$.*

Proof. We begin by showing that u must be bounded at least on one side. Suppose on the contrary that $\limsup_{x \to 0} u = +\infty$ and $\liminf_{x\to 0} u = -\infty$. Since u is continuous except possibly at 0, this implies that in every sphere about 0 there must be at least one point where $u = 0$. Hence there exists a sequence of points x_m, $m = 1, 2, \dots,$ such that

$$u(x_m) = 0 , \qquad x_m \to 0 .$$

We now assert that $u(x)$ is uniformly bounded on the spheres $r = |x_m|$. For let x be a point on one of these spheres. By hypothesis there exists a constant $N > 0$ such that

$$|\nabla u| \leqq N/r .$$

Hence

$$|u(x)| = \left| \int_{x_m}^{x} \frac{du}{ds} \, ds \right| \leqq 2\pi N$$

the path of integration lying on the sphere $r=|x|=|x_m|$. The assertion is thereby proved. But this implies, because of the maximum principle, that u is bounded throughout $0<r\leqq|x_1|$, which contradicts the assumption $\limsup u=+\infty$.

Having shown that u is bounded on one side, say $u\geqq -M$, we can continue as in the proof of Theorem 2. We must now apply Lemma 1d rather than Lemmas 1a or 1b, but otherwise the argument is unchanged. This completes the proof of Theorem 4.

An analogous result is true when the singular point is at infinity, but will not be stated formally.

5. Behavior of positive solutions at an isolated singularity.

In this section we present some analogues of Bôcher's theorem for the general elliptic equation (1). The background for these results has already been discussed in the Introduction.

Theorem 5. *Suppose $n=2$. Let the equation*

$$Lu=0 \tag{24}$$

possess a solution $V(x)$ in the punctured disc S_0 with the property $V\to\infty$ as $r\to 0$. Then every positive solution of (24) in S_0 has the form

$$u=\text{Const.}\,V+w,$$

where Const. $\geqq 0$ and w is of class C^1 in the closed disc $0\leqq r\leqq r_0$.

It is evident that w is a solution of (24) in S_0 and that the representation of u in terms of V is unique.

A solution of $Lu=0$ which becomes positively infinite as $r\to 0$ is ordinarily called a *fundamental solution*, and, as is well-known, if the coefficients are Hölder continuous the equation always possesses a fundamental solution which is asymptotic to $\log 1/r$. It is a consequence of Theorem 5 that two fundamental solutions, whatever the smoothness properties of the coefficients, must be positive multiples of one another, up to a C^1 solution. The asymptotic nature of a fundamental solution is therefore an inherent property of the particular equation. This result is given substance by the existence of equations having fundamental solutions of quite different asymptotic behavior (§ 7); it is only when the coefficients are at least Hölder (or Dini) continuous that the asymptotic behavior stabilizes to $\log 1/r$ (Theorem 6).

Proof. It is sufficient to carry through the proof in the special case $c \leqq 0$, for the general case may be reduced to this one by a change of dependent variables, as in Theorem 2.

Now let a constant $V_0 > 0$ be chosen so large that the function $V^* = V - V_0$ is non-positive on $r = r_0$. Also let α be the largest number such that $u - \alpha V^* \geqq 0$ in S_0. This number exists since the set of all values α for which $u - \alpha V^* \geqq 0$ is closed, bounded above, and contains at least $\alpha = 0$. We set $W \equiv u - \alpha V^* \geqq 0$.

For a function $f(x)$ defined in S_0 we adopt the notation

$$\overline{f}(r) = \max_{|x|=r} f(x) \quad , \quad \underline{f}(r) = \min_{|x|=r} f(x) .$$

We assert, to begin with, that

$$\underline{W} = o(\overline{V}), \qquad \text{as } r \to 0 . \tag{25}$$

For suppose to the contrary that there exists an $\varepsilon \to 0$ and a sequence of numbers r_m, $r_m \to 0$, such that

$$\underline{W}(r_m) \geqq \varepsilon \overline{V}(r_m) , \qquad m = 1, 2, \ldots . \tag{26}$$

On any circle $r = r_m$ we have by (26) that $W \geqq \varepsilon V^*$, while on $r = r_0$ $W - \varepsilon V^* \geqq 0$. Since $L(W - \varepsilon V^*) = (\alpha + \varepsilon) V_0 c \leqq 0$, it follows from the maximum principle that

$$W - \varepsilon V^* \geqq 0 \qquad \text{in } r_m \leqq r \leqq r_0 .$$

But r_m may be arbitrarily small, hence

$$W - \varepsilon V^* \geqq 0 \qquad \text{in } S_0 .$$

This inequality may be rewritten as $u - (\alpha + \varepsilon) V^* \geqq 0$, which violates the definition of α as the largest number such that $u - \alpha V^* \geqq 0$. This contradiction establishes (25).

Observe now that W is a positive solution of the equation

$$LW = \alpha V_0 c ,$$

where $\alpha V_0 c$ is bounded, say $|\alpha V_0 c| \leqq B'$. Also $\underline{W}(r)$ and $\overline{W}(r)$ are the values of W at two points on the circle $|x| = r$. Therefore, according to Lemma 1a, there exists a constant K such that

$$\overline{W}(r) \leqq K[\underline{W}(r) + r^2 B'] , \qquad 0 < r \leqq \mathrm{Min}(1 , \tfrac{1}{2} r_0) .$$

For the same reasons

$$\overline{V}(r) \leqq K\underline{V}(r), \qquad 0<r\leqq \mathrm{Min}(1, \tfrac{1}{2}r_1),$$

where it is assumed that r_1 is so small that $V>0$ in $0<r<r_1$. The last two inequalities together with the order of magnitude relation (25) imply

(28) $$W = o(V) \qquad \text{as } r\to 0.$$

It follows that

(29) $$u-\alpha V = o(V) \qquad \text{as } r\to 0.$$

We can infer at once from (29) that $u-\alpha V$ is bounded (the argument is that of Theorem 1), and then from Theorem 2 that $u-\alpha V$ is continuous.

To show further that $u-\alpha V$ is in C^1 we apply the theory of C^1 solutions of (24), as developed by Bers and Nirenberg [2, 3]. The solutions considered there have continuous first derivatives and "strong" second derivatives satisfying the equation almost everywhere. This work establishes the maximum principle for C^1 solutions and proves the existence of C^1 solutions of the Dirichlet problem. We now follow a standard argument in proving that $w=u-\alpha V$ is itself a C^1 solution in the closure of S_0. Let w_1 be the C^1 solution of (24) in $0\leqq r\leqq r_0$, such that $w_1=w$ on $r=r_0$. Then $w-w_1$ is bounded in S_0, is a C^1 solution except possibly at $r=0$, and therefore satisfies the extended maximum principle in S_0 (as is seen by following the procedure of § 2 with V as comparison function). This implies $w\equiv w_1$, and the final assertion of the theorem is proved.

Theorem 6. *Let the coefficients of the equation*

(30) $$Lu = 0 \qquad (c\leqq 0)$$

be uniformly Hölder continuous in $0\leqq r\leqq r_0$. Then every positive solution of (30) in S_0 has the form

(31) $$u = \mathrm{Const.}\, G + w, \quad \mathrm{Const.} \geqq 0,$$

where G is the Green's function for the sphere $0\leqq r\leqq r_0$ with singularity at $r=0$, and w is the (unique) solution of the Dirichlet problem in $0\leqq r\leqq r_0$ that coincides with u on $r=r_0$.

Proof. As in Theorem 5 we define α to be the largest number such that $u-\alpha G\geqq 0$ in S_0. (The existence of the Green's function G is guaranteed by known existence theorems for equations with Hölder continuous coefficients). We set $w \equiv u - \alpha G \geqq 0$. It follows as in Theorem 5 that

(25′) $$\underline{w} = o(\overline{G}) \qquad \text{as } r \to 0 .$$

Since the coefficients of L are supposed Hölder continuous we may apply Lemma 1c to any positive solution of $Lu = 0$. Therefore, as in Theorem 5, we may derive

(28′) $$w = o(G) \qquad \text{as } r \to 0 .$$

According to the theorem on removable singularities [7], w is a solution of (30) throughout $0 \leqq r \leqq r_0$. Since clearly $w = u$ on $r = r_0$, this completes the proof of Theorem 6.

For general values of c, not obeying $c \leqq 0$, the explicit representation (31) can no longer be obtained, at least by the present route. We can still assert, however, that any two positive solutions in S_0 are multiples of one another (up to a solution in the closure of S_0) and asymptotic to $\log 1/r$. The proof of this fact is the same as in Theorem 5.

Finally we note that certain equations, falling in the category of neither Theorem 5 nor Theorem 6, can still be treated by the preceding methods. For example, it can be shown, although the details are omitted here, that every positive solution of

$$(r^m u_i)_i = 0 , \qquad m < n - 2 ,$$

in S_0, has the form

$$u = \text{Const.}\, r^{-p} + w , \quad p = n - m - 2 ,$$

where w is continuous in $0 \leqq r \leqq r_0$.

6. Equations of divergence type.

In this section we treat elliptic equations with divergence structure,

(32)$_n$ $$Du = (a_{ik} u_i)_k = 0$$

and, more generally, equations of the form

(33)$_n$ $$Du + b_i u_i = 0 .$$

The domain of definition of the coefficients and of u will be the punctured sphere, S_0 (or $r \geqq r_0$), and unless specifically mentioned *no continuity properties of the coefficients are assumed* at $r = 0$ (or $r = \infty$). For sake of simplicity we suppose the functions u are in C^2, that the coefficients a_{ik} have bounded first derivatives in closed subdomains of

S_0 and that the coefficients b_i are bounded and measurable. The assumptions on u and a_{ik} are made to allow the formal steps in the proofs of the following theorems to proceed with most dispatch. However, it is not difficult to see, after suitable reformulation, that basically the same proofs and results hold for C^1 solutions when the coefficients a_{ik} are only bounded and measurable in S_0. We prove first the maximum principle for $(32)_n$ in S_0, an unusual feature of the result being its validity for non-uniformly as well as uniformly elliptic equations.

Theorem 7. *Let u be a non-constant function satisfying the inequality $Du \geqq 0$ in the punctured sphere S_0. In S_0 let the operator D be elliptic, but not necessarily uniformly elliptic, and assume the a_{ik} are bounded, and that* $a_{ik} = a_{ki}$. We set $M = \max_{r=r_0} u$. Then if

$$u = \begin{cases} O(|\log r|^{1-\delta}), & n = 2 \\ O(r^{2-n+\delta}), & n > 2, \quad \text{as } r \to 0, \end{cases} \tag{34}$$

for some $\delta > 0$, it follows that $u < M$ in $0 < r < r_0$, and furthermore, $\limsup_{r \to 0} u < M$.

If $n = 2$ and the domain of definition is $r \geqq r_0$, the same conclusion holds when $r = 0$ is replaced by $r = \infty$ in the hypotheses.

Proof. We shall show first that $u \leqq M$ in S_0. For suppose to the contrary that $u > M$ at some point in S_0, and let G be any open component of the set of points in which $u > M$. The function $v \equiv u - M$ is positive in G, vanishes on the boundary of G (except perhaps at $r = 0$) and is a solution of (32).

We set $\varepsilon = \delta/n$, and apply the divergence theorem to the expression $(v^\varepsilon a_{ik} v_i)_k$ in the region G_ρ consisting of the points in G for which $r > \rho$. Because of the vanishing of v on the boundary of G, we obtain

$$\int_{G_\rho} (v^\varepsilon a_{ik} v_i)_k \, dx = \int_{S_\rho} v^\varepsilon a_{ik} v_i n_k \, dS, \quad (dx = dx_1 \dots dx_n)$$

where the surface integral on the right is extended over the portion S_ρ of the spherical boundary of G_ρ on which $r = \rho$, and the n_k are the direction cosines of the outward normal. Expanding the left side and taking account of the hypothesis $Du \geqq 0$, we have

$$(35)\qquad \varepsilon \int_{G_\rho} v^{\varepsilon-1} a_{ik} v_i v_k \, dx \leq \int_{S_\rho} v^{\varepsilon} a_{ik} v_i n_k \, dS .$$

If, for some $\rho > 0$, $G_\rho = G$, then clearly the volume integral vanishes and, since its integrand is positive definite, $v \equiv 0$ in G. This is impossible, so $G_\rho \neq G$ and the singular point $r = 0$ lies on the boundary of G. The condition $v > 0$ now leads to a contradiction by the following argument.

Schwarz's inequality for symmetric, definite forms gives

$$(36)\qquad \int_{S_\rho} v^{\varepsilon} a_{ik} v_i n_k \, dS \leq \int_{S_\rho} v^{\varepsilon} (a_{ik} v_i v_k)^{1/2} (a_{ik} n_i n_k)^{1/2} \, dS$$
$$\leq K_1 \int_{S_\rho} v^{\varepsilon} (a_{ik} v_i v_k)^{1/2} \, dS ,$$

where $K_1 \geq (a_{ik} n_i n_k)^{1/2}$. Also, proceeding formally,[9]

$$(37)\qquad \int_{S_\rho} v^{\varepsilon} (a_{ik} v_i v_k)^{1/2} \, dS = \int_{S_\rho} v^{(\varepsilon+1)/2} v^{(\varepsilon-1)/2} (a_{ik} v_i v_k)^{1/2} \, dS$$
$$\leq \left(\int_{S_\rho} v^{\varepsilon+1} \, dS \int_{S_\rho} v^{\varepsilon-1} a_{ik} v_i v_k \, dS \right)^{1/2}$$

and hence, from (35), (36), (37), we obtain

$$(38)\qquad \left(\int_{G_\rho} v^{\varepsilon-1} a_{ik} v_i v_k \, dx \right)^2 \leq K_2 \int_{S_\rho} v^{\varepsilon+1} \, dS \int_{S_\rho} v^{\varepsilon-1} a_{ik} v_i v_k \, dS ,$$

where $K_2 = K_1^2 / \varepsilon^2$. Consider now the integral $\int_{S_\rho} v^{\varepsilon+1} \, dS$: From the hypothesis (34) concerning the growth of u, and from our choice of ε, we have that

$$\int_{S_\rho} v^{\varepsilon+1} \, dS = O(\rho^{(2-n+\delta)(\varepsilon+1)} \rho^{n-1}) = O(\rho^{1+\varepsilon}), \quad \text{if } n > 2 ,$$

and

$$\int_{S_\rho} v^{\varepsilon+1} \, dS = O[|\log \rho|^{(1-\delta)(\varepsilon+1)} \rho] = O(|\log \rho|^{1-\varepsilon} \rho), \quad \text{if } n = 2 .$$

Hence (38) becomes, for some constant K_3,

9. In this proof we disregard certain details of rigor, such as that necessitated by the singularity of $v^{\varepsilon-1}$ on the boundary. However, it is not difficult to modify the proof so that all the details are in order.

$$(39) \qquad \left(\int\limits_{G_\rho} v^{\varepsilon-1} a_{ik} v_i v_k \, dx\right)^2 \leq K_3 R(\rho) \int\limits_{S_\rho} v^{\varepsilon-1} a_{ik} v_i v_k \, dS ,$$

where $R(\rho) = \rho^{1+\varepsilon}$ if $n > 2$ and $R(\rho) = \rho\,|\log \rho|^{1-\varepsilon}$ if $n = 2$.
Now the integral

$$Q(\rho) = \int\limits_{G_\rho} v^{\varepsilon-1} a_{ik} v_i v_k \, dx = \int\limits_{\rho}^{\rho_0} d\rho \int\limits_{S_\rho} v^{\varepsilon-1} a_{ik} v_i v_k \, dS$$

(ρ_0 = outer radius of G_ρ) is a positive monotonically decreasing function of ρ, and

$$\frac{dQ}{d\rho} = -\int\limits_{S_\rho} v^{\varepsilon-1} a_{ik} v_i v_k \, dS .$$

Hence (39) becomes the differential inequality

$$(40) \qquad -\frac{1}{Q^2}\frac{dQ}{d\rho} \geqq \frac{1}{K_3 R(\rho)} .$$

If this is integrated from $\rho = \rho_1$ to $\rho = \rho_2$ ($\rho_1 < \rho_2 < \rho_0$), we find

$$\frac{1}{Q(\rho_2)} - \frac{1}{Q(\rho_1)} \geqq K_4(\rho_1^{-\varepsilon} - \rho_2^{-2}) \qquad \text{if } n > 2 ,$$

or

$$\geqq K_4(|\log \rho_1|^\varepsilon - |\log \rho_2|^\varepsilon) \qquad \text{if } n = 2 ,$$

where $K_4 = (\varepsilon K_3)^{-1}$. If now $\rho_1 \to 0$ and ρ_2 remains fixed, the right member becomes infinite while the left is bounded by $1/Q(\rho_2)$. This contradiction implies that the initial assumption, namely $u > M$ at some point in S_0, is untenable. Hence $u \leqq M$ in S_0.

To infer the sharper conclusions, $u < M$ in $0 < r < r_0$ and $\limsup\limits_{r \to 0} < M$, we observe that under our assumptions concerning the coefficients, $Du \geqq 0$ can be written as

$$Du = a_{ik} u_{ik} + a_{ik,k} u_i \geqq 0 \qquad (a_{ik,k} = \partial a_{ik}/\partial x_k),$$

where the $a_{ik,k}$ are bounded in closed subregions of S_0. Since Du is elliptic in S_0 it is furthermore uniformly elliptic in closed subregions. Consequently u satisfies the strong form of the maximum principle in subregions of S_0. From this fact and the previously proved inequality $u \leqq M$ follow the desired conclusions, $u < M$ in $0 < r < r_0$ and $\limsup\limits_{r \to 0} u < M$. The proof of the final conclusion of the theorem (the maximum principle for exterior domains when $n = 2$) requires only trivial changes in the preceding arguments.

Remarks: 1) The use of differential inequalities and similar methods to obtain estimates concerning uniformly elliptic equations have usually been based on the Dirichlet integral (or slight modifications of it). However, the above proof succeeds in treating the non-uniformly elliptic case precisely because of the use (in Q) of the quadratic form $a_{ik} v_i v_k$, which may become semi-definite at $r = 0$, rather than the customary Σv_i^2 of the Dirichlet integral. This device is due to R. Finn. (10)

2) It is clear from the proof that the conclusion $u \leqq M$ (but not necessarily $u < M$) holds even when the elliptic operator D becomes parabolic at a set of measure zero in S_0.

3) If the coefficient matrix a_{ik} is not symmetric, but instead is uniformly elliptic, then the same proof applies. In this case inequality (36) is obtained from:

$$a_{ik} v_i n_k \leqq [(\sum_{i,k} a_{ik} n_k)^2]^{1/2} (\Sigma v_i^2)^{1/2} \leqq K'_1 (\Sigma v_i^2)^{1/2} \leqq K''_1 (a_{ik} v_i v_k)^{1/2},$$

where $K''_1 = K'_1 \lambda^{1/2}$, and λ is the ellipticity modulus. Also, the hypothesis of uniform ellipticity allows one to obtain an extended maximum principle for equations $(33)_n$ with first order terms.

Theorem 7 has as immediate consequence the following analogue of Bernstein's theorem [1].

Let u be a bounded solution in the entire plane of the equation

$$(a u_x + b u_y)_x + (b u_x + c u_y)_y = 0, \tag{41}$$

or, more generally, of the system,

$$\begin{aligned} v_y &= a u_x + b u_y, \\ -v_x &= b u_x + c u_y, \end{aligned} \tag{42}$$

where $ac - b^2 > 0$, and a, b, c are bounded and measurable. Then u is a constant. (11)

Proof. Since the strong maximum principle holds for these equations, even under the weak assumptions on the coefficients, (cf. [12], [2]), a non-

10. See [5], which states several results based on this idea, although the details are not included.

11. Essentially the same result is stated by Finn [5], and is mentioned here for completeness.

constant u cannot have its maximum or minimum at a finite point, and the result follows from Theorem 7. (A slight modification in the first steps of the proof is required to take care of the system (42) rather than the single equation (41)). Of course, the theorem holds as well when $u = O(\log r)^{1-\delta}$ as $r \to \infty$.

Theorem 7 is an example of the stronger type of result effected by the divergence structure. As example 1 of § 7 illustrates, this result does not hold for equations of the form $(1)_2$ even when the equation is uniformly elliptic with continous coefficients at the singular point. However, in the latter case a bounded solution at least has a limit at the singular point, whereas for non-uniformly elliptic equations of divergence type Theorem 7 cannot be thus sharpened to yield a limit at the point; (cf. Example 2 of § 7). To obtain stronger results we must therefore confine attention to uniformly elliptic equations. *In the following the operator D will be assumed uniformly elliptic without further mention.*

Let us first consider (33) when $n = 2$. The following theorems are analogues of Theorems 2 (for $n = 2$) and 5.

Theorem 8. *Let $u(x)$ be a solution of the equation*

$$(33)_2 \qquad Du + b_i u_i = 0,$$

in the punctured disc, S_0: $0 < r \leqq r_0$, and suppose $u \geqq 0$. Then u tends to a limit (possibly $+\infty$) as $r \to 0$, and if the limit is finite, it is approached Hölder continuously. The same conclusion is true as $r \to \infty$ if the domain of definition is $r \geqq r_0$, and if $b_i = O(1/r)$.

Theorem 9. *Let the equation $(33)_2$ possess a solution $V(x)$ in the punctured disc, $S_0: 0 < r \leqq r_0$, with the property that $V \to \infty$ as $r \to 0$. Then every positive solution u of $(33)_2$ in S_0 has the form*

$$u = \text{const.}\, V + w,$$

where const. $\geqq 0$, and w is a solution in S_0, Hölder continuous at $r = 0$.

Proof. The proof of Theorem 8 is patterned exactly after that of Theorem 2 ($n = 2$), except that certain simplifications can be made since now $c = 0$. We recall that the essential step in the proof rests on the use of Lemma 1a, this lemma being in turn a direct consequence of the Harnack inequality 1. For equations of the form $(33)_2$ such a Harnack inequality has been proved (by elementary means) by Bers and Nirenberg [3], and hence an analogue of Lemma 1a is true also for these equations. The proof of the existence of a limit now follows as in Theorem 2. That the limit, if finite, is approached Hölder continuously is proved by a slight modification of the type of estimates used by Morrey [12] and Nirenberg [14] in establishing Hölder continuity in other connections.[12]

The proof of the theorem for $r \to \infty$ proceeds in the same way as for finite points. The argument proving Hölder continuity is essentially the same at infinity as at finite points when $n = 2$.

Turning now to Theorem 9, we need only observe that since Lemma 1a holds for $(33)_2$ the proof follows that of Theorem 5, with only a few changes (simplifications) due to the fact that now $c \equiv 0$.

With regard to limit theorems for $n > 2$, the situation is somewhat similar for both $(1)_n$ and $(32)_n$ in that the existing proofs require some continuity of the coefficients at the singular point, as well as uniform ellipticity. We mention the following result of L. Nirenberg,[13] which bears on this problem.

Let the coefficients of the uniformly elliptic equation,

$$(32)_n \qquad Du = (a_{ik} u_i)_k = 0\,, \qquad n > 2\,,$$

be continuous at $r = 0$, and let u be a bounded solution in $0 < r \leqq r_0$. Then u approaches a limit Hölder continuously at $r = 0$ (with any Hölder exponent < 1). If u is a bounded solution in $r \geqq r_0$ and the coefficients of D approach limits at infinity, then $u \to u_0$ as $r \to \infty$, and $|u - u_0| = O(r^{-\alpha})$, where $\alpha > \frac{n}{2} - 1$.

12. These references on Hölder continuity do not prove *as such* that a bounded (or continuous) solution of $(33)_2$ must be Hölder continuous at a singular point, but a simple adaptation of their estimates to an annulus enclosing the point shows first that the Dirichlet integral of u over any disc containing the point as center is finite, and then that the Dirichlet integral has a Hölder growth with respect to the radius of the disc. From this the desired Holder continuity of the solution follows at once by Morrey's lemma.

13. To appear in a subsequent publication.

The same theorem holds if the solution is allowed certain growth properties, specifically, $u = O[r^{(1-n+\delta)/2}]$ as $r \to 0$, and $u = O(r^{\alpha})$, $\alpha < \frac{1}{2}$, as $r \to \infty$.

7. Examples.

In this section we discuss two examples bearing on the sharpness of some of the preceding theorems.

1. Consider the equation,

$$Lu = A_{ik} u_{ik} = 0\,, \tag{43$_n$}$$

where

$$A_{ik} = \delta_{ik} + g(r)\frac{x_i x_k}{r^2}\,, \qquad \delta_{ik} = \begin{cases} 1, & i = k\,, \\ 0, & i \neq k\,. \end{cases} \tag{44}$$

A simple computation shows that for every $g(r)$ this equation has a solution $u = u(r)$ satisfying the ordinary differential equation,

$$\frac{u''}{u'} = \frac{1-n}{r(1+g)}\,, \tag{45}$$

and hence given by the formula,

$$u(r) = \int^{r} \left[\exp\left(\int^{r} \frac{1-n}{1+g(\rho)}\frac{d\rho}{\rho}\right)\right] dr\,. \tag{46}$$

Since

$$\operatorname{Min}(1\,, 1+g)\,\Sigma\xi_i^2 \leqq A_{ik}\xi_i\xi_k \leqq \operatorname{Max}(1\,, 1+g)\,\Sigma\xi_i^2\,,$$

the equation is uniformly elliptic with modulus λ in any region in which

$$\lambda^{-1} \leqq 1 + g \leqq \lambda\,, \qquad \lambda = \text{Const.} > 0\,. \tag{47}$$

When $n = 2$, (43) becomes

$$(1 + g(r)\cos^2\vartheta)\,u_{xx} + 2g(r)\sin\vartheta\cos\vartheta\, u_{xy} + (1 + g(r)\sin^2\vartheta)\,u_{yy} = 0\,, \tag{48}$$

and a solution $u = u(r)$, corresponding to

$$g(r) = -\frac{2}{2 + \log r}\,, \qquad r > 0\,; \qquad g(0) = 0\,, \tag{49}$$

is

$$u(r) = a + b/\log r\,. \tag{50}$$

Since $g \to 0$ as $r \to 0$, it is evident that (48) is uniformly elliptic near $r = 0$; in particular, (47) is satisfied for $\lambda = 3$ in the circle $0 \leqq r \leqq e^{-3}$. The coefficients of (48) are analytic in this circle, except at $r = 0$ where they are continuous. Thus for $a=1$, $b=3$ in (50), $u=1+3/\log r$ is a solution of the uniformly elliptic equation (48), (49) in $0<r\leqq e^{-3}$; this solution is singular at the origin, and has the further properties that it vanishes on $r=e^{-3}$, is positive for $0<r<e^{-3}$, and has the limit value 1 as $r\to 0$.

This example illustrates the following general remarks: 1) If only continuity is demanded of the coefficients, the maximum principle (in the sense of § 2) does not hold, even for bounded solutions; for uniformly elliptic equations of the form $(43)_2$. 2) A bounded solution of such an equation, although it will always have a limit at the singular point (Theorem 2), need not be Hölder or Dini continuous at the point. 3) Equations $(43)_2$ with only continuous coefficients will not in general possess a fundamental solution, or any positively infinite solution; for the existence of such a solution implies (by the methods of § 2) that bounded solutions satisfy the maximum principle. 4) The theorem on removable singularities does not hold for equations $(43)_2$ with only continuous coefficients. 5) If the domain of definition in $(43)_2$ is $r\geqq r_0$ and the singular point is taken at infinity, the remarks corresponding to 1) and 2) hold equally well, since the same example works for exterior domains, say in $r\geqq 2$.

By inserting into (44) other functions g varying in modulus of continuity from that of (49) to a Dini modulus, one can obtain a variety of possible solutions $u(r)$ with different orders of growth at $r=0$.[14] The function $g(r)=\dfrac{2}{\log r-2}$, for example, gives rise to the solution $u=\log^3(1/r)$ which tends to ∞ as $r\to 0$ but not in the "expected" manner ($\sim\log 1/r$).

For $n>2$, let g be the function

$$g(r)=-\frac{1}{1+(n-1)\log r}, \qquad r>0, \qquad g(0)=0.$$

Evidently $(43)_n$ is uniformly elliptic in $0\leqq r\leqq e^{-1}$, and has continuous coefficients at $r=0$. The solution given by (46) is, within a constant factor,

$$u(r)=\frac{r^{2-n}}{\log r}(1+\varepsilon(r)), \qquad \varepsilon(r)\to 0 \quad \text{as} \quad r\to 0.$$

14. The possible growths at $r=0$ are easily studied by inserting solutions into (45) and then computing the corresponding g.

Thus $(43)_n$ has a solution that is $o(r^{2-n})$ at $r=0$, but at the same time is not $o(r^{2-n+\delta})$ for any $\delta>0$. From this it is apparent that Theorem 1′ cannot be essentially sharpened, and that a fundamental solution of growth r^{2-n} does not in general exist for equations $(43)_n$ having only continuous coefficients. Finally if

$$g=\frac{(n-2)\log r-2}{\log r+2}, \quad r>0, \quad g(0)=n-2,$$

then exactly as in the first special case noted,

$$(1-n)/(1+g)=-1-2/\log r,$$

and again the solution given by (46) is

$$u=a+b/\log r.$$

Thus even for $n>2$ it is possible for a bounded solution to have an isolated singularity where the maximum principle does not hold. In this example (43) is uniformly elliptic in $0\leq r\leq e^{-3}$ but the coefficients A_{ik} are discontinuous at $r=0$.

The same example, when considered for large r, shows that a uniformly elliptic equation with discontinuous coefficients at infinity may have a solution which approaches a limit at infinity with less than a Dini modulus. This is in contrast with the result stated in Theorem 3′ for continuous coefficients and in the remark following Theorem 3′ for coefficients having sufficiently small oscillation at infinity.

It is of interest to see what solutions of the form $u(r)$ result from inserting the coefficients (44) into equations of divergence type. It is again a simple computation to check that, corresponding to each $g(r)$, the equation

$$(51) \qquad (A_{ik}u_i)_k=0, \quad A_{ik}=\delta_{ik}+g(r)\frac{x_i x_k}{r^2},$$

has a solution

$$u(r)=\int^{r}\frac{1}{1+g(\rho)}\frac{d\rho}{\rho^{n-1}}.$$

It is clear that any $g(r)$, whether continuous or not at $r=0$, and satisfying (47), yields solutions of the form

$$\log r(1+\varepsilon(r)), \quad n=2, \quad \text{and} \quad r^{2-n}(1+\varepsilon(r)), \quad n>2,$$

where $\varepsilon(r)\to 0$ as $r\to 0$. Thus the solutions have the same order of growth as the expected fundamental solutions, and do not reveal any unusual behavior on the part of the divergence equation (51), in contrast with (43).

2. The following example appears in Finn [6], p. 394. In the equation

$$(\rho u_x)_x + (\rho u_y)_y = 0\,, \tag{52}$$

let

$$\rho(x\,,y) = r^{(1-\alpha^2)} \exp\left(-r^{-\alpha}/\alpha^2\right), \qquad \alpha > 0\,.$$

For this value of ρ the equation (52) is elliptic, except at $r = 0$ where $\rho \to 0$, and satisfies the hypothesis of Theorem 7. A solution in $0 < r < \infty$ is

$$u = (1 + r^{\alpha}) \sin \vartheta\,.$$

This function is bounded in the finite part of the plane but has no limit at $r = 0$, thereby demonstrating that Theorem 8 cannot in general be sharpened to provide a limit at the singular point when the equation is non-uniformly elliptic.

REFERENCES

1. S. Bernstein, Ueber ein geometrisches Theorem und seine Anwendung auf die partiellen Differentialgleichungen vom elliptischen Typus, *Math. Zeit.* 26 (1927), pp. 551—558.

2. L. Bers and L. Nirenberg, On a representation theorem for linear elliptic systems with discontinuous coefficients and its applications, *Convegno Internazionale sulle Equazioni Derivate Parziali*, 1954, pp. 111—140.

3. L. Bers and L. Nirenberg, On linear and non-linear boundary value problems in the plane, same as preceding, pp. 141—167.

4. M. Bôcher, Singular points of functious which satisfy partial differential equations of elliptic type, *Bull. Amer. Math. Soc.* 9 (1903), pp. 455—465.

5. R. Finn, Sur quelques généralisations du théorème de Picard, *C. R. Acad. Sci.* Paris 235 (1952), pp. 596—598.

6. R. Finn, Isolated singularities of solutions of non-linear partial differential equations, *Trans. Amer. Math. Soc.* 75 (1953), pp. 385—404.

7. M. Gevrey, Sur certaines propriétés des fonctions harmoniques et leur extension aux équations aux dérivées partielles, *C. R. Acad. Sci.* Paris 183 (1926), pp. 546—548.

8. M. Gevrey, Sur une généralisation du principe des singularités positives de M. Picard, *C. R. Acad. Sci.* Paris 211 (1940), pp. 581—584.

9. P. Hartman and A. Wintner, On the local behavior of solutions of non-parabolic partial differential equations, I, II, III, *Amer. J. of Math.* 75 (1953), pp. 449—476; 76 (1954), pp. 351—361; 77 (1955), pp. 453—474.

10. E. Hopf, Elementare Betrachtungen ueber die Lösungen partieller Differentialgleichungen zweiter Ordnung vom elliptischen Typus, *Sitzungsberichte Preuss. Akad. Wiss.* 19 (1927), pp. 147—152.

11. E. H o p f, Bemerkungen zu einem Satze von S. Bernstein aus der Theorie der elliptischen Differentialgleichungen, *Math. Zeit.* 29 (1928), pp. 744—745.

12. C. B. M o r r e y, On the solutions of quasi-linear elliptic partial differential equations, *Trans. Amer. Math. Soc.* 43 (1938), pp. 126—166.

13. C. B. M o r r e y, Second order elliptic systems of differential equations, in Contributions to the theory of partial differential equations, *Ann. of Math. Studies* No. 33, Princeton, 1954, pp. 101—159.

14. L. N i r e n b e r g, On non-linear elliptic partial differential equations and Holder continuity, *Comm. on pure and appl. Math.* 6 (1953), pp. 103—156.

15. J. S e r r i n, On the Harnack inequality for linear elliptic equations, *J. d'Analyse Math.* (preceding in this issue).

Indiana University and
University of Minnesota

(Received January 6, 1956)

Arch. Ration. Mech. Anal. 20 (1965), 163–169.

Reprint I.5.[ii]

Removable Singularities of Solutions of Elliptic Equations. II

JAMES SERRIN

It is well known that solutions of the minimal surface equation cannot possess isolated non-removable singularities. This result, first proved by BERS, has since been shown to hold for several large classes of elliptic partial differential equations, namely those of minimal surface type and those arising from a bounded conservation law depending on the gradient of the dependent variable. A different direction in which BERS' result can be generalized was indicated in a recent paper of mine [9, p. 288], the question there being raised as to how large a singular set can be allowed in the removable singularity theorem. Although my results were not quite sharp, they did lead to the conjecture that the single isolated singular point of BERS' theorem could be replaced by singularities occupying a set of vanishing $(n-1)$-dimensional Hausdorff measure. Recently NITSCHE (for two independent variables) and somewhat later DE GIORGI & STAMPACCHIA (for $n \geqq 2$ independent variables) proved that this is indeed the case. More precisely, they obtained the following result: *if u is a solution of the minimal surface equation defined in a domain $D-Q$, where D is open and Q is a compact subset of D having $(n-1)$-dimensional Hausdorff measure zero, then u can be given values on Q so that the resulting function is a solution of the minimal surface equation in all of D.* NITSCHE in fact observed that when $n=2$ the singularities can be allowed to extend to the boundary of D; *cf.* the concluding remarks of Section 2.

In this paper we shall prove the same result for every elliptic equation arising from a bounded conservation law in which only the gradient of the dependent variable appears explicitly. More generally, we shall consider quasi-linear equations which can be written in the divergence form

$$\operatorname{div} \mathscr{A}(x, u, u_x) = \mathscr{B}(x, u, u_x). \tag{1}$$

Here $\mathscr{A}$ is a given continuous vector function of the variables x, u, u_x, $\mathscr{B}$ is a given scalar function of the same variables, and u_x denotes the gradient of the dependent variable $u = u(x) = u(x_1, \ldots, x_n)$. The functions $\mathscr{A}(x, u, p)$ and $\mathscr{B}(x, u, p)$ are required to satisfy the conditions

$$|\mathscr{A}(x, u, p)| \leqq a \tag{2}$$

and

$$|\mathscr{B}(x, u, p)| \leqq f, \tag{3}$$

where a is a positive constant and f is a measurable function in the Lebesgue class L_n. Conditions of monotonicity and ellipticity will also be imposed, but it is convenient to defer their statement until later. (It may be observed in passing that the assumption $a = \text{const} > 0$ can be replaced by the weaker condition that a belong to a suitable Lebesgue class, provided the singular set is at the same time reduced in size.)

Our specific conclusions are stated in Theorems 1, 2, and 3. Theorem 3 is concerned with the general equation (1), while Theorems 1 and 2 deal with the special case

$$\operatorname{div} \mathscr{A}(u_x) = \mathscr{B}(u), \tag{4}$$

where stronger results can be obtained. These theorems apply particularly to the non-parametric Euler-Lagrange equation of any regular parametric variational problem. More specifically, both the minimal surface equation and the equation for surfaces of constant mean curvature are covered by the result of Theorem 2. For the minimal surface equation this yields precisely the results of NITSCHE and DE GIORGI & STAMPACCHIA, while for surfaces of constant mean curvature we obtain the following interesting geometric conclusion:

Let S be a surface of class C^2 defined over a domain $D-Q$ in E^2, where D is open and Q is a compact set of 1-dimensional Hausdorff measure zero. Suppose the mean curvature of S is constant. Then S can be continuously extended to a surface of constant mean curvature over all of D.

Our results include the removable singularity theorems of references [1], [2], [3], [4], and [8]. The boundedness theorems of [5], [6] are not covered, however, since the equations treated there cannot necessarily be written in divergence form with a bounded conservation law. To conclude the introduction, it is worth noting that inequalities (2), (3) arise by setting $\alpha = 1$ in the general conditions of reference [9]. In fact, with the help of Theorem 3 of this paper, and with obvious changes in Theorems 1, 3, and 4 of [9], the general theory given there carries over to the case $\alpha = 1$ in a smooth, continuous way. The only exception to this rule is the Harnack inequality (Theorem 5 of [9]), whose form when $\alpha = 1$ is known only in certain special cases.

1. The special equation div $\mathscr{A}(u_x) = \mathscr{B}(u)$

In this section, as well as in the remaining parts of the paper, a function u will be called a *solution* of equation (1) in a domain $\mathscr{D}$ if it is strongly differentiable* in D and if

$$\int (\varphi_x \cdot \mathscr{A} + \varphi \mathscr{B})\, dx = 0 \tag{5}$$

for any continuously differentiable function $\varphi = \varphi(x)$ with compact support in $\mathscr{D}$. If u is also continuous in D, it will be called a *continuous solution*, and so forth. In view of assumption (2), it is evident that if u is a solution of (1), then (5) holds not only for continuously differentiable functions φ, but in fact for any strongly differentiable function φ with compact support in $\mathscr{D}$.

* A function u is called strongly differentiable in a domain $\mathscr{D}$ if it is locally summable in $\mathscr{D}$ and has distribution derivatives u_x which are themselves locally summable in $\mathscr{D}$.

In the following theorem we shall be concerned with the special case of equation (1) in which $\mathscr{A}=\mathscr{A}(p)$, $\mathscr{B}=\mathscr{B}(u)$, and

$$(6)\qquad \begin{aligned}(p-q)\cdot\{\mathscr{A}(p)-\mathscr{A}(q)\}&>0 \qquad p\neq q,\\ (u-v)\cdot\{\mathscr{B}(u)-\mathscr{B}(v)\}&\geqq 0 \qquad u\neq v.\end{aligned}$$

Condition (6) implies that there exists a constant $\kappa>0$ such that

$$(7)\qquad p\cdot\{\mathscr{A}(p)-\mathscr{A}(0)\}\geqq\kappa(|p|-1).$$

It is clearly enough to prove this for $|p|>1$. For any real number ϑ, $0<\vartheta<1$, we have

$$(p-\vartheta p)\cdot\{\mathscr{A}(p)-\mathscr{A}(\vartheta p)\}>0;$$

thus follows easily

$$p\cdot\{\mathscr{A}(p)-\mathscr{A}(0)\}>p\cdot\{\mathscr{A}(\vartheta p)-\mathscr{A}(0)\}.$$

Setting $\vartheta=|p|^{-1}$ and $\vec{n}=p/|p|$ then yields

$$p\cdot\{\mathscr{A}(p)-\mathscr{A}(0)\}\geqq\kappa\,|p|$$

where $\kappa=\min\vec{n}\cdot\{\mathscr{A}(\vec{n})-\mathscr{A}(0)\}$. From (6) and the fact that $\mathscr{A}$ is continuous, it is evident that $\kappa>0$, completing the proof.

Theorem 1. *Let u be a continuous solution of the equation*

$$(8)\qquad \operatorname{div}\mathscr{A}(u_x)=\mathscr{B}(u)$$

in a domain $D-Q$, where D is open and Q is a compact subset of D having $(n-1)$-dimensional Hausdorff measure zero. Suppose that (6) *holds, and furthermore that*

$$(9)\qquad |\mathscr{A}(u_x)|\leqq a,\qquad |\mathscr{B}(u)|\leqq f$$

where a is a positive constant, and f is a measurable function of class L_n in D. Then u can be defined on Q so that the resulting function is a continuous solution of (9) *in all of D.*

Remarks. Condition (9) calls for some explanation and clarification. Specifically, these inequalities are to be understood in the sense that for the particular solution $u=u(x)$ under consideration, the functions $\mathscr{A}(u_x)$ and $\mathscr{B}(u)$ are respectively bounded by a constant and by a function of class L_n, or, what comes to the same thing, that $\mathscr{A}(u_x)$ is in L_∞ and $\mathscr{B}(u)$ is in L_n. This is of course slightly weaker than conditions (2) and (3), but should cause no confusion.

It should be emphasized that (9) may hold independently of the particular solution considered, simply as a consequence of the structure of $\mathscr{A}(p)$ and $\mathscr{B}(u)$. This is the most striking case of the theorem in that it places *no* a priori restriction on the behavior of the solution near the singular set. The minimal surface equation is a case in point, for here

$$\mathscr{A}(p)=p/\sqrt{1+|p|^2}$$

and this expression is always less than one in magnitude. Another related example is that of surfaces of constant mean curvature, where

$$\mathscr{A}(p)=p/\sqrt{1+|p|^2},\qquad \mathscr{B}(u)=2H=\text{constant}.$$

12*

In any case, it is the boundedness of the expression $\mathscr{A}(u_x)$, rather than the particular behavior of the solution near the singular set, which is the relevant consideration.

Proof of Theorem 1. We may clearly assume that u is continuous on the boundary of D, and that D itself is bounded. Moreover, by subtracting $\mathscr{A}(0)$ from the function $\mathscr{A}(p)$, we can suppose without loss of generality that $\mathscr{A}(0)=0$.

Now let M be the maximum value of u on the boundary of D. Let k be a non-negative number greater than M, whose particular value will be determined later. Also choose any number $l>k$. We define the new function*

$$\bar{u}=\begin{cases} k & \text{if} \quad u\leqq k, \\ u & \text{if} \quad k<u<l, \\ l & \text{if} \quad u\geqq l, \end{cases}$$

and set

$$\varphi=(\bar{u}-k)\bar{\eta},$$

where $\bar{\eta}=\bar{\eta}(x)$ is a smooth function vanishing in a neighborhood of the singular set Q and such that $0\leqq\bar{\eta}\leqq 1$ elsewhere. Because of our choice of k, it is apparent that φ has compact support in $D-Q$. We observe, moreover, that φ is non-zero only in the set E where $u>k$, and that φ is strongly differentiable.

Using (7), (9), and the fact that $u_x=\bar{u}_x$ almost everywhere in the set where $\bar{u}_x\neq 0$, we find

$$\varphi_x\cdot\mathscr{A}(u_x)+\varphi\,\mathscr{B}(u)\geqq\bar{\eta}\,\kappa(|\bar{u}_x|-1)-a\,\bar{\eta}_x(\bar{u}-k)-f\,\bar{\eta}(\bar{u}-k), \qquad \text{a.e.} \tag{10}$$

Thus from (5) follows (since φ has compact support in D)

$$\kappa\,\|\bar{\eta}\,\bar{u}_x\|_1\leqq\kappa\,|D|+a(l-k)\,\|\bar{\eta}_x\|_1+\|f\|_n\,\|\bar{\eta}(\bar{u}-k)\|_{n/(n-1)},$$

where $\|\cdot\|_p$ denotes the L_p norm over the set D and where HÖLDER's inequality was applied to the final term. Moreover, the norm $\|f\|_n$ clearly need be taken only over the set E, where $\bar{u}-k\neq 0$. Now by SOBOLEV's inequality

$$\|\bar{\eta}(\bar{u}-k)\|_{n/(n-1)}\leqq K\{\|\bar{\eta}\,\bar{u}_x\|_1+(l-k)\,\|\bar{\eta}_x\|_1\},$$

the constant K depending only on n. If k is chosen large enough, then $K\,\|f\|_n\leqq\frac{1}{2}\,\kappa$. With this choice of k we easily obtain from the two preceding display lines

$$\|\bar{\eta}(\bar{u}-k)\|_{n/(n-1)}\leqq 2K(l-k)(a\,\kappa^{-1}+1)\,\|\bar{\eta}_x\|_1+2K\,|D|. \tag{11}$$

Since Q is of $(n-1)$-dimensional Hausdorff measure zero, it is also of 1-capacity zero (*cf.* [9], Section 7, or [11]). Setting $\bar{\eta}=\bar{\eta}^{(\nu)}$ as in [9], Lemma 8, we have $\bar{\eta}^{(\nu)}\to 1$ almost everywhere, while $\|\bar{\eta}_x^{(\nu)}\|_1\to 0$. Thus (11) implies

$$\|\bar{u}-k\|_{n/(n-1)}\leqq 2K\,|D|.$$

Finally, letting l tend to infinity, we obtain

$$\|u\|_{n/(n-1),E}\leqq 2K\,|D|+k\,|E|^{(n-1)/n}.$$

* The double truncation of u to form the function $\bar{u}$ is fundamental to the proof. This important technique is essentially due to R. FINN [3].

In the same way, one can prove that the negative part of u has bounded norm. Thus, in particular, u must be contained in the Lebesgue class $L_{n/(n-1)}(D-Q)$.

We can now apply Theorem 10′ of [9]*. It follows that u can be defined on Q so that the resulting function is a solution of (8) in all of D; at this stage, in fact, any specification of u on the set Q is allowable, inasmuch as we are not yet dealing with a continuous solution in D. To complete the proof, then, it must be shown that the values of u on Q can be assigned so that u is *continuous* in D.

Now under the hypothesis (6) the difference of two solutions of (8) satisfies a maximum principle. Moreover, since $\mathscr{A}$ und $\mathscr{B}$ are independent of x, both $u(x)$ and $u(x+h)$ are solutions of (8). It follows that for suitably small h

$$|u(x+h)-u(x)| \leqq \omega(h), \tag{12}$$

where $\omega(t)$ is a modulus of continuity for the solution on the boundary of D (recall that u was assumed to be continuous there). Actually, since we are here dealing with a *weak* solution in D, the estimate (12) holds only at almost all points x in D. A modification of u on a set of measure zero, however, establishes (12) for all x in D; since this modification cannot affect the values of u on $D-Q$, where it is already continuous, we see that by suitably assigning values to u on Q, it becomes a continuous solution of (8) in all of D. This completes the proof.

If we slightly strengthen the hypotheses on the functions $\mathscr{A}$ and $\mathscr{B}$, it is possible to show that the extension of u is necessarily of class C^2 and that u becomes a classical solution in all of D. In particular, suppose $\mathscr{A}$ and $\mathscr{B}$ are both of class C^2, with $\mathscr{B}_u \geqq 0$ and

$$\mathscr{A}_{i,k}\,\xi_i\,\xi_k > 0 \qquad (\xi \neq 0)$$

where $\mathscr{B}_u = \partial\mathscr{B}/\partial u$ and $\mathscr{A}_{i,k} = \partial\mathscr{A}_i/\partial p_k$. Then the following result holds.

Theorem 2. *Let u be a twice continuously differentiable solution of the equation*

$$\operatorname{div} \mathscr{A}(u_x) = \mathscr{B}(u)$$

in a domain $D-Q$, where D is open and Q is a compact subset of D of $(n-1)$-dimensional Hausdorff measure zero. Suppose that the functions $\mathscr{A}$ and $\mathscr{B}$ satisfy the conditions just stated, and also that

$$|\mathscr{A}(u_x)| \leqq a, \qquad |\mathscr{B}(u)| \leqq f,$$

where a is a positive constant, and f is a measurable function of class L_n in D. Then u can be defined on Q so that the resulting function is a twice continuously differentiable solution in all of D.

Proof. The hypotheses on $\mathscr{A}$ and $\mathscr{B}$ obviously imply (6). Therefore by Theorem 1, u can be extended to be a continuous solution of (8) in all of D. Since u is continuously differentiable in $D-Q$, inequality (12) shows that the resulting

* The hypotheses of Theorem 10′ are slightly stronger than those in the present case. In particular it was there assumed that $|\mathscr{B}| \leqq f$, with f in $L_{n+\varepsilon}$, while here we have only $f \in L_n$. Theorem 10′ remains valid under the weaker assumption $f \in L_n$, however, as follows by modifying the proof so that the norm $\|f\|_n$ is suitably small when taken over the support of η. Finally, the fact that our hypothesis (9) is slightly weaker than (2) makes no essential difference in the applicability of Theorem 10′.

function, which we shall also call u, satisfies a uniform Lipschitz condition in D. (We assume as in the proof of Theorem 1 that D is bounded, and that u is smooth on the boundary of D.) From this result, and the fact that $\mathscr{A}$ is of class C^2, it follows that

$$\lambda|\xi|^2 \leqq \mathscr{A}_{i,k}(u_x)\,\xi_i\,\xi_k \leqq \Lambda|\xi|^2 \qquad \text{a.e.}$$

for appropriate positive constants λ and Λ. This being the case, standard estimates of LADYZHENSKAYA & URALTSEVA, or MORREY, imply that u must be of class $C^{1+\alpha}$ in D. Consequently (using the standard theory of linear equations and the fact that both $\mathscr{A}$ and $\mathscr{B}$ are of class C^2) the gradient u_x must be of class $C^{1+\alpha}$ in D. Hence $u \in C^{2+\alpha}$, and the proof is completed.

Remark. The hypothesis that u is twice differentiable in $D-Q$ is stronger than necessary for the conclusion to hold. In particular, according to the argument just given, if u is even Lipschitz continuous in $D-Q$, then it is twice differentiable there.

2. The equation div $\mathscr{A}(x, u, u_x)=\mathscr{B}(x, u, u_x)$

Here we shall treat the general equation div $\mathscr{A}(x, u, u_x)=\mathscr{B}(x, u, u_x)$ under the hypotheses (2) and (3). The monotonicity and ellipticity conditions of Section 1 are here to be replaced by the assumption

$$(13) \qquad p\cdot\mathscr{A}(x, u, p) \geqq |p| - d\,|u| - g$$

where d and f are measurable functions in $L_{n+\varepsilon}$ [note that (13) neither implies nor is implied by ellipticity of (1)]. With the present assumptions our conclusions are somewhat weaker than before, in that we are unable to guarantee a *continuous* extension of u to the singular set Q. Nevertheless, the resulting function remains a (weak) solution in all of D.

Theorem 3. *Let u be a solution of the equation*

$$\operatorname{div}\mathscr{A}(x, u, u_x)=\mathscr{B}(x, u, u_x)$$

in a domain $D-Q$, where D is open and Q is a compact subset of D of $(n-1)$-dimensional Hausdorff measure zero. Suppose that conditions (2), (3), *and* (13) *hold. Then u can be defined on Q so that the resulting function is a solution in all of D.*

Proof. As in the proof of Theorem 1, it will be enough to show that u is in the Lebesgue class $L_{n/(n-1)}(D-Q)$.

We begin by noting that u must be bounded on any compact subset of $D-Q$, this following exactly as in the proof of Theorem 1 of reference [*9*]. We can therefore suppose without loss of generality that D is bounded, and that u is uniformly bounded in some neighborhood of the boundary of D. Let M be an upper bound for u in the neighborhood of the boundary of D. Then proceeding exactly as in the proof of Theorem 1, we find in place of (10)

$$\varphi_x\cdot\mathscr{A}+\varphi\,\mathscr{B} \geqq \bar\eta(|\bar u_x|-d\,|\bar u|-g)-a\,\bar\eta_x(\bar u-k)-(\bar u-k)\,\bar\eta\,f, \qquad \text{a.e.}$$

Writing

$$|\bar u|=(\bar u-k)+k$$

and using (5), we find easily

$$\|\bar{\eta}\,\bar{u}_x\|_1 \leqq \|\bar{\eta}(k\,d+g)\|_1 + a(l-k)\,\|\bar{\eta}_x\|_1 + \|d+f\|_n\,\|\bar{\eta}(\bar{u}-k)\|_{n/(n-1)}$$

where the norm of $d+f$ need be taken only over the set E where $u>k$. Then, following the steps in the proof of Theorem 1, we are led to

$$\|\bar{u}-k\|_{n/(n-1)} \leqq 2K\,\|k\,d+g\|_1,$$

provided k is chosen so large that $2K\,\|d+f\|_{n,E} \leqq 1$. The rest of the proof is the same as Theorem 1, with the exception that the final paragraph of that proof can be omitted.

Concluding remarks. In two dimensions, the hypothesis that Q is a compact subset of D is unnecessary. This follows from the fact (noted by NITSCHE, *cf.* [8]) that any point of D can be enclosed by a smooth curve in D disjoint from Q. A similar construction need not hold in more than two dimensions, as one sees easily by taking Q to be a one dimensional continuum extending to the boundary of D. Whether our results hold in more than two dimensions without the condition that Q is a compact subset of D remains an open question. It may be remarked, however, that if u is contained in *any* Lebesgue class L_δ over $D-Q$, where $\delta>0$, then Q is removable even if it extends to the boundary of D, *cf.* [9], Theorem 10′.

I wish finally to thank Professor JOHANNES C. C. NITSCHE for several very stimulating conversations in regard to the contents of this paper.

References

[1] BERS, L., Isolated singularities of minimal surfaces. Ann. Math. **53**, 364—386 (1951).
[2] DE GIORGI, E., & G. STAMPACCHIA, Sulle singolarità eliminabili delle ipersuperficie minimali. (To appear.)
[3] FINN, R., Isolated singularities of solutions of nonlinear partial differential equations. Trans. Amer. Math. Soc. **75**, 385—404 (1953).
[4] FINN, R., On partial differential equations (whose solutions admit no isolated singularities). Scripta Math. **26**, 107—115 (1961).
[5] JENKINS, H., Super-solutions for quasi-linear elliptic equations. Arch. Rational Mechanics and Analysis **16**, 402—410 (1963).
[6] JENKINS, H., On the behavior of solutions of quasi-linear elliptic equations. (To appear Trans. Amer. Math. Soc.)
[7] MORREY, C.B., Jr., Multiple integral problems in the calculus of variations and related topics. Ann. Scuola Norm. Sup. Pisa **14**, 1—62 (1960).
[8] NITSCHE, J.C.C., Über ein verallgemeinertes Dirichletsches Problem für die Minimalflächengleichungen und hebbare Unstetigkeiten ihrer Lösungen. Math. Ann. **158**, 203—214 (1965).
[9] SERRIN, J., Local behavior of solutions of quasi-linear equations. Acta Mathematica **111**, 247—302 (1964).
[10] SERRIN, J., Removable singularities of solutions of elliptic equations. Arch. Rational Mechanics and Analysis **17**, 67—76 (1964).
[11] WALLIN, H., A connection between α-capacity and L^p-classes of differentiable functions. Ark. Mat. **5**, 331—341 (1964).

School of Mathematics
Institute of Technology
University of Minnesota
Minneapolis

(Received June 19, 1965)

Am. J. Math. 88 (1966), 258–272. **Reprint I.5.[iii]**

ISOLATED SINGULARITIES OF SOLUTIONS OF LINEAR ELLIPTIC EQUATIONS.*

By JAMES SERRIN and H. F. WEINBERGER.

---

1. Introduction. It was shown by De Giorgi [1] and Nash [6] that every solution of the self-adjoint uniformly elliptic second order linear equation in divergence form,

$$(1) \qquad (a^{ij}u_{x^i})_{x^j}=0, \qquad (a^{ij}=a^{ji})$$

is Hölder continuous. The Hölder exponent μ depends only on the ellipticity constant λ, where λ is defined by

$$(2) \qquad \lambda^{-1}\,|\,\xi\,|^2 \leqq a^{ij}\xi_i\xi_j \leqq \lambda\,|\,\xi\,|^2.$$

More recently Moser [5] showed that positive solutions of (1) obey a Harnack inequality. Using this inequality, Royden [7] and Littman, Stampacchia, and Weinberger [3] independently proved that in $n \geqq 3$ dimensions a solution which approaches plus infinity at the origin is bounded above and below by positive multiples of $|\,x\,|^{2-n}$, and that a fundamental solution actually exhibiting this behavior exists.

Recently one of the present authors [8] has investigated the singularities of solutions of a class of non-linear equations which includes (1) as a special case. It was shown there that every solution u of (1) in the punctured domain $0<|\,x\,|<R$ which is $O(|\,x\,|^{2-n})$ must be of the form $u=ag+w$, where a is a constant, g is any particular solution which behaves like $|\,x\,|^{2-n}$, and w is a regular solution of (1) in $|\,x\,|<R$.

Analogously, it was shown [9] that a bounded solution of (1) in $|\,x\,|>R$ must be of the form $u=u_\infty+O(|\,x\,|^{2-n})$, where u_∞ is a constant. See also [5], p. 589 (footnote). Moreover, the existence of a solution $G(x)$ such that $G(x)\approx|\,x\,|^{2-n}$ at infinity is easily established by the procedures which give the existence of a fundamental singularity at the origin (cf. [8]). Here we use the notation $f(x)\approx|\,x\,|^{2-n}$ to denote the fact that there are positive constants a and b such that $a\,|\,x\,|^{2-n}\leqq f\leqq b\,|\,x\,|^{2-n}$.

Received March 4, 1965.

* This research was supported by the United States Air Force Office of Scientific Research through grant no. AF OSR 883-65 and by the National Science Foundation through grant no. GP3904.

The purpose of the present paper is to refine and generalize these results concerning the behavior of a solution at an isolated singularity. It is shown that there exists a positive constant δ depending only on the ellipticity constant λ, such that if u is a solution for $0 < |x| < R$, and if

$$\limsup |x|^{n-2+\delta} u \leqq 0,$$

$|x| \to 0$, then $u = ag + w$, where $g \approx |x|^{2-n}$ and w is a regular solution in $|x| < R$. In other words, a singular solution other than a fundamental solution must approach both plus and minus infinity at a rate definitely faster than a fundamental solution.

Similarly, there exist a constant κ depending only on λ such that if u is a solution at infinity and $\limsup |x|^{-\kappa} u \leqq 0$, $|x| \to \infty$, then

$$u = u_0 + bG + 0(|x|^{2-n-\nu})$$

where G is a particular solution with $G \approx |x|^{2-n}$, and ν depends only on λ.

Thus any solution which does not approach a limit at infinity must go to both plus and minus infinity at least as fast as $|x|^{\kappa}$ on some sets of points. On the other hand, a solution that goes to a limit more rapidly than $|x|^{2-n}$ must go like $|x|^{2-n-\nu}$. The relation between the constant b and the outflow of u is discussed in Section 7.

In Section 3 we introduce an inversion which is the natural generalization of the Kelvin transformation for the Laplace equation. It takes solutions of equation (1) into solutions of another uniformly elliptic equation of the same form. In particular, solutions in the punctured domain $0 < |x| < R$ go into solutions at infinity, and conversely. This transformation allows us to obtain theorems about behavior at infinity from results concerning behavior at the origin, and vice versa. It is applied to yield our main theorems in Sections 4 and 5.

The above inversion transformation requires the symmetry hypothesis $a^{ij} = a^{ji}$. In Section 6 we present some similar results which can be derived without this assumption. In Section 7 we introduce the concept of outflow and use it to obtain more specific results concerning the behavior of a solution at infinity. The result of Theorem 8, which relates asymptotic behavior, the strong maximum principle at infinity, and outflow should be particularly noted. We conclude the paper with some remarks concerning the solvability of the Dirichlet problem for (1) in exterior domains with prescribed outflow, or with prescribed limiting value at infinity.

For the sake of simplicity we have assumed $n \geqq 3$ in the preceding discussion, and also in the proofs to follow. All our results remain valid when $n=2$, though of course the solutions $g(x)$ and $G(x)$ must in this case behave like $\log|x|$.

By a solution of (1) in a domain Ω we shall always mean a continuous function $u(x)$ having strong derivatives u_x which are locally of class L^2, such that

$$\int a^{ij}u_{x^i}\phi_{x^j}dx=0$$

for any continuously differentiable function ϕ with compact support in Ω. The summation convention will be used throughout, with sums running from 1 to n on each index.

2. An oscillation theorem. Let $u=u(x)$, $x=(x^1,\cdots,x^n)$, be a solution of (1) in an exterior domain, say $|x|>R$. For any $\sigma>R$ we define $M(\sigma)$ and $m(\sigma)$ to be the maximum and minimum of u on the sphere $|x|=\sigma$. Then the following result holds.

THEOREM 1. *Either u tends to a limit u_∞ as x tends to infinity, or else for all suitably large σ we have the relations*

$$M(\sigma)\geqq A\sigma^\kappa \text{ and } m(\sigma)\leqq -A\sigma^\kappa,$$

where $A>0$ and κ is a positive number depending only on the ellipticity constant λ. If $n\geqq 3$ then the limiting value u_∞ is finite.

Remark. In [5] it is shown that if $M(\sigma)$ is increasing and $m(\sigma)$ is decreasing, then $M(\sigma)-m(\sigma)$ grows at least like a power of σ. The present result provides a strengthening of this conclusion.

Proof of Theorem 1. According to [9], Lemma 11, if u is bounded on one side then it tends to a limit u_∞ as $|x|$ tends to infinity; when $n\geqq 3$ the limit u_∞ is finite, while for $n=2$ it may be infinite. We may thus assume in what follows that u becomes both positively and negatively unbounded as x tends to infinity. By the maximum principle, it is clear that $M(\sigma)$ can have no more than one relative minimum for $\sigma>R$, while similarly $m(\sigma)$ can have no more than one relaltive maximum. Then since $M(\sigma)\to\infty$ and $m(\sigma)\to-\infty$, there exists some value σ_0 such that M is increasing and m decreasing for $\sigma>\sigma_0$. It may furthermore be assumed that $M>0$ and $m<0$ for $\sigma>\sigma_0$.

We apply the Harnack principle to the solution $u-m(2\sigma)$ in the annulus $\sigma/2<|x|<2\sigma$, where $\sigma>2\sigma_0$. (Clearly $u-m(2\sigma)>0$ in this region). There results easily

$$M(\sigma)-m(2\sigma)\leqq C(m(\sigma)-m(2\sigma))$$

where $C > 1$ depends only on λ. The same argument applied to $M(2\sigma) - u$ leads to

$$M(2\sigma) - m(\sigma) \leqq C(M(2\sigma) - M(\sigma)).$$

Adding the above two inequalities and transposing, we obtain

$$M(2\sigma) - m(2\sigma) \geqq [M(\sigma) - m(\sigma)](C+1)/(C-1).$$

By iteration, we get

$$M(2^\nu\sigma) - m(2^\nu\sigma) \geqq [M(\sigma) - m(\sigma)][(C+1)/(C-1)]^\nu, \quad \nu = 1, 2, \cdots.$$

We conclude (cf. [5], p. 580) that

$$M(\sigma) - m(\sigma) \geqq K\sigma^\kappa, \qquad \sigma > 2\sigma_0,$$

where

$$\kappa = \log_2[(C+1)/(C-1)].$$

By the first two of the above inequalities,

$$(C-1)m(2\sigma) \leqq -[M(\sigma) - Cm(\sigma)] \leqq -[M(\sigma) - m(\sigma)],$$
$$(C-1)M(2\sigma) \geqq CM(\sigma) - m(\sigma) \geqq M(\sigma) - m(\sigma).$$

Thus $m(\sigma) \leqq -A\sigma^\kappa$ and $M(\sigma) \geqq A\sigma^\kappa$, where $A = K/(C-1)$, and $\sigma > 4\sigma_0$.

The following corollary of Theorem 1 restates the result in a form more useful for our purposes.

COROLLARY. *There exists a constant* $\kappa > 0$, *depending only on the ellipticity constant* λ, *such that if* u *is a solution of* $(a^{ij}u_{x^i})_{x^j} = 0$, *and either*

$$\liminf_{|x|\to\infty} |x|^{-\kappa}u \leqq 0 \text{ or } \limsup_{|x|\to\infty} |x|^{-\kappa}u \geqq 0,$$

then u *approaches a limit at infinity.*

Theorem 1 states that if a solution $u(x)$ of (1) in $n \geqq 3$ dimensions is to go to infinity, it must go at least as fast as $|x|^\kappa$ for some definite κ. In the case of Laplace's equation $\kappa = 1$. The following example shows that, in general, κ depends on λ. The equation

$$(3) \qquad \{(1+\alpha^2)^{-\frac{1}{2}}(\delta_{ij} + \alpha^2 x^i x^j / |x|^2)u_{x^i}\}_{x^j} = 0$$

is uniformly elliptic with the ellipticity constant $\lambda = (1+\alpha^2)^{\frac{1}{2}}$. It has the solution

$$u = x^1|x|^{-q}, \qquad q = \tfrac{1}{2}\{n - [(n-2)^2 + 4(n-1)(1+\alpha^2)^{-1}]^{\frac{1}{2}}\}$$

which goes to infinity at the rate $|x|^{1-q}$. Hence

$$\kappa(\lambda) \leqq \tfrac{1}{2}\{[(n-2)^2+4\lambda^{-2}(n-1)]^{\frac{1}{2}}-(n-2)\}.$$

For large λ this number is arbitrarily near zero. Hence we cannot choose κ independent of λ.

If the coefficients a^{ij} tend to a limit at infinity, a simple change of variables reduces the problem to the case where λ is arbitrarily near one. In this case Finn and Gilbarg showed (for $n \leqq 6$) that one can take $\kappa = [n-1]^{\frac{1}{2}} - \frac{1}{2}(n-2)$.

3. A Kelvin transformation. In this section we establish a generalization of the Kelvin inversion transformation for Laplace's equation. This transformation allows us to take a uniformly elliptic equation of the form (1) defined in a neighborhood of infinity into another equation of the same form defined in a punctured neighborhood of the origin, and vice versa. We first prove the following lemma:

LEMMA. *Let u and w be two solutions of the elliptic equation*

$$(a^{ij}u_{x^i})_{x^j} = 0$$

in a domain Ω, with $w > 0$. Let $y^k = y^k(x^1, \cdots, x^n)$, $k = 1, \cdots, n$, be a continuously differentiable one-to-one coordinate transformation with non-vanishing Jacobian $J = \det(\partial y^k/\partial x^i)$ and inverse $x^i = x^i(y^1, \cdots, y^n)$. Then the function

$$v(y) = u(x(y))/w(x(y))$$

is a solution of the elliptic equation

$$(\bar{a}^{kl}v_{y^k})_{y^l} = 0 \tag{4}$$

in the image domain Ω', where

$$\bar{a}^{kl} = \frac{w^2}{J} a^{ij} \frac{\partial y^k}{\partial x^i} \frac{\partial y^l}{\partial x^j}. \tag{5}$$

Proof. Let $\phi(y)$ be a continuously differentiable function having compact support in Ω'. Then

$$\int \bar{a}^{kl}\phi_{y^k}v_{y^l}\, dy = \int (w^2/J) a^{ij} (\partial y^k/\partial x^i)(dy^l/\partial x^j)\phi_{y^k}v_{y^l}J\, dx.$$

Since the chain rule and the product rule hold for strong derivatives, we have

$$\begin{aligned}\int \bar{a}^{kl}\phi_{y^k}v_{y^l}\, dy &= \int w^2 a^{ij}\phi_{x_i}(u/w)_{x^j}\, dx \\ &= \int a^{ij}\phi_{x_i}(wu_{x^j} - uw_{x^j})\, dx \\ &= \int a^{ij}[(\phi w)_{x^i}u_{x^j} - (\phi u)_{x^i}w_{x^j}]\, dx.\end{aligned}$$

(The symmetry of a^{ij} was used in deriving the last expression.) Since ϕw and ϕu are L^2 strongly differentiable functions with compact support in Ω and u and w are solutions of (1), it follows that

$$\int \bar{a}^{kl}\phi_{y^k}v_{y^l}\,dy = 0.$$

Hence v is a solution of (4).

Suppose now that u is a solution of (1) in the punctured domain $0 < |x| < R$, and that the equation is uniformly elliptic there:

$$\lambda^{-1}|\xi|^2 \leqq a^{ij}\xi_i\xi_j \leqq \lambda|\xi|^2 \tag{6}$$

for all real vectors ξ. Then there is a positive solution $g(x)$ in $0 < |x| < R$ such that

$$|x|^{2-n} \leqq g(x) \leqq c|x|^{2-n} \qquad (n \geqq 3). \tag{7}$$

where c is a positive constant depending only on λ. For example, g can be taken to be a multiple of the Green's function for $|x| < R+1$ with singularity at the origin [3], [8].

We let $w(x) = c^{-\frac{1}{2}}g(x)$, and introduce the inversion $y^k = x^k/|x|^2$. This transformation takes the punctured domain into the exterior domain $|y| > R^{-1}$. We have

$$\partial y^k/\partial x^i = |x|^{-2}(\delta_{ik} - 2x^ix^k/|x|^2), \qquad J = |x|^{-2n},$$

whence

$$\bar{a}^{kl} = c^{-1}(g|x|^{n-2})^2a^{ij}(\delta_{ik} - 2x^ix^k/|x|^2)(\delta_{jl} - 2x^jx^l/|x|^2).$$

The matrix $\delta_{ik} - 2x^ix^k/|x|^2$ is unitary. Therefore it follows from (6) that

$$(c\lambda)^{-1}|\eta|^2 \leqq \bar{a}^{kl}\eta_k\eta_l \leqq c\lambda|\eta|^2.$$

Thus the equation (4) satisfied by

$$v(y) = u(y/|y|^2)/g(y/|y|^2)$$

is again uniformly elliptic, and its ellipticity constant $\lambda' = c\lambda$ depends only on the ellipticity constant λ of (1). We have proved

THEOREM 2. *If u is a solution of the uniformly elliptic equation* (1) *in $0 < |x| < R$, then*

$$v(y) = u(y/|y|^2)/g(y/|y|^2)$$

is a solution of the uniformly elliptic equation (4) *in $|y| > R^{-1}$, and the* *ellipticity constant* λ' *of* (4) *depends only on the ellipticity constant* λ *of* (1).

Similarly, on using a solution $G(x)$ of (1) which satisfies

$$|x|^{2-n} \leqq G(x) \leqq C\,|x|^{2-n} \qquad (n \geqq 3) \tag{8}$$

with C depending only on λ, we find:

THEOREM 3. *If u is a solution of the uniformly elliptic equation* (1) *in* $|x| > R$, *then*

$$v(y) = u(y/|y|^2)/G(y/|y|^2)$$

is a solution of the uniformly elliptic equation (4) *in* $0 < |y| < R^{-1}$, *and the ellipticity constant* λ' *of* (4) *depends only on the ellipticity constant* λ *of* (1).

Remark. By the same proofs with $w \equiv 1$ we can show that Theorems 2 and 3 still hold in two dimensions. In fact, $\lambda = \lambda'$ in this case.

4. Behavior at an isolated singularity. A solution $g(x)$ of (1) in a punctured neighborhood of the origin will be called a *fundamental solution* if $g \approx |x|^{2-n}$, or $g \approx -\log|x|$ if $n = 2$. In the same way, a solution G at infinity with $G \approx |x|^{2-n}$ (or $G \approx \log|x|$) will also be called a fundamental solution.

THEOREM 4. *There exists a positive number δ depending only on the elliptic constant λ, with the following property: If u is a solution of the self-adjoint uniformly elliptic equation* (1) *in the domain* $0 < |x| < R$ *and if* $\limsup |x|^{n-2+\delta} u \leqq 0$ (*or* $\liminf |x|^{n-2+\delta} u \geqq 0$), $x \to 0$, *then*

$$u = ag + w \tag{9}$$

where a is a constant, g is any given fundamental solution, and w is a solution of (1) *in the whole disk* $|x| < R$.

Proof. We first prove the theorem for a fixed fundamental solution g satisfying (7). By Theorem 2 the function

$$v(y) = u(y/|y|^2)/g(y/|y|^2)$$

satisfies equation (4), which is uniformly elliptic with ellipticity constant $\lambda' = c\lambda$. By (7) the limit superior of $|y|^{-\delta} v$ has the same sign as that of $|x|^{n-2+\delta} u$, for any $\delta > 0$.

We choose δ to be the constant κ of Theorem 1 corresponding to the ellipticity constant $\lambda' = c\lambda$. Then, by Theorem 1, v must be bounded in a neighborhood of infinity. It follows that $u(x)/g(x)$ is bounded near the origin. Hence by Theorems 14 and 15 of [8] u is of the form (9). Since c depends only on λ and δ depends only on $\lambda' = c\lambda$, δ depends only on λ. Thus the theorem is proved for our particular function g.

Now let $\bar{g}$ be any fundamental solution. By what has already been shown, $\bar{g} = \tilde{a}g + \tilde{w}$ where $\tilde{a} > 0$. Then $g = (\bar{g} - \tilde{w})/\tilde{a}$, and hence for any u satisfying the hypotheses of the theorem we have $u = ag + w = \bar{a}\bar{g} + \bar{w}$. Since we can obviously omit the bars on the right side, this completes the proof.

Remark. For $n = 2$ an application of Theorem 1 to $u(y/|y|^2)$ shows that $u(x)$ is bounded either above or below near the origin. It follows from [8, Theorem 14] that u has the form (9).

Theorem 4 states that a function with an unremovable isolated singularity different from the fundamental solution g must go to both plus and minus infinity at least like g times $|x|^{-\delta}$. In the case of Laplace's equation, $\delta = 1$.

We again examine the equation

$$\{(1+\alpha^2)^{-\frac{1}{2}}(\delta_{ij} + \alpha^2 x^i x^j/|x|^2)u_{x^i}\}_{x^j} = 0,$$

where α is a constant. As we have seen, $\lambda = (1+\alpha^2)^{\frac{1}{2}}$. The function

(10) $\quad u = x^1 |x|^{-q}$, where $q = \frac{1}{2}\{n + [(n-2)^2 + 4(n-1)(1+\alpha^2)^{-1}]^{\frac{1}{2}}\}$,

is a solution. Hence

$$\delta(\lambda) \leqq \tfrac{1}{2}\{[(n-2)^2 + 4\lambda^{-2}(n-1)]^{\frac{1}{2}} - (n-2)\}.$$

This number can be made arbitrarily close to zero by choosing λ large enough. Hence it is not possible to choose δ independent of the ellipticity constant λ.

5. Behavior at infinity. We shall now extend the result of Theorem 1 to obtain a sharper asymptotic estimate.

THEOREM 5. *Let $\kappa = \kappa(\lambda)$ be the constant of Theorem 1. Let u be a solution of the self-adjoint uniformly elliptic equation (1) for $|x| > R$. There exists an exponent ν depending only on λ, such that if* $\limsup |x|^{-\kappa}u \leqq 0$ (*or* $\liminf |x|^{-\kappa}u \geqq 0$), $x \to \infty$, *then*

$$u(x) = u_\infty + bG(x) + 0(|x|^{2-n-\nu}), \tag{11}$$

where u_∞ and b are constants, and G is any given fundamental solution at infinity. Moreover the last term satisfies a strong maximum principle at infinity.

(We say that a function u satisfies a strong maximum principle at infinity if u tends to a limit u_∞ as x tends to infinity and $u - u_\infty$ either changes sign or is identically zero in every neighborhood of infinity.)

Proof. We first consider a particular G satisfying (8). By Theorems 1 and 3 the function

$$v(y) = u(y/|y|^2)/G(y/|y|^2)$$

satisfies the hypotheses of Theorem 4. Therefore it is of the form

$$v(y) = ag(y) + w(y) \tag{12}$$

where g is any fundamental solution of the transformed equation (4). By putting $u \equiv 1$ in Theorem 3, we see that $G(y/|y|^2)^{-1}$ is a fundamental solution of (4). Hence

$$v(y) = u_\infty G(y/|y|^2)^{-1} + w(y)$$

where u_∞ is a constant, and $w(y)$ is a solution of (4) for $|y| < R$. Now w is Hölder continuous with exponent ν depending only on the ellipticity constant $\lambda' = c\lambda$ of (4), which, in turn, depends only on λ. Hence,

$$\begin{aligned} v(y) &= u_\infty G(y/|y|^2)^{-1} + w(0) + \{w(y) - w(0)\} \\ &= u_\infty G(y/|y|^2)^{-1} + w(0) + 0(|y|^\nu). \end{aligned}$$

Returning to $u = Gv$ yields (11) for our particular G.

Now let $\bar{G}$ be any fundamental solution. By what has already been shown

$$\bar{G} = bG + 0(|x|^{2-n-\nu}),$$

and hence for any u satisfying the hypotheses of the theorem

$$u = u_\infty + \bar{b}\bar{G} + 0(|x|^{2-n-\nu}).$$

To complete the proof, it is only necessary to observe that the final term is of the form

$$G(x) \cdot (w'(y) - w'(0))$$

for some regular solution $w'(y)$ of (4); since $w'(y) - w'(0)$ satisfies the maximum principle near $y = 0$, the last statement of the theorem is clearly valid.

Remark. For $n = 2$ we find by Theorem 1, 3 and 4 that $u(y/|y|^2) = ag(y) + w(y)$. By Theorem 3 we can let $g(y) = G(y/|y|^2)$. Inverting again yields (11). Of course, u_∞ is no longer the limit of u when $b \neq 0$.

In the special case when $u - u_\infty = o(|x|^{2-n})$, Theorem 5 states that $u - u_\infty = 0(|x|^{2-n-\nu})$ and that u satisfies a strong maximum principle at infinity. The latter part of this result is due to Finn and Gilbarg [2, Section 12], and holds even when (2) is replaced by the weaker condition

$$0 < a^{ij}\xi_i\xi_j \leqq \lambda \mid \xi \mid^2, \qquad \xi \neq 0.$$

For Laplace's equation, clearly $\nu = 1$. The solution (10) of equation (3) shows, however, that

$$\nu(\lambda) \leqq \tfrac{1}{2}\{[(n-2)^2 + 4\lambda^{-2}(n-1)]^{\frac{1}{2}} - (n-2)\},$$

which approaches zero as $\lambda \to \infty$. Thus ν cannot be chosen independent of λ.

6. Non-self-adjoint equations. In this section we shall obtain some related results concerning the behavior of solutions of (1) in the neighborhood of infinity, without making the restrictive assumption $a^{ij} = a^{ji}$.

We first remark that the Harnack inequality has been proved without the symmetry assumption [8]. Therefore, Theorem 1 holds in the non-self-adjoint case. We must, however, add the inequality $\mid a^{ij} \mid \leqq \lambda$ to the definition of the ellipticity constant λ.

We now investigate the behavior of solutions at infinity.

THEOREM 6. *Let u be a solution of* (1) *in the domain* $\mid x \mid > R$, *and suppose that* $\limsup \mid x \mid^{-\kappa} u \leqq 0$, *where κ is the constant of Theorem* 1. *Then the representation*

$$u = u_\infty + bG + o(\mid x \mid^{2-n}) \tag{13}$$

is valid, where u_∞ and b are constants, and G is any given fundamental solution at infinity.

Remark. The conclusion of Theorem 6 is weaker than that of Theorem 5 in that the last term is now only known to be $o(\mid x \mid^{2-n})$ rather than $O(\mid x \mid^{2-n-\nu})$. We shall show in the next section that it again satisfies a strong maximum principle at infinity.

Proof of Theorem 6. By Theorem 1, u tends to a finite limit u_∞ as x tends to infinity ($n \geqq 3$). We may assume $u_\infty = 0$ without loss of generality in obtaining the representation (13). Since $G > 0$ there exists a constant A such that

$$\mid u \mid < AG \text{ on } \mid x \mid = 2R. \tag{14}$$

Now consider the set of numbers β such that

$$u - \beta G > 0 \tag{15}$$

in some neighborhood of infinity. By (14) and the maximum principle it is evident that this set contains $\beta = -A$, but does not contain $\beta = A$. Let b denote the least upper bound of the values β in this set, and put

$$w = u - bG.$$

We assert that $w = o(G)$ as x tends to infinity.

Indeed, suppose that the assertion is not true. Then there exists an $\epsilon > 0$ and a sequence of points x_m tending to infinity, such that either

$$w(x_m) > \epsilon G(x_m) \tag{16}$$

or

$$w(x_m) < -\epsilon G(x_m). \tag{17}$$

By the definition of b it is clear that

$$u - (b - \epsilon/2) G > 0 \text{ near infinity,}$$

whence the relation (17) cannot be true. It remains to be shown that (16) also cannot hold.

Let θ be a small positive number which will be determined later, and put

$$w' \equiv u - (b - \epsilon\theta) G.$$

By the definition of b, the function w' is positive in some neighborhood of infinity. Moreover, if (16) holds,

$$w'(x_m) = w(x_m) + \theta\epsilon G(x_m) > (1 + \theta)\epsilon G(x_m).$$

By the Harnack inequality applied to the solution w' in the annulus $|x_m|/2 \leqq |x| \leqq 2|x_m|$, one has for all sufficiently large m,

$$w'(x) > [(1+\theta)\epsilon/C] G(x_m), \qquad |x| = |x_m|, \tag{18}$$

where C is the Harnack constant already used in the proof of Theorem 1. A second application of the Harnack inequality similarly establishes that

$$G(x) < CG(x_m), \qquad |x| = |x_m|. \tag{19}$$

Combining (18) and (19) we have for all sufficiently large m

$$w'(x) > [(1+\theta)\epsilon/C^2] G(x) \text{ on } |x| = |x_m|.$$

Hence by the maximum principle

$$w' > [(1+\theta)\epsilon/C^2] G(x)$$

near infinity. This may be written in the form

$$u - \{b + [(1+\theta)/C^2 - \theta]\epsilon\} G > 0. \tag{20}$$

Now choose $\theta < 1/(C^2 - 1)$. Then $(1+\theta)C^{-2} - \theta > 0$, and inequality (20) violates the definition of b as the least upper bound of the numbers β such that (15) holds in some neighborhood of infinity. This contradiction establishes the assertion $w = o(G)$.

Since $G \approx |x|^{2-n}$, it follows that

$$u = bG + w$$

where $w = o(|x|^{2-n})$. Since we have assumed that $u_\infty = 0$, the theorem is proved.

Remark. When $n = 2$, we observe that the limit value of u may be infinite, but in any case by Theorem 1 and [9, Theorem 2] we have

$$u = O(\log |x|).$$

Hence the consideration beginning with relation (15) can easily be carried out, with the conclusion that $u = bG + w$, where $w = o(\log |x|)$. It then follows (for example, from [9] or from Lemmas 1 and 2 of Section 7) that w has a finite limit.

7. Outflow. We shall now relate our results to the outflow of the solution u. If u is continuously differentiable, we define the outflow

$$K[u] = \int_{|x|=R} a^{ij} u_{x^i} x^j / R \, dS$$

where $|x| = R$ is any sphere such that u is a solution in $|x| \geqq R$. In general, we define the outflow by the formula

$$K[u] = \int a^{ij} u_{x^i} \psi_{x^j} \, dx$$

where ψ is any strongly differentiable continuous function whose support lies in the set $|x| > R$, and which is identically one in some neighborhood of infinity.

Since the difference of two such functions ψ has compact support, and since u is a solution of (1), it follows that $K[u]$ is independent of the particular ψ which is chosen. Since the support of ψ_x is finite, $K[u]$ is finite. The two definitions for $K[u]$ given above are easily seen to be in agreement when u is smooth.

The following two lemmas relate the outflow to the behavior at infinity. The first lemma states that a solution at infinity which does not satisfy the strong maximum principle there (in particular, a fundamental solution G) has non-zero outflow. The second lemma says that any solution which is $o(G)$ has zero outflow.

LEMMA 1. *Let v be a solution of* (1) *in a domain which contains a neighborhood* $|x| \geqq R$ *of infinity. Let v have a limit v_∞ (possibly $-\infty$) as $|x| \to \infty$, and let $v \geqq v_\infty$. Then either $v \equiv v_\infty$ or $K[v] < 0$.*

Proof. Suppose v is not constant. By the Harnack principle $v > v_\infty$, and hence there are constants A and B such that $v > A > B > v_\infty$ for $|x| = R$. Let

$$\psi = \begin{cases} 0 & \text{when } v > A \\ (A-v)/(A-B) & \text{when } B \leqq v \leqq A \\ 1 & \text{when } v < B. \end{cases}$$

Then by definition

$$K[v] = -(A-B)^{-1} \int_{B \leqq v \leqq A} a^{ij} v_{x^i} v_{x^j}\, dx < 0$$

since v_x cannot vanish almost everywhere on the set $B \leqq v \leqq A$.

LEMMA 2. *Let w be a solution at infinity of the uniformly elliptic equation* (1). *If $w = o(|x|^{2-n})$ (or $o(\log|x|)$ if $n = 2$), then $K[w] = 0$.*

Proof. Let G be a fundamental solution satisfying (8). We observe that $K[u]$ is a linear functional, so that for any constant α

$$K[w + \alpha G] = K[w] + \alpha K[G]. \tag{21}$$

When $\alpha > 0$, $w + \alpha G$ is again a fundamental solution at infinity since $w = o(|x|^{2-n})$. Hence by Lemma 1, $K[w + \alpha G] < 0$ when $\alpha > 0$. Similarly, $K[w + \alpha G] > 0$ when $\alpha < 0$. Therefore we see from (21) that $K[w] = 0$, which is the required conclusion.

When $n = 2$, Lemma 1 shows that $K[G] > 0$ and we must only reverse the inequalities to obtain the same result.

By combining Lemmas 1 and 2 with Theorems 6 and 5, we obtain the following two theorems:

THEOREM 7. *Let u be a solution at infinity of* (1) *with* $\limsup |x|^{-\kappa} u \leqq 0$. *Then there exists a constant u_∞ such that if G is any fundamental solution, then*

$$u = (K[u]/K[G])\, G + u_\infty + o(|x|^{2-n}).$$

If $a^{ij} = a^{ji}$, the last term becomes $O(|x|^{2-n-\nu})$.

THEOREM 8. *The following three statements about a bounded solution u at infinity are equivalent:*

(a) $u - u_\infty = o(|x|^{2-n})$,

(b) *u satisfies a strong maximum principle at infinity,*

(c) $K[u] = 0$.

As we have remarked earlier, the conclusion (a) $\Rightarrow$ (b) is due to Finn and Gilbarg [2].

8. The exterior Dirichlet problem. We conclude the paper with some remarks concerning the solvability of the Dirichlet problem for (1) in exterior domains.

Let D be a given open region containing a neighborhood of infinity and having a smooth internal boundary ∂D. A continuous function $\psi(x)$ is prescribed on ∂D. We wish to find a solution u of (1) in D which continuously takes on the values $\psi(x)$ on the boundary ∂D.

Naturally one must specify some further conditions at infinity in order to guarantee uniqueness of the solution. If $n = 2$ we require that the solution should be bounded in D (by Theorem 7 this involves the requirement $K[u] = 0$). If $n \geqq 3$ we may specify the solution in either of two ways:

1) The solution should tend to an assigned limit u_∞ as x tends to infinity, or

2) The solution is bounded, and $K[u]$ is assigned.

We shall show in these cases that the exterior problem is well set. Moreover, when $n \geqq 3$ all bounded solutions of (1) taking on the boundary values $\psi(x)$ are realized as u_∞ ranges over the real numbers, or as $K[u]$ ranges over the real numbers.

When $n = 2$ the existence of a bounded solution of the Dirichlet problem is easily established by forming sequences of solutions defined in $D \cap \{|x| < m\}$, as in reference [4]. The uniqueness then follows from the strong maximum principle of Theorem 8.

When $n \geqq 3$ one can find a bounded solution $\bar{u}$ exactly as in the case $n = 2$. In order to find a solution with the appropriate limit at infinity, or the appropriate value of $K[u]$ one may add to $\bar{u}$ another solution which is zero on the internal boundary. In particular, it is clear that there exists a solution $H(x)$ in D such that $H = 0$ on D and $H \to 1$ as $|x| \to \infty$. Let

$$u = \bar{u} + \alpha H,$$

where α remains to be chosen. Then

$$u_\infty = \bar{u}_\infty + \alpha,$$
$$K[u] = K[\bar{u}] + \alpha K[H].$$

By Lemma 1, $K[H] \neq 0$, so that we can choose α to give any preassigned value to either u_∞ or $K[u]$.

If u_∞ is assigned, uniqueness follows from the ordinary maximum principle. If $K[u]$ is assigned, uniqueness follows from the strong maximum principle of Theorem 8.

Finally, since every bounded solution has a finite limit at infinity and a finite outflow, it is clear that all bounded solutions of the Dirichlet problem are obtained as the limit value u_∞ in 1) ranges over the real numbers, or else as the outflow $K[u]$ in 2) ranges over the real numbers.

UNIVERSITY OF MINNESOTA.

REFERENCES.

[1] E. De Giorgi, "Sulla differenziabilità e l'analiticità delle estremali degli integrali multipli regolari," *Memorie dell' Accademia delle Scienze de Torino*, Ser. 3a, 3 (1957), pp. 25-43

[2] R. Finn and D. Gilbarg, "Three-dimensional subsonic flows and asymptotic estimates for elliptic partial differential equations," *Acta Mathematica*, vol. 98 (1957), pp. 265-296.

[3] W. Littman, G. Stampacchia and H. Weinberger, "Regular points for elliptic equations with discontinuous coefficients," *Annali della Scuola Normale Superiore di Pisa*, Ser. III, 17 (1963), pp. 45-79.

[4] N. Meyers and J. Serrin, "The exterior Dirichlet problem for second order elliptic partial differential equations," *Journal of Mathematics and Mechanics*, vol. 9 (1960), pp. 513-538.

[5] J. Moser, "On Harnack's theorem for elliptic differential equations," *Communications on Pure and Applied Mathematics*, vol. 14 (1961), pp. 577-591.

[6] J. Nash, "Continuity of solutions of parabolic and elliptic equations," *American Journal of Mathematics*, vol. 80 (1958), pp. 931-953.

[7] H. Royden, "The growth of a fundamental solution of an elliptic divergence structure equation," *Studies in Mathematical Analysis and Related Topics; Essays in honor of G. Polya*, pp. 333-340, Stanford University Press, Stanford, 1962.

[8] J. Serrin, "Local behavior of solutions of quasi-linear equations," *Acta Mathematica*, vol. 111 (1964), pp. 247-302.

[9] ———, "Singularities of solutions of nonlinear equations," *Nonlinear Differential Equations* (*Proceedings of Symposia on Pure Mathematics*, vol. 17), pp. 68-88, American Mathematical Society, Providence, R. I., 1965.

Commun. Pure Appl. Math. 39 (1986), 379–399. **Reprint I.5.[iv]**

Nonexistence Theorems for Singular Solutions of Quasilinear Partial Differential Equations

WEI-MING NI AND JAMES SERRIN
University of Minnesota

1. Introduction

In previous papers (see [5], [6]), we have studied the semilinear Poisson equation

$$\Delta u + f(u) = 0 \tag{1.1}$$

and its quasilinear analogue

$$\operatorname{div}(A(|Du|)Du) + f(u) = 0, \tag{1.2}$$

where Δ is the n-dimensional Laplacian and $A(p)$, $f(u)$ are given functions (it will be supposed, except in Section 5, that $A \in C^1[0, \infty)$ and $f \in C[0, \infty)$ with $f(0) = 0$). We have obtained a class of nonexistence results for ground states of equations (1.1) and (1.2), a ground state being a positive classical solution in $\mathbb{R}^n$ which tends to zero at ∞.

The main purpose of the present paper is to study *singular ground states* of equations (1.1) and (1.2). By a *singular ground state* we mean a positive classical solution in $\mathbb{R}^n \setminus \{0\}$ which tends to zero at ∞ and tends to ∞ at the origin. We shall derive several nonexistence theorems for singular ground states as well as for the corresponding singular Dirichlet problems for balls. (We actually prove a stronger result, namely, we establish nonexistence for positive radial solutions in $\mathbb{R}^n \setminus \{0\}$ which tends to 0 at ∞ *without imposing any condition on u near the singularity*. An analogous result also holds true for balls.)

The nonlinearities $f(u)$ we consider here for equation (1.1) are superlinear, that is, roughly speaking, $f(u) \geqq Cu^q$ for u near ∞, where $q > 1$. A simple but typical example which is included is the celebrated nonlinearity $f(u) = -u + u^q$ for $u \geqq 0$. For ground states it is well known that the existence range for q is $1 < q < (n+2)/(n-2)$ while the nonexistence range is $q \geqq (n+2)/(n-2)$. For singular ground states, the story is incomplete. We are able to establish nonexistence for $q \geqq (n+2)/(n-2)$, but the existence question remains open.

Our method of proof has two ingredients, a generalized Pohozaev-type identity (which will be recalled later) and *a priori* estimates for positive radial (local) solutions near the singularity. *A priori* estimates for solutions of (1.1) near an isolated singularity have been the subject of long and active research interest.

Communications on Pure and Applied Mathematics, Vol. XXXIX 379–399 (1986)
 CCC 0010-3640/86/030379-21$04.00

However for "strong" nonlinearities such as the ones considered here, little work appears in the literature. Nevertheless, we are able to prove that if u is a positive radial solution of (1.1) in $\Omega \setminus \{0\}$ with $u \to \infty$ as $r \to 0$, where Ω is a neighborhood of the origin and $f(u) \geqq Cu^q$ for u near ∞, then

$$u(r) \leqq \frac{C}{r^{2/(q-1)}} \quad \text{for} \quad r \quad \text{near} \quad 0.$$

Once this is established, a somewhat intricate argument involving an application of the generalized Pohozaev identity (see (1.4) below) gives nonexistence in the case $q \geqq (n+2)/(n-2)$.

The method described above carries over to equation (1.2), if $A(p)$ is bounded and positive in $(0, \infty)$. This covers, in particular, the mean curvature operator where

$$A(p) = (1+p^2)^{-1/2}, \qquad p > 0.$$

For the degenerate Laplacian

$$A(p) = p^{m-2}, \qquad p > 0,$$

where $1 < m < n$, our method continues to apply although the corresponding exponent is now $[(m-1)n + m)]/(n-m)$ instead of $(n+2)/(n-2)$.

Before turning to the main body of our work, we recall for the reader's convenience a version of the generalized Pohozaev identity proved in [5].

PROPOSITION. *Let $u = u(r)$ be a radial solution of equation* (1.2) *with domain $r \in [R_1, R_2)$, where $0 \leqq R_1 < R_2 \leqq \infty$. Then for any real constant a the following identity holds*:

$$\begin{aligned} &\frac{d}{dr}\left[r^n\left\{\int_0^p \rho E\,d\rho + F(u) + aA(p)\frac{uu'}{r}\right\}\right] \\ (1.3)\qquad &= r^{n-1}\left[n\int_0^p \rho E\,d\rho + nF(u) - auf(u) + (a+1-n)A(p)p^2\right], \end{aligned}$$

where $p = |u'| = |du/dr|$ and

$$E(p) = A(p) + p\frac{dA}{dp}, \qquad F(u) = \int_0^u f(t)\,dt.$$

For equation (1.1), $A \equiv 1$; therefore (1.3) takes the simpler form

$$\begin{aligned} &\frac{d}{dr}\left[r^n\left\{\frac{1}{2}u'^2 + F(u) + a\frac{uu'}{r}\right\}\right] \\ (1.4)\qquad &= r^{n-1}\left[nF(u) - auf(u) + (a+1-\tfrac{1}{2}n)u'^2\right]. \end{aligned}$$

The paper is organized as follows. In Section 2, we derive the asymptotic estimates described above. In Section 3, we establish nonexistence results for equation (1.1), while in Section 4, we deal with the quasilinear equation (1.2). In Section 5, we treat the case of the degenerate Laplacian, where $A(p) = |p|^{m-2}$.

2. Asymptotic Behavior near a Singularity

In this section we derive some asymptotic behavior theorems for positive radial solutions of equations (1.1) and (1.2) near an isolated singularity at $r = 0$. As was mentioned in the introduction, our primary concern is the "super-critical" case (i.e., for equation (1.1) if $f(u) \sim u^q$ at $u = \infty$ then $q \geqq (n+2)/(n-2)$, while for equation (1.2) analogously $q \geqq [(1-c)n+1]/(n-1)$, see Section 4). We shall first consider the special case (1.1), and then indicate the necessary modifications for the general case (1.2).

THEOREM 2.1. *Let u be a positive radial solution of* (1.1) *in* $0 < r \leqq r_0$ *with u tending to ∞ as $r \to 0$. If*

$$f(u) \geqq \text{Pos. Const. } u^q \quad \textit{for large} \quad u,$$

where q is a constant > 1, *then for* $0 < r \leqq r_0$ *we have*

$$u(r) \leqq \frac{C}{r^{2/(q-1)}} \tag{2.1}$$

for some positive constant C.

Proof: For a radial solution $u = u(r)$, we write (1.1) in the form

$$(r^{n-1}u')' + r^{n-1}f(u) = 0. \tag{2.2}$$

Since $u \to \infty$ as $r \to 0$, we may assume without loss of generality that $f(u) \geqq 0$ for $u = u(r)$, $0 < r \leqq r_0$. Thus

$$(r^{n-1}u')' \leqq 0 \quad \text{for} \quad 0 < r \leqq r_0,$$

that is, $r^{n-1}u'$ is decreasing in $(0, r_0)$. Since $u \to \infty$ as $r \to 0$, we conclude that $u' < 0$ in $(0, r_0)$.

Integrating (2.2) from $\bar{r}$ to r (where $0 < \bar{r} < r < r_0$), we obtain

$$r^{n-1}u'(r) = \bar{r}^{n-1}u'(\bar{r}) - \int_{\bar{r}}^{r} s^{n-1}f(u(s))\,ds.$$

Therefore

$$r^{n-1}u'(r) < -\int_{\bar{r}}^{r} s^{n-1}f(u(s))\,ds$$

for all $0 < \bar{r} < r$. Then letting $\bar{r} \to 0$ we obtain

$$r^{n-1}u'(r) < -\int_0^r s^{n-1}f(u(s))\,ds \leqq -C\int_0^r s^{n-1}u^q(s)\,ds,$$

and the second inequality now follows from our hypothesis on f. Also, since u is decreasing near $r = 0$, we find that

$$r^{n-1}u'(r) < -Cu^q(r)\int_0^r s^{n-1}\,ds = -Cr^n u^q(r),$$

which in turn leads to

$$u'(r) \leqq -Cru^q(r) \quad \text{for} \quad 0 < r \leqq r_0. \tag{2.3}$$

Dividing both sides by u^q and integrating from 0 to r yields the desired estimate (2.1).

In addition to (2.1), we shall also need some control on the derivative u' near $r = 0$ in order to apply the generalized Pohozaev identity. However, what we need is rather weak and may be derived directly from (2.1).

LEMMA 2.1. *Under the hypotheses of Theorem* 2.1, *there exists a sequence* $r_k \downarrow 0$ *such that*

$$|u'(r_k)| \leqq C/r_k^{1+2/(q-1)}, \qquad C > 0. \tag{2.4}$$

Proof: The assertion is equivalent to the statement

$$\limsup_{r\to 0} r^{1+2/(q-1)}u'(r) > -\infty.$$

Suppose this is false. Then

$$\lim_{r\to 0} r^{1+2/(q-1)}u'(r) = -\infty,$$

that is, for all $M > 0$ there exists a value r_M such that

$$r^{1+2/(q-1)}u'(r) \leqq -M$$

for all $0 < r < r_M$. Integrating this from r to r_M, we get

$$u(r_M) - u(r) = \int_r^{r_M} u'(s)\,ds \leqq \int_r^{r_M} -Ms^{-(1+2/(q-1))}\,ds$$

$$= \tfrac{1}{2}M(q-1)\left(r_M^{-2/(q-1)} - r^{-2/(q-1)}\right).$$

Let $r = \frac{1}{2}r_M = r_m$. Then we obtain

$$u(r_m) \geqq CMr_m^{-2/(q-1)},$$

where C is a positive constant depending only on q. Since $r_m \downarrow 0$ as $m \to \infty$, this contradicts (2.1) and the proof is complete.

Remark. If one assumes further that $f(u) \leqq Cu^q$ for large u, then it is possible to obtain a more satisfactory estimate for u' near $r = 0$, namely

$$|u'(r)| \leqq C/r^{1+2/(q-1)}$$

for all $0 < r < r_0$. However such a hypothesis is too strong for our purposes.

We now turn to the quasilinear equation (1.2). *We shall assume in the rest of this section that A is positive and bounded in* $(0, \infty)$, *with say,* $A(p) \leqq \Lambda$.

The conclusions of Theorem 2.1 and Lemma 2.1 then remain true for the quasilinear equation (1.2). Indeed, in this case, equation (2.2) is replaced by

$$(r^{n-1}Au')' + r^{n-1}f(u) = 0.$$

As before, we observe that $u' < 0$ near $r = 0$ by the positivity of A. Then, following the proof of Theorem 2.1, instead of (2.3) we arrive at

$$Au'(r) \leqq -Cru^q(r) \quad \text{in} \quad (0, r_0).$$

Since $A \leqq \Lambda$ and $u' < 0$, this gives, for $r \in (0, r_0)$,

$$u'(r) \leqq -cru^q(r)/\Lambda.$$

Thus estimate (2.1) follows as before. For Lemma 2.1, one only has to observe that (2.1) is the sole ingredient used in the proof.

3. Nonexistence Results for Equation (1.1)

In this section we shall present several nonexistence results concerning singular radial solutions of equation (1.1) both for balls and for $\mathbb{R}^n$. Our method is close to the one used in Section 4 of [6], except we must now pay attention to the behavior of the solution u near the origin. We assume throughout this section that $n > 2$. Our first result is the following

THEOREM 3.1. *Suppose $f(u)$ is positive for u sufficiently large. If there exists $\alpha \geqq (n+2)/(n-2)$ such that*

$$(\alpha + 1)F(u) \leqq uf(u) \tag{3.1}$$

for all $u > 0$, *then the singular Dirichlet problem*

$$(3.2)\qquad \begin{aligned} &\Delta u + f(u) = 0 \quad in \quad \Omega \setminus \{0\}, \\ &u = 0 \quad on \quad \partial\Omega, \qquad u > 0 \quad in \quad \Omega, \end{aligned}$$

where Ω *is a ball* B_R *of radius* R *in* $\mathbb{R}^n$, *does not possess any radial solutions.*

Remarks. Integrating (3.1) one sees that

$$(3.3)\qquad F(u) \geqq Cu^{1+\alpha}$$

for u sufficiently large, where C is a positive constant. Substituting this back to (3.1) yields

$$(3.4)\qquad f(u) \geqq Cu^{\alpha}$$

for u sufficiently large. (Here and in what follows we continue to use the letter C as a generic positive constant, whose particular value depends on the context and may even change within the proof of a given theorem.)

Note that no behavior of u near the (possible) singularity 0 is imposed. Our first step towards the proof is to derive such behavior for "local" solutions of (1.1). (By a "local solution" of (1.1) we mean a solution u of (1.1) in $\Omega' \setminus \{0\}$ without any boundary conditions, where Ω' is an arbitrary neighborhood of 0.) The following result is therefore of independent interest.

LEMMA 3.1. *Let* $u > 0$ *be a radial solution of equation* (1.1) *in* $\Omega' \setminus \{0\}$ *where* Ω' *is an arbitrary neighborhood of the origin. If* $f(u) \geqq -C$ *for all* $u > 0$, *where* C *is an arbitrary but fixed positive constant, then* u *must be either a classical solution of* (1.1) *in* Ω' *or* u *tends to* ∞ *at the origin.*

Proof: Suppose u is not a classical solution of (1.1) in the entire neighborhood Ω'. Then u must be unbounded in Ω'. For if u is bounded in Ω', then by standard arguments one sees that u is actually a distributional solution of (1.1) in the entire Ω'. Then, elliptic regularity theory implies that u is classical in the entire domain Ω', a contradiction. Alternately, the result follows from Corollary 2 in Section 7 of [5].

Next, we claim that $u(r) \to \infty$ as $r \to 0$. In fact, suppose for the sake of contradiction that

$$\liminf_{r \to 0} u(r) < \infty.$$

Then there exist two sequences $\{s_i\}, \{t_i\}$ such that $s_1 > t_1 > s_2 > t_2 > \cdots \to 0$ and u assumes local maxima at s_i, local minima at t_i; moreover, $u(s_i) \to \infty$ while $u(t_i)$ remain bounded. Since $f(u) \geqq -C$, we have $\Delta u \leqq C$. By the maximum principle we conclude that

$$u(t_i) \geqq \min(u(s_i), u(s_{i+1})) - \bar{C},$$

where $\overline{C}$ is a constant independent of i, a contradiction.

Therefore, for a positive radial solution u of equation (1.1) in $\Omega' \setminus \{0\}$ which is not smooth at the origin, u must tend to ∞ as $r \to 0$ and $u' < 0$ near $r = 0$. Hence estimate (2.1) holds by virtue of Theorem 2.1.

The proof of Theorem 3.1 now depends on several preliminary lemmas which we shall obtain first. It will be assumed throughout that the hypotheses of Theorem 3.1 are in force and that u (for eventual contradiction) is a positive radial solution of (3.2) which tends to a limit (possibly infinite) as $r \to 0$. (The fact that (3.2) does not have any *regular* solutions is established in [5] but is not needed here.)

LEMMA 3.2. *For all sufficiently small r, we have*

$$r^n\left[\tfrac{1}{2}u'^2(r) + F(u(r)) + \frac{n}{\alpha+1}\frac{uu'}{r}\right] \geqq C^2, \tag{3.5}$$

where C is some positive constant.

Proof: Integrating (1.4) from r to R, we obtain

$$\begin{aligned} &\tfrac{1}{2}R^n u'^2(R) - r^n\left[\tfrac{1}{2}u'^2(r) + F(u(r)) + \frac{n}{\alpha+1}\frac{uu'}{r}\right] \\ &\quad = \int_r^R \left[nF(u) - \frac{n}{\alpha+1}uf(u) + \left(\frac{n}{\alpha+1} + 1 - \tfrac{1}{2}n\right)u'^2\right]s^{n-1}\,ds, \end{aligned} \tag{3.6}$$

where we put $a = n/(\alpha+1)$. The right-hand side of (3.6) is nonpositive and decreases as $r \to 0$. If the right-hand side of (3.6) is not identically zero, our conclusion follows easily. If it is identically zero for all $r > 0$, it is not hard to see that α then must be $(n+2)/(n-2)$ and the strict equality in (3.1) holds for all u in the range of the solution $(0, u_0)$, which in turn implies that

$$F(u) = Cu^{1+\alpha} \quad \text{and} \quad f(u) = Cu^{\alpha}$$

for all u in $(0, u_0)$. In that case, $u'(R) < 0$ by the Hopf boundary point lemma (or by uniqueness of the initial value problem for ordinary differential equations) and we are done.

LEMMA 3.3. *Set* $\gamma = 2/(\alpha - 1)$ *and* $w = r^\gamma u$. *Then for all sufficiently small r we have either*

$$\text{(i)} \qquad w' \geqq 0$$

or

$$\text{(ii)} \qquad w' \leqq 0.$$

Proof: It is easy to see that

$$
\begin{aligned}
&u = r^{-\gamma} w, \\
(3.7)\qquad &u' = -\gamma r^{-\gamma-1} w + r^{-\gamma} w', \\
&u'' = \gamma(\gamma+1) r^{-\gamma-2} w - 2\gamma r^{-\gamma-1} w' + r^{-\gamma} w'';
\end{aligned}
$$

thus (1.1) becomes

$$
(3.8)\qquad w'' + \frac{n-1-2\gamma}{r} w' + r^{\gamma}\left[f(u) - \frac{\gamma(n-2-\gamma)}{r^2} u\right] = 0.
$$

We claim that w has no local minimum when r is so small that (3.5) holds. Indeed, at a local minimum of w we would have

$$
w' = 0, \quad w'' \geqq 0 \quad \text{and} \quad u' = -\gamma r^{-\gamma-1} w = -\gamma u/r.
$$

Hence by (3.5), at such a point

$$
(\alpha+1) F(u) \geqq \left(n - \tfrac{1}{2}\gamma(\alpha+1)\right)\gamma u^2/r^2.
$$

By (3.8) wc have

$$
u f(u) \leqq \gamma(n-2-\gamma)\frac{u^2}{r^2}.
$$

Since $n - \frac{1}{2}\gamma(\alpha+1) > (n-2-\gamma)$, this contradicts (3.1), and our assertion is proved. This in turn implies that either (i) or (ii) must hold for all r sufficiently small.

LEMMA 3.4. *The function w tends to a finite non-negative limit L as $r \to 0$. Moreover, there is a sequence $r_m \to 0$ such that*

$$
r_m w'(r_m) \to 0 \quad \textit{and} \quad r_m^2 w''(r_m) \to 0.
$$

Proof: From (3.4) and Theorem 2.1, we see that w is bounded near $r = 0$. Thus Lemma 3.3 implies that w tends to a finite limit $L \geqq 0$ as $r \to 0$. Moreover, it is evident that either

$$
\liminf_{r\to 0} r w' = 0 \qquad \text{(case (i) in Lemma 3.3)}
$$

or

$$
\limsup_{r\to 0} r w' = 0 \qquad \text{(case (ii) in Lemma 3.3)}.
$$

(Otherwise an easy integration would show that $|w|$ becomes infinite as $r \to 0$.) In case (i) either $\varphi = rw'$ decreases monotonically to zero or $\varphi = rw'$ has a sequence of local minima along which it tends to zero. If the second possibility occurs, we take exactly these points as the r_m. Then

$$\varphi(r_m) = r_m w'(r_m) \to 0,$$

$$r_m \varphi'(r_m) = r_m^2 w''(r_m) + r_m w'(r_m) = 0,$$

and the lemma follows at once. If the first possibility occurs, we have

$$\varphi = rw' \to 0, \qquad \varphi' < 0$$

as $r \to 0$ so that

$$\limsup_{r \to 0} r\varphi' = 0,$$

(otherwise φ tends to negative infinity at $r = 0$, as in the earlier discussion of w) and our lemma again follows.

In case (ii) we proceed similarly. Either rw' increases monotonically to zero as $r \to 0$ or there is a sequence of local maxima along which rw' tends to zero. In either case the existence of the required sequence $\{r_m\}$ follows exactly as before.

We may now complete the proof of Theorem 3.1. From (3.5) and (3.1), we have

$$2F(u) \leqq \frac{2}{\alpha + 1} uf(u) = \frac{2}{\alpha + 1}\left(-uu'' - \frac{n-1}{r} uu'\right)$$

and

$$r^n\left[u'^2 - \frac{2}{\alpha + 1} uu'' + \frac{2}{\alpha + 1}\frac{uu'}{r}\right] \geqq C^2 > 0, \tag{3.9}$$

for all small $r > 0$. Now, choose $r = r_m$ and let $m \to \infty$. Then (3.7) and Lemma 3.4 imply that

$$\begin{aligned} r_m^{\gamma} u(r_m) &\to L, \\ r_m^{1+\gamma} u'(r_m) &\to -\gamma L, \\ r_m^{2+\gamma} u''(r_m) &\to \gamma(\gamma + 1) L. \end{aligned} \tag{3.10}$$

Rewrite the left-hand side of (3.9) as follows:

$$r^{n-2\gamma-2}\left[r^{2+2\gamma} u'^2 - \frac{2}{1+\alpha} r^{\gamma} u \cdot r^{2+\gamma} u'' + \frac{2}{\alpha + 1} r^{\gamma} u \cdot r^{1+\gamma} u'\right].$$

As $m \to \infty$, we see that the term inside the bracket tends to the limit

$$\gamma^2 L^2 - \frac{2}{1+\alpha}\gamma(\gamma+1)L^2 - \frac{2}{\alpha+1}\gamma L^2 = -\gamma\frac{2}{1+\alpha}L^2 \leqq 0.$$

The exponent $n - 2\gamma - 2 \geqq 0$ since $\alpha \geqq (n+2)/(n-2)$, and we therefore conclude that the left-hand side of (3.9) when evaluated at $r = r_m$ tends to a nonpositive limit as $m \to \infty$ while the right-hand side is a fixed positive constant. This is clearly a contradiction and the proof of Theorem 3.1 is complete.

COROLLARY. *Assume that the hypotheses of Theorem* 3.1 *hold, and further that f is Lipschitz continuous on* $[0, \infty)$. *Then the singular Dirichlet problem*

$$\Delta u + f(u) = 0 \quad in \quad \Omega \setminus \{0\},$$
(3.11)
$$u = 0 \quad on \quad \partial\Omega, \qquad u > 0 \quad near \quad x = 0,$$

does not possess any positive solutions nor any radial solutions.

This corollary follows from the above theorem and the fact that *a positive solution of* (3.11) *must be radial.* This may be proved by suitably adapting the arguments in [2] (we omit the details here).

The following example shows that our conditions in Theorem 3.1 and the corollary are the best possible.

Consider the case $f(u) = u^q$, $u > 0$. In [4] it is proved that for $1 < q < (n+2)/(n-2)$ there are infinitely many positive radial solutions of (3.2) and (3.11); while for $q \geqq (n+2)/(n-2)$, there are no solutions of (3.2) or (3.11) (Theorem 2 in [4]). Our results may be considered as generalizations of Theorem 2(ii) in [4]. It is interesting to note that in [4], the proofs make no use of Pohozaev-type results.

We now turn to nonexistence results in $\mathbb{R}^n$.

THEOREM 3.2. *Suppose $f(u)$ is positive for u sufficient large and that* (3.1) *holds for all $u > 0$, where $\alpha \geqq (n+2)/(n-2)$ is a constant. If $f \leqq 0$ in $(0, \varepsilon)$ for some $\varepsilon > 0$, then equation* (1.1) *does not possess any radial solutions in $\mathbb{R}^n \setminus \{0\}$ which tends to* 0 *at infinity and remains positive in a neighborhood of the origin.*

Remark. Once again note that no assumption on the asymptotic behavior of (possible) solutions near the (possible) isolated singularity (the origin) is imposed.

Proof: To begin with, we may assume that $u > 0$ for all $r > 0$. Otherwise there would exist an $R > 0$ such that $u > 0$ for $0 < r < R$ and $u(R) = 0$, contradicting Theorem 3.1. Then using Proposition 5.1 of [5] we have (the condition $u'(0) = 0$ in that result is not essential)

$$\lim_{r \to \infty} r^{n-1}u'(r) \quad \text{exists (finite)}$$

and

$$\lim_{r\to\infty} r^{2(n-1)}F(u(r)) = 0.$$

From here on the proof is similar to that of Theorem 3.1, so we can be brief. First, Lemma 3.1 still applies. For Lemma 3.2 we again integrate (1.4) from r to R, and then let R tend to infinity. By the asymptotic results established above, we then get, with $a = n/(\alpha+1)$,

$$(3.12)\qquad \begin{aligned} &-r^n\left[\tfrac{1}{2}(u'^2(r)) + F(u(r)) + \frac{n}{\alpha+1}\frac{uu'}{r}\right] \\ &= \int_r^\infty \left[nF(u) - \frac{n}{\alpha+1}uf(u) + \left(\frac{n}{\alpha+1} + 1 - \tfrac{1}{2}n\right)u'^2\right]s^{n-1}\,ds. \end{aligned}$$

The integrand on the right-hand side of (3.12) is strictly negative for all $r > 0$. Indeed otherwise we would have, as in the proof of Lemma 3.2,

$$f(u) = Cu^\alpha, \qquad \alpha = (n+2)/(n-2),$$

for all u in the range of the solution, say $(0, u_0)$ with u_0 possibly infinite. But $f(u) \leqq 0$ near 0 so that $C = 0$. Thus $f(u) = 0$ on the range of u. This is impossible if $u_0 = \infty$ since f is assumed positive for u large. Hence u_0 is finite and u must be a bounded solution of the Laplace equation on $\mathbb{R}^n \setminus \{0\}$, which is impossible by the Liouville theorem.

It follows that (3.5) holds for all $r \leqq 1$, say, and the rest of the proof of Theorem 3.1 then carries over without any change.

COROLLARY. *Suppose $f(u)$ is positive for u sufficiently large and* (3.1) *holds for all $u > 0$ where $\alpha \geqq (n+2)/(n-2)$ is a constant. If $f'(0) < 0$, then equation* (1.1) *does not possess any positive solutions in $\mathbb{R}^n \setminus \{0\}$ which tend to ∞ at 0 and tend to 0 at ∞.*

This corollary is a direct consequence of Theorem 3.2 and the fact that *any positive solution of* (1.1) *in $\mathbb{R}^n \setminus \{0\}$ which tends to 0 at ∞ and tends to ∞ at 0 must be radial* under the hypotheses of the corollary. Although this fact is not stated explicitly in [3], the arguments there may be modified in an obvious way (by making use of (3.4) and Theorem 2.1 above) to yield the required conclusion.

Remark. Theorem 3.2 and the corollary yield new results even in the simple case $f(u) = -u + u^q$, where $q \geqq (n+2)/(n-2)$.

4. Quasilinear Equations

Most of the results in the previous section admit generalizations to the more general equation (1.2) under appropriate assumptions on A.

Throughout this section we assume that c is a positive constant satisfying the condition

$$\int_0^p \rho A \, d\rho \geqq cAp^2 \tag{4.1}$$

for all $p \geqq 0$. Observe that (4.1) is equivalent to

$$\int_0^p \rho E \, d\rho \leqq (1-c)Ap^2 \tag{4.2}$$

for all $p \geqq 0$. For various choices of c under different circumstances we refer the reader to [5], Section 2. We shall also assume throughout this section that $n > 1/c$, $\Lambda \geqq A(p) > 0$ for all $p > 0$, and

$$\int_0^p \rho E \, d\rho > 0 \quad \text{for all} \quad p > 0.$$

There are two different types of generalizations of results in the previous section. We present the simpler one first.

THEOREM 4.1. *Suppose*

$$f(u) \geqq Cu^q, \qquad F(u) \leqq Cu^{1+\bar{q}}, \tag{4.3}$$

for u sufficiently large, where q, $\bar{q}$ are positive exponents satisfying the condition

$$q \leqq \bar{q} < \tfrac{1}{2}(q-1)n - 1. \tag{4.4}$$

Assume also that there exists a constant α such that

$$(\alpha+1)F(u) \leqq uf(u) \tag{4.5}$$

holds for all $u > 0$, where α satisfies

$$\alpha \geqq \frac{(1-c)n+1}{cn-1}. \tag{4.6}$$

Then the singular Dirichlet problem

$$\begin{aligned} &\operatorname{div}(A(|Du|)Du) + f(u) = 0 \quad in \quad \Omega\setminus\{0\}, \\ &u = 0 \quad on \quad \partial\Omega, \qquad u > 0 \quad in \quad \Omega\setminus\{0\}, \end{aligned} \tag{4.7}$$

where Ω is a ball B_R in $\mathbb{R}^n$, does not possess any radial solutions.

Remark. Again, no asymptotic condition on u is imposed near the (possible) isolated singularity 0. Incidentally, the conclusion of Theorem 4.1 still holds if we replace the condition "$u > 0$ in $\Omega \setminus \{0\}$" in (4.7) by "$u > 0$ in a deleted neighborhood of the origin".

To prepare our proof, we first present the following

LEMMA 4.1. *Suppose* (4.5) *and* (4.6) *hold for all* $u > 0$. *Then* (4.7) *does not possess any bounded radial solutions.*

Proof: Suppose u is a bounded radial solution of (4.7). We integrate (1.3) from r to R, and let $a = n/(\alpha + 1)$, giving ($p(r) \equiv |u'(r)|$)

$$R^n \int_0^{p(R)} \rho E \, d\rho - r^n \left[\int_0^{p(r)} \rho E \, d\rho + F(u(r)) + aA(p(r))u'(r)u(r)/r \right]$$
(4.8)
$$= \int_r^R \left[nF(u) - auf(u) + n \int_0^{p(s)} \rho E \, d\rho + (a + 1 - n)Au'^2(s) \right] s^{n-1} \, ds.$$

Note that the right-hand side of (4.8) is less than or equal to

$$\int_r^R [nF(u) - uf(u) + (a + 1 - cn)Au'^2(s)] s^{n-1} \, ds,$$

which is nonpositive and decreasing as $r \to 0$. Next we assert that the left-hand side of (4.8) tends to the non-negative limit

$$R^n \int_0^{p(R)} \rho E \, d\rho \tag{4.9}$$

along a sequence $r_k \to 0$. Indeed, the term $r^n F(u(r)) \to 0$ as $r \to 0$ since u is bounded. Since

$$|r^{n-1} Au'(r)| = \left| - \int_0^r s^{n-1} f(u(s)) \, ds \right| \leqq Cr^n \to 0 \quad \text{as} \quad r \to 0,$$

the term $aAu'ur^{n-1} \to 0$ as $r \to 0$. Finally, to estimate the term

$$r^n \int_0^{p(r)} \rho E \, d\rho \leqq (1 - c) r^n \Lambda u'^2(r),$$

we distinguish two cases. First, there exists a sequence $r_k \to 0$ such that $u'(r_k) = 0$ for all k. In this case our assertion follows readily. Second, u is monotone in $0 < r < \varepsilon$, for some $\varepsilon > 0$. Then

$$\liminf_{r \to 0} r^n u'^2(r) = 0.$$

Otherwise, there exists a constant $C > 0$ such that $r^n u'^2(r) \geqq C > 0$ for all

$r \in (0, \varepsilon')$, for some $\varepsilon' > 0$. Integrating, we have for $r \leqq \varepsilon'$

$$|u(0) - u(r)| = \int_0^r |u'(s)|\, ds \geqq C \int_0^r \frac{1}{s^{n/2}}\, ds = \infty$$

since $n \geqq 2$. This is a contradiction, and the assertion is thus proved.

If the right-hand side of (4.8) is not identically zero, it must then be negative, which contradicts (4.9). If the right-hand side is identically zero, then the equality holds in (4.2) and $c = \frac{1}{2}$. Moreover, $\alpha = (n+2)/(n-2)$, $f(u) = Cu^{\alpha}$ and $A(p) \equiv$ positive constant. Therefore $u'(R) < 0$ by the Hopf boundary point lemma and (4.9) is then positive, again a contradiction, and the proof is complete.

LEMMA 4.2. *Suppose* $u > 0$ *is a radial solution of* (1.2) *in* $\Omega' \setminus \{0\}$ *with*

$$\limsup_{r \to 0} u(r) = \infty,$$

where Ω' *is an arbitrary neighborhood of the origin. If* (4.5), (4.6) *hold for all* $u > 0$ *and* $f(u) > 0$ *near* $u = \infty$, *then*

$$\lim_{r \to 0} u(r) = \infty.$$

Proof: Suppose on the contrary that

$$\liminf_{r \to 0} u(r) < \infty.$$

Then there exist two sequences $\{s_i\}, \{t_i\}$ such that $s_1 > t_1 > s_2 > t_2 > \cdots \to 0$ and u assumes local maxima at s_i, local minima at t_i; moreover, $u(s_i) \to \infty$ and $\{u(t_i)\}$ remains bounded. Integrating (1.3) from t_i to s_i with the obvious choice $a = n/(\alpha + 1)$, we see that

$$s_i^n F(u(s_i)) \leqq t_i^n F(u(t_i)).$$

This gives rise to a contradiction for i large, in view of (3.3) (which also follows from the present hypotheses).

We now prove Theorem 4.1. In view of Lemmas 4.1 and 4.2, it suffices to show that (4.7) does not possess any positive radial solution which tends to ∞ at the origin. Suppose for the sake of contradiction that there is such a solution u. Estimates (2.1) and (2.4) hold also for this solution u, as was noted at the end of Section 2. Integrating (1.3) from r_k to R (where r_k appeared in Lemma 2.1), we obtain (4.8) with r replaced by r_k. Again, as in the proof of Lemma 4.1, $a = n/(\alpha + 1)$ and the right-hand side of (4.8) tends to a limit $L \leqq 0$. We also claim that the left-hand side of (4.8) tends to (4.9) as $k \to \infty$. Assuming this for the moment, we see that if $L < 0$, we have derived a contradiction. If $L = 0$, then arguing as in the proof of Lemma 4.1, we conclude $u'(R) < 0$ and (4.9) is then positive, also a contradiction. Therefore it only remains to show that the left-hand side of (4.8) approaches (4.9) as $k \to \infty$. We proceed as follows.

Since $A \leqq \Lambda < \infty$, we deduce from (4.2) that

$$\left| r_k^n \int_0^{|u'(r_k)|} \rho E \, d\rho \right| \leqq (1 - c)\Lambda u'^2(r_k) r_k^n \leqq C r_k^{n - 2[1 + 2/(q-1)]} \to 0$$

as $r_k \to 0$ since $n - 2[1 + 2/(q - 1)] > 0$ because $q > (n + 2)/(n - 2)$, the latter condition being entailed by (4.4). For the term $r_k^n F(u(r_k))$, we apply conditions (4.3), (4.4) and Theorem 2.1, so that

$$r^n F(u(r)) \leqq C r^n u^{1+\bar{q}}(r) \leqq C r^{n-(1+\bar{q})2/(q-1)} \to 0$$

as $r \to 0$. The last term can be estimated in a similar way,

$$\left| r_k^{n-1} u(r_k) u'(r_k) \right| \leqq C r_k^{n-1-(2/(q-1))-1-2/(q-1)} \to 0$$

as $r_k \to 0$, and our assertion follows.

We now turn to results concerning the entire space. Lemma 4.1 has the following counterpart.

LEMMA 4.3. *Suppose* (4.5) *and* (4.6) *hold for all* $u > 0$. *Assume also* $f \leqq 0$ *in* $(0, \varepsilon)$ *for some* $\varepsilon > 0$. *Then equation* (1.2) *does not possess any bounded radial solution in* $\mathbb{R}^n \setminus \{0\}$ *which is positive near the origin and tends to zero at infinity.*

Proof: Since this proof is very similar to that of Lemma 4.1, we shall be brief.

Suppose u is a bounded radial solution of (1.2) in $\mathbb{R}^n \setminus \{0\}$ which tends to 0 at ∞. Integrating (1.3) from r to R with $a = n/(\alpha + 1)$, we have

$$\begin{aligned} R^n &\left[\int_0^{p(R)} \rho E \, d\rho + F(u(R)) + aAu(R)u'(R)/R \right] \\ &- r^n \left[\int_0^{p(r)} \rho E \, d\rho + F(u(r)) + aAu(r)u'(r)/r \right] \\ &= \int_r^R \left[nF(u) - auf(u) + n \int_0^{p(s)} \rho E \, d\rho + (a + 1 - n) A u'^2(s) \right] s^{n-1} \, ds. \end{aligned}$$

Let $r = r_k$, where $\{r_k\}$ is the sequence we chose in the proof of Lemma 4.1. Hence the right-hand side is nonpositive and eventually tends to a negative limit as $k \to \infty$ and $R \to \infty$ since $f \leqq 0$ in $(0, \varepsilon)$. The left-hand side tends to 0, as in the proof of Lemma 4.1 (the first term on the left-hand side tends to 0 as $R \to \infty$

in view of the asymptotic behavior for u developed in Theorem 5.1 of [5]). This contradiction completes the proof of the lemma.

THEOREM 4.2. *Suppose* (4.3) *and* (4.4) *hold for u near* ∞, *and that* (4.5) *and* (4.6) *hold for all* $u > 0$. *Assume also* $f \leqq 0$ *in* $(0, \varepsilon)$ *for some* $\varepsilon > 0$. *Then equation* (1.2) *does not possess any radial solution in* $\mathbb{R}^n \setminus \{0\}$ *which is positive near the origin and tends to* 0 *at infinity.*

Proof: By Lemmas 4.2, 4.3 and Theorem 4.1, it suffices to prove that (1.2) does not have any positive radial solution in $0 < r < \infty$ with $u \to \infty$ as $r \to 0$ and $u \to 0$ as $r \to \infty$.

Suppose $u = u(r)$ is a positive radial solution of (1.2) with $u \to 0$ as $r \to \infty$ and $u \to \infty$ as $r \to 0$. As in the proof of Theorem 4.1, we integrate (1.3) from r_k to R (R being arbitrary at this point) to obtain

$$R^n\left[\int_0^{p(R)} \rho E\,d\rho + F(u(R)) + aA(p(R))u(R)u'(R)/R\right]$$

$$-r_k^n\left[\int_0^{p(r_k)} \rho E\,d\rho + F(u(r_k)) + aA(p(r_k))u(r_k)u'(r_k)/r_k\right]$$

$$\leqq \int_{r_k}^{R}\left[nF(u) - auf(u) + (a+1-cn)Au'^2(r)\right]r^{n-1}\,dr.$$

Letting $k \to \infty$, we see by the arguments used in the proof of Theorem 4.1 that

$$R^n\left[\int_0^{p(R)} \rho E\,d\rho + F(u(R)) + aA(p(R))u(R)u'(R)/R\right]$$

$$\leqq \int_0^{R}\left[nF(u) - auf(u) + (a+1-cn)Au'^2\right]r^{n-1}\,dr.$$

By Theorem 5.1 of [5] the left-hand side tends to zero as R tends to ∞. Thus we have

$$\int_0^{\infty}\left[nF(u) - auf(u) + (a+1-cn)Au'^2\right]r^{n-1}\,dr \geqq 0.$$

Choosing $a = n/(\alpha + 1)$, a contradiction arises exactly as in the proof of Theorem 3.2, and the theorem is proved.

Remark. Our hypothesis on A does not even require the operator $\operatorname{div}(A(|Du|)Du)$ to be elliptic, since ellipticity of this operator is characterized by $A > 0$, $E > 0$ for all $p > 0$.

An example which satisfies all of the above conditions on A is the *generalized mean curvature operator*

$$\operatorname{div}\Big[\big(1+|Du|^2\big)^{(m/2)-1}Du\Big],$$

where $1 \leqq m \leqq 2$. In this case, an easy computation shows that we can take $c = \frac{1}{2}$, and the condition (4.6) becomes the usual one $\alpha \geqq (n+2)/(n-2)$. Therefore, applying Theorems 4.1 and 4.2 to the nonlinearity $f(u) = -u + u^q$, $q > 1$, yields nonexistence in the range $q > (n+2)/(n-2)$. In the remaining case $1 < q \leqq (n+2)/(n-2)$, nothing is known.

In the rest of this section, we shall present generalizations of Theorems 3.1 and 3.2 in a different direction, requiring the condition $dA/dp \geqq 0$.

THEOREM 4.3. *Suppose* $0 \leqq dA/dp \leqq c/p$ *for all* $p > 0$ *in addition to the hypotheses listed at the beginning of the section. Suppose further that* $f(u)$ *is positive for* u *sufficiently large and that*

$$uf(u) \geqq sF(u) \tag{4.10}$$

for all $u > 0$, *where* s *satisfies*

$$s \geqq n/(cn-1). \tag{4.11}$$

Then the singular Dirichlet problem (4.7) *does not possess any radial solutions.*

Remark. Before turning to the proof it is useful to consider some relations between the values c and s which follow from our conditions. In particular, since $dA/dp \geqq 0$ we have

$$\int_0^p \rho A\, d\rho \leqq A(p)\int_0^p \rho\, d\rho = \tfrac{1}{2}p^2 A(p).$$

Hence by (4.1) necessarily $c \leqq \frac{1}{2}$. Now put

$$\beta = 2n - (n-2)s$$

and observe that $\beta \leqq 0$ by (4.11) and the fact that $c \leqq \frac{1}{2}$. In addition

$$s > 1/c \geqq 2,$$

again because of (4.11) and the fact that $c \leqq \frac{1}{2}$. Finally, let

$$t = s(2-\beta)/(2cs-\beta).$$

Then $t \leqq s$ because $\beta \leqq 0$ and $s \geqq 1/c$. It follows in particular that

$$uf(u) \geqq tF(u) \tag{4.12}$$

for all u sufficiently large.

We remark again that no *a priori* asymptotic behavior is required of the solution u near the (possible) isolated singularity at 0.

Proof: The proof uses essentially the ideas of Theorem 3.1. We shall therefore be brief and only indicate the modifications.

First, observe that due to Lemmas 4.1 and 4.2 it suffices to prove that (4.7) does not possess any positive radial solutions which tend to ∞ at 0. Suppose for the sake of contradiction that u is such a solution of (4.7), we first establish the analogue of (3.5), for all sufficiently small r,

$$(4.13)\quad r^n\left[\int_0^{p(r)}\rho E\,d\rho + F(u(r)) + \frac{n}{s}A(p(r))u'(r)u(r)/r\right] \geqq C > 0.$$

We proceed as in the proof of Theorem 4.1. Integrating (1.3) from r to R, we obtain (4.8). Set $a = n/s$. The arguments in the proof of Theorem 4.1 then give (4.13) with the help of (4.10) and (4.11).

Next, setting $\gamma = 2/(s-2)$ and $w = r^\gamma u$, we claim that for all sufficiently small r either $w' \geqq 0$ or $w' \leqq 0$. Following the proof of Lemma 3.3, with (3.8) replaced by

$$Ew'' + [(n-1)A - 2\gamma E]\frac{w'}{r}$$
$$+r^\gamma\left\{f(u) - \gamma[(n-1)A - (\gamma+1)E]\frac{u}{r^2}\right\} = 0,$$

it is apparent that at a local minimum we must have

$$uf(u) \leqq [(n-1)A - (\gamma+1)E]\gamma u^2/r^2.$$

On the other hand, from (1.3), (4.13) and (4.2), at a local minimum,

$$F(u) > \left[\frac{n}{s} - \gamma(1-c)\right]\gamma A u^2/r^2.$$

Combining the above inequalities and using the relation $E \geqq A > 0$, we derive (note that $n > s\gamma(1-c)$ since $\beta \leqq 0$ and $s > 2$)

$$uf(u) < \{s(n-\gamma-2)/[n - \gamma s(1-c)]\}F(u) = tF(u),$$

which contradicts (4.12) if w has a local minimum near $r = 0$.

Now the proof of Lemma 3.4 shows that w tends to a finite limit $L \geqq 0$ as $r \to 0$ and that there is a sequence $r_m \to 0$ with

$$r_m w'(r_m) \to 0 \quad\text{and}\quad r_m^2 w''(r_m) \to 0.$$

To complete the proof of the theorem, we start with (4.13). By (4.2) and (4.12) we have at $r = r_m$ (and r_m sufficiently small) in the case $L > 0$

$$(4.14)\quad \begin{aligned} 0 < C &\leqq r^n\left[(1-c)Au'^2 - \frac{1}{t}Euu'' + \frac{Auu'}{r}\left(\frac{n}{s} - \frac{n-1}{t}\right)\right] \\ &\leqq r^{n-2(1+\gamma)}\left\{r^{2(1+\gamma)}\left[(1-c)u'^2 - \frac{1}{t}uu'' + \frac{uu'}{r}\left(\frac{n}{s} - \frac{n-1}{t}\right)\right]\right\}A. \end{aligned}$$

Since by (3.10) both u and u'' are non-negative for sufficiently small r_m and since $E \geqq A$ by assumption. Again by (3.10), as $m \to \infty$ the term inside the braces tends to the limit

$$\gamma\left((1-c)\gamma + \frac{n-\gamma-2}{t} - \frac{n}{s}\right)L^2,$$

which is zero since $t = s(n-\gamma-2)/[n-\gamma s(1-c)]$. This gives a contradiction since $n \geqq 2(1+\gamma)$ (by $\beta \leqq 0$).

The case $L = 0$ may be handled in a similar fashion by showing that each of the following terms

$$r^{2(1+\gamma)}Au'^2, \quad r^{2(1+\gamma)}Euu'' \quad \text{and} \quad r^{2(1+\gamma)-1}Auu'$$

tends to zero as $r_m \to 0$ (the boundedness of E is used here, which follows from our hypotheses on A). Then again (4.14) gives rise to a contradiction and the proof is complete.

For the entire space, we have the following analogous result.

THEOREM 4.4. *Under the hypotheses of Theorem* 4.3 *if we assume further that* $f \leqq 0$ *in* $(0, \varepsilon)$ *for some* $\varepsilon > 0$, *then equation* (1.2) *does not possess any radial solution in* $\mathbb{R}^n \setminus \{0\}$ *which is positive near the origin and tends to* 0 *at infinity.*

Proof: By Lemma 4.2 and 4.3 and Theorem 4.3 it suffices to show that equation (1.2) does not have any positive radial solution in $\mathbb{R}^n \setminus \{0\}$ which tends to ∞ at the origin and tends to 0 at infinity. The rest of the proof is then almost identical to that of Theorem 4.3 except we need the asymptotic behavior of solutions at infinity (see Theorem 5.1 in [5]) to establish (4.13). The argument is essentially contained in the proof of Theorem 4.2, and therefore we shall not repeat it here. (Note that the right-hand side of (4.8) is not identically zero because $f \leqq 0$ in $(0, \varepsilon)$ for some $\varepsilon > 0$.)

5. The Generalized Laplacian

The *generalized Laplacian* is given by

$$(5.1)\qquad A(p) = p^{m-2},$$

where $1 < m < n$. The corresponding nonlinear equation (1.2) takes the form

$$\operatorname{div}(|Du|^{m-2}Du) + f(u) = 0. \tag{5.2}$$

When $m = 2$, this is equation (1.1), but otherwise the equation is quasilinear, and degenerate or singular (depending on whether $m > 2$ or $m < 2$) when $Du = 0$; moreover it does not satisfy the hypotheses in Section 4 except in the case $m = 2$. Nevertheless, due to the homogeneity of A, corresponding results can be obtained by following the procedures in Sections 2, 3, and 4. Since all the proofs in this section are exact analogues of previous ones, we shall leave them to the reader.

THEOREM 5.1. *Let u be a positive radial solution of* (5.2) *in* $0 < r \leqq r_0$ *and* $u \to \infty$ *as* $r \to 0$. *If* $f(u) \geqq Cu^q$ *at* $u = \infty$ *for some constants* $C > 0$, $q > 1$, *then for* $0 < r \leqq r_0$

$$u(r) \leqq C/r^{m/(q-m+1)}. \tag{5.3}$$

LEMMA 5.1. *Under the hypotheses of Theorem* 5.1, *there exists a sequence* $r_k \to 0$ *such that*

$$|u'(r_k)| \leqq C/r_k^{(m/(q-m+1))+1}. \tag{5.4}$$

To apply our generalized Pohozaev identity (1.3), we first observe that

$$\int_0^p \rho E \, d\rho = \left(1 - \frac{1}{m}\right) A p^2$$

for all $p \geqq 0$, whence (1.3) takes the simpler form

$$\frac{d}{dr}\left[r^n\left\{|u'|^m\left(1 - \frac{1}{m}\right) + F(u) + auu'|u'|^{m-2}/r\right\}\right]$$

$$= r^{n-1}\left[nF(u) - auf(u) + (a + 1 - n/m)|u'|^m\right].$$

THEOREM 5.2. *Suppose that $f(u)$ is positive for u sufficiently large. Let s be a constant, put $\beta = mn - (n - m)s$ and assume that $\beta \leqq 0$. Suppose further that*

$$uf(u) \geqq sF(u) \tag{5.5}$$

for all $u > 0$ *where s satisfies*

$$s \geqq mn/(n - m). \tag{5.6}$$

Then the singular Dirichlet problem

$$\begin{aligned} &\operatorname{div}(|Du|^{m-2}Du) + f(u) = 0 \quad in \quad \Omega\setminus\{0\}, \\ &u = 0 \quad on \quad \partial\Omega, \qquad u > 0 \quad in \quad \Omega\setminus\{0\}, \end{aligned} \tag{5.7}$$

where Ω is a ball in $\mathbb{R}^n$, does not possess any radial solutions.

THEOREM 5.3. *Under the hypotheses of Theorem 5.2, if we assume further that $f \leqq 0$ in $(0, \varepsilon)$ for some $\varepsilon > 0$, then equation (5.2) does not possess any radial solution in $\mathbb{R}^n \setminus \{0\}$ which is positive near the origin and tends to 0 at infinity.*

The asymptotic behavior at $r = \infty$ of a positive radial solution is needed here; these results are derived in [1] and in [5].

Acknowledgment. The authors were supported in part by NSF Grants DMS 8200033A01 and DMS 8403233.

Bibliography

[1] Franchi, B., Lanconelli, E., and Serrin, J., *Uniqueness of positive solutions of quasilinear equations in* $\mathbb{R}^n$, in preparation.
[2] Gidas, B., Ni, W.-M., and Nirenberg, L., *Symmetry and related properties via the maximum principle*, Comm. Math. Phys. 68, 1979, pp. 209–243.
[3] Gidas, B., Ni, W.-M., and Nirenberg, L., *Symmetry of positive solutions of nonlinear elliptic equations in* $\mathbb{R}^n$, Advances in Math. Suppl. Studies 7A, 1981, pp. 369–402.
[4] Ni. W.-M., and Sacks, P., *Singular behavior in nonlinear parabolic equations*, Trans. Amer. Math. Soc., 287, 1985, pp. 657–671.
[5] Ni, W.-M., and Serrin, J., *Non-existence theorems for quasilinear partial differential equations*, Rendiconti. Circolo Mat. Palermo, Centenary Supplement, to appear.
[6] Ni, W.-M., and Serrin, J., *Existence and non-existence theorems for ground states for quasilinear partial differential equations*, to appear, Accad. Naz. dei Lincei.

Received March, 1985.
Revised July, 1985.

I.6. Liouville Theorems

Alberto Farina

Mathematics Subject Classification (2010). 35j60, 35B53.
Keywords. Nonlinear Partial differential equations, Liouville theorems.

The classical Liouville theorem states that any bounded real harmonic function defined on the Euclidean space $\mathbb{R}^N$ must be constant. As a generalization of this property, by the Liouville theorem, we shall mean the assertion that any bounded entire solution (i.e., defined over $\mathbb{R}^N$) of an elliptic, possibly nonlinear, partial differential equation is identically constant. Besides the elegance and the implicit beauty of these kinds of rigidity results, the Liouville theorems play a key role in the study of solutions of general nonlinear boundary value problems. In particular, they are crucial to deriving a priori estimates on solutions [3, 17], to obtain the existence of solutions [4] and also to study the regularity properties of weak solutions in arbitrary domains [24]. Due to their importance, there has been a large number of works devoted to Liouville-type theorems. Among them, two contributions of James Serrin play a *prominent* part.

In the pioneering work [i] below, J. Serrin was able to generalize the standard Liouville theorem for Laplace's equation to quasilinear elliptic equations. The simplest case [1], illustrating both the importance and the generality of the results obtained in this work, can be stated in the following way

Theorem 1. *Let f be a function of class C^1 in its variables. Let u be an entire solution of*

$$\Delta u = f(u, \nabla u) \quad in \quad \mathbb{R}^N, \tag{1}$$

where $\frac{\partial f}{\partial u} \geq 0$. Suppose that both u and ∇u are bounded. Then u must be constant.

When f depends only on u, the same conclusion holds true under the sole assumption that $u(x) = o(|x|)$ as $|x| \to +\infty$.

Theorem 1 is sharp, as witness the Laplace equation, which admits non-constant affine solutions on $\mathbb{R}^N$. Also the monotonicity assumption on f cannot be dropped, since $u(x) = \sin(x_1)$ is a bounded entire solution of the equation $\Delta u = -u$.

To the best of our knowledge, this is essentially the first contribution concerning Liouville theorems in a general nonlinear setting and in any dimension. The proof is based on pointwise gradient bounds. More precisely, J. Serrin derives from equation (1) (possibly after a clever change of dependent variable) a differential inequality satisfied by the function $|\nabla u|^2$ and then, via the Maximum Principle, he deduces the desired

[1]The methods developed also apply to quasilinear elliptic equations of the form $a_{ij}(\nabla u)\partial_{ij} u = f(u, \nabla u)$, with smooth and bounded coefficients a_{ij}, encompassing thus the important case of the equation of prescribed mean curvature.

P. Pucci et al. (eds.), *James Serrin. Selected Papers: Volume 1*, Contemporary Mathematicians, DOI 10.1007/978-3-0348-0845-3_6, © Springer Basel 2014

estimates. The idea of dealing with pointwise bounds (which may be traced back to S.N. Bernstein [5] in the two-dimensional case) has then been extended and modified in several ways, and used for a detailed classification of entire solutions in many cases of interest, including degenerate and/or singular quasilinear equations in divergence form (see, among the others, [7, 12, 16, 23, 25, 26, 30, 33, 34] and the references therein). In particular, instead of looking at the differential inequality satisfied by $|\nabla u|^2$ only, it has become relevant to look at a more general variational inequality involving the analogue of the Hamiltonian (or Lagrangian) function in the dynamical system framework. The pointwise estimates obtained in this way are therefore somewhat more precise, since they better take into account the particular features of the nonlinearities involved. They may be seen as generalizations of the Conservation of Energy Principle in the PDE setting.

In 2002, J. Serrin published the beautiful and impressive work [ii] below. This article, written in collaboration with H. Zou, has been published in Acta Mathematica and it is a major contribution in the area of nonlinear Liouville theorems and, more generally, to the study of qualitative properties of the solutions to the generalized Lane–Emden equation. A detailed description of this contribution would go beyond the scope of the present article. Nevertheless, I cannot resist the pleasure to state the following beautiful and deep result

Theorem 2.

(a) **(Universal a priori estimate)** *Assume $N > m > 1$ and let $\Omega \not\equiv \mathbb{R}^N$ be a domain of $\mathbb{R}^N$. Let $u \in C^1(\Omega)$ be a non-negative distribution solution of the generalized Lane–Emden equation*

$$-\mathrm{div}(|\nabla u|^{m-2}\nabla u) = u^{p-1} \qquad in \qquad \Omega. \tag{2}$$

Suppose $m < p < \frac{Nm}{N-m}$ then there exists a universal constant

$$C = C(N, m, p) > 0$$

(independent of both Ω and u) such that

$$u(x) \leq C[\mathrm{dist}\,(x, \partial\Omega)]^{-\frac{m}{p-m}} \qquad \forall x \in \Omega.$$

(b) **(Liouville theorem)** *Assume $N \geq 1$ and let $u \in C^1(\mathbb{R}^N)$ be a non-negative distribution solution of the generalized Lane–Emden equation (2) on the entire Euclidean space $\mathbb{R}^N$.*

Suppose $1 < m < p < \frac{Nm}{N-m}$ ($= +\infty$ if $N \leq m$). Then, $u \equiv 0$.

The above theorem greatly extends the celebrated results of Gidas and Spruck [17, 18] (which hold only in the semilinear case $m = 2$) to quasilinear equations of the form (2) and once again, amazingly, the results are optimal. The Liouville theorem stated in part (b) is seen as a consequence of the universal a priori estimate obtained in part (a) (actually being the extreme case where the domain is the entire space $\mathbb{R}^N$), which in turn is derived from a Harnack-type theorem and an accurate application of the test functions technique. This point of view and the methods developed in the paper provide a unified approach to the study of local properties of non-negative solutions, to classifications results, and to Liouville-type theorems for equations involving the m-Laplace operator. Its influence on further developments in the area of nonlinear PDE's is attested to by the large number of published articles, as well as by the numerous topics of research stimulated since its first appearance. In this respect we mention (and the list is very far from being exhaustive): nonlinear Liouville-type theorems (possibly in unbounded domains of various geometry) [8, 12, 19, 22], a priori and singularity estimates for local solutions [6, 10, 29, 32, 36], monotonicity and symmetry results [9, 11, 13], regularity issues [9, 20], existence and uniqueness/multiplicity results for quasilinear problems in bounded domains [1, 21, 32, 36], as well as their extensions to the anisotropic case [35], to fully nonlinear equations [2, 27], or to manifolds [28]. To conclude, we wish to quote the recent works [10, 14, 15, 31]. They are devoted to Liouville theorems for very general weak elliptic quasilinear equations and inequalities, and all of them are closely related to Serrin's works considered in the present article. They are based on the method of test functions and on comparison principles, two topics in which James Serrin is incontestably a *Master*.

References

[1] Aftalion A., Pacella F., Uniqueness and nondegeneracy for some nonlinear elliptic problems in a ball. *J. Differential Equations* 195 (2003), no. 2, 380–397.

[2] Armstrong S.N., Sirakov B., Liouville results for fully nonlinear elliptic equations with power growth nonlinearities. *Ann. Sc. Norm. Super. Pisa Cl. Sci.* 10 (2011), 711–728.

[3] Bahri A., Lions P.-L., Solutions of superlinear elliptic equations and their Morse indices. *Comm. Pure Appl. Math.* 45 (1992), no. 9, 1205–1215.

[4] Berestycki H., Capuzzo-Dolcetta I., Nirenberg L., Superlinear indefinite elliptic problems and nonlinear Liouville theorems. *Topol. Methods Nonlinear Anal.*, 4 (1994), no. 1, 59–78.

[5] Bernstein S.N., Sur la généralisation du problème de Dirichlet.(French) *Math. Ann.*, 69 (1910), no. 1, 82–136.

[6] Bidaut-Véron M-F., Borghol R., Véron L., Boundary Harnack inequality and a priori estimates of singular solutions of quasilinear elliptic equations. *Calc. Var. Partial Differential Equations* 27 (2006), no. 2,159–177.

[7] Caffarelli L., Garofalo N., Segála F., A gradient bound for entire solutions of quasi-linear equations and its consequences. *Comm. Pure Appl. Math.* 47 (1994), no. 11, 1457–1473.

[8] Damascelli L., Farina A., Sciunzi B., Valdinoci E., Liouville results for m-Laplace equations of Lane–Emden–Fowler type. *Ann. Inst. H. Poincaré Anal. Non Linéaire* 26 (2009), no. 4, 1099–1119.

[9] Damascelli L., Sciunzi B., Regularity, monotonicity and symmetry of positive solutions of m-Laplace equations. *J. Differential Equations*, 206 (2) 483–515, 2004.

[10] D'Ambrosio L., Mitidieri E., A priori estimates, positivity results, and nonexistence theorems for quasilinear degenerate elliptic inequalities. *Adv. Math.* 224 (2010), no. 3, 967–1020.

[11] Du Y., Guo Z., Symmetry for elliptic equations in a half-space without strong maximum principle. *Proc. Roy. Soc. Edinburgh Sect. A*, 134 (2), pp. 259–269, 2004.

[12] Farina A., Liouville-type theorems for elliptic problems. In *Handbook of differential equations: stationary partial differential equations. Vol. IV, Handb. Differ. Equ.*, 61–116. Elsevier/North-Holland, Amsterdam, 2007.

[13] Farina A., Montoro L, Sciunzi B., Monotonicity and one-dimensional symmetry for solutions of $-\Delta_p u = f(u)$ in half-spaces. *Calc. Var. Partial Differential Equations* 43 (2012), 123–145.

[14] Farina A., Serrin J., Entire solutions of completely coercive quasilinear elliptic equations. *J. Differential Equations* 250 (2011), 4367–4408.

[15] Farina A., Serrin J., Entire solutions of completely coercive quasilinear elliptic equations II. *J. Differential Equations* 250 (2011), 4409–4436.

[16] Farina A., Valdinoci E., Pointwise estimates and rigidity results for possibly singular and degenerate elliptic PDEs with general nonlinearities. *ESAIM Control Optim. Calc. Var.* 19 (2013), 616–627.

[17] Gidas B., Spruck J., A priori bounds for positive solutions of nonlinear elliptic equations. *Comm. Partial Differential Equations* 6 (1981), no. 8, 883–901.

[18] Gidas B., Spruck J., Global and local behavior of positive solutions of nonlinear elliptic equations. *Comm. Pure Appl. Math.* 34 (1981), no. 4, 525–598.

[19] Kondratiev V., Liskevich V., Sobol, Z., Positive solutions to semi-linear and quasi-linear elliptic equations on unbounded domains. In *Handbook of differential equations: stationary partial differential equations. Vol. VI,* 177–267, *Handb. Differ. Equ.*, Elsevier/North-Holland, Amsterdam, 2008.

[20] Koch G., Nadirashvili N., Seregin G.A., Šverák V., Liouville theorems for the Navier–Stokes equations and applications. *Acta Math.* 203 (2009), no. 1, 83–105.

[21] Mercuri C., Willem M., A global compactness result for the p-Laplacian involving critical nonlinearities. *Discrete Contin. Dyn. Syst.* 28 (2010), no. 2, 469–493.

[22] Mitidieri E., Pohozaev S.I., Towards a unified approach to nonexistence of solutions for a class of differential inequalities. *Milan J. Math.* 72 (2004), 129–162.

[23] Modica L., A gradient bound and a Liouville theorem for nonlinear Poisson equations. *Comm. Pure Appl. Math.* 38 (1985), no. 5, 679–684.

[24] Nečas J., Introduction to the theory of nonlinear elliptic equations. *Teubner-Texte zur Mathematik [Teubner Texts in Mathematics]*, 52. BSB B. G. Teubner Verlagsgesellschaft, Leipzig, 1983. 202 pp.

[25] Payne L.E., Some remarks on maximum principles. *J. Analyse Math.* 30 (1976), 421–433.

[26] Peletier L. A., Serrin J., Gradient bounds and Liouville theorems for quasilinear elliptic equations. *Ann. Scuola Norm Sup. Pisa Cl. Sci.* (4) 5 (1978), no. 1, 65–104.

[27] Phuc, N.C., Verbitsky, I.E., Quasilinear and Hessian equations of Lane–Emden type. *Ann. of Math.* (2) 168 (2008), no. 3, 859–914.

[28] Pigola S., Rigoli M., Setti A.G., A Liouville-type result for quasi-linear elliptic equations on complete Riemannian manifolds. *J. Funct. Anal.* 219 (2005), no. 2, 400–432.

[29] Polácik P., Quittner P., Souplet P., Singularity and decay estimates in superlinear problems via Liouville-type theorems. I. Elliptic equations and systems. *Duke Math. J.* 139 (2007), no. 3, 555–579.

[30] Pucci P., Serrin J., The maximum principle. *Progress in Nonlinear Differential Equations and their Applications*, 73. Birkhäuser Verlag, Basel, 2007.

[31] Pucci P., Serrin J., A remark on entire solutions of quasilinear elliptic equations. *J. Differential Equations* 250 (2011), no. 2, 675–689.

[32] Ruiz D., A priori estimates and existence of positive solutions for strongly nonlinear problems. *J. Differential Equations* 199 (2004), no. 1, 96–114

[33] Serrin J., Entire solutions of quasilinear elliptic equations. *J. Math. Anal. Appl.* 352 (2009), no. 1, 3–14.

[34] Sperb, R.P., Maximum principles and their applications. *Mathematics in Science and Engineering*, 157. Academic Press, Inc. [Harcourt Brace Jovanovich, Publishers], New York-London, 1981.

[35] Vétois J., A priori estimates for solutions of anisotropic elliptic equations. *Nonlinear Anal.* 71 (2009), no. 9, 3881–3905.

[36] Zou H., A priori estimates and existence for quasi-linear elliptic equations. *Calc. Var. Partial Differential Equations* 33 (2008), no. 4, 417–437.

Alberto Farina
Université de Picardie Jules Verne
LAMFA, CNRS UMR 6140
Amiens, France

Proc. Lond. Math. Soc. (3) 24 (1972), 348–366. **Reprint I.6.[i]**

ENTIRE SOLUTIONS OF NONLINEAR POISSON EQUATIONS

By JAMES SERRIN

[Received 29 April 1971—Revised 14 May 1971]

We shall be concerned here with the classical Poisson partial differential equation

$$\Delta u = f(x, u, \mathrm{D}u)$$

on the Euclidean space $\mathbf{R}^n$. Here f is a given function of class C^1 in its variables, $\mathrm{D}u$ denotes the gradient vector of the solution $u = u(x)$, and Δ is the Laplace operator. By an *entire solution* we mean a solution which is defined and of class C^3 on all of $\mathbf{R}^n$. The simplest case of our results can be stated in the following way.

THEOREM. *Let u be an entire solution of the differential equation*

(1) $$\Delta u = f(\mathrm{D}u).$$

Suppose that both u and Du *are bounded. Then $u \equiv$ constant.*

In geometric terms, equation (1) may be interpreted as saying (for example) that the charge density in an electric field is a function of the electric force alone. In this case, the result of the theorem is that if both the force and the potential are bounded over all space, then in fact the field must vanish. Conversely, if the charge density or the force do not vanish, then either the potential or the force must become unbounded. We shall see later that in the circumstances of the theorem one can replace the boundedness condition on u by the weaker assumption that $u = o(\log d)$ as d becomes large (here d denotes the distance from some fixed origin of coordinates).

When the function f depends also on x and u it is necessary to add some further condition on its structure. The simplest case of this is the following

THEOREM. *Let u be an entire solution of the differential equation*

(2) $$\Delta u = f(u, \mathrm{D}u),$$

where $\partial f/\partial u \geqslant 0$. Suppose that both u and Du *are bounded. Then $u \equiv$ constant.*

Without further hypotheses on f we cannot allow here the weaker assumption $u = o(\log d)$. On the other hand, in the interesting special case when f depends only on u one can drop the condition that Du be bounded and even allow u to be $o(d)$ as d becomes large. The restriction that u be $o(d)$ cannot be removed, however, since even the Laplace equation admits linear entire solutions.

The results stated above are immediate corollaries of the main Liouville-type theorems of §§ 2 and 3. Moreover, as we shall show in § 4, the method of proof applies to much more general second-order elliptic differential operators than the Laplacian. To indicate somewhat the scope which may be allowed, we note a further result somewhat analogous to the Bernstein theorem for the minimal surface equation.

THEOREM. *Let u be an entire solution of the differential equation*

$$(1+|\mathrm{D}u|^2)\Delta u - \frac{\partial u}{\partial x_i}\frac{\partial u}{\partial x_j}\frac{\partial^2 u}{\partial x_i \partial x_j} = f(\mathrm{D}u), \tag{3}$$

and suppose that both u and $\mathrm{D}u$ are bounded. Then $u \equiv$ constant.

Since equation (3) relates directly to the mean curvature of the surface $x_{n+1} = u(x)$, one obtains the following geometric interpretation. Let S be a non-parametric surface of class C^3 defined over $\mathbf{R}^n$, whose mean curvature is a function of the gradient vector. Then, if both the gradient and the ordinate of the surface are bounded, the surface is a plane.† We can replace here the condition that u be bounded by the weaker assumption that u is $o(\log d)$ as d becomes large; alternatively f can be allowed to depend on x and u provided further conditions are added.

In the results stated above the conclusion $u \equiv$ constant implies that $f(u, p)$ must satisfy

$$f(c, 0) = 0$$

for at least one real value c. In fact such a condition is *necessary* if we are to discuss entire solutions which are $o(d)$ as d becomes large. This is shown by the following proposition (we consider only the Laplace operator for simplicity).

Let u be an entire solution of the equation

$$\Delta u = f(u, \mathrm{D}u), \tag{4}$$

where $\partial f/\partial u \geqslant 0$. Suppose that $u = o(d)$ as $d \to \infty$. Then $f(c, 0) = 0$ for every c in the range of u.

To prove this, we build on an idea of Redheffer ([**6**]). Let P be an arbitrary point of $\mathbf{R}^n$, and let c be the value of u at P. Taking P as origin of coordinates, consider for $\varepsilon > 0$ the function

$$u - \varepsilon h - c + 2\varepsilon, \quad \text{where} \quad h = \sqrt{(1+|x|^2)}.$$

This is positive at P, but negative for sufficiently large values of x in view of the order condition on u. Consequently it takes a positive maximum

† A similar result for the minimal surface equation was found by Moser ([**4**]).

value at some point Q; hence at Q

$$u \geqslant \varepsilon h + c - 2\varepsilon \geqslant c - 2\varepsilon, \quad \mathrm{D}u = \varepsilon\,\mathrm{D}h, \quad \Delta(u - \varepsilon h) \leqslant 0.$$

Therefore, using the condition $\partial f/\partial u \geqslant 0$, we find at Q

$$f(c - 2\varepsilon, \varepsilon\,\mathrm{D}h) \leqslant f(u, \varepsilon\,\mathrm{D}h) = f(u, \mathrm{D}u) = \Delta u \leqslant \varepsilon\Delta h.$$

By direct calculation

$$|\mathrm{D}h| \leqslant 1, \quad \Delta h = \ \leqslant n.$$

Thus, letting $\varepsilon \to 0$ in (4), there results

$$f(c, 0) \leqslant 0.$$

Similarly, considering the function $u + \varepsilon h - c - 2\varepsilon$, we find that $f(c, 0) \geqslant 0$, and the proposition is proved.

As a corollary, if $\partial f/\partial u \geqslant 0$, the only entire solutions of $\Delta u = f(u)$ which are $o(d)$ are constants (a remark noted earlier). Indeed in this case the proposition implies that $f(u) = 0$ over the range of the solution; hence $\Delta u = 0$, and the result follows from the standard Liouville theorem for Laplace's equation.

As noted previously, the main results are proved in §§ 2–4. Section 5 treats the case where f depends on x as well as on u and p. The first section of the paper contains the derivation of an identity which is fundamental to the later proofs; although this identity is entirely elementary, it is at the same time rather complicated and we must apologize to the reader for the care necessary to follow the calculations.

Other Liouville-type theorems have been obtained by Bernstein ([**1**]), Redheffer ([**6**]), Gilbarg and Serrin ([**3**]), and Bohn and Jackson ([**2**]). The paper of Redheffer in particular presents a number of interesting theorems for the differential inequality $\Delta u \geqslant f(u, |\mathrm{D}u|)$ and contains a bibliography of earlier works (Redheffer's results, however, do not apply to the *equation* $\Delta u = f(u, \mathrm{D}u)$ unless f has a very specific form; in particular it would be necessary for $\partial f/\partial p$ to be zero at $p = 0$ and, for $n > 2$, the second differences of f with respect to p would have to become unbounded as $p \to 0$).

1. It will be convenient to begin with some simple matters of notation. If $u = u(x)$ is a function of class C^3, we let

$$\mathrm{D}u = (u_1, \ldots, u_n) = \left(\frac{\partial u}{\partial x_1}, \ldots, \frac{\partial u}{\partial x_n}\right)$$

be its gradient vector, and similarly denote by $\mathrm{D}^2 u$ the matrix of its second derivatives $u_{ij} = \partial^2 u/\partial x_i\,\partial x_j$.

Let $A(u,p)$ be an $n \times n$ matrix whose components $A_{ij}(u,p)$ are continuously differentiable functions of the variables u and $p = (p_1, \ldots, p_n)$, for all real values of u and p. Similarly let $f(u,p)$ be a continuously differentiable function of u and p. Now consider the differential equation

$$A_{ij}(u, \mathrm{D}u)u_{ij} = f(u, \mathrm{D}u), \tag{5}$$

which for convenience will be written in the compressed form

$$A\,\mathrm{D}^2u = f. \tag{5$'$}$$

In the following sections we shall interpret A as the identity matrix, so that (5) becomes $\Delta u = f(u, \mathrm{D}u)$. For the moment, however, it is valuable to retain greater generality in the operator A.

Our first object is to derive from (5) a corresponding differential equation† for the function

$$w = |\mathrm{D}u|^2.$$

To this end, let us apply the differential operator $u_k(\partial/\partial x_k)$ to (5); with an obvious notation for the partial derivatives of A and f, this yields easily

$$\begin{aligned} 0 &= u_k \frac{\partial}{\partial x_k}[A_{ij}u_{ij} - f] \\ &= A_{ij}u_k u_{ijk} + \left(\frac{\partial A_{ij}}{\partial u}u_{ij} - \frac{\partial f}{\partial u}\right)u_k u_k + \left(\frac{\partial A_{ij}}{\partial p_l}u_{ij} - \frac{\partial f}{\partial p_l}\right)u_k u_{lk}. \end{aligned}$$

Now $w = u_k u_k$, so that $w_i = 2u_k u_{ki}$ and $w_{ij} = 2u_k u_{kij} + 2u_{ik}u_{jk}$. Consequently the preceding identity may be rewritten in the form

$$A_{ij}w_{ij} = 2A_{ij}u_{ik}u_{jk} - 2\left(\frac{\partial A_{ij}}{\partial u}u_{ij} - \frac{\partial f}{\partial u}\right)w - \left(\frac{\partial A_{ij}}{\partial p_l}u_{ij} - \frac{\partial f}{\partial p_l}\right)w_l.$$

At this point it is useful to introduce the two differential operators

$$\delta \equiv -\frac{\partial}{\partial u} + r$$

and

$$\delta_2 \equiv \frac{\partial}{\partial p} + t = \left(\frac{\partial}{\partial p_1} + t_1, \ldots, \frac{\partial}{\partial p_n} + t_n\right),$$

r and $t = (t_1, \ldots, t_n)$ being respectively a given scalar function and a given vector function of the variables u and p. Using (5) once more, the last identity can then be written

$$A\,\mathrm{D}^2w = 2\,\mathrm{D}^2uA\,\mathrm{D}^2u + 2(\delta A\,\mathrm{D}^2u - \delta f)w - (\delta_2 A\,\mathrm{D}^2u - \delta_2 f)\cdot \mathrm{D}w.$$

† This method occurs first in the work of Bernstein, and has since been extended and developed in papers of Leray, Ladyzhenskaya, and others.

Note in particular that A and f are functions of u and $p = \mathrm{D}u$ and that the operators δ and δ_2 apply only to the first symbol directly following them.

Using a technique of Oleinik and Kruzhkov (**[5]**), we next consider the function

$$z = \zeta w, \quad \zeta = \left(1 - \frac{|x|^2}{d^2}\right)^2$$

in the domain $\Sigma = \{|x| \leqslant d\}$. Since

$$w = \frac{z}{\zeta}, \quad \mathrm{D}w = \frac{\mathrm{D}z}{\zeta} - \frac{\mathrm{D}\zeta}{\zeta^2}z, \quad \mathrm{D}^2 w = \frac{\mathrm{D}^2 z}{\zeta} - 2\frac{\mathrm{D}\zeta\,\mathrm{D}z}{\zeta^2} - \left(\frac{\mathrm{D}^2\zeta}{\zeta^2} - 2\frac{\mathrm{D}\zeta\,\mathrm{D}\zeta}{\zeta^3}\right)z,$$

it is evident that z satisfies the relation

$$\begin{aligned} A\,\mathrm{D}^2 z &= 2\zeta\,\mathrm{D}^2 u A\,\mathrm{D}^2 u + 2(\delta A\,\mathrm{D}^2 u - \delta f)z \\ &\quad + (\delta_2 A\,\mathrm{D}^2 u - \delta_2 f)\cdot\frac{\mathrm{D}\zeta}{\zeta}z + \left(\frac{A\,\mathrm{D}^2\zeta}{\zeta} - 2\frac{\mathrm{D}\zeta A\,\mathrm{D}\zeta}{\zeta^2}\right)z + M\cdot\mathrm{D}z, \end{aligned} \tag{6}$$

where M is a continuous function on Σ and $\mathrm{D}\zeta A\,\mathrm{D}\zeta$ denotes the quadratic form on A generated by $\mathrm{D}\zeta$.

While the identity (6) is fundamental in what follows, it is in fact of importance to express it as a differential inequality in which the quantity $\mathrm{D}^2 u$ is absent. To this end, we suppose (as will be the case in the applications) that the matrix A is positive definite. Then letting $\lambda = \lambda(u, p)$ be the least eigenvalue of A, it follows that

$$\mathrm{D}^2 u A\,\mathrm{D}^2 u \geqslant \lambda \|\mathrm{D}^2 u\|^2.$$

On the other hand, by Cauchy's inequality,

$$\begin{aligned} 2|\delta A\,\mathrm{D}^2 u|z &\leqslant \lambda\zeta\|\mathrm{D}^2 u\|^2 + \frac{\|\delta A\|^2 z^2}{\lambda\zeta} \\ &= \lambda\zeta\|\mathrm{D}^2 u\|^2 + \frac{|p|^2\|\delta A\|^2}{\lambda}z, \\ |\delta_2 A\,\mathrm{D}^2 u| \times \left|\frac{\mathrm{D}\zeta}{\zeta}\right| z &\leqslant \lambda\zeta\|\mathrm{D}^2 u\|^2 + \frac{\|\delta_2 A\|^2}{4\lambda\zeta}\left|\frac{\mathrm{D}\zeta}{\zeta}\right|^2 z^2 \\ &= \lambda\zeta\|\mathrm{D}^2 u\|^2 + \frac{|p|^2\|\delta_2 A\|^2}{4\lambda}\left|\frac{\mathrm{D}\zeta}{\zeta}\right| z, \end{aligned}$$

where p is used as a replacement variable for $\mathrm{D}u$.

Finally, by direct calculation,

$$|\mathrm{D}\zeta| \leqslant \frac{4}{d}\zeta^{\frac{1}{2}}, \quad \|\mathrm{D}^2\zeta\| \leqslant \frac{4n}{d^2}.$$

Inserting the preceding inequalities into (6), we obtain the important relation

$$(7) \qquad A\,\mathrm{D}^2 z \geqslant M\cdot \mathrm{D}z - 2Nz \quad \text{in } \Sigma$$

with

$$(8) \qquad N = \delta f + \frac{|p|^2 \| \delta A \|^2}{2\lambda} + 2|\delta_2 f| \frac{|p|}{d\sqrt{z}} + 2\left\{(n+8)\|A\| + \frac{|p|^2 \|\delta_2 A\|^2}{\lambda}\right\} \frac{|p|^2}{d^2 z},$$

using at the last stage the fact that $\zeta = z/w = z/|p|^2$.

In the following sections (7) will be used only after an initial change of dependent variables. Consider in particular a new variable $\bar{u}$, related to u through the transformation

$$u = \varphi(\bar{u}), \quad \varphi' > 0.$$

One checks without difficulty that, if u satisfies (5), then $\bar{u}$ satisfies the equation

$$\bar{A}\,\mathrm{D}^2 \bar{u} = \bar{f}$$

with

$$(9) \qquad \bar{A} = A, \quad \bar{f} = \frac{|\bar{p}|}{|p|}(f + \omega E)$$

and

$$(9') \qquad \omega = -\frac{\varphi''}{\varphi'^2}, \quad E = pAp,$$

where the arguments of A and f are u and $p = \varphi'(\bar{u})\bar{p} = \varphi'(\bar{u})\,\mathrm{D}\bar{u}$. Note that the symbol $\bar{p}$ is used here as a replacement variable for $\mathrm{D}\bar{u}$ in the same way as p is used for $\mathrm{D}u$.

It follows at once from the preceding work (specifically (7) and (8)) that the function

$$\bar{z} = \zeta \bar{w} = \zeta |\mathrm{D}\bar{u}|^2$$

satisfies the inequality

$$(10) \qquad A\,\mathrm{D}^2 \bar{z} \geqslant \bar{M}\cdot \mathrm{D}\bar{z} - 2\bar{N}\bar{z},$$

where $\bar{N}$ denotes the expression N evaluated for the barred variables. To obtain $\bar{N}$ explicitly, observe that

$$|\bar{p}|\bar{\delta} = |p|\delta + \omega\delta_1 |p|, \quad \delta_1 \equiv p_l \frac{\partial}{\partial p_l} + s,$$

$$|\bar{p}|\delta_2 = |p|\delta_2$$

(here one uses $\bar{r} = |\bar{p}|^{-1}|p|(r + \omega(s+1))$, $\bar{t} = |\bar{p}|^{-1}|p|t$). Consequently

$$|\bar{p}|\bar{\delta}\bar{A} = |p|\delta A + \omega\delta_1(|p|A), \quad |\bar{p}|\delta_2\bar{A} = |p|\delta_2 A, \quad \delta_2\bar{f} = \delta_2 f + \omega\delta_2 E$$

and

$$\delta\bar{f} = \delta(f+\omega E)+\omega\delta_1(f+\omega E)$$
$$= \omega^2\delta_1 E+\omega(\delta E+\delta_1 f)+\delta f-\frac{\omega'}{\varphi'}E.$$

Finally $|\bar{p}|^2/\bar{z} = |p|^2/z$ since $z = \varphi'^2\bar{z}$. Substitution of these expressions into the barred form of (8) yields the fundamental relation

$$E^{-1}\bar{N} \leqslant \alpha\omega^2+2\beta\omega+\gamma-\frac{\omega'}{\varphi'}$$

$$(11)\qquad +\frac{2}{E}|\delta_2 f+\omega\delta_2 E|\frac{|p|}{d\sqrt{z}}+\frac{2}{E}\left\{(n+8)\|A\|+\frac{|p|^2\|\delta_2 A\|^2}{\lambda}\right\}\frac{|p|^2}{d^2z},$$

the coefficients α, β, γ being defined by

$$\alpha = \frac{1}{E}\left\{\delta_1 E+\frac{\|\delta_1(|p|A)\|^2}{\lambda}\right\},\quad 2\beta = \frac{1}{E}\{\delta E+\delta_1 f\},\quad \gamma = \frac{1}{E}\left\{\delta f+\frac{\|\delta(|p|A)\|^2}{\lambda}\right\}$$

(in these formulae it should be emphasized again that p is a substitution variable for Du).

2. We now turn to the main results of the paper, recalling that the solutions u under consideration are of class C^3 on the entire Euclidean space $\mathbf{R}^n$, and that the function f in Poisson's equation is of class C^1 in all its variables.

THEOREM 1. *Let u be an entire solution of the equation*

$$\Delta u = f(u, \mathrm{D}u),$$

where $\partial f/\partial u \geqslant 0$. Suppose that u and Du *are bounded. Then $u \equiv$ constant.*

Proof. Let P be an arbitrary point of $\mathbf{R}^n$ which we shall hold fixed throughout, and let $\varepsilon > 0$ be a given small number. It will be shown that

$$|\mathrm{D}u| \leqslant 2\varepsilon \quad \text{at } P,$$

and this (since ε and P are arbitrary) will complete the proof.

We introduce a new dependent variable $\bar{u}$ by means of the transformation

$$u = \varphi(\bar{u}),\quad \varphi' > 0.$$

Choosing P as origin of coordinates and putting

$$\zeta = 1-\left(\frac{|x|^2}{d^2}\right)\quad (d > 0),$$

it then follows from (10) that the function $\bar{z} = \zeta |D\bar{u}|^2$ satisfies the differential inequality

$$\Delta \bar{z} \geqslant \bar{M} \cdot D\bar{z} - 2\bar{N}\bar{z} \quad \text{in } \Sigma; \tag{12}$$

moreover, using (9′) and (11), we have $E = |p|^2$ and

$$|p|^{-2}\bar{N} \leqslant \alpha\omega^2 + 2\beta\omega + \gamma - \frac{\omega'}{\varphi'}$$

$$+ 2\left\{\frac{|\delta_2 f|}{|p|} + (2 + |p| \times |t|)|\omega|\right\}\frac{1}{d\sqrt{z}}$$

$$+ 2\{(n+8)\sqrt{n} + n|p|^2|t|^2\}\frac{1}{d^2 z}, \tag{13}$$

where

$$\alpha = (s+2) + n(s+1)^2, \quad 2\beta = |p|^{-2}\delta_1 f, \quad \gamma \leqslant 0$$

(to obtain (13) we have put $r = 0$ in the differential operator δ, and have used the hypotheses of the theorem to establish that $\delta f = -\partial f/\partial u \leqslant 0$).

By hypothesis u and Du are bounded on $\mathbf{R}^n$, whence

$$|u| \leqslant m, \quad |Du| \leqslant l$$

for appropriate constants m and l. Now let

$$a = 2\sup \alpha, \quad b = 2\sup(1 + |\beta|),$$

the suprema being taken over $|u| \leqslant m$ and $\varepsilon \leqslant |p| \leqslant l$. (Here we take $s = 0, t = 0$, though in later proofs this will not always be the case.) Choose $u = \varphi(\bar{u})$ according to the inverse relation

$$\bar{u} = \int_{-2m}^{\varphi - m} \frac{dv}{(2 - \exp(2bv))^{1/a}}. \tag{14}$$

Since u varies in the range $[-m, m]$, the integral is clearly well defined and $\varphi'(\bar{u}) > 0$. Putting $v = \varphi - m$ ($\leqslant 0$) one finds that

$$\varphi' = (2 - \exp(2bv))^{1/a}, \quad \omega = \frac{2b}{a}\frac{\exp(2bv)}{2 - \exp(2bv)},$$

$$\frac{\omega'}{\varphi'} = \frac{8b^2}{a}\frac{\exp(2bv)}{(2 - \exp(2bv))^2} = a\omega^2 + 2b\omega.$$

Therefore, for $|u| \leqslant m$, $\varepsilon \leqslant |p| \leqslant l$, there is a constant B, depending only on ε, m, l, and the function f, such that

$$|p|^{-2}\bar{N} \leqslant -\tfrac{1}{2}a\omega^2 - b\omega + B\left(\frac{1+\omega}{d\sqrt{z}} + \frac{1}{d^2 z}\right)$$

$$\leqslant -\frac{1}{n}\exp(-4bm) + \tilde{B}\left(\frac{1}{d\sqrt{z}} + \frac{1}{d^2 z}\right) \quad \left(\tilde{B} = \left(1 + 2\frac{b}{a}\right)B\right), \tag{15}$$

since $\omega \leqslant 2b/a$, $a = 2(2+n)$, and $b \geqslant 2$.

We assert that

$$|\mathrm{D}\bar{u}| \leqslant \varepsilon' = \frac{\varepsilon}{\min \varphi'} \quad \text{at } P. \tag{16}$$

Indeed, suppose that the contrary holds. Then, since $\zeta = 1$ at P, we would have

$$\bar{z} > \varepsilon'^2 \quad \text{at } P.$$

Consequently (since $\bar{z} = 0$ when $|x| = d$), there must be a point Q in Σ where $\bar{z}$ has a maximum value exceeding ε'^2. At Q we have then (recall that $z = \varphi'^2\bar{z}$)

$$\varepsilon = \varepsilon' \min \varphi' \leqslant \sqrt{\bar{z}} \min \varphi' \leqslant \sqrt{z}$$

and moreover, since $z = \zeta |\mathrm{D}u|^2 \leqslant |\mathrm{D}u|^2$,

$$|u| \leqslant m, \quad \varepsilon \leqslant |\mathrm{D}u| \leqslant l.$$

Also at Q

$$\Delta \bar{z} \leqslant 0, \quad \mathrm{D}\bar{z} = 0.$$

Consequently, using (12) and (15),

$$0 \leqslant \bar{N} \leqslant l^2 \left\{ -\frac{1}{n} \exp(-4bm) + \tilde{B}\left(\frac{1}{d\varepsilon} + \frac{1}{d^2\varepsilon^2}\right) \right\} \quad \text{at } Q,$$

which gives an obvious contradiction for sufficiently large d. Thus (16) is proved.

It now follows that at P

$$|\mathrm{D}u| = \varphi' |\mathrm{D}\bar{u}| \leqslant \varepsilon' \max \varphi' = \varepsilon \frac{\max \varphi'}{\min \varphi'} \leqslant 2\varepsilon.$$

This completes the proof of Theorem 1.

THEOREM 2. *Let u be an entire solution of the equation*

$$\Delta u = f(u, \mathrm{D}u),$$

where $\partial f/\partial u \geqslant 0$. Assume also that $\partial f(u,p)/\partial p$ is bounded (independently of u) when p lies in any compact subset of $\mathbf{R}^n \backslash \{0\}$.

Then, if $\mathrm{D}u$ *is bounded and $u = o(\log d)$ as $d \to \infty$, we have $u \equiv$ constant.*

Proof. We proceed exactly as in the proof of Theorem 1, except that, since now u can become unbounded, one must replace m in the definition (14) of φ by

$$m' = \sup_{\Sigma} |u|.$$

Clearly $m' = o(\log d)$ according to hypothesis. Therefore (15) yields further, for $|u| \leqslant m'$, $\varepsilon \leqslant |p| \leqslant l$, and for suitably large d,

$$|p|^{-2}\bar{N} \leqslant -\frac{1}{n\sqrt{d}} + \tilde{B}\left(\frac{1}{d\sqrt{z}} + \frac{1}{d^2 z}\right).$$

The remaining part of the proof is the same as before.

It is clear that the boundedness condition on $\partial f(u,p)/\partial p$ could be avoided provided that the growth restriction on u is strengthened appropriately. Rather than pursue this, however, we wish now to show that in certain circumstances the growth restriction on u can be weakened to the optimal order $o(d)$.

For this purpose, we say that the function f satisfies *condition* H if, for any pair of positive numbers ε and l, there exists a corresponding number C such that

$$\left|\frac{\partial f(u,p)}{\partial p}\right| \leqslant C|f|$$

when $\varepsilon \leqslant |p| \leqslant l$, $-\infty < u < \infty$. With this hypothesis, we have

THEOREM 3. *Let u be an entire solution of the equation*

$$\Delta u = f(u, \mathrm{D}u),$$

where $\partial f/\partial u \geqslant 0$ and f satisfies condition H. *Suppose that* Du *is bounded and that $u = o(d)$ as $d \to \infty$. Then $u \equiv$ constant.*

Proof. As in the proof of Theorem 1, we consider the differential equation (12) satisfied by $\bar{z}$. We take $r = 0$ and then choose the functions s and t so that

$$\delta_1 f = \delta_2 f = 0.$$

This requires in particular that

$$s = -\frac{p_i}{f}\frac{\partial f}{\partial p_i}, \quad t_i = -\frac{1}{f}\frac{\partial f}{\partial p_i}.\dagger$$

Since we are assuming that condition H is satisfied, both s and t are bounded in the range $\varepsilon \leqslant |p| \leqslant l$, $-\infty < u < \infty$, where l denotes the supremum of $|\mathrm{D}u|$.

We now find (see (13)) that

$$|p|^{-2}\bar{N} \leqslant \alpha\omega^2 - \frac{\omega'}{\varphi'} + 2(2+|p|\times|t|)|\omega|\frac{1}{d\sqrt{z}} + 2\{(n+8)\sqrt{n}+n|p|^2|t|^2\}\frac{1}{d^2 z},$$

where $\alpha = (s+2)+n(s+1)^2$. Putting $a = 2\sup\alpha$, the supremum being taken for $\varepsilon \leqslant |p| \leqslant l$ and $-\infty < u < \infty$, let us choose $u = \varphi(\bar{u})$ according to the relation

$$\bar{u} = \int_{-m'}^{\varphi} \frac{dv}{(3m'-v)^{1/a}},$$

† At any point where $f = 0$ we define $s = t = 0$.

where $m' = \sup_\Sigma |u|$. Since u varies in the range $[-m', m']$ when x is in Σ, the integral is well defined and $\varphi'(\bar{u}) > 0$. Moreover,

$$\varphi' = (3m' - \varphi)^{1/a}, \quad \omega = \frac{1}{a}(3m' - \varphi)^{-1}, \quad \frac{\omega'}{\varphi'} = a\omega^2.$$

Thus, for $|u| \leqslant m'$, $\varepsilon \leqslant |p| \leqslant l$, there is a constant B such that

$$\begin{aligned}|p|^{-2}\bar{N} &\leqslant -\tfrac{1}{2}a\omega^2 + B\left(\frac{\omega}{d\sqrt{z}} + \frac{1}{d^2 z}\right)\\ &= -\frac{1}{2a(3m'-u)^2} + B\left(\frac{1}{a(3m'-u)d\sqrt{z}} + \frac{1}{d^2 z}\right)\\ &\leqslant -\frac{1}{32am'^2} + B\left(\frac{1}{2am'd\sqrt{z}} + \frac{1}{d^2 z}\right) = \frac{1}{32am'^2}\left\{-1 + \tilde{B}\left(\frac{m'}{d\sqrt{z}} + \frac{m'^2}{d^2 z}\right)\right\},\end{aligned}$$

where $\tilde{B}$ is another constant $(= 32aB)$.

The remaining part of the proof is now the same as before, once we note that $m' = o(d)$ and $\max\varphi'/\min\varphi' \leqslant 2$.

The Liouville-type theorems which we have obtained so far all rely on the assumption that the gradient Du is bounded. If by some additional argument we are able to show that the gradient *must* be bounded, then of course the previous results yield a Liouville theorem of the usual type. Such an additional result obviously must take into account the asymptotic behaviour of the function $f(u,p)$ for large p, and accordingly it becomes natural to consider this aspect of the problem separately from the part which has already been treated. We shall turn our attention to this matter in the following section.

3. We say that the function $f(u,p)$ satisfies *condition* H_∞ if

$$\frac{\partial f(u,p)}{\partial p} = O\left(\frac{f(u,p)}{|p|}\right)$$

as $p \to \infty$, uniformly in u. In particular this condition holds if there exist positive numbers C and L such that

$$\left|\frac{\partial f(u,p)}{\partial p}\right| \leqslant C\frac{|f(u,p)|}{|p|} \quad \text{for } |p| \geqslant L, \quad -\infty < u < \infty.$$

The following result is now valid.

THEOREM 4. *Let u be an entire solution of the equation*

$$\Delta u = f(u, \mathrm{D}u),$$

where $\partial f/\partial u \geqslant 0$ and f satisfies condition H_∞. Suppose that $u = o(d)$ as $d \to \infty$. Then Du *is bounded.*

Proof. This is exactly the same as the proof of Theorem 3, except that one replaces ε by L and the range $\varepsilon \leqslant |p| \leqslant l$ by $|p| \geqslant L$. This being done, the condition $|D\bar{u}| \geqslant L' = L/\min\varphi'$ at P then leads in the usual way to a contradiction for sufficiently large d. Hence we find that

$$|Du| \leqslant \max\varphi' |D\bar{u}| \leqslant 2L$$

at P, completing the proof.

Condition H_∞ is not the most general one for which the boundedness of the gradient can be proved. We say that $f(u,p)$ satisfies *condition* I_∞ if there exists a decomposition of f,

$$f = f_1 + f_2, \tag{17}$$

such that

$$\frac{\partial f_1}{\partial p} = O\left(\frac{f_1}{|p|}\right); \quad \frac{\partial f_2}{\partial p}, \frac{f_2}{|p|} = O(|p|) \tag{18}$$

as $p \to \infty$, uniformly in u. Similarly we say that $f(u,p)$ satisfies the *restricted condition* I_∞ if (18) holds as $p \to \infty$ for any bounded set of values u.

THEOREM 5. *Let u be an entire solution of the equation*

$$\Delta u = f(u, Du),$$

where $\partial f/\partial u \geqslant 0$ and f satisfies condition I_∞. Suppose also that $u = o(\log d)$ as $d \to \infty$. Then $|Du|$ is bounded.

If f satisfies only the restricted condition I_∞, then the conclusion remains true for bounded solutions u.

Proof. After introducing the new variable $\bar{u}$ and putting

$$r = 0, \quad s = -\frac{p_i}{f_1}\frac{\partial f_1}{\partial p_i}, \quad t_i = -\frac{1}{f_1}\frac{\partial f_1}{\partial p_i},$$

we find that $\bar{z}$ satisfies (12) with

$$|p|^{-2}\bar{N} \leqslant \alpha\omega^2 + 2\beta\omega - \frac{\omega'}{\varphi'} + 2\left\{\frac{1}{|p|}\left|\frac{\partial f_2}{\partial p}\right| + \frac{|t| \times |f_2|}{|p|} + (2 + |p| \times |t|)|\omega|\right\}\frac{1}{d\sqrt{z}}$$

$$+ 2\{(n+8)\sqrt{n} + n|p|^2|t|^2\}\frac{1}{d^2 z}$$

and

$$\alpha = (s+2) + n(s+1)^2, \quad 2\beta = |p|^{-2}\delta_1 f_2.$$

By condition I_∞ the functions s, $|p|t$, $|p|^{-1}\partial f_2/\partial p$, and $|p|^{-2}f_2$ are bounded for $|p| \geqslant L$, say, and all values of u. This being shown, the remaining argument is the same as in Theorems 1 and 2, with ε replaced by L and the range $\varepsilon \leqslant |p| \leqslant l$ replaced by $|p| \geqslant L$.

It should be remarked that the condition $\partial f/\partial u \geqslant 0$ in Theorems 4 and 5 can easily be replaced with the weaker hypothesis

$$\limsup_{p\to\infty} \frac{1}{|p|^2} \frac{\partial f(u,p)}{\partial u} \geqslant 0.$$

We omit the details.

Liouville theorems

By combining the previous results we can obtain various Liouville theorems in which no hypothesis is placed on the behaviour of the gradient of the solution.

We shall say that the function $f(u,p)$ satisfies *condition* H_0 if, for every $\varepsilon > 0$, there exists a corresponding number C such that

$$|p| \times \left|\frac{\partial f}{\partial p}\right| \leqslant C|f|$$

for $|p| \geqslant \varepsilon$, $-\infty < u < \infty$. Similarly we shall say that $f(u,p)$ satisfies *condition* I_0 if, for every $\varepsilon > 0$, there exists a decomposition of f,

$$f = f_1 + f_2,$$

such that

$$|p| \times \left|\frac{\partial f_1}{\partial p}\right| \leqslant C|f_1|, \quad |p| \times \left|\frac{\partial f_2}{\partial p}\right| + |f_2| \leqslant C|p|^2$$

for $p \geqslant \varepsilon$, $-\infty < u < \infty$. Finally, we say that f satisfies the *restricted condition* I_0 if f satisfies I_0 for any bounded set of values u.

THEOREM. *Let u be an entire solution of the equation*

$$\Delta u = f(u, \mathrm{D}u)$$

with $\partial f/\partial u \geqslant 0$. Suppose that any one of the following three conditions holds.

(i) *f satisfies the restricted condition I_0 and u is bounded.*

(ii) *f satisfies condition I_0 and $u = o(\log d)$ as $d \to \infty$.*

(iii) *f satisfies condition H_0 and $u = o(d)$ as $d \to \infty$.*

Then $u \equiv$ constant.

Proof. Case (i) follows from Theorems 1 and 5. Case (iii) follows from Theorems 3 and 4 (or in fact directly from Theorem 4 since we can take $L = \varepsilon$). Case (ii) follows from Theorem 5 by taking $L = \varepsilon$ arbitrarily small.

4. The methods developed in the preceding sections can be applied also to elliptic differential equations of the form

$$A(\mathrm{D}u)\,\mathrm{D}^2u = f(u, \mathrm{D}u).$$

For simplicity we shall confine our discussion to the analogues of Theorems 1, 2, and 3, it being supposed throughout that the gradient of the solution is bounded, say $|\mathrm{D}u| \leqslant l$. Since A is positive definite its least eigenvalue $\lambda(p)$ satisfies

$$\lambda(p) \geqslant \kappa \quad \text{when } |p| \leqslant l$$

for some positive constant κ; hence $E = pAp \geqslant \kappa |p|^2$. Moreover, for some appropriate constant K,

$$\|\partial A/\partial p\|,\ \|A\| \leqslant K \quad \text{when } |p| \leqslant l.$$

Finally we note that

$$\delta A = \delta E = 0,$$

since $r = 0$ in the operator δ. The arguments of § 2 now apply almost word for word, the only necessary changes being due to the fact that the values κ and K will enter the relations. We have therefore proved the following

THEOREM 6. *The results of Theorems* 1 *to* 3 *remain unchanged if we replace the equation* $\Delta u = f(u, \mathrm{D}u)$ *by* $A(\mathrm{D}u)\,\mathrm{D}^2u = f(u, \mathrm{D}u)$, *where* $A(p)$ *is a continuously differentiable positive definite matrix in the variable* p.

As a special case of this result, consider the matrix

$$A(p) = (1 + |p|^2)I - pp,$$

where I is the identity and pp denotes the dyadic $p_i p_j$. This yields the third theorem stated in the introduction. Concerning bounds on $|\mathrm{D}u|$, results analogous to Theorems 4 and 5 are given in [**7**].

5. Here we shall indicate how the previous results can be modified in order to treat the more general equation

$$\Delta u = f(x, u, \mathrm{D}u),$$

where, as always, it is assumed that f is of class C^1 in all its variables.

The calculations of § 1 apply equally when the matrix A and the function f depend on x, except that one must now put

$$\delta \equiv -\left(\frac{\partial}{\partial u} + \frac{p_i}{|p|^2}\frac{\partial}{\partial x_i}\right) + r.$$

Here of course r can additionally depend on x, as is also the case for the multipliers s and t in the operators δ_1 and δ_2.

By applying Cauchy's inequality with a free constant, it is also not hard to see that (11) holds with the terms

$$\|\delta(|p|A)\|^2, \quad \|\delta_1(|p|A)\|^2, \quad \|\delta_2 A\|^2$$

multiplied respectively by the additional factors

$$\tfrac{1}{4}(1+\theta), \qquad \frac{2}{\theta}, \qquad \frac{2}{\theta},$$

where θ is any constant between 0 and 1. Moreover, θ can be chosen zero if $\delta_1(|p|A)$ and $\delta_2 A$ vanish.

This being the case, it follows that, when $A = I$, the coefficient γ in (11) can be written in the form

$$|p|^2\gamma = -\left(\frac{\partial f}{\partial u}+\frac{p_i}{|p|^2}\frac{\partial f}{\partial x_i}\right)+rf+\tfrac{1}{4}(1+\theta)n|p|^2r^2.$$

Choosing

$$r = -\frac{2f}{(1+\theta)n|p|^2} \tag{19}$$

(rather than $r = 0$), one finds that

$$|p|^2\gamma = -\left(\frac{\partial f}{\partial u}+\frac{p_i}{|p|^2}\frac{\partial f}{\partial x_i}+\frac{f^2}{(1+\theta)n|p|^2}\right).$$

In consequence it is possible to generalize the earlier theorems by allowing f to depend on x, u, p and by replacing the condition $\partial f/\partial u \geqslant 0$ by a weaker restriction. We state the results analogous to Theorems 2 and 3.

THEOREM 7. *Let u be an entire solution of the equation*

$$\Delta u = f(x, u, \mathrm{D}u),$$

where

$$\frac{\partial f}{\partial u}+\frac{1}{|p|^2}\left(p_i\frac{\partial f}{\partial x_i}+\frac{f^2}{n}\right) \geqslant 0 \quad (p \neq 0).$$

Assume that $f(x,u,p)$ and $\partial f(x,u,p)/\partial p$ are bounded (independently of x and u) when p lies in any compact subset of $\mathbf{R}^n\backslash\{0\}$. Suppose also that $\mathrm{D}u$ is bounded and $u = o(\log d)$ as d tends to ∞. Then $u \equiv$ constant.

For the proof we choose r as above, and $s = -1$, $t = 0$. Then

$$\delta_1(|p|A) = (s+1)I = 0, \quad \delta_2 A = 0.$$

Hence we can take $\theta = 0$ in (11), the remainder of the proof being the same as for Theorem 2.

THEOREM 8. *Let u be an entire solution of the equation*

$$\Delta u = f(x, u, \mathrm{D}u),$$

where

$$\frac{\partial f}{\partial u}+\frac{1}{|p|^2}\left(p_i\frac{\partial f}{\partial x_i}+\mu\frac{f^2}{n}\right)\geqslant 0 \quad (p\neq 0)$$

for some $\mu < 1$. *Assume that* f *satisfies hypothesis* H (*see* § 2) *with the constant* C *also independent of* x. *Suppose also that* Du *is bounded and that* $u = o(d)$ *as* d *tends to infinity. Then* $u \equiv$ *constant.*

Proof. This is the same as that of Theorem 3, except that, with the new choice (19) for r, one must now put

$$s = -\frac{p_i}{f}\frac{\partial f}{\partial p_i}+\frac{2}{(1+\theta)n}$$

in order to assure $2\beta = (1/E)(\delta E+\delta_1 f) = 0$. Since this choice of s makes $\delta_1(|p|A) \neq 0$ we can no longer take $\theta = 0$; this accounts for the appearance of the constant μ in the statement of the theorem.

Even though the paper is predominantly concerned with non-linear problems it is also of interest to consider the linear equation

$$\Delta u + a(x)\cdot Du = 0, \tag{20}$$

where $a(x)$ is a bounded vector of class C^1. In this case, suppose that the matrix

$$aa+n\frac{\partial a}{\partial x}$$

(in an obvious notation) is non-negative definite. Then, if u is an entire solution satisfying $u = o(\log d)$ as $d \to \infty$, we have $u \equiv$ constant.

To see this, we first compute

$$p_i\frac{\partial f}{\partial x_i}+\frac{f^2}{n} = \frac{\partial a_j}{\partial x_i}p_ip_j+\frac{a_ia_j}{n}p_ip_j.$$

This will be non-negative if $aa+n\,\partial a/\partial x$ is non-negative definite. Moreover, since

$$f(x,p) = a\cdot p, \quad \frac{\partial f}{\partial p} = a,$$

one can apply the argument of Theorem 5 with $f_1 = 0$, $s = -1$, $t = 0$ to show that Du must be bounded. Hence Theorem 7 applies. The only other Liouville theorem known to the writer concerning equations of the type (20) is due to Gilbarg and Serrin ([**3**]). In their work, entire solutions of (20) which are $o(d)$ are shown to be constant provided that $a(x) = O(1/d)$. While this result does not require differentiability of the vector $a(x)$ it nevertheless places heavy restrictions on its asymptotic behaviour, and consequently does not apply even to constant coefficient equations. (On the other hand, the result of [3] does apply to equations whose principal part is much more general than the Laplacian.)

We note finally that results analogous to Theorems 4, 5, and 6 can also be obtained when f depends on x, u, and p.

6. We present here several remarks auxiliary to our earlier results, and also an application to parabolic equations.

1. The condition that u be of class C^3 can be omitted if it is assumed that A and f are in the Hölder class $C^{1+\mu}$ in u and p, for some $\mu > 0$. Indeed in this case if $u \in C^2(\mathbf{R}^n)$ then $A(x,u,Du)$ and $f(x,u,Du) \in C^1(\mathbf{R}_n)$ so that, by the Schauder theory, we find that $u \in C^{2+\mu}(\mathbf{R}^n)$. But then $A(x,u,Du)$ and $f(x,u,Du) \in C^{1+\mu}(\mathbf{R}^n)$. Therefore, once again by the Schauder theory, $u \in C^{3+\mu}(\mathbf{R}^n)$ as required.

2. Even if the coefficient γ in (13) does not vanish by assumption, one can still carry through the arguments provided the oscillation of u is suitably small.

3. It should be noted that the proofs do not actually require the differentiability of f at $p = 0$. Thus the results apply for example when $f = |p|^\alpha$, $0 < \alpha < 1$, even though $\partial f/\partial p$ does not exist at $p = 0$.

4. Finally the methods developed here can be applied to the heat equation $u_t = \Delta u - f(x,t,u,Du)$ and to other parabolic equations of the form $u_t = A(Du)\,D^2u - f(x,t,u,Du)$. The derivations of § 1 in fact carry over directly, with the important exception that one must put $r = t = 0$, $s = -1$ throughout. Indeed, taking

$$\zeta = \left(1 - \frac{|x|^2}{d^2}\right)^2 \left(1 - \frac{t^2}{d_1{}^2}\right)$$

one obtains finally the differential inequality

$$-\bar{z}_t + A\,D^2\bar{z} \geqslant \bar{M}\cdot D\bar{z} - 2\bar{N}z, \tag{10'}$$

where now

$$E^{-1}\bar{N} \leqslant \{\ \} + \frac{1}{E}\frac{|p|^2}{d_1 z}, \tag{11'}$$

the braces denoting the expression already determined in (11).

This being the case, it is apparent that the proofs of Theorems 1 and 2 carry over immediately, while Theorem 5 also applies if we require $f_1 = a\cdot p$ in (17). In this work we must of course apply the maximum principle in the typical framework of parabolic problems. Correspondingly we shall mean by an *entire solution* a solution which is defined and of class C^3 on some domain $\mathbf{R}^n \times (-\infty, T]$. The following results now hold.

THEOREM 1′. *Let u be an entire solution of the equation*

$$u_t = \Delta u - f(u, \mathrm{D}u),$$

where $\partial f/\partial u \geqslant 0$ and u and $\mathrm{D}u$ are bounded. Then $u \equiv$ constant.

THEOREM 2′. *Let u be an entire solution of the equation*

$$u_t = \Delta u - f(u, \mathrm{D}u),$$

where $\partial f/\partial u \leqslant 0$ and both f and $\partial f/\partial p$ are bounded when p is bounded, uniformly in u. Suppose that $\mathrm{D}u$ is bounded and that $u = o(\log h)$ as $h \to \infty$, where $h = \sqrt{(d^2 + d_1)}$ is the 'parabolic distance' from some fixed origin. Then $u \equiv$ constant.

Proof. As in the earlier sections of the paper if follows that $\mathrm{D}u \equiv$ constant. Hence $u = u(t)$ satisfies

$$\frac{du}{dt} = -f(u, 0) \quad (-\infty < t \leqslant T). \tag{21}$$

It must be shown that the only solutions of (21) which are bounded, or $o(\log|t|)$ as $t \to \infty$, are constants. To this end, we use once more the hypothesis that $\partial f/\partial u \geqslant 0$. Put $\tau = -t$ and consider the transformed equation

$$\frac{du}{d\tau} = f(u, 0) \quad (-T \leqslant \tau < \infty).$$

Put $c = u(T)$. If $f(c, 0) = 0$, then $u \equiv c$. If $f(c, 0) > 0$, then quite clearly

$$\frac{du}{d\tau} \geqslant f(c, 0) \quad (-T \leqslant \tau < \infty),$$

whence $u \geqslant f(c, 0)\tau$. But this gives in the original variables

$$u \geqslant c + f(c, 0)(T - \tau)$$

as $t \to -\infty$, contradicting the growth restriction on u. In the same way $f(c, 0) < 0$ is impossible. This completes the proof.

THEOREM 5′. *Let u be an entire solution of the equation*

$$u_t = \Delta u - f(u, \mathrm{D}u),$$

where $\partial f/\partial u \geqslant 0$ and f satisfies the condition I_∞ (with $f_1 = a \cdot p$). Suppose that $u = o(\log h)$ as $h \to \infty$. Then $\mathrm{D}u$ is bounded.

Similar theorems hold when f satisfies the restricted condition I_∞ and the conditions I_0; finally there are obvious generalizations to equations with non-linear elliptic parts as in § 4.

The results of § 5 also apply to parabolic equations, though only in the weaker form resulting when $r = 0$. For example, corresponding to Theorem 7 one has the following parabolic analogue.

THEOREM 7′. *Let u be an entire solution of the equation*

$$u_t = \Delta u - f(x, t, u, \mathrm{D}u),$$

where

$$\frac{\partial f}{\partial u} + \frac{p_i}{|p|^2}\frac{\partial f}{\partial x_i} \geqslant 0 \quad (p \neq 0).$$

Assume that $f(x,t,u,p)$ and $\partial f(x,t,u,p)/\partial p$ are bounded (independently of x, t, and u) when p lies in any compact subset of $\mathbf{R}^n \backslash \{0\}$. Suppose also that $\mathrm{D}u$ is bounded and that $u = o(\log h)$ as h tends to infinity. Then $u \equiv u(t)$. Moreover, if

$$f \cdot \left(f\frac{\partial f}{\partial u} - \frac{\partial f}{\partial t}\right) \geqslant 0 \quad \textit{when } p = 0,$$

then $u \equiv$ constant.

This result includes Theorem 2′ by taking f independent of x and t. A similar generalization of Theorem 5′ can also be given, leading to a Liouville theorem of standard form.

REFERENCES

1. S. N. BERNSTEIN, 'Über ein geometrisches Theorem und seine Anwendung auf die partiellen Differentialgleichungen vom elliptischen Typus', *Math. Z.* 26 (1927) 551–58.
2. E. BOHN and L. K. JACKSON, 'The Liouville theorem for a quasilinear elliptic partial differential equation', *Trans. Amer. Math. Soc.* 104 (1962) 392–97.
3. D. GILBARG and J. B. SERRIN, 'On isolated singularities of solutions of second order elliptic differential equations', *J. Analyse Math.* 4 (1954–6) 165–74.
4. J. MOSER, 'On Harnack's theorem for elliptic differential equations', *Comm. Pure Appl. Math.* 14 (1961) 577–91.
5. O. A. OLEINIK and S. N. KRUZHKOV, 'Quasi-linear second order parabolic equations with many independent variables', *Uspehi Mat. Nauk* 16 (1961) 115–56. (Translated in *Russian Math. Surveys* 16 (1961) 105–46.)
6. R. M. REDHEFFER, 'On the inequality $\Delta u \geqslant f(u, |\operatorname{grad} u|)$', *J. Math. Anal. Appl.* 1 (1960) 277–99.
7. J. SERRIN, 'Gradient estimates for nonlinear elliptic and parabolic partial differential equations', *Symposium on Nonlinear Functional Analysis* (University of Wisconsin Press, 1972).

University of Minnesota

Acta. Math. 189 (2002), 79–142. **Reprint I.6.[ii]**

Pages 116–137 have been removed for the sake of brevity

Cauchy–Liouville and universal boundedness theorems for quasilinear elliptic equations and inequalities

by

JAMES SERRIN and HENGHUI ZOU

University of Minnesota
Minneapolis, MN, U.S.A.

University of Alabama
Birmingham, AL, U.S.A.

Contents

1. Introduction

In 1844, in a short note [8] in the Comptes Rendus de l'Académie des Sciences, Paris, Augustin Cauchy published the first statement of what is now known as the Liouville theorem for bounded analytic functions:

Any bounded entire function of a single complex variable must be constant.

This classical theorem generalizes at once to real harmonic functions on $\mathbf{R}^n$ which are bounded only on one side:

Let $n\geqslant 2$ and let u be a real harmonic function on $\mathbf{R}^n$, bounded either from above or below. Then u must be constant.

When the dimension $n=2$, one can even consider superharmonic functions (e.g., satisfying the inequality $\Delta u\leqslant 0$):

Let u be a real superharmonic function on $\mathbf{R}^2$, bounded from below. Then u is constant.

Proofs can be found for example in [21, pp. 111 and 130]. More recently, the Liouville theorem was further generalized to solutions of quasilinear elliptic equations (Serrin [25]):

Let u be an entire solution of the equation

$$\Delta u + f(u, \nabla u) = 0 \quad \text{in } \mathbf{R}^n.$$

Suppose that $\partial f/\partial u \leqslant 0$ and that both u and ∇u are bounded. Then u must be constant.

Under further assumptions, it can be shown [20], [26] that ∇u is necessarily bounded on all $\mathbf{R}^n$. Using this fact, one gets a standard Liouville theorem for bounded solutions. See also Caffarelli, Garofalo and Segala [6] and references therein.

Still other Liouville theorems have been obtained for non-negative solutions of the Lane–Emden equation

$$\Delta u + u^{p-1} = 0, \quad p > 1 \tag{1.1}$$

(note that the previous result does not cover (1.1), since u^{p-1} is *increasing* for $u>0$). We first state a beautiful and deep result of Gidas and Spruck.

THEOREM 1 (Gidas and Spruck [12]). *Assume $n>2$. Let u be a non-negative solution of (1.1) in $\mathbf{R}^n$ with $2 \leqslant p < 2n/(n-2)$ (Sobolev number for $\mathbf{R}^n$). Then $u \equiv 0$.*

A striking fact about this result is that it *fails* for any $p \geqslant 2n/(n-2)$. For example, when $p = 2n/(n-2)$, we have the Emden solution

$$u(x) = \left(C \frac{\varkappa}{\varkappa^2 + |x|^2} \right)^{(n-2)/2} \tag{1.2}$$

where $\varkappa > 0$ is a parameter and $C = C(n) = \sqrt{n(n-2)}$. This solution also shows that when $n>2$ there are *non-constant* bounded superharmonic functions.

A second marvelous result concerning equation (1.1) is due to Bidaut-Veron.

THEOREM 2 (Bidaut-Veron [3]). *Let $n \geqslant 2$ and let u be a non-negative solution of (1.1) on an exterior domain. Suppose $2 < p < 2(n-1)/(n-2)$ ($=\infty$ if $n=2$). Then $u \equiv 0$.*(1)

Again this result *fails* for any $p > 2(n-1)/(n-2)$. For example, for p in this range we have the singular solution

$$u(x) = C|x|^{-2/(p-2)}, \quad C = (p-2)^{-2/(p-2)} \left[2(n-2) \left(p - \frac{2(n-1)}{n-2} \right) \right]^{1/(p-2)}, \tag{1.3}$$

(1) In [3] this result was actually given for the more general equation $\Delta_m u + u^{p-1} = 0$, under related restrictions on the parameters m, p; see also Theorem I below of Bidaut-Veron and Pohozaev.

The first author to observe the importance of the exponent $p = 2(n-1)/(n-2)$ is R. H. Fowler, this exponent being equivalent to the special value $\sigma = -2$ in Fowler's work. Cf. *Quart. J. Pure Appl. Math.*, 45 (1914), 289–350, and *Quart. J. Math. Oxford Ser.*, 2 (1931), 259–288.

defined in the exterior domain $\mathbf{R}^n\setminus\{0\}$.

The purpose of the present paper is to extend the above considerations to non-homogeneous degenerate elliptic equations of the form

$$\Delta_m u+f(u)=0, \quad u\geqslant 0,\, x\in\Omega, \tag{1.4}$$

where Ω is a domain (connected open set) in $\mathbf{R}^n$, $n\geqslant 2$, and

$$\Delta_m u=\operatorname{div}(|\nabla u|^{m-2}\nabla u)$$

is the well-known m-Laplace operator, $m>1$. Equation (1.4) arises in many nonlinear phenomena, for instance, in the theory of quasi-regular and quasi-conformal mappings, see [17], [22], [29], and in mathematical modeling of non-Newtonian fluids, see [2], [9], [14], [15] for a discussion of the physical background. The equation also has a large and well-known theoretical literature, some of which will be particularly discussed below.

A function $u\in C^1(\Omega)$ is said to be a *weak solution* of (1.4) if

$$-\int |\nabla u|^{m-2}\nabla u\cdot\nabla\phi+\int f(u)\phi=0 \quad \text{for all } \phi\in C_0^\infty(\Omega). \tag{1.5}$$

We shall assume throughout the paper that $f(u)$ is a non-negative function in $C([0,\infty))\cap C^1((0,\infty))$. Then a strong maximum principle holds for equation (1.4), in the sense that all non-negative non-trivial solutions must be strictly positive, see Lemma 2.1 below. In what follows we shall always be concerned with weak solutions, without further mention.(2)

The first main goal of the paper is to consider Liouville-type results *for the degenerate equation* (1.4), *and also for continuously differentiable (or even* $W^{1,m}_{\mathrm{loc}}(\Omega)\cap C(\Omega)$*) weak solutions of the differential inequalities*

$$-\Delta_m u\geqslant u^{p-1}, \quad u\geqslant 0,\, x\in\Omega, \tag{1.6}$$

with $p>1$, *and*

$$-\Delta_m u\geqslant 0, \quad u\geqslant 0,\, x\in\Omega. \tag{1.6'}$$

Our second principal purpose is to derive universal a priori estimates *for solutions of* (1.4) *and* (1.6), including, in particular, the generalized Lane–Emden equation $\Delta_m u+u^{p-1}=0$, that is, (1.4) with $f(u)=u^{p-1}$. By using the word "universal" here, we mean that our bounds are not only *independent* of any given solution under consideration but also *do not require, or assume, any boundary conditions whatsoever.* We are not aware of any previous results of this type for equation (1.4) with $f(u)\geqslant 0$, with the exception of the a priori estimates obtained by Gidas and Spruck for solutions of (1.1) in the neighborhood of an isolated singularity, and a result of Dancer for the same case.

(2) It is possible to use an even weaker definition for weak solutions. That is, one needs to require only that $u\in W^{1,m}_{\mathrm{loc}}(\Omega)\cap L^\infty_{\mathrm{loc}}(\Omega)$. By classical results, however, the two definitions are equivalent (see the references in §8 below).

THEOREM 3 (Dancer [34, Lemma 1]). *Assume* $n>2$ *and* $2<p<2n/(n-2)$. *Let* u *be a non-negative solution of the canonical equation* (1.1) *in a domain* $\Omega\not\equiv\mathbf{R}^n$. *Then for every* $x\in\Omega$ *we have*

$$u(x)\leqslant C(n,p)[\operatorname{dist}(x,\partial\Omega)]^{-2/(p-2)}.$$

In particular, u *is bounded on any compact subset* Ω' *of* Ω, *the bound being independent of the solution.*

The range $2<p<2n/(n-2)$ *and the exponent* $2/(p-2)$ *are both optimal in view of the special solutions* (1.2) *and* (1.3).

The relevance of a universal boundedness theorem can be immediately illustrated by Theorem 3. Indeed, Theorem 1 is a direct corollary of Theorem 3, since $\operatorname{dist}(x,\partial\Omega)$ can be chosen arbitrarily large when the solution is defined on all $\mathbf{R}^n$. The relation between Theorem 3 and the Liouville Theorem 1 can also be considered in a deeper way. That is, they both provide *upper bounds* for non-negative solutions, with Theorem 1 being the extreme case where the domain is all of $\mathbf{R}^n$ and the upper bound becomes zero, the smallest value it could have. In still other terms, Theorem 3 provides a continuous embedding of the Liouville theorem for (1.1) in a family of results for an expanding sequence of *bounded* domains. Theorem 3 is a special case of Theorem IV below, which in turn is contained in Theorems 4.1 and 4.2.

Returning to the general equation (1.4), when $n>m$ we define

$$m_*=\frac{m(n-1)}{n-m}>0,$$

the *lower critical exponent*, and

$$m^*=\frac{nm}{n-m}>0,$$

the *critical exponent for Sobolev embedding*. We say that f is *subcritical* if $n>m$ and there exists a number $1<\alpha<m^*$ such that

$$f(u)\geqslant 0,\quad (\alpha-1)f(u)-uf'(u)\geqslant 0,\quad \text{for } u>0. \tag{1.7}$$

Note in particular that the function $f(u)=u^{p-1}$ is subcritical when $1<p<m^*$. A domain Ω is called *exterior* if $\Omega\supset\{|x|>R>0\}$ for some $R>0$. An important recent Liouville-type result is the following

THEOREM I (Bidaut-Veron and Pohozaev [4, Theorems 3.3 (iii) and 3.4 (ii)]). *Let* Ω *be an exterior domain. Then the differential inequality* (1.6) *has only the trivial solution* $u\equiv 0$, *provided* $p\in(1,m_*]$ *when* $n>m$, *or* $p\in(1,\infty)$ *when* $n=m$.

The result also applies when $n<m$, as follows.

THEOREM I$'$. *Let Ω be an exterior domain and let $n<m$. Then the differential inequality* (1.6) *has only the trivial solution $u\equiv 0$, provided $p\in(1,\infty)$.*

When the domain Ω is the *entire space* $\mathbf{R}^n$, rather than simply an exterior set, Theorem I can be extended to a larger range of exponents. The full result is as follows.

THEOREM II. *Let $\Omega=\mathbf{R}^n$. Then the following conclusions hold.*

(a) *Let $u(x)$ be a non-negative solution of $\Delta_m u=0$ (if $n>m$), or of $\Delta_m u\leqslant 0$ (if $n\leqslant m$). Then u is constant.*

(b) *Suppose either $n=2$ and $m>\frac{1}{4}(1+\sqrt{17})$, or $n\in[3,2m)$, $m>\frac{3}{2}$. Assume that f is subcritical. Then every solution of* (1.4) *is constant.*

(c) *Assume that f is subcritical and that there exists $p>m$ such that*

$$f(u)\geqslant u^{p-1} \tag{1.8}$$

for sufficiently large u. Then (1.4) *has only the trivial solution $u\equiv 0$. The same conclusion holds if $p\in(1,m]$, provided $\alpha\leqslant m$ in* (1.7).

(d) *If $n>m$ and $p\in(1,m_*]$, then the differential inequality* (1.6) *has only the trivial solution $u\equiv 0$.*

Remarks. When $n\leqslant m$ the inequality $-\Delta_m u\geqslant 0$ has the (bounded) positive non-constant solution $u=1-1/|x|$ on the exterior domain $\{|x|>1\}$, which indicates the necessity of considering (a) on the entire space $\mathbf{R}^n$. Notice also that the result of (b), when it is applicable, is stronger than (c), and that (d) overlaps with (b) and (c), both cases being of independent interest.

Case (d) is of course an immediate consequence of Theorem I; it was first proved by Mitidieri and Pokhozhaev [18].

The special case $m=2$ of Theorem II (Laplace operator) is important enough to be stated as a separate result, especially in order to compare our results with those of Gidas and Spruck. We consider particularly the cases (b), (c), since for the Laplacian case (a) is classical while (d) is a special case of the Mitidieri–Pokhozhaev theorem.

THEOREM 4. *Let $m=2$ and $\Omega=\mathbf{R}^n$. Then the following conclusions hold.*

(b) *Let $n=3$ and assume that f is subcritical. Then every solution of* (1.4) *is constant.*

(c) *Suppose $n\geqslant 4$. Assume that f is subcritical and that* (1.8) *holds with $p>2$. Then* (1.4) *has only the trivial solution $u\equiv 0$. The same conclusion holds if $p\in(1,2]$, provided $\alpha\leqslant 2$ in* (1.7).

Case (b) is due to Gidas and Spruck [12] under the additional assumption $f(u)>0$ for $u>0$, the conclusion then of course being that $u\equiv 0$. The first statement of case (c) is similarly due to Gidas and Spruck, see [12, Theorem 6.1]. The second statement of case (c) is new.

Gidas and Spruck have conjectured (in view of case (b)) that the extra condition (1.8) in case (c) may be unnecessary. We are inclined to doubt this, since even in the more general case of Theorem II, case (b), the required condition is $n<2m$, that is, $n<4$ when $m=2$. If, however, one treats solutions which are also bounded above, then their conjecture is essentially true.

THEOREM III. *Let $\Omega=\mathbf{R}^n$ with $n>m$. Assume that f is subcritical and that $f(u)>0$ for all $u>0$. Then every bounded solution of* (1.4) *is trivial.*

Theorems I–III are sharp, in the sense of the following corollaries.(3)

COROLLARY I. *Let Ω be an exterior domain. Then the differential inequality* (1.6) *has a non-trivial solution if and only if $m\in(1,n)$ and $p>m_*$.*

The "only if" part follows from Theorem I. On the other hand, for $\Omega=\mathbf{R}^n\backslash\{0\}$, one readily verifies when $p>m_*$ that (1.6) has a positive singular solution $C'|x|^{-m/(p-m)}$, where

$$C'=\left(\frac{m}{p-m}\right)^{m/(p-m)}\left[\frac{n-m}{m}(p-m_*)\right]^{1/(p-m)},$$

this being the exact analogue of the solution (1.3) when $m=2$.

COROLLARY II. *Let Ω be the entire space $\mathbf{R}^n$. Then:*

(i) *The inequality $-\Delta_m u\geqslant 0$ has a non-constant positive solution in $\mathbf{R}^n$ if and only if $n>m$.*

(ii) *Assume $n>m$. Then the generalized Lane–Emden equation $\Delta_m u+u^{p-1}=0$ has a positive solution in $\mathbf{R}^n$ if and only if $p\geqslant m^*$.*

(iii) *Assume $n>m$. Then the differential inequality $-\Delta_m u\geqslant u^{p-1}$ has a positive solution in $\mathbf{R}^n$ if and only if $p>m_*$.*

The "only if" part of (i) and of (iii) are an immediate consequence of Theorem II (a) and II (d), respectively. Similarly, (b), (c) of Theorem II imply that when $n>m$ and $p<m^*$ all solutions of the generalized Lane–Emden equation in $\mathbf{R}^n$ must be constant, and hence zero, which is the "only if" part of (ii).

(3) Some of the above Liouville theorems have previously been established in [19] for *radially symmetric* solutions. If one knew a priori that solutions of (1.4) when $\Omega=\mathbf{R}^n$ were *necessarily* radially symmetric, then of course Theorem II would follow at once. Such an approach, however, seems an unlikely possibility in any kind of generality; moreover when $p\geqslant m^*$ it is not even true that all solutions are radially symmetric (see [32]).

On the other hand, direct calculation shows that the inequality $-\Delta_m u \geqslant 0$ has a (bounded) positive non-constant solution in $\mathbf{R}^n$ of the form

$$C(1+|x|^{m/(m-1)})^{-(n-m)/m},$$

which gives the "if" part for (i). For (ii), existence is a special case of Theorem 6.4 of [19].

Finally, by direct calculation, the differential inequality (1.6) has a (bounded) positive solution in $\mathbf{R}^n$ of the form (see Remark 4 in Mitidieri and Pokhozhaev [18])

$$u(x) = C'(\varkappa + |x|^{m/(m-1)})^{-(m-1)/(p-m)},$$

where C' is the coefficient given in the proof of Corollary I. (In fact, one finds explicitly that $-\Delta_m u = u^{p-1} + \varkappa c u^{(mp-2m+1)/(m-1)}$ for an appropriate constant $c = c(n,m,p) > 0$.) This yields existence for (iii), and the proof is complete.

COROLLARY III. *Let $\Omega = \mathbf{R}^n$. Assume $n > m$. Then the generalized Lane–Emden equation has a bounded positive solution on Ω if and only if $p \geqslant m^*$, and* (1.6) *has a bounded positive solution if and only if $p > m_*$.*

This is a special case of Corollary II. Two model nonlinearities may be noticed here,

$$f(u) = u^{p-1} + u^{s-1}, \quad f(u) = \frac{u^{p-1}}{1+u^t},$$

where $1 < s \leqslant p < m^*$ and $t > 0$. Both nonlinearities are subcritical as one easily checks. For the first, Theorem II shows that the only solution of (1.4) on $\mathbf{R}^n$ is $u \equiv 0$. For the second, Theorem II (b) applies, but not Theorem II (c). Nevertheless, since $f(u) > 0$ for $u > 0$, by Theorem III the only *bounded* solution of (1.4) for this nonlinearity on $\mathbf{R}^n$ is again $u \equiv 0$.

Turning to the second principal goal of the paper, we have the following *universal a priori estimate* (see §5 for other related results).

THEOREM IV. *Let $\Omega \subset \mathbf{R}^n$ and assume $n > m$. Then the following conclusions hold.*

(a) *Let u be a non-negative weak solution of the two-sided differential inequality*

$$u^{p-1} - u^{m-1} \leqslant -\Delta_m u \leqslant \Lambda(u^{p-1}+1), \quad x \in \Omega, \tag{1.9}$$

where $\Lambda > 1$ and $m < p < m_*$. Then there exists a constant $C = C(n,m,p,\Lambda) > 0$ such that for all $x \in \Omega$

$$u(x) \leqslant C R^{-m/(p-m)}, \tag{1.10}$$

where $R = \min(1, \operatorname{dist}(x, \partial\Omega))$. If the additive terms u^{m-1} and 1 are dropped from (1.9), *then* (1.10) *is satisfied with $R = \operatorname{dist}(x, \partial\Omega)$.*

(b) *Let u be a solution of* (1.4). *Suppose that f is subcritical and that, for some $\Lambda>1$ and $p>m$, it satisfies the power-like condition*

$$u^{p-1} \leqslant f(u) \leqslant \Lambda(u^{p-1}+1). \tag{1.11}$$

Then (1.10) *holds with $C=C(n,m,p,\alpha,\Lambda)>0$ and $R=\min(1,\operatorname{dist}(x,\partial\Omega))$. If, instead of* (1.11),

$$u^{p-1} \leqslant f(u) \leqslant \Lambda u^{p-1}, \tag{1.12}$$

then (1.10) *holds with $R=\operatorname{dist}(x,\partial\Omega)$.*

The inequality (1.10) yields *absolute bounds for non-negative solutions on any compact subdomain of their domain Ω of definition, the constant C being independent of any particular solution under consideration.* We note also that the range $m<p<m_*$ for case (a) and $m<p<m^*$ for case (b), and the exponent $m/(p-m)$ in (1.10), are each optimal; see the discussion in §5.

It is interesting to ask about the size of the *singular set* of a solution u which is defined over some domain Ω. Certainly it cannot consist of the entire boundary of Ω, since u is superharmonic. On the other hand, can one estimate in some way the Hausdorff dimension or the Hausdorff measure of this set? Here, a tentative conjecture is that the Hausdorff dimension of a singular set on $\partial\Omega$ must be less than or equal to $n-m(p-1)/(p-m)$; see Mazzeo and Pacard [16] for the case $m=2$, and also Veron [31, pp. 242–254].

Remark. In (1.9), one might wish to study the apparently more general left-hand side, $\lambda u^{p-1}-\mu u^{m-1}$. The constants λ and μ can however be reduced to 1 by simple rescaling; thus the special form of the left-hand side of (1.9) involves no loss of generality. The same remark obviously applies to later formulations of the principal conditions on f.

Theorem IV has useful implications for the asymptotic behavior of solutions near isolated singularities, see the corollary below.

Corollary IV. *Let u be a solution of* (1.4), *where f is subcritical. Then*:

(i) *Suppose that* (1.11) *holds, and let $\Omega=B_1(0)\setminus\{0\}$. Then there exists a constant $C=C(n,m,p,\Lambda)>0$ such that*

$$u(x) \leqslant C|x|^{-m/(p-m)} \tag{1.13}$$

for all $x\in\Omega$.

(ii) *Suppose that* (1.12) *is satisfied, and let Ω be an exterior domain. Then* (1.13) *holds for all sufficiently large $|x|$.* (*Note that if $p\leqslant m_*$ then Theorem* I *gives a stronger result.*)

The asymptotic behavior (1.13) near an isolated singularity (and further extensions of this) were established for the Laplace operator ($m=2$) by Gidas and Spruck [12] and by Bidaut-Veron and Veron [5].

Theorem IV also implies the following non-existence theorem.

COROLLARY IV′. *Let the hypotheses of Theorem* IV (a), *or* (b), *hold. Then the Dirichlet problem for* (1.9), *or* (1.4), *with data* $u\geqslant M$ *on* $\partial\Omega$ *has no solution if* M *is sufficiently large.*

Proof. The result follows at once from from (1.10) together with the fact that solutions are m-superharmonic and so satisfy $u\geqslant M$ in Ω.

The a priori estimates given in Theorem IV are obtained from the following Harnack-type theorem, itself of independent interest, which will be proved in §4; see Theorems 4.1 and 4.2.

THEOREM V. *Let* R *and* x_0 *be such that* $B_R\equiv B_R(x_0)\subset B_{2R}(x_0)\subset\Omega$, *and assume* $n>m$. *Then we have the following conclusions.*

(a) *Let* u *be a non-negative weak solution of the differential inequality* (1.9). *Then for every* $R_0>0$ *there exists* $C=C(n,m,p,\Lambda,R_0)>0$ *such that*

$$\sup_{B_R} u\leqslant C\inf_{B_R} u \tag{1.14}$$

provided $R\leqslant R_0$.

If the terms u^{m-1} *and* 1 *are dropped in* (1.9), *then* (1.14) *holds with* $C=C(n,m,p,\Lambda)$ *and with no further restriction on* R.

(b) *Let* u *be a solution of* (1.4), *where* f *is subcritical. Suppose either* $n=2$ *and* $m>\frac{1}{4}(1+\sqrt{17})$, *or* $n\in[3,2m)$, $m>\frac{3}{2}$. *Then* (1.14) *holds with* $C=C(n,m,\alpha)>0$.

(c) *Let* u *be a solution of* (1.4), *where* f *is subcritical, and suppose that* (1.12) *is satisfied for some* $p>m$. *Then* (1.14) *holds with* $C=C(n,m,p,\alpha,\Lambda)$.

The case $n\leqslant m$ can also be treated, see Theorem 4.3.

Some remarks on the proof methods are worthwhile. First, Theorems I′, II (a) and IV (a) are relatively elementary, with the exception that the last two cases require an application of the classical Harnack inequality for quasilinear equations (Serrin [23]).

The proofs for Theorem II (b), (c), Theorem IV (b) and Theorem V (b), (c) additionally rely on an important integral inequality for solutions of equation (1.4), see Proposition 6.1. For the Laplace case $m=2$ this result is (essentially) due to Gidas and Spruck [12]. When $m\geqslant 2$, Proposition 6.1 is proved by direct calculation, using as a key element an unusual nonlinear vector field ω, see (6.10), and also, at one point, a delicate interchange of order of differentiation. (For $m=2$ the interchange is elementary because of the smoothness of solutions of (1.4) in this case; the loss of C^2 regularity is at the heart of the difficulty otherwise.)

When $1<m<2$ the proof of the inequality is still more technical, the difficulty again being due to the degeneracy of the m-Laplace operator and the loss of smoothness of solutions. As part of the derivation, we have been led to an improved regularity result for solutions of (1.4), extending the "classical" theory of [10], etc. This result is of independent interest and is, to the best of the authors' knowledge, new. It is worth remarking as well that, when $1<m<2$, it appears to be impossible to obtain a strict analogue of the Gidas–Spruck identity for the case $m=2$. Rather, we employ related integral inequalities, which, fortunately, seem at least as useful as the identity itself.

The paper is organized as follows. Chapter I contains the proofs of Theorems I–V and of several further results of a more special nature; see the Table of Contents for the specific content of these sections. Chapter II is devoted to the proof of Proposition 6.1, that is, the general integral inequality (6.1). In particular, in §§6 and 7 we prove the inequality respectively for the case $m\geqslant 2$ and for the more delicate range $1<m<2$. Finally §8 contains our regularity results for solutions of (1.4); see the main Proposition 8.1.

It may seem paradoxical that so much effort must be devoted to the integral inequality for (1.4), in view of the fact that it is applied only at one point in Chapter I. On the other hand, on this application alone stands or falls the entire structure for functions $f(u)$ whose growth rate in the variable u exceeds the power m_*-1 $(=n(m-1)/(n-m))$; for the case of the Laplace operator, in particular, a growth rate exceeding the "classical" power $n/(n-2)$. This being the case, the further effort seems more than worthwhile, even if it is lengthy and difficult. Moreover, this aspect of the theory makes abundantly clear the great difference between the ranges $(1, m_*]$ and (m_*, m^*) of the variable p in the equation $\Delta_m u + u^{p-1} = 0$.

It almost goes without saying that much of the work in the paper can be expected to carry over to more general operators and to nonlinearities f depending on x and ∇u as well as on u.

The second author wishes to thank E. DiBenedetto and D. Andreucci for useful discussions at the beginning of this work. The authors also wish to thank the referee for carefully reading the manuscript and, in particular, for pointing out to us the importance of reference [16].

Chapter I

2. Liouville theorems I

In this section, we prove the Liouville Theorem I′, and also, for completeness, Theorem I. Our proof of the latter result was obtained independently of the work of Bidaut-Veron and Pohozaev, and in some respects depends on different ideas.

The letter C will be used throughout to denote a generic positive constant, which may vary from line to line and only depends on arguments inside the parentheses or which are otherwise clear from the context.

We begin with a series of lemmas. The first is the well-known strong maximum principle.

LEMMA 2.1. *Let u be a weak solution of (1.6′). Then either $u \equiv 0$ or $u > 0$ on Ω.*

Lemma 2.1 is a consequence of the weak Harnack inequality (see Lemma 3.2 below).

LEMMA 2.2. *Let u and v be continuous functions in the Sobolev space $W^{1,m}_{\mathrm{loc}}(\Omega)$ which satisfy the distribution inequality*

$$\Delta_m u - \Delta_m v \leqslant 0 \tag{2.1}$$

in a domain Ω of $\mathbf{R}^n$. Suppose that $u \geqslant v$ on $\partial\Omega$, in the sense that the set $\{u - v + \varepsilon \leqslant 0\}$ has compact support in Ω for every $\varepsilon > 0$. Then $u \geqslant v$ in Ω.

Lemma 2.2 is a well-known comparison lemma; its proof can be omitted.

LEMMA 2.3. *Suppose $\{|x| > R > 0\} \subset \Omega$. Let u be a positive weak solution of the inequality*

$$\Delta_m u \leqslant 0, \quad x \in \Omega. \tag{2.2}$$

Then there exists a constant $C = C(m, n, u, R) > 0$ such that

$$u(x) \geqslant C|x|^{-(n-m)/(m-1)} \tag{2.3}$$

provided $n > m$, while

$$\liminf_{x \to \infty} u(x) > 0 \tag{2.4}$$

if $n \leqslant m$.

Proof. First assume $n > m$. Define

$$K = R^{(n-m)/(m-1)} \min_{|x|=2R} u(x) > 0$$

and

$$v(x) = K|x|^{-(n-m)/(m-1)}.$$

Then v is a fundamental solution of $-\Delta_m v=0$, so by applying Lemma 2.2 in the domain $|x|>2R$ we get (2.3).

To prove (2.4), let $S>R$ and define

$$v(x) = \frac{\ln(4S/|x|)}{\ln(2S/R)}$$

for x in Ω. Observe that

$$-\operatorname{div}(|\nabla v|^{m-2}\nabla v) \leqslant 0, \quad x \in \Omega,$$

while also $v=1$ when $|x|=2R$; $v=\frac{1}{2}$ when $|x|=2(2RS)^{1/2}$; and $|v|=0$ when $|x|=4S$.

Since $u>0$ in Ω, we have $\varepsilon \equiv \inf_{|x|=2R} u>0$. Hence by the weak comparison principle, we get $u \geqslant \varepsilon v$ for $2R<|x|<4S$. But then $u(x) \geqslant \frac{1}{2}\varepsilon$ for $|x| \leqslant 2(2RS)^{1/2}$. Letting S tend to infinity now yields $\liminf_{x\to\infty} u(x) \geqslant \frac{1}{2}\varepsilon>0$, as required.

We next give an integral estimate for solutions of (1.6) in a domain Ω of $\mathbf{R}^n$. Here and in the sequel, by $B_R=B_R(x_0)$ we shall mean a ball of radius R and center x_0, such that the corresponding ball $B_{2R}(x_0)$ of radius $2R$ is contained in the domain Ω.

LEMMA 2.4. *Let u be a weak solution of* (1.6) *in Ω for some $p>m$ and let $R>0$. Then for all $\gamma \in (0, p-1)$ there exists a constant $C=C(n,m,p,\gamma)>0$ such that*

$$\int_{B_R} u^\gamma \leqslant CR^{n-m\gamma/r}, \tag{2.5}$$

where $r=p-m>0$. Similarly, for all $\mu \in (0, m(p-1)/p)$ there exists $C(n,m,p,\mu)>0$ such that

$$\int_{B_R} |\nabla u|^\mu \leqslant CR^{n-p\mu/r}. \tag{2.6}$$

Remark. The inequality (2.5) is due (in a slightly different form) to Bidaut-Veron and Pohozaev [4, Lemma 2.5]; see also Mitidieri and Pokhozhaev [18]. We include the proof for completeness and also for later reference.

Proof. It can be assumed without loss of generality that $u>0$ in Ω, since otherwise $u\equiv 0$ by the strong maximum principle, and (2.5) and (2.6) are trivially satisfied.

Now let ξ be a radially symmetric C^2 cut-off function on the double unit ball $B_2(0)$, namely,

(1) $\xi \equiv 1$ for $|x|<1$; $0 \leqslant \xi \leqslant 1$ for $|x| \geqslant 1$,

(2) ξ has compact support in $B_2(0)$,

and, without loss of generality,

(3) $|\nabla\xi|\leqslant 2$, $|\nabla^2\xi|\leqslant c$, where c is a suitable constant, e.g., $c=4\sqrt{n}$ suffices.

For $k>m$ to be determined later and $d=p-1-\gamma>0$, take $\phi=[\xi(|x-x_0|/R)]^k u^{-d}>0$ as a test function in the weak form of (1.6). (Test functions involving negative powers of the solution have been classically used at least since the work of Moser in the 1950's.) This gives at once

$$d\int \xi^k u^{\gamma-p}|\nabla u|^m+\int \xi^k u^\gamma \leqslant \int u^{-d}\mathbf{u}\cdot\nabla\xi^k, \tag{2.7}$$

where $\mathbf{u}=|\nabla u|^{m-2}\nabla u$. Write

$$|\nabla\xi^k|=k\xi^{k-1}|\nabla\xi|\leqslant \xi^k(1\cdot 2k/R\xi)$$

and $1=u^{-(m-1)/m}\cdot u^{(m-1)/m}$. Then by Young's inequality with the exponent pair $m/(m-1)$, m (and the usual trick(4)) one finds

$$\left|\int u^{-d}\mathbf{u}\cdot\nabla\xi^k\right|\leqslant \frac{d}{2}\int \xi^k u^{\gamma-p}|\nabla u|^m+CR^{-m}\int \xi^{k-m}u^{\gamma-r},$$

where $r=p-m>0$. In turn, by (2.7),

$$\frac{d}{2}\int \xi^k u^{\gamma-p}|\nabla u|^m+\int \xi^k u^\gamma\leqslant CR^{-m}\int \xi^{k-m}u^{\gamma-r}, \tag{2.8}$$

the constant C depending on m,k,p,γ.

We can now prove (2.5). First suppose $\gamma>r$. Letting

$$k=m\frac{\gamma}{r}>m$$

and applying the Young inequality to the right side of (2.8), using the exponent pair

$$\frac{\gamma}{\gamma-r},\quad \frac{\gamma}{r},$$

we obtain (again with the usual trick)

$$CR^{-m}\int \xi^{k-m}u^{\gamma-r}\leqslant \frac{1}{2}\int \xi^k u^\gamma+C(m,p,\gamma)R^{-m\gamma/r}\int \xi^0.$$

(4) Namely, to use a small coefficient multiplying one of the terms of the inequality at the expense of a larger coefficient for the other, that is,

$$ab\leqslant \varepsilon a^q+\frac{1}{\varepsilon^{1/(q-1)}}b^{q/(q-1)}$$

for any $a,b>0$ and exponents q, $q/(q-1)>1$.

Consequently, (2.8) yields

$$\frac{d}{2}\int \xi^k u^{\gamma-p}|\nabla u|^m + \frac{1}{2}\int \xi^k u^\gamma \leqslant C(n,m,p,\gamma)R^{n-m\gamma/r}. \tag{2.9}$$

Condition (2.5) now follows at once. The special case $\gamma=r$, on the other hand, is immediately obvious from (2.8).

Finally, if $\gamma<r$ we apply Hölder's inequality with exponents r/γ and $r/(r-\gamma)$ to obtain

$$\int_{B_R} u^\gamma \leqslant C\left(\int_{B_R} u^r\right)^{\gamma/r} R^{n(1-\gamma/r)}.$$

Since (2.5) is already known to hold for the exponent $\gamma=r$, this gives the required conclusion for all γ.

To get (2.6), note that $\mu<m$, and write

$$\int |\nabla u|^\mu \leqslant \left(\int u^{\gamma-p}|\nabla u|^m\right)^{\mu/m}\left(\int u^{\bar\gamma}\right)^{1-\mu/m},$$

where $\gamma\in(0,p-1)$ and $\bar\gamma=(p-\gamma)\mu/(m-\mu)$. Since $\mu\in(0,m(p-1)/p)$, we have $\bar\gamma\in(0,p-1)$ provided γ is suitably near $p-1$. Hence one can apply (2.5) to the second integral on the right. On the other hand, by (2.9) the first integral is bounded by $CR^{n-m\gamma/r}$. Combining these estimates and simplifying then gives (2.6). This completes the proof of the lemma.

Remark. For $1<p<m_*$ one can give an improved version of (2.5), see Lemma 4.1.

Now we are ready to prove the first Liouville theorem.

Proof of Theorem I. Let u be a positive weak solution of (1.6), where Ω is an exterior domain containing $\{|x|>R\}$. Take a sequence of points $\{x^j\}\subset\mathbf{R}^n$ such that $|x^j|>3R$ and $x^j\to\infty$.

Consider first the case $n\leqslant m$, $p>m$. In (2.5), take $\gamma=r=p-m<p-1$ to obtain

$$\int_B u^r \leqslant C|x^j|^{n-m}, \quad j=1,2,...,$$

where $B=B_{|x^j|/4}(x^j)$ (note that $B_{|x^j|/2}(x^j)\subset\Omega$) and $C=C(n,m,p)$. Then

$$\min_B u^r \leqslant \frac{1}{|B|}\int_B u^r \leqslant C|x^j|^{-m}. \tag{2.10}$$

But this contradicts (2.4) as $x^j\to\infty$. Hence there are no everywhere positive solutions u, so from the strong maximum principle we get $u\equiv0$, as required.

Next suppose $n>m$ and $p\in(m,m_*)$. Let u be a positive weak solution of (1.6). As in the previous case the inequality (2.10) is valid. Consequently there exists $y^j\in B$ such that

$$u^r(y^j)=\min_B u^r\leqslant C|x^j|^{-m}\leqslant\left(\tfrac{5}{4}\right)^m C|y^j|^{-m}$$

(since $\frac{3}{4}|x^j|<|y^j|<\frac{5}{4}|x^j|$). On the other hand, recalling that $n>m$ and using (2.3), we have

$$u(y^j)\geqslant C|y^j|^{-(n-m)/(m-1)}$$

for some $C>0$. This yields an immediate contradiction, since $y^j\to\infty$ and

$$\frac{m}{r}>\frac{n-m}{m-1}\quad\text{when } p<m_*.$$

It now follows as before that $u\equiv 0$, as required.

For the case $n>m$, $p=m_*$, we need an auxiliary lemma.

LEMMA 2.5. *Let u be a weak solution of (1.6) on $\Omega\supset\{|x|>R_0>0\}$, with $p>m$. Also let $\mu\in(0,m(p-1)/p)$. Then there exist a constant $C=C(m,n,p,\mu)>0$ and an increasing sequence $\{R_j\}\to\infty$ as $i\to\infty$, such that*

$$\int_{S^{n-1}}|\nabla u(R_j,\theta)|^\mu\,d\theta\leqslant CR_j^{-p\mu/r},\quad j=1,2,\ldots,\tag{2.11}$$

where $d\theta$ is the surface area differential on S^{n-1} and $r=p-m$.

Proof. We first show that for $R>R_0$

$$\int_{B_{3R}(0)\setminus B_{2R}(0)}|\nabla u|^\mu\leqslant CR^{n-p\mu/r}.\tag{2.12}$$

To see this, notice that one can cover the set $B_{3R}(0)\setminus B_{2R}(0)$ by a finite number of balls $B_{2R}(y_j)$ with $|y_j|=3R$. Thus (2.12) follows immediately from (2.6) by a covering argument.

Now take a sequence of positive integers $\{K_j\}\to\infty$ such that $2K_{j+1}>3K_j$. Then (2.12) implies

$$\int_{2K_j}^{3K_j}t^{n-1}\int_{S^{n-1}}|\nabla u(t,\theta)|^\mu\,d\theta\,dt=\int_{B_{3K_j}(0)\setminus B_{2K_j}(0)}|\nabla u|^\mu\leqslant CK_j^{n-p\mu/r}.$$

Hence by the mean-value theorem for integrals, there exists $R_j\in(2K_j,3K_j)\to\infty$ such that

$$K_jR_j^{n-1}\int_{S^{n-1}}|\nabla u(R_j,\theta)|^\mu\,d\theta\leqslant CK_j^{n-p\mu/r},$$

and (2.11) follows immediately.

We can now finish the proof for the case $p=m_*$. Integrate (1.6) over $B_{R_j}\backslash B_{R_0}$ to obtain

$$\int_{B_{R_j}\backslash B_{R_0}} u^{m_*-1}\,dx \leqslant -\int_{\partial B_{R_j}} \mathbf{u}\cdot\nu_1\,d\theta - \int_{\partial B_{R_0}} \mathbf{u}\cdot\nu_2\,d\theta, \tag{2.13}$$

where ν_1 and ν_2 are the unit outer normal vectors (this is easily justified for the weak form of (1.6)). The left-hand side of (2.13) tends logarithmically to infinity as $j\to\infty$ since

$$u^{m_*-1} \geqslant C|x|^{-(n-m)(m_*-1)/(m-1)} = C|x|^{-n},$$

by (2.3). Clearly

$$\left|\int_{\partial B_{R_0}} \mathbf{u}\cdot\nu_2\,d\theta\right| \leqslant C$$

for some $C>0$. Using (2.11) with $\mu=m-1$ and $p=m_*>m$, one can also bound the first term on the right-hand side of (2.13) as follows:

$$\begin{aligned}\left|\int_{\partial B_{R_j}} \mathbf{u}\cdot\nu_1\,d\theta\right| &\leqslant C(n)R_j^{n-1}\int_{S^{n-1}} |\nabla u(R_j,\theta)|^{m-1}\,d\theta\\ &\leqslant CR_j^{n-1}R_j^{-(m-1)m_*/(m_*-m)} = C\end{aligned}$$

for some $C>0$ independent of j as $j\to\infty$. This contradicts (2.13), and the proof is complete.

It remains to take up the case $1<p\leqslant m$. This will be done with the help of three lemmas.

LEMMA 2.6. *Let $n>1$, $m>1$. There exists $R_m>0$ such that the equation*

$$\Delta_m v + v^{m-1} = 0 \tag{2.14}$$

has a positive radial solution $v_m(|x|)$ in the ball $|x|<R_m$, with $v_m=0$ on $|x|=R_m$ and $v_m(0)=1$.

Lemma 2.6 is well-known. A proof can for example be given by the shooting method and use of Theorem 6.2 (i) of Ni and Serrin [19].

The function v_m plays the role of an "eigenfunction" for equation (2.14). Since (2.14) is homogeneous in v, clearly any multiple of v_m is also a solution in $|x|<R_m$ with zero boundary data, while moreover any translation of v_m is equally a solution.

LEMMA 2.7. *Let $g(s)$, $s>0$, be a positive function, with $\inf_{s>s_0} g(s)>0$ for any $s_0>0$. Suppose that u is a non-negative solution of the inequality*

$$-\Delta_m u \geqslant g(u) \tag{2.15}$$

in an exterior domain Ω. Then $\liminf_{x\to\infty} u(x)=0$.

Proof. An easy calculation shows that the function

$$w(x)=\frac{1}{n^{1/(m-1)}}\,\frac{m-1}{m}\,|x|^{m/(m-1)}$$

satisfies $\Delta_m w=1$.

Now suppose for contradiction that $\liminf_{x\to\infty} u(x)=\varepsilon>0$, and let y^j be a sequence in Ω tending to ∞ as $j\to\infty$, such that $\lim_{j\to\infty} u(y^j)=\varepsilon$. Define $\gamma=\inf_{s>\varepsilon/2} g(s)$, so by (2.15) and the conditions on g,

$$-\Delta_m u(x) \geqslant g(u(x)) > \gamma$$

whenever $|x|$ is suitably large. The function

$$w_\varepsilon(x)=2\varepsilon-\gamma w(x-y^j)$$

is positive and satisfies $-\Delta_m w_\varepsilon=\gamma$ when $x\in B_{R_\varepsilon}(y^j)$, with $w_\varepsilon=0$ when $|x-y^j|=R_\varepsilon$ for some appropriate constant R_ε. Clearly $B_{R_\varepsilon}(y^j)\subset\Omega$ if $|y^j|$ is large enough. Hence by the weak comparison principle we get $u\geqslant w_\varepsilon$ in $B_{R_\varepsilon}(y^j)$ for all suitably large j. In turn $u(y^j)\geqslant w_\varepsilon(y^j)=2\varepsilon$, an obvious contradiction if $|y^j|$ is sufficiently large. The lemma is proved.

Remark. An immediate consequence of Lemma 2.3 and Lemma 2.7 is that if $n\leqslant m$, then the only possible solution of (2.15) in an exterior domain is $u\equiv 0$.

LEMMA 2.8. *If $1<p\leqslant m$, then the only solution of* (1.6) *in an exterior domain Ω is $u\equiv 0$.*

Proof. Suppose for contradiction that $u\not\equiv 0$ is a solution of (1.6) in Ω. Then $u>0$ in Ω by Lemma 2.1. Let $y\in\Omega$ be such that the ball $B_{R_m}(y)$ is contained in Ω and $u(y)\leqslant 1$. (The second condition is possible because of Lemma 2.7.)

Clearly there is some constant $c\in(0,1]$ such that $u\geqslant cv_m>0$ in $B_{R_m}(y)$ while also $u=cv_m$ at some point in $B_{R_m}(y)$. In turn it is not hard to see that, for any sufficiently small constant $\varepsilon>0$, there exists a (non-empty) domain D_ε strictly contained in $B_{R_m}(y)$, such that

$$cv_m > u-\varepsilon \quad \text{in } D_\varepsilon$$

and

$$cv_m \leqslant u-\varepsilon \quad \text{in } B_{R_m}(y)\backslash D_\varepsilon.$$

Obviously for $x\in D_\varepsilon$,

$$\begin{aligned}\Delta_m(cv_m)-\Delta_m(u-\varepsilon) &= \Delta_m(cv_m)-\Delta_m u \\ &\geqslant -(cv_m)^{m-1}+u^{p-1} \geqslant -(cv_m)^{p-1}+u^{p-1} \geqslant 0,\end{aligned} \tag{2.16}$$

since $1<p\leqslant m$, $cv_m\leqslant v_m\leqslant 1$ and $cv_m\leqslant u$. By the weak comparison principle Lemma 2.2 we then find $cv_m\leqslant u-\varepsilon$ in D_ε, a contradiction.

This proves Lemma 2.8, and so completes the proof of Theorem I.

Remark. Lemma 2.8 applies also to the more general inequality

$$-\Delta_m u = \min(u^{p-1}, u^{m-1}),$$

the proof being essentially the same, up to a simple and easily seen modification of the second line of (2.16).

The argument used to prove Theorem I no longer works when $p>m_*$. In fact, when $p\in(m_*,\infty)$, $n>m$, it is easy to verify directly that the equation

$$\Delta_m u+u^{p-1}=0$$

(and accordingly also (1.6)) has solutions of the form $C|x-x_0|^{-m/(p-m)}$ in exterior domains.

3. Liouville theorems II

In this section we shall prove the (more difficult) results of Theorem II. To this end, it will be critically important to use the generalized Gidas–Spruck inequality (Theorem 6.1 below) together with the assumption that f is "subcritical", in the sense that (1.7) holds for some $\alpha\in(1,m^*)$. The following lemma, extending the range of Lemma 2.4, is the key to the discussion.

LEMMA 3.1. *Let u be a positive weak solution of* (1.4) *with $n>m$, and let*

$$R>0, \quad d\in(0,1), \quad k>2m.$$

Suppose also that f is subcritical, with

$$1<\alpha<m^*-\frac{m^*-1}{m_*}d. \tag{3.1}$$

Then there exists a positive constant $C=C(n,m,\alpha,d,k)$ such that

$$\int \xi^k f^2(u)u^{2-d-m_*} \leqslant CR^{-2m}\int \xi^{k-2m}u^{\sigma-d}, \tag{3.2}$$

where $\sigma=2m-m_*$ and $\xi=\xi(|x-x_0|/R)$ is a scaled cut-off function on the ball B_{2R}, as in the proof of Lemma 2.4.

Proof. Let $A, \hat{A}, B, D, \widehat{D}$ be the coefficients in (6.2). By our assumptions on d and α we have $B>0$ and

$$\delta = \frac{n-1}{n}\left(m^* - \alpha - \frac{m^*-1}{m_*}d\right) > 0.$$

In turn, by (1.7),

$$Af(u)+\hat{A}uf'(u) = \delta f(u) + \frac{n-1}{n}[(\alpha-1)f(u)-uf'(u)] \geqslant \delta f(u).$$

Taking $\phi=[\xi(|x-x_0|/R)]^k$, one readily sees from (6.1) that

$$\begin{aligned}\delta\int \xi^k f(u)u^{1-d-m_*}|\nabla u|^m + B\int \xi^k u^{-d-m_*}|\nabla u|^{2m} \\ \leqslant \int u^{2-d-m_*}\{\mathbf{u}\nabla^2(\xi^k)\mathbf{u}\} + \int u^{1-d-m_*}\{Duf(u)+\widehat{D}|\nabla u|^m\}\mathbf{u}\cdot\nabla(\xi^k).\end{aligned} \tag{3.3}$$

Using Young's inequality with the respective exponent pairs

$$\left(m, \frac{m}{m-1}\right), \quad \left(2m, \frac{2m}{2m-1}\right), \quad \left(m, \frac{m}{m-1}\right),$$

we may bound the terms on the right side of (3.3) as follows:

$$\left|\int u^{2-d-m_*}\{\mathbf{u}\nabla^2(\xi^k)\mathbf{u}\}\right| \leqslant \frac{B}{2}\int \xi^k u^{-d-m_*}|\nabla u|^{2m} + C\int u^{\sigma-d}\xi^{(1-m)k}|\nabla^2\xi^k|^m,$$

$$\left|\int \widehat{D}u^{1-d-m_*}|\nabla u|^m\mathbf{u}\cdot\nabla(\xi^k)\right| \leqslant \frac{B}{2}\int \xi^k u^{-d-m_*}|\nabla u|^{2m} + C\int u^{\sigma-d}\xi^{(1-2m)k}|\nabla\xi^k|^{2m},$$

$$\begin{aligned}\left|\int Df(u)u^{2-d-m_*}\mathbf{u}\cdot\nabla(\xi^k)\right| \leqslant \frac{\delta}{2}\int \xi^k f(u)u^{1-d-m_*}|\nabla u|^m \\ + C\int f(u)u^{1-d+m-m_*}\xi^{(1-m)k}|\nabla\xi^k|^m,\end{aligned}$$

where $B=(1-d)(m-1)d/m$ and $C=C(n,m,d,\delta)$. Hence, with the help of the estimates $|\xi|\leqslant 1$, $|\nabla\xi|\leqslant 2/R$ and $|\nabla^2\xi|\leqslant c/R^2$, we get from (3.3)

$$\int \xi^k f(u)u^{1-d-m_*}|\nabla u|^m \leqslant a\left(R^{-2m}\int \xi^{k-2m}u^{\sigma-d} + R^{-m}\int \xi^{k-m}f(u)u^{1+m-d-m_*}\right), \tag{3.4}$$

where the coefficient a depends only on n, m, d, δ, k.

Note that the condition $k>2m$ is used here to make the integrals in (3.4) well-defined.

With the crucial estimate (3.4) in hand, we can now turn to the main conclusion (3.2). The first step in its derivation is to take $\phi=\xi^k f(u)u^{2-d-m_*}$ as a test function in the weak form of (1.4); this is allowable since u is assumed to be positive. One thereby obtains

$$\int \xi^k f^2(u)u^{2-d-m_*} = \int \xi^k (f(u)u^{2-d-m_*})'|\nabla u|^m + k\int \xi^{k-1} f(u)u^{2-d-m_*}\mathbf{u}\cdot\nabla\xi. \quad (3.5)$$

The Young inequality with exponents $m/(m-1)$ and m yields the following estimate for the second integral on the right of (3.5):

$$\left|\int \xi^{k-1} f(u)u^{2-d-m_*}\mathbf{u}\cdot\nabla\xi\right| \leqslant \int \xi^k f(u)u^{1-d-m_*}\{|\nabla u|^m + (2u/[R\xi])^m\}.$$

Moreover, the first integral can be controlled with the help of (1.7), namely

$$\begin{aligned}(f(u)u^{2-d-m_*})' &= u^{1-d-m_*}[(2-d-m_*)f(u)+uf'(u)]\\ &\leqslant u^{1-d-m_*}[\alpha+1-d-m_*]f(u) \leqslant \frac{n}{n-m}f(u)u^{1-d-m_*}\end{aligned}$$

since $\alpha < m^*$. Thus (3.5) gives

$$\begin{aligned}\int \xi^k f^2(u)u^{2-d-m_*} \leqslant \left(\frac{n}{n-m}+k\right)\int \xi^k f(u)u^{1-d-m_*}|\nabla u|^m\\ +2^m kR^{-m}\int \xi^{k-m} f(u)u^{1+m-d-m_*}.\end{aligned}$$

Eliminating the first integral on the right by using (3.4) now gives

$$\int \xi^k f^2(u)u^{2-d-m_*} \leqslant a_1 R^{-2m}\int \xi^{k-2m}u^{\sigma-d} + a_2 R^{-m}\int \xi^{k-m} f(u)u^{1+m-d-m_*}, \quad (3.6)$$

where $a_1 = (k+n/(n-m))a$, $a_2 = a_1 + 2^m k$.

On the other hand, by the Cauchy inequality,

$$a_2 R^{-m}\int \xi^{k-m} f(u)u^{1+m-d-m_*} \leqslant \frac{1}{2}\int \xi^k f^2(u)u^{2-d-m_*} + \frac{1}{2}a_2^2 R^{-2m}\int \xi^{k-2m}u^{\sigma-d}.$$

Using this to eliminate the second integral on the right in (3.6) then yields (3.2), with the constant $C = 2a_1 + a_2^2$.

We also need the following weak Harnack inequality, due to Trudinger [28].

LEMMA 3.2. *Let $-\Delta_m u \geqslant 0$ and $u \geqslant 0$ in Ω. Then for all $\gamma \in (0, m_* - 1)$ and $R > 0$, there exists a constant $C = C(n, m, \gamma) > 0$ such that*

$$\min_{x\in B_R} u(x) \geqslant CR^{-n/\gamma}\|u\|_{L^\gamma(B_{2R})}. \quad (3.7)$$

Proof. The function u satisfies the hypotheses of Theorem 1.2 of [27], with

$$a_0 = 1, \quad b_0 = 0; \quad a_i(x) = b_i(x) \equiv 0, \quad i \geqslant 1,$$

and (3.7) then follows immediately from Theorem 1.2 of [28].(5)

The following estimate is a consequence of Lemmas 3.1 and 3.2.

LEMMA 3.3. *Let u be a positive weak solution of (1.4) in Ω. Assume that f is subcritical and $n \in (m, 2m^2 - m)$. Let*

$$\varrho = \min\left(2, \frac{2n(m-1)}{2m^2 - m - n}\right).$$

Then for all $q \in (0, \varrho)$ there exists $C = C(n, m, \alpha, q) > 0$ such that

$$\|fu^{1-m}\|_{L^q(B_R)} \leqslant CR^{n/q-m}, \tag{3.8}$$

where $B_R = B_R(x) \subset B_{4R} \subset \Omega$.

Proof. Choose d so small that (3.1) is valid. Then using Hölder's inequality with exponents $2/q$ and $2/(2-q)$, together with (3.2), we bound

$$\begin{aligned} \int_{B_R} (fu^{1-m})^q &= \int_{B_R} u^{\sigma-d}(fu^{1-m})^q u^{d-\sigma} \\ &\leqslant \left(\int_{B_R} f^2(u) u^{2-d-m_*}\right)^{q/2} \left(\int_{B_R} u^{q(d-\sigma)/(2-q)}\right)^{(2-q)/2} \\ &\leqslant \left(CR^{-2m} \int_{B_{2R}} u^{\sigma-d}\right)^{q/2} \left(\int_{B_R} u^{q(d-\sigma)/(2-q)}\right)^{(2-q)/2}. \end{aligned} \tag{3.9}$$

There are now two cases.

(i) $n \in (m, 2m-1]$. Because $n \leqslant 2m-1$ one sees at once that

$$\sigma \leqslant 0 \quad \text{and} \quad \varrho = \frac{2n(m-1)}{2m^2 - m - n}.$$

(5) The weak Harnack inequality is also a direct consequence of earlier arguments in [23]. In particular, one may apply the proof given in §3 of [23], restricting however to the case $\beta < 0$ because u obeys only the super-solution inequality $-\operatorname{div}(|\nabla u|^{m-2}\nabla u) \geqslant 0$. This means that Case I (p. 265 of [23]) can be omitted, and the iteration of (35) in Case II must be terminated at the first point p_ν where $p_\nu = m + \beta_\nu - 1 \geqslant m - 1$. But then $p_{\nu-1} < m-1$, while $p_\nu = \varkappa p_{\nu-1}$, where $\varkappa = m^*/m = n/(n-m)$.

In turn, by adjusting p'_0 appropriately (see [23, top of p. 268]), we can take $p_{\nu-1}$ arbitrarily near $m-1$. Thus the relation (40), p. 268, holds with $\max u$ replaced by $\Phi(\gamma, 2)$, where γ is any exponent less than $(m-1)\varkappa = m_* - 1$. The remainder of the argument on p. 268 of [23] (that is, Case III with $\beta < 1 - m < 0$) continues to apply, and accordingly we reach the display line immediately after (41), with $\max u$ replaced by $\Phi(\gamma, 2)$, exactly the conclusion (3.7) of Lemma 3.2.

Then, using the fact that $q<\varrho\leqslant 2$, we obtain by direct calculation

$$-\frac{\sigma q}{2-q}\in[0,m_*-1).$$

Choose $d>0$ so small that

$$\gamma=\frac{q(d-\sigma)}{2-q}\in(0,m_*-1).$$

Since $\sigma-d<0$, by Lemma 3.2 we have

$$\begin{aligned}\int_{B_{2R}}u^{\sigma-d}&\leqslant\int_{B_{2R}}\left[CR^{-n/\gamma}\left(\int_{B_{4R}}u^\gamma\right)^{1/\gamma}\right]^{\sigma-d}\\&=CR^{2n/q}\left(\int_{B_{4R}}u^{q(d-\sigma)/(2-q)}\right)^{-(2-q)/q}.\end{aligned}\tag{3.10}$$

Combining (3.9) and (3.10) yields (3.8).

(ii) $n\in(2m-1,2m^2-m)$. For this case one checks in a straightforward way that $\sigma\in(0,m_*-1)$. Choose $d>0$ so small that

$$\gamma=\sigma-d\in(0,m_*-1).$$

Since $q(d-\sigma)/(2-q)<0$, we may now (symmetrically) apply the argument of (3.10) to the *second* integral in (3.9), and the conclusion follows as before.

Now we are ready to prove our second Liouville theorem.

Proof of Theorem II. (a) First part. Let u be a non-negative solution of $\Delta_m u=0$ in $\mathbf{R}^n$. By subtracting an appropriate constant, we can assume without loss of generality that $\inf_{\mathbf{R}^n}u=0$. We must show that $u\equiv 0$.

By the Harnack inequality, Theorem 5 of [23],(6) we have

$$\max_{B_R(x_0)}u\leqslant C(n,m)\min_{B_R(x_0)}u.$$

Letting $R\to\infty$ we get

$$\sup_{\mathbf{R}^n}u\leqslant C(n,m)\inf_{\mathbf{R}^n}u=0,$$

which is the required conclusion.

Second part. As before, we can assume that $\inf_{\mathbf{R}^n}u=0$, and must show $u\equiv 0$. But if $u\not\equiv 0$, then by the strong maximum principle u is everywhere positive in $\mathbf{R}^n$. Hence (2.4) gives $\inf_{x\to\infty}u(x)>0$, so u must have a zero minimum at some finite point y, an immediate contradiction. Case (a) is proved.

(6) See also Lemma 4.2 below, in the special case $\hat{c}=\hat{d}=\hat{f}=0$.

Remark. The Liouville theorem of the first part of case (a) is probably known, but we do not have a direct reference.

(b) By the second part of case (a), we can assume without loss of generality that $n>m$. Also, as previously, it is enough to consider positive solutions u. It is not hard to check that the hypothesis $n<2m^2-m$ of Lemma 3.3 holds in the present case, hence (3.8) is valid for any $q\in(0,\varrho)$. A short calculation verifies that $n/m<\varrho$ (recall that $1<n/m<2$ by assumption). Thus there exists an exponent $q\in(n/m,\varrho)$. With such a choice of q, by letting $R\to\infty$ in (3.8) we get

$$\|fu^{1-m}\|_{L^q(\mathbf{R}^n)}=0.$$

It follows that $f(s)=0$ for all values s in the range of the solution $u(x)$. Thus in turn $\Delta_m u\equiv 0$ in $\mathbf{R}^n$, whence by the Liouville theorem for the m-Laplacian (first part of case (a)) we get $u(x)\equiv\text{Const.}$, as required.

Before proving case (c) it is convenient to give a simple lemma. (Recall that $\sigma=2m-m_*$.)

LEMMA 3.4. *If either*

$$n=2,\ 1<m<\tfrac{4}{3}\quad or\quad n\geqslant 3,\ n\geqslant 2m-\tfrac{1}{2}, \tag{3.11}$$

then

(i) $\sigma>m/n$,

(ii) $(\alpha-m)n<(2\alpha-m_*-m/n)m$ *for all* $\alpha\in[m,m^*]$.

Proof. It is easy to check that (i) is equivalent to $(n-m)^2>m^2-m$, which is satisfied in either case of (3.11). Similarly, (ii) is equivalent to

$$n(n-m)(n-2m)\alpha<m(n^3-2mn^2+m^2). \tag{3.12}$$

If $n\geqslant 2m$ it is enough to verify (3.12) when $\alpha=m^*=mn/(n-m)$, as is easily done. On the other hand, if $n=2$ or $n<2m$ then the worst case of (3.12) occurs when $\alpha=m$, that is, we must verify

$$nm(n-m)(n-2m)<m(n^3-2mn^2+m^2).$$

This reduces to $(n-m)^2>m^2-m$, and as before is satisfied in either case of (3.11). This completes the proof.

We can now return to the proof of the theorem.

(c) If either $n=2$, $m\geqslant\frac{4}{3}$ or $n\in[3,2m-\frac{1}{2})$, then the result of case (b) applies, so we are done. Thus without loss of generality we can assume that either

$$n=2,\ m<\tfrac{4}{3}\quad\text{or}\quad n\geqslant 3,\ n\geqslant 2m-\tfrac{1}{2}.$$

Now suppose for contradiction that u is a positive solution of (1.4) on $\mathbf{R}^n$, with f being subcritical. By (1.8) there is some $u_0>0$ such that $f(u)\geqslant u^{p-1}$ for $u\geqslant u_0$. By integrating (1.7) we get, for $u\geqslant u_0$,

$$f(u)\leqslant \frac{f(u_0)}{u_0^{\alpha-1}}u^{\alpha-1}.$$

Comparing the last lines it follows that $p\leqslant\alpha$. Again by integrating (1.7), we find for $u\leqslant u_0$

$$f(u)\geqslant \frac{f(u_0)}{u_0^{\alpha-1}}u^{\alpha-1}\equiv \lambda u^{\alpha-1}, \tag{3.13}$$

where $\lambda\geqslant u_0^{p-\alpha}>0$.

Consider first the case when $p>m$, so that $m<p\leqslant\alpha<m^*$. Define

$$\sigma=2m-m_*,\quad \tau=2p-m_*,\quad \nu=2\alpha-m_*.$$

Then clearly

$$\tau-\sigma=2(p-m)>0,\quad \nu-\sigma=2(\alpha-m)\geqslant 2(p-m)>0.$$

Also by Lemma 3.4 (i),

$$\sigma-d>0,\quad \tau-d>0 \tag{3.14}$$

provided d is fixed less than m/n.

Let $k>2m$. From (3.2), with the positive constant C rewritten as C_1, we have

$$\int \xi^k f^2(u)u^{2-m_*-d}\leqslant C_1R^{-2m}\int \xi^{k-2m}u^{\sigma-d}$$

(since $\alpha<m^*$, the hypothesis (3.1) of Lemma 3.1 is satisfied if d is made even smaller, if necessary).

Now, for $u\geqslant u_0$, by Young's inequality with the exponent pair

$$\frac{\tau-d}{\sigma-d},\quad \frac{\tau-d}{\tau-\sigma},$$

we get (since $f(u)\geqslant u^{p-1}$)

$$(\xi R)^{-2m}u^{\sigma-d}\leqslant \frac{1}{2C_1}f^2(u)u^{2-m_*-d}+C_2(\xi R)^{-m(\tau-d)/(p-m)}; \tag{3.15}$$

while for $u\leqslant u_0$, similarly (since $f(u)\geqslant\lambda u^{\alpha-1}$)

$$(\xi R)^{-2m}u^{\sigma-d}\leqslant \frac{1}{2C_1}f^2(u)u^{2-m_*-d}+C_3(\xi R)^{-m(\nu-d)/(\alpha-m)} \tag{3.16}$$

for appropriate constants C_2 and C_3. Combining the previous lines yields (with the constant $C=2C_1\max(C_2, C_3)$)

$$\int_{B_R} f^2(u)u^{2-m_*-d} \leqslant C\max[R^{n-m(\tau-d)/(p-m)}, R^{n-m(\nu-d)/(\alpha-m)}] \tag{3.17}$$

provided that k is taken suitably large (greater than both $m(\tau-d)/(p-m)$ and $m(\nu-d)/(\alpha-m)$; in fact $k=2mp/(p-m)$ suffices for all purposes.(7))

We assert that

$$n-m\frac{\tau-d}{p-m} \leqslant n-m\frac{\nu-d}{\alpha-m} < 0. \tag{3.18}$$

Indeed the first inequality can be rewritten in the form $(\alpha-p)(\sigma-d)\geqslant 0$, which is clearly satisfied in view of (3.14). The second (strict) inequality, on the other hand, follows at once from Lemma 3.4 (ii), since $d<m/n$.

Let $R\to\infty$ in the inequality (3.17). In view of (3.18) there results

$$\int_{\mathbf{R}^n} f^2(u)u^{2-m_*-d} = 0.$$

But this is impossible because, as we have seen, $f(u)>0$ for all $u>0$.

There remains the case $p\leqslant m$. Here by assumption also $\alpha\leqslant m$. Now by hypothesis $f(u)\geqslant u^{p-1}$ for $u\geqslant u_0$. Also (3.13) gives $f(u)\geqslant\lambda u^{\alpha-1}$ for $u\leqslant u_0$. Hence

$$-\Delta_m u \geqslant \min(\lambda u^{\alpha-1}, u^{p-1}) \geqslant l\min(u^{m-1}, u^{p-1}),$$

where $l=\min(\lambda, 1)$. After a trivial change of scale, it now follows from the remark after the proof of Lemma 2.8 that $u\equiv 0$.

(d) This is an immediate consequence of Theorem I.

Remark. The proof of parts (b), (c) give upper bounds for various norms of u in terms of the radius of the ball B_R, see (3.8) and (3.17); these can be considered as another type of universal a priori estimate.

Proof of Theorem III. Let M be the upper bound for the solution u, that is, $0<u(x)\leqslant M$ in $\mathbf{R}^n$. As in (3.13), there holds

$$f(u) \geqslant \frac{f(M)}{M^{\alpha-1}}u^{\alpha-1} \equiv \hat{\lambda}u^{\alpha-1}$$

(7) For its interest, one may note that

$$C_2 = (2C_1)^{(\sigma-d)/(\tau-d)}, \quad C_3 = (2C_1/\lambda^2)^{(\sigma-d)/(\nu-d)}.$$

for $u \leqslant M$. Here it should be noted that $\hat{\lambda}>0$, since by assumption $f(u)>0$ for all $u>0$.

Arguing as in the proof of case (c), but without the necessity for introducing the constant τ or deriving the inequality (3.15), we obtain (3.16), this now being valid for all x in $\mathbf{R}^n$. Inequality (3.17) is therefore replaced simply by

$$\int_{B_R} f^2(u) u^{2-m_*-d} \leqslant 2C_1C_3 R^{n-m(\nu-d)/(\alpha-m)},$$

and the proof is completed with the help of (3.18) exactly as before.

4. The Harnack inequality

Here we establish Harnack inequalities for weak solutions of the equation (1.4) and the inequality (1.6), see Theorems 4.1–4.3.

As previously, $B_R = B_R(x_0)$ denotes a ball centered at x_0 with radius R, such that the corresponding ball $B_{2R}(x_0)$ of radius $2R$ is contained in Ω. Our arguments will be restricted throughout to such "admissible" balls B_R. Moreover, the letter C denotes a generic positive constant, which may vary from line to line and only depends on the arguments inside the parenthesis.

The following result is an extension and generalization of Lemma 2.4.

LEMMA 4.1. (i) *Assume that $m<p<m_*$ in Lemma 2.4. Then* (2.5) *holds for all $\gamma \in (0, m_*-1)$, and* (2.6) *for all $\mu \in (0, n(m-1)/(n-1))$.*

(ii) *Assume $p>m$, and let u be a non-negative weak solution of the differential inequality*

$$-\Delta_m u \geqslant u^{p-1} - u^{m-1} - |\nabla u|^{m-1} \quad \text{in } \Omega.$$

Then there exists a constant $C = C(n, m, p, \gamma) > 0$ such that

$$\int_{B_R} u^\gamma \leqslant C R^{n-m\gamma/r} + C R^n, \tag{4.1}$$

for all $\gamma \in (0, m_*-1)$ provided $m<p<m_*$, and for all $\gamma \in (0, p-1)$ provided $m_* \leqslant p < m^*$.

Proof. (i) We proceed as in the proof of Lemma 2.4. To begin with, for $k>m$ we have

$$\nabla(\xi^{k/m} u^{(\gamma-r)/m}) = \xi^{k/m} \frac{\gamma-r}{m} u^{(\gamma-p)/m} \nabla u + \frac{k}{m} \xi^{k/m-1} u^{(\gamma-r)/m} \nabla \xi.$$

With the help of the elementary inequality $|x+y|^m \leqslant 2^{m-1}(|x|^m + |y|^m)$, an easy calculation then leads to (recall $|\nabla\xi|\leqslant 2/R$)

$$d\int|\nabla(\xi^{k/m}u^{(\gamma-r)/m})|+\left(\frac{2(\gamma-r)}{m}\right)^m\int\xi^k u^\gamma$$
$$\leqslant\left(\frac{2(\gamma-r)}{m}\right)^m\left[\frac{d}{2}\int\xi^k u^{\gamma-p}|\nabla u|^m+\int\xi^k u^\gamma\right]+\left(\frac{4k}{m}\right)^m\frac{d}{2}R^{-m}\int\xi^{k-m}u^{\gamma-r}$$
$$\leqslant C(m,k,p,\gamma)R^{-m}\int\xi^{k-m}u^{\gamma-r}$$

by virtue of (2.8). Then exactly as in the derivation of (2.9) from (2.8), we find that

$$\int|\nabla(\xi^{k/m}u^{(\gamma-r)/m})|^m\leqslant CR^{n-m\gamma/r}$$

for $\gamma\in(r,p-1)$. Then by the Sobolev inequality,

$$\int[\xi^{k/m}u^{(\gamma-r)/m}]^{m^*}\leqslant CR^{(n-m\gamma/r)m^*/m}.$$

Rewriting this by setting $\bar\gamma=m^*(\gamma-r)/m$, we obtain exactly (2.5) with γ replaced by $\bar\gamma$. But from the condition $\gamma\in(r,p-1)$ then follows $\bar\gamma\in(0,m_*-1)$, which is the first result.

The second is then obtained by following the derivation of (2.6), but using at the final step the result just shown.

Remark. We have not in fact used the condition $p<m_*$ in this argument. However, if $p\geqslant m_*$ then the original restriction $\gamma<p-1$ is weaker than or equivalent to $\gamma<m_*-1$. This can be restated alternately, that (2.5) holds for any $\gamma\in(0,\max(p-1,m_*-1))$.

(ii) The proof of Lemma 2.4 carries over without difficulty, once one notes that the inequality (2.7) continues to hold, but with the addition of two further terms

$$\int\xi^k u^{\gamma-r}+\int\xi^k u^{-d}|\nabla u|^{m-1}$$

on the right side. In turn, we derive (2.8) essentially as before, but now with an added term

$$C\int\xi^k u^{\gamma-r}$$

on the right-hand side. Finally, again essentially as before, by Young's inequality one gets (2.9) with an additional term CR^n on the right side. The rest of the proof then follows exactly as in Lemma 2.4 and the previous case (i).

LEMMA 4.2. *Assume* $\Omega\subset\mathbf{R}^n$ *and* $n>m$. *Let* u *be a non-negative weak solution (in* $W^{1,m}_{\mathrm{loc}}\cap C$*) of the two-sided inequality*

$$|\Delta_m u|\leqslant \hat{d}u^{m-1}+\hat{f}, \tag{4.2}$$

where $\hat{d}$ *and* $\hat{f}$ *are non-negative measurable functions on* Ω. *Let* $q\in(n/m, n/(m-1))$. *Then for every* $R>0$ *for which* B_R *is admissible, there exists a constant* C, *depending only on the parameters*

$$n,\ m,\ q,\ R^{m-n/q}\|\hat{d}\|_{L^q(B_{2R})},$$

such that

$$\sup_{B_R} u\leqslant C\big(\inf_{B_R} u+R^{m-n/q}\|\hat{f}\|_{L^q(B_{2R})}\big). \tag{4.3}$$

With slight change of notation, this is exactly the special case $c=0$ of Theorem 5 of [23], after restriction to the operator $\mathcal{A}(p)=|p|^{m-2}p$, see §1 of [23]. The factor $R^{m-n/q}$ is just that required to rescale to the unit ball B_1 in the proof of Theorem 5; see p. 263 of [23].

Remark. Lemma 4.2 also holds when $n=m$, and even when $n<m$ provided $q=1$; see §1, relation (8), and §6 of [23].

We can now prove our first Harnack inequality. It will be assumed unless otherwise stated that $n>m$. All balls B_R are assumed to be admissible, in the sense noted at the beginning of the section.

THEOREM 4.1. (a) *Suppose* $n>m$ *and* $p\in(m,m_*)$. *Let* u *be a non-negative weak solution of the differential inequality*

$$u^{p-1}\leqslant -\Delta_m u\leqslant \Lambda u^{p-1}\quad \textit{in } \Omega, \tag{4.4}$$

for some constant $\Lambda>1$. *Then there is a constant* $C=C(n,m,p,\Lambda)>0$ *such that*

$$\sup_{x\in B_R} u(x)\leqslant C\inf_{x\in B_R} u(x). \tag{4.5}$$

(b) *If* $m<p\leqslant s<m_*$ *and* (4.4) *is replaced by*

$$u^{p-1}-u^{m-1}-|\nabla u|^{m-1}\leqslant -\Delta_m u\leqslant \Lambda(u^{s-1}+u^{m-1}+|\nabla u|^{m-1})\quad \textit{in } \Omega, \tag{4.6}$$

then (4.5) *holds with* $C=C(n,m,p,s,\Lambda,R)>0$. *The constant* C *may become arbitrarily large if* $R\to 0$ *(when* $p<s$*) or if* $R\to\infty$.

Proof. (a) First assume $B_{4R}\subset\Omega$. We shall apply the Harnack inequality, Lemma 4.2. In the present case, in view of (4.2) and (4.4) we can take $\hat{c}=\hat{f}=0$ and

$$\hat{d}=\hat{d}(x)=\Lambda u^{p-m}.$$

We require an estimate of the norm $I\equiv\|\hat{d}(x)\|_{L^q(B_{2R})}$ for some $q\in(n/m,n/(m-1))$.

By the left side of (4.4) and by (2.5) with B_R replaced by B_{2R}, we get

$$I^q\leqslant\Lambda^q\int_{B_{2R}}u^{(p-m)q}\leqslant C\Lambda^q R^{n-mq(p-m)/(p-m)}=CR^{n-mq}$$

provided q satisfies the principal condition $\gamma=(p-m)q<p-1$. Since $p<m_*$, however, it is easy to see that one can choose (and fix) a value $q\in(n/m,n/(m-1))$ so that this condition holds. Hence $I\leqslant C\Lambda R^{n/q-m}$ and in turn $R^{m-n/q}I\leqslant C\Lambda$.

Lemma 4.2 now gives the conclusion (4.5), but under the additional assumption $B_{4R}\subset\Omega$. By a chaining argument, increasing C appropriately, but still dependent only on n,m,p,Λ, one can replace B_{4R} by B_{2R}. This finishes the proof for case (a).

(b) Again it will be assumed to begin with that $B_{4R}\subset\Omega$. From (4.2) and (4.6) we have $\hat{f}=0$ and

$$\hat{c}=\Lambda,\quad \hat{d}=\Lambda(u^{s-m}+1).$$

Therefore, besides an estimate for the norm I we shall also need to bound the norm $J=\|\hat{c}\|_{L^{q'}(B_{2R})}$ for some $q'\in(n,\infty)$. First,

$$I^q\leqslant\Lambda^q\int_{B_{2R}}(u^{s-m}+1)^q\leqslant 2^{q-1}\Lambda^q\int_{B_{2R}}(u^{(s-m)q}+1)=2^{q-1}\Lambda^q\int_{B_{2R}}(u^\gamma+1),\qquad(4.7)$$

where $\gamma\equiv(s-m)q$. By choosing q near enough to n/m, and recalling that $s<m_*$, we have $\gamma<s-1<m_*-1$. Hence, with the help of Lemma 4.1 (ii) and the left-hand inequality of (4.6), it follows that

$$I^q\leqslant C\Lambda^q(R^{n-m\gamma/(p-m)}+R^n)=C\Lambda^q(R^{n-mq(s-m)/(p-m)}+R^n),$$

where $C=C(n,m,p,s)$. Moreover, $J=\Lambda\big(\int_{B_{2R}}\big)^{1/q'}=C\Lambda R^{n/q'}$. In turn (with the appropriate rescaling factors)

$$R^{m-n/q}I\leqslant C\Lambda(R^{-(s-p)/(p-m)}+R)^m,\quad R^{1-n/q}J=C\Lambda R.\qquad(4.8)$$

Lemma 4.2 now yields the required conclusion, since the ball B_{4R} can always be replaced by B_{2R}. The proof is complete.

Remark. When $s=p$ the first estimate of (4.8) reduces to $R^{m-n/q}I\leqslant C\Lambda(1+R)^m$. In turn the coefficient $C=C(n,m,p,s,\Lambda,R)$ in (4.5) becomes $C=C(n,m,p,\Lambda,R)$, of course remaining bounded as $R\to0$.

THEOREM 4.2. *Let u be a solution of* (1.4), *with $n>m$. Then:*

(a) *If f is subcritical and either $n=2$ and $m>\left(1+\frac{1}{4}\sqrt{17}\right)$ or $n\in[3,2m)$, $m>\frac{3}{2}$, then*

$$\sup_{B_R} u \leqslant C \inf_{B_R} u \tag{4.9}$$

for some constant $C=C(n,m,\alpha)$.

(b) *Suppose that f is subcritical, and that for some $p>m$*

$$u^{p-1} \leqslant f(u) \leqslant \Lambda u^{p-1} \quad \textit{for all } u>0. \tag{4.10}$$

Then (4.9) *holds with $C=C(n,m,p,\alpha,\Lambda)$.*

(c) *Let the conditions of case* (b) *hold, but with* (4.10) *replaced by*

$$u^{p-1} \leqslant f(u) \leqslant \Lambda(u^{p-1}+u^{m-1}) \quad \textit{for all } u>0. \tag{4.11}$$

Then (4.9) *is satisfied with $C=C(n,m,p,\alpha,\Lambda,R)$. The constant C remains bounded as $R\to 0$.*

Proof. (a) As in the proof of Theorem 4.1 we shall apply Lemma 4.2. By (1.4) and (4.2) one can take $\hat{c}=\hat{f}=0$ and

$$\hat{d}=\hat{d}(x)=u^{1-m}f(u).$$

We must estimate the norm $I\equiv\|\hat{d}(x)\|_{L^q(B_{2R})}$ for some $q\in(n/m,n/(m-1))$.

As in the proof of Theorem II (b) we can choose q so that $q\in(n/m,\varrho)$, and even more so that $q\in(n/m,n/(m-1))$. Then by Lemma 3.3, we find

$$I\leqslant CR^{n/q-m},$$

and the Harnack inequality follows exactly as in Theorem 4.1 (a).

(b) Before proceeding with the main proof, we note that necessarily $p\leqslant\alpha$, as follows by integration of (1.7); see (3.13).

Now utilizing the left side of (4.10) in (3.2) gives

$$\int \xi^k u^{\tau-d} \leqslant CR^{-2m}\int \xi^{k-2m}u^{\sigma-d}, \tag{4.12}$$

where $\tau=2p-m_*$, $\sigma=2m-m_*$ and d is chosen so that (3.1) is satisfied.

If $n=2$ and $m\geqslant\frac{4}{3}$, or if $n\geqslant 3$ and $n\leqslant 2m-\frac{1}{2}$, the previous case (a) applies and we are done. It is therefore enough to consider the ranges $n=2$, $m<\frac{4}{3}$ and $n\geqslant 3$, $n>2m-\frac{1}{2}$.

Lemma 3.4 shows that $\sigma > m/n$ in both. We may of course suppose that $d < m/2n$, so that $\sigma - d > m/2n$. Clearly $\tau - d > \sigma - d$ since $p > m$. Then from the Young inequality with exponents

$$\frac{\tau-d}{\sigma-d}, \quad \frac{\tau-d}{\tau-\sigma},$$

we find (with the usual trick)

$$CR^{-2m}\int \xi^{k-2m}u^{\sigma-d} \leqslant \frac{1}{2}\int \xi^k u^{\tau-d} + CR^{n-(\tau-d)m/(p-m)},$$

since $\tau - \sigma = 2(p-m)$. Note also that we must take k suitably large, say $k = 2pm/(p-m)$; see the analogous derivation of (3.17) in the previous section. In turn, with the aid of (4.12),

$$\int_{B_{2R}} u^{\tau-d} \leqslant CR^{n-(\tau-d)m/(p-m)}, \tag{4.13}$$

where we have replaced B_R by B_{2R} and used the fact that $B_{4R} \subset \Omega$.

We assert that

$$(p-m)\frac{n}{m} < \tau - \frac{m}{n}. \tag{4.14}$$

In fact, since $m < p \leqslant \alpha$, therefore also $p \in (m, m^*)$. Hence (4.14) is exactly the result of Lemma 3.4 (ii) with α replaced by p.

We are now able to estimate the norm I. First, by the right side of (4.10) there results

$$I^q \leqslant \Lambda^q \int_{B_{2R}} u^{(p-m)q}.$$

By (4.14) and the fact that $d < m/2n$, one can choose $q \in (n/m, n/(m-1))$ so that $(p-m)q < \tau - d$. Then by Hölder's inequality,

$$I^q \leqslant C\Lambda^q R^{n(1-(p-m)q/(\tau-d))}\left(\int_{B_{2R}} u^{\tau-d}\right)^{(p-m)q/(\tau-d)} \leqslant C\Lambda^q R^{n-mq}$$

by (4.13). The rest of the proof is as before.

(c) The inequality (4.13) follows exactly as in case (b). On the other hand, by the right side of (4.11),

$$I^q \leqslant \Lambda^q \int_{B_{2R}} (u^{p-m}+1)^q \leqslant 2^{q-1}\Lambda^q \int_{B_{2R}} (u^{(p-m)q}+1).$$

Hence, exactly as in case (b), one gets $I \leqslant C\Lambda(R^{n/q-m} + R^{n/q})$. With the rescaling factor of Lemma 4.2, we then find

$$R^{m-n/q} I \leqslant C\Lambda(1+R^m).$$

The Harnack inequality (4.9) now follows as previously, except of course the coefficient C now depends on R.

Harnack inequalities can also be given when $n \leqslant m$, the case earlier left aside.

THEOREM 4.3. *Let* $n \leqslant m$. *Then:*

(a) *Assume the hypotheses of Theorem* 4.1 (a), *except that the condition* $p \in (m, m_*)$ *is replaced by* $p \in (1, \infty)$, *that is,* $m_* = \infty$. *Then* (4.5) *is valid with* $C = C(n, m, p, \Lambda)$.

(b) *Assume the hypotheses of Theorem* 4.1 (b), *with* $m < p \leqslant s < \infty$. *Then* (4.5) *holds with* $C = C(n, m, p, s, \Lambda, R)$.

Proof. Everything is the same as in the proof of Theorem 4.1, with the exception that the exponents q, q' must be chosen as in the remark after the proof of Lemma 4.2. Such a choice is obviously possible, and the proof is complete.

Remark. Since q is here subject to a weaker condition than in the previous case $n > m$, it is no longer necessary to have the upper bound $p < m_*$ or $\alpha < m^*$.

Comment on the form of the coefficient C *in* (4.9). This constant arises in a complicated way, depending on Lemmas 3.1, 3.2, 3.3, 4.2, as well as on the coefficient on the right side of (4.13). By following the proof, however, it is not hard to see that the corresponding coefficients C can become unbounded only when one or another of the following limits occur:

$$d \to 0; \quad m \to n; \quad q \to \varrho; \quad q \to n/m; \quad 2m^2 - m \to n; \quad p \to m$$

(omitting the trivial limits $m, n \to 1$, $m, n \to \infty$ and $\Lambda \to \infty$). Moreover, $d \to 0$ only when $\alpha \to m^*$; $q \to \varrho$ or $q \to n/m$ only when $m \to n$ or $m \to \frac{1}{2}n$ (in the proof of Theorem 4.2 (a)); and $2m^2 - m \to n$ only when $n = 2$ and $m \to \frac{1}{4}(1 + \sqrt{17})$.

In turn, the coefficient $C = C(n, m, \alpha)$ in case (a) can become unbounded only if $\alpha \to m^*$, or $n = 2$ and $m \to \frac{1}{4}(1 + \sqrt{17})$, or $n \geqslant 3$ and m approaches either n or $\frac{1}{2}n$.

Similarly, the coefficient $C = C(n, m, p, \alpha, \Lambda)$ in case (b) can become unbounded only if $\alpha \to m^*$, or $p \to m$, or $m \to n$ (and of course $\Lambda \to \infty$).

5. Universal a priori estimates

In this section, we shall establish the universal a priori estimate Theorem IV, as well as several other related conclusions. The notation will be adapted from the previous section, in particular, $B_R = B_R(x_0) \subset B_{2R} = B_{2R}(x_0) \subset \Omega$.

Proof of Theorem IV. (a) By (1.9), Lemma 4.1 (ii) implies that there exists a constant $C = C(m, n, p) > 0$ such that

$$\int_{B_R} u^r \leqslant C(R^{n-m} + R^n)$$

where $r=p-m$. Therefore

$$\inf_{B_R} u^r \leqslant \frac{1}{|B_R(x)|} \int_{B_R} u^r \leqslant C(1+R^{-m}).$$

In turn,

$$\inf_{B_R} u \leqslant C(1+R^{-m/r}). \tag{5.1}$$

On the other hand, again by (1.9), a simple modification of Theorem V (a) implies that, for $R\leqslant 1$, there exists a constant $C=C(m,n,p,\Lambda)>0$ such that for $x\in B_R$,

$$u(x) \leqslant \sup_{B_R} u \leqslant C\big(\inf_{B_R} u+1\big). \tag{5.2}$$

The conclusion (1.10) now follows at once from (5.1) and (5.2).

To prove the second part of Theorem IV (a), note first that in this case (5.1) is now valid without the additional term $R^{-m/r}$ and also without the restriction $R\leqslant 1$. Similarly, in view of the second part of Theorem V (a), the inequality (5.2) holds with no restriction on R and without the additive term 1. The required conclusion then follows as before.

Remark. A similar result can also be given for the differential inequality (4.6).

(b) We first prove (1.10) under the assumption (1.11). This being the case, we need to apply Lemma 4.2 with

$$\hat{c} \equiv 0, \quad \hat{d} = \Lambda u^{p-m}, \quad \hat{f} \equiv \Lambda.$$

Proceeding exactly as in Theorem 4.2 (b), we deduce that there exists $C=C(n,m,p)>0$ such that

$$\|\hat{d}\|_{L^q(B_{2R})} \leqslant C\Lambda R^{n/q-m}$$

for some $q\in(n/m, n/(m-1))$. Therefore, by (4.3), since $\|\hat{f}\|_{L^q(B_{2R})}=C\Lambda R^{n/q}$ we get

$$\sup_{B_R} u \leqslant C\big(\inf_{B_R} u+R^m\big) \tag{5.3}$$

for some $C=C(n,m,p,\Lambda)>0$.

On the other hand, by (1.11) and Lemma 2.4, we have

$$\inf_{B_R} u^{p-m} \leqslant \frac{1}{|B_R(x)|} \int_{B_R} u^{p-m} \leqslant CR^{-m}$$

for some $C=C(n,m,p)>0$, that is,

$$\inf_{B_R} u \leqslant CR^{-m/(p-m)}. \tag{5.4}$$

Now (1.10) follows immediately from (5.3) and (5.4).

Next suppose that (1.12) holds in place of (1.11). Then (5.3) is valid without the additional term ΛR^m since $\hat{f}\equiv 0$. Therefore in (1.10) one has $R=\min(x,\partial\Omega)$, and the proof is complete.

We say that the exponent $m/(p-m)$ in (1.10) is optimum, if for any $\delta>0$ there exist a domain Ω and a solution u of (1.4) in Ω such that

$$u(x)\geqslant C\delta^{-m/(p-m)} \tag{5.5}$$

for some $x\in\Omega$ satisfying $\operatorname{dist}(x,\partial\Omega)=\delta$, where the constant $C=C(n,m,p)$ is independent of the solution as well as the domain Ω.

When $f=u^{p-1}$ and $p\in(m_*,m^*)$, the singular solution of (1.4) defined in the proof of Corollary I (see the Introduction) immediately reveals that $m/(p-m)$ is optimum, by placing the singularity $x=0$ on the boundary. In fact, we have the following more complete result.

PROPOSITION 5.1. *Assume that $n>m$, $p\in(m,m^*)$ and $f=u^{p-1}$. Then the exponent $m/(p-m)$ is optimum for* (1.10).

Proof. Let $m>1$ and $p>m$. Consider the initial value problem

$$\begin{cases} (|u'|^{m-2}u')'+u^{p-1}=0, & u>0,\ t>0,\\ u(0)=u_0>0, \\ u'(0)=0. \end{cases} \tag{IVP}$$

Local existence and uniqueness are well known, with $u'<0$ for $t>0$. Furthermore, since $(|u'|^{m-2}u')'<0$, it is easy to see that the solution can be continued, still with $u'<0$, as long as $u>0$; see for example the Appendix of [19]. Define

$$T_0=\sup\{T>0\,|\,u(t)>0 \text{ for } t\in[0,T)\}.$$

Then since $p<m^*$ we get from [16, Theorem 6.2]

$$T_0<\infty,\quad u(T_0)=0.$$

Therefore

$$((-u')^{m-1}(t))'=u^{p-1}(t),\quad t\in(0,T_0]. \tag{5.6}$$

Multiply (5.6) by u' and integrate from 0 to $t\in(0,T_0]$ to obtain

$$-\frac{m-1}{m}(-u')^m=\frac{1}{p}(u^p(t)-u_0^p).$$

It follows that

$$-C\frac{u'}{u_0}\cdot\left(1-\frac{u^p(t)}{u_0^p}\right)^{-1/m}=u_0^{(p-m)/m}, \tag{5.7}$$

where

$$C=\left(\frac{p(m-1)}{m}\right)^{1/m}.$$

Integrating (5.7) once more from 0 to t, we get

$$tu_0^{(p-m)/m}=\int_{u(t)/u_0}^{1}(1-s^p)^{-1/m}\,ds,$$

that is,

$$u_0=\left(C\int_{u(t)/u_0}^{1}(1-s^p)^{-1/m}\,ds\right)^{m/(p-m)}\cdot t^{-m/(p-m)}.$$

In particular,

$$u_0=KT_0^{-m/(p-m)}, \tag{5.8}$$

where

$$K=K(m,p)=\left(C\int_0^1(1-s^p)^{-1/m}\,ds\right)^{m/(p-m)}>0.$$

For $\delta>0$, let $B=B_\delta$ be the ball centered at 0 with radius δ, and let $u=u(t)$ be the unique solution of (IVP) with $u_0=K\delta^{-m/(p-m)}>0$. By (5.8), for this solution we have $T_0=\delta$. It follows that

$$v(x)=v(x_1,x_2,\dots,x_n)=u(x_1)$$

is a solution of (1.4) which is defined and positive in B, and of course depends only on the variable $x=x_1$. Clearly $\operatorname{dist}(0,\partial B)=\delta$, while $u(0)=u_0=K\delta^{-m/(p-m)}>0$. This is just (5.5), and the proof is complete.

It is also interesting to know for what values of p the estimate (1.10) holds at all. Again, we consider the pure power $f=u^{p-1}$ for simplicity.

PROPOSITION 5.2. *Let $n>m$. Then* (1.10) *fails to hold for* (1.4) *if $p=m$ or $p\geqslant m^*$. Similarly* (1.10) *fails for* (1.6) *if $p=m$ or $p>m_*$.*

Proof. For $p=m$, the equation (1.4) is homogeneous. Clearly a multiple of a non-trivial non-negative solution is still a solution, which can have an arbitrarily large maximum value, whence (1.10) cannot hold. On the other hand, when $p\geqslant m^*$, it is well-known (e.g., Theorem 6.4 of [19]) that (1.4) has positive entire solutions on the entire space $\mathbf{R}^n$, whose maximum value can be arbitrarily large. Again, (1.10) cannot hold.

For (1.6), we observe that when $p>m_*$ the solution given in the proof of Corollary II (iii) takes arbitrarily large values $C'\varkappa^{-(m-1)/(p-m)}$ when $x=0$, so again (1.10) fails. The proof is complete.

Remarks for the supercritical range $p\geqslant mn/(m-n)$. Proposition 5.2 indicates that the optimum range of p for the global estimate (1.10) to hold for (1.1) is (m,m^*). Yet a corresponding local version (1.13) near an isolated singularity may continue to hold outside this range. Indeed, for $m=2$ and $p=2n/(n-2)$, the estimate (1.13) was established for solutions of (1.1) by Caffarelli, Gidas and Spruck, see [6, Theorem 1.1, p. 272], though there the constant C might depend on the solution itself.

For a non-removable singularity at the origin, the estimate can be strengthened to show that C is independent of the solution for sufficiently small $|x|$ (see [6, Theorem 1.2, p. 273]). Of course, similar estimates hold on exterior domains via the Kelvin transform (always assuming $m=2$).

On the other hand, when $n>m+1$ and $p\geqslant m(n-1)/(n-m-1)$, (1.13) need not hold for an *exterior* domain, in contrast to case (ii) of Corollary IV. To see this, we observe that u^{p-1} is supercritical for the dimension $n-1$ if $n>m+1$ and $p\geqslant(n-1)m/(n-m-1)$. Thus (1.1) has a positive radial solution $u_0(r)$ on $\mathbf{R}^{n-1}$ (e.g., Theorem 6.4 of [19]). Put

$$u(x)=u(x',x_n)=u_0(|x'|).$$

Obviously u is also a solution of (1.1) on $\mathbf{R}^n$, but (1.13) does not hold in an exterior domain since u is constant (>0) along the direction x_n.

Whether the local estimate (1.13) holds for (1.4) in $B_1(0)\backslash\{0\}$ remains open even for the pure power $f=u^{p-1}$ when $p\in[m^*,(n-1)m/(n-m-1))$, except $m=2$ and $p=2n/(n-2)$.

Finally, it is interesting to note that, for a positive solution u of (1.4) on $\mathbf{R}^n$, the estimate (1.13) for large $|x|$ is equivalent to the radial symmetry of u, provided $m=2$ and $p\in(2n/(n-2),2(n-1)/(n-3))$. In fact, (1.13) plainly holds if u is radially symmetric. On the other hand, (1.13) for large $|x|$ implies that u must be radially symmetric with respect to some point $x_0\in\mathbf{R}^n$ by Theorem 1.1, p. 48, of [33], provided $p\in(2n/(n-2),2(n-1)/(n-3))$.

Chapter II

6. A general integral inequality I

In this section we shall establish an important integral inequality for solutions u of (1.4), generalizing the Gidas–Spruck identity for solutions of (1.4) in the case $m=2$. The following agreements will be used throughout. Boldface lower and upper case letters respectively denote vector and matrix quantities; when $\mathbf{x}$ is a vector and $\mathbf{A}$ a matrix, by $\{\mathbf{xAx}\}$ we mean the quadratic form $\mathbf{x}\cdot(\mathbf{Ax})$, where standard matrix multiplication is always understood and $\mathbf{x}\cdot\mathbf{y}$ means the inner product of the vectors $\mathbf{x}$ and $\mathbf{y}$.

For a weak solution $u\in C^1(\Omega)$ of (1.4), we introduce the *centrally important vector field*

$$\mathbf{u}=|\nabla u|^{m-2}\nabla u, \tag{6.1}$$

where in the usual way it is understood that $\mathbf{u}=0$ when $\nabla u=0$ (recall $m>1$). (The simple notation $\mathbf{u}$ in (6.1) may not carry a clear indication of the meaning of this vector; alternative notations suggest themselves but all seem cumbersome in view of the many appearances of this vector in later places. Consequently we retain the indicated notation, and simply remind the reader again of the central importance of the vector $\mathbf{u}$.) Recall that

$$m_*=\frac{m(n-1)}{n-m},\qquad m^*=\frac{mn}{n-m}.$$

Then we have the following principal result.

PROPOSITION 6.1. *Suppose $m\in(1,n)$. Let u be a positive weak solution of* (1.4) *and $\phi\in C_0^\infty(\Omega)$ a non-negative test function.*

Then for any $d\in\mathbf{R}$ we have

$$\begin{aligned}\int u^{1-m_*-d}\{Af(u)+\hat{A}uf'(u)\}|\nabla u|^m\phi+B\int u^{-m_*-d}|\nabla u|^{2m}\phi\\ \leqslant\int u^{1-m_*-d}\{Duf(u)+\widehat{D}|\nabla u|^m\}\mathbf{u}\cdot\nabla\phi+\int u^{2-m_*-d}\{\mathbf{u}\nabla^2\phi\mathbf{u}\},\end{aligned} \tag{6.2}$$

where $\nabla^2\phi$ is the Hessian of ϕ and

$$\begin{aligned}A=\frac{n-1}{n}\left(1-\frac{d}{m_*}\right)(m^*-1),\quad \hat{A}=-\frac{n-1}{n},\quad B=\frac{m-1}{m}d(1-d),\\ D=-\frac{n+1}{n},\quad \widehat{D}=\left(2-\frac{m_*}{m}\right)-\left(2-\frac{1}{m}\right)d.\end{aligned} \tag{6.3}$$

The vector $\mathbf{u}$ is nonlinear in ∇u when $m\neq 2$; this makes the proof of Proposition 6.1 more delicate than that for the linear case treated by Gidas and Spruck. In addition, the proofs when $m\geqslant 2$ and $1<m<2$ are distinctly different, with the latter requiring extreme care.

Before turning to the main proof, it is convenient to introduce some further notation. Let

$$\Omega_{\text{cr}}=\{x\in\Omega\mid\nabla u(x)=0\}$$

On the other hand, from (8.13) and another use of Lemma 6.1, we get $|\mathbf{U}|\leqslant|\nabla u|^{-l}z$.

Consequently, by the Cauchy–Schwarz inequality and (8.13),

$$\begin{aligned}-\xi(\mathbf{u}_\varepsilon\mathbf{U})\cdot\nabla\xi &\leqslant \xi(|\nabla u|_\varepsilon|\nabla u|)^{-l}|\mathbf{U}|\cdot|\mathbf{u}_\varepsilon|(|\nabla u|_\varepsilon|\nabla u|)^{l}|\nabla\xi| \\ &\leqslant \frac{1}{2}(m-1)^2\xi^2|\nabla u|_\varepsilon^{m-2}z^2+\frac{1}{2(m-1)^2}|\nabla u|_\varepsilon^{2(m-1)}|\nabla\xi|^2.\end{aligned}$$

Thus by (8.16),

$$\frac{1}{2}(m-1)^2\int\xi^2|\nabla u|_\varepsilon^{m-2}z^2\leqslant\int A(x)|\nabla\xi|^2+\int B(x)\xi^2,$$

where

$$A(x)=\frac{1}{2(m-1)^2}|\nabla u|_\varepsilon^{2(m-1)},\quad B(x)=|f'(u)|\cdot|\nabla u|^m$$

are in $L^\infty_{\text{loc}}(\Omega)$. In turn

$$\int_{B_R}|\nabla u|_\varepsilon^{m-2}z^2\leqslant\text{Const.},$$

the constant being *independent* of ε (since $|\nabla u|_\varepsilon\leqslant 1+|\nabla u|$ in the expression for $A(x)$). By the Lebesgue monotone convergence theorem, we now get

$$\int_{B_R}|\nabla u|^{m-2}z^2\leqslant\text{Const.},$$

which by (8.13) is exactly the statement that $|\nabla u|^{m-2}|\nabla^2 u|$ is in $L^2(\widehat{\Omega}\backslash\Omega_{\text{cr}})$. This completes the proof of Proposition 8.1.

9. Historical note: Cauchy and Liouville, a question of priority

Augustin Cauchy was the first person to publish the result now known as Liouville's theorem (see [8]). The contribution of Joseph Liouville is an interesting and tangled story, worth recounting in some detail. A few weeks before Cauchy's note appeared, Liouville announced to the academy his first results for doubly-periodic functions, for which he is justly famous (*C. R. Acad. Sci. Paris*, 19 (1844), 1262). This announcement includes, without proof, a weak version of Cauchy's theorem, namely the statement that a *doubly-periodic* holomorphic function must be constant. Cauchy was entirely aware of the relation of his result to that of Liouville, as he writes (*C. R. Acad. Sci. Paris*, 19 (1844), 1379), "If one considers separately the case of doubly-periodic functions, one recovers the special theorem regarded with reason, by one of our honorable associates, as particularly applicable to the theory of elliptic functions."

Three years later, Liouville gave a series of informal lectures on his theory for F. Joachimsthal and C. W. Borchardt; these lectures, containing the previously cited weak version of Cauchy's result, but with no reference to Cauchy, were transcribed and edited by Borchardt and (much later) published in *J. Reine Angew. Math.*, 88 (1880), 277–310. Outside of the first announcement and one later note (see below), this is the entire published record of Liouville's work; surprisingly it does not contain Liouville's own proof, but instead an alternate discussion due to Borchardt.

In 1851 Cauchy again wrote explicitly that his work of 1844 "furnished the fundamental principle invoked by M. Liouville for doubly-periodic functions" and went on to restate his result of 1844 (see *C. R. Acad. Sci. Paris*, 32 (1851), 452–454; *Œuvres complètes*, Ire série, tome XI, 373–376). At about the same time, Liouville delivered a carefully written course of lectures at the Collège de France on doubly-periodic functions, containing a relatively simple proof of his doubly-periodic theorem, but again not citing Cauchy's contribution.

Liouville was clearly much concerned with what he considered his priority to the doubly-periodic result, for in 1855 (*J. Math. Pures Appl.*, 20, 201–208) he republished his 1844 remarks together with a later comment of 1851 containing much the same material; indeed he even went on to refer explicitly to his lectures at the Collège de France "in the second semester of the year 1850–1851". This degree of concern almost certainly stems from the remarkable fact that near the end of his mathematical notebook for the year 1844 he had written the following *"Remarque d'analyse"*:

*Soit $f(z)$ une fonction bien déterminée de z. Si le module de $f(z)$ ne dépasse jamais M, on a $f(z)$=*Constante.

Since it is evident that he understood the function $f(z)$ to be given on the *entire* complex plane, this is clearly the general result! There follows a one-line proof sketch, which however can only be considered tentative. From internal evidence it seems highly likely that these words were written prior to the announcement of 1844, that Liouville then devoted his effort to finding a proof of the doubly-periodic result, and, upon finding a (difficult) demonstration, then reported this (but only this) result to the academy. He never afterwards referred to the "Remarque".

Liouville saw the utility and centrality of the doubly-periodic theorem for elliptic function theory, but in his preoccupation with this he missed the elegance and beauty of the main result. Cauchy on the other hand immediately understood its importance, as have all subsequent writers. Clearly disappointed at the turn of events, Liouville at no time thereafter ever made reference to Cauchy's theorem. The irony is that nevertheless it is Liouville's name which has become attached to the theorem.

The first modern proof of the main theorem (still found in texts today, based on the Taylor expansion) is apparently due to Briot and Bouquet in the first edition of their monograph *Théorie des fonctions elliptiques*, Paris, 1859. They refer neither to Cauchy nor to Liouville for the result. Very curiously, the first published attribution of the theorem to Liouville (to our knowledge) occurs in the *second edition*, Paris, 1875, pp. 202–203. In the preface (p. iv) they cite Liouville's lectures of 1851 as the source for their attribution, though as we have noted above, their proof is neither Cauchy's nor Liouville's but essentially the modern one based on Taylor's series.

A few years later, E. Borel, *Leçons sur les fonctions entières*, Paris, 1900, p. 2, gives the result back to Cauchy, though without citing a source. Finally, Whittaker and Watson in *A Course of Modern Analysis*, Cambridge, 1902, explicitly call the result "Liouville's theorem", again without citation; in their second edition, however, Cambridge, 1915, while still naming the result Liouville's theorem, they specifically attribute it to Cauchy and cite the 1844 reference at the beginning of this note. On rare occasions a more modern treatise on complex analysis still refers to Cauchy, e.g., works of Copson, 1935; Dinghas, 1961; Sansone and Gerretson, 1962. But by this time the die has been cast.

For simplicity throughout the present work, we have continued the customary practice of attaching Liouville's name to results in which a non-negative solution of an elliptic equation is shown to be constant.

(We are indebted to Professor Edgar Reich for his aid in locating some of the sources cited above, and for helpful discussions of the various historical issues involved. We also thank Fabienne Queyroux of the Bibliothéque, Institut de France, for her help in locating Liouville's notebooks and manuscripts. We cite also J. Lützen's *Joseph Liouville 1809–1882: Master of Pure and Applied Mathematics*, Springer-Verlag, 1990, where there is an extended and interesting discussion of some of the material above, though the presentation is partly marred by championship of Liouville's priority claims.)

References

[1] Acerbi, E. & Fusco, N., Regularity for minimizers of non-quadratic functionals: the case $1<p<2$. *J. Math. Anal. Appl.*, 140 (1989), 115–135.

[2] Astarita, G. & Marrucci, G., *Principles of Non-Newtonian Fluid Mechanics*. McGraw-Hill, 1974.

[3] Bidaut-Veron, M.-F., Local and global behavior of solutions of quasilinear equations of Emden–Fowler type. *Arch. Rational Mech. Anal.*, 107 (1989), 293–324.

[4] Bidaut-Veron, M.-F. & Pohozaev, S., Nonexistence results and estimates for some nonlinear elliptic problems. *J. Anal. Math.*, 84 (2001), 1–49.

[5] Bidaut-Veron, M.-F. & Veron, L., Nonlinear elliptic equations on compact Riemannain manifolds and asymptotics of Emden equations. *Invent. Math.*, 106 (1991), 489–539.

[6] CAFFARELLI, L., GAROFALO, N. & SEGALA, F., A gradient bound for entire solutions of quasi-linear equations and its consequences. *Comm. Pure Appl. Math.*, 47 (1994), 1457–1473.

[7] CAFFARELLI, L., GIDAS, B. & SPRUCK, J., Asymptotic symmetry and local behavior of semilinear elliptic equations with critical Sobolev growth. *Comm. Pure Appl. Math.*, 42 (1989), 271–297.

[8] CAUCHY, A., Mémoire sur les fonctions complémentaires. *C. R. Acad. Sci. Paris*, 19 (1844), 1377–1384; also *Œuvres complètes*, Ire série, tome VIII, 378–383.

[9] DÍAZ, J. I., *Nonlinear Partial Differential Equations and Free Boundaries, Vol. 1. Elliptic Equations.* Pitman Res. Notes Math. Ser., 106. Pitman, Boston, MA, 1985.

[10] DIBENEDETTO, E., $C^{1+\alpha}$ local regularity of weak solutions of degenerate elliptic equations. *Nonlinear Anal.*, 7 (1983), 827–850.

[11] EVANS, L. C., A new proof of local $C^{1,\alpha}$ regularity for solutions of certain degenerate elliptic P.D.E. *J. Differential Equations*, 45 (1982), 356–373.

[12] GIDAS, B. & SPRUCK, J., Global and local behavior of positive solutions of nonlinear elliptic equations. *Comm. Pure Appl. Math.*, 34 (1981), 525–598.

[13] LEWIS, J. L., Regularity of the derivatives of solutions to certain degenerate elliptic equations. *Indiana Univ. Math. J.*, 32 (1983), 849–858.

[14] MARTINSON, L. K. & PAVLOV, K. B., The effect of magnetic plasticity in non-Newtonian fluids. *Magnit. Gidrodinamika*, 2 (1969), 69–75.

[15] — Unsteady shear flows of a conducting fluid with a rheological power flow. *Magnit. Gidrodinamika*, 3 (1970), 5869–5875.

[16] MAZZEO, R. & PACARD, F., A construction of singular solutions for a semilinear elliptic equation using asymptotic analysis. *J. Differential Geom.*, 44 (1996), 331–370.

[17] MIKLJUKOV, V. M., On the asymptotic properties of subsolutions of quasilinear equations of elliptic type and mappings with bounded distortion. *Mat. Sb. (N.S.)*, 111 (1980), 42–66 (Russian).

[18] MITIDIERI, È. & POKHOZHAEV, S. I., The absence of global positive solutions to quasilinear elliptic inequalities. *Dokl. Akad. Nauk*, 359 (1998), 456–460; English translation in *Dokl. Math.*, 57 (1998), 250–253.

[19] NI, W.-M. & SERRIN, J., Existence and non-existence theorems for ground states of quasilinear partial differential equations: The anomalous case. *Atti Convegni Lincei*, 77 (1986), 231–257.

[20] PELETIER, L. A. & SERRIN, J., Gradient bounds and Liouville theorems for quasilinear elliptic equations. *Ann. Scuola Norm. Sup. Pisa Cl. Sci (4)*, 5 (1978), 65–104.

[21] PROTTER, M. H. & WEINBERGER, H. F., *Maximum Principles in Differential Equations.* Springer-Verlag, New York, 1984 (reprinted).

[22] RESHETNYAK, YU. G., Index boundedness condition for mappings with bounded distortion. *Siberian Math. J.*, 9 (1968), 281–285.

[23] SERRIN, J., Local behavior of solutions of quasi-linear equations. *Acta Math.*, 111 (1964), 247–302.

[24] — Isolated singularities of solutions of quasi-linear equations. *Acta Math.*, 113 (1965), 219–240.

[25] — Entire solutions of nonlinear Poisson equations. *Proc. London Math. Soc. (3)*, 24 (1972), 348–366.

[26] — Liouville theorems for quasilinear elliptic equations, in *Atti del Convegno Internazionale sui Metodi Valutativi nella Fisica-Matematica* (Rome, 1975), pp. 207–215. Accad. Naz. Lincei, Quaderno 217. Accad. Naz. Lincei, Rome, 1975.

[27] TOLKSDORF, P., Regularity for a more general class of quasilinear elliptic equations. *J. Differential Equations*, 51 (1984), 126–150.

[28] TRUDINGER, N. S., On Harnack type inequalities and their application to quasilinear elliptic equations. *Comm. Pure Appl. Math.*, 20 (1967), 721–747.

[29] UHLENBECK, K., Regularity for a class of non-linear elliptic systems. *Acta Math.*, 138 (1977), 219–240.

[30] URAL'TSEVA, N. N., Degenerate quasilinear elliptic systems. *Zap. Nauchn. Sem. Leningrad. Otdel. Mat. Inst. Steklov. (LOMI)*, 7 (1968), 184–222 (Russian).

[31] VERON, L., *Singularities of Solutions of Second Order Quasilinear Equations*. Pitman Res. Notes Math. Ser., 353. Longman, Harlow, 1996.

[32] ZOU, H., Slow decay and the Harnack inequality for positive solutions of $\Delta u+u^p=0$ in $\mathbf{R}^n$. *Differential Integral Equations*, 8 (1995), 1355–1368.

[33] — Symmetry of positive solutions of $\Delta u+u^p=0$ in $\mathbf{R}^n$. *J. Differential Equations*, 120 (1995), 46–88.

[34] DANCER, E. N., Superlinear problems on domains with holes of asymptotic shape and exterior problems. *Math. Z.*, 229 (1998), 475–491.

JAMES SERRIN
Department of Mathematics
University of Minnesota
Minneapolis, MN 55455
U.S.A.
serrin@math.umn.edu

HENGHUI ZOU
Department of Mathematics
University of Alabama at Birmingham
Birmingham, AL 35294
U.S.A.
zou@math.uab.edu

Received May 5, 2000

Filomena Pacella, Rome, 2010.

Moshe Marcus and James Serrin, Technion University, Haifa, Israel, 2010.

I.7. Symmetry of Solutions – Moving Plane Method

Filomena Pacella

Mathematics Subject Classification (2010). 35J60, 35B05, 35J65.
Keywords. Nonlinear elliptic equations, symmetry problem, Hopf's boundary point lemma.

i. A symmetry problem in potential theory

This paper appeared in 1971, and can be considered a pioneering work on the theory of symmetry of solutions of elliptic problems. It concerns the following easily formulated question about an overdetermined problem:

Let Ω be a bounded smooth connected open domain in $\mathbb{R}^n$ and suppose that there exists a function $u = u(x)$ satisfying the Poisson equation

$$-\Delta u = 1 \quad in \ \ \Omega$$

together with the boundary conditions

$$u = 0\,, \quad u_\nu = \ constant \ on \ \ \partial\Omega$$

where ν is the outer normal to $\partial\Omega$. Must then Ω be a ball?

Serrin answered this question affirmatively and also showed that the solution must be radially symmetric and equal to $\frac{b^2-r^2}{2n}$, where b is the radius of the ball and r denotes the distance from the center.

Like many other results of Serrin, this one has important consequences in applications. For example for a viscous incompressible fluid moving in straight parallel streamlines through a straight pipe of given cross sectional form Ω it states that "the tangential stress on the pipe wall is the same at all points of the wall if and only if the pipe has a circular cross section". Another applications was given in the linear theory of torsion of a solid straight bar of cross section Ω in which case Serrin's theorem states that "when a solid straight bar is subject to torsion, the magnitude of the resulting traction which occurs at the surface of the bar is independent of position if and only if the bar has a circular cross section". It also applies to the study of a liquid rising in a straight capillary tube of cross section Ω.

From a mathematical point of view, in spite of its simple formulation, the question, and hence the result, is important in understanding how the boundary conditions can determine the shape of the domain and of the solution.

P. Pucci et al. (eds.), *James Serrin. Selected Papers: Volume 1*, Contemporary Mathematicians,
DOI 10.1007/978-3-0348-0845-3_7, © Springer Basel 2014

Equally important and innovative was Serrin's proof of this result. It is, in fact, in this paper that the, by now classical, "moving plane method" was introduced into the theory of partial differential equations. It had previously been used by A.D. Alexandroff in differential geometry.

This efficient and elegant method is based on the moving parallel planes[1] up to a critical position and showing, via the maximum principle, that the solution is symmetric about the limiting plane. Since then, this method has been used, mainly to prove monotonicity and symmetry results, in a large variety of differential equation problems. To understand its importance, it is enough to think of the famous symmetry result of Gidas, Ni and Nirenberg.

Another important tool used by Serrin in his proof is an extension of Hopf's boundary point Lemma to domains with corners, which he derived in the same paper. This result is known as Serrin's lemma.

Finally, the paper also contains extensions of the above result to more general elliptic equations or to more general boundary conditions involving the mean curvature of the boundary surface $\partial\Omega$, which lead to other consequences in applications.

ii. Symmetry of ground states of quasilinear elliptic equations

This paper addresses the problem of radial symmetry of *non-negative* solutions of elliptic equations of the general form

$$\operatorname{div}(A(|Du|)Du) + f(u) = 0 \qquad \text{in } \mathbb{R}^n, \quad n \geq 2,$$

under the ground state condition

$$u(x) \to 0 \quad \text{as} \quad |x| \to \infty$$

Two cases have to be considered:

a) the "regular" case when $[tA(t)]' > 0$ for all $t \geq 0$,

b) the "singular" case when either $A(t)$ is undefined at $t = 0$ or fails to satisfy a) at $t = 0$.

The first case includes, in particular, the mean curvature operator, while a relevant example for the second one is the m-Laplace operator, when $m > 1$.

On the nonlinearity $f(u)$ only the following hypotheses are assumed:

i) f is continuous for $u \geq 0$ and locally Lypschitz continuous for $u > 0$,

ii) $f(0) = 0$ and f is nonincreasing near zero.

Serrin–Zou's symmetry results are the following:

Case a) Any non-negative C^1-weak ground state solution whose (open) support is connected must be radially symmetric about some point and radially decreasing.

Case b) Any non-negative C^1-weak ground state solution with only one critical point where it is positive must be radially symmetric about that point and radially decreasing.

These results are nontrivial extensions of previous theorems, including the classical result by Gidas, Ni and Nirenberg for the Laplace operator ($A(t) \equiv 1$). Among the significant improvements of Serrin–Zou's results are weaker requirements on the nonlinearity $f(u)$ which permit the existence of non-negative solutions with zeros. This is very

[1]by moving parallel planes we mean the motion of a single plane with the property that the plane remains parallel to its initial orientation

important, since it can be proved that solutions with compact support in $\mathbb{R}^n$ do exist. Indeed, necessary and sufficient conditions on $f(u)$ can be given (as shown in a paper by Pucci and Serrin) for such phenomenon to appear. This is related to the celebrated paper of Vazquez for the m-Laplace operator. As a consequence, Serrin–Zou's theorems show that the condition of positivity used in several previous papers can be dropped without affecting the conclusion.

For Case b) some counterexamples, involving the m-Laplace operator, with $m > 2$, are presented to show that some assumption on the critical set of the solution is needed in order to get the radial symmetry. On the other hand, when the operator is singular at $t = 0$ (which, for the m-Laplace operator corresponds to the assumption $1 < m < 2$) the hypothesis on the critical set of the non-negative solution can be replaced by a connectivity assumption. It was shown by Damascelli, Pacella and Ramaswamy that if $1 < m < 2$ and the solution is positive, then no connectivity condition is needed.

Finally Serrin–Zou's approach also applies to more general nonlinearities and even allows to consider singular ground states.

Filomena Pacella
Dipartimento di Matematica
University of Roma "Sapienza"

Arch. Ration. Mech. Anal. 43 (1971), 304–318. **Reprint I.7.[i]**

A Symmetry Problem in Potential Theory

JAMES SERRIN

The following problem has been posed by Professor R. L. FOSDICK. Let Ω be a bounded open connected domain in the Euclidean space R^n having a smooth boundary $\partial\Omega$. Suppose there exists a function $u=u(x)=u(x_1, \ldots, x_n)$ satisfying the Poisson differential equation

$$\Delta u = -1 \quad \text{in } \Omega \tag{1}$$

together with the boundary conditions

$$u=0, \quad \frac{\partial u}{\partial n} = \text{constant} \quad \text{on } \partial\Omega. \tag{2}$$

Must Ω then be a ball? We shall show here that the answer is affirmative, and that u must have the specific form $(b^2-r^2)/2n$ where b is the radius of the ball and r denotes distance from its center. The precise result is as follows.

Theorem 1. *Let Ω be a domain whose boundary is of class C^2. Suppose there exists a function $u \in C^2(\bar{\Omega})$ satisfying conditions* (1) *and* (2). *Then Ω is a ball and u has the specific form noted above.*

The proof of this result is given in Section 1; in Section 3 we give various generalizations to elliptic differential equations other than (1). Before turning to the detailed arguments it will be of interest to discuss the physical motivation for the problem itself.

Consider a viscous incompressible fluid moving in straight parallel streamlines through a straight pipe of given cross sectional form Ω. If we fix rectangular coordinates in space with the z axis directed along the pipe, it is well known that the flow velocity u is then a function of x, y alone satisfying the Poisson differential equation (for $n=2$)

$$\Delta u = -A \quad \text{in } \Omega$$

where A is a constant related to the viscosity and density of the fluid and to the rate of change of pressure per unit length along the pipe. Supplementary to the differential equation one has the adherence condition

$$u=0 \quad \text{on } \partial\Omega.$$

Finally, the tangential stress per unit area on the pipe wall is given by the quantity $\mu\, \partial u/\partial n$, where μ is the viscosity. Our result states that *the tangential stress on the pipe wall is the same at all points of the wall if and only if the pipe has a circular cross section.*

Exactly the same differential equation and boundary condition arise in the linear theory of torsion of a solid straight bar of cross section Ω; see [3] pp. 109–119. Theorem 1 then states that, *when a solid straight bar is subject to torsion, the magnitude of the resulting traction which occurs at the surface of the bar is independent of position if and only if the bar has a circular cross section.*

A more sophisticated example occurs in the case of a liquid rising in a straight capillary tube of cross section Ω. The function $u(x, y)$ describing the upper surface of the liquid satisfies the differential equation

$$(1+u_y^2)u_{xx}-2u_x u_y u_{xy}+(1+u_x^2)u_{yy}=\kappa u(1+u_x^2+u_y^2)^{3/2} \tag{3}$$

where κ is a certain positive constant; also the requirement that the wetting angle γ be constant leads to the boundary condition

$$\frac{\partial u}{\partial n}=-\cot\gamma=\text{constant} \quad \text{on } \partial\Omega \tag{4}$$

($\vec{n}$ being the inner normal direction). By the theorem of Section 3 it then follows that *provided the wetting angle γ is different from $\pi/2$, a liquid will rise to the same height at each point of the wall of a capillary tube if and only if the tube has circular cross section.* (When $\gamma=\pi/2$, the unique solution of (3), (4) is $u\equiv 0$ for *any* cross sectional form of the tube.)

In the final section of the paper we show that our results can be applied to somewhat more general boundary conditions than (2). A curious consequence of this generalization is the following result. Consider a viscous fluid flowing in straight streamlines through a straight pipe whose cross section is non-circular. Then there must be two points P and Q on the wall, such that the curvature of the wall is greater at P than at Q but the tangential stress is greater at Q than at P. A similar result also holds for the torsion problem.

It may be noted in conclusion that (2) constitutes Cauchy data on the boundary surface for the elliptic equation (1). It is of course well known that such data is generally overdetermined with regard to solving (1) in a given domain Ω; in light of this remark, our results provide a concrete example where the overdetermined nature of the condition can be rigorously analyzed.

1. Proof of Theorem 1

Let T_0 by a hyperplane in R^n not intersecting the domain Ω. We suppose this plane to be continuously moved normal to itself to new positions, until ultimately it begins to intersect Ω. From that moment onward, at each stage of the motion the resulting plane T will cut off from Ω a cap $\Sigma(T)$: that is, $\Sigma(T)$ will be that portion of Ω which lies on the same side of T as the original plane T_0.

For any cap $\Sigma(T)$ thus formed, we let $\Sigma'(T)$ be its reflection in T. Evidently $\Sigma'(T)$ will be contained in Ω at the beginning of the process; and indeed as T advances into Ω *the resulting cap $\Sigma'(T)$ will stay within Ω at least until one of the following two events occurs:*

(*i*) $\Sigma'(T)$ becomes internally tangent to the boundary of Ω at some point P not on T, or

(*ii*) T reaches a position where it is orthogonal to the boundary of Ω at some point Q.

We denote the plane T when it reaches either one of these positions by T'.

We now assert that Ω must be symmetric about T'. In fact, if this assertion is proved the theorem follows immediately. To see this, we observe that for *any given direction* in R^n there would then be a plane T' with normal in that direction such that Ω is symmetric about T'. Moreover, according to the construction Ω would have to be simply connected. But the only simply connected domains which have the symmetry property just noted are balls. Having thus proved that Ω is a ball we say that the function $(b^2-r^2)/2n$ is then the unique solution of the given boundary value problem ([2], pages 68–69).

In proving the assertion, we introduce a new function v defined in $\Sigma' \equiv \Sigma'(T')$ by the formula

(5) $$v(x)=u(x') \quad (x\in\Sigma')$$

where x' is the reflected value of x across T'. Evidently v satisfies the differential equation

$$\Delta v=-1 \quad \text{in } \Sigma'$$

and the boundary conditions

$$\begin{aligned} & v=u && \text{on } \partial\Sigma'\cap T' \\ v=0, \quad & \frac{\partial v}{\partial n}=\text{constant} && \text{on } \partial\Sigma'\cap \mathrm{Comp}(T'), \end{aligned}$$

the constant being the same as in (2).

Since Σ' is contained in Ω by construction, we may consider the function $u-v$ in Σ'. Evidently

(6) $$\Delta(u-v)=0 \quad \text{in } \Sigma'$$

and

(7) $$\begin{aligned} u-v&=0 \quad \text{on } \partial\Sigma'\cap T' \\ u-v&\geqq 0 \quad \text{on } \partial\Sigma'\cap \mathrm{Comp}(T'); \end{aligned}$$

the latter condition is a consequence of the fact that $u>0$ in Ω.* If we apply the strong version of the maximum principle ([2], page 53), it is easy to see from (6) and (7) that either

(8) $$u-v>0 \quad \text{at all interior points of } \Sigma'$$

or else $u\equiv v$ in Σ'. In the latter case it is clear that the reflected cap Σ' must coincide with that part of Ω on the same side of T' as Σ'; that is, Ω must be symmetric about T'.

* For otherwise u would have a minimum at some interior point of Ω, which is impossible since $\Delta u=-1$.

To complete the proof of the theorem it must therefore be shown that (8) is impossible. Suppose first that we are in case (*i*), that is, Σ' is internally tangent to the boundary of Ω at some point P not on T'. Then $u-v=0$ at P. Consequently, using (6), (8) and the boundary point version of the maximum principle ([2], page 65), we conclude that

$$\frac{\partial}{\partial n}(u-v)>0 \quad \text{at } P.$$

This however contradicts the fact that $\partial u/\partial n=\partial v/\partial n=$constant at P. Hence (8) is impossible in case (*i*).

In case (*ii*) the situation is more complicated, for even though $u-v=0$ at Q the boundary point version of the maximum principle does not apply (this is because Q is a right angled corner of Σ' and the requisite internally tangent ball [see [2], page 65] is not available). Consequently we must proceed in an alternate fashion. It will be shown (*a*) that $u-v$ has a zero of second order at Q and then (*b*) a contradiction will be obtained from a more delicate version of the boundary point maximum principle.

(*a*) By hypothesis the boundary of Ω is of class C^2. Consider a rectangular coordinate frame with origin at Q, the x_n axis being directed along the inward normal to $\partial\Omega$ at Q, and the x_1 axis being normal to T'. In this frame we can represent the boundary of Ω locally by the equation

$$x_n=\phi(x_1, \dots, x_{n-1}), \qquad \phi\in C^2.$$

Since u is in $C^2(\bar{\Omega})$ the condition $u=0$ on $\partial\Omega$ can then be expressed as a twice differentiable identity

$$u(x_1, \dots, x_{n-1}, \phi)\equiv 0. \tag{9}$$

Similarly, the boundary condition $\partial u/\partial n=$constant$=c$ on $\partial\Omega$ can be written as an identity,

$$\frac{\partial u}{\partial x_n}-\sum_1^{n-1}\frac{\partial u}{\partial x_k}\frac{\partial \phi}{\partial x_k}=c\left\{1+\sum_1^{n-1}\left(\frac{\partial\phi}{\partial x_k}\right)^2\right\}^{1/2} \tag{10}$$

where x_n is to be replaced throughout by $\phi(x_1, \dots, x_{n-1})$.

At this stage some simple notation will be convenient; thus

$$u_i=\frac{\partial u}{\partial x_i} \quad (i=1, \dots, n-1); \qquad u_n=\frac{\partial u}{\partial x_n}.$$

Differentiating (9) with respect to x_i, $i=1, \dots, n-1$, we now obtain

$$u_i+u_n\phi_i=0. \tag{11}$$

Evaluating this at Q, where $\phi_i=0$, we find

$$u_i=0, \qquad u_n=c \quad (\text{at } Q).$$

Next differentiating (11) with respect to x_j, $j=1, \dots, n-1$, and evaluating at Q yields

$$u_{ij}+c\phi_{ij}=0 \quad (\text{at } Q)$$

where $u_{ij}=\partial^2 u/\partial x_i\,\partial x_j$. Lastly, differentiating (10) with x_i and evaluating at Q gives

$$u_{ni}=0 \quad (\text{at } Q).$$

Since also $u_{nn}=-\sum_1^{n-1} u_{ii}-1=c\Delta\varphi-1$ at Q, we have accordingly determined all the first and second derivatives of u at Q.

Since the reflected cap Σ' lies inside Ω, it is not hard to see that the second derivatives of ϕ must also satisfy

$$\phi_{1l}=0 \quad \text{at } Q, \qquad l=2,\dots,n-1.$$

Taking these relations into account, and observing that (5) implies

$$v(x_1, x_2, \dots, x_n)=u(-x_1, x_2, \dots, x_n),$$

we find that the first and second derivatives of u and v agree at Q. This completes the proof of (*a*).

Turning now to (*b*), we require the following preliminary result.

Lemma 1. *Let D^* be a domain with C^2 boundary and let T be a plane containing the normal to ∂D^* at some point Q. Let D then denote the portion of D^* lying on some particular side of T.*

Suppose that w is of class C^2 in the closure of D and satisfies

$$\Delta w\leqq 0 \quad in\ D,$$

while also $w\geqq 0$ in D and $w=0$ at Q. Let $\vec{s}$ be any direction at Q which enters D non-tangentially. Then either

$$\frac{\partial w}{\partial s}>0 \quad or \quad \frac{\partial^2 w}{\partial s^2}>0 \quad at\ Q$$

unless $w\equiv 0$.

The proof of this result will be given in Section 2. Assuming that the lemma holds, we may apply it to the function $w=u-v$ in Σ'. Since $w>0$ there, and $w=0$ at Q, this yields

$$\frac{\partial(u-v)}{\partial s}>0 \quad \text{or} \quad \frac{\partial^2(u-v)}{\partial s^2}>0 \quad \text{at } Q$$

contradicting the fact that both u and v have the same first and second partial derivatives at Q. This completes the proof of the theorem.

2. Proof of Lemma 1

Let K_1 be a ball which is internally tangent to D^* at Q, and which touches the boundary of D^* only at Q. Such a ball exists by virtue of the fact that the boundary of D^* is of class C^2.

Construct a ball K_2 with center at Q and radius $\frac{1}{2}r_1$ where r_1 is the radius of K_1. Finally let $K'=K_1\cap K_2\cap D$. Now define the auxiliary function

$$z=z(x)=x_1\left(e^{-\alpha r^2}-e^{-\alpha r_1^2}\right)$$

where α is a positive constant to be determined (we have chosen coordinates with origin at the center of K_1 and with T being the plane $x_1=0$; moreover it can be assumed that D is on the side of T where $x_1>0$). It is clear that

$$z>0 \quad \text{in } K', \qquad z=0 \quad \text{on } \partial K_1 \text{ and on } T \tag{12}$$

(at this stage we mean by K' the *interior* of the region described, i.e., K_1, K_2 and D are taken to be open sets). We compute

$$\Delta z = 2\alpha x_1 e^{-\alpha r^2}\{2\alpha r^2-(n+2)\}.$$

Now $r \geqq \frac{1}{2} r_1$ in K' so that by choosing α suitably large, say $\alpha=(n+2)\, r_1^{-2}$, we obtain $\Delta z>0$ in K'.

Now suppose w is not identically zero in D. Then by the strong maximum principle we have $w>0$ in D. We consider the part of the boundary of K' lying on ∂K_2. This set intersects the boundary of D only on the plane T. Moreover the intersection set lies at a finite distance from the corners of D. By virtue of the boundary point lemma ([2], page 65), therefore, it is not hard to see that there exists a constant $\varepsilon>0$ such that

$$w \geqq \varepsilon x_1 \quad \text{on } \partial K' \cap \partial K_2 .$$

Moreover

$$w \geqq 0 \quad \text{on } \partial K' \cap \partial K_1 \text{ and } \partial K' \cap T.$$

On the other hand, it is clear that

$$z \leqq x_1 \quad \text{on } \partial K' \cap \partial K_2 . \tag{13}$$

Consequently, by use of (12), the function $w-\varepsilon z$ is non-negative on the entire boundary of K', and is zero at Q. Moreover

$$\Delta(w-\varepsilon z)=\Delta w-\varepsilon \Delta z<0$$

in K'. By the maximum principle, therefore, $w-\varepsilon z>0$ in K'. Hence at Q, where $w-\varepsilon z=0$, we have either

$$\frac{\partial(w-\varepsilon z)}{\partial s}>0 \quad \text{or} \quad \frac{\partial^2(w-\varepsilon z)}{\partial s^2} \geqq 0 .$$

Since by direct calculation

$$\frac{\partial z}{\partial s}=0, \quad \frac{\partial^2 z}{\partial s^2}>0 \quad \text{at } Q,$$

this completes the proof of the lemma.

3. More General Elliptic Equations

If the proof of Theorem 1 is re-examined, one finds that the properties of the Poisson equation which were applied are the following.

(A) It is invariant to the symmetric substitution $x \to x'$.

(B) The second partial derivatives u_{nn} in an arbitrary rectangular coordinate frame can be determined in terms of the remaining second partial derivatives.

(C) The difference of two solutions obeys the strong maximum principle, and

(D) The difference of two solutions obeys both boundary point versions of the maximum principle.

Now conditions (A) and (B) are satisfied by any nonlinear *elliptic* differential equation of the form

$$a\,\Delta u + h\,u_i u_j u_{ij} = f \tag{14}$$

provided that a, f and h are functions only of u and $|p|$, where $p=(u_1, \ldots, u_n)$ denotes the gradient vector of the solution.* To see this, note that (14) is invariant under coordinate rotations as well as under substitutions $(x_1, \ldots, x_n) \to (-x_1, \ldots, x_n)$ which change the sign of a single coordinate.

On the other hand, the strong maximum principle does *not* hold for differences of solutions of (14) in the form which was applied in the proof of Theorem 1. Nevertheless, a somewhat restricted version does remain valid. Suppose in fact that the functions a, f and $h\,p_i\,p_j$ are continuously differentiable in u and p, and that u and v are two solutions in a domain D such that $u-v \geqq 0$ in D and $u-v=0$ at some interior point of D. Then according to a theorem of E. HOPF ([1], pages 149–150) we have $u \equiv v$.

In the proof of Theorem 1 the maximum principle was used only to show that either (8) holds or that $u \equiv v$ in Σ'. This conclusion can be reached by an alternate argument, however, using only the restricted version of the maximum principle.

Deferring this proof for a moment, it remains only to consider the two boundary point versions of the maximum principle. In this regard, the standard boundary point lemma ([2], page 65) applies immediately to the difference of solutions of (14) since this difference is zero at the point P. (The argument is the same as for the restricted version of the maximum principle above.) Finally, the required genralization of Lemma 1 needed for application to the difference of solutions of (14) will be proved in the following section.

We may thus turn our attention to showing that either (8) holds or else $u \equiv v$ in Σ'. For this purpose we shall make an additional assumption about the behavior of the solution, namely**

$$u > 0 \quad \text{in } \Omega. \tag{15}$$

(We note that this assumption applies automatically in the case of Theorem 1, as a consequence of (1), (2).) Letting c denote the constant which occurs in the boundary condition (2), we assert to begin with that either $c>0$ or else $\partial^2 u/\partial n^2 > 0$ at each point of the boundary. Thus suppose $c=0$. Then, according to the formulas given in part (*a*) of the previous proof, if we introduce at any fixed point P of the boundary a special coordinate frame with x_n axis directed along the inner normal to the boundary, all the first and second derivatives of u vanish at P except $\partial^2 u/\partial n^2 = u_{nn}$, and this derivative has the value

$$\frac{f(0,0)}{a(0,0)} = d$$

* From here onward we adopt the standard convention that repeated indices are to be summed from 1 to n.

** Recall that Ω is assumed to be open so that (15) is not in conflict with the given boundary condition (2).

according to (14). (Note that $a>0$ for all values of u and p because (14) is elliptic.)

Since $c=0$ by supposition, the inequality $d<0$ is obviously contradictory to (15); it remains to show that $d=0$ also cannot occur. In such a case, however, we would have $f(0,0)=0$. Then the function $\bar{u}\equiv 0$ would be a solution of (14), and correspondingly the solution u under consideration could be regarded as the *difference* of two solutions. Then in view of (15) we could apply the standard boundary point lemma to infer that $\partial u/\partial n>0$ at every point of the boundary, contradicting the original supposition that $c=0$. Hence $d>0$.

From what has just been proved it is now clear that u increases monotonically as one enters Ω along any non-tangential direction $\vec{s}$, for some positive distance so into the domain. Moreover, s_0 can be chosen independent of position on $\partial\Omega$, depending only on a bound for $\vec{s}$ away from the tangent direction.

Let us now return to the opening stages of the proof of Theorem 1. By what has been shown it is clear that immediately after T has penetrated Ω not only will $\Sigma'(T)$ be contained in Ω but also

$$u>v \quad \text{at interior points of } \Sigma'(T). \tag{16}$$

Here we construct v in $\Sigma'(T)$ in exactly the same way as the previous construction of v in $\Sigma'(T')$.

We assert that (16) persists as T advances into Ω, for all positions of T prior to T'. Suppose in fact that there is some instant where (16) fails prior to T reaching T', Then there would be a position T'' of T such that either

$$u\geqq v \quad \text{in } \Sigma'(T''), \qquad u=v \quad \text{at some interior point of } \Sigma'(T'') \tag{17}$$

or

$$\begin{aligned} &u>v && \text{at interior points of } \Sigma'(T'') \\ &\frac{\partial u}{\partial n}=\frac{\partial v}{\partial n} && \text{at some interior point of } T''\cap\Omega. \end{aligned} \tag{18}$$

If (17) holds we reach a contradiction with the restricted version of the maximum principle, and if (18) holds, with the boundary point principle. Thus the assertion is proved.

It follows by continuity that when T reaches T' we have

$$u\geqq v \quad \text{at interior points of } \Sigma'.$$

By applying the restricted maximum principle once more we see that either the strict inequality holds or else $u\equiv v$. But this is what we originally set out to demonstrate. Summing up, we have proved the following result.

Theorem 2. *Let Ω be a domain whose boundary is of class C^2. Suppose there exists a function $u\in C^2(\bar{\Omega})$ satisfying the elliptic differential equation*

$$a(u,|p|)\Delta u+h(u,|p|)u_i u_j u_{ij}=f(u,|p|) \quad \text{in } \Omega \tag{19}$$

where a, f and $h p_i p_j$ are continuously differentiable functions of u and p (here $p=(u_1, \ldots, u_n)$ denotes the gradient vector of u). Suppose also that

$$u>0 \quad \text{in } \Omega \tag{20}$$

and that u satisfies the boundary conditions

$$u=0, \quad \frac{\partial u}{\partial n}=\text{constant} \quad on\ \partial\Omega. \tag{2}$$

Then Ω must be a ball and u is radially symmetric.

Remarks 1. It is clear that condition (20) can be replaced by the alternate assumption that $u<0$ in Ω. Also (20) can be deleted from the statement of the theorem if it is assumed that f is never zero. To see this, note that if $f<0$, say, then u cannot take on an interior minimum value, so that the boundary condition $u=0$ on $\partial\Omega$ implies $u>0$ at interior points of Ω. In particular, one has $f=-1$ for equation (1), explaining why it was unnecessary to make the explicit assumption $u>0$ in Theorem 1.

2. Equation (3) describing the upper surface of a liquid in a capillary tube (i.e. a liquid under the combined influence of gravity and surface tension) is a special case of (19). The result stated at the end of the introduction is therefore a consequence of the following remark. *Let u be a solution of* (3), (4) *in Ω, such that $u=\alpha=constant$ on $\partial\Omega$. Suppose that $\gamma\neq\pi/2$. Then $u\neq\alpha$ in Ω.* Proof. Assume first $0<\gamma<\pi/2$. Then by (4), since solutions of (3) can take on neither a positive maximum nor a negative minimum in Ω, we have $\alpha>0$ and $0\leqq u<\alpha$ in Ω. Similarly if $\gamma>\pi/2$ then $\alpha<0$ and $\alpha<u\leqq 0$ in Ω.

A further class of equations to which Theorem 2 is applicable can be obtained from regular variational problems of the form

$$\delta\int F(u,|p|)\,dx=0. \tag{21}$$

Indeed if we assume that $F(u, t)$ is three times differentiable, and that $F'(u,0)=F'''(u,0)=0$ for consistency, then the Euler-Lagrange equation for (21) can be written in the form (19) with

$$a=1, \quad h=|p|^{-2}\left(\frac{|p|F''}{F'}-1\right), \quad f=\frac{|p|}{F'}(F_u-|p|F'_u)$$

where primes here denote differentiation with respect to the second argument of F.

3. Under the hypotheses of the theorem we cannot conclude any more about the form of u then that it must be a function of the radial distance r from the center of Ω. If we write $u=u(r)$ and use primes to denote differentiation with respect to r, then one obtains the differential equation

$$a(u,|u'|)\left(u''+\frac{n-1}{r}u'\right)+h(u,|u'|)u'^2u''=f(u,|u'|), \tag{22}$$

for $0\leqq r\leqq b$ with the end conditions

$$u'(0)=0, \quad u(b)=0. \tag{23}$$

The ellipticity of (20) implies

$$a(u,|u'|)+h(u,|u'|)|u'|^2>0$$

so that in any case (22) is non-singular and can be written in the form

$$(22') \qquad u'' = F(r, u, u').$$

To make further progress it may be assumed that a and h are independent of u and that $\partial f/\partial u \geqq 0$, as is the case for the equation describing a capillary surface. Then standard arguments show that there cannot be more than one solution h. There would be some interest in pursuing the discussion of equation (22) further, as JOHNSON and PERKO have done in the case of the capillary equation (Arch. Rational Mech. Analysis, vol. 29).

4. It is of interest to observe that without some assumption of the type described in the preceeding remark, equation (22) may have more than one solution (or even infinitely many solutions) satisfying the end conditions (23). In consequence, the property of having unique solutions corresponding to given boundary data is *not* a prerequisite for Theorem 2 to hold for a given equation.

4. The Boundary Point Lemma at a Corner

Here we shall prove a generalization of Lemma 1 suitable for application to non-linear elliptic equations. We then use this result to obtain the conclusion

$$(24) \qquad \frac{\partial(u-v)}{\partial s} > 0 \quad \text{or} \quad \frac{\partial^2(u-v)}{\partial s^2} > 0 \quad \text{at } Q$$

which is needed, just as in the case of Theorem 1, for completing the proof of Theorem 2.

Lemma 2. *Let D be a domain of the type described in Lemma 1. Suppose that w is of class C^2 in the closure of D and satisfies the elliptic differential inequality*

$$Lw = a_{ij}(x)\, w_{ij} + b_i(x)\, w_i \leqq 0 \qquad \text{in } D$$

where the coefficients are uniformly bounded. We assume that the matrix a_{ij} is uniformly definite

$$(25) \qquad a_{ij}(x)\, \xi_i \xi_j \geqq \kappa |\xi|^2 \quad (\kappa = \text{constant} > 0),$$

and that

$$(26) \qquad |a_{ij} \xi_i \eta_j| \leqq K(|\xi \cdot \eta| + |\xi| \cdot |d|) \quad (K = \text{constant} > 0)$$

where $\xi = (\xi_1, \ldots, \xi_n)$ is an arbitrary real vector, $\eta = (\eta_1, \ldots, \eta_n)$ is the unit normal to the plane T, and d is the distance from T.

Suppose also that $w \geqq 0$ in D and $w = 0$ at Q. Then either

$$\frac{\partial w}{\partial s} > 0 \quad \text{or} \quad \frac{\partial^2 w}{\partial s^2} > 0 \quad \text{at } Q$$

unless $w \equiv 0$, where $\vec{s}$ is any direction at Q which enters D non-tangentially.

Proof. We proceed in the same way as in the proof of Lemma 1, except that the comparison function z and the ball K_2 must be chosen with greater care.

As in the proof of Lemma 1, we introduce the ball K_1, and then construct K_2 to have center at Q and radius θr_1, where $\theta \leqq 1/2$ is a constant to be determined.

Then in $K' = K_1 \cap K_2 \cap D$ we define the auxiliary function

$$z = z(x) = \{e^{-\alpha(x_1-r_1)^2} - e^{-\alpha r_1^2}\} \cdot \{e^{-\alpha r^2} - e^{-\alpha r_1^2}\}$$

where α is also to be determined. With the choice of coordinates as before, it is clear that

$$z > 0 \quad \text{in } K', \qquad z = 0 \quad \text{on } \partial K_1 \text{ and on } T.$$

We compute

$$\begin{aligned} Lz = {} & e^{-\alpha r^2}(e^{-\alpha(x_1-r_1)^2} - e^{-\alpha r_1^2}) \cdot \{4\alpha^2 a_{ij} x_i x_j - 2\alpha[a_{ii} + b_i x_i]\} \\ & + e^{-\alpha(x_1-r_1)^2}(e^{-\alpha r^2} - e^{-\alpha r_1^2}) \cdot \{4\alpha^2 a_{11}(x_1-r_1)^2 - 2\alpha[a_{11} + b_1(x_1-r_1)]\} \\ & + 8\alpha^2 e^{-\alpha r^2} e^{-\alpha(x_1-r_1)^2} \cdot (x_1-r_1) a_{1j} x_j. \end{aligned}$$

Because of the ellipticity condition (25)

$$a_{ij} x_i x_j \geqq \kappa r^2 \geqq \tfrac{1}{4}\kappa r_1^2 \quad \text{in } K'$$

and, for the same reason, $a_{11}(x_1-r_1)^2 \geqq \frac{1}{4}\kappa r_1^2$ in K'. Moreover by (26),

$$|a_{1j} x_j| = |a_{ij} \eta_i x_j| \leqq K(|x_1| + |x_1|)$$

since in the present case $\eta = (1, 0, \ldots, 0)$. Thus

$$|(x_1-r_1) a_{ij} x_j| \leqq 2 x_1 r_1 K \quad \text{in } K'.$$

Finally we observe that by the mean value theorem

$$e^{-\alpha(x_1-r_1)^2} - e^{-\alpha r_1^2} \geqq 2\alpha(1-\theta) r_1 e^{-\alpha r_1^2} x_1 \geqq \alpha x_1 r_1 e^{-2\alpha\theta r_1^2 - \alpha(x_1-r_1)^2}.$$

Inserting these inequalities into the earlier expression for Lz, and using the fact that the terms $[a_{ii} + b_i x_i]$ and $[a_{11} + b_1(x_1-r_1)]$ are bounded, we find for large α

$$\begin{aligned} Lz \geqq {} & \alpha^2 x_1 r_1 e^{-\alpha(r^2 + (x_1-r_1)^2)} \cdot \{(\alpha\kappa r_1^2 - B) e^{-2\alpha\theta r_1^2} - 16K\} \\ & + \alpha e^{-\alpha(x_1-r_1)^2}(e^{-\alpha r^2} - e^{-\alpha r_1^2})\{\alpha\kappa r_1^2 - B\} \end{aligned}$$

where B is an appropriate constant. By choosing $\theta = 1/\alpha$ and then α suitably large it is clear that we can make the quantities in braces positive. That is, we have now constructed a function z such that $Lz > 0$ in K'.

The remaining part of the proof is the same as for Lemma 1, since both the maximum principle and the boundary point lemma apply to the elliptic operator L in the same way as to the Laplace operator ([2], pages 61 and 65). The only slight difference to be noted is that (13) must now be replaced by

(13′) $$z \leqq 2\alpha r_1 x_1,$$

which follows easily from the mean value theorem and the definition of z. Consequently the function $w - \varepsilon z$ of Lemma 1 must be replaced by $w - (\varepsilon/2\alpha r_1) z$ but otherwise the argument is left unchanged. This completes the demonstration of Lemma 2.

To complete the proof of Theorem 2 we must show that the final argument in the proof of Theorem 1 can be carried over to the more general case of equation (19). Thus let u and v be, respectively, the original solution and the reflected solution in the region Σ'. It has already been established that either $u-v>0$ or $u\equiv v$ in Σ'. We must show in the former case that (24) holds (the plane $T=T'$ being in the position indicated by case (*ii*) of the proof of Theorem 1). We begin by obtaining an appropriate second order differential equation for the difference function $u-v$.

Since both u and v satisfy (19) we have

$$a[u]\Delta u+h[u]u_i u_j u_{ij}=f[u]$$

$$a[v]\Delta v+h[v]v_i v_j v_{ij}=f[v]$$

where $a[u]\equiv a(u,|p|)$ and similarly for the other square brackets. Differencing these equations yields easily

$$\{a[u]+a[v]\}\Delta(u-v)+\{h[u]u_i u_j+h[v]v_i v_j\}(u-v)_{ij}$$
$$+\{a[u]-a[v]\}\Delta(u+v)+\{h[u]u_i u_j-h[v]v_i v_j\}(u+v)_{ij}=2\{f[u]-f[v]\}.$$

Now by the mean value theorem of multidimensional calculus

$$a[u]-a[v]=\left(\frac{\partial a}{\partial u}\right)_0(u-v)+\left(\frac{\partial a}{\partial p_i}\right)_0(u-v)_i$$

with similar expressions for $h[u]u_i u_j-h[v]v_i v_j$ and $f[u]-f[v]$. Thus if we define

$$a_{ij}(x)=\{a[u]+a[v]\}\delta_{ij}+\{h[u]u_i u_j+h[v]v_i v_j\}$$

it follows from the above identity that

$$a_{ij}(x)(u-v)_{ij}+b_i(x)(u-v)_i+c(x)(u-v)=0 \tag{27}$$

where $b_i(x)$ and $c(x)$ are certain bounded functions.

Here the matrix a_{ij} is uniformly definite:

$$a_{ij}\xi_i\xi_j\geqq\kappa|\xi|^2$$

since both expressions $a[u]\delta_{ij}+h[u]u_i u_j$ and $a[v]\delta_{ij}+h[v]v_i v_j$ have this property (recall that equation (19) is elliptic). Consider next the expression $a_{ij}\xi_i\eta_j$ for which the estimate (26) must be established. We have by computation

$$a_{ij}\xi_i\eta_j=\{a[u]+a[v]\}\xi\cdot\eta+\{h[u](p\cdot\xi)(p\cdot\eta)+h[v](q\cdot\xi)(q\cdot\eta)\} \tag{28}$$

where $p=(u_1,\ldots,u_n)$ and $q=(v_1,\ldots,v_n)$. Now v is the reflection of u in the plane $T=T'$. Hence $u=v$ on T, and moreover (in the usual rectangular coordinate frame centered at Q with x_1 axis normal to T)

$$p=(u_1,u_2,\ldots,u_n),\qquad q=(-u_1,u_2,\ldots,u_n)\quad\text{on } T.$$

Thus on T there holds

$$|p|=|q|,\qquad a[u]=a[v],\qquad h[u]=h[v];$$

and, since $\eta=(1,0,\dots,0)$,

$$\begin{aligned} I &\equiv h[u](p\cdot\xi)(p\cdot\eta)+h[v](q\cdot\xi)(q\cdot\eta)\\ &=h[u]\,u_1(p-q)\cdot\xi=2h[u]\,u_1^2\xi_1=2h[u]\,u_1^2\xi\cdot\eta. \end{aligned}$$

By continuity it then follows that off T

$$|I|\leqq M(|\xi\cdot\eta|+|\xi|\cdot|x_1|) \tag{29}$$

for some constant M. Noting that also $\{a[u]+a[v]\}\leqq M'$ for some M', we have finally from (28) and (29)

$$|a_{ij}\xi_i\eta_j|\leqq K(|\xi\cdot\eta|+|\xi|\cdot|x_1|)\qquad (K=M+M')$$

as required.

The differential equation (27) for $u-v$ is not quite in the form to which Lemma 2 applies. Hence as in the proof of the restricted maximum principle we make HOPF's substitution $w=(u-v)\,e^{\beta x_1}$. Then obviously

$$a_{ij}(x)\,w_{ij}+\tilde b_i(x)\,w_i+\tilde c(x)\,w=0$$

where $\tilde b_i=b_i-2\beta a_{1i}$, $\tilde c=a_{11}\beta^2-b_1\beta+c$. Since $u-v\geqq 0$ by assumption it follows that for β sufficiently large

$$a_{ij}(x)\,w_{ij}+\tilde b_i(x)\,w_i\leqq 0.$$

Consequently Lemma 2 may be applied directly to w (note that w satisfies the hypotheses of Lemma 2 since $u-v=0$ at Q); we find therefore

$$\frac{\partial w}{\partial s}>0\quad\text{or}\quad\frac{\partial^2 w}{\partial s^2}>0\quad\text{at } Q$$

since $w\not\equiv 0$. But at Q

$$\frac{\partial(u-v)}{\partial s}=\frac{\partial w}{\partial s},\qquad \frac{\partial^2(u-v)}{\partial s^2}=\frac{\partial^2 w}{\partial s^2}-2\beta s_1\frac{\partial w}{\partial s},$$

so that also

$$\frac{\partial(u-v)}{\partial s}>0\quad\text{or}\quad\frac{\partial^2(u-v)}{\partial s^2}>0\quad\text{at } Q.$$

Thus the difference $u-v$ satisfies (24), and the proof of Theorem 2 is complete.

5. A Different Boundary Condition

The previous results can be extended to more general boundary conditions than (2) without changing the basic method. We let H denote the *mean curvature* of the boundary surface $\partial\Omega$, chosen so that H is positive when the surface is "convex." Analytically, if a portion of the boundary is locally represented by the equation

$$x_n=\phi(x_1,\dots,x_{n-1})$$

with the x_n axis directed into Ω, then on this part

$$H=\frac{(1+\phi_k\phi_k)\,\Delta\phi-\phi_i\phi_j\phi_{ij}}{(n-1)(1+\phi_k\phi_k)^{3/2}}. \tag{30}$$

We now replace (2) by the more general condition

$$u=0, \quad \frac{\partial u}{\partial n}=c(H) \quad \text{on } \partial\Omega \tag{31}$$

where c is a continuously differentiable non-decreasing function of H. Then the following result holds

Theorem 3. *The conclusions of Theorems* 1 *and* 2 *remain valid under the more general boundary condition* (31).

Proof. It will be enough to show how the proof of Theorem 1 needs to be modified to cover the more general boundary condition (31), and then to make several comments concerning the further modification required for the case of equation (19).

For the generalization of Theorem 1, then, we proceed as in the initial steps of the proof given in Section 1 to show that either (8) holds or $u\equiv v$ in Σ'. To prove then that (8) is impossible, suppose first that case (*i*) holds in the definition of Σ'. Then $u-v=0$ at P, and by the boundary point version of the maximum principle we conclude as before that

$$\frac{\partial}{\partial n}(u-v)>0 \quad \text{at } P. \tag{32}$$

On the other hand, if H is the mean curvature of $\partial\Omega$ at P and H' is the mean curvature of $\partial\Sigma'$ at P, then $H\leqq H'$ since $\Sigma'\subset\Omega$. Consequently

$$\frac{\partial u}{\partial n}=c(H)\leqq c(H')=\frac{\partial v}{\partial n}$$

which contradicts (32). Hence (8) is impossible in case (*i*).

In case (*ii*) we must further analyze the second partial derivatives of u and v at Q. For this purpose we assume for simplicity that the boundary of Ω is of class C^3, though by taking more care one can weaken this assumption. Fixing coordinates as before, we obtain then

$$u_i=0, \quad u_n=c, \quad u_{ij}+c\,\phi_{ij} \quad (\text{at } Q)$$

where i, j range from 1 to $n-1$. Moreover, differentiating (10) and evaluating at Q gives

$$u_{ni}=\frac{dc}{dH}H_i \quad (\text{at } Q)$$

since c is no longer constant. Also from (30),

$$H_i=(\Delta\varphi)_i/(n-1) \quad (\text{at } Q).$$

Since v is the reflection of u, it now follows that all the first and second partial derivatives of these functions agree at Q, with the possible exception of u_{1n} and v_{1n}, which have the values

$$u_{1n}=-v_{1n}=\frac{dc}{dH}H_1 \quad (\text{at } Q). \tag{33}$$

Now Σ' is contained in Ω; continuing the analysis of this situation (which has already yielded the relations $\phi_{1l}=0$ at Q), we find that the third partial derivatives of ϕ at Q must satisfy the condition

$$\varepsilon\{\phi_{lm1}\zeta_l\zeta_m+\phi_{111}\}\leqq 0 \quad (\varepsilon=\operatorname{sign} x_1 \text{ in } \Sigma')$$

where $\zeta=(\zeta_2, \ldots, \zeta_{n-1})$ is an arbitrary real vector and the indices l and m are to be summed from 2 to $n-1$. It follows that

$$\phi_{111}\leqq 0, \quad \varphi_{ll1}\leqq 0 \quad (\text{at } Q) \tag{34}$$

since Σ' may be assumed to lie in the space $x_1>0$. Clearly (34) implies

$$H_1=(\Delta\varphi)_1/(n-1)\leqq 0.$$

Thus by (33), recalling that $dc/dH\geqq 0$, we find

$$u_{1n}\leqq 0, \quad v_{1n}\geqq 0 \quad (\text{at } Q).$$

But these are the only (possibly) unequal partial derivatives of u, v at Q. Since $u>v$ in Σ', it follows that in fact $u_{1n}=v_{1n}=0$, at Q. Hence all the first and second partial derivatives of u and v agree at Q. The remainder of the proof is the same as before.

The argument for the elliptic equation (19) proceeds almost exactly as it did earlier in Section 2, though of course taking into account the modifications in the proof of Theorem 1 which were described above.

The only slight change required involves the demonstration that u increases monotonically as one enters Ω along any non-tangential direction $\vec{s}$. The difficulty is that one might have $c(H)=0$ at some points of the boundary and $c(H)>0$ at others (of course $c(H)$ can never be negative at any boundary point because $u>0$ in Ω). In any case, just as in Section 3 one can show that at any point where $\partial u/\partial n=0$ all the first and second partial derivatives of u vanish except for $\partial^2 u/\partial n^2$, and this is positive. A not too difficult compactness argument then yields the required monotonicity property of u near the boundary. This completes the proof of Theorem 3.

References

1. HOPF, E., Elementare Bemerkungen über die Lösungen partieller Differentialgleichungen zweiter Ordnung vom elliptischen Typus. Berlin, Sber. Preuss. Akad. Wiss. **19**, 147–152 (1927).
2. PROTTER, M. H., & H. F. WEINBERGER, Maximum Principles in Differential Equations. New Jersey: Prentice-Hall 1967.
3. SOKOLNIKOFF, I. S., Mathematical Theory of Elasticity. New York: McGraw-Hill 1956.

Department of Mathematics
University of Minnesota
Minneapolis

(Received May 4, 1971)

Arch. Ration. Mech. Anal. 148 (1999), 265–290. **Reprint I.7.[ii]**

Pages 280–285 have been removed for the sake of brevity.

Symmetry of Ground States of Quasilinear Elliptic Equations

JAMES SERRIN & HENGHUI ZOU

Abstract

We consider the problem of radial symmetry for *non-negative* continuously differentiable weak solutions of elliptic equations of the form

$$\mathrm{div}(A(|Du|)Du) + f(u) = 0, \quad x \in \boldsymbol{R}^n, \quad n \geqq 2, \tag{1}$$

under the ground state condition

$$u(x) \to 0 \text{ as } |x| \to \infty. \tag{2}$$

Using the well-known moving plane method of ALEXANDROV and SERRIN, we show, under suitable conditions on A and f, that all ground states of (1) are radially symmetric about some origin O in $\boldsymbol{R}^n$. In particular, we obtain radial symmetry for compactly supported ground states and give sufficient conditions for the positivity of ground states in terms of the given operator A and the nonlinearity f.

1. Introduction

We consider the problem of radial symmetry for non-negative solutions of elliptic equations of the form

$$\mathrm{div}(A(|Du|)Du) + f(u) = 0, \quad x \in \boldsymbol{R}^n, \quad n \geqq 2, \tag{1.1}$$

under the ground state condition

$$u(x) \to 0 \text{ as } |x| \to \infty. \tag{1.2}$$

This problem was first considered by GIDAS, NI & NIRENBERG [10] for the case of the Laplace operator ($A(t) \equiv 1$), that is, for the equation

$$\Delta u + f(u) = 0, \quad x \in \boldsymbol{R}^n,$$

and for C^1 functions $f(u)$ which are (essentially) of class $C^{1+\gamma}$ near $u = 0$ and satisfy $f(0) = 0$, $f'(0) < 0$. Under the additional restriction that solutions are *positive* in $\boldsymbol{R}^n$, they showed that corresponding ground states are radially symmetric about some origin O in $\boldsymbol{R}^n$. Their work depends crucially on the moving plane method of ALEXANDROV and SERRIN (as also does that in the present paper), as well as upon appropriate asymptotic estimates for the behavior of ground states as $|x| \to \infty$.

Here we extend their result in three main and comprehensive directions — first, by asking of the function $f(u)$ only the following properties,

(i) f is continuous for $u \geqq 0$ and locally Lipschitz continuous for $u > 0$, and
(ii) $f(0) = 0$ and f is non-increasing on some interval $0 \leqq u < \delta$;

second, by replacing the Laplace operator with appropriate quasilinear operators for which (1.1) is locally uniformly elliptic for any vector $Du \in \boldsymbol{R}^n$; and third by treating degenerate operators.

In particular, our conclusions cover as special cases the Laplace operator, the mean curvature operator

$$A(t) = \frac{1}{\sqrt{1+t^2}} \qquad \text{(see Theorem 1}'\text{)},$$

and the degenerate Laplace operator

$$A(t) = |t|^{m-2}, \; m > 1 \qquad \text{(Theorems 2, 3)}.$$

The importance of (1.1) in applications is discussed in [3], including the case when f is not Lipschitz continuous at $u = 0$.

In view of the generality of the hypotheses, it is necessary to deal carefully with the meaning of solutions of (1.1). To be specific, we understand a *ground state* of (1.1) to be a non-negative, non-trivial C^1 distribution (weak) solution of (1.1) on $\boldsymbol{R}^n$, which satisfies condition (1.2).

We can now describe our main results, first considering the case of *regular operators* A.

Theorem 1. *Let $f(u)$ satisfy conditions* (i) *and* (ii) *above. Suppose $A(t)$ is of class $C^{1,1}([0,\infty))$ and has the property*

$$\Omega'(t) > 0 \quad \textit{for all } t \geqq 0, \tag{1.3}$$

where $\Omega(t) = tA(t)$. Then any ground state of (1.1) *whose (open) support is connected must be radially symmetric about some origin O in $\boldsymbol{R}^n$, and the corresponding radial function $u(r)$ obeys $u'(r) < 0$ for all $r > 0$ such that $u(r) > 0$.*

Condition (1.3) guarantees the ellipticity of equation (1.1); see Section 2. It is immediate that both the Laplacian and the mean curvature operator satisfy (1.3), as well as operators arising from regular variational problems.

In Theorem 1 it is explicitly understood that ground states need not be everywhere positive, this being reflected in the final statement of the theorem, that $u'(r) < 0$ *only* on the open support of u.

Under appropriate conditions on the nonlinearity $f(u)$ the ground state is necessarily positive and the connectivity condition in Theorem 1 can be dropped, but in other cases this condition is necessary for the validity of the theorem. The precise behavior to be expected is described in the next results.

Proposition 1. *Let the hypotheses of Theorem 1 be satisfied, and let* $F(t) = \int_0^t f(s)ds$. *Suppose that* u *is a ground state of* (1.1). *Then* u *is positive on* $\boldsymbol{R}^n$ *if and only if*[1]

$$\int_0 \frac{dt}{|F(t)|^{1/2}} = \infty, \tag{1.4}$$

while conversely u *has compact support in* R^n *if and only if*

$$\int_0 \frac{dt}{|F(t)|^{1/2}} < \infty. \tag{1.5}$$

This result, which combines ideas due to VAZQUEZ [20] and CORTÁZAR, ELGUETA & FELMER [3], is proved in [17]. In view of Theorem 1, the first part of the proposition has the following corollary.

Corollary. *Suppose the hypotheses of the first paragraph of Theorem 1 hold, and let* f *be locally Lipschitz continuous for* $u \geqq 0$. *Then ground states of* (1.1) *are necessarily positive, radially symmetric about some origin* O, *and satisfy* $u'(r) < 0$ *for all* $r > 0$.

Under the a priori assumption that the ground states in question are *everywhere positive* this result is due to LI & NI [14] (in fact their work also covers regular fully nonlinear equations). The conclusion of course also generalizes the Gidas-Ni-Nirenberg theorem, and shows even more that *the condition of positivity in that result can be dropped without affecting the conclusion.* See also [8] and [13], and the remarks at the end of Section 3.

A second corollary of the proposition concerns the case when (1.5) applies:

Corollary. *Suppose that* (1.5) *holds, and let* u *be a ground state of* (1.1) *with connected open support. Then the support of* u *is a ball and* u *is radially symmetric about the center.*

The assumption that the open support of u be connected *cannot be omitted when* (1.5) *holds, or even weakened to the requirement that the (closed) support of* u *be connected*, since this would allow the possibility of multibump, non-radial ground states (even though each bump, in isolation from the others, would be radially symmetric about some origin).

Theorem 1′. *Let the hypotheses of the first paragraph of Theorem 1 hold, and let* f *be locally Hölder continuous for* $u \geqq 0$. *Assume also that* (1.5) *is satisfied. Then any ground state of* (1.1) *has (open) support on a finite number of open balls in* $\boldsymbol{R}^n$, *on each of which it is radially symmetric about the center of the ball.*

[1] If $f(u) \equiv 0$ for all sufficiently small u, then the integral (1.4) is understood to be divergent. The same convention is used later for the integral in (1.8).

Theorem 1′ generalizes Theorem 1.2 of [3], which deals in particular with the Laplace operator Δu and the special nonlinearity $f(u) = -u^q + u^p$, $0 < q < 1 < p$; see also [4].

For the degenerate Laplace operator, condition (1.3) fails at $t = 0$ when $m \neq 2$, and Theorem 1 must accordingly be modified. In the next result we turn our attention to such exceptional operators, namely those which either are undefined at $t = 0$ or fail to satisfy (1.3) at $t = 0$. Here it is necessary to add further conditions on the critical set of u before symmetry can be expected.

Theorem 2. *Suppose that $f(u)$ satisfies conditions* (i) *and* (ii), *and assume $A(t)$ is of class $C^{1,1}((0,\infty))$ such that*

$$\Omega'(t) > 0 \quad for\ t > 0, \qquad \Omega(t) \to 0 \quad as\ t \to 0. \tag{1.6}$$

Let u be a ground state of (1) *with the property that the set*

$$\{x \in \boldsymbol{R}^n \mid u(x) > 0,\ Du(x) = 0\} \tag{1.7}$$

contains exactly one point O. Then u is radially symmetric about O, and the corresponding function $u(r)$ satisfies $u'(r) < 0$ for all $r > 0$ such that $u(r) > 0$.

Critical point conditions, as in Theorem 2, are well-known for the symmetry problem for the degenerate Laplace operator; see [1, 2, 5]. In the generality of Theorem 2 they are moreover *necessary*, as shown by the counter-examples in Section 6.

We also have the following complementary result to Theorem 2, in which the function

$$H(t) = t\Omega(t) - \int_0^t \Omega(s)ds, \quad t > 0,$$

plays a vital part. Note that $H'(t) > 0$ for all $t > 0$, in view of (1.6), and that $H(0) \to 0$ as $t \to 0$. Clearly then $H(t) > 0$ for all $t > 0$.

Proposition 2. *Let the hypotheses of the first sentence of Theorem 2 be satisfied, and let u be a ground state of* (1.1). *Then u is positive on $\boldsymbol{R}^n$ if*

$$\inf_{t\to 0} \frac{H(t)}{t\Omega(t)} > 0, \qquad \int_0 \frac{dt}{H^{-1}(|F(t)|)} = \infty, \tag{1.8}$$

while conversely it has compact support in R^n if

$$\int_0 \frac{dt}{H^{-1}(|F(t)|)} < \infty. \tag{1.9}$$

Proposition 2 is proved in [17] under the additional restriction that $\inf_{t\to 0} H(t)/t\Omega(t) > 0$. That this condition is not in fact necessary has been shown in recent work of PUCCI & SERRIN. We note also that Proposition 1 is a special case of Proposition 2; that is, when $\Omega'(t) > 0$ for $t \geqq 0$, we get $\Omega(t) = \Omega'(0)t + o(t)$ for small t, so also $H(t) = \{\frac{1}{2}\Omega'(0)\}t^2 + o(t^2)$. Thus the integral conditions (1.8) and (1.9) reduce to (1.4) and (1.5).

For the degenerate Laplace operator $A(t) = t^{m-2}, m > 1$, we have $t\Omega(t) = t^m$ and $H(t) = \frac{m-1}{m}t^m$, so (1.8) reduces to the condition

$$\int_0 \frac{dt}{|F(t)|^{1/m}} = \infty. \tag{1.10}$$

For functions $f(u)$ which are locally Lipschitz continuous on $u \geqq 0$ this condition is satisfied for all $m \leqq 2$, and hence corresponding ground states are positive. When $m > 2$, there is a converse result:

Corollary. *Suppose* $\sup_{u\to 0} f(u)/u < 0$ *(for example,* $f'(0) < 0$*). Let* u *be a corresponding ground state for the degenerate Laplace operator with* $m > 2$*, such that the set* (1.7) *contains exactly one point. Then* (1.9) *holds and* u *has support in an open ball in* $\boldsymbol{R}^n$.

When the operator A is *singular* at $t = 0$ the result of Theorem 2 can be improved.

Theorem 3. *Let the hypotheses of the first sentence of Theorem 2 be satisfied. Suppose also that* $\Omega'(t) \to \infty$ *as* $t \to 0$*. Let* u *be a ground state of* (1.1) *with the property that the (open) support of* Du *is connected. Then* u *must be radially symmetric about some origin* O *in* $\boldsymbol{R}^n$*, and the corresponding function* $u(r)$ *obeys* $u'(r) < 0$ *for all* $r > 0$ *such that* $u(r) > 0$.

An important example for Theorem 3 is the degenerate Laplace operator $A(t) = t^{m-2}$ with $1 < m < 2$, a case also considered in the following paper of Damascelli, Pacella & Ramaswamy [6]. In particular, using special properties of the degenerate Laplace operator, but without requiring the connectivity condition of Theorem 3, these authors show that corresponding *positive* ground states are radially symmetric about some center in $\boldsymbol{R}^n$.

Remark. Once one is able to show that ground states are radially symmetric, a great deal is known about their uniqueness as functions of r and their asymptotic behavior as $r \to \infty$. For these issues, we refer the reader to [3, 9, 16] and the references noted therein.

In conclusion, it is worth emphasizing that by allowing *non-negative* ground states into consideration we obtain an altogether richer theory than that for which ground states are a priori assumed to be positive, for besides the virtue of additional generality one moreover avoids undue restrictions on the behavior of the nonlinearity $f(u)$ near $u = 0$; see Propositions 1 and 2. Moreover, compactly supported ground states, which arise in this context, can even be considered as solutions of the *free boundary problem* to find solutions of (1.1) which are *positive* on a domain B and have the property that $u(x),\ Du(x) \to 0$ as $x \to \partial B$.

Theorems 1, 1′ and 2 will be proved in Sections 2 and 3. Theorem 3 will be proved in Sections 4 and 5. In Section 6, we present two counter-examples of *non-symmetric* ground states.

Finally, we remark that our approach applies to more general nonlinearities $f(r, u, |Du|)$ and that we also can treat singular ground states. With appropriately corresponding assumptions, results similar to those above can be obtained in both these cases. For singular ground states, in particular, we give an outline of the required details in Section 7.

2. Preliminaries

Consider the quasilinear elliptic equation

$$\operatorname{div}(A(|\nabla u|)\nabla u) + f(u) = 0, \qquad x \in \boldsymbol{R}^n. \tag{2.1}$$

We are interested in C^1 non-negative weak solutions u in $\boldsymbol{R}^n$ such that

$$\lim_{|x|\to\infty} u(x) = 0, \tag{2.2}$$

the so-called *ground state* condition.

Throughout the paper we treat the following two separate cases for the operator A, that is, *regular* operators and *degenerate* operators. Our assumptions are as follows.

1. *Regular case.* A is in $C^{1,1}([0,\infty))$. We let

$$\Omega(t) = tA(t), \quad t \geqq 0,$$

and suppose that

$$\Omega'(t) > 0 \quad \text{for } t \geqq 0. \tag{2.3}$$

2. *Degenerate case.* A is of class $C^{1,1}((0,\infty))$. Here $\Omega(t)$ is only defined for $t > 0$, and correspondingly we may suppose only that

$$\Omega'(t) > 0 \quad \text{for } t > 0.$$

Additionally, it is then assumed that

$$\Omega(t) \to 0 \quad \text{as } t \to 0,$$

a condition which is of course automatic in the regular case.

Finally, the function f is assumed to satisfy the conditions (i), (ii) in the introduction.

In the regular case, it is evident from (2.3) that both $A(t)$ and $\Omega(t)$ are *positive* for $t > 0$. Clearly also, by L'Hôpital's rule,

$$A(0) = \lim_{t\to 0} A(t) = \Omega'(0) > 0.$$

When A is degenerate we obviously continue to have $A(t)$ and $\Omega(t)$ positive for $t > 0$, though A itself remains undefined at 0. Note moreover that

$$\lim_{p\to 0} A(|p|)p = 0,$$

where $p = (p_1, p_2, ..., p_n)$, so that with the obvious meaning that $A(|Du|)Du = 0$ when $Du = 0$ the weak form of (2.1) remains valid even for degenerate operators A.

Let $p = (p_1, \dots, p_n)$. Then for regular operators A we have

$$\frac{\partial}{\partial p_j}(A(|p|)p_i) = A(|p|)\delta_{ij} + |p|A'(|p|)\frac{p_i p_j}{|p|^2} \tag{2.4}$$

for all $p \in \boldsymbol{R}^n$, while for degenerate A the same calculation applies provided $p \neq 0$.

Let the right-hand side of (2.4) be denoted by $a_{ij}(p)$.

Lemma 2.1. *For regular operators A the matrix $a_{ij}(p)$ is Lipschitz continuous and positive-definite for all $p \in \boldsymbol{R}^n$. When A is degenerate, the same result holds provided $p \neq 0$.*

Proof. By linear algebra, for $p = (p_1, p_2, ..., p_n) \neq 0$ the only non-zero eigenvalue of the matrix $\{p_i p_j |p|^{-2}\}$ is 1, with eigenvector p; in fact we have

$$\lambda_1 = 1, \qquad \lambda_2 = \cdots = \lambda_n = 0.$$

In turn, $a_{ij}(p)$ has the n eigenvalues

$$\lambda_1 = A(|p|) + |p|A'(|p|) = \Omega'(|p|), \qquad \lambda_2 = \cdots = \lambda_n = A(|p|),$$

and so is positive-definite. When $p = 0$ and A is regular, we have $a_{ij}(0) = A(0)\delta_{ij}$, so again a_{ij} is positive-definite.

Lipschitz continuity is obvious from the basic hypotheses for both regular and degenerate operators. □

Lemma 2.2. *If A is regular, then any C^1 non-negative weak solution u of* (2.1) *is of class C^2 in the neighborhood of points y where $u(y) > 0$. For degenerate operators, the same result is valid provided that $Du(y) \neq 0$.*

In view of Lemma 2.1 and condition (i) for the nonlinearity f, this is an immediate consequence of [12, Theorems 6.2, 6.3, pp. 282–283].

In any domain in which u is of class C^2 we can clearly write (2.1) in the classical form

$$a_{ij}(Du)u_{ij} + f(u) = 0, \tag{2.5}$$

where $u_{ij} = \partial^2 u/\partial x_i \partial x_j$. By Lemma 2.1 one sees that (2.5) is locally uniformly elliptic.

In what follows it will be convenient to introduce the function

$$\lambda(s) \stackrel{\text{def}}{=} \inf_{0<t\leqq s} \{A(t), \Omega'(t)\}, \qquad s > 0.$$

Clearly λ is well-defined for both regular and degenerate operators, and is positive and non-increasing. We have the following monotonicity theorem.

Lemma 2.3. *Let A be either regular or degenerate, and suppose that $p \neq q$. Then*

$$\{A(|p|)p - A(|q|)q\} \cdot \{p - q\} > 0$$

and, more specifically,

$$\{A(|p|)p - A(|q|)q\} \cdot \{p - q\} \geqq \lambda(\sigma)|p - q|^2,$$

where $\sigma = \max(|p|, |q|)$.

Proof. By an obvious interpolation identity we have

$$(A(|p|)p - A(|q|)q) \cdot (p - q) = \int_0^1 a_{ij}(p_\theta)(p_i - q_i)(p_j - q_j)\, d\theta$$

where $p_\theta = \theta p + (1 - \theta)q$. (*Note.* The integrand may have a singularity when $\theta p + (1 - \theta)q = 0$, which however can occur at most at a single value θ_0 of θ. On the other hand, since $a_{ij}(p)$ is positive-definite for $p \neq 0$, it is not hard to see that the identity is valid even in this exceptional case — e.g., delete an interval $(\theta_0 - \varepsilon, \theta_0 + \varepsilon)$ from the integration domain, calculate the resulting integral and then let $\varepsilon \to 0$.) The first conclusion now follows from the fact that the integrand is continuous and positive (in θ) except for at most one point.

To obtain the second conclusion, it is enough to note that, with the possible exception of a single value θ,

$$a_{ij}(p_\theta)\xi_i\xi_j \geqq \min\{A(|p_\theta|),\ \Omega'(|p_\theta|)\}|\xi|^2 \geqq \lambda(|p_\theta|)|\xi|^2$$

(since the eigenvalues of $a_{ij}(p)$ are $A(|p|)$ and $\Omega(|p|)$; see the proof of Lemma 2.1), and

$$|p_\theta| \leqq \sigma = \max(|p|, |q|),$$

where of course $\sigma > 0$.

Now let u, v be two non-negative weak solutions of (2.1) defined in a half space Σ in $\mathbf{R}^n$. Put $w = u - v$. Then by differencing the classical equations (2.5) satisfied by u and v, we obtain the principal differential equation

$$a_{ij}(p)w_{ij} + (a_{ij}(p) - a_{ij}(q))v_{ij} + c(x)w = 0 \tag{2.6}$$

in Σ, where p stands for Du and q for Dv, and

$$c(x) = \begin{cases} \frac{f(u(x)) - f(v(x))}{u(x) - v(x)}, & u(x) \neq v(x), \\ 0, & u(x) = v(x) \end{cases}.$$

By Lemma 2.2 and the remarks immediately afterward, it is clear for regular operators A that equation (2.6) is well-defined and locally uniformly elliptic in any open subset of Σ where $u,\ v \neq 0$, while for degenerate operators it is necessary also to have $Du,\ Dv \neq 0$.

Because a_{ij} is Lipschitz continuous, one also has

$$(a_{ij}(p) - a_{ij}(q))v_{ij} = b_i(x)w_i,$$

where $b_i(x)$ is a locally bounded function in the domains of validity indicated above, while of course $c(x)$ is locally bounded in any open set in which either u or v is nonzero. Thus (2.6) can finally be written in the standard quasilinear elliptic form

$$a_{ij}(p)w_{ij} + b_i(x)w_i + c(x)w = 0. \qquad \square \tag{2.7}$$

3. Symmetry: Proof of Theorems 1, 2

Let u be a C^1 non-negative weak solution of (2.1), and write

$$\operatorname{supp}_+ u = \{x \in \boldsymbol{R}^n \mid u(x) > 0\},$$

the *open support* of u, and

$$Z = \{x \in \boldsymbol{R}^n \mid u(x) > 0,\ Du(x) = 0\},$$

the set of critical points of u on its open support.

For $\gamma \in \boldsymbol{R}$ and $x = (x_1, \ldots, x_n)$, we define

$$\Sigma = \Sigma_\gamma = \{x \in \boldsymbol{R}^n \mid x_1 < \gamma\}$$

and denote by Γ the hyperplane

$$\Gamma = \Gamma_\gamma = \{x \in \boldsymbol{R}^n \mid x_1 = \gamma\}.$$

Let x^γ be the *reflection* in Γ of a point x in $\boldsymbol{R}^n$, that is,

$$x^\gamma = (2\gamma - x_1, x_2, \ldots, x_n),$$

and similarly let Ω^γ be the reflection in Γ of a set Ω in $\boldsymbol{R}^n$,

$$\Omega^\gamma = \{x^\gamma \mid x \in \Omega\}.$$

For $x \in \overline{\Sigma}$, put

$$v(x) = v_\gamma(x) = u(x^\gamma), \qquad w(x) = u(x) - v(x).$$

Then by the reflection invariance of equation (2.1), it follows that v as well as u is a solution of (2.1) in Σ, of class C^1 in $\overline{\Sigma}$. We observe also that $w = 0$ on Γ, and that, furthermore, if $Du(x) = 0$ at some point x in Γ, then also $Dv(x) = Dw(x) = 0$.

We next obtain a strong comparison lemma and a Hopf boundary lemma for the function w, which will be crucial in our moving plane procedure.

Lemma 3.1. *Let u be a C^1 non-negative weak solution of* (2.1) *and let $A(t)$ be regular (that is, of class $C^{1,1}([0, \infty))$ such that* (1.3) *is satisfied). Assume that* $\operatorname{supp}_+ u$ *is connected and suppose that*

$$w \geqq 0, \qquad w \not\equiv 0 \qquad in\ \Sigma.$$

Then

$$w > 0 \quad in\ \Sigma \cap \operatorname{supp}_+ u, \qquad \frac{\partial w}{\partial x_1} < 0 \quad on\ \Gamma \cap \operatorname{supp}_+ u. \tag{3.1}$$

Proof. Suppose for contradiction that $(3.1)_1$ fails, that is, there exists $y \in \Sigma \cap \operatorname{supp}_+ u$ with $w(y) = 0$. By (2.7) we can write

$$a_{ij}(x)w_{ij} + b_i(x)w_i + c(x)w = 0 \tag{3.2}$$

as in Section 2. Clearly (3.2) is uniformly elliptic in a neighborhood of y since $v(y) = u(y) > 0$. For the same reason, the coefficients b_i are uniformly bounded near y, and finally $c(x)$ is also uniformly bounded near y.

Put $c(x) = c_+(x) + c_-(x)$, where $c_+ \geqq 0$ and $c_- \leqq 0$ are respectively the positive and negative parts of c. Then (3.2) can be rewritten as a uniformly elliptic differential inequality

$$a_{ij}(x)w_{ij} + b_i(x)w_i + c_-(x)w = -c_+(x)w \leqq 0 \tag{3.3}$$

for x near y. The strong maximum principle [11, Theorem 3.5, p. 35], then implies that

$$w \equiv w(y) = 0$$

in a neighborhood of y.

Let Y be the (open) component of $\Sigma \cap \operatorname{supp}_+ u$ which contains y. Then by a chaining argument one immediately concludes that w vanishes in Y. We claim in fact that Y coincides with $\Sigma \cap \operatorname{supp}_+ u$.

To see this, we note first the relations

$$u = v > 0 \ \text{ in } Y, \qquad u > 0 \ \text{ in } Y^\gamma$$

(where the second inequality follows by reflection), and

$$u = v = 0 \ \text{ on } \partial Y \backslash \Gamma, \qquad u = 0 \ \text{ on } \partial Y^\gamma \backslash \Gamma.$$

There are now two cases to consider.

First, suppose that $u = v = 0$ at all points of ∂Y. It is clear that Y is a component of $\operatorname{supp}_+ u$. Then, by reflection, Y^γ is a second component of $\operatorname{supp}_+ u$. This is impossible, however, since $\operatorname{supp}_+ u$ is assumed to be connected.

In the remaining case, the set $Y_\Gamma = \partial Y \cap \operatorname{supp}_+ u$ is non-empty, in which case we must have $Y_\Gamma \subset \Gamma$. Consider the set

$$Y^* = Y \cup Y^\gamma \cup Y_\Gamma.$$

It is not difficult to see that each point of Y_Γ has a neighborhood consisting entirely of points in Y^*. Hence Y^* is open, and $u > 0$ in Y^*. Let $x \in \partial Y^*$. Obviously either $x \in \partial Y \backslash \Gamma$, or $x \in \partial Y^\gamma \backslash \Gamma$ or $x \in \Gamma$. In the first two cases we have $u(x) = 0$. In the last, we see easily that x is also in ∂Y. On the other hand $x \notin \operatorname{supp}_+ u$, for otherwise $x \in Y_\Gamma$, and x is an interior point of Y^*. Hence $u(x) = 0$. We have thus shown that $u = 0$ on ∂Y^*.

It is easy to see that Y^* is connected. It follows therefore that Y^* is itself a component of $\operatorname{supp}_+ u$. Since $\operatorname{supp}_+ u$ is connected by hypothesis, this means that Y^* *coincides* with $\operatorname{supp}_+ u$. Consequently,

$$Y = \Sigma \cap Y^* = \Sigma \cap \operatorname{supp}_+ u,$$

as claimed.[2]

This being shown, we conclude that $w \equiv 0$ on $\Sigma \cap \mathrm{supp}_+ u$. But also (since $w \geqq 0$, $v \geqq 0$ and $u = 0$ on $\Sigma \backslash \mathrm{supp}_+ u$) we have $w = u = v = 0$ on $\Sigma \backslash \mathrm{supp}_+ u$. Hence $w \equiv 0$ in Σ, contradicting the principal assumption of the lemma. This proves $(3.1)_1$.

Next, let $y \in \Gamma \cap \mathrm{supp}_+ u$. Then $w(y) = 0$, $v(y) = u(y) > 0$ and so w again satisfies the uniformly elliptic differential inequality (3.3) for $x \in \Sigma$ near y. Moreover $w > 0$ in $\Sigma \cap \mathrm{supp}_+ u$ by $(3.1)_1$. Thus the boundary point lemma [11, Lemma 3.4, p. 34] yields

$$\frac{\partial w}{\partial x_1}(y) < 0,$$

which completes the proof. □

Lemma 3.2. *Let u be a C^1 non-negative weak solution of* (2.1) *and let $A(t)$ be degenerate. Assume that* $\mathrm{supp}_+ u$ *is connected, and moreover let*

$$Z \subset \Sigma. \tag{3.4}$$

Then the conclusion of Lemma 3.1 continues to hold.

Proof. This is essentially the same as for Lemma 3.1, requiring only slight modifications due to the degeneracy of A at the origin. Suppose for contradiction that $(3.1)_1$ fails, that is, there exists $y \in \Sigma \cap \mathrm{supp}_+ u$ with $w(y) = 0$. Obviously $Dw(y) = 0$ because $w(y) = 0$ is a minimum on the open set Σ. Moreover $v(y) = u(y) > 0$. Hence by (3.4) we have $Dv(y) \neq 0$, and in turn

$$Du(y) = Dv(y) \neq 0.$$

Consequently w satisfies the uniformly elliptic differential inequality (3.3) near y, and the first part of the proof of Lemma 3.1 thus carries over immediately.

Next we prove $(3.1)_2$. Suppose for contradiction that it fails. Then there exists $y \in \Gamma \cap \mathrm{supp}_+ u$ such that

$$\frac{\partial w}{\partial x_1}(y) = 0.$$

Therefore, since $w = 0$ on Γ, we have $Dw(y) = 0$ and $v(y) = u(y) > 0$. Hence by (3.4) again, there results $Du(y) = Dv(y) \neq 0$. Therefore the proof of Lemma 3.1 carries over, yielding

$$\frac{\partial w}{\partial x_1}(y) < 0,$$

an immediate contradiction. This completes the proof. □

Lemma 3.3. *Let u be a C^1 non-negative weak solution of* (2.1) *and let $A(t)$ be either regular or degenerate. Suppose there exists $y \in \Sigma$ such that $w(y) < 0$. Then there exists $\xi \in \Sigma$ with the properties*

$$w(\xi) < 0, \qquad v(\xi) > \delta,$$

where $\delta > 0$ is defined in condition (ii).

[2] The same point is discussed in [3, Lemma 3.2] with a more involved proof.

Proof. Choose $\varepsilon > 0$ so small that $w(y)+\varepsilon < 0$. Then, since w vanishes at infinity and on Γ, clearly

$$w_\varepsilon = \min\{w + \varepsilon, 0\}$$

is non-positive and has compact support in Σ. By the distribution meaning of solutions, taking the Lipschitzian function w_ε as test function one gets

$$\int_\Sigma [A(|Du|)Du - A(|Dv|)Dv]Dw_\varepsilon = \int_\Sigma [f(u) - f(v)]w_\varepsilon. \tag{3.5}$$

Clearly the left-hand side of (3.5) is positive by the first part of Lemma 2.3, since $Dw_\varepsilon \equiv Dw$ and $Dw \not\equiv 0$ in the set $\{w + \varepsilon < 0\}$, while otherwise $Dw_\varepsilon = 0$ (a.e.).

We claim that there exists $\xi \in \mathrm{supp}_+ (-w_\varepsilon)$ such that $v(\xi) > \delta$. Otherwise, when $w + \varepsilon < 0$, there holds

$$0 \leqq u < v - \varepsilon < \delta,$$

so

$$f(u) - f(v) \geqq 0$$

since $f(u)$ is non-increasing for $u < \delta$. It follows that the right-hand side of (3.5) is non-positive, a contradiction. □

Now we are ready to prove Theorems 1 and 2.

Proof of Theorem 1. *Step 1.* There exists $\gamma_0 > 0$ such that, for $\gamma \geqq \gamma_0$,

$$w \geqq 0 \qquad \text{in } \Sigma_\gamma. \tag{3.6}$$

Indeed, choose γ_0 so that $u(x) \leqq \delta$ for $x_1 \geqq \gamma_0$, which can be done in view of the ground state condition (1.2). Then if (3.6) fails, there would exist y in Σ_γ with $w(y) < 0$. In this case, by Lemma 3.3 there must also be some ξ in Σ_γ with $v(\xi) > \delta$, and in turn $u(\xi^\gamma) > \delta$, contradicting the choice of γ_0.

Step 2. There exists $\gamma \in \boldsymbol{R}$ such that $w \ngeqq 0$ in Σ_γ. Indeed, let $y_0 \in \boldsymbol{R}^n$ be fixed, and choose ν (suitably large negatively) so that

$$u(x) < u(y_0) \tag{3.7}$$

whenever $x \in \Sigma_\nu$, which can be done in view of the ground state condition (1.2). Clearly $y_0 \notin \Sigma_\nu$. Thus there is $y \in \Sigma_\nu$ such that $y^\nu = y_0$. In turn

$$w(y) = u(y) - v_\nu(y) = u(y) - u(y_0) < 0$$

by (3.7). Hence $w \ngeqq 0$ in Σ_γ when $\gamma = \nu$, as required.

Step 3. Let $[\mu, \infty)$ be the maximal (obviously closed) interval for which (3.6) holds for all γ in $[\mu, \infty)$. That such an interval exists follows at once from the result of Step 2.

We show that $w \equiv 0$ in Σ_μ. Suppose the contrary. By Lemma 3.1, one then has

$$w > 0 \text{ in } \Sigma_\mu \cap \mathrm{supp}_+ u, \quad \frac{\partial w}{\partial x_1} < 0 \text{ on } \Gamma_\mu \cap \mathrm{supp}_+ u. \tag{3.8}$$

We now derive a contradiction. Since $[\mu, \infty)$ is maximal, there exist sequences $\{\gamma_i\}$ and $\{x^i\}$ such that (in an obvious notation)

$$\gamma_i \in (-\infty, \mu), \quad x^i \in \Sigma^i = \Sigma_{\gamma_i}, \quad w^i(x^i) < 0, \qquad i = 1, 2, \ldots, \tag{3.9}$$

where $w^i = u - v^i$ and $\lim_{i\to\infty} \gamma_i = \mu$.

By Lemma 3.3 we can assume without loss of generality that $v^i(x^i) > \delta$. In turn, it is not hard to see that the sequence $\{x^i\}$ must be bounded. Moreover, there can be no subsequence which approaches any point ξ in $\overline{\Sigma}_\mu$ where $u(\xi) = 0$. In fact, otherwise, by (3.9) and the assumption that $w \geqq 0$ in Σ_μ, we would have $w(\xi) = 0$. But then $v(\xi) = 0$, an impossibility since $v^i(x^i) > \delta$.

Consequently there is a subsequence of $\{x^i\}$ tending either to a point y in $\Sigma_\mu \cap \operatorname{supp}_+ u$, or to a point z in $\Gamma_\mu \cap \operatorname{supp}_+ u$.

The first case cannot occur, for then $w(y) = 0$ as above, contradicting $(3.8)_1$. For the remaining case, since $w^i = 0$ on Γ_{γ_i}, there must exist for each i a point z^i on the line segment joining x^i to the point $(\gamma_i, x_2^i, \ldots, x_n^i)$ on Γ_{γ_i}, such that

$$\frac{\partial w^i}{\partial x_1}(z^i) \geqq 0.$$

Therefore

$$\frac{\partial w}{\partial x_1}(z) \geqq 0$$

since $z^i \to z$, and $w^i \to w$ in C^1 as $i \to \infty$. This contradicts $(3.8)_2$, so also this case is impossible.

It follows that (3.8) cannot occur and u must be symmetric with respect to the hyperplane $x_1 = \mu$. Moreover, since $\partial w/\partial x_1 = 2\partial u/\partial x_1$ on any hyperplane Γ, the proof shows clearly that

$$\frac{\partial u}{\partial x_1}(x) < 0 \tag{3.10}$$

in $\{x_1 > \mu\} \cap \operatorname{supp}_+ u$. In view of the rotational invariance of (2.1), the same argument of course applies for any assigned direction of the x_1-axis in $\boldsymbol{R}^n$.

The proof is now concluded with the help of

Lemma 3.4. *Let u be a nonnegative C^1 function on $\boldsymbol{R}^n$ satisfying the ground state condition* (1.2). *Suppose that corresponding to each direction $\boldsymbol{e}$ in $SO(n-1)$ there exists a hyperplane $\Gamma_{\boldsymbol{e}}$, with normal $\boldsymbol{e}$, having the properties that $u(x)$ is symmetric across $\Gamma_{\boldsymbol{e}}$ and*

$$Du(x)\cdot \boldsymbol{e} \leqq 0 \qquad \text{whenever } (x - x_0)\cdot \boldsymbol{e} > 0 \text{ for some } x_0 \in \Gamma_{\boldsymbol{e}}. \tag{3.11}$$

Then u is radially symmetric about some origin O in $\boldsymbol{R}^n$, and the corresponding radial function $u(r)$ satisfies $u'(r) \leqq 0$.

Proof. We determine an origin O by the intersection of n orthogonal hyperplanes

$$\Gamma_{e_1}, \Gamma_{e_2}, \ldots, \Gamma_{e_n}.$$

By (3.11) it is evident that O is a maximum point of u. Now let $\boldsymbol{e} \neq \boldsymbol{e}_1, \boldsymbol{e}_2, \ldots, \boldsymbol{e}_n$. We assert that $\Gamma_{\boldsymbol{e}}$ contains O.

If not, then by (3.11) it is easy to see that $u \equiv \text{constant} = u(O)$ on the line segment from O to the nearest point of $\Gamma_{\boldsymbol{e}}$, say $\bar{O}$. By successive reflections in the planes Γ_{e_i} and $\Gamma_{\boldsymbol{e}}$ one sees easily that $u \equiv u(O)$ on a line segment of length $4|O\bar{O}|$ parallel to $O\bar{O}$ and with center at O. By further similar reflections, we get $u \equiv u(O)$ on segments through O of lengths $8|O\bar{O}|$, $16|O\bar{O}|$, etc. But then $u \nrightarrow 0$ as $|x| \to \infty$, which is impossible.

This being shown, to complete the proof of the lemma it is enough to verify that if $x, y \in \boldsymbol{R}^n$ with $|x| = |y| > 0$, then $u(x) = u(y)$. But in this case, let $\boldsymbol{k}$ be the unit vector with direction $\boldsymbol{x} - \boldsymbol{y}$ (we assume $x \neq y$) and $\Gamma_{\boldsymbol{k}}$ the corresponding hyperplane. Since $|x| = |y|$, it is clear that $\Gamma_{\boldsymbol{k}}$ bisects the line segment joining x and y. Hence $u(x) = u(y)$ by the reflection property of $\Gamma_{\boldsymbol{k}}$, as required. □

To finish the proof of Theorem 1, it still must be shown that $u'(r) < 0$ for all $r > 0$ such that $u(r) > 0$. This is obvious however by applying (3.10) to the radial function $u(r)$. Theorem 1 is thus complete.

Note. Lemma 3.4 is given under slightly weaker hypotheses than are actually needed for Theorems 1 and 2, but the general form will be useful later for the proof of Theorem 3.

Proof of Theorem 2. This is the same as for Theorem 1, except that we can use Lemma 3.2 instead of Lemma 3.1. In this case, moreover, the center point O must be the unique positive critical point of u, which will be taken as the origin of coordinates. We note additionally that $\text{supp}_+ u$ can have at most one component (i.e., $\text{supp}_+ u$ is connected) since otherwise Z would contain maximum points in each component, contradicting the hypothesis that it contains exactly one critical point.

Step 1. Exactly the same as that of Theorem 1.

Step 2. Since u has only one maximum point, the origin O, it is clear that $w \ngeqq 0$ in Σ_γ whenever $\gamma < 0$.

Step 3. Let $[\mu, \infty)$ be the maximal (obviously closed) interval for which (3.6) holds for all γ in $[\mu, \infty)$. By Step 2, we have $\mu \geqq 0$. We want to show that $\mu = 0$.

If $\mu > 0$, then $Z \subset \Sigma_\mu$, so by Lemma 3.2 either (3.8) holds or $w \equiv 0$ in Σ_μ. Obviously the latter cannot occur since then $u(x) \equiv u(x^\mu)$, from which we would infer at once that there are critical points of u in the set $\{x_1 > 0\} \cap \text{supp}_+ u$, for example at O^μ, contrary to assumption. The former case also cannot occur (hence $\mu = 0$), by a virtually unchanged argument from that used in proving Theorem 1, but applying Lemma 3.2 instead of Lemma 3.1 since $Z \subset \Sigma_\mu$.

Since we have shown that $\mu = 0$, it now follows that $w \geqq 0$ in Σ_0. Moreover, the proof shows that $\partial u/\partial x_1 < 0$ in $\{x_1 > 0\} \cap \text{supp}_+ u$. Reversing the x_1-axis, one sees that $w \geqq 0$ for the correspondingly reversed half-plane. By reflection this yields $w \leqq 0$ in the original half-space Σ_0, and so finally $w = 0$ in Σ_0. That is, u must be symmetric with respect to the hyperplane $x_1 = 0$.

Since the direction of the x_1-axis is arbitrary, as remarked at the end of the proof of Theorem 1, the last part of the symmetry Lemma 3.4 shows that u must be radially symmetric, of course now with respect to the origin O.

That $u'(r) < 0$ when $r > 0$ and $u(r) > 0$ also follows exactly as before. This completes the proof of Theorem 2. □

Proof of Theorem 1′. We first observe that, for functions f which are Hölder continuous at $u = 0$ and Lipschitz continuous for $u > 0$, any corresponding non-negative weak solution of (2.1) is of class $C^2(\boldsymbol{R}^n)$. (One first shows that $u \in C^{1,\alpha}_{\text{loc}} \cap W^{2,2}_{\text{loc}}$; see Theorem 1 on page 127 and Proposition 1 on page 129 of [19], or Theorem 2 on page 829 of [7], and then applies Theorem 6.3 on page 283 of [12].)

Now let u be a ground state for (1.1), and let Ω be any component of $\text{supp}_+ u$. We define

$$u^* = \begin{cases} u(x), & x \in \Omega, \\ 0, & \text{otherwise.} \end{cases}$$

Clearly u^* is of class C^2 in $\overline{\Omega}$ with $u = Du = 0$ on $\partial\Omega$; hence it is of class $C^2(\boldsymbol{R}^n)$ provided also $D^2 u = 0$ on $\partial\Omega$.

But if $y \in \partial\Omega$, then at this point we have $a_{ij}u_{ij} = -f(u(y)) = -f(0) = 0$ by condition (ii). On the other hand, since y is a minimum of u, either $D^2u(y) = 0$ or there are directions $\boldsymbol{n}$ at y such that $\partial^2 u/\partial n^2 > 0$. The second alternative obviously stands in contradiction with $a_{ij}u_{ij} = 0$. Thus $D^2u(y) = 0$ as required.

The function u^* is obviously a ground state of (1.1), while also $\text{supp}_+ u^* = \Omega$ is connected. Hence, recalling that condition (1.5) is assumed to hold, by Theorem 1′ we see that u^* is radially symmetric and its support is an open ball. It follows now that the support of u consists of a set of open balls on each of which u is radially symmetric about the center.

Finally, there can be at most a *finite* number of such balls, since at any center O' (where u takes a local maximum value) we would have $a_{ij}u_{ij} \leqq 0$, so $f(u(O')) \geqq 0$. However, the case $f(u(O')) = 0$ cannot occur because of uniqueness of the radial initial-value problem. Thus $f(u(O')) > 0$ and $u(O') > \delta$ by (ii). Accordingly, if $\text{supp}_+ u$ had an infinite number of components, then either $u \nrightarrow 0$ as $|x| \to \infty$, and u would *not* be a ground state, or there would be an accumulation point of centers, equally impossible. □

Remark. Theorem 1 can be given a simpler proof when (1.4) holds. Indeed, once it is known that u is positive (Proposition 1), it is no longer necessary to avoid points where $u = 0$ in carrying out the argument of Lemma 3.1. That is, the conclusion (3.1) can then be obtained much more immediately, and even in the stronger form $w > 0$ in Σ, $\partial w/\partial x_1 < 0$ on Γ.

In addition, the integrand in the proof of Lemma 2.3 is never singular, further simplifying the argument, while, finally, the concluding proof of radial symmetry can be shortened (as indicated above) since (3.11) can be used with the strict inequality sign.

$u(r_0) > 0$, then the set $\{Du = 0\}$ contains the sphere $\{|x| = r_0\}$. Therefore we must have $Du(x) \equiv 0$ for $|x| \leqq r_0$, for otherwise, the (open) support of Du would be disconnected, contradicting the main assumption of the theorem. In turn $u \equiv u(r_0)$ for $r \leqq r_0$, and $f(u(r_0)) = 0$ from the equation. Recalling that f is Lipschitz continuous at the value $u(r_0) > 0$, and applying Proposition A2 in the Appendix of [9] to the singular operator A, now shows that $u \equiv u(r_0)$ for all $r \geqq 0$, which is absurd. This completes the proof of Theorem 3.

6. Counter-Examples

In this section, we give two counter-examples showing that the hypothesis of Theorem 2, namely that the set (1.7) contains only a single point, cannot be removed or even significantly weakened.

Example 1. We construct a non-symmetric ground state with compact support. Let $m > 1$ and $s \geqq 0$ be real. For $a \in (0, \min\{1/2, (m-1)/m\})$ to be determined later, we put

$$k = m - 1 - am > 0.$$

Let $r = |x|$ be the radial distance, and define

$$u(x) = u(r) = \begin{cases} 1, & r \leqq s, \\ [1 - (r-s)^{1/a}]^{1/a}, & s \leqq r < s+1, \\ 0, & r \geqq s+1, \end{cases}$$

Since $1/a > 2$, it is clear that $u \in C^{2,\alpha}(\boldsymbol{R}^n)$ for some $\alpha > 0$, and

$$u(s) = 1, \quad u'(s) = u''(s) = 0, \quad u(s+1) = u'(s+1) = u''(s+1) = 0,$$

where $'$ is the derivative with respect to the radius r. Moreover,

$$r = (1 - u^a)^a + s, \quad r \in [s, s+1].$$

By direct calculation, it follows that for $r \in [s, s+1]$

$$u'(r) = -a^{-2}[1 - (r-s)^{1/a}]^{(1-a)/a}(r-s)^{(1-a)/a} = -a^{-2}[u(1-u^a)]^{1-a} < 0.$$

A further calculation then gives

$$(|u'(r)|^{m-1})' = -(k+a)a^{-2m}[u(1-u^a)]^k[1 - (a+1)u^a].$$

Hence, for $r \in [s, s+1]$,

$$\begin{aligned} - \div (|Du|^{m-2}Du) &= (|u'(r)|^{m-1})' + \frac{n-1}{r}|u'(r)|^{m-1} \\ &= -(k+a)a^{-2m}[u(1-u^a)]^k[1-(a+1)u^a] \\ &\quad q + (n-1)a^{2(1-m)}[u(1-u^a)]^{k+a}[(1-u^a)^a + s]^{-1}. \end{aligned}$$

Therefore u satisfies

$$-\mathrm{div}(|Du|^{m-2}\nabla u) = f(u), \quad x \in \boldsymbol{R}^n, \tag{6.1}$$

where

$$\begin{aligned} f(u) = &-(k+a)a^{-2m}[u(1-u^a)]^k[1-(a+1)u^a] \\ &+(n-1)a^{2(1-m)}[u(1-u^a)]^{k+a}[(1-u^a)^a + s]^{-1} \end{aligned}$$

for $u \in [0, 1]$ and is identically zero elsewhere. In particular, if $m \in (1, 2]$, it is easy to see that f is locally Hölder continuous for $u \geqq 0$, with exponent $\alpha = k > 0$, while if $m > 2$, then by taking a sufficiently small one finds that f is at least of class $C^{1,\alpha}$ for some $\alpha > 0$. Finally

$$f'(u) \approx -ka^{-2m}(1-a)(m-1)u^{k-1} < 0 \quad \text{for small } u > 0.$$

Thus we have constructed a (classical) $C^{2,\alpha}$ solution $u(r)$ of the degenerate Laplace equation (6.1), where f is Hölder continuous and, if $m > 2$ and a is suitably small, at least of class $C^{1,\alpha}$. Moreover, $u(x)$ has compact support, f is decreasing for small $u > 0$ and

$$F(u(0)) > 0, \quad f(u(0)) = 0, \quad u(0) = 1.$$

Now we are going to construct the counter-example. For $s \geqq 0$, let $f_s(u)$ be the function and u_s the solution of (6.1) constructed above. Fix $s > 3$. Choose

$$x_0 \neq 0, \qquad |x_0| < 1.$$

and define

$$\hat{u}(x) = \begin{cases} u_0(|x - x_0|) + 1, & |x - x_0| < 1, \\ u_s(|x|), & \text{elsewhere.} \end{cases}$$

Clearly $\hat{u}$ is of class $C^{2,\alpha}$ by our construction. Moreover, by direct verification, $\hat{u}$ is a classical non-symmetric solution of the degenerate Laplace equation (6.1), with

$$f(u) = \begin{cases} f_s(u), & u \in [0, 1], \\ f_0(u-1), & u \in [1, 2], \\ 0, & u > 2. \end{cases}$$

Plainly f is of class C^α if $m \in (1, 2)$ and at least of class $C^{1,\alpha}$ if $m > 2$ and a is suitably small. It is evident finally that $\hat{u}$ has compact support.

When $m > 2$, this construction provides a counterexample to Theorem 2 when the set (1.7) contains more than one point. Moreover, condition (1.7) cannot be replaced by the simple hypothesis that $\text{supp}_+ u$ is connected (as in Theorem 1), since this condition is satisfied by $\hat{u}$.

Conversely, when $m = 2$ (Laplace operator), the example shows that *for the validity of Theorem 1 it is* **not** *enough simply to have f locally Hölder continuous for $u \geqq 0$;* see condition (i) in the Introduction.

Example 2. This example is a non-symmetric *positive* ground state. In analogy to the earlier construction, for $a > 0$, $s \geqq 0$ and $m > 1$, and also with

$$\frac{1}{a} > 2, \frac{m}{m-1}, \frac{n-m}{m-1},$$

we define

$$u(x) = u(r) = \begin{cases} 1, & r \leqq s, \\ [1 + (r-s)^{1/a}]^{-1}, & r \geqq s, \end{cases}$$

where r is the radius. Clearly $u \in C^{2,\alpha}(\boldsymbol{R}^n)$ for some $\alpha > 0$, with

$$u(s) = 1, \quad u'(s) = u''(s) = 0,$$

and

$$u(r) > 0; \quad r = u^{-a}(1-u)^a + s \quad \text{for } r > s.$$

By direct calculation, it follows that for $r > s$

$$u'(r) = -a^{-1}[1 + (r-s)^{1/a}]^{-2}(r-s)^{(1-a)/a} = -a^{-1}u^{a+1}(1-u)^{1-a} < 0,$$

$$(|u'(r)|^{m-1})' = (m-1)a^{-m}u^l(1-u)^k[2u - (a+1)],$$

where

$$k = m - 1 - am > 0, \quad l = m - 1 + am > 0.$$

As before, u is a classical solution of the equation

$$-\operatorname{div}(|Du|^{m-2}Du) = f(u) \tag{6.2}$$

for $r > 0$, where

$$\begin{aligned} f(u) = (|u'(r)|^{m-1})' &= (m-1)a^{-m}u^l(1-u)^k[2u-(a+1)] \\ &+ (n-1)a^{1-m}u^l(1-u)^{k+a}[(1-u)^a + su^a]^{-1} \end{aligned}$$

for $u \in [0, 1]$, and is identically zero elsewhere. Moreover $u(x)$ is strictly positive on $\boldsymbol{R}^n$ and

$$f'(u) \approx -la^{-m}[(m-1)-(n-m)a]u^{l-1} < 0 \quad \text{for small } u > 0.$$

Thus as before, f is Hölder continuous for $u \geqq 0$ and decreasing for small $u > 0$; furthermore, a proper choice of a obviously allows higher differentiability when $m > 2$.

Finally, a construction similar to that above yields a strictly positive non-symmetric solution $\hat{u}$ of (6.2). Of course, the remarks following the first counter-example continue to apply.

7. Singular Ground States

Let O be an arbitrary point in $\boldsymbol{R}^n$, which can be taken as the origin of coordinates. For convenience we put $\boldsymbol{S} = \boldsymbol{R}^n \backslash \{O\}$. A non-negative function u which is of class C^1 on $\boldsymbol{S}$ is called a *singular ground state* of (1.1) if

(1) u is a weak solution of (1.1) on $\boldsymbol{S}$,
(2) u satisfies (1.2), and
(3) $\lim_{x\to 0} u(x) = \infty$.

The previous symmetry results continue to be valid for singular ground states, provided their formulations are appropriately modified. To this end, we make the following further definitions (these are slightly different from those given above, but no confusion should result),

$$\operatorname{supp}_+ u = \{x \in \boldsymbol{S} \mid u(x) > 0\},$$

$$Z = \{x \in S \mid u(x) > 0, \ Du(x) = 0\}.$$

The following results then hold.

Theorem 1 (Singular case). *Let the hypotheses of Theorem 1 be satisfied. Then any singular ground state of* (1.1) *such that* $\text{supp}_+ u$ *is connected must be radially symmetric about* O. *The corresponding radial function* $u(r)$ *obeys* $u'(r) < 0$ *for all* $r > 0$ *such that* $u(r) > 0$.

Theorem 2 (Singular case). *Suppose that the hypotheses of the first sentence of Theorem 2 are satisfied, and let* u *be a singular ground state of* (1.1) *such that the set* Z *is empty. Then the conclusion of Theorem 2 holds.*

Theorem 3 (Singular case). *Theorem 3 holds for singular ground states of* (1.1).

We prove only the result of Theorem 1, the other theorems being treated analogously. The notation of Section 3 will be carried over, with the additional definition

$$\Sigma' = \Sigma'_\gamma = \Sigma_\gamma \backslash \{O\}.$$

Lemma 3.1 is then easily seen to hold for singular ground states, provided that in the statement Σ is replaced by Σ' and $\gamma > 0$ (recall that O is taken as origin). Similarly, Lemma 3.3 remains valid when $\gamma > 0$ and Σ is replaced by Σ', the proof being unchanged since w_ε has compact support in Σ'.

Turning to the main proof of Theorem 1, we see that Step 1 clearly carries over unchanged, while obviously Step 2 applies for any $\gamma < 0$.

We assert that $\mu = 0$ in Step 3. Otherwise, if $\mu > 0$, we reach a contradiction exactly as before. (Note that the sequence $\{x^i\}$ necessarily has the property that $x^i \in \Sigma'_{\gamma_i}$, since w^i is undefined at O.) It follows that $w \geqq 0$ in Σ_0. Then by the reflection procedure at the end of the proof of Theorem 2 we obtain $w \equiv 0$ in Σ_0. Theorem 1 now follows at once, where of course $u(r) \to \infty$ as $r \to 0$.

Results corresponding to the auxilliary conclusions in the introduction can be left to the interested reader.

Acknowledgements. The research of H. Zou was supported in part by an NSF Grant DMS-9622937 and an Alabama EPSCoR grant.

References

1. Badiale, M. & E. Nabana, *A note on radiality of solutions of p-Laplacian equation*, Applicable Anal., **52** (1994), 35–43.
2. Berestycki, H. & L. Nirenberg, *On the method of moving planes and the sliding method*, Boletim da Soc. Brasileira de Mat. Nova Ser. **22** (1991), 1–37.
3. Cortázar, C., M. Elgueta & P. Felmer, *On a semilinear ellptic problem in* $\mathbf{R}^n$ *with a non-Lipschizian nonlinearity*, Adv. in Diff. Equations, **1** (1996), 199–218.
4. Cortázar, C., M. Elgueta & P. Felmer, *Symmetry in an elliptic problem and the blow-up set of a quasilinear heat equation*, Comm. Part. Diff. Eqs., **21** (1996), 507–520.
5. Damascelli, L., *Comparison theorems for some quasilinear degenerate elliptic operators and applications to symmetry and monotonicity results*, to appear, Ann. Inst. H. Poincaré.
6. Damascelli, L., F. Pacella & M. Ramaswamy, *Symmetry of ground states of p-Laplace equations via the moving plane method*, Arch. Rational Mech. Anal. **148** (1999), 291–316

7. DIBENEDETTO, E., $C^{1+\alpha}$ *local regularity of weak solutions of degenerate elliptic equations*, Nonlinear Anal., **7** (1983), no.8, 827–850.
8. FRANCHI, B. & E. LANCONELLI, *Radial symmetry of the ground states for a class of quasilinear elliptic equations*, in 'Nonlinear diffusion equations and their equilibrium states, Springer-Verlag, New York, 1988.
9. FRANCHI, B., E. LANCONELLI & J. SERRIN, *Existence and uniqueness of nonnegative solutions of quasilinear equations in R^n*. Adv. Math., **118** (1996), 177–243.
10. GIDAS, B., W.-M. NI & L. NIRENBERG, *Symmetry of positive solutions of nonlinear elliptic equations in R^n*, Adv. Math. Supp. Stud., **7A** (1981), 369–403.
11. GILBARG, D. & N. TRUDINGER, *Elliptic Partial Differential Equations of Second Order*, 2nd edition, Springer-Verlag, 1983.
12. LADYZHENSKAYA, O. A. & N. N. URAL'TSEVA, *Linear and Quasilinear Elliptic Equations*, Academic Press, New York, 1968.
13. LI, C.-M., *Monotonicity and symmetry of solutions of fully nonlinear elliptic equations on unbounded domains*, Comm. Part. Diff. Eqs., **16** (1991), 585–615.
14. LI, Y. & W.-M. NI, *Radial symmetry of positive solutions of nonlinear elliptic equations in R^n*, Comm. Part. Diff. Eqs., **18** (1993), 1043–1054.
15. PELETIER, L. A. & J. SERRIN, *Uniqueness of non-negative solutions of semilinear equations in $\mathbf{R}^n$*, J. Diff. Eqs., **61** (1986), 380–397.
16. PUCCI, P. & J. SERRIN, *Uniqueness of ground states for quasilinear elliptic operators*, Indiana Univ. Math. J., (1998).
17. PUCCI, P., J. SERRIN & H. ZOU, *A strong maximum principle for quasilinear elliptic equations*, to appear, J. Math. Pures Appl., 1999.
18. SERRIN, J., *A symmetry problem in potential theory*, Arch. Rational Mech. Anal., **43** (1971), 304–318.
19. TOLKSDORF, P., *Regularity for a more general class of quasilinear elliptic equations*, J. Diff. Eqs., **51** (1984), 126–150.
20. VAZQUEZ, J.-L., *A strong maximum principle for some quasilinear elliptic equations*, Appl. Math. Opt., **112** (1984), 191–202.
21. ZIEMER, W., *Weakly Differentiable Functions*, Springer-Verlag, New York, 1989.

Department of Mathematics
University of Minnesota, Minneapolis

and

Department of Mathematics
University of Alabama at Birmingham

(Accepted September 21, 1998)

Stanislav Pohozaev, Patrizia Pucci, James Serrin and Enzo Mitidieri, Trieste, Italy, 1995.

Ermanno Lanconelli, James Serrin and Bruno Franchi, Bologna, Italy, 1996.

Filippo Gazzola and James Serrin, Frontenac State Park, MN, 1998.

I.8. Ground States – Existence

Filippo Gazzola

Mathematics Subject Classification (2010). 35B33, 35J60, 46E35.
Keywords. Ground states, quasilinear elliptic equations, radial solutions.

1. Ground states

Let $f \in C^1(\mathbb{R})$ and consider the following autonomous semilinear elliptic equation

$$-\Delta u = f(u) \qquad \text{in } \mathbb{R}^n. \tag{1}$$

Physically meaningful solutions to (1) have finite energy and, if they exist, one is interested in the least energy solution. In many significant cases, least energy solutions exist not only, but they are positive and radially symmetric with respect to some point (the origin). With a slight abuse, mathematicians call a positive radial solution to (1) a **ground state**.

Starting with the paper of Strauss [27], equation (1) has been studied intensively. Berestycki–Lions [2, 3] used variational methods to prove the existence of finite energy solutions to (1). Radial solutions to (1) solve an ODE and a *shooting method* can be used, see [4]. In this setting a major contribution was given by James Serrin, in particular with the seminal papers [17, 18]. When f is a positive power-like function, Ni–Serrin [18][1] prove existence and nonexistence results for ground states to

$$\operatorname{div}\big[A(|Du|)Du\big] + f(u) = 0 \qquad \text{in } \mathbb{R}^n$$

where $A \in C^1(\mathbb{R}^+)$ satisfies $\int_0^t sA(s)\,ds \geq ct^2A(t)$ for some $c > 0$ so that the p-Laplacian and the mean curvature operator are included. A crucial tool for these results is a Pohožaev-type variational identity [18, Section 2] which also gives information on the behavior of solutions (when they exist!). If $f(u) = u^p$ then [18] yields sharp existence results for ground states exhibiting the critical exponents for which existence fails. These results were extended to more general elliptic operators and sources f by Serrin and his coauthors [9, 12, 13, 14, 19, 20, 21, 22, 23, 24] and by many other mathematicians. Among others we mention [6, 7, 8, 11, 10, 15, 16].

Another interesting contribution for a slightly different equation is the paper by Peletier–Serrin–Zou [20]. It was motivated by previous work of Chipot–Weissler [5] who were interested in blow-up phenomena for the parabolic equation

$$u_t = \Delta u - |\nabla u|^q + \lambda |u|^{p-1}u, \qquad \lambda > 0, \qquad \text{for } (x,t) \in \mathbb{R}^n \times (0,\infty). \tag{2}$$

[1] Quite surprisingly, the paper by Ni–Serrin [18] does not appear in the database of MathSciNet.

P. Pucci et al. (eds.), *James Serrin. Selected Papers: Volume 1*, Contemporary Mathematicians,
DOI 10.1007/978-3-0348-0845-3_8,

The source term $|u|^{p-1}u$ tends to cause blow-up while the gradient term $-|\nabla u|^q$ has a damping effect against blow-up. So, it appears natural to study the competition between these two terms for different values of the parameters. A crucial role for the study of blow-up phenomena for (2) is played by stationary solutions, or ground states. Peletier–Serrin–Zou [20] study the one-dimensional case, that is

$$u'' - |u'|^q + \lambda u^p = 0 \qquad \text{on } \mathbb{R}. \tag{3}$$

Their main results are existence and uniqueness of solutions to (3) which vanish at infinity under suitable combined assumptions on p, q, and λ. The results in [20] gave rise to several subsequent developments, see in particular [1, 25, 26].

References

[1] J.P. Bartier, *Global behavior of solutions of a reaction-diffusion equation with gradient absorption in unbounded domains.* Asymptot. Anal. **46** (2006), 325–347.

[2] H. Berestycki, P.L. Lions, *Nonlinear scalar field equations. I. Existence of a ground state.* Arch. Ration. Mech. Anal. **82** (1983), 313–345.

[3] H. Berestycki, P.L. Lions, *Nonlinear scalar field equations. II. Existence of infinitely many solutions.* Arch. Ration. Mech. Anal. **82** (1983), 347–375.

[4] H. Berestycki, P.L. Lions, L.A. Peletier, *An ODE approach to the existence of positive solutions for semilinear problems in* $\mathbb{R}^N$. Indiana Univ. Math. J. **30** (1981), 141–157.

[5] M. Chipot, F. Weissler, *Some blow-up results for a nonlinear parabolic equation with a gradient term.* SIAM J. Math. Anal. **20** (1989), 886–907.

[6] D.G. De Figueiredo, J.V. Gonçalves, O.H. Miyagaki, *On a class of quasilinear elliptic problems involving critical exponents.* Commun. Contemp. Math. **2** (2000), 47–59.

[7] A. Ferrero, F. Gazzola, *On subcriticality assumptions for the existence of ground states of quasilinear elliptic equations.* Adv. Diff. Eq. **8** (2003), 1081–1106.

[8] A. Ferrero, F. Gazzola, *Asymptotic behavior of ground states of quasilinear elliptic problems with two vanishing parameters. III.* J. Diff. Eq. **198** (2004), 53–90.

[9] B. Franchi, E. Lanconelli, J. Serrin, *Existence and uniqueness of nonnegative solutions of quasilinear equations in* $\mathbb{R}^n$. Adv. Math. **118** (1996), 177–243.

[10] M. García-Huidobro, R. Manásevich, E. Mitidieri, C. Yarur, *Existence and nonexistence of positive singular solutions for a class of semilinear elliptic systems.* Arch. Ration. Mech. Anal. **140** (1997), 253–284.

[11] F. Gazzola, *Critical exponents which relate embedding inequalities with quasilinear elliptic problems.* In: Dynamical systems and differential equations (Wilmington, NC, 2002). Discrete Contin. Dyn. Syst. 2003, suppl., 327–335.

[12] F. Gazzola, B. Peletier, P. Pucci, J. Serrin, *Asymptotic behavior of ground states of quasilinear elliptic problems with two vanishing parameters. II.* Ann. Inst. H. Poincaré ANL **20** (2003), 947–974.

[13] F. Gazzola, J. Serrin, *Asymptotic behavior of ground states of quasilinear elliptic problems with two vanishing parameters.* Ann. Inst. H. Poincaré ANL **19** (2002), 477–504.

[14] F. Gazzola, J. Serrin, M. Tang, *Existence of ground states and free boundary problems for quasilinear elliptic operators.* Adv. Diff. Eq. **5** (2000), 1–30.

[15] E. Mitidieri, S. Pohožaev, *Towards a unified approach to nonexistence of solutions for a class of differential inequalities.* Milan J. Math. **72** (2004), 129–162.

[16] E. Montefusco, P. Pucci, *Existence of radial ground states for quasilinear elliptic equations.* Adv. Diff. Eq. **6** (2001), 959–986.

[17] W.M. Ni, J. Serrin, *Nonexistence theorems for quasilinear partial differential equations.* Rend. Circ. Mat. Palermo **2** Suppl. No. 8 (1985), 171–185.

[18] W.M. Ni, J. Serrin, *Existence and nonexistence theorems for ground states for quasilinear partial differential equations. The anomalous case.* Accad. Naz. Lincei, Atti dei Convegni **77** (1986), 231–257.

[19] L. Peletier, J. Serrin, *Ground states for the prescribed mean curvature equation.* Proc. Amer. Math. Soc. **100** (1987), 694–700.

[20] L.A. Peletier, J. Serrin, H. Zou, *Ground states of a quasilinear equation.* Differential Integral Equations **7** (1994), 1063–1082.

[21] P. Pucci, J. Serrin, *Uniqueness of ground states for quasilinear elliptic operators.* Indiana Univ. Math. J. **47** (1998), 501–528.

[22] J. Serrin, M. Tang, *Uniqueness of ground states for quasilinear elliptic equations.* Indiana Univ. Math. J. **49** (2000), 897–923.

[23] J. Serrin, H. Zou, *Existence and nonexistence results for ground states of quasilinear elliptic equations.* Arch. Ration. Mech. Anal. **121** (1992), 101–130.

[24] J. Serrin, H. Zou, *Symmetry of ground states of quasilinear elliptic equations.* Arch. Ration. Mech. Anal. **148** (1999), 265–290.

[25] P. Souplet, *Recent results and open problems on parabolic equations with gradient nonlinearities.* Electron. J. Differential Equations No. 10 (2001), 19 pp. (electronic).

[26] P. Souplet, S. Tayachi, *Blowup rates for nonlinear heat equations with gradient terms and for parabolic inequalities.* Colloq. Math. **88** (2001), 135–154.

[27] W.A. Strauss, *Existence of solitary waves in higher dimensions.* Comm. Math. Phys. **55** (1977), 149–162.

Filippo Gazzola
Dipartimento di Matematica
Politecnico di Milano
e-mail: `filippo.gazzola@polimi.it`

Accad. Naz. Lincei, Atti dei Convegni, 77 (1986), 231–257. **Reprint I.8.[i]**

WEI–MING NI* and JAMES SERRIN*

EXISTENCE AND NONEXISTENCE THEOREMS FOR GROUND STATES OF QUASILINEAR PARTIAL DIFFERENTIAL EQUATIONS. THE ANOMALOUS CASE

In an earlier paper [6] we have considered a class of non–existence theorems for *ground state solutions* of the partial differential equations

$$\Delta u + f(u) = 0 \tag{1.1}$$

and

$$\operatorname{div}(A(|Du|)\,Du) + f(u) = 0\,, \tag{1.2}$$

where $f(u)$ and $A(p)$ are given functions, and $f(0) = 0$. (By a ground state we mean a positive solution u in R^n, $n \geq 2$, which tends to zero at ∞).

Our earlier paper was devoted primarily to the situation when $f(u) < 0$ for u positive and small, which one may call the *normal case.* Since it is well known that ground states cannot exist when $f'(0) > 0$, there remains the delicate but interesting possibility that $f'(0) = 0$ and $f(u) \geq 0$ near $u = 0$. We shall call this the *anomalous case.*

Because of the intricacy of this case we find it necessary to make some fairly explicit assumptions concerning the asymptotic behavior of the function $f(u)$ for u near 0. We prove first that if

$$f(u) \geq \text{Pos. Const. } u^q,$$

for u positive and small, where q is a positive constant less than or equal to $n/(n-2)$, then (1.1) can admit no radial solution $u = u(r)$ which is positive for $r \geq r_0$ and tends to zero as $r = |x| \to \infty$. Of course, this result also shows that no ground state exists when $f(u) \geq$ Pos. Const. u^q and $q \leq n/(n-2)$.

If $q > n/(n-2)$ then positive radial solutions of (1.1) are possible in exterior domains $r \geq r_0 > 0$, as shown by the example

$$u = \text{Pos. Const. } r^{-2/(q-1)} \quad \text{when } f(u) = u^q$$

(the particular constant which is necessary here is of no immediate importance). On the other hand, when $f(u) = u^q$ and

$$\frac{n}{n-2} < q < \frac{n+2}{n-2}$$

(*) University of Minnesota - School of Mathematics - 127 Vincent Hall - 206 Church Street S.E. - MINNEAPOLIS, Minnesota 55455 (U.S.A.).

then there are no ground states open to (1.1), a result which can be obtained by scaling arguments. In Section 3 we generalize this result to a large class of functions $f(u)$ which obey the inequality $f(u) \geq$ Pos. Const. u^q for u small.

The converse situation occurs when $f(u) > 0$ satisfies the condition

$$(\alpha + 1) F(u) \leq uf(u) \tag{1.3}$$

for some constant $\alpha \geq (n+2)/(n-2)$; here we have put

$$F(u) = \int_0^u f(t)\, dt.$$

(Upon integration the condition (1.3) yields $F(u) \leq$ Const. $u^{1+\alpha}$ for small u; then (1.3) is compatible with the requirement $f(u) \geq$ Pos. Const. u^q only when $q \geqslant \alpha$). In Theorem 3.2 we show that when (1.3) holds there *always* are ground states of (1.1), a conclusion obtained under slightly stronger conditions in [2] and [5].

The results described above in many ways constitute a reversal of the conclusions obtained earlier for the normal case. That is, in the anomalous case it is exponents *greater* than $(n+2)/(n-2)$ which make ground states possible, while for the normal case it is exponents *less* than $(n+2)/(n-2)$ which serve this purpose (see [6], Theorems 3.2 and 3.3, and [1]).

The case $f(u) \sim$ Pos. Const. $u^{(n+2)/(n-2)}$ lies at the borderline of existence and non-existence. Here small perturbations can seriously affect the situation. To illustrate this, observe that if

$$f(u) = u^{(n+2)/(n-2)} + \lambda u^t \tag{1.4}$$

where $\lambda > 0$ and $t > (n+2)/(n-2)$, then it is always possible to find ground states (Theorem 3.2), while when

$$f(u) = u^{(u+2)/(n-2)} - \lambda u^t \tag{1.5}$$

it is not. The latter result can be obtained as a fairly easy corollary of Theorem 5.13 of [3]. In Theorem 4.1 we give another proof of this result, and generalize it as well to a wider class of functions f.

From a heuristic point of view the existence of ground states corresponding to the function (1.4) is a reflection of the fact that the "exponent" here can be considered slightly larger than $(n+2)/(n-2)$, while conversely in the case (1.5) the "exponent" can be considered slightly smaller than $(n+2)/(n-2)$.

It is one of the main purposes of this paper to extend these results to the general quasilinear equation (1.2). For simplicity, we shall consider classical solutions of (1.2) and thus assume throughout that $A(p)$ *is continuously differentiable for* $p > 0$. Then, when $A(p)$ is positive for $p > 0$ and bounded near $p = 0$ we prove direct analogues of both the existence and non-existence theorems of Section 3. Quite surprisingly, the exponent $n/(n-2)$ shows up exactly as before as the borderline constant for non-existence of positive solutions in exterior domains (see the remarks at the end of section 2). The generality of results which are possible here can perhaps best be illustrated by noting that Theorems 5.1 and 5.2 apply even to the mean curvature equation

$$\operatorname{div}\left(\frac{Du}{\sqrt{1+|Du|^2}}\right)+f(u)=0\,,$$

while at the same time the hypotheses even include certain equations which are *not* elliptic.

In the final section we consider the degenerate elliptic equation

$$\operatorname{div}\,(|\,Du\,|^{m-2}\,Du)+f(u)=0\,,$$

where m is a constant greater than 1. Here it is possible to find analogues for every result earlier obtained for equation (1.1). These are given in Section 6 without proof.

Before turning to the proofs, we also call the reader's attention to the asymptotic estimate obtained in Theorem 2.2, namely that if

$$f(u)\geq \text{Pos. Const. } u^q$$

for u near 0, where q is a constant greater than $n/(n-2)$, then any positive radial solution of (1.1) - or of (1.2) if $A(p)$ is positive and bounded - which is defined for $r\geq r_0$ and tends to zero as $r\to\infty$ must have the asymptotic behavior

$$u=0(r^{-2/(q-1)})\quad\text{as } r\to\infty.$$

Our methods are predicated throughout on straightforward partial integrations together with application of a generalized Pohozaev identity obtained in our first paper ([6], section 2). For convenience we record this identity in the following form.

Proposition. *Let $u=u(r)$ be a radial solution of equation* (1.2) *with domani $0\leq r<\infty$. Then for any real constant a we have the identity*

$$\begin{aligned}&\frac{d}{dr}\left[r^n\left\{\int_0^p \rho\,E\,d\rho+F(u)+aA(p)\frac{uu'}{r}\right\}\right]\\ (1.6)\qquad &=r^{n-1}\left[n\int_0^p \rho\,E\,d\rho+(a+1-n)\,A(p)\,p^2+nF(u)-auf(u)\right].\end{aligned}$$

Here $u'=du/dr$, $p=|\,u'\,|$, and

$$E(p)=A(p)+p\,dA(p)/dp.$$

For the Poisson equation (1.1) it is clear that $A\equiv E\equiv 1$, so that (1.6) takes the simpler form

$$(1.7)\;\frac{d}{dr}\left[r^n\left\{\frac{1}{2}|\,u'\,|^2+F(u)+a\frac{uu'}{r}\right\}\right]=r^{n-1}\left[nF(u)-auf(u)+(a+1-\frac{n}{2})|\,u'\,|^2\right].$$

This relation will be the principal tool needed in Sections 3 and 4. In Sections 5 and 6 we shall require the general result (1.6).

Note. Equation (1.1) in the radial case takes the form

$$u'' + \frac{n-1}{r} u' + f(u) = 0.$$

Making the change of variables $y = r^{n-2} u(r)$ and $t = r^{n-2}$, one finds that the function $t \to y(t)$ satisfies the generalized Emden Fowler equation

$$\ddot{y} + t^{\alpha} f(y / t) = 0, \tag{1.8}$$

where dots denote differentiation with respect to t and where $\alpha = (4 - n) / (n - 2)$. This equation is the subject of an extensive literature, in which some elements of our results (especially for the special case $f(u) = u^q$ when (1.8) reduces to the Emden–Fowler equation $\ddot{y} + t^{\alpha - q} y^q = 0$) can be found. * At the same time, because of the technical nature of this literature it is not generally an easy task to translate the conclusions to our circumstances. Even more, these studies are in the main concerned with the behavior of solutions for large t (large r) rather than with the globally regular solutions considered here. Thus on the whole it has seemed best to treat equation (1.1) directly without resort to the Emden equation. Finally, our interest is equally in the quasilinear equation (1.2), and for this case there is in fact no literature corresponding to the Emden equation; indeed, it is precisely the methods which we use for the *direct* treatment of (1.1) which turn out to be crucial in the corresponding study of (1.2).

2. *Asymptotic Behavior for the Anomalous Case.*

In this section we present several preliminary results which can be obtained directly without using the general identities (1.6) or (1.7).

Since we shall be dealing throughout with positive solutions of (1.1) or (1.2) it will be tacitly assumed that any conditions stated on the function $f(u)$ are always intended to hold only for the range $u > 0$.

Our first result shows that in general one cannot have radial ground states of (1.1) in the anomalous case unless $n > 2$. The proof of this result, which is a simple variant of the demonstration of Theorem 2.1 below, can be left to the reader.

THEOREM 2.0. *Suppose that $n = 2$ (or $n \leq 2$ if n is considered as a real parameter), and assume that $f \geq 0$ for all u near* 0. *Then there can be no positive radial solutions of* (1.1) *defined for $r \geq r_0$ and tending to zero as $r \to \infty$.*

In what follows we can therefore suppose that $n > 2$, a condition which we shall tacitly suppose throughout the sequel, unless otherwise mentioned.

(*) Fowler's work on the Emden equation, Quarterly J. Mathematics, 1931, is still interesting, though his results in Table I are partly in error. A recent publication, carriyng the flavor of much modern work, is M. Kwong and J. Wong, Proc. Amer. Math. Soc. 1983.

The following two theorems, which concern the asymptotic behavior at ∞ of radial solutions of (1.1), are essentially due to McLeod [4] and Ni [5]. We include their proofs here both for the sake of completeness as well as to make clear our later extensions to the quasilinear case (1.2).

THEOREM 2.1. *Suppose that $f(u) \geq 0$ for u near 0. Let u be a positive radial solution of* (1.1) *defined for $r \geq r_0$ and tending to zero as $r \to \infty$. Then*

$$u \geq \text{Pos. Const. } r^{2-n} \quad \textit{and} \quad u' \leq \text{Neg. Const. } r^{1-n}$$

for all sufficiently large r.

Proof. For a radial solution $u = u(r)$ we can write (1.1) in the form

$$(r^{n-1} u')' + r^{n-1} f = 0 . \tag{2.1}$$

Since $u \to 0$ as $r \to \infty$ one can suppose without loss of generality that $f(u) \geq 0$ for $u = u(r)$, $r \geq r_0$. Hence in turn

$$(r^{n-1} u')' \leq 0 \text{ for } r \geq r_0$$

and $r^{n-1}u'$ is a decreasing function. Since u' obviously cannot be everywhere non-negative, it follows that

$$r^{n-1} u' \to \text{negative limit (possibly } -\infty) \tag{2.2}$$

as $r \to \infty$, hence $u' < 0$ for all suitably large r.

In particular for all large r, we have

$$r^{n-1} u' \leq -\mathrm{C}$$

where C is some positive constant. Integrating this relation from any fixed value r to ∞ yields $u \geq [\mathrm{C} / (n-2)]\, r^{2-n}$, completing the proof.

THEOREM 2.2. *Suppose that $f(u) \geq$ Pos. Const. u^q for u near 0, where q is a positive constant.*

If $q \leq n / (n-2)$ there can exist no positive radial solution of (1.1) *defined for $r \geq r_0$ and tending to zero as $r \to \infty$.*

If $q > n / (n-2)$ then any positive radial solution of (1.1) *which is defined for $r \geq r_0$ and tends to zero as $r \to \infty$ has the asymptotic behavior*

$$u = 0\,(r^{-2/(q-1)}) \quad \textit{as } r \to \infty.$$

Proof. Let $u = u(r)$ be a positive radial solution defined for $r \geq r_0$ and tending to zero as $r \to \infty$. Without loss of generality we can suppose that, for $r \geq r_0$,

$$f(u) \geq \lambda u^q , \quad u' < 0 , \tag{2.3}$$

where $\lambda > 0$. Integrating (2.1) from r_0 to r gives

$$r^{n-1} u'(r) - r_0^{n-1} u'(r_0) = - \int_{r_0}^{r} s^{n-1} f(u(s))\, ds .$$

Hence by (2.3)

$$\int_{r_0}^{r} s^{n-1} f(u(s))\,ds \geq \int_{r_0}^{r} \lambda s^{n-1} [u(s)]^q\,ds$$

$$\geq \lambda [u(r)]^q \int_{r_0}^{r} s^{n-1}\,ds = \frac{\lambda}{n}(r^n - r_0^n)\,[u(r)]^q,$$

where we have used the fact that $u(s) > u(r)$ for $s < r$. Choosing $r \geq 2r_0$ and noting again that $u'(r_0) < 0$, it now follows that

$$-r^{n-1} u'(r) > \frac{\lambda}{n} n(1 - 2^{-n})\, r^n [u(r)]^q,$$

that is

$$(2.4) \qquad -u^{-q} u' > Cr \quad (r \geq 2r_0)$$

where $C = \lambda(1 - 2^{-n})/n$.

Integrating (2.4) from $s_0 = 2r_0$ to r $(> s_0)$ we get

$$\frac{1}{q-1}\{[u(r)]^{1-q} - [u(s_0)]^{1-q}\} \geq \frac{1}{2} C(r^2 - s_0^2),$$

provided $q \neq 1$ (if $q = 1$, the proof is similar).

When $q < 1$ an immediate contradiction results by letting $r \to \infty$. If $q > 1$ then for all sufficiently large r we have

$$(2.5) \qquad u^{q-1} \leq \frac{3}{(q-1)C} r^{-2}.$$

When $q > n/(n-2)$ this is the required estimate.

Next suppose that $1 < q < n/(n-2)$. In this case (2.5) contradicts the conclusion of Theorem 2.1, namely that $u \geq$ Pos. Const. r^{2-n}. Hence no solution of the type under consideration can exist. It remains to show the same result when $q = n/(n-2)$.

In this case, using Theorem 2.1 and the hypothesis $f(u) \geq \lambda u^q = \lambda u^{n/(n-2)}$, we see from Theorem 2.1 that

$$f(u) \geq \text{Pos. Const. } r^{-n}$$

for all sufficiently large r. Hence by (2.1)

$$(r^{n-1} u')' = -r^{n-1} f \leq -Cr^{-1}$$

where $C > 0$. By integration from s to r, where s is suitably large and $r > s$, we get

$$r^{n-1} u'(r) - s^{n-1} u'(s) \leq C \log(s/r).$$

It follows that $r^{n-1} u'$ tends to *negative infinity* as $r \to \infty$. By the argument used at the end of the proof of Theorem 2.1 this in turn shows that $r^{n-2} u \to \infty$ as $r \to \infty$. On the other hand, since $q = n/(n-2)$ in the present case, the inequality (2.5) shows that $u = 0(r^{n-2})$. The resulting contradiction establishes that there can be no solutions of the type under consideration even when $q = n/(n-2)$, completing the proof of the theorem.

Remark. The example $u = \text{Pos. Const. } r^{-2/(q-1)}$ when $f(u) = u^q$, noted in the introduction, shows that the asymptotic estimate $u = 0(r^{-2/(q-1)})$ in the second part of the theorem is best possible.

The above results also remain valid for the quasilinear equation (1.2) *provided that* $A(p)$ *is continuously differentiable for* $p > 0$, *bounded for* p *near* 0, *and satisfies*

$$A(p) > 0 \qquad \textit{for all } p > 0. \tag{2.6}$$

This assumption is a natural one, and is much weaker than the requirement that (1.2) be elliptic.

Indeed let us suppose that $u = u(r)$ is a positive radial solution of (1.2) defined for $r \geq r_0$ and tending to zero as $r \to \infty$. Then in place of (2.1) we now have the corresponding equation

$$(r^{n-1} u' A(|u'|))' + r^{n-1} f = 0 \tag{2.7}$$

(see [6], equation (2.2) and (2.3)). If $f \geq 0$ for u near 0 it then follows that $r^{n-1}u' A(|u'|)$ is ultimately decreasing, and in turn, since $A > 0$, that

$$r^{n-1} u' A(|u'|) \leq -C \tag{2.8}$$

for sufficiently large r. Consequently $u' < 0$ for all sufficiently large r.

We show next that in fact $u' \to 0$ as $r \to 0$. Writing (1.2) in the alternate form (see [6], equation (2.2))

$$E(|u'|) u' u'' + (n-1) A(|u'|) \frac{u'^2}{r} + f(u) u' = 0,$$

we obtain by integration from r_1 to $r > r_1$

$$\int_{p_1}^{p(r)} \rho E(\rho) d\rho + (n-1) \int_{r_1}^{r} A \frac{u'^2}{r} dr + F(u) - F(u_1) = 0 \tag{2.9}$$

where $u_1 = u(r_1)$, $p_1 = |u'(r_1)|$ and $p(r) = |u'(r)|$.

Now define

$$H(p) = \int_0^p \rho E(\rho) d\rho, \quad p \geq 0.$$

Since $E = A + pA_p$ we have also

$$H(p) = p^2 A(p) - \int_0^p \rho A(\rho) d\rho \tag{2.10}$$

so that H is continuous for $p \geq 0$ and $H(0) = 0$. Moreover one cannot have $H(p) \leq 0$ for all small p, as can be seen by treating (2.10) as a differential inequality for $\int \rho A d\rho$ and using the boundedness of A near $p = 0$.

Let $\varepsilon > 0$ be fixed arbitrarily. Then there exists a value $\varepsilon' < \varepsilon$ such that $h = H(\varepsilon') > 0$. Choose $\delta, 0 < \delta < \varepsilon'$, so that

$$H(p) < h/2 \quad \text{for } p \in (0, \delta).$$

Now let $r \to \infty$ in (2.9). Since $A(p) > 0$ it follows that the second integral is convergent (possibly to positive infinity), and in turn

$$H(p) \to \text{limit (possibly negative infinity)},$$

where $p = p(r) = |u'(r)|$.

Suppose this limit is greater than $h/2$. Then $|u'(r)| \geq \delta$ for all sufficiently large r, which is impossible. Hence the limit is $\leq h/2$. Let $\bar{r}$ be so large that $H(p) < h$ for all $r \geq \bar{r}$. Then since $p(r)$ is continuous, and certainly has some values $< \varepsilon'$ for r tending to infinity, we see that in fact $p < \varepsilon'$ for all $r \geq \bar{r}$. But $\varepsilon' < \varepsilon$ and ε in turn is arbitrary, whence we obtain the desired conclusion $u' \to 0$ as $r \to \infty$.

Now return to the inequality (2.8). Because $A(p)$ is locally bounded and $u' \to 0$ as $r \to \infty$ we have

$$A(|u'|) \leq \Lambda, \quad r \geq r_0,$$

for some positive constant Λ. Consequently

$$r^{n-1} u' \leq -C/\Lambda$$

for all sufficiently large r, this being enough to obtain the conclusions of Theorems 2.0 and 2.1.

For Theorem 2.2, in place of (2.4) we find easily that

$$-u^q u' A(|u'|) > Cr$$

for sufficiently large r. Thus also $-u^q u' > (C/\Lambda) r$ and the rest of the proof is the same as before.

3. *Existence and Non-existence Theorems for the Anomalous Case*

Here we shall apply the identity (1.6) to obtain both existence and non-existence theorems for ground states of (1.1) when $f(u) > 0$ for u near 0. As noted in Section 2 one can always suppose that $n > 2$.

THEOREM 3.1. *Suppose that*

$$f(u) \geq \text{Pos. Const. } u^q, \quad F(u) \leq \text{Pos. Const. } u^{\bar{q}+1} \tag{3.1}$$

for all sufficiently small u, where q and $\bar{q}$ are positive exponents satisfying the condition

$$\frac{q-1}{2} n - 1 < \bar{q} \leq q. \tag{3.2}$$

Suppose also that for some constant $\alpha \leq (n+2)/(n-2)$ we have

$$(3.3) \qquad (\alpha + 1)\, F(u) \geq u f(u)$$

for all $u > 0$. *Then there cannot exist any positive radial solutions of equation* (1.1) *in* R^n *which tend to zero as* $r \to \infty$.

Remark. By integrating the condition (3.3), considered as a differential inequality for the function F, we find that $F(u) \geq$ Pos. Const. $u^{1+\alpha}$ for all sufficiently small u, whence the constant α cannot be less than $\bar{q}$. Note also that condition (3.2) entails the restriction $q < (n+2)/(n-2)$.

Proof of Theorem 3.1. We can assume that $q > n/(n-2)$, for otherwise the result follows at once from the first part of Theorem 2.2.

Now suppose for contradiction that $u = u(r)$ is a positive radial solution of (1.1) in R^n which tends to zero as $r \to \infty$. By integrating the identity (1.7) from 0 to r we obtain

$$r^n \left(\frac{1}{2} |u'|^2 + F(u) + \frac{auu'}{r} \right)$$
$$= \int_0^r [n F(u) - a u f(u) + (a + 1 - \frac{n}{2}) |u'|^2] s^{n-1} ds.$$

By Theorem 2.1 we have $u' < 0$ for all sufficiently large values of r. Thus if we choose $a = n/(\alpha+1)$ in the above relation and use (3.3) we obtain, for sufficiently large r,

$$(3.4) \qquad r^n \left[\frac{1}{2} |u'|^2 + F(u) \right] \geq C^2$$

where C is some positive constant. [Since $\alpha \leq (n+2)/(n-2)$ we have $a + 1 - n/2 \geq 0$. Therefore the only way in which C could be zero would be to have $\alpha = (n+2)/(n-2)$ together with equality in (3.3) — at least for sufficiently small u. But in this case integration of (3.3) yields F = Pos. Const. $u^{1+\alpha}$ and $f =$ Pos. Const. u^α, contradicting the hypothesis (3.1)].

Now by $(3.1)_2$ and the second part of Theorem 2.2 we have, for sufficiently large r,

$$F(u) \leq \bar{\lambda}\, u^{\bar{q}+1} \leq \text{Const.}\, r^{-2(\bar{q}+1)/(q-1)}.$$

Hence by $(3.2)_1$

$$F(u) = o(r^{-n}) \quad \text{as } r \to \infty.$$

It now follows from (3.4) that $|u'| \geq C r^{-n/2}$ for all suitably large r, and in turn by integration

$$(3.5) \qquad u \geq \frac{2C}{n-2} r^{1-n/2},$$

valid for all suitably large r. Since $q < (n+2)/(n-2)$ this contradicts the asymptotic behavior stated for u in the second part of Theorem 2.2, completing the proof.

THEOREM 3.2. *If* $f(u) > 0$ *for* u *near* 0, *and there exists* $\alpha \geq (n+2)/(n-2)$ *such that*

$$(\alpha + 1) F(u) \leq u f(u) \tag{3.6}$$

for all sufficiently small $u > 0$, *then there is at least one positive radial solution of* (1.1) *in* R^n *which tends to zero as* $r \to \infty$.

Remarks. The hypotheses of Theorem 3.2 are satisfied in particular by any function $f(u)$ of the form

$$f(u) = \text{Pos. Const. } u^q + g(u),$$

where $q > (n+2)/(n-2)$ and $g(u) = o(u^q)$ for u near 0. Another case of interest when the conditions hold is

$$f(u) = \text{Pos. Const. } u^{(n+2)/(n-2)} + g(u),$$

where $g(u) = o(u^{(n+2)/(n-2)})$ and $o < \frac{2n}{n-2} G(u) \leq u\, g(u)$ for u near 0.

Theorem 3.2 fails if the condition $f > 0$ for u near zero is replaced by the condition $f(u) \geq 0$ for u near zero. The function $f(u) \equiv 0$ provides an easy counterexample; a more interesting one, where $f \not\equiv 0$ and $f \geq 0$ for all $u > 0$, is given by

$$f(u) = \begin{cases} 0, & 0 \leq u \leq \delta \\ (u - \delta)^{(n+2)/(n-2)}, & u > \delta, \end{cases}$$

where δ is a positive constant. In fact, if $u(0) \leq \delta$ the corresponding radial solution is $u \equiv u(0)$, while if $u(0) > \delta$ the radial solution is found simply by adding δ to the solution (3.8) below and putting $u(0) = \delta + \mu^{(n-2)/2}$.

Proof of Theorem 3.2. We use the shooting method to establish existence. In particular, according to the arguments developed in [2], section III, and [5], section 4, it is enough to show that, for any suitably small positive constant u_0, the Dirichlet problem

$$\begin{aligned} \Delta u + f(u) = 0 \quad & \text{in } \Omega \\ u = 0 \text{ on } \partial\Omega, \quad & u(0) = u_0, \end{aligned} \tag{3.7}$$

where Ω is a ball, cannot possess any positive radial solution. But any such solution has $u' < 0$ for $r > 0$, so that $0 < u(r) \leq u_0$. Applying the identity (1.6) exactly as in the proof of Theorem 3.1 in [6], we obtain a contradiction when u_0 is sufficiently small. Therefore no such solution can exist, and the theorem is proved.

We have in fact proved somewhat more than stated, namely that there is a positive radial ground state corresponding to all sufficiently small assigned values $u(0) = u_0$.

The special case $f(u) = u^{(n+2)/(n-2)}$ is included in Theorem 3.2. Here, as is well known, the solution has the particular form

$$u = \left(\frac{n\,(n-2)\,\mu}{n\,(n-2) + \mu^2 r^2} \right)^{(n-2)/2} \tag{3.8}$$

where the central value $u(0) = \mu^{(n-2)/2}$ can be assigned arbitrarily.

4. *The critical case*

In the previous section we have obtained a non–existence theorem for ground states of (1.1) when condition (3.3) holds with $\alpha \leq (n + 2 / (n - 2)$ and condition (3.1) holds with $q < (n + 2) / (n - 2)$. Conversely if (3.6) holds with $\alpha \geq (n + 2) / (n - 2)$ then ground states must exist.

The question arises as to whether non–existence holds when

$$f(u) \geq \text{Pos. Const. } u^{(n+2)/(n-2)} \quad (u \text{ small})$$

and (3.3) is satisfied with $\alpha = (n + 2) / (n - 2)$. A particular case in point is the function

$$f(u) = \text{Pos. Const. } u^{(n+2)/(n-2)} - \text{Pos. Const. } u^t, \tag{4.1}$$

where $t > (n + 2) / (n - 2)$. (If the coefficient of u^t is positive, rather than negative as here, then Theorem 3.2 supplies the existence of ground states.)

The following theorem answers this question under fairly mild conditions on the nonlinearity $f(u)$, and in particular shows that (4.1) does *not* admit a corresponding ground state. It can be considered the analogue of Theorem 3.1 for the limiting case $q = (n + 2) / (n - 2)$.

THEOREM 4.1 *Suppose that for all sufficiently small* $u > 0$ *we have*

$$uf(u) \geq 2F(u) > 0\,. \tag{4.2}$$

Let $l = (n + 2) / (n - 2)$ *and assume that*

$$uf(u) \leq (l + 1)\, F(u) \quad \textit{for all } u > 0\,, \tag{4.3}$$

with the strict inequality holding for some sequence of values u tending to zero. Then (1.1) *does not admit any positive radial solutions in* R^n *which tend to zero as* $r \to \infty$.

The theorem is false if equality holds in (4.3) for all sufficiently small u. Indeed in this case we have $f(u) = \text{Pos. Const. } u^l$ for all sufficiently small u, and corresponding ground states, given by (3.8), in fact exist.

COROLLARY. *Suppose that*

$$f(u) = \text{Pos. Const. } u^l - g(u)\,,$$

where $g(u) = o(u^l)$ *for* u *near* 0 *and* $(l + 1)\,G(u) \leq u\,g(u)$. *Then* (1.1) *does not admit any positive radial solutions in* R^n *which tend to zero as* $r \to \infty$ *(unless* $g(u) \equiv 0$ *for* u *near* 0).

This corollary should be compared whith the remark following Theorem 3.2 (note that $2n / (n - 2) = l + 1$).

The proof of Theorem 4.1 depends on several preliminary lemmas, which we shall obtain first. It will be supposed throughout that the hypotheses of Theorem 4.1 are in force, and that u (for eventual contradiction) is a positive radial solution of (1.1) in R^n which tends to zero as $r \to \infty$.

LEMMA 4.1. *For all sufficiently large r we have*

$$r^n \left[\frac{1}{2} |u'|^2 + F(u) + \frac{n-2}{2} \frac{uu'}{r} \right] \geq C^2 , \tag{4.4}$$

where C is some positive constant.

Proof. Put $a = (n-2)/2$ in (1.7) and integrate from 0 to r. Then, using condition (4.3) as in the first part of the proof of Theorem 3.1, we obtain (4.4).

LEMMA 4.2. *Let $\gamma = (n-2)/2$, and introduce the new variable $w = r^\gamma u$. Then for all sufficiently large r we have either*

$$\text{(i) } w' \geq 0 \quad \textit{or} \quad \text{(ii) } w' \leq 0 .$$

Proof. By direct calculation

$$\begin{aligned} u &= r^{-\gamma} w \\ u' &= -\gamma r^{-\gamma-1} w + r^{-\gamma} w' \\ u'' &= \gamma(\gamma+1) r^{-\gamma-2} w - 2\gamma r^{-\gamma-1} w' + r^{-\gamma} w'' . \end{aligned} \tag{4.5}$$

and so from (1.1)

$$w'' + \frac{w'}{r} + r^\gamma \left\{ f(u) - \gamma^2 \frac{u}{r^2} \right\} = 0 . \tag{4.6}$$

We shall show that w can have no local minimum when r is so large that both (4.2) and (4.4) hold. Indeed at a local minimum of w we would have

$$w' = 0 , \quad u' = -\gamma r^{-\gamma-1} w = -\gamma u / r , \quad w'' \geq 0 .$$

Hence, by (4.4), at such a point (with r sufficiently large)

$$\frac{1}{2} \gamma^2 \frac{u^2}{r^2} + F(u) - \gamma^2 \frac{u^2}{r^2} > 0 ,$$

while by (4.6) we would also have

$$f(u) - \gamma^2 \frac{u}{r^2} \leq 0.$$

Combining the last two inequalities we see that, at the local minimum,

$$2F(u) - uf(u) > 0 .$$

Since this contradicts the assumption (4.2) it is accordingly impossible for w to have a local minimum for r sufficiently large. This shows in turn that either (i) or (ii) must hold for all sufficiently large r.

LEMMA 4.3. *The function w tends to a finite non-negative limit* L *as* $r \to \infty$. *Moreover, there is a sequence of points* $\{r_m\}$, $m = 1, 2, \ldots$, *such that* $r_m \to \infty$ *as* $m \to \infty$ *and*

$$r_m w'(r_m) \to 0 , \quad r_m^2 w''(r_m) \to 0 .$$

Proof. By (4.2) we have $F(u) > 0$ for small u. Then, considering (4.3) as a differential inequality for the function F (recall that $f = F'$), we find that

$$F(u) \geq \text{Pos. Const. } u^{1+l}$$

for u small. The assumption (4.2) then yields

$$f(u) \geq \text{Pos. Const. } u^l ,$$

whence we can apply Theorem 2.2, with $q = l = (n+2)/(n-2)$ to obtain

$$u = 0\,(r^{1/2(2-n)}) \quad \text{as } r \to \infty.$$

Hence w is bounded, and consequently, in either case (i) or (ii) of Lemma 4.2, we have $w \to$ finite non-negative limit.

Moreover it is evident that

$$\liminf_{r\to\infty} r w' = 0 \quad \text{(case (i) in Lemma 4.2)}$$

or

$$\limsup_{r\to\infty} r w' = 0 \quad \text{(case (ii) in Lemma 4.2).}$$

(Otherwise an easy integration would show that $|w|$ becomes infinite at least logarithmically). In case (i) either $r w'$ decreases monotonically to zero or $\varphi = r w'$ has a sequence of *local minima* along which it tends to zero. If the second of these possibilities holds we take exactly these points as the r_m. Then

$$\varphi(r_m) = r_m w'(r_m) \to 0$$

$$\varphi'(r_m) = r_m w''(r_m) + w'(r_m) = 0$$

and the lemma follows at once. In the first possibility, we have

$$\varphi = rw' \to 0 \quad \text{as } r \to \infty$$

with $\varphi' < 0$, so that

$$\limsup_{r\to\infty} r \varphi' = 0$$

(otherwise φ tends logarithmically to negative infinity, as in the earlier discussion of the function w). This proves the lemma in case (i).

In case (ii) we proceed similarly. Either $r w'$ increases monotonically to zero or there is a sequence of *local maxima* along which $r w'$ tends to zero. In either case the existence of the required sequence $\{r_m\}$ follows exactly as before.

We may now complete the proof of Theorem 4.1. Suppose for contradiction that u is a positive radial solution of (1.1) which tends to zero as $r \to \infty$. By hypothesis, for sufficiently large r we have

$$2F(u) \leq uf(u) = u\left(-u'' - \frac{n-1}{r}u'\right).$$

Combining this with (4.4) it follows that, for sufficiently large r,

$$r^n\left\{|u'|^2 - uu'' - \frac{uu'}{r}\right\} \geq 2C^2, \quad C > 0. \tag{4.7}$$

On the other hand, by (4.5) and Lemma 4.3 we have

$$\begin{aligned} &r^\gamma u \to L, \\ &r_m^{1+\gamma} u'(r_m) \to -\gamma L \\ &r_m^{2+\gamma} u''(r_m) \to \gamma(\gamma+1) L. \end{aligned} \tag{4.8}$$

Putting $r = r_m$ in (4.7) and letting $m \to \infty$ yields the contradiction $0 \geq 2C^2$, completing the proof.

Two of the main ideas in this proof, namely the monotonicity result of Lemma 4.2 and the existence of the sequence $\{r_m\}$ in Lemma 4.3, are drawn from a paper of Ding and Ni [3].

The argument leading to Theorem 4.1 can also be used to obtain a related result, complementing Theorem 3.1 in an interesting and unusual way.

THEOREM 4.2 *Let s be a constant, put*

$$\beta = n - (n-2)s, \quad t = s(1-\beta)/(s-\beta)$$

and assume that $0 < \beta < 1$, or equivalently $0 < t < 1$. Suppose further that

$$uf(u) \geq 2tF(u) > 0 \tag{4.9}$$

for all sufficiently small u and

$$uf(u) \leq 2sF(u) \tag{4.10}$$

for all $u > 0$. Then (1.1) *does not possess any positive radial solution in* R^n *which tend to zero as $r \to \infty$.*

Remarks. Theorem 4.1 can be considered the limiting case where $2s = l + 1 = 2n/(n-2)$, $\beta = 0$, $t = 1$. On the other hand, integration of (4.10) and use of (4.9) leads to

$$f(u) \geq \text{Pos. Const. } u^{2s-1} \tag{4.11}$$

for all u near 0 (see the first part of the proof of Lemma 4.3, noting also that $F(u) > 0$ for u near 0). Thus, with $2s = q + 1$, we may view (4.9) as an alternative hypothesis to $(3.1)_2$ in the non-existence Theorem 3.1.

The inequality $\beta \geqslant 1$ is equivalent to $2s - 1 \leqslant n / (n-2)$. In this case, in view of (4.11) and Theorem 2.2 no ground state can exist.

Proof of Theorem 4.2. We only outline the ideas, since they closely parallel those in the proof of Theorem 4.1.

To begin with, using condition (4.10) as in the proof of Lemma 4.1, we obtain in place of (4.4) the relation

$$r^n \left[\frac{1}{2} |u'|^2 + F(u) + \frac{n}{2s} \frac{uu'}{r}\right] \geq C^2 \tag{4.11}$$

where C is a positive constant.

As in Lemma 4.2 we now introduce the new variable

$$w = r^{\gamma} u$$

where, however, $\gamma = 1 / (s-1)$, rather than $(n-2)/2$ as before. Then, exactly as in the proof of Lemma 4.2, it follows from (4.9) and (4.11) that ultimately either $w' \geq 0$ or $w' \leq 0$.

Because of Theorem 2.2 it is evident that w tends to a non-negative limit L as $r \to \infty$. The remaining conclusions of Lemma 4.3 then carry over without difficulty, and the proof finally is completed exactly as before.

5. *Quasilinear operators*

The results of Section 3 have direct extensions to the case of ground states of the quasilinear equation (1.2). We shall state the results in question and outline the necessary proofs, in most cases being brief in view of the similarity of the arguments to those already given for the case of the Laplace operator.

THEOREM 5.1. *Suppose that* $A(p)$ *is positive for all* $p > 0$ *and bounded for* p *near* 0. *Assume also that* $f(u)$ *satisfies the conditions* (3.1) *and* (3.2) *of Theorem* 3.1. *Finally suppose that*

$$(\alpha + 1) F(u) \geq uf(u) \tag{5.1}$$

for all $u > 0$, *where*

$$\alpha \leq \frac{(1-d)n+1}{dn-1} \qquad (n > 1/d) \tag{5.2}$$

and $d < 1$ *is a constant such that*

$$\int_0^p \rho A \, d\rho \leq dp^2 A(p), \quad p > 0. \tag{5.3}$$

Then there cannot exist positive radial solutions of (1.2) *in* R^n *which tend to zero as* $r \to \infty$.

Remark. If $A_p \geq 0$ we can obviously take $d = 1/2$ and condition (5.2) then becomes the same as in Theorem 3.1.

As in Theorem 3.1 there is no need to consider values $q \leq n/(n-2)$, since by Theorem 2.2 (see the discussion at the end of Section 2) no ground states can possibly exist in this case. On the other hand, by integration of (5.1) we find that $F(u) \geq$ Pos. Const. $u^{\alpha+1}$ for small u. Hence, using (3.1) and (3.2) there results

$$\alpha \geq \bar{q} > \frac{q-1}{2} n - 1 > \frac{2}{n-2}$$

when $q > n/(n-2)$. Consequently, from (3.3) it is only for values $d < (n-1)/n$ that Theorem 5.1 produces a stronger result than Theorem 2.2.

Proof of Theorem 5.1. Let $u = u(r)$ be a positive radial solution of (1.2) in R^n which tends to zero as $r \to \infty$, which for contradiction we shall suppose exists.

To begin with, since $E = A + p A_p$ it follows at once that

$$\int_0^p \rho E \, d\rho = p^2 A(p) - \int_0^p \rho A \, d\rho$$
$$\geq (1-d) p^2 A(p)$$

by (5.3). Then as in the proof of Theorem 3.1 we integrate the identity (1.6) from 0 to r and use (5.3), to obtain

$$r^n \left[\int_0^p \rho E \, dp + F(u) + a \frac{A(p) u u'}{r} \right] \tag{5.4}$$
$$\geq \int_0^r s^{n-1} [n F(u) - a u f(u) + (a + 1 - dn) A(p) p^2] \, ds .$$

Using the conditions (5.1) and (5.2) and setting $a = n/(\alpha+1)$, it is apparent that the right hand side of this inequality is non-negative. By Theorem 2.1, which of course holds equally for the quasilinear equation (1.2) as noted at the end of Section 2, we have $u' < 0$ for all sufficiently large r. Hence from (5.4) we get, for all sufficiently large r,

$$r^n \left[\int_0^p \rho E \, d\rho + F(u) \right] \geq C^2 \tag{5.5}$$

where C is a positive constant. [The only way in which C could be zero would be to have equality in all three of the conditions (5.1), (5.2) and (5.3) - in the case of (5.1) at least for sufficiently small u, and in the case of (5.3) at least for sufficiently small p. This implies by integration that both

$$f(u) = \text{Pos. Const. } u^\alpha \tag{5.6}$$

for small $u > 0$, and

$$A(p) = \text{Pos. Const. } p^{-2+1/d}$$

for small $p > 0$. By assumption the function $A(p)$ is bounded for small $p > 0$, so this gives $d \leq 1/2$, whence in turn

$$\alpha = \frac{(1-d)n+1}{dn-1} \geq \frac{n+2}{n-2}.$$

Condition (5.6) now stands in contradiction with $(3.1)_1$ since by (3.2) we must have $q < (n+2)/(n-2)$. Hence the constant C must be positive].

From the discussion at the end of Section 2 we know that $u' \to 0$ as $r \to \infty$. Consequently, using the boundedness of $A(p)$ near $p = 0$, we find that, for some constant $\Lambda > 0$,

$$\int_0^p \rho E \, d\rho = p^2 A(p) - \int_0^p \rho A \, d\rho \tag{5.7}$$
$$\leq p^2 A(p) \leq \Lambda p^2, \quad p \text{ small}.$$

Combining this with (5.5) yields, for all sufficiently large r,

$$r^n \{\Lambda |u'|^2 + F(u)\} \geq C^2.$$

The rest of the proof is now the same as for Theorem 3.1.

Remark. The proof of Theorem 5.1, and also subsequently Theorems 5.2, 5.3 and 5.4, rests heavily on the Pohozaev type identity (1.6). The careful observer might notice, however, that the hypotheses of Theorem 5.1 are so weak as to call into question the validity of (1.6). In particular we have assumed only that $A(p)$ is continuously differentiable for $p > 0$, that $A(p) > 0$, and that $A(p)$ is bounded near zero. Consequently it is not immediately obvious that (1.6) can be maintained. In fact, however, (5.7) serves to establish the existence of the integral $\int_0^p \rho E \, d\rho$ (see also (2.10)); even more, this relation shows that the *derivative* of the integral is precisely $p\,E(p)$, even when $p = 0$, provided we *define* $p\,E(p)$ to be zero when $p = 0$. With this in hand, it is then easy to check the validity of (1.6) by direct calculation, exactly as in [6], Section 2.

THEOREM 5.2. *Suppose that*

$$E(p) > 0 \quad \textit{for all sufficiently small } p > 0 \tag{5.8}$$

and

$$p\,A(p) \to 0 \quad \textit{as } p \to 0. \tag{5.9}$$

Assume also that $f(u)$ is continuous and

$$0 < (\alpha + 1) F(u) \leq u f(u)$$

for all u near 0, where

$$\alpha \geq \frac{(1-c)n+1}{cn-1} \quad (n > 1/c)$$

and c is a positive constant such that

$$\int_0^p \rho \, A \, d\rho \geq c p^2 A(p) \quad \textit{for small } p > 0 . \tag{5.10}$$

Then there exists at least one positive radial solution of (1.2) *in* R^n *which tends to zero as* $r \to \infty$.

Proof. We use the method of [2] and [5], exactly as in the proof of Theorem 3.2. The details, however, are more delicate than before and require considerable additional discussion.

We first show that if the value $u_0 = u(0)$ in the quasilinear analogue of the problem (3.7) is chosen sufficiently small, then for any eventual positive solution $u = u(r)$ there holds

$$-\delta < u'(r) < 0 \text{ for } r > 0 , \tag{5.11}$$

where $\delta > 0$ is fixed so that

$$A(p) > 0 , \quad E(p) > 0$$

for $0 < p < \delta$. To see that such a value δ exists, we observe first that $p\,A(p) - -p_1\,A(p_1) = \int_{p_1}^{p} E(\rho)\,d\rho$ when $p > p_1 > 0$ (recall that $(p\,A)_p = E$). By (5.8) and (5.9) we can let $p_1 \to 0$, so that

$$p\,A(p) = \int_0^p E(\rho)\,d\rho .$$

Thus, since $E(p) > 0$ for small p, it follows that also $A(p) > 0$ for small p.

Now put (the integral obviously exists by what has just been shown)

$$H(p) = \int_0^p \rho \, E(\rho)\,d\rho , \quad p > 0,$$

set $h = H(\delta) > 0$ and let u_0 be chosen so that

$$f(u) > 0 \text{ for } 0 < u \leq u_0 , \quad F(u_0) \leq h .$$

We shall show that (5.11) then holds for the eventual positive solution $u(r)$. To begin with, from (2.6) the expression

$$\psi(r) = r^{n-1} u' \, A(|u'|)$$

is strictly decreasing as long as $0 < u(r) \leq u_0$ (since $f > 0$ in this range). Because $\psi(0) = 0$ it follows that $\psi(r)$ is negative for sufficiently small $r > 0$. Thus in turn $u' A(|u'|)$ is negative for sufficiently small r. Hence $u' \neq 0$ for small r. Moreover, since $u'(0) = 0$, we have $|u'| < \delta$ for small r and so $A(|u'|) > 0$. Consequently for small r we have both $u' < 0$ and $u' > -\delta$, that is, the inequality (5.11) holds at least initially.

Now consider any range $0 < r < \bar{r}$ in which (5.11) holds. Since $u' < 0$ it follows that $u(r) < u_0$. Hence $f(u) > 0$ and $\psi(r)$ is decreasing. Conseuqently $\psi(\bar r) < 0$, and in turn $u'(\bar r)\,A(|u'(\bar r)|) < 0$. This obviously can happen only if $u'(\bar r) \neq 0$. But then $u'(\bar r) < 0$ by continuity. On the other hand, by (2.9) with $r_0 = 0, r_1 = \bar r$,

$$H(p(\bar r)) + (n+1)\int_0^{\bar r} A(|u'|)\frac{u'^2}{r}\,dr + F(\bar u) - F(u_0) = 0.$$

Clearly $\bar u = u(\bar r) < u_0$ so that $F(\bar u) > 0$. Moreover $0 < |u'| < \delta$ for $0 < r < \bar r$ so $A(|u'|) > 0$ and the integral is positive. Consequently

$$H(p(\bar r)) < F(u_0) \leq h.$$

Thus $p(\bar r) \neq \delta$, whence $u'(\bar r) > -\delta$ by continuity.

We have therefore shown that if the inequality (5.11) holds for an interval $0 < r < \bar r$ it also holds for the interval $0 < r \leqslant \bar r$. Accordingly (5.11) must hold for all $r > 0$ for the eventual solution, and our assertion is proved. In particular, the proof shows equally that $-\delta < u' < 0$ at the outer radius $r = R$ of the original domain Ω, where $u(R) = 0$. Applying the identity (1.6), then, exactly as in the proof of Theorem 3.1 in [6], we obtain a contradiction (this is where (5.10) is used). Therefore no solution of (3.7) can exist.

To complete the proof it must be shown that if u_0 is chosen as indicated above, then the corresponding positive radial solution of (1.2) is defined for *all* r, $0 \leq r < \infty$, and tends to zero as $r \to \infty$.

The first part of this requirement, namely that the solution is defined for all $r \geq 0$, follows from the proof already given. More precisely, by the principal theorem of the Appendix (Section 7), conditions (5.8) and (5.9) guarantee the *existence* of a radial solution $u = u(r)$ of (1 .2) with initial value $u(0) = u_0$, at least for all sufficiently small values of r. Moreover, *this solution can fail to be continuable to all positive values of r only if u' were to attain values outside the interval $(-\delta, 0)$ or if u were to reach the value* $u = 0$, and by what has already been shown this cannot happen.

Next let

$$\lim_{r\to\infty} u(r) = l \tag{5.12}$$

where we suppose for contradiction that $l > 0$. Exactly as in the argument at the end of Section 2 we find that $u' \to 0$ as $r \to \infty$ (in fact, the hypothesis $(5.8)_2$ implies $H(p) > 0$ so the proof can even be slightly shortened). Consequently, from the equation

$$Eu'' + (n-1)A\frac{u'}{r} + f(u) = 0$$

we see that, for all r sufficiently large,

$$\begin{aligned} Eu'' &= -(n-1)A(|u'|)\frac{u'}{r} - f(u) \\ &\leq -f(l) + 2\varepsilon(r) \end{aligned}$$

where $\varepsilon(r) \to 0$ as $r \to \infty$. Since $l > 0$ we have $f(l) > 0$ and so, again if r is suitably large,

$$\mathrm{E}u'' \leq -\tfrac{1}{2} f(l).$$

Since $\mathrm{E} > 0$ for r sufficiently large, it follows finally that $u'' < 0$ for large r. But in combination with the fact that $u' < 0$ this clearly stands in contradiction to condition (5.12). Therefore $l = 0$, and the proof is complete.

The proof of Theorem 5.2 shows in fact that if $f(u) > 0$ for all $u > 0$, and $\mathrm{A}(p) > 0$, $\mathrm{E}(p) > 0$ for all $p > 0$, then *radial ground states exist corresponding to arbitrarily assigned values* $u(0)$ *satisfying the condition*

$$\mathrm{F}(u(0)) \leq \int_0^\infty \rho\,\mathrm{E}(\rho)\,d\rho\,. \tag{5.13}$$

In particular if the integral is divergent then ground states exist for all assigned values $u(0)$.

For the case of the mean curvature equation

$$\operatorname{div} \frac{\mathrm{D}u}{\sqrt{1+|\mathrm{D}u|^2}} + f(u) = 0 \tag{5.14}$$

we have non-existence of ground states when $f(u) \geq$ Pos. Const. u^q for $q \leq n/(n-2)$, by Theorem 2.2, and existence when

$$0 < (\alpha + 1)\,\mathrm{F}(u) \leq uf(u)$$

for some $\alpha \geq (n+2)/(n-2)$, by Theorem 5.2 (here $c = 1/2$). In particular, existence holds when $f(u) =$ Pos. Const. u^α and $\alpha \geq (n+2)/(n-2)$. (More precisely, in view of (5.13) ground state solutions of the mean curvature equation (5.14) exist for the function $f(u) = \lambda u^q$, with $q \geq (n+2)/(n-2)$, whenever $u(0) \leq ((q+1)/\lambda)^{1/(q+1)}$). It is an open question whether solutions exist when $f(u) =$ Pos. Const. u^q and $n/(n-2) < q < (n+2)/(n-2)$, though it appears unlikely.

For the quasilinear equation (1.2) there are also direct analogues of Theorems 4.1 and 4.2 provided that we assume $d\mathrm{A}/dp \geq 0$.

THEOREM 5.3. *Suppose that* $d\mathrm{A}/dp \geq 0$ *for* $p > 0$. *and that* c *is a positive constant with the property* (5.10). *Furthermore, assume that*

$$uf(u) \geq \frac{1}{c}\,\mathrm{F}(u) > 0 \quad \textit{for all } u \textit{ near } 0,$$

and that

$$(l+1)\,\mathrm{F}(u) \geq uf(u) \quad \textit{for all } u > 0,$$

where $l = (n+2)/(n-2)$, *with the strict inequality holding for some sequence of values* u *tending to zero. Then* (1.2) *does not admit any positive radial solutions in* R^n *which tend to zero as* $r \to \infty$.

The proof is almost identical to that of Theorem 4.1, with the exception that in place of (4.6) we have the corresponding equation

$$Ew'' + [(n-1)A - 2\gamma E]\frac{w'}{r} + r^{\gamma}\left\{f(u) - \gamma[(n-1)A - (\gamma+1)E]\frac{u}{r^2}\right\} = 0.$$

The hypothesis $dA/dp \geq 0$ moreover is used in three different places. First, to show that A is bounded near $p = 0$, so that Theorem 2.2 can be applied, second to show that $A \leq E$, a relation which enables Lemma 4.2 to be carried over to the present case, and third to show that $d = 1/2$ in (5.3), which gives the conclusion corresponding to Lemma 4.1.

THEOREM 5.4. *Suppose that* $dA/dp \geq 0$ *for* $p > 0$ *and that* (5.10) *holds. Let s be a constant, put*

$$\beta = n - (n-2)s, \quad t = s(1-\beta)/(2cs - \beta)$$

and assume that $0 < \beta < 1$, *or equivalently* $0 < t < 1/2c$. *Suppose further that* (4.9) *and* (4.10) *hold. Then* (1.2) *does not possess any positive radial solutions in* R^n *which tend to zero as* $r \to \infty$.

The proof is essentially the same as for Theorem 4.2, once one takes into account the remarks following Theorem 5.3.

6. *The degenerate Laplacian*

Here we consider the equation

$$\operatorname{div}(|Du|^{m-2}Du) + f(u) = 0 \tag{6.1}$$

where m is a constant, $1 < m < n$. When $m = 2$ this is precisely the original problem (1.1), but otherwise (6.1) is both quasilinear and degenerate (when $Du = 0$). In the present case

$$A(p) = |p|^{m-2}.$$

Consequently Theorem 5.2 applies with $c = 1/m$ and Theorems 2.0 - 2.2 5.1, 5.3 and 5.4 apply when $m \geq 2$. However, in the latter cases the results are not best possible since the precise nature of $A(p)$ for p near 0 is not taken into account.

Accordingly it is the purpose of this section to show that each of the results previously obtained for the Poisson equation (1.1) has an immediate generalization to equation (6.1). Because the procedures are essentially the same as our earlier ones we shall only state the conclusions, leaving the proofs to the reader.

THEOREM 6.1. *Suppose that* $f(u) \geq 0$ *for* u *near* 0. *Let* u *be a positive radial solution of* (6.1) *defined for* $r > r_0$ *and tending to zero as* $r \to \infty$. *Then*

$$u \geq \text{Pos. Const. } r^{(m-n)/(m-1)}, \quad u' \leq \text{Neg. Const. } r^{-(n-1)/(m-1)}$$

for all sufficiently large r.

The argument used to prove this result can also be used to show that, when $n \leq m$ and $f(u) \geq 0$ for all u near 0, there can exist no positive radial solutions of (6.1) which tend to zero as $r \to \infty$.

THEOREM 6.2. *Suppose that* $f(u) \geq$ Pos. Const. u^q *for* u *near* 0, *where* q *is a positive constant.*

If $q \leq (m-1)n/(n-m)$ *there can exist no positive radial solution of* (6.1) *defined for* $r \geq r_0$ *and tending to zero as* $r \to \infty$.

If $q > (m-1)n/(n-m)$ *then any positive radial solution of* (6.1) *which is defined for* $r \geq r_0$ *and tends to zero as* $r \to \infty$ *has the asymptotic behavior*

$$u = 0\,(r^{-m/(q-m+1)})$$

as $r \to \infty$.

The exponent $(m-1)n/(n-m)$ is best possible in the first part of Theorem 6.2, as shown by the example

$$f(u) = u^q, \quad u = \text{Pos. Const. } r^{-m/(q-m+1)}\,.$$

This example also shows that the asymptotic estimate in the second part of the theorem is best possible.

THEOREM 6.3. *Suppose that, for all sufficiently small* u,

$$f(u) \geq \text{Pos. Const. } u^q\,,\ F(u) \leq \text{Pos. Const. } u^{\bar{q}+1}\,,$$

where q *and* $\bar{q}$ *are positive exponents satisfying the condition*

$$\frac{q-m+1}{m}n - 1 < \bar{q} \leq q\,.$$

Suppose also that for some $\alpha \leq ((m-1)n+m)/(n-m)$ *we have*

$$(6.2) \qquad (\alpha+1)\,F(u) \geq uf(u)$$

for all $u > 0$. *Then there cannot exist any positive radial solutions of* (6.1) *in* R^n *which tend to zero as* $r \to \infty$.

Remarks. By integrating the condition (6.2) we find that $F(u) \geq$ Const. $u^{1+\alpha}$ for small u, whence the constant α cannot be less than $\bar{q}$. One may also notice that the limiting constant

$$\alpha + 1 = \frac{mn}{n-m}$$

is precisely the Sobolev exponent for the space $W^{1,m}$ in n dimensions. This observation in turn is related to the fact that (6.1) itself is the Euler–Lagrange equation corresponding to the variational problem

$$\delta \int \left\{ \frac{1}{m} \mid Du \mid^m - F(u) \right\} dx = 0\,.$$

THEOREM 6.4. *If* $f(u) > 0$ *for* u *near* 0, *and there exists*

$$\alpha \geq \frac{(m-1)n+m}{n-m}$$

such that $(\alpha+1)\,F(u) \le u f(u)$ *for all sufficiently small* $u > 0$, *then there is at least one positive radial solution of* (6.1) *in* R^n *which tends to zero as* $r \to \infty$.

THEOREM 6.5. *Suppose that for all sufficiently small* $u > 0$ *we have*

$$uf(u) \ge mF(u) > 0\,.$$

Suppose also that

$$(l+1)\,F(u) \ge uf(u) \quad \textit{for all } u > 0\,,$$

where $l = ((m-1)\,n + m)\,/\,(n-m)$ *and where the strict equality does not hold in any interval* $(0\,,\varepsilon)$. *Then* (6.1) *does not admit any positive radial solutions which tend to zero as* $r \to \infty$.

The condition that strict equality not hold in any interval $(0\,,\varepsilon)$ is necessary for the validity of the theorem, as shown by the example

$$f(u) = u^l\,.$$

For this function the equation (6.1) has the corresponding positive radial solution

$$u = \left(\frac{(n-m)\mu}{(n-m) + n^{1/(m-1)}\,(\mu r)^{m/(m-1)}}\right)^{(n-m)/m},$$

where the central value $u(0) = \mu^{(n-m)/m}$ can be assigned arbitrarily.

THEOREM 6.6. *Let* s *be a constant, put*

$$\beta = n - (n-m)\,s\,, \qquad t = s\,(1-\beta)\,/\,(s-\beta)\,,$$

and assume that $0 < \beta < 1$, *or equivalently* $0 < t < 1$. *Suppose further that*

$$u\,f(u) \ge mt\;F(u) > 0$$

for all sufficiently small u *and*

$$u\,f(u) \le ms\;F(u)$$

for all $u > 0$. *Then* (6.1) *does not possess any positive radial solutions in* R^n *which tend to zero as* $r \to \infty$.

Theorem 4.1 arises as the limiting case

$$m = 2\quad,\ 2\,s = l + 1 = 2\,n\,/\,(n-2)\quad,\ \beta = 0\quad,\ t = 1\,,$$

while Theorem 6.5 similarly arises when

$$ms = l + 1 = m\,n\,/\,(n-m)\quad,\ \beta = 0\quad,\ t = 1\,.$$

7. *Appendix. The initial value problem*

Here we study the initial value problem for radial solutions of (1.2), namely

$$\text{(7.1)}\quad \begin{aligned} &E(|u'|)\,u'' + (n-1)\,A(|u'|)\frac{u'}{r} + f(u) = 0\,, r > 0\,,\\ &u(0) = u_0 > 0\,. \end{aligned}$$

We suppose throughout that $f(u)$ is continuous. In addition, as stated in the introduction, the function $A(p)$ is assumed continuously differentiable for $p > 0$, and of course $E(p) = A(p) + p\,dA(p)/dp$.

PROPOSITION 1. *Suppose that* $E(p) > 0$ *for small* $p > 0$ *and that* $p\,A(p) \to 0$ *as* $p \to 0$. *Then* (7.1) *has a continuously differentiable solution* $u = u(r)$ *on some interval* $0 \le r < R$. *Moreover, for any such solution we have* $u'(0) = 0$, *provided* $A(p) > 0$ *for all* p.

Proof. Put $\psi(p) = p\,A(p)$. Then $\psi'(p) = E(p) > 0$ for small $p > 0$ so that ψ is increasing. Moreover $\psi(p)$ is continuous for all $p > 0$ when we set $\psi(0) = 0$. It is clear that any eventual solution of (7.1), for small r, must be a fixed point of the operator

$$\text{(7.2)}\qquad T[u](r) = u_0 - \int_0^r \varphi\left(\int_0^t f(u(s))\left(\frac{s}{t}\right)^{n-1} ds\right) dt\,,$$

where $\sigma \to \varphi(\sigma)$ is an odd function, coinciding for $\sigma \ge 0$ with the inverse of $\psi(p)\,, p \ge 0$. Naturally φ is defined and continuous for all small σ, with $\varphi(0) = 0$.

Consider the space X of continuous functions on $[0\,, R]$, with the uniform topology. We put

$$C = \{u \in X : \| u(r) - u_0 \| < \tfrac{1}{2} u_0\}\,.$$

For $u \in C$ and R suitably small, say $R \le R_0$, the value

$$\int_0^t f(u(s))\left(\frac{s}{t}\right)^{n-1} ds\,,\quad 0 \le t \le R\,,$$

lies in the domain of φ. Thus for $R \le R_0$ it is evident that T is a compact continuous map of the convex set C into X (use the Arzelà–Ascoli theorem). Moreover, if R is made even smaller if necessary, we have $T[C] \subset C$.

Thus by Schauder's fixed point theorem, T possesses a fixed point u in C. That u is continuously differentiable for $r \ge 0$ follows at once from the representation (7.2). This proves the first part of the theorem.

The second part of the theorem is the assertion that if $u(r)$ is a continuously differentiable solution of (7.1) for some interval $0 \le r < R$, then $u'(0) = 0$. To prove this we write (7.1) in the form

$$(r^{n-1} u'\,A(|u'|))' = -r^{n-1} f(u)\,.$$

By integration from r to $\bar{r} < R$ we get

$$\bar{r}^{n-1} u'(\bar{r})\,A(|u'(\bar{r})|) - r^{n-1} u'\,A(|u'|) = -\int_r^{\bar{r}} s^{n-1} f(u(s))\,ds\,.$$

Since the right hand integrand is bounded it is clear that

$$r^{n-1}\, u'\, A(|u'|) \to \text{limit}$$

as $r \to 0$. Moreover, since $u'\, A(|u'|)$ must be bounded near $r = 0$ by the hypothesis that u' is continuous, it is evident that the limit must be zero. Consequently we have in fact

$$r^{n-1}\, u'\, A(|u'|) = -\int_0^r s^{n-1} f(u(s))\, ds$$

and in turn

$$u'\, A(|u'|) = 0(r)$$

as $r \to 0$. Since $A(p) > 0$ for all $p > 0$ it follows at once that $u'(0) = 0$.

Remark. Because of the generality of the hypotheses above, it is technically necessary to use the notion of weak solution. In particular, the solutions provided by the proposition actually satisfy equation (7.1) only in the weak sense that

$$r^{n-1}\, u'\, A(|u'|) = -\int^r s^{n-1} f(u(s))\, ds\,.$$

They are of class C^2 and satisfy (7.1) in the classical sense *in the neighborhood of any point r where $u'(r) \neq 0$.* [If $E(0) = \lim_{p \to 0} E(p)$ exists and is *non-zero*, then the solution is everywhere classical]. This observation is ultimately at the heart of the proof of Theorem 5.2: since $-\delta < u'(r) < 0$ for $r > 0$ the solution constructed there is in fact *twice continuously differentiable* for $r > 0$ and satisfies (7.1) in the classical sense.

Solutions of the initial value problem obtained under the minimal hypotheses above need not be unique. To obtain uniqueness it is necessary to strengthen the hypotheses sufficiently in order to apply the contraction mapping theorem. In particular, this would require that the function f be locally Lipschitz continuous and that, say, $E(0)$ exists and is positive, or $A(p) \equiv$ Pos. Const. p^{m-2}, $m > 1$, for small p.

The proof of the second part of Proposition 1 actually gives a related and slightly deeper result, which we give here for its independent interest.

PROPOSITION 2. *Suppose that $u = u(r)$ is a bounded radial solution of* (1.2) *for $0 < r < R$. Assume also that $A(p) > 0$ for all $p > 0$ and that*

$$\text{Pos. Const.} \leq p^2 A(p) \leq \text{Pos. Const. } p^n$$

for all sufficiently large p. Then u' is continuously differentiable for $0 \leq r < R$ and $u'(0) = 0$. Moreover either

$$\lim_{p \to 0} p\, A(p) = 0 \quad \text{or} \quad \lim_{p \to 0} p\, A(p) = \infty\,.$$

Proof. As above, we find easily that

$$r^{n-1} u' \mathrm{A}(| u' |) \to \text{limit}$$

as $r \to 0$. If the limit is positive then it is clear from the condition $\mathrm{A}(p) > 0$ for $p > 0$ that either

$$u' \to 0 \text{ or } u' \to \pm \infty \text{ as } r \to 0 .$$

In the latter case, using the behavior of $\mathrm{A}(p)$ for large p one gets

$$| u' |^{n-1} \geq \text{Pos. Const. } r^{1-n}$$

for small r, which is impossible if u is to remain bounded. Hence $u' \to 0$. Of course this case can only happen if $\lim_{p \to 0} p \mathrm{A}(p) = \infty$.

Next, if the limit is zero, we find as before that

$$u' \mathrm{A}(| u' |) = 0(r) \text{ as } r \to 0 .$$

Again it is necessary that $u' \to 0$ or $u' \to \pm \infty$. In the latter case, using the the behavior of $\mathrm{A}(p)$ for large p, we get

$$| u' |^{-1} \leq \text{Pos. Const. } r$$

which is again impossible. Hence again $u' \to 0$ as $r \to 0$. Moreover in this case $\lim p \mathrm{A}(p) = 0$, completing the proof.

Corollary 1. *Let the hypotheses of Proposition 2 hold. If $\mathrm{E}(p) > 0$ for small $p > 0$, so that (7.1) is elliptic for p near 0, then necessarily $p \mathrm{A}(p) \to 0$ as $p \to 0$.*

For the case of the degenerate Laplacian, $\mathrm{A}(p) = | p |^{m-2}$. Hence $p^2 \mathrm{A}(p) = | p |^m$ and the asymptotic conditions for large p are fulfilled. In addition $p \mathrm{A}(p) \to 0$ as $p \to 0$ (recall that $1 < m < n$) so that we are in the usual elliptic case. For Theorem 5.2, moreover, the corollary shows that condition (5.9) is necessary for the existence of a solution.

Corollary 2. *Let the hypotheses of Proposition 2 hold.* If $\mathrm{E}(p)$ *is defined and positive for all $p \geq 0$ then u is a twice continuously differentiable classical solution of (7.1), with*

$$\lim_{r \to 0} \frac{u'(r)}{r} = u''(0) = -\frac{1}{n\mathrm{E}(0)} f(u_\circ) .$$

REFERENCES

[1] Berestycki, H. and Lions, P.-L. – *Nonlinear scalar field equations, I. Existence of ground states.* « Arch. Rational Mech. Anal. » *82*, 313–345 (1983).

[2] Berestycki, H., Lions, P.-L., and Peletier, L. A., – *An O.D.E. approach to the existence*

of positive solution for semilinear problems in R^n. « Indiana Univ. Math. J. » *30*, 141–157 (1981).

[3] DING, W.-Y. and NI, W.-M. – *On the elliptic equation* $\Delta u + K u^{(n+2)/(n-2)} = 0$ *and related topics.* « Duke Math. J. » *52* (1985), 485-506.

[4] MCLEOD, K. – *Uniqueness of positive radial solutions of semi-linear elliptic equations in* R^n. Doctoral thesis, University of Minnesota, 1984.

[5] NI, W.-M. – *On the elliptic equation* $\Delta u + K(x) u^{(n+2)/(n-2)} = 0$, *its generalizations and applications in geometry.* « Indiana Univ. Math. J. » *31*, 493–529 (1982).

[6] NI, W.-M. and SERRIN, J. – *Non-existence theorems for quasilinear partial differential equations.* To appear, « Rend. Circolo. Mat. Palermo » Suppl. 5 (1986).

Differ. Integral Equ. 7 (1994), 1063–1082. **Reprint I.8.[ii]**

GROUND STATES OF A QUASILINEAR EQUATION

L.A. PELETIER
Mathematical Institute, Leiden University, Leiden, The Netherlands

J. SERRIN
School of Mathematics, University of Minnesota, Minneapolis, MN 55455

HENGHUI ZOU
Department of Mathematics, Northwestern University, Evanston, IL 60201

To the memory of Peter Hess

1. Introduction. In 1989, Chipot and Weissler [2] proposed the equation

$$u_t = \Delta u - |\nabla u|^q + \lambda |u|^{p-1}u, \quad \lambda > 0 \tag{1.1}$$

for the study of blow-up phenomena for parabolic equations. The term $|u|^{p-1}u$ in this equation has the tendency to cause blow-up, whilst the term $-|\nabla u|^q$ just opposes this tendency. It was the competition between these terms, for different values of the parameters p and q, which they studied. Whereas [2] was mainly focused on the study of (1.1) in bounded domains in $\mathbb{R}^N$, Alfonso and Weissler [1] and Serrin and Zou [4] studied (1.1) in unbounded domains. In particular, in [4] a very extensive analysis was given of nonnegative radially symmetric equilibrium solutions of (1.1) in $\mathbb{R}^N$ – often referred to as *ground-states* when $u(x) \to 0$ as $x \to \infty$ – in different regions of the (p, q) parameter space and for $N > 1$.

In this paper we study the existence, uniqueness and asymptotic behavior of ground states when $N = 1$. Thus, we consider the problem

$$\text{(I)} \qquad \begin{cases} u'' - |u'|^q + \lambda u^p = 0 & \text{on} \quad \mathbb{R} & (1.2\text{a}) \\ u \geq 0, \; u \not\equiv 0 & \text{on} \quad \mathbb{R} & (1.2\text{b}) \\ u(x) \to 0 & \text{as} \quad x \to \pm\infty. & (1.2\text{c}) \end{cases}$$

About the parameters p and q we assume initially merely that they are positive.

It was shown in [2] that if

$$p > 1, \quad q > \frac{2p}{p+1}, \tag{1.3}$$

Received September 1993.

then Problem (I) has no solution, whilst if

$$p > 1, \quad q = \frac{2p}{p+1},$$

then Problem (I) can be solved if and only if

$$\lambda \leq \frac{(2p)^p}{(p+1)^{2p+1}}.$$

Here we extend these results to the parameter range $0 < p \leq 1$ and also give an alternative proof of non-existence of ground states for the parameter values (1.3).

The main effort of the paper will therefore be devoted to investigating the existence, uniqueness and asymptotic properties of ground states for the remaining parameter values

$$0 < p < \infty \quad \text{and} \quad q < \frac{2p}{p+1}. \tag{1.4}$$

In this analysis it is convenient to consider separately the special case of ground states u which satisfy the following additional conditions:

$$\text{(i)} \quad u(0) > 0, \quad u'(0) = 0$$
$$\text{(ii)} \quad xu'(x) < 0 \ \text{when} \ u(x) > 0 \ \text{and} \ x \neq 0.$$

We refer to these as *restricted ground states*. Since (1.2a) is invariant under the change $x \to -x$, and solutions of the Cauchy problem are unique, it is clear that restricted ground states are symmetric about $x = 0$. Hence by (ii) a restricted ground state is either everywhere positive or else its support is a finite interval.

With these conventions understood, we have the following principal conclusions.

Theorem A. *Suppose $q > 1$, so also $p > 1$. Then there exists a unique positive restricted ground state $u_0(x)$ with the asymptotic behavior*

$$u(x) \sim B|x|^{-(2-q)/(q-1)} \quad \text{as} \quad |x| \to \infty, \tag{1.5a}$$

where

$$B = \frac{1}{2-q}\Big(\frac{1}{q-1}\Big)^{(2-q)/(q-1)} \tag{1.5b}$$

and a one-parameter family of positive restricted ground states $u(x)$ with the asymptotic behavior

$$u(x) \sim A|x|^{-q/(p-q)} \quad \text{as} \quad |x| \to \infty, \tag{1.6a}$$

where

$$A = \Big(\frac{q}{p-q}\Big)^{q/(p-q)}\lambda^{-1/(p-q)}. \tag{1.6b}$$

The central values $u(0)$ in the latter case cover exactly the parameter range $(0, u_0(0))$.

Theorem B. *Suppose $q < \min(p, 1)$. Then there exists a unique restricted ground state $u_0(x)$ with compact support, and a one-parameter family of positive restricted ground states $u(x)$ with the asymptotic behavior* (1.5).

The central values $u(0)$ in the latter case cover exactly the parameter range $(0, u_0(0))$.

Theorem C. *Suppose $p < q$, so also $q < 1$. Then there exists a one parameter family of restricted ground states with compact support. Their central values $u(0)$ cover exactly a parameter range* $(0, \gamma_0]$.

Theorem A continues to hold for $q = 1$, with the exception that (1.5) should be replaced by

$$e^x u(x) \to \text{const.} > 0 \qquad \text{as } x \to \infty. \tag{1.7}$$

Here the constant depends only on p and λ. Similarly on the line segment $p = q$, $0 < q < 1$, the results correspond to Theorem B except that the asymptotic behavior of the positive restricted ground states is again replaced by $e^{\kappa x}u(x) \to$ const. $>$ 0 as $x \to \infty$, where $\kappa = \lambda^{1/p}$.

Beyond the set of restricted ground states given by Theorems A, B and C, we note explicitly (see Theorem 4 in Section 3) that there are no further restricted ground states for the parameter range (1.4).

It is clear from the above discussion that there are exactly four different types of restricted ground states possible in the parameter range (1.4), namely positive ground states with one of the three behaviors (1.5), (1.6) and (1.7), and compact support ground states. As we shall see in Section 3 compact support ground states also have slightly different behaviors in the parameter regions $q < \min(1, p)$ and $q > p$.

Positive ground states with the behavior (1.6) are called *slow* ground states, while ground states with the behavior (1.5) or (1.7) are called *fast* ground states.

Theorems A, B and C will be proved by means of a shooting technique in which we study solutions $u(x, \gamma)$ of equation (1.2a) satisfying the initial value problem

$$u(0) = \gamma \quad \text{and} \quad u'(0) = 0, \qquad \gamma > 0. \tag{1.8}$$

It will be shown that a family of values γ can be found for which the corresponding solutions satisfy

$$u(x) \geq 0 \text{ for all real } x;\ \ xu'(x) < 0 \text{ whenever } xu(x) \neq 0;\ \ \lim_{x\to\pm\infty} u(x) = 0 \tag{1.9}$$

and so are restricted ground states. Since solutions of the initial value problem (1.8) are symmetric with respect to the origin, it will be enough to consider the range $x \geq 0$. Note moreover that γ is then the central value of the corresponding ground state.

Because (1.2a) is autonomous, it is convenient to study solutions of the initial value problem (1.8), (1.9) in the phase plane. By an asymptotic analysis of orbits near the origin, we establish the asymptotic behavior of the slow and fast ground states and the uniqueness of the latter; see Section 2. The existence of restricted ground states is then proved in Section 3.

Turning to the structure of ground states which are not restricted by the conditions (i) and (ii), we have the following result.

Theorem D. *Let $u = u(x)$ be a ground state for* (1.1) *when $N = 1$. Then either u is a translation to a new origin x_0 of a positive restricted ground state, or there exists a finite or countable set of bounded disjoint intervals $\{I_k\}$ such that*

$$u(x) \equiv 0 \quad \text{for} \quad x \notin \cup I_k,$$

and the function

$$w_k(x) = \begin{cases} u(x) & \text{if} \quad x \in I_k \\ 0 & \text{if} \quad x \notin I_k \end{cases}$$

is a translation of a restricted ground state with compact support.

Conversely, all functions constructed in this way by translation and superposition are ground states of (1.1).

Proof. Let u be a ground state of (1.1) when $N = 1$. If u is everywhere positive, then there exists a point $x_0 \in \mathbb{R}$ such that $u'(x_0) = 0$. By translation of x_0 to the origin we obtain a positive ground state $\hat{u}$ which moreover is easily seen to satisfy the condition $x\hat{u}'(x) < 0$ for $x \neq 0$. Hence, $\hat{u}$ is a positive restricted ground state as required.

If on the other hand u is not everywhere positive, we define the set

$$I = \{x \in \mathbb{R} : u(x) > 0\}.$$

Since I is nonempty and open, it can be decomposed into a countable family of disjoint open intervals $\{I_k\}$. Note that $u = 0$ and $u' = 0$ at the (finite) endpoints of each I_k.

Now consider any interval $I_k = (a, b)$ where a or b, but not both, may be infinite. As before, there must be at least one point $x_0 \in (a, b)$ where $u'(x_0) = 0$. Translating x_0 to the origin, it follows again that $\hat{u}'(x) < 0$ for $x \in (0, b - x_0)$ and $\hat{u}'(x) > 0$ for $x \in (a - x_0, 0)$. Hence the function

$$\hat{w}(x) = \begin{cases} \hat{u}(x) & \text{if } x \in (a - x_0, b - x_0) \\ 0 & \text{otherwise} \end{cases}$$

is a restricted ground state. Since at least one of the points $a - x_0$ and $b - x_0$ is finite, it follows by symmetry that both of them must be finite. Hence $\hat{w}$ also has compact support. This completes the proof of the first part of the theorem.

The converse statement is obvious from the autonomy of (1.2a).

From Theorems A and D we see that when (1.4) holds, and $q > 1$, every ground state is positive. Theorems B and D imply that if $q < \min(p, 1)$, then ground states are either everywhere positive or compactly supported on a *finite* number of congruent intervals, on which the functions w_k are all translations of w_1, and Theorems C and D imply that if $p < q$, there are no positive ground states, although the support $\cup \bar{I}_k$ may still be infinite.

It is worth noting that for any *fixed* parameter pair (p, q) which satisfies (1.4) there exists an a priori bound for the absolute value norm of a corresponding ground state.

A point worth making is that the analysis of ground states in the phase plane can be interpreted as the description of solutions of first order equations of quasi-Ricatti type

$$v\frac{dv}{du} = \lambda v^q - u^p,$$

see (2.2), or after the change of variables $v^2 \to y$, $u \to t$,

$$\dot{y} = 2(\lambda y^{q/2} - t^p).$$

When $q = 4$ in particular we obtain the explicit Ricatti equation

$$\dot{y} = \lambda y^2 - 2t^p$$

(of course this value of q lies outside the parameter range (1.4)).

Another related equation is

$$\ddot{u} + \frac{n-1}{r}\dot{u} + \frac{1}{r^2}u^p = 0, \qquad \dot{} = \frac{d}{dr}, \tag{1.10}$$

essentially the Matukuma equation with the term $u^p/(1+r^2)$ replaced by u^p/r^2. Thus the asymptotic behavior of positive solutions as $r \to \infty$ can be expected to be the same, at the first order. If we make the change of variables $x = (n-2)\log r$, then (1.10) becomes

$$u'' + u' + (n-2)^{-2}u^p = 0, \qquad ' = \frac{d}{dx}. \tag{1.11}$$

Positive solutions of (1.10) necessarily approach zero as $r \to \infty$, and therefore satisfy $\dot{u} < 0$ for all sufficiently large r, say $r \geq r_0 > 0$. In turn for $x \geq x_0$ we have $u' < 0$, and (1.11) becomes just

$$u'' - |u'| + \lambda u^p = 0 \tag{1.12}$$

with $\lambda = (n-2)^{-2}$, that is (1.2a) with $q = 1$.

We can thus infer the asymptotic behavior of solutions of (1.10) for $p > 1$ from that of (1.12). Using the remark just after Theorem C, we then see that either

$$r^{n-2}u(r) \to \text{ constant } > 0 \quad \text{as } r \to \infty$$

(the constant depending only on p and n, and there being exactly one solution with this property), or

$$u(r) \sim \left(\frac{n-2}{p-1}\,\frac{1}{\log r}\right)^{1/(p-1)}.$$

This result can be compared with the expansion (1.4), (1.5) in Li [3] where $n = 3$.

In conclusion, we note that when $q \neq 2p/(p+1)$, we may set $\lambda = 1$ by scaling the variables according to

$$y = \lambda^{-(q-1)/\{p(2-q)-q\}}x \quad \text{and} \quad w(y) = \lambda^{(2-q)/\{p(2-q)-q\}}u(x).$$

We shall accordingly often set $\lambda = 1$ in the following analysis.

2. Preliminaries. Suppose u is a solution of (1.2a) on $[0, \infty)$, which satisfies (1.8) and (1.9). Then an easy argument shows that $u(x) \to 0$ as $x \to \xi$, where $\xi > 0$ may be finite or infinite. Thus the solutions we are interested in correspond to orbits (u, v) of the first order system

$$\begin{cases} u' = -v \\ v' = u^p - v^q, \end{cases} \tag{2.1}$$

in the first quadrant $Q = \{u > 0, \ v \geq 0\}$, that is, maximally continued solutions in Q. In particular, we seek orbits which start (with $x = 0$) on the positive u-axis at points $(\gamma, 0)$ and converge through Q to the origin, the only equilibrium point. The isoclines $\Gamma_u = \{u' = 0\}$ and $\Gamma_v = \{v' = 0\}$ are

$$\Gamma_u : v = 0 \quad \text{and} \quad \Gamma_v : v = u^\alpha, \qquad \alpha = \frac{p}{q}. \tag{2.2}$$

It is clear that orbits must cross Γ_v, and then stay above Γ_v.

As long as p and q are greater than or equal to 1, the right hand sides of (2.1) are C^1, so that the initial value problem with

$$u(0) = \gamma \quad \text{and} \quad v(0) = 0 \tag{2.3}$$

has a unique solution in Q, which moreover depends continuously on γ. The same is also true when p or q is less than 1. To see this, we write (2.1) in the domain $\{0 \leq v < u^\alpha\}$ in the form

$$\frac{du}{dv} = \frac{v}{v^q - u^p},$$

and note that the right hand side is *continuous in* v and C^1 in u near $u = \gamma$. It now follows at once that the trajectory is unique, and since $dv/dt = \gamma^p$ when $t = 0$, also the orbit is unique.

To determine the behavior of orbits near the origin, we introduce the family of curves

$$C_\lambda = \{u > 0, \ v = \lambda u^\alpha\}, \quad \lambda > 0$$

(the parameter λ here is not to be confused with that in (1.2)).

Lemma 2.1. *For every* $\lambda > 1$ *there exists a number* $u_\lambda > 0$ *such that the region*

$$\Omega_\lambda = \{0 < u < u_\lambda,\ u^\alpha < v < \lambda u^\alpha\}$$

is invariant.

Proof. Since orbits cannot leave Ω_λ, either through the isocline Γ_v or through the line $\{u = u_\lambda\}$, it remains to inspect the vectorfield on C_λ. There we have

$$\left.\frac{dv}{du}\right|_{C_\lambda} = \left(\lambda^{q-1} - \frac{1}{\lambda}\right) u^{p-\alpha},$$

whilst the tangent to C_λ has slope

$$T = \lambda\alpha u^{\alpha-1}.$$

Orbits cannot leave Ω_λ through C_λ when

$$\left.\frac{dv}{du}\right|_{C_\lambda} > T;$$

that is, when

$$\left(\lambda^{q-2} - \frac{1}{\lambda^2}\right)\frac{q}{p} > u^{2\alpha-1-p}. \tag{2.4}$$

By (1.4)

$$2\alpha = \frac{2p}{q} > p + 1,$$

so that the exponent in (2.4) is positive. Thus, when we set

$$u_\lambda^{2\alpha-1-p} = \left(\lambda^{q-2} - \frac{1}{\lambda^2}\right)\frac{q}{p},$$

we find that when $u < u_\lambda$ orbits cannot leave Ω_λ (see Figure 1).

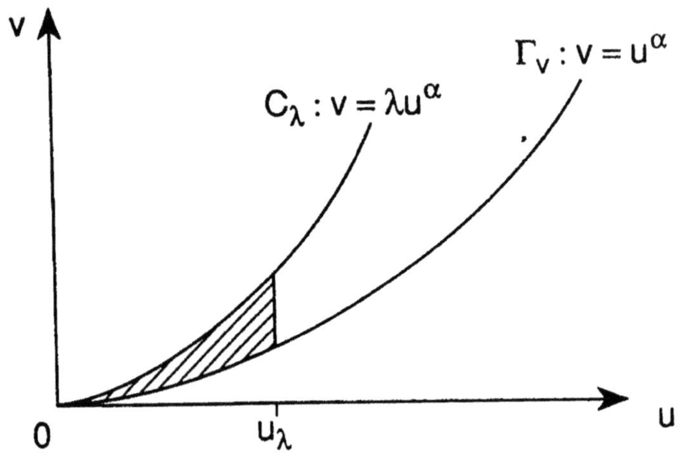

FIGURE 1. THE INVARIANT SET Ω_λ IS SHADED.

Define
$$\beta = \frac{1}{2-q}.$$
Note that (1.4) implies that
$$0 < \beta < \alpha.$$

Lemma 2.2. *Let ξ be the supremum of values x along an orbit (u, v), the possibility $\xi = \infty$ being admitted. Then there exists a nonnegative constant v_0, such that*
$$(u(x), v(x)) \to (0, v_0) \quad \textit{as} \quad x \to \xi.$$

This follows by inspection of the vector field in the phase plane.

If $v_0 > 0$, then v_0 is a continuous function of the initial value γ. This is obvious if p and q are greater than or equal to 1, by the uniqueness of the Cauchy problem. Otherwise, we can write (2.1) in the domain $\{0 \le u < v^{1/\alpha}\}$ in the form
$$\frac{dv}{du} = \frac{v^q - u^p}{v}.$$
The right hand side is now *continuous in u* and C^1 in v, and the argument preceding Lemma 2.1 applies.

Next we turn to the case $v_0 = 0$.

Lemma 2.3. *Suppose $v_0 = 0$. Then along the orbit (u, v) we have*

(*a*)
$$v < \left(\frac{u}{\beta}\right)^{\beta}.$$

Moreover, either

(*b*)
$$\lim_{x\to\xi} \frac{v(x)}{u^{\alpha}(x)} = 1$$

or

(*c*)
$$\lim_{x\to\xi} \frac{v(x)}{u^{\beta}(x)} = \beta^{-\beta},$$

where $\alpha = p/q$ and $\beta = 1/(2-q)$.

Proof. Clearly we have
$$\frac{dv}{du} = \frac{v^q - u^p}{v} < v^{q-1}.$$
Integrating with respect to u over $(0, u)$ and using Lemma 2.2, we obtain (a).

We now prove (b) and (c). Since any orbit must cross Γ_v, it follows at once that
$$\liminf_{x\to\xi} \frac{v(x)}{u^{\alpha}(x)} \ge 1.$$

Suppose now that

$$\liminf_{x\to\xi} \frac{v(x)}{u^\alpha(x)} = \ell < \infty.$$

Then it follows easily from Lemma 2.1 that

$$\lim_{x\to\xi} \frac{v(x)}{u^\alpha(x)} = \ell.$$

Hence, by l'Hôpital's rule,

$$\begin{aligned}\ell = \lim_{x\to\xi} \frac{v'(x)}{\alpha u^{\alpha-1}(x)u'(x)} &= \frac{1}{\alpha}\lim_{x\to\xi} \frac{v^{q-1}(x)}{u^{\alpha-1}(x)}\Big(1 - \frac{u^p}{v^q}\Big)\\ &= \frac{1}{\alpha}\Big(\ell^{q-1} - \frac{1}{\ell}\Big) \lim_{x\to\xi} u^{1-(\alpha/\beta)}(x).\end{aligned}$$

If $\ell \neq 1$, then the limit on the right hand side exists, and is infinite because $1-(\alpha/\beta) < 0$. Thus $\ell = \infty$, a contradiction. Therefore $\ell = 1$.

Next suppose that

$$\lim_{x\to\xi} \frac{v(x)}{u^\alpha(x)} = \infty.$$

Then, using l'Hôpital's rule again, we find that

$$\begin{aligned}\lim_{x\to\xi} \frac{v^{1/\beta}(x)}{u(x)} &= \frac{1}{\beta}\lim_{x\to\xi} \frac{v^{1-q}(x)v'(x)}{u'(x)} = \frac{1}{\beta}\lim_{x\to\xi} v^{-q}(x)\{v^q(x) - u^p(x)\}\\ &= \frac{1}{\beta}\lim_{x\to\xi}\Big\{1 - \Big(\frac{u^\alpha(x)}{v(x)}\Big)^q\Big\} = \frac{1}{\beta}.\end{aligned}$$

This completes the proof.

In the following lemma we shall show that only one orbit enters the origin, such that $v \sim (u/\beta)^\beta$. This orbit will be called *exceptional.*

Lemma 2.4. *There exists at most one orbit which enters the origin and is such that*

$$\lim_{x\to\xi} \frac{v(x)}{u^\beta(x)} = \beta^{-\beta}.$$

Proof. Because $u' = -v < 0$, we may use u as an independent variable. Setting $w = v^2$, we then obtain the equation

$$w' = 2w^{q/2} - 2u^p \qquad (' = d/du).$$

Since we are interested in solutions which approach zero according to

$$w(u) \sim \beta^{-2\beta}u^{2\beta} \quad \text{as} \quad u \to 0,$$

we introduce the new variables

$$t = -\log u \quad \text{and} \quad y(t) = \left(\frac{\beta}{u}\right)^{2\beta} w(t).$$

This yields the problem

$$\begin{cases} y' = f(y) + Ae^{-\nu t} & (' = d/dt) & (2.5a) \\ y(t) \to 1 \text{ as } t \to \infty, & & (2.5b) \end{cases}$$

where

$$f(y) = 2\beta(y - y^{q/2}), \qquad A = 2\beta^{2\beta} \tag{2.6}$$

and

$$\nu = \frac{2p - q(p+1)}{2-q} > 0. \tag{2.7}$$

We first prove an asymptotic estimate for $y(t)$.

Lemma 2.5. *Let $y(t)$ be a solution of Problem* (2.5). *Then*

$$y(t) = 1 - \frac{A}{1+\nu}e^{-\nu t} + \mathcal{O}(e^{-2\nu t}) \quad \textit{as} \quad t \to \infty.$$

Proof. By Lemma 2.3(a) and the fact that $v_0 = 0$, we have $0 < y < 1$. Also it is easy to check that $f(y) < 0$ for $0 < y < 1$, so from (2.5a) follows

$$y' < Ae^{-\nu t}.$$

Integrating from t to ∞ then yields

$$1 - y(t) < \frac{A}{\nu}e^{-\nu t}. \tag{2.8}$$

Next we write

$$y = 1 - \eta \quad \text{and} \quad f(y) = -\eta + g(\eta).$$

Then because $f(y)$ is smooth near $y = 1$, and $f(1) = 0$, $f'(1) = 1$, we have

$$g(\eta) = \mathcal{O}(\eta^2) \quad \text{and} \quad g'(\eta) = \mathcal{O}(\eta) \quad \text{as} \quad \eta \to 0. \tag{2.9}$$

We obtain for η the differential equation

$$\eta' - \eta = -Ae^{-\nu t} - g(\eta),$$

which in turn gives the integral equation

$$\eta(t) = \frac{A}{\nu+1}e^{-\nu t} + e^t \int_t^\infty g(\eta(s))e^{-s}ds. \tag{2.10}$$

Observe that by (2.8) and (2.9) the last term in (2.10) is $\mathcal{O}(e^{-2\nu t})$. This completes the proof of Lemma 2.5.

We are now ready to prove the uniqueness. Suppose y_1 and y_2 are solutions of Problem (2.5). Then the corresponding functions η_1 and η_2 both satisfy (2.10) and by Lemma 2.5 their difference $z = |\eta_1 - \eta_2|$ satisfies

$$z(t) \le Ce^t \int_t^\infty e^{-(\nu+1)s} z(s)\,ds \quad \text{and} \quad z(t) \to 0 \text{ as } t \to \infty,$$

where C is some positive constant. From this inequality we deduce, as in the proof of Gronwall's lemma, that

$$z(t) \le C\exp\Big(t + \frac{Ce^{-\nu t}}{\nu}\Big) \cdot \int_T^\infty e^{-(\nu+1)s} z(s)\,ds$$

for any $T > t$. Letting $T \to \infty$ then gives

$$z(t) \equiv 0.$$

Thus (2.5) can have no more than one solution, and in turn at most one orbit can enter the origin such that $v \sim (u/\beta)^\beta$.

Lemma 2.5 allows us to refine the local behavior of the exceptional orbit near the origin.

Corollary 2.6. *Let (u, v) correspond to the exceptional orbit entering the origin. Then*

$$v(u) = \Big(\frac{u}{\beta}\Big)^\beta \Big\{1 - \frac{\beta^{2\beta}}{1+\nu} u^\nu + \mathcal{O}(u^{2\nu})\Big\} \quad as \quad u \to 0.$$

We conclude this section with an improved estimate for those orbits which approach the origin like $v \sim u^\alpha$.

Lemma 2.7. *Suppose (u, v) corresponds to an orbit entering the origin, which is not exceptional. Then*

$$v(u) = u^\alpha \Big\{1 + \frac{p}{q^2} u^\mu + o(u^\mu)\Big\} \quad as \quad u \to 0,$$

where $\mu = 2(p/q) - (p+1)$.

Proof. Set

$$t = -\log u \quad \text{and} \quad z(t) = u^{-2\alpha} v^2(u). \tag{2.11}$$

Then

$$z' = 2\alpha z - 2e^{\mu t}(z^{q/2} - 1) \quad \text{and} \quad z(t) \to 1 \text{ as } t \to \infty,$$

and we wish to know the second term in the expansion for $z(t)$ as $t \to \infty$.

Inspection of the phase plane shows that orbits must eventually lie above the isocline $\{z' = 0\}$, so that

$$\liminf_{t\to\infty} e^{\mu t}\{z(t) - 1\} \geq \frac{2\alpha}{q}. \tag{2.12}$$

We assert that

$$\limsup_{t\to\infty} e^{\mu t}\{z(t) - 1\} \leq \frac{2\alpha}{q}. \tag{2.13}$$

Once we have proved (2.13) we can conclude in view of (2.12) that

$$z(t) = 1 + \frac{2\alpha}{q}e^{-\mu t} + o(e^{-\mu t}) \quad \text{as} \quad t \to \infty,$$

which readily yields the estimate we set out to prove.

Suppose to the contrary that (2.13) does not hold and that there exists an $\varepsilon > 0$ and an unbounded sequence (t_k) such that

$$e^{\mu t}\{z(t) - 1\} \geq \frac{2\alpha}{q} + \varepsilon \tag{2.14}$$

along this sequence. Whenever (2.14) holds, and t is large enough, we deduce from (2.11) after an elementary computation that

$$z'(t) < -\frac{1}{2}q\varepsilon. \tag{2.15}$$

Thus (2.14) cannot hold on an interval (T, ∞), but only in an unbounded sequence of intervals. Therefore we may choose the points t_k so that

$$\left(e^{\mu t_k}\{z(t_k) - 1\}\right)' = 0 \quad \text{or} \quad z'(t_k) + \mu\{z(t_k) - 1\} = 0.$$

Because $z(t) \to 1$ as $t \to \infty$, this is again incompatible with (2.15). This completes the proof of Lemma 2.7.

3. Main results. Here we obtain the general behavior of restricted ground states of Problem (I). First we consider the orbits of (2.1) introduced in the previous section.

Theorem 1. *Suppose* $q < 2p/(p+1)$. *Then there exists a unique positive number* γ_0 *such that* (i) *the orbit* (u_γ, v_γ) *of* (2.1) *starting at* $(\gamma, 0)$ *converges to* $(0, 0)$ *as* $x \to \xi$ *whenever* $\gamma \leq \gamma_0$, *and* (ii) *crosses the* v*-axis at some finite positive value* v_0 *whenever* $\gamma > \gamma_0$.

Moreover, we have

$$\lim_{x\to\xi} \frac{u^p(x)}{v^q(x)} = 1 \qquad \textit{if} \quad \gamma \in (0, \gamma_0) \tag{3.1}$$

and

$$\lim_{x\to\xi} \frac{u(x)}{v^{2-q}(x)} = \frac{1}{2-q} \qquad \text{if } \gamma = \gamma_0. \tag{3.2}$$

Proof. Consider the sets

$$I_0 = \{\gamma \in (0,\infty) : (u_\gamma, v_\gamma) \text{ eventually enters one of the invariant domains } \Omega_\lambda\}$$
$$I_1 = \{\gamma \in (0,\infty) : (u_\gamma, v_\gamma) \text{ eventually enters the domain } \{v > (u/\beta)^\beta\}\}.$$

Here $\lambda > 1$. We shall show below that I_0 and I_1 are open and disjoint, that I_0 contains all sufficiently small γ, and that I_1 contains all sufficiently large γ.

This will prove the theorem. Indeed, since two open nonempty disjoint sets cannot cover $(0,\infty)$, it follows that there is at least one γ, say $\gamma = \gamma_0$, which is in neither I_0 nor I_1. On the other hand, any orbit which never enters the set $\{v > (u/\beta)^\beta\}$ certainly approaches $(0,0)$ as $x \to \xi$; see Lemma 2.2. Thus if the orbit also does not enter any domain Ω_λ, we clearly must be in case (c) of Lemma 2.3. By Lemma 2.4 there is *at most one* value γ which is in neither I_0 nor I_1. Consequently there is *exactly one* value $\gamma = \gamma_0$ which is neither in I_0 nor in I_1. Hence I_0 and I_1 must be intervals, separated by the single point γ_0, and the corresponding orbit for γ_0, since it falls under case (c) of Lemma 2.3, satisfies (3.2).

Now suppose $\gamma \in I_0$. Then since Ω_λ is invariant it follows from Lemma 2.2 that $(u_\gamma, v_\gamma) \to (0,0)$ as $x \to \xi$, proving (i); moreover from Lemma 2.3 it is case (b) which occurs, proving (3.1).

Finally, if $\gamma \in I_1$ then by Lemma 2.3 (a) we must have $v_0 \neq 0$. Thus the orbit (u_γ, v_γ) exits $\overline{Q}$ at $(0, v_0)$, as required for (ii).

The following steps now complete the proof.

1. *I_0 and I_1 are open.* This is immediate from the continuous dependence of solutions on the initial Cauchy data established in Section 2.

2. *I_0 and I_1 are disjoint.* This is obvious from our earlier discussion, since $v_0 = 0$ for I_0 and $v_0 \neq 0$ for I_1.

3. *I_0 contains all sufficiently small values γ.* Let $\lambda > 1$ be fixed, say $\lambda = 2$. Then for any $\gamma \leq u_\lambda$, the orbit crosses the isocline Γ_v and consequently (since along the orbit we have $u(x) < u_\lambda$) enters Ω_λ.

4. *I_1 contains all sufficiently large values γ.* The curves $v = u^\alpha$ and $v = (u/\beta)^\beta$ cross at a *single point* (μ, μ^α) with $\mu = \beta^{\beta/(\beta-\alpha)}$. Let $y > \mu$ be fixed and consider the line

$$L : v = -au + b$$

where a, b are positive numbers, chosen so that L intersects the curve $v = (u/\beta)^\beta$ at the point (y, z). If we write $z = \rho y^\alpha$, then clearly,

$$\rho = \frac{y^{\beta-\alpha}}{\beta^\beta} < \frac{\mu^{\beta-\alpha}}{\beta^\beta} = 1$$

since $0 < \beta < \alpha$. See Figure 2.

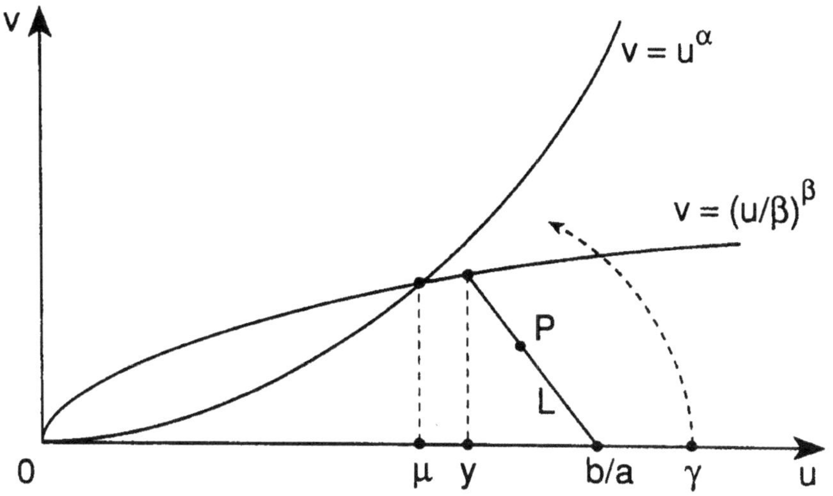

FIGURE 2. ORBITS WITH $\gamma \in I_1$.

Let $\Gamma = (u_\gamma(x), v_\gamma(x))$ be an orbit with $\gamma > b/a$, so that the initial point $(\gamma, 0)$ is to the right of L. Suppose Γ intersects L at some point $P = (u, v)$ with $u > y$, $0 < v < z$. Then at P we must have

$$\frac{dv_\gamma}{du_\gamma} = \frac{u^p - v^q}{-v} \geq -a \tag{3.3}$$

since $-a$ is the slope of L. On the other hand, if we set

$$a = \frac{1-\rho^q}{\rho} y^{\alpha(q-1)}$$

we find that $u^p - v^q > av$ at P, which contradicts (3.3). Hence with this choice of a, and with b chosen as above, the orbit Γ cannot intersect L at P, and in turn therefore crosses the curve $v = (u/\beta)^\beta$. Thus $\gamma \in I_1$, which completes the proof.

Having determined the structure of orbits of (2.1) in the preceeding theorem, we can now translate these results to solutions of (I). In particular if $(u(x), v(x))$ is an orbit of (2.1) starting at the point $(\gamma, 0)$, it is clear that $u(x)$ is then a solution for all $x \in [0, \xi)$ of the initial value problem (1.2a), (1.8). If $\gamma \leq \gamma_0$ and $\xi = \infty$ then the solution, when reflected to all of $\mathbb{R}$, obviously constitutes a restricted ground state with central value γ. If $\gamma \leq \gamma_0$ and $\xi < \infty$ then clearly

$$u(\xi) = u'(\xi) = 0$$

and in turn $u(x)$ can be continued as a solution to values $x > \xi$ by setting $u(x) \equiv 0$ for $x > \xi$. Thus such a solution (when reflected to all real x) yields a compact support ground state.

Finally, if $\gamma > \gamma_0$ then the orbit $(u(x), v(x))$ reaches the u-axis at a point $(0, v_0)$ with $v_0 > 0$ and with ξ finite. The corresponding solution $u(x)$ thus reaches zero at $x = \xi$, with a negative derivative, and accordingly cannot represent a ground state.

We turn finally to the asymptotics of $u(x)$ in terms of the principal variable x. Recall that

$$q < \frac{2p}{p+1}$$

throughout the discussion.

Theorem 2. *Suppose $u(x)$ is a restricted ground state with central value $u(0) < \gamma_0$.* (a) *If $0 < p < q$ then $u(x)$ has compact support. Moreover if supp $u(x) = [0, \xi]$ then*

$$\lim_{x\to\xi} \frac{u(x)}{(\xi - x)^{q/(q-p)}} = \left(\frac{q-p}{q}\right)^{q/(q-p)}. \tag{3.4}$$

(b) *If $p \geq q$ then $u(x)$ is positive for all $x > 0$, and*

$$\begin{cases} x^{q/(p-q)}u(x) \to \left(\dfrac{q}{p-q}\right)^{q/(p-q)} & (p > q) \\ e^{x}u(x) \to \text{const.} > 0 & (p = q) \end{cases} \tag{3.5}$$

as $x \to \infty$.

Proof. (a) Along the orbit $(u(x), v(x))$ we have $v > u^\alpha$ for all sufficiently small u, say $u \leq \varepsilon$. Hence for $u \leq \varepsilon$

$$x = \int_\gamma^u \frac{du}{u'} = \int_u^\gamma \frac{du}{v} < \text{const.} + \int_u^\varepsilon \frac{du}{u^\alpha} < \infty,$$

the integral being convergent since $\alpha < 1$. Thus x remains bounded as the orbit approaches $(0, 0)$. Consequently $\xi < \infty$ and $u(x)$ has compact support. The asymptotic behavior at $x = \xi$ is found from l'Hôpital's rule by the calculation

$$\lim_{x\to\xi} \frac{u^{(q-p)/q}(x)}{\xi - x} = (\alpha - 1)\lim_{x\to\xi} u^{-\alpha}(x)u'(x) = (1-\alpha)\lim_{x\to\xi}\frac{v(x)}{u^\alpha(x)} = 1 - \alpha,$$

according to Theorem 1.

(b) From (3.1) it follows that $v < 2u^\alpha$ for all sufficiently small u, say $u \leq \varepsilon$. Hence for $u \leq \varepsilon$

$$x = \text{const.} + \int_u^\varepsilon \frac{du}{v} > \text{const.} + \frac{1}{2}\int_u^\varepsilon \frac{du}{u^\alpha}$$

and the latter integral diverges as $u \to 0$ since $\alpha \geq 1$. Thus $\xi = \infty$.

When $p > q$, we have by l'Hôpital's rule

$$\lim_{x\to\infty} xu^{(p-q)/q}(x) = \lim_{x\to\infty} \frac{x}{u^{1-\alpha}(x)} = \frac{1}{1-\alpha} \lim_{x\to\infty} \frac{1}{u^{-\alpha}(x)u'(x)}$$
$$= \frac{1}{\alpha-1} \lim_{x\to\infty} \frac{u^{\alpha}(x)}{v(x)} = \frac{1}{\alpha-1},$$

as required.

When $p = q$ we proceed in a similar manner. In this case we have $\alpha = 1$, and so from Lemma 2.7, in which now $\mu = 1 - p > 0$, we obtain

$$x = \int_u^\gamma \frac{du}{v} = \int_u^\gamma \frac{du}{u}[1 + \mathcal{O}(u^\mu)] = \log\frac{\gamma}{u} + \text{constant}.$$

This proves the second part of (3.5).

Theorem 3. *Let $u(x)$ be a restricted ground state with central value $u(0) = \gamma_0$.*
(a) *If $q < 1$ then $u(x)$ has compact support. Moreover*

$$\lim_{x\to\xi} \frac{u(x)}{(\xi - x)^{(2-q)/(1-q)}} = \frac{1}{2-q}(1-q)^{(2-q)/(1-q)}. \tag{3.6}$$

(b) *If $q \geq 1$ then $u(x)$ is positive for all $x > 0$ and*

$$\begin{cases} x^{(2-q)/(q-1)}u(x) \to \dfrac{1}{2-q}\left(\dfrac{1}{q-1}\right)^{(2-q)/(q-1)} & (q > 1) \\ e^x u(x) \to \textit{const.} > 0 & (q = 1) \end{cases} \tag{3.7}$$

as $x \to \infty$.

Proof. (a) From (3.2) we obtain as in part (b) of Theorem 2, that

$$x < \text{const.} + \frac{2}{(2-q)^{1/(2-q)}} \int_u^\varepsilon \frac{du}{u^{1/(2-q)}} < \infty$$

since $q < 1$. Thus u has compact support. The formula (3.6) is then obtained exactly as in part (a) of Theorem 2.

To prove part (b), we observe from Lemma 2.3(a) that $v < (u/\beta)^\beta$ along the orbit, so

$$x = \int_u^\gamma \frac{du}{v} > \beta^\beta \int_u^\gamma \frac{du}{u^\beta}.$$

The latter integral diverges since $q \geq 1$ and so $\beta \geq 1$; consequently $u(x) > 0$ for all $x \geq 0$. The required asymptotic behavior (3.7) when $q > 1$ is obtained from l'Hôpital's rule as in part (b) of Theorem 2.

When $q = 1$, we need a slightly different argument. Then $\beta = 1$ and we obtain from Corollary 2.6 that

$$x = \int_u^\gamma \frac{du}{v} = \int_u^\gamma \frac{du}{u}[1 + \mathcal{O}(u^\nu)].$$

Here, $\nu = p - 1 > 0$ because $q = 1$ implies that $p > 1$. Thus, we obtain

$$x = \log\frac{\gamma}{u} + \text{constant}$$

from which the assertion follows.

Theorem 4. *If $\gamma > \gamma_0$ then the solution $u(x)$ of the initial value problem* (1.8) *reaches zero (with a negative derivative) at some finite value $x = \xi$*

This was already proved in the earlier discussion, prior to the statement of Theorem 2.

4. The parameter range $q \geq 2p/(p+1)$. Here we prove that there exist no ground states when

$$q > \frac{2p}{p+1} \tag{4.1}$$

and determine for which values of λ there exist ground states when p and q lie on the critical line

$$q = \frac{2p}{p+1}. \tag{4.2}$$

Because λ cannot be scaled away when (p, q) lies on the critical line, we shall retain λ throughout this section.

The following lemma is important in the proof of non-existence, giving the asymptotic behavior of positive ground states when $q \geq 2p/(p+1)$, assuming that they exist at all.

Lemma 4.1. *Suppose that $u(x)$ is a positive ground state of Problem* (I) *with*

$$p > 0, \qquad q \geq \frac{2p}{p+1}.$$

Then necessarily $p \geq 1$. Moreover
(a) *If $p > 1$, then $u = \mathcal{O}(x^{-2/(p-1)})$ and $u' = \mathcal{O}(x^{-(p+1)/(p-1)})$ as $x \to \infty$.*
(b) *If $p = 1$, then $u = \mathcal{O}(e^{-cx})$ and $u' = \mathcal{O}(e^{-cx})$ as $x \to \infty$ with $c = c(\lambda, p) > 0$;*

Proof. We follow ideas of [4] and set

$$S(x) = \lambda u^p - 2|u'|^{2p/(p+1)}.$$

Then by direct calculation and use of equation (1.2a), we find that $S'(\rho) < 0$ for any sufficiently large value $x = \rho$ where $S = 0$. Thus there are two alternatives, namely,
(i) S is ultimately negative, or
(ii) S is ultimately positive.
If (i) occurs, we have (recall $u' < 0$)

$$-u' > cu^{(p+1)/2}, \qquad c = (\lambda/2)^{(p+1)/2p}$$

for suitably large x. When $p < 1$ integration of this yields an immediate contradiction, while when $p \geq 1$ we obtain the first parts of (a) and (b).

To prove the second part of (a) we can argue exactly as in the proof of $(6.4)_2$ on page 121 of [4], using the second of the two cases and setting $n = 1$.

To prove the second part of (b) we observe that

$$|u'(x)| = \int_x^\infty u'' = \int_x^\infty |u'|^q + \mathcal{O}(e^{-cx}).$$

Because $u'(x) \to 0$ as $x \to \infty$, it follows that $|u'|^q \leq |u'| = -u'$ for x sufficiently large and so

$$|u'(x)| \leq u(x) + \mathcal{O}(e^{-cx}) = \mathcal{O}(e^{-cx}).$$

If (ii) occurs, then for suitably large x, say $x \geq x_1$, we have

$$|u'|^{2p/(p+1)} < \frac{\lambda}{2}u^p \tag{4.3}$$

so that

$$u'' + \frac{\lambda}{2}u^p < 0, \qquad x \geq x_1.$$

Then by integration from x_1 to x ($\geq 2x_1$)

$$|u'(x)| = |u'(x_1)| - \int_{x_1}^x u''(t)\,dt > \frac{\lambda}{2}\int_{x/2}^x u^p(t)\,dt > \frac{\lambda}{4}xu^p(x),$$

since $u(x)$ is decreasing. Another integration yields an immediate contradiction if $p < 1$.

When $p > 1$ integration gives $u(x) \leq \text{Const.}\ x^{-2/(p-1)}$ so that the first part of (a) holds. A similar integration also yields the first part of (b). The second part of (a) and (b) then follows at once from (4.3).

Lemma 4.2. *Suppose that*

$$p \geq 1 \quad \textit{and} \quad q > \frac{2p}{p+1}.$$

Then for any $\varepsilon > 0$ there exists a constant $\rho > 0$ such that

$$-u'(x) \geq cx^{-\varepsilon}, \qquad x \geq \rho, \tag{4.4}$$

where c is a positive constant.

Proof. First consider $p > 1$. Define

$$\delta = \frac{q(p+1) - 2p}{p-1} > 0.$$

By part (a) of Lemma 4.1 we have ultimately

$$|u'|^{q-1} \leq cx^{-1-\delta}$$

(note that here and in what follows we use c as a generic positive constant). Therefore, given $\varepsilon > 0$ we can choose ρ_0 large enough so that (1.2a) gives

$$u'' + \frac{\varepsilon}{x} u' < 0, \qquad x \geq \rho_0.$$

Thus $(x^{\varepsilon} u')$ is ultimately decreasing, and since u' is negative it follows that $-u' \geq cx^{-\varepsilon}$ for all large x.

If $p = 1$ and $q > 1$ the same argument applies, since from case (b) of Lemma 4.1 we find for any $\delta > 0$ and for all sufficiently large x, that

$$|u'|^{q-1} \leq cx^{-1-\delta}.$$

Hence again (4.4) holds also for this case.

Theorem 5. *Problem* (I) *admits no ground states if*

$$p > 0 \quad \text{and} \quad q > \frac{2p}{p+1} \tag{4.1}$$

while if $q = 2p/(p+1)$ *ground states exist if and only if*

$$p > 0 \quad \text{and} \quad \lambda \leq \lambda_p = \frac{(2p)^p}{(p+1)^{2p+1}}.$$

Proof. We first consider *positive* ground states when (4.1) holds. By Lemma 4.1 we must then have $p \geq 1$, so that (4.4) holds. Putting $\varepsilon = \frac{1}{2}$ and integrating (4.4) over (ρ, x), we find that $u(x) \to -\infty$ as $x \to \infty$, which is absurd. Thus there cannot be positive ground states when (4.1) holds. As for compact support ground states when (4.1) holds, their nonexistence for the same range follows from Theorem 6 of [4] (the proof there works for all $n \geq 1$).

To prove the second half of the theorem we use ideas and results of [2]. In this case we can scale the equation so that $u(0) = 1$. We denote the solution of (1.2a), with initial data $u(0) = 1$ and $u'(0) = 0$ by $u(x, \lambda)$, write

$$\xi(\lambda) = \sup\{x > 0 : u(\cdot, \lambda) > 0 \text{ on } [0, x)\},$$

and define the sets

$$S^+ = \{\lambda > 0 : \xi(\lambda) = \infty, \text{ or if } \xi(\lambda) < \infty \text{ then } u'(\xi(\lambda), \lambda) = 0\},$$
$$S^- = \{\lambda > 0 : \xi(\lambda) < \infty \text{ and } u'(\xi(\lambda), \lambda) < 0\}.$$

Plainly, for every $\lambda \in S^+$ the solution $u(x, \lambda)$, continued by zero beyond $\xi(\lambda)$ when $\xi(\lambda) < \infty$, is a restricted ground state.

It is shown in [2] that $(0, \lambda_p) \subset S^+$ and $(\lambda_p, \infty) \subset S^-$ (Actually, the proof is only given for $p > 1$, but it can easily be modified to apply for $0 < p \leq 1$ as well).

For $\lambda \in \mathcal{S}^-$ it is evident that the limit value v_0 in Lemma 2.2 is positive. Hence by the remark following Lemma 2.2 and the scaling properties of the equation, it follows that $\mathcal{S}^-$ is open. Because $\mathcal{S}^+ \cup \mathcal{S}^- = (0, \infty)$ it follows that $\mathcal{S}^+$ must be closed, so that $\lambda_p \in \mathcal{S}^+$. This completes the proof.

REFERENCES

[1] L. Alfonsi and F.B. Weissler, Blow up in $\mathbf{R}^n$ for a parabolic equation with a damping nonlinear gradient term, In "Nonlinear Diffusion Equations and Their Equilibrium States 3," (Eds. N.G. Lloyd, W.M. Ni, L.A. Peletier and J. Serrin), pp. 1–20, Birkhäuser, Boston, 1992.

[2] M. Chipot and F.B. Weissler, *Some blowup results for a nonlinear parabolic equation with a gradient term*, SIAM J. Math. Anal., 20 (1989), 886–907.

[3] Yi Li, *On the positive solutions of the Matukuma equation*, Duke. Math. J., 70 (1993), 575–589.

[4] J. Serrin and H. Zou, *Existence and non-existence results for ground states of quasilinear elliptic equations*, Arch. Rational Mech. Anal., 121 (1992), 101–130.

Giovanni Leoni, Carnegie Mellon University, Pittsburgh, PA, 2012.

Henghui Zou, Djairo de Figueredo and James Serrin, Niagara Falls, Canada, 1996.

James Serrin and Moxun Tang, Niagara Falls, Canada, 1996.

I.9. Ground States – Uniqueness

Moxun Tang

Mathematics Subject Classification (2010). 35J92, 35J93, 35J25, 34B18.
Keywords. Ground states; uniqueness; separation technique.

1. The uniqueness theorems

A central question in the study of the semilinear elliptic equation

$$\Delta u + f(u) = 0 \quad \text{in} \quad \mathbb{R}^n, \qquad n > 1, \tag{1.1}$$

has been understanding how the number of solutions correlates with the qualitative behavior of the nonlinearity $f(u)$. In the 1983 paper [i] below, Peletier and Serrin discussed the uniqueness of positive classical solutions of (1.1) which tend to zero at ∞, the so-called ground states, when f is negative near zero. They assumed, with a slight modification, the following condition:

A. f is locally Lipschitz continuous on $[0, \infty)$; $f(0) = 0$ and $f(u) < 0$ in $(0, b)$ for some $b > 0$.

Under this condition, the integral $F(u) = \int_0^u f(s)ds$ is clearly negative in $(0, b)$. Peletier and Serrin observed that $F(u(0)) > 0$ for any radial ground state. Hence a necessary condition for the existence of a radial ground state is that

$$\beta = \inf\{u : F(u) > 0\} < \infty. \tag{1.2}$$

Paper [i] contains a uniqueness theorem. We present a slightly improved version of this theorem, which is found in the 1986 paper [4] of the same authors.

Theorem 1. [i], [4] *Let condition* **A** *hold, and suppose that* $f(u)/(u - \beta)$ *is non-increasing for all* $u \in (\beta, \infty)$ *where* $f(u) > 0$. *Then* (1.1) *cannot have more than one ground state.*

In 1996, Franchi, Lanconelli, and Serrin [ii] extended this remarkable uniqueness theorem to a large class of quasilinear elliptic equations of the form

$$\operatorname{div}[A(|Du|)Du] + f(u) = 0,$$

with A subject to the ellipticity conditions $A(p) > 0$ and $[pA(p)]' > 0$ for $p > 0$ and some further inequalities. A large class of special cases is given by the equation

$$\operatorname{div}\left(\frac{|Du|^{m-2}Du}{(1 + |Du|^2)^{\kappa/2}}\right) + f(u) = 0 \quad \text{in } \mathbb{R}^n, \qquad n \geq 2, \ \kappa > 0, \ m \geq 1 + \kappa; \ \text{or } \kappa = 0, \ m > 1. \tag{1.3}$$

P. Pucci et al. (eds.), *James Serrin. Selected Papers: Volume 1*, Contemporary Mathematicians, DOI 10.1007/978-3-0348-0845-3_9, © Springer Basel 2014

This applies in particular to the m-Laplacian $\Delta_m u = \operatorname{div}(|Du|^{m-2}Du)$, and the mean curvature operator $\operatorname{div}\left(Du/(1+|Du|^2)^{1/2}\right)$, which were not previously treated in the ground state literature. Theorem A of [ii] proves the existence of at least one radially symmetric ground state under the minimal hypothesis

H. f is continuous on $[0,\infty)$, $f(0)=0$, and β defined by (1.2) is positive.

The following important uniqueness theorem was proved on pages 213–215 of [ii].

Theorem 2. [ii] *Let condition* **H** *hold. Suppose that there exists $\gamma \in (\beta,\infty]$ such that f is locally Lipschitz continuous, positive, and non-increasing in (β,γ). Then (1.3) has at most one radial ground state, with $||u||_\infty < \gamma$.*

In paper [iii] below, Serrin and the Commentator considered the case in which f is superlinear at infinity, which was not covered by papers [i] and [ii]. The paper focused on the m-Laplace equation

$$\Delta_m u + f(u) = 0 \quad \text{in} \quad \mathbb{R}^n, \qquad n > m > 1, \tag{1.4}$$

under the basic assumption

C. f is continuous and $f(u) \le 0$ on $(0,b]$ for some $b>0$; $f(u)>0$ on (b,γ) for some $b<\gamma\le\infty$.

They proved

Theorem 3. [iii] *Let condition* **C** *hold, and suppose that $f \in C^1(b,\gamma)$, and $g(u) = uf'(u)/f(u)$ is non-increasing on (b,γ). Then (1.4) admits at most one radial ground state, with $||u||_\infty < \gamma$.*

It is noted that if $\gamma<\infty$, and $f(u)<0$ for $\gamma<u<\infty$, then there can be no ground states with $||u||_\infty \ge \gamma$. Under the conditions of Theorem 3, it was also shown in [iii] that the uniqueness result is valid for radial solutions of the homogeneous Dirichlet–Neumann free boundary problem

$$\Delta_m u + f(u) = 0, \qquad n > m > 1,$$
$$u > 0 \ \text{ in } B_R, \quad u = \frac{\partial u}{\partial \nu} = 0 \ \text{ on } \partial B_R,$$

where B_R is an open ball in $\mathbb{R}^n$ with radius $R>0$.

Although Theorems 1–3 do not produce conditions which are both necessary and sufficient for the uniqueness of ground states, they contain the weakest sufficient conditions for the m-Laplace equation found so far. Moreover, Theorem 3 *completely* covers the important canonical nonlinearities

$$f(u) = -u^p + u^q, \quad \text{or} \quad f(u) = -u^s - \lambda u^p + \mu u^q, \quad 0<s<p<q;\ \lambda,\ \mu>0.$$

On the other hand, the canonical example

$$f(u) = -u + \lambda u^p + u^q, \quad 1<p<q,\ \lambda>0,$$

does not satisfy the conditions of any of Theorems 1–3. Indeed in this case it has recently been shown that there exist parameter values such that (1.1) admits at least three radial ground states [1].

A more striking example of non-uniqueness of ground states was given by Peletier and Serrin [i]. Let $f(u)$ be a function satisfying the conditions of Theorem 1. Then (1.1) has a unique ground state $u(|x|)$ with $u(0)>\beta$. Let a be an arbitrary number in $(\beta, u(0))$. In their elegant construction, they first defined a continuous function f_1 by letting $f_1 \equiv f$

in $[0, a-\epsilon]$ for a sufficiently small $\epsilon > 0$, f_1 linear in $[a-\epsilon, a]$ with $f_1(a) = 0$, and $f_1 \equiv 0$ for $u > a$. The new function f_1 again satisfies the condition of Theorem 1, so (1.1) with f replaced by f_1 admits a ground state $u_1(|x|)$ with $u_1(0) < a$. They then defined a second continuous function f_2 by letting $f_2 \equiv f_1$ in $[0, u_1(0)]$, f_2 linear in $(u_1(0), a+\epsilon)$ with $f_2(a+\epsilon) = f(a+\epsilon)$, and $f_2 \equiv f$ for $u \geq a+\epsilon$. Let f be replaced by f_2 in (1.1). It is clear that $u_1(|x|)$ is again a ground state of (1.1). Furthermore, they proved that there is another ground state $u_2(|x|)$ with $u_2(0)$ close to $u(0)$. Since f_2 satisfies the condition A, this non-uniqueness example demonstrates the sharpness of the uniqueness theorem, and indicates that it is, in general, impossible to find a set of conditions which is both sufficient and necessary for the uniqueness of ground states.

The proof of Theorem 1 is based on what Peletier and Serrin [i] called the separation technique. This technique consists of establishing the following three properties of radial solutions of (1.1).

1. If $u_1(r)$ and $u_2(r)$ are radial ground states of (1.1) and $u_1(0) < u_2(0)$, then there is a $\rho > \beta$ such that $u_1(\rho) = u_2(\rho)$.
2. If $u_1(r)$ and $u_2(r)$ are radial ground states of (1.1) and $u_1(0) < u_2(0)$, and if $u_1(\rho) = u_2(\rho)$ for some $\rho > \beta$, then $u_1(r) \equiv u_2(r)$.
3. If $u_1(r)$ and $u_2(r)$ are radial ground states of (1.1) and $u_1(0) = u_2(0)$, then $u_1(r) \equiv u_2(r)$.

These three properties clearly imply uniqueness. Furthermore, Property 3 is a standard uniqueness theorem for a singular initial value problem. To prove Properties 1 and 2, they first showed that $u_1(r)$ and $u_2(r)$ are decreasing functions of r, and then worked with their inverse functions $r_1(u)$ and $r_2(u)$. The rationale behind this innovative approach lies on the following important discovery.

The Maximum Principle. [i] *Let condition* **A** *hold. Then the difference* $t(u) = r_1(u) - r_2(u)$ *cannot assume a positive local minimum value, or a negative local maximum value, in* $(0, u_1(0))$.

It is interesting to note that there is no similar result for the difference $u_1(r) - u_2(r)$.

Paper [ii] extended the separation technique to the class of semilinear equations treated therein. The proof requires the introduction of an unusual differential identity. For the special case of the equation

$$\left[u'/\sqrt{1+u'^2}\,\right]' + f(u) = 0$$

for a radial solution for the quasilinear equation with the minimal surface operator, the identity is

$$\frac{d}{dr}\left[r\left(\frac{p^2}{1+p^2} + \frac{2F(u)}{\sqrt{1+p^2}} - F^2(u)\right)^{1/2(n-1)}\right]$$
$$= \left(\frac{p^2}{1+p^2} + \frac{2F(u)}{\sqrt{1+p^2}} - F^2(u)\right)^{-(n-3/2)/(n-1)} \left(\frac{p^2+2}{\sqrt{1+p^2}}\, F(u) - F^2(u)\right)$$

where $p = |u'(r)|$.

The proof of Theorem 3 combines these techniques with some ideas from a paper of Erbe and Tang [2], as well as a Pohozaev-type identity due to Ni and Serrin [3], and Pucci and Serrin [5].

References

[1] J. DÁVILA, M. DEL PINO AND I. GUERRA, Non-uniqueness of positive ground states of nonlinear Schrödinger equations, *Proc. London Math. Soc.* **106** (2013), 318–344.

[2] L. ERBE AND M. TANG, Uniqueness theorems for positive solutions of quasilinear elliptic equations in a ball, *J. Diff. Eqs.* **138** (1997), 351–379.

[3] W.-M. NI AND J. SERRIN, Nonexistence theorems for quasilinear partial differential equations, *Rend. Circolo Mat. Palermo (Centenary Supplement)* **8** (1985), 171–185.

[4] L. A. PELETIER AND J. SERRIN, Uniqueness of non-negative solutions of semilinear equations in $\mathbb{R}^N$, *J. Diff. Eqs.* **61** (1986), 380–397.

[5] P. PUCCI AND J. SERRIN, A general variational identity, *Indiana Univ. Math. J.* **35** (1986), 681–703.

Moxun Tang
Department of Mathematics
Michigan State University
East Lansing
Michigan, 48824, USA
e-mail: tang@math.msu.edu

J. Differ. Equations 61 (1986), pp. 380–397. **Reprint I.9.[i]**

Uniqueness of Non-negative Solutions of Semilinear Equations in $\mathbb{R}^n$

L. A. PELETIER

Department of Mathematics, University of Leiden, Leiden, The Netherlands

AND

JAMES SERRIN

Department of Mathematics, University of Minnesota, Minneapolis, Minnesota

Received March 12, 1984; revised January 7, 1985

In an earlier paper [6] we discussed the uniqueness of positive classical solutions $u(x)$ of the problem

$$\text{(I)} \quad \begin{aligned} &\Delta u + f(u) = 0 && \text{in} \quad \mathbb{R}^n \\ &u(x) \to 0 && \text{as} \quad x \to \infty, \end{aligned}$$

in which $n > 1$ and $x = (x_1, ..., x_n)$, under the following basic assumptions on the function f:

A1. f is defined and locally Lipschitz continuous on $(0, \infty)$.

A2. The integral $F(u) = \int_0^u f(s)\, ds$ exists for all $u > 0$ and is positive at some value $\delta > 0$.

By **A2** both of the quantities

$$\alpha = \inf\{u > 0 \colon f(u) > 0\}, \qquad \beta = \inf\{u > 0 \colon F(u) > 0\}$$

are well-defined (clearly $0 \leqslant \alpha \leqslant \beta$). It was shown that Problem (I) has at most one positive radial solution if f additionally satisfies the hypotheses:

H. $\lim_{u \to 0}(f(u)/u) = -m$ for some positive number m.

380

0022-0396/86 $3.00

S. The function

$$u \mapsto \frac{f(u)}{u-\beta}$$

is nonincreasing for all $u \in (\beta, \infty)$ where $f(u) > 0$.

Conditions **A1** and **A2** are quite mild, **A2** in fact being necessary for the existence of a solution. There are, however, no intrinsic reasons for assuming **H**, and it is the object of this paper to weaken this assumption considerably. We shall replace it by

H*. One of the following conditions holds:

(i) $n = 2$ and $\beta > 0$,

(ii) $n > 2$ and $\alpha > 0$.

Hypothesis **H** of course implies $\alpha > 0$, so that **H*** obviously is a considerably weaker condition than **H**. In particular, we note that **H*** allows the possibility $f \equiv 0$ for u near 0.

It is known that there can be no positive radial solution of (I) when

$$\int_0 \frac{du}{\sqrt{F(u)}} < \infty;$$

see [1] and Section 4. On the other hand, even though (I) cannot have positive radial solutions in this case, it *can* have solutions which are non-negative but not identically zero. Such behavior is interesting in itself; in order to include it in our considerations we shall henceforth allow not just the class of positive solutions but also those which are non-negative and non-trivial.

So that the equation be well-defined when $u = 0$ we naturally add the condition:

A3. $\lim_{u \to 0} f(u) = 0$ and $f(0) = 0$.

(In general, any limit other than 0 is incompatible with the condition $u(x) \to 0$ as $x \to \infty$.)

In work of Gidas, Ni, and Nirenberg [2, 3] it is shown that when **H** holds and $f \in C^{1+\varepsilon}$ near $u = 0$, then any positive solution of (I) is radially symmetric about some center of symmetry. In the generality of our present assumptions, a corresponding result is not yet known (and may be quite difficult to verify). At the same time, radially symmetric solutions of (I) are an important class of solutions in their own right, and in many cases they are he *only* possible global solutions. Consequently, it is reasonable to study the uniqueness of radially symmetric solutions for their own sake.

In the sequel we shall consider, then, the following radial problem associated with (I):

$$\text{(II)}\qquad \begin{aligned} &u''+((n-1)/r)\,u'+f(u)=0, && r>0, && (1)\\ &u\geqslant 0, && r\geqslant 0, && (1')\\ &u'(0)=0, \qquad \lim_{r\to\infty} u(r)=0, && && (1'') \end{aligned}$$

where u is a C^2 function of the radial variable r on $[0, \infty)$, with $u \not\equiv 0$. We shall assume throughout that f is a given function satisfying the natural conditions **A1**, **A2**, **A3**. For simplicity we shall not refer to these conditions in the statements of our results, it being tacitly understood, however, that they are always present.

Our purpose is to show that Problem (II) has at most one solution under the hypotheses **H*** and **S**. Thus we considerable generalize the results of [6] both by extending the class of solutions considered and by weakening the restrictions which are placed on f near $u=0$.

The general approach is as follows. We say that a solution of Problem (II) is of class $\mathscr{C}$ if

$$\lim_{r\to\infty} r^{n-1}u'(r) \text{ exists (finite).} \tag{2}$$

This condition, which is crucial to our deliberations, is easily shown to follow from condition **H***(ii); see Lemma 5, part (i). We then prove

THEOREM 1. *Suppose $\beta>0$, and let u and v be two different solutions of Problem* (II). *Assume moreover either that $n=2$ or that both u and v are of class $\mathscr{C}$. Then the graphs of u and v must intersect at some point (R, U) with $U>0$.*

The proof of this theorem is modelled somewhat along the lines of the corresponding result in [6], but requires considerably more care because we can no longer rely on the strong hypothesis **H**. An interesting set of lemmas used in the proof concerns the *horizontal* separations of solution graphs (see Section 2). We subsequently obtain (see Section 3) a strengthened version of Theorem 1.

THEOREM 2. *Let the hypotheses of Theorem* 1 *hold. Then the graphs of u and v must intersect at some point (R, U) with $U>\beta$.*

The final step in the proof consists in showing that if condition **S** holds then the graphs of the solutions cannot intersect at any point (R, U) for which $U>\beta$. Together with Theorem 2 this yields the following conclusions.

THEOREM 3. *Let conditions* **H*** *and* **S** *hold. Then Problem* II *cannot have more than one solution.*

THEOREM 4. *Suppose* $\beta > 0$. *If condition* **S** *holds, there can be at most one solution of Problem* II *which is of class* $\mathscr{C}$.

The number n in Problem (II) denotes the number of space dimensions, and was assumed to be an integer greater than 1. If we drop this interpretation, and assume merely that n is a parameter in the range $(1, \infty)$, we find that Theorem 1 continues to hold while Theorems 2 and 3 require the extra condition $n \geqslant \frac{3}{2}$. For values of n in the range $(1, \frac{3}{2})$ the conclusion of Theorem 2 continues to hold however if we replace β by α. If we correspondingly strengthen condition **S** be replacing β by α this gives the following result.

Suppose $n \in (1, \frac{3}{2})$ *and let* **S** *hold with* β *replaced by* α. *Then Problem* (II) *cannot have more than one solution if* $\alpha > 0$.

Problems in which f fails to satisfy **H** occur frequently in applications. As an example we mention a model arising in population dynamics, due to Gurtin and MacCamy [4], describing the spread of biological populations. If u denotes the population density, one is led to the diffusion equation

$$u_t = \Delta(u^m) + u(1-u)(u-a) \qquad (m > 1, 0 < a < 1).$$

Let $\tilde{u}$ be an equilibrium solution of this equation. Since $\tilde{u}$ is a density, we may expect $\tilde{u} \geqslant 0$. Then $v = \tilde{u}^m$ satisfies Eq. (1) in which $f(v) \sim -av^{1/m}$ near $v = 0$, and hence, since $m > 1$, f satisfies condition **H*** but not **H**.

The outline of the paper is as follows. In Section 1 we give some general properties of solutions of Problem II, quoting from [6] when convenient. In Section 2 we consider various properties of solutions of class $\mathscr{C}$, including the main separation lemmas and Theorem 1. Section 3 contains the proofs of Theorems 2, 3, and 4.

Other uniqueness theorems for global solutions of Problems (I) and (II), which replace condition **S** by various alternate restrictions on f, have been obtained by K. McLeod and Serrin [5].

1. GENERAL BEHAVIOUR OF SOLUTIONS

We begin by deriving some basic identities.

LEMMA 1. *Let* $u = u(r)$ *be a solution of Eq.* (1) *on a finite interval* $[r_0, r_1] \subset \bar{\mathbb{R}}^+$ *and let* $u > 0$ *on* (r_0, r_1). *Then*

$$\left\{\frac{1}{2}u'(r_1)^2+F(u_1)\right\}-\left\{\frac{1}{2}u'(r_0)^2+F(u_0)\right\}=-(n-1)\int_{r_0}^{r_1}u'(r)^2\frac{dr}{r}, \quad (3)$$

where $u_i=u(r_i)$, $i=0, 1$.

Proof. Multiply Eq. (1) by u', and integrate over (r_0, r_1).

LEMMA 2. *Let u be a strictly monotone solution of Eq.* (1) *on a finite interval* $[r_0, r_1]\subset\bar{\mathbb{R}}^+$. *Let* $r(u)$ *be the inverse of* $u(r)$. *Then*

$$\text{(i)}\quad \tfrac{1}{2}r_1^{2n-2}u'(r_1)^2-\tfrac{1}{2}r_0^{2n-2}u'(r_0)^2=\int_{u_1}^{u_0}r^{2n-2}(u)\,f(u)\,du \quad (4)$$

$$\text{(ii)}\quad r_1^{2n-2}\{\tfrac{1}{2}u'(r_1)^2+F(u_1)\}-r_0^{2n-2}\{\tfrac{1}{2}u'(r_0)^2+F(u_0)\}$$

$$=2(n-1)\int_{r_0}^{r_1}r^{2n-3}F(u(r))\,dr, \quad (5)$$

where $u_i=u(r_i)$, $i=0, 1$.

Proof. Multiply Eq. (1) by r^{n-1}. Then we can write it as

$$(r^{n-1}u')'+r^{n-1}f(u)=0.$$

Next we multiply by $r^{n-1}u'$ and integrate over (r_0, r_1). There results

$$\tfrac{1}{2}r_1^{2n-2}u'(r_1)^2-\tfrac{1}{2}r_0^{2n-2}u'(r_0)^2=-\int_{r_0}^{r_1}r^{2n-2}f(u)u'\,dr. \quad (6)$$

Changing the integration variable to u in the integral yields (i). Since $f(u)u'=dF(u)/dr$, an integration by parts gives

$$\int_{r_0}^{r_1}r^{2n-2}f(u)u'\,dr=r^{2n-2}F(u(r))\Big|_{r_0}^{r_1}-2(n-1)\int_{r_0}^{r_1}r^{2n-3}F(u(r))\,dr. \quad (7)$$

Putting (6) and (7) together, we obtain (ii).

Next, we shall extend these identities to solutions of Problem (II). We first recall a lemma from [6].

LEMMA 3. *Let u be a solution of Problem* (II). *Then* $u'(r)\leqslant 0$ *for all* $r>0$. *Equality holds at a point* $\bar{r}$ *if and only if* $u(\bar{r})=0$.

Remark. Lemma 3 was proved in [6] only for positive solutions of (1). The conclusion there was stronger: $u'<0$. However, the argument given there can easily be generalized to non-negative solutions. Since $u'=0$ if $u=0$, the conclusion can only be $u'\leqslant 0$.

COROLLARY. *Let u be a solution of Problem* (II). *Then either* $u>0$ *on* $[0, \infty)$, *or* $u>0$ *on* $[0, a)$ *and* $u=0$ *on* $[a, \infty)$ *for some* $a>0$.

LEMMA 4. *Let u be a non-trivial solution of Problem* (II). *Then*

$$\text{(i)}\quad \lim_{r\to\infty} u'(r)=0,$$

$$\text{(ii)}\quad \frac{1}{2}u'(r^2)+F(u(r))=(n-1)\int_r^\infty u'(s)^2\frac{ds}{s},\qquad r\geqslant 0. \tag{8}$$

Proof. In view of the preceding Corollary, we may assume that $u>0$ on $[0, a)$ for some maximal $a>0$ which may be finite or infinite. Suppose $a<\infty$. Then (i) is obvious since $u'=0$ on $[a, \infty)$ and (ii) follows by setting $r_1=a$ in (3).

Next suppose $a=\infty$. We let $r_1\to\infty$ in (3). The right-hand side converges to some negative limit, or to $-\infty$. This implies that $\frac{1}{2}u'(r_1)^2+F(u(r_1))$ converges, though possibly to $-\infty$. However, as $r_1\to\infty$, $u(r_1)\to 0$ and hence $F(u(r_1))\to 0$. Thus $\frac{1}{2}u'(r_1)^2$ converges (clearly not to $-\infty$, and so, equally clearly) to some non-negative limit l^2. Since $u(r_1)\to 0$ as $r_1\to\infty$, we can only have $l=0$. This proves part (i) and moreover yields the relation

$$\lim_{r\to\infty}\{\tfrac{1}{2}u'(r)^2+F(u(r))\}=0. \tag{9}$$

Part (ii) now follows at once if we take $r_0\geqslant 0$ in (3), let $r_1\to\infty$, and then use (9).

Remark. Let u be a solution of Problem (II). If we set $r=0$ in Lemma 4(ii) we obtain

$$F(u(0))=(n-1)\int_0^\infty u'(r)^2\frac{dr}{r}>0.$$

Since $F\leqslant 0$ on $(0, \beta)$, this implies that $u(0)>\beta$.

LEMMA 5. *Let u be a solution of Problem* (II). *Then*

(i) *if* $\alpha>0$, *there exists a non-negative number L such that*

$$\lim_{r\to\infty} r^{n-1}u'(r)=-L \tag{10}$$

and

$$\lim_{r\to\infty} r^{2(n-1)}F(u(r))=0; \tag{11}$$

(ii) *if $\beta > 0$, there exists a non-negative number L_1 such that*

$$\lim_{r \to \infty} r^{2(n-1)}\{\tfrac{1}{2}u'(r)^2 + F(u(r))\} = \tfrac{1}{2}L_1^2 \tag{12}$$

and

$$r^{2(n-1)}\{\tfrac{1}{2}u'(r)^2 + F(u(r))\} = \tfrac{1}{2}L_1^2 - 2(n-1)\int_r^\infty t^{2n-3}F(u(t))\,dt. \tag{13}$$

If $n \leqslant 2$ then $L_1 = 0$.

Proof. If u vanishes for all sufficiently large r then (i) and (ii) follow immediately with $L = L_1 = 0$ (for (13), let $r_1 \to \infty$ in (5)). Thus let $u > 0$ on $[0, \infty)$.

(i) Since $u(r) \to 0$ as $r \to \infty$, and $f(u) \leqslant 0$ for $0 < u \leqslant \alpha$, we have $f(u(r)) \leqslant 0$ for r large. Hence, if we let $r_1 \to \infty$ in (4) the integral will eventually decrease monotonically. On the other hand, using (4) again, the integral is bounded below: for all $r_1 > r_0 \geqslant 0$

$$\int_{u_1}^{u_0} r^{2n-2}(u)\,f(u)\,du \geqslant -\tfrac{1}{2}r_0^{2(n-1)}u'(r_0)^2,$$

where $u_i = u(r_i)$, $i = 0, 1$. Hence the integral in (4) converges to some finite limit as $r_1 \to \infty$, from which (10) follows at once.

By l'Hôpital's rule

$$\lim_{r \to \infty} r^{2(n-1)}\left\{\frac{1}{2}u'(r)^2 + F(u(r))\right\} = \lim_{r \to \infty} \frac{\tfrac{1}{2}u'(r)^2 + F(u(r))}{r^{-2(n-1)}}$$

$$= \lim_{r \to \infty} \frac{u'u'' + f(u)u'}{-2(n-1)\,r^{-2n+1}}$$

$$= \frac{1}{2}\lim_{r \to \infty} r^{2(n-1)}u'(r)^2 \qquad \text{(using (1))}$$

$$= \frac{1}{2}L^2 \qquad \text{(by (10))}.$$

Thus, again in view of (10),

$$\lim_{r \to \infty} r^{2(n-1)}F(u(r)) = 0.$$

(ii) The proof of (12) is almost identical to that of (10), except that we use the identity (5). To prove that the integral is bounded below as $r_1 \to \infty$, we now use (8) to obtain $\frac{1}{2}u'(r)^2 + F(u(r)) > 0$ for $r \geqslant 0$.

To show that $L_1 = 0$ when $n \leqslant 2$, suppose for contradiction that L_1 were positive in this case. Then by (12), since $F \leqslant 0$ for u near 0, we have

$$-r^{n-1}u' \geqslant \tfrac{1}{8}L_1 > 0$$

for all sufficiently large r. A simple integration then shows that $u \to \infty$ as $r \to \infty$, an obvious contradiction. The proof is therefore completed.

Remark. If $\alpha = 0$ but $\beta > 0$, Lemma 5 does not give an estimate for the behavior of either $u(r)$ or $u'(r)$ as $r \to \infty$. On the other hand, if we assume that u is of class $\mathscr{C}$, with

$$\lim_{r \to \infty} r^{n-1}u'(r) = -L \tag{14}$$

then $L_1 = L$ and

$$\lim_{r \to \infty} r^{2(n-1)}F(u(r)) = 0 \tag{15}$$

as can be seen from an argument similar to the one used to prove (11).

2. Separation of Solution Graphs

Let $u(r)$ and $v(r)$ be solutions of Problem (II), and let $u(r) \not\equiv v(r)$. By Lemma 3 and its corollary $u'(r) < 0$ and $v'(r) < 0$ when $u(r) > 0$ and $v(r) > 0$. Hence their inverses $r(u)$ and $s(u)$ are respectively well-defined for u in $(0, u(0))$ and u in $(0, v(0))$. Note that $u(0) \neq v(0)$, for otherwise the two solutions would be identical.

We begin with two preliminary lemmas.

LEMMA 6. *The function $r(u) - s(u)$ can have at most a finite number of zeros if one of the following conditions is satisfied*:

(i) $\alpha > 0$, $n > 1$;

(ii) $\beta > 0$, $n \geqslant \frac{3}{2}$.

Proof. The proof of part (i) is essentially the same as that in [6], so we omit it.

To prove part (ii), we observe, as in [6], that the points of intersection are isolated, and thus, that if there exists an infinite number of them, they can be enumerated, say $u_1 > u_2 > u_3 \ldots$, and $u_k \to 0$ as $k \to \infty$. Let η_0 be a particular zero with $\eta_0 \leqslant \beta$ and set $r_0 = r(\eta_0)$. We can assume without loss of generality that $r > s$ in some right neighborhood of $u = \eta_0$.

If there exists a zero of $r - s$ on $(0, \eta_0)$, there must exist a number $\eta_1 \in (0, \eta_0)$ such that

$$r > s,\ r' < s' \text{ on } (\eta_1, \eta_0) \qquad \text{and} \qquad r'(\eta_1) = s'(\eta_1). \tag{16}$$

Set $r_1 = r(\eta_1)$ and $s_1 = s(\eta_1)$. If we apply the identity (5) for u on (r_0, r_1) and for v on (r_0, s_1) and then subtract we obtain

$$(r_1^{2n-2} - s_1^{2n-2})\{\tfrac{1}{2}u'(r_1)^2 + F(\eta_1)\} - \tfrac{1}{2}r_0^{2n-2}\{u'(r_0)^2 - v'(r_0)^2\}$$
$$= -\int_{\eta_1}^{\eta_0} \{r(u)^{2n-2} - s(u)^{2n-2}\}' F(u)\, du. \tag{17}$$

Since $0 > u'(r_0) > v'(r_0)$ we find, in view of Lemma 4(ii), that the left-hand side of (17) is positive. For $u \in (\eta_1, \eta_0)$ we have

$$\{r(u)^{2n-2} - s(u)^{2n-2}\}' = (2n-2)\{r^{2n-3}(r-s)' + (r^{2n-3} - s^{2n-3})s'\} < 0,$$

by (16) and the fact that $n \geqslant \frac{3}{2}$. Remembering that $(\eta_1, \eta_0) \subset (0, \beta)$, and hence that $F(u) \leqslant 0$ in (17), it follows that the right-hand side of (17) is non-positive.

Thus, there cannot be a zero of $r - s$ on $(0, \eta_0)$, and hence the number of zeros of $r - s$ must be finite.

Remark. The proof of Lemma 6(ii) yields in addition our first result on the horizontal separation of solution graphs.

Let u and v be solutions of Eq. (1) which intersect at a point (r_0, η_0). *If* $\beta > 0$, $\eta_0 \in (0, \beta]$, *and* $n \geqslant \frac{3}{2}$ *we have*

$$r'(\eta_0) < s'(\eta_0) \textit{ implies } r' < s' \textit{ on } (0, \eta_0], \tag{18}$$

If $1 < n < \frac{3}{2}$, *then* (18) *still holds if* $\alpha > 0$ *and* $\eta_0 \in (0, \alpha]$.

The conclusion for $1 < n < \frac{3}{2}$ follows when we use (4) instead of (5) in the proof of Lemma 6.

In the next set of lemmas we further consider the function $r(u) - s(u)$. Since the difference $r - s$ can have at most a finite number of zeros necessarily either $r - s > 0$ or $r - s < 0$ for all u sufficiently small. Without loss of generality we can always suppose the former to occur.

LEMMA 7. *Suppose* $r(u) - s(u) > 0$ *one some interval I. Then* $r(u) - s(u)$ *can have at most one critical point on I. Moreover, this critical point is a strict maximum.*

Since the proof of Lemma 7 in [6] is slightly more complicated than necessary, we give here a simpler version.

Proof of Lemma 7. We have

$$u_r = 1/r_u, \qquad u_{rr} = -r_{uu}/r_u^3.$$

Hence $r(u)$ satisfies the equation

$$r_{uu} - \frac{n-1}{r} r_u^2 - f(u) r_u^3 = 0.$$

A similar equation can be derived for $s(u)$. At a critical point we have $r > s$ and $r_u = s_u < 0$, so that by subtraction

$$(r-s)_{uu} = (n-1)\left(\frac{1}{r} - \frac{1}{s}\right) r_u^2 < 0.$$

Thus $r - s$ can only have a strict maximum.

We next consider solutions u and v satisfying condition $\mathscr{C}$ in the Introduction, with $n > 2$. We write

$$\lim_{r \to \infty} r^{n-1} u'(r) = -L, \qquad \lim_{r \to \infty} r^{n-1} v'(r) = -M.$$

LEMMA 8. *Suppose u and v satisfy condition $\mathscr{C}$, with $n > 2$. If $r(u) > s(u)$ for $0 < u < u_0$, then $L \geqslant M$.*

Proof. Suppose to the contrary that $L < M$. By l'Hôpital's rule

$$\frac{u(r)}{r^{2-n}} \to \frac{L}{n-2}, \quad \frac{v(r)}{r^{2-n}} \to \frac{M}{n-2} \qquad \text{as} \quad r \to \infty.$$

Hence, if we subtract, there results

$$r^{n-2}\{u(r) - v(r)\} \to \frac{L-M}{n-2} < 0$$

which contradicts the assumption that $r > s$ in $(0, u_0)$.

LEMMA 9. *Suppose u and v satisfy condition $\mathscr{C}$, with $n > 2$. Then if $\beta > 0$ and $r > s$ for $0 < u < u_0$ we have*

$$\{r(u) - s(u)\}' < 0 \qquad \textit{for} \quad 0 < u < u_0.$$

Proof. By Lemma 8 we have $L \geqslant M$. We now distinguish two cases: (i) $L > M$ and (ii) $L = M$.

(i) $L > M$. Choose $\varepsilon = \frac{1}{3}(L - M)$. Then we find, as in the proof of Lemma 8, that for r sufficiently large

$$u(r) > \frac{L-\varepsilon}{n-2} r^{-(n-2)} = u^*(r)$$

$$v(r) < \frac{M+\varepsilon}{n-2} r^{-(n-2)} = v^*(r).$$

Let r^* and s^* be the respective inverses of u^* and v^*. Then for u sufficiently small

$$r(u) - s(u) > r^*(u) - s^*(u) = \left\{ \left(\frac{L-\varepsilon}{n-2}\right)^{1/(n-2)} - \left(\frac{M+\varepsilon}{n-2}\right)^{1/(n-2)} \right\} u^{-1/(n-2)}.$$

Since $L - \varepsilon > M + \varepsilon$ we conclude that $r(u) - s(u) \to \infty$ as $u \to 0$, whence, by Lemmma 7, $\{r(u) - s(u)\}' < 0$ for $u \in (0, u_0)$.

(ii) $L = M$. By Lemma 4

$$\frac{1}{2} u'(r(u))^2 + F(u) = -(n-1) \int_0^u \frac{u'(r(t))}{r(t)} \, dt \tag{19}$$

and similarly for v, with r replaced by s. If we subtract the identity for v from that for u and write

$$a(u) = |u'(r(u))| \qquad \text{and} \qquad b(u) = |v'(s(u))|$$

we obtain (recall that $u', v' < 0$)

$$\begin{aligned} a(u)^2 - b(u)^2 &= 2(n-1) \int_0^u \left\{ \frac{a(t)}{r(t)} - \frac{b(t)}{s(t)} \right\} dt \\ &< 2(n-1) \int_0^u \frac{a(t) - b(t)}{r(t)} \, dt \end{aligned} \tag{20}$$

because $s < r$ on $(0, u_0)$.

Define

$$w(u) = \int_0^u \frac{a-b}{r} \, dt,$$

so that (20) becomes, with $' = d/du$,

$$(a+b) \, r w' < 2(n-1) w.$$

Integrating this inequality leads to

$$w(u) < w(\bar{u}) \exp\left\{2(n-1)\int_{\bar{u}}^{u} \frac{dt}{(a+b)\,r}\right\} \tag{21}$$

for $0 < \bar{u} < u < u_0$.

Assertion. $w(u) < 0$ for $0 < u < u_0$.

Assuming the assertion for the moment, we can readily complete the proof. Indeed if $w < 0$, then $a < b$ by (20), that is,

$$|u'(r(u))| < |v'(s(u))|, \qquad 0 < u < u_0,$$

or, in turn $\{r(u) - s(u)\}' < 0$ for $0 < u < u_0$, which was to be proved.

By Lemma 7 either $r(u) - s(u)$ is everywhere decreasing on $0 < u < u_0$ or $r(u) - s(u)$ is increasing for u near 0. In the first case $a(u) - b(u) < 0$ on $0 < u < u_0$, and in the second $a(u) - b(u) > 0$ for u near 0. Hence, by the definition of $w(u)$, either $w < 0$ for $0 < u < u_0$, and we are done, or $w(u) > 0$ for all u sufficiently small, say $0 < u < \delta$. Suppose then for contradiction that the latter case holds. Choose $\bar{u} \in (0, \delta)$. Then $w(\bar{u}) > 0$ and, by (21),

$$\begin{aligned} w(u) &< w(\bar{u}) \exp\left\{2(n-1)\int_{\bar{u}}^{u} \frac{dt}{ar}\right\} \\ &= w(\bar{u}) \exp\left\{2(n-1)\int_{\bar{r}}^{r} \frac{u'\,dr}{-u'r}\right\}, \end{aligned}$$

where $\bar{r} = r(\bar{u})$ and $0 < \bar{u} < u < u_0$. Thus $w(u) < w(\bar{u}) \exp\{2(n-1)\log(\bar{r}/r)\}$, or

$$r^{2(n-1)} w(u) < \bar{r}^{2(n-1)} w(\bar{u}), \qquad 0 < \bar{u} < u < u_0. \tag{22}$$

Since $b > 0$ we have from the definition of $w(u)$, for any $u > 0$,

$$\begin{aligned} r^{2(n-1)} w(u) &< r^{2(n-1)} \int_0^u \frac{a}{r}\,du \\ &= r^{2(n-1)} \int_r^{\infty} \frac{u'^2}{r}\,dr \\ &= \frac{r^{2(n-1)}}{n-1}\left\{\frac{1}{2}u'(r)^2 + F(u(r))\right\} \end{aligned}$$

by Lemma 4. Now from the remark after Lemma 5 (here we use the condition $\beta > 0$ and the fact that u is of type $\mathscr{C}$) there results

$$\limsup_{u \to 0} r^{2(n-1)} w(u) \leqslant \frac{L^2}{2(n-1)}. \tag{23}$$

Next, let $\bar{u} \to 0$ in (22) and use relation (23) (with the dummy variable u replaced by $\bar{u}$). This give

$$r^{2(n-1)} w(u) \leqslant \limsup_{\bar{u} \to 0} \bar{r}^{2(n-1)} w(\bar{u}) \leqslant \frac{L^2}{2(n-1)}, \qquad 0 < u < u_0.$$

If $L = 0$ this implies $w(u) \leqslant 0$ on $(0, u_0)$, contradicting the assumption that $w(u) > 0$ on $(0, \delta)$. Thus $w(u) < 0$ on $(0, u_0)$ in this case.

If $L > 0$, we need a more delicate estimate of the behavior of $r^{2(n-1)} w(u)$ as $u \to 0$. By l'Hôpital's rule we have (recall that $w(u) \to 0$ as $u \to 0$)

$$\begin{aligned} \lim_{u \to 0} r^{2(n-1)} w(u) &= \lim_{u \to 0} \frac{\int_0^u ((a-b)/r)\, dt}{r^{-2(n-1)}} \\ &= \lim_{u \to 0} \frac{(a-b) u'}{-2(n-1) r^{-2(n-1)}} \\ &= -\frac{1}{2(n-1)} \lim_{r \to \infty} (r^{n-1} u')^2 \lim_{u \to 0} \left\{ \frac{v'(s(u))}{u'(r(u))} - 1 \right\} \\ &= -\frac{L^2}{2(n-1)} \left\{ \lim_{u \to 0} \frac{v'}{u'} - 1 \right\} \end{aligned}$$

assuming the last limit to exist.

Because $r^{n-1} u'(r) \to -L$ and $r^{n-2} u(r) \to L/(n-2)$ (see Lemma 8), and since $L > 0$, we see that

$$\lim_{u \to 0} \frac{u'(r(u))}{u^{(n-1)/(n-2)}} = \lim_{r \to \infty} \frac{r^{n-1} u'(r)}{(r^{n-2} u(r))^{(n-1)/(n-2)}} = -L \left(\frac{L}{n-2} \right)^{-(n-1)/(n-2)}.$$

Similarly, since $L = M > 0$,

$$\lim_{u \to 0} \frac{v'(s(u))}{u^{(n-1)/(n-2)}} = -L \left(\frac{L}{n-2} \right)^{-(n-1)/(n-2)}.$$

Consequently

$$\lim_{u \to 0} \frac{v'(s(u))}{u'(r(u))} = 1$$

and so we find

$$\lim_{u \to 0} r^{2(n-1)} w(u) = 0.$$

The rest of the proof is the same as in the case $L = 0$.

Lemma 9 is crucial for the proof of Theorem 1 when u and v are of class $\mathscr{C}$. The corresponding result when $n \leqslant 2$ is the following

LEMMA 10. *Suppose $r > s$ for $0 < u < u_0$. Then if $\beta > 0$ and $n \leqslant 2$ we have*

$$\{r(u) - s(u)\}' < 0 \qquad \textit{for} \quad 0 < u < u_0.$$

Proof. By Lemma 5(ii) we have

$$r^{2(n-1)}\{\tfrac{1}{2}|u'|^2 + F(u)\} \to 0 \qquad \text{as} \quad r \to \infty.$$

The result now follows exactly as in the proof of Lemma 9, part (ii), since the condition (23) yields $\lim \bar{r}^{2(n-1)} w(\bar{u}) = 0$. Note in particular that the final step in the proof of Lemma 9 (the possibility $L > 0$) is not needed here.

The proof of Theorem 1 now follows as a simple consequence of Lemmas 9 and 10. Indeed suppose u and v are solutions of Problem (II) of class $\mathscr{C}$ which do not intersect at any positive value of u. Without loss of generality we may assume that $u > v$ as long as $u > 0$, and hence, that $r(u) - s(u) > 0$ on $(0, v(0)]$. By Lemma 9,

$$(r - s)' < 0 \qquad \text{for} \quad u \in (0, v(0)).$$

On the other hand,

$$(r - s)' \to \infty \qquad \text{as} \quad u \to v(0),$$

because $s'(u) \to -\infty$ and $r'(u)$ remains bounded as $u \to v(0)$. This contradiction shows that the graphs of u and v must intersect at some positive u. For $n \leqslant 2$ the proof is the same except that we use Lemma 10 instead of Lemma 9.

3. PROOFS OF THEOREMS 2, 3 AND 4

Let u and v be two different solutions of Problem (II), both of class $\mathscr{C}$, and let their last point of intersection (for $u > 0$) be (R, U). By Lemma 6 and Theorem 1 this point exists. By (ii) the Remark following Lemma 5 we have

$$R^{2n-2}\{\tfrac{1}{2}u'(R)^2+F(U)\}=\tfrac{1}{2}L^2-2(n-1)\int_R^{\infty} t^{2n-3}F(u(t))\,dt;$$

of course the same identity holds for v, provided L is replaced by M. Thus we obtain upon subtraction

$$\tfrac{1}{2}R^{2n-2}\{v'(R)^2-u'(R)^2\}+\tfrac{1}{2}(L^2-M^2)=-\int_0^U \Psi'(u)\,F(u)\,du, \tag{24}$$

where $\Psi(u)=r^{2n-2}-s^{2n-2}$.

Suppose, without loss of generality, that $r>s$ on $(0, U)$. Then $|v'(R)|>|u'(R)|$ and, by Lemma 8, $L\geqslant M$. Hence the left-hand side of (24) is positive.

From Lemma 9 we deduce that

$$\Psi'=2(n-1)\{(r^{2n-3}-s^{2n-3})\,r'+s^{2n-3}(r-s)'\}<0 \qquad \text{on} \quad (0, U)$$

for $n\geqslant\frac{3}{2}$. Therefore, if $U\leqslant\beta$, then $F(u)\leqslant 0$ on the whole of $(0, U)$ and the right-hand side of (24) is non-positive. Remembering that the left-hand side is positive, we conclude that $U>\beta$. This completes the proof of Theorem 2 for solutions of class $\mathscr{C}$. When $n\leqslant 2$ the argument is essentially the same except that $L_1=M_1=0$ by Lemma 5(ii), and Lemma 10 must be used at the final stage instead of Lemma 9.

The proofs of Theorems 3 and 4 are now completed as in [6]. One shows that the graph of two solutions cannot intersect above the line $u=\beta$ if **S** is satisfied [6, Lemma 10]. Since, by Theorem 2, the graphs of two solutions must necessarily intersect above the line $u=\beta$, the proof is complete.

Finally, let us consider the case $n\in(1,\frac{3}{2})$. If we let $r_1\to\infty$ in (4) and set $r_0=R$, we obtain from Lemma 5(i)

$$\tfrac{1}{2}R^{2n-2}u'(R)^2-\tfrac{1}{2}L^2=-\int_0^U r(u)^{2n-2}f(u)\,du.$$

For v one obtains a similar expression with L replaced by M. Subtraction yields

$$\tfrac{1}{2}R^{2n-2}\{v'(R)^2-u'(R)^2\}+\tfrac{1}{2}(L^2-M^2)=\int_0^U \Psi(u)\,f(u)\,du.$$

Assuming again that $r>s$ on $(0, U)$ we see that the left-hand side is positive and that the right-hand side is negative if $U\leqslant\alpha$. Thus $U>\alpha$. The argument for the case $n\in(1,\frac{3}{2})$ is then completed as before, with β in **S** replaced by α, and $\alpha>0$.

4. Compact Support

We denote by $u(r)$ a decreasing solution of Eq. (1) in $[R, \infty)$ with $U = u(R) > 0$ and $u(r) \to 0$ as $r \to \infty$. Let $r(u)$ denote as usual the inverse of $u(r)$.

LEMMA 11. *If* $U \in (0, \beta)$ *then*

$$r(u) \leqslant r(U) + 2^{-1/2} \int_u^U |F(s)|^{-1/2} \, ds, \qquad 0 < u \leqslant U.^{1} \tag{25}$$

Proof. By Lemma 4

$$\tfrac{1}{2}u'(r)^2 + F(u(r)) \geqslant 0 \qquad \text{for} \quad r \geqslant R$$

or, because $u(r) \in [0, \beta)$ and hence $F(u(r)) \leqslant 0$,

$$\tfrac{1}{2}u'(r)^2 \geqslant |F(u(r))| \qquad \text{for} \quad r \geqslant R.$$

Since $u' \leqslant 0$ this implies that

$$u'(r) \leqslant -2^{1/2} |F(u(r))|^{1/2}. \tag{26}$$

By the corollary to Lemma 3 there exists a number $a \leqslant \infty$ such that $u(r) > 0$ when $r < a$ and $u(r) = 0$ when $r \geqslant a$. Thus $R < a$, and we obtain from (26)

$$|F(u(r))|^{-1/2} u'(r) \leqslant -2^{1/2} \qquad \text{for} \quad R \leqslant r < a.$$

Integrating this inequality from R to $r \in (R, a)$ we obtain

$$-\int_u^U |F(s)|^{-1/2} \, ds \leqslant -2^{1/2}[r(u) - R],$$

or

$$r(u) \leqslant R + 2^{-1/2} \int_u^U |F(s)|^{-1/2} \, ds.$$

THEOREM 5. *Let* u *be a solution of Problem* II. *The condition*

$$\int_0 |F(s)|^{-1/2} \, ds < \infty \tag{27}$$

[1] If the integral fails to exist, then (25) holds vacuously.

is sufficient for u to have compact support. If $\alpha > 0$, *this condition is also necessary.*

Proof. That the condition (27) is sufficient for u to have compact support is an immediate consequence of (25).

To prove that (27) is necessary we assume that u has compact support $[0, a]$ with $a < \infty$.

By Lemma 4 we have

$$u'(r)^2 \leqslant 2\,|F(u)(r))| + c\int_r^a u'(s)^2\,ds, \qquad a-\delta < r < a,$$

where $c \equiv 2(n-1)/(a-\delta)$ and $\delta \in (0, a)$ will be chosen later. Applying Gronwall's lemma we deduce that

$$u'(r)^2 \leqslant 2\,|F(u(r))| + 2c\int_r^a |F(u(s))|\,e^{c(s-r)}\,ds. \tag{28}$$

Now we choose δ so small that $u(r) \in (0, \alpha)$ for $a-\delta < r < a$. Then

$$\frac{d}{dr}\,|F(u(r))| = -f(u(r))\,u'(r) \leqslant 0$$

and hence $|F(u(s))| \leqslant |F(u(r))|$. Using this in (28) we obtain

$$u'(r)^2 \leqslant 2\,|F(u(r))| \left\{1 + c\int_r^a e^{c(s-r)}\,ds\right\} \leqslant 2\,|F(u(r))|\,e^{c\delta}. \tag{29}$$

To complete the proof we proceed as in the proof of Lemma 11. Dividing by $|F|$ in (29) and taking the square root, we obtain

$$-|F(u(r))|^{-1/2}u'(r) \leqslant 2^{1/2}e^{\delta(n-1)/(a-\delta)}.$$

This yields, upon integration over $(a-\delta, a-\varepsilon)$, $0 < \varepsilon < \delta$,

$$\int_{u(a-\varepsilon)}^{u(a-\delta)} |F(u)|^{-1/2}\,du \leqslant 2^{1/2}\delta e^{\delta(n-1)/(a-\delta)}.$$

Because $u(a-\varepsilon) \to 0$ as $\varepsilon \to 0$, we conclude that

$$\int_0 |F(u)|^{-1/2}\,du < \infty,$$

which was to be proved.

Acknowledgment

This work was supported by a grant from the Netherlands Organization for the Advancement of Pure Research (ZWO) and by a grant from the National Science Foundation.

References

1. Ph. Benilan, H. Brezis, and M. G. Crandall, A semilinear elliptic equation in $L^1(\mathbb{R}^N)$, *Ann. Scuola Norm Sup. Pisa* **2** (1975), 523–555.
2. B. Gidas, W.-M. Ni, and L. Nirenberg, Symmetry and related properties via the maximum principle, *Comm. Math. Phys.* **68** (1979), 209–243.
3. B. Gidas, W.-M. Ni, and L. Nirenberg, Symmetry of positive solutions nonlinear elliptic equations in $\mathbb{R}^n$, *Math. Anal. Appl. Adv. in Math. Suppl. Stud.* **7A** (1981), 369–402.
4. M. E. Gurtin and R. C. MacCamy, On the diffusion of biological populations, *Math. Biosci.* **33** (1977), 35–49.
5. K. McLeod and J. Serrin, Uniqueness of the ground state solution for $\Delta u + f(u) = 0$, *Proc. Nat. Acad. Sci. USA* **78** (1981), 6592–6595.
6. L. A. Peletier and J. Serrin, Uniqueness of positive solutions of semilinear equations in $\mathbb{R}^n$, *Arch. Rational Mech. Anal.* **81** (1983), 181–197.

Advances in Math. 118 (1996), 177–243. **Reprint I.9.[ii]**

Pages 226–237 have been removed for the sake of brevity.

Existence and Uniqueness of Nonnegative Solutions of Quasilinear Equations in R^n

BRUNO FRANCHI AND ERMANNO LANCONELLI

Dipartimento di Matematica, Università di Bologna, Italy

AND

JAMES SERRIN

School of Mathematics, University of Minnesota, Minneapolis, Minnesota 55455

CONTENTS

0. INTRODUCTION

We shall consider the quasilinear elliptic equation

$$\operatorname{div}\{A(|Du|)\, Du\} + f(u) = 0, \qquad x \in R^n, \quad n \geqslant 2, \tag{1}$$

where $A(p)$ is a real positive continuous function defined for all $p > 0$ and $f(u)$ is a real continuous function defined for all $u \geqslant 0$. Our principal concern will be with ground states of (1), namely non-negative non-trivial continuously differentiable solutions $u = u(x)$ which tend to zero as the point x approaches infinity.

The special case where $A \equiv 1$, corresponding to the Laplace operator, is well-known and takes the classical form

$$\Delta u + f(u) = 0. \tag{2}$$

0001-8708/96 $18.00

When the function $f(u)$ is of class $C^{1+\varepsilon}$ for u near zero, and when $f(0)=0$, $f'(0)<0$, it has been shown by Gidas, Ni & Nirenberg that ground states of (2) are necessarily radially symmetric. In addition, under further appropriate conditions on f, the existence of radially symmetric ground states has been established by Strauss, Coleman, Glazer & Martin, Berestycki & Lions, Atkinson & Peletier, and Kaper & Kwong, while the uniqueness of such ground states is further treated in papers of Coffman, Cortázar *et al.*, Peletier & Serrin, McLeod & Serrin, Kaper & Kwong, Kwong, Kwong & Zhang and Yanagida.

Ground states for the general equation (1) can in most cases be expected to be radially symmetric, and we shall accordingly restrict our discussion to that case. More specifically, we shall establish both existence and uniqueness of *radially symmetric ground states of* (1) under appropriate conditions on the operator $A(p)$ and the function $f(u)$, and we shall also obtain a number of qualitative results concerning the behavior of solutions. Finally, occasionally at the expense of somewhat longer proofs, we shall always deal with the weakest hypotheses which we can treat.

Our results apply in particular to the important cases of the degenerate Laplace operator $A(p)=p^{m-2}$, $m>1$, the mean curvature operator $A(p)=(1+p^2)^{-1/2}$, and more generally to $A(p)=(1+p^2)^{-s/2}\,p^{m-2}$, where m and s are positive constants which satisfy suitable conditions (see below). Equation (1) can also be viewed as the Euler–Lagrange equation corresponding to the variational problem

$$\delta \int \{G(|Du|)-F(u)\}\, dx=0, \tag{3}$$

where

$$f(u)=\frac{dF}{du}, \qquad F(u)=\int_0^u f(t)\, dt,$$

and

$$A(p)=\frac{G_p(p)}{p}, \qquad G(p)=\int_0^p \rho A(\rho)\, d\rho.$$

Section 1 deals with general properties of radial solutions and radial ground states of (1), using ideas introduced by Peletier & Serrin in [PS1] and [PS2]. In Section 2, we give our existence results. The approach here is based on shooting methods, partially following work of Berestycki, Lions & Peletier and of Peletier & Serrin. It is also possible to develop variational methods to obtain existence, as in the works of Berestycki & Lions and of

Strauss; for several results in this direction, see the works of Citti and Kichenassamy & Smoller. Further work on existence and non-existence of solutions of (1), particularly for the case of the mean curvature operator, can be found in [NS1], [NS2], [PS3], and [APS]. In addition some related existence results for Monge–Ampère operators are proved in [F].

Section 3 is concerned with the uniqueness question for radial ground states. The essential tools for this purpose are the monotone separation theorem (see [PS1] and [PS2] for the case $A \equiv 1$) and a new identity for radial solutions of (1). In general form, this identity appears in Lemma 3.1.3. For the mean curvature operator $A(p) = 1/\sqrt{1+p^2}$, in particular, it reduces to the striking relation

$$\frac{d}{dr}\left\{ r\left(\frac{p^2}{1+p^2} + \frac{2F(u)}{\sqrt{1+p^2}} - F^2(u)\right)^{1/2(n-1)}\right\}$$
$$= \left(\frac{p^2}{1+p^2} + \frac{2F(u)}{\sqrt{1+p^2}} - F^2(u)\right)^{-(n-3/2)/(n-1)} \left(\frac{p^2+2}{\sqrt{1+p^2}} F(u) - F^2(u)\right),$$

where $u = u(r)$ is a radial solution of (1) for this operator, and $p = |u'(r)|$. Another remarkable formula occurs for the degenerate Laplace operator $A(p) = p^{m-2}$; namely for this case, radial solutions of (1) satisfy the identity

$$\frac{d}{dr}\left\{ r\left(p^m + \frac{m}{m-1} F(u)\right)^{(m-1)/m(n-1)}\right\}$$
$$= \frac{m}{m-1}\left(p^m + \frac{m}{m-1} F(u)\right)^{-(n-2+1/m)/(n-1)} F(u).$$

(Of course the above formulas are applicable only when they are meaningful, that is when $(1+p^2)^{-1/2} - F(u) \in (0, 1)$ for the first, and $p^m + (m/(m-1)) F(u) > 0$ for the second.)

We turn now to a more detailed statement of our assumptions and main conclusions. Throughout the paper we shall assume, without further comment, the following hypotheses concerning the function $f(u)$ and the operator $A(p)$:

(H1) f is continuous on $[0, \infty)$ and $f(0) = 0$,

(H2) A is continuous on $(0, \infty)$ and $pA(p) \rightarrow$ as $p \rightarrow 0$,

(H3) $pA(p)$ is strictly increasing for $p > 0$.

The continuity assumptions in (H1) and (H2) are essentially minimal, and indeed one of the purposes of the paper is just that, to work with minimal hypotheses. The conditions on f at $u = 0$ and on A as $p \rightarrow 0$ are required in order for a ground state to exist. Assumption (H3) is simply the condition that (1) be weakly elliptic or alternately that the integrand in (3) be strictly convex in Du. Note also that (H2) and (H3) imply $A>0$ for $p>0$.

It is easy to check that conditions (H2) and (H3) are satisfied both for the Laplace operator and for the degenerate Laplace operator $A=p^{m-2}$, $m>1$. They also hold for the mean curvature operator and even for the general operator $(1+p^2)^{-s/2}\,p^{m-2}$ when $s\geqslant 0$, $m>1$, $m\geqslant s+1$.

In order to state our main *existence theorems*, we first define

$$H(p)=p^2A(p)-\int_0^p \rho A(\rho)\,d\rho \quad \text{for} \quad \rho>0, \quad H(0)=0, \tag{4}$$

or, equivalently, with

$$\Omega(p)=pA(p) \quad \text{for} \quad p>0, \quad \Omega(0)=0,$$

and with $p(\Omega)$ denoting the inverse of $\Omega(p)$,

$$H(p)=\int_0^{\Omega(p)} p(\Omega)\,d\Omega. \tag{5}$$

If $A(p)$ arises from the Euler–Lagrange equation of the variational integrand $G(p)$, as in (3), then $H(p)$ becomes the Legendre transform of $G(p)$, that is

$$H(p)=pG'(p)-G(p). \tag{6}$$

It is clear from (5) and (H3) that $\Omega(p)$ and $H(p)$ are strictly increasing and positive for $p>0$. In turn we can define

$$\Omega(\infty)=\lim_{p\to\infty}\Omega(p), \qquad H(\infty)=\lim_{p\to\infty}H(p) \quad \text{(possibly infinite)}.$$

For the operator $A=p^{m-2}$ we have in particular $H=((m-1)/m)\,p^m$ and $\Omega(\infty)=H(\infty)=\infty$, while, for the mean curvature operator, $H=1-1/\sqrt{1+p^2}$ and $H(\infty)=1$.

Our principal conclusion concerning existence, Theorem 1 in Section 2, can now be stated.

THEOREM A. *Suppose that there exist constants $\beta>0$ and $\gamma>\beta$ (γ possibly infinite) such that the following conditions hold*:

(a) $F(u)<0$ *for* $0<u<\beta$; $\quad F(\beta)=0$,

(b) $f(u)>0$ *for* $\beta\leqslant u<\gamma$; $\quad f(\gamma)=0$ *if* $\gamma<\infty$,

(c) $\max_{[0,\beta]} |F(u)| + F(\gamma) < H(\infty)$ *if* $H(\infty) < \infty$,

(d) $\liminf_{u\to\infty} \dfrac{H^{-1}(F(u))}{u} = 0$ *if* $H(\infty) = F(\gamma) = \infty$, $\Omega(\infty) < \infty$,

(e) $\liminf_{u\to\infty} \dfrac{H^{-1}(F(u))}{u} < \infty$ *if* $H(\infty) = \Omega(\infty) = F(\gamma) = \infty$.

Then there exists a radially symmetric ground state $u = u(r)$ *of* (1), *with central value* $u(0) \in (\beta, \gamma]$ *if* $\gamma < \infty$, *or* $u(0) \in (\beta, \infty)$ *if* $\gamma = \infty$. *Moreover* $u(r) \leqslant 0$ *for all* $r \geqslant 0$.

In the statement of this theorem for the case $\gamma = \infty$ we define $F(\gamma) = \lim_{u\to\infty} F(u)$, which certainly exists though possibly being infinite. Clearly $F(\gamma) = \infty$ can occur only in the case $\gamma = \infty$. Note also that H^{-1} is well-defined since H is strictly increasing, and that $H(\infty) = \infty$ whenever $\Omega(\infty) = \infty$.

As we shall see from Proposition 1 below, conditions (a), (b), and (c), while not quite necessary for solutions to exist, are nevertheless nearly necessary. On the other hand, neither smoothness nor even Lipschitz continuity is required for the function $f(u)$ in this proposition. Clearly, if $F(\gamma) < \infty$ and $H(\infty) = \infty$ then only the conditions (a), (b) are needed. The appearance of the value $H(\infty)$ in (c) and (d), and $\Omega(\infty)$ in (e) is of interest. For the Laplace operator or the degenerate Laplace operator $H(\infty)$ and $\Omega(\infty)$ are both infinite, so conditions (c), (d) and (e) are not needed if $\gamma < \infty$, though, of course, (e) must be used when $F(\gamma) = \infty$. On the other hand, for the mean curvature operator we have $H(\infty) = 1$. Therefore, in consequence of (c), the function $F(u)$ must be quite restricted in order to obtain existence; see also Proposition 1 below.

For the case of the Laplace operator condition (e) takes the form

$$\liminf_{u\to\infty} \frac{F(u)}{u^2} < \infty$$

and for the degenerate Laplace operator

$$\liminf_{u\to\infty} \frac{F(u)}{u^m} < \infty.$$

Even for the Laplace operator this condition for existence when $\gamma = F(\infty) = \infty$ seems to be new. In fact, for the Laplace operator in the case

$$\liminf_{u\to\infty} f(u) > 0,$$

(so also $\gamma = F(\gamma) = \infty$) it was shown in [KK4] that existence holds if (a), (b) and

$$\text{(e}'\text{)} \quad \frac{f(u)}{u-\beta} \text{ is nonincreasing for } u > \beta$$

are satisfied, and f is Lipschitz continuous on $(0, \infty)$. This result clearly falls as a special case under (e) in Theorem A.

Various properties of ground state solutions of (1) are derived in Section 1, several of which are worth noting here. We first observe that radial solutions $u = u(r)$ of (1) satisfy the differential equation

$$(A(|u'|)\, u')' + \frac{n-1}{r} A(|u'|)\, u' + f(u) = 0, \qquad r > 0.$$

PROPOSITION 1. *Let $u = u(r)$ be a radially symmetric ground state of* (1), *under the hypotheses* (H1)–(H3) *on $f(u)$ and $A(p)$. Then*

$$\text{(i)} \quad f(u(0)) \geqslant 0$$

$$\text{(ii)} \quad F(u(0)) > 0$$

$$\text{(iii)} \quad u'(r) \leqslant 0 \qquad \text{for} \quad 0 < r < \infty$$

$$\text{(iv)} \quad u'(r) \to 0 \qquad \text{as} \quad r \to \infty.$$

If $F(u) \leqslant 0$ for $0 < u < \beta$ then also

$$\text{(v)} \quad \max_{[0,\beta]} |F| < H(\infty).$$

If $F(u) \leqslant 0$ at every point u (in the range of a solution) where $f(u) = 0$, then equality cannot hold in (i), *while* (iii) *can be improved to $u'(r) < 0$ for all $r > 0$ such that $0 < u(r) < u(0)$.*

The asymptotic behavior (iv) can be considerably strengthened in case conditions (A2) and (F1) below are satisfied; see Lemmas 3.2.1 and 3.2.2.

By condition (iii), if $u(\bar{r}) = 0$ for some $\bar{r} > 0$ then $u(r) \equiv 0$ for $r > \bar{r}$, so that the solution has compact support. The next result gives a necessary and sufficient condition for this situation to occur (see Section 1.3).

PROPOSITION 2. *Assume that constant $c > 1$ and $\delta > 0$ exist such that*

$$f(u) \leqslant 0 \quad \text{for } 0 < u < \delta, \qquad c \int_0^p \rho A(\rho)\, d\rho \leqslant p^2 A(p) \quad \text{for } 0 < p < \delta.$$

Then radial ground states of (1) *have compact support if and only if*

$$\int_0 \frac{1}{H^{-1}(|F(t)|)}\,dt < \infty.$$

The final section of the paper concerns the *uniqueness* of radially symmetric ground states. The situation here is quite delicate, and accordingly we shall require further restrictions on the function $f(u)$ and the operator $A(p)$:

(A1′) $p^3\Omega(p)\,\Omega'(p)$ is increasing (see Section 3, condition (A1), for a more general hypothesis),

(A2) $p^{2-m}A(p) \to 1$ as $p \to 0$, for some $m > 1$,

(A3) $\Omega'(p) \geqslant [\Omega(p)]^{\mu}$ for all p near 0 and some $\mu \in [0, 2)$,

(F1) $F(u) < 0$ for $0 < u < \beta$, $F(\beta) = 0$,

(F2) $f(u)$ is positive and non-increasing for $\beta < u < \gamma$,

(F3) $f(u)$ is locally Lipschitz continuous on (β, γ).

THEOREM B. *Assume* $n \geqslant 2$ *and that* $A(p)$ *is of class* C^1 *for* $p > 0$. *Suppose also that the hypotheses* (A1′), (A2), (A3) *and* (F1), (F2), (F3) *are satisfied. Then equation* (1) *cannot have more than one radially symmetric ground state* $u = u(r)$ *such that* $u(0) < \gamma$.

THEOREM C. *Let the hypotheses of Theorem* B *hold, except that*

(i) $\mu = 0$ *in* (A3);

(ii) $\gamma < \infty$ *in* (F2), (F3);

(iii) (β, γ) *is replaced by* $(\beta, \gamma]$ *in* (F3).

Then equation (1) *cannot have more than one radially symmetric ground state such that* $u(0) \leqslant \gamma$.

The requirement (F2) in Theorem B is fairly strong, and under appropriate conditions on the operator $A(p)$ can be significantly weakened. In particular if A satisfies (A1′)–(A3) and also

(A2′) $p^{2-\nu}A$ is non-decreasing on $(0, \infty)$, where $\nu > 1$ is constant,

then it is enough to suppose that f satisfies (F1) and (F3) and, in place of (F2), the weaker condition

(F2′) $f(u)/(u-\beta)^{\nu-1}$ is positive and non-increasing for $\beta < u < \gamma$.

Note that (A2′), (F2′) reduce simply to (F2) when $\nu \to 1$.

For the Laplace operator ($A \equiv 1$) we can take $m=2$ in (A2) and $\mu=0$ in (A3), and also $\nu=2$ in (A2′). Then (F2′) becomes condition (S) in [PS1]. Conditions (A2′) and (A3) also are obeyed by the degenerate Laplace operator $A \equiv p^{m-2}$, with $\nu=m$ and with $\mu=0$ for $1<m\leqslant 2$ and $\mu=(m-2)/(m-1)$ for $m>2$. For the mean curvature operator one can take $m=2$ in (A2) and $\mu=0$ in (A3) exactly as for the Laplace operator. On the other hand (A2′) is not satisfied for any $\nu>1$. Consequently (F2′) cannot be directly used in this case (but see the footnote in the introductory part of Section 3). Finally, condition (A1′) holds for the degenerate Laplace operator for all $m>1$ and $n \geqslant 2$, and also for the mean curvature operator—see the introduction to Section 3.

The fact that f in Theorem B needs to be Lipschitz continuous *only on* (β, γ) is worth noting, both because the natural expectation is that if Lipschitz continuity is needed at all it should be for the whole range of values $(0, \gamma)$ which a solution might encounter, and also because in previous uniqueness results for equation (2) Lipschitz continuity or even continuous differentiability were required at all values $u \in (0, \gamma)$. Thus Theorem B is, even for equation (2), a generalization of earlier work. Similar comments apply also for Theorem C.

Even after giving up both hypothesis (A3) and the Lipschitz continuity of f, a result of considerable interest still persists, describing the structure of the set of all ground states when uniqueness fails.

PROPOSITION 3. *Assume* $n \geqslant 2$ *and that* $A(p)$ *is of class* C^1 *with* $\{pA(p)\}'>0$ *for* $p>0$. *Suppose hypotheses* (A1′), (A2), (F1) *and* (F2) *are satisfied* (*or, in place of* (F2), *the conditions* (A2′) *and* (F2′)). *If* u *and* v *are radially symmetric ground states of* (1) *with* $u(0), v(0) \leqslant \gamma$, *then necessarily*

$$u(0) = v(0).$$

Moreover if $u \not\equiv v$ *then* $u(r) \neq v(r)$ *for all* $r>0$ *such that* $u(r) \in (0, \gamma)$. (*If* u *and* v *have compact support, then also* $\inf\{r>0; u(r)=0\} \neq \inf\{r>0; v(r)=0\}$.)

In other words, even if the radial ground state itself is not unique for a given function $f(u)$, *its central value is*. Note moreover that if either u or v does *not* have compact support then $u(r) \neq v(r)$ for all $r>0$. Of course, if one adds to the hypotheses of Proposition 3 the assumption that the Cauchy problem for the equation satisfied by radial solutions of (1) has only a unique solution, then the ground state is unique. This happens in particular for Theorems B and C above.

Under the hypotheses of Theorem B, and with the help of Proposition 3, it is clear that the only way there could be *more than one* radial ground state of (1) with central value $u(0)$ in $(0, \gamma]$, $\gamma < \infty$, is that there are *no radial ground states* with $u(0) \in (0, \gamma)$ and *at least* two radial ground states with $u(0) = \gamma$.

This remark is of particular importance for the degenerate Laplace operator when $m > 2$, since Theorem C cannot be applied in this case because assumption (i) fails.

The examples of the Laplace operator, the degenerate Laplace operator and the mean curvature operator are sufficiently important to be considered separately. By combining our previous conclusions of existence and uniqueness we can state the following results for these cases.

COROLLARY 1. *Suppose $n \geqslant 2$. Assume conditions* (a), (b) *of Theorem* A *are satisfied, and that, for some constant $m > 1$,*

(c) *$f(u)/(u-\beta)^{m-1}$ is non-increasing for $\beta < u < \gamma$.*

There exists a radially symmetric ground state $u = u(r)$ of the equation

$$\operatorname{div}\{|Du|^{m-2} Du\} + f(u) = 0$$

with the following properties (*see Theorem A*):

$$\text{if } \quad \gamma = \infty, \quad \text{then } \quad u(0) \in (\beta, \infty),$$
$$\text{if } \quad \gamma < \infty, \quad \text{then } \quad u(0) \in (\beta, \gamma].$$

Suppose also that f is Lipschitz continuous on (β, γ). Then if $\gamma = \infty$ the solution is unique, and if $\gamma < \infty$ the solution is unique in the class where $u(0) < \gamma$.

If $f(u) \leqslant 0$ for small u, then the solution is positive for all $r \geqslant 0$ if and only if

$$\int_0 |F(t)|^{-1/m}\, dt = \infty.$$

COROLLARY 2. *Let the preceding hypotheses hold, except that* (c) *is replaced by*

(c′) *f is non-increasing for $\beta < u < \gamma$*

(c″) $\max\limits_{[0, \beta]} |F(u)| + F(\gamma) < 1.$

Then there exists a radially symmetric ground state $u = u(r)$ of the equation

$$\operatorname{div}\left(\frac{Du}{\sqrt{1+|Du|^2}}\right) + f(u) = 0$$

with the same properties as in Corollary 1.

If $f(u) \leqslant 0$ for small u, then the solution is positive for all $r \geqslant 0$ if and only if

$$\int_0 |F(t)|^{-1/2}\, dt = \infty.$$

In both corollaries the solutions have the general properties listed in Propositions 1 and 2. Also, in the first corollary, when $m=2$, and in the second corollary, *all* ground states are necessarily radially symmetric if $f \in C^{1+\varepsilon}$ near $u=0$ and $f'(0)<0$. This follows at once from the symmetry results of Gidas, Ni & Nirenberg and from work of Franchi & Lanconelli.

Corollary 1 also shows the possibility in certain cases of ground states which are *not* radially symmetric, a point first raised by Kichenassamy & Smoller. That is, should a *compact support* radial ground state exist, then two or more of these, with translated origins and disjoint supports, would still constitute a ground state, but clearly not one which is radially symmetric. All the more, in such a case the family of non-radially symmetric ground states is non-denumerable. To obtain *existence* of a compact support ground state it is enough by Corollary 1 that the function $f(u)$ satisfy the condition

$$\int_0 |F(t)|^{-1/m}\, dt < \infty. \tag{7}$$

In particular, in the natural case when $f'(0)$ exists and is negative, condition (7) becomes simply $m>2$, as noted in [KS], Section 1 C (the argument there does not, however, quite conclusively demonstrate the existence of the requisite compact support ground states, since the final hypothesis of Theorem 1.2 is not easy to verify). Finally, remarkably enough, condition (7) can hold even for the Laplace operator, for example whenever

$$f(u) \leqslant -\text{const.}\, u^{1-\varepsilon}, \qquad \varepsilon > 0,$$

for small u, see [KK4]. This relation on course is not covered by the Gidas–Ni–Nirenberg theorem, since $f \notin \mathbf{C}^{1+\sigma}([0,\infty))$.

For the existence results of both Corollaries 1 and 2 we emphasize again that there is no need to assume that $f(u)$ is Lipschitz continuous, even though such an assumption is certainly necessary for uniqueness. At the same time, if the hypothesis of Lipschitz continuity is dropped in either of the above corollaries, the partial uniqueness result of Proposition 3 still remains true, and in particular the central value $u(0)$ is still unique. The weakening of the Lipschitz condition to the subinterval (β, γ), and the complete removal of this hypothesis in Proposition 3, are of course improvements of previous results even for the case of the Laplace operator.

A detailed table of contents is given at the beginning of the paper. Chapters II and III, respectively covering existence and uniqueness, can be read independently of each other. Chapter I, on the other hand, is fundamental for all further results.

Since we shall be dealing throughout with radially symmetric solutions of (1) it is possible to consider the dimension n simply as a real parameter greater than 1; we shall consistently follow this point of view in the sequel.

Some particular cases of the above results were announced earlier in two papers [FLS1] and [FLS2].

1. General Theory

1.1. *Preliminary Results*

A radially symmetric ground state $u=u(r)$ of equation (1) can be considered as a solution of the problem

$$\begin{gathered}(Au')' + \frac{n-1}{r} Au' + f(u) = 0, \qquad r>0 \\ u'(0)=0,\ u \geqslant 0 \quad \text{for } r>0; \qquad u \to 0 \quad \text{as } r\to\infty; \qquad u \not\equiv 0,\end{gathered} \tag{$*$}$$

where $A = A(|u'|)$ and n is the dimension, which for our purposes may be considered any real number greater than 1.

As noted in the Introduction, we shall always maintain the hypotheses (H1)–(H3) without further comment. Moreover, to avoid confusion *we shall specifically define the function Au' to be* 0 *whenever* $u'=0$.

Because of the possible singularity of A when $u'=0$ (which occurs in particular for the degenerate Laplace operator, and equally because we do not require the operator $A(p)$ itself to be differentiable) it is necessary to be precise concerning the meaning of a solution of the differential equation in (*), namely

$$(Au')' + \frac{n-1}{r} Au' + f(u) = 0, \qquad r>0. \tag{1.1.1}$$

Here we shall treat *classical* solutions, with the precise meaning that $u \in C^1([0,\infty))$ with $u'(0)=0$, and $w = Au' \in C^1((0,\infty))$.

In this section we shall prove some simple but important identities for non-negative classical solutions of the equation (1.1.1). For the moment we shall not assume that $u\to 0$ as $r\to\infty$. Obvious specializations of these results also hold when the domain $[0,\infty)$ for u is replaced by some subinterval of $[0,\infty)$.

LEMMA 1.1.1. *If u is a classical solution of* (1.1.1), *then* $w=Au'\in C^1([0,\infty))$ *and*

$$w(r)=-\int_0^r\left(\frac{t}{r}\right)^{n-1}f(u(t))\,dt, \qquad w(0)=0, \qquad w'(0)=-\frac{1}{n}f(u(0)).$$

Proof. Obviously w satisfies the equation

$$w'+\frac{n-1}{r}w+f(u)=0, \qquad r>0, \tag{1.1.2}$$

and $w(0)=0$ because $u'(0)=0$. Writing (1.1.2) in the form

$$(r^{n-1}w)'=-r^{n-1}f(u), \tag{1.1.3}$$

an easy integration now yields

$$r^{n-1}w(r)=-\int_0^r t^{n-1}f(u(t))\,dt,$$

whence also

$$w'(r)=-f(u(r))+\frac{n-1}{r^n}\int_0^r t^{n-1}f(u(t))\,dt, \qquad r>0.$$

By l'Hopital's rule the right hand side approaches $-f(u(0))/n$ as $r\to 0$, completing the proof.

Note further that

$$\lim_{r\to 0}\frac{w(r)}{r}=w'(0)=-\frac{1}{n}f(u(0))$$

so in particular w/r is bounded as $r\to 0$.

LEMMA 1.1.2. *Suppose u is a classical solution of* (1.1.1). *Then for any pair of numbers r_0, $r\geqslant 0$ we have*

$$H(p)+F(u)=H(p_0)+F(u_0)-(n-1)\int_{r_0}^r Au'^2\frac{dr}{r}, \tag{1.1.4}$$

where $H(p)$ is defined by (4),

$$F(u)=\int_0^u f(t)\,dt$$

is the primitive of f vanishing at $u=0$, and $u=u(r)$, $p=|u'(r)|$, $u_0=u(r_0)$, $p_0=|u'(r_0)|$.

Proof. We note first that $Au'^2/r = (w/r)\,u'$, so this quantity is locally integrable on $[0, \infty)$. Now put

$$I(r) = H(p) + F(u) + (n-1)\int_{r_0}^{r} Au'^2 \frac{dr}{r}.$$

We must show that $dI/dr = 0$ when $r > 0$. But

$$\frac{dF}{dr} = f(u)\,u', \qquad \frac{d}{dr}\int_{r_0}^{r} Au'^2 \frac{dr}{r} = \frac{Au'^2}{r},$$

and by (5)

$$\frac{dH}{dr} = \frac{d}{dr}\int_0^{\Omega(p)} p(\Omega)\,d\Omega = u'w' = u'(Au')'$$

since $\Omega(p) = w$ sign u', and so $d\Omega/dr = (dw/dr)$ sign u'. Hence, because u satisfies (1.1.1) when $r > 0$,

$$\frac{d}{dr}I(r) = \left((Au')' + \frac{n-1}{r}Au' + f(u)\right)u' = 0.$$

LEMMA 1.1.3. *Suppose $n > 1$ and let u be a classical solution of* (1.1.1). *If $\bar{r}$ ($\geqslant 0$) is a critical point of u, then either*

$$u(r) \leqslant u(\bar{r}) \text{ for } r > \bar{r}, \qquad \text{and} \qquad f(u(\bar{r})) \geqslant 0$$

or

$$u(r) \geqslant u(\bar{r}) \text{ for } r > \bar{r}, \qquad \text{and} \qquad f(u(\bar{r})) \leqslant 0.$$

Proof. Let $\bar{r}$ be a critical point of u. For contradiction, we suppose that there exist $r_1, r_2 \geqslant \bar{r}$ such that $u(r_1) > u(\bar{r})$ and $u(r_2) < u(\bar{r})$. Then we have $u(\bar{r}) = u(\tilde{r})$ for some $\tilde{r}$ between r_1 and r_2. Obviously u is non-constant on $[\bar{r}, \tilde{r}]$. Now, from (1.1.4) with $r_0 = \bar{r}$ and $r = \tilde{r}$, we obtain

$$H(\tilde{p}) + (n-1)\int_{\bar{r}}^{\tilde{r}} \frac{Au'^2}{r}\,dr = 0.$$

Hence since $n > 1$ and because $H > 0$, $A > 0$ for $p > 0$ we get $u' \equiv 0$ on $[\bar{r}, \tilde{r}]$. This contradiction proves that $u(r) \leqslant u(\bar{r})$ for every $r > \bar{r}$ or $u(r) \geqslant u(\bar{r})$ for every $r > \bar{r}$. In the first case it is easy to see that $u'(\bar{r}) = 0$, $Au' = w(\bar{r}) = 0$, $w'(\bar{r}) = (Au')' \leqslant 0$. Hence by (1.1.2), or by the relation $w'(0) = -f(u(0))/n$ if $\bar{r} = 0$, we have $f(u(\bar{r})) \geqslant 0$. In the same way it follows in the second case that $f(u(\bar{r})) \leqslant 0$.

1.2. *Behavior of Solutions*

In this section we shall obtain some elementary but useful results concerning classical solutions of (*). We maintain always the hypotheses (H1)–(H3).

LEMMA 1.2.1. *If u is a classical solution of (*) then*

$$u' \to 0 \qquad as \quad r \to \infty$$

and

$$H(p) + F(u) = (n-1) \int_r^\infty \frac{Au'^2}{\rho} \, d\rho \geqslant 0 \tag{1.2.1}$$

$$(n-1) \int_0^\infty \frac{Au'^2}{r} \, dr = F(u(0)). \tag{1.2.2}$$

Proof. Let $r \to \infty$ in (1.1.4). Since $F(u)$ then tends to zero (recall $u \to 0$ as $r \to \infty$), and the integral tends to some limit, also $H(p)$ tends to a limit. Hence $p \to$ limit, which must necessarily be zero since $u \to 0$ as $r \to \infty$. This proves the first part of the lemma. For (1.2.1), we again let $r \to \infty$ in (1.1.4), and then for (1.2.2) we set $r = 0$ in (1.2.1).

In view of (1.2.2) it is clear that $F(u(0)) > 0$ and so a fortiori $u(0) > 0$. Consequently if there exists a solution of (*) then necessarily

$$B = \{u > 0 \mid F(u) > 0\} \neq \varnothing$$

In the sequel we shall put $\beta = \inf B$. The lemma shows that $u(0) > \beta$.

PROPOSITION 1.2.2. *A necessary condition for Problem (*) to have a solution is that*

$$\max_{[0,\beta]} |F| < H(\infty). \tag{1.2.3}$$

Proof. Put $r_0 = 0$ in (1.1.4); since $H(p)$ is strictly increasing we get

$$F(u(0)) - F(u) = H(p) + (n-1) \int_0^r \frac{Au'^2}{r} \, dr$$

$$< H(\infty) + (n-1) \int_0^\infty \frac{Au'^2}{r} \, dr = H(\infty) + F(u(0)),$$

by (1.2.2). Since the range of $u(r)$ strictly includes $(0, \beta]$, the result follows at once.

LEMMA 1.2.3. *Let u be a classical solution of* (*). *If* $u(a)=0$ *for some* $a>0$ *then* $u\equiv 0$ *for* $r\geqslant a$.

Proof. Clearly $u'(a)=0$. Consequently (1.2.1) with $r=a$ yields

$$\int_a^\infty \frac{Au'^2}{r}\,dr=0.$$

Hence $u'\equiv 0$ and $u\equiv 0$ for $r\geqslant a$.

LEMMA 1.2.4. *Let u be a classical solution of* (*). *Then* $r=0$ *is a maximum of u and* $u'(r)\leqslant 0$ *for* $r\geqslant 0$.

Proof. By Lemma 1.1.3, it is evident that the critical point $r=0$ must be a maximum of u for otherwise the condition $u\to 0$ at infinity would be impossible.

Suppose that $u'(\rho)>0$ for some $\rho>0$. Since $u(\rho)\leqslant u(0)$ by what was just shown, it follows that there must be some minimum point $\bar{r}$ contained in $(0,\rho)$. But then $u(r)\geqslant u(\bar{r})$ for $r>\bar{r}$. If $u(\bar{r})>0$ it would be impossible to have $u\to 0$ as $r\to\infty$. If $u(\bar{r})=0$ then $u\equiv 0$ for $r\geqslant\bar{r}$ and so $u'(\rho)=0$, which is again a contradiction.

Even more, by using Lemma 1.1.3 we get

COROLLARY 1.2.5. *If u is a classical solution of* (*) *then* $f(u(0))\geqslant 0$.

The situation is somewhat simpler when f has the property that $F(u)\leqslant 0$ at all points u (in the range of a solution) for which $f(u)=0$ (this is the case, for example, when the conditions (a), (b) noted in the Introduction are satisfied and $u(0)\in(0,\gamma]$). To begin with, the conclusion of Corollary 1.2.5 can be strengthened to $f(u(0))>0$. Indeed if $f(u(0))=0$, then $F(u(0))\leqslant 0$, contradicting (1.2.2).

The possibility $u'(\rho)=0$, $u(\rho)>0$ for some $\rho>0$ can also be ruled out in Lemma 1.2.4. For if this happens, then by the result of Lemma 1.2.4 we must have $u(r)\geqslant u(\rho)$ for $r<\rho$ and $u(r)\leqslant u(\rho)$ for $r>\rho$. Consequently $w'(\rho)$ can be neither positive nor negative. Hence $w'(\rho)=0$. That is, $w(\rho)=w'(\rho)=0$ and so $f(u(\rho))=0$ by (1.1.2). But in this case again $F(u(\rho))\leqslant 0$. Hence by (1.2.1) we get (since $u'(\rho)=0$)

$$F(u(\rho))=(n-1)\int_\rho^\infty \frac{Au'^2}{r}\,dr=0$$

so that $u'(r)\equiv 0$, $u(r)\equiv u(\rho)$ for $r\geqslant\rho$, which contradicts the condition $u\to 0$ as $r\to\infty$. We state this result as

PROPOSITION 1.2.6. *Let u be a classical solution of* (*). *Suppose that* $F(u)\leqslant 0$ *at all values* $u\leqslant u(0)$ *for which* $f(u)=0$. *Then we have*

(i) $u'(r) < 0$ *for* $r > 0$ *as long as* $u(r) > 0$,

(ii) $f(u(0)) > 0$.

1.3. *Compact Support*

In this section we prove a sufficient condition and a related necessary condition for a solution of (*) to have compact support.

For the Laplace operator the result of Proposition 1.3.1 was obtained by Peletier and Serrin [PS2], as well as a conclusion closely related to Proposition 1.3.2.

PROPOSITION 1.3.1. *Let* u *be a classical solution of* (*). *If* $\beta > 0$ *and*

$$\int_0 \frac{1}{H^{-1}(|F(s)|)}\, ds < \infty, \tag{1.3.1}$$

then u *has compact support.*

Proof. Let R be such that $0 \leqslant u(r) < \beta$ for $r \geqslant R$. Then by (1.2.1) we have $p = |u'| > 0$ at all $r \geqslant R$ for which $u(r) > 0$. Hence by Lemma 1.2.3 either $u \equiv 0$ for all sufficiently large r or $u > 0$ and $u' < 0$ for $r \geqslant R$. In the first case we are done. Otherwise, again by (1.2.1),

$$u'(r) < -H^{-1}(|F(u(r))|) \qquad \text{for} \quad r \geqslant R.$$

Hence

$$\int_{u(r)}^{u(R)} \frac{ds}{H^{-1}(|F(s)|)} > r - R \qquad \text{for} \quad r \geqslant R.$$

Since $u(r) \to 0$ as $r \to \infty$, this inequality implies

$$\int_0^{u(R)} \frac{ds}{H^{-1}(|F(s)|)}\, ds = \infty. \tag{1.3.2}$$

The assertion is proved.

The next result is a partial converse.

PROPOSITION 1.3.2. *Assume that there exist constants* $c > 1$, $\delta > 0$ *and an increasing function* Φ: $[0, \delta) \to \mathbb{R}$ *with* $\Phi(0) = 0$ *such that*

(i) $|F(u)| \leqslant \Phi(u)$ *for every* $u \in (0, \delta)$

(ii) $c \int_0^p \rho A(\rho)\, d\rho \leqslant p^2 A(p)$ *for* $p \in (0, \delta)$.

Let u be a classical solution of (*). *Then* $u(r)>0$ *for every* $r>0$ *if*

$$\int_0 \frac{ds}{H^{-1}(\Phi(s))}=\infty. \tag{1.3.3}$$

Remarks. (a) Condition (i) is satisfied with $\Phi=|F|$ if there exists $\delta>0$ such that $f(u)\leqslant 0$ for $0\leqslant u\leqslant\delta$.

(b) Condition (ii) holds with $c=m$ for the degenerate Laplacian ($A(p)=p^{m-2}$, $m>1$) and with any $c\in(1,m)$ for the generalized mean curvature operator ($A(p)=(1+p^2)^{-s/2}p^{m-2}$, $m>1$).

(c) Condition (ii) is equivalent to $G(p)\leqslant$ const. $H(p)$; see relations (3) and (6) in the Introduction.

Proof of Proposition 1.3.2. For contradiction we assume $u(r)=0$ for some $r>0$. From Lemmas 1.2.3 and 1.2.4 there exists a constant $a>0$ such that $u'(r)\leqslant 0$ and $0<u(r)\leqslant u(0)$ for $0<r<a$, and $u(r)\equiv 0$ for $r\geqslant a$. Then by (1.2.1)

$$H(p)=-F(u)+(n-1)\int_r^a Au'^2\frac{d\rho}{\rho}, \qquad r<a.$$

Now by (4) and hypothesis (ii)

$$p^2A(p)=H(p)+\int_0^p \rho A(\rho)\,d\rho\leqslant H(p)+\frac{1}{c}p^2A(p),$$

so

$$p^2A(p)\leqslant\frac{c}{c-1}H(p), \tag{1.3.4}$$

provided of course that $p\in(0,\delta)$. It follows that if r_0 is sufficiently close to a then

$$H(p)\leqslant|F(u)|+\frac{(n-1)\,c}{c-1}\int_r^a H(p)\frac{d\rho}{\rho}\leqslant\Phi(u)+c_1\int_r^a H(p)\,d\rho$$

for $r_0<r<a$, where $c_1=(n-1)\,c/(c-1)\,r_0$.

Applying Gronwall's inequality we obtain

$$H(p)\leqslant\Phi(u)+c_1\int_r^a \Phi(u(t))\,e^{c_1(t-r)}\,dt.$$

Now observe that $t \to \Phi(u(t))$ is decreasing, because Φ is increasing and u is decreasing. Thus for $r_0 < r < a$, we have

$$H(p) \leqslant \Phi(u)\left(1 + c_1 \int_r^a e^{c_1(t-r)}\, dt\right) = e^{c_1(a-r)}\Phi(u) \leqslant e^{c_1(a-r_0)}\Phi(u) = c_2\Phi(u).$$

with $c_2 > 1$. This inequality in turn yields

$$-u' \leqslant H^{-1}(c_2\Phi(u)), \qquad r_0 < r < a.$$

By the following Lemma 1.3.3 there exists a constant $c_3 > 0$ such that

$$H^{-1}(c_2 q) \leqslant c_3 H^{-1}(q) \qquad \text{for} \quad 0 < q < \Phi(u_0), \quad u_0 = u(r_0).$$

Therefore

$$\frac{-u'}{H^{-1}(\Phi(u))} \leqslant c_3, \qquad r_0 < r < a.$$

Integrating this inequality on (r_1, r_2), $r_0 < r_1 < r_2 < a$, we obtain

$$\int_{u(r_2)}^{u(r_1)} \frac{ds}{H^{-1}(\Phi(s))} \leqslant c_3(r_2 - r_1).$$

When $r_2 \to a$ this yields

$$\int_0^{u(r_1)} \frac{ds}{H^{-1}(\Phi(s))} \leqslant c_3(a - r_1).$$

Because $u(r_1) > 0$, we have obtained a contradiction with the hypothesis (1.3.3). The proof is complete.

LEMMA 1.3.3. *If the hypothesis* (ii) *of the preceding proposition is satisfied, then for every fixed* $\theta > 1$ *and* $q_0 > 0$, *there exists a positive constant d such that*

$$H^{-1}(\theta q) \leqslant d H^{-1}(q) \qquad \textit{for every} \quad q \in (0, q_0).$$

Remark. For the special case $A(p) = p^{m-2}$, $m > 1$, we have $H(p) = (m/(m-1))\, p^m$ so that

$$\frac{H^{-1}(\theta q)}{H^{-1}(q)} = \left(\frac{\theta q}{q}\right)^{1/m} = \theta^{1/m}$$

so we can take $d = \theta^{1/m}$. For a general function A the proof is more complicated.

Proof. Put

$$\varphi(p) = \int_0^p \rho A(\rho)\, d\rho.$$

For every $p \in (0, \delta)$ we have by hypothesis (ii)

$$\varphi(p) \leqslant \frac{1}{c} p^2 A(p) = \frac{1}{c} p\varphi'(p),$$

or equivalently, putting $q = \varphi(p)$,

$$\frac{(\varphi^{-1})'(q)}{\varphi^{-1}(q)} \leqslant \frac{1}{cq} \qquad \text{if} \quad 0 < q < \varphi(\delta).$$

An integration on $(q, \lambda q)$, $\lambda > 1$, yields

$$\log \frac{\varphi^{-1}(\lambda q)}{\varphi^{-1}(q)} \leqslant \frac{1}{c} \log \lambda \qquad \text{if} \quad 0 < q < \frac{\varphi(\delta)}{\lambda}. \tag{1.3.5}$$

On the other hand, using (ii) and (1.3.4)

$$\varphi(p) \leqslant \frac{1}{c} p^2 A(p) \leqslant \frac{H(p)}{c-1} = c' H(p), \qquad c' = \frac{1}{c-1}.$$

Therefore, putting $q = \varphi(p)/c'$ we get

$$H^{-1}(q) \leqslant \varphi^{-1}(c'q) \qquad \text{if} \quad 0 < q < \varphi(\delta)/c'. \tag{1.3.6}$$

Moreover, since $pA(p)$ is positive and increasing,

$$\varphi(p) \geqslant \int_{p/2}^p \rho A(\rho)\, d\rho \geqslant \left(\frac{p}{2}\right)^2 A\left(\frac{p}{2}\right) \geqslant H\left(\frac{p}{2}\right),$$

by (4). Hence, putting $q = \varphi(p)$,

$$H^{-1}(q) \geqslant \tfrac{1}{2} \varphi^{-1}(q). \tag{1.3.7}$$

Then for $0 < q < \varphi(\delta)/c'\theta$ we have from (1.3.6) and (1.3.7)

$$\frac{H^{-1}(\theta q)}{H^{-1}(q)} \leqslant 2 \frac{\varphi^{-1}(c'\theta q)}{\varphi^{-1}(q)}.$$

If $c'\theta \leqslant 1$ then the last ratio is $\leqslant 1$, while if $c'\theta > 1$ it is $\leqslant (c'\theta)^{1/c}$ by (1.35). On the other hand $H^{-1}(\theta q)/H^{-1}(q)$ is bounded for $\varphi(\delta)/c'\theta \leqslant q < q_0$. Thus in all cases $H^{-1}(\theta q)/H^{-1}(q)$ is bounded, which completes the proof.

COROLLARY. *Assume* $\beta > 0$ *and that* (i) *in Proposition* 1.3.2 *holds. Suppose also that* $p^{2-m}A(p) \to 1$ *as* $p \to 0$ *for some* $m > 1$. *Then* $u(r) > 0$ *for every* $r > 0$ *if*

$$\int_0 \frac{ds}{(\Phi(s))^{1/m}} = \infty$$

and only if

$$\int_0 \frac{ds}{|F(s)|^{1/m}} = \infty.$$

Proof. From (4) and L'Hopital's rule we get $p^{-m}H(p) \to 1 - 1/m$ as $p \to 0$, and the results follow at once from Propositions 1.3.1 and 1.3.2.

If $F(u)$ is monotone for all sufficiently small u, then by taking $\Phi(s) = |F(s)|$ we get necessary and sufficient conditions for compact support.

2. EXISTENCE

In this chapter, we consider the existence question for Problem (*).

THEOREM I. *Assume that, for suitable constants* $0 < \beta < \gamma$, $\gamma \leqslant \infty$,

(a) $F(u) < 0$ *for* $0 < u < \beta$; $\quad F(\beta) = 0$,

(b) $f(u) > 0$ *for* $\beta \leqslant u < \gamma$; $\quad f(\gamma) = 0$ *if* $\gamma < \infty$,

(c) $F(\gamma) + \max_{[0,\beta]} |F| < H(\infty)$ *if* $H(\infty) < \infty$,

(d) $\liminf_{u \to \infty} \dfrac{H^{-1}(F(u))}{u} = 0$ *if* $H(\infty) = F(\gamma) = \infty$ *and* $\Omega(\infty) < \infty$,

(e) $\liminf_{u \to \infty} \dfrac{H^{-1}(F(u))}{u} < \infty$ *if* $\Omega(\infty) = H(\infty) = F(\gamma) = \infty$.

Then (*) *has a solution with* $u(0) \in (\beta, \gamma]$ *and* $u' \leqslant 0$.

Remark. If $\Omega(\infty) = \infty$ then necessarily $H(\infty) = \infty$. We will in fact see in the sequel that

$$\lim_{p \to \infty} \frac{H(p)}{\Omega(p)} = \infty \qquad \text{if } \Omega(\infty) = \infty$$

(see (2.1.15)). We note explicitly that condition (d) is non-empty. Indeed it is possible to have $\Omega(\infty)<\infty$ and $H(\infty)=\infty$, as can be seen from the example

$$A(p)=\frac{1}{1+\sqrt{1+p^2}},$$

since in this case

$$H(p)=\frac{1}{2}\left[\frac{p^2}{1+\sqrt{1+p^2}}-\log\frac{1+\sqrt{1+p^2}}{2}\right].$$

2.1. *Existence for Regular Operators*

In this section we prove Theorem I under the following extra conditions,

$$\begin{aligned}&A \text{ is of class } C^1 \text{ in } [0,\infty) \text{ and } \Omega'=pA'+A>0 \text{ for all } p\geqslant 0;\\ &f \text{ is Lipschitz continuous in } [0,\gamma] \text{ if } \gamma<\infty \text{ (in } [0,\infty) \text{ if } \gamma=\infty).\end{aligned} \tag{2.11}$$

Even this special case of Theorem I is new and of interest of itself.

Consider the Cauchy problem

$$\begin{aligned}&(Av')'+\frac{n-1}{r}Av'+f(v)=0, \qquad r>0\\ &v(0)=\xi, \qquad v'(0)=0\end{aligned} \tag{2.1.2}$$

with $\xi\in[\beta,\gamma)$. The existence of a unique solution of this Cauchy problem, at least in some neighborhood of $r=0$, follows from Propositions A.1 and A.2 in the Appendix (here we make strong use of (2.1.1)). For convenience we shall write $w=Av'$, so that $w(0)=0$ and $w'(0)=-f(\xi)/n<0$ (as in Lemma 1.1.1). Thus $v'<0$ and $v<\xi$ in some interval to the right of zero. The solution can be continued either for all $r>0$, with $v(r)>0$ and $v'(r)<0$, or else reaches a first point $r=R$ where $v(R)=0$, $v'(R)\leqslant 0$ or where $v'(R)=0$ and $v(R)>0$. To prove this, note first that, by (1.1.4) and the fact (see (b)) that $F(u)$ is positive and increasing for $u>\beta$,

$$H(|v'|)\leqslant F(\xi)-F(v)\leqslant F(\xi)+\max_{[0,\beta]}|F|=M(\xi) \tag{2.1.3}$$

as long as v exists and $0\leqslant v\leqslant\xi$. If $H(\infty)=\infty$, or by (c) if $H(\infty)<\infty$, we have $M<H(\infty)$. Hence (2.1.3) implies

$$|v'|\leqslant H^{-1}(M) \tag{2.1.4}$$

as long as v exists and $0\leqslant v\leqslant\xi$. This being shown, the result now follows by the standard continuation theory of ordinary differential equations. In what follows we shall assume, except where otherwise explicitly stated, that v is continued *exactly* until a first point is reached where either $v'=0$, $v>0$ or $v=0$. If no such point occurs then we assume v is continued for all $r>0$ (of course with $v'<0$).

Now put

$$I^+ = \{\xi \in [\beta, \gamma) \mid \inf v > 0\}$$

$$I^- = \{\xi \in [\beta, \gamma) \mid v(R) = 0 \text{ for some finite } R > 0\}.$$

Clearly I^+ and I^- are disjoint. We shall show that both the sets I^+ and I^- are open in $[\beta, \gamma)$ and non-empty. Once this is done the theorem follows at once, for then $I^+ \cup I^-$ cannot cover $[\beta, \gamma)$ and so there exists some $\bar{\xi} \in [\beta, \gamma)$ which is neither in I^+ nor I^-. The corresponding solution v is decreasing, but neither reaches $v=0$ nor remains bounded from zero. Consequently it must exist for all $r>0$ and be a (non-compact support) solution of (*). The proof, then, consists of the following four steps:

Step 1. I^- is open,

Step 2. I^+ is open,

Step 3. I^+ is non-empty,

Step 4. I^- is non-empty.

Step 1. I^- is open. Let $\bar{\xi} \in I^-$. Then the solution $\bar{v}$ vanishes at $r=R$, and by the uniqueness of the Cauchy problem for smooth functions A and f, necessarily $\bar{v}'(R)<0$. The solution can be extended to values $\bar{v}<0$ by defining $f(u) \equiv 0$ for $u<0$. Then for all nearby values of ξ the solution also must reach points where $v<0$. Clearly for these values of ξ we have $v'(r)<0$ for $r>0$. Consequently all such values are in I^-, and I^- is open.

Step 2. I^+ is open. Let $\bar{\xi} \in I^+$. Then for the corresponding solution $\bar{v}$ either there exists $R>0$ such that $\bar{v}(R)>0$ and $\bar{v}'(R)=0$, or $\bar{v}'(r)<0$ on $0<r<\infty$ and $\lim_{r\to\infty} v(r)>0$. If $\bar{v}$ has a vanishing derivative at $r=R$, then necessarily $\bar{w}(R)=0$, $\bar{w}'(R)>0$ (if $\bar{w}'(R)=0$ we would have $f(\bar{v}(R))=0$ and $\bar{v} \equiv \bar{v}(R)$ by uniqueness). Hence $\bar{v}$ can be continued to larger values of r with $\bar{v}'>0$. Then for all nearby values of ξ the solution also can be continued to reach values $v'>0$ *before* reaching $v=0$. In turn these solutions must have some (possibly different) value $R>0$ such that $v(R)>0$, $v'(R)=0$, and $v'(r)<0$ for $0<r<R$. Consequently all such values are in I^+.

These remains the case that $\bar{v}'(r)<0$ on $0<r<\infty$, with $\lim_{r\to\infty} \bar{v}(r)=l>0$. As in the proof of Lemma 1.2.1, it is clear that $\bar{v}'(r)\to 0$ as $r\to\infty$. Hence $\bar{w}(r)\to 0$ as $r\to\infty$ and so by (1.1.2) also $\bar{w}'(r)\to$ limit, necessarily zero, as $r\to\infty$. Hence $f(l)=0$ by (1.1.2).

Since $f(u)>0$ for $\beta\leqslant u<\gamma$, and since $0<l<\xi<\gamma$, it follows that $0<l<\beta$ and so $F(l)<0$ by (a).

By (1.1.4) applied to $\bar{v}$, with $r_0=0$ and $r_1\to\infty$, we get

$$(n-1)\int_0^\infty \frac{\bar{A}\bar{v}'^2}{r}\,dr+F(l)-F(\bar{\xi})=0$$

where $\bar{A}=A(|\bar{v}'|)$. Choose $\bar{R}>0$ so large that

$$(n-1)\int_{\bar{R}}^\infty \frac{\bar{A}\bar{v}'^2}{r}\,dr<\frac{1}{4}\,|F(l)|.$$

In turn, choose $\delta>0$ so that when $\xi\in[\beta,\gamma)$ and $|\xi-\bar{\xi}|<\delta$ we have

(1) $|F(\xi)-F(\bar{\xi})|<\frac{1}{4}\,|F(l)|$,

(2) the corresponding solution v exists at least for $0\leqslant r\leqslant\bar{R}$ and satisfies $v(r)>0$ when $0\leqslant r\leqslant\bar{R}$,

(3) $(n-1)\left|\int_0^{\bar{R}}\frac{Av'^2-\bar{A}\bar{v}'^2}{r}\,dr\right|<\frac{1}{4}\,|F(l)|$.

Conditions (2) and (3) are possible because of the continuous dependence of solutions of the Cauchy problem on the initial conditions, together with the fact that w/r is uniformly bounded (see Proposition A3 in the Appendix) and v' is uniformly bounded (by (2.1.4)).

Now consider any $\xi\in[\beta,\gamma)$ with $|\xi-\bar{\xi}|<\delta$, and let v be the corresponding solution of the Cauchy problem. For any $r\geqslant\bar{R}$ for which v is defined, we have by (1.1.4) with $r_0=0$,

$$\begin{aligned}F(\xi)-F(v(r))&\geqslant(n-1)\int_0^r\frac{Av'^2}{r}\,dr\geqslant(n-1)\int_0^{\bar{R}}\frac{Av'^2}{r}\,dr\\&\geqslant(n-1)\int_0^{\bar{R}}\frac{\bar{A}\bar{v}'^2}{r}\,dr-\frac{1}{4}\,|F(l)|\\&\geqslant(n-1)\int_0^\infty\frac{\bar{A}\bar{v}'2}{r}\,dr-\frac{1}{2}\,|F(l)|\\&=F(\bar{\xi})-F(l)-\frac{1}{2}\,|F(l)|\geqslant F(\xi)+\frac{1}{4}\,|F(l)|\end{aligned}$$

Hence $-F(v(r))\geqslant\frac{1}{4}|(F(l)|$ and therefore of course $\xi\in I^+$, possibly with $v'(R)=0$, $v(R)>0$ at some value $R>\bar{R}$. This completes Step 2.

Step 3. $I^+ \neq \varnothing$. We shall show that $\beta \in I^+$. Let v denote the corresponding solution of the Cauchy problem. Applying (1.1.4) with $r_0 = 0$ there results (since $F(\beta) = 0$)

$$H(|v'(r)|) + (n-1) \int_0^r \frac{Av'^2}{r} \, dr + F(v(r)) = 0. \tag{2.1.5}$$

Thus $F(v(r)) < 0$ and so $v(r) > 0$ for all r for which v is defined.

If $v'(R) = 0$, $v(R) > 0$ for some $R > 0$ then obviously $\beta \in I^+$. Otherwise $v' < 0$ and $v > 0$ for all $r > 0$, and $v \to l$, $v' \to 0$ as $r \to \infty$ (as before). Letting $r \to \infty$ in (2.1.5) yields

$$F(l) = -(n-1) \int_0^\infty \frac{Av'^2}{r} \, dr < 0$$

so that $l > 0$. Thus again $\beta \in I^+$, completing the proof of Step 3.

Step 4. $I^- \neq \varnothing$. Consider any $\xi \in [\beta, \gamma)$. The corresponding solution v is defined on an interval $[0, R)$, with $R = R_\xi$ possibly infinite, and

$$\lim_{r \to R} v = m \geqslant 0, \qquad v'(r) < 0 \qquad \text{for} \quad r \in (0, R).$$

We assert to begin with that $f(m) \leqslant 0$. Indeed, if $m > 0$ and $R < \infty$ then clearly $v(R) = m$, $w(R) = 0$ and $w'(R) > 0$. Hence by (1.1.2) we have $f(m) < 0$. If $m = 0$ or $R = \infty$ then $f(m) = 0$ as in Step 2. It follows next from (b) that $m < \beta$ and $F(m) \leqslant 0$.

Let $\bar{\beta} \in (\beta, \gamma)$ be fixed. Suppose $\xi > \bar{\beta}$, and define $\bar{R} = \bar{R}(\xi) \in (0, R)$ by

$$v(\bar{R}) = \bar{\beta} \qquad (\text{note } \bar{\beta} > \beta > m).$$

Now suppose $\xi \notin I^-$. Then $\lim_{r \to R} v'(r) = 0$ (since the case $m = 0$, $R < \infty$ does not occur). Put $\bar{M} = \bar{M}(\xi) = \sup_{[\bar{R}, R]} |v'| = |v'(R_0)|$ where $R_0 = R_0(\xi) \in [\bar{R}, R)$. The identity (1.1.4) in $[R_0, R]$ (or in $[R_0, \infty)$ if $R = \infty$) gives

$$\begin{aligned} H(\bar{M}) &= F(m) - F(v(R_0)) + (n-1) \int_{R_0}^R Av'^2 \frac{d\rho}{\rho} \\ &\leqslant \sup_{[0, \beta]} |F| + (n-1) \frac{\Omega(\bar{M})}{\bar{R}} \int_{R_0}^R (-v') \, d\rho \\ &\leqslant \sup_{[0, \beta]} |F| + (n-1) \frac{\Omega(\bar{M})}{\bar{R}} \bar{\beta}. \end{aligned} \tag{2.1.6}$$

On the other hand, the same identity in $[\bar{R}, R]$ gives

$$F(\bar{\beta})=F(m)-H(|v'(\bar{R})|)+(n-1)\int_{\bar{R}}^{R} Av'^2\frac{d\rho}{\rho}\leqslant(n-1)\frac{\Omega(\bar{M})}{\bar{R}}\bar{\beta}. \quad (2.1.7)$$

Defining

$$C=C(\Omega)=\max\left\{\Omega(2),\frac{\Omega(2)}{\Omega(2)-\Omega(1)}\right\} \quad (2.1.8)$$

we assert that

$$\Omega(p)\leqslant C(1+H(p)) \quad (2.1.9)$$

for every $p>0$. The relation (2.1.9) is obvious if $0<p\leqslant 2$. On the other hand, if $p\geqslant 2$,

$$H(p)\geqslant\int_1^p \rho\, d\Omega(\rho)\geqslant\Omega(p)-\Omega(1)\geqslant\Omega(p)\left(1-\frac{\Omega(1)}{\Omega(2)}\right).$$

as required. Now by (2.1.6) and (2.1.9) with $p=\bar{M}$,

$$\Omega(\bar{M})\left(1-C\frac{(n-1)\bar{\beta}}{\bar{R}}\right)\leqslant C(1+\sup_{[0,\beta]}|F|) \quad (2.1.10)$$

We can now prove the following important

LEMMA 2.1.1. *Suppose*

$$\bar{R}(\xi)>2C\frac{(n-1)\bar{\beta}}{\bar{m}}, \quad (2.1.11)$$

where

$$\bar{m}=\min\{1, F(\bar{\beta})/(1+\sup_{[0,\beta]}|F|)\}. \quad (2.1.12)$$

Then $\xi\in I^-$.

Proof. If $\xi\notin I^-$, then by (2.1.10), (2.1.11) and (2.1.7) we get

$$\Omega(\bar{M})<2C(1+\sup_{[0,\beta]}|F|)<\frac{F(\bar{\beta})\,\bar{R}}{(n-1)\,\bar{\beta}}\leqslant\Omega(\bar{M})$$

which is contradiction.

When γ is finite, the existence of a value ξ satisfying (2.1.11) is an obvious consequence of the continuous dependence of v on ξ. Indeed, in this case $v \to \gamma$ as $\xi \to \gamma$ uniformly on every bounded subset of $[0, \infty)$, and so there exists $\xi_0 \in [0, \gamma)$ such that (2.1.11) holds. Thus $I^- \neq \emptyset$ and Theorem I for the regular case and for $\gamma < \infty$ is proved.

Next suppose that $\gamma = \infty$ and either (c) or (d) holds. From (1.1.4) with $r_0 = 0$, we obtain for any $\xi \in [\beta, \infty)$

$$H(|v'(r)|) < F(\xi) \tag{2.1.13}$$

for every r in $[0, \bar{R}]$, because of course $v(r) \geqslant \bar{\beta}$ when $r \in [0, \bar{R}]$ and so $F(v(r)) > 0$. Therefore, since

$$\frac{\xi - \bar{\beta}}{\bar{R}} = \frac{v(0) - v(\bar{R})}{\bar{R}} \leqslant \sup_{[0, \bar{R}]} |v'|$$

and because $F(\xi) < F(\infty) < H(\infty)$ in case (c), and $H(\infty) = \infty$ in case (d), we have

$$\bar{R} \geqslant \frac{\xi - \bar{\beta}}{H^{-1}(F(\xi))}. \tag{2.1.14}$$

Hence, by conditions (c) or (d), we can choose $\xi_0 \in [\beta, \infty)$ such that $\bar{R} = \bar{R}(\xi_0)$ is so large that (2.1.11) holds. Thus $\xi_0 \in I^-$ by the lemma.

The case (e).

Assume finally that $\gamma = \infty$ and (e) holds. In this case, the proof that $I^- \neq \emptyset$ is quite long and will be divided into several steps. First, we need the following result:

If $\Omega(\infty) = \infty$, then

$$\lim_{p \to \infty} \frac{H(p)}{\Omega(p)} = \infty. \tag{2.1.15}$$

In fact, let $k > 0$ be fixed. For every fixed $p > k$ we obtain from (5)

$$H(p) \geqslant \int_k^p \rho \, d\Omega(\rho) \geqslant k(\Omega(p) - \Omega(k)).$$

Then

$$\liminf_{p \to \infty} \frac{H(p)}{\Omega(p)} \geqslant k \liminf_{p \to \infty} \left(1 - \frac{\Omega(k)}{\Omega(p)}\right) \geqslant k$$

for every $k > 0$. This proves (2.1.15).

In addition, a crucial role will be played by the following generalized Hadamard inequality; the standard Hadamard inequality is in fact the special case $A \equiv 1$, $\Omega \equiv p$, $H = \frac{1}{2}p^2$ (our attention was drawn to Hadamard's inequality by the work of Kaper and Kwong [KK4]).

LEMMA 2.1.2. *Let u be a continuously differentiable function defined on an interval $I = [r_0, R)$, $0 \leqslant r_0 < R \leqslant \infty$. Suppose $\Omega(\infty) = \infty$, $w = u'A(|u'|) \in \mathbf{C}^1(I)$, $u' < 0$ and $u'(r) \to 0$ as $r \to R$. Then*

$$H(\Omega^{-1}(M)) \leqslant DS. \tag{2.1.16}$$

where

$$M = \sup_I |w|, \qquad S = \sup_I |w'|, \qquad D = u(r_0) - u(R).$$

Proof. Let η be a point on I such that $|w(\eta)| = M$. For every $r \in I$, $r > \eta$, we have

$$S(r-\eta) \geqslant \int_\eta^r w'(s)\, ds = w(r) - w(\eta) = M - |w(r)|,$$

and hence

$$|w(r)| \geqslant M - S(r-\eta).$$

This inequality implies that $R \geqslant \eta + M/S$ and

$$|u'| \geqslant \Omega^{-1}(M - S(r-\eta)) \qquad \text{if} \quad \eta < r < \eta + M/S.$$

Thus we obtain

$$\begin{aligned} D &\geqslant \int_\eta^{\eta+M/S} |u'(s)|\, ds \geqslant \int_\eta^{\eta+M/S} \Omega^{-1}(M - S(r-\eta))\, dr \\ &\geqslant \int_0^M \Omega^{-1}(t) \frac{dt}{S} = \frac{1}{S}\int_0^{\Omega^{-1}(M)} p\, d\Omega(p) = \frac{H(\Omega^{-1}(M))}{S}, \end{aligned}$$

and the assertion is proved.

We can now go back to the solution v of the Cauchy problem (2.1.2). First we prove the following assertion.

Let $c^* > 0$ be such that $H^{-1}(F(\xi))/\xi < c^*$ and put $r^* = (2c^*)^{-1}$. Then

$$v(r^*) \geqslant \tfrac{1}{2}\xi. \tag{2.1.17}$$

In fact, as long as $v(r) > \beta$, from (1.1.4) with $r_0 = 0$ we obtain $H(p) \leqslant F(\xi)$, and hence $p \leqslant H^{-1}(F(\xi)) \leqslant c^*\xi$. Then

$$v(r) = \xi + \int_0^r v'(s)\, ds \geqslant \xi - c^*\xi r = \xi(1 - c^* r).$$

Hence (2.1.17) follows.

We are now able to prove an a priori estimate for v' on the interval $[\bar{R}, R)$, where $\bar{R}$ and R have been defined at the beginning of this step. More precisely, we have the following result.

Suppose $\xi \notin I^-$ *and let* $d > 0$ *be such that*

$$H(p) > 2(n-1)\,\bar{\beta} c^* \Omega(p) + \bar{\beta} \sup_{[0,\beta]} |f| \qquad \textit{for} \quad p > d \tag{2.1.18}$$

(*note that* d *exists, by* (2.1.15)). *Then if* $\xi \geqslant 2\bar{\beta}$ *and* $H^{-1}(F(\xi))/\xi < c^*$ *we have*

$$\sup_{[\bar{R}, R)} |v'| \leqslant d. \tag{2.1.19}$$

Indeed, from the equation (1.1.2)

$$w' + \frac{n-1}{r} w + f(u) = 0$$

we have

$$S \leqslant \frac{n-1}{\bar{R}} M + \sup_{[0,\beta]} |f|, \tag{2.1.20}$$

where we have used the notation of Lemma 2.2.2 (with $r_0 = \bar{R}$). Note that

$$D = v(\bar{R}) - v(r) \leqslant v(\bar{R}) = \bar{\beta}.$$

Hence, by Lemma 2.2.1, which clearly applies since $\xi \notin I^-$, we have

$$\bar{\beta} S \geqslant DS \geqslant H(\Omega^{-1}(M)).$$

Moreover $\bar{\beta} \leqslant \xi/2 \leqslant v(r^*)$, so that $\bar{R} \geqslant r^* = (2c^*)^{-1}$; thus by (2.1.20),

$$\bar{\beta} S \leqslant \bar{\beta}\{2(n-1)\, Mc^* + \sup_{[0,\beta]} |f|\}.$$

Then, by (2.1.18) (with $p = \Omega^{-1}(M)$), we must have

$$\Omega^{-1}(M) \leqslant d,$$

and hence (2.1.19) follows since $M = \Omega(\sup |v'|)$.

The last step is to show that if $\xi \notin I^-$ then

$$\frac{1}{2}\xi \leqslant \bar{\beta} + \Omega^{-1}\left(\left(\Omega(2d) + \frac{1}{(n-1)\,c^*}\sup_{[0,\,\beta]}|f|\right)\exp(2(n-1)\,c^*\bar{R})\right)\bar{R}. \tag{2.1.21}$$

To prove (2.1.21), put $N = 2\sup_{[0,\,\beta]}|f|$ and consider the Cauchy problem

$$\begin{gathered}(Au')' + \frac{n-1}{r^*}Au' - N = 0, \qquad r^* \leqslant r \leqslant \bar{R}\\ u(\bar{R}) = v(\bar{R}), \qquad u'(\bar{R}) = 2v'(\bar{R})\end{gathered} \tag{2.1.22}$$

An easy comparison argument proves that

$$v \leqslant u \qquad \text{on} \quad r^* < r < \bar{R}. \tag{2.1.23}$$

In fact $v = u$ at $r = r_1$ and $v < u$ in a left neighborhood of $\bar{R}$ (since $u'(r_1) = 2v'(r_1) < v'(r_1) < 0$). For contradiction suppose (2.1.23) does not hold; then there exists a point $\bar{r} \in (r^*, \bar{R})$ such that

$$u(\bar{r}) > v(\bar{r}), \qquad u'(\bar{r}) = v'(\bar{r}) < 0, \qquad u''(\bar{r}) \leqslant v''(\bar{r}).$$

Consequently at $r = \bar{r}$ we have

$$(Au')' \leqslant (Av')', \qquad \frac{n-1}{r^*}Au' \leqslant \frac{n-1}{\bar{r}}Av', \qquad -N < f(v).$$

From the equation in (2.1.22) these inequalities imply

$$(Av')' + \frac{n-1}{\bar{r}}Av' + f(v) > 0,$$

an obvious contradiction.

Now define

$$z = A(|u'|)\,u',$$

so that

$$z' + \frac{n-1}{r^*}z - N = 0, \qquad z(\bar{R}) = -\Omega(2\,|v'(\bar{R})|).$$

Then, if we put

$$U=\left[\Omega(2\,|v'(\bar{R})|)+\frac{r^*}{n-1}N\right]\exp((n-1)\,\bar{R}/r^*), \tag{2.1.24}$$

we have $|z|\leqslant U$ and hence

$$|u'|\leqslant\Omega^{-1}(U). \tag{2.1.25}$$

If we integrate (2.1.25) on the interval $[r,\bar{R}]$ there results

$$|u(r)-u(\bar{R})|\leqslant\Omega^{-1}(U)\,\bar{R},$$

which implies

$$u(r)\leqslant u(\bar{R})+\Omega^{-1}(U)\,\bar{R}=v(\bar{R})+\Omega^{-1}(U)\,\bar{R},\qquad r\in[r^*,\bar{R}]. \tag{2.1.26}$$

Combining (2.1.26) and (2.1.23) we get

$$v(r^*)\leqslant v(\bar{R})+\Omega^{-1}(U)\,\bar{R}=\bar{\beta}+\Omega^{-1}(U)\,\bar{R}. \tag{2.1.27}$$

Then from (2.1.17), (2.1.27), (2.1.24) and (2.1.29) we obtain (2.1.21).

Lemma 2.1.3. *Suppose*

$$\frac{H^{-1}(F(\xi))}{\xi}<c^* \tag{2.1.28}$$

and

$$\frac{\xi}{2}>\bar{\beta}+\frac{2(n-1)\,C\bar{\beta}}{\bar{m}}\,\Omega^{-1}\left(U_1\exp\frac{4(n-1)^2\,Cc^*\bar{\beta}}{\bar{m}}\right), \tag{2.1.29}$$

where $U_1=\Omega(2d)+(1/((n-1)\,c^*))\sup_{[0,\beta]}|F|$. *Then* $\xi\in I^-$.

Proof. If $\xi\notin I^-$ then by Lemma (2.1.1) we get

$$R(\xi)\leqslant\frac{2(n-1)\,C\bar{\beta}}{\bar{m}}$$

(here $\bar{\beta}$ is any fixed number greater than β, while the numbers C, $\bar{m}$ and d are defined respectively by (2.1.8), (2.1.12) and (2.1.18)). Then (2.1.21) contradicts (2.1.29).

To complete the proof of Theorem I in the case when A and f are regular it is now enough to take

$$c^*=2\liminf_{u\to\infty}\frac{H^{-1}(F(u))}{u}$$

and then to choose ξ_0 (sufficiently large) so that (2.1.28) and (2.1.29) are satisfied. Then $\xi_0 \in I^-$ and I^- is not empty.

Remark. The solution v we have just obtained has the property

$$\beta < v(0) < \gamma. \tag{2.1.30}$$

Moreover when $\gamma = \infty$ and either (c) or (d) holds, we can choose $\Gamma > \bar{\beta}$ such that

$$\frac{H^{-1}(F(\Gamma))}{\Gamma - \bar{\beta}} < \frac{\bar{m}}{2(n-1)\bar{\beta}C} \tag{2.1.31}$$

where $\bar{m}$ is defined by (2.1.12). Then by (2.1.14) the inequality (2.1.11) is satisfied with $\bar{R} = \bar{R}(\Gamma)$. Hence $\Gamma \in I^-$. Analogously, in the case (e) if we choose Γ such that

$$\begin{aligned} &\frac{H^{-1}(F(\Gamma))}{\Gamma} < c^* \\ &\frac{\Gamma}{2} > \bar{\beta} + \frac{2(n-1)\,C\bar{\beta}}{\bar{m}}\,\Omega^{-1}(U_1 \exp(4(n-1)^2\, Cc^*\bar{\beta}/\bar{m})). \end{aligned} \tag{2.1.32}$$

Then by Lemma 2.1.3 we have $\Gamma \in I^-$. Thus in all cases, recalling that $\beta \in I^+$, this implies that there exists $\xi \in (\beta, \Gamma)$ such that $\xi \notin I^+ \cup I^-$. The corresponding solution of the Cauchy problem (2.1.2) is then defined for all $r > 0$ and is a solution of (*) having the property

$$\beta < v(0) < \Gamma. \tag{2.1.33}$$

Obviously, we also obtain a solution of (*) verifying (2.1.33) if in Theorem I we assume either (2.1.31) and (2.1.32) holds for some values $\beta < \bar{\beta} < \Gamma$ instead of either hypothesis (d) or (e), respectively.

2.2. *Existence for General Operators*

In this section we complete the proof of Theorem I without the extra conditions on A and f assumed in Section 2.1. The importance of this step lies not only in the minimal continuity behavior required of A and f, but also in the fact that degenerate operators can be considered. For example, the operator $A(p) \equiv |p|^{m-2}$, $m > 2$, fails to have $\Omega' > 0$ for $p = 0$ and thus does not satisfy (2.1.1) even though it is covered by Theorem I. Our proof is not entirely simple, but this is perhaps to be expected in view of the generality of the conclusion.

In what follows we may suppose without loss of generality that f is extended to have values for all $u \in \mathbb{R}$, with $f(u) \equiv 0$ for $u \leqslant 0$. We may also suppose without loss of generality that $f(u) \equiv 0$ for $u \geqslant \gamma$ if $\gamma < \infty$. Con-equently $F(u) = 0$ for $u \leqslant 0$ and $F(u) = F(\gamma)$ for $u \geqslant \gamma$ if $\gamma < \infty$.

For every $\varepsilon > 0$ we define

$$\Omega_\varepsilon(p) = \int_{\mathbb{R}} \Omega(p + \varepsilon t)\, J(t)\, dt + \varepsilon p, \qquad p \in \mathbb{R}$$

and

$$F_\varepsilon(u) = \int_{\mathbb{R}} F(u + \varepsilon(t-1))\, J(t)\, dt, \qquad u \in \mathbb{R}$$

where $\Omega(t) = tA(|t|)$, $t \in \mathbb{R}$, and J is a C^∞ function on $\mathbb{R}$ such that $0 \leqslant J \leqslant 1$, $J(t) = 0$ if $|t| \geqslant 1$, $J(t) = J(|t|) > 0$ if $|t| < 1$, and $\int_{\mathbb{R}} J(t)\, dt = 1$.

Clearly $\Omega_\varepsilon(p)$ is an odd C^∞ function on $\mathbb{R}$, with $\Omega'_\varepsilon(p) > 0$ and $\Omega_\varepsilon(p) \nearrow \infty$ as $p \to \infty$. If we put

$$A_\varepsilon(p) = \frac{\Omega_\varepsilon(p)}{p} \quad \text{if } p \neq 0, \qquad A_\varepsilon(0) = \lim_{p \to 0} \frac{\Omega_\varepsilon(p)}{p},$$

then A_ε is a C^∞ function which verifies the hypotheses (H2) and (H3) (see Introduction) and satisfies (2.1.1). In addition we have

$$\frac{d}{du} F_\varepsilon(u) = \int_{\mathbb{R}} f(u + \varepsilon(t-1))\, J(t)\, dt = \int_{\mathbb{R}} f(u - \varepsilon + \varepsilon t)\, J(t)\, dt \equiv f_\varepsilon(u);$$

clearly f_ε is a C^∞ function and $f_\varepsilon(0) = 0$. Moreover, since $F_\varepsilon(0) = 0$,

$$F_\varepsilon(u) = \int_0^u f_\varepsilon(t)\, dt.$$

It is easy to show that, for every $\varepsilon > 0$ sufficiently small, there exist $\gamma_\varepsilon \leqslant \infty$ and $\beta_\varepsilon \in (0, \gamma_\varepsilon)$, such that $F_\varepsilon(u) < 0$ for $0 < u < \beta_\varepsilon$, $F_\varepsilon(\beta_\varepsilon) = 0$, $f_\varepsilon(u) > 0$ for $\beta_\varepsilon \leqslant u < \gamma_\varepsilon$. Moreover $\beta_\varepsilon \to \beta$ as $\varepsilon \to 0$, while

if $\gamma < \infty$, then $\gamma_\varepsilon < \infty$ and $f_\varepsilon(u) \equiv 0$ for $u > \gamma_\varepsilon$;

if $\gamma = \infty$, then $\gamma_\varepsilon = \infty$.

Finally, if we put

$$H_\varepsilon(p) = \int_0^p \rho\, d\Omega_\varepsilon(\rho) \equiv p\Omega_\varepsilon(p) - \int_0^p \Omega_\varepsilon(\rho)\, d\rho,$$

then since $\Omega_\varepsilon \to \Omega$ as $\varepsilon \to 0$ uniformly on every compact subset of $\mathbb{R}$, we have $H_\varepsilon \to H$ as $\varepsilon \to 0$ uniformly on every compact subset of $\mathbb{R}$. Moreover when $H(\infty)=\infty$ it is clear that also $H_\varepsilon(p)\nearrow\infty$ as $p\to\infty$, that is $H_\varepsilon(\infty)=\infty$.

Now consider the problem

$$(A_\varepsilon u'_\varepsilon)+\frac{(n-1)}{r}A_\varepsilon u'_\varepsilon+f_\varepsilon(u_\varepsilon)=0,\qquad r>0$$
$$u'_\varepsilon(0)=0,\ u_\varepsilon(r)\geqslant 0 \text{ for } r>0;\qquad u_\varepsilon(r)\to 0 \text{ as } r\to\infty;\qquad u_\varepsilon\neq 0. \tag{$*$}_\varepsilon$$

If $\gamma_\varepsilon<\infty$, i.e. the case $\gamma<\infty$, then by the results proved in Section 2.1, this problem has a positive solution u_ε, such that

$$\beta_\varepsilon<u_\varepsilon(0)<\gamma_\varepsilon \tag{2.2.1}$$

for every $\varepsilon>0$ sufficiently small.

Next suppose $\gamma_\varepsilon=\infty$, and so of course $\gamma=\infty$. In this case, by conditions (c), (d) or (e), there exist constants $\Gamma>\bar\beta>\beta$ such that either (2.1.31) or (2.1.32) holds. Of course also $\bar\beta>\beta_\varepsilon$ for every $\varepsilon>0$ small enough.

Now let $C_\varepsilon=C(\Omega_\varepsilon)$ as in (2.1.8) and let $\bar m_\varepsilon=\bar m(F_\varepsilon)$ as in (2.1.12). If either condition (c) or (d) holds, then clearly there exists $\varepsilon_0>0$ such that

$$\frac{H_\varepsilon^{-1}(F_\varepsilon(\Gamma))}{\Gamma-\bar\beta}<\frac{\bar m_\varepsilon}{2(n-1)\,\bar\beta C_\varepsilon}$$

for every ε in $(0,\varepsilon_0)$, where Γ has been defined in (2.1.31). Therefore, from the principal remark at the end of Section 2.1, the solution u_ε clearly has the property

$$\beta_\varepsilon<u_\varepsilon(0)<\Gamma,\qquad \varepsilon\in(0,\varepsilon_0). \tag{2.2.2}$$

If condition (e) holds, then the proof of the analogous a priori estimate requires some further argument since in (2.1.32) the further quantities c^* and d also appear. Choose now $\bar d>0$ such that the following slightly stronger version of (2.1.18) holds:

$$H(p)>4(n-1)\,\bar\beta c^*\Omega(p)+\bar\beta\sup_{[0,\,\beta]}|f|\qquad\text{for}\quad p>\bar d.$$

Then, if $p>2\bar d$, so that $p+\varepsilon t>\bar d$ for $|t|<1$ and ε small, we have

$$\begin{aligned}H_\varepsilon(p)&=p\Omega_\varepsilon(p)-\int_0^p\Omega_\varepsilon(\rho)\,d\rho\\&=p\left\{\int_{\mathbb R}\Omega(p+\varepsilon t)\,J(t)\,dt+\varepsilon p\right\}-\int_0^p\left(\int_{\mathbb R}\Omega(\rho+\varepsilon t)\,J(t)\,dt+\varepsilon\rho\right)d\rho\\&=\int_{\mathbb R}\left\{p\Omega(p+\varepsilon t)-\int_0^p\Omega(\rho+\varepsilon t)\,d\rho\right\}J(t)\,dt+\frac{1}{2}\varepsilon p^2\end{aligned}$$

$$=\int_{\mathbb{R}}\left\{(p+\varepsilon t)\,\Omega(p+\varepsilon t)-\int_{\varepsilon t}^{p+\varepsilon t}\Omega(\rho)\,d\rho\right.$$

$$\left.-\varepsilon t\Omega(p+\varepsilon t)\right\}J(t)\,dt+\frac{1}{2}\varepsilon p^2$$

$$=\int_{\mathbb{R}}\left\{H(p+\varepsilon t)-\varepsilon t\Omega(p+\varepsilon t)+\int_0^{\varepsilon t}\Omega(\rho)\,d\rho\right\}J(t)\,dt+\frac{1}{2}\varepsilon p^2$$

$$\geqslant\int_{\mathbb{R}}\left\{(4(n-1)\,\bar\beta c^*-\varepsilon)\,\Omega(p+\varepsilon t)+\bar\beta\sup_{[0,\,\beta_\varepsilon]}|f_\varepsilon|\right\}J(t)\,dt+\frac{1}{2}\varepsilon p^2$$

$$\geqslant 3(n-1)\,\bar\beta c^*\Omega_\varepsilon(p)+\bar\beta\sup_{[0,\,\beta_\varepsilon]}|f_\varepsilon|\,\frac{1}{2}\varepsilon p(p-6(n-1)\,\bar\beta c^*)$$

$$\geqslant 2(n-1)\,\bar\beta c^*\Omega_\varepsilon(p)+\bar\beta\sup_{[0,\,\beta_\varepsilon]}|f_\varepsilon|$$

if ε is small. Hence we can repeat our previous arguments to show that (2.2.2) holds for ε small and with an obviously modified value for Γ (i.e. with d replaced by $2\bar d$ in (2.1.32)).

We claim that there exists a positive constant M such that

$$-M\leqslant u_\varepsilon'(r)\leqslant 0,\qquad r\geqslant 0,\tag{2.2.3}$$

for every ε in $(0,\varepsilon_0)$. The inequality $u_\varepsilon'\leqslant 0$ follows from Lemma 1.2.4. On the other hand, if $\gamma=\infty$ then by (1.1.4) and (2.2.2) we have

$$H_\varepsilon(|u_\varepsilon'(r)|)\leqslant F_\varepsilon(u_\varepsilon(0))-F_\varepsilon(u_\varepsilon(r))\leqslant F_\varepsilon(\Gamma)+\sup_{[0,\,\beta_\varepsilon]}|F_\varepsilon|.$$

Hence, using condition (c) in case $H(\infty)<\infty$, we have $H_\varepsilon(|u_\varepsilon'|)<H(\infty)$ for every sufficiently small ε in $(0,\varepsilon_0)$. Choose $\bar M$ so that

$$H_\varepsilon(|u_\varepsilon'|)\leqslant\bar M<H(\infty).$$

Thus $u_\varepsilon'\geqslant -H_\varepsilon^{-1}(\bar M)$. Clearly $H_\varepsilon^{-1}(\bar M)\to H^{-1}(\bar M)$ as $\varepsilon\to 0$, and inequality (2.2.3) therefore follows with $M=H^{-1}(\bar M)+1$, changing, if necessary, the choice of ε_0. Next suppose $\gamma<\infty$. In this case from (1.1.14) and (2.2.1) we get

$$H_\varepsilon(|u'(r)|)\leqslant F_\varepsilon(\gamma_\varepsilon)+\sup_{[0,\,\beta_\varepsilon]}|F|$$

and (2.2.3) is proved as before.

It now follows from the Arzelá-Ascoli theorem that, for a suitable sequence $\varepsilon_j \downarrow 0$ we have $u_{\varepsilon_j} \to u$ uniformly on every compact subset of $[0, \infty)$. Therefore, because (see Lemma 1.1.1)

$$u'_\varepsilon A_\varepsilon(|u'_\varepsilon|) = -\int_0^r f_\varepsilon(u_\varepsilon(\rho)) \left(\frac{\rho}{r}\right)^{n-1} d\rho, \tag{2.2.4}$$

we see that

$$\Omega_{\varepsilon_j}(|u'_{\varepsilon_j}|) = |u'_{\varepsilon_j}| \, A_{\varepsilon_j}(|u'_{\varepsilon_j}|) \to g(r) \qquad \text{for all} \quad r \geqslant 0, \tag{2.2.5}$$

where

$$g(r) = \int_0^r f(u(\rho)) \left(\frac{\rho}{r}\right)^{n-1} d\rho.$$

Moreover, since $\Omega_\varepsilon(p) \to \Omega(p)$ uniformly on compact subsets of $p \geqslant 0$, it follows that

$$u'_{\varepsilon_j} \to -\Omega^{-1}(g(r)) \qquad \text{for all} \quad r \geqslant 0.$$

Hence from the relation

$$u_\varepsilon(r) = u_\varepsilon(0) + \int_0^r u'_\varepsilon(\rho)\, d\rho$$

together with (2.2.3), (2.2.5) and Lebesgue's dominated convergence theorem we get

$$u(r) = u(0) - \int_0^r \Omega^{-1}(g(\rho))\, d\rho;$$

thus u is continuously differentiable and a classical solution of (1.1.1). In turn of course (2.2.5) and (2.2.4) supply the relations

$$u'_{\varepsilon_j} \to u' \qquad \text{and} \qquad \Omega_{\varepsilon_j}(|u'_{\varepsilon_j}|) \to \Omega(|u'|) \qquad \text{for all} \quad r \geqslant 0. \tag{2.2.6}$$

Obviously $u \geqslant 0$, $u' \leqslant 0$ and

$$\beta < u(0) \leqslant \gamma \qquad (\leqslant \Gamma < \infty \text{ if } \gamma = \infty);$$

note that the possibility $u(0) = \beta$ is ruled out by Lemma 1.2.1.

It remains only to be shown that $u \to 0$ as $r \to \infty$. Put $l = \lim_{r \to \infty} u(r)$. As in Step 2 of the proof in Section 2.1 we find easily that $u' \to 0$ as $r \to \infty$ and $f(l) = 0$. Of course also $0 \leqslant l \leqslant \gamma$ so that necessarily either

$$0 \leqslant l < \beta \qquad \text{or} \qquad l = \gamma,$$

the second possibility occurring only in the case $\gamma<\infty$. Now

$$0\leqslant\int_R^\infty A_\varepsilon u_\varepsilon'^2\frac{d\rho}{\rho}\leqslant\frac{A_\varepsilon(M)\,M}{R}\int_R^\infty -u_\varepsilon'(\rho)\,d\rho=\frac{A_\varepsilon(M)\,M}{R}\,u_\varepsilon(R)$$
$$\leqslant\frac{2A(M)\,M\gamma}{R}\qquad\left(\text{or }\frac{2A(M)\,M\Gamma}{R}\quad\text{if }\gamma=\infty\right)\tag{2.2.7}$$

for every $R>0$.

On the other hand, by (1.2.2) applied to the solution u_ε,

$$F_\varepsilon(u_\varepsilon(0))=(n-1)\int_0^\infty A_\varepsilon u_\varepsilon'^2\frac{dr}{r}.$$

Letting $\varepsilon=\varepsilon_j\to 0$ and using (2.2.3), (2.2.4) to show that the integrand is uniformly bounded, together with (2.2.6) and (2.2.7) and the dominated convergence theorem, then gives

$$F(u(0))=(n-1)\int_0^\infty Au'^2\frac{dr}{r}.\tag{2.2.8}$$

In the case $\gamma<\infty$, if $l=\gamma$ then $u\equiv\gamma$, because u is decreasing function with $u(0)\leqslant\gamma$. In turn (2.2.8) gives

$$F(\gamma)=0,$$

which however is impossible since (a), (b) imply $F(\gamma)>0$. Thus in all cases $0\leqslant l<\beta$. Finally, because $u(r)\to l$ and $u'(r)\to 0$ as $r\to\infty$, we obtain from (1.1.4)

$$(n-1)\int_0^\infty Au'^2\frac{d\rho}{\rho}=F(u(0))-F(l).$$

From this result and (2.2.8) it now follows that $F(l)=0$, and so $l=0$ by hypothesis (a). This completes the proof.

3. Uniqueness

In this section, we prove our main uniqueness results. We shall require throughout that the operator $A(p)$ be *continuously differentiable* on $(0,\infty)$ rather than simply continuous as in the hypothesis (H2). We recall from the Introduction the definition

$$\Omega(p)=pA(p),\quad p>0;\qquad \Omega(0)=0.$$

By hypothesis (H3) it is clear that $\Omega(p)>0$ for $p>0$ and $\Omega'(p)\geqslant 0$. We shall strengthen this in the sequel, requiring henceforth that

(H3′) $\Omega(p)>0,\quad E(p)=\Omega'(p)>0$ for $p>0$.

In terms of the variational integrand $G(p)$ in the Introduction, these are exactly the conditions $G'(p)$, $G''(p)>0$ for $p>0$.

THEOREM I. *Suppose $n>1$, and assume also the following hypotheses:*

(A1) *$S(l,p)>0$ for $0<l<p$, where*

$$S(l,p)=M(l)-M(p)+\frac{n-2}{n-1}N(l)(N(l)-N(p))$$

and

$$M(p)=1/p^3\Omega\Omega', \qquad N(p)=1/p\Omega.$$

(A2) *$p^{2-m}A(p)\to 1$ as $p\to 0$, for some constant $m>1$,*

(A3) *$\Omega'(p)\geqslant\{\Omega(p)\}^{\mu}$ for all $p>0$ near zero, where μ is a constant in* $[0,2)$,

(F1) *$F(u)\leqslant 0$ for $0<u<\beta$, $F(\beta)=0$,*

(F2) *f is positive and non-increasing on (β,γ), where $\gamma\in(\beta,\infty]$,*

(F3) *f is locally Lipschitz continuous on (β,γ).*

Then $(*)$ has at most one solution u such that $u(0)<\gamma$.

THEOREM II. *Let the hypotheses of Theorem* I *hold, with the following exceptions:*

(i) *$\mu=0$ in condition* (A3);

(ii) *$\gamma<\infty$ in conditions* (F2) *and* (F3);

(iii) *(β,γ) is replaced by $(\beta,\gamma]$ in condition* (F3).

Then $(*)$ has at most one solution u such that $u(0)\leqslant\gamma$.

THEOREM III. *Suppose the hypotheses of Theorem* I (*or Theorem* II) *hold, with the exception that condition* (F2) *is weakened to*

(F2′) *The function $u\to f(u)/(u-\beta)^{v-1}$ is positive and non-increasing for $\beta<u<\gamma$, where v is a constant $\geqslant 1$.*

Assume also

(A2′) $p^{2-\nu}A$ *is non-decreasing on* $(0, \infty)$.[1]

Then (*) *has at most one solution such that* $u(0) < \gamma$ (*or* $u(0) \leqslant \gamma$).

If $\gamma = \infty$ the condition $u(0) < \gamma$ is satisfied for any solution, and so can be omitted from the statement of Theorem I. Similarly the condition $u(0) \leqslant \gamma$ can be dropped in Theorem II if f satisfies the condition $f(u) \leqslant 0$ for $u > \gamma$. Indeed in this case, if $u(0) > \gamma$ then (1.1.4) gives a contradiction when applied to the interval $(0, r_0)$ where $u(r) \in (\gamma, u(0))$.

When conditions (A3) and (F3) are dropped from Theorems I–III, we still obtain results of interest, as stated in Proposition 3 in the Introduction.

Condition (A2′) together with (A2) forces $\nu \leqslant m$. In the particular case $A(p) = p^{m-2}$ condition (A2) is trivially satisfied, while (A2′) holds with $\nu = m$; similarly (A3) is valid for $\mu = 1$.

Conditions (A2) and (A3) concern the behavior of $A(p)$ and $A'(p)$ near $p = 0$, whereas (A1) has a global character. Some simpler sufficient conditions can be formulated in order that (A1) holds; in particular we have the following result.

LEMMA 3.0. *The hypothesis* (A1) *is satisfied if either of the following conditions holds*:

(i) $n > 2$ *and* $M + \frac{1}{2}\frac{n-2}{n-1} N^2$ *is non-increasing*;

(ii) $n < 2$ *and* $M + \frac{n-2}{n-1} N^2$ *is non-increasing*;

(iii) $n = 2$ *and* M *is decreasing.*

Proof. Write $\Psi(p) = M(p) + ((n-2)/(n-1))\, N^2(p)$. Then

$$S(l, p) = \Psi(l) - \Psi(p) + \frac{n-2}{n-1} N(p)(N(l) - N(p))$$

and the conclusion for case (ii) follows at once since N is decreasing (by (H3)). Condition (i) is proved in essentially the same way; that is, putting

$$\Psi^*(p) = M(p) + \frac{1}{2}\frac{n-2}{n-1} N^2(p)$$

[1] When $F(\gamma) < \infty$ the interval $(0, \infty)$ can be replaced by a smaller set $(0, p_0)$, where p_0 is defined by $H(p_0) = F(\gamma)$. Indeed by (1.2.1) and (1.2.2), when $u(r) > \beta$,

$$H(p) < H(p) + F(u) \leqslant F(u(0)) < F(\gamma),$$

so $p < p_0$.

we get for $0<l<p$,

$$S(l,p)=\Psi^*(l)-\Psi^*(p)+\frac{1}{2}\frac{n-2}{n-1}\{N(l)-N(p)\}^2$$
$$>\Psi^*(l)-\Psi^*(p)\geqslant 0$$

since $n>2$. Condition (iii) is obvious.

It is clear that (i) and (iii) are most simply satisfied when M is decreasing. This in turn is implied by condition (A1′) in the Introduction, as is the condition $\Omega'(p)>0$ for $p>0$.

Remark. For the operator

$$A(p)=(1+p^2)^{-s/2}p^{m-2}, \qquad s\geqslant 0,$$

we have

$$\Omega(p)=(1+p^2)^{-s/2}p^{m-1},$$
$$E(p)=\Omega'(p)=(1+p^2)^{-s/2-1}p^{m-2}\{(m-1)+(m-1-s)p^2\},$$
$$M(p)=(1+p^2)^{s+1}p^{-2m}\{(m-1)+(m-1-s)p^2\}^{-1},$$
$$N^2(p)=(1+p^2)^s p^{-2m}.$$

Clearly (H3′) holds if and only if $m\geqslant 1+s$, or $m>1$ if $s=0$. In this case, moreover, it is easy to check that M is decreasing. Hence (A1) and (A1′) are satisfied for $n\geqslant 2$.

When $1<n<2$ we write

$$\Psi(p)=M(p)-\frac{1}{m-1}N^2(p)+\left(\frac{n-2}{n-1}+\frac{1}{m-1}\right)N^2(p).$$

Now

$$M(p)-\frac{1}{m-1}N^2(p)=\frac{s}{m-1}\frac{(1+p^2)^s}{p^{2(m-1)}}\frac{1}{(m+1)+(m-1-s)p^2}.$$

This quantity is non-increasing provided

$$s\geqslant 0, \qquad m\geqslant s+1 \text{ (or } m>1 \text{ if } s=0).$$

Hence $\Psi(p)$ itself is non-increasing and (A1) holds if we suppose also that

$$\frac{n-2}{n-1}+\frac{1}{m-1}\geqslant 0, \qquad \text{that is} \quad n\geqslant 2-\frac{1}{m}.$$

(It is not hard to see that this condition is best possible).

The following three sections contain introductory material, including a discussion of the asymptotic behavior of solutions in Section 3.2 and a monotone separation theorem in Section 3.3.

Theorems I–III are almost immediate corollaries of Proposition 3 in the Introduction (with assumption (A1′) being replaced by (A1) and also with $n>1$ rather than $n\geqslant 2$). Consequently our principal effort can be devoted to proving Proposition 3. The proofs of Theorems I and II are then given at the end of Section 3.4, and the proof of Theorem III in Section 3.5. Theorems A, B, C of the Introduction are the special cases of Theorems I–III when $n\geqslant 2$ and (A1) is replaced by the stronger condition (A1′).

The main Theorems I–III require the satisfaction of conditions (A1) and (A2). These conditions can be avoided provided that other hypotheses are suitably strengthened. This result is given in Theorem IV of Section 6. In the final section of the paper we add several remarks concerning the exterior Dirichlet problem.

3.1. *Preliminary Results.*

Denote by $l=l(u,p)$ the positive function determined implicitly by the equation

$$H(l)=H(p)+F(u)$$

on the domain

$$P=\{(u,p)\colon u,p>0,\ 0<H(p)+F(u)<H(\infty)\}.$$

Obviously l is well-defined and strictly positive since H is increasing. Moreover (recall that $H'(p)=pE(p)$ by (4) and that $E(p)>0$ for $p>0$).

$$\frac{\partial l}{\partial u}=\frac{f(u)}{lE(l)},\qquad \frac{\partial l}{\partial p}=\frac{pE(p)}{lE(l)}. \tag{3.1.1}$$

We also define $K(u,p)$ on P by the formula

$$K(u,p)=\frac{(lA(l))^{\alpha-1}}{l}\{l^2A(l)-p^2A(p)\},\qquad \alpha=1/(n-1).$$

Then the following result holds.

LEMMA 3.1.1. *Suppose $(u,p)\in P$. Then*

(i) $\dfrac{\partial}{\partial p}\{lA(l)\}>0;$

(ii) $\dfrac{\partial}{\partial p}\left\{\dfrac{K(u,p)}{p}\right\}>0$ *if and only if* $S(l,p)>0.$

Moreover, if $F(u)\leqslant 0$, *then*

$$\text{(iii)}\quad K(u,p)\leqslant 0.$$

Proof. The first relation is obvious since

$$\frac{\partial}{\partial p}\{lA(l)\}=\frac{\partial}{\partial p}\Omega(l)=E(l)\frac{pE(p)}{lE(l)}=\frac{pE(p)}{l}. \tag{3.1.2}$$

The third result also follows at once since $p^2A(p)$ is increasing and $l\leqslant p$ when $F(u)\leqslant 0$. Finally $l=p$ and $K=0$ if and only if $F(u)=0$.

We turn to the second conclusion. By direct calculation, keeping in mind (3.1.1) and (3.1.2), we have

$$\begin{aligned}\frac{\partial}{\partial p}\left\{\frac{K(u,p)}{p}\right\}&=\frac{\partial}{\partial p}\left\{(\Omega(l))^{\alpha-1}\left(\frac{\Omega(l)}{p}-\frac{\Omega(p)}{l}\right)\right\}\\&=(\alpha-1)(\Omega(l))^{\alpha-2}\frac{pE(p)}{l}\left(\frac{\Omega(l)}{p}-\frac{\Omega(p)}{l}\right)\\&\quad+(\Omega(l))^{\alpha-1}\left(-\frac{\Omega(l)}{p^2}+\frac{\Omega(p)}{l^2}\frac{pE(p)}{lE(l)}\right)\\&=(\Omega(l))^{\alpha}\,p\Omega(p)\,E(p)\,S(l,p),\end{aligned}$$

and (ii) is proved.

We next introduce an important identity (Lemma 3.1.3) for solutions of (*). We begin by showing that the solution u is twice differentiable whenever $u'\neq 0$.

LEMMA 3.1.2. *Suppose* $u'\neq 0$ *at some point* $r>0$. *Then at this point* u'' *exists and satisfies* (1.1.1) *in the form*

$$E(p)\,u''+\frac{n-1}{r}A(p)\,u'+f(u)=0\qquad\textit{where}\quad p=|u'|. \tag{3.1.3}$$

Proof. Suppose $u'(r)<0$. By virtue of Lemma 1.1.1

$$u'(r)=-\Omega^{-1}\left(\int_0^r\left(\frac{t}{r}\right)^{n-1}f(u(t))\,dt\right) \tag{3.1.4}$$

Since A is continuously differentiable and $\Omega' = E$, it follows that Ω^{-1} is differentiable and $(\Omega^{-1})' = 1/E$, when $p > 0$. Hence from (3.1.4) we see that u'' exists and

$$u'' = -\frac{1}{E(p)}\left\{f(u(r)) - \int_0^r \frac{n-1}{r}\left(\frac{t}{r}\right)^{n-1} f(u(t))\,dt\right\}$$

$$= \frac{1}{E(p)}\left\{-f(u) + \frac{n-1}{r}\,\Omega(p)\right\},$$

as required. The proof when $u' > 0$ is similar.

LEMMA 3.1.3. *If u is a solution of (*) such that $u(r) > 0$ and $F(u(r)) \leqslant 0$ on (r_0, r_1) we have*

$$\text{(i)}\quad (u(r), |u'(t)|) \in P \qquad \textit{for any} \quad r \in (r_0, r_1).$$

Moreover, if we write $p = p(r) = |u'(r)|$ and $l = l(r) = l(u(r), p(r))$, then

$$\text{(ii)}\quad r(lA(l))^{\alpha}\big|_{r_0}^{r_1} = \int_{r_0}^{r_1} K(u, p)\,dr, \qquad \alpha = \frac{1}{n-1}. \tag{3.1.5}$$

Proof. (i) It is obvious from (1.2.1) that $H(p) + F(u) > 0$ on (r_0, r_1). Since $F(u) \leqslant 0$ we get in turn that $p > 0$ and $H(p) + F(u) \leqslant H(p) < H(\infty)$. Thus (i) follows from the definition of P.

(ii) If $r \in (r_0, r_1)$ we have

$$\frac{d}{dr}\{r(lA(l))^{\alpha}\} = (lA(l))^{\alpha} + \alpha r(lA(l))^{\alpha-1} E(l)\left\{\frac{\partial l}{\partial u}u' + \frac{\partial l}{\partial p}p'\right\}$$

$$= (lA(l))^{\alpha} + \alpha r(lA(l))^{\alpha-1} u'\{f(u) + E(p)\,u''\}\,l^{-1}$$

$$= \frac{(lA(l))^{\alpha-1}}{l}\{l^2A(l) - p^2A(p)\} = K(u, p),$$

where (3.1.1) was used at the second step and equation (3.1.3) at the third (recall that $p > 0$ in (r_0, r_1) by (i)). Integrating over (r_0, r_1) completes the proof.

3.2. *Asymptotic Behavior.*

We can now obtain the following important results concerning the asymptotic behavior of a solution of (*).

LEMMA 3.2.1. *Suppose* (F1) *holds and let u be a solution of* (*). *Then*

(i) *if* $u>0$ *for all* $r>0$ *there exists* $\lambda \geqslant 0$ *such that*

$$r^{n-1} l A(l) \to \lambda \qquad \text{as} \quad r \to \infty, \tag{3.2.1}$$

(ii) $\displaystyle \lim_{r\to\infty} r^{n-1} \int_r^\infty A u'^2 \frac{d\rho}{\rho} = 0.$

Proof. (i) By Lemma 3.1.3 we get

$$(r^{n-1} l(r)\, A(l(r)))^\alpha = (R^{n-1} l(R)\, A(l(R)))^\alpha + \int_R^r K(u, p)\, d\rho \tag{3.2.2}$$

for $r>R$, where R is so large that $u(r)<\beta$ for $r>R$; see (F1).

The right hand side of (3.2.2) goes to a limit $\bar{\lambda}$ as $r\to\infty$, since $K(u,p)\leqslant 0$ by Lemma 3.1.2 (ii). Obviously $\bar{\lambda}\in[-\infty,\infty)$. On the other hand, the left hand side is $\geqslant 0$. Hence $\bar{\lambda}\geqslant 0$ and there exists $\lambda\geqslant 0$ such that (3.2.1) holds. Note that this proof shows also that $K(u,p)\in L^1((r,\infty))$ for any $r>0$.

(ii) It is enough to consider the case when $u>0$ for all $r>0$. By Lemma 1.2.1 and the definition of l we have

$$(n-1)\int_r^\infty A u'^2 \frac{d\rho}{\rho} = H(p) + F(u) = H(l).$$

On the other hand, by equation (4) in the introduction,

$$r^{n-1} H(l) = r^{n-1}\left(l^2 A(l) - \int_0^l \rho A(\rho)\, d\rho\right) \leqslant r^{n-1} l^2 A(l) \to 0$$

since $l\to 0$ $(0<l<p)$ and $r^{n-1} l A(l)\to\lambda$ as $r\to\infty$.

In the proof of the following lemma we use an ingenious idea of Kaper & Kwong [KK2].

LEMMA 3.2.2. *Suppose hypotheses* (A2) *and* (F1) *are satisfied and let u be a positive solution of* (*). *Then, with* λ *defined as in* (3.2.1), *we have*

(i) $\lambda=0$ *when* $n\leqslant m$;

(ii) *when* $n>m$

$$r^{(n-m)/(m-1)} u(r) \to \frac{m-1}{n-m} \lambda^{1/(m-1)} \qquad \text{as} \quad r\to\infty.$$

Proof. (i) Since $l \to 0$ and $r^{n-1} l A(l) \to \lambda$ as $r \to \infty$, from (A2) it follows that

$$r^{(n-1)/(m-1)} l = (r^{n-1} l A(l))^{1/(m-1)} (l^{2-m} A(l))^{-1/(m-1)} \to \lambda^{1/(m-1)} \qquad (3.2.3)$$

as $r \to \infty$. Therefore since $0 < l < p$ for large r,

$$\liminf_{r \to \infty} r^{(n-1)/(m-1)} p(r) \geqslant \lambda^{1/(m-1)}.$$

Because $p = p(r) = |u'(r)| = -u'(r)$ is integrable on $[0, \infty)$, the last inequality implies $\lambda = 0$ if $n \leqslant m$.

(ii) The proof for the case $n > m$ lies much deeper. By (3.2.3), l is integrable on (R, ∞), where $R > 0$ is such that $u(r) < \beta$ for every $r > R$. Moreover $F(u(r)) < 0$ for $r > R$ by (F1), so that also

$$H(p(r)) = H(l(r)) + |F(u(r))| \qquad \text{for} \quad r > R.$$

In particular, because H is strictly increasing we obtain

$$l < p = H^{-1}(H(l) + |F(u)|) \qquad \text{for} \quad r > R. \qquad (3.2.4)$$

Now, from (4) and (A2) it follows easily that $p^{-m} H(p) \to (m-1)/m$ as $p \to 0$. Thus since $m > 1$ we see that for every $\varepsilon > 0$ there exists $r_\varepsilon \geqslant R$ such that

$$\begin{aligned} p &< \left(1 + \frac{\varepsilon}{2}\right) \left\{ \frac{m}{m-1} H(p) \right\}^{1/m} \\ &= \left(1 + \frac{\varepsilon}{2}\right) \left\{ \frac{m}{m-1} (H(l) + |F(u)|) \right\}^{1/m} \\ &\leqslant (1+\varepsilon) \left\{ l^m + \frac{m}{m-1} |F(u)| \right\}^{1/m} \\ &\leqslant (1+\varepsilon) \left(l + \left(\frac{m}{m-1} |F(u)| \right)^{1/m} \right) \end{aligned} \qquad (3.2.5)$$

on (r_ε, ∞). Therefore we obtain from (3.2.4) and (3.2.5)

$$\begin{aligned} \int_r^\infty l(\rho)\, d\rho &< \int_r^\infty p(\rho)\, d\rho \\ &< (1+\varepsilon) \left(\int_r^\infty l(\rho)\, d\rho + \text{const.} \int_r^\infty |F(u(\rho))|^{1/m}\, d\rho \right) \end{aligned} \qquad (3.2.6)$$

for every $r > r_\varepsilon$.

We estimate the second integral on the right side by using Hölder's inequality, namely

$$\int_r^\infty |F(u)|^{1/m}\, d\rho$$

$$\leqslant \left(\int_r^\infty B(l)\,|F(u)|\, d\rho\right)^{1/m} \cdot \left(\int_r^\infty (B(l))^{-1/(m-1)}\, d\rho\right)^{1-1/m}, \quad (3.2.7)$$

where $B(l)=((lA(l))^{\alpha-1}/l$. Since $l<p$ and $r>r_\varepsilon$,

$$|F(u)| = H(p)-H(l) = p^2A(p)-l^2A(l) - \int_0^p \rho A(\rho)\, d\rho + \int_0^l \rho A(\rho)\, d\rho$$

$$\leqslant p^2A(p)-l^2A(l) = \frac{|K(u,p)|}{B(l)}.$$

Then, from the fact that $K(u,p)\in L^1$ we get

$$B(l)\, F(u)\in L^1. \quad (3.2.8)$$

On the other hand, by (3.2.1), (3.2.3), and a direct computation using L'Hopital's rule, the quantity

$$r^{(n-m)/(m-1)}\left(\int_r^\infty (B(l))^{-1/(m-1)}\, d\rho\right)^{1-1/m}$$

tends to a finite limit as $r\to\infty$. In fact it is enough to show that

$$\frac{B(l)^{-1/(m-1)}}{r^{-m(n-m)/(m-1)^2-1}}$$

tends to a limit, or equivalently that

$$\frac{B(l)}{r^{m(n-m)/(m-1)+(m-1)}}$$

tends to a limit. But

$$B(l)=[lA(l)]^{(2-n)/(n-1)}/l$$

while by (3.2.1) and (3.2.3)

$$r^{n-1}lA(l)\to \text{limit}, \qquad lr^{(n-1)/(m-1)}\to \text{limit}.$$

The conclusion now follows immediately, since

$$\frac{m(n-m)}{m-1}+(m-1)=(n-2)+\frac{n-1}{m-1}.$$

From (3.2.7) and (3.2.8) we then get

$$r^{(n-m)/(m-1)}\int_r^\infty |F(u)|^{1/m}\,d\rho \to 0 \qquad \text{as} \quad r\to\infty. \tag{3.2.9}$$

Finally, again by (3.2.3) and L'Hospital's rule,

$$r^{(n-m)/(m-1)}\int_r^\infty l\,d\rho \to \frac{m-1}{n-m}\lambda^{1/(m-1)}. \tag{3.2.10}$$

From (3.2.6), (3.2.9) and (3.2.10) we obtain the assertion since $\int_r^\infty p\,d\rho = u(r)$.

3.3. *Monotone Separation Theorems*

Our purpose is to study the separation properties of pairs of solutions of Problem (*).

Let u and v be two solutions of (*), which we assume throughout to satisfy the conditions

$$u(0)\leqslant\gamma, \qquad v(0)\leqslant\gamma.$$

By Lemma 1.2.6, in view of (F1) and (F2), it is clear that

$$u'(r)<0 \textit{ whenever } r>0 \textit{ and } u(r)>0;$$

$$v'(r)<0 \textit{ whenever } r>0 \textit{ and } v(r)>0.$$[2]

Thus both u and v possess inverses on the domain where they are positive. We denote by r and s the inverses of u and v, defined respectively on the intervals $(0, u(0))]$ and $(0, v(0)]$. Note also the principal fact that

$$u(0) \textit{ and } v(0) \textit{ must exceed } \beta,$$

by virtue of Lemma 1.2.1 and (F1).

LEMMA 3.3.1. *Assume that* (F1), (F2) *hold. If* $r(u)-s(u)>0$ *on some open interval* I *of the domain of* r *and* s, *then* $r(u)-s(u)$ *can have at most*

[2] To be precise in the application of Lemma 1.2.6, the reader may note that condition (F1) is used to obtain $u'(r)<0$ when $u\leqslant\beta$ while condition (F2) is used when $u\geqslant\beta$.

one critical point on I. Moreover if such a point exists it must be a strict maximum.

Proof. By Lemma 3.1.2, equation (1.1.1) can be written in the form (3.1.3). Then it is immediately verified that the function $r=r(u)$ satisfies the equation

$$E\left(\left|\frac{1}{r_u}\right|\right)r_{uu}-\frac{n-1}{r}A\left(\left|\frac{1}{r_u}\right|\right)r_u^2-r_u^3f(u)=0$$

for $u\in I$ (clearly $r>0$ and $r_u<0$). A similar equation also holds for $s=s(u)$.

Now suppose $r-s$ has a critical point in I. Then $r_u=s_u<0$ and $r>s>0$ at this point. Consequently by subtraction we find

$$(r-s)_{uu}=(n-1)\frac{A}{E}\left(\frac{1}{r}-\frac{1}{s}\right)r_u^2<0$$

which proves the assertion.

Lemma 3.3.2. *Suppose* (A1) *and* (F1) *are satisfied. If* $r-s$ *has two zeros in* $(0,\beta]$, *say* ξ_0 *and* ξ_1, *then* $r-s=0$ *for all* u *between* ξ_0 *and* ξ_1.

Proof. Suppose the conclusion is false. Then without loss of generality we can assume that $\xi_0<\xi_1$ and $r-s>0$ for all $u\in[\xi_0,\xi_1]$.

In the open interval (ξ_0,ξ_1) the difference $r-s$ can have at most one critical point by Lemma 3.3.1, and at the same time at least one, since it vanishes at the endpoints. If follows that there exists a point $\xi_2\in(\xi_0,\xi_1)$ such that $r_u=s_u$ at $u=\xi_2$, while

$$r_u<s_u \qquad \text{for} \quad u\in(\xi_2,\xi_1).$$

Now put $r_1=r(\xi_1)$, $r_2=r(\xi_2)$, $l_1=l(r_1)$, $l_2=l(r_2)$. By (3.1.5) we have

$$r_2(l_2A(l_2))^\alpha-r_2(l_2A(l_2))^\alpha=\int_{r_1}^{r_2}K(u,p)\,dr=\int_{\xi_2}^{\xi_1}K(u,p)\frac{du}{p},$$

where p denotes the function $p(r(u))=|u'(r(u))|$.

A similar identity naturally holds for the solution v. In analogy with the preceding notation, we denote by s_1,s_2,m_1,m_2 the quantities relative to v corresponding to the quantities r_1,r_2,l_1,l_2 for u.

Subtracting the two identities, and using the fact that $r_1=s_1$ and $l_2=m_2$ (because $r_u(\xi_2)=s_u(\xi_2)$), we obtain

$$\begin{aligned}(r_2-s_2)\{l_2A(l_2)\}^\alpha&-r_1\{(l_1A(l_1))^\alpha-(m_1A(m_1))^\alpha\}\\&=\int_{\xi_2}^{\xi_1}\left(\frac{K(u,p)}{p}-\frac{K(u,q)}{q}\right)du,\end{aligned}\tag{3.3.1}$$

where, by analogy with the above notation, $q = q(s(u)) = |v'(s(u))|$. Note particularity that $p > q$ for $u \in (\xi_2, \xi_1)$ since $r_u < s_u$.

Now let

$$\frac{\partial}{\partial p}\left(\frac{K(u, p)}{p}\right) = D(u, p).$$

Then by the mean value theorem, for any fixed $u \in (\xi_2, \xi_1)$,

$$\frac{K(u, p)}{p} - \frac{K(u, q)}{q} = D(u, \tilde{p})(p - q)$$

where $\tilde{p} \in (p, q)$. Let $\tilde{l} = l(u, \tilde{p})$, see Section 3.1. By the definition of l and the fact that $H(\cdot)$ is strictly increasing it is clear that $\tilde{l} \leqslant \tilde{p}$, with $\tilde{l} = \tilde{p}$ if and only if $F(u) = 0$ (recall $F(u) \leqslant 0$ for $u \in (\xi_2, \xi_1)) \subset (0, \beta]$.

By Lemma 3.1.2 (ii) we have $D(u, \tilde{p}) > 0$ if $S(\tilde{l}, \tilde{p}) > 0$, so by (A1) also $D(u, \tilde{p}) > 0$ if $\tilde{l} < \tilde{p}$. Thus

$$\frac{K(u, p)}{p} - \frac{K(u, q)}{q} < 0 \qquad \text{if } \tilde{l} < \tilde{p}.$$

On the other hand, if $\tilde{l} = \tilde{p}$ then $D(u, \tilde{p}) = S(\tilde{l}, \tilde{p}) = 0$ and $K(u, p)/p = K(u, q)/q$. Hence the right hand side of (3.3.1) is non-positive.

But also $r_2 > s_2$ and $l_1 \leqslant m_1$ because $r_u \leqslant s_u$ at $u = \xi_1$. Thus the left hand side of (3.3.1) is strictly positive, a contradiction.

LEMMA 3.3.3. *Suppose hypotheses* (F1) *and* (F2) *are satisfied. If* $r(u) - s(u) > 0$ *on an interval* $I = (0, a)$, *then* $(r - s)' < 0$ *on* I.

Proof. By Lemma 3.3.1 either $r(u) - s(u)$ is everywhere decreasing on I, or $r(u) - s(u)$ is increasing for u near zero. In the first case we are done, so let us assume for contradiction that $r_u - s_u > 0$ in some interval $0 < u < \delta$. As in the previous lemma, we put $r = r(u)$, $p = p(u) = |u'(r(u))|$. Then equation (1.2.1) becomes

$$H(p) + F(u) = (n - 1) \int_0^u \frac{Ap}{r}\, dt, \qquad 0 < u < \delta.$$

A similar identity holds for the solution v, replacing p by $q = q(u) = |v'(s(u))|$.

Now writing $A = A(p)$, $B = A(q)$ and subtracting the two identities we get

$$H(p) - H(q) = (n - 1) \int_0^u \left(\frac{Ap}{r} - \frac{Bq}{s}\right) dt. \tag{3.3.2}$$

Define

$$\varphi = \varphi(u) = \int_0^u \left(\frac{Ap}{r} - \frac{Bq}{s} \right) dt.$$

Since $s_u < r_u < 0$ on $(0, \delta)$ we have $p > q$ for $0 < u < \delta$. Hence, because H is strictly increasing, (3.3.2) gives

$$(n-1)\,\varphi = H(p) - H(q) > 0, \qquad 0 < u < \delta. \tag{3.3.3}$$

Also, since $r > s$,

$$\begin{aligned} qs\varphi' = qs\left(\frac{Ap}{r} - \frac{Bq}{s}\right) &< q(Ap - Bq) = q\int_q^p d(\rho A(\rho)) \\ &< \int_q^p \rho d(\rho A(\rho)) = H(p) - H(q) = (n-1)\,\varphi, \end{aligned}$$

by (5).

Now we integrate the inequality

$$\frac{\varphi'}{\varphi} < \frac{(n-1)}{qs},$$

obtaining for $0 < u < \bar{u} < \delta$,

$$\log \frac{\varphi(\bar{u})}{\varphi(u)} < (n-1)\int_u^{\bar{u}} \frac{1}{qs}\, dt = (n-1)\int_{s(\bar{u})}^{s(u)} \frac{ds}{s} = (n-1) \log \frac{s(u)}{s(\bar{u})}.$$

Hence

$$(s(\bar{u}))^{n-1}\, \varphi(\bar{u}) < (s(u))^{n-1}\, \varphi(u). \tag{3.3.4}$$

On the other hand, since $r > s$ and $Bq > 0$, we have

$$0 < (s(u))^{n-1}\, \varphi(u) < (r(u))^{n-1}\, \varphi(u) < (r(u))^{n-1} \int_0^u \frac{Ap}{r}\, dt = r^{n-1} \int_r^\infty Au'^2 \frac{d\rho}{\rho}.$$

Thus by Lemma 3.2.1 (ii), it follows that $(s(u))^{n-1}\, \varphi(u) \to 0$ as $u \to 0$. From (3.3.4) we then deduce that $\varphi(\bar{u}) \leqslant 0$, which contradicts (3.3.3).

3.4. *Uniqueness Theorem* I

In this section we shall prove Theorem I. We first obtain several preliminary results, of interest in themselves.

Consider the initial value problem

$$\begin{aligned} &(Au')' + \frac{n-1}{r} Au' + f(u) = 0, \qquad r > 0 \\ &u(0) = \xi > 0, \qquad u'(0) = 0 \end{aligned} \tag{P}$$

for a classical solution u (see Section 1.1).

PROPOSITION A1. *Suppose the hypotheses* (H1)–(H3) *are satisfied. Then problem* (P) *has a classical solution in a neighborhood of the origin.*

Proof. Put $\Omega(p) = pA(|p|)$ for $p \neq 0$, $\Omega(0) = 0$.[5] Existence of a solution is proved by applying the Schauder fixed point theorem to the operator

$$(Tu)(r) = \xi - \int_0^r \Omega^{-1} \left(\int_0^\rho f(u(t)) \left(\frac{t}{\rho} \right)^{n-1} dt \right) d\rho;$$

see [NS].

PROPOSITION A2. *Suppose that A is of class C^1 on $(0, \infty)$ and that $f(u)$ is Lipschitz continuous for $u \in J$, where J is a subinterval of $(0, \infty)$ containing ξ. Assume also that the derivative of the (increasing) function Ω is bounded from zero on every bounded subset of $(0, \infty)$. Then the solution of* (P) *is unique as long as it exists and remains in J.*

Proof. Suppose that u and v are two different solutions of (P) whose values lie in J. Then the function

$$\omega = \Omega(u') - \Omega(v')$$

is a solution of the Cauchy problem

$$\omega' + \frac{n-1}{r}\omega = \psi(r), \qquad \omega(0) = 0,$$

where $\psi(r) = f(v(r)) - f(u(r))$. It follows that

$$|\omega(r)| = \int_0^r \psi(t) \left(\frac{t}{r} \right)^{n-1} dt \leqslant \frac{r}{n} \sup_{[0, r]} |\psi(t)|. \tag{1}$$

On the other hand, since f is Lipschitz continuous on J, we obtain, for appropriate values r_0, $M > 0$,

$$|\psi(r)| \leqslant M\, |u(r) - v(r)| \qquad \text{for} \quad r \in [0, r_0].$$

[5] This extends the domain of $\Omega(p)$ from $p \geqslant 0$ to all $p \in \mathbf{R}$.

Hence recalling that $u(0)=v(0)$, the principal hypothesis of the proposition yields

$$|\psi(r)| \leqslant M \int_0^r |u'(s)-v'(s)|\, ds \leqslant \frac{M}{k} \int_0^r |\omega(s)|\, ds, \tag{2}$$

where $k>0$ is the infimum of the derivative of $\Omega(p)$ on the set $(0, p_0)$, and p_0 is a bound for $|u'(r)|$ and $|v'(r)|$ on the interval $[0, r_0]$; here we recall that both u' and v' are zero when $r=0$, and are also continuous.

Combining (1) and (2) yields

$$|\omega(r)| \leqslant \frac{1}{n} \frac{M}{k} r \int_0^r |\omega(s)|\, ds$$

for $r \leqslant r_1$. It now follows from Gronwall's inequality that $\omega \equiv 0$ for $r \leqslant r_1$. Consequently $\Omega(u') \equiv \Omega(v')$, and $u' \equiv v'$ for $r \leqslant r_1$. With the initial point $r=r_0$ replaced by $r=\rho>0$, for an appropriate value ρ, the same proof can be reapplied as often as necessary to give uniqueness of any continuation of the solution whose values lie in J.

Proposition A3. *Suppose that the hypotheses of Proposition* A2 *are satisfied. Then solutions of problem* (P) *depend continuously on the initial data* ξ.

Proof. We proceed in essentially the same way as in the demonstration of Proposition A2. Let $u(0)=\xi$, $v(0)=\xi+h$. Then on any compact subset of values r for which the solutions are defined, we have

$$|u(r)-v(r)| \leqslant h + \int_0^r |u'(s)-v'(s)|\, ds$$

(it can be assumed that $h>0$). Consequently (2) can be replaced by

$$|\psi(r)| \leqslant M\left(h + \frac{1}{k}\int_0^r |\omega(s)|\, ds\right),$$

with the constants M, k as before. In turn

$$|\omega(r)| \leqslant \frac{1}{n} Mr\left(h + \frac{1}{k}\int_0^r |\omega(s)|\, ds\right),$$

so by Gronwall's inequality

$$|\omega(r)| \leqslant \frac{M}{n} rh \exp\left(\frac{M}{2nk} r^2\right).$$

This gives

$$|u'-v'| \leqslant \frac{M}{nk} rh \exp\left(\frac{M}{2nk} r^2\right),$$

and

$$|u-v| \leqslant h \exp\left(\frac{M}{2nk} r^2\right),$$

completing the proof.

Proposition A2 does not apply to the important case when $A(p)=p^{m-2}$ and $m>2$, since then $\Omega'(p)=(m-1)\,p^{m-2}$ is not bounded from zero on bounded subsets of $(0,\infty)$. Therefore it is important to have the following extension of its validity.

PROPOSITION A4. *Suppose A is of class C^1 on $(0,\infty)$ and that $f(u)$ is non-vanishing and Lipschitz continuous for $u \in J$, where J is a subinterval of $(0,\infty)$ containing ξ. Assume also that $\Omega'(p)>0$ for $p \neq 0$ and*

$$\Omega'(p) \geqslant |\Omega(p)|^{\mu} \tag{3}$$

for all sufficiently small $p \neq 0$, where μ is a fixed exponent in $[0,2)$. Then the solution of (P) *is unique as long as it exists and remains in J.*

Condition (3) clearly holds when $A(p)=p^{m-2}$, with $\mu=0$ if $m \leqslant 2$ and $\mu=1$ (and $p \leqslant m-1$) if $m>2$. It is also satisfied when $A(p)=(1+p^2)^{-s/2}\,p^{m-2}$ for $s \geqslant 0$, $m>1$, as one easily checks. We are indebted to L. Veron for the idea of the proof.

Proof of Proposition A4. It can be assumed that $f(u)>0$ for $u \in J$, the opposite case being treated similarly. We proceed now as in the proof of Proposition A.2, until relation (2). Here the estimate

$$|u'(s)-v'(s)| \leqslant \frac{1}{k}\,|\omega(s)|, \qquad s \in (0, r_0)$$

must be replaced by

$$|u'(s)-v'(s)| \leqslant \frac{1}{\inf \Omega'(\tilde{p})}\,|\omega(s)|, \qquad s \in (0, r_0), \tag{4}$$

where the infimum is taken over all intermediate values $\tilde{p}$ between $|u'(s)|$ and $|v'(s)|$. Note, for this estimate, that necessarily $u'(r)$, $v'(r)$ are negative for $r>0$ because $f(u)>0$ when $u \in J$, see Lemma 1.1.1 or the proof of Proposition A1.

Obviously, for any solution of (P)

$$\Omega(u') = -\int_0^r f(u(t))(t/r)^{n-1}\, dt.$$

Therefore supposing without loss of generality that $|u'(s)| \leqslant |v'(s)|$ in (4), we find that

$$\Omega(\tilde{p}) > \Omega(|u'|) = -\int_0^s f(u(t))(t/s)^{n-1}\, dt \geqslant \frac{s}{2n} f(u(0)) = \frac{s}{2n} f(\xi),$$

provided that r_0 is chosen even smaller, if necessary, so that $f(u(r)) \geqslant \frac{1}{2} f(u(0))$ for $0 \leqslant r \leqslant r_0$. Using (3) and the fact that $\mu \geqslant 0$, we get

$$\Omega'(\tilde{p}) \geqslant \left(\frac{s}{2n} f(\xi)\right)^{\mu}, \qquad s \in (0, r_0),$$

and of course the same holds when $|v'(s)| \leqslant |u'(s)|$. Hence from (4) and the first inequality of (2) follows

$$|\psi(r)| \leqslant M \int_0^r \left(\frac{2n}{sf(\xi)}\right)^{\mu} |\omega(s)|\, ds.$$

In turn, using (1) for $r \in (0, r_0)$ there holds

$$|\omega(r)| \leqslant Cr \int_0^r \frac{|\omega(s)|}{s^{\mu}}\, ds,$$

where $C = Mn^{\mu-1}(2/f(\xi))^{\mu}$.

One can now apply Gronwall's lemma (for this purpose, it is convenient to replace ω by ω/r), with the conclusion

$$\left|\frac{\omega(r)}{r}\right| \leqslant C \int_0^t \frac{|\omega(s)|}{s^{\mu}}\, ds \cdot \exp\left(C \int_t^r \frac{ds}{s^{\mu-1}}\right)$$

for any $t \in (0, r)$. Of course $|\omega(s)/s|$ is bounded on $(0, r_0)$ by (1). Hence, recalling that $\mu < 2$ and letting $t \to 0$, we get $\omega(r) \equiv 0$ for $r \in (0, r_0)$. The procedure can of course be repeated as often as necessary, proving the theorem.

Remark. The condition that $f(u(0)) > 0$ is essential for the validity of Proposition A.4. Indeed, consider the equation

$$((u')^3)' + \frac{n-1}{r}(u')^3 + 8(1-u) = 0,$$

that is, the case $A(p)=p^2$ and $f(u)=8(1-u)$. If we take $\xi=1$, then $f(\xi)=0$, and the solution of the Cauchy problem (P) is in fact not unique, there being at least the three solutions,

$$u\equiv 1, \qquad u=1-at^2, \qquad u=1+at^2,$$

where $a=\sqrt{1/(n+2)}$.

One can restore uniqueness in case $f(\xi)=0$ by appropriately strengthening the Lipschitz condition at $u=\xi$, but (for once) we shall not pursue this further.

References

[AP] F. V. Atkinson and L. A. Peletier, Ground states of $-\Delta u=f(u)$ and the Emden–Fowler equation, *Arch. Rational Mech. Anal.* **93** (1986), 103–127.

[APS] F. V. Atkinson, L. A. Peletier, and J. Serrin, Ground states for the prescribed mean curvature equation: the supercritical case, *in* "Nonlinear Diffusion Equations and Their Equilibrium States" (W. M. Ni, L. A. Peletier, and J. Serrin, Eds.), Springer-Verlag, New York/Berlin, 1988.

[BLP] H. Berestycki, P. L. Lions, and L. A. Peletier, An O.D.E. approach to the existence of positive solutions for semilinear problems in $\mathbb{R}^n$, *Indiana Univ. Math. J.* **30** (1981), 141–157.

[BL] H. Berestycki and P. L. Lions, Nonlinear scalar field equations, I. Existence of a ground state, *Arch. Rational Mech. Anal.* **82** (1983), 313–345.

[C] G. Citti, Positive solutions of a quasilinear degenerate elliptic equation in $\mathbb{R}^n$, *Rend. Circ. Mat. Palermo* (2) **35** (1986), 364–375.

[CC] C. V. Coffman, Uniqueness of the ground state solution for $\Delta u-u+u^3=0$ and a variational characterization of other solutions, *Arch. Rational Mech. Anal.* **46** (1972), 81–95.

[CEF] C. Cortázar, M. Elgueta, and P. Felmer, Uniqueness of positive solutions of $\Delta(u)+f(u)=0$ in $\mathbb{R}^N$, preprint, 1995.

[CGM] S. Coleman, V. Glazer, and A. Martin, Actions minima among solutions to a class of Euclidean scalar field equations, *Comm. Math. Phys.* **58** (1978), 211–221.

[DS] J. I. Diaz and J. E. Saa, Uniqueness of solutions of nonlinear diffusion equations, *C. R. Acad. Sci. Paris* **305** (1987), 521–524.

[F] B. Franchi, Global solutions for a class of Monge–Ampére equations, *in* "Nonlinear Diffusion Equations and Their Equilibrium States, 3" (W. M. Ni, L. A. Peletier, and J. Serrin, Eds.), Birkhäuser, Boston/Basel/Berlin, 1992.

[FL] B. Franchi and E. Lanconelli, Radial symmetry of the ground states for a class of quasilinear elliptic equations, *in* "Nonlinear Diffusion Equations and their Equilibrium States" (W. M. Ni, L. A. Peletier, and J. Serrin, Eds.), Springer-Verlag, New York/Berlin, 1988.

[FLS1] B. Franchi, E. Lanconelli, and J. Serrin, Esistenza e unicitá degli stati fondamentali per equazioni ellittiche quasilinear, *Atti Accad. Naz. dei Lincei, Rend. Cl. Sci. Fis. Mat. Natur. (8)* **79** (1985), 121–126.

[FLS2] B. Franchi, E. Lanconelli, and J. Serrin, Existence and uniqueness of ground state solutions of quasilinear elliptic equations, *in* "Nonlinear Diffusion Equations and their Equilibrium States" (W. M. Ni, L. A. Peletier, and J. Serrin, Eds.), Springer-Verlag, New York/Berlin, 1988.

[GNN] B. GIDAS, W. M. NI, AND L. NIRENBERG, Symmetry and related properties of nonlinear elliptic equations in $\mathbf{R}^n$, *Adv. Math. Studies* **74** (1981), 369–402.

[K] M. K. KWONG, Uniqueness of positive solutions of $\Delta u - u + u^p = 0$ in $\mathbf{R}^n$, *Arch. Rational Mech. Anal.* **105** (1989), 243–266.

[KK1] H. G. KAPER AND M. K. KWONG, Uniqueness of nonnegative solutions of semilinear elliptic equations, *in* "Nonlinear Diffusion Equations and their Equilibrium States" (W. M. Ni, L. A. Peletier, and J. Serrin, Eds.), Springer-Verlag, New York/Berlin, 1988.

[KK2] H. G. KAPER AND M. K. KWONG, Uniqueness results for some nonlinear initial and boundary value problems, *Arch. Rational Mech. Anal.* **102** (1988), 45–56.

[KK3] H. G. KAPER AND M. K. KWONG, Uniqueness for a class of nonlinear initial value problems, *J. Math. Anal. Appl.* **130** (1988), 467–473.

[KK4] H. G. KAPER AND M. K. KWONG, Free boundary problems for Emden–Fowler equations, *Differential Integral Equations* **3** (1990), 353–362.

[KS] S. KICHENASSAMY AND J. SMOLLER, On the existence of radial solutions of quasilinear elliptic equations, *Nonlinearity* **3** (1990), 677–694.

[KZ] M. K. KWONG AND L. ZHANG, Uniqueness of the positive solutions of $\Delta u + f(u) = 0$, *Differential Integral Equations* **4** (1991), 583–599.

[MLS] K. MCLEOD AND J. SERRIN, Uniqueness of solutions of semilinear Poisson equations, *Proc. Nat. Acad. Sci. USA* **78** (1981), 6592–6595.

[NS] W. M. NI AND J. SERRIN, Existence and nonexistence theorems for ground states of quasilinear partial differential equations, the anomalous case, *Rome, Accad. Naz. dei Lincei, Atti dei Convegni* **77** (1986), 231–257.

[PS1] L. A. PELETIER AND J. SERRIN, Uniqueness of positive solutions of semilinear equations in $\mathbb{R}^n$, *Arch. Rational Mech. Anal.* **81** (1983), 181–197.

[PS2] L. A. PELETIER AND J. SERRIN, Uniqueness of nonnegative solutions of semilinear equations in $\mathbb{R}^n$, *J. Differential Equations* **61** (1986), 380–397.

[PS3] L. A. PELETIER AND J. SERRIN, Ground states for the prescribed mean curvature equation, *Proc. Amer. Math. Soc.* **100** (1987), 694–700.

[PuS] P. PUCCI AND J. SERRIN, A general variational identity, *Indiana U. Math.* **35** (1986), 681–703.

[ST] W. A. STRAUSS, Existence of solitary waves in higher dimensions, *Comm. Math. Phys.* **55** (1977), 149–162.

[Y] E. YANAGIDA, Uniqueness of positive radial solutions of $\Delta u + g(r)u + h(r)u^p = 0$ in $\mathbb{R}^n$, *Arch. Rational Mech. Anal.* **115** (1991), 257–274.

Indiana Univ. Math. J. 49 (2000), 897–923. **Reprint I.9.[iii]**

Uniqueness of Ground States for Quasilinear Elliptic Equations

JAMES SERRIN & MOXUN TANG

ABSTRACT. We give a simple elementary condition for the uniqueness of ground states of quasilinear elliptic equations. Several open problems are thus resolved and previous work is significantly extended, even for the classical scalar field equation.

1. INTRODUCTION

We consider the uniqueness of *ground states* of the quasilinear elliptic equation

$$\Delta_m u + f(u) = 0 \ \text{ in } \mathbb{R}^n, \quad n > m > 1, \tag{1.1}$$

where $\Delta_m u = \operatorname{div}(|Du|^{m-2}Du)$ is the degenerate m-Laplace operator. Here by a ground state we mean a non-negative non-trivial C^1 distribution solution of (1.1) which tends to zero at ∞. We shall prove

Theorem 1. (1.1) *admits at most one radial ground state if, for some* $b > 0$,

(H1) f *is continuous on* $(0,\infty)$, *with* $f(u) \le 0$ *on* $(0,b]$ *and* $f(u) > 0$ *for* $u > b$;
(H2) $f \in C^1(b,\infty)$, *with* $g(u) = uf'(u)/f(u)$ *non-increasing on* (b,∞).

For the important canonical nonlinearity

$$f(u) = -u^p + u^q, \quad p < q, \tag{1.2}$$

(H1) and (H2) are satisfied with $b = 1$, since

$$g'(u) = -\frac{(q-p)^2 u^{q-p-1}}{(u^{q-p}-1)^2} < 0, \quad u > 1.$$

A further model nonlinearity (1.4) is discussed in detail in Theorem 4 below.

In addition to the ground state problem, we can also consider uniqueness of radial solutions of the homogeneous Dirichlet-Neumann free boundary problem

$$(1.3) \qquad \begin{aligned} &\Delta_m u + f(u) = 0, \quad n > m > 1, \\ &u > 0 \quad \text{in } B_R, \quad u = \frac{\partial u}{\partial n} = 0 \quad \text{on } \partial B_R, \end{aligned}$$

where B_R is an open ball in $\mathbb{R}^n$ with radius $R > 0$.

Theorem 2. *The Dirichlet-Neumann free boundary problem* (1.3) *admits at most one radial solution if conditions* (H1), (H2) *hold.*

To present our next result, we introduce several important constants. By (H2), we can define $q = \lim_{u\to\infty} g(u)$; clearly $-\infty \le q < \infty$. By a Gronwall type argument one sees that $\lim_{u\downarrow b} g(u) \to \infty$, since $f(b) = 0$. Write

$$m^* = \frac{(m-1)n + m}{n - m},$$

the Sobolev critical number. When $q = \lim_{u\to\infty} g(u) < m^*$ we define

$$\eta = \sup\{s \mid s > b,\ g(s) \ge m^*\},$$

in which case $b < \eta < \infty$. When $q \ge m^*$ we let $\eta = \infty$.

The behavior of f near zero is of importance to our results for existence of radial ground states. It is natural to assume further

$$\text{(H3)} \qquad \int_0 |f(s)|ds < \infty \quad \text{and} \quad F(u) = \int_0^u f(s)ds < 0 \quad \text{for } u \text{ small.}$$

This condition is very mild; in fact for (1.2) it requires only that $p > -1$. Under (H3), we can define

$$\beta = \sup\{s \mid F(s) \le 0\}.$$

Clearly $\beta > b$, and $F(\beta) = 0$ when $\beta < \infty$. In the following considerations we shall identify two mutually exclusive situations:

(1) REGULAR CASE. f is continuous on $[0, \infty)$; clearly (H1) implies that $f(0) \le 0$;
(2) SINGULAR CASE. f cannot be extended as a continuous function to $[0, \infty)$; we leave the value of f at 0 undefined.

Note that the nonlinearity (1.2) is regular if $p \ge 0$, and singular if $p < 0$.

Theorem 3. *Let* (H1)-(H3) *hold. If* $q < m^*$ *and* $\beta < \infty$, *then either* (1.1) *has a unique radial ground state, or* (1.3) *has a unique radial solution. More precisely, we have*

(i) *When* f *is continuous at zero and* $f(0) = 0$, *there is a unique radial ground state* u. *Furthermore, when* $\int_0 |F(s)|^{-1/m} ds = \infty$, u *is positive and* (1.3) *has no radial solution; when the integral is finite,* u *has compact support and accordingly is the unique radial solution of* (1.3) *for exactly one* $R > 0$.
(ii) *When* f *is singular and* $\limsup_{s\downarrow 0} f(s) = 0$, *there is a unique positive radial ground state* u *provided* $\int_0 |F(s)|^{-1/m} ds = \infty$, *and a unique radial solution* u *of* (1.3) *when the integral is finite.*
(iii) *When* $\limsup_{s\downarrow 0} f(s) < 0$, *there are no ground states, and* (1.3) *has a unique radial solution* u.

Moreover, in all cases, $u'(r) < 0$ *for all* $r > 0$ *such that* $u(r) > 0$, *and* $\|u\|_\infty > \max\{\eta, \beta\}$.

Conversely, if $q \geq m^*$ *or* $\beta = \infty$, *then* (1.1) *has no radial ground state and* (1.3) *has no radial solution.*

When (H1)-(H3) are satisfied, a necessary and sufficient condition for existence, and also uniqueness, of ground states is that $q < m^*$ and $\beta < \infty$. Another immediate consequence of Theorem 3 is the following result:

Corollary. *Let* $f(u)$ *be given by* (1.2) *with* $p > -1$. *If* $q \geq m^*$, *then* (1.1) *has no ground state and* (1.3) *has no radial solution. If* $q < m^*$, *then*

(i) *When* $p \geq m - 1$, *there exists a unique radial ground state which is positive on* $\mathbb{R}^n$, *while* (1.3) *has no radial solution for any* $R > 0$.
(ii) *When* $0 < p < m - 1$, *there exists a unique radial ground state which has compact support on* $\mathbb{R}^n$ *and is also the unique radial solution of* (1.3) *for exactly one* $R > 0$.
(iii) *When* $-1 < p \leq 0$, *there exists no radial ground state, and* (1.3) *has a unique radial solution for exactly one* $R > 0$.

For the range $1 < m \leq 2$, $m - 1 \leq p \leq 1$, $m - 1 < q < m^*$, uniqueness for (1.2) was obtained by Citti [Ci]. Additionally, for the range $0 < p \leq m^* - 1$, $p < q < m^*$, Pucci and Serrin [PuS2] obtained the uniqueness results of the corollary (even for more general operators), but they left open the question whether uniqueness holds in the entire range $-1 < p < q < m^*$, which is of course resolved above.

We remark that when $f(u) = -u^p + u^q$, $p > q > m^*$, there are infinitely many ground states (see [T1]) and the solution structure is completely different.

Theorems 1-3 can be extended to the case when $f(u)$ vanishes at some point in (b, ∞), that is, when for some constant $\gamma > b$ we have the following conditions:

(H1′) f is continuous on $(0, \gamma)$, with $f(u) \leq 0$ on $(0, b]$, $f(u) > 0$ for $b < u < \gamma$, and $f(\gamma) = 0$;
(H2′) $f \in C^1(b, \gamma)$, with $g(u) = uf'(u)/f(u)$ non-increasing on (b, γ).

Theorem 1′. *Let* (H1′) *and* (H2′) *hold. Then* (1.1) *admits at most one radial ground state, and* (1.3) *admits at most one radial solution, with* $\|u\|_\infty < \gamma$.

An analogue of Theorem 3 also can be established under conditions (H1′) and (H2′), if we take into account the fact that in this case $\lim_{u \uparrow \gamma} g(u) = -\infty < m^*$. The details can easily be supplied, following the ideas of Theorem 3.

It is of interest to apply the above results to the model function

$$f(u) = -u^t - \lambda u^p + \mu u^q, \tag{1.4}$$

where

$$0 < t < p,\ q; \quad p \neq q; \quad \lambda,\ \mu > 0,$$

a nonlinearity which appears importantly in several physical applications, particularly with $t = 1$ (see [AD], [S]). Another related nonlinearity, occurring in population density problems, is

$$f(u) = u^t(u - b)(\gamma - u), \quad t > 0,\ 0 < b < \gamma. \tag{1.4′}$$

In the following two theorems we address these cases.

Theorem 4. *Let f obey* (1.4) *and suppose that $m \leq 2$ if $p > q$. Then there can be at most one radial ground state, which exists if and only if one of the following is valid:*

(i) $p < q < m^*$;
(ii) $p > q$, $\lambda < \lambda_0(\mu)$, *where* $\lambda_0(\mu) > 0$ *is a constant depending on* μ *(and on* p, q, t).

The ground state is positive in $\mathbb{R}^n$ *if* $t \geq m - 1$, *and has compact support if* $0 < t < m - 1$.

Theorem 4′. *Let f obey* (1.4′) *and suppose* $m \leq 2$. *Then there can be at most one radial ground state, which exists if and only if*

$$b < \frac{1+t}{3+t}\gamma. \tag{1.5}$$

Condition (H2) can be weakened without affecting the uniqueness results of Theorems 1-3, though at the price of a somewhat more complicated formulation. In fact, it is enough if $g(u)$ has the following properties:

(a) $g(u) \geq m^*$ for $b < u \leq \eta$;
(b) $g(u)$ is non-increasing and $m-1 < g(u) < m^*$ for $\eta < u < \kappa$, for some $\kappa \in (\eta, \infty]$;
(c) $g(u) \leq m-1$ for $u \geq \kappa$. (If $\kappa = \infty$, then condition (c) is empty.)

This extension is discussed in the second remark at the end of Section 4.

A very general existence theorem for radial ground states of the classical semilinear equation $(m = 2)$

$$\Delta u + f(u) = 0, \quad n > 2, \tag{1.6}$$

was given by Berestycki and Lions [BL], where the nonlinearities (1.2) and (1.4) were also treated as canonical examples. Their result gives existence of ground states for (1.2), (1.6) when $p = 1$ and $1 < q < (n+2)/(n-2)$, and also for (1.4), (1.6) in all three cases (i), (ii), (iii) in Theorem 4 above. Non-existence for (1.2), (1.6) in the supercritical case $q \geq (n+2)/(n-2)$, and for (1.4) when $p \leq 2^* \leq q$, follows from the well-known Pohozaev identity; non-existence for (1.4) in the range $2^* < p < q$ was left open in [BL], but was later proved by Ni and Serrin [NS1]. They also generalized the results of [BL] to the case of the quasilinear equation (1.1).

Berestycki and Lions asked, as a major open problem in [BL], which nonlinearities f admit uniqueness of ground states for the semilinear elliptic equation (1.6) with $n > 2$? This problem has been extensively studied, the first result being due to Coffman [C], who proved uniqueness for the model case (1.2) of (1.6) with $p = 1$, $q = 3$, and $n = 3$. Later, McLeod and Serrin [MS] improved Coffman's method to include more general functions f, in particular the cases $p = 1$, $1 < q \leq n/(n-2)$ of (1.2) with $n = 3, 4$. Based on the initial work of Coffman [C] and its extension by McLeod and Serrin [MS], and employing the Sturm comparison theory for second order linear differential equations, Kwong [K] proved uniqueness for (1.2), (1.6) with $p = 1$ and $1 < q < (n+2)/(n-2)$, a proof simplified and generalized later in [KZ, Ci, CL, M, Y]. In all these cases the nonlinearity was assumed to be differentiable on $[0, \infty)$. We note in particular that (H2) was introduced in [KZ] for $m = 2$, and in [Ci] for the range $1 < m \leq 2$; their proofs, however, need a number of further conditions on the behavior of $f(u)$ for u near 0 and for u large.

A different approach was initiated by Peletier and Serrin [PS1, 2], where they introduced a key Monotone Separation Theorem. Combining the comparison argument of Coffman and Kwong and the separation technique of Peletier and Serrin, Cortázar, Elgueta, and Felmer [CEF1] extended the previously cited work for (1.2), (1.6) to the more general case $0 < p < 1 < q < (n+2)/(n-2)$, where indeed the function f is not even Lipschitz continuous at 0. More recently, they obtained an even further result, asserting that (1.6) has at most one radial ground state if (H1) (H2) are valid; if f is continuous at zero with $f(0) = 0$; and if $f(u) \leq f'(u)(u-b)$ for all $u > b$, a type of convexity assumption, see [CEF2]. These conditions are substantially weaker than those cited above and in fact apply to (1.2) for any $p > 0$. On the other hand, due to the use of Sturmian theory, their proof is restricted to the semilinear case (1.6) only. Also, the convexity assumption for large u - which is unnecessary as shown by Theorem 1 of this paper - prevents the application of their results to case (iii) of Theorem 4. Our results thus provide simpler sufficient conditions for the uniqueness of ground states than those noted above (even for the semilinear case (1.6)), as well as clarifying the structure of radial ground states for the nonlinearity (1.4).

The method of [PS1], [PS2] was continued by Franchi, Lanconelli, and Serrin [FLS], in which equation (1.6) is generalized to (1.1) and even to further quasilinear operators. The recent paper [PuS2], noted earlier, further extended this work, with results which overlap those of the present paper. See also the comments in Section 6 below.

We present some preliminary results in the next section, and several technical lemmas in Section 3. The proof of Theorems 1, 2 is given in Section 4, and that of Theorems 3, 4, 4′ in Section 5. Some further remarks are made in Section 6. The proof of Theorems 1, 2 given here is self-contained and uses only elementary calculus, together with standard uniqueness theorems for the initial value problem for ordinary differential equations.

2. General Behavior of Solutions

Let $u = u(r)$ be a radial ground state of (1.1) or a radial solution of the free-boundary problem (1.3). Then u is a C^1 solution of the initial value problem

$$(2.1)\qquad (|u'|^{m-2}u')' + \frac{n-1}{r}|u'|^{m-2}u' + f(u) = 0, \quad r > 0,$$
$$u(0) = \alpha > 0, \quad u'(0) = 0,$$

for some $\alpha > 0$. Equation (2.1) can be rewritten in the form

$$(2.2)\qquad (r^{n-1}u'|u'|^{m-2})' = -r^{n-1}f(u),$$

which will be useful in a number of later arguments.

Proposition 2.1. *Let condition* (H1) *be satisfied. If u is a radial ground state, or a solution of* (1.3), *then $\alpha = u(0) > b$ and $u'(r) < 0$ as long as $u(r) > 0$ and $r > 0$.*

Under a slightly weaker condition than (H1), but assuming additionally that $\int_0 |f(s)|\,ds < \infty$, this result was given in [FLS], Proposition 1.2.6. With (H1) satisfied, but with no other restrictions placed on the behavior of f near 0, the result is new; our proof uses an argument slightly different than the usual energy function analysis.

Proof. We first show that if $u'(r_0) = 0$ for any $r_0 \geq 0$ at which $u(r_0) > 0$, then $u(r) \leq u(r_0)$ for all $r > r_0$. Indeed, if this were not true, then there would be some $\sigma > r_0$ such that $u(\sigma) = u(r_0)$, and u is not a constant over (r_0, σ). Consider the modified energy function

$$\bar{E}(r) = \frac{m-1}{m}|u'(r)|^m + \int_{u(r_0)}^{u(r)} f(s)\,ds, \quad r \in [r_0, \sigma].$$

Clearly, $\bar{E}(r_0) = 0$ and $\bar{E}(\sigma) \geq 0$. On the other hand, a differentiation yields

$$\frac{d\bar{E}(r)}{dr} = -\frac{n-1}{r}|u'(r)|^m,$$

implying that $\bar{E}$ is non-increasing in $[r_0, \sigma]$. Since moreover u' is not identically zero on $[r_0, \sigma]$, there results $\bar{E}(r_0) > \bar{E}(\sigma)$, giving a contradiction.

When $u'(r_0) = 0$, we set $u(r_0) = u_0$ (> 0); thus $u(r) \leq u_0$ for $r > r_0$ by what was shown above.

We assert that $0 < u_0 \leq b$ is impossible. In fact, integrating (2.2) from r_0 to r gives, since $u'(r_0) = 0$,

$$r^{n-1}u'|u'|^{m-2} = -\int_{r_0}^{r} s^{n-1} f(u(s))\,ds. \tag{2.3}$$

If $u_0 \leq b$, then the right side is non-negative for $r > r_0$ by virtue of (H1). Hence $u'(r) \geq 0$ for $r > r_0$, which is clearly absurd. The assertion is proved, and so $u'(r) = 0$ cannot occur when $0 < u(r) \leq b$.

Since $u(0) = \alpha$ and $u'(0) = 0$, it then follows as well that $\alpha > b$.

Finally, let r_b be the unique number such that $u(r) > b$ on $[0, r_b)$ and $u(r_b) = b$. By integrating (2.2) from 0 to r we obtain, since $u'(0) = 0$,

$$r^{n-1}u'|u'|^{m-2} = -\int_{0}^{r} s^{n-1} f(u(s))\,ds. \tag{2.4}$$

It is evident from (H1) that the integral on the right side is positive for $0 < r < r_b$, so in turn $u'(r) < 0$ in this interval. Since by the assertion above we have $u'(r) \neq 0$ whenever $0 < u(r) \leq b$, we conclude that $u'(r_b) < 0$ and thus $u'(r) < 0$ as long as $0 < u(r) < b$. In summary, $u'(r) < 0$ as long as $u(r) > 0$ and $r > 0$. This completes the proof. □

Let $J = (0, R)$ be the maximum interval on which $u(r) > 0$ (and so also $u'(r) < 0$ by Proposition 2.1). Then $R = \infty$ when u is positive in the whole space $\mathbb{R}^n$, and $R < \infty$ otherwise. In the latter case, if u is a ground state, we understand that it is extended to all $r > 0$ by setting $u(r) \equiv 0$ for $r \geq R$ (it is worth noting that this situation can occur only when f is regular and $f(0) = 0$).

Since $u'(r) < 0$ on J, it follows that the inverse of $u(r)$, denoted by $r = r(u)$, is well-defined and is strictly decreasing on $(0, \alpha)$. By (2.1) it is easy to see that

$$(m-1)r'' = \frac{n-1}{r}r'^2 + f(u)|r'|^m r', \qquad ' = \frac{d}{du}. \tag{2.5}$$

(In the sequel, the superscript $'$ denotes either d/du or d/dr, depending on the context. This should not be a cause for confusion.)

Proposition 2.2. *Let* (H1) *hold. If* u *is a radial ground state or a solution of* (1.3), *then*

$$\lim_{u\to 0} \frac{r^{(n-1)/(m-1)}(u)}{|r'(u)|} = \lim_{r\to R} r^{(n-1)/(m-1)}|u'(r)| = \mu, \tag{2.6}$$

$$\lim_{u\to 0} u r^{(n-m)/(m-1)}(u) = \lim_{r\to R} r^{(n-m)/(m-1)}u(r) = \nu, \tag{2.7}$$

for some numbers $0 \leq \mu, \nu < \infty$.

Proof. Obviously (2.6) and (2.7) are true with $\mu = \nu = 0$, when u is a radial ground state with compact support or a radial solution of (1.3). If u is a positive ground state, then by (H1) and (2.2) the quantity $r^{n-1}|u'|^{m-1}$ is ultimately non-increasing, and so approaches a finite, non-negative number, implying (2.6) for some $0 \leq \mu < \infty$. By L'Hopital's rule, since $n > m$,

$$\lim_{r\to\infty} r^{(n-m)/(m-1)}u(r) = \frac{m-1}{n-m}\mu \equiv \nu,$$

and so (2.7) follows. □

The next result is well known. We give the proof for completeness.

Proposition 2.3. $r^{-1}|u'(r)|^{m-1} \to f(\alpha)/n$ *as* $r \downarrow 0$.

Proof. By (2.2), for all $r > 0$,

$$r^{-1}|u'|^{m-1} = r^{-n}\int_0^r s^{n-1} f(u(s))\, ds.$$

The required limit is now obtained from L'Hopital's rule. □

Lemma 2.4. *Let condition* (H1) *hold. Suppose u is a radial ground state or a radial solution of* (1.3), *and define*

$$A(u) = (n-m)u + (m-1)\frac{r(u)}{r'(u)}.$$

Then $A(b) \leq 0$ and $A(u) \to 0$ as $u \downarrow 0$.

Proof. Using (2.6) and recalling that $n > m$, we find $ru'(r) \to 0$ as $r \to R$, implying $\lim_{u\downarrow 0} A(u) = 0$. Moreover, differentiation and use of (2.4) and (H1) yields

$$A'(u) = -f(u)rr'|r'|^{m-2} \leq 0, \quad 0 < u < b, \tag{2.8}$$

and so $A(b) \leq 0$. □

Let a be an arbitrary real number, and define for $0 < u < \alpha$,

$$P(u) = auw(u) + r^n\left[\frac{m-1}{m}\frac{1}{|r'|^m} + \frac{a}{n}uf(u)\right], \quad w(u) = \frac{r^{n-1}}{r'|r'|^{m-2}}. \tag{2.9}$$

Then we have the following identity, a variant of a result of Ni, Pucci, and Serrin [NS1, NS2, PuS1].

Proposition 2.5. *Let* (H1) *and* (H2) *be satisfied. If u is a radial ground state or a radial solution of* (1.3), *then*

$$P(u) = \left(a+1-\frac{n}{m}\right)\int_\alpha^u w(v)\,dv + \frac{a}{n}\int_\alpha^u r^n f(v)\left(g(v) - \frac{n-a}{a}\right)dv, \tag{2.10}$$

where $\alpha = u(0)$ and $b \leq u < \alpha$.

Proof. Using the definition of w in (2.9) and A in Lemma 2.4, we can rewrite $P(u)$ in the form

$$P(u) = \left(a+1-\frac{n}{m}\right)uw + \frac{1}{m}wA + \frac{a}{n}r^n uf(u). \tag{2.11}$$

With the help of (2.8) and the relation

$$w'(u) = -r^{n-1}r'f(u), \tag{2.12}$$

see (2.2), we find by differentiating (2.11) with respect to u that

$$P'(u) = \left(a+1-\frac{n}{m}\right)w + \frac{a}{n}r^n f(u)\left(g(u) - \frac{n-a}{a}\right).$$

Integrating this identity over (u, α) gives (2.10), since $P(\alpha) = 0$. □

Remark. If $a \le n/m - 1$, then from (2.11) and Lemma 2.4 follows (since $w < 0$ and $f(b) = 0$)

$$P(b) \ge \frac{w(b)A(b)}{m} \ge 0. \tag{2.13}$$

Proposition 2.6. *Let* (H1) *and* (H2) *be satisfied. If u is a radial ground state or a radial solution of* (1.3), *then* $u(0) = \alpha > \eta$.

In particular, if $q \ge m^*$, then there can be no radial ground state or radial solution of the free boundary problem (1.3).

Proof. Suppose for contradiction that $b < \alpha \le \eta$. Then $g(u) \ge m^*$ and $f(u) > 0$ for $u \in (b, \alpha)$, while as noted earlier $g(u) \to \infty$ as $u \downarrow b$. Let $a = n/m - 1$. Then applying (2.10) and (H2) we obtain

$$P(b) = \left(\frac{1}{m} - \frac{1}{n}\right) \int_\alpha^b r^n f(u)(g(u) - m^*)\, du < 0,$$

which contradicts (2.13). The final statement follows, since $\eta = \infty$ when $q \ge m^*$. □

3. SEPARATION LEMMAS

Let u_1 and u_2 be two distinct ground states of (1.1), or two distinct radial solutions of (1.3). Let $\alpha_i = u_i(0)$; then by Proposition 2.6 we have $\alpha_i > \eta$, $i = 1, 2$, and of course by the definition of η, also $\eta \ge b$. Let $r_i = r_i(u)$ be the inverse of $u_i(r)$. Then $r_i = r_i(u)$ is defined on $(0, \alpha_i)$ with $r_i(\alpha_i) = 0$.

Proposition 3.1. *Under the above assumptions, and assuming that* (H1) *holds, the relations*

$$u_1(\bar{r}) = u_2(\bar{r}) > 0, \quad u_1'(\bar{r}) = u_2'(\bar{r}), \quad \bar{r} > 0, \tag{3.1}$$

cannot occur. Moreover, the initial values α_1, α_2 cannot be equal.

Proof. Suppose for contradiction that u_1 and u_2 are distinct ground states of (1.1) or distinct solutions of (1.3), and that (3.1) holds. The inverses $r_1(u)$, $r_2(u)$ each satisfy the equation (2.5) for $0 < u < \min(\alpha_1, \alpha_2)$. But the right side of (2.5) is Lipschitz continuous in the dependent variables r, r' (for $r > 0$), and by (H1) continuous in the independent variable u. Condition (3.1) moreover implies

$$r_1(u) = r_2(u) = \bar{r} > 0, \quad r_1'(u) = r_2'(u)$$

at the "initial" value $\bar{u} = u_1(\bar{r}) = u_2(\bar{r}) > 0$. Hence by the standard uniqueness theory for ordinary differential equations, we find $r_1(u) \equiv r_2(u)$ for $0 < u <$

$\min(\alpha_1, \alpha_2)$, and in turn $\alpha_1 = \alpha_2$. It follows that the corresponding intervals J_1 and J_2 for u_1 and u_2 are also the same, which contradicts the assumption that u_1 and u_2 are distinct (either as ground states of (1.1), or as solutions of (1.3)).

Suppose next that $\alpha_1 = \alpha_2 = \alpha$. Then, as noted in Proposition 2.1, we have $\alpha > b$. Since $f(u)$ is Lipschitz continuous for $u > b$, see (H2), it then follows from the uniqueness of the initial value problem for (2.1) (see for example, [NS2], or Proposition A4 of [FLS]) that

$$u_1(r) \equiv u_2(r) > 0$$

on some interval $0 \le r < \delta$. But then by what has been shown above, we get $u_1 \equiv u_2$ and $J_1 = J_2$, again contradicting the assumption that u_1 and u_2 are distinct. □

We shall suppose for definiteness throughout the rest of the paper that

$$\eta < \alpha_1 < \alpha_2.$$

The following auxiliary functions related to r_1 and r_2 are then well defined on $(0, \alpha_1]$:

$$t(u) = r_1(u) - r_2(u), \qquad s(u) = \frac{r_1(u)}{r_2(u)}, \tag{3.2}$$

$$T(u) = \frac{r_1(u)}{r_1'(u)} - \frac{r_2(u)}{r_2'(u)}, \quad S(u) = \frac{w_1(u)}{w_2(u)}, \tag{3.3}$$

where $w_i(u)$ is the function $w(u)$ defined in (2.9) with $r(u)$ replaced by $r_i(u)$; here we note particularly that $w_2(u) < 0$ on $(0, \alpha_1]$ since $\alpha_2 > \alpha_1$.

As a first step in the proof we recall the following maximum principle, due to Peletier and Serrin [PS1, PS2] for the Laplace operator, and, for more general operators, to Franchi, Lanconelli, and Serrin [FLS].

Proposition 3.2. *Let condition* (H1) *be satisfied. The difference function* $t(u) = r_1(u) - r_2(u)$ *cannot assume a positive local minimum value* (*or a negative local maximum value*) *in* $(0, \alpha_1)$.

Proof. If $u_c \in (0, \alpha_1)$ is a critical point of t, then by (2.5)

$$(m-1)t''(u_c) = (n-1)r_1'^2(u_c)\left(\frac{1}{r_1(u_c)} - \frac{1}{r_2(u_c)}\right), \tag{3.4}$$

from which the principle follows at once. □

Lemma 3.3. (Tang [T2]) *Let* (H1) *hold. If* $s'(u_m) = 0$ *for some* $u_m \in (0, \alpha_1)$, *then*

$$s''(u_m) = \frac{f(u_m)r_1'}{(m-1)r_2}(|r_1'|^m - |r_2'|^m). \tag{3.5}$$

Proof. Note that $s'(u) = (r_1'r_2 - r_2'r_1)/r_2^2$. Thus (3.5) follows by a direct differentiation with the help of (2.5). □

Lemma 3.4. (Erbe and Tang [ET]) *Let* (H1) *hold. Then* $S'(u)$ *and* $t'(u)$ *have the same sign in the interval* (b, α_1).

Proof. By (2.12) we have $w'(u) = -r^{n-1}r'f(u)$. Applying the quotient rule gives

$$S'(u) = \left(\frac{r_1}{r_2}\right)^{n-1}\left(\frac{r_2'}{r_1'}\right)^{m-1} f(u)(|r_2'|^m - |r_1'|^m), \tag{3.6}$$

leading to the desired result, since $t'(u) = r_1' - r_2' = |r_2'| - |r_1'|$ and $f(u) > 0$ for $u > b$. □

Lemma 3.5. *Let* (H1) *be satisfied. We have* $T_0 = \lim_{u\downarrow 0} T(u) = 0$, *and*

$$T(u) = \frac{A_1(u) - A_2(u)}{(m-1)} = -\frac{r_2^2}{r_1'r_2'}s'(u), \tag{3.7}$$

$$T'(u) = \frac{f(u)}{m-1}(r_1|r_1'|^{m-1} - r_2|r_2'|^{m-1}). \tag{3.8}$$

Proof. That $T_0 = 0$ follows from Lemma 2.4. The identities (3.7) and (3.8) are obtained by direct differentiation with the help of (2.5). □

Proposition 3.6. (Monotone Separation Theorem) *Let* (H1) *hold. Then* $t(u)\cdot t'(u) < 0$ *on* $(0, b]$.

This result was first established by Peletier and Serrin [PS1, PS2] for the semilinear case, as a major part of their separation technique; for more general elliptic operators, it was proved by Franchi, Lanconelli and Serrin (see Lemma 3.6.5, Lemma 3.6.7 of [FLS]). For completeness, we give (a slightly simplified) proof here.

Proof of Proposition 3.6. We first show that $t(u)$ has at most one zero in $(0,b]$. Otherwise we can pick two numbers τ_1, $\tau_2 \in (0,b]$, $\tau_1 < \tau_2$, such that $t(\tau_1) = t(\tau_2) = 0$ and without loss of generality $t(u) > 0$ for $\tau_1 < u < \tau_2$; there results, for some $\tau \in (\tau_1, \tau_2)$,

$$t'(\tau) = 0, \; t'(u) < 0 \quad \text{for } \tau < u \le \tau_2.$$

Thus by (3.6) and (H1) one finds $S'(u) \ge 0$ on (τ, τ_2), and by (3.3)

$$S(\tau) = s^{n-1}(\tau) > 1, \quad S(\tau_2) = \left| \frac{r_2'(\tau_2)}{r_1'(\tau_2)} \right|^{m-1} < 1,$$

giving a contradiction.

Hence we may assume without loss of generality that for some $\delta \le b$

$$t(u) = r_1 - r_2 > 0 \quad \text{for } 0 < u \le \delta, \tag{3.9}$$

By the maximum principle of Proposition 3.2, $t(u)$ can have at most one critical point, necessarily a strict maximum, in $(0,\delta)$. Thus we have either

$$t'(u) = r_1' - r_2' < 0 \quad \text{for } 0 < u \le \delta, \tag{3.10}$$

or for some $0 < \delta_1 \le \delta$,

$$t'(u) = r_1' - r_2' > 0 \quad \text{for } 0 < u < \delta_1. \tag{3.11}$$

We show that (3.11) is impossible; for this purpose we assume for contradiction that (3.9) and (3.11) hold, and put

$$B(u) = \frac{1}{|r_1'(u)|^m} - \frac{1}{|r_2'(u)|^m}.$$

Then $B(u) > 0$ for $0 < u < \delta_1$. By Proposition 2.2 it is clear that $r_1^{(n-1)/(m-1)}/|r_1'|$ tends to a finite limit, and $1/|r_1'|$ tends to zero, as $u \downarrow 0$. We thus find, using (3.9), that

$$\limsup_{u\downarrow 0} r_2(u)^{n-1} B(u) \le \limsup_{u\downarrow 0} r_1(u)^{n-1} B(u) \le \lim_{u\downarrow 0} \frac{r_1(u)^{n-1}}{|r_1'(u)|^m} \tag{3.12}$$

$$= \lim_{u\downarrow 0} \frac{1}{|r_1'(u)|} \left[\frac{r_1^{(n-1)/(m-1)}}{|r_1'|} \right]^{m-1} = 0.$$

By (2.5)

$$B'(u) = \ell\left[\frac{1}{r_1|r_1'|^{m-1}} - \frac{1}{r_2|r_2'|^{m-1}}\right], \quad \ell = \frac{m(n-1)}{m-1}. \tag{3.13}$$

Now from Young's inequality follows

$$\frac{1}{|r_1'|^{m-1}|r_2'|} \le \frac{m-1}{m}\frac{1}{|r_1'|^m} + \frac{1}{m}\frac{1}{|r_2'|^m}.$$

Hence by (3.9) and (3.13) we get

$$\frac{r_2B'}{|r_2'|} < \ell\left[\frac{1}{|r_1'|^{m-1}} - \frac{1}{|r_2'|^{m-1}}\right]\frac{1}{|r_2'|} \le (n-1)\left[\frac{1}{|r_1'|^m} - \frac{1}{|r_2'|^m}\right],$$

that is

$$\frac{B'}{B} < (n-1)\frac{|r_2'|}{r_2}.$$

By integration, we find for any $\bar{u} \in (u, \delta_1)$,

$$r_2(\bar{u})^{n-1}B(\bar{u}) < r_2(u)^{n-1}B(u).$$

But then, by taking the limit as u tends to 0,

$$r_2(\bar{u})^{n-1}B(\bar{u}) \le \limsup_{u\downarrow 0} r_2(u)^{n-1}B(u) \le 0$$

by (3.12). This gives a contradiction, since $B(\bar{u}) > 0$ for $0 < \bar{u} < \delta_1$.

Thus (3.10) is true under the assumption (3.9), and so $t > 0$, $tt' < 0$ is valid on $(0,\delta]$. If $tt' < 0$ were not true on the whole interval $(0,b]$, then there would be a number $\delta_2 \in (\delta, b]$ such that at $u = \delta_2$ one of the following holds:

$$t > 0,\ t' = 0,\ t'' \ge 0; \quad \text{or} \quad t = 0,\ t' = 0; \quad \text{or} \quad t = 0,\ t' < 0.$$

The first of these cannot occur because of (3.4), see Proposition 3.2, and the second also cannot occur in view of Proposition 3.1. The remaining possibility shows by (3.2), (3.3) that $T(\delta_2) > 0$. But $T'(u) \le 0$ over $(0,\delta_2)$ by (3.8), when we recall that $f(u) \le 0$ for $u < b$. This leads to $T_0 \ge T(\delta_2) > 0$, contradicting the fact that $T_0 = 0$. The proof is completed. □

To state the next result, we recall from the introduction the definition $q = \lim_{u\to\infty} g(u)$ and the limit relation $\lim_{u\downarrow b} g(u) = \infty$. Put

$$\kappa = \sup\{s \mid s > b,\ g(s) \ge m-1\}.$$

Then $b < \kappa < \infty$ provided $q < m-1$, while $\kappa = \infty$ otherwise; if (H2′) holds, then $b < \kappa < \gamma$. By definition $g(\kappa) = m-1$ when κ is finite.

The proof of the next result is similar to that of Proposition 2.1 of [T3].

Lemma 3.7. *Suppose* (H1), (H2) *hold and that* $q < m-1$. *Let* $\alpha_2 > \alpha_1 > \kappa$. *Then* $u_1(r) < u_2(r)$ *as long as* $u_1(r) \geq \kappa$.

Proof. By (H2), $g(u) < m-1$ for $u > \kappa$. Then

$$\frac{d}{du}\left(\frac{f(u)}{u^{m-1}}\right) = u^{-m} f(u)[g(u) - (m-1)] < 0, \quad u > \kappa.$$

Thus $f(u)/u^{m-1}$ is strictly decreasing for $u > \kappa$. Then with the aid of Proposition 2.3 we find

$$\lim_{r\downarrow 0} \frac{u_2}{u_2'}\left(\frac{u_1}{u_2}\right)' = \lim_{r\downarrow 0}\left(\frac{u_1'}{u_2'} - \frac{u_1}{u_2}\right) = \left(\frac{f(\alpha_1)}{f(\alpha_2)}\right)^{1/(m-1)} - \frac{\alpha_1}{\alpha_2} > 0,$$

since $\alpha_2 > \alpha_1 > \kappa$. Hence $(u_1/u_2)' < 0$ for r sufficiently small.

Let r_κ be the unique value of r such that $u_1(r_\kappa) = \kappa$. We claim that

$$\left(\frac{u_1}{u_2}\right)' < 0 \quad \text{for } 0 < r < r_\kappa. \tag{3.14}$$

Otherwise there would exist $r_0 \in (0, r_\kappa)$ such that

$$\left(\frac{u_1}{u_2}\right)' < 0, \quad r \in (0, r_0); \quad \left(\frac{u_1}{u_2}\right)' = 0 \quad \text{at } r = r_0. \tag{3.15}$$

Now by direct calculation, using (2.1) and the fact that $(u_1/u_2)' = 0$ at $r = r_0$, we find that

$$\begin{aligned}
\left(\frac{u_1}{u_2}\right)'' &= \frac{1}{u_2^2}(u_1'' u_2 - u_1 u_2'') \\
&= \frac{1}{(m-1)u_2^2}(-f(u_1)u_2|u_1'|^{2-m} + f(u_2)u_1|u_2'|^{2-m}) \\
&= \frac{1}{(m-1)u_2} f(u_2)|u_1'|^{2-m}\left(\left(\frac{u_1}{u_2}\right)^{m-1} - \frac{f(u_1)}{f(u_2)}\right),
\end{aligned}$$

at $r = r_0$. But $u_1/u_2 = \alpha_1/\alpha_2 < 1$ at $r = 0$, so by (3.15) we get $u_1/u_2 < 1$ at $r = r_0$. Hence, again using the fact that $f(u)/u^{m-1}$ is strictly decreasing for $u > \kappa$, and that $u_2 > u_1 > \kappa$ at r_0, we obtain

$$\left(\frac{u_1}{u_2}\right)'' < 0 \quad \text{at } r = r_0.$$

This clearly contradicts (3.15), proving our claim. But now from (3.14) follows, as above,

$$\frac{u_1(r)}{u_2(r)} < 1, \quad 0 < r \leq r_K,$$

which is exactly the desired result. □

4. PROOF OF THEOREMS 1 AND 2

We assume that (H1) and (H2) are satisfied. As in Section 3, let u_1 and u_2 be two distinct ground states of (1.1), or two distinct radial solutions of (1.3). We shall show that this cannot in fact occur, which will prove Theorems 1 and 2.

We observe that if $t(u) = r_1(u) - r_2(u) < 0$ over the entire interval $(0, \alpha_1)$, then Proposition 3.2, together with the fact that $t(\alpha_1) < 0$ and $t'(\alpha_1) = -\infty$, yields $t'(u) < 0$ on $(0, \alpha_1)$; thus $t(u) \cdot t'(u) > 0$ on $(0, b)$, contradicting Lemma 3.6.

There remains the much more difficult case that $t(u)$ has one or more zeros in $(0, \alpha_1)$, which by Lemma 3.6 must necessarily lie in the interval (b, α_1).

Case I. First let us suppose that $t'(u) < 0$ for $b < u < \alpha_1$; then t has exactly one zero $u_I \in (b, \alpha_1)$, and so

(4.1) $t(u) > 0 \quad \text{for } u \in (0, u_I); \quad t(u_I) = 0; \quad t(u) < 0 \quad \text{for } u \in (u_I, \alpha_1).$

Now $T_0 = 0$ by Lemma 3.5, while also, by (3.8), $(4.1)_1$, (H1), and Lemma 3.6,

$$T'(u) \leq 0 \quad \text{for } 0 < u \leq b.$$

Hence we find

$$T(b) \leq 0. \tag{4.2}$$

Then using the identity (2.11) for both the solutions u_1 and u_2, together with the definition $(3.3)_2$ of $S(u)$ and (3.7), we get

$$\begin{aligned} P_1(b) - S(b)P_2(b) &= \frac{1}{m}[w_1(b)A_1(b) - S(b)w_2(b)A_2(b)] \\ &= \frac{m-1}{m} w_1(b)T(b) \geq 0. \end{aligned} \tag{4.3}$$

We note that (4.3) is valid for *any* real number a in (2.9).

Clearly $s(b) = r_1(b)/r_2(b) > 1$, since $t(b) = r_1(b) - r_2(b) > 0$. Using the relation

$$S(u) = [s(u)]^{n-1} \cdot \left(\frac{|r_2'(u)|}{|r_1'(u)|}\right)^{m-1},$$

together with $t'(b) = |r_2'(b)| - |r_1'(b)| < 0$, we find

$$S(b) < [s(b)]^{n-1} < [s(b)]^n. \tag{4.4}$$

Also by (4.2) and (3.7) we have

$$s'(b) \geq 0. \tag{4.5}$$

Hence $|r_2'(b)/r_1'(b)| \geq r_2(b)/r_1(b) = 1/s(b)$ and so

$$S(b) = [s(b)]^{n-1} \cdot \left(\frac{|r_2'(b)|}{|r_1'(b)|}\right)^{m-1} \geq [s(b)]^{n-m} > 1. \tag{4.6}$$

By Lemma 3.3 and the assumption that $t'(u) = |r_2'| - |r_1'| < 0$, $s(u)$ takes a maximum value at any critical point in (b, α_1). Thus $s(u)$ can have at most one critical point in this interval.

Recall that $s(u_I) = 1$ and $s(b) > 1$. If $s'(b) > 0$, then necessarily $s'(u)$ vanishes at some point, say, $u = u_m$ in (b, u_I), such that

$$s'(u) > 0 \quad \text{for } b < u < u_m; \quad s'(u) < 0 \quad \text{for } u_m < u < \alpha_1. \tag{4.7}$$

If $s'(b) = 0$, then either $s'(u) < 0$ over (b, α_1), that is, $u_m = b$, or (4.7) holds.

In all cases, by (4.4) and (4.6)-(4.7) we can specify a number $u_e \in (u_m, u_I)$ such that

$$[s(u_e)]^n = S(b);$$

thus $s(u_e) < s(b)$ by (4.4), and necessarily $s'(u_e) < 0$ and u_e is unique. It is then not hard to check that

$$s^n > S(b) \quad \text{for } b < u < u_e, \quad s^n < S(b) \quad \text{for } u_e < u < \alpha_1. \tag{4.8}$$

To continue, we distinguish two subcases:

▷ SUBCASE (i). $u_e \geq \eta$. Let a be given by

$$\frac{n-a}{a} = g(u_e), \quad \text{that is,} \quad a = \frac{n}{1 + g(u_e)}. \tag{4.9}$$

We claim that $g(u_e) \geq m - 1$. This is obvious if $q \geq m - 1$. Otherwise $q < m - 1$, in which case by Lemma 3.7 it is clear that $u_I < \kappa$. Then by (H2) $g(u_e) \geq g(u_I) \geq g(\kappa) = m - 1$, and the claim is shown. But also, since $u_e \geq \eta$, we have $g(u_e) \leq g(\eta) = m^*$ by (H2). Thus by (4.9)

$$\frac{n-m}{m} \leq a \leq \frac{n}{m}. \tag{4.10}$$

By (H2) again,

$$(4.11)\quad g(u) \geq \frac{n-a}{a} \quad \text{for } b < u < u_e, \quad g(u) \leq \frac{n-a}{a} \quad \text{for } u_e < u < \alpha_2.$$

Now applying the identity (2.10) we get

$$\begin{aligned}(4.12)\qquad & P_1(b) - S(b)P_2(b) \\ &= \left(a+1-\frac{n}{m}\right)\int_{\alpha_1}^{b} w_2(u)[S(u)-S(b)]\,du \\ &+ \frac{a}{n}\int_{\alpha_1}^{b} f(u)\left(g(u)-\frac{n-a}{a}\right)r_2^n(u)[s^n(u)-S(b)]\,du \\ &+ \left(a+1-\frac{n}{m}\right)S(b)\int_{\alpha_1}^{\alpha_2} w_2(u)\,du \\ &+ \frac{a}{n}S(b)\int_{\alpha_1}^{\alpha_2} f(u)\left(g(u)-\frac{n-a}{a}\right)r_2^n(u)\,du \\ &\equiv I_1+I_2+I_3+I_4.\end{aligned}$$

Notice that $w_2(u) < 0$ on (b, α_2). Also $S(u) < S(b)$ for $u \in (b, \alpha_1)$, since $S'(u) < 0$ by Lemma 3.4 and the assumption that $t'(u) < 0$. Hence $I_1 \leq 0$ by (4.10). By (4.8), (4.10), and (4.11) we find that $I_2 \leq 0$; indeed, $I_2 < 0$ since $g(u) \to \infty$ as $u \downarrow b$. Finally, one can easily check that $I_3 \leq 0$ and $I_4 \leq 0$ by (4.11) and $u_e < \alpha_1 < \alpha_2$. Hence $I_1 + I_2 + I_3 + I_4 < 0$, contradicting (4.3).

▷ SUBCASE (ii). $u_e < \eta$. Since $s'(u_e) < 0$, by (4.7) one finds that $u_m < \eta < \alpha_1$ (recall from Proposition 2.6 that $\eta < \alpha_1$). Hence

$$(4.13)\qquad s(u) > s(\eta) \quad \text{for } b < u < \eta, \quad s(u) < s(\eta) \quad \text{for } \eta < u < \alpha_1.$$

By the definition of η

$$(4.14)\qquad g(u) \geq m^* \quad \text{for } b < u < \eta, \quad g(u) < m^* \quad \text{for } \eta < u < \alpha_2.$$

Now taking $a = -1 + n/m > 0$, and applying (2.10), (4.13), and (4.14), we obtain

$$\begin{aligned}(4.15)\qquad & P_1(b) - s^n(\eta)P_2(b) \\ &= \left(\frac{1}{m}-\frac{1}{n}\right)\int_{\alpha_1}^{b} f(u)[g(u)-m^*]r_2^n(u)(s^n(u)-s^n(\eta))\,du \\ &+ s^n(\eta)\left(\frac{1}{m}-\frac{1}{n}\right)\int_{\alpha_1}^{\alpha_2} f(u)[g(u)-m^*]r_2^n(u)\,du < 0.\end{aligned}$$

Since $u_e \in (u_m, \eta)$, we have $s^n(\eta) < s^n(u_e) = S(b)$. Using the fact that $P_2(b) \geq 0$, see (2.13), the inequality (4.15) then gives

$$P_1(b) - S(b)P_2(b) < 0,$$

again contradicting (4.3). *Thus the assumption of Case I cannot occur.*

Case II. $t'(u)$ vanishes at some point in (b, α_1). Then there is a first point $u_c \in (b, \alpha_1)$ where

$$t'(u_c) = 0 \quad \text{and} \quad t'(u) < 0 \quad \text{for } u \in (u_c, \alpha_1). \tag{4.16}$$

Clearly $t(u_c) \neq 0$ by Proposition 3.1, since u_1 and u_1 are assumed to be distinct. Using Proposition 3.2 one finds that

$$t(u_c) > 0 \quad \text{and so} \quad s'(u_c) > 0. \tag{4.17}$$

Thus using (2.9), $(3.3)_2$, and the fact that $f(u_c) > 0$ yields, for any $a > 0$,

$$\begin{aligned} (4.18)\quad & P_1(u_c) - S(u_c)P_2(u_c) \\ &= \left[\frac{m-1}{m}\frac{1}{|r_1'(u_c)|^m} + \frac{a}{n}u_c f(u_c)\right](r_1^n(u_c) - S(u_c)r_2^n(u_c)) > 0, \end{aligned}$$

since $S(u_c) = [s(u_c)]^{n-1} < [s(u_c)]^n$ by (4.16)-(4.17). Using exactly the arguments in Case I we find, after replacing b with u_c, that (4.5)-(4.7) are still valid with the same definition of u_m. Now redefine $u_e \in (u_m, u_I)$ by

$$s^n(u_e) = S(u_c).$$

Then repeating the above argument for the subcase $u_e \geq \eta$, with b replaced by u_c, we derive, as in (4.12),

$$P_1(u_c) - S(u_c)P_2(u_c) < 0,$$

contradicting (4.18). Hence $u_e \geq \eta$ is impossible.

In case $u_e < \eta$, we easily show, as in (4.15) with $a = -1 + n/m$, that

$$P_1(u_c) - s^n(\eta)P_2(u_c) < 0.$$

Now by (2.10) and the fact that $g(u) \geq m^*$ over (b, u_c) since $u_c < \eta$, we obtain

$$P_2(u_c) \geq P_2(u_c) + \left(\frac{1}{m} - \frac{1}{n}\right)\int_{u_c}^{b} f(u)[g(u) - m^*]r_2^n(u)\,du = P_2(b).$$

Hence $P_2(u_c) \geq 0$ by (2.13). Also, using the fact that $u_e \in (u_m, \eta)$, we have $s^n(\eta) < S(u_c)$, so finally

$$P_1(u_c) - S(u_c)P_2(u_c) < 0,$$

again contradicting (4.18). *Thus Case II also cannot occur,* and the proof is complete. □

If (H1′), (H2′) hold and α_1, α_2 are in the finite interval (b, γ), we can repeat the above proof essentially without change to obtain Theorem 1′.

Remark. Checking through the preceding proofs, one sees that in Proposition 2.5 we have used only the fact that $g(u) \geq m^*$ for $b < u < \eta$; in Lemma 3.7 only the fact that $g(u) \leq m - 1$ for $u \geq \eta$; and in the proof of Theorems 1 and 2 only the further fact that g is non-increasing in the interval $\eta < u < \kappa$. Since these are the only places where the main condition (H2) was in fact applied, it follows that *this condition can be weakened to the form stated in the introduction (after Theorem 4′), without affecting our uniqueness results.*

5. PROOF OF THEOREMS 3 AND 4

Proof of Theorem 3. If u is a radial ground state or a radial solution of the free boundary problem (1.3), then $u(0) > \eta$ by Proposition 2.6, yielding at once $\eta < \infty$, that is, $q < m^*$. We show next that $\beta < \infty$. Indeed, differentiating the energy function

$$E(r) = \frac{m-1}{m}|u'(r)|^m + F(u(r)), \quad r \in J,$$

one finds that

$$\frac{dE(r)}{dr} = -\frac{n-1}{r}|u'(r)|^m.$$

Thus E is strictly decreasing in J, implying

$$F(u(0)) = E(0) > \lim_{r \to R} E(r) = 0.$$

Hence $u(0) > \beta$ and of course $\beta < \infty$. This proves $\|u\|_\infty > \max\{\eta, \beta\}$ and also shows that the condition η, $\beta < \infty$ is necessary for the existence of u.

In what follows, we assume $\beta < \infty$ and $q < m^*$.

To show the existence of a radial ground state or a radial solution of (1.3), we shall apply Theorem 1 of [GST]. Thus we must prove

$$\limsup_{s \to \infty} Q(s_2)\left(\frac{s^{m-1}}{f(s_1)}\right)^{n/m} = \infty, \tag{5.1}$$

for an arbitrary pair of numbers s_1, s_2 in $[s/2, s]$, say, where

$$Q(s) = nmF(s) - (n-m)sf(s). \tag{5.2}$$

To verify (5.1), we fix a number $\varepsilon > 0$ such that

$$\varepsilon < \frac{(n-m)(m^* - q)}{n}.$$

Choose $\eta' > b$ so large that for all $s > \eta'$

$$q \leq g(s) < q + \varepsilon,$$

or equivalently

$$\frac{q}{s} \leq \frac{f'(s)}{f(s)} < \frac{q+\varepsilon}{s}.$$

Then integrating over (η', s) yields

$$\frac{1}{c}s^q \leq f(s) < cs^{q+\varepsilon}, \quad s > \eta', \tag{5.3}$$

for some constant $c > 0$. Furthermore, for $s > \eta'$ we have $f(s) > 0$ and

$$\begin{aligned} Q'(s) &= (n-m)[m^* - g(s)]f(s) \\ &> (n-m)(m^* - q - \varepsilon)f(s) > m\varepsilon f(s). \end{aligned} \tag{5.4}$$

There are now two cases, $q \geq -1$ and $q < -1$ (including $q = \infty$). In the first case, from (5.3), (5.4) we find $Q'(s) \geq \text{Const.}\, s^q$, whence by integration

$$Q(s) \geq \text{Const.}\, s^{q+1}, \quad s > \eta', \tag{5.5}$$

(or $Q(s0) \geq \text{Const.} \log s$ if $q = -1$). Hence, for any pair of numbers s_1, s_2 in $[s/2, s]$, $s > 2\eta'$, we have by (5.3) and (5.5)

$$Q(s_2)\left(\frac{s^{m-1}}{f(s_1)}\right)^{n/m} \geq \text{Const.}\, s^{\xi}, \quad \xi = \frac{n-m}{m}\left(m^* - q - \frac{n\varepsilon}{n-m}\right) > 0$$

(or $\text{Const.} \log s (s^{m+\varepsilon})^{m/n}$ if $q = -1$). This verifies (5.1).

In the second case, when $q < -1$, we observe that $F(s)$ is strictly increasing for $u > b$, so $F(s) \geq$ Pos. Const. for all large s. But then by (5.2) it follows that

$$Q(s) \geq \text{Pos. Const.} \tag{5.6}$$

for large s, and (5.1) is easily seen to hold.

(In [GST] it is also required that $Q(s)$ should be bounded below and non-negative for suitably large s. The latter follows from (5.5) and (5.6), and the former is then obvious from the definition of Q, since $f(s) \le 0$ for $u < b$ and $F(s)$ is bounded below near $u = 0$.)

To complete the proof, note that uniqueness follows from Theorems 1 and 2, while the remaining results can be obtained from Theorem 1 of [GST]. □

We now give a simple lemma which will be needed below.

Lemma 5.1. *Suppose that f is continuous on $(0, \infty)$ and*

$$0 \le f(u) \le k(\gamma - u)^{m-1} \quad \text{for } u \in (\gamma - \varepsilon, \gamma]$$

for some positive constants k, γ, and ε. Then (1.1) *has no radial ground states, and* (1.3) *no radial solutions, with $u(0) = \gamma$.*

Proof. As in the first part of the proof of Proposition 2.1, any ground state with $u(0) = \gamma$ must satisfy $u \le \gamma$ for all r. Then from (2.4) and the main hypothesis we get $u'(r) \le 0$ and

$$|u'|^{m-1} \le \frac{k}{r^{n-1}} \int_0^r s^{n-1}(\gamma - u(s))^{m-1}\, ds$$

for all sufficiently small r, say $0 \le r < r_0$. But then, since $\gamma - u(s) \le \gamma - u(r)$ for $0 \le s \le r$, we get

$$|u'(r)|^{m-1} \le \frac{kr}{n}(\gamma - u(r))^{m-1}, \quad r < r_0,$$

that is

$$|u'(r)| \le \text{Const.}\, (\gamma - u(r)), \quad r < r_0.$$

Integrating from 0 to r then gives

$$\gamma - u(r) \le \text{Const.} \int_0^r (\gamma - u(s))\, ds, \quad r < r_0.$$

Thus by the Gronwall lemma, $\gamma - u(r) \equiv 0$ for $0 \le r < r_0$, and by an obvious continuation also $\gamma - u(r) \equiv 0$ for all r, which is impossible for a ground state. Hence $u(0) = \gamma$ cannot occur, and the proof is complete. □

Proof of Theorem 4. When $p < q$ an easy analysis of (1.4) shows that $f(u)$ has a unique positive zero b, and that (H1) holds.

Similarly, when $p > q$ either $f(u) \le 0$ over $(0, \infty)$, in which case (1.1) admits no ground states, or $f(u)$ has two positive zeros, say at b and γ. Then clearly $f(u) > 0$ for $b < u < \gamma$, so (H1′) holds. Moreover, in this case it is easy to see (since $f(u) < 0$ for $u > \gamma$) that there can be no ground state with $u(0) > \gamma$. Also, since $m \leq 2$ and f is Lipschitz continuous at $u = \gamma$, by Lemma 5.1 there can be no ground state with $u(0) = \gamma$. Hence necessarily $\|u\|_\infty < \gamma$.

Uniqueness. For $u > b$ (or for $b < u < \gamma$ when $p > q$) we have $f(u) > 0$, and thus can write

$$g'(u) = \frac{\varphi(u)}{u[f(u)]^2}.$$

But $\mu u^q > u^t + \lambda u^p$ when $f(u) > 0$. Hence for $u > b$ (or $b < u < \gamma$), there holds

$$\begin{aligned}\varphi(u) &= \lambda(p-t)^2 u^{p+t} - \mu u^q[\lambda(p-q)^2 u^p + (q-t)^2 u^t] \\ &< \lambda(p-t)^2 u^{p+t} - (u^t + \lambda u^p)[\lambda(p-q)^2 u^p + (q-t)^2 u^t] \\ &= -[\lambda(p-q)u^p - (q-t)u^t]^2 \leq 0.\end{aligned}$$

Thus (H2) (or (H2′) when $p > q$) holds and uniqueness is an immediate consequence of Theorems 1 or 1′.

Existence. When $p < q < m^*$, existence follows directly from Theorem 3. When $p > q$, we apply Theorem 1 (case (C1)) of [GST]. The required hypothesis (1.6) of [GST] is easily seen to hold for the nonlinearity(1.4′), since $m \leq 2$. Thus the existence of a ground state u with $u(0) < \gamma$ is equivalent to the existence of a number $\bar{u} \in (b, \gamma)$ such that $F(\bar{u}) > 0$, that is, $b < \beta < \gamma$. As observed in [BL], see also [AD], such a number $\bar{u} \in (b, \gamma)$ can be found if and only if $\lambda < \lambda_0$, where $\lambda_0 = \lambda_0(\mu) > 0$ is a constant depending on μ.

The last part of the theorem is a consequence of Theorem 3, case (i), together with an easy calculation (cf. the Corollary of Theorem 3 in the introduction). □

Proof of Theorem 4′. Uniqueness for the model (1.4′) is immediate, since (1.4′) can be written in the form (1.4) with $p = t + 2$, $q = t + 1$, and appropriate constants λ, μ.

To obtain existence, we first calculate that

$$F(u) = -\frac{u^{t+3}}{t+3} + (b+\gamma)\frac{u^{t+2}}{t+2} - b\gamma\frac{u^{t+1}}{t+1}.$$

In turn

$$F(\gamma) = \frac{\gamma^{t+2}}{t+2}\left\{\frac{\gamma}{t+3} - \frac{b}{t+1}\right\} > 0$$

when (1.5) is satisfied. As in the final part of the proof of Theorem 4, it follows that $b < \beta < \gamma$, so existence is again a consequence of Theorem 1 (case (C1)) of [GST]. □

6. FURTHER REMARKS

6.1. Conditions (H1) *and* (H2). Condition (H1) is essentially a minimum requirement for uniqueness of radial ground states. On the other hand, condition (H2) is certainly not a necessary condition, though it is weaker than those assumed in [K, KZ, CC, CEF2, M, MS].

Another uniqueness result, which neither includes nor is included in our Theorem 1, was recently found by Pucci and Serrin [PuS2]. By using the separation technique of Peletier and Serrin [PS1, PS2] and an identity of Erbe and Tang [ET], they showed that equation (1.1) (and also more general elliptic equations) admit at most one radial ground state, provided that

(a) $f \in C^1(0,\infty)$, and is continuous at zero with $f(0) = 0$;
(b) $f(u) < 0$ for $0 < u < b$ and $f(u) > 0$ for $u > b$; and
(c) $(F(u)/f(u))' \geq 1/m - 1/n$.

For the case $n > m$, the results of Theorem 1 are in many ways simpler to apply than those of [PuS2], and moreover include the model cases (1.2) and (1.4), which, as noted in the introduction, are not completely covered by [PuS2].

6.2. The case of the n–Laplacian in $\mathbb{R}^n$. The proofs in the present paper rely extensively on the assumption $n > m$ and cannot be extended easily to values $n \leq m$. On the other hand, the conclusions in [PuS2] remain applicable when $n \leq m$, condition (c) then reducing simply to

(c′) $(F(u)/f(u))' \geq 0$.

As shown in [PuS3], when $n \leq m$, conditions (a), (b), (c′) are satisfied by the nonlinearity

$$f(u) = u^{q-1}[-\ell u^q + (\exp(u^q) - 1 - u^q)], \quad \ell > 0,\ \tfrac{1}{2} < q \leq 1,\ n \geq 2.$$

This function is at the present time the only known example for which the n–Laplacian in $\mathbb{R}^n$ has a unique radial ground state. We mention that in [GMSTY] we proved the existence of radial ground states of the n–Laplacian for a large class of functions f, including the model case

$$\Delta_n u + (-u^{p_1} + u^{p_2})\exp(u^q) = 0 \quad \text{in } \mathbb{R}^n,$$
$$n > 1,\ 0 < p_1 < p_2,\ 0 < q \leq n/(n-1).$$

We conjecture that ground states for this model are also unique.

6.3. Uniqueness in a ball. Our method may work for studying uniqueness of radial solutions of the Dirichlet problem

$$(6.1)\qquad \Delta_m u + f(u) = 0, \quad n > m > 1, \quad u > 0 \text{ in } B_R, \quad u = 0 \text{ on } \partial B_R,$$

where B_R is an open ball in $\mathbb{R}^n$ with radius $R > 0$. If (H1′) holds, i.e., $f(u)$ has two positive zeros b and γ, $b < \gamma$, or if (H2) holds and $q < m - 1$, then (6.1) admits at least two solutions for large R (see [T1, T2]). Thus the conditions of Theorem 1 are not sufficient for uniqueness of radial solutions of (6.1). Nevertheless, we conjecture that (6.1) admits exactly one radial solution when f is the model nonlinear function (1.2). To deal with more general nonlinearities, it may be necessary to have a stronger condition than (H1) on the interval $(0, b)$.

6.4. Symmetry of ground states. The classical paper of Gidas, Ni, Nirenberg [GNN] showed that positive ground states of the semilinear equation (1.6) are necessarily radially symmetric about some origin O, provided that f is of class $C^{1+\alpha}$ in $[0, \varepsilon)$, with $f(0) = 0$, $f'(0) < 0$. Turning to equation (1.1), the results are more subtle. First, if $\int_0 |F(s)|^{-1/m} ds = \infty$ and $1 < m \le 2$, one can again assert that ground states are symmetric. When $m > 2$, or if the integral is finite, symmetry is known only under the fairly strong condition that the ground state in question has only a single critical point, see [SZ] and references therein.

Under conditions for which it is known that ground states are necessarily radially symmetric, the present results show that ground states (whether they are initially assumed to be symmetric or not) are *unique up to simple translations of the origin.*

Acknowledgement. We wish to thank Patrizia Pucci for her careful reading of the first version of this paper, and her suggestions for improvement.

REFERENCES

[AD] D. ANDERSON & G. DERRICK, *Stability of time dependent particle like solutions in nonlinear field theories*, J. Math. Phys. **11** (1970), 1336-1346.

[AD′] D. ANDERSON & G. DERRICK, *Stability of time dependent particle like solutions in nonlinear field theories*, J. Math. Phys. **12** (1971), 945-952.

[BL] H. BERESTYCKI & P.-L. LIONS, *Non-linear scalar field equations. I, Existence of a ground state, II, Existence of infinitely many solutions*, Arch. Rational Mech. Anal. **82** (1983), 313-375.

[CL] C.-C. CHEN & C.-S. LIN, *Uniqueness of the ground state solution of* $\Delta u \; fu \; 0$ *in* $\mathbb{R}^n$, $n \geq 3$, Comm. Partial Diff. Eqs. **16** (1991), 1549-1572.

[Ci] G. CITTI, *A uniqueness theorem for radial ground states of the equation* $\Delta_p u \; fu$ 0, Boll. Un. Mat. Ital. B **7** (1993), 283-310.

[C] C.V. COFFMAN, *Uniqueness of the ground state solution for* $\Delta u - u \; u^3 \;\; 0$ *and a variational characterization of other solutions*, Arch. Rational Mech. Anal. **46** (1972), 81-95.

[CEF1] C. CORTZAR, M. ELGUETA & P. FELMER, *On a semilinear elliptic problem in* $\mathbb{R}^N$ *with a non-Lipschitzian non-linearity*, Advances in Diff. Eqs. **1** (1996), 199-218.

[CEF2] ———, *Uniqueness of positive solutions of* $\Delta u\ fu\quad 0$ *in* $\mathbb{R}^N$, $N \geq 3$, Arch. Rational Mech. Anal. **142** (1998), 127-141.

[ET] L. ERBE & M. TANG, *Uniqueness theorems for positive solutions of quasilinear elliptic equations in a ball*, J. Diff. Eqs. **138** (1997), 351-379.

[FLS] B. FRANCHI, E. LANCONELLI & J. SERRIN, *Existence and uniqueness of non-negative solutions of quasilinear equations in* $\mathbb{R}^n$, Advances in Math. **118** (1996), 177-243.

[GMSTY] M. GARCA-HUIDOBRO, R. MANSEVICH, J. SERRIN, M. TANG & C. YARUR, *Ground states and free boundary value problems for the n-Laplacian in n dimensional space*, J. Funct. Anal **172** (2000), 177–201.

[GST] F. GAZZOLA, J. SERRIN & M. TANG, *Existence of ground states and free boundary problems for quasilinear elliptic operators*, Advances in Diff. Eqs **5** (2000), 1–30.

[GNN] B. GIDAS, W.M. NI & L. NIRENBERG, *Symmetry of positive solutions of nonlinear elliptic equations in* $\mathbb{R}^n$, Adv. in Math. Suppl. Stud. **7a** (1981), 369-402.

[K] M.K. KWONG, *Uniqueness of positive solutions of* $\Delta u - u\ u^p\quad 0$ *in* $\mathbb{R}^n$, Arch. Rational Mech. Anal. **105** (1989), 243-266.

[KZ] M.K. KWONG & L. ZHANG, *Uniqueness of the positive solution of* $\Delta u\ fu\quad 0$ *in an annulus*, Diff. Int. Eqs. **4** (1991), 583-596.

[M] K. MCLEOD, *Uniqueness of positive radial solutions of* $\Delta u\ fu\quad 0$ *in* $\mathbb{R}^n$, *II*, Tran. Amer. Math. Soc. **339** (1993), 495-505.

[MS] K. MCLEOD & J. SERRIN, *Uniqueness of positive radial solutions of* $\Delta u\ fu\quad 0$ *in* $\mathbb{R}^n$, Arch. Rational Mech. Anal. **99** (1987), 115-145.

[NS1] W.-M. NI & J. SERRIN, *Nonexistence theorems for quasilinear partial differential equations*, Rend. Circolo Mat. Palermo (Centenary Supplement) **8** (1985), 171-185.

[NS2] ———, *Existence and nonexistence theorems for ground states for quasilinear partial differential equations. The anomalous case*, Rome, Accad. Naz. dei Lincei, Atti dei Connegni **77** (1986), 231-257.

[PS1] L.A. PELETIER & J. SERRIN, *Uniqueness of positive solutions of semilinear equations in* $\mathbb{R}^N$, Arch. Rational Mech. Anal. **81** (1983), 181-197.

[PS2] ———, *Uniqueness of non-negative solutions of semilinear equations in* $\mathbb{R}^N$, J. Diff. Eqs. **61** (1986), 380-397.

[PuS1] P. PUCCI & J. SERRIN, *A general variational identity*, Indiana Univ. Math. J. **35** (1986), 681-703.

[PuS2] ———, *Uniqueness of ground states for quasilinear elliptic operators*, Indiana Univ. Math. J. **47** (1998), 501-528.

[PuS3] ———, *Uniqueness of ground states for quasilinear elliptic equations in the exponential case*, Indiana Univ. Math. J. **47** (1998), 529-539.

[SZ] J. SERRIN & H. ZOU, *Symmetry of ground states of quasilinear elliptic equations*, Arch. Rational Mech. Anal **148** (1999), 265–290.

[S] W. STRAUSS, *Existence of solitary waves in higher dimensions*, Comm. Math. Phys. **55** (1977), 149-162.

[T1] M. TANG, *Existence and uniqueness of fast decay entire solutions of quasilinear elliptic equations*, J. Diff. Eqs **164** (2000), 155–179.

[T2] ———, *Uniqueness and global structure of positive radial solutions for quasilinear elliptic equations*, Comm. Partial Diff. Eqs. (to appear).

[T3] ———, *Uniqueness of positive radial solutions for n-Laplacian Dirichlet problems*, to appear, Proc. Roy. Soc. Edinburgh Sect. A.

[Y] E. YANAGIDA, *Uniqueness of positive radial solutions of* Δu *gru hru*^{p} 0 *in* $\mathbb{R}^n$, Arch. Rational Mech. Anal. **115** (1991), 257-274.

School of Mathematics
University of Minnesota
Minneapolis, MN 55455 U. S. A.
EMAIL: serrin@math.umn.edu (James Serrin)
EMAIL: mtang@math.umn.edu (Moxun Tang)

KEYWORDS: quasilinear elliptic equations, scalar field equation.

Submitted: November 4th, 1999, revised: June 27th, 2000.

I.10. Ground States – Special Results

Lambertus A. Peletier

Mathematics Subject Classification (2010). 35B05; 35B33; 35B40; 35B60; 35J66.

Keywords. Ground states, bifurcation theory, nonlinear eigenvalue problems, minimal surface operator.

Paper [i] below is the third in a series of papers [APS1], [AP] and [APS2] in which the notion of a *Ground State* was extended to equations involving the minimal surface operator instead of the Laplace operator, i.e., equations such as

$$\operatorname{div}\left(\frac{\nabla u}{\sqrt{1+|\nabla u|^2}}\right) + u^p = 0, \quad u > 0 \qquad x \in \mathbf{R}^N \ (N > 2). \tag{1}$$

If $p > (N+2)/(N-2)$ then it is known that this equation has a radially symmetric ground state with any prescribed central value $u(0) = c$, provided $c > 0$ is small enough. Specifically it was shown that if $c \leq c_1 \stackrel{\text{def}}{=} (p+1)^{1/(p+1)}$ then radial ground states exist [NS] and if $c \geq c_2 \stackrel{\text{def}}{=} (4N^2p)^{1/(p+1)}$ they do not exist [S].

As c increases beyond the value c_1 it was found that the ground state eventually ceases to exist for some finite $c_\infty \in (c_1, c_2)$ because $|u'|$ blows up at some finite radius, when c approaches c_∞ from below. This was the reason for introducing the notion of a *Generalized Ground State* in [APS1].

Such a ground state is a multivalued function $u(r)$, which is single valued for small and for large values of r, and can be represented by a function

$$r = r(u), \qquad 0 < u \leq c, \tag{2}$$

endowed with the properties: $r(u) \to \infty$ as $u \searrow 0$ and $r(u) \to 0$ as $u \nearrow c$.

In [AP], the second paper in this series of three, estimates were obtained for the location of the first point (r, u) – counting from the center – where $u'(r) = -\infty$, i.e., where the graph of the *multi-valued* function $u(r)$ has a vertical tangent.

As c increases further, numerical computations showed that the graph of $u(r)$ exhibits an increasing number of vertical points, and the location of these points was seen to possess an elegant regularity. Let us denote the vertical points by $P_m = (u_m, r_m)$, where $m = 1, 2, 3, \ldots, M$, and points have been ordered so that $c > u_1 > u_2 > \cdots > u_M$. Thus, P_1 is the first vertical point, counting from the centre, a maximum of $r(u)$, P_2 the second vertical point, a minimum, and so on. Clearly P_m is a maximum of the function $r(u)$ if m is odd and a minimum if m is even. In [APS1] it was proved that if $p \geq 1$, then for any $c > 0$ the number of vertical points is finite and that they have the following

P. Pucci et al. (eds.), *James Serrin. Selected Papers: Volume 1*, Contemporary Mathematicians,
DOI 10.1007/978-3-0348-0845-3_10, © Springer Basel 2014

properties:

$$\frac{N-1}{u_m^p} < r(u_m) < \frac{N}{u_m^p} \quad (m \text{ odd}) \qquad \text{and} \qquad r(u_m) < \frac{N-1}{u_m^p} \quad (m \text{ even}). \tag{3}$$

(See also the elegant monograph by Finn [F]).

Paper [i], which is [APS2], focusses on the location of a range of vertical points, especially when the central value c is large and the number of vertical points is also large. It uses appropriately chosen *Delaunay surfaces* [D] to derive sharp asymptotic estimates.

Paper [ii] below is concerned with the Dirichlet problem

$$-\Delta u = \lambda u + u^p, \quad u > 0 \quad \text{in} \quad \Omega \subset \mathbf{R}^N; \qquad u|_{\partial\Omega} = 0, \qquad (\lambda \geq 0,\ p > 1,\ N > 2), \tag{4}$$

in which Ω is a bounded star-shaped domain. This seminal problem, which was introduced in1983 by Brezis and Nirenberg [BN], has attracted a great deal of attention, and proven to possess a very rich structure. For convenience we put $\Omega = B_1$, the unit ball in $\mathbf{R}^N$. In [BN] the focus was primarily on the question of existence for different values of N and λ, when p is equal to the critical Sobolev exponent $p_N \stackrel{\text{def}}{=} (N+2)/(N-2)$. For instance, it was shown that when $p = p_N$, then if $N \geq 4$ and $\lambda = 0$, Problem (4) does not have a solution, but if $0 < \lambda < \lambda_1$, where λ_1 is the principal eigenvalue of the Laplacian on B_1, it does. If $N = 3$ the corresponding range of λ was shown to be $\frac{1}{4}\pi^2 < \lambda < \lambda_1 = \pi^2$.

Many subsequent studies have approached this problem from the perspective of *Bifurcation theory*, the relevant parameter being λ or p. Of particular interest are problems about what one might call *the edge of existence*. A typical example of this *genre* analyses the behaviour of a solution u_λ of Problem (4) as λ approaches the endpoints of the existence interval $\lambda_0 < \lambda < \lambda_1$, where $\lambda_0 = 0$ if $N \geq 4$, and $\lambda_0 = \frac{1}{4}\pi^2$ when $N = 3$.

Since the point $(\lambda, u) = (\lambda_1, 0)$ is a bifurcation point, a branch Γ of positive solutions emanates from this point and is unbounded in the half-strip $\{(\lambda, u) : \lambda_0 < \lambda < \lambda_1,\ u > 0\}$. As $\lambda \to \lambda_0$ this branch approaches the half-line $\{(\lambda, u) : \lambda = \lambda_0,\ u > 0\}$ asymptotically.

The character of the bifurcation branch Γ does not change much if p becomes *supercritical*, i.e., $p > p_N$. It still bifurcates from the point $(\lambda_1, 0)$ and tends to infinity in a half-strip. In paper [ii] below the properties of the branch Γ are investigated when the power $p \to \infty$. It is shown that the width of the half-strip tends to zero and that the strip eventually contracts around the vertical line emanating from the bifurcation point: $\{(\lambda, u) : \lambda = \lambda_1,\ u > 0\}$. The solutions on the branch acquire a characteristic shape: If we put $u(0) = c$, then when $c \leq 1$ the solution converges to $c \times \varphi_1$, where φ_1 is the eigenfunction of the Laplace operator normalized so that $\varphi_1(0) = 1$, and when $c > 1$, the solution approaches φ_1, except near the origin, where it contracts to a *needle* with height $c - 1$ on top of φ_1.

References

[AP] F.V. Atkinson and L.A. Peletier, *Bounds for vertical points of solutions of prescribed mean curvature type equations I*, Proc. Royal Soc. Edinb. **112A** (1989) 15–32.

[APS1] F.V. Atkinson, L.A. Peletier and J. Serrin, *Ground states for the prescribed mean curvature equation: The supercritical case*, In *Nonlinear Diffusion Equations and their Equilibrium States* (eds. W.-M. Ni, L.A. Peletier & J. Serrin), pp. 51–74, MSRI Publications 12/13, Springer, 1988.

[APS2] F.V. Atkinson, L.A. Peletier and J. Serrin, *Estimates for vertical points of solutions of prescribed mean curvature type equations II*, Asymptotic Anal. **5** (1992), 283–310.

[BN] H. Brezis and L. Nirenberg, *Positive solutions of nonlinear elliptic equations involving critical Sobolev exponents*, Comm. Pure Appl. Math. **XXXVI** (1983) 437–477.

[D] C.E. Delaunay, *Sur la surface de révolution dont la courbure moyenne est constante*, J. Math. Pures Appl. **6** (1841) 309–315.

[F] R. Finn, *Equilibrium Capillary Surfaces*, Springer-Verlag, New York, 1986.

[NS] W.-M. Ni and J. Serrin, *Existence and nonexistence theorems for ground states of quasilinear partial differential equations. The anomalous case*, Accad. Naz. Lincei Convegni del Lincei **77** (1986) 231–257.

[S] J. Serrin, *Positive solutions of the a prescribed mean curvature equation*, In "Calculus of Variations and Partial Differential Equations", Lecture Notes in Mathematics #1340. Springer, 1988, 248–255.

Lambertus A. Peletier
Mathematical Institute
Leiden University
PB 9512
NL-2300 RA Leiden, The Netherlands
e-mail: `peletier@math.LeidenUniv.nl`

Asymptotic Anal. 5 (1992), 283–310. **Reprint I.10.[i]**

Pages 289–308 have been removed for the sake of brevity.

Asymptotic Analysis 5 (1992) 283–310
North-Holland

Estimates for vertical points of solutions of prescribed mean curvature type equations II

F.V. Atkinson
Department of Mathematics, University of Toronto, Toronto M5S 1A1, Ontario, Canada

L.A. Peletier
Mathematical Institute, Leiden University, Leiden, The Netherlands

J. Serrin
School of Mathematics, University of Minnesota, Minneapolis, MN, USA

Received 30 January 1991

Abstract

Atkinson, F.V., L.A. Peletier and J. Serrin, Estimates for solutions of mean curvature equations, Asymptotic Analysis 5 (1992) 283–310.

Radially symmetric solutions $u(r)$ of the prescribed mean curvature equation involving a source term $f(u)$ may exhibit vertical points. In this paper we study the asymptotic properties of these vertical points when f is positive and increasing and the parameter $\varepsilon = f'/f^2$ becomes small. Our results are based on a sharp asymptotic estimate for Delaunay surfaces as the distance of the vertical points nearest to the symmetry axis $r = 0$ tends to zero.

1. Introduction

In this paper we continue our study of the qualitative properties of radial solutions of the Prescribed Mean Curvature Equation

$$\operatorname{div}\left(\frac{\nabla u}{\sqrt{1+|\nabla u|^2}}\right) + f(u) = 0. \tag{1.1}$$

It is well known that the graph of the solution $u(r, \gamma)$ of (1.1) with central value $u(0, \gamma) = \gamma$, may exhibit vertical points [1–3,6]. The first such vertical point, counting from the axis, was studied in detail in [1] under the following assumptions on the function f:

(H1) $f \in C^2(c, \infty)$,

(H2) $f > 0,\ f' > 0,\ f'' \geqslant 0$ on (c, ∞),

(H3) $\dfrac{f'(u)}{f^2(u)} \to 0$ as $u \to \infty$,

(H4) $\dfrac{f''(u)}{f(u)f'(u)} \to 0$ as $u \to \infty$,

where c is some real number.

0921-7134/92/$05.00

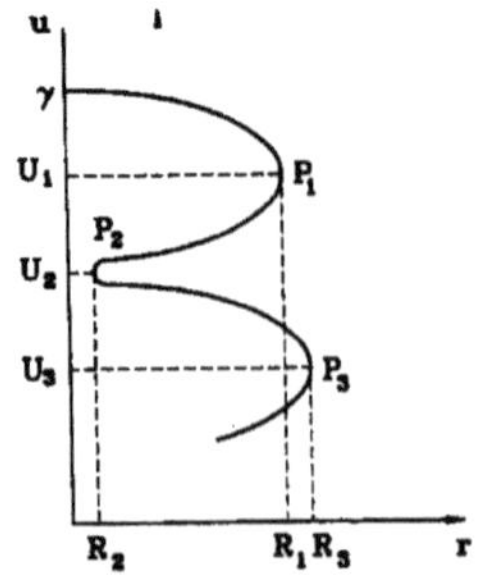

Fig. 1. The graph of $u(r, \gamma)$.

It was established in [6] and [2] that for sufficiently large values of γ, the graph of $u(r, \gamma)$ has a sequence of isolated vertical points. We shall denote them by $P_i = (R_i, U_i)$, P_1 being the first vertical point, P_2 the second, and so forth, which are met when we proceed along the graph starting from the axis $r = 0$ (see Fig. 1). Moreover, it was shown that for $k = 1, 2, \dots$

$$\frac{N-1}{f(U_k)} < R_k < \frac{N}{f(U_k)} \quad (k \text{ odd}), \qquad R_k < \frac{N-1}{f(U_k)} \quad (k \text{ even}). \tag{1.2}$$

In [1] we obtained an asymptotic estimate for the location (R_1, U_1) of the first vertical point as $\gamma \to \infty$. It is the purpose of this paper to extend this estimate to subsequent vertical points (R_k, U_k), $k = 2, 3, \dots$. Our main result will be stated in the following theorem; just as for the first point P_1, the quantity

$$\varepsilon(\gamma) = \frac{f'(\gamma)}{f^2(\gamma)}$$

will play a key role in the estimates. Note that in view of hypothesis (H3), $\varepsilon(\gamma) \to 0$ as $\gamma \to \infty$.

Theorem A. *Suppose f satisfies* (H1)–(H4). *Then for any integer* $k \geqslant 1$,

$$U_k = \gamma - k\frac{N}{f(\gamma)}[1 + \mathrm{O}(\varepsilon^\beta)] \quad as\ \gamma \to \infty, \tag{1.3)$_k$$$

where

$$\beta = \min\left\{\frac{1}{2}, \frac{1}{N-1}\right\}.$$

If k is odd

$$R_k = \frac{N}{f(\gamma)}[1 + kN(1 - \alpha_N)\varepsilon + \mathrm{o}(\varepsilon)] \quad as\ \gamma \to \infty, \qquad (1.4)_k$$

and if k is even

$$R_k = \frac{N}{f(\gamma)}(kN\alpha_N\varepsilon)^{1/(N-1)}[1 + \mathrm{o}(1)] \quad as\ \gamma \to \infty, \qquad (1.5)_k$$

where

$$\alpha_N = \int_0^1 \frac{s^{N+1}\,\mathrm{d}s}{\sqrt{1-s^2}} = \frac{\sqrt{\pi}}{2}\frac{\Gamma(N/2+1)}{\Gamma(N/2+3/2)}.$$

We note particularly that $(1.3)_k$ and $(1.4)_k$ were proved for $k = 1$ in [1]. In fact, in this case it was even possible to take $\beta = 1$.

The factor

$$R_0 = \frac{N}{f(\gamma)}$$

which appears in the estimates of Theorem A is precisely the reciprocal mean curvature at the starting point $(0, \gamma)$. If we normalize $\gamma - U_k$ and R_k with respect to R_0 and write

$$\xi_k = \frac{\gamma - U_k}{R_0}, \qquad \eta_k = \frac{R_k}{R_0},$$

if follows from Theorem A that

$$\xi_k = k + \mathrm{O}(\varepsilon^\beta),$$

$$\eta_k = \begin{cases} 1 + kN(1 - \alpha_N)\varepsilon + \mathrm{o}(\varepsilon) & k \text{ odd}, \\ (kN\alpha_N\varepsilon)^{1/(N-1)}[1 + \mathrm{o}(1)] & k \text{ even}. \end{cases}$$

Thus, when we omit the error terms and make this normalization, the vertical points lie on the curves

$$\eta = 1 + N(1 - \alpha_N)\varepsilon\xi \quad \text{outer points},$$

$$\eta^{N-1} = N\alpha_N\varepsilon\xi \quad \text{inner points}.$$

Note that the curve of outer points is linear and that when $N = 2$ the curve of inner points is also linear.

As examples we consider the cases $N = 2$, 3 and 4. Since

$$\alpha_2 = \tfrac{2}{3}, \ \alpha_3 = \frac{3\pi}{16}, \ \alpha_4 = \tfrac{8}{15},$$

we find

	inner points	outer points
if $N = 2$:	$\eta = \frac{4}{3}\varepsilon\xi$	$\eta = 1 + \frac{2}{3}\varepsilon\xi,$
if $N = 3$:	$\eta^2 = \frac{9\pi}{16}\varepsilon\xi$	$\eta = 1 + \left(3 - \frac{9\pi}{16}\right)\varepsilon\xi,$
if $N = 4$:	$\eta^3 = \frac{32}{15}\varepsilon\xi$	$\eta = 1 + \frac{28}{15}\varepsilon\xi.$

In Fig. 2 we give some numerical results for $f(u) = \lambda u^q$ when $\lambda = 5$; here $\varepsilon(\gamma) = q/(5\gamma^{q+1})$.

The case $N = 2$ is particularly instructive, for one sees at once that the slope of the inner envelope is exactly one-half the slope of the outer envelope, whatever the values of ε or γ. This is clearly borne out by Fig. 2(a). We note as well that the inner and the outer envelopes which are shown in Fig. 2(a) are generated directly from the asymptotic formulae of Theorem A, rather than by simple curve matching; thus the accurate fit, which is shown confirms and verifies the results of Theorem A.

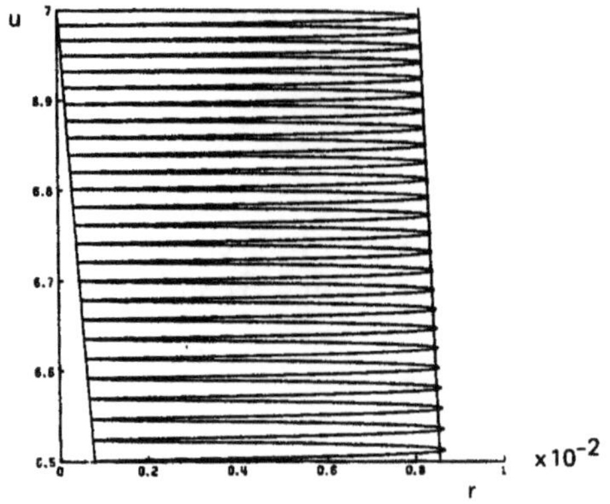

Fig. 2(a). $N = 2$, $q = 2$, $\lambda = 5$, $\gamma = 7$, $\varepsilon = 0.0012$.

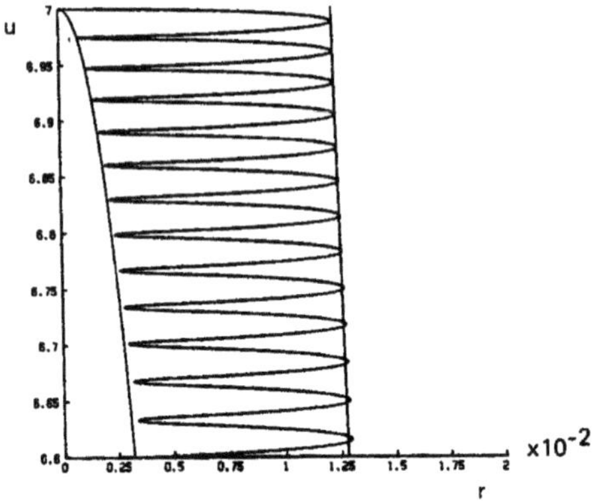

Fig. 2(b). $N = 3$, $q = 2$, $\lambda = 5$, $\gamma = 7$, $\varepsilon = 0.0012$.

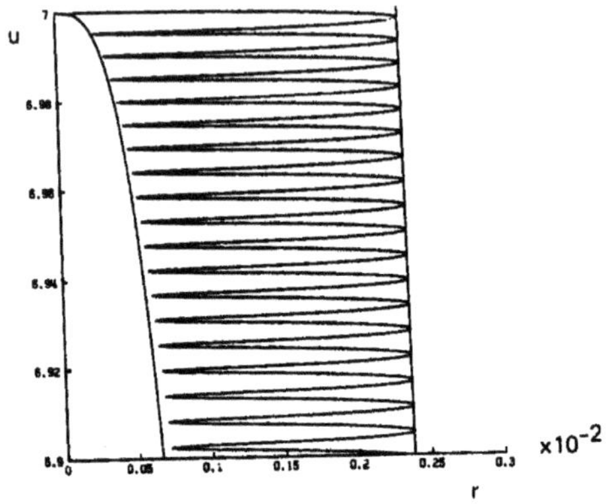

Fig. 2(c). $N = 4$, $q = 3$, $\lambda = 5$, $\gamma = 7$, $\varepsilon = 0.00025$.

As we explained in [2], the function $u(r)$, representing the segment of the graph between the points P_k and P_{k+1}, is a solution of the equation

$$\left(\frac{r^{N-1}u'}{\sqrt{1+u'^2}}\right)' = (-1)^{k-1}r^{N-1}f(u), \qquad \begin{cases} R_{k+1} < r < R_k & \text{if } k \text{ is odd} \\ R_k < r < R_{k+1} & \text{if } k \text{ is even,} \end{cases} \tag{1.6}$$

with the following properties at the endpoints

$$u(R_l) = U_l, \quad \lim_{r \to R_l} |u'(r)| = +\infty \qquad l = k, k+1. \tag{1.7}$$

To estimate R_k and U_k, we regard $u(r)$ as the solution of the initial value problem for (1.6) with data (1.7) given at the point P_{k-1}. We start this process at P_1, about which we already know the estimates $(1.3)_1$ and $(1.4)_1$ and so obtain estimates for P_2. These results are then used to obtain estimates for P_3, which, in turn, are used to estimate P_4 and so on.

The main idea involved in estimating P_{k+1} from information about P_k will be to compare $u(r)$ between P_k and P_{k+1} with solutions $v(r)$ of the constant mean curvature problem $(k = 1, 2, \ldots)$

$$\left(\frac{r^{N-1}v'}{\sqrt{1+v'^2}}\right)' = (-1)^{k-1}r^{N-1}H, \qquad \begin{cases} r < R_k & \text{if } k \text{ is odd} \\ r > R_k & \text{if } k \text{ is even,} \end{cases} \tag{1.8a}$$

$$v(R_k) = U_k, \qquad (-1)^k v'(R_k) = \infty, \tag{1.8b}$$

where the (positive) values to be used for H will be chosen appropriately at each step. The graphs of these solutions clearly represent segments of *Delaunay surfaces* corresponding to the mean curvature H/N; see [3–5].

Specifically, between P_1 and P_2 the behaviour of $u(r)$ is captured if we choose comparison solutions v^+ and v^- corresponding to the values

$$H = H^+ = f(U_1) \quad \text{and} \quad H = H^- = f\left(U_1 - \frac{K}{H^+}\right) \tag{1.9}$$

for some suitable positive constant K. The local existence of the solutions v^+ and v^- can be established thanks to (1.2) with $k = 1$. Let their maximal intervals of existence be (R_2^+, R_2) and (R_2^-, R_2). Using (1.2) again, we show that $R_2^\pm > 0$ and that

$$(v^\pm)'(R_2^\pm) = \infty,$$

that is, the Delaunay surfaces represented by $v^\pm$ do not intersect the vertical u-axis.

The fact that f is strictly increasing yields the following relations:

$$R_2^+ < R_2 < R_2^-, \tag{1.10}$$

together with the comparison estimates

$$u(r) < v^+(r) \quad \text{for } R_2 \leqslant r < R_1, \tag{1.11}$$

and

$$u(r) > v^-(r) \quad \text{for } R_2^- \leqslant r < R_1, \tag{1.12}$$

as shown in Fig. 3.

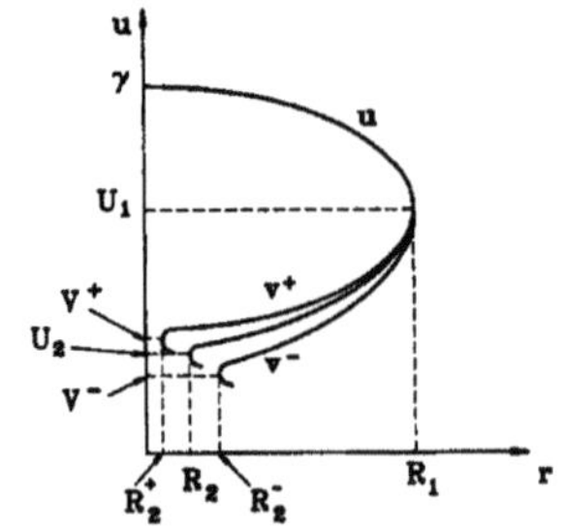

Fig. 3. The comparison functions v^+ and v^-.

The functions v^+ and v^-, as well as their end points R_2^+ and R_2^-, can be computed explicitly and enable us to obtain from (1.10)–(1.12) the following estimates (Lemma 2.5 and Theorem 2.7)

$$U_2 = U_1 - \frac{N}{f(\gamma)}[1 + \mathrm{O}(\omega(\varepsilon))] \quad \text{as } \gamma \to \infty \tag{1.13}$$

and

$$\left(\frac{R_2}{R_1}\right)^{N-1} = \mathrm{O}(\varepsilon) \quad \text{as } \gamma \to \infty, \tag{1.14}$$

where

$$\omega(x) = \begin{cases} x^\alpha & \text{if } N > 2, \\ x \log \dfrac{1}{x} & \text{if } N = 2, \end{cases} \tag{1.15}$$

in which $\varepsilon = f'(\gamma)/f^2(\gamma)$ and $\alpha = 1/(N-1)$.

Finally to improve (1.14) we integrate the differential equation (1.6) over the entire interval (R_2, R_1). Remembering that $u' = \infty$ at both end points, this yields the relation

$$x^{N-1} = 1 - R_1^{1-N} \int_{R_2}^{R_1} r^{N-1} f(u(r))\, \mathrm{d}r,$$

with $x = R_2/R_1$. Using (1.13) and (1.14) then gives (see Theorem 3.1)

$$\left(\frac{R_2}{R_1}\right)^{N-1} = 2N\alpha_N \varepsilon + \mathrm{o}(\varepsilon) \quad \text{as } \gamma \to \infty.$$

This analysis, which is carried out in Sections 2 and 3, proves the case $k = 2$ of Theorem A. For larger values of k we use a similar procedure to obtain the corresponding conclusions of Theorem A. This is done in Sections 4 and 5.

Finally we note that the asymptotic results obtained in this paper rely on an asymptotic estimate for a certain integral, which also arises in the description of Delaunay surfaces [3–5]. This estimate is obtained in the Appendix and is interesting in its own right.

For convenience, we shall frequently write

$$f_0 = f(\gamma) \quad \text{and} \quad f_k = f(U_k), \ k = 1, 2, \ldots.$$

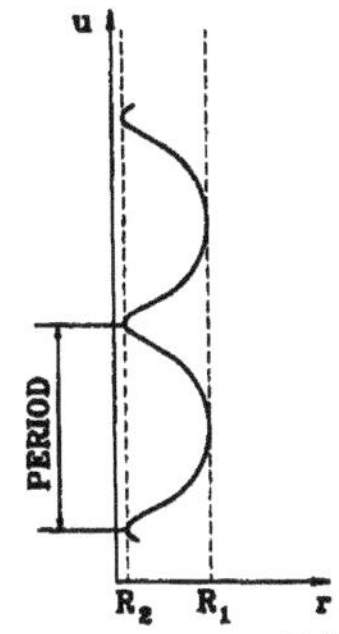

Fig. 4. Delaunay surface of revolution.

with a similar result for J_{32}. The other integrals remain in force. Hence, if $N=2$,

$$J(x)=1+x\log\left(\frac{1}{x}\right)+\mathrm{O}(x^{2/N}).$$

Remark. The integral $J(x)$ is closely connected with the Delaunay surface of revolution [4] (see Fig. 4).

In fact, if the mean curvature is taken to be constant, H/N, then the Delaunay surface is represented by $u=u(r)$, where

$$u'(r)=\frac{I(r)}{\sqrt{1-I^2(r)}},$$

$$I(r)=r^{1-N}\left\{R_1^{N-1}-\int_r^{R_1}s^{N-1}H\,\mathrm{d}s\right\},$$

and R_1 denotes the outer radius. Putting

$$t=\frac{H}{N}r \quad\text{and}\quad x=\left(\frac{R_1H}{N}\right)^{N-1}\left(1-\frac{R_1H}{N}\right),$$

we can write

$$I=xt^{1-N}+t.$$

Thus since $I=1$ at both the outer radius $r=R_1$ and the inner radius $r=R_2$, it follows that the *period* of the Delaunay surface is precisely $2NJ(x)/H$.

The asymptotic behaviour of $J(x)$ as $x\to 0$ thus corresponds to finding the period of the surface as $R_1\to N/H$, that is as the outer radius approaches the radius of the sphere of mean curvature H/N. Finally, to ascertain the inner radius R_2 in terms of R_1 we need only notice that the second (small) root of

$$xt^{1-N}+t=1$$

is given by

$$t=x^{1/(N-1)}\left(1+\frac{1}{N-1}x^{1/(N-1)}+\cdots\right).$$

Thus,

$$\frac{R_2}{R_1} = \left(1 - \frac{R_1 H}{N}\right)^{\alpha} \left\{1 + \frac{1}{N-1}\left(1 - \frac{R_1 H}{N}\right)^{\alpha} + \cdots \right\}$$

and

$$\frac{\text{Period}}{2R_1} = \begin{cases} 1 + \left(1 - \frac{R_1 H}{N}\right) \log \frac{1}{1 - R_1 H/N} + \cdots & \text{if } N = 2 \\ 1 + \alpha K \left(1 - \frac{R_1 H}{N}\right)^{\alpha} \cdots & \text{if } N > 2, \end{cases}$$

$$= \begin{cases} 1 + \frac{R_2}{R_1} \log \frac{R_1}{R_2} + \cdots & \text{if } N = 2 \\ 1 + \alpha K \frac{R_2}{R_1} + \cdots & \text{if } N > 2. \end{cases}$$

Acknowledgements

It is a pleasure to thank Ruud Schotting of the Technological University Delft for the numerical work involved in Figure 2. This research was partially supported by NATO Grant 86/151, EEC Grant SC1-0019-C(TT) and Grant A3979 from the National Sciences and Engineering Council of Canada.

References

[1] F.V. Atkinson and L.A. Peletier, Bounds for vertical points of solutions of prescribed mean curvature type equations I, *Proc. Roy. Soc. Edinb. 112A* (1989) 15–32.
[2] F.V. Atkinson, L.A. Peletier and J. Serrin, Ground states for the prescribed mean curvature equation: the supercritical case, in: S.-M. Ni, L.A. Peletier and J. Serrin, eds., *Nonlinear Diffusion Equations and their Equilibrium States*, MSRI Conference Proceedings (Springer, New York, 1989) 51–74.
[3] P. Concus and R. Finn, The shape of a pendent drop, *Philos. Trans. Roy. Soc. London 292* (1979) 307–340.
[4] C.E. Delaunay, Sur la surface de révolution dont la courbure moyenne est constante, *J. Math. Pures Appl. 6* (1841) 309–315.
[5] R. Finn, *Equilibrium Capillary Surfaces* (Springer, New York, 1986).
[6] J. Serrin, Positive solutions of a prescribed mean curvature equation, Lecture Notes Math. 1340 (Springer, New York, 1989) 248–255.

Indiana Univ. Math. J. 43 (1994), 585–609. **Reprint I.10.[ii]**

Pages 289–308 have been removed for the sake of brevity.

A Bifurcation Problem at a Singular Limit

F. MERLE, L. A. PELETIER & J. SERRIN

1. Introduction. We consider the Dirichlet problem

$$\text{(I)} \qquad \begin{cases} -\Delta u = \lambda u + u^p & \text{in } B_1, \\ u > 0 & \text{in } B_1, \\ u = 0 & \text{on } \partial B_1, \end{cases}$$

in which B_1 is the unit ball in $\mathbf{R}^N$ $(N > 2)$ and λ and p are positive constants. We shall be interested in the existence of solutions (λ, u) when p is "super critical", that is

$$p > p_N = \frac{N+2}{N-2},$$

and, more specifically, in the case when p tends to infinity.

It is well known and easily demonstrated that Problem (I) has no solutions if $\lambda \geq \mu$, where μ is the principal eigenvalue of the Laplacian on B_1 with Dirichlet boundary conditions, and that (see below) solutions *do exist* in a left neighborhood of $\lambda = \mu$. On the other hand, it was shown by Brézis and Nirenberg [BN] and Pucci and Serrin [PS] that there exists a positive lower bound for the values of λ for which Problem (I) can have a solution. They proved in fact that if

$$\text{(1.1)} \qquad \lambda \leq \lambda^*(p, N) = \mu \frac{N-2}{N} \frac{p - p_N}{p-1},$$

then Problem (I) cannot have a solution. We note particularly that, as p tends to infinity, $\lambda^*(p, N)$ tends to a limit which is strictly less than μ.

By classical global bifurcation theory [R] there exists for every fixed $p > p_N$ a continuum $\mathcal{C}$ of solutions (λ, u) of (I) which emanates from the bifurcation point $(\mu, 0)$, and is unbounded in $\mathbf{R} \times C(B_1)$. Thus, since solutions are possible only when $\lambda^* < \lambda < \mu$, there must exist for every $\gamma > 0$, a solution (λ_p, u_p) of (I) such that $|u_p| = \gamma$, where $|\cdot|_\infty$ denotes the L^∞-norm in B_1. By the radial symmetry of B_1 and the positivity of u_p, the famous result of Gidas, Ni and Nirenberg [GNN] implies that u_p is also radially symmetric, so that $u_p(0) = \gamma$.

In this paper we shall be interested in the behavior of λ_p and u_p as $p \to \infty$. We shall prove the following theorem.

Theorem A. *Let $\gamma > 0$ be fixed and, for any given $p > p_N$, let (λ_p, u_p) be a solution of Problem* (I) *such that $|u_p| = \gamma$. Then*

(a)
$$\lim_{p\to\infty} \lambda_p = \mu$$

uniformly with respect to γ. Here μ is the principal eigenvalue of the Laplacian in B_1 with Dirichlet boundary data. Let φ be the corresponding eigenfunction, normalized so that $\varphi(0) = 1$. Then for any fixed $x \neq 0$,

$$u_p(x) \to \begin{cases} \gamma\varphi(x) & \text{if } \gamma < 1, \\ \varphi(x) & \text{if } \gamma \geq 1, \end{cases} \quad \text{as } p \to \infty.$$

Moreover,

$$u_p \to \begin{cases} \gamma\varphi & \text{if } \gamma < 1, \\ \varphi & \text{if } \gamma \geq 1, \end{cases} \quad \text{as } p \to \infty \text{ in } H^1(B_1).$$

Remark. The pointwise convergence is uniform over $\bar{B}_1$ when $\gamma \leq 1$ and uniform over compact subsets of $\bar{B}_1 \setminus \{0\}$ when $\gamma > 1$.

If $\gamma < 1$, the result remains true in *general* bounded domains.

It follows from the theorem that as p tends to infinity, the *maximal* interval $I_p = (0, \hat{\lambda}_p)$ for which there exist *no solutions for any* $\lambda \in I_p$, expands to the full interval $(0, \mu)$. This implies that the lower bound for λ given in (1.1) is not the best possible one.

In addition, when $\gamma = u_p(0) > 1$, the solution u_p clearly must develop a *needle* near the origin and a neck at the level $u_p = 1$ (see Fig.1).

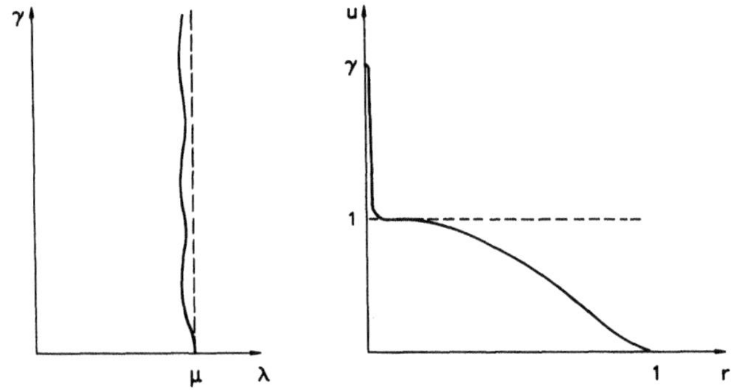

FIGURE 1: THE BIFURCATION DIAGRAM AND THE GRAPH OF u_p FOR LARGE p.

For the description of the shape of the needle, and of the neck at its base, we need to distinguish two regions: an *outer* region for the sides of the needle and its base, and an *inner* region for the tip.

The outer region can be described conveniently in terms of the transformed variables

$$\rho = \sqrt{p-1}\,r \quad \text{and} \quad w_p(\rho) = (p-1)\log u_p(r), \tag{1.2}$$

where $r = |x|$. We show that

$$w_p(\rho) \to \bar{w}(\rho) \quad \text{as } p \to \infty \tag{1.3}$$

uniformly on compact sets in $(0, \infty)$. Here $\bar{w}$ is the (singular) solution of the problem

$$\begin{cases} -\Delta w = \mu + e^w & \text{in } \mathbf{R}^N \setminus \{0\}, \\ w(0) = \infty. \end{cases} \tag{1.4}$$

Thus, for large values of p the function $\bar{w}$ provides a good description of the sides and the base of the needle. It is interesting to note that

$$\bar{w}(\rho) \sim \log\left(\frac{2N-4}{\rho^2}\right) \quad \text{as } \rho \to 0.$$

For the inner region we scale the radial variable once more and set

$$\eta = \rho\gamma^{(p-1)/2} \quad \text{and} \quad w_p(\rho) = W_p(\eta). \tag{1.5}$$

Note that

$$W_p(0) = (p-1)\log\gamma \to \infty \text{ as } p \to \infty \quad \text{and} \quad W_p'(0) = 0,$$

when $\gamma > 1$. We show in Section 5 that there exists a function $\bar{W}(\eta)$ such that

$$W_p(\eta) - W_p(0) \to \bar{W}(\eta) \quad \text{as } p \to \infty \tag{1.6}$$

uniformly on compact sets in $[0, \infty)$. Here $\bar{W}$ has the properties

$$\bar{W}(0) = 0, \quad \bar{W}'(0) = 0 \quad \text{and} \quad \bar{W}(\eta) \sim \log\left(\frac{2N-4}{\eta^2}\right) \quad \text{as } \eta \to \infty. \tag{1.7}$$

The outer and inner asymptotic behavior do not quite connect here, the outer behavior being valid for $\rho \geq \varepsilon$ ($\varepsilon > 0$ arbitrary but fixed) and the inner for $\eta \leq M$ (M arbitrary but fixed). For p large this leaves a gap, namely the interval

$$M\gamma^{-(p-1)/2} \leq \rho \leq \varepsilon,$$

in which the asymptotic behavior is more complicated and which we leave aside in the present paper.

The proof of our results is based on a shooting technique. Thus we consider the initial value problem

$$\text{(II)} \qquad \begin{cases} u'' + \dfrac{N-1}{r}u' + f(u) = 0, \quad 0 < r < 1, & (1.8) \\ u(0) = \gamma, \quad u'(0) = 0, & (1.9) \end{cases}$$

where

$$(1.10) \qquad f(u) = \lambda u + u^p.$$

For any fixed $\gamma > 0$, Problem (II) has a unique solution $u_p(r)$ in a right neighbourhood of the origin, which is decreasing and can be continued until it vanishes. As we observed before, there will be at least one value λ_p of λ such that

$$(1.11) \qquad u_p(r) > 0 \quad \text{for } 0 \leq r < 1 \quad \text{and} \quad u_p(1) = 0,$$

so that u_p corresponds to a radial solution of Problem (I).

In the analysis of u_p as p tends to infinity, it makes a great difference whether $\gamma \leq 1$ or $\gamma > 1$. If $\gamma \leq 1$, then $u_p(r) < 1$ for all $0 < r \leq 1$ and so the contribution of the term u^p in $f(u)$ tends to zero as $p \to \infty$. On the other hand if $\gamma > 1$, then $u_p(r)$ takes on values both greater and smaller than 1 and the situation is much more delicate.

In Section 2 we shall deal with the simpler case $\gamma < 1$; in Sections 3, 4 and 5 we study the asymptotics when $\gamma > 1$; and in Section 6 we prove the main result when $\gamma \geq 1$. Finally, in Section 7 we show that the convergence of λ_p to μ is uniform with respect to γ.

Since u_p is radially symmetric and decreasing in r we can define its inverse function $r_p(u)$ on $[0,\gamma]$. It is useful to note the following global upper bound

$$(1.12) \qquad r_p(u) < \sqrt{\mu}\, u^{-(p-1)/2} \quad \text{for all } 0 < u \leq \gamma.$$

The proof of (1.12) is really quite elementary. We write

$$g(s) = \lambda + s^{p-1}.$$

Then, for any $a \in (0,\gamma)$, we consider the solution z of the problem

$$\begin{cases} z'' + \dfrac{N-1}{r}z' + g(a)z = 0, \\ z(0) = 1, \quad z'(0) = 0. \end{cases}$$

It is obvious by direct scaling that $z(r)$ is an eigenfunction for the ball of radius R, where

$$R^2 = \frac{\mu}{g(a)} = \frac{\mu}{\lambda + a^{p-1}}. \tag{1.13}$$

Because $g(s)$ is increasing, a standard Sturm Comparison argument implies that

$$r_p(a) < R. \tag{1.14}$$

Since $\lambda > 0$, the desired bound now follows from (1.13) and (1.14).

2. The case $\gamma < 1$**.** Suppose (λ_p, u_p) is a solution of Problem (I). When we multiply the differential equation by u_p and integrate over B_1, we obtain

$$\int |\nabla u_p|^2 = \lambda_p \int u_p^2 + \int u_p^{p+1},$$

where the integrals are taken over B_1. Using Poincaré's inequality we can conclude that

$$\mu \int u_p^2 \leq \lambda_p \int u_p^2 + \int u_p^{p+1}$$

and hence, since $u \leq \gamma$ in B_1, that

$$\mu - \gamma^{p-1} < \lambda_p < \mu. \tag{2.1}$$

Thus, since $\gamma < 1$, it follows that

$$\lambda_p \to \mu \quad \text{as } p \to \infty,$$

the convergence being uniform with respect to $\gamma \in [0, 1-\delta]$ for any $\delta \in (0,1)$.

In addition, since the family $\{u_p\}$ is bounded in $W^{2,\infty}$, it follows from a standard compactness argument, and the fact that the functions u_p are radial, that there exists a sequence $\{u_{p_k}\}$, where $p_k \to \infty$ as $k \to \infty$, and a function $v \in C^1(B_1)$ such that

$$u_{p_k} \to v \quad \text{as} \quad k \to \infty \quad \text{in} \quad C^1(B_1). \tag{2.2}$$

For any $\zeta \in C_0^\infty(B_1)$ we have

$$\int (\nabla\zeta \cdot \nabla u_p - \lambda_p \zeta u_p - \zeta u_p^p) = 0 \tag{2.3}$$

which yields, when we pass to the limit along the subsequence $\{p_k\}$,

$$\int (\nabla\zeta \cdot \nabla v - \mu\zeta v) = 0.$$

Hence,

$$\Delta v + \mu v = 0 \quad \text{in } B_1$$

and so

$$v = \gamma\varphi.$$

Since φ is unique, the entire family $\{u_p\}$ converges to $\gamma\varphi$ as $p \to \infty$.

Thus we have shown the following result.

Theorem 2.1. *Let* $\gamma < 1$. *Then the family* $\{(\lambda_p, u_p)\}$ *has the properties*

(a) $$\lambda_p \to \mu \quad \text{as } p \to \infty,$$

(b) $$u_p \to \gamma\varphi \quad \text{as } p \to \infty,$$

uniformly on compact subsets of $B_1 \setminus \{0\}$, *and*

(c) $$u_p \to \gamma\varphi \quad \text{as} \quad p \to \infty \quad \text{in} \quad C^1(B_1).$$

Remark. Actually the same result holds when $\gamma \le 1$. For this we rely first on the main result of Section 7, that $\lambda_p \to \mu$ uniformly for $\gamma > 0$. Moreover, the result (2.2) then holds for all $\gamma \le 1$. We must finally show when $\gamma = 1$ that $v = \varphi$. By (2.3) and the fact that $u_p \to v$ we get

$$\left| \int_{B_1} (\nabla\zeta \cdot \nabla v - \mu\zeta v) \right| \le \lim_{p\to\infty} \int_{B_1} \zeta u_p^p.$$

But from Lemma 6.4, (6.14) and the fact that $r_p = 0$ when $\gamma = 1$, the right hand limit is zero. Hence $v = \varphi$.

3. The case $\gamma > 1$. When $\gamma > 1$ we reformulate the problem through a series of transformations. As a first step, we write

(3.1) $$\rho = \sqrt{p-1}\, r \quad \text{and} \quad w(\rho) = (p-1)\log u(r).$$

Then Problem (II) becomes

(III) $$\begin{cases} w'' + \dfrac{N-1}{\rho} w' + \dfrac{1}{p-1} w'^2 + \lambda + e^w = 0, & (3.2) \\[2ex] w(0) = (p-1)\log\gamma, \quad w'(0) = 0. & (3.3) \end{cases}$$

Following an idea of Tartar (see [MiPu]), we transform (3.2) to a second order equation which becomes autonomous when $\lambda = 0$. Set

(3.4) $$t = \log\rho, \quad \tilde{w}(t) = w(\rho), \quad y(t) = \tilde{w}'(t).$$

Then $\rho = e^t > 0$ and

(3.5) $$w'(\rho) = \frac{\tilde{w}'(t)}{\rho} = e^{-t} y(t),$$

(3.6) $$w''(\rho) = \{e^{-t} y(t)\}' e^{-t} = e^{-2t}\{y'(t) - y(t)\}.$$

and so we obtain from (3.2)

$$y' + (N-2)y + \frac{1}{p-1}y^2 + (\lambda + e^{\tilde{w}})e^{2t} = 0. \tag{3.7}$$

Differentiation yields, when we use (3.7) to eliminate $\tilde{w}$,

$$y'' + \{N-4-(1-\vartheta)y\}y' - y(y+2)\left(N-2+\frac{1}{2}\vartheta y\right) - \lambda y e^{2t} = 0, \tag{3.8}$$

where

$$\vartheta = \frac{2}{p-1}.$$

Because $u'(r) < 0$, we conclude from (3.1) that $w'(\rho) < 0$, and hence from (3.4) that $y(t) < 0$.

To determine the behavior of $y(t)$ as $t \to \infty$, we recall that

$$u(r) \to \gamma, \quad u'(r) \to 0, \quad u''(r) \to -\frac{1}{N}(\lambda\gamma + \gamma^p) \quad \text{as } r \to 0. \tag{3.9}$$

Now

$$y(t) = \tilde{w}'(t) = (p-1)\frac{ru'(r)}{u(r)};$$

hence if we use the limits from (3.9) we obtain

$$e^{-2t}y(t) = \frac{u'(r)}{r}\frac{1}{u(r)} \to -\frac{1}{N}(\lambda + \gamma^{p-1}) \quad \text{as } t \to -\infty. \tag{3.10}$$

When we introduce the new variable

$$z(t) = -(\lambda + e^{\tilde{w}(t)})e^{2t} = -(\lambda + u^{p-1})\cdot(p-1)r^2, \tag{3.11}$$

we obtain from (3.7) the following system in the (y,z)-plane:

$$\text{(IV)} \qquad \begin{cases} y' = -(N-2)y + z - \dfrac{1}{2}\vartheta y^2, \\ z' = (y+2)z + \lambda y e^{2t}. \end{cases}$$

If $\vartheta = 0$ and $\lambda = 0$, this system reduces to the autonomous system

$$\text{(IV}'\text{)} \qquad \begin{cases} y' = -(N-2)y + z, \\ z' = (y+2)z. \end{cases}$$

4. The reduced system. In this section we establish some properties of the reduced system (IV′). Its critical points are the origin (0,0) and the point $P = \big(-2, -2(N-2)\big)$ in the third quadrant. An elementary analysis shows that the origin is a *saddle point* and that P is either a *stable node* (if $N \geq 10$) or a *stable spiral point* (if $3 \leq N \leq 9$). Specifically, the eigenvalues λ_1 and λ_2 and the corresponding eigenvectors v_1 and v_2 for (0,0) are

$$\begin{aligned} \lambda_1 &= -(N-2), & v_1 &= (1,0), \\ \lambda_2 &= 2, & v_2 &= (1,N), \end{aligned}$$

whilst at $\big(-2, -2(N-2)\big)$ the eigenvalues are

$$\lambda_{1,2} = \frac{1}{2}\big\{-(N-2) \pm \sqrt{(N-2)(N-10)}\,\big\}.$$

Note, moreover, that the y-axis is invariant.

For our purposes we are particularly interested in the orbit Γ, which leaves the origin and enters the third quadrant.

Lemma 4.1. *The orbit Γ connects the two critical points through the third quadrant.*

Proof. Let

$$\Gamma = \big\{\big(\bar{y}(t), \bar{z}(t)\big) : t \in \mathbf{R}\big\}.$$

By construction, $\big(\bar{y}(t), \bar{z}(t)\big) \to (0,0)$ as $t \to -\infty$. We need to show that $\big(\bar{y}(t), \bar{z}(t)\big) \to P$ as $t \to +\infty$.

By inspecting the vector field we find that the set

$$\Sigma = \{(x,y) : -2N/(N-2) < y < z/N,\ -2N < z < 0\}$$

is invariant and that Γ enters Σ from the origin. By means of standard arguments we can show that Γ either enters P or goes around P an infinite number of times and converges either to a periodic orbit around P or to P itself. We shall show that there are no periodic orbits around P in Σ so that the assertion is proved.

It is convenient to shift the origin to P and write

$$y = -2 + \eta, \quad z = -2(N-2) + \zeta. \tag{4.1}$$

In terms of the new variables η and ζ, we obtain the system

$$\begin{cases} \eta' = -(N-2)\eta + \zeta, \\ \zeta' = \eta\{-2(N-2) + \zeta\}. \end{cases}$$

Suppose Γ_0 is a periodic orbit around P and suppose it passes through the point $(0,a)$ on the vertical ζ-axis, where $0 < a < 2(N-2)$. We shall compare the forward and the backward component of this orbit. Since $\eta' = a > 0$ at the initial point, the graph $\{(\eta,\zeta^+(\eta)) : \eta > 0\}$ of the forward component satisfies the equation

$$\frac{d\zeta}{d\eta} = \frac{\eta\{-2(N-2)+\zeta\}}{-(N-2)\eta+\zeta} = h^+(\eta,\zeta).$$

Going backward, η becomes negative. Thus, to compare this component with the forward one we reflect it with respect to the vertical axis, replacing η by $-\eta$. The thus reflected graph $\{(\eta,\zeta^-(\eta)) : \eta > 0\}$ of the backward component satisfies the equation

$$\frac{d\zeta}{d\eta} = \frac{\eta\{-2(N-2)+\zeta\}}{(N-2)\eta+\zeta} = h^-(\eta,\zeta).$$

Since

$$h^+(\eta,\zeta) < h^-(\eta,\zeta) \quad \text{for } \eta > 0 \quad \text{and} \quad (N-2)\eta < \zeta < 2(N-2),$$

it follows that

$$\zeta^+(\eta) < \zeta^-(\eta) \quad \text{when } \zeta^+(\eta) > (N-2)\eta.$$

Hence, if η^+ and η^- are the zeros of

$$\zeta^\pm(\eta) = \pm(N-2)\eta,$$

we can conclude that

$$\eta^+ < \eta^-.$$

Starting from a point on the negative vertical axis, we obtain a similar result. Thus the orbit does not return to the same point on the isocline $\ell = \{y' = 0\}$ and, therefore, cannot be periodic, a contradicton. □

The asymptotic stability of the point P will play a key role in the subsequent analysis. In the next lemma we supply a second order polynomial in the local coordinates η and ζ defined in (4.1), which is a local Lyapunov function.

Lemma 4.2. *The function*

$$E(\eta,\zeta) = \frac{3}{2}(N-2)\eta^2 + \zeta^2 - \eta\zeta$$

is a local positive definite Lyapunov function.

Proof. Writing $E = E(t)$ along orbits, we obtain upon differentiation that

$$\begin{aligned} E'(t) &= 3(N-2)\eta\eta' + 2\zeta\zeta' - \eta'\zeta - \eta\zeta' \\ &= -(N-2)(3N-8)\eta^2 - \zeta^2 - \eta^2\zeta + 2\eta\zeta^2. \end{aligned} \tag{4.2}$$

Hence, for $\eta^2 + \zeta^2$ sufficiently small, $E'(t) < 0$, since $N \geq 3$.

A simple computation shows that

$$E(t) \geq \frac{1}{2}(\eta^2 + \zeta^2) \quad \text{for all } (\eta, \zeta) \in \mathbf{R}^2. \qquad \square$$

5. Asymptotic properties of (y_p, z_p). After the preparatory work of Sections 3 and 4, we return to the analysis of the asymptotic behavior of the eigenvalue λ_p, and the corresponding solution u_p, or the corresponding orbit (y_p, z_p), as $p \to \infty$. In particular, we shall describe how, when $\gamma > 1$, the orbit (y_p, z_p) converges to the connecting orbit $(\bar{y}, \bar{z})$ of the reduced autonomous system (IV′) defined in Section 3.

Let us define

$$s = t + \frac{p-1}{2}\log\gamma, \quad \tilde{y}(s) = y(t), \quad \tilde{z}(s) = z(t).$$

For simplicity in printing we omit the tilde and frequently drop the subscript p. We begin with some uniform estimates for (y_p, z_p).

Lemma 5.1. *There exist constants s_0 and p_0 such that*

$$\frac{1}{2N} < -y_p(s)e^{-2s} < \frac{3}{2N},$$

(5.1) *for all $s < s_0$ and $p > p_0$.*

$$\frac{3}{4} < -z_p(s)e^{-2s} < \frac{5}{4},$$

Proof. Define the functions

$$f(s) = -z(s)e^{-2s} \quad \text{and} \quad g(s) = \{-Ny(s) + z(s)\}e^{-2s}. \tag{5.2}$$

We show that constants s_0 and p_0 can be found so that

$$\frac{3}{4} < f(s) < \frac{5}{4} \quad \text{and} \quad |g(s)| < \frac{1}{4} \quad \text{for all } s < s_0 \text{ and } p > p_0. \tag{5.3}$$

The desired inequalities (5.1) then follow at once.

For fixed p, we find from (3.10), (3.11) and (3.1) that

$$\left.\begin{aligned} y(t)e^{-2t} &\to -\frac{1}{N}(\lambda+\gamma^{p-1}),\\ z(t)e^{-2t} &\to -(\lambda+\gamma^{p-1}), \end{aligned}\right\} \quad \text{as } t\to-\infty,$$

and hence

$$\left.\begin{aligned} y(s)e^{-2s} &\to -\frac{1}{N}(1+\lambda\gamma^{1-p}),\\ z(s)e^{-2s} &\to -(1+\lambda\gamma^{1-p}), \end{aligned}\right\} \quad \text{as } s\to-\infty.$$

Therefore,

$$f(s)\to 1+\lambda\gamma^{1-p} \quad\text{and}\quad g(s)\to 0 \qquad \text{as } s\to-\infty, \tag{5.5}$$

so that for fixed p, (5.1) is satisfied for sufficiently large negative values of s.

Suppose (5.1) first fails to hold when $s=s_p$. Then, since

$$\begin{aligned} f' &= yze^{-2s}-\lambda y\gamma^{1-p},\\ g'+Ng &= \frac{N}{2}\vartheta y^2e^{-2s}+yze^{-2s}+\lambda y\gamma^{1-p}, \end{aligned} \tag{5.6}$$

we have as long as $s<s_p$,

$$|f'(s)|<c_0e^{2s} \quad\text{and}\quad |g'(s)+Ng(s)|<c_0e^{2s},$$

where c_0 is a positive constant which does not depend on p. Hence, when we integrate over $(-\infty,s)$ and keep $s<s_p$, we obtain in view of (5.5),

$$|f(s)-(1+\lambda\gamma^{1-p})|<\frac{c_0}{2}e^{2s} \quad\text{and}\quad |g(s)|<\frac{c_0}{N+2}e^{2s}. \tag{5.7}$$

Now choose

$$s_0=\frac{1}{2}\log\left(\frac{1}{4c_0}\right).$$

Then, by (5.7),

$$|f(s)-1|<\frac{1}{8}+\lambda\gamma^{1-p} \quad\text{and}\quad |g(s)|<\frac{1}{4(N+2)} \quad \text{if } s<s_0.$$

Since (5.1) fails if

$$\text{either}\quad |f(s_p)-1|=\frac{1}{4} \quad\text{or}\quad |g(s_p)|=\frac{1}{4},$$

it follows that $s_p>s_0$ when $\lambda\gamma^{1-p}<\frac{1}{8}$, which is the case when $p>p_0=1+(\log 8\lambda/\log\gamma)$. □

We are now equipped to use a compactness argument to establish the limiting behavior of (y_p, z_p) as $p \to \infty$. We denote by $(\bar{y}, \bar{z})$ the unique solution of the reduced system (IV′) which connects the two critical points (0,0) and $(-2, -2(N-2))$ and is so normalised that

$$\bar{y}(s)e^{-2s} \to -\frac{1}{N} \quad \text{and} \quad \bar{z}(s)e^{-2s} \to -1 \qquad \text{as } s \to -\infty, \tag{5.8}$$

where we recall that the eigenvalue and eigenvector in question at (0,0) are $\lambda_2 = 2$ and $v_2 = (1, N)$.

Lemma 5.2. *We have*

$$(y_p, z_p) \to (\bar{y}, \bar{z}) \quad \text{as } p \to \infty$$

uniformly on $(-\infty, S]$, *where* S *may be any real number.*

Proof. By Lemma 5.1, the family $\{(y_p, z_p) : s \le s_0,\ p > p_0\}$ is uniformly bounded. Moreover,

$$\begin{cases} y' = -(N-2)y + z - \dfrac{y^2}{p-1}, \\ z' = (y+2)z + \lambda y e^{2s-(p-1)\log\gamma}, \end{cases} \tag{5.9}$$

so that (y_p', z_p') are uniformly bounded for $s \le s_0$, $p > p_0$. Hence, by the Arzela-Ascoli theorem, we can extract a sequence, still denoted by $\{(y_p, z_p)\}$, which converges to a pair (y_0, z_0) as $p \to \infty$ uniformly on compact sets in $(-\infty, s_0]$. By Lemma 5.1, the convergence is seen to be even uniform over the entire interval $(-\infty, S]$. Passing to the limit in the system (5.9), we find that (y_0, z_0) solves the reduced system (IV′) for $s \le s_0$.

To prove that (y_0, z_0) satisfies the normalization (5.8) we fix $\varepsilon > 0$ and choose $s_\varepsilon \le s_0$ so that $\frac{1}{2}c_0 e^{-2s_\varepsilon} \le \varepsilon$. Then, by (5.7),

$$|-z_p(s)e^{-2s} - (1 + \lambda\gamma^{1-p})| < \varepsilon \quad \text{for all } s < s_\varepsilon \text{ and } p > p_0.$$

If we now let p tend to infinity, we find that

$$|-z_0(s)e^{-2s} - 1| < \varepsilon \quad \text{for all } s < s_\varepsilon,$$

so that

$$z_0(s)e^{-2s} \to -1 \quad \text{as } s \to -\infty. \tag{5.10}$$

Proceeding similarly with the estimate for g in (5.7), we find that

$$\{-Ny_0(s)+z_0(s)\}e^{-2s} \to 0 \quad \text{as } s \to -\infty,$$

so that in view of (5.10),

$$y_0(s)e^{-2s} \to -\frac{1}{N} \quad \text{as } s \to -\infty. \tag{5.11}$$

Thus, using (5.10) and (5.11), we conclude that $(y_0, z_0) = (\bar{y}, \bar{z})$ for $s \leq s_0$.

By continuous dependence on the initial data and the differential equation, the convergence can be extended to compact sets in $(-\infty, S]$ for any $S \in \mathbf{R}$.

Finally, since $(\bar{y}, \bar{z})$ is unique, the entire family $\{(y_p, z_p)\}$ converges to $(\bar{y}, \bar{z})$ as $p \to \infty$. □

Remark. From Lemmas 5.1 and 5.2 we can deduce the stronger result

$$e^{-\nu s}\big(y_p(s), z_p(s)\big) \to e^{-\nu s}\big(\bar{y}(s), \bar{z}(s)\big) \quad \text{as } s \to -\infty \tag{5.12}$$

uniformly on $(-\infty, S]$ as $p \to \infty$ when $\nu < 2$.

To translate this into a result for the function $W_p(\eta)$ introduced in (1.5) to describe the shape of the needle near the centre, we note that

$$W_p'(\eta) = w_p'(\rho)\gamma^{-(p-1)/2} = e^{-t}y_p(t)\gamma^{-(p-1)/2} = e^{-s}\tilde{y}_p(s).$$

As before, we omit the tilde for the rest of the section. Define

$$\bar{W}'(\eta) = e^{-s}\bar{y}(s) \quad \text{and} \quad \bar{W}(0) = 0. \tag{5.13}$$

Then, by (5.8) and the relation $\eta = e^s$,

$$\bar{W}'(\eta) \to 0 \quad \text{as } \eta \to 0.$$

Moreover, for any bounded $\eta > 0$, (5.12) with $\nu = 1$ implies that

$$W_p(\eta) - W_p(0) = \int_0^\eta W_p'(\zeta)\,d\zeta \to \bar{W}(\eta) \quad \text{as } p \to \infty. \tag{5.14}$$

To determine the asymptotic behaviour of $\bar{W}(\eta)$ as $\eta \to \infty$, we use the behaviour of z_p as $p \to \infty$. We can write (3.11) as

$$\exp\{W_p(\eta) - (p-1)\log\gamma\} - \lambda\gamma^{-(p-1)} = -z_p(\log\eta)\eta^{-2}. \tag{5.15}$$

Hence, when we let $p \to \infty$ and keep η fixed, we obtain in the limit, using (5.14),

$$e^{\bar{W}(\eta)} = -\bar{z}(\log \eta)\eta^{-2}.$$

If we now let $\eta \to \infty$, we deduce from Lemma 4.1 that

$$\bar{W}(\eta) \sim \log\left(\frac{2N-4}{\eta^2}\right) \quad \text{as } \eta \to \infty.$$

In the remainder of this section we focus on the asymptotics for the outer region of the needle, away from the tip, and thus return to the variable t. It will be convenient to shift the origin now to the critical point $\big(-2,-2(N-2)\big)$, as we did in Section 4. Thus, we again set

$$y = -2+\eta, \quad z = -2(N-2)+\zeta,$$

and we obtain for the pair (η,ζ):

$$\text{(V)} \qquad \begin{cases} \eta' = -(N-2)\eta + \zeta - \dfrac{(2-\eta)^2}{p-1}, \\ \zeta' = \eta\{-2(N-2)+\zeta\} - \lambda(2-\eta)e^{2t}. \end{cases}$$

Let

$$\bar{y}(s) = -2+\bar{\eta}(s), \quad \bar{z}(s) = -2(N-2)+\bar{\zeta}(s).$$

Then, as we saw in Section 4,

$$(5.16) \qquad \big(\bar{\eta}(s),\bar{\zeta}(s)\big) \to (0,0) \quad \text{as } s \to \infty.$$

For any $c > 0$, define the set

$$\Sigma_c = \{(\eta,\zeta) : E(\eta,\zeta) < c^2\},$$

where the function E was given in Lemma 4.2. Since E is homogeneous of second order and positive, there exist positive numbers σ_- and σ_+ such that

$$B_{\sigma_-} \subset \Sigma_1 \subset B_{\sigma_+},$$

where B_d denotes the ball with radius d centered at the origin.

Now fix $\varepsilon \in (0,\sigma_-^2/6\sigma_+^3)$. Then, by (5.16), there exists a time s_1 such that

$$\big(\bar{\eta}(s_1),\bar{\zeta}(s_1)\big) \in B_{\varepsilon\sigma_-/2}$$

and hence, by Lemma 5.2, there exists a parameter value p_1 such that

$$\big(\eta(s_1),\zeta(s_1)\big) \in B_{\varepsilon\sigma_-} \quad \text{if } p > p_1.$$

We showed in Section 4 that the function E is a local Lyapunov function for the autonomous system. If (η,ζ) is an orbit for the perturbed system (V), we find that

$$\frac{d}{dt}E(\eta(t),\zeta(t)) = -(N-2)(3N-8)\eta^2 - \zeta^2 - \eta^2\zeta + 2\eta\zeta^2 + R_1 + R_2,$$

where

$$R_1 = \frac{1}{p-1}\{3(N-2)\eta - \zeta\} \quad \text{and} \quad R_2 = -\lambda(2\zeta-\eta)(2-\eta)e^{2t}.$$

Note that there exists a constant $C > 0$ such that

$$|R_1| \le \frac{Cd}{p-1} \quad \text{and} \quad |R_1| \le Cde^{2t},$$

where we have set $d = \sqrt{\eta^2+\zeta^2}$. Therefore,

$$\frac{d}{dt}E\big(\eta(t),\zeta(t)\big) \le -d^2 + 3d^3 + Cd\left(\frac{1}{p-1} + e^{2t}\right).$$

Thus, if we choose $p_\varepsilon > p_1$ and T_ε so large that

$$\frac{1}{p_\varepsilon - 1} < \frac{\varepsilon\sigma_-}{4C} \quad \text{and} \quad e^{-2T_\varepsilon} < \frac{\varepsilon\sigma_-}{4C},$$

then, if $p > p_\varepsilon$ and $t < -T_\varepsilon$,

$$\frac{d}{dt}E\big(\eta(t),\zeta(t)\big) \le -\frac{1}{2}(\varepsilon\sigma_-)^2 + 3(\varepsilon\sigma_+)^3 < 0 \quad \text{if } \varepsilon\sigma_- \le d \le \varepsilon\sigma_+,$$

because $\varepsilon < \sigma_-^2/6\sigma_+^3$.

Now, by the definition of s_1 the orbit is inside the ball $B_{\varepsilon\sigma_-}$ when $s = s_1$, so that $E \le \varepsilon^2$ at that instant. It follows that $E \le \varepsilon^2$ on the further orbit, at least until it leaves $B_{\varepsilon\sigma_+}$. But this cannot happen, since at the point of leaving we should have both $E < \varepsilon^2$ and $E \ge \varepsilon^2$. This means that

$$\big(\eta(t),\zeta(t)\big) \in \Sigma_\varepsilon \quad \text{if } s_1 - \tfrac{1}{2}(p-1)\log\gamma < t < -T_\varepsilon \text{ and } p > p_\varepsilon,$$

or, in terms of the old variables, since $\Sigma_\varepsilon \subset B_{\varepsilon\sigma_+}$,

$$\big|\big(y(t),z(t)\big) - \big(-2,-2(N-2)\big)\big| \le \varepsilon\sigma_+$$

if

$$s_1 - \frac{1}{2}(p-1)\log\gamma < t < -T_\varepsilon \quad \text{and} \quad p > p_\varepsilon.$$

In turn, this implies that

$$\limsup_{p\to\infty}\big|\big(y(t),z(t)\big) - \big(-2,-2(N-2)\big)\big| \le \varepsilon\sigma_+ \quad \text{if } t < -T_\varepsilon.$$

Since ε can be chosen arbitrarily small, we have proved:

Lemma 5.3. *We have*

$$\lim_{t\to-\infty}\limsup_{p\to\infty}\big|\big(y_p(t),z_p(t)\big)-\big(-2,-2(N-2)\big)\big|=0. \tag{5.17}$$

We conclude this section with the observation that the critical point $\big(-2,-(N-2)\big)$ in the (y,z)-plane corresponds to the singular function

$$w^*(\rho)=\log\left(\frac{2N-4}{\rho^2}\right)\quad\text{or}\quad u^*(r)=\left(\frac{2N-4}{p-1}\right)^{1/(p-1)}r^{-2/(p-1)}.$$

6. Proof of the main results. With Lemmas 5.2 and 5.3 in hand, we can now establish the limiting profile of the functions $w_p(\rho)$ on compact sets in $\mathbf{R}^+$ and thus, in turn, obtain estimates for the radius $r_p(1)$, where $u_p=1$, and for the slope of u_p at $r_p(1)$. This finally enables us to determine the limit in the outer region $(r_p(1),1)$. As we shall see, the limiting profile near the origin will be a solution of the singular problem

$$\text{(III}'\text{)}\qquad\begin{cases} w''+\dfrac{N-1}{\rho}w'+\lambda+e^w=0,\\ w(\rho)\to\infty\quad\text{as }\rho\to0.\end{cases}$$

Since λ_p is restricted to a compact interval which does not contain the origin, we may assume that λ_p converges to a positive number $\bar\lambda$ by taking an appropriate subsequence.

Lemma 6.1. *There exists a singular solution $\bar w_{\bar\lambda}$ of Problem* (III$'$) *such that, along a subsequence of values $p\to\infty$, we have*

$$w_p(\rho)\to\bar w_{\bar\lambda}(\rho)$$

uniformly in C^1 on compact intervals in $\mathbf{R}^+$.

Proof. Let us translate our results about $\big(y_p(t),z_p(t)\big)$ to results about $w_p(\rho)$. As usual, we shall often omit the subscript p.

From Lemma 5.3 there exist constants P and T sufficiently large and a constant C such that

$$\big|\big(y(t),z(t)\big)\big|\le C\qquad\text{for all } p>P \text{ and } t<-T.$$

This means in terms of ρ and w that

$$\rho|w'(\rho)|\le C, \tag{6.1}$$

$$\rho^2e^{w(\rho)}\le C, \tag{6.2}$$

for all $p>P$ and $\rho<\rho_1$,

where $\rho_1 = e^{-T}$. By the differential equation (3.2), we have

$$\rho^2 w'' = -\rho^2 e^w - (N-1)\rho w' - \frac{1}{p-1}(\rho w')^2 - \rho^2 \lambda, \tag{6.3}$$

so that (6.1) and (6.2) imply that

$$\rho^2 |w''(\rho)| \le C \quad \text{for all } p > p_0 \text{ and } \rho < \rho_1. \tag{6.4}$$

□

To obtain a bound for $|w(\rho)|$ we still need to supplement (6.2) with a bound from below. As a first step to such a bound we prove:

Lemma 6.2. *There exist positive constants p_1 and $\rho_2 < \rho_1$ such that*

$$\rho_2^2 e^{w_p(\rho_2)} \ge N-2 \quad \text{for all } p > p_1.$$

Proof. Translating (5.17) to ρ and w, we obtain

$$\lim_{\rho\to 0} \limsup_{p\to\infty} |\rho w'(\rho) + 2| = 0 \tag{6.5}$$

and

$$\lim_{\rho\to 0} \limsup_{p\to\infty} |\rho^2 e^{w(\rho)} - 2(N-2)| = 0. \tag{6.6}$$

Therefore, there exists a radius $\rho_3 > 0$ such that

$$\limsup_{p\to\infty} |\rho^2 e^{w(\rho)} - 2(N-2)| \le N-2 \quad \text{if } \rho \le \rho_3,$$

so that

$$\liminf_{p\to\infty} \rho^2 e^{w(\rho)} \ge \frac{3}{2}(N-2) \quad \text{if } \rho \le \rho_3,$$

from which the assertion follows. □

We now return to the proof of Lemma 6.1. From (6.1), (6.2) and Lemma 6.2, we see by integration over (ρ, ρ_2) that there exist constants C, p^* and ρ^* such that if $p > p^*$ and $\rho < \rho^*$,

$$|w_p(\rho)| \le C\{1 + \log(1/\rho)\}, \quad \rho|w_p'(\rho)| \le C, \quad \rho^2|w_p''(\rho)| \le C. \tag{6.7}$$

Thus, by a standard compactness argument, we conclude that for a subsequence of values $p \to \infty$,

$$w_p(\rho) \to \bar{w}_{\bar{\lambda}}(\rho)$$

uniformly in C^1 on compact intervals of $(0,\rho^*)$. By passing to the limit in equation (3.2), we find that $\bar{w}_{\bar{\lambda}}$ is a solution of

$$w'' + \frac{N-1}{\rho}w' + \lambda + e^w = 0 \quad \text{on } (0,\rho^*),$$

and from (6.6) we conclude that

$$\rho^2 e^{\bar{w}_{\bar{\lambda}}(\rho)} \to 2(N-2) \quad \text{as } \rho \to 0.$$

Thus, $\bar{w}_{\bar{\lambda}}$ is a solution of Problem (III′) with $\lambda = \bar{\lambda}$.

From the fact that

$$0 < e^{w_p(\rho)} < e^{w_p(\rho^*)} \quad \text{for } \rho \geq \rho^*$$

follows that we can continue w_p for all $\rho > 0$, and by standard continuity arguments we conclude that

$$w_p(\rho) \to \bar{w}_{\bar{\lambda}}(\rho) \quad \text{as } p \to \infty$$

uniformly in C^1 on compact intervals in $\mathbf{R}^+$. This completes the proof of Lemma 6.1. □

In our next step we study the asymptotic behavior of $r_p = r_p(1)$ and $u_p'(r_p)$ as $p \to \infty$.

Lemma 6.3. *There exist constants $\rho_0 > 0$ and $c_0 > 0$ such that along a subsequence of values of p*

$$\text{(a)} \quad r_p \sim \frac{\rho_0}{\sqrt{p-1}}, \qquad \text{(b)} \quad u_p'(r_p) \sim -\frac{c_0}{\sqrt{p-1}}, \qquad \text{as } p \to \infty.$$

Proof. We first show that any solution $w(\rho)$ of Problem (III′) has a zero, which, since w is decreasing, must necessarily be the only one. Suppose to the contrary that $w(\rho) > 0$ for all $\rho > 0$. Then $e^w > 1$ and so

$$w'' + \frac{N-1}{\rho}w' < -1.$$

If we multiply this inequality by ρ^{N-1} and integrate over (δ,ρ) for some $\delta > 0$, we obtain

$$w'(\rho) < -\frac{\rho}{N}\left\{1 - \left(\frac{\delta}{\rho}\right)^N\right\},$$

which yields upon integration over (δ,ρ),

$$w(\rho) < C(N,\delta) - \frac{\rho^2}{2N},$$

where $C(N,\delta)$ is some constant. It follows that $w(\rho)$ must eventually become negative, contrary to our assumption.

Thus every solution of Problem (III′) has a unique zero. We denote the zero of $\bar{w}_{\bar{\lambda}}$ by ρ_0. From Lemma 6.1 we conclude that $w_p(\rho)$ also has a zero ρ_p for p large enough and that

$$\rho_p \to \rho_0 \quad \text{and} \quad w_p'(\rho_p) \to \bar{w}_{\bar{\lambda}}'(\rho_0) \quad \text{as } p \to \infty. \tag{6.8}$$

This yields, for $r_p = r_p(1)$ and $u_p'(r_p)$,

$$\sqrt{p-1}\, r_p \to \rho_0 \quad \text{as } p \to \infty \tag{6.9}$$

and

$$\sqrt{p-1}\, u_p'(r_p) = w_p'(\rho_p) \to \bar{w}_{\bar{\lambda}}'(\rho_0) \quad \text{as } p \to \infty. \tag{6.10}$$

□

It now remains to study the behavior of u_p as $p \to \infty$ on the interval $(r_p, 1)$. We first establish an upper and a lower bound.

Lemma 6.4. *Let $\gamma \geq 1$. Then, for p sufficiently large,*

$$\left.\begin{aligned} &\alpha(r) - Ap^{-1} < u_p(r) < \beta(r) + Bp^{-1}, \\ &\alpha'(r) - Cp^{-1} < u_p'(r) < 0, \end{aligned}\right\} \quad \text{for } r_p < r < 1,$$

where A, B and C are positive constants and

$$\alpha(r) = 1 - \frac{\mu+1}{2N} r^2,$$

$$\beta(r) = 1 - \frac{\lambda^*}{N-2} \int_0^r s\alpha(s) \left\{1 - \left(\frac{s}{r}\right)^{N-2}\right\} ds,$$

where λ^* is the positive lower bound for λ.

Proof. Suppose that $\gamma > 1$. We write (1.7) as

$$(r^{N-1}u')' = -r^{N-1}(\lambda u + u^p). \tag{6.11}$$

Since $u' < 0$, we have $u < 1$ on $(r_p, 1)$. Remembering that $\lambda < \mu$, we find that

$$(r^{N-1}u')' > -(\mu+1)r^{N-1}.$$

When we integrate this inequality over (r_p, r) and then divide by r^{N-1}, we obtain

$$u'(r) > \left(\frac{r_p}{r}\right)^{N-1} u'(r_p) - \frac{\mu+1}{N}r \quad \text{for } r_p < r < 1. \tag{6.12}$$

One more integration over (r_p, r) now yields

$$u(r) > 1 - \frac{\mu+1}{2N}r^2 + \frac{r_p u'(r_p)}{N-2} \quad \text{for } r_p < r < 1. \tag{6.13}$$

Since by Lemma 6.3 the last term in (6.13) is $O(p^{-1})$ as $p \to \infty$, and $u > 0$, the lower bound follows.

To obtain an upper bound, we use the lower bound which we write as

$$u(r) > \alpha(r) - \delta(p),$$

where $\delta(p) = O(p^{-1})$ as $p \to \infty$. From (6.11) we now obtain

$$(r^{N-1}u')' = -\lambda^* r^{N-1}\{\alpha(r) - \delta(p)\}$$

which yields, after an elementary computation,

$$\begin{aligned} u(r) &< 1 - \frac{\lambda^*}{N-2}\int_{r_p}^{r} s\{\alpha(s) - \delta(p)\}\left\{1 - \left(\frac{s}{r}\right)^{N-2}\right\} ds \\ &< 1 - \frac{\lambda^*}{N-2}\int_{0}^{r} s\alpha(s)\left\{1 - \left(\frac{s}{r}\right)^{N-2}\right\} ds + O(p^{-1}) \quad \text{as } p \to \infty, \end{aligned}$$

where the estimates from Lemma 6.3 have been used.

The lower bound for u_p' follows from (6.12) and Lemma 6.3 and the upper bound is known.

If $\gamma = 1$, then $r_p = r_p(1) = 0$ and $u_p'(r_p) = 0$. Thus, in this case $u(r) > \alpha(r)$ for $0 < r < 1$ and we may set $\delta(r)$ equal to zero. The desired upper bound is now immediate. This completes the proof. □

From the expression for $\beta(r)$ we see that

$$\beta(r) = 1 - \frac{\lambda^*}{2N}r^2 + O(r^4) \quad \text{as } r \to 0. \tag{6.14}$$

We are now in a position to prove the main result.

Theorem 6.5. *Let $\gamma \geq 1$. Then the family $\{(\lambda_p, u_p)\}$ has the properties*

$$\text{(a)} \qquad \lambda_p \to \mu, \qquad\qquad \textit{as } p \to \infty$$

$$\text{(b)} \qquad u_p \to \varphi,$$

uniformly on compact subsets of $B_1 \setminus \{0\}$. Also,

$$\text{(c)} \qquad u_p \to \varphi \quad \text{as } p \to \infty \text{ in } H^1(B_1).$$

Proof. Fix $r_0 \in (0,1)$. Then, by Lemmas 6.3 and 6.4, there exist numbers p_0 and δ_0 such that

$$\text{(6.15)} \qquad r_p < r_0 \quad \text{and} \quad u_p(r) < 1 - \delta_0 \quad \text{on } [r_0, 1] \quad \text{if} \quad p > p_0.$$

By (6.12), u_p' is uniformly bounded on $[r_0, 1]$ and by (1.3) and (6.15), u_p'' and u_p''' are uniformly bounded on $[r_0, 1]$. Hence, by compactness we can extract a subsequence, again denoted by $\{u_p\}$, and a function v such that

$$u_p \to v \quad \text{as } p \to \infty \text{ in } C^2,$$

uniformly on $[r_0, 1]$. By letting r_0 tend to zero, we thus find a sequence for which (6.16) holds uniformly on every compact set in $(0, 1]$.

Taking the limit in the differential equation (1.3), we find that

$$\text{(6.17)} \qquad v'' + \frac{N-1}{r} v' + \bar{\lambda} v = 0, \qquad \text{for } 0 < r < 1,$$

and taking the limit in Lemma 6.4, we obtain

$$\text{(6.18)} \qquad \alpha(r) \leq v(r) \leq \beta(r) \quad \text{and} \quad \alpha'(r) \leq v'(r) \leq 0 \qquad \text{for } 0 < r < 1,$$

so that, in view of (6.14),

$$\text{(6.19)} \qquad v(r) \to 1 \quad \text{and} \quad v'(r) \to 0 \quad \text{as } r \to 0.$$

Finally, since $u_p(1) = 0$ for all p, we have

$$\text{(6.20)} \qquad v(1) = 0.$$

The properties (6.17)–(6.20) of v imply that

$$\bar{\lambda} = \mu \quad \text{and} \quad v = \varphi.$$

Since μ and φ are unique, the entire family $\{(\lambda_p, u_p)\}$ converges as required to (μ, φ) as $p \to \infty$.

To prove (c) we need to show that

$$\int_{B_1} |\nabla(u_p - \varphi)|^2 \to 0 \quad \text{as } p \to \infty.$$

We have, omitting the subscript p for convenience,

$$-\Delta(u - \varphi) = \lambda u - \mu\varphi + u^p$$

and so

$$\begin{aligned}\int_{B_1} |\nabla(u_p - \varphi)|^2 &= \int_{B_1} \{\lambda(u-\varphi) + (\lambda - \mu)\varphi + u^p\}(u - \varphi) \\ &< \lambda \int_{B_1} (u-\varphi)^2 + (\lambda - \mu) \int_{B_1} \varphi(u-\varphi) + \int_{B_1} u^{p+1} \\ &= J_1 + J_2 + J_3.\end{aligned}$$

Plainly, as p tends to infinity, $J_1 \to 0$ by (b) and $J_2 \to 0$ by (a) and (b). Thus, it remains to show that

$$J_3 = \int_{B_1} u^{p+1} \to 0 \quad \text{as } p \to \infty. \tag{6.21}$$

Fix $\varepsilon > 0$ and write for some $\delta > 0$ to be selected,

$$\int_{B_1} u^{p+1} = \int_{B_\delta} u^{p+1} + \int_{B_1 \setminus B_\delta} u^{p+1}. \tag{6.22}$$

To estimate the first term on the right in (6.22), we recall the upper bound (6.2), which reads in terms of r and u,

$$u^{p-1}(r) \le \frac{\mu}{r^2} \quad \text{for } 0 < r \le 1.$$

Hence

$$\int_{B_\delta} u^{p+1} \le \mu^{(p+1)/(p-1)} \omega_N \int_0^\delta r^{-2\{(p+1)/(p-1)\}+N-1}\, dr,$$

where ω_N denotes the surface area of the unit ball. Note that

$$N - 2\frac{p+1}{p-1} = \frac{N-2}{p-1}\left(p - \frac{N+2}{N-2}\right) > 0,$$

and so

$$\int_{B_\delta} u^{p+1} \le \mu^{(p+1)/(p-1)} \omega_N \delta^{N-2-\{4/(p-1)\}}.$$

Therefore, there exists a $\delta_\varepsilon > 0$ such that

$$\int_{B_{\delta_\varepsilon}} u^{p+1} < \frac{1}{2}\varepsilon \quad \text{if} \quad p > \frac{N+2}{N-2} + \varepsilon = p'_\varepsilon.$$

By the upper bound of Lemma 6.4, there exists a $p''_\varepsilon > (N+2)/(N-2)$ such that

$$\int_{B_1 \setminus B_{\delta_\varepsilon}} u^{p+1} < \frac{1}{2}\varepsilon \quad \text{if} \quad p > p''_\varepsilon.$$

Hence

$$\int_{B_1} u^{p+1} < \varepsilon \quad \text{if} \quad p > p_\varepsilon = \max\{p'_\varepsilon, p''_\varepsilon\},$$

which proves (6.21).

This completes the proof of Theorem 6.5. □

7. Uniform convergence. In Theorems 2.1 and 6.5 we have proved that for each fixed $\gamma > 0$, $\lambda_p \to \mu$ as $p \to \infty$. In this section we shall show that this convergence is uniform with respect to $\gamma > 0$.

Theorem 7.1. *Let* $(\lambda_p(\gamma), u_p(\gamma))$ *be a solution of Problem* (I) *such that* $|u_p| = \gamma$. *Then*

$$\sup_{\gamma > 0} |\lambda_p(\gamma) - \mu| \to 0 \quad \text{as } p \to \infty.$$

Proof. Suppose to the contrary that there exist infinite sequences $\{p_n\}$ and $\{\gamma_n\}$ such that

$$\lim_{n\to\infty} |\lambda_{p_n}(\gamma_n) - \mu| > 0.$$

Without loss of generality, it is enough to distinguish three cases for contradiction

$$\begin{aligned} &\text{(a)} \quad (p_n - 1)\log \gamma_n \to +\infty, && \gamma_n > 1; \\ &\text{(b)} \quad (p_n - 1)\log \gamma_n \to -\infty, && \gamma_n < 1; \\ &\text{(c)} \quad (p_n - 1)\log \gamma_n \to c \in \mathbf{R}, && \gamma_n \to 1. \end{aligned}$$

In case (a) we can proceed as in Sections 5 and 6 to show that $\lambda_{p_n}(\gamma_n) \to \mu$, a contradiction. In case (b) we proceed as in Section 2 to establish a contradiction when the sequence $\{\gamma_n\}$ is bounded away from 1. Otherwise, we can suppose that $\gamma_n \to 1$ and proceed as in case (c), but with $c = -\infty$. Thus, case (c) remains.

To obtain a contradiction in case (c) we need an upper and a lower bound for u_p as in Lemma 6.4. □

Lemma 7.2. *Suppose*

$$(p-1)\log\gamma \to c \in (-\infty,\infty) \quad \text{as } p \to \infty.$$

Then

$$\tilde{\alpha}(r) + o(1) < u_p(r) < \tilde{\beta}(r) + o(1) \quad \text{as } p \to \infty,$$

where

$$\tilde{\alpha}(r) = 1 - \frac{\mu + e^c}{2N} r^2$$

and $\tilde{\beta}$ is defined in terms of $\tilde{\alpha}$ as β was defined in terms of α in Lemma 6.4.

Proof. From (6.11) we conclude that

$$(r^{N-1}u')' > -\gamma(\mu + \gamma^{p-1}) \quad \text{for } 0 < r < 1,$$

which yields upon integration over $(0,r)$ and division by r^{N-1},

$$u'(r) > -\frac{\gamma}{N}(\mu + \gamma^{p-1})r.$$

One more integration over $(0,r)$ gives

$$\begin{aligned} u(r) &> \gamma - \frac{\gamma}{2N}(\mu + \gamma^{p-1})r^2 \\ &> 1 - \frac{\mu + e^c}{2N} r^2 + o(1) \\ &= \tilde{\alpha}(r) + o(1) \quad \text{as } p \to \infty. \end{aligned}$$

As in the proof of Lemma 6.4, we derive from this lower bound the desired upper bound.

Proceeding now as in the proof of Theorem 6.5, we conclude that $\lambda_{p_n}(\gamma_n) \to \mu$, a contradiction.

This completes the proof of Theorem 7.1. □

REFERENCES

[BN] H. BREZIS & L. NIRENBERG, *Positive solutions of nonlinear elliptic equations involving critical Sobolev exponents*, Comm. Pure Appl. Math. **XXXVI** (1983), 437–477.

[GNN] B. GIDAS, W.-M. NI & L. NIRENBERG, *Symmetry and related properties via the maximum principle*, Comm. Math. Phys. **68** (1979), 209–243.

[JL] D. D. JOSEPH & T. S. LUNDGREN, *Quasilinear Dirichlet problems driven by positive sources*, Arch. Rational Mech. Anal. **49** (1973), 241–269.

[MP] F. MERLE & L. A. PELETIER, *Positive solutions of elliptic equations involving supercritical growth*, Proc. Roy. Soc. Edinburgh **118A** (1991), 49–63.

[MiPu] F. MIGNOT & J.-P. PUEL, *Solution radiale singulière de* $-\Delta u = \lambda e^u$, C. R. Acad. Sci. Paris **307** (1988), 379–382.

[P] S. I. POHOZAEV, *Eigenfunctions of the equation* $\Delta u + \lambda f(u) = 0$, Dokl. Akad. Nauk. SSSR **165** (1965), 36–39 (in Russian), Sov. Math. **6** (1965), 1408–1411.

[PS] P. PUCCI & J. SERRIN, *A general variational identity*, Indiana Univ. Math. J. **35** (1986), 681–703.

[R] P. H. RABINOWITZ, *Variational methods for nonlinear eigenvalue problems*, Indiana Univ. Math. J. **23** (1974), 729–754.

This work was partially supported by a grant from the Netherlands Organisation for Scientific Research and EEC Grant SC1-0019-C(TT).

F. MERLE
Departément de Mathématiques
Université de Cergy-Pontoise
Avenue du Parc 8 le Campus
95033 Cedex France

L. A. PELETIER
Mathematical Institute
Leiden University
PB9512
2300 RA Leiden, Netherlands

J. SERRIN
Department of Mathematics
University of Minnesota
Minneapolis, Minnesota 55455-0100

Received: July 14, 1992; Revised: January 7, 1994.

James Serrin and Haim Brezis, Minneapolis, MN, 2002.

J. Bryce McLeod, James Serrin, Haim Brezis and Lambertus Peletier, 2000.

I.11 Variational Identities and their Applications

Hans-Christoph Grunau

Mathematics Subject Classification (2010). 35J66; 35J35, 49K20.
Keywords. Noether's theorem, semilinear polyharmonic problems, critical growth, uniqueness results, critical dimensions.

Variational identities

What today we call "Noether's theorem" [13] was proved in 1918 and has been well known for a long time, particularly to physicists. Roughly speaking, it says that symmetries of variational problems give rise to conservation laws, i.e., identities between certain quantities related to the variational integral. For a profound exposition and for putting this into a very general framework one may consult Olver's monograph [14].

Apparently, a first application of identities or, more generally, of inequalities deduced from invariance or, more generally, from monotonicity properties of variational integrals in the context of nonexistence results for *nonlinear elliptic boundary value problems* was given by Pohožaev [15]. There, among others, the boundary value problem

$$\begin{cases} -\Delta u = |u|^{p-1}u & \text{in } \Omega, \\ u = 0 & \text{on } \partial\Omega \end{cases} \tag{1}$$

was studied. By making use of the test functions u and $x \cdot \nabla u$, what was later called a "Pohožaev"- or a "Rellich-type-identity" (cf. also [20] for the case $p = 1$) was deduced. For strictly star-shaped bounded domains $\Omega \subset \mathbb{R}^n$ one obtains in the supercritical case $p > (n+2)/(n-2)$ immediately that $u(x) \equiv 0$ is the unique classical solution of (1). In the critical case $p = (n+2)/(n-2)$ one finds first that $\frac{\partial u}{\partial \nu} = 0$ on $\partial\Omega$ and in order to conclude further that $u(x) \equiv 0$ one has to apply Hopf's strong maximum principle (to exclude existence of strictly positive solutions) or the unique continuation principle in the general case (after having extended u by 0 outside Ω as a solution to the differential equation in (1)). Pohožaev generalised these results in [16] to quasilinear equations of arbitrary order. However, the above-mentioned extension trick is not applicable here. More restrictive assumptions are needed in the case of higher-order equations of critical growth in order to prove the nonexistence of nontrivial solutions.

The work [17] (see pp. 667–689 below) of Patrizia Pucci and James Serrin addresses these topics and puts them now into a general and systematic context by uncovering the relations to Noether's theorem. The authors comment on the corresponding key identity that "its wide usefulness in the theory of elliptic partial differential equations" "has not

DOI 10.1007/978-3-0348-0845-3_11,

previously been emphasized", see [17, p. 683, lines 13, 14]. They introduced "Pohožaev-type" variational identities as a basic tool into the theory of nonlinear elliptic equations and have thereby inspired a huge amount of subsequent research. Let us just put a spotlight on a few works, the selection of which is just due to the author's taste and knowledge.

One strand of research is concerned with the question how essential it is for the validity of nonexistence results for (super-) critical problems that the domain is star-shaped, see, e.g., [1, 5]. Reichel [19] developed and analysed the notion of conformally contractible domains as a generalisation of starshapedness, which may be most suitable for applying Noether's approach and deducing nonexistence results (i.e., uniqueness of the trivial solution).

Some uniqueness results for geometric equations, which were previously proven in an ad-hoc fashion, can also be shown with the help of Pohožaev–Pucci–Serrin identities. One nice example is Wente's uniqueness result [21] for the homogeneous Dirichlet problem for the prescribed mean curvature equation. For a proof in this spirit, see [19, pp. 137/138]. For Willmore graphs over strictly star-shaped domains, a related uniqueness result was recently proved by Dall'Acqua [4].

Another strand is concerned, e.g., with the derivation of Liouville-type theorems for entire solutions (that is, solutions, which are defined in the whole space) of semilinear elliptic equations of subcritical growth, see, e.g., [11]. Moreover, "Pohožaev–Pucci–Serrin-type" identities have quite often proved to be a useful tool in showing qualitative and quantitative properties of solutions.

A further strand will be discussed in what follows in some more detail.

Critical dimensions in polyharmonic critical growth problems

The main concern of Patrizia Pucci's and James Serrin's work [18] (see pp. 690–718 below) is the following semilinear polyharmonic eigenvalue problem

$$\begin{cases} (-\Delta)^K u = \lambda u + |u|^{s-1}u & \text{in } B, \\ D^\alpha u|_{\partial B} = 0 & \text{for } |\alpha| \leq K-1. \end{cases} \tag{2}$$

Here $K \in \mathbb{N}$, $B = B_1(0) \subset \mathbb{R}^n$ $(n > 2K)$ denotes the unit ball, $s = (n+2K)/(n-2K)$ is the critical exponent with respect to the embedding $W_0^{K,2} \hookrightarrow L^{s+1}$, $\lambda \in \mathbb{R}$ is an eigenvalue parameter. If $K = 1$, this is the celebrated Brézis–Nirenberg eigenvalue problem (see [3]).

It follows from the variational identities mentioned before that, due to the critical growth of the nonlinearity, a necessary condition for a nontrivial solution to (2) to exist is $\lambda > 0$ if $K = 1$ (see [15]) and $\lambda \geq 0$ if $K \geq 2$ (see [16, 17]). For $K \geq 2$ the case $\lambda = 0$ is more difficult and it was only recently that Lazzo and Schmidt [10] proved the nonexistence of nontrivial radial solutions in this case. One should observe that by a sort of Gidas–Ni–Nirenberg result of Berchio, Gazzola, and Weth [2] any nonnegative solution to (2) is radially symmetric.

However, the partial loss of compactness due to the critical growth becomes manifest in (2) in much more refined nonexistence results. In the work [18] under consideration, Pucci and Serrin call a dimension n *critical* with respect to the Dirichlet problem (2), if there exists a positive number $\Lambda_* > 0$ such that a necessary condition for a nontrivial radial solution to (2) to exist is that $\lambda > \Lambda_* > 0$. Brézis and Nirenberg [3] discovered that $n = 3$ is critical in this sense for the second-order problem, i.e., when $K = 1$. Pucci and Serrin proved in [18] by making use of refined test functions – when compared with those which were used in [16, 17] – that, for any K, the dimension $n = 2K + 1$ is critical. One should observe that there is so far no obvious guess or general method to construct test functions, which uncover critical dimensions.

If $K = 2$, one finds from [6, 18] that the critical dimensions are precisely $n = 5, 6, 7$. Motivated by the biharmonic results and inspired by their intuition concerning a general existence theory, Pucci and Serrin **conjecture** in [18] that the critical dimensions for (2) are precisely $n = 2K + 1, \dots, 4K - 1$. Although this conjecture is quite convincing and although it initiated quite some subsequent research a proof is in general still missing. Concerning partial results one may find detailed information and references in the book [7]. A weakened variant of this conjecture – restricted to positive solutions – is proved in [2, 8].

Pucci's and Serrin's work [18] has inspired a lot of further research, e.g., for polyharmonic operators with different boundary conditions, critical dimensions in different problems, attempts to explain the critical dimension phenomenon in terms of properties of the corresponding Green functions or in terms of refined optimal Sobolev inequalities with remainder terms. More detailed information and references can be found in [9, 12] or in the book [7]. It should be mentioned that to the author's knowledge nothing is known in the critical dimensions about nonexistence of *any* nontrivial solution to (2) in the range $\lambda \in [0, \Lambda_*)$. A further interesting open problem is to determine the optimal value of Λ_* for $K \geq 2$.

References

[1] A. Bahri, J.M. Coron, On a nonlinear elliptic equation involving the critical Sobolev exponent: the effect of the topology of the domain, *Comm. Pure Appl. Math.* **41**, 253–294 (1988).

[2] E. Berchio, F. Gazzola, T. Weth, Radial symmetry of positive solutions to nonlinear polyharmonic Dirichlet problems, *J. Reine Angew. Math.* **620**, 165–183 (2008).

[3] H. Brézis, L. Nirenberg, Positive solutions of nonlinear elliptic equations involving critical Sobolev exponents, *Comm. Pure Appl. Math.* **36**, 437–477 (1983).

[4] A. Dall'Acqua, Uniqueness for the homognenous Dirichlet Willmore boundary value problem, *Ann. Global Anal. Geom.* **42**, 411–420 (2012).

[5] E.N. Dancer, A note on an equation with critical exponent, *Bull. London Math. Soc.* **20**, 600–602 (1988).

[6] D.E. Edmunds, D. Fortunato, E. Jannelli, Critical exponents, critical dimensions and the biharmonic operator, *Arch. Rational Mech. Anal.* **112**, 269–289 (1990).

[7] F. Gazzola, H.-Ch. Grunau, G. Sweers, *Polyharmonic boundary value problems*, Positivity preserving and nonlinear higher order elliptic equations in bounded domains, Lecture Notes in Mathematics **1991**, Springer-Verlag, Berlin etc. 2010.

[8] H.-Ch. Grunau, On a conjecture of P. Pucci and J. Serrin, *Analysis* **16**, 399–403 (1996).

[9] E. Jannelli, The role played by space dimension in elliptic critical problems, *J. Differential Equations* **156**, 407–426 (1999).

[10] M. Lazzo, P.G. Schmidt, Nonexistence criteria for polyharmonic boundary-value problems, *Analysis (Munich)* **28**, 449–460 (2008).

[11] E. Mitidieri, A Rellich type identity and applications, *Comm. Partial Differential Equations* **18**, 125–151 (1993).

[12] E. Mitidieri, On the definition of critical dimension, unpublished manuscript, 1993.

[13] E. Noether, Invariante Variationsprobleme, *Nachr. Königl. Gesellschaft Wissensch. Göttingen*, Math.-Phys. Klasse 235–257 (1918). English translation in: *Tansport Theory and Statistical Physics* **1**, 186–207 (1971).

[14] P.J. Olver, *Applications of Lie groups to differential equations*, second edition, Graduate Texts in Mathematics **107**, Springer-Verlag, New York etc., 1993.

[15] S.I. Pohožaev, Eigenfunctions of the equation $\Delta u + \lambda f(u) = 0$, *Soviet. Math. Dokl.* **6**, 1408–1411 (1965). Russian original: *Dokl. Akad. Nauk SSSR*, **165**, 36–39 (1965).

[16] S.I. Pohožaev, On the eigenfunctions of quasilinear elliptic problems, Math. USSR Sbornik **11**, 171–188 (1970). Russian original: *Mat. Sb. (N.S.)*, **82 (124)**, 192–212 (1970).

[17] P. Pucci, J. Serrin, A general variational identity, *Indiana Univ. Math. J.* **35**, 681–703 (1986).

[18] P. Pucci, J. Serrin, Critical exponents and critical dimensions for polyharmonic operators, *J. Math. Pures Appl.* **69**, 55–83 (1990).

[19] W. Reichel, *Uniqueness theorems for variational problems by the method of transformation groups*, Lecture Notes in Mathematics **1841**, Springer-Verlag, Berlin etc. (2004).

[20] F. Rellich, Darstellung der Eigenwerte von $\Delta u + \lambda u = 0$ durch ein Randintegral, *Math. Z.* **46**, 635–636 (1940).

[21] H. Wente, The differential equation $\Delta x = 2H(x_u \wedge x_v)$ with vanishing boundary values, *Proc. Amer. Math. Soc.* **50**, 131–137 (1975).

Hans-Christoph Grunau
Institut für Analysis und Numerik
Otto von Guericke-Universität Magdeburg
D-39016 Magdeburg, Germany
e-mail: `hans-christoph.grunau@ovgu.de`

Indiana Univ. Math. J. 35 (1986), 681–703

Reprint I.11.[i]

A General Variational Identity

PATRIZIA PUCCI & JAMES SERRIN

In 1965 Pohožaev discovered an important identity for solutions of the Dirichlet problem

$$(1) \qquad \Delta u + f(u) = 0 \quad \text{in } \Omega, \qquad u = 0 \quad \text{on } \partial\Omega.$$

This identity enabled him to show that problem (1) has no nontrivial solutions when Ω is a bounded star-shaped domain in $\mathbf{R}^n$ and $f = f(u)$ is a continuous function on $\mathbf{R}$ satisfying the condition

$$(n-2)uf(u) - 2nF(u) > 0 \qquad \text{for } u \neq 0,$$

where F denotes the primitive $F(u) = \int_0^u f(t)\,dt$ of f.

Here we shall obtain a corresponding identity for extremals of the general variational problem

$$(2) \qquad \delta \int_\Omega \mathcal{F}(x,u,Du)\,dx = 0,$$

see formula (4) below. This result can be applied similarly to demonstrate that no nontrivial extremals u can exist in a star-shaped domain Ω when $u = 0$ on $\partial\Omega$ and the function $\mathcal{F} = \mathcal{F}(x,u,p)$ satisfies appropriate conditions (see Theorem 1). For the special case $\mathcal{F}(x,u,p) = G(p) - F(x,u)$ the results are considerably easier to state, and a number of immediately ensuing applications are given in §2.

In §3 we apply our identity to an eigenvalue problem, generalizing a result noted in [1]. The case of unbounded domains is treated in §4. Here two essentially different situations arise, the first when $\Omega = \mathbf{R}^n$, the second when Ω has the structure of a "half-space"; cf. Esteban and Lions [2].

The existence of a general identity for the variatonal problem (2) leads one to expect a similar identity for vector-valued extremals, mapping Ω into $\mathbf{R}^N$, and for higher order problems. This expectation indeed can be realized, as we show in the final section. The conclusions are new even for the system

$$\Delta u^i + f_i(u) = 0, \qquad i = 1,\ldots,N,$$

and for the semilinear pluriharmonic equation

$$(-\Delta)^K u - f(u) = 0,$$

with $K \geq 2$. A particular result in this direction, again generalizing Pohožaev, is the following.

Suppose that $n > 2K$, where K is a positive integer. Let $u \in C^{2K}(\Omega) \cap C^{2K-1}(\overline{\Omega})$ be a solution of the equation

$$(-\Delta)^K u - \lambda u - u|u|^{s-1} = 0$$

in a star-shaped domain Ω, satisfying $u = Du = \ldots = D^{K-1}u = 0$ on $\partial\Omega$. If

$$\lambda < 0 \quad \text{and} \quad s \geq \frac{n+2K}{n-2K}$$

or if

$$\lambda \leq 0 \quad \text{and} \quad s > \frac{n+2K}{n-2K},$$

then $u \equiv 0$.

§1. The main identity. Throughout this section Ω will denote a smooth, bounded domain in $\mathbf{R}^n$. We let $\nu = \nu(x) = (\nu_1, \ldots, \nu_n)(x)$ be the outer normal to $\partial\Omega$ at x. The domain will be called *star-shaped* if there exists a point $x_0 \in \Omega$ such that the line segment $\overline{x_0 x}$ is contained in Ω whenever $x \in \Omega$. If Ω is star-shaped with respect to the origin, it is easy to see that $x \cdot \nu \geq 0$ on $\partial\Omega$ and that $x \cdot \nu > 0$ on some subset of $\partial\Omega$ of positive measure; with the present definition, the inequality $x \cdot \nu > 0$ need not hold a.e. on $\partial\Omega$, as simple examples show.

We consider integrands $\mathcal{F} = \mathcal{F}(x,u,p)$, $p = (p_1, \ldots, p_n)$, which are of class C^1 on the domain $\overline{\Omega} \times \mathbf{R} \times \mathbf{R}^n$, and for which, additionally, the vector function

$$\mathcal{F}_p(x,u,p) = \left(\frac{\partial \mathcal{F}}{\partial p_1}, \ldots, \frac{\partial \mathcal{F}}{\partial p_n}\right)(x,u,p)$$

is of class C^1 on $\overline{\Omega} \times \mathbf{R} \times \mathbf{R}^n$. For definiteness we shall treat extremals of (2) which are of class $C^2(\Omega)$, and so satisfy the Euler–Lagrange equation

$$\text{(3)} \qquad \operatorname{div}\{\mathcal{F}_p(x,u,Du)\} = \mathcal{F}_u(x,u,Du), \qquad x \in \Omega,$$

where $Du = (\partial u/\partial x_1, \ldots, \partial u/\partial x_n)$ and $\mathcal{F}_u = \partial \mathcal{F}/\partial u$. We shall also write $\mathcal{F}_{x_i} = \partial \mathcal{F}/\partial x_i$ and $\mathcal{F}_{p_i} = \partial \mathcal{F}/\partial p_i$. Finally, we assume for convenience and without loss of generality that $\mathcal{F}(x,0,0) = 0$ in Ω. We now have the following result.

Proposition 1. *Let $u \in C^2(\Omega)$ be a solution of the Euler–Lagrange equation (3), and let a and h be, respectively, scalar and vector–valued functions of class $C^1(\Omega)$. Then the following relation holds in Ω,*

$$\frac{\partial}{\partial x_i}\left[h_i\mathcal{F}(x,u,Du) - h_j\frac{\partial u}{\partial x_j}\mathcal{F}_{p_i}(x,u,Du) - au\mathcal{F}_{p_i}(x,u,Du)\right]$$

$$= \frac{\partial h_i}{\partial x_i}\mathcal{F}(x,u,Du) + h_i\mathcal{F}_{x_i}(x,u,Du)h_i - \left(\frac{\partial u}{\partial x_j}\frac{\partial h_j}{\partial x_i} + u\frac{\partial a}{\partial x_i}\right)\mathcal{F}_{p_i}(x,u,Du)$$

$$-a\left(\frac{\partial u}{\partial x_i}\mathcal{F}_{p_i}(x,u,Du) + u\mathcal{F}_u(x,u,Du)\right),$$

where repeated indices i and j are understood to be summed from 1 to n.

The proof is obtained by direct computation, with (3) being used to eliminate the term $\operatorname{div}\{\mathcal{F}_p(x,u,Du)\}$.

From an abstract point of view, results analogous in form to that of Proposition 1 can be found in the literature of general Noetherian theory, see for example [7], Chapter 4. To the authors' knowledge, however, the specific form of the identity in Proposition 1 has not previously been emphasized, nor its wide usefulness in the theory of elliptic partial differential equations.

From now on we consider solutions $u \in C^2(\Omega) \cap C^1(\overline{\Omega})$ of (3) with $u = 0$ on $\partial\Omega$. Clearly then $\partial u/\partial x_i = (\partial u/\partial\nu)\nu_i$ on $\partial\Omega$, so

$$h_j\frac{\partial u}{\partial x_j}\mathcal{F}_{p_i}(x,u,Du)\nu_i = \frac{\partial u}{\partial x_i}\mathcal{F}_{p_i}(x,u,Du)h_j\nu_j \qquad \text{on } \partial\Omega.$$

We now integrate the identity of Proposition 1 over Ω and use the condition $u = 0$ on $\partial\Omega$. Assuming that a and h are in $C^1(\Omega) \cap C(\overline{\Omega})$, it follows from the divergence theorem that

$$(4) \quad \oint_{\partial\Omega}\left[\mathcal{F}(x,0,Du) - \frac{\partial u}{\partial x_i}\mathcal{F}_{p_i}(x,0,Du)\right](h\cdot\nu)\,ds$$

$$= \int_\Omega\left\{\mathcal{F}(x,u,Du)\operatorname{div} h + h_i\mathcal{F}_{x_i}(x,u,Du) - \left[\frac{\partial u}{\partial x_j}\frac{\partial h_j}{\partial x_i} + u\frac{\partial a}{\partial x_i}\right]\mathcal{F}_{p_i}(x,u,Du)\right.$$

$$\left.-a\left[\frac{\partial u}{\partial x_i}\mathcal{F}_{p_i}(x,u,Du) + u\mathcal{F}_u(x,u,Du)\right]\right\}dx,$$

a principal identity which is fundamental for our purposes.

If we take $\mathcal{F}(x,u,p) = \frac{1}{2}|p|^2 - F(u)$, the corresponding Euler–Lagrange equation is just (1). In this case, the choice $h(x) = x$ and $a(x) = (n-2)/2$

gives precisely the inequality discovered by Pohožaev, that is,

$$-\frac{1}{2}\oint_{\partial\Omega}|Du|^2(x\cdot\nu)\,ds=\int_\Omega\left[\frac{n-2}{2}uf(u)-nF(u)\right]dx.$$

In passing, we observe that the special case of this identity when $f(u)=\frac{1}{2}\lambda u^2$ was already obtained by Rellich in 1940.

Now putting $h(x)=x$ and $a(x)=\text{constant}$ in (4), we get correspondingly

$$\begin{aligned}(5)\qquad &\oint_{\partial\Omega}\left[\mathcal{F}(x,u,Du)-\frac{\partial u}{\partial x_i}\mathcal{F}_{p_i}(x,0,Du)\right](x\cdot\nu)\,ds\\ &=\int_\Omega\Big[n\mathcal{F}(x,u,Du)+x_i\mathcal{F}_{x_i}(x,u,Du)-au\mathcal{F}_u(x,u,Du)\\ &\qquad\qquad -(a+1)\frac{\partial u}{\partial x_i}\mathcal{F}_{p_i}(x,u,Du)\Big]\,dx.\end{aligned}$$

This leads to the following *exact generalization* of Pohožaev's non-existence theorem.

Theorem 1. *Assume Ω is bounded and star-shaped with respect to the origin. Suppose also that*

$$(6)\qquad p_i\mathcal{F}_{p_i}(x,0,p)-\mathcal{F}(x,0,p)\geq 0\qquad \textit{for all } x\in\partial\Omega \textit{ and } p\in\mathbf{R}^n$$

and that there exists a real number a such that

$$(7)\qquad n\mathcal{F}(x,u,p)+x_i\mathcal{F}_{x_i}(x,u,p)-au\mathcal{F}_u(x,u,p)-(a+1)p_i\mathcal{F}_{p_i}(x,u,p)\geq 0$$

for all $(x,u,p)\in\Omega\times\mathbf{R}\times\mathbf{R}^n$. Assume finally that either $u=0$ or $p=0$ whenever the equality in (7) holds. Then the variational equation (3) has no nontrivial solution $u\in C^2(\Omega)\cap C^1(\overline{\Omega})$ which vanishes on $\partial\Omega$.

Proof. Any solution of the variational equation (3) which vanishes on the boundary of Ω must satisfy (5). By virtue of (6), the left-hand side of (5) is non-positive (since $x\cdot\nu\geq 0$ on $\partial\Omega$). On the other hand, by (7) the right-hand side of (5) is positive, unless either $u=0$ or $p=0$ at each point of Ω. It follows at once that $u\equiv 0$ in Ω, completing the proof.

Condition (6) is closely related to the weak ellipticity of the Euler–Lagrange equation (3) and the Weierstrass condition for the variational problem (2). In particular, if the latter is satisfied, then for all $p, \overline{p} \in \mathbf{R}^n$ we have

$$\mathcal{F}(x,u,\overline{p}) - \mathcal{F}(x,u,p) - (\overline{p}_i - p_i)\mathcal{F}_{p_i}(x,u,p) \geq 0.$$

Setting $u = 0$ and $\overline{p} = 0$ gives

$$-\mathcal{F}(x,0,p) + p_i \mathcal{F}_{p_i}(x,0,p) \geq 0,$$

which is precisely (6). The same conclusion can of course be reached from the weak ellipticity of (3).

When both (6) and (7) hold, it is not difficult to show that both $\mathcal{F}_u(x,0,0)$ and $\mathcal{F}_p(x,0,0)$ must vanish. In particular, $u \equiv 0$ is then a solution of the Euler–Lagrange equation (3), and the result of Theorem 1 is that this is the only solution in Ω which vanishes on the boundary.

Remark. The same methods can be applied to modified variational equations of the form

$$\operatorname{div}\{\mathcal{F}_p(x,u,Du)\} = \mathcal{F}_u(x,u,Du) + \mathcal{G}(x,u,Du).$$

In this case the fundamental identity of Proposition 1 is changed only by adding to the right–hand side the expression

$$-\left[h_j \frac{\partial u}{\partial x_j} + au\right] \mathcal{G}(x,u,Du).$$

All further results can similarly be modified to take into account this additional term, the exact statements being left to the reader.

§2. **Examples.** A case of special interest occurs when

$$\mathcal{F}(x,u,p) = G(p) - F(x,u),$$

with $F(x,0) = G(0) = 0$ for all $x \in \Omega$. The Euler–Lagrange equation is then simply

$$\operatorname{div}\{G_p(Du)\} + f(x,u) = 0,$$

where $f(x,u) = F_u(x,u)$. Condition (6) is guaranteed by the *convexity* of $G = G(p)$, while (7) becomes

$$nG(p) - (a+1)p_i G_{p_i}(p) \geq nF(x,u) - auf(x,u) + x_i F_{x_i}(x,u).$$

For this to be satisfied for all (x,u,p) requires that the inequalities

$$nG(p) - (a+1)p_i G_{p_i}(p) \geq 0 \qquad \text{for all } p \in \mathbf{R}^n, \tag{8}$$

$$auf(x,u) - nF(x,u) - x_i F_{x_i}(x,u) \geq 0 \qquad \text{for all } (x,u) \in \Omega \times \mathbf{R} \tag{9}$$

hold for some constant $a \in \mathbf{R}$. That is, *the growth of G with respect to p must be directly related to the growth of F with respect to (x,u).*

Two examples are relevant here:

$$G(p) = \frac{1}{m}|p|^m, \qquad m > 1,$$

the degenerate Dirichlet operator, see [6]; and

$$G(p) = \sqrt{1+|p|^2} - 1,$$

the area integrand. In the first case (8) reduces to $a \leq (n-m)/m$ and in the second $a \leq (n-2)/2$. Suppose in turn that $F = F(u)$ and that there exists $\alpha > 0$ such that

$$uf(u) - \alpha F(u) > 0 \qquad \text{for } u \neq 0,$$

as in the case $F(u) = \lambda u^2 + |u|^\alpha$, $\lambda < 0$, $\alpha > 2$. Then for the integrand $G(p) = |p|^m/m$ we can satisfy (8), (9), and hence (7), provided that

$$\frac{1}{\alpha} \leq \frac{1}{m} - \frac{1}{n}, \qquad a = \frac{n}{\alpha}. \tag{10}$$

In turn, no nontrivial solution of (3) which vanishes on $\partial\Omega$ can exist. For the area integrand the corresponding condition is

$$\frac{1}{\alpha} \leq \frac{1}{2} - \frac{1}{n}, \qquad a = \frac{n}{\alpha}. \tag{11}$$

Suppose $1 \leq s < n$ and let $r = r(s)$ be the critical exponent for the Sobolev space $W^{1,s}(\Omega)$, that is, $1/r = 1/s - 1/n$. Then (10) becomes simply $\alpha \geq r(m)$ and (11) reduces to $\alpha \geq r(2)$. Analogous results for the special case when Ω is a ball were recently found in [6], while other related results will be given later in §3.

A further example of importance is the semilinear elliptic integrand

$$\mathcal{F}(x,u,p) = \frac{1}{2}a_{ij}(x,u)p_i p_j - F(x,u),$$

where (a_{ij}) is a given symmetric positive definite matrix. We assume of course that a_{ij} and F are of class C^1 on $\overline{\Omega} \times \mathbf{R}$ and that $F(x,0) = 0$. Condition (6) in Theorem 1 is automatically satisfied, while (7) leads to the requirement that the matrix

$$\left((n-2(a+1))a_{ij}+x_k\frac{\partial a_{ij}}{\partial x_k}-u\,\frac{\partial a_{ij}}{\partial u}\right)$$

be non–negative definite for some constant a and that

$$auf(u)-nF(u)-x_k\,\frac{\partial F}{\partial x_k}>0 \qquad \text{for } u\neq 0.$$

When $a_{ij}(x,u)=\varphi(x,u)\delta_{ij}$, the definiteness condition simplifies to

$$(n-2(a+1))\varphi+x_k\,\frac{\partial\varphi}{\partial x_k}-u\,\frac{\partial\varphi}{\partial u}\geq 0.$$

To see the specific effect of this type of semilinearity, we consider particularly the cases $\varphi=|x|^2$ and $\varphi=u^2$. In the first, the definiteness condition becomes

$$a\leq\frac{n}{2},$$

which is weaker than the condition $a\leq(n-2)/2$ obtained before. On the other hand, in the second it becomes

$$a\leq\frac{n-4}{2},$$

which is more stringent.

Remark. The hypothesis made in Theorem 1, that equality in (7) implies either $u=0$ or $p=0$, may not always be applicable. For example, when

$$\mathcal{F}(x,u,p)=\frac{1}{2}|p|^2-|u|^{\ell+1}, \qquad \ell=\frac{n+2}{n-2}, \quad n>2,$$

it is necessary to choose $a=(n-2)/2$, in which case the left–hand side of (7) is identically zero.

For examples of this kind an alternative method is required to complete the proof. There are two possibilities here, each with its own advantages.

First, one may strengthen (6) by requiring that equality holds if and only if $p=0$, that is, by invoking the strong Weierstrass condition. Then, as in the proof of Theorem 1, we obtain the conclusion $Du=0$ at each point of $\partial\Omega$ where $x\cdot\nu>0$. Thus on some non–empty subset of $\partial\Omega$ we have $u=Du=0$. *If equation* (3) *now admits a unique continuation theorem, then we get $u\equiv 0$ in Ω, as required.*

This approach applies most simply in the case of radial solutions, for then the Euler–Lagrange equation (3) becomes an ordinary differential equation and unique continuation is nothing more than uniqueness of the initial value problem.

Second, one may deal with the special Dirichlet problem $u = 0$ on $\partial\Omega$, $u > 0$ in Ω. With (6) strengthened as before, we then have $Du = 0$ on a non-empty subset of $\partial\Omega$. Hence, *if the Hopf Boundary Point Lemma applies, then a contradiction is immediately reached and so no positive solution exists.*

A sufficient condition for the applicability of the boundary point lemma is that (3) can be written as a non-degenerate linear elliptic equation. In particular, from (3) follows

$$\mathcal{F}_{p_ip_j}\frac{\partial^2 u}{\partial x_i \partial x_j} + \mathcal{F}_{p_iu}\frac{\partial u}{\partial x_i} + \mathcal{F}_{p_ix_i} - \mathcal{F}_u = 0, \tag{12}$$

where summation from 1 to n over repeated indices is understood. Assuming that $u = 0$ is a solution of (12), we clearly have

$$\mathcal{F}_{p_ix_i}(x,0,0) - \mathcal{F}_u(x,0,0) = 0.$$

Suppose the matrix

$$\left(\mathcal{F}_{p_ip_j}(x,u,p)\right)$$

is positive definite for all $(x,p) \in \overline{\Omega} \times \mathbf{R}^n$ and all $u \geq 0$, and that

$$\mathcal{F}_{p_ix_i}(x,u,p) - \mathcal{F}_u(x,u,p)$$

is of class C^1 in the variables u and p, $u \geq 0$. Then, *for any given nonnegative solution* $u = u(x)$, equation (12) can be written in the form

$$a_{ij}(x)\frac{\partial^2 u}{\partial x_i \partial x_j} + b_i(x)\frac{\partial u}{\partial x_i} + c(x)u = 0$$

with $(a_{ij}(x))$ continuous and positive definite and $b_i(x)$, $c(x)$ bounded. Application of the general boundary point lemma given below now shows that any positive solution of equation (3) with $u = 0$ on $\partial\Omega$ satisfies $\frac{\partial u}{\partial \nu} < 0$ at each point of $\partial\Omega$, where ν is the outer normal to $\partial\Omega$. The required contradiction is thus established.

Boundary Point Lemma. *Let u satisfy the inequality*

$$a_{ij}(x)\,\frac{\partial^2 u}{\partial x_i\,\partial x_j} + b_i(x)\,\frac{\partial u}{\partial x_i} + c(x)u \geq 0$$

in a domain Ω in which (a_{ij}) is uniformly positive definite. Suppose that $u \geq 0$ in Ω and that $u = 0$ at a boundary point x_0, where x_0 lies on the boundary of a ball K in Ω. If u and and Du are continuous at x_0, then

$$\frac{\partial u}{\partial \nu} < 0 \qquad \text{at } x_0$$

unless $u \equiv 0$.

Proof. Suppose $u > 0$ in Ω. Choose $R > 0$ so small that

$$\tilde{c}(x) = (R^2 - |x - x_0|^2)\,c(x) - 2(x_i - x_{0i})\,b_i(x) - 2a_{ii}(x) \leq 0$$

in $\Omega \cap B$, where B denotes the open ball of radius $R/2$ and center x_0. The substitution

$$u = (R^2 - |x - x_0|^2)w$$

now gives the relation

$$(R^2 - |x - x_0|^2)\,a_{ij}(x)\,\frac{\partial^2 w}{\partial x_i \partial x_j} + d_i(x)\,\frac{\partial w}{\partial x_i} + \tilde{c}(x)w \geq 0$$

and $w > 0$ in $\Omega \cap B$, $w = 0$ at x_0. By the standard boundary point lemma (cf. [9], page 65) it follows that

$$\frac{\partial w}{\partial \nu} < 0 \qquad \text{at } x_0,$$

so that in turn

$$\frac{\partial u}{\partial \nu} = R^2\,\frac{\partial w}{\partial \nu} < 0 \qquad \text{at } x_0,$$

as required.

If $u = 0$ at some point in Ω, a similar argument leads to a contradiction, unless $u \equiv 0$ in Ω. This completes the proof.

§3. A bifurcation problem. Here we show how an argument of Rabinowitz [10] and Brezis and Nirenberg [1] concerning non–existence of solutions of radial eigenvalue problems can be extended to bounded star–shaped regions and more general operators. For simplicity we consider only the special integrand

$$\mathcal{F}(x,u,p) = G(p) - \frac{\lambda}{m}|u|^m - \Phi(u), \qquad G(0) = \Phi(0) = 0,$$

and the corresponding bifurcation problem

$$(13)\qquad \begin{aligned} &\operatorname{div}\{G_p(Du)\} + \lambda u|u|^{m-2} + \varphi(u) = 0, \qquad u \not\equiv 0, \quad \text{in } \Omega, \\ &u = 0 \qquad \text{on } \partial\Omega, \end{aligned}$$

where $m > 1$ is a real number and $\varphi = \Phi'$. Brezis and Nirenberg treat specifically the case $G(p) = |p|^2/2$, $\varphi(u) = u|u|^{s-1}$, $m = 2$, $s > 1$ and $n \geq 3$.

Assuming that $u\colon \Omega \to \mathbf{R}$ is a solution of (13) and that Ω is bounded and star–shaped with respect to the origin, it follows from (5) that

$$(14)\qquad \begin{aligned} \oint_{\partial\Omega} \left[G(Du) - \frac{\partial u}{\partial x_i} G_{p_i}(Du)\right](x\cdot\nu)\, ds &= \int_\Omega \left[\lambda\left(a - \frac{n}{m}\right)|u|^m + au\varphi(u) - n\Phi(u)\right. \\ &\qquad \left. + nG(Du) - (a+1)\frac{\partial u}{\partial x_i} G_{p_i}(Du)\right] dx. \end{aligned}$$

We assume

$$(15)\qquad p_i G_{p_i}(p) - G(p) \geq 0,$$

$$(16)\qquad nG(p) - (a+1)p_i G_{p_i}(p) \geq k|p|^m,$$

for all $p \in \mathbf{R}^n$ and for some constants $a < n/m$ and $k > 0$. Moreover, we suppose that

$$(17)\qquad au\varphi(u) - n\Phi(u) \geq 0.$$

Then from (14)–(17) there results

$$0 \geq \int_\Omega \left[\lambda\left(a - \frac{n}{m}\right)|u|^m + k|Du|^m\right] dx.$$

Let λ_1 be the Poincaré constant for the domain Ω, namely the maximal coefficient λ_1 such that

$$\lambda_1 \int_\Omega |\psi|^m \, dx \le \int_\Omega |D\psi|^m \, dx \qquad \text{when } \psi = 0 \text{ on } \partial\Omega. \tag{18}$$

It now follows at once that

$$\lambda\left(a - \frac{n}{m}\right) + \lambda_1 k \le 0,$$

that is,

$$\lambda \ge \lambda_1 \frac{mk}{n - ma}.$$

This proves the following result.

Theorem 2. *Suppose that* (15)–(18) *hold and that* Ω *is a bounded star-shaped domain. Then problem* (13) *has no solution when*

$$\lambda < \lambda_1 \frac{mk}{n - ma}.$$

Remark. If either the unique continuation theorem can be applied, or $u > 0$ in Ω and the boundary point lemma holds, the assertion of Theorem 2 can be improved to the form

$$\lambda \le \lambda_1 \frac{mk}{n - ma}.$$

The second possibility applies in particular when $G(p) = \frac{1}{2}|p|^2$ and $\varphi(u) \in C^1(\mathbf{R}_0^+)$.

For the case treated in [1] we have

$$a = \frac{n}{s+1}, \qquad m = 2,$$

$$k = \frac{n}{2} - (a+1) = \frac{n-2}{2(s+1)}\left(s - \frac{n+2}{n-2}\right),$$

so that the non-existence condition for positive solutions in Ω becomes

$$\lambda \le \lambda_1 \frac{n-2}{n} \frac{s - \dfrac{n+2}{n-2}}{s-1}, \qquad s > \frac{n+2}{n-2}, \quad n > 2,$$

a result obtained in [1] for the case when Ω is a ball and $u = u(|x|)$.

When $G(p) = |p|^m/m$ and $\varphi(u) = u|u|^{s-1}$, we take

$$a = \frac{n}{s+1}$$

$$k = \frac{n}{m} - (a+1) = \frac{n-m}{m(s+1)}\left[s - \frac{(m-1)n+m}{n-m}\right],$$

leading to the non–existence condition

$$\lambda \le \lambda_1 \frac{n-m}{m} \frac{s - \dfrac{(m-1)n+m}{n-m}}{s-(m-1)},$$

provided $s > [(m-1)n+m]/(n-m)$, $n > m$.

§4. The unbounded case. Here we extend the results of §1 to the case of unbounded domains Ω. We treat separately the cases $\Omega = \mathbf{R}^n$, and Ω an unbounded proper subset of $\mathbf{R}^n$.

Suppose first that $\Omega = \mathbf{R}^n$. In this case the condition $u = 0$ on $\partial\Omega$ must be replaced by an appropriate decay condition at infinity in order to be able to carry out the earlier arguments. In particular, letting

$$\mathcal{H}(x,u,p) = |x|\,\mathcal{F}(x,u,p) - \frac{1}{|x|}(x_j p_j + au)x_i \mathcal{F}_{p_i}(x,u,p),$$

where a is constant, it is apparent that the proof of Theorem 1 carries over intact provided only that

$$\liminf_{R\to\infty} \oint_{\partial B_R} \mathcal{H}(x,u,Du)\,ds \le 0, \tag{19}$$

where B_R denotes the ball of radius R about the origin. This proves the following theorem.

Theorem 3. *Let a be a real number and let* (7) *hold for all $(x,u,p) \in \mathbf{R}^n \times \mathbf{R} \times \mathbf{R}^n$, the equality being satisfied only if $u = 0$ or $p = 0$. Then the variational equation* (3) *has no nontrivial solution on $\mathbf{R}^n$ satisfying the limit condition* (19).

Condition (19) can be verified under appropriate L^p restrictions on u and Du, as indicated by the following proposition.

Proposition 2. *Suppose that $u \in C^2(\mathbf{R}^n)$ is such that*

$$\mathcal{F}(x,u,Du)^+, \quad |Du| \cdot |\mathcal{F}_p(x,u,Du)|, \quad \frac{|u|}{|x|+1} |\mathcal{F}_p(x,u,Du)| \in L^1(\mathbf{R}^n).$$

Then (19) *holds.*

Proof. Clearly

$$\oint_{\partial B_R} \mathcal{H}(x,u,Du)\, ds \le (R+1) \oint_{\partial B_R} \left[\mathcal{F}(x,u,Du)^+ + \left(|Du| + \frac{|a|}{R+1} |u| \right) |\mathcal{F}_p(x,u,Du)| \right] ds$$

for all $R > 0$. Hence

$$\int_0^\infty \oint_{\partial B_R} \mathcal{H}(x,u,Du)\, ds \frac{dR}{R+1} \le \int_{\mathbf{R}^n} \left[\mathcal{F}(x,u,Du)^+ + \left(|Du| + \frac{|a|}{R+1} |u| \right) |\mathcal{F}_p(x,u,Du)| \right] dx < \infty$$

and the result follows at once.

In some circumstances Proposition 2 can be improved. Suppose, for example, that the conditions

$$u, \ |Du| \to 0 \qquad \text{as } |x| \to \infty$$

are imposed on the solution. Assume moreover that the Weierstrass condition holds and that $\mathcal{F}$ satisfies the "natural" conditions

$$\mathcal{F}(x,u,0) \le 0(u^2) \quad \text{uniformly as } u \to 0, \ |x| \to \infty$$

$$\mathcal{F}_p(x,u,p) = 0(|p|) \quad \text{uniformly as } u, \ |p| \to 0, \ |x| \to \infty.$$

Then (19) holds provided that $u \in C^2(\mathbf{R}^n) \cap W^{1,2}(\mathbf{R}^n)$. The proof is essentially the same as before, once one observes that

$$\mathcal{F}(x,u,p) \le p_i \mathcal{F}_{p_i}(x,u,p) + \mathcal{F}(x,u,0)$$

by the Weierstrass condition.

Next consider the case when Ω is unbounded but not all of $\mathbf{R}^n$, with $\partial\Omega$ smooth. There are two possibilities which can be treated.

I. *$\partial\Omega$ is star-shaped, say with respect to the origin.*

The discussion here is essentially the same as in the case $\Omega = \mathbf{R}^n$, leading to the following result.

Theorem 4. *Let a be a real number, let (6) hold, and suppose (7) is satisfied for all $(x,u,p) \in \Omega \times \mathbf{R} \times \mathbf{R}^n$, with equality only if $u = 0$ or $p = 0$. Then the variational equation (3) has no nontrivial solution vanishing on $\partial\Omega$ and satisfying the limit condition*

$$\liminf_{R\to\infty} \oint_{\partial B_R \cap \Omega} \mathcal{H}(x,u,Du)\, ds \le 0. \tag{20}$$

A sufficient condition that (20) be satisfied can be obtained in direct analogy with Proposition 2. We leave the details to the reader.

II. *$\partial\Omega$ is star-shaped with respect to infinity.*

More precisely, following Esteban and Lions [2], we assume that there exists a fixed vector $X \in \mathbf{R}^n$ such that

$$\nu(x)\cdot X \ge 0, \qquad \nu(x)\cdot X \not\equiv 0 \text{ on } \partial\Omega. \tag{21}$$

The development here will be based directly on Proposition 1, with the choices $h(x) = X$ and $a(x) = 0$ in Ω. Assuming that $u = 0$ on $\partial\Omega$, we then obtain by integration over $\Omega \cap B_R$,

$$\begin{aligned} &\oint_{\partial\Omega\cap B_R} \left[\mathcal{F}(x,0,Du) - \frac{\partial u}{\partial x_i}\mathcal{F}_{p_i}(x,0,Du)\right](X\cdot\nu)\, ds \\ &\qquad + \oint_{\partial B_R\cap\Omega} [\mathcal{F}(x,u,Du)(X\cdot\nu) - (X\cdot Du)(\mathcal{F}_p(x,u,Du)\cdot\nu)]\, ds \\ &= \int_{\Omega\cap B_R} \mathcal{F}_x(x,u,Du)\cdot X\, dx, \end{aligned} \tag{22}$$

where of course $\nu(x) = x/R$ on ∂B_R. From (22) one may read off various non-existence theorems. For definiteness, and because it is the principal case, let us suppose that $\mathcal{F}$ is independent of x so that the integral on the right-hand side vanishes. The following result then holds (cf. also [2], Theorem I.1).

Theorem 5. *Suppose that* (6) *holds and that* $\mathcal{F}$ *is independent of* x. *Then the variational equation* (3) *has no nontrivial solution when the following conditions hold:*

(a) $u = 0$ *on* $\partial\Omega$;

(b) Ω *satisfies the Esteban–Lions property* (21);

(c) $\liminf_{R\to\infty} \oint_{\partial B_R\cap\Omega} [\mathcal{F}(u,Du)(X\cdot\nu) - (X\cdot Du)(\mathcal{F}_p(u,Du)\cdot\nu)]\, ds \le 0$;

(d) *Equation* (3) *admits a unique continuation theorem.*

Remark. If we add the condition $u > 0$ in Ω, then (d) can be replaced by

(d′) *The Hopf Boundary Point Lemma applies to* (3).

This follows exactly as in the Remark at the end of §2.

A sufficient condition that (c) hold is that $u \in C^2(\Omega)$ and

$$\frac{1}{|x|+1}\mathcal{F}(x,u,Du)^+, \quad \frac{1}{|x|+1}|Du|\cdot|\mathcal{F}_p(x,u,Du)| \in L^1(\Omega),$$

as can be seen from the proof of Proposition 2.

§5. Vector–valued extremals and higher–order problems. In view of the results of §1 we can expect an identity similar to that in Proposition 1 to hold for vector–valued extremals and higher–order variational problems. First consider vector functions $u = (u^1, \ldots, u^N)\colon \Omega \to \mathbf{R}^N$, and, for simplicity, the first–order problem

$$\delta\int_\Omega \mathcal{F}(x,u,Du)\,dx = 0, \qquad \mathcal{F}(x,0,0) = 0, \tag{23}$$

where now Du denotes the matrix $(\partial u^k/\partial x_i)$, $i = 1,\ldots,n$, $k = 1,\ldots,N$. The corresponding Euler–Lagrange system is simply

$$\operatorname{div}\{\mathcal{F}_{p^k}(x,u,Du)\} = \mathcal{F}_{u^k}(x,u,Du), \qquad k = 1,\ldots,N, \tag{24}$$

the notation being clear. We now have the following proposition.

Proposition 3. *Let $u \in C^2(\Omega)$ be a solution of the Euler–Lagrange system (24), and let a and h be, respectively, scalar and vector-valued functions of class $C^1(\Omega)$. Then the following identity holds in Ω,*

$$\frac{\partial}{\partial x_i}\left[h_i \mathcal{F}(x,u,Du) - \left(h_j \frac{\partial u^k}{\partial x_j} + a u^k\right)\mathcal{F}_{p_i^k}(x,u,Du)\right]$$

$$= \frac{\partial h_i}{\partial x_i}\mathcal{F}(x,u,Du) + h_i \mathcal{F}_{x_i}(x,u,Du) - \left(\frac{\partial u^k}{\partial x_j}\frac{\partial h_j}{\partial x_i} + u^k \frac{\partial a}{\partial x_i}\right)\mathcal{F}_{p_i^k}(x,u,Du)$$

$$-a\left(\frac{\partial u^k}{\partial x_i}\mathcal{F}_{p_i^k}(x,u,Du) + u^k \mathcal{F}_{u^k}(x,u,Du)\right),$$

where the indices i,j are understood to be summed from 1 to n and the index k from 1 to N.

The proof is again by direct computation, with (24) being used to eliminate the term $\operatorname{div}\{\mathcal{F}_{p^k}(x,u,Du)\}$. Choosing $h(x) = x$ and $a(x) =$ constant and integrating over Ω, we obtain the following principal result.

Theorem 6. *Assume Ω is bounded and star-shaped with respect to the origin. Suppose also that*

$$p_i^k \mathcal{F}_{p_i^k}(x,0,p) - \mathcal{F}(x,0,p) \geq 0 \tag{25}$$

for all $x \in \Omega$ and all $p = (p_i^k)$ having the dyadic form $p_i^k = C^k \xi_i$, $C \in \mathbf{R}^N$, $\xi \in \mathbf{R}^n$. Assume finally that there exists a real number a such that

$$n\mathcal{F}(x,u,p) + x_i \mathcal{F}_{x_i}(x,u,p) - a u^k \mathcal{F}_{u^k}(x,u,p) - (a+1)p_i^k \mathcal{F}_{p_i^k}(x,u,p) \geq 0 \tag{26}$$

for all $(x,u,p) \in \Omega \times \mathbf{R}^N \times \mathbf{R}^{n\times N}$, the equality holding only when either $u = 0$ or $p = 0$. Then the variational system (24) has no nontrivial solution $u \in C^2(\Omega) \cap C^1(\overline{\Omega})$ which vanishes on $\partial\Omega$.

The proof is the same as for Theorem 1, once we observe that $\partial u^k/\partial x_i = (\partial u^k/\partial \nu)\nu_i$ on $\partial\Omega$ because u vanishes there.

The inequality (25) is related to the Weierstrass condition for the variational problem (23). In particular, this condition takes the form (see Graves [3]),

$$\mathcal{F}(x,u,\bar{p}) - \mathcal{F}(x,u,p) - (\bar{p}-p)_i^k \, \mathcal{F}_{p_i^k}(x,u,p) \geq 0,$$

for all $(x,u) \in \Omega \times \mathbf{R}^N$ and all $p,\bar{p} \in \mathbf{R}^{n\times N}$ satisfying $(\bar{p}-p)_i^k = C^k \xi_i$, $C \in \mathbf{R}^N$, $\xi \in \mathbf{R}^n$. If we set $u=0$, $\bar{p}=0$, this gives precisely (25).

A particular case of interest is the integrand

$$\mathcal{F}(x,u,p) = \frac{1}{2}\sum_{k=1}^{N} |p^k|^2 - F(u), \qquad F(0)=0.$$

Condition (25) then holds automatically, while condition (26) applies if

$$a u^k \frac{\partial F}{\partial u^k} - nF(u) > 0 \qquad \text{for } u \neq 0$$

for some constant $a \leq (n-2)/2$ and $n > 2$. In this form, we have a direct generalization of Pohožaev's Theorem to systems. For the special case $F(u) = -|u|^2 + |u|^{s+1}$, the condition for non-existence becomes the familiar inequality

$$s \geq \frac{n+2}{n-2}, \qquad n > 2.$$

Remark. Special cases of Proposition 3 have been remarked in the theory of finite elastostatics; the integrand $\mathcal{F}$ here represents the stored energy of a solid and depends only on the variable Du. In this setting Knowles & Sternberg [5] have obtained the case $h_i = \delta_{i\ell}$, $a = 0$ and A. E. Green [4] the case $h_i = x_i$, $a = -1$. The special example of a quadratic integrand $\mathcal{F}(Du)$ is also noted in both papers.

As an example of a higher-order integrand, we consider the variational problem

$$\text{(27)} \qquad \delta \int_\Omega \mathcal{F}(x,u,Du,D^2u)\,dx = 0,$$

where for simplicity u is scalar and $\mathcal{F} = \mathcal{F}(x,u,p,r)$ is of class C^3 in its variables. Without loss of generality we may assume that $\mathcal{F}(x,0,0,0) \equiv 0$ and that $\mathcal{F}_{r_{ij}} = \mathcal{F}_{r_{ji}}$. The latter condition can be realized, if necessary, by writing $\frac{1}{2}(r_{ij}+r_{ji})$ in place of r_{ij} in the function $\mathcal{F}$. The corresponding Euler–Lagrange equation, for solutions $u \in C^4(\Omega)$, is now

$$\text{(28)} \qquad \frac{\partial^2}{\partial x_i \partial x_j}\mathcal{F}_{r_{ij}}(x,u,Du,D^2u) - \frac{\partial}{\partial x_i}\mathcal{F}_{p_i}(x,u,Du,D^2u) + \mathcal{F}_u(x,u,Du,D^2u) = 0.$$

We have the following principal identity.

Proposition 4. *Let $u \in C^4(\Omega)$ be a solution of the Euler–Lagrange equation (28) and let a and h be, respectively, scalar and vector-functions of class $C^2(\Omega)$. Then the following identity holds in Ω,*

$$\frac{\partial}{\partial x_i}\left[h_i\mathcal{F} - \left(h_k\frac{\partial u}{\partial x_k}+au\right)\left(\mathcal{F}_{p_i} - \frac{\partial}{\partial x_j}\mathcal{F}_{r_{ij}}\right) - \frac{\partial}{\partial x_j}\left(h_k\frac{\partial u}{\partial x_k}+au\right)\mathcal{F}_{r_{ij}}\right]$$

$$= \frac{\partial h_i}{\partial x_i}\mathcal{F} + h_i\mathcal{F}_{x_i} - au\,\mathcal{F}_u - \left[\frac{\partial h_j}{\partial x_i}\frac{\partial u}{\partial x_j} + \frac{\partial}{\partial x_i}(au)\right]\mathcal{F}_{p_i}$$

$$-\left[\frac{\partial^2 h_k}{\partial x_i\partial x_j}\frac{\partial u}{\partial x_k} + 2\frac{\partial h_k}{\partial x_i}\frac{\partial^2 u}{\partial x_j\partial x_k} + \frac{\partial^2}{\partial x_i\partial x_j}(au)\right]\mathcal{F}_{r_{ij}},$$

where $\mathcal{F}$ and its derivatives are evaluated at (x,u,Du,D^2u).

The special case $h(x) = x$ and $a(x) =$ constant in Ω is considerably simpler, giving the identity

$$\frac{\partial}{\partial x_i}\left[x_i\mathcal{F} - \left(x_k\frac{\partial u}{\partial x_k}+au\right)\left(\mathcal{F}_{p_i} - \frac{\partial}{\partial x_j}\mathcal{F}_{r_{ij}}\right) - \frac{\partial}{\partial x_j}\left(x_k\frac{\partial u}{\partial x_k}+au\right)\mathcal{F}_{r_{ij}}\right]$$
$$= n\mathcal{F} + x_i\mathcal{F}_{x_i} - au\,\mathcal{F}_u - (a+1)\frac{\partial u}{\partial x_i}\mathcal{F}_{p_i} - (a+2)\frac{\partial^2 u}{\partial x_i\partial x_j}\mathcal{F}_{r_{ij}}, \tag{29}$$

again with $\mathcal{F}$ and its derivatives evaluated at (x,u,Du,D^2u). As in Section 1, this identity leads to the following result.

Theorem 7. *Assume that Ω is bounded and star–shaped with respect to the origin. Suppose also that*

$$r_{ij}\mathcal{F}_{r_{ij}}(x,0,0,r) - \mathcal{F}(x,0,0,r) \geq 0, \qquad x \in \Omega, \tag{30}$$

for all matrices $r = (r_{ij})$ such that $r_{ij} = C\xi_i\xi_j$ with $C \in \mathbf{R}$ and $\xi \in \mathbf{R}^n$. Finally, let

$$n\mathcal{F} + x_i\mathcal{F}_{x_i} - au\,\mathcal{F}_u - (a+1)p_i\mathcal{F}_{p_i} - (a+2)r_{ij}\mathcal{F}_{r_{ij}} \geq 0, \tag{31}$$

for all $(x,u,p,r) \in \Omega \times \mathbf{R} \times \mathbf{R}^n \times \mathbf{R}^{n\times n}$, with equality only when either $u = 0$

or $p = 0$ or $r = 0$. Then the variational equation (28) *has no nontrivial solutions $u \in C^4(\Omega) \cap C^3(\bar{\Omega})$ such that $u = Du = 0$ on $\partial\Omega$.*

Proof. When $u = Du = 0$ on $\partial\Omega$, we have

$$\frac{\partial^2 u}{\partial x_i \partial x_j} = \frac{\partial^2 u}{\partial \nu^2} \nu_i \nu_j \qquad \text{on } \partial\Omega,$$

and so in turn

$$x_k \frac{\partial^2 u}{\partial x_j \partial x_k} \mathcal{F}_{r_{ij}} \nu_i = \frac{\partial^2 u}{\partial x_i \partial x_j} \mathcal{F}_{r_{ij}} x_k \nu_k .$$

Integrating (29) over Ω, applying the divergence theorem to the left–hand side and finally using the boundary conditions, we are led immediately to the required conclusion.

Condition (30) is again related to the Weierstrass condition for the problem (27), namely,

$$\mathcal{F}(x,u,p,\bar{r}) - \mathcal{F}(x,u,p,r) - (\bar{r} - r)_{ij} \mathcal{F}_{r_{ij}}(x,u,p,r) \geq 0,$$

for all $r, \bar{r}$ such that $(\bar{r} - r)_{ij} = C\xi_i\xi_j$, with $C \in \mathbf{R}$ and $\xi \in \mathbf{R}^n$ (see [3]). Setting $u = p = \bar{r} = 0$, we get precisely (30).

As an application of Theorem 7, suppose that

$$\mathcal{F}(x,u,p,r) = \frac{1}{2}\left(\sum_{i=1}^{n} r_{ii}\right)^2 - F(u),$$

with $F(0) = 0$, in which case (28) takes the biharmonic form

$$\Delta^2 u - f(u) = 0,$$

where $f = F'$. The condition of non–existence then becomes

$$auf(u) - nF(u) > 0 \qquad \text{for } u \neq 0,$$

for some constant $a \leq (n-4)/2$ and $n > 4$. When

$$F(u) = -\frac{1}{2}u^2 + \frac{1}{s+1}|u|^{s+1},$$

that is, for the equation $\Delta^2 u + u - u|u|^{s-1} = 0$, this reduces to the condition $s \geq (n+4)/(n-4)$.

Analogous identities for third and higher-order variational problems are somewhat more complicated to state. The general principle can be understood clearly, however, if we consider a third-order integrand $\mathcal{F} = \mathcal{F}(x,u,p,r,s)$, the notation being obvious. The appropriate identity is then obtained by expanding the expression

$$\frac{\partial}{\partial x_i}\Bigg[h_i\mathcal{F} - \left[h_\ell\frac{\partial u}{\partial x_\ell}+au\right]\left[\mathcal{F}_{p_i} - \frac{\partial}{\partial x_j}\mathcal{F}_{r_{ij}} + \frac{\partial^2}{\partial x_j\,\partial x_k}\mathcal{F}_{s_{ijk}}\right] - \frac{\partial}{\partial x_j}\left[h_\ell\frac{\partial u}{\partial x_\ell}+au\right]\left[\mathcal{F}_{r_{ij}} - \frac{\partial}{\partial x_k}\mathcal{F}_{s_{ijk}}\right] - \frac{\partial^2}{\partial x_j\,\partial x_k}\left[h_\ell\frac{\partial u}{\partial x_\ell}+au\right]\mathcal{F}_{s_{ijk}}\Bigg],$$

where $\mathcal{F}$ and its derivatives are evaluated at (x,u,Du,D^2u,D^3u).

To express the corresponding formulae for a general K-th order integrand, it is convenient to use multi-index notation. Let $D_i = \partial/\partial x_i$ and $\alpha = (\alpha_1,\dots,\alpha_n)$, where the α_i are non-negative integers, and define the operator

$$D^\alpha = D_1^{\alpha_1}\cdots D_n^{\alpha_n}.$$

Thus the order of D^α is $|\alpha| = \alpha_1+\cdots+\alpha_n$. Since we are assuming that $\mathcal{F}$ is symmetric in its arguments r_{ij}, as well as in the variables corresponding to all higher derivatives of u, it is convenient to define

$$\mathcal{F}_{r^\alpha} = C_\alpha\mathcal{F}_{r_{ij\cdots k}}$$

for $r^\alpha = r_i r_j\cdots r_k$ and $C_\alpha = |\alpha|!/\alpha_1!\alpha_2!\cdots\alpha_n!$.

The Euler-Lagrange equation for a K-th order integrand $\mathcal{F}$ is now simply

$$\sum_{|\alpha|\le K}(-1)^{|\alpha|}D^\alpha\mathcal{F}_{r^\alpha} = 0.$$

In turn, the divergence expression necessary for the principal identity takes the form

$$D_i\left\{h_i\mathcal{F} - \sum_{|\alpha+\beta|\le K-1}(-1)^{|\beta|}\frac{C_\alpha C_\beta}{C_\gamma}D^\alpha v\,D^\beta\mathcal{F}_{r^\gamma}\right\},$$

where $v = h \cdot Du + au$, the multi-index γ is defined by

$$r^\gamma = r^{\alpha+\beta} r_i ,$$

and we sum over i from 1 to n. Of course it is now necessary that h and a be of class $C^K(\Omega)$. A straightforward, though somewhat long, calculation gives the divergence expansion as

$$\mathcal{F} \operatorname{div} h + h \cdot \mathcal{F}_x - \sum_{|\alpha| \le K} \mathcal{F}_{r^\alpha} \{ D^\alpha (h \cdot D + a) - (h \cdot D) D^\alpha \} u .$$

The term in braces involves derivatives of u only up to the order $|\alpha|$, thus providing the required identity. In particular, if $h = x$ and $a =$ constant, it is easy to see that

$$D^\alpha (h \cdot D + a) - (h \cdot D) D^\alpha = (a + |\alpha|) D^\alpha .$$

The corresponding *integral identity* for K–th order integrands, when $u = Du = \ldots = D^{K-1} u = 0$ on $\partial\Omega$ and $h(x) = x$, $a(x) =$ constant in Ω, is easily found. Reverting to our original notation, we obtain the following formula for an integrand $\mathcal{F}(x,u,p,r,\ldots,t)$,

$$\oint_{\partial\Omega} \Big[\mathcal{F}(x,0,\ldots,0,t) - t \cdot \mathcal{F}_t(x,0,\ldots,0,t) \Big] (x \cdot \nu) \, ds$$
$$= \int_\Omega \Big[n\mathcal{F} + x \cdot \mathcal{F}_x - au \mathcal{F}_u - (a+1) p \cdot \mathcal{F}_p - \cdots - (a+K) t \cdot \mathcal{F}_t \Big] \, dx ,$$

where $p, \ldots, t$ respectively stand for the first derivatives up to the K–th derivatives of u. This identity allows a number of non–existence theorems to be read off with ease.

For example, the integrand

$$\mathcal{F} = \begin{cases} \frac{1}{2} |D\Delta^{(K-1)/2} u|^2 - F(u) & \text{for } K \text{ odd} \\ \frac{1}{2} |\Delta^{K/2} u|^2 - F(u) & \text{for } K \text{ even,} \end{cases}$$

where K is a positive integer, has the corresponding Euler–Lagrange equation

$$(-\Delta)^K u - f(u) = 0. \tag{32}$$

When K is even, the preceding identity takes the form

$$-\frac{1}{2}\oint_{\partial\Omega}|\Delta^{K/2}u|^2(x\cdot\nu)\,ds$$
$$=\int_\Omega\left[\left(\frac{n}{2}-a-K\right)|\Delta^{K/2}u|^2+auf(u)-nF(u)\right]dx,$$

with a similar relation of course holding when K is odd. This gives the following result.

Theorem 8. *Suppose $n>2K$. Let $u\in C^{2K}(\Omega)\cap C^{2K-1}(\overline{\Omega})$ be a solution of* (32) *in a star-shaped region Ω, with $u=Du=\ldots=D^{K-1}u=0$ on $\partial\Omega$. If*

$$auf(u)-nF(u)>0 \qquad \text{for } u\neq 0$$

for some $a\leq(n-2K)/2$, then $u\equiv 0$ in Ω.

In particular, when

$$F(u)=\frac{\lambda}{2}u^2+\frac{1}{s+1}|u|^{s+1},$$

we obtain the result stated at the end of the introduction.

With the exception of those results based on use of the unique continuation theorem or the boundary point lemma, the conclusions of Sections 2, 3, and 4 carry over with only minor changes to vector–valued extremals and higher–order integrands. We leave the statement of these results to the interested reader.

Acknowledgment. The first author wishes to thank the Department of Mathematics of the University of Minnesota for its hospitality during the period when this work was carried out, and also the Gruppo Nazionale di Analisi Funzionale a sue Applicazioni of the Consiglio Nazionale delle Ricerche for its support.

References

[1] H. Brezis & L. Nirenberg, *Positive solutions of nonlinear elliptic equations involving critical Sobolev exponents,* Comm. Pure Appl. Math. **36** (1983), 437–477.

[2] M. J. Esteban & P. L. Lions, *Existence and non-existence results for semilinear elliptic problems in unbounded domains,* Proc. Royal Soc. Edinburgh **93–A** (1982), 1–14.

[3] L. M. Graves, *The Weierstrass condition for multiple integral variational problems involving higher derivatives,* Proc. Amer. Math. Soc. **11** (1960), 750–756.

[4] A. E. Green, *On some general formulae in finite elastostatics,* Archive Rational Mech. Anal. **50** (1973), 73–80.

[5] J. K. KNOWLES & E. STERNBERG, *On a class of conservation laws in linearized and finite elasticity*, Archive Rational Mech. Anal. **44** (1972), 187–211.
[6] W.-M. NI & J. SERRIN, *Non-existence theorems for quasilinear partial differential equations*, Suppl. Rend. Circ. Mat. Palermo (to appear).
[7] P. OLVER, *Applications of Lie Groups to Differential Equations*, Springer-Verlag, New York, 1986 (to appear).
[8] S. I. POHOŽAEV, *Eigenfunctions of the equation* $\Delta u+\lambda f(u)=0$, Soviet Math. Dokl. **6** (1965), 1408–1411 (English transl., Dokl. Akad. Nauk. SSRR **165** (1965), 33–36).
[9] M. H. PROTTER & H. F. WEINBERGER, *Maximum Principles in Differential Equations*, Prentice-Hall, Englewood Cliffs, New Jersey, 1967.
[10] P. RABINOWITZ, *Some global results for nonlinear eigenvalue problems*, J. Funct. Anal. **7** (1971), 487–513.
[11] F. RELLICH, *Darstellung der Eigenwerte von* $\Delta u+\lambda u=0$ *durch ein Randintegral*, Math. Z. **46** (1940), 635–636.

P. PUCCI
Dipartimento di Matematica
Universitá degli Studi
06100 Perugia, Italy

J. SERRIN
Department of Mathematics
University of Minnesota
Minneapolis, Minnesota 55455

Received September 26, 1985.

J. Math. Pures Appl. 69 (1990), pp. 55–83. **Reprint I.11.[ii]**

CRITICAL EXPONENTS AND CRITICAL DIMENSIONS FOR POLYHARMONIC OPERATORS

By Patrizia PUCCI and James SERRIN

Dedicated to Jacques-Louis Lions on the occasion of his sixtieth birthday.

Introduction

In the well known work of Brezis and Nirenberg on positive solutions of semilinear elliptic equations involving critical exponents there occurs a remarkable phenomenon, relating to the dimension $n=3$ and the possibility of solving the Dirichlet problem in a bounded domain. The purpose of this paper is to show that the same phenomenon also appears in the study of higher order elliptic boundary value problems.

More specifically, for the representative problem

$$\text{(I)} \qquad \begin{aligned} -\Delta u &= \lambda u + u^s \quad \text{in } B \\ u &= 0 \qquad\quad \text{on } \partial B, \end{aligned}$$

where λ is a real constant, $s=(n+2)/(n-2)$ is the critical Sobolev exponent, and B is the unit ball in $\mathbf{R}^n$, $n>2$, it was shown in [1] that the conditions for existence of a positive radial solution are surprisingly different when $n=3$ and when $n\geqq 4$. When $n\geqq 4$, problem (I) has a positive solution if and only if $\lambda\in(0, \lambda_1)$, where λ_1 denotes the first eigenvalue of $-\Delta$ in B with homogeneous Dirichlet boundary conditions. On the other hand, when $n=3$ the range $(0, \lambda_1)$ is no longer correct, but must instead be replaced by the *smaller interval* $(\lambda_1/4, \lambda_1)$. The lower bound for the set of values λ for which a positive solution can exist is therefore *greater than zero* precisely in the case $n=3$.

Here we treat various polyharmonic problems analogous to (I). For example, for the model biharmonic problem

$$\text{(II)} \qquad \begin{cases} \Delta^2 u = \lambda u + u^s & \text{in } B\subset\mathbf{R}^n \\ u = Du = 0 & \text{on } \partial B, \end{cases}$$

where $s=(n+4)/(n-4)$, $n>4$, the conditions for existence of a positive radial solution in dimensions $n=5, 6, 7$ correspond to the earlier condition for the case $n=3$, while the

JOURNAL DE MATHÉMATIQUES PURES ET APPLIQUÉES. – 0021-7824/1990/01 55 29/$ 4.90/

situation when $n \geqq 8$ corresponds to the Laplace case for $n \geqq 4$. The appearance of more than one critical dimension for the biharmonic operator, rather than a single critical dimension as in the case of the Laplace operator, indicates a further remarkable and unexpected feature of the phenomenon discovered by Brezis and Nirenberg.

The demonstration of the lower bound $\lambda_1/4$ given by Brezis and Nirenberg relies crucially on an identity – (1.37) in their paper – having implicit roots in Noetherian theory. We shall adopt a similar outlook here, examining for polyharmonic operators the consequences of various analogous identities of variational type. These identities in turn are most easily obtained as special cases of a general result established by the authors in [5].

To introduce our results, we first state a refinement of a central conclusion of [5] for the general polyharmonic problem

$$(\mathrm{III})_\Omega \qquad \begin{cases} (-\Delta)^K u = f(u) & \text{in } \Omega \\ u = \mathrm{D}u = \ldots = \mathrm{D}^{K-1} u = 0 & \text{on } \partial\Omega. \end{cases}$$

Here Ω is a bounded star-shaped domain in $\mathbf{R}^n$, $n > 2K$, and

$$f(u) = \lambda u + g(u).$$

This refinement, given formally in Theorem 8 of Section 3, asserts that, if $u\, g(u) \geqq 0$ and if furthermore

$$(\alpha+1)\, G(u) \leqq u\, g(u)$$

for same real $\alpha \geqq s$, $s = (n+2K)/(n-2K)$, then a lower bound for λ for the existence of non-trivial solutions of $(\mathrm{III})_\Omega$ is

$$(\mathrm{IV}) \qquad \lambda \geqq \frac{2}{s+1}\, \frac{\alpha-s}{\alpha-1}\, \lambda_K.$$

Here $G(u) = \int_0^u g(t)\, dt$ and λ_K is the first eigenvalue of the polyharmonic operator $(-\Delta)^K$ in Ω with Dirichlet boundary conditions. The existence condition (IV) reduces precisely to the typical relation $\lambda \geqq 0$ when α takes the limiting value s; for greater values of α the right hand side is positive.

Should g be a pure power, say $g(u) = u\, |u|^{p-1}$, then naturally we take $\alpha = p$. This gives the condition $p \geqq s$ for the validity of (IV) and in turn the existence condition

$$\lambda \geqq \frac{2}{s+1}\, \frac{p-s}{p-1}\, \lambda_K.$$

The number

$$s = \frac{n+2K}{n-2K}$$

is *critical:* it is exactly at this value of the exponent p that (IV) becomes valid. In alternate terms, it is exactly at this value of p that the variational integral for problem $(\mathrm{III})_\Omega$ loses compactness, as follows from the Sobolev-Kondrachov embedding theorem, that is $W_0^{K,2}(\Omega) \overset{\text{comp}}{\subset} L^{p+1}(\Omega)$ fails exactly when $p \geqq s$.

Returning to the strictly radial case — when $\Omega = B$ — we have the following main result, formally stated in Theorem 5. Let u be a radially symmetric solution of problem $(\mathrm{III})_B$ in the case $n=2K+1$, and suppose that $u\,g(u) \geqq 0$ and

$$(s+1)\,G(u) \leqq u\,g(u).$$

Then the condition $\lambda \geqq 0$ given by (IV) can be replaced by the stronger restriction

$$\lambda > \left(2K - \frac{1}{2}\right)\lambda_{K-1}.$$

This result corresponds closely with the conclusion obtained by Brezis and Nirenberg for the Laplace case. Indeed, if we take $K=1$, $n=2K+1=3$, and $s=5$, and also $ug(u)=u^6=6\,G(u)$, then our necessary condition for existence becomes $\lambda > 3/2$, while they have $\lambda > \lambda_1/4 = \pi^2/4$. That our bound is not best possible is due to the complexity of treating the problem when $K \geqq 2$.

We shall say that the dimension $n=2K+1$ is *critical*, in view of the fact that the natural existence condition $\lambda \geqq 0$ can in *this dimension* be replaced by $\lambda >$ Pos. Const.

It is worth observing finally that if $g(u)>0$ for $u>0$, then a *necessary condition for the existence of non-negative, non-trivial solutions of problem* $(\mathrm{III})_B$ *is that*

$$\lambda < \lambda_K.$$

This follows, exactly as in the Laplace case, from the positivity of the first eigenfunction of $(-\Delta)^K$ in B, the latter result having been established by P. L. Lions in a recent paper [4].

Since the eigenvalues λ_K are not generally tabulated, it is worth expressing the main estimate above in more accessible terms. In this respect the following eigenvalue inequality, obtained in [6], is useful:

$$\lambda_K \geqq \begin{cases} \lambda_1 (\lambda_1 \mu_2)^m, & K=2m+1 \quad \text{odd} \\ (\lambda_1 \mu_2)^m, & K=2m \quad \text{even,} \end{cases}$$

where λ_1 and μ_2 are respectively the first eigenvalue of the Laplace operator in B with Dirichlet boundary conditions and the first non-zero eigenvalue for the radial Laplace operator in B with Neumann boundary conditions. Since $\lambda_1 = (j_{(n-2)/2})^2$ and $\mu_2 = (j_{n/2})^2$, it is clear that the eigenvalues λ_K increase rapidly with the order of the equation and with the dimension n. For example, when $K=3$ and $n=2K+1=7$ we already have $\lambda_K \geqq 1622$.

For its intrinsic curiosity, consider the case $K=5$. Here the critical dimension is $n=2K+1=11$ and the critical exponent is $s=21$. This yields the equation

$$\Delta^5 u+\lambda u+u^{21}=0$$

and the necessary condition for the existence of a non-trivial radial solution of the Dirichlet problem in the unit ball in $\mathbb{R}^{11}$,

$$\lambda \geqq 9.5\lambda_4 \geqq 9.5\,(j_{9/2})^4\,(j_{11/2})^4 \approx 326\cdot 10^6,$$

a very large number indeed.

When $K=2$ the results are more extensive, as we have noted above. Our most general conclusion in this direction is Theorem 3′, where we show for the biharmonic operator that the dimensions $n=5, 6, 7$ are critical, or, considering n as a real parameter, the values $4<n<8$. In particular, if $u\,g(u)\geqq 0$ and

$$(s+1)\,G(u)\leqq u\,g(u),$$

then a lower bound for λ for the existence of non-trivial radial solutions of problem $(\mathrm{III})_B$ is

$$\lambda>\frac{n}{4}(n^2-16)(8-n), \qquad 4<n<8.$$

It seems almost certain that as the order K increases there should be increasingly large intervals of critical dimensions. The computational difficulties in verifying this appear extreme. Nevertheless, with the results for $K=2$ in mind, and taking into account the variational techniques introduced by Brezis and Nirenberg in [1], we conjecture that for any $K\geqq 1$ the critical dimensions n are *precisely* those in the range

$$2K<n<4K,$$

while for $n\geqq 4K$ problem $(\mathrm{III})_B$ – at least in the case $g(u)=u\,|u|^{s-1}$ – has a positive radial solution if and only if $\lambda\in(0, \lambda_K)$. Recently we have learned from Edmunds, Fortunato and Jannelli [2] that our conjecture is valid for $K=2$; that is, for $n\geqq 8$ positive radial solutions corresponding to the case $g(u)=u\,|u|^{s-1}$ exist for all $\lambda\in(0, \lambda_K)$.

The outline of the paper is as follows. In Section 1 we consider the simplest model case for the polyharmonic problem $(\mathrm{III})_B$, namely

$$f(u)=\lambda u+u\,|u|^{s-1}, \qquad s=\frac{n+2K}{n-2K}.$$

We begin by reformulating, for the model case of polyharmonic operators, the general identity given for higher order variational integrals on pages 700-701 of [5]. The particular results of Section 1, as well as those of later sections, then follow by suitably specializing this polyharmonic identity to radially symmetric functions of various types.

In Section 2 we treat the case of the biharmonic operator, both in the model case $f(u)=\lambda u+u|u|^{s-1}$ and in the more general algebraic case

$$f(u)=\lambda u|u|^{q-1}+u|u|^{s-1}, \qquad 1\leq q<s.$$

It can be expected that even for $q>1$ there is some range of values for which each of the dimensions $n=5, 6, 7$ remains critical, particularly since in [1], where $n=3$ and $s=5$, it was shown that the dimension 3 stays critical for all $q\in[1, 3]$. This expectation is indeed correct, as shown in Theorem 4, with the following ranges:

$$\begin{array}{lll} n=5, & s=9, & q\in[1, 5.77] \\ n=6, & s=5, & q\in[1, 3) \\ n=7, & s=11/3, & q\in[1, 5/3); \end{array}$$

we observe that the upper limiting value for q is exactly $s-2$ when $n\geqq 6$.

Finally in Section 3 we turn to the general situation when

$$f(u)=\lambda u+g(u)$$

or when

$$f(u)=\lambda u|u|^{q-1}+g(u), \qquad q\geqq 1.$$

The computations in the latter cases are direct extensions of those in the earlier parts of the paper; they have been deferred to the end in order not to obscure the main ideas presented in Sections 1 and 2. We remark that the case of general functions $g(u)$ does not seem to have been treated previously, even in the case $K=1$ of the Laplace operator.

1. The general polyharmonic operator

We are concerned with the semilinear polyharmonic problem

$$(1) \qquad \begin{cases} (-\Delta)^K u-f(u)=0 & \text{in } B, \\ u=Du=\ldots=D^{K-1}u=0 & \text{on } \partial B. \end{cases}$$

where B denotes the open unit ball of $\mathbb{R}^n$, K is a positive integer and $n>2K$. The function f we consider is defined by

$$(2) \qquad f(u)=\lambda u+u|u|^{s-1}$$

where $s=(n+2K)/(n-2K)$ is the critical exponent of problem (1) in the sense of the Sobolev embedding theorem.

We are interested in solutions of (1) of class $C^{2K}(\bar{B})$. These of course are also extremals of the variational problem whose integrand $\mathscr{F}$ is given by

$$
\mathscr{F}=\begin{cases} \frac{1}{2}|D\Delta^m u|^2 - F(u), & \text{for } K=2m+1 \text{ odd} \\ \frac{1}{2}|\Delta^m u|^2 - F(u), & \text{for } K=2m \text{ even,} \end{cases} \tag{3}
$$

where

$$
F(u)=\int_0^u f(t)\,dt=\frac{\lambda}{2}u^2+\frac{|u|^{s+1}}{s+1}.
$$

This fact allows us to use the identities established in [5], and in particular shows that solutions $u=u(x)$ of problem (1) have the property

$$
\int_B \mathscr{H}\,dx=0 \tag{4}
$$

where the function $\mathscr{H}$ is computed according to the formula

$$
\mathscr{H}=\mathscr{F}\operatorname{div}h+auf(u)-\Delta^m u\{\Delta^m(h\cdot Du+au)-h\cdot D(\Delta^m u)\} \tag{5}
$$

when $K=2m$ is even, and according to the formula

$$
\mathscr{H}=\mathscr{F}\operatorname{div}h+auf(u)-D\Delta^m u\cdot\{D\Delta^m(h\cdot Du+au)-(h\cdot D)(D\Delta^m u)\} \tag{6}
$$

when $K=2m+1$ is odd. In these formulas a and h are, respectively, scalar and vector-valued functions of class $C^K(\bar{B})$, which can be arbitrarily assigned subject only to the condition that $h=0$ on ∂B. Finally D denotes the differential operator $(\partial/\partial x_1, \ldots, \partial/\partial x_n)$ and x the vector $(x_1, \ldots, x_n)$.

It is our main purpose to consider radial solutions u of (1). We first specialize (5) and (6) by appropriately choosing the functions a and h in order to obtain more immediately useful identities. To this end we require a series of lemmas.

Lemma 1. — *If v is any function of class $C^{2m+1}(B)$, $m=1, 2, \ldots$, then the identities*

$$
\Delta^m(x\cdot Dv)-x\cdot D(\Delta^m v)=2m\Delta^m v, \tag{7}
$$

$$
\Delta^m(r^2 v)=r^2\Delta^m v+4mx\cdot D\Delta^{m-1}v+2m[n+2(m-1)]\Delta^{m-1}v \tag{8}
$$

hold in B.

Proof. — We proceed by induction. It is easily seen that (7) holds for $m=1$. If it is true for $m=1, \ldots, i-1$, $i \geq 2$, then

$$\begin{aligned}\Delta^i(x\cdot Dv)-x\cdot D(\Delta^i v)&=\Delta(\Delta^{i-1}(x\cdot Dv))-x\cdot D(\Delta^i v)\\ &=2(i-1)\Delta^i v+x\cdot D(\Delta^i v)+2\Delta^i v-x\cdot D(\Delta^i v)=2i\Delta^i v,\end{aligned}$$

which proves (7). For $m=1$ it is easily checked that

$$\Delta(r^2 v)=r^2\Delta v+2Dr^2\cdot Dv+(\Delta r^2)v=r^2\Delta v+4x\cdot Dv+2nv.$$

Suppose that (8) holds for $m=1, \ldots, i-1$, $i \geq 2$. Using (8) for $m=i-1$, then (8) again for $m=1$ with v replaced by $\Delta^{i-1}v$, and finally (7) for $m=1$ with v replaced $\Delta^{i-2}v$, we obtain

$$\begin{aligned}\Delta^i(r^2 v)&=\Delta(\Delta^{i-1}(r^2 v))=\Delta(r^2\Delta^{i-1}v)+4(i-1)\Delta(x\cdot D\Delta^{i-2}v)\\ &\quad+2(i-1)[n+2(i-2)]\Delta^{i-1}v=r^2\Delta^i v+4x\cdot D\Delta^{i-1}v+2n\Delta^{i-1}v+4(i-1)x\cdot D\Delta^{i-1}v\\ &\quad+8(i-1)\Delta^{i-1}v+2(i-1)[n+2(i-2)]\Delta^{i-1}v\\ &=r^2\Delta^i v+4ix\cdot D\Delta^{i-1}v+2i[n+2(i-1)]\Delta^{i-1}v.\end{aligned}$$

The proof is complete.

LEMMA 2. — *Let v be a radial function on B, of appropriate smoothness that the derivatives below are well-defined. Put*

$$v'=v'(r)=\frac{dv}{dr}.$$

Then

$$x\cdot Dv=rv'(r),\qquad |Dv|^2=v'(r)^2. \tag{9}$$

Moreover if $v=0$ on ∂B then

$$\int_B rvv'\,dx=-\frac{n}{2}\int_B v^2\,dx; \tag{10}$$

if $Dv=0$ on ∂B then

$$\int_B v'(\Delta v)'\,dx=-\int_B|\Delta v|^2\,dx \tag{11}$$

$$\int_B rv'\Delta v\,dx=\frac{n-2}{2}\int_B|Dv|^2\,dx; \tag{12}$$

and if $Dv=D^2v=0$ on ∂B then

$$\int_B rv''(\Delta v)'\,dx=\frac{n-2}{2}\int_B|\Delta v|^2\,dx. \tag{13}$$

Proof. — The first result is obvious since $Dv=v'(r)x/r$. Now if $v=0$ on ∂B, then letting $\omega(n)$ be the surface measure of the unit ball in $\mathbb{R}^n$, we have

$$\int_B rvv'\,dx=\omega(n)\int_0^1 vv'r^n\,dr=\frac{1}{2}\omega(n)\int_0^1 (v^2)'r^n\,dr=-\frac{n}{2}\omega(n)\int_0^1 v^2r^{n-1}\,dr,$$

which gives (10). If $Dv=0$ on ∂B, then

$$\int_B v'(\Delta v)'\,dx=\int_B Dv\cdot D\Delta v\,dx=-\int_B |\Delta v|^2\,dx$$

by Green's formula. Moroever

$$\int_B rv'\Delta v\,dx=\int_B rv'\left(v''+\frac{n-1}{r}v'\right)dx=-\frac{n}{2}\int_B (v')^2\,dx+(n-1)\int_B (v')^2\,dx$$

by (10), proving (12). Finally, if $Dv=D^2v=0$ on ∂B, then

$$\int_B rv''(\Delta v)'\,dx=\int_B r\left(\Delta v-\frac{n-1}{r}v'\right)(\Delta v)'\,dx=-\frac{n}{2}\int_B |\Delta v|^2\,dx+(n-1)\int_B |\Delta v|^2\,dx$$

by (10) and (11), and the lemma is proved.

We shall use the simple relations (9) repeatedly in the sequel without further mention.

LEMMA 3. — *Let v be a radial function of class $C^{2m+1}(B)$, $m=1,2,\ldots$ Then*

$$\begin{aligned} I\,v &\equiv \Delta^m(r^2x\cdot Dv)-r^2x\cdot D\Delta^m v\\ &=2m\{3r^2\Delta^m v+r(6m-n-2)(\Delta^{m-1}v)'+2(m-1)(2m+n-2)\Delta^{m-1}v\}. \end{aligned}$$

Proof. — Using formula (8) of Lemma 1 with v replaced by $x\cdot Dv$, we have

$$\Delta^m(r^2x\cdot Dv)=r^2\Delta^m(x\cdot Dv)+4mx\cdot D\Delta^{m-1}(x\cdot Dv)+2m[n+2(m-1)]\Delta^{m-1}(x\cdot Dv).$$

Moreover

$$\Delta^m(x\cdot Dv)=x\cdot D\Delta^m v+2m\Delta^m v$$

and

$$\Delta^{m-1}(x\cdot Dv)=x\cdot D\Delta^{m-1}v+2(m-1)\Delta^{m-1}v$$

by virtue of formula (7) in Lemma 1. Hence, using the fact that v is radial, we find

$$\begin{aligned}
\mathrm{I}\,v &= 2mr^2\Delta^m v + 4mr\{r(\Delta^{m-1}v)' + 2(m-1)\Delta^{m-1}v\}' \\
&\quad + 2m[n+2(m-1)]\{r(\Delta^{m-1}v)' + 2(m-1)\Delta^{m-1}v\} \\
&= 2mr^2\Delta^m v + 4mr^2\left\{\Delta^m v - \frac{n-1}{r}(\Delta^{m-1}v)'\right\} \\
&\quad + 2mr[6m+n-4](\Delta^{m-1}v)' + 4m(m-1)[n+2(m-1)]\Delta^{m-1}v \\
&= 2m\{3r^2\Delta^m v + r(6m-n-2)(\Delta^{m-1}v)' + 2(m-1)(2m+n-2)\Delta^{m-1}v\},
\end{aligned}$$

as stated.

LEMMA 4. — *Suppose that v is a radial function of class $C^{2m+2}(B)$, $m=1,2,\ldots$ Then*

$$x\cdot D(x\cdot D\Delta^m v) - x\cdot(x\cdot D)D\Delta^m v = r(\Delta^m v)' \tag{14}$$

and analogously

$$x\cdot D((r^2x\cdot D)\Delta^m v) - x\cdot(r^2x\cdot D)D\Delta^m v = 3r^3(\Delta^m v)'. \tag{15}$$

Proof. — The left hand sides of (14) and (15) are, respectively, $(x\cdot Dx\cdot D)\Delta^m v$ and $(x\cdot D(r^2x)\cdot D)\Delta^m v$.

We are now ready to evaluate the integral $\int_B \mathscr{H}\,dx$ for radial solutions u of problem (1) and for the particular pair of functions

$$a(x) = a(r) = \alpha + \beta r^2, \qquad h(x) = (1-r^2)x,$$

where α and β are real numbers which will be determined later. Note here that $h(x) = D\varphi(r)$ where $\varphi(r) = (1/2)r^2 - (1/4)r^4$.

For the sake of clarity we discuss the cases (5) and (6) separately. Consider first the case when K is even and $m = K/2$. Let u be a radial solution of (1), so that in particular, by standard theory of ordinary differential equations, u is of class $C^{2K}(\bar{B})$. Then from (4) and (5) we obtain

$$\begin{aligned}
\int_B \{F(u)\Delta\varphi - auf(u)\}\,dx &= \frac{n}{2}\int_B |\Delta^m u|^2\,dx - \frac{n+2}{2}\int_B |\Delta^m u|^2 r^2\,dx \\
&\quad - \int_B (\Delta^m u)\{\Delta^m[(1-r^2)x\cdot Du + (\alpha+\beta r^2)u] - (1-r^2)x\cdot D\Delta^m u\}\,d
\end{aligned} \tag{16}$$

since $\Delta\varphi = n-(n+2)r^2$. For simplicity we denote the expression on the left hand-side of (16) by Φ and the third integral on the right hand-side of (16) by Σ.

By virtue of formula (7) in Lemma 1 there results

$$\Sigma=(2m+\alpha)\int_B |\Delta^m u|^2\,dx+\beta\int_B (\Delta^m u)\,\Delta^m(r^2 u)\,dx$$
$$-\int_B (\Delta^m u)\{\Delta^m(r^2 x\cdot Du)-(r^2 x\cdot D)\,\Delta^m u\}\,dx=\sigma_1+\sigma_2+\sigma_3.$$

Using formula (8) in Lemma 1 we find

$$\sigma_2=\beta\int_B |\Delta^m u|^2 r^2\,dx+4\beta m\int_B (\Delta^m u)(\Delta^{m-1}u)'\,r\,dx$$
$$+2\beta m[n+2(m-1)]\int_B (\Delta^m u)(\Delta^{m-1}u)\,dx$$
$$=\beta\int_B |\Delta^m u|^2 r^2\,dx-4\beta m^2\int_B |D\,\Delta^{m-1}u|^2\,dx$$

where at the last step we have applied (12) with $v=\Delta^{m-1}u$ to the second integral and Green's formula $\int_B \psi\Delta\psi\,dx=-\int_B |D\psi|^2\,dx$ to the third. By means of Lemma 3, we obtain also

$$-\sigma_3=6m\int_B |\Delta^m u|^2 r^2\,dx+2m(6m-n-2)\int_B \Delta^m u(\Delta^{m-1}u)'\,r\,dx$$
$$+4m(m-1)(2m+n-2)\int_B (\Delta^m u)(\Delta^{m-1}u)\,dx$$
$$=6m\int_B |\Delta^m u|^2 r^2\,dx-m[n^2-2(m+2)n+4(2m^2-m+1)]\int_B |D\,\Delta^{m-1}u|^2\,dx,$$

by virtue of (12) and Green's formula. Combining σ_1, σ_2, σ_3 and (16) yields

$$\Phi=\left(\frac{n}{2}-2m-\alpha\right)\int_B |\Delta^m u|^2\,dx+\left(6m-\beta-\frac{n+2}{2}\right)\int_B |\Delta^m u|^2 r^2\,dx$$
$$+\{4\beta m^2-m[n^2-2(m+2)n+4(2m^2-m+1)]\}\int_B |D\,\Delta^{m-1}u|^2\,dx.$$

Now choosing

$$\alpha=\frac{n}{2}-K \qquad\text{and}\qquad \beta=3K-\frac{n+2}{2}.$$

we find

(17) $$c\int_B |D\Delta^{m-1}u|^2\,dx+\int_B \{auf(u)-F(u)\Delta\varphi\}\,dx=0,$$

where

(18) $$c=\frac{K}{2}(2K+n-2)(2K-n+2).$$

On the other hand, since $s+1=2n/(n-2K)$ an easy calculation using (2) yields

(19) $$\begin{aligned} auf(u)-F(u)\Delta\varphi &= \lambda u^2\left[\frac{n}{2}-K-\frac{n}{2}+\left(3K-\frac{n+2}{2}+\frac{n+2}{2}\right)r^2\right]\\ &\quad+|u|^{s+1}\left[\frac{n}{2}-K-\frac{n}{s+1}+\left(3K-\frac{n+2}{2}+\frac{n+2}{s+1}\right)r^2\right]\\ &= -\lambda K u^2(1-3r^2)+2K\frac{n-1}{n}r^2|u|^{s+1}. \end{aligned}$$

Combining (17) and (19) gives our principal identity for the case of even K:

(20) $$c\int_B |D\Delta^{m-1}u|^2\,dx-\lambda K\int_B u^2\,dx+3\lambda K\int_B u^2r^2\,dx+2K\frac{n-1}{n}\int_B |u|^{s+1}r^2\,dx=0.$$

Next suppose that u is a regular radial solution of problem (1) when K is odd and $m=(K-1)/2$. Using (4) and (6) and taking a and h as before in the form

$$a(x)=a(r)=\alpha+\beta r^2,\qquad h(x)=(1-r^2)x=D\varphi(r),$$

we find with the help of (9) that

(21) $$\begin{aligned} \Phi &= \frac{n}{2}\int_B |D\Delta^m u|^2\,dx-\frac{n+2}{2}\int_B |D\Delta^m u|^2 r^2\,dx\\ &\quad-\int_B (\Delta^m u)'\{\Delta^m(\alpha+\beta r^2)u\}'\,dx\\ &\quad-\int_B \frac{1}{r}(\Delta^m u)'\{x\cdot D\Delta^m(x\cdot Du)-x\cdot(x\cdot D)D\Delta^m u\}\,dx\\ &\quad+\int_B \frac{1}{r}(\Delta^m u)'\{x\cdot D\Delta^m(r^2x\cdot Du)-x\cdot(r^2x\cdot D)D\Delta^m u\}\,dx\\ &\equiv \tau_1+\tau_2+\tau_3+\tau_4+\tau_5. \end{aligned}$$

Using formula (8) in Lemma 1, we obtain (if $m \geq 1$)

$$-\tau_3 = \alpha \int_B [(\Delta^m u)']^2 \, dx + \beta \int_B (\Delta^m u)' \{ r^2 \Delta^m u + 4mr (\Delta^{m-1} u)'$$

$$+ 2m[n+2(m-1)] \Delta^{m-1} u \}' \, dx = \alpha \int_B |D \Delta^m u|^2 \, dx + \beta \int_B |D \Delta^m u|^2 r^2 \, dx$$

$$+ \beta \int_B (\Delta^m u)' \{ 2r \Delta^m u + 2m(n+2m)(\Delta^{m-1} u)' + 4mr (\Delta^{m-1} u)'' \} \, dx$$

$$= \alpha \int_B |D \Delta^m u|^2 \, dx + \beta \int_B |D \Delta^m u|^2 r^2 \, dx - \beta (4m^2 + 4m + n) \int_B |\Delta^m u|^2 \, dx,$$

where at the last step we have used (10), (11) and (13) with, respectively, $v = \Delta^m u$, $v = \Delta^{m-1} u$ and $v = \Delta^{m-1} u$. The same formula holds if $m = 0$, by direct checking.

Next, from (7) of Lemma 1 and (14) of Lemma 4 we derive

$$-\tau_4 = \int_B \frac{1}{r} (\Delta^m u)' \{ x \cdot D(x \cdot D \Delta^m u) + 2mx \cdot D \Delta^m u - x \cdot D(x \cdot D \Delta^m u) + r (\Delta^m u)' \} \, dx$$

$$= (2m+1) \int_B |D \Delta^m u|^2 \, dx.$$

By virtue of Lemma 3 and (15) of Lemma 4 it follows that

$$\tau_5 = \int_B (\Delta^m u)' \{ r^2 x \cdot D \Delta^m u + 6mr^2 \Delta^m u$$

$$+ 2mr(6m-n-2)(\Delta^{m-1} u)' + 4m(m-1)(2m+n-2) \Delta^{m-1} u \}' \, dx$$

$$+ \int_B (\Delta^m u)' \{ 3r^2 (\Delta^m u)' - (r^2 x \cdot D \Delta^m u)' \} \, dx = 3(2m+1) \int_B |D \Delta^m u|^2 r^2 \, dx$$

$$+ 2m \int_B (\Delta^m u)' \{ 6r \Delta^m u + [4m^2 - 2m + 2 + n(2m-3)] (\Delta^{m-1} u)'$$

$$+ r(6m-n-2)(\Delta^{m-1} u)'' \} \, dx$$

$$= 3(2m+1) \int_B |D \Delta^m u|^2 r^2 \, dx - m[n^2 - 2mn + 8m(m+1)] \int_B |\Delta^m u|^2 \, dx,$$

by (10), (11) and (13). Combining the evaluations of τ_3, τ_4, τ_5 and (21), we obtain

$$\begin{aligned}\Phi=&\left[\frac{n}{2}-\alpha-(2m+1)\right]\int_B |D\Delta^m u|^2\,dx\\ &+\left[3(2m+1)-\beta-\frac{n+2}{2}\right]\int_B |D\Delta^m u|^2 r^2\,dx\\ &+\{\beta(4m^2+4m+n)-m[n^2-2mn+8m(m+1)]\}\int_B |\Delta^m u|^2\,dx.\end{aligned}$$

Therefore, if we take as before

$$\alpha=\frac{n}{2}-K \quad \text{and} \quad \beta=3K-\frac{n+2}{2},$$

there results

$$c\int_B |\Delta^m u|^2\,dx+\int_B \{\alpha u f(u)-F(u)\Delta\varphi\}\,dx=0 \tag{22}$$

where again

$$c=\frac{K}{2}(2K+n-2)(2K-n+2). \tag{23}$$

Observe that (17) and (22) are perfectly analogous, the constant c being the same whether K is even or odd. We now put

$$\mathscr{G}_K[u]=\frac{1}{2}\begin{cases}|D\Delta^m u|^2 & \text{for } K=2m+1 \text{ odd}\\ |\Delta^m u|^2 & \text{for } K=2m \text{ even,}\end{cases} \tag{24}$$

(so that the integrand (3) takes the form $\mathscr{F}=\mathscr{G}_K-F$) and state the result just proved.

THEOREM 1. — *If u is a radial solution of problem* (1), (2), *then u satisfies*

$$A\int_B \mathscr{G}_{K-1}[u]\,dx-\lambda\int_B u^2\,dx+3\lambda\int_B u^2 r^2\,dx+2\frac{n-1}{n}\int_B |u|^{s+1} r^2\,dx=0 \tag{25}$$

with

$$A=(2K+n-2)(2K-n+2).$$

Our next goal is to use Theorem 1 to obtain a lower bound for the values of λ for which non-trivial radial solutions of the problem (1), (2) can exist. As we shall see, the situation in this regard is strongly dependent on the dimension n, as was originally noted by Brezis and Nirenberg in the case K = 1.

To begin with, in [5] we have shown that the unique solution of problem (1), (2) for $\lambda<0$ is $u\equiv 0$. Hence, independent of Theorem 1, a lower bound for existence for all dimensions is $\lambda=0$. Turning to Theorem 1, it is evident that (25) can supply an improved lower bound for λ only in case the coefficient A is positive. Since $A>0$ only when $n<2K+2$ and since the basic condition corresponding to the critical exponent s is that $n>2K$, we see that Theorem 1 can be applied exactly in the range

$$2K<n<2K+2. \tag{26}$$

The only (integer) dimension obeying the last inequality is $n=2K+1$, for which

$$A=4K-1. \tag{27}$$

For $K=1$ and $n=2K+1=3$, Theorem 1 then yields

$$\left(\frac{3}{2}-\lambda\right)\int_B u^2\,dx<0$$

if $\lambda\geq 0$ and $u\not\equiv 0$. Hence for the Laplace operator in three dimensions there is no non-trivial radial solution unless $\lambda>3/2$.

A slightly better bound was obtained by Brezis and Nirenberg, namely $\lambda>\pi^2/4$, at least for the family of positive solutions. (It appears, however, that their proof applies also to radial solutions, whether positive or not.) Of course, our estimate can also be extended to arbitrary positive solutions, whether radial or not, by application of the well-known symmetry theorems of Serrin [7] and of Gidas, Ni and Nirenberg [3].

The bound $\lambda>\pi^2/4$ is optimal for the class of positive solutions, or of radial solutions, a further result of Brezis and Nirenberg. For solutions which are neither positive nor radial we are unaware of any corresponding result.

For $K=2$ we denote by λ_1 the first eigenvalue of $-\Delta$ in B with homogeneous Dirichlet data. Here the Poincaré inequality holds, namely

$$\lambda_1\int_B |v|^2\,dx\leq\int_B |Dv|^2\,dx \quad \text{when } v=0 \text{ on } \partial B. \tag{28}$$

For $n=2K+1=5$, by combining (25) and (28) we see that necessarily $(1/2)A\lambda_1-\lambda<0$ in order for a non-trivial radial solution to exist. But $\lambda_1=j_{3/2}^2$ where $j_{3/2}$ is the first positive zero of the Bessel function $J_{3/2}(r)$. Hence for the biharmonic operator it is clear that for $\lambda\leq(7/2)j_{3/2}^2\approx 70.67$ the only radial solution possessed by problem (1), (2) is $u\equiv 0$.

For the general polyharmonic operator, the analogue of (28) is

$$\lambda_K\int_B |v|^2\,dx\leq 2\int_B \mathcal{G}_K[v]\,dx$$

where $v=Dv=\ldots=D^{K-1}v=0$ on ∂B, and λ_K is the first eigenvalue of the polyharmonic operator $(-\Delta)^K$ in B with homogeneous Dirichlet boundary conditions. Thus problem

(1), (2) has no non-trivial radial solutions for $n=2K+1$ provided that

$$\lambda \leqq \left(2K-\frac{1}{2}\right)\lambda_{K-1}.$$

This proves:

THEOREM 2. — *Suppose $n=2K+1$. In order for* (1) *to have a non-trivial radial solution when $f(u)$ is given by* (2) *it is necessary that*

$$\lambda > \left(2K-\frac{1}{2}\right)\lambda_{K-1}, \tag{29}$$

where λ_K is the first eigenvalue of the Kth order polyharmonic operator in B with homogeneous Dirichlet boundary conditions.

If the solution is non-negative in B then also it is necessary that $\lambda<\lambda_k$.

Only the last statement requires proof. In fact, by a recent result of P. L. Lions [4] we know that the first eigenfunction φ_K corresponding to λ_K is radial and of constant sign in B. Suppose in particular that $\varphi_K>0$ in B. If u were a positive solution of (1), (2), then

$$\begin{aligned}\lambda_K \int_B u\varphi_K\,dx &= (-1)^K \int_B u\Delta^K\varphi_K\,dx = (-1)^K \int_B \varphi_K\Delta^K u\,dx\\ &= \lambda \int_B u\varphi_K\,dx + \int_B u|u|^{s-1}\varphi_K\,dx > \lambda \int_B u\varphi_K\,dx.\end{aligned}$$

Thus $\lambda<\lambda_K$ as required. Note that this argument is independent of the dimension, indeed depending only on the positivity of the first eignefunction φ_K and the positivity of the second term on the right hand side of (2).

The upper bound on λ in Theorem 2 is optimal; on the other hand the lower bound (27), while of real interest because of its simple and explicit form, is not.

Since the eigenvalues λ_K are not generally tabulated, it is worthwhile giving another form of Theorem 2 in which λ_K does not appear explicitly. The following corollary holds.

COROLLARY. — *The inequality* (29) *in Theorem* 2 *can be replaced by*

$$\lambda > \left(2K-\frac{1}{2}\right)\begin{cases}(\lambda_1\mu_2)^m, & K=2m+1 \text{ odd}\\ \lambda_1(\lambda_1\mu_2)^{m-1}, & K=2m \quad \text{even,}\end{cases} \tag{29'}$$

where λ_1 is as before, and μ_2 is the first non-zero eigenvalue for the radial Laplace operator in B with homogeneous Neumann boundary conditions.

The corollary follows at once from the eigenvalue inequality

$$\lambda_K \geq \begin{cases}\lambda_1(\lambda_1\mu_2)^m, & K=2m+1 \text{ odd}\\ (\lambda_1\mu_2)^m, & K=2m \quad \text{even}\end{cases}$$

proved in [6].

In particular, recalling that $n=2K+1$, we have

$$\lambda_1=(j_{K-1/2})^2, \qquad \mu_2=(j_{K+1/2})^2, \tag{30}$$

where j_l is the first positive zero of the Bessel function $J_l(r)$. (Note that here l is half and odd integer so that J_l can be expressed in terms of elementary functions.) The values (30), when inserted in (29'), give the following necessary conditions for existence for problem (1), (2):

$K=1$, $n=3$, $\lambda>3/2$;

$K=2$, $n=5$, $\lambda>70.67$;

$K=3$, $n=7$, $\lambda>8{,}921$;

$K=4$, $n=9$, $\lambda>1{,}197{,}386$;

$K=5$, $n=11$, $\lambda>326{,}292{,}082$.

Further values are easily computed from Bessel function tables but become increasingly large ($K=19$, $\lambda>(37.5)(592)^{18}\approx 31\cdot 10^{50}$). Note moreover that these are all lower estimates for λ, and that lower bounds given by (29), themselves not optimal, would be appreciably larger.

Remark. — All the properties established in this section continue to hold when the exponent s in (2) is supercritical, namely satisfies

$$s\geqq(n+2K)/(n-2K).$$

Indeed, the main identities (17) and (22) are unchanged, while the coefficients of $|u|^{s+1}$ and $|u|^{s+1}r^2$ in (19) are still nonnegative, as required for our argument. To be specific,

$$\frac{n}{2}-K-\frac{n}{s+1}\geqq\frac{n-2K}{2}-\frac{n-2K}{2}=0.$$

and $3K-(n+2)/2+(n+2)/(s+1)\geqq 0$ since, in particular, $3K-(n+2)/2\geqq 0$ when $n\leqq 4K$. Theorems 1 and 2 then hold provided (25) is replaced by the obvious analogue

$$\begin{aligned}2c\int_B \mathscr{G}_{K-1}[u]\,dx-\lambda K\int_B u^2\,dx+3\lambda K\int_B u^2r^2\,dx\\ +\left(\frac{n-2K}{2}-\frac{n}{s+1}\right)\int_B|u|^{s+1}\,dx+\left(3K-\frac{n+2}{2}+\frac{n+2}{s+1}\right)\int_B|u|^{s+1}r^2\,dx=0.\end{aligned} \tag{25'}$$

In Section 3 we shall see that the same conclusions extend to considerably more general functions $f(u)$ than those noted above.

2. The biharmonic case

We consider here the semilinear biharmonic problem

(31) $$\begin{cases} (-\Delta)^2 u - f(u) = 0 & \text{in B} \\ u = \mathrm{D}u = 0 & \text{on } \partial\mathrm{B}, \end{cases}$$

where B denotes the open unit ball of $\mathbf{R}^n$ and $n>4$. The function f we treat first is defined as before by

(32) $$f(u) = \lambda u + u|u|^{s-1},$$

where $s=(n+4)/(n-4)$ is the critical exponent of (31) in the sense of the Sobolev embedding theorem. As in the previous section we are interested in radial solutions of (31), (32), which of course are of class $C^4(\bar{\mathrm{B}})$.

We consider formulas (4) and (5) of Section 1, taking $K=2$ and $h=\mathrm{D}\varphi$ where φ and a are assumed to be radial and such that $\mathrm{D}\varphi=0$ on $\partial\mathrm{B}$. (The explicit choice of φ will be decided later.) Then if u is a radial solution of problem (31) we obtain

(33) $$\begin{aligned} \Phi &\equiv \int_{\mathrm{B}} \{ \mathrm{F}(u)\Delta\varphi - auf(u)\}\,dx \\ &= \frac{1}{2}\int_{\mathrm{B}} |\Delta u|^2 \Delta\varphi\,dx - \int_{\mathrm{B}} \Delta u \{\Delta(\mathrm{D}\varphi\cdot\mathrm{D}u) - \mathrm{D}\varphi\cdot\mathrm{D}\Delta u\}\,dx - \int_{\mathrm{B}} \Delta u\,\Delta(au)\,dx. \end{aligned}$$

Here $\Delta(au) = a\Delta u + 2a'u' + u\Delta a$; since $\mathrm{D}\varphi\cdot\mathrm{D}u = \varphi'u'$ we also have

$$\begin{aligned} \Delta(\mathrm{D}\varphi\cdot\mathrm{D}u) - \mathrm{D}\varphi\cdot\mathrm{D}\Delta u &= \varphi'\Delta(u') + 2\varphi''u'' + u'\Delta(\varphi') - \varphi'(\Delta u)' \\ &= 2\varphi''\Delta u + u'\left(\varphi''' - \frac{n-1}{r}\varphi'' + \frac{n-1}{r^2}\varphi'\right). \end{aligned}$$

Consequently (33) becomes

(34) $$\begin{aligned} \Phi &= \int_{\mathrm{B}} |\Delta u|^2 \left(\frac{1}{2}\Delta\varphi - 2\varphi'' - a\right) dx \\ &\quad - \int_{\mathrm{B}} \Delta u \left\{ u'\left(\varphi''' - \frac{n-1}{r}\varphi'' + \frac{n-1}{r^2}\varphi' + 2a'\right)\right\} dx - \int_{\mathrm{B}} u\,\Delta u\,\Delta a\,dx. \end{aligned}$$

We choose a and φ so that

$$\frac{1}{2}\Delta\varphi - 2\varphi'' - a = 0,$$

or in other words

(35) $$a = -\frac{1}{2}\left(3\varphi'' - \frac{n-1}{r}\varphi'\right).$$

The first integral on the right hand side of (34) now vanishes, so (34) takes the form

$$\Phi = 2\int_B u'\varphi'''\Delta u\,dx - \int_B u\,\Delta u\,\Delta a\,dx. \tag{36}$$

Next

$$\Delta u^2 = 2u\,\Delta u + 2|Du|^2$$

and

$$\Delta(u^2\Delta a) = u^2\Delta^2 a + 2\,Du^2\cdot D\Delta a + \Delta u^2\,\Delta a = u^2\Delta^2 a + 2\,Du^2\cdot D\Delta a + 2(u\,\Delta u + |Du|^2)\,\Delta a.$$

By Green's formula, the divergence theorem and the given boundary conditions, it follows also that

$$\int_B Du^2\cdot D\Delta a\,dx = -\int_B u^2\Delta^2 a\,dx, \qquad \int_B \Delta(u^2\Delta a)\,dx = 0.$$

Hence

$$\int_B u\,\Delta u\,\Delta a\,dx = \frac{1}{2}\int_B u^2\Delta^2 a\,dx - \int_B |Du|^2\Delta a\,dx.$$

Moreover we have

$$\begin{aligned}2\int_B u'\varphi'''\Delta u\,dx &= \int_B\left(2u'u''\varphi''' + 2\frac{n-1}{r}u'^2\varphi'''\right)dx\\ &= \omega(n)\int_0^1 (u'^2)'\varphi''' r^{n-1}\,dr + 2\int_B u'^2\frac{n-1}{r}\varphi'''\,dx\\ &= -\int_B u'^2\varphi^{IV}\,dx + \int_B u'^2\frac{n-1}{r}\varphi'''\,dx = -\int_B |Du|^2\left(\varphi^{IV} - \frac{n-1}{r}\varphi'''\right)dx.\end{aligned}$$

Hence the principal identity (36) can be written

$$\Phi = \int_B |Du|^2\left(\Delta a - \varphi^{IV} + \frac{n-1}{r}\varphi'''\right)dx - \frac{1}{2}\int_B u^2\Delta^2 a\,dx, \tag{37}$$

where by (32)

$$\Phi = \int_B \{F(u)\Delta\varphi - auf(u)\}\,dx = \int_B\left[\frac{1}{2}\lambda u^2\Delta\varphi + \frac{1}{s+1}|u|^{s+1}\Delta\varphi - a\lambda u^2 - a|u|^{s+1}\right]dx.$$

Now choose

$$\varphi = \varphi(r) = \frac{1}{2}r^2 - \frac{1}{b+2}r^{b+2},$$

where $b>0$ is constant, in which case we have

$$a(r)=\frac{n-4}{2}+\frac{3b+4-n}{2}r^b$$

by (35). It follows in turn that $\Delta\varphi=n-(n+b)r^b$ and

$$\Delta a-\varphi^{\text{IV}}+\frac{n-1}{r}\varphi'''=\left\{\frac{b}{2}(3\,b+4-n)(b+n-2)+b(b+1)(b-n)\right\}r^{b-2}$$
$$=\frac{b}{2}(5b^2+4n-n^2-8)r^{b-2}=\frac{b}{2}[(n+2)(6-n)+5(b^2-4)]\,r^{b-2};$$
$$\frac{1}{2}\Delta^2 a=\frac{b}{4}(b-2)(b-2+n)(b-4+n)(3b+4-n)r^{b-4};$$
$$a-\frac{1}{2}\Delta\varphi=-2\varphi''=-2[1-(b+1)r^b];$$
$$a-\frac{1}{s+1}\Delta\varphi=\frac{n-4}{2}-\frac{n}{s+1}+\left(\frac{3b+4-n}{2}+\frac{b+n}{s+1}\right)r^b$$
$$=2b\frac{n-1}{n}r^b \qquad \text{for} \quad s=(n+4)/(n-4).$$

Thus the main identity (37) can be written in the final form

$$(38)\quad c_1\int_{\mathrm{B}}|\mathrm{D}u|^2 r^{b-2}\,dx+c_2\int_{\mathrm{B}}u^2 r^{b-4}\,dx-2\lambda\int_{\mathrm{B}}u^2(1-c_3 r^b)\,dx+c_4\int_{\mathrm{B}}|u|^{s+1}r^b\,dx=0,$$

where

$$c_1=\frac{b}{2}[(n+2)(6-n)+5(b^2-4)],$$
$$c_2=-\frac{b}{4}(b-2)(b-2+n)(b-4+n)(3b+4-n),$$
$$c_3=b+1>0,$$
$$c_4=2b\frac{n-1}{n}>0.$$

Clearly all the integrals in (38) are convergent since $b>0$, $n>4$.

It is interesting to note that if $b=2$, hence $c_2=0$, we recover the result given in Section 1 for the special case $\mathrm{K}=2$.

We now observe that, since $u=0$ on $\partial\mathrm{B}$,

$$\int_{\mathrm{B}}u^2 r^{b-4}\,dx=\omega(n)\int_0^1 u^2 r^{b-5+n}\,dr=-\frac{2\omega(n)}{b-4+n}\int_0^1 u u' r^{b-4+n}\,dr$$
$$=-\frac{2}{b-4+n}\int_{\mathrm{B}}u u' r^{b-3}\,dx=\frac{2}{b-4+n}\left|\int_{\mathrm{B}}u u' r^{b-3}\,dx\right|,$$

since $b>0$ and $n>4$. By the Cauchy-Schwarz inequality

$$\left|\int_B u u' r^{b-3}\,dx\right| \leqq \left(\int_B u^2 r^{b-4}\,dx\right)^{1/2} \left(\int_B |u'|^2 r^{b-2}\,dx\right)^{1/2}.$$

It therefore follows that

$$\int_B |Du|^2 r^{b-2}\,dx \geqq \frac{(b-4+n)^2}{4} \int_B u^2 r^{b-4}\,dx. \tag{39}$$

Substituting (39) into (38) we obtain

$$\left[\frac{(b-4+n)^2}{4} c_1 + c_2\right] \int_B u^2 r^{b-4}\,dx - 2\lambda \int_B u^2 (1 - c_3 r^b)\,dx + c_4 \int_B |u|^{s+1} r^b\,dx \leqq 0, \tag{40}$$

provided $c_1 \geqq 0$. Since $b>0$, we have $c_1 \geqq 0$ if and only if

$$b \geqq \left(\frac{n^2-4n+8}{5}\right)^{1/2}; \tag{41}$$

in what follows we shall always assume that this condition holds.

The coefficient

$$\mu = \frac{(b-4+n)^2}{4} c_1 + c_2$$

in (40) is strictly positive if and only if $n-4<b<n$. Indeed,

$$\begin{aligned} \mu &= \frac{b}{8}(b-4+n)\left[(b-4+n)(5b^2+4n-n^2-8) + 2(2-b)(b-2+n)(3b+4-n)\right] \\ &= \frac{b}{8}(b-4+n)(b+4-n)(n+b)(n-b), \end{aligned} \tag{42}$$

and the assertion follows since $b>0$ and $n>4$.

As a first application, we consider (40) in the special case when $4<n<8$ a $b=4$. Since (41) is surely satisfied, this gives

$$\mu \int_B u^2\,dx - 2\lambda \int_B u^2 (1 - 5r^4)\,dx + 8\frac{n-1}{n} \int_B |u|^{s+1} r^4\,dx \leqq 0, \tag{43}$$

where

$$\mu = \frac{n}{2}(n^2-16)(8-n) > 0.$$

We recall again that for $\lambda \leqq 0$ the unique solution of problem (31), (32) is $u \equiv 0$. Hence let us assume that $\lambda > 0$. In this case if u is a non-zero radial solution, then u must satisfy (43) and λ must be strictly greater than $\mu/2$, that is

$$\lambda > \frac{n}{4}(n^2-16)(8-n), \qquad 4 < n < 8.$$

We have thus shown the following

THEOREM 3. — *Let $4 < n < 8$ and consider the Dirichlet problem*

$$\begin{aligned} &\Delta^2 u = \lambda u + u|u|^{s-1} && \textit{in } \mathrm{B}, \qquad s = (n+4)/(n-4), \\ &u = \mathrm{D}u = 0 && \textit{on } \partial\mathrm{B}, \end{aligned}$$

where B is the open unit ball of $\mathbb{R}^n$. Then a necessary condition for this problem to have a non-zero radial solution is that

$$\lambda > \frac{n}{4}(n^2-16)(8-n).$$

Remarks. — For the case $n=5$, noted at the end of the previous section, Theorem 3 gives the bound $\lambda > 33.75$, not nearly as good as before. For $n=6$ we have $\lambda > 60$ and for $n=7$ the condition is $\lambda > 57.75$. These cases, of course, could not be treated at all before.

After completing this manuscript, we learned that Edmunds, Fortunato and Jannelli have shown that when $K=2$ and $n \geqq 8$ there exist positive radial solutions of (31)-(32) for all $\lambda \in (0, \lambda_1)$. Thus the range $4 < n < 8$ in Theorem 3 is *best possible* in the sense that there is no larger set of n for which *non-existence* can be established for λ in some neighborhood of zero.

We now consider problem (31) with the function f defined by

(44) $$f(u) = \lambda u|u|^{q-1} + u|u|^{s-1}, \quad 1 \leqq q < s, \qquad s = (n+4)/(n-4).$$

Arguing as above, we see that for $b > 0$ and $n > 4$ the main identity (38) takes the form

(45) $$\begin{aligned} &c_1 \int_{\mathrm{B}} |\mathrm{D}u|^2 r^{b-2}\,dx + c_2 \int_{\mathrm{B}} u^2 r^{b-4}\,dx \\ &\qquad - \lambda \int_{\mathrm{B}} \left(\frac{n}{q+1} - \frac{n-4}{2} - c_5 r^b\right)|u|^{q+1}\,dx + c_4 \int_{\mathrm{B}} |u|^{s+1} r^b\,dx = 0, \end{aligned}$$

where c_1, c_2, c_4 are as before and

(46) $$c_5 = \frac{3b+4-n}{2} + \frac{b+n}{q+1}.$$

Thus, as before, if (41) holds, in other words if $c_1 \geqq 0$, then a non-zero radial solution u of problem (31), with f given by (44), must satisfy (45) and hence, by (39), the inequality

$$(47)\qquad \mu\int_B u^2 r^{b-4}\,dx-\lambda\int_B\left(\frac{n}{q+1}-\frac{n-4}{2}-c_5 r^b\right)|u|^{q+1}\,dx+c_4\int_B |u|^{s+1} r^b\,dx\leqq 0.$$

We now use (47) to obtain a positive lower bound for λ for appropriate values of $q\geqq 1$. In particular, by Young's inequality there holds

$$|u|^{q-1}\leqq C(r^{b-4}+r^b|u|^{s-1}),$$

where C is a positive constant (depending on q and s), provided that b has the interpolation value

$$(48)\qquad b=4\frac{s-q}{s-1}.$$

Thus if (41) holds, then from (47) and this choice of b we get

$$\lambda\left(\frac{n}{q+1}-\frac{n-4}{2}\right)\int_B |u|^{q+1}\,dx>\mu\int_B u^2 r^{b-4}\,dx+c_4\int_B |u|^{s+1} r^b\,dx\geqq\frac{\operatorname{Min}(\mu,\, c_4)}{C}\int_B |u|^{q+1}\,dx$$

provided that $\mu>0$ and $c_5>0$. By (42) and (46) the last two conditions hold if $n-4<b<n$. That $b<n$ is obvious from (48). On the other hand the condition $b>n-4$ is easily checked to be equivalent to

$$(49)\qquad q<s-2.$$

Since $q\geqq 1$ this condition can hold only if $n<8$. We note also that

$$\frac{n}{q+1}-\frac{n-4}{2}>\frac{(n-4)^2}{8}>0$$

when $q<s-2$.

It remains to verify condition (41) when (48) and (49) hold. In particular, an easy calculation shows that all the required conditions are satisfied when

$$(50)\qquad q\begin{cases}\leqq s-2\sqrt{\dfrac{n^2-4n+8}{5(n-4)^2}} & \text{for } 4<n<6\\ <s-2 & \text{for } 6\leqq n<8.\end{cases}$$

We have proved:

THEOREM 4. – *Assume $4<n<8$ and consider the Dirichlet problem*

$$(51)\qquad \begin{cases}\Delta^2 u=\lambda u|u|^{q-1}+u|u|^{s-1} & \text{in B}\\ u=Du=0 & \text{on } \partial B,\end{cases}$$

where $q \geqq 1$ and $s = (n+4)/(n-4)$. Suppose also that (50) *holds. If this problem admits a non-trivial radial solution, then $\lambda > \lambda_0$ where λ_0 is a positive number depending only on n, q, and s.*

Remarks. — Condition (50) can also be written in the form $q \leqq q(n)$. The graph of this relation is shown in Figure 1. The conclusion of Theorem 4 applies to any pair (n, q) in the shaded region.

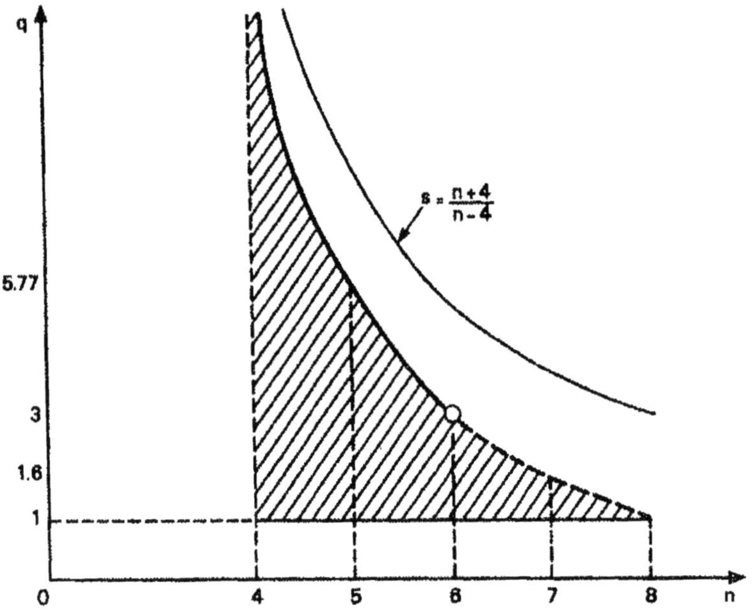

Fig. 1. — Allowed exponents q in (44) for the biharmonic operator. The values on the solid lines and in the shaded region are allowed. The values on the dotted lines cannot be taken by q.

The case $n=6$ yields $1 \leqq q < 3$, $s=5$, which corresponds closely with the Brezis-Nirenberg case $n=3$ for the Laplace equation, when $1 \leqq q \leqq 3$ and $s=5$. The fact that we have not obtained the case $q=3$ is due to the complication of the calculations, for a logarithmic term would then have to be inserted into the formula for φ.

The case $n=5$, corresponding to the discussion in Section 1, gives the condition $1 \leqq q \leqq 5.77$, $s=9$.

The values of n in Theorems 3 and 4 are of course integers, corresponding to the setting of equation (31) in $\mathbb{R}^n$. By treating the *radial* version of the biharmonic operator from the outset, one could, as in the case of the radial Laplace operator, consider n to be a real parameter. The results of Theorems 3 and 4 thus can be extended without change to all real parameters n in the range (4,8). The same remark presumably applies even to the results of Section 1 when $K > 2$. This, however, would require first the proof of the identities (4)-(6) for the radial operator $\Delta = (\ )'' + ((n-1)/r)(\ )'$ with real parameter n, an extension which seems to be calculationally difficult for larger values of K.

3. More general function $f(u)$

In place of condition (2) we consider functions f of the form

$$f(u)=\lambda u+g(u) \tag{52}$$

where $g\in C(\mathbb{R})$. Writing

$$G(u)=\int_0^u g(t)\,dt,$$

we obtain from (17) and (22) the following generalization of (25),

$$\begin{aligned}\mathrm{AK}\int_B \mathscr{G}_{K-1}[u]\,dx-\lambda K\int_B u^2\,dx+3\lambda K\int_B u^2r^2\,dx\\ +\int_B\left\{\left[\frac{n}{2}-K+\left(3K-\frac{n+2}{2}\right)r^2\right]ug(u)\right.\\ \left.-[n-(n+2)r^2]\,G(u)\right\}dx=0\end{aligned} \tag{53}$$

where $A=(2K+n-2)(2K-n+2)$.

Let us suppose that $n>2K$ and

$$(n-2K)\,ug(u)-2nG(u)\geqq 0,\qquad ug(u)\geqq 0, \tag{54}$$

or equivalently

$$(s+1)\,G(u)\leqq ug(u),\qquad ug(u)\geqq 0.$$

In turn, for $n\leqq 6K-2$ we have

$$\begin{aligned}\frac{1}{2}[6K-(n+2)]\,ug(u)+(n+2)\,G(u)&\geqq\left[\frac{n(6K-(n+2))}{n-2K}+(n+2)\right]G(u)\\ &=4K\frac{n-1}{n-2K}\,G(u)\geqq 0.\end{aligned}$$

Consequently, exactly as in Section 1 we obtain the inequality

$$A\int_B \mathscr{G}_{K-1}[u]\,dx-\lambda\int_B u^2\,dx\leqq 0$$

and thus the conclusions of Section 1 follow unchanged. This proves:

THEOREM 5. — *Suppose* $n=2K+1$ *and assume that* (54) *holds. Then problem* (1), (52) *has no non-trivial radial solution when*

$$\lambda \leqq \left(2K-\frac{1}{2}\right)\lambda_{K-1}.$$

If the solution is assumed to be non-negative in B *then also there is no solution when* $\lambda \geqq \lambda_K$.

Remark. — If $g(u)$ is a pure power, $g(u)=u|u|^{p-1}$, say, then (54) forces

$$p \geqq \frac{n+2K}{n-2K},$$

that is, p must be critical or supercritical. Even for the case $K=1$ this provides a generalization of the original non-existence theorem of Brezis and Nirenberg to supercritical exponents.

Next consider the case

$$f(u)=\lambda u|u|^{q-1}+g(u), \qquad q \geqq 1, \tag{55}$$

and assume that

$$(n-2K)\,ug(u)-2nG(u)\geqq 0. \qquad G(u)\geqq \text{Pos. Const. } |u|^{p+1} \tag{56}$$

with $q<p$. Then we obtain as above, for $2K<n\leqq 6K-2$,

$$AK\int_B \mathscr{G}_{K-1}[u]\,dx+\lambda\left(\frac{n}{2}-K-\frac{n}{q+1}\right)\int_B |u|^{q+1}\,dx+A'\int_B |u|^{p+1} r^2\,dx \leqq 0$$

where $A=(2K+n-2)\,(2K-n+2)$ and A′ is a positive constant [*see* the calculation following (54)]. Clearly

$$\frac{n}{2}-K-\frac{n}{q+1}<0$$

provided that $q<(n+2K)/(n-2K)$. Moreover, by iterating the calculation yielding inequality (39) and using the equivalence of the norms $\int_B |D^K u|^2\,dx$ and $\int_B \mathscr{G}_K[u]\,dx$, we find that for some positive constant depending only on K and n,

$$\int_B \mathscr{G}_K[u]\,dx \geqq \text{Pos. Const.} \int_B u^2 r^{-2K}\,dx,$$

provided only that $u=Du=\ldots=D^{K-1}u=0$ on ∂B. Now by Young's inequality

$$|u|^{q-1}\leqq C\,(r^{-2\delta}+r^2|u|^{p-1})$$

where $d=(q-1)/(p-q)$. Consequently, if $d \leqq K-1$ then we obtain a positive lower bound on λ for a solution of (1), (55) to exist. The condition $d \leqq K-1$ is equivalent to

$$q \leqq p-\frac{1}{K}(p-1),$$

which thus proves

THEOREM 6. — *Suppose that $n=2K+1$ and that there exists a non-trivial radial solution u of* (1), (55). *Suppose also that* (56) *holds, and that*

$$(57) \qquad 1 \leqq q \leqq p-\frac{p-1}{K}, \qquad q<4K+1.$$

Then necessarily $\lambda>\lambda_0$, where λ_0 is a positive constant depending only on p, q, K and the constant in (56).

If $p=s=(n+2K)/(n-2K)$ then (57) takes the precise form $1 \leqq q<s-4$. For the particular case of the biharmonic operator, when $n=2K+1=5$ and $s=9$ this is a slightly stronger restriction than that given by Theorem 4 (namely $1 \leqq q \leqq 5.77$).

Turning to a more detailed treatment of the biharmonic case, with $f(u)$ now given by (52), we see that (43) must be replaced by

$$\mu \int_B u^2\,dx-2\lambda \int_B u^2(1-5r^4)\,dx + \int_B \left\{\left(\frac{n-4}{2}+\frac{16-n}{2}r^4\right) u\,g(u)-[n-(n+4)r^4]\,G(u)\right\} dx \leqq 0.$$

Suppose that $n>4$ and that

$$(58) \qquad (n-4)\,u\,g(u)-2n\,G(u) \geqq 0, \qquad u\,g(u) \geqq 0,$$

or equivalently $(s+1)\,G(u) \leqq u\,g(u)$, $u\,g(u) \geqq 0$. In turn

$$\frac{1}{2}(16-n)\,u\,g(u)+(n+4)\,G(u) \geqq \left(\frac{n(16-n)}{n-4}+(n+4)\right) G(u)=16\,\frac{n-1}{n-4}\,G(u) \geqq 0$$

so that the following theorem holds.

THEOREM 3'. — *The result of Theorem 3 holds exactly as before, provided that* (58) *is satisfied.*

Finally consider the case

$$(55) \qquad f(u)=\lambda u|u|^{q-1}+g(u), \qquad q \geqq 1.$$

The argument then proceeds as in the last part of Section 2, with (47) being replaced by

$$\mu \int_B u^2 r^{b-4}\,dx - \lambda \int_B \left(\frac{n}{q+1} - \frac{n-4}{2} - c_5 r^b\right) |u|^{q+1}\,dx$$
(59)
$$+ \int_B \left\{\left(\frac{n-4}{2} + \frac{3b+4-n}{2} r^b\right) u\,g(u) - [n-(n+b)r^b]\,G(u)\right\} dx \leqq 0.$$

provided that (41) holds.

Again we suppose that (58) is satisfied, whence in turn (if $3b+4-n \geqq 0$)

(60) $$\frac{1}{2}(3b+4-n)\,u\,g(u) + (n+b)\,G(u) \geqq 4b\frac{n-1}{n-4}\,G(u) \geqq 0.$$

Now assume additionally that

(61) $$G(u) \geqq \text{Pos. Const.}\,|u|^{p+1}, \qquad q<p.$$

In this case, by (60) and (61) we can write (59) in the form

$$\mu \int_B u^2 r^{b-4}\,dx - \lambda \int_B \left(\frac{n}{q+1} - \frac{n-4}{2} - c_5 r^b\right) |u|^{q+1}\,dx + \mu' \int_B |u|^{p+1} r^b\,dx \leqq 0,$$

for some positive constant μ', provided of course that (41) holds.

The remaining argument of Section 2 can now be applied, with

$$b = 4\frac{p-q}{p-1}, \qquad 4<n<8.$$

We must check that $\mu>0$, that (41) is satisfied, that $c_5 \geqq 0$, and that

(62) $$\frac{n}{q+1} - \frac{n-4}{2} > 0.$$

Condition (62) immediately requires

(63) $$q < \frac{n+4}{n-4}.$$

Next, by (46) and (63).

$$c_5 = \frac{3b+4-n}{2} + \frac{b+n}{q+1} > 2b\frac{n-1}{n} > 0,$$

as required. The condition $\mu>0$ is equivalent, as in Section 2, to $n-4<b<n$. The condition $n-4<b$ is equivalent to

(64) $$q < p - \frac{n-4}{4}(p-1),$$

while the condition $b<n$ follows directly from the facts that $b\leqq 4$ (since $q\geqq 1$) and $n>4$. Note also that $3b+4-n\geqq 0$ when (64) holds. (Condition (64) also forces $n<8$ because $p>q\geqq 1$.)

It remains to examine condition (41). This gives easily

$$q\leqq p-\frac{1}{4}\delta(p-1), \tag{65}$$

where $\delta=[(n^2-4n+8)/5]^{1/2}$. When $4<n<6$ we have $\delta>n-4$, while $\delta\leqq n-4$ when $6\leqq n<8$. Thus (65) is stronger than (64) when $n<6$ and weaker when $n\geqq 6$.

We thus obtain the conditions

$$1\leqq q<\frac{n+4}{n-4}, \tag{66}$$

and

$$q\begin{cases}\leqq p-\dfrac{1}{4}\delta(p-1) & \text{for } 4<n<6\\ <p-\dfrac{n-4}{4}(p-1) & \text{for } 6\leqq n<8.\end{cases} \tag{67}$$

This gives the following result.

THEOREM 7. — *Suppose* $4<n<8$ *and that* (66) *and* (67) *hold. Then problem* (1) *for* $K=2$ *(biharmonic operator), with* $f(u)$ *obeying* (55), (58) *and* (61), *has no non-trivial radial solution when* $\lambda\leqq\lambda_0$, *where* λ_0 *is a positive number depending only on* p, q, n *and the constant in* (61).

Remark. — For the case $n=6$ conditions (66) and (67) reduce to the simple form

$$q<\begin{cases}(p+1)/2, & 5\leqq p\leqq 9\\ 5, & p>9.\end{cases}$$

Final remarks. — The conclusions of Section 3 apply only to the critical dimensions 5, 6, 7 for the biharmonic case and $2K+1$ for the polyharmonic case. As we noted earlier, however, when $f(u)$ is given by (2) a universal lower bound for existence in any dimension is $\lambda=0$; see [5]. In [5] we also studied the *supercritical* case for the Laplace operator. The argument given there (see pages 690-691) can be carried over directly to the case of polyharmonic operators, and yields a companion result to Theorems 5, 6 and 7. In particular from the identity appearing on page 702 — with the coincidence $n/a=\alpha+1$ — we obtain

THEOREM 8. — *Consider problem* (1) *with* $n>2K$ *and*

$$f(u)=\lambda u+g(u), \tag{52}$$

where

(68) $$(\alpha+1)\, G(u) \leqq u\, g(u)$$

for some $\alpha \geqq s$, $s=(n+2K)/(n-2K)$. *Then a necessary condition for a non-trivial solution to exist for an arbitrary bounded star-shaped domain* Ω *is that*

$$\lambda \geqq \frac{2}{s+1}\,\frac{\alpha-s}{\alpha-1}\,\lambda_K,$$

where λ_K *is the first eigenvalue of* $(-\Delta)^K$ *in* Ω *with Dirichlet boundary conditions.*

When $g(u)$ is a pure power, say $u|u|^{p-1}$, it is easy to check that (68) holds only when $\alpha \leqq p$, which in turn gives the condition $p \geqq s$. Similar results can be obtained when $f(u)$ has the form (55).

Acknowledgment

The first author is a member of *Gruppo Nazionale di Analisi Funzionale e sue Applicazioni* of the *Consiglio Nazionale delle Ricerche*, Rome, Italy, and this research has been partly supported by the Italian *Ministero della Pubblica Istruzione.*

REFERENCES

[1] H. Brezis and L. Nirenberg, *Positive solutions of nonlinear elliptic equations involving critical Sobolev exponents*, (*Comm. Pure Appl. Math.*, Vol. 36, 1983, pp. 437-477).

[2] D. E. Edmunds, D. Fortunato and E. Jannelli, *Critical exponents, critical dimensions and biharmonic operator*, (to appear, *Arch. Rational Mech. Anal.*).

[3] B. Gidas, W. M. Ni and L. Nirenberg, *Symmetry and related properties via the maximum principle* (*Comm. Math. Phys.*, Vol. 68, 1979, pp. 209-243).

[4] P. L. Lions, *The concentration-compactness principle in the calculus of variations. The limit case. Part I* (*Revista Matematica Iberoamericana*, Vol. 1, 1985, pp. 145-201).

[5] P. Pucci and J. Serrin, *A general variational identity*, (*Indiana Univ. Math. J.*, Vol. 35, 1986, pp. 681-703).

[6] P. Pucci and J. Serrin, *Remarks on the first eigenspace for polypharmonic operators* (*Atti Sem. Mat. Fis. Univ. Modena*, Vol. 36, 1988, pp. 107-117; see also *Proc. of the Focused Research Program—Spectral Theory and Boundary Value Problems*, Argonne National Laboratory, Report ANL-87-26, Vol. 3, 1989, pp. 135-145, H. G. Kaper, M. K. Kwong and A. Zettl Eds.).

[7] J. Serrin, *A symmetry problem in potential theory*, (*Arch. Rational Mech. Anal.*, Vol. 43, 1971, pp. 304-318).

(Manuscrit reçu en mai 1988.)

Patrizia Pucci,
Dipartimento di Matematica "G. Vitali",
Università di Modena Italy.

James Serrin,
Department of Mathematics,
University of Minnesota, Minneapolis

I.12. The Maximum Principle

Juan Luis Vázquez

Mathematics Subject Classification (2010). 35J15; 35J62; 35J70: 35J92; 35J93.

Keywords. Nonlinear elliptic partial differential equations, quasilinear elliptic inequalities, maximum principles, strong maximum principle, compact support principle.

This is a commentary on two of the most famous papers by James Serrin in collaboration with Patrizia Pucci in the area of the Maximum Principle, and more precisely on the Strong Maximum Principle. It is widely recognized that Maximum Principles are among the most powerful and widely used analytic tools in the study of second-order linear and nonlinear elliptic and parabolic equations. Moreover, the absence of such principles in other types of equations and systems leads to enormous extra difficulties in constructing theories. On the other hand, the kind of maximum principle called strong, associated with the work of Eberhard Hopf in the 1920s and later, is a more specialized tool, but nevertheless it has come to be a cornerstone of the qualitative study of elliptic partial differential equations, both linear and nonlinear. Early interest of Prof. Serrin in the Maximum Principle is shown in his work [1]. Since 1984, Profs. Pucci and Serrin have collaborated to produce more than 40 papers in different topics of the theory of partial differential equations. Ten of them, starting with the paper [3] below, which was written together with H. Zou in 1999, dealt with the Maximum Principle. A different type of contribution is contained in the 2006 paper [6]. In 2007 they published the famous book [7] "Maximum principles for elliptic partial differential equations"; Chapter 5 of that book contains the last word on the topic at this time.

1. Let us review the contents of [3] below. It starts by mentioning that the strong maximum principle (SMP) had been established by Vázquez in [10] for equations of the form

$$-\Delta u + f(u) = g(x) \tag{1}$$

where f is assumed to be a real, nondecreasing function with $f(0) = 0$, and $g(x) \geq 0$ a.e. The SMP says that *any nonnegative weak solution u of* (1) *defined in a domain D of* $\mathbb{R}^N$ *which vanishes at some point $x_0 \in D$ must vanish everywhere in D.* The criterion in [10] says that SMP holds if an integral condition is satisfied by f. Indeed, the condition is the divergence at 0 of the integral $\int_0^r F^{-1/2}(t)\,dt$, where $F(t) = \int_0^t f(s)ds$. For instance, if $f(t) = t^r$, then the SPM holds if $r \geq 1$, while when $f(t) = t\,|\log(t)|^r$ for small $t > 0$, it holds if $r \leq 2$. The sharpness of these conditions is seen from the construction of explicit examples when the integral converges.

P. Pucci et al. (eds.), *James Serrin. Selected Papers: Volume 1*, Contemporary Mathematicians,
DOI 10.1007/978-3-0348-0845-3_12, © Springer Basel 2014

This is then generalized by replacing the Laplacian operator in the equation by the so-called p-Laplacian $\Delta_p(u) = \nabla \cdot (|\nabla u|^{p-2}\nabla u)$ with $p > 1$. The condition for the SMP becomes $\int_0^r F^{-1/p}(t)\,dt = \infty$.

In [3], Pucci, Serrin, and Zhou address the strong maximum principle and a compact support principle for a wide class of elliptic inequalities, generalizing such previous results. Consider a domain D of $\mathbf{R}^n$, $n \geq 2$, and the differential inequality

$$\nabla \cdot (A(|Du|)Du) - f(u) \leq 0, \quad u \geq 0, \tag{2}$$

where A is a continuous function in $(0, \infty)$ such that $\Omega(t) \equiv tA(t)$ is strictly increasing and $\Omega(t) \to 0$ as $t \to 0$; f is a continuous function in $[0, \infty)$, $f(0) = 0$, and f is nondecreasing on the interval $[0, \delta)$ for some $\delta > 0$. They define $H(t) = \int_0^{\Omega(t)} \Omega^{-1}(s)ds$ for $t \geq 0$, and $F(u) = \int_0^u f(s)ds$, and suppose that either $f(s) = 0$ for $s \in [0, \tau)$, $\tau > 0$ or

$$\int_0^\delta ds/H^{-1}(F(s)) = \infty. \tag{$*$}$$

Under these conditions the strong maximum principle is satisfied in the above-defined sense; i.e., whenever u is a nonnegative solution of (2) with $u(x_0) = 0$ for some $x_0 \in D$, then $u \equiv 0$ in D. The concept of solution used in this result is C^1 weak solution. Theorem 1.1.1 of [7] shows that a slight improvement of condition $(*)$ is a necessary and sufficient condition for the SMP.

If the SMP is not valid, there are nonnegative solutions which vanish at some points but not identically. A typical situation encountered in the literature is the equation $\Delta u + f(u) = h$, where h is, e.g., a bounded function with compact support in D. The question is then to determine if the solution of the problem with suitable boundary conditions is also compactly supported. This is called the compact support principle.

The problem with the above general setting is that the validity of the property may depend on the distance of the support of h from the boundary of D. In order to avoid such difficulty, the approach of [3] is to consider the compact support principle on exterior domains. It assumes that D contains the exterior of a ball in $\mathbf{R}^n$, $f(s) > 0$ for $s \in (0, \delta)$ and the integral defined in $(*)$ is finite. Then, if u is a solution of the reverse inequality to (2) and $u(x) \to 0$ as $|x| \to \infty$, it must either vanish identically or else it must have compact support in D. See Theorem 1.1.2 of [7]. A precedent of these results is the work of Diaz and collaborators. See [2].

These results are extended to the case of fully quasilinear inequalities of the form

$$\nabla \cdot (a^{ij}(x)A(|Du|)D_j u) - B(x, u, Du) \leq 0 \qquad [\text{resp.} \geq 0] \tag{3}$$

under the hypotheses: $(a^{ij}(x))$, $i, j = 1, \ldots, n$ is a continuously differentiable, symmetric coefficient matrix on D, which is uniformly elliptic; A is continuously differentiable in $(0, \infty)$, $\Omega(t)$ is defined as before, and there exists a positive constant $\kappa > 0$ such that $B(x, u, p) \leq \kappa\Omega(|p|) + f(u)$ for $x \in D$, $u \geq 0$, and $|p| < 1$.

Finally, the authors obtain comparison principles which generalize previous ones. There have been numerous extensions of those results. For instance, they apply to equations in non-divergence form, as proved by Tahraoui in [9]

2. Let us review the contents of [6] below. The inequalities under consideration take the form

$$\nabla \cdot \mathbf{A}(x, t, u, Du) + \mathbf{B}(x, t, u, Du) \geq 0 \tag{4}$$

in a bounded domain $\Omega \in \mathbb{R}^N$ with precise structure conditions on $\mathbf{A}$ and $\mathbf{B}$. In particular, $\mathbf{A}$ is elliptic with linear growth in Du. The interest is in proving that solutions will be a priori bounded above if $u \leq 0$ on $\partial\Omega$ and the structure conditions are made precise in a suitable way. One important example in this theory is the mean curvature equation

$$\nabla \cdot \left(Du/\sqrt{1+|Du|^2} \right) = n\, H(x)\,. \tag{5}$$

Very precise a priori bounds are obtained both for the particular model and a more general class of inequalities. The relation with the classical literature on the subject is discussed at length. The above SMP results hold true for these mean curvature operators.

References

[1] Serrin, J.; On the strong maximum principle for quasilinear second order differential inequalities. *J. Functional Analysis* 5 (1970), pp. 184–193.

[2] Diaz, J.I.; Nonlinear partial differential equations and free boundaries. Vol. I. Elliptic equations. Research Notes in Mathematics, 106. Pitman (Advanced Publishing Program), Boston, MA, 1985. vii + 323 pp.

[3] Pucci, P., Serrin, J., Zou, H.; A strong maximum principle and a compact support principle for singular elliptic inequalities. *J. Math. Pures Appl.* 78 (1999), pp. 769–789.

[4] Pucci, P., Serrin, J.; A note on the strong maximum principle for elliptic differential inequalities. *J. Math. Pures Appl.* (9) 79 (2000), no. 1, 57–71

[5] Pucci, P., Serrin, J.; The strong maximum principle revisited. *J. Differential Equations* 196 (2004), no. 1, pp. 1–66.

[6] Pucci, P., Serrin, J.; A maximum principle for mean-curvature type elliptic inequalities. *J. Eur. Math. Soc.* (*JEMS*) 8 (2006), pp. 389–398.

[7] Pucci, P., Serrin, J.; The maximum principle. Progress in Nonlinear Differential Equations and their Applications, 73. Birkhäuser Verlag, Basel, 2007. x + 235 pp. ISBN: 978-3-7643-8144-8

[8] Pucci, P., Serrin, J.; Maximum principles for elliptic partial differential equations. Handbook of differential equations: stationary partial differential equations. Vol. IV, 355–483, Handb. Differ. Equ., Elsevier/North-Holland, Amsterdam, 2007.

[9] Tahraoui, R.; Strong maximum principle for general quasilinear problems in non-divergence form. *Nonlinear Anal.* 65 (2006), no. 10, 1955–1976.

[10] Vazquez, J.-L.; A strong maximum principle for some quasilinear elliptic equations. *Appl. Math. Optim.* 12 (1984), pp. 191–202.

Juan Luis Vázquez
Dpto. de Matemáticas
Univ. Autónoma de Madrid
E-28049 Madrid, Spain
e-mail: `juanluis.vazquez@uam.es`.

J. Math. Pures Appl. 69 (1999), pp. 769–789. **Reprint I.12.[i]**

A STRONG MAXIMUM PRINCIPLE AND A COMPACT SUPPORT PRINCIPLE FOR SINGULAR ELLIPTIC INEQUALITIES

Patrizia PUCCI [a], **James SERRIN** [b], **Henghui ZOU** [c]

[a] *Department of Mathematics, University of Perugia*
[b] *Department of Mathematics, University of Minnesota*
[c] *Department of Mathematics, University of Alabama at Birmingham*

Received 8 March 1999

ABSTRACT. – Vazquez in 1984 established a strong maximum principle for the classical m-Laplace differential inequality

$$\Delta_m u - f(u) \leqslant 0,$$

where $\Delta_m u = \operatorname{div}(|Du|^{m-2}Du)$ and $f(u)$ is a non-decreasing continuous function with $f(0) = 0$. We extend this principle to a wide class of singular inequalities involving quasilinear divergence structure elliptic operators, and also consider the converse problem of compact support solutions in exterior domains. © Elsevier, Paris

1. Introduction

In a classical paper, Vazquez established a strong maximum principle for the differential inequality

$$\Delta u - f(u) \leqslant 0 \tag{1.1}$$

in a domain D of $\mathbb{R}^n$, where f is a non-decreasing continuous function in $\mathbb{R}$ with $f(0) = 0$. The essential assumption required is that

$$\int_0^1 \frac{\mathrm{d}s}{\sqrt{F(s)}} = \infty, \tag{1.2}$$

where $F(u) = \int_0^u f(s)\,\mathrm{d}s$, the case $f(u) \equiv 0$ on $[0, \tau)$, $\tau > 0$, being included in (1.2) by agreement. More precisely, he showed under these conditions that *any non-negative solution of* (1.1) *in D which vanishes at some point of D must vanish everywhere in D.*

Vazquez also extended this result to the m-Laplace inequality:

$$\Delta_m u - f(u) \leqslant 0, \quad m > 1, \tag{1.3}$$

provided that condition (1.2) is replaced by

$$\int_0^1 \frac{ds}{[F(s)]^{1/m}} = \infty, \tag{1.4}$$

though the details were not fully carried out.

Conditions (1.2) and (1.4) are known to be necessary for the validity of the strong maximum principle for the inequalities (1.1) and (1.3), see respectively [2] and [5, theorem 1.4]. Indeed in that result more is proved. Consider the general divergence structure inequality:

$$\operatorname{div}\{A(|Du|)Du\} - f(u) \leqslant 0 \tag{1.5}$$

under the ellipticity condition that $\Omega(t) = tA(t)$ is continuously differentiable in $(0, \infty)$ with $\Omega'(t) > 0$ for $t > 0$ and $\Omega(t) \to 0$ as $t \to 0$. Then it was shown that the strong maximum principle *cannot hold* when

$$\int_0^1 \frac{ds}{H^{-1}(F(s))} < \infty, \tag{1.6}$$

where $H = H(t)$ is the strictly increasing function on $[0, \infty)$ defined by $H(0) = 0$ and

$$H(t) = t^2 A(t) - \int_0^t sA(s)\,ds, \quad t > 0. \tag{1.7}$$

For example, the case of the m-Laplace operator $A(t) = t^{m-2}$, $m > 1$, yields

$$H(t) = \frac{m-1}{m} t^m,$$

so that (1.6) becomes exactly the converse of (1.4).

Taking this discussion into account, we are led to the conjecture that *the strong maximum principle should hold for* (1.5) *precisely if*

$$\int_0^1 \frac{ds}{H^{-1}(F(s))} = \infty. \tag{1.8}$$

This conjecture was partially answered in the affirmative in [7]. Our first main result is to provide a more complete resolution of the conjecture, see Theorem 1 in Section 2. In so doing we simultaneously weaken the ellipticity condition on $\Omega(t)$, requiring only that it be continuous and strictly increasing on $(0, \infty)$ and that it approach zero as $t \to 0$.

For the m-Laplace operator, condition (1.8) reduces to Vazquez's condition (1.4). On the other hand, for the minimal surface operator $A(t) = 1/\sqrt{1+t^2}$ we have $H(t) = 1 - 1/\sqrt{1+t^2} \approx \frac{1}{2}t^2$ as $t \to 0$, so that (1.8) is equivalent to (1.2), that is, the same condition which is required for the Laplace operator.

Inequality (1.5), in the special case of equality, is the Euler–Lagrange equation for the canonical variational problem

$$\delta \int_D \{G(|Du|) + F(u)\}\,dx = 0,$$

where $F, G \in C^1[0,\infty)$ with $F(0) = G(0) = F'(0) = G'(0) = 0$, and where G is strictly convex in $[0,\infty)$. The relation with (1.5) is attained by defining $A(t) = G'(t)/t$ for $t > 0$. It is worth noting that

$$H(t) = tG'(t) - G(t)$$

is then the Legendre transform of the integrand G. In particular, for a regular integrand, namely whenever $G \in C^2[0,\infty)$ with $G''(t) > 0$ for all $t \geqslant 0$, we have $H(t) \approx \frac{1}{2}G''(0)t^2$ as $t \to 0$, and condition (1.8) reduces exactly to (1.2), again the same as for the Laplace operator.

In Theorem 2 we consider the case when (1.6) holds. Under this condition, it is shown that *any non-negative solution of the converse inequality*

$$\operatorname{div}\{A(|Du|)Du\} - f(u) \geqslant 0 \tag{1.9}$$

in the exterior of a ball in $\mathbb{R}^n$, *which approaches* 0 *as* $|x| \to \infty$, *must vanish for all sufficiently large* $|x|$, *that is, must either vanish identically or have compact support.*

In Theorem 3 and its Corollary we formalize the result noted above, that (1.8) is a necessary condition for the strong maximum principle to hold for (1.5).

The results described above can be extended to a wider class of differential inequalities by replacing $\operatorname{div}\{A(|Du|)Du\}$ by the more general operator $\operatorname{div}\{a^{ij}(x)A(|Du|)D_ju\}$ and $f(u)$ by $B(x,u,Du)$, where $a^{ij}(x)$ is a positive definite symmetric matrix on D and where B satisfies a condition of the form

$$B(x,u,p) \leqslant \text{Const.}\,|p|\,A(|p|) + f(u), \tag{1.10}$$

for $x \in D$, $u \geqslant 0$ and all $p \in \mathbb{R}^n$ with $|p|$ sufficiently small (reverse the inequality sign for the compact support principle!). *These extensions are the second main purpose of the paper*; see Theorems 1′ and 2′ in Section 4.

An important prototype is the equation

$$\Delta_m u - |Du|^p - f(u) = 0, \quad m > 1,\ p > 0. \tag{1.11}$$

Since $A(t) = t^{m-2}$ for this case, condition (1.10) then requires $p \geqslant m-1$; that is, the strong maximum principle holds for (1.11) when $p \geqslant m-1$ and f obeys (1.8). On the other hand, when $p \in (0, m-1)$ the strong maximum principle can fail, even when f obeys (1.8), e.g., the C^1 function $u(x) = C|x|^k$ satisfies

$$\Delta_m u - |Du|^p = 0, \tag{1.12}$$

where

$$k = 1 + \frac{1}{s}, \quad \frac{1}{C} = k\left[(m-1)\frac{1}{s} + n - 1\right]^{1/s}, \quad s = m - 1 - p > 0$$

(for $m = 2$, this example is due to G. Barles, G. Diaz and J.I. Diaz). It is of further interest in connection with this example that the compact support principle fails when $p > m-1$, e.g., the function $u(x) = L|x|^{-l}$ satisfies (1.12) with $l = |1 + 1/s| > 0$ provided that:

$$p < \min\left(m, \frac{n(m-1)}{n-1}\right), \quad \frac{1}{L} = l\left|(m-1)\frac{1}{s} + n - 1\right|^{1/s}.$$

In the next section we state the main hypotheses on the operator A and the nonlinearity f, and give our main results for the canonical models (1.5) and (1.9). Their proofs are contained in

Section 3, while in Section 4 we consider the case of fully quasilinear inequalities

$$\operatorname{div}\{a^{ij}(x)A(|Du|)D_j u\} - B(x,u,Du) \leqslant 0 \ (\geqslant 0).$$

It is worth adding that the comparison principles in Section 5 are of interest in themselves, as generalizations of the well known comparison Theorem 10.7 of [9] [1] to degenerate quasilinear equations.

Part I. Canonical inequalities

2. Main results

Consider the differential inequalities

$$\operatorname{div}\{A(|Du|)Du\} - f(u) \leqslant 0, \quad u \geqslant 0, \tag{2.1}$$

and

$$\operatorname{div}\{A(|Du|)Du\} - f(u) \geqslant 0, \quad u \geqslant 0, \tag{2.2}$$

in a domain D, possibly unbounded, of $\mathbb{R}^n$, $n \geqslant 2$. Here we assume without further mention the following conditions on $A = A(t)$ and $f = f(u)$:

(A1) $A \in C(0,\infty)$,

(A2) $t \mapsto tA(t)$ is strictly increasing in $(0,\infty)$ and $tA(t) \to 0$ as $t \to 0$;

(F1) $f \in C[0,\infty)$,

(F2) $f(0) = 0$ and f is non-decreasing on some interval $[0,\delta)$, $\delta > 0$.

By a *solution* of (2.1) or (2.2) in D we mean a non-negative function $u \in C^1(D)$ which satisfies (2.1) or (2.2) in the distribution sense.

We define $F(u) = \int_0^u f(s)\,\mathrm{d}s$, $u > 0$. Also, with the notation $\Omega(t) = tA(t)$ when $t > 0$, and $\Omega(0) = 0$, we introduce the function:

$$H(t) = t\Omega(t) - \int_0^t \Omega(s)\,\mathrm{d}s, \quad t \geqslant 0. \tag{2.3}$$

Letting $\Omega^{-1}(\omega)$ be the inverse of the strictly increasing function $\Omega(t)$, then from Stieltjes integration it is easy to see that

$$H(t) = \int_0^{\Omega(t)} \Omega^{-1}(\omega)\,\mathrm{d}\omega, \quad t \geqslant 0. \tag{2.4}$$

Therefore H is strictly increasing on $[0,\infty)$.

We can now state our first main result.

THEOREM 1 (Strong maximum principle). – *Suppose*

$$\liminf_{t\to 0} \frac{H(t)}{t\Omega(t)} > 0, \tag{2.5}$$

[1] Theorem 9.5 in the first edition.

and either $f(s) \equiv 0$ *for* $s \in [0, \tau)$, $\tau > 0$, *or*

$$\int_0^\delta \frac{\mathrm{d}s}{H^{-1}(F(s))} = \infty. \tag{2.6}$$

If u *is a solution of* (2.1) *with* $u(x_0) = 0$ *for some* $x_0 \in D$, *then* $u \equiv 0$ *in* D.

Remark. – Condition (2.5) is perhaps unexpected. It is however automatically satisfied for the canonical example of the m-Laplacian, and for any variational problem with regular integrand G, as follows from the comments in the introduction. At the same time, (2.5) is *not* a consequence of conditions (A1), (A2), as shown by the example

$$A(t) = \left(t \log \frac{1}{t}\right)^{-1} \left[1 + \left(\log \frac{1}{t}\right)^{-1}\right],$$

for which (A1), (A2) are satisfied but not (2.5).

Note added in proof. – We have recently shown by using a more delicate version of Lemma 2 (in Section 3) that (2.5) is in fact not needed for Theorem 1.

It is well-known for linear equations that there is a close relation between the strong maximum principle and the standard boundary point lemma. The same relation holds true here, see in particular Corollary 1′ in Section 4.

In the next result we consider the situation when the integral in (2.6) is convergent. Here the appropriate hypothesis is that u "vanishes" at ∞, rather than at some finite point $x_0 \in D$.

THEOREM 2 (Compact support principle). – *Suppose* $f(s) > 0$ *for* $s \in (0, \delta)$ *and*

$$\int_0^\delta \frac{\mathrm{d}s}{H^{-1}(F(s))} < \infty. \tag{2.7}$$

Let D *be unbounded, with* $\{x \in \mathbb{R}^n\colon |x| > R\} \subset D$ *for some* $R > 0$. *If* u *is a solution of* (2.2) *in* D, *with* $u(x) \to 0$ *as* $|x| \to \infty$, *then there exists* $R_1 \geqslant R$ *such that* $u(x) \equiv 0$ *for* $|x| > R_1$.

Theorem 2 also applies when f satisfies the alternative conditions:

(F1)′ $f \in C(0, \infty)$,

(F2)′ f is a maximal graph with $f(0) = 0$ and $\liminf_{u \to 0} f(u) > 0$ (or $+\infty$);

see the remark following the proof of Theorem 2 in the next section.

Theorem 2 is essentially due to Redheffer [11, Theorem 4], by specializing his results to the matrix $a_{ij} = A(|p|)\delta_{ij} + |p|A'(|p|)p_i p_j/|p|^2$ which arises by expansion of the divergence term in (2.2). This specialization requires, however, two assumptions which are not needed here, first that the operator A be of class $C^1(0, \infty)$, and second, that the solutions in consideration should be of class C^2 at points of D where $Du \neq 0$. [In the proof of Theorem 4 of [11] it is not evident that an appropriate comparison principle can be applied without the further assumption that the nonlinearity f be non-decreasing for small $u > 0$ – that is, for the validity of Theorem 4 of [11] this additional assumption, which is exactly (F2) above, seems to be required as well.] For the special case of the degenerate Laplacian, see also [6].

If Theorem 2 were an exact analogue of Theorem 1, the conclusion would be that $u \equiv 0$ in D, but this would be incorrect since (2.2) admits non-trivial compact support solutions under assumption (2.7). We formalize this statement in the following:

THEOREM 3. – *Suppose that* $f > 0$ *in* $(0, \delta)$. *Let* $D_R = \{x \in \mathbb{R}^n\colon\ |x| > R\}$. *Then if* R *is sufficiently large, there is a non-trivial solution of* (2.1) *in* D_R, *with the equality sign, such that* $u(x) \to 0$ *as* $|x| \to \infty$.

This result has the important consequence,

COROLLARY 1. – *Condition* (2.6) *is necessary for the strong maximum principle to be valid for* (2.1).

Proof. – Suppose (2.6) fails, with $f > 0$ in $(0, \delta)$, and let u be the solution obtained in Theorem 3 for $D = D_R$. By Theorem 2, since (2.7) now holds, u vanishes for $|x| > R_1 \geqslant R$. But $u \not\equiv 0$, which violates the strong maximum principle. [Another proof of the Corollary will be given in the next section.] □

COROLLARY 2. – *Suppose* (2.5) *holds. Then condition* (2.7) *is necessary for the compact support principle to be valid for* (2.2).

Proof. – Suppose (2.7) fails and let u be the solution obtained in Theorem 3 for $D = D_R$. By Theorem 1, since both (2.5) and (2.6) now hold, it is clear that $u > 0$ in D_R, but this violates the compact support principle. □

3. Proofs

3.1. Preliminary lemmas

For the proof of Theorem 1 we require several preliminary results.

LEMMA 1. –

(i) *For any constant* $\sigma \in [0, 1]$ *we have*:

$$F(\sigma u) \leqslant \sigma F(u), \qquad u \in [0, \delta).$$

(ii) *Let* $w = w(r)$ *be of class* $C^1(r_0, r_1)$ *with* $w'(r) \geqslant 0$, *and assume that* $\Omega \circ w'$ *is also of class* $C^1(r_0, r_1)$. *Then* $H \circ w'$ *is of class* $C^1(r_0, r_1)$, *and*

$$\{H(w'(r))\}' = w'(r)\{\Omega(w'(r))\}' \quad \text{in } (r_0, r_1).$$

To obtain (i), observe that $\sigma f(\sigma u) \leqslant \sigma f(u)$ for $u \in (0, \delta)$, since f is non-decreasing. Integrating this relation from 0 to u yields the result.

On the other hand, (ii) is an immediate consequence of (2.4).

LEMMA 2. – *Suppose conditions* (2.5) *and* (2.6) *are satisfied Then for any numbers* k, $R > 0$ *and* $\varepsilon \in (0, \delta)$ *the ordinary differential equation,*

$$\{H(|v'|)\}' + \frac{k}{r} H(|v'|) - f(v)v' = 0, \tag{3.1}$$

has a C^1 *solution* $v = v(r)$ *in the interval* $\frac{1}{2}R \leqslant r \leqslant R$, *with*

$$v(R) = 0, \qquad v'(R) = -\alpha < 0 \tag{3.2}$$

and

$$0 < v < \varepsilon, \qquad -1 < v' < 0 \quad \text{in } [R/2, R) \tag{3.3}$$

provided α *is sufficiently small.*

Proof. – By a standard Schauder fixed point argument (see [8, Proposition A1]) there exists a *maximal interval* $[R_0, R]$, $\frac{1}{2}R \leqslant R_0 < R$, for which a C^1 solution $v = v(r)$ of (3.1), (3.2) exists and satisfies

$$0 \leqslant v \leqslant \varepsilon, \qquad -1 \leqslant v' \leqslant 0 \quad \text{in } [R_0, R]. \tag{3.4}$$

Step 1. We assert that $v > 0$ and $|v'| > \alpha$ in $[R_0, R)$. To see this, multiply (3.1) by r^k to get

$$\{r^k H(|v'|)\}' = r^k f(v) v' \leqslant 0 \quad \text{in } [R_0, R]. \tag{3.5}$$

Therefore by integration we obtain

$$H(|v'(r)|) \geqslant \left(\frac{R}{r}\right)^k H(\alpha) \geqslant H(\alpha) > 0$$

and the claim follows since H is strictly increasing.

Step 2. Clearly from (3.5) we have:

$$\{r^k H(|v'(r)|)\}' - R^k\{F(v(r))\}' \geqslant 0 \quad \text{for } r \in [R_0, R].$$

Integrating from r to R yields, since $F(0) = 0$,

$$R^k H(\alpha) - r^k H(|v'(r)|) + R^k F(v(r)) \geqslant 0,$$

and in turn

$$H(|v'(r)|) \leqslant \left(\frac{R}{r}\right)^k \{F(v(r)) + H(\alpha)\}.$$

Therefore, finally, since $r \geqslant R_0 \geqslant \frac{1}{2}R$, we obtain:

$$|v'(r)| \leqslant H^{-1}(2^k\{F(v(r)) + H(\alpha)\}) \tag{3.6}$$

in $[R_0, R]$.

Step 3. We complete the proof of the Lemma. Consider first the case when $f(u) \equiv 0$ in some interval $[0, \tau)$. Without loss of generality, we can assume that $\varepsilon \leqslant \tau$. Then by (3.6) we get $|v'(r)| \leqslant H^{-1}(2^k H(\alpha))$ in $[R_0, R]$. Hence for $\alpha > 0$ sufficiently small, $|v'(r)| < \max\{1, 2\varepsilon/R\}$. Therefore $-1 < v' < 0$, $0 < v < \varepsilon$ on $[R_0, R)$, from which we conclude that $R_0 = R/2$ for the maximal interval in which (3.4) holds. This completes the proof of the Lemma for the first case.

The principal case, when $f(u) > 0$ for $0 < u < \delta$, is more delicate. Let $\sigma = 2^{-(k+1)}$; without loss of generality we can assume that

$$\varepsilon < \sigma\delta, \qquad F(\varepsilon) < \sigma H(1). \tag{3.7}$$

Let $\alpha_0 > 0$ be fixed such that

$$H(\alpha_0) < F(\varepsilon). \tag{3.8}$$

For $\alpha < \alpha_0$ we define $\hat{\varepsilon}(\alpha) = F^{-1}(H(\alpha))$. Clearly $\hat{\varepsilon}(\alpha) < \varepsilon$ by (3.8). Now for any fixed $\alpha \in (0, \alpha_0]$ and for any $r \in [R_0, R]$ we have either

$$v(r) > \hat{\varepsilon}(\alpha) \quad \text{or} \quad v(r) \leqslant \hat{\varepsilon}(\alpha) < \varepsilon.$$

In the first case we claim that if α is sufficiently small then also $v(r) < \varepsilon$. Indeed in this case it is clear that $F(v(r)) > F(\hat{\varepsilon}(\alpha)) = H(\alpha)$, so that by (3.6)

$$|v'(r)| < H^{-1}\left(\frac{1}{\sigma} F\big(v(r)\big)\right).$$

Clearly there exists $r_0 \in (r, R)$ where $v(r_0) = \hat{\varepsilon}(\alpha)$. Integrating from r_0 to r then yields

$$\int_{\hat{\varepsilon}(\alpha)}^{v(r)} \frac{\mathrm{d}s}{H^{-1}(F(s)/\sigma)} < \frac{R}{2};$$

making the substitution $s = \sigma t$ and using Lemma 1(i), we then get (since $t \leqslant \varepsilon/\sigma < \delta$ when $r \in [R_0, R]$)

$$\int_{\hat{\varepsilon}(\alpha)/\sigma}^{v(r)/\sigma} \frac{\mathrm{d}t}{H^{-1}(F(t))} < 2^k R.$$

The claim now follows from (2.6) and the fact that $\hat{\varepsilon}(\alpha) \to 0$ as $\alpha \to 0$. In summary, whenever α is sufficiently small we have

$$v(r) < \varepsilon \quad \text{for } r \in [R_0, R];$$

also by (3.6)–(3.8)

$$|v'(r)| \leqslant H^{-1}\big(2^k\{F(\varepsilon) + H(\alpha_0)\}\big) < H^{-1}\big(2^k\{\sigma H(1) + \sigma H(1)\}\big) = 1.$$

Consequently $R_0 = R/2$ and the Lemma is proved.

LEMMA 3 (Weak comparison principle). – *Let u and v be respective solutions of* (2.1) *and* (2.2) *in a bounded domain D. Suppose also that u and v are continuous in $\overline{D}$, with $v < \delta$ in D and $u \geqslant v$ on ∂D. Then $u \geqslant v$ in D.*

Proof. – Let $w = u - v$ in $\overline{D}$. If the conclusion fails, then there exists a point $x_1 \in D$ such that $w(x_1) < 0$. Fix $\varepsilon > 0$ so small that $w(x_1) + \varepsilon < 0$. Consequently, since $w \geqslant 0$ on ∂D it follows that the function $w_\varepsilon = \min\{w + \varepsilon, 0\}$ is non-positive and has compact support in D. By the distribution meaning of solutions, taking the Lipschitzian function w_ε as test function, we get

$$\int_D \big\{A(|Du|)Du - A(|Dv|)Dv\big\} Dw_\varepsilon \leqslant \int_D \big\{f(v) - f(u)\big\} w_\varepsilon. \tag{3.9}$$

The left hand side of (3.9) is positive due to the strict monotonicity of $tA(t)$ and the fact that $Dw_\varepsilon \equiv Dw \not\equiv 0$ when $w + \varepsilon < 0$, while otherwise $Dw_\varepsilon = 0$ (a.e.).

Moreover, when $w + \varepsilon < 0$ there holds $0 \leqslant u < v - \varepsilon$ $(< \delta)$; hence $f(v) - f(u) \geqslant 0$ since $f(s)$ is non-decreasing for $s < \delta$. Thus the right hand side of (3.9) is non-positive, a contradiction. This completes the proof of the Lemma.

Remark. – Lemma 3 is essentially the result of theorem 10.7 of [9], with $\mathbf{A}$ independent of z and B independent of p, and with the differentiability of $\mathbf{A}$ replaced by a strict convexity condition; see also Theorem 4′ in Section 5.

Now we are ready to prove Theorem 1. We first show that the function $v(x) = v(r)$, $r = |x|$, where v is given by Lemma 2, satisfies in $E_R = \{x \in \mathbb{R}^n\colon\ R/2 \leqslant |x| \leqslant R\}$ the differential inequality (2.2). This is a consequence of the calculation:

$$\begin{aligned} v'\big[\operatorname{div}\{A(|Dv|)Dv\} - f(v)\big] &= v'\big[\operatorname{div}\{A(|v'|)v'x/r\} - f(v)\big] \\ &= |v'|\left[\{\Omega(|v'|)\}' + \frac{(n-1)}{r}\Omega(|v'|)\right] - v'f(v) \\ &= \{H(|v'|)\}' + \frac{(n-1)}{r}|v'|\,\Omega(|v'|) - v'f(v) \\ &\leqslant \{H(|v'|)\}' + kH(|v'|)/r - v'f(v) = 0, \end{aligned}$$

where Lemma 1 has been used at the third step, while at the last step we have applied (2.5) and the fact that $|v'| < 1$, with the constant

$$k = (n-1)\sup_{|t|\leqslant 1} \frac{t\Omega(t)}{H(t)}.$$

Therefore, since $v' < 0$, the function v is a solution of (2.2) in E_R.

The rest of the proof is exactly the same as in the standard demonstration of the strong maximum principle (see [9, proof of theorem 3.5 on page 35]), since the comparison function v satisfies the conditions, see [9, proof of lemma 3.4 on page 34]:

(i) $v > 0$ in E_R,
(ii) $v = 0$ when $|x| = R$,
(iii) $\partial v/\partial \nu = v' < 0$ when $|x| = R$, where ν is the outer normal to ∂E_R,
(iv) $v < \varepsilon$ when $|x| = R/2$,

where $\varepsilon, R > 0$ can be taken arbitrarily small and the origin of coordinates can be chosen arbitrarily in D. Note that the use of the weak maximum principle (corollary 3.2 of [9]) is here replaced by application of Lemma 3.

This completes the proof of Theorem 1.

3.2. Proof of Theorem 2

By (2.7) we can define

$$C = \int_0^\delta \frac{\mathrm{d}s}{H^{-1}(F(s))} < \infty.$$

Introduce $w = w(r)$, $0 \leqslant r < C$, by:

$$r = \int_{w(r)}^{\delta} \frac{\mathrm{d}s}{H^{-1}(F(s))}. \tag{3.10}$$

Differentiation gives

$$-\frac{w'(r)}{H^{-1}(F(w(r)))} = 1 \quad \text{for } 0 \leqslant r < C,$$

that is, w is of class $C^1[0, C)$, with $0 < w \leqslant \delta$, $w' < 0$, and $H(|w'|) = F(w)$. Hence also, from Lemma 1,

$$-\{\Omega(|w'|)\}' = f(w). \tag{3.11}$$

Obviously $w(r) \to 0$, $w'(r) \to 0$ as $r \to C$. Therefore, by defining $w(r) \equiv 0$ for $r \geqslant C$, it is clear that w becomes a C^1 solution of (3.11) in $[0, \infty)$.

Now let u be the solution of (2.2) in D given in the statement of the theorem, with $u(x) \to 0$ as $|x| \to \infty$. Clearly there exists $R_0 \geqslant R$ such that $u(x) < \delta$ if $|x| \geqslant R_0$. Define $v(x) = w(|x| - R_0)$ for any $x \in D_0 = \{x \in \mathbb{R}^n\colon |x| > R_0\}$. Consequently, for $x \in D_0$, and $r = |x|$, we have

$$\operatorname{div}\{A(|Dv|)Dv\} - f(v) = -\{\Omega(|v'|)\}' - \frac{(n-1)}{r}\Omega(|v'|) - f(v) \leqslant 0$$

in view of (3.11) and the fact that $\Omega \geqslant 0$. Since $0 \leqslant u(x) < \delta = v(x)$ on ∂D_0, and since $u(x)$, $v(x) \to 0$ as $|x| \to \infty$, we can apply the comparison principle Lemma 3 (with the roles of u and v being interchanged) to obtain $0 \leqslant u(x) \leqslant v(x)$ in D_0. In particular $u(x) = 0$ when $|x| \geqslant R_1 = R_0 + C$, as required.

Remarks. –

(1) The last sentence of the proof gives an a priori estimate for the support of u.
(2) Theorem 2 is closely related to the results in [11], see also the remarks after the statement of Theorem 2 in Section 2. The proof we have given is in fact not different in its underlying ideas from those in [2,4,6,11,13], the principal improvements here being the direct approach, the generality of the solution class, and the clarification of the method.
(3) When conditions (F1)′, (F2)′ are assumed, rather than (F1), (F2), we can transform the vertical segment of f at $u = 0$ into a linear segment with finite slope, thus arriving at a function $\overline{f} \leqslant f$ satisfying (F1) and (F2). But then every solution of (2.2) remains a solution of (2.2) with f replaced by $\overline{f}$, and the result of Theorem 2 continues to apply.[2]

A second proof of Corollary 1. – Suppose (2.6) fails, that is (2.7) holds. We can then introduce the function $w = w(r)$, defined on $[0, \infty)$, as in the proof of Theorem 2. Let $u(x) = w(x_n)$ for any $x \in \mathbb{R}^n_+ = \{x \in \mathbb{R}^n\colon x_n > 0\}$. By (3.11), u is obviously a solution of (2.1), *with the equality sign*, in the domain $D = R^n_+$. Clearly $u(0, \ldots, 0, C) = w(C) = 0$ and at the same time $u \not\equiv 0$ in D. Hence the strong maximum principle fails.

This result also yields a direct and simple answer to the unique continuation question for the equation $\operatorname{div}\{A(|Du|)Du\} - f(u) = 0$, that is, the function $u(x) = w(x_n)$ shows that a solution in a domain D may vanish in a subdomain without vanishing throughout D. Another counterexample to unique continuation was given in the proof of Corollary 1 in Section 2.

3.3. Proof of Theorem 3

Let the function $\tilde{f}$ be defined on $[0, \infty)$ by:

$$\tilde{f}(u) = \begin{cases} f(u), & 0 \leqslant u \leqslant \delta, \\ -u + f(\delta) + \delta, & \delta \leqslant u \leqslant \beta, \\ u + \tilde{f}(\beta) - \beta, & u \geqslant \beta, \end{cases}$$

[2] A similar argument can be used also for maximal monotone graphs f, see [13].

where $\beta > \delta$ is chosen so that

$$\int_0^\beta \tilde{f}(s)\,\mathrm{d}s = 0.$$

Clearly $\tilde{f}(\gamma) = 0$, where $\gamma = \beta - \tilde{f}(\beta) > \beta$.

Case 1. Suppose that $H(t) \to \infty$ as $t \to \infty$. Then from Theorem A of [8, page 180], there exists a radially symmetric non-negative, non-trivial C^1 solution of

$$\operatorname{div}\{A(|Du|)Du\} - \tilde{f}(u) = 0 \quad \text{in } \mathbb{R}^n,$$

such that $u(x) \to 0$ as $|x| \to \infty$. This solution moreover has the property that $u'(r) \leqslant 0$ for $r \geqslant 0$, and $u(0) \in (\beta, \gamma)$.

Let $R > 0$ be so large that $u(x) = \delta$ when $|x| = R$. Thus $u(x) \leqslant \delta$ when $|x| \geqslant R$. Consequently u is a solution of

$$\operatorname{div}\{A(|Du|)Du\} - f(u) = 0 \quad \text{in } D_R,$$

and u does not vanish identically in D_R. This completes the proof of Case 1.

Case 2. Assume next that $H(\infty) < \infty$. Then the above argument requires the further condition:

(3.12) $$h = \max_{[0,\beta]} \tilde{F}(u) - \tilde{F}(\gamma) < H(\infty)$$

(see case (c) of Theorem A in [8]). It is not hard to see that

$$h = 2\int_0^\delta f(s)\,\mathrm{d}s + f(\delta)^2.$$

Therefore, in order to verify (3.12) it is enough, if necessary, to replace δ by a sufficiently small number $\tilde{\delta} \in (0, \delta)$ in the construction of $\tilde{f}$. The rest of the proof is now the same as in Case 1.

Remark. – It is the authors' belief that Theorem 3 can be proved more simply and directly, without applying the work of [8].

Part II. Fully quasilinear inequalities

4. The strong maximum principle and the compact support principle

4.1. Let D be a domain in $\mathbb{R}^n$. Let $\{a^{ij}(x)\}$, $i, j = 1, \ldots, n$, be a continuously differentiable, symmetric coefficient matrix on D, which is uniformly elliptic in the sense that

$$a^{ij}(x)\xi_i\xi_j \geqslant \lambda|\xi|^2, \quad x \in D,\ \xi \in \mathbb{R}^n,$$

for some positive number λ. Moreover, let $B(x, u, p)$ be a continuous function on $D \times \mathbb{R}_0^+ \times \mathbb{R}^n$.

We shall suppose without further mention that the operator $A = A(t)$ satisfies the following strengthened versions of (A1), (A2), namely

(A1)′ $A \in C^1(0, \infty)$,

(A2)′ $\Omega'(t) > 0$ for $t > 0$, and $\Omega(t) \to 0$ as $t \to 0$,

(A3)′ condition (2.5) holds;
we also continue to assume that the nonlinearity f obeys (F1) and (F2).

The following direct extension of Theorem 1 now holds.

THEOREM 1′ (Strong maximum principle). – *Consider the differential inequality*[3]

$$\operatorname{div}\{a^{ij}(x)A(|Du|)D_j u\} - B(x,u,Du) \leqslant 0, \quad u \geqslant 0,\ x \in D. \tag{4.1}$$

Assume that there exists a constant $\kappa > 0$ such that:

$$B(x,u,p) \leqslant \kappa\Omega(|p|) + f(u) \tag{4.2}$$

for $x \in D$, $u \geqslant 0$, and all $p \in \mathbb{R}^n$ with $|p| < 1$. Suppose fina ly that either $f(s) \equiv 0$ for $s \in [0,\tau)$, $\tau > 0$, or else (2.6) *holds.*

If u is a C^1 solution of (4.1) *with $u(x_0) = 0$ for some $x_0 \in D$, then $u \equiv 0$ in D.*

The semilinear case $A(t) = 1$ was treated by Vazquez [13, theorem 4]. (Note that in [13] the condition (C4) should include the assumption $d(x) > 0$.) For earlier work on linear inequalities, see [3]; and for non-singular quasilinear inequalities with $B(x,u,p)$ Lipschitz continuous in u, p, see [12].

Proof. – The proof is similar to that of Theorem 1. We first show that for any origin O in D the function $v(x) = v(r)$, $r = |x|$, where v is given by Lemma 2, satisfies the differential inequality

$$\operatorname{div}\{a^{ij}(x)A(|Du|)D_j u\} - \kappa\Omega(|Du|) - f(u) \geqslant 0, \quad u \geqslant 0, \tag{4.3}$$

in $E_R = \{x \in \mathbb{R}^n\colon R/2 \leqslant |x| \leqslant R\}$. Let

$$\Lambda = \text{max eigenvalue of } \{a^{ij}(x)\} \text{ in } E_R, \qquad a = \max\bigl|D_i a^{ij}(x)\bigr| \text{ in } E_R.$$

It is easy to see that

$$D_i\left(a^{ij}(x)\frac{x_j}{r}\right) = \bigl(D_i a^{ij}(x)\bigr)\frac{x_j}{r} + \frac{a^{ij}}{r}\left(\delta_{ij} - \frac{x_i x_j}{r^2}\right),$$

so

$$\max_{E_R}\left|D_i\left(a^{ij}(x)\frac{x_j}{r}\right)\right| \leqslant a + \frac{n-1}{r}\Lambda. \tag{4.4}$$

By a trivial change of scale we can suppose without loss of generality that $\lambda = 1$. Then calculations similar to those in the proof of Theorem 1 yield (since $-1 < v' < 0$):

$$\begin{aligned}
&v'\bigl[\operatorname{div}\{a^{ij}(x)A(|Dv|)D_j v\} - \kappa\Omega(|Dv|) - f(v)\bigr]\\
&\quad\leqslant a^{ij}(x)\frac{x_i x_j}{r^2}|v'|\{\Omega(|v'|)\}' + \left(a + \kappa + \frac{n-1}{r}\Lambda\right)|v'|\,\Omega(|v'|) - v' f(v)\\
&\quad\leqslant a^{ij}(x)\frac{x_i x_j}{r^2}\bigl[\{H(|v'|)\}' + k[\Lambda + R(a+\kappa)]H(|v'|)/r - v' f(v)\bigr] = 0,
\end{aligned}$$

where at the last steps we have used (A3)′ and Lemma 2 (replacing k with $k[\Lambda + R(a+\kappa)]$). Therefore, again since $v' < 0$, the function v is a solution of (4.3) in E_R.

[3] By $\operatorname{div}\mathbf{A}^i$ we mean $D_i(\mathbf{A}^i)$, with the obvious summation.

We next require a comparison lemma corresponding to Lemma 3, but applying to the more general inequality (4.1).

LEMMA 4 (Comparison principle). – *Let u and v be respectively solutions of* (4.1) *and* (4.3) *in a bounded domain D. Suppose that $|Du| + |Dv| > 0$ in D; that u and v are continuous in $\overline{D}$; and that*

$$v < \delta \quad \text{in } D, \qquad u \geqslant v \quad \text{on } \partial D.$$

Then $u \geqslant v$ in D.

Lemma 4 has been stated in the context of the inequalities (4.1) and (4.3), but the result immediately extends to the inequalities treated in theorem 10.7 of [9]; see Section 5. The main point of Lemma 4 is that if $|Du| + |Dv| > 0$ in D, then just as for Lemma 3 *it is not necessary to have ellipticity at the value $p = 0$.* We defer the proof of Lemma 4 until Section 5.

The rest of the proof of Theorem 1′ is now the same as before, the only change being that at the last step we rely on Lemma 4 instead of Lemma 3.

The following boundary lemma is immediate as a by-product of Theorem 1′. It will be important for the proof of Theorem 2′ below.

COROLLARY 1′ (Boundary lemma). – *Let $x_0 \in \partial D$ and suppose that D satisfes an interior sphere condition at x_0. Assume that the conditions of Theorem* 1′ *hold.*

If u is a C^1 solution of (4.1) *in $\overline{D}$, with $u > 0$ in D and $u = 0$ at x_0, then $\partial u/\partial \nu < 0$ at x_0, where ν is the outer normal to ∂D at x_0.*

Proof. – We proceed essentially as in Theorem 1′. By the interior sphere condition there exist $y \in D$ and $R > 0$ such that $B = B_R(y) \subset D$ and $x_0 \in \partial B$. Let v be the solution of (4.3) given in Lemma 2 and put $w(x) = v(|x - y|)$. Then as in the proof of Theorem 1′ it follows that

$$u(x) \geqslant w(x) \quad \text{in } B_R(y)\backslash B_{R/2}(y)$$

provided that ε is sufficiently small. This completes the proof of the Corollary, since $\partial w/\partial \nu = v'(R) < 0$.

4.2. We next consider the corresponding compact support principle, when condition (2.7) rather than (2.6) is satisfied. Let D be unbounded, with $\{x \in \mathbb{R}^n\colon |x| > R\} \subset D$ for some $R > 0$, and consider the inequality

$$\operatorname{div}\{a^{ij}(x)A(|Du|)D_j u\} - B(x, u, Du) \geqslant 0, \quad u \geqslant 0, \tag{4.5}$$

in D, where for some constant $\kappa > 0$,

$$B(x, u, p) \geqslant -\kappa\Omega(|p|) + f(u) \tag{4.6}$$

for $x \in D$, $u \geqslant 0$, and all $p \in \mathbb{R}^n$ with $|p| < 1$.

THEOREM 2′ (Compact support principle). – *Let* (4.6) *hold. Suppose $f(s) > 0$ for $s \in (0, \delta)$ and let* (2.7) *be satisfied If u is a solution of* (4.5) *in D with $u(x) \to 0$ as $|x| \to \infty$, then there exists $R_1 \geqslant R$ such that $u(x) \equiv 0$ for $|x| > R_1$.*

Remark. – For the importance of the lower bound (4.6), see the counterexample after the proof.

Proof. – The proof is almost the same as that of Theorem 2, except that we need to reconstruct the comparison function v. For simplicity, we shall assume as before that $\lambda = 1$.

Fix $\sigma \in (0,1)$. By using Lemma 1(i) and the strict monotonicity of H^{-1}, we get:

$$\int_0^\delta \frac{ds}{H^{-1}(\sigma F(s))} \leqslant \int_0^\delta \frac{ds}{H^{-1}(F(\sigma s))} = \frac{1}{\sigma}\int_0^{\delta\sigma} \frac{ds}{H^{-1}(F(s))} < \infty$$

by (2.7). It follows that $\gamma \in (0,\delta)$ can be chosen so that $F(\gamma) \leqslant H(1)$ and

$$0 < C = \int_0^\gamma \frac{ds}{H^{-1}(\sigma F(s))} \leqslant 1. \tag{4.7}$$

Introduce $w = w(r)$, $0 \leqslant r < C$, by

$$r = \int_{w(r)}^{\gamma} \frac{ds}{H^{-1}(\sigma F(s))}.$$

Differentiation gives

$$-\frac{w'(r)}{H^{-1}(\sigma F(w(r)))} = 1 \quad \text{for } 0 \leqslant r < C,$$

that is, w is of class $C^1[0,C)$, with $0 < w \leqslant \gamma$, $-1 < w' < 0$, $w'' > 0$, and $H(|w'|) = \sigma F(w)$. Hence also, from Lemma 1,

$$-\{\Omega(|w'|)\}' = \sigma f(w). \tag{4.8}$$

Obviously $w(r) \to 0$, $w'(r) \to 0$ as $r \to C$. Therefore, by defining $w(r) \equiv 0$ for $r \geqslant C$, it is clear that w becomes a C^1 solution of (4.8) in $[0,\infty)$.

For any $R_0 > 0$ and $D_0 = \{x \in \mathbb{R}^n\colon |x| > R_0\}$, let us define $v(x) = w(|x| - R_0)$. Then, for $x \in D_0$ and $r = |x|$ we have, as in the earlier calculations in the proof of Theorem 1′:

$$\begin{aligned}
&\operatorname{div}\{a^{ij}(x)A(|Dv|)D_j v\} + \kappa\Omega(|v'|) - f(v)\\
&\quad\leqslant -a^{ij}(x)\frac{x_i x_j}{r^2}\{\Omega(|v'|)\}' + \left(a + \kappa - \frac{n-1}{r}\right)\Omega(|v'|) - f(v)\\
&\quad\leqslant a^{ij}(x)\frac{x_i x_j}{r^2}\sigma f(w) + (a+\kappa)k\frac{H(|v'|)}{|v'|} - f(v)\\
&\quad\leqslant KH(|v'|)/|v'| - (1 - \Lambda\sigma)f(v);
\end{aligned} \tag{4.9}$$

the steps in obtaining (4.9) are as follows:

First line: when $|x| > R_0 + C$ we have $v \equiv 0$ so there is nothing to show, while for $|x| \leqslant R_0 + C$ there exist constants $a = a_0 > 0$ and $\Lambda = \Lambda_0 > 0$ such that

$$|D_i a^{ij}(x)| \leqslant a, \quad a^{ij}(x)\xi_i\xi_j \leqslant \Lambda|\xi|^2$$

and finally, $a^{ij}(x)(\delta_{ij} - x_i x_j/r^2) \geqslant n-1$ (since $\lambda = 1$).

Second line: use (A3)′, (4.8), and the fact that $\Omega \geqslant 0$.

Third line: use (4.6) and put $K = (a+\kappa)k$.

Next we want to show that, for $\sigma > 0$ small enough,

$$KH(|v'|) \leqslant (1 - \Lambda\sigma) f(v)|v'|. \tag{4.10}$$

Since f is non-decreasing for $s < \delta$, there holds $F(v) \leqslant vf(v)$. Moreover, since $v' < 0$ and $v'' > 0$,

$$v(|x|) = \int_{C+R_0}^{|x|} v'(s)\,\mathrm{d}s \leqslant C|v'(|x|)| \leqslant |v'(|x|)|,$$

since $C \leqslant 1$. By taking $\sigma = 1/(K + \Lambda)$, it now follows that [4]

$$\frac{K\sigma}{1 - \Lambda\sigma} \cdot \frac{F(v)}{f(v)} \leqslant v \leqslant |v'|.$$

Therefore

$$KH(|v'|) = K\sigma F(v) \leqslant (1 - \Lambda\sigma) f(v)|v'|,$$

which is (4.10). Combining (4.9) and (4.10) we have finally

$$\operatorname{div}\{A(|Dv|)Dv\} + \kappa\Omega(|v'|) - f(v) \leqslant KH(|v'|)/|v'| - (1 - \Lambda\sigma) f(v) \leqslant 0. \tag{4.11}$$

By assumption, since $u(x) \to 0$ as $|x| \to \infty$, we can choose $R_0 > R$ so that $u(x) < \gamma$ in D_0, where γ is given in (4.7). We wish to show that $u \leqslant v$ as in the proof of Theorem 2, where v is the comparison function above, satisfying (4.11). For this purpose it is not possible to resort directly to Lemma 4, since $Dv \equiv 0$ for large $|x|$ while Du is unrestricted. Accordingly we use an indirect argument.

Define $\omega = v - u$ in D_0. We claim that $\omega \geqslant 0$. If this is not the case, then

$$\omega_0 = \inf_{D_0} \omega < 0$$

and we shall reach a contradiction. Note first that $\omega = \gamma - u > 0$ when $|x| = R_0$, and that $\omega(x) \to 0$ as $|x| \to \infty$; hence the infimum of ω must be attained at some point y in D_0.

Let $R_1 = R_0 + C$, where C is given by (4.7), and define

$$\hat{D} = \{R_0 < |x| < R_1\}, \qquad D_1 = \{|x| > R_1\}.$$

Then $D_0 = \hat{D} \cup \partial D_1 \cup D_1$, so exactly the following three cases can occur:

(1) The infimum of ω is attained in D_1.

(2) The infimum of ω is not attained in D_1, but is reached at a point on ∂D_1.

(3) The infimum of ω is not attained in $\overline{D}_1$, but is reached in $\hat{D}$.

In Case 1, let the infimum be attained at x_0. For x in D_1, define $z(x) = -u(x) - \omega_0$. Then since $v \equiv 0$ in D_1, we see that $z = \omega - \omega_0 \geqslant 0$ has a zero minimum at x_0. Moreover,

$$\operatorname{div}\{a^{ij}(x)A(|Dz|)D_j z\} - B_1(x, z, Dz) \leqslant 0, \quad z \geqslant 0$$

[4] Clearly $\sigma < 1$ since $\Lambda \geqslant \lambda = 1$ and $K > 0$.

in D_1, where (using (4.6) and the fact that $f(u) > 0$ when $u < \gamma$)

$$B_1(x, z, Dz) = -B(x, u, Du) \leqslant \kappa\Omega(|Du|) - f(u) \leqslant \kappa\Omega(|Dz|).$$

That is, B_1 satisfies (4.2) with $f \equiv 0$. Hence by the strong maximum principle (Theorem 1′) applied to the domain D_1 we obtain $z \equiv 0$. Thus $u \equiv -\omega_0 > 0$ in D_1, which is impossible since $u(x) \to 0$ as $|x| \to \infty$.

In Case 2, let the infimum of ω be reached at x_0 on ∂D_1. In this case, obviously $z > 0$ in D_1 while $z = 0$ at x_0 (we can of course consider z as a C^1 function on $\overline{D}_1$). Then, since D_1 clearly satisfies an interior sphere condition at x_0, the boundary Lemma (Corollary 1′) gives $\partial z/\partial \nu < 0$ at x_0. But this is also impossible, because $Dz = D\omega = 0$ at x_0.

In Case 3, necessarily $v - u = \omega > \omega_0$ in $\overline{D}_1$, while as noted earlier $v - u > 0$ when $|x| = R_0$. Thus $v - u > \omega_0$ on the boundary of the domain $\hat{D}$ and of course $u < \delta$ and $Dv \neq 0$ in $\hat{D}$. This is exactly the case of Lemma 4 for $D = \hat{D}$ (with the roles of u and v interchanged). The conclusion is that $v - u > \omega_0$ in $\hat{D}$ (not $v - u \geqslant 0$!). But this contradicts the condition of Case 3 that $\omega = v - u$ attains its infimum ω_0 in $\hat{D}$.

We have thus shown that all three cases lead to a contradiction. Consequently $\omega \geqslant 0$ in D_0, that is $v \geqslant u$. In turn, $u \equiv 0$ for $|x| > R_1$, which completes the proof of the Theorem.

Here we give a counterexample showing the importance of the lower bound condition (4.6). Consider the inequality

$$\Delta_m u + |Du|^p - u^q \geqslant 0, \quad m > 1,\ p, q > 0. \tag{4.12}$$

Clearly, conditions (2.7) and (4.6) are satisfied if and only if $p \geqslant m - 1$ and $q < m - 1$. The compact support principle then holds for (4.12). On the other hand, for any $p \in (0, m-1)$ we can take $p < q < m - 1$. One easily checks that (4.12) then has positive solutions $u = \text{const.}\,|x|^{-l}$ on $\{|x| > R\}$ for l and R large. Hence the compact support principle fails even though condition (2.7) is fulfilled!

4.3. The inequalities (4.1) and (4.5) do not involve the variable u in the divergence term, a restriction which in some cases can be overcome. Confining the discussion for simplicity to the strong maximum principle, we consider the inequalities:

$$\operatorname{div}\{a^{ij}(x, u)A(|Du|)D_j u\} - B(x, u, Du) \leqslant 0, \quad u \geqslant 0, \tag{4.13}$$

and

$$\operatorname{div}\{a^{ij}(x)c(u)A(|Du|)D_j u\} - B(x, u, Du) \leqslant 0, \quad u \geqslant 0. \tag{4.14}$$

For (4.13), assume that $a^{ij}(x, u)$ is continuously differentiable in x and u, with

$$a^{ij}(x, u)\xi_i\xi_j \geqslant \lambda|\xi|^2, \quad x \in D,\ u \in [0, \infty),\ \xi \in \mathbb{R}^n,$$

for some positive number λ, and that $B(x, u, p) \leqslant f(u)$, i.e., $\kappa = 0$ in (4.2). The proof of Theorem 1′ then carries over essentially unchanged up to the application of Lemma 4. This Lemma, however, is no longer applicable since the variable u appears in the divergence term. On the other hand, we can use theorem 10.7(ii) of [9] to obtain a corresponding comparison theorem which can be applied (see the remark after Theorem 4′ in Section 5). Here it is crucial that $\hat{B}(x, u, p)$ (which in the present case is to be identified with $f(u)$) is independent of the gradient variable p. In summary, when (2.5) and (2.6) hold, one obtains the strong maximum principle for (4.13) exactly as for (4.1).

For (4.14), assume that $c(u)$ is positive and continuously differentiable for $u \in [0, \delta)$, while $B(x, u, p)$ continues to satisfy (4.2). We rewrite (4.14) in the form

$$\operatorname{div}\{a^{ij}(x)A(|Du|)D_j u\} - \tilde{B}(x, u, Du) \leqslant 0, \quad u \geqslant 0, \tag{4.15}$$

where, for some constant $\tilde{\kappa} \geqslant \kappa$,

$$\begin{aligned}\tilde{B}(x, u, p) &= -a^{ij}(x)\left(\frac{c'(u)}{c(u)}\right)A(|p|)p_i p_j + \frac{B(x, u, p)}{c(u)} \\ &\leqslant \tilde{\kappa}\Omega(|p|) + \text{const.}\, f(u)\end{aligned}$$

for $0 \leqslant u < \delta$ and $|p| < 1$. Clearly Theorem 1′ applies to (4.15), so again the strong maximum principle is valid when (2.5) and (2.6) are satisfied.

5. A comparison theorem for singular divergence form operators

Consider the differential inequalities

$$\operatorname{div}\{\hat{A}(x, Du)\} - \hat{B}(x, u, Du) \leqslant 0, \quad u \geqslant 0, \tag{5.1}$$

$$\operatorname{div}\{\hat{A}(x, Dv)\} - \hat{B}(x, v, Dv) \geqslant 0, \quad v \geqslant 0, \tag{5.2}$$

in a possibly unbounded domain $D \subset \mathbb{R}^n$. Let the vector function

$$\hat{A}(x, p) : D \times \mathbb{R}^n \to \mathbb{R}^n$$

be continuous in $D \times \mathbb{R}^n$ and continuously differentiable with respect to p for $p \neq 0$ in $\mathbb{R}^n$ and let

$$\hat{B}(x, z, p) : D \times \mathbb{R} \times \mathbb{R}^n \to \mathbb{R}$$

be continuous in $D \times \mathbb{R} \times \mathbb{R}^n$ and continuously differentiable with respect to p for $p \neq 0$ in $\mathbb{R}^n$. Suppose moreover that $\hat{A}$ is uniformly elliptic in the sense that the matrix $\{D_j \hat{A}^i(x, p)\}$ is positive definite for $x \in D$ and $p \neq 0$ in $\mathbb{R}^n$. Finally assume that $\hat{B}(x, z, p)$ is non-decreasing in the variable z. Then the following comparison principle holds:

THEOREM 4 (Comparison principle). – *Let u and v be respective solutions of* (5.1) *and* (5.2) *in D. Suppose that u and v are continuous in $\overline{D}$, that $|Du| + |Dv| > 0$ in D, and that $u \geqslant v$ on ∂D.*[5] *Then $u \geqslant v$ in D.*

This is exactly theorem 10.7(i) of [9] with the exception that the functions $\hat{A}$ and $\hat{B}$ are allowed to be singular at $p = 0$, this being compensated by the additional condition $|Du| + |Dv| > 0$ in D. [We have written $\hat{A}$, $\hat{B}$ here, rather A, B as in [9], in order to avoid confusion with earlier notation in the paper.]

[5] If D is unbounded, the boundary condition is understood to include the limit relation

$$\liminf_{|x| \to \infty} \{u(x) - v(x)\} \geqslant 0.$$

Proof. – Suppose for contradiction that the theorem is false. Put $w(x) = u(x) - v(x)$, whence

$$\bar{\varepsilon} = -\inf_{x \in D} w(x) > 0.$$

By the boundary condition, clearly there exists $x_0 \in D$ such that $w(x_0) = -\bar{\varepsilon}$.

For $\varepsilon \in (0, \bar{\varepsilon})$, plainly $w_\varepsilon = \min(w + \varepsilon, 0)$ has compact support in D. Then, as in the proof of Lemma 3, one has

$$\begin{aligned} \int_D \{\hat{A}(x, Du) - \hat{A}(x, Dv)\} Dw_\varepsilon &\leqslant \int_\Sigma \{\hat{B}(x, v, Dv) - \hat{B}(x, u, Du)\} w_\varepsilon \\ &\leqslant \int_\Sigma \{\hat{B}(x, u, Dv) - \hat{B}(x, u, Du)\} w_\varepsilon, \end{aligned} \tag{5.3}$$

where

$$\Sigma = \Sigma_\varepsilon = \operatorname{supp} w_\varepsilon$$

is a compact subset of D, and where at the last step of (5.3) we have used the facts that $w_\varepsilon \leqslant 0$ and $u \leqslant v$ in Σ, and that $\hat{B}$ is non-decreasing in the variable z.

As in [9, proof of theorem 10.7], define $u_t = tv + (1-t)u$, $t \in [0, 1]$ and

$$\hat{a}^{ij}(x) = \int_0^1 D_{p_j} A^i(x, Du_t)\,\mathrm{d}t, \qquad \hat{c}^i(x) = \int_0^1 D_{p_i} B(x, u(x), Du_t)\,\mathrm{d}t.$$

Then (5.3) can be written

$$\int_D \hat{a}^{ij}(x) D_i w_\varepsilon D_j w_\varepsilon \leqslant \int_\Sigma \hat{c}^i(x) D_i w_\varepsilon w_\varepsilon. \tag{5.4}$$

At this point we wish to proceed as in the proof of theorem 8.1 in [9], but this requires further preliminary argument since $\hat{A}$ and $\hat{B}$ are now allowed to be singular at $p = 0$.

We claim that if $\varepsilon \in (0, \bar{\varepsilon})$ is sufficiently close to $\bar{\varepsilon}$ then

$$|Du_t| \geqslant \text{const.} > 0 \quad \text{in } \Sigma. \tag{5.5}$$

To prove (5.5), let

$$d = \frac{1}{6} \min\{|Du| + |Dv|\} \quad \text{in } \hat{\Sigma} = \operatorname{supp} w_{\bar{\varepsilon}/2} \subset D.$$

Note particularly that $d\ (> 0)$ is independent of ε.

Obviously $Du - Dv = Dw = 0$ on the closed subset $E = \{x \in D\colon w(x) = -\bar{\varepsilon}\}$ of Σ. Moreover, distance$(E, \partial\Sigma) \to 0$ as $\varepsilon \to \bar{\varepsilon}$. Hence by continuity, $|Du - Dv| < d$ in Σ provided ε is suitably near $\bar{\varepsilon}$. In particular, for such values of ε we find easily that

$$\min\{|Du|, |Dv|\} \geqslant 2d$$

and so

$$|Du_t| = |tDu + (1-t)Dv| \geqslant |Dv| - t|Du - Dv| \geqslant d > 0 \quad \text{in } \Sigma,$$

which is just (5.5).

This being shown, we now have by the principal hypotheses on the functions $\hat{A}$ and $\hat{B}$,

$$\hat{a}^{ij}(x)\xi_i\xi_j \geqslant \lambda|\xi|^2, \quad \xi \in \mathbb{R}^n; \qquad |c^i(x)| \leqslant \lambda^{-1}$$

for all $x \in \Sigma$ and all ε suitably near $\bar{\varepsilon}$, where λ is an appropriate positive constant (independent of ε). Hence from (5.4) it follows that

$$\int_D |Dw_\varepsilon|^2 \leqslant \lambda^{-2} \int_\Sigma |Dw_\varepsilon||w_\varepsilon|. \tag{5.6}$$

Let $\Gamma = \Gamma_\varepsilon = \{\bar{\varepsilon} - \varepsilon < w_\varepsilon < 0\}$. Then $Dw_\varepsilon = 0$ a.e. on $\Sigma \setminus \Gamma$, so the integral on the right side of (5.6) can equally be taken over the set Γ.

Using the Hölder and Sobolev inequalities, we now obtain (see the proof of theorem 8.1 in [9], 2nd edition)

$$|\Gamma| \geqslant C\lambda^{2n}, \tag{5.7}$$

where C depends only on the dimension n. On the other hand, $\Gamma \to \emptyset$ as $\varepsilon \to \bar{\varepsilon}$, a contradiction to (5.7). This completes the proof.

When $\hat{B}$ is independent of p, Theorem 4 continues to hold without assuming that $\hat{A}$ is differentiable with respect to p. This gives the following generalization of Lemma 3:

THEOREM 4′ (Comparison principle). – *Let the hypotheses of Theorem 4 hold, except for the condition that $\hat{A}$ is differentiable with respect to p. Suppose that $\hat{B}$ is independent of p and that $\hat{A}$ is strictly monotone, that is*

$$\{\hat{A}(x,p) - \hat{A}(x,q)\}(p-q) > 0, \quad \text{when } p \neq q.$$

Then $u \geqslant v$ in D.

This follows at once from (5.3), exactly as in the proof of Lemma 3.

Remarks. – A related case, again under the condition that $\hat{B}$ is independent of p but allowing $\hat{A}$ to depend on u (continuously differentiably when $p \neq 0$), continues to yield the result of Theorem 4; see theorem 10.7(ii) of [9] together with (5.5).

If in Theorems 4 and 4′ we add the hypothesis that $v < \delta$, then the monotonicity of $\hat{B}$ is needed only in the interval $0 < z < \delta$; see the proof of Lemma 3.

We can now give the:

Proof of Lemma 4. – First observe that, for some constant $\kappa > 0$, the functions u and v respectively satisfy the inequalities

$$\operatorname{div}\{a^{ij}(x)A(|Du|)D_j u\} - \kappa\Omega(|Du|) - f(u) \leqslant 0,$$

and

$$\operatorname{div}\{a^{ij}(x)A(|Dv|)D_j v\} - \kappa\Omega(|Dv|) - f(v) \geqslant 0$$

in D. [The first inequality is a consequence of (4.1), (4.2) together with the continuity of $B(x,z,p)$.] Thus we can apply Theorem 4 (together with the remark above). Lemma 4 then follows immediately from the identifications

$$\hat{A}^i(x,p) = a^{ik}(x)A(|p|)p_k, \qquad \hat{B}(x,z,p) = \kappa\Omega(|p|) + f(z),$$

provided we show that the matrix $\{D_{p_j}\hat{A}^i(x,p)\}$ is positive definite for $p \neq 0$. But

$$D_{p_j}\hat{A}^i(x,p) = a^{ik}(x)b^{kj}(p),$$

where

$$b^{kj}(p) = A(|p|)\delta_{kj} + |p|\,A'(|p|)\,\frac{p_i p_k}{|p|^2}, \qquad p \neq 0.$$

The matrix $\{b^{kj}(p)\}$ has eigenvalues $A(|p|)$ (repeated $n-1$ times) and $\Omega'(|p|)$. By assumption (A2)′ we have $\Omega'(|p|) > 0$ for $p \neq 0$, while also

$$A(|p|) = \Omega(|p|)/|p| > 0, \quad \text{for } p \neq 0,$$

since Ω is increasing and $\Omega(0) = 0$. Hence $\{b^{ij}\}$ is positive definite for $p \neq 0$. Because $\{a^{ij}(x)\}$ is assumed positive definite in D, it now follows that $\{D_{p_j}\hat{A}^i(x,p)\}$ is positive definite for $x \in D$ and $p \neq 0$, completing the proof.

Acknowledgement

This work is partially supported by CNR, Progetto Strategico *Modelli e Metodi per la Matematica e l'Ingegneria*, by the Italian *Ministero della Università e della Ricerca Scientifca e Tecnologica* under the auspices of *Gruppo Nazionale di Analisi Funzionale e sue Applicazioni* of the CNR and by the Project *Metodi Variazionali ed Equazioni Differenziali Non Lineari*; and by NSF Grant DMS-9622937 and an Alabama EPSCoR grant.

REFERENCES

[1] G. BARLES, G. DIAZ and J.I. DIAZ, Uniqueness and continuum of foliated solutions for a quasilinear elliptic equation with a non-Lipschitz nonlinearity, *Comm. Partial Differential Equations* 17 (1992) 1037–1050.

[2] P. BENILAN, H. BREZIS and M. CRANDALL, A semilinear equation in $L^1(\mathbb{R}^n)$, *Ann. Scuola Norm. Sup. Pisa* 4 (1975) 523–555.

[3] M. CHICCO, Principio di massimo forte per sottosoluzioni di equazioni ellitiche di tipo variazionale, *Boll. Un. Mat. Italiana Ser. 2*, 22 (1967) 368–372.

[4] C. CORTÁZAR, M. ELGUETA and P. FELMER, On a semilinear elliptic problem in $\mathbb{R}^n$ with a non-Lipschitzian nonlinearity, *Adv. in Diff. Equations* 1 (1996) 199–218.

[5] J.I. DIAZ, *Nonlinear Partial Differential Equations and Free Boundaries*, Research Notes in Mathematics 106, 1985.

[6] J.I. DIAZ and M.A. HERRERO, Estimates on the support of the solutions of some nonlinear elliptic and a parabolic problems, *Proc. Roy. Soc. Edinburgh* 89-A (1981) 249–258.

[7] J.I. DIAZ, J.E. SAA and U. THIEL, Sobre la ecuación de curvatura media prescrita y otras ecuaciones cuasilineales elípticas con soluciones anulándose localmente, *Rev. Un. Mat. Argentina* 35 (1990) 175–206.

[8] B. FRANCHI, E. LANCONELLI and J. SERRIN, Existence and uniqueness of nonnegative solutions of quasilinear equations in $\mathbb{R}^n$, *Adv. in Math.* 118 (1996) 177–243.

[9] D. GILBARG and N. TRUDINGER, *Elliptic Partial Differential Equations of Second Order*, 2nd edn, Springer-Verlag, 1983.

[10] B. KAWOHL and N. KUTEV, Strong maximum principle for semicontinuous viscosity solutions of nonlinear partial differential equations, *Arch. Math.* 70 (1998) 470–478.

[11] R. REDHEFFER, Nonlinear differential inequalities and functions of compact support, *Trans. Amer. Math. Soc.* 220 (1976) 133–157.
[12] J. SERRIN, On the strong maximum principle for quasilinear second order differential inequalities, *J. Funct. Anal.* 5 (1970) 184–193.
[13] J.-L. VAZQUEZ, A strong maximum principle for some quasilinear elliptic equations, *Appl. Math. Optim.* 12 (1984) 191–202.

(JEMS) 8 (2006), 389–398. **Reprint I.12.[ii]**

J. Eur. Math. Soc. 8, 389–398

James Serrin

A maximum principle for mean-curvature type elliptic inequalities

Dedicated to Antonio Ambrosetti on the occasion of his sixtieth birthday

Received October 26, 2005

Abstract. Consider the divergence structure elliptic inequality

$$\operatorname{div}\{\boldsymbol{A}(x,u,Du)\}+B(x,u,Du)\geq 0 \tag{1}$$

in a bounded domain $\Omega\subset\mathbb{R}^n$. Here

$$\boldsymbol{A}(x,z,\boldsymbol{\xi}):K\to\mathbb{R}^n,\qquad B(x,z,\boldsymbol{\xi}):K\to\mathbb{R},\qquad K=\Omega\times\mathbb{R}^+\times\mathbb{R}^n,$$

and $\boldsymbol{A}$, B satisfy the following conditions:

$$\langle\boldsymbol{A}(\boldsymbol{\xi}),\boldsymbol{\xi}\rangle\geq|\xi|-c(x)z-a(x),\qquad|\boldsymbol{A}(x,z,\boldsymbol{\xi})|\leq\text{const},\qquad B(x,z,\boldsymbol{\xi})\leq b(x),$$

for all $(x,z,\boldsymbol{\xi}))\in K$, where $a(x)$, $b(x)$, $c(x)$ are given non-negative functions. Our interest is in the validity of the maximum principle for solutions of (1), that is, the statement that *any solution which satisfies $u\leq 0$ on $\partial\Omega$ must be a priori bounded above in Ω.*

This question arises, in particular, when one is interested in the mean curvature equation

$$\operatorname{div}\frac{Du}{\sqrt{1+|Du|^2}}=nH(x).$$

Keywords. Quasilinear elliptic inequalities, maximum principles, mean curvature equation

Introduction

We are concerned with the divergence structure elliptic inequality

$$\operatorname{div}\{\boldsymbol{A}(x,u,Du)\}+B(x,u,Du)\geq 0 \tag{1}$$

in a bounded domain $\Omega\subset\mathbb{R}^n$. Here

$$\boldsymbol{A}(x,z,\boldsymbol{\xi}):K\to\mathbb{R}^n,\qquad B(x,z,\boldsymbol{\xi}):K\to\mathbb{R},\qquad K=\Omega\times\mathbb{R}^+\times\mathbb{R}^n,$$

and $\boldsymbol{A}$, B satisfy the following structure conditions:

$$|\boldsymbol{A}(x,z,\boldsymbol{\xi})|\leq\text{const},\qquad\langle\boldsymbol{A}(\boldsymbol{\xi}),\boldsymbol{\xi}\rangle\geq|\xi|-c(x)z-a(x),\qquad B(x,z,\boldsymbol{\xi})\leq b(x), \tag{2}$$

J. Serrin: Department of Mathematics, University of Minnesota, Minneapolis, MN, USA;
e-mail: serrin@math.umn.edu

Mathematics Subject Classification (2000): Primary 35J15; Secondary 35J70

for all $(x, z, \boldsymbol{\xi})) \in K$, where $a(x)$, $b(x)$, $c(x)$ are given non-negative functions. Our interest is in the validity of the maximum principle for solutions of (1), that is, the statement that *any solution of* (1) *which satisfies* $u \leq 0$ *on* $\partial\Omega$ *must be a priori bounded above in* Ω.

This question arises, in the first instance, from the example of the mean curvature inequality

$$\operatorname{div}\frac{Du}{\sqrt{1+|Du|^2}} \geq nH(x), \tag{3}$$

and secondly from a result of Gilbarg and Trudinger ([1, Theorem 10.10]) for the more general case of inequality (1). In particular, Gilbarg and Trudinger prove the following result.

Let $u \in C(\bar{\Omega}) \cap C^1(\Omega)$ *be a distribution solution of the inequality* (1), *where* $\boldsymbol{A}$, B *satisfy conditions* (2) *with* $a(x)$, $b(x)$, $c(x)$ *constants, written simply* a, b, c.[1] *Suppose* $u \leq 0$ *on* $\partial\Omega$ *and that*

$$(b+c)|\Omega|^{1/n} < 1/S$$

where $S = 1/n\omega_n^{1/n}$ *is the Sobolev constant for the space* $W_0^{1,1}(\mathbb{R}^n)$ *and* ω_n *denotes the measure of the unit ball in* $\mathbb{R}^n$. *Then*

$$u \leq Ca \quad \text{in } \Omega,$$

with C *a constant depending only on* n, b, c *and* $|\Omega|$.

This result can be applied immediately to the mean curvature equation, though for purposes of comparison with our results, we shall defer this until later.

Gilbarg and Trudinger assert as well (without proof) that the same conclusion holds when $u \in W_0^{1,1}(\Omega)$ and the coefficients $a(x)$, $b(x)$, $c(x)$ in (2) are in $L^n(\Omega)$. It seems to me, however, that the space $L^n(\Omega)$ is too weak and should be replaced by $L^q(\Omega)$ for some $q > n$. In fact, we have the following result.

Theorem 1. *Let* $u \in W_{\text{loc}}^{1,1}(\Omega)$ *be a distribution solution of the inequality* (1) *where* $\boldsymbol{A}$, B *satisfy conditions* (2) *with* $a(x)$, $b(x)$, $c(x)$ *in* $L^q(\Omega)$ *for some* $q > n$. *Then if*[2]

$$\|b+c\|_q\,|\Omega|^{1/n-1/q} < 1/S \tag{4}$$

we have

$$u \leq C\,\|a\|_q|\Omega|^{1/n-1/q}/\delta \quad \text{in } \Omega, \tag{5}$$

where (explicitly)

$$C = C_1(n) = 3S\left(3\left[\frac{n}{n-1}\right]^{n-1}\right)^{nq/(q-n)} \tag{6}$$

and $\delta \in (0,1)$ *is any constant such that* (4) *holds with the right side replaced by* $(1-\delta)S$.

(For the record, $(n/(n-1))^{n-1}$ has the universal bound e for all $n \geq 2$.)

---

[1] Alternatively, $a(x)$, $b(x)$, $c(x)$ can be bounded by constants, which without confusion can equally be called a, b, c.

[2] By $\|\cdot\|_q$ we mean $\|\cdot\|_{L^q(\Omega)}$.

If $a(x)$, $b(x)$, $c(x)$ are bounded, the conclusion can be strengthened. It is convenient here to introduce the *effective radius* $\mathcal{R}$ of Ω by the definition

$$|\Omega| = \omega_n \mathcal{R}^n.$$

We can then replace (4) by

$$(b+c)\mathcal{R} < n, \tag{4'}$$

and (5) by

$$u \le Ca\mathcal{R}/\delta \tag{5'}$$

where now

$$C = C_2(n) = \frac{2}{n}\left(3\left[\frac{n}{n-1}\right]^{n-1}\right)^n. \tag{6'}$$

Condition (4) in Theorem 1 can in fact be replaced by the weaker restriction

$$\int_\Omega \left(\frac{b+c}{n}\right)^n < \omega_n, \tag{7}$$

but at the cost that the conclusion (5) arises in the less precise form

$$u \le C(n, q, |\Omega|; \|b+c\|_n, \|b+c\|_q)\|a\|_q, \tag{8}$$

with no simple expression for C in terms of the listed parameters. Moreover, $C \to \infty$ not only as $q \to n$ but also as $\|b+c\|_q \to \infty$, even when $\|b+c\|_n$ is finite.

Theorem 1 and the extensions given above will be proved in Section 1. For the explicit case of the mean curvature inequality (3), the conclusions take the following form.

Theorem 2. *Let $u \in W^{1,1}_{\mathrm{loc}}(\Omega)$ be a distribution solution of the mean curvature inequality (3), with $u \le 0$ on $\partial\Omega$. Assume that $H^- \in L^q(\Omega)$ for some $q > n$. Then if*

$$\|H^-\|_q \, |\Omega|^{1/n-1/q} < \omega_n^{1/n} \tag{9}$$

we have

$$u \le C\mathcal{R}/\delta \tag{10}$$

where

$$C = C_3(n) = \frac{1}{n}\left(3\left[\frac{n}{n-1}\right]^{n-1}\right)^{nq/(q-n)}. \tag{11}$$

If $H(x)$ is bounded below, say $H(x) \ge -H_0$, $H_0 > 0$, the conclusion can be strengthened, with (9) *replaced by*

$$H_0\mathcal{R} < 1 \tag{9'}$$

and (11) *by*

$$C = C_4(n) = \frac{0.6}{n}\left(3\left[\frac{n}{n-1}\right]^{n-1}\right)^n. \tag{11'}$$

Corresponding to condition (7), it is actually enough to assume only that

$$\|H^-\|_n^n < \omega_n, \tag{12}$$

though the conclusion again takes the less explicit form

$$u \leq C(n, q, |\Omega|; \|H\|_n, \|H\|_q). \tag{13}$$

In the canonical case $n = 2$, $H(x) = -H_0$ and $\delta = 1/2$ the estimate (10) can be replaced by (see Section 3)

$$u \leq 15\mathcal{R}. \tag{14}$$

Of course this is not too accurate for the case of balls, or even convex sets, where optimal estimates can be obtained, at least for C^1 solutions, by using the standard maximum principle and spherical caps as comparison functions. See also the comments in Section 3.

Theorem 2 appears to be new. We emphasize in particular that the solution class consists of weakly differentiable functions, that Ω is unrestricted except to have finite measure, and finally that (11), (11′) and (14) are all explicit.

It remains an open question, at least to the present author, whether the space $L^q(\Omega)$ for the coefficients can be replaced by $L^n(\Omega)$, as asserted by Gilbarg and Trudinger (see the comments at the foot of page 276 and the top of page 277 of [1]). This is, in fact, true when the solution is in the strong class $C^2(\Omega)$ (or even $W^{2,n}_{\mathrm{loc}}(\Omega)$), as proved by the same authors (Corollary 10.6), but again it is an open question whether the smoothness condition can be removed.

The reader may question whether the restrictions (9) or (9′) are necessary for the validity of (10). In fact, for the specific case $H(x) \geq -H_0 =$ negative constant, no solution at all is possible if Ω contains a ball $B = B_R$ of radius R and $H_0 \geq 1/R$. To see this, take a hemispherical cap $v = v(x)$ over the ball, of radius exactly R. The graph $z = v(x)$ over B has mean curvature $-1/R$, so that

$$\operatorname{div}\left(\frac{Dv}{\sqrt{1+|Dv|^2}}\right) + n/R = 0 \quad \text{in } B. \tag{15}$$

By a vertical translation of $v(x)$, either up or down, it can be arranged that the cap is tangent to the graph $z = u(x)$ at some point $x_0 \in B$, while also $v(x) \leq u(x)$ in B. The translated cap of course continues to solve (15).

We can now apply a beautiful theorem of Eberhard Hopf ([2, Satz 3′]; see also [3, Theorem 2.5]) concerning *tangent solutions* of quasilinear elliptic equations, to conclude that in this case $u \equiv v$ in B, which is obviously impossible.

A simple consequence is that if Ω is a ball of radius R and $H(x) = \text{constant} = -H$, then the condition $HR \leq 1$ is necessary for the mere existence of a solution, while $HR < 1$ is sufficient to provide an upper bound on the solution in terms of its boundary values.

Section 1 is devoted to the proof of Theorem 1, and Section 2 to the proof of Theorem 2. In Section 3 we add some final remarks.

1. Proof of Theorem 1

For completeness we first recall that a function $u \in L^1_{\mathrm{loc}}(\Omega)$ is a *weak distribution solution* of (1) in Ω if $\boldsymbol{A}(\cdot, u, Du)$ is measurable in Ω; $B(\cdot, u, Du) \in L^1_{\mathrm{loc}}(\Omega)$; and

$$\int_\Omega \langle \boldsymbol{A}(x, u, Du), D\varphi \rangle \leq \int_\Omega B(x, u, Du)\varphi \tag{16}$$

for all $\varphi \in C^1(\Omega)$ such that $\varphi \geq 0$ in Ω and $\varphi \equiv 0$ near $\partial\Omega$. (In view of the condition $|\boldsymbol{A}| \leq$ const in (2), this inequality is meaningful.)

In order to have a sufficiently large supply of test functions φ for later purposes, the following lemma is crucial.

Lemma 1. *Let $\psi : \mathbb{R} \to \mathbb{R}_0^+$ be a non-decreasing continuous function such that $\psi(t) = 0$ for $t \in (-\infty, \ell]$ and $\psi \in C^1[\ell, m] \cap C^1[m, \infty)$, with possible corners at $t = \ell$ and $t = m$, $\ell \leq m$, and with ψ' uniformly bounded. If $u \in W^{1,1}_{\mathrm{loc}}(\Omega)$ is a distribution solution of (1) such that $u \leq 0$ on $\partial\Omega$, then (16) is valid for $\varphi = \psi(u)$, in the sense that*

$$\int_\Omega \langle \boldsymbol{A}(x, u, Du), D\varphi \rangle \leq \int_\Omega [B(x, u, Du)]^+ \varphi. \tag{17}$$

A proof can be found in [4, Section 3.1].

The proof of Theorem 1 is based on two main lemmas.

Lemma 2. *Let $w \in W^{1,1}_{\mathrm{loc}}(\Omega)$ be a solution of inequality (1), and suppose the coefficients $a(x)$, $b(x)$, $c(x)$ in (2) are in the Lebesgue space $L^q(\Omega)$ for some $q > n$. Assume that $|\Omega| = 1$ and*

$$w \geq k \quad \text{in } \Omega, \qquad w = k \quad \text{on } \partial\Omega,$$

with $k = \alpha S \|a\|_q$, $\alpha > 0$ and $S = 1/n\omega_n^{1/n}$. Then $w \in L^\infty(\Omega)$ and

$$w \leq \left(K \left[\frac{n}{n-1} \right]^{n-1} \right)^{nq/(q-n)} \|w\|_1, \qquad K = e^{1/e} + 1/\alpha + S\|b + c\|_q. \tag{18}$$

(In the case of bounded coefficients we can take more simply $K = 1 + 1/\alpha + S(b + c)$.)

Proof. Step 1. This relies on a Moser-type iteration argument. We argue by induction, showing that if $w \in W^{1,1}_{\mathrm{loc}}(\Omega) \cap L^r(\Omega)$ for some $r \geq 1$, then

$$w \in W^{1,1}_{\mathrm{loc}}(\Omega) \cap L^{\kappa r}(\Omega),$$

where $\kappa = n/(n-1)$. For $r = 1$ one obviously has $w \in W^{1,1}_{\mathrm{loc}}(\Omega) \cap L^1(\Omega)$, since $w = k$ on $\partial\Omega$. Hence by induction

$$w \in W^{1,1}_{\mathrm{loc}}(\Omega) \cap L^\varrho(\Omega) \qquad \text{for all } \varrho \geq 1.$$

Let ℓ, m be fixed, with $k < \ell < m$ (ultimately we take $\ell \to k$ and $m \to \infty$). For all real numbers $r \geq 1$, define

$$\psi(t) = \begin{cases} 0 & \text{if } k \leq t \leq \ell, \\ t^r - \ell^r & \text{if } \ell < t < m, \\ m^r - l^r & \text{if } t \geq m; \end{cases} \tag{19}$$

$$v(t) = \begin{cases} \ell^r & \text{if } k \leq t \leq \ell, \\ t^r & \text{if } \ell < t < m, \\ m^r & \text{if } t \geq m. \end{cases} \tag{20}$$

The truncation of $\psi(t)$ and $v(t)$ when $t \geq m$ makes $\psi(t)$ and $v(t)$ piecewise smooth with corners only at $t = \ell$ and $t = m$.

By Lemma 1 it is clear that $\varphi = \psi(w)$ can serve as a test function for (1) in Ω. In particular by (17) we have

$$\int_\Omega \langle \boldsymbol{A}(x, w, Dw), D\varphi \rangle \leq \int_\Omega [B(x, w, Dw)]^+ \varphi, \tag{21}$$

where $D\varphi = \mathbf{0}$ in $\Omega \setminus \Gamma$ and

$$D\varphi = r w^{r-1} Dw \quad \text{in } \Gamma, \tag{22}$$

with $\Gamma = \{x \in \Omega : \ell < w(x) < m\}$. For $x \in \Gamma$,

$$r w^{r-1} |Dw| = |Dv|, \qquad v = v(w) = w^r.$$

Hence by the structure conditions (2), the inequality (21) leads to

$$\|Dv\|_1 \leq r \int_\Omega (b + c + a/k) v. \tag{23}$$

where we have used the fact that (23) is trivially satisfied on the set $\Omega \setminus \Gamma$. (The integrals in (23) are well defined, since $v \leq m^r$ and $Dv \in L^1(\Omega)$. Even more, $k > 0$ unless $a \equiv 0$.)

Step 2. By Hölder's inequality and the fact that $k = \alpha S \|a\|$ we obtain from (23)

$$\|Dv\|_1 \leq r(\|b + c\|_q + 1/\alpha S) \|v\|_1^{1-n/q} \|v\|_{n/(n-1)}^{n/q}.$$

On the other hand, by Sobolev's inequality (recall $v \equiv \ell^r$ near $\partial\Omega$) it follows that $\|v - \ell^r\|_{n/(n-1)} \leq S\|Dv\|_1$. But also $\|\ell^r\|_{n/(n-1)} = \|\ell^r\|_1$ since $|\Omega| = 1$. Therefore

$$\|v\|_{n/(n-1)} \leq \|v - \ell^r\|_{n/(n-1)} + \|\ell^r\|_{n/(n-1)} \leq S\|Dv\|_1 + \|\ell^r\|_1 \leq S\|Dv\|_1 + \|v\|_1.$$

since $v \geq \ell^r$ in Ω. Combining the previous two displayed inequalities now gives

$$\|v\|_{n/(n-1)} \leq Srd \|v\|_1^{1-n/q} \|v\|_{n/(n-1)}^{n/q} + \|v\|_1, \tag{24}$$

where $d = \|b + c\|_q + 1/\alpha S$.

Application of Young's inequality with the exponent pair $(q/(q-n),\ q/n)$ shows that

$$Srd\|v\|_1^{1-n/q}\|v\|_{n/(n-1)}^{n/q} \le \left(1-\frac{n}{q}\right)(Srd)^{q/(q-n)}\|v\|_1 + \frac{n}{q}\|v\|_{n/(n-1)}.$$

Thus (24) implies (!)

$$\|v\|_{n/(n-1)} \le [(Srd)^{q/(q-n)} + q/(q-n)]\|v\|_1 \le [Srd + e^{1/e}]^{q/(q-n)}\|v\|_1.$$

Step 3. It is not hard to check that

$$\|w^r\|_{\kappa,\Gamma'} \le \|v\|_\kappa = \|v\|_{n/(n-1)}, \qquad \Gamma' = \{x \in \Omega : k \le w(x) < m\}$$

with $\kappa = n/(n-1)$. Similarly

$$\|v\|_1 \le \|w^r\|_1 + (\ell^r - k^r)\,|\Omega \setminus \Gamma| \le \|w^r\|_1 + \ell^r - k^r.$$

Hence the concluding inequality of Step 2 can be rewritten

$$\|w^r\|_{\kappa,\Gamma'} \le [Kr]^{q/(q-n)}\{\|w^r\|_1 + \ell^r - k^r\},$$

with $K = 1/\alpha + e^{1/e} + S\|b+c\|_q$. Letting $\ell \to k$ gives $\|w^r\|_{\kappa,\Gamma'} \le [Kr]^{q/(q-n)}\|w^r\|_1$. Finally, taking $m \to \infty$ and using the monotone convergence theorem, there follows

$$\|w\|_{\kappa r}^r \le [Kr]^\nu \|w\|_r^r, \qquad \nu = q/(q-n). \tag{25}$$

The right side of (25) is finite by the induction assumption $w \in L^r(\Omega)$. Thus also $\|w\|_{\kappa r}$ is finite, completing the induction step and consequently showing that $w \in L^\varrho(\Omega)$ for all $\varrho \ge 1$.

Step 4. The inequality (25) holds for all $r \ge 1$. Taking first $r = 1$, we get

$$\|w\|_\kappa \le K^\nu \|w\|_1.$$

Next, take $r = \kappa$ so that

$$\|w\|_{\kappa^2} \le (K\kappa)^{\nu/\kappa}\,\|w\|_\kappa \le [(K\kappa)^{1/\kappa}\,K]^\nu\,\|w\|_1 = [K^{1+1/\kappa}\,\kappa^{1/\kappa}]^\nu\,\|w\|_1.$$

Continuing in this way, with r successively equal to 1, κ, κ^2, etc., we get

$$\|w\|_{\kappa^j} \le [K^\Sigma \kappa^{\Sigma'}]^\nu \|w\|_1, \tag{26}$$

where

$$\Sigma = \Sigma_j = \sum_{i=0}^{j-1}\frac{1}{\kappa^i}, \qquad \Sigma' = \Sigma_j' = \sum_{i=1}^{j-1}\frac{i}{\kappa^i}.$$

The series Σ converges to $\kappa/(\kappa-1) = n$ as $j \to \infty$, and similarly Σ' converges to $\kappa/(\kappa-1)^2 = n(n-1)$. Thus letting $j \to \infty$ in (26) gives

$$\|w\|_\infty \le \left(K\left[\frac{n}{n-1}\right]^{n-1}\right)^{nq/(q-n)}\|w\|_1. \tag{27}$$

If a, b, c are bounded, we can effectively suppose $q = \infty$, so that (24) reduces to

$$\|v\|_\kappa \leq [Srd + 1]\|v\|_1.$$

Hence in this case one can take $K = 1 + 1/\alpha + S(b + c)$. The proof is complete.

Lemma 3. *Let the hypotheses of Lemma 2 hold with the exception that*

$$\|b + c\|_n \leq (1 - \delta)/S \tag{28}$$

where $0 < \delta < 1$. *Then*

$$\|w\|_1 \leq (1 + \alpha)S\|a\|_q/\delta. \tag{29}$$

Proof. As in the proof of Lemma 2, but using *only* the case $r = 1$, we obtain corresponding to (23)

$$\|Dv\|_1 \leq \int_\Omega [(b + c)v + a].$$

In turn, by Hölder's inequality,

$$\|Dv\|_1 \leq \|b + c\|_n \, \|v\|_{n/(n-1)} + \|a\|_1 \leq \frac{1 - \delta}{S}\|v\|_{n/(n-1)} + \|a\|_1 \tag{30}$$

by (28).

Next by Sobolev's inequality, as before,

$$\|v\|_{n/(n-1)} \leq \|v - \ell\|_{n/(n-1)} + \|\ell\|_{n/(n-1)} \leq S\|Dv\|_1 + \ell.$$

Using (30), this gives

$$\|v\|_{n/(n-1)} \leq (1 - \delta)\|v\|_{n/(n-1)} + S\|a\|_1 + \ell.$$

After transposing, one can take $\ell \to k$ and $m \to \infty$. Then as in the proof of Lemma 2 there follows

$$\|w\|_{n/(n-1)} \leq (1 + \alpha)S\|a\|_q/\delta \tag{31}$$

since $k = \alpha S\|a\|$. The required conclusion (29) is now obvious.

Proof of Theorem 1. First take $|\Omega| = 1$. Define $w = u^+ + k$. Then $w \geq k$ in Ω and $w = k$ on $\partial\Omega$, so that Lemmas 2 and 3 apply. In particular by (4) and (18),

$$K = e^{1/e} + 1/\alpha + S\|b + c\| \leq e^{1/e} + 1/\alpha + 1 - \delta. \tag{32}$$

We are free to choose α as we wish. For bounded coefficients take, say, $\alpha = 1$, and replace $e^{1/e}$ by 1. Then $1 + \alpha = 2$ and $K \leq 2 + 1/\alpha = 3$. Otherwise take $\alpha = 2$, giving $1 + \alpha = 3$ and $K \leq 3$.

The conclusions (5) and (5′) for the case $|\Omega| = 1$ now follow at once from (18) and (29). The general result is then obtained by scaling.

These estimates can be compared with the case $p = 1$ of Theorem 10.10 in [1]. A boundary condition $u \leq M$ on $\partial\Omega$ can be handled by the change of variable $\tilde{u} = u - M$ and replacing a by $a + Mc$.

2. The mean curvature inequality

Consider the mean curvature inequality

$$\operatorname{div}\left(\frac{Du}{\sqrt{1+|Du|^2}}\right) \geq nH(x). \tag{33}$$

Putting $A(\xi) = \xi/\sqrt{1+|\xi|^2}$, $B = -nH(x)$, we see that $|A| < 1$ and

$$\langle A(\xi), \xi \rangle \geq |\xi| - a, \qquad a = \sqrt{\frac{5\sqrt{5}-11}{2}} \sim 0.3002831.$$

[This is an easy exercise in differential calculus. In fact, the best value for a is

$$\max_{0 \leq t < \infty} (t - t^2/\sqrt{1+t^2}).$$

By elementary calculus the maximum occurs at $t_0 = \sqrt{(\sqrt{5}-1)/2} \sim 0.78614$, and in turn $t_0 - t_0^2/\sqrt{1+t_0^2}$ takes the value a given above.]

Thus (33) satisfies the structure conditions (2), with a as given, $c = 0$ and $b = -nH^{-1}(x)$. From Theorem 1 we then get

$$u \leq \left(3\left[\frac{n}{n-1}\right]^{n-1}\right)^{nq/(q-n)} S|\Omega|^{1/n}/\delta \tag{34}$$

provided $\|H^-\|_q |\Omega|^{1/n-1/q} \leq (1-\delta)\omega_n^{1/n}$. Rewriting this in terms of the *effective radius* $\mathcal{R}$ of Ω, it becomes

$$u \leq \frac{1}{n}\left(3\left[\frac{n}{n-1}\right]^{n-1}\right)^{nq/(q-n)} \mathcal{R}/\delta,$$

which is just (10)–(11). In the same way we obtain (10)–(11′). This proves Theorem 2.

In the canonical case $n = 2$, $H(x) \geq -H_0$ and $\delta = 1/2$ the estimate (10)–(11) can be somewhat improved. In fact, from (32) we can take $K = 3/2 + 1/\alpha$. Then with $\alpha = 1.1$, $(n/(n-1))^{n-1} = 2$, we see from (18), (29) that $u/\mathcal{R} = 2.1 \cdot (5.3/1.1)^2 a \sim 14.6$, as required for (14).

Remark. If $u \leq 0$ on $\partial\Omega$ and $H = 0$ one expects the conclusion $u \leq 0$ in Ω. This however cannot be obtained by the present approach since the constant $a \sim 0.3002831$ acts as an inhomogeneous term in the structure conditions (2).

3. Remarks

The a priori estimates (10)–(11′) for the supremum of a solution of (3) in Ω can be supplemented, as we have noted above, by the result of Corollary 10.6 in [1] as well as by simple bounds based on spherical caps (or even the upper part of cylindrical tubes). Each of these results has its own separate merit, depending on given smoothness assumptions on u, or on the geometry of the domain Ω.

A final esimate, obtained also by Gilbarg and Trudinger ([1, p. 409]) can also be noted. In particular, assume that the boundary $\partial\Omega$ of Ω is of class $C^{2,\alpha}$ (no smoothness conditions on $\partial\Omega$ were required in the previous results). Suppose also that the mean curvature H' of the boundary satisfies

$$H'(y) \geq \frac{n}{n-1} H_0, \tag{35}$$

where $H(x) \geq -H_0 =$ negative constant. Then the following upper bound holds.

Let u be a solution of (3) *of class $C^1(\Omega)$. Then if $u \leq 0$ on $\partial\Omega$ we have*

$$u \leq (e^{\mu D} - 1)/\mu \quad \text{in } \Omega \tag{36}$$

where $\mu = 1 + nH_0$ and $D =$ diam Ω.

Gilbarg and Trudinger in fact assume that $H(x) \equiv$ const, that $u \in C^2(\Omega)$, and that u solves (3) with equality, but it is not hard to see that their proof implies the more general result stated here.

When Ω is a ball of radius R, condition (35) becomes $1/R \geq (n/(n-1))H_0$, that is, $H_0 R \leq (n-1)/n$, a condition somewhat stronger than (9′). Since obviously $D = 2R$ for balls, (36) gives

$$u \leq (e^{2(1+nH_0)R} - 1)/(1 + nH_0). \tag{37}$$

If R is small the bound is essentially $u \leq 2R$, which is quite good. On the other hand, for large R clearly (37) is not nearly as good as (10) or (14).

Whatever else, it should be kept in mind (i) that condition (35) is exactly the boundary curvature condition necessary for the existence of solutions of the Dirichlet problem for the constant mean curvature equation for arbitrary continuous data, while (ii) the estimate (36) does not apply at all in the generality of Theorem 1.

Acknowledgements. The author wishes to thank Professor Patrizia Pucci for her help and for her kind interest in this work.

References

[1] Gilbarg, D., Trudinger, N.: Elliptic Partial Differential Equations of Second Order. 2nd ed., Springer (1983) Zbl 0562.35001 MR 0737190

[2] Hopf, E.: Elementare Bemerkungen über die Lösungen partieller Differentialgleichungen zweiter Ordnung vom elliptischen Typus. Sitz. Preussischen Akad. Wissenschaften **1927**, 147–152 JFM 53.0454.02

[3] Pucci, P., Serrin, J., The strong maximum principle revisited. J. Differential Equations **196**, 1–66 (2004); Erratum, J. Differential Equations **207**, 226–227 (2005) Zbl pre02033000MR2100819

[4] Pucci, P., Serrin, J.: The maximum principle. In: Handbook of Differential Equations, M. Chipot and P. Quittner (eds.), Elsevier (2006)

Olga Ladyzenskaya and David Gilbarg, Warsaw, Poland, 1988.

I.13 Dead Cores

J.I. Díaz

Mathematics Subject Classification (2010). Primary 35J60; Secondary 35R35.

Keywords. Dead core, semilinear elliptic problems, free boundary problem, compact support solutions.

The paper [i] below deals with several qualitative properties of the solutions of the quasilinear elliptic partial differential equations

$$\operatorname{div}\{A(|Du|)Du\} = f(u) \quad \text{in } \Omega, \tag{1}$$

where A and f satisfy some structure conditions which do not imply their differentiability. Most of the paper is devoted to the "monotone case" in which $f \in C(\mathbb{R})$, $f(0) = 0$, f is nondecreasing on $\mathbb{R}$, and $f(u) > 0$ for $u > 0$. This "monotone case" is also considered (for a more special diffusion term) in the presence of some weight coefficients, including for example the equations of Matukuma or Batt–Faltenbacher–Horst type. For the sake of the presentation, only nonnegative solutions are considered. A final short section considers also the "nonmonotone case", in which f is assumed to be nondecreasing only near 0.

Here Ω is an open and connected set in $\mathbb{R}^n$, $n \geq 1$, not necessarily bounded. We follow the usual terminology of the chemical engineering literature by saying that a solution u which is strictly positive on $\partial\Omega$ has a dead core if u vanishes on a set of positive measure. The main difficulty in studying this problem lies in obtaining the existence and behavior of the free boundary defined as the boundary of the dead core. This behavior had been studied previously in connection with other problems for the case of unbounded domains. It is difficult to be absolutely precise at this point, but it seems that this type of free boundary was first considered, in the class of second-order elliptic equations, by H. Brezis and G. Stampacchia in a series of papers, starting in 1973 (see references in [4]) on the "classical" problem of the flow past a given profile with prescribed velocity at infinity. In this special case, the location of the free boundary associated to an auxiliary semilinear problem leads to some sharp estimates on the maximum velocity. The study of solutions with compact support on unbounded domains was extended later to more general equations. The compactness of the support of the solution of some variational inequalities (of obstacle type) formulated in terms of the multivalued semilinear equation $Lu + \beta(u) \ni \psi(x)$ in Ω, where L is a linear second-order elliptic operator and β is a maximal monotone graph in $\mathbb{R}^2$ such that $0 \in \beta(0)$ was proved in [3]. Other nonlinear variational problems having a solution with compact support were studied earlier by other authors (J. Auchmuty and R. Beals, L. Berkowitz and H. Pollard, and R. Redheffer). The auxiliary problems in those case involved higher-order semilinear equations.

P. Pucci et al. (eds.), *James Serrin. Selected Papers: Volume 1*, Contemporary Mathematicians,
DOI 10.1007/978-3-0348-0845-3_13, © Springer Basel 2014

It seems important to point out that this type of free boundary was studied almost twenty years before in an entirely different framework: the solutions of some degenerate quasilinear parabolic equations such as the porous medium equation. The compactness of the support of the solution $u(t,\cdot)$, for any fixed $t > 0$ when $u(0,\cdot)$ has compact support was proved for the first time in the literature in 1958 by O.A. Oleinik, A.S. Kalashnikov and Chzou Yui Lin. Some special solutions were considered even earlier by G.I. Barenblatt, Ya.B. Zeldovich and later by many other distinguished specialists. This relevant class of parabolic problems, attracted the attention of other authors too numerous to mention here.

In 1975, Ph. Benilan, H. Brezis and M.G. Crandall [2] studied the support of the solution of the semilinear equation $-\Delta u + \beta(u) \ni \psi$ in $\mathbb{R}^N$ considering the case in which the monotone term $\beta(u)$ is not necessarily multivalued at the origin as assumed in the above-mentioned papers. They proved that the necessary and sufficient condition on β in order to get a solution with compact support is given by the integrability at the origin of a suitable improper integral. Starting in 1978, this criterion was extended to the case of quasilinear problems of the type $-\text{div }(|\nabla u|^{p-2}\,\nabla u) + \beta(u) \ni \psi,\ p > 1$ in several papers by J.I. Díaz and M.A. Herrero (see, *e.g.*, [7]), to a new criterion which now applies to Lipschitz functions $\beta(u)$ if $p > 2$.

The above results were extended later in many directions in the literature. For instance, the study of the semilinear elliptic equation but now on a bounded domain Ω (and with $\psi = 0$ on Ω and $u = 1$ on the boundary) was studied by C. Bandle, R. Sperb and I. Stakgold. They showed that the same criterion of [2] is, in some sense, the necessary and sufficient condition on β for the formation of an internal free boundary (the boundary of the *dead core*). The general idea of the construction of super and subsolutions vanishing at suitable points of the domain (and without requiring the condition Ω unbounded) was first carried out in 1984 by J.I. Díaz and J. Hernández for the case of the dead core problem. It was later extended systematically to the class of general quasilinear equations of the type (1) in the 1985 monograph [5]. Some further results were published by J.I. Díaz, J.E. Saa and U. Thiel in 1990. These types of results attained their maximum elegance and deepness with the 2006 paper by Pucci and Serrin [i].

They improve all the previous papers in the literature by using only arguments which can be considered today as "elementary" in the theory of elliptic partial differential equations: in the style of the famous 1927 paper by E. Hopf. Their presentation is highly pedagogical. They study the optimality of their main assumptions. For instance, the above-mentioned criteria of [2] and [7] now become

$$\int_{0+} \frac{ds}{H^{-1}(F(s))} < +\infty,$$

where

$$H(\rho) = \int_0^{\Phi(\rho)} \Phi^{-1}(s)ds,$$

with

$$\Phi(\rho) = \rho A(\rho) \quad \text{and} \quad F(u) = \int_0^u f(s)ds.$$

The criterion on the balance between the data and the domain (see Section 1.2b of [5]) was here made sharper for the special case in which Ω is a ball and the boundary datum is a constant. Several steps of their study (perhaps the finest and newest parts of the paper) have some interest by themselves. For instance Theorem 4.1 proves the strict separation among solutions outside their dead cores, even if the strong maximum principle can not be applied directly.

The short section on the "nonmonotone case" presents some special solutions (labeled here, for the first time, as *burst within the core solutions*) which can be built, like a puzzle, due to the lack of uniqueness of solutions. The literature on nonmonotone equations leading to this type of free boundary is also quite long (see, *e.g.*, [8], several previous papers by Serrin and coauthors and the exposition of [6]), but never before has the problem been treated in the generality considered in this paper. The deepness and clarity of the techniques seem to make it possible to obtain a bifurcation diagram with respect to a chosen parameter.

Finally, we briefly mention that some appropriate energy methods can be applied in studying this type of free boundary, even for higher-order equations or systems of equations (see, *e.g.*, the monograph [1]).

References

[1] S.N. Antontsev, J.I.Díaz and S.I. Shmarev, *Energy Methods for Free Boundary Problems*, Birkhäuser, Boston, 2002.

[2] Ph. Benilan, H. Brezis and M.G. Crandall, *A semilinear equation in* $L^1(R^N)$, Ann. Scuola Norm. Sup. Pisa **4**, 2 (1975), 523–555.

[3] H. Brezis, *Solutions of variational inequalities with compact support.* Uspekhi Mat. Nauk., **129**, (1974) 103–108.

[4] H. Brezis and G. Stampacchia, *The Hodograph Method in Fluid-Dynamics in the Light of Variational Inequalities*, Arch. Rat. Mech.Anal., **61**, (1976), 1–18.

[5] J.I. Díaz, *Nonlinear Partial Differential Equations and Free Boundaries.* Research Notes in Mathematics, 106, Pitman, London 1985.

[6] J.I. Díaz, J. Hernández and F.J. Mancebo, Branches of positive and free boundary solutions for some singular quasilinear elliptic problems, J. Math. Anal. Appl., **352** (2009) 449–474.

[7] J.I. Díaz and M.A. Herrero, *Estimates on the support of the solutions of some non linear elliptic and parabolic problems.* Proc. Roy. Soc. Edinburgh, **98A** (1981) 249–258.

[8] A. Friedman and D. Phillips, *The free boundary of a semilinear elliptic equation*, Trans. Amer. Math. Soc. (1984), 282, 153–182.

J.I. Díaz
Departamento de Matemática Aplicada
Facultad de Matemáticas, UCM
Plaza de las Ciencias 3
ES 28040 Madrid, Spain
e-mail: `diaz.racefyn@insde.es`

SIAM J. Math. Anal. 38 (2006), 259–278 **Reprint I.13.[i]**

SIAM J. MATH. ANAL.
Vol. 38, No. 1, pp. 259–278

DEAD CORES AND BURSTS FOR QUASILINEAR SINGULAR ELLIPTIC EQUATIONS*

PATRIZIA PUCCI† AND JAMES SERRIN‡

Abstract. We consider divergence structure quasilinear singular elliptic partial differential equations on domains of $\mathbb{R}^n$ and show that there exist solutions with dead cores and, furthermore, solutions which involve both a dead core and *bursts* within the core. The results are obtained under appropriate monotonicity conditions on both the nonlinearity and the elliptic operator. Important special cases treated here are the p-Laplace and the mean curvature operators.

We also study related problems for p-Laplace equations with weights, which include the Matukuma equation as a prototype.

While it is usually thought that dead cores arise due to loss of smoothness of the underlying equation, we show by examples that they can occur equally for analytic p-Laplace equations.

Key words. quasilinear singular elliptic equations, dead cores

AMS subject classifications. Primary, 35J15; Secondary, 35J70

DOI. 10.1137/050630027

1. Introduction. We consider quasilinear elliptic partial differential equations having the canonical divergence structure

$$\operatorname{div}\{A(|Du|)Du\} = f(u) \quad \text{in } \Omega. \tag{1.1}$$

Here Ω is a domain (connected open set) in $\mathbb{R}^n$, $n \geq 1$, and Du denotes the vector gradient of the given function $u = u(x)$, $x \in \Omega$. *Unless otherwise stated explicitly, we assume throughout the paper the following conditions on the operator $A = A(\varrho)$ and the nonlinearity $f = f(u)$:*

(A1) $A \in C(\mathbb{R}^+)$, $\mathbb{R}^+ := (0, \infty)$;
(A2) $\varrho \mapsto \varrho A(\varrho)$ is strictly increasing in $\mathbb{R}^+$ and $\varrho A(\varrho) \to 0$ as $\varrho \to 0$;
(F1) $f \in C(\mathbb{R})$; and
(F2) $f(0) = 0$, f is nondecreasing on $\mathbb{R}$, and $f(u) > 0$ for $u > 0$.

Condition (A2) is a minimal requirement for ellipticity of (1.1). Furthermore, it allows both singular and degenerate behavior of the operator A at $\varrho = 0$, that is, at critical points of u. We emphasize that no assumptions of differentiability are made on either A or f.

We also study the related elliptic equation

$$\operatorname{div}(g(|x|)|Du|^{p-2}Du) = h(|x|)f(u) \quad \text{in } \Omega, \quad p > 1, \tag{1.2}$$

where g, $h : \mathbb{R}^+ \to \mathbb{R}^+$ are radial functions of class $C^1(\mathbb{R}^+)$ and where the general elliptic operator A is replaced by the p-Laplacian function $A(\varrho) = \varrho^{p-2}$. The celebrated Matukuma equation is a prototype for (1.2).

*Received by the editors April 26, 2005; accepted for publication December 2, 2005; published electronically April 21, 2006. This research was supported by the Project Metodi Variazionali ed Equazioni Differenziali Non Lineari.

http://www.siam.org/journals/sima/38-1/63002.html.

†Dipartimento di Matematica e Informatica, Università degli Studi di Perugia, Via Vanvitelli 1, 06123 Perugia, Italy (pucci@dipmat.unipg.it).

‡ Department of Mathematics, University of Minnesota, Minneapolis, MN 55455 (serrin@math.umn.edu).

By a *classical solution* (or a classical distribution solution) of (1.1) or (1.2) in Ω we mean a function $u \in C^1(\Omega)$ which satisfies (1.1) or (1.2) in the distribution sense.

With the notation $\Phi(\varrho) = \varrho A(\varrho)$ when $\varrho > 0$, and $\Phi(0) = 0$, we introduce the function

$$H(\varrho) = \varrho\Phi(\varrho) - \int_0^{\varrho} \Phi(s)ds, \qquad \varrho \geq 0.$$

This function is easily seen to be strictly increasing, as follows from the inequality

$$\varrho_1\Phi(\varrho_1) - \varrho_0\Phi(\varrho_0) > (\varrho_1 - \varrho_0)\Phi(\varrho_1) > \int_{\varrho_0}^{\varrho_1} \Phi(s)ds$$

when $\varrho_1 > \varrho_0 \geq 0$. Alternatively, monotonicity follows from the representation

$$H(\varrho) = \int_0^{\Phi(\varrho)} \Phi^{-1}(s)ds, \qquad \varrho \geq 0,$$

this being a consequence of the Stieltjes formula $H(\varrho) = \int_0^{\varrho} s\, d\Phi(s)$.

For the Laplace operator, that is, when (1.1) takes the classical form

$$\Delta u = f(u),$$

we have $A(\varrho) \equiv 1$ and $H(\varrho) = \frac{1}{2}\varrho^2$. Similarly, for the degenerate p-Laplace operator, here denoted by Δ_p, $p > 1$, we have $A(\varrho) = \varrho^{p-2}$ and $H(\varrho) = \varrho^p/p'$, while for the mean curvature operator, one has $A(\varrho) = 1/\sqrt{1+\varrho^2}$ and $H(\varrho) = 1 - 1/\sqrt{1+\varrho^2}$. In the last example, note the anomalous behavior $\Phi(\infty) = H(\infty) = 1$, a possibility which requires extra care and will be treated and discussed separately.

It is also worth observing that (1.1) is precisely the Euler–Lagrange equation for the variational integral

$$I[u] = \int_\Omega \{\mathcal{G}(|Du|) + F(u)\}dx, \qquad F(u) = \int_0^u f(s)ds,$$

where $\mathcal{G}$ and A are related by $A(\varrho) = \mathcal{G}'(\varrho)/\varrho$, $\varrho > 0$. In this case $H(\varrho) = \varrho\mathcal{G}'(\varrho) - \mathcal{G}(\varrho)$, the pre-Legendre transform of $\mathcal{G}$. Similarly, the variational integral for (1.2) is given by

$$I[u] = \int_\Omega \left\{g(|x|)\frac{|Du|^p}{p} + h(|x|)F(u)\right\} dx,$$

where now $H(\varrho) = \varrho^p/p'$ and p' is the Hőlder conjugate of p.

An elliptic equation or inequality is said to have a *dead core solution* u in some domain $\Omega \subset \mathbb{R}^n$ provided that there exists an open subset Ω_1 with compact closure in Ω, called the *dead core* of u, such that

$$u \equiv 0 \quad \text{in } \Omega_1, \qquad u > 0 \quad \text{in } \Omega \setminus \overline{\Omega_1}.$$

The condition $u > 0$ could be replaced by $u \not\equiv 0$, but for definiteness (and physical reality) we prefer the condition as stated. By the strong maximum principle (see Theorem 1.1 of [7, 8]) a nonnegative solution of (1.1) or (1.2) can have a dead core only if

$$\int_{0+} \frac{ds}{H^{-1}(F(s))} < \infty, \tag{1.3}$$

where H^{-1} is the inverse of H. *We assume that condition* (1.3) *holds throughout the sequel, except for Theorems* 1.2, 1.3, 6.2, *and* 6.3.

The equation $\Delta u = u^q$, for example, allows dead cores only if $0 < q < 1$. Actually condition (1.3) is not only necessary but also sufficient for the existence of solutions with dead cores. More specifically, we have the following main result for (1.1).

THEOREM 1.1. *Suppose* $\Phi(\infty) = H(\infty) = \infty$. *Assume the dead core condition* (1.3) *holds and let* u *be a solution of* (1.1), *with* $0 \le u(x) \le m$ *on* $\partial\Omega$ *for some positive constant* m. *Then the following properties are valid:*

(a) $0 \le u < m$ *in* Ω.

(b) *Assume that*

$$C = \int_0^\infty \frac{ds}{H^{-1}(F(s)/n)} < \infty, \tag{1.4}$$

and let B_R *be a ball with radius* $R \ge C$, *compactly contained in* Ω. *Then* u *has a dead core in* Ω *for all* $m > 0$.

(c) *If* B *is any ball compactly contained in* Ω, *then* $u \equiv 0$ *in* B *provided that* $m > 0$ *is suitably small.*

Equation (1.2) allows a corresponding dead core result, which however we defer until section 6.

A more refined version of Theorem 1.1 can be obtained when $\Omega = B_R$, where B_R is any open ball in $\mathbb{R}^n$, $n \ge 1$, of radius $R > 0$. Until explicitly noted in section 9, we continue to assume that $\Phi(\infty) = H(\infty) = \infty$.

THEOREM 1.2. *The problem*

$$\begin{cases} \operatorname{div}\{A(|Du|)Du\} = f(u) & \text{in } B_R, \\ u = m > 0 \quad \text{on } \partial B_R, & u \in C(\overline{B_R}), \end{cases} \tag{1.5}$$

admits a unique classical (distribution) solution u, *necessarily radial. Moreover* $u = u(r) = u(r,m)$ *is of class* $C^1[0,R]$ *and satisfies* $u \ge 0$, $u' \ge 0$ *in* $[0,R]$ *and* $u'(0) = 0$, *where* $' = d/dr$.

Finally, at any $r > 0$ *where* $u(r,m) > 0$ *we have also* $u'(r,m) > 0$.

It follows from Theorem 1.2 that the solution $u(\cdot\,, m)$ must be one of the following three types:

1. $u > 0$ in B_R.
2. $u(0,m) = 0$ and $u'(r,m) > 0$ when $r > 0$.
3. There exists $S \in (0,R)$ such that $u \equiv 0$ in B_S and $u'(r,m) > 0$ when $r > S$.

That is, in case 3 the solution u of (1.5) has a dead core B_S. The solution $u = u(\cdot\,, m)$ of (1.5) has further properties of interest, given in the next result.

THEOREM 1.3. *The function* $u = u(\cdot\,, m)$ *is continuous and nondecreasing in the variable* $m\,(> 0)$, *and* $u < m$ *in* B_R.

The following theorem gives an important relation between the value m and dead cores solutions of (1.5).

THEOREM 1.4. *Let* $u(\cdot\,, m)$ *be the unique solution of* (1.5). *Then either* $u(\cdot\,, m)$ *has a dead core for all* $m > 0$, *or there is a unique (finite) number*

$$m = m_0 = m_0(R) > 0$$

for which a solution $u_0 = u_0(r) = u_0(r\,, m_0)$ *of* (1.5) *in* B_R *exists, with the properties that*

(i) $u_0(0) = 0$;

(ii) $u(0,m) > 0$ *for every* $m > m_0$*; and*
(iii) $u(\cdot,m)$ *has a dead core for every* $0 < m < m_0$.

For convenience we define $m_0 = m_0(R)$ to be ∞ when $u(0,m) = 0$ for all $m > 0$. The examples

$$\Delta u = (\operatorname{sign} u)\sqrt{|u|}, \tag{1.6}$$

$$\Delta_4 u = u \tag{1.7}$$

are particularly interesting as illustrations of the main theorems above. Indeed, both of these are included in the canonical case

$$\Delta_p u = u|u|^{q-1}, \qquad p > 1, \qquad q > 0,$$

for which $F(u) = |u|^{q+1}/(q+1)$. Here the dead core condition (1.3) reduces exactly to

$$0 < q < p - 1.$$

For these special cases, we search for u_0 in the form $c\,r^k$, c, $k > 0$. Then from (1.5) one finds

$$k = \frac{p}{p-1-q}, \qquad c = k^{-k/p'}(n+kq)^{-k/p}, \qquad m_0 = c\,R^k. \tag{1.8}$$

For the case (1.6) we have $p = 2$, $q = 1/2$, $k = 4$, so that

$$m_0 = \frac{1}{(n+2)^2}\left(\frac{R}{2}\right)^4,$$

while $p = 4$, $q = 1$, $k = 2$ for (1.7), and, in turn,

$$m_0 = \frac{R^2}{2\sqrt{2(n+2)}},$$

which reduces exactly to $m_0 = R^4/400$ and $m_0 = R^2/2\sqrt{10}$ when $n = 3$. In particular for the unit radius $R = 1$ we obtain, respectively, the unexpectedly small numbers $m_0 = 0.00125$ and $m_0 \cong 0.158$.

Equation (1.7), when written in full for $n = 2$, has the form

$$|Du|^2\Delta u + 2u_x^2 u_{xx} + 4u_x u_y u_{xy} + 2u_y^2 u_{yy} = u,$$

which is analytic in all its variables. Thus dead core behavior is not due simply to a lack of smoothness in the basic equation. In fact (1.7) is an analytic partial differential equation, elliptic except at the singular point $Du = 0$, which has a nonanalytic solution.

As a final example, consider the equation

$$\Delta u = (\operatorname{sign} u)\sqrt{|u|} + u|u|^2.$$

Here $A(\varrho) \equiv 1$, $H^{-1}(\varrho) = \sqrt{2\varrho}$, and $F(u) = \frac{2}{3}|u|^{3/2} + \frac{1}{4}|u|^4$. Then

$$C = \sqrt{\frac{n}{2}}\int_0^\infty \frac{ds}{\sqrt{(2/3)s^{3/2} + s^4/4}} < \infty.$$

By numerical calculation $C \cong 6.4334$ if $n = 2$. Therefore, by the results of section 4 we have $m_0 = \infty$ whenever $R \geq 7$. In particular, for the problem

$$\begin{cases} \Delta u = (\operatorname{sign} u)\sqrt{|u|} + u|u|^2 & \text{in } B_7 \subset \mathbb{R}^2, \\ u = m > 0 & \text{on } \partial B_7, \end{cases}$$

a dead core occurs for all $m > 0$. This result also follows without recourse to numerical calculation, since one can write, when $n = 2$,

$$C = \left(\frac{9}{2}\right)^{1/5} \int_0^\infty \frac{dt}{\sqrt{t^{3/2} + t^4}} < \left(\frac{9}{2}\right)^{1/5} \left\{ \int_0^1 \frac{dt}{\sqrt{t^{3/2}}} + \int_1^\infty \frac{dt}{\sqrt{t^4}} \right\} = 5(4.5)^{1/5} \cong 6.75.$$

The case $n = 3$ can be treated in the same way, with $C \cong 7.879$, but here the radius $R = 7$ should be replaced by $R = 8$.

In a related paper [1] Bandle and Vernier-Piro also studied the dead core problem for the weighted equation (1.2). Because of the different assumptions on the weights g and h made there it is hard to compare the two papers. A further dead core theorem concerning p-regular equations (see section 11 of [7, 8]) was given by Diaz and Veron [3]. Again the assumptions are different enough to make it difficult to compare the results.

Sperb [10] considers similar dead core problems for the particular special case of the Laplace equation without weights, that is, $A \equiv 1$, $g \equiv 1$, and $h \equiv 1$. He estimates the critical value m_0 for more general domains than balls, but only for the homogeneous case $f(u) = \text{Const}\, u|u|^{q-1}$, $0 < q < 1$. For balls B_R his estimate is weaker than our exact result (1.8). Similarly his estimates for the size of dead cores apply to more general domains than balls, but again are weaker than ours in the latter case.

Theorems 1.4 for the general equation (1.1) and 7.2 for the weighted p-Laplace equation (1.2) seem to capture and extend many of the ideas of these earlier papers.

2. Proof of Theorem 1.2. *Existence of a radial solution u of* (1.5), *with $u \geq 0$, $u' \geq 0$, and $u'(0) = 0$.* For the purpose of this proof only, we shall redefine f so that $f(v) = f(m)$ for all $v \geq m$, and $f(v) = 0$ when $v \leq 0$. This will not affect the conclusion of the theorem, since clearly any ultimate solution u of (1.5), with $u \geq 0$, $u' \geq 0$ in $[0, R]$, satisfies $0 \leq u \leq m$.

We shall make use of the Leray–Schauder fixed point theorem, as in Proposition 4.1 of [7, 8]. Denote by X the Banach space $X = C[0, R]$, endowed with the usual norm $\|\cdot\|_\infty$, and let $\mathcal{T}$ be the mapping from X to X defined pointwise for all $w \in X$ by

$$\mathcal{T}[w](r) = m - \int_r^R \Phi^{-1}\left(s^{1-n} \int_0^s t^{n-1} f(w(t))dt\right) ds, \qquad r \in [0, R]. \tag{2.1}$$

Clearly $\mathcal{T}[w](R) = m$. Also

$$\mathcal{T}[w]'(r) = \Phi^{-1}\left(r^{1-n} \int_0^r t^{n-1} f(w(t))dt\right), \qquad r \in (0, R]. \tag{2.2}$$

Obviously $\mathcal{T}[w]'$ is continuous and nonnegative in $(0, R]$, since $0 \leq f(w) \leq f(m)$ for all $w \in X$. Moreover $r^{1-n} \int_0^r t^{n-1} f(w(t))dt$ tends to zero as $r \to 0^+$. Therefore $\mathcal{T}[w]'(r)$ approaches 0 as $r \to 0^+$, since $\Phi(0) = 0$, and in turn $\mathcal{T}[w] \in C^1[0, R]$ with $\mathcal{T}[w]'(0) = 0$.

We claim that if w is a fixed point of $\mathcal{T}$ in X, then $w(0) \geq 0$. Otherwise $w(0) < 0$ and $w(R) = m > 0$. Thus there exists a first point $r_0 \in (0, R)$ such that $w(r) < 0$ in $[0, r_0)$ and $w(r_0) = 0$. Consequently $f(w(r)) = 0$ in $[0, r_0]$ and so $w' \equiv 0$ for $r \in [0, r_0]$ by (2.2). Hence $w(r_0) = w(0) < 0$, which is impossible, proving the claim.

Define the homotopy $\mathcal{H} : X \times [0, 1] \to X$ by

$$\mathcal{H}[w, \sigma](r) = \sigma m - \int_r^R \Phi^{-1}\left(\sigma s^{1-n} \int_0^s t^{n-1} f(w(t))dt\right) ds. \tag{2.3}$$

By the above argument, any fixed point $w_\sigma = \mathcal{H}[w_\sigma, \sigma]$ is of class $C^1[0, R]$ and has the properties $w_\sigma \geq 0$, $w'_\sigma \geq 0$ in $[0, R]$, and $w_\sigma(R) = \sigma m$. Additionally, by (2.2) we find that $\Phi(w'_\sigma) \in C^1[0, R]$, and then from (2.1) that w_σ is a classical (distribution) solution of the problem

$$\begin{cases} [r^{n-1}\Phi(w'_\sigma(r))]' - \sigma r^{n-1} f(w_\sigma(r)) = 0 & \text{in } (0, R], \\ w'_\sigma(0) = 0, \quad w_\sigma(R) = \sigma m. \end{cases} \tag{2.4}$$

In turn, it is evident that any function w_1 which is a fixed point of $\mathcal{H}[w, 1]$ (that is $w_1 = \mathcal{H}[w_1, 1]$) is a nonnegative radial distribution solution of problem (1.5), in $B_R \setminus \{0\}$, with $w'(0) = 0$ and $w' \geq 0$ in $[0, R]$.

Since $f > 0$ for $u > 0$ it follows equally from (2.4) that the final statement of the theorem is valid.

We assert that such a fixed point $w = w_1$ exists. We shall use Browder's version of the Leray–Schauder theorem for this purpose (see Theorem 11.6 of [5]).

To begin with, obviously $\mathcal{H}[w, 0] \equiv 0$ for all $w \in X$; that is, $\mathcal{H}[w, 0]$ maps X into the single point $w_0 = 0$ in X. (This is the first hypothesis required in the application of the Leray–Schauder theorem.)

We show next that $\mathcal{H}$ is compact from $X \times [0, 1]$ into X. First, $\mathcal{H}$ is continuous on $X \times [0, 1]$. Indeed, let $w_j \to w$, $\sigma_j \to \sigma$, $(w_j, \sigma_j) \in X \times [0, 1]$. Then in (2.3) clearly $\sigma_j f(w_j) \to \sigma f(w)$, since the modified function f is continuous on $\mathbb{R}$. Hence $\mathcal{H}[w_j, \sigma_j] \to \mathcal{H}[w, \sigma]$, as required.

Next let $(w_k, \sigma_k)_k$ be a bounded sequence in $X \times [0, 1]$. It is clear from (2.2) that

$$\|\mathcal{H}[w_k, \sigma_k]'\|_\infty \leq \Phi^{-1}(Rf(m)/n). \tag{2.5}$$

As an immediate consequence of the Ascoli–Arzelà theorem $\mathcal{H}$ then maps bounded sequences into relatively compact sequences in X, so $\mathcal{H}$ is compact.

To apply the Leray–Schauder theorem it is now enough to show that there is a constant $M > 0$ such that

$$\|w\|_\infty \leq M \quad \text{for all } (w, \sigma) \in X \times [0, 1], \quad \text{with } \mathcal{H}[w, \sigma] = w. \tag{2.6}$$

Let (w, σ) be a pair of type (2.6). But, as observed above, one has $w \geq 0$, $w' \geq 0$, so that $\|w\|_\infty = w(R) \leq \sigma m \leq m$. Thus we can take $M = m$ in (2.6).

The Leray–Schauder theorem therefore implies that the mapping $\mathcal{T}[w] = \mathcal{H}[w, 1]$ has a fixed point $w \in X$ as asserted, which is the required solution of (1.5) in $B_R \setminus \{0\}$. *The fixed point $u = w$ is a C^1 distribution solution of* (1.5) *in B_R*. The proof is standard. Let $\varphi \in C^1_c(B_R)$. We have to show that

$$\int_{B_R} A(|Du|)Du \cdot D\varphi \, dx = -\int_{B_R} f(u)\varphi \, dx.$$

To this end let $\psi = \varphi k_\varepsilon$, $0 < 2\varepsilon < R$, where

$$k_\varepsilon(x) = \begin{cases} 0 & \text{for} \quad |x| \le \varepsilon, \\ 1 & \text{for} \quad |x| \ge 2\varepsilon, \end{cases}$$

and such that $k_\varepsilon \in C^1(\mathbb{R}^n)$, $0 \le k_\varepsilon \le 1$ in $\mathbb{R}^n$, $|Dk_\varepsilon(x)| \le 2/\varepsilon$ for all x with $\varepsilon \le |x| \le 2\varepsilon$. Consequently, using ψ as a test function in $B_R \setminus \{0\}$, we get

$$\int_{B_R\setminus B_{2\varepsilon}} A(|Du|)Du \cdot D\varphi\, dx + \int_{B_{2\varepsilon}\setminus B_\varepsilon} A(|Du|)Du \cdot (k_\varepsilon D\varphi + \varphi Dk_\varepsilon)dx$$
$$= -\int_{B_R\setminus B_{2\varepsilon}} f(u)\varphi dx - \int_{B_{2\varepsilon}\setminus B_\varepsilon} f(u)k_\varepsilon \varphi dx.$$

Now

$$\left|\int_{B_{2\varepsilon}\setminus B_\varepsilon} A(|Du|)Du \cdot (k_\varepsilon D\varphi + \varphi Dk_\varepsilon)dx\right| \le \sup_{B_{2\varepsilon}} \left\{\Phi(|Du|) \cdot \left[|D\varphi| + \frac{2}{\varepsilon}|\varphi|\right]\right\} \cdot |B_{2\varepsilon}|$$
$$= o(\varepsilon^{n-1})$$

since $Du(0) = \mathbf{0}$, and Φ is continuous at $\varrho = 0$ by (A2). Moreover

$$\left|\int_{B_{2\varepsilon}\setminus B_\varepsilon} f(u)k_\varepsilon \varphi\, dx\right| \le \text{Const}\, \varepsilon^n.$$

Letting $\varepsilon \to 0$ we get the required conclusion.

Uniqueness of C^1 distribution solutions of (1.5). This is an immediate consequence of the weak comparison principle given in Theorem 5.4 of [7, 8].

THEOREM 2.1 (weak comparison principle [7, 8]). *Assume* (A1), (A2), (F1), *and* (F2) *are satisfied. Let u and v be, respectively, classical solutions of*

$$\operatorname{div}\{A(|Du|)Du\} - f(u) \ge 0 \quad \textit{and} \quad \operatorname{div}\{A(|Dv|)Dv\} - f(v) \le 0$$

in a bounded domain Ω; that is, u is a subsolution and v is a supersolution of (1.1). *Suppose also that u and v are continuous in $\overline{\Omega}$, with $u \le v$ on $\partial\Omega$. Then $u \le v$ in Ω.*

3. Proof of Theorem 1.3. *Monotonicity.* This follows from the weak comparison principle, as above.

Continuity. Let $0 < m_1 < m_2$ and write $u_1(r) = u(r, m_1)$ and $u_2(r) = u(r, m_2)$. We show that

$$0 \le u_2(r) - u_1(r) \le m_2 - m_1, \qquad r \in [0, R]. \tag{3.1}$$

By (2.1), for all $r \in [0, R]$,

$$u_2(r) = m_2 - \int_r^R \Phi^{-1}\left(s^{1-n}\int_0^s t^{n-1} f(u_2(t))dt\right) ds,$$
$$u_1(r) = m_1 - \int_r^R \Phi^{-1}\left(s^{1-n}\int_0^s t^{n-1} f(u_1(t))dt\right) ds.$$

Then by subtraction

$$u_2(r) - u_1(r) = m_2 - m_1$$
$$- \int_r^R \left[\Phi^{-1}\left(s^{1-n}\int_0^s t^{n-1} f(u_2(t))dt\right) - \Phi^{-1}\left(s^{1-n}\int_0^s t^{n-1} f(u_1(t))dt\right)\right] ds.$$

The function Φ^{-1} is strictly increasing by (A2) and f is nondecreasing by (F2). Therefore, since $u_1 \le u_2$ in $[0,R]$ by monotonicity, one sees that the quantity in square brackets above is nonnegative, and (3.1) is proved.

Proof that $u < m$ in B_R. By (2.1) it is enough to show that

$$I = \int_r^R \Phi^{-1}\left(s^{1-n}\int_0^s t^{n-1} f(u(t))dt\right) ds > 0 \quad \text{for } r \in [0,R).$$

Clearly $u > 0$ in some interval $(r_0, R]$ with $r_0 \ge 0$, and in turn $f(u(s)) > 0$ in $(r_0, R]$ by (F2). Therefore

$$I \ge \int_{\max\{r_0,r\}}^R \Phi^{-1}\left(s^{1-n}\int_{r_0}^s t^{n-1} f(u(t))dt\right) ds > 0,$$

as required. □

4. The critical value m_0. We begin with a preliminary result, of interest in itself.

THEOREM 4.1. *If $u_1 = u(\cdot, m_1)$ has a dead core B_{S_1}, then $u_2 = u(\cdot, m_2)$, where $m_2 < m_1$, has a dead core B_{S_2}, with $S_2 > S_1$. Similarly, if either $u_1(0) > 0$ or $u_1(0) = 0$ and $u_1(r) > 0$ for $r \in (0,R]$, then $u_2 > u_1$ in B_R when $m_2 > m_1$.*

Proof. To prove the first part of the lemma, assume for contradiction that $m_2 < m_1$, and either $u_2(r) > 0$ in $(0,R]$, or $0 < S_2 \le S_1$. In the first of these cases the solutions u_1 and u_2 must cross at some point $r_0 \in (S_1, R)$. Then applying Theorem 2.1 we find that $u_1 \equiv u_2$ in $[0, r_0]$, which is an obvious contradiction since $u_2(r) > 0$ on $(0, r_0]$, while $u_2 \equiv u_1 \equiv 0$ in $[0, S_1]$. The next case $0 < S_2 < S_1$ leads to a contradiction in the same way.

The remaining case, when $S = S_2 = S_1 > 0$, needs more care. For $\varepsilon \in (0,R)$ define

$$u_\varepsilon(r) = \begin{cases} 0, & r \in [0,\varepsilon], \\ u_1(r-\varepsilon), & r \in (\varepsilon, R]. \end{cases}$$

If $\varepsilon > 0$ is suitably small then one has $m_1 > u_\varepsilon(R) > m_2 = u_2(R)$, while at the same time

$$u_2(S+\varepsilon) > 0 = u_1(S) = u_\varepsilon(S+\varepsilon). \tag{4.1}$$

Thus there is a point $r_0 \in (S+\varepsilon, R)$ where u_ε and u_2 cross.

We assert that u_ε is a supersolution of (1.1) in the annulus $B_R \setminus \overline{B_\varepsilon}$. Indeed in this set we have

$$\begin{aligned} \operatorname{div}\{A(|Du_\varepsilon|)Du_\varepsilon\} - f(u_\varepsilon) &= \{A(|u_\varepsilon'|)u_\varepsilon'\}' + \frac{n-1}{r}A(|u_\varepsilon'|)u_\varepsilon' - f(u_\varepsilon) \\ &= \left(\frac{n-1}{r} - \frac{n-1}{r-\varepsilon}\right)\Phi(u_1'(r-\varepsilon)) \\ &= -\varepsilon\frac{n-1}{r(r-\varepsilon)}\Phi(u_1'(r-\varepsilon)) \le 0. \end{aligned} \tag{4.2}$$

Observing that $u_2(0) = u_\varepsilon(0) = 0$, we can then apply the comparison principle, Theorem 2.1, in B_{r_0}. Therefore $u_2 \le u_\varepsilon$ in $[0, r_0]$, which contradicts (4.1) at the point $r = S + \varepsilon$ and completes the first part of the proof.

To obtain the second part of the theorem, assume for contradiction that $u_2(0) = u_1(0)$ when $m_2 > m_1$. Of course by Theorem 1.2 we have $u_1'(r) > 0$ for $r \in (0, R]$. Define

$$\tilde{u}_\varepsilon(r) = \begin{cases} u_2(0), & r \in [0, \varepsilon], \\ u_2(r - \varepsilon), & r \in (\varepsilon, R], \end{cases}$$

where ε is chosen so small that $m_2 > \tilde{u}_\varepsilon(R) > m_1 = u_1(R)$. On the other hand $u_1(\varepsilon) > \tilde{u}_\varepsilon(\varepsilon) = 0$. Hence there is a crossing point $r_0 \in (\varepsilon, R)$ where $u_1(r_0) = \tilde{u}_\varepsilon(r_0)$. As before u_ε is a supersolution of (1.1) in B_R so that $u_1 \leq \tilde{u}_\varepsilon$ in B_{r_0} by Theorem 2.1. Therefore, $u_1 \equiv \tilde{u}_\varepsilon \equiv 0$ in $[0, \varepsilon]$, which is impossible, since $u_1(r) > 0$ for all $r \in (0, R]$.

That $u_2 > u_1$ in all B_R now follows at once, since otherwise u_2 and u_1 would cross at some value $r = r_0$ in which case comparison would lead to the absurd result $u_2 \equiv u_1$ in B_{r_0}. □

Proof of Theorem 1.4. For the purpose of this proof, we suppose that there is some $m > 0$ for which $u(0, m) > 0$.

Existence of u_0. Define

$$m_0 = \inf\{m > 0 \ : \ u(0, m) > 0\}. \tag{4.3}$$

We assert first that $m_0 > 0$. Let $\bar{m} > 0$ be so small that

$$C_{\bar{m}} = \int_0^{\bar{m}} \frac{ds}{H^{-1}(F(s)/n)} < R, \tag{4.4}$$

which of course is possible by assumption (1.3); see Lemma 3.2 of [7, 8]. Define $v(r) = w(r - S)$, $r \in [S, R]$, $S = R - C_{\bar{m}}$, where w is the function constructed in the dead core Lemma 7.1 of [7, 8], with $\sigma = 1/n$ and $C = C_{\bar{m}}$. We assert that v is a supersolution of (1.1) in the set $B_R \setminus \overline{B}_S$. In fact

$$\operatorname{div}\{A(|Dv|)Dv\} = [\Phi(v')]' + \frac{n-1}{r}\Phi(v') \leq \left\{1 + \frac{n-1}{r}(r - S)\right\}\sigma f(v)$$

by (iii) and (iv) of Lemma 7.1 of [7, 8]. Thus

$$\operatorname{div}\{A(|Dv|)Dv\} \leq \left\{1 - \frac{n-1}{nr}S\right\} f(v) \leq f(v),$$

as required. Then, since $v(S) = v'(S) = 0$ and $v(R) = \bar{m}$, by defining v to be zero on B_S, the extended function v is a supersolution of (1.1) in B_R. By the comparison principle, Theorem 2.1, we find that $u(\cdot, \bar{m}) \equiv 0$ in B_S. Therefore $m_0 \geq \bar{m} > 0$ by (4.3) and the claim is proved.

Next, if (i) would be false, then $u_0(0) > 0$ and by Theorem 1.3 also $u(0, m) > 0$ for all $m > 0$ sufficiently near m_0, which would contradict (4.3). Property (ii) is again a direct consequence of the definition (4.3) of m_0 and Theorem 1.3. Finally if there is $m \in (0, m_0)$ such that the corresponding solution $u(\cdot, m)$ of (1.5) has no dead core, then $u(0, m) \geq 0$ and $u(r, m) > 0$ for $r \in (0, R]$. Thus by Theorem 4.1, with $m_1 = m$ and $m_2 = m_0$, we get $u_0(0) > u(0, m) \geq 0$, contradicting (i) and proving (iii).

Uniqueness of u_0. Suppose both m_0 and $\bar{m}_0$ have the properties (i)–(iii) of the theorem. Then $u_0(0) = u_0(0, m_0) = 0$ by (i), while $u(0, m) > 0$ when $m > \bar{m}_0$ by (ii). Hence $m_0 \leq \bar{m}_0$. Similarly $\bar{m}_0 \leq m_0$. Therefore $\bar{m}_0 = m_0$, as desired.

The case $m_0 = \infty$. If every solution of (1.5) is such that $u(0, m) = 0$ for all $m > 0$, then $u(\cdot, m)$ has a dead core for all $m > 0$. Otherwise there would be $\bar{m} > 0$ for which $u(0, \bar{m}) = 0$ and $u(r, \bar{m}) > 0$ for $r \in (0, R]$. Hence $u(0, m) > 0$ for $m > \bar{m}$ by Theorem 4.1, contradicting the assumption. This also justifies the earlier agreement that $m_0 = \infty$ in this case. □

Remark. In summary, if m_0 is finite and $m > m_0$, then the solution $u = u(\cdot, m)$ of (1.5) is positive; namely, $u(r, m) > 0$ for all $r \in [0, R]$. On the other hand, if $m < m_0 \leq \infty$ then the solution $u = u(\cdot, m)$ of (1.5) has a dead core $B_S \subset B_R$, $0 < S < R$.

5. The size of a dead core and proof of Theorem 1.1. Recall the assumption that $\Phi(\infty) = H(\infty) = \infty$, and let

$$C = \int_0^\infty \frac{ds}{H^{-1}(F(s)/n)}. \tag{5.1}$$

Clearly $0 < C \leq \infty$ since the integral is convergent at 0 by (1.3) and by Lemma 3.2 of [7, 8] with $\sigma = 1/n$. Of course the integral can possibly diverge at ∞.

THEOREM 5.1. *We have*

$$m_0 = \infty \quad \text{if} \quad C < \infty \quad \text{and} \quad R \geq C, \tag{5.2}$$

while

$$m_0 \geq \bar{m} \quad \text{if} \quad R < C, \tag{5.3}$$

where $\bar{m}$ *is defined by the relation*

$$R = \int_0^{\bar{m}} \frac{ds}{H^{-1}(F(s)/n)}.$$

Proof. The proof of (5.2) is essentially the same as the proof of the first part of Theorem 1.4, the only exception being that $C_{\bar{m}}$ given by (4.4) is replaced by C.

To obtain (5.3), we define $v(r) = w(r)$ as in the proof of Theorem 1.4, with $S = 0$, $\sigma = 1/n$, and $C = R$. Then by Lemma 7.1 of [7, 8] we have $v(0) = v'(0) = 0$, $v(R) = w(R) = \bar{m}$, while v is a supersolution of (1.1) in B_R. It follows by comparison that $0 \leq u(r, \bar{m}) \leq v(r)$. Hence $u(0, \bar{m}) = v(0) = 0$, and in turn from the definition (4.3) of m_0 we get $m_0 \geq \bar{m}$, as required in (5.3). □

THEOREM 5.2. *Let* $m < m_0$, *so that a dead core exists by Theorem* 1.4, (iii). *In particular the solution* $u = u(\cdot, m)$ *satisfies*

$$u \equiv 0 \quad \text{in } B_S \subset B_R,$$

where

$$R - \int_0^m \frac{ds}{H^{-1}(F(s)/n)} < S < R.$$

If $R \geq C$, *then for all* $m > 0$ *one has*

$$R - C < S < R.$$

Proof. The proof is the same as the first part of the proof of Theorem 1.4. □

Remark. For any $\varepsilon > 0$, if m is suitably small (depending on ε) we have $R - \varepsilon < S < R$.

Proof of Theorem 1.1. (a) That $u \geq 0$ follows from Theorem 2.1 by comparing the given solution u with the trivial solution 0.

The constant function m is a supersolution of (1.1), so that again by Theorem 2.1 we have $u \leq m$ in Ω. In fact $u < m$ in Ω. To see this, let y be any point of Ω and B a ball in Ω centered at y. Let $v(\cdot\,, m)$ be the radial solution of (1.1) in B constructed in Theorem 1.2, with $v(|x-y|\,, m) = m$ for $x \in \partial B$. Therefore $u(x) \leq m = v(|x-y|\,, m)$ for $x \in \partial B$, and in turn $u(x) \leq v(|x-y|\,, m) < m$ for $x \in B$ by the final part of Theorem 1.3.

(b) This is a direct consequence of Theorem 5.1.

(c) Clearly there exists $R > 0$ such that $\overline{B} \subset B_R \subset\subset \Omega$, with B and B_R centered at the same point of Ω. By (a) we know that $u < m$ on ∂B_R. Denote by $R - \varepsilon$ the radius of B; then by comparison, together with the remark after Theorem 5.2, we have $u \equiv 0$ in B when $m > 0$ is suitably small. □

6. The equation $\operatorname{div}\{g(|x|)|Du|^{p-2}Du\} = h(|x|)f(u)$. Consider the quasilinear singular elliptic equation

$$\operatorname{div}(g(|x|)|Du|^{p-2}Du) = h(|x|)f(u) \qquad \text{in } \Omega, \qquad p > 1, \tag{6.1}$$

where $g,\ h : \mathbb{R}^+ \to \mathbb{R}^+$ are radial functions of class $C^1(\mathbb{R}^+)$, and Ω is a domain of $\mathbb{R}^n$, $n \geq 1$, containing the origin. Prototypes of (6.1), with nontrivial functions g, h, are given, for example, by equations of Matukuma type and equations of Batt–Faltenbacher–Horst type. More precisely, the Matukuma-type equation is given by

$$\Delta_p u = \frac{f(u)}{1 + r^\sigma}, \qquad r = |x|, \qquad \sigma > 0, \tag{6.2}$$

where Δ_p denotes the p-Laplace operator, $p > 1$, and where also $g(|x|) \equiv 1$, $h(|x|) = 1/(1 + r^\sigma)$. A second example is the equation

$$\Delta_p u = \frac{r^\sigma}{(1 + r^{p'})^{\sigma/p'}} \cdot \frac{f(u)}{r^{p'}}, \qquad \sigma > 0, \tag{6.3}$$

where now $g(|x|) \equiv 1$, $h(|x|) = r^{\sigma - p'}/(1 + r^{p'})^{\sigma/p'}$.

All these equations are discussed in detail in section 4 of [6], as special cases of the main example[1]

$$\begin{gathered} \operatorname{div}(r^k|Du|^{p-2}Du) = r^\ell \left(\frac{r^s}{1 + r^s}\right)^{\sigma/s} f(u), \\ k \in \mathbb{R}, \qquad \ell \in \mathbb{R}, \qquad s > 0, \qquad \sigma > 0. \end{gathered} \tag{6.4}$$

In particular, in [6] conditions on the exponents were found so that, under appropriate behavior of the nonlinearity f, radial ground states for (6.2)–(6.4) are unique.

We shall also be interested in the radial version of (6.1), when Ω is a ball B_R *centered at* 0 with radius $R > 0$, namely,

$$[a(r)|u'|^{p-2}u']' = b(r)f(u) \qquad \text{in } (0, R), \tag{6.5}$$

[1] In [6] the equation (6.1) was written in the form

$$\operatorname{div}(g(|x|)|Du|^{p-2}Du) + h(|x|)f(u) = 0.$$

The two versions are reconciled by replacing f by $-f$.

where, with obvious notation,

$$a(r) = r^{n-1}g(r), \quad b(r) = r^{n-1}h(r). \tag{6.6}$$

Motivated by the case $a(r) = b(r) = r^{n-1}$ in [6], the functions a and b are assumed to be such that (6.5) *can be transformed* by the change of variables $r \mapsto t(r)$,

$$t(r) = \int_0^r [b(s)/a(s)]^{1/p}\, ds, \qquad r \geq 0, \tag{6.7}$$

$t: \mathbb{R}_0^+ \to \mathbb{R}_0^+$, $t(0) = 0$, to the form

$$[q(t)|v_t|^{p-2}v_t]_t = q(t)f(v), \tag{6.8}$$

where

$$q(t) = [a(r(t))]^{1/p}[b(r(t))]^{1/p'}, \qquad t > 0. \tag{6.9}$$

This requires, in particular, the following conditions on the coefficients:

$$a,\ b > 0, \qquad a,\ b \in C^1(\mathbb{R}^+), \qquad (b/a)^{1/p} \in L^1[0, R]. \tag{A3}$$

We shall ask as well that the transformed equation (6.8) be compatible with the structure:

$$q \in C^1(\mathbb{R}^+), \quad q > 0 \quad \text{in } \mathbb{R}^+; \tag{Q1}$$

$$\text{there is} \quad \delta > 0 \quad \text{such that} \quad q_t > 0 \quad \text{in} \quad (0, \delta). \tag{Q2}$$

By (Q2) the weight $q(t)$ has a finite limit as $t \to 0^+$ and in turn also $\bar{q}(r) = q(t(r))$ is bounded as $r \to 0^+$ by (6.7). Hence $b = (b/a)^{1/p}\bar{q} = O((b/a)^{1/p})$ as $r \to 0^+$ by (6.9), and so by (A3)

$$b \in L^1[0, R] \quad \text{and} \quad h \in L^1_{\text{loc}}(B_R). \tag{6.10}$$

Definition. A classical solution of (6.1) is a nonnegative function u of class $C(\Omega) \cap C^1(\Omega \setminus \{0\})$, which is a distribution solution of (6.1) in Ω, of course also with

$$g|Du|^{p-1} \in L^1_{\text{loc}}(\Omega). \tag{6.11}$$

THEOREM 6.1 (weak comparison principle). *Let u and $\tilde{u}$ be, respectively, classical super- and subsolutions of (6.1) in a bounded domain Ω. Suppose also that u and $\tilde{u}$ are continuous in $\overline{\Omega}$, with $u \geq \tilde{u}$ on $\partial\Omega$. Then $u \geq \tilde{u}$ in Ω.*

Proof. We follow the proof of Theorem 5.4 of [7, 8].

Let $w = u - \tilde{u}$ in $\overline{\Omega}$. If the conclusion fails, then there exists a point $x_1 \in \Omega$ such that $w(x_1) < 0$. Fix $\varepsilon > 0$ so small that $w(x_1) + \varepsilon < 0$. Consequently, since $w \geq 0$ on $\partial\Omega$ it follows that the function $w_\varepsilon = \min\{w + \varepsilon,\ 0\}$ is nonpositive and has compact support in Ω. By the distribution meaning of solutions, taking the Lipschitz continuous function w_ε as a test function, we get

$$\int_\Omega g(|x|)\{|Du|^{p-2}Du - |D\tilde{u}|^{p-2}D\tilde{u}\}Dw_\varepsilon \leq \int_\Omega h(|x|)\{f(\tilde{u}) - f(u)\}w_\varepsilon. \tag{6.12}$$

The left-hand side of (6.12) is positive since $p > 1$ and $g(|x|) > 0$ for $x \in \Omega \setminus \{0\}$ and $Dw_\varepsilon \equiv Dw = Du - D\tilde{u} \not\equiv \mathbf{0}$ when $w + \varepsilon < 0$, while otherwise $Dw_\varepsilon = \mathbf{0}$ (a.e.).

Moreover, when $w + \varepsilon < 0$ there holds $0 \le u < \tilde{u} - \varepsilon$; hence $f(\tilde{u}) - f(u) \ge 0$ since f is nondecreasing by (F2). Thus the right-hand side of (6.12) is nonpositive, since $h(|x|) > 0$ for $x \in \Omega \setminus \{0\}$, which is a contradiction. □

In the rest of this section we continue to denote by B_R the open ball *centered at* 0 with radius $R > 0$, that is, $B_R = B(0, R)$. Our first result is the analogue of Theorem 1.2, where now we use $A(\varrho) = \varrho^{p-2}$, $\Phi(\varrho) = \varrho^{p-1}$, consistent with section 1. In this case of course $\Phi(\infty) = H(\infty) = \infty$ and $\Phi^{-1}(\tau) = \tau^{1/(p-1)}$.

THEOREM 6.2. *The problem*

$$\begin{cases} \operatorname{div}\{g(|x|)|Du|^{p-2}Du\} = h(|x|)f(u) & \text{in } B_R, \\ u = m > 0 \quad \text{on } \partial B_R, \quad u \in C(\overline{B_R}), \end{cases} \tag{6.13}$$

admits a unique classical solution u, necessarily radial.

Moreover $u = u(r) = u(r, m)$ *is of class* $C^1[0, R]$ *and satisfies* $u \ge 0$, $u' \ge 0$ *in* $[0, R]$ *and* $u'(0) = 0$, *where* $' = d/dr$. *Finally, at any* $r > 0$ *where* $u(r, m) > 0$ *we have also* $u'(r, m) > 0$.

It follows from Theorem 6.2 that the solution $u(\cdot, m)$ must be one of the following three types:

1. $u > 0$ in B_R.
2. $u(0, m) = 0$ and $u'(r, m) > 0$ when $r > 0$.
3. There exists $S \in (0, R)$ such that $u \equiv 0$ in B_S and $u'(r, m) > 0$ when $r > S$.

That is, in case 3 the solution u of (1.5) has a dead core B_S.

Proof. Existence of a radial solution u of (6.13), *with* $u \ge 0$, $u' \ge 0$, *and* $u'(0) = 0$. This will be accomplished by first showing that (6.8) has a solution $v = v(t)$ on $[0, T]$, $T = t(R)$, with $v \ge 0$, $v_t \ge 0$, and $v_t(0) = 0$, where $t = t(r)$ is given by (6.7).

Indeed, by following the proof of Theorem 1.2 almost word for word, including the redefinition of f, but with the exception that r^{n-1} is replaced by $q(t)$, one obtains[2] an appropriate fixed point $w = w(t) \in X = C[0, T]$ satisfying

$$w(t) = m - \int_t^T \left(\frac{1}{q(s)} \int_0^s q(\tau) f(w(\tau)) d\tau \right)^{1/(p-1)} ds, \qquad t \in [0, T]. \tag{6.14}$$

Clearly $w(T) = m$ and

$$w_t(t) = \left(\frac{1}{q(t)} \int_0^t q(\tau) f(w(\tau)) d\tau \right)^{1/(p-1)}, \qquad t \in (0, T]. \tag{6.15}$$

Obviously w_t is continuous and nonnegative in $(0, T]$, since $0 \le f(w) \le f(m)$ for all $w \in X$. Moreover by (Q2)

$$0 < \frac{1}{q(t)} \int_0^t q(\tau) f(w(\tau)) d\tau \le t f(m) \to 0 \quad \text{as} \quad t \to 0^+.$$

[2]In view of (Q2), the bound $Rf(m)/n$ in (2.5), holding for the standard weight r^{n-1}, should be replaced by

$$L = \begin{cases} Rf(m) & \text{if } 0 < R \le \delta, \\ R\,\dfrac{\max_{t\in[\delta,R]} q(t)}{\min_{t\in[\delta,R]} q(t)} f(m) & \text{if } \delta < R. \end{cases}$$

Therefore $w_t(t)$ approaches 0 as $t \to 0^+$ by (6.15). In turn we have $w \in C^1[0,T]$ with $w_t(0) = 0$.

We claim moreover that $w(0) \geq 0$. Otherwise $w(0) < 0$ and $w(T) = m > 0$. Thus there exists a first point $t_0 \in (0,T)$ such that $w(t) < 0$ in $[0,t_0)$ and $w(t_0) = 0$. Consequently $f(w(t)) = 0$ in $[0,t_0]$ and so $w_t \equiv 0$ for $t \in [0,t_0]$ by (6.15). Hence $w(t_0) = w(0) < 0$, which is impossible, proving the claim.

The function $v = w$ is the desired solution of (6.8) in $[0,T]$, and in turn $u(r) = v(t(r))$ is a C^1 solution of (6.5) on $(0,R]$.

The function $u(|x|) = v(t(|x|))$ *is a* C^1 *distribution solution of* (6.13) *in* B_R. First we show (6.11) for $\Omega = B_R$. Recall that $v \in C^1[0,T]$ and $v_t(0) = 0$, while also, as shown above, q is bounded as $t \to 0^+$. Then using the change of variables (6.7) we get

$$a(r)|u'(r)|^{p-1} = q(t)|v_t(t)|^{p-1} \to 0 \qquad \text{as} \quad r,\ t \to 0^+.$$

Therefore, recalling (6.6), it follows that (6.11) holds in B_R for $u(|x|) = v(t(|x|))$. The rest of the proof is standard, as in the last part of the proof of Theorem 1.2. Thus u is a C^1 distribution solution of (6.13) in B_R.

Uniqueness of C^1 *distribution solutions of* (6.13). This is an immediate consequence of the weak comparison principle given in Theorem 6.1. □

The solution $u = u(\cdot\,,m)$ of (6.13) has further properties of interest, given in the next result.

THEOREM 6.3. *The function* $u = u(\cdot\,,m)$ *is continuous and nondecreasing in the variable* m, *and* $u < m$ *in* B_R.

Proof. Monotonicity. This follows from the weak comparison principle Theorem 6.1.

Continuity. Using the notation of the proof of Theorem 1.3, we show that if $0 < m_1 < m_2$, then

$$0 \leq v_2(t) - v_1(t) \leq m_2 - m_1 \qquad \text{on } [0,T]. \tag{6.16}$$

By (6.14) we have for all $t \in [0,T]$

$$v_2(t) - v_1(t) = m_2 - m_1$$
$$- \int_t^T \left[\left(\frac{1}{q(s)} \int_0^s q(\tau) f(v_2(\tau))d\tau \right)^{1/(p-1)} - \left(\frac{1}{q(s)} \int_0^s q(\tau) f(v_1(\tau))d\tau \right)^{1/(p-1)} \right] ds,$$

and (6.16) now follows exactly as in the proof of Theorem 1.3. Consequently

$$0 \leq u_2(r) - u_1(r) = v_2(t(r)) - v_1(t(r)) \leq m_2 - m_1 \qquad \text{for } r \in [0,R],$$

as required.

Proof that $u < m$ *in* B_R. It is equivalent to show the same property for the corresponding fixed point solution v of (6.9). By (6.14) it is enough to prove that

$$I = \int_t^T \left(\frac{1}{q(s)} \int_0^s q(\tau) f(v(\tau))d\tau \right)^{1/(p-1)} ds > 0 \quad \text{for } t \in [0,T).$$

Clearly $v > 0$ in some interval $(t_0, T]$ with $t_0 \geq 0$, and in turn $f(v(t)) > 0$ in $(t_0,T]$ by (F2). Therefore

$$I \geq \int_{\max\{t_0,\,t\}}^T \left(\frac{1}{q(s)} \int_{t_0}^s q(\tau) f(v(\tau))d\tau \right)^{1/(p-1)} ds > 0,$$

as required. □

7. The critical value m_0 for (6.1). We now turn to the dead core theorem for (6.1), analogous to Theorem 1.4 for (1.1). In the present case we have $H^{-1}(\tau) = (p'\tau)^{1/p}$ and the dead core condition (1.3) becomes

$$\int_{0+} \frac{ds}{[F(s)]^{1/p}} < \infty. \tag{7.1}$$

In what follows we assume also that

$$\frac{q_t}{q} \quad \text{is nonincreasing on } \mathbb{R}^+; \tag{Q3}$$

$$\limsup_{t\to 0+} \frac{tq_t(t)}{q(t)} < \infty. \tag{Q4}$$

THEOREM 7.1. *If $u_1 = u(\cdot, m_1)$ has a dead core B_{S_1}, then $u_2 = u(\cdot, m_2)$, $m_2 < m_1$, has a dead core B_{S_2}, with $S_2 > S_1$. On the other hand, if either $u_1(0) > 0$ or $u_1(0) = 0$ and $u_1(r) > 0$ for $r \in (0, R]$, then $u_2 > u_1$ in B_R when $m_2 > m_1$.*

Proof. It is equivalent to show the same result with u_1 and u_2 replaced by $v_1 = v_1(t) = u_1(r(t))$ and $v_2 = v_2(t) = u_2(r(t))$, $t \in [0, T]$, respectively. To do this we can repeat the proof of Theorem 4.1 almost word for word, of course with u_1 and u_2 replaced by v_1 and v_2, and also with the following necessary changes: (i) replace (4.2) by

$$\begin{aligned}\frac{1}{q(t)}\{[q(t)|v_{\varepsilon,t}|^{p-2}v_{\varepsilon,t}]_t - q(t)f(v_\varepsilon)\} &= [|v_{\varepsilon,t}|^{p-2}v_{\varepsilon,t}]_t + \frac{q_t(t)}{q(t)}|v_{\varepsilon,t}|^{p-2}v_{\varepsilon,t} - f(v_\varepsilon)\\ &= \left[\frac{q_t(t)}{q(t)} - \frac{q_t(t-\varepsilon)}{q(t-\varepsilon)}\right][v_{1,t}(t-\varepsilon)]^{p-1},\end{aligned}$$

which is nonpositive in view of (Q3) and the fact that $v_{1,t} \geq 0$, that is,

$$[q(t)|v_{\varepsilon,t}|^{p-2}v_{\varepsilon,t}]_t - q(t)f(v_\varepsilon) \leq 0;$$

and (ii) replace the use of Theorem 2.1 in B_{r_0} by the use of the radial analogue of Theorem 6.1 in $[0, r_0)$. □

The following theorem gives an important relation between the value m and dead core solutions of (6.13).

THEOREM 7.2. *Let $u(\cdot, m)$ be the unique solution of (6.13). Then either $u(\cdot, m)$ has a dead core for all $m > 0$, or there is a unique (finite) number*

$$m = m_0 = m_0(R) > 0$$

for which a solution $u_0 = u_0(r) = u_0(r, m_0)$ of (6.13) in B_R exists, with the properties that

(i) *$u_0(0) = 0$;*
(ii) *$u(0, m) > 0$ for every $m > m_0$;*
(iii) *$u(\cdot, m)$ has a dead core for every $0 < m < m_0$.*

For convenience we define $m_0 = m_0(R)$ to be ∞ when all solutions of (6.13) are such that $u(0, m) = 0$ for all $m > 0$.

Before proving Theorem 7.2, it is useful to give a preliminary lemma. For this and later purposes we introduce the constant $M \geq 1$ by

$$\sup_{0<t<T} \frac{tq_t(t)}{q(t)} = M - 1, \tag{7.2}$$

which is well defined by (Q1), (Q2), and (Q4).

LEMMA 7.3. *Any nonnegative solution u of* (6.1) *in B_R which is suitably small on ∂B_R has a dead core.*

Proof. Let $\bar{m} > 0$ be so small that

$$D_{\bar{m}} = \int_0^{\bar{m}} \frac{ds}{[F(s)/M]^{1/p}} = M^{1/p} \int_0^{\bar{m}} \frac{ds}{[F(s)]^{1/p}} < T, \qquad T = t(R), \tag{7.3}$$

which of course is possible by assumption (7.1), where $T = t(R)$ and t is given in (6.7). Define $\tilde{v}(t) = w(t - \tilde{S})$, $t \in [\tilde{S}, T]$, $\tilde{S} = T - D_{\bar{m}}$, where w is the function constructed in the dead core Lemma 7.1 of [7, 8], with $\sigma = 1/M$. By analogy with (7.4) of [7, 8] the function $\tilde{v}$ is a supersolution of (6.8) in the interval $[\tilde{S}, T]$; that is, for $t \in (\tilde{S}, T)$

$$\begin{aligned} \{(\tilde{v}_t(t))^{p-1}\}_t + \frac{q_t(t)}{q(t)}(\tilde{v}_t(t))^{p-1} - f(\tilde{v}(t)) &\le \left\{\sigma\left[1 + (M-1)\frac{t - \tilde{S}}{t}\right] - 1\right\} f(\tilde{v}(t)) \\ &\le (\sigma M - 1) f(\tilde{v}(t)) = 0, \end{aligned} \tag{7.4}$$

where we have used (iii) and (iv) of the dead core Lemma 7.1 of [7, 8]. Of course also $\tilde{v}(\tilde{S}) = \tilde{v}_t(\tilde{S}) = 0$ and $\tilde{v}(T) = \bar{m}$. Let $\tilde{u}(x) = \tilde{v}(t(|x|))$, $x \in B_R \setminus \overline{B_S}$, where $\tilde{S} = t(S)$. Then

$$\operatorname{div}(g(|x|)|D\tilde{u}|^{p-2}D\tilde{u}) - h(|x|)f(\tilde{u}) \le 0 \quad \text{in } B_R \setminus \overline{B_S}.$$

Moreover defining $\tilde{u}$ to be zero on B_S, the extended function $\tilde{u}$ is a supersolution in B_R of (6.1). Let the given solution u of (6.1) in B_R have nonnegative boundary values less than or equal to m. By the weak comparison Theorem 6.1 we get $u \le \tilde{u}$ whenever $m \le \bar{m}$. Therefore $u \equiv 0$ in B_S. This completes the proof. □

Proof of Theorem 7.2. *Existence of* u_0. Suppose that there is some $m > 0$ for which $u(0, m) > 0$ and define as before

$$m_0 = \inf\{m > 0 \ : \ u(0, m) > 0\}, \tag{7.5}$$

so that $m_0 > 0$ by Lemma 7.3.

The rest of the proof is exactly the same as the proof of Theorem 1.4, with the exception that Theorems 1.3 and 4.1 are replaced by the corresponding Theorems 6.3 and 7.1.

Uniqueness of u_0. The proof is exactly as for the corresponding result in Theorem 1.4.

The case $m_0 = \infty$. If every solution of (6.1) in B_R is such that $u(0, m) = 0$ for all $m > 0$, then $u(\cdot, m)$ has a dead core for all $m > 0$. Otherwise there would be $\bar{m} > 0$ for which $u(0, \bar{m}) = 0$ and $u(r, \bar{m}) > 0$ for $r \in (0, R]$. Hence $u(0, m) > 0$ for $m > \bar{m}$ by Theorem 7.1, contradicting the assumption. This also justifies the earlier agreement that $m_0 = \infty$ in this case. □

Remark. In summary, if m_0 is finite and $m > m_0$, then the solution $u = u(\cdot, m)$ of (6.13) is positive; namely, $u(r, m) > 0$ for all $r \in [0, R]$. On the other hand, if $m < m_0 \le \infty$ then the solution $u = u(\cdot, m)$ of (6.13) has a dead core $B_S \subset B_R$, $0 < S < R$.

8. The size of a dead core and the case of general domains for (6.1). Recall the assumption that $\Phi(\infty) = H(\infty) = \infty$, and let

$$D = \int_0^{\infty} \frac{ds}{[F(s)/M]^{1/p}}, \tag{8.1}$$

where $M \geq 1$ is the number given in (7.2). Clearly $0 < D \leq \infty$ since the integral is convergent at 0 by (7.1) and by Lemma 3.2 of [7, 8] with $\sigma = 1/M \leq 1$, but possibly divergent at ∞. The next two results are the analogues of Theorems 5.1 and 5.2.

THEOREM 8.1. *We have*

$$m_0 = \infty \quad \text{if} \quad D < \infty \quad \text{and} \quad R \geq r(D), \tag{8.2}$$

where r is the inverse function of change of variable t given in (6.7), *while*

$$m_0 \geq \bar{m} \qquad \text{if } R < r(D), \tag{8.3}$$

where $\bar{m}$ *is given by the relation* $R = r(D_{\bar{m}})$.

Proof. The proof of (8.2) is essentially the same as that of Lemma 7.3, the only exception being that $D_{\bar{m}}$ is replaced by D, and R by $T = t(R)$.

To obtain (8.3) one again follows the outline of the proof of Lemma 7.3. Now, however, in addition to replacing $D_{\bar{m}}$ by D, we take $\tilde{S} = S = 0$. One thus obtains (7.4) for $\tilde{v}$, with $\tilde{v}(0) = \tilde{v}_t(0) = 0$ and $\tilde{v}(T) = \bar{m}$. Hence $\tilde{u}(0) = \tilde{u}'(0) = 0$ and $\tilde{u}(R) = \bar{m}$, where $\tilde{u}(r) = \tilde{v}(t(r))$. The conclusion is that $0 \leq \tilde{u}(r, \bar{m}) \leq u(r, \bar{m})$, by virtue of the weak comparison Theorem 6.1. Hence $\tilde{u}(0, \bar{m}) = u(0, \bar{m}) = 0$, and in turn from the definition (7.5) of m_0 we get $m_0 \geq \bar{m}$, as required in (8.3). □

Remark. It is interesting to calculate the function $t \mapsto r(t)$ for the classical Matukuma equation, in which $g \equiv 1$ and $h(r) = 1/(1 + r^2)$, $p = 2$. One finds from (6.7) that $t(r) = \operatorname{arcsinh} r$, so that $r(t) = \sinh t$. Thus in this case (8.2) becomes

$$m_0 = \infty \quad \text{if } D < \infty \quad \text{and} \quad R \geq \sinh D,$$

with a similar relation for (8.3).

THEOREM 8.2. *Let* $m < m_0$, *so that a dead core exists in* B_R *by Theorem* 7.2. *In particular*

$$u \equiv 0 \qquad \text{in } B_S \subset B_R,$$

where $T = t(R)$, $\tilde{S} = t(S)$, *and*

$$T - \int_0^m \frac{ds}{[F(s)/M]^{1/p}} < \tilde{S} < T. \tag{8.4}$$

If $R \geq r(D)$, *then for all* $m > 0$ *one has*

$$R - r(D) < S < R.$$

Proof. This is the same as the proof of (8.2). □

Remark. Let $\varepsilon > 0$. Then if m is suitably small (depending on ε) we have $T - \varepsilon < \tilde{S} < T$ by (8.4).

The next result is the analogue for (6.1) of Theorem 1.1 for (1.1) and for general domains Ω. Let $0 \in \Omega$. As before we denote by B_R the open ball *centered at* 0 with radius $R > 0$, and we let

$$\overline{R} = \sup\{R > 0 \ : \ B_R \subset \Omega\}.$$

THEOREM 8.3. *Assume the dead core condition* (7.1) *holds and let u be a solution of* (6.1), *with* $0 \leq u(x) \leq m$ *on* $\partial\Omega$ *for some positive constant m. Then the following properties are valid:*

(a) $0 \le u \le m$ *in* Ω *and* $0 \le u < m$ *in* $B_{\overline{R}}$.
(b) *If* $\overline{R} \ge r(D)$, *with* D *given in* (8.1), *then* u *has a dead core in* Ω *for all* $m > 0$.
(c) *Let* B_R *be compactly contained in* Ω. *Then* $u \equiv 0$ *in* B_R *provided that* $m > 0$ *is suitably small.*

Proof of Theorem 8.3. (a) That $u \ge 0$ follows by comparison (Theorem 6.1) of the given solution u with the trivial solution 0. Also the constant function m is a supersolution of (6.1), so that again by comparison $u \le m$ in Ω.

Moreover $u < m$ in $B_{\overline{R}}$. To see this, let $\overline{u}(\cdot, m)$ be the radial solution of (6.1) in $B_{\overline{R}}$ constructed in Theorem 6.2. Clearly $u(x) \le m = \overline{u}(|x|, m)$ for $x \in \partial B_{\overline{R}}$, and in turn $u(x) \le \overline{u}(|x|, m) < m$ for $x \in B_{\overline{R}}$ by the final part of Theorem 6.2.

(b) This is a direct consequence of Theorem 8.1.

(c) Let $\varepsilon > 0$ be so small that $B_R \subset B_{R+\varepsilon} \subset \Omega$. In analogy with part (a), let $\overline{u}(\cdot, m)$ be the solution of (6.13) with B_R replaced by $B_{R+\varepsilon}$. Then $u(x) \le \overline{u}(|x|, m) = m$ for $x \in \partial B_{R+\varepsilon}$. By the remark after Theorem 8.2 we have $\overline{u} \equiv 0$ in B_R when $m > 0$ is suitably small, and in turn $u \equiv 0$ in B_R by the weak comparison Theorem 6.1. □

9. The case $\Phi(\infty) < \infty$. This is the case, for example, for the mean curvature operator noted in the introduction, in which $\Phi(\infty) = H(\infty) = 1$. Here the proof of the critical Theorem 1.2 requires the modification that the parameter m in (1.5) should be restricted so that

$$f(m) + \Phi(m) < \left(\frac{R}{R+1}\right)^{n-1} \Phi(\infty) \qquad \text{and} \qquad F(m) \le nH(\infty); \tag{9.1}$$

see Proposition 4.1 of [7, 8]. Denote by m_∞ the supremum of all $m > 0$ satisfying (9.1).

Then the main results stated in the introduction remain true provided that the condition $m < m_\infty$ is assumed in all the statements. For instance we have the following analogue of Theorem 1.1.

THEOREM 9.1. *Assume the dead core condition* (1.3) *holds and let* u *be a solution of* (1.1), *with* $0 \le u(x) \le m$ *on* $\partial\Omega$ *for some positive constant* $m < m_\infty$. *Then the following properties are valid:*
(a) $0 \le u < m$ *in* Ω.
(b) *Assume that* $C = \int_0^{m_\infty} \frac{ds}{H^{-1}(F(s)/n)} < \infty$, *and let* B_R *be a ball with radius* $R \ge C$, *compactly contained in* Ω. *Then* u *has a dead core in* Ω *for all* $m \in (0, m_\infty)$.
(c) *If* B *is any ball compactly contained in* Ω, *then* $u \equiv 0$ *in* B *provided that* $m > 0$ *is suitably small.*

It is not hard to show that if $\Phi(\infty) = \infty$ then necessarily $H(\infty) = \infty$, but it is possible to have $\Phi(\infty) < \infty$ and $H(\infty) = \infty$, as shown by $A(\varrho) = 1/(1 + \sqrt{1+\varrho^2})$, with corresponding

$$H(\varrho) = \frac{1}{2}\left[\frac{\varrho^2}{1+\sqrt{1+\varrho^2}} - \log\frac{1+\sqrt{1+\varrho^2}}{2}\right].$$

In this example $\Phi(\infty) = 1$, while $H(\infty) = \infty$.

The case $H(\infty) < \infty$ for unrestricted $m > 0$ was treated by Siegel in [9].

10. Bursts. Here we assume (F1) and apply the previous theory to problems with nonlinearities f with (F2) replaced by the following:

(F3) f is nondecreasing in $(0,\delta)$, $F(u) > 0$ on $(0,\beta)$, with $F(\beta) = 0$, and there is $\gamma \in (\beta, \infty]$ such that the following conditions hold:
(i) $f(u) < 0$ on $[\beta, \gamma)$, $f(\gamma) = 0$ if $\gamma < \infty$;
(ii) $\max_{[0,\beta]} F(u) + |F(\gamma)| < H(\infty)$ if $H(\infty) < \infty$;
(iii) $\liminf_{u\to\infty} \frac{H^{-1}(|F(u)|)}{u} = 0$ if $H(\infty) = |F(\gamma)| = \infty$ and $\Phi(\infty) < \infty$;
(iv) $\liminf_{u\to\infty} \frac{H^{-1}(|F(u)|)}{u} < \infty$ if $H(\infty) = \Phi(\infty) = |F(\gamma)| = \infty$.

In (F3) for the case $\gamma = \infty$ we define $F(\gamma) = \lim_{u\to\infty} F(u)$, which certainly exists though possibly being $-\infty$. Clearly $F(\gamma) = -\infty$ can occur only if $\gamma = \infty$. As noted above, $H(\infty) = \infty$ whenever $\Phi(\infty) = \infty$.

For such functions f Franchi, Lanconelli, and Serrin proved in Theorem A of [4] that (1.1) *admits a nonnegative radial ground state* $u = u(r)$, *with central value* $u(0) = \alpha \in (\beta, \gamma]$ *if* $\gamma < \infty$, *or* $u(0) = \alpha \in (\beta, \infty)$ *if* $\gamma = \infty$; *moreover* $u'(r) \le 0$ *for all* $r \ge 0$. *Furthermore, since the dead core condition* (1.3) *is assumed to hold, then the solution is compactly supported.*

Remarks. Clearly, if $F(\gamma) > -\infty$ and $H(\infty) = \infty$ then only condition (i) is needed. For the Laplacian operator and the p-Laplacian degenerate operator $H(\infty) = \infty$ and $\Phi(\infty) = \infty$, so conditions (ii), (iii), and (iv) are not needed if $\gamma < \infty$, though (iv) must be used when $|F(\gamma)| = \infty$. On the other hand, for the mean curvature operator $H(\infty) = 1$, and so in consequence of (ii), the function F must be quite restrictive to apply Theorem A of [4] in order to get existence of radially symmetric ground states for (1.1).

Consider next the Dirichlet problem

$$\begin{cases} \operatorname{div}\{A(|Du|)Du\} = f(u) & \text{in } B_R \subset \mathbb{R}^n, \quad n \ge 2, \\ u = \delta & \text{on } \partial B_R, \end{cases} \tag{10.1}$$

under the dead core condition (1.3) for functions f satisfying (F1) and (F3). If $H(\infty) = \infty$, define

$$C' = \int_0^\delta \frac{ds}{H^{-1}(F(s)/n)} < \infty,$$

by (1.3) and (F1). Then there exists, according to Theorems 1.2 and 5.2, a solution of (10.1) which has a dead core B_S, with $R - S < C'$. In other words the main equation (1.1) has two nonnegative solutions—the compact support *burst* shown above by Theorem A of [4], and the dead core solution just derived by Theorem 1.2.

These two solutions can clearly be superposed, even if problem (10.1) is nonlinear, to obtain another combined solution of (10.1), with a dead core $B_S \setminus B_T$, $T < S < R$, and a symmetrical *burst* at its center.

In particular, the Dirichlet problem (10.1), with the *loop* nonlinearity f verifying (F1) and (F3), neither has a *unique* solution nor obeys the ordinary maximum principle.

Since (1.1) in problem (10.1) is translation invariant, it is evident that the *burst* need not be centered in the ball B_R. Even more, if R is made many times larger, one can place multiple bursts into original dead cores.

The main open problem arising in the study of dead cores is the question of stability, particularly in dynamic situations for parabolic time dependent problems of the type, e.g.,

$$u_t = \operatorname{div}\{A(|Du|)Du\} - f(u), \qquad u = u(t,x), \qquad (t,x) \in \mathbb{R}^+ \times \Omega.$$

Acknowledgments. The paper is based on a minicourse given by P. Pucci in September 2002, in Grado, Italy, at the workshop Stationary and Evolution Problems, supported by the *GNAMPA* of the Istituto Nazionale di Alta Matematica "F. Severi," and on the lecture of J. Serrin given in June 2003 in the series Lezioni Leonardesche, organized by the Mathematics departments of the two universities in Milan and of the Politechnic of Milan.

REFERENCES

[1] C. Bandle and S. Vernier-Piro, *Estimates for solutions of quasilinear problems with dead cores*, Z. Angew. Math. Phys., 54 (2003), pp. 815–821.
[2] J. Batt, W. Faltenbacher, and E. Horst, *Stationary spherically symmetric models in stellar dynamics*, Arch. Ration. Mech. Anal., 93 (1986), pp. 159–183.
[3] J.I. Diaz and L. Veron, *Local vanishing properties of solutions of elliptic and parabolic quasilinear equations*, Trans. Amer. Math. Soc., 290 (1985), pp. 787–814.
[4] B. Franchi, E. Lanconelli, and J. Serrin, *Existence and uniqueness of nonnegative solutions of quasilinear equations in* $\mathbb{R}^n$, Adv. Math., 118 (1996), pp. 177–243.
[5] D. Gilbarg and N. Trudinger, *Elliptic Partial Differential Equations of Second Order*, 2nd ed., Springer-Verlag, New York, 1983.
[6] P. Pucci, M. García-Huidobro, R. Manásevich, and J. Serrin, *Qualitative proprieties of ground states for singular elliptic equations with weights*, Ann. Mat. Pura Appl. (4), 185 (2006), pp. 205–243.
[7] P. Pucci and J. Serrin, *The strong maximum principle revisited*, J. Differential Equations, 196 (2004), pp. 1–66.
[8] P. Pucci and J. Serrin, *Erratum: "The strong maximum principle revisited,"* J. Differential Equations, 207 (2004), pp. 226–227.
[9] D. Siegel, *Height estimates for capillarity surfaces*, Pacific J. Math., 88 (1980), pp. 471–515.
[10] R. Sperb, *Some complementary estimates in the dead core problem*, in Nonlinear Problems in Applied Mathematics. In Honor of Ivar Stakgold on His 70th Birthday, T.S. Angell, et al., eds., Philadelphia, 1996, PA, pp. 217–224.

I.14. Pathological Solutions

Haim Brezis

Mathematics Subject Classification (2010). 35J62, 35B45, 35B65, 35D30 35J70.
Keywords. Divergence elliptic equation, weak solutions.

A well-known result from 1957, due to De Giorgi and Nash (see [3]), asserts that if $u \in H^1(\Omega)$ satisfies

$$\sum_{ij} \int_\Omega a_{ij} \frac{\partial u}{\partial x_i} \frac{\partial \varphi}{\partial x_j} = 0 \quad \forall \varphi \in C_c^\infty(\Omega), \tag{1}$$

where the coefficients $a_{ij} \in L^\infty(\Omega)$ are uniformly elliptic, then $u \in C^{0,\alpha}_{\text{loc}}(\Omega)$ for some $\alpha > 0$, with a corresponding estimate. In particular

$$\|u\|_{L^\infty(\omega)} \leq C\|u\|_{H^1(\Omega)}, \quad \forall \omega \quad \text{with } \overline{\omega} \subset \Omega. \tag{2}$$

Since (1) makes sense for functions $u \in W^{1,1}(\Omega)$, one may wonder whether the De Giorgi–Nash estimate (or the weaker form (2)) is still valid when H^1 is replaced by $W^{1,1}$. In his 1964 paper (paper [i] below) J. Serrin produced a striking example showing that such a stronger version of the De Giorgi–Nash estimate fails: Given any $p < 2$, J. Serrin constructed a function $u \in W^{1,p}(\Omega)$ in a ball $\Omega \subset \mathbb{R}^N, N \geq 2$, satisfying (1) for some elliptic coefficients $a_{ij} \in L^\infty$, and such that $u \notin L^\infty_{\text{loc}}(\Omega)$. In the same paper he proposed the following:

Conjecture (J. Serrin, 1964). *Assume that the coefficients a_{ij} are Hölder and uniformly elliptic. Assume that $u \in W^{1,1}(\Omega)$ satisfies* (1)*. Then u is "classical", i.e., $u \in H^1_{\text{loc}}(\Omega)$* (and therefore u is Hölder by De Giorgi–Nash).

A partial answer to Serrin's conjecture was given in 1971 by two former PhD students of J. Serrin, R. Hager and J. Ross:

Theorem 1 ([4])**.** *Assume that the coefficients a_{ij} are Hölder and uniformly elliptic. Assume that u belongs to $W^{1,p}(\Omega)$ for some $p > 1$, and satisfies* (1)*. Then $u \in H^1_{\text{loc}}(\Omega)$.*

H. Brezis gave a complete solution of Serrin's conjecture and also weakened the first assumption in the theorem of Hager–Ross. The results were announced in [1] and the detailed proofs are presented in [2] as an Appendix in a paper by A. Ancona (who, in turn, used this result while answering a question raised by H. Brezis and A. Ponce).

Theorem 2 ([1], [2])**.** *Assume that the coefficients a_{ij} are continuous in $\overline{\Omega}$ and uniformly elliptic. Assume that u belongs to $W^{1,p}(\Omega)$ for some $p > 1$, and satisfies* (1)*. Then u belongs to $W^{1,q}_{\text{loc}}(\Omega)$ for every $q < \infty$. Moreover*

$$\|u\|_{W^{1,q}(\omega)} \leq C\|u\|_{W^{1,p}(\Omega)} \quad \forall \omega \quad \text{with } \overline{\omega} \subset \Omega, \tag{3}$$

P. Pucci et al. (eds.), *James Serrin. Selected Papers: Volume 1*, Contemporary Mathematicians,
DOI 10.1007/978-3-0348-0845-3_14,

where C depends only on ω, Ω, the ellipticity constant, p, q, and the modulus of continuity of the a_{ij}'s.

Theorem 3 ([1], [2]). *Assume that the coefficients a_{ij} are Dini-continuous, i.e.,*

$$|a_{ij}(x) - a_{ij}(y)| \leq \gamma(|x-y|) \quad \forall x, y \in \Omega, \qquad \textit{with} \int_0^1 \frac{\gamma(t)}{t} dt < \infty$$

(e.g., a_{ij} are Hölder). Assume that u belongs to $W^{1,1}(\Omega)$ and satisfies (1). Then $u \in C^1(\Omega)$ and moreover

$$\|Du\|_{L^\infty(\omega)} \leq C\|u\|_{W^{1,1}(\Omega)} \quad \forall \omega \quad \textit{with } \overline{\omega} \subset \Omega, \tag{4}$$

where C depends only on ω, Ω, and on the Dini-modulus of continuity of the a_{ij}'s.

Here are several comments showing that the results of Theorems 2 and 3 are sharp and more subtle than they seem! Under the assumptions of Theorem 3 the following estimate holds, for every $p < \infty$,

$$\|Du\|_{L^p(\omega)} \leq C\|u\|_{W^{1,1}(\Omega)}, \quad \forall \omega \quad \text{with } \overline{\omega} \subset \Omega \tag{5}$$

where C depends only on ω, Ω, p and the *modulus of continuity* of the a_{ij}'s (but *not* on the Dini-modulus of continuity!). Estimate (5) suggested that the answer to Serrin's conjecture might still be positive when the a_{ij}'s are merely continuous. H. Brezis raised that question in [2]. It turns out that the answer is negative! T. Jin, V. Maz'ya and J. Van Schaftingen [5] have constructed an ingenious example of a function $u \in W^{1,1}(\Omega)$ satisfying (1) for some elliptic coefficients $a_{ij} \in C^0$, such that Du does *not* belong even to the class $L \log L$! They also answered another question raised by H. Brezis: they found a function $\tilde{u}$ which belongs to $W^{1,p}(\Omega)$ $\forall p < \infty$, which satisfies (1) for some elliptic coefficients $a_{ij} \in C^0$, and such that $D\tilde{u} \notin$ BMO. Putting all those facts together one has now a clear picture of the "smoothing effects" for elliptic equations in divergence form with continous coefficients.

References

[1] H. Brezis, *On a conjecture of J. Serrin*, Rend Lincei Mat. Appl. **19** (2008), 335–338.

[2] H. Brezis, *Solution of a conjecture by J. Serrin,* Appendix in the paper by A. Ancona, *Elliptic operators, conormal derivatives and positive parts of functions*, J. Funct. Anal., **257** (2008), 2147–2158.

[3] E. De Giorgi, *Sulla differenziabilità e l'analiticità delle estremali degli integrali multipli regolari* Mem. Accad. Sci. Torino, **3** (1957), 25–43.

[4] R. Hager and J. Ross, *A regularity theorem for linear second order elliptic divergence equations,* Ann. Sc. Norm. Sup. Pisa, **23** (1971), 283–290.

[5] T. Jin, V. Maz'ya and J. Van Schaftingen, *Pathological solutions to elliptic problems in divergence form with continuous coefficients*, C. R. Acad. Sci. **347** (2009), 773–778.

Haim Brezis
Rutgers, The State University
Department of Mathematics
Piscataway, NJ 08854, USA
e-mail: brezis@math.rutgers.edu

Ann. Sc. Normale Sup. Pisa, Sci. Fis. Math. (3) 18 (1964), 385–387. **Reprint I.14.[i]**

PATHOLOGICAL SOLUTIONS OF ELLIPTIC DIFFERENTIAL EQUATIONS

by JAMES SERRIN

In this note we consider the simplest kind of divergence structure elliptic equation, namely

$$\Sigma \frac{\partial}{\partial x_i}\left(a_{ij}\frac{\partial u}{\partial x_j}\right) = 0 \tag{1}$$

where the coefficients a_{ij} are bounded measurable functions of $x = (x_1, \dots, x_n)$ satisfying the ellipticity condition

$$\lambda \xi^2 \leq a_{ij}\, \xi_i\, \xi_j \leq \Lambda \xi^2, \quad \lambda, \Lambda > 0). \tag{2}$$

Because of the general assumptions made on the coefficients it is natural that the equation be interpreted in a weak sense. In particular, let $u = u(x)$ be a function having strong derivatives $\partial u/\partial x_i$ which are locally summable over a domain D. Then u may be called a *weak solution* of (1) over D if

$$\int a_{ij}\frac{\partial u}{\partial x_j}\frac{\partial \Phi}{\partial x_i}\, dx = 0 \tag{3}$$

for any continuously differentiable function $\Phi = \Phi(x)$ with compact support in D.

In developing the theory of equations (1) or (3) it is generally assumed that the derivatives $\partial u/\partial x_i$ are locally of class L_2 in D, and this additional requirement is accordingly built into the definition of weak solutions. This being done, one can then prove that solutions are locally bounded, have a local maximum principle, and that the Dirichlet problem for

---

Pervenuto alla Redazione il 3 Aprile 1964.

smooth data on smooth boundaries has one and only one solution. If, however, one does not introduce this additional requirement, the definition of weak solution still remains meaningful as we have given it. In fact, the above definition is in a sense the most « natural » one, in that it requires of the function $u(x)$ just exactly enough to make relation (3) well defined.

Nevertheless, the generality allowed in this definition is too great to allow a comprehensive theory to be developed. We shall show here that there exist weak solutions in the sense defined above that are neither locally bounded nor have a local maximum principle; and that, moreover, the Dirichlet problem need not have unique solutions in the class considered. In addition, the violation of local boundedness will be shown to hold even in the class of solutions whose derivatives are locally in $L_{1+\mu}$, where μ is any fixed number less than one. Thus the usual requirement that the derivatives be in L_2 forms an important and essential part of the theory.

Consider now equation (1) with the particular coefficients

$$a_{ij} = \delta_{ij} + (a-1)\frac{x_i x_j}{r^2}, \qquad (r = |x|),$$

where a is a constant greater than one. It is easily checked that these coefficients are bounded and that (2) holds with $\lambda = 1$, $\Lambda = a$. Moreover one finds that the function

$$u = x_1 r^{1-n-\varepsilon}$$

is a classical solution in $|x| > 0$ provided a and ε are related by

$$a = \frac{n-1}{\varepsilon(\varepsilon + n - 2)}, \qquad 0 < \varepsilon < 1.$$

We shall now show that u is in fact a weak solution throughout E^n. First of all, since $u = 0(r^{2-n-\varepsilon})$, $u_x = 0(r^{1-n-\varepsilon})$, it is easily verified that u is strongly differentiable with $u_x \in L_\beta$ for any $\beta < n/(n+\varepsilon-1)$. Moreover, if Φ is any continuously differentiable function with compact support in E^n, then obviously

$$\int a_{ij}\frac{\partial u}{\partial x_j}\frac{\partial \Phi}{\partial x_i}dx = \lim_{\varrho\to 0}\int_{r>\varrho} a_{ij}\frac{\partial u}{\partial x_j}\frac{\partial \Phi}{\partial x_i}dx$$

$$= -\lim_{\varrho\to 0}\oint_{r=\varrho}\Phi\, a_{ij}\frac{\partial u}{\partial x_j}\frac{x_i}{r}ds.$$

But by direct calculation

$$-\oint_{r=\varrho} \Phi\, a_{ij} \frac{\partial u}{\partial x_j} \frac{x_i}{r}\, ds = \frac{\omega_n (n-1)}{\varepsilon} \varrho^{-1-\varepsilon} \oint_{r=\varrho} \Phi\, x_1\, d\omega.$$

Since Φ is continuously differentiable the last integral is easily seen to be $0(\varrho^2)$ as $\varrho \to 0$, whence combining these results it follows that

$$\int a_{ij} \frac{\partial u}{\partial x_j} \frac{\partial \Phi}{\partial x_i}\, dx = 0.$$

This proves that u is a weak solution in E^n.

Clearly the function u is neither locally bounded nor has a local maximum principle. It can be used, moreover, to show that the Dirichlet problem need not have unique solutions. In fact, let v be the unique weak solution of (1) with L_2 derivatives, taking the same values on $r = 1$ as the function u. Then $u - v$ has zero data on $r = 1$, but is not identically zero. This shows clearly that the Dirichlet problem can have two solutions corresponding to the same data, provided we give up the requirement that these solutions have L_2 derivatives.

In case $n = 2$, we have $u_x \in L_\beta$ for any $\beta < 2/(1+\varepsilon)$, whence by choosing ε sufficiently near zero it is clear that any relaxation of the requirement that derivatives be in L_2 will lead to difficulties. When $n > 2$ the function

$$u(x) = x_1 (x_1^2 + x_2^2)^{-(1+\varepsilon)/2}$$

may be thought of as a solution (by descent) of another uniformly elliptic equations whose coefficients a_{11}, a_{12}, a_{22} are those associated with the case $n = 2$ above. Since this solution is not locally bounded, the proof of our assertions is complete.

Remarks. Although the above example shows that in general it is reasonable to require solutions to have strong derivatives which are locally in L_2, nevertheless, for any given equation (1) it can be shown that regularity properties remain valid in a somewhat more extended class of functions, depending in particular on the moduli of ellipticity [1]. From this point of view, the above example shows that there are definite limits beyond which one cannot go without losing the outlines of the theory.

In conclusion, we should like to conjecture that if the coefficients a_{ij} are Hölder continuous, then any weak solution as defined here must be in fact a classical solution.

---

[1] N. G. Meyers, *An L^p estimate for the gradient of solutions of second order elliptic divergence structure equations*. Ann. Scuola Norm. Pisa XVII (1963), pp. 189-206.

I.15 Lane–Emden Systems

Enzo Mitidieri

Mathematics Subject Classification (2010). 35J60, 34A34, 35B05, 35B40.

Keywords. Nonlinear elliptic systems, nonexistence, Liouville-type theorem, Lane–Emden conjecture.

Elliptic Hamiltonian systems are an interesting and challenging topic in the theory of partial differential equations. The first contributions on this subject were obtained in the early nineties. In these regards, the two papers which are the subject of this commentary, represent a breakthrough from different point of view.

1. Non-existence of positive solutions of Lane–Emden systems

There are two main kind of problems that we shall focus on during the course namely, existence and non-existence of positive solutions for Lane–Emden system. The latter is a special Hamiltonian system of the type:

$$\Delta u + v^p = 0, \quad \Delta v + u^q = 0, \; x \in \mathbb{R}^n, \tag{1}$$

where $n \geq 3$ and $p, q > 0$. This problem can be thought of as a natural extension of the Lane–Emden equation,

$$\Delta w + w^p = 0, \; x \in \mathbb{R}^n. \tag{2}$$

Equation (2) is strongly related to Sobolev inequality. Indeed it is well known, see [5], that (2) admits a positive solution if and only if p is on or above the critical Sobolev exponent, i.e., $q \geq \dfrac{n+2}{n-2}$.

On the other hand, system is related to a special case of the bilinear form of the Hardy–Littlewood–Sobolev inequality. See [6] for details. The corresponding *conformal invariant* condition on (p, q) for (1) is given by

$$\frac{1}{p+1} + \frac{1}{q+1} = \frac{n-2}{n},$$

the so-called *critical hyperbola*. Hence, it is a natural question to conjecture that (1) posses a positive solution if and only the pair (p, q) is on or above the *critical hyperbola*.

The main contribution, among others, of the authors was the proof of this conjecture if $n = 3$ for polynomially bounded solutions as well as the complete classification of the solution set for positive radial solutions for (1). This result is of fundamental importance for establishing existence of positive solutions for the Dirichlet problem in

P. Pucci et al. (eds.), *James Serrin. Selected Papers: Volume 1*, Contemporary Mathematicians,
DOI 10.1007/978-3-0348-0845-3_15, © Springer Basel 2014

a bounded domain via the blow-up procedure (see [4]) for a large class of elliptic systems not necessarily of variational type. Recently in [7] the restriction to polynomially bounded solutions has been removed.

In higher dimensions, the problem is very difficult since the available tools that are employed in the scalar case, do not seem to apply for this class of systems. However some partial contributions in this direction are known. We quote here an earlier results obtained in [2] and the recent one contained in the important paper [10] where the complete conjecture has been established if $n = 4$.

Obviously there are other related problems of interest. In this direction we mention the study of the corresponding systems associated to the full Hardy–Littlewood–Sobolev inequality and the complete classification of existence and non existence of global solutions for the Cauchy problem associated to the corresponding parabolic or hyperbolic systems in bounded or unbounded domains contained in $\mathbb{R}^n$.

This is a promising area of research for which we strongly believe that the original ideas contained in the paper that we are commenting will reveal all its potential.

2. Existence of positive entire solutions of elliptic Hamiltonian systems

An important related problem is to prove the existence (and non-existence) of positive solutions for general elliptic Hamiltonian systems of the type,

$$\Delta u + H_v(u,v) = 0, \;\; \Delta v + H_v(u,v) = 0, \; x \in \mathbb{R}^n. \tag{3}$$

Here the Hamiltonian function H satisfies, together with its derivatives, some natural regularity, positivity and superlinearity assumptions (see the author's paper for the precise hypotheses). The important point here is that, thought the problem has a strongly indefinite variational structure, the possible solutions do not necessarily belong to a precise energy space. This complicates the analysis. However, the authors, by restricting their interest to positive radially symmetric solutions, were able to prove existence by using a delicate and innovative two-dimensional shooting method.

To our knowledge this was an absolutely novelty for these kind of problems. The method employed is strictly related to a suitable form of a variational identity proved in [8]. One other very important aspect of this contribution is that the authors were able to handle also the case when H_u, H_v were not Lipschitz continuous functions at the values $u = 0$ and $v = 0$. This is indeed a very delicate point since in this case one loose continuous dependence on the data, a fundamental aspect of the application of the shooting method!

During the last fifteen years, the impact of this paper in the literature on the field was twofold. On one hand it has stimulated the study of existence of solutions for general quasilinear elliptic Hamiltonian system, see [1]. On the other hand, the nonexistence problem for generalized forms of Lane–Emden systems started to become and interesting and motivated research topic. In these respects we mention only the very recent contributions related to the so-called Hènon–Lane–Emden systems. See [3, 9] and the references therein.

References

[1] M.F. Bidaut-Veron, H. Giacomini; A new dynamical approach of Emden-Fowler equations and systems, Adv. Differential Equations **15** (2010), no. 11-12, 1033–1082.

[2] J. Busca, R. Manásevich, A Liouville-type theorem for Lane–Emden system, Indiana Univ. Math. J. **51** (2002) 37–51.

[3] M. Fazly, N. Ghoussoub, On the Hènon–Lane–Emden conjecture, arXiv:1107.5611, 2011.

[4] B. Gidas, J. Spruck, A priori bounds for positive solutions of nonlinear elliptic equations, Comm. Partial Differential Equations **6** (1981) 883–901.

[5] B. Gidas, J. Spruck, Global and local behavior of positive solutions of nonlinear elliptic equations, Comm. Pure Appl. Math. **34** (1981) 525–598.

[6] E.H. Lieb, M. Loss, Analysis. Second edition. Graduate Studies in Mathematics, 14. American Mathematical Society, Providence, RI, 2001. xxii + 346 pp.

[7] P. Polàcǐk, P. Quittner, Ph. Souplet, Singularity and decay estimates in superlinear problems via Liouville-type theorems. Part I: Elliptic systems, Duke Math. J. **139** (2007) 555–579.

[8] P. Pucci, J. Serrin, A general variational identity, Indiana Univ. Math. J. **35** (1986) 681–703.

[9] Q.P. Phan, Liouville-type theorems and bounds of solutions for Hardy–Hènon elliptic systems, preprint July 2011.

[10] P. Souplet, The proof of the Lane–Emden conjecture in four space dimensions. Adv. Math. **221** (2009), no. 5, 1409–1427.

Enzo Mitidieri
Dipartimento di Matematica e Informatica
Università degli Studi di Trieste
Via Valerio 12/1
I-34127 Trieste, Italy
e-mail: mitidieri@units.it

Differ. Integral Equ. 9 (1996), 635–653. **Reprint I.15.[i]**

Pages 646–651 have been removed for the sake of brevity.

NON-EXISTENCE OF POSITIVE SOLUTIONS OF LANE-EMDEN SYSTEMS

JAMES SERRIN
School of Mathematics, University of Minnesota, Minneapolis, MN 55455

HENGHUI ZOU*
Department of Mathematics, University of Alabama at Birmingham, Birmingham, AL 35294

1. Introduction. In this paper we consider (component-wise) positive solutions of the weakly coupled system

$$\begin{aligned} \Delta u + v^p &= 0, \\ \Delta v + u^q &= 0, \end{aligned} \qquad x \in \mathbb{R}^n, \tag{I}$$

where $p, q > 0$ and $n \geq 3$ is the dimension of the space, and are concerned with the question of non-existence of such solutions. This system arises in chemical, biological and physical studies, and has been investigated by several authors, see for example [3, 8, 9] and references therein.

The system (I) is a natural extension of the celebrated Lane-Emden equation, and we thus refer to it as the Lane–Emden system. The Lane-Emden equation

$$\Delta u + u^p = 0, \qquad x \in \mathbb{R}^n,\ n > 2,\ p > 1 \tag{II}$$

has been extensively studied, going back to the pioneering work of Fowler (cf. [4], and the recent paper [12]). It is well-known that the Sobolev exponent

$$l = \frac{n+2}{n-2}$$

serves as the dividing number for existence and non-existence of solutions of (II), that is, equation (II) admits non-negative, non-trivial solutions if and only if $p \geq l$, see [4] and [5].

It is natural to ask if there exists a corresponding *dividing curve* in the pq—plane for the Lane-Emden system, that is, a curve with the property that (I) admits positive solutions if and only if (p, q) is on or above the curve.

Mitidieri [9] showed that (I) does not have any positive *radial* solutions if

$$\frac{1}{p+1} + \frac{1}{q+1} > \frac{n-2}{n}, \qquad p, q > 1,$$

Received for publication February 1996.
*Research supported in part by NSF grant DMS-9418779, by a grant from Alabama EPSCoR and a faculty research grant from University of Alabama at Birmingham, Birmingham.
AMS Subject Classifications: 35J60, 35B40.

a result which was improved in [13] to be valid for $p, q > 0$. On the other hand, a concentrated - compactness argument (cf. [7]) implies that (I) admits a positive radial solution when (p, q) is on the hyperbola

$$\frac{1}{p+1}+\frac{1}{q+1}=\frac{n-2}{n}. \tag{1.1}$$

This suggests that the hyperbola (1.1) is the dividing curve for existence and nonexistence for (I). The purpose of the paper is to justify this assertion in a number of important cases, see Corollaries 1.1 and 1.2. We also note that condition (1.1) becomes exactly $p = l$ when $p = q$, as for the Lane–Emden equation.

We call a point (p, q) in the first quadrant of the pq–plane *critical, subcritical* or *supercritical* if it is respectively on, below or above the hyperbola (1.1). Also we say that a function w on $\mathbb{R}^n$ has algebraic growth at infinity if there exists a constant K such that

$$|w(x)| \leq \text{Const. } |x|^K, \qquad |x| > 1. \tag{1.2}$$

Throughout the paper we write vector-valued functions using bold face type, e.g., $\mathbf{u}(x) = (u(x), v(x))$, and we say that $\mathbf{u}$ is positive if both components are positive. We shall establish the following non-existence results concerning positive solutions of (I).

Theorem 1.1. *Let $n = 3$ and suppose that (p, q) is subcritical with respect to the hyperbola* (1.1). *Then* (I) *does not admit any non-negative and non-trivial solutions* $\mathbf{u}$ *with algebraic growth at infinity.*

Since by definition *ground states* of (I) tend to zero at infinity, *Theorem* 1.1 *excludes their existence for subcritical exponents when $n = 3$*, see Corollary 1.1.

Theorem 1.2. *The system* (I) *does not admit any non-negative and non-trivial solutions* $\mathbf{u}$, *provided either $pq \leq 1$ or*

$$pq > 1 \quad \textit{and} \quad \max\left\{\frac{2(p+1)}{pq-1}, \frac{2(q+1)}{pq-1}\right\} \geq n-2. \tag{1.3}$$

Remark. When $p, q > 1$, Theorem 1.2 was obtained in [9]. Condition (1.3) corresponds exactly to the relation $p \leq n/(n-2)$ when $p = q$.

The proofs depend heavily on asymptotic estimates at infinity. For the proof of Theorem 1.2, this amounts to obtaining estimates for spherical averages of solutions. Such an argument is no longer valid when the point (p, q) lies above the curve given by (1.3), as in the case of Theorem 1.1. A subtler approach is to apply a Pohozaev-type identity together with estimates for the solutions themselves.

There are various difficulties in obtaining such estimates, partly due to the imbalance between the two exponents p and q. This makes it generally impossible to apply techniques used in treating single equations. For instance, in contrast to the case of the Lane-Emden equation for which solutions have fast decay $|x|^{2-n}$ at infinity when $p = l$, the components u and v of solutions of (I) cannot both have fast decay everywhere on the critical hyperbola.

Problem (I) admits infinitely many positive radial solutions when (p, q) is critical or supercritical with respect to the hyperbola (1.1), see [14]. Combining the results of [9, 13] and Theorem 1.1 with this conclusion, we obtain the following corollaries showing that (1.1) is the dividing curve for (I) for radial solutions, and for all solutions with algebraic growth when $n = 3$.

Corollary 1.1. *Suppose $n = 3$. Then the hyperbola given by* (1.1) *is the dividing curve for positive solutions of* (I) *with algebraic growth at infinity, and, in particular, for ground states.*

Corollary 1.2. *The hyperbola given by* (1.1) *is the dividing curve for positive radial solutions of* (I).

General functions $f(v)$ and $g(u)$, replacing v^p and u^q, can also be treated in a similar manner, with corresponding existence and non-existence results under suitable conditions on f and g. We refer the reader to [13, 14] for details.

The outcomes could be quite different if there were special homogeneous relations between the exponents. For example, Bidaut-Veron and Raoux [1] considered the system

$$\begin{aligned} \Delta u + |x|^\sigma u^q v^{p+1} &= 0, \\ \Delta v + |x|^\sigma u^{q+1} v^p &= 0, \end{aligned} \qquad u, v > 0, \tag{III}$$

where $p,\ q > 0$. Estimates near isolated singularities were obtained when

$$p + q + 1 < \frac{n + 2 + 2\sigma}{n - 2} \tag{1.4}$$

by an argument similar to that used in [5]. The duality between u and v plays a major role here and methods used to treat single equations remain applicable. Moreover, for comparable n-component systems (with appropriate symmetry, and with (III) a special case), the moving plane method shows non-existence and symmetry of positive solutions when the nonlinearity is, respectively, subcritical or critical with respect to (1.4); see [11].

This approach also can be used for the Lane–Emden system in certain cases. Indeed when both exponents p and q are less than or equal to l, the moving plane method applies with the aid of a Kelvin transform. This shows in particular that the system (I) has no positive solutions if $p \le l$ and $q \le l$ with at most one equality holding, and that all solutions must be radially symmetric when $p = q = l$; see [3].

As a final remark, we observe that one can combine Theorem 1.1 with a blow-up argument to get a priori estimates for positive solutions of a large class of semilinear elliptic systems on bounded domains, especially systems without variational structure, see the forthcoming paper [16].

The organization of the paper is as follows. In Section 2, we present some preliminaries and our main estimates. In Section 4, appropriate asymptotic estimates are obtained for the case $n = 3$. Section 3 contains the proof of Theorem 1.2, and Section 5 the proof of Theorem 1.1.

2. Preliminaries and principal asymptotic estimates. Consider the weakly coupled system

$$\begin{cases} \Delta u + v^p = 0, \\ \Delta v + u^q = 0, \end{cases} \qquad x \in \mathbb{R}^n, \tag{I}$$

where $p, q > 0$ and $n \geq 3$ is an integer. We are interested in non-existence of non-negative and non-trivial C^2-solutions.

As mentioned in the introduction, throughout the paper we shall write vector-valued functions in the form $\mathbf{u}(x) = (u(x), v(x))$, and say that $\mathbf{u}$ is positive if both components u and v are positive. Generic constants will be denoted by c, M etc.; these depend on the arguments explicitly indicated, but may vary from line to line.

We begin with several preliminary results. The first is a standard comparison lemma.

Lemma 2.1. *Suppose that $w \geq 0$ is non-trivial and satisfies*

$$\Delta w \leq 0, \qquad x \in \mathbb{R}^n. \tag{2.1}$$

Then there exists a positive constant $c = c(w)$ such that

$$w(x) \geq c|x|^{2-n}, \qquad |x| \geq 1.$$

Proof. By the strong maximum principle, $w > 0$ (we take this granted without mention in the sequel). Take

$$c = \min_{|x|=1} w(x) > 0,$$

and put

$$w_1(x) = c|x|^{2-n}, \qquad |x| > 1.$$

Clearly $w - w_1$ is non-negative on the boundary of the set $\Omega = \mathbb{R}^n \backslash B_1$. Moreover

$$\liminf_{|x|\to\infty} (w - w_1) \geq 0$$

and

$$\Delta(w - w_1) \leq 0, \qquad x \in \Omega.$$

Therefore $w - w_1$ is non-negative in Ω by the maximum principle.

Lemma 2.2. *Suppose that $w \geq 0$ is non-trivial and satisfies* (2.1). *Then for $\eta \in C_0^\infty(\mathbb{R}^n)$ and $\gamma < 1$, there exists a constant $c = c(\gamma) > 0$ such that*

$$\int_{\mathbb{R}^n} \eta^2 |Dw|^2 w^{\gamma-2} \leq c \int_{\mathbb{R}^n} |D\eta|^2 w^\gamma. \tag{2.2}$$

Proof. Multiplying (2.1) by $\phi(x) = w^{\gamma-1}(x)\eta^2 \in C_0^\infty(\mathbb{R}^n)$ and integrating over $\mathbb{R}^n$, we obtain

$$(1-\gamma)\int_{\mathbb{R}^n} \eta^2 |Dw|^2 w^{\gamma-2} - 2\int_{\mathbb{R}^n} \eta w^{\gamma-1} Dw \cdot D\eta \leq 0.$$

It follows by the Schwarz inequality that

$$\begin{aligned}\int_{\mathbb{R}^n} \eta^2 |Dw|^2 w^{\gamma-2} &\leq \frac{2}{1-\gamma}\int_{\mathbb{R}^n} \eta|D\eta||Dw|w^{\gamma-1}\\ &\leq \frac{2}{1-\gamma}\left(\int_{\mathbb{R}^n} \eta^2|Dw|^2 w^{\gamma-2}\right)^{1/2}\cdot\left(\int_{\mathbb{R}^n}|D\eta|^2 w^{\gamma}\right)^{1/2}\\ &\leq \frac{1}{2}\int_{\mathbb{R}^n}\eta^2|Dw|^2w^{\gamma-2} + \frac{2}{(1-\gamma)^2}\int_{\mathbb{R}^n}|D\eta|^2 w^{\gamma},\end{aligned}$$

as required ($c = 4/(1-\gamma)^2$).

For $w \in C(\mathbb{R}^n)$, denote the spherical average of w by

$$\overline{w}(r) = \frac{1}{\omega_n}\int_{S^{n-1}} w(r,\theta), \qquad r > 0,$$

where (r, θ) are spherical-coordinates and $\omega_n = |S^{n-1}|$ is the area of the unit sphere in $\mathbb{R}^n$.

Lemma 2.3. *Suppose that $w \geq 0$ is non-trivial and satisfies* (2.1). *Then for $a \in (0, 1]$, we have $(\overline{w^a})' \leq 0$. In particular, $\overline{w^a}$ is a non-increasing function of r.*

Proof. For $a \in \mathbb{R}$, it follows by direct calculation and (2.1) that

$$\Delta(w^a) + a(1-a)w^{a-2}|\nabla w|^2 \leq 0.$$

Taking the average over S^{n-1} yields, by a standard calculation,

$$\left(\overline{w^a}\right)'' + \frac{n-1}{r}\left(\overline{w^a}\right)' + a(1-a)\overline{w^{a-2}|\nabla w|^2} \leq 0, \qquad r > 0,$$

with $'$ denoting differentiation with respect to r. In turn

$$(\overline{w^a})'' + \frac{n-1}{r}(\overline{w^a})' \leq 0,$$

since $0 < a \leq 1$. Thus $r^{n-1}(\overline{w^a})'$ is non-increasing. But this function vanishes as $r \to 0$, so that $r^{n-1}(\overline{w^a})' \leq 0$, and in turn $(\overline{w^a})' \leq 0$ and $\overline{w^a}$ is non-increasing.

We next state a useful version of the Poincaré-Sobolev embedding lemma.

Lemma 2.4. *Let k be a positive integer and $s > 1$. Suppose $w \in W^{k,s}(\Omega)$, where Ω is a smooth bounded connected domain in $\mathbb{R}^n$. Then there exists a positive number $C = C(n, s, k, \Omega)$ such that*

$$|w|_{L^\infty(\Omega)} \leq C|D^k w|_{L^s(\Omega)} + |w_\Omega|,$$

provided $ks > n$, and

$$|w|_{L^{ns/(n-ks)}(\Omega)} \leq C|D^k w|_{L^s(\Omega)} + |\Omega|^{(n-ks)/ns}|w_\Omega|,$$

provided $ks < n$, where

$$w_\Omega = \frac{1}{|\Omega|}\int_\Omega w.$$

When Ω is a compact smooth connected manifold in $\mathbb{R}^n$ the conclusions still hold, but with the conditions $ks > n$ and $ks < n$ respectively replaced by $ks > \dim(\Omega)$ and $ks < \dim(\Omega)$.

Proof. Let Ω be a smooth bounded connected domain. From [15], Corollary 4.2.3, $L(w) = T(w) = w_\Omega$ and $p = s$, $m = k+1$, we obtain (replacing k by $k-1$)

$$|w - w_\Omega|_{L^{s^*}(\Omega)} \le C|D^k w|_{L^s(\Omega)};$$

here one uses also that $T(1) = 1$ and that $\|T\|$ depends on Ω. The required estimates now follow from the triangle inequality.

The same conclusion obviously continues to hold (by a partition of unity argument) when Ω is a compact smooth connected manifold.

We remark that the constant C in Lemma 2.4 is invariant under geometric similarity (scaling, rotation and translation etc.) for the case when $ks < n$.

Lemma 2.5. (i) *Suppose that $w \ge 0$ is non-trivial and satisfies* (2.1). *Also assume that there exist $\gamma \in (0,1)$ and $\mu > 0$ such that*

$$\overline{w^\gamma}(r) \le M r^{-\mu}. \tag{2.3}$$

Then for any $R > 0$ and $\sigma > 1$ we have

$$\int_{B_{\sigma R}\setminus B_R} |Dw|^2 w^{\gamma-2} \le cMR^{n-2-\mu}$$

and

$$\int_{B_{\sigma R}\setminus B_R} w^{\gamma n/(n-2)} \le cM^{n/(n-2)} R^{n-\mu n/(n-2)},$$

where $c = c(n,\gamma,\sigma)$.

(ii) *If moreover $\gamma \in (0,(n-2)/n]$, then also*

$$\overline{w^{\gamma n/(n-2)}}(r) \le cM^{n/(n-2)} r^{-\mu n/(n-2)}.$$

Proof. Choose

$$\eta(x) \equiv \begin{cases} 0, & |x| > 2\sigma R \text{ or } |x| < R/2, \\ 1, & R < |x| < \sigma R, \end{cases} \qquad |\nabla\eta(x)| \le c/R \tag{2.4}$$

in Lemma 2.2. Then, recalling that $\gamma < 1$, we get

$$\begin{aligned}\int_{B_{\sigma R}\setminus B_R} |D(w^{\gamma/2})|^2 &= \frac{\gamma^2}{4}\int_{B_{\sigma R}\setminus B_R} |Dw|^2 w^{\gamma-2} \le \frac{c}{R^2}\int_{B_{2\sigma R}\setminus B_{R/2}} w^\gamma \\ &= \frac{c}{R^2}\int_{R/2}^{2\sigma R} r^{n-1}\overline{w^\gamma}(r)\,dr \le \frac{cM}{R^2}\int_{R/2}^{2\sigma R} r^{n-1-\mu}\,dr = cMR^{n-2-\mu}\end{aligned}$$

by (2.3) (note that the final step is valid for arbitrary μ). This gives the first result of (i), with the constant c clearly depending also on σ.

Next we apply Lemma 2.4 with

$$k = 1, \quad s = 2, \quad \Omega = B_{\sigma R} \backslash B_R$$

and with w replaced by $w^{\gamma/2}$. Since $ks < n$ in the present case, this gives, in view of the comment after the proof of Lemma 2.4,

$$J = |w^{\gamma/2}|^2_{L^{2n/(n-2)}(\Omega)} \leq c|Dw^{\gamma/2}|^2_{L^2(\Omega)} + 2|\Omega|^{(n-2)/n}|w^{\gamma/2}|^2_{L^1(\Omega)}$$

where $c = c(n, \sigma)$. By the Hölder inequality

$$|w^{\gamma/2}|^2_{L^1(\Omega)} \leq cR^n \int_{B_{\sigma R}\backslash B_R} w^\gamma \leq cMR^{2n-\mu}$$

as in the last steps of the preceding calculation. Combining the previous lines then yields

$$J \leq cMR^{n-2-\mu}.$$

The second result of (i) now follows from the calculation

$$\int_\Omega w^{\gamma n/(n-2)} = J^{n/(n-2)} \leq (cMR^{n-2-\mu})^{n/(n-2)} = cM^{n/(n-2)}R^{n-\mu n/(n-2)}. \tag{2.5}$$

We now prove (ii). Since $\overline{w^{\gamma n/(n-2)}}(r)$ is non-increasing by Lemma 2.3 and the hypothesis $\gamma n/(n-2) \leq 1$, it follows that

$$\int_\Omega w^{\gamma n/(n-2)} = \omega_n \int_R^{\sigma R} r^{n-1}\overline{w^{\gamma n/(n-2)}}(r)\,dr \geq \frac{(\sigma^n - 1)\omega_n}{n}\overline{w^{\gamma n/(n-2)}}(\sigma R) \cdot R^n.$$

Hence

$$\overline{w^{\gamma n/(n-2)}}(\sigma R) \leq cR^{-n}\int_\Omega w^{\gamma n/(n-2)} \leq cM^{n/(n-2)}(\sigma R)^{-\mu n/(n-2)},$$

where we have used (2.5) at the last step. This completes the proof of the lemma, when we put $r = \sigma R$.

The previous lemmas concerned solutions of (2.1). We now turn to the main problem of positive solutions of (I).

Lemma 2.6. *Let* $\mathbf{u}=(u, v)$ *be a positive solution of* (I). *Then*

$$U_1'' + \frac{n-1}{r}U_1' + V_1^p \leq 0, \qquad \textit{for } 0 < p_1 \leq p, \tag{2.6}$$

and

$$V_2'' + \frac{n-1}{r} V_2' + U_2^q \le 0, \qquad \text{for } 0 < q_1 \le q, \tag{2.7}$$

where

$$U_1(r) = \left(\overline{u^{p_1/p}}\right)^{p/p_1}, \qquad V_1(r) = \left(\overline{v^{p_1}}\right)^{1/p_1};$$

and

$$U_2(r) = \left(\overline{u^{q_1}}\right)^{1/q_1}, \qquad V_2(r) = \left(\overline{v^{q_1/q}}\right)^{q/q_1}.$$

Proof. We only prove (2.6) for $p_1 < p$, the case (2.7) being the same and the case with equality being trivial. For $a \in \mathbb{R}$, it follows by direct calculation and $(\mathrm{I})_1$ that

$$\Delta(u^a) + a(1-a)u^{a-2}|\nabla u|^2 + a u^{a-1} v^p = 0.$$

Taking the average over S^{n-1} yields

$$\left(\overline{u^a}\right)'' + \frac{n-1}{r}\left(\overline{u^a}\right)' + a(1-a)\overline{u^{a-2}|\nabla u|^2} + a\overline{v^p u^{a-1}} = 0, \qquad r > 0. \tag{2.8}$$

By the Hölder inequality, we have

$$\begin{aligned}\overline{v^{p_1}} &= \frac{1}{\omega_n}\int_{S^{n-1}} v^{p_1} \le \frac{1}{\omega_n}\left(\int_{S^{n-1}} v^p u^{-1/p'}\right)^{p_1/p} \cdot \left(\int_{S^{n-1}} u^{p_1/p}\right)^{1/p'} \\ &= \left(\overline{u^{p_1/p}}\right)^{1/p'} \cdot \left(\overline{v^p u^{-1/p'}}\right)^{p_1/p},\end{aligned}$$

where $p' = p/(p-p_1)$ is the conjugate of p/p_1, that is,

$$\overline{v^p u^{a-1}} \ge V_1^p \cdot \left(\overline{u^a}\right)^{1-p/p_1}$$

with $a = p_1/p \in (0,1)$. Therefore from (2.8)

$$\left(\overline{u^a}\right)'' + \frac{n-1}{r}\left(\overline{u^a}\right)' + a(1-a)\overline{u^{a-2}|\nabla u|^2} + aV_1^p \cdot \left(\overline{u^a}\right)^{1-p/p_1} \le 0, \qquad r > 0.$$

Multiply by $\left(\overline{u^a}\right)^{p/p_1 - 1}$ and rewrite in terms of $U_1(r) = \left(\overline{u^a}\right)^{1/a}$ to obtain after a short calculation

$$U_1'' + \frac{n-1}{r} U_1' + \frac{1-a}{a^2}\left[a^2\overline{u^a} \cdot \overline{u^{a-2}|\nabla u|^2} - \left(\overline{u^a}\right)'^2\right] U_1^{1-2a} + V_1^p \le 0. \tag{2.9}$$

By the Schwarz inequality, setting $u' = \partial u/\partial r$,

$$\begin{aligned}\left(\overline{u^a}\right)'^2 &= \left[\left(\frac{1}{\omega_n}\int_{S^{n-1}} u^a\right)'\right]^2 = \left(\frac{a}{\omega_n}\int_{S^{n-1}} u' u^{a-1}\right)^2 \\ &\le \frac{a^2}{\omega_n^2}\int_{S^{n-1}} u'^2 u^{a-2} \cdot \int_{S^{n-1}} u^a \le a^2\overline{u^a} \cdot \overline{u^{a-2}|\nabla u|^2}.\end{aligned}$$

Thus (2.6) follows from (2.9), and the proof of the lemma is complete.

Lemma 2.7. *Suppose $z = z(r) > 0$ satisfies*

$$z'' + \frac{n-1}{r} z' + \phi(r) \le 0, \qquad r > 0,$$

with ϕ non-negative and non-increasing, and z' bounded for r near 0. Then

$$z(r) \ge c r^2 \phi(r),$$

where $c = c(n)$.

Proof. Clearly

$$-z'(r) \ge r^{1-n} \int_0^r s^{n-1} \phi(s) ds \ge r^{1-n} \int_0^{r/2} s^{n-1} \phi(s) ds \ge \frac{r}{2^n n} \phi(r/2).$$

Integrating from r to $2r$ then yields

$$z(r) \ge z(2r) + \frac{1}{2^n n} \int_r^{2r} s \phi(s/2) ds \ge \frac{r^2}{2^n n} \phi(r),$$

as required.

For $pq > 1$, put

$$\alpha = \frac{2(p+1)}{pq-1} > 0, \qquad \beta = \frac{2(q+1)}{pq-1} > 0. \tag{2.10}$$

Then we have the following proposition.

Proposition 2.1. *If $pq = 1$, then* (I) *admits no positive solutions. When $pq > 1$ there exists a positive constant $M = M(p, q, n)$ such that*

$$\overline{u}(r) \le M r^{-\alpha}, \qquad \overline{v}(r) \le M r^{-\beta} \quad \textit{for} \quad r > 0. \tag{2.11}$$

Proof. We first consider $p,\ q \ge 1$. Taking the average of (I) on S^{n-1} and using Jensen's inequality, we have

$$\overline{u}'' + \frac{n-1}{r} \overline{u}' + \overline{v}^p \le 0, \qquad \overline{v}'' + \frac{n-1}{r} \overline{v}' + \overline{u}^q \le 0, \qquad r > 0.$$

Since $\overline{u}$ and $\overline{v}$ are non-increasing by Lemma 2.3 with $a = 1$, we get from Lemma 2.7

$$\overline{u}(r) \ge c r^2 \overline{v}^p(r), \qquad \overline{v}(r) \ge c r^2 \overline{u}^q(r). \tag{2.12}$$

We immediately obtain a contradiction if $pq = 1$ ($p = q = 1$). Otherwise, solving the inequalities, we obtain (2.11).

Next, without loss of generality, assume $pq \geq 1$ and $q < 1$. Clearly from (I)

$$\overline{v}'' + \frac{n-1}{r}\overline{v}' + \overline{u^q} = 0.$$

Using (2.6) with $1 = p_1 < p$, and noticing that $\overline{u^q}$ $(q < 1)$ and $\overline{v}$ are non-increasing functions by Lemma 2.3, we again obtain from Lemma 2.7

$$U_1(r) = \left(\overline{u^{1/p}}\right)^p \geq cr^2\overline{v}^p, \qquad \overline{v}(r) \geq cr^2\overline{u^q}. \tag{2.13}$$

But from the Hölder inequality $\left(\overline{u^{1/p}}\right)^p \leq \left(\overline{u^q}\right)^{1/q}$, since $q \geq 1/p$. It follows that

$$(\overline{u^q})^{1/q} \geq cr^2\overline{v}^p, \qquad \overline{v}(r) \geq r^2\overline{u^q},$$

again a contradiction if $pq = 1$. Thus we assume $pq > 1$ and solve the inequalities to get

$$\overline{u^q}(r) \leq Mr^{-\alpha q}, \quad \overline{v}(r) \leq Mr^{-\beta}. \tag{2.14}$$

We now require Lemma 2.5. Because $q \in (0, 1)$ it is clear that there exists a positive number $\gamma_0 \leq q$ and an integer $N \geq 1$ such that

$$\left(\frac{n}{n-2}\right)^N \gamma_0 = 1.$$

Define

$$\gamma_i = \left(\frac{n}{n-2}\right)^i \gamma_0, \qquad i = 0, 1, 2, \dots,$$

so

$$0 < \gamma_0 < \gamma_1 < \cdots < \gamma_{N-1} = \frac{n-2}{n} \quad \text{and} \quad \gamma_N = 1.$$

Now from (2.14) and the Hölder inequality

$$\overline{u^{\gamma_0}}(r) \leq \left(\overline{u^q}(r)\right)^{\gamma_0/q} \leq (Mr^{-\alpha q})^{\gamma_0/q} \leq Mr^{-\alpha\gamma_0}.$$

Hence from Lemma 2.5 (ii), we get by iteration (with $\gamma = \gamma_i$ and $\mu = \mu_i = \alpha\gamma_i$)

$$\overline{u^{\gamma_i}}(r) \leq Mr^{-\alpha\gamma_i}, \qquad i = 1, 2, \dots, N \tag{2.15}$$

since $\gamma_i \in (0, (n-2)/n]$ for $i = 1, 2, \dots, N-1$. Taking $i = N$ in (2.15) completes the proof of the proposition.

Corollary 2.1. *Suppose that $pq > 1$ and that* **u** *is a positive solution of* (I). *Then*

$$\int_{B_R} u^q \leq cR^{n-q\alpha}, \qquad \int_{B_R} v^p \leq cR^{n-p\beta} \tag{2.16}$$

where $c = c(p, q, n)$.

Proof. By (I) and (2.11),

$$\int_{B_R} u^q \leq cR^{n-2}\int_R^{2R} r^{1-n}\int_{B_r} u^q = -cR^{n-2}\int_R^{2R} r^{1-n}\int_{B_r}\Delta v$$
$$= -cR^{n-2}\int_R^{2R}\overline{v}' \leq cR^{n-2}\overline{v}(R) \leq cMR^{n-2-\beta} = cR^{n-q\alpha},$$

since $2 + \beta = q\alpha$, which yields $(2.16)_1$. Similarly one obtains $(2.16)_2$.

Corollary 2.2. *Suppose that $pq > 1$ and that* $\mathbf{u}$ *is a positive solution of* (I). *Then for* $\gamma \in (0,1)$, $R > 0$ *and* $\sigma > 1$, *there exists a constant* $c = c(\gamma, \sigma) > 0$ *such that*

$$\int_{B_{\sigma R}\setminus B_R} |Du|^2 u^{\gamma-2} \le cR^{n-2-\gamma\alpha}, \qquad \int_{B_{\sigma R}\setminus B_R} |Dv|^2 v^{\gamma-2} \le cR^{n-2-\gamma\beta}. \tag{2.17}$$

Moreover, for any $\delta < n/(n-2)$ *there exists a constant* $c = c(\delta, \sigma) > 0$ *such that*

$$\int_{B_{\sigma R}\setminus B_R} u^\delta \le cR^{n-\delta\alpha}, \qquad \int_{B_{\sigma R}\setminus B_R} v^\delta \le cR^{n-\delta\beta}. \tag{2.18}$$

Proof. Since $\gamma \in (0,1)$, we have $\overline{u^\gamma} \le \overline{u}^\gamma \le cr^{-\gamma\alpha}$ by Hölder's inequality and $(2.11)_1$. This is just (2.3) with $w = u$ and $\mu = \alpha\gamma$. Hence from Lemma 2.5 (i) we get $(2.17)_1$, and we derive $(2.17)_2$ similarly.

To obtain (2.18), choose $\gamma = \delta(n-2)/n$ and use the second part of Lemma 2.5 (i).

3. Non-existence I. In this section, we prove two non-existence results for positive solutions of (I). Consider the following region in the first quadrant of the (p,q)—plane,

$$\Sigma_2 = \{p,\ q > 0,\ pq > 1 : n-2 \le \max\{\alpha, \beta\}\},$$

where α and β are given by (2.10). We first show non-existence in Σ_2.

Theorem 3.1. *The system* (I) *does not admit any non-negative and non-trivial solutions* $\mathbf{u}$ *if* $(p,q) \in \Sigma_2$.

Remark. When $p,\ q \ge 1$, Theorem 3.1 was proved in [9].

Proof. Suppose for contradiction that (I) admits a non-negative and non-trivial solution $\mathbf{u}$. Then $\mathbf{u}$ must be strictly positive by the maximum principle. Without loss of generality, we assume in what follows that $p \ge q$. Then necessarily $p > 1, \alpha \ge \beta$ and $\alpha \ge n-2$. Moreover,

$$n - p\beta = n - (2+\alpha) = n - 2 - \alpha \le 0,$$

with equality holding if and only if $\alpha = n-2$. Thus by (2.16),

$$\int_{B_R} v^p \le cR^{n-p\beta} \to 0 \quad \text{as} \quad R \to \infty$$

when $\alpha > n-2$. It follows that $u \equiv v \equiv 0$ when $n-2 < \alpha$, which is a contradiction.

Next suppose $\alpha = n-2$. Then by $(2.12)_2$ and Lemma 2.1 we have, when $q \ge 1$,

$$\overline{v} \ge cr^2\,(\overline{u})^q \ge cr^{2-(n-2)q}, \qquad r \ge 1,$$

and, by $(2.13)_2$ and Lemma 2.1, when $q < 1$,

$$\overline{v} \ge cr^2\overline{u^q} \ge cr^{2-(n-2)q}, \qquad r \ge 1,$$

We shall show that the right hand side of (5.2) tends to zero as $j \to \infty$. Indeed, by Theorem 4.1

$$I_1 = R_j^3 \int_{S^2} v^{p+1} \le cR_j^{\epsilon(p+1)-\sigma}, \qquad I_2 = R_j^3 \int_{S^2} u^{q+1} \le cR_j^{\epsilon(q+1)-\sigma},$$

$$|I_3| \le R_j^3 \int_{S^2} |u'v'| \le cR_j^{2\epsilon-\sigma}, \qquad |I_4| \le R_j \int_{S^2} |\nabla_\theta u \nabla_\theta v| \le cR_j^{2\epsilon-\sigma},$$

and

$$|I_5| \le R_j^2 \int_{S^2} |u'v| \le cR_j^{2\epsilon-\sigma}, \qquad |I_6| \le R_j^2 \int_{S^2} |uv'| \le cR_j^{2\epsilon-\sigma}.$$

Taking

$$\epsilon < \min\left\{2, \frac{\sigma}{p+1}, \frac{\sigma}{q+1}\right\}$$

(so also $2\epsilon < \sigma$ since $pq > 1$), we infer that

$$|I| \le I_1 + I_2 + |I_3| + |I_4| + |I_5| + |I_6| \to 0 \qquad \text{as } j \to \infty.$$

This implies that $u \equiv v \equiv 0$, which contradicts our assumption and completes the proof.

We remark that Theorem 5.1 excludes the existence of ground states of (I) for subcritical exponents, since every ground state tends to zero at infinity. Combining this with the existence result in [14], we obtain the following corollary.

Corollary 5.1. *Suppose $n = 3$. Then* (I) *has a positive solution with algebraic growth at infinity if and only if (p, q) is critical or supercritical. In particular,* (I) *has a ground state if and only if (p, q) is critical or supercritical.*

REFERENCES

[1] M. Bidaut-Veron and T. Raoux, *Asymptotics of solutions of some nonlinear elliptic systems*, preprint.

[2] L. Caffarelli, B. Gidas, and J. Spruck, *Asymptotic symmetry and local behavior of semilinear elliptic equations with critical Sobolev growth*, Comm. Pure Appl. Math. XLII (1989), 271–297.

[3] D.G. de Figueiredo and P.L. Felmer, *A Liouville type theorem for elliptic systems*, Ann. Scuola Norm. Sup. Pisa, 21 (1994), 387–397.

[4] R.H. Fowler, *Further studies of Emden's and similar differential equations*, Quart. J. Math, Oxford Series, 2 (1931), 259–288.

[5] B. Gidas and J. Spruck, *Global and local behavior of positive solutions of nonlinear elliptic equations*, Comm. Pure Appl. Math., 34 (1981), 525–598.

[6] D. Gilbarg and N.S. Trudinger, "Elliptic Partial Differential Equations of Second Order," Springer-Verlag, New York, 1983 (second-edition).

[7] P.L. Lions, *The concentration-compactness principle in the calculus of variations, Part 1*, Rev. Mat. Iberoamer 1 (1985), 145–201.

[8] E. Mitidieri, *A Rellich type identity and applications*, Comm. P.D.E., 18 (1993), 125–151.

[9] E. Mitidieri, *Non–existence of positive solutions of semilinear elliptic systems in $\mathbb{R}^n$*, Quaderno Matematico, 285 (1992).

[10] P. Pucci and J. Serrin, *A general variational identity*, Indiana J. Math., 35 (1986), 681–703.

[11] P. Pucci, J. Serrin, and H. Zou, *A note on positive solutions of a semilinear elliptic systems*, unpublished note.

[12] J. Serrin and H. Zou, *Classification of positive solutions of quasilinear elliptic equations*, Topological Methods in Nonlinear Analysis, 3 (1994), 55–68.

[13] J. Serrin and H. Zou, *Non-existence of positive solutions of semilinear elliptic systems*, Discourses in Mathematics and its Applications 3, Department of Mathematics, Texas A&M University, College Station, Texas, 1994, pp. 55–68.

[14] J. Serrin and H. Zou, *Existence of positive solutions of the Lane-Emden system*, submitted for publication.

[15] W.P. Ziemer, "Weakly Differentiable Functions," Springer-Verlag, New York, 1989.

[16] H. Zou, *Existence of positive solutions for semilinear elliptic systems without variational structure*, in preparation.

Commun. Partial Differ. Equations 23 (1998), 577–599. **Reprint I.15.[ii]**

EXISTENCE OF POSITIVE ENTIRE SOLUTIONS OF ELLIPTIC HAMILTONIAN SYSTEMS

James Serrin
School of Mathematics
University of Minnesota, Minneapolis

Henghui Zou*
Department of Mathematics
University of Alabama at Birmingham, Birmingham

1 Introduction

In recent papers [8, 9] we have considered (component-wise) positive solutions of the weakly coupled system

$$\begin{aligned} \Delta v + f(u) &= 0, \\ \Delta u + g(v) &= 0, \end{aligned} \qquad x \in \mathbf{R}^n, \tag{1.1}$$

*Research supported in part by NSF grants DMS-9418779 and DMS-9622937, by a grant from Alabama EPSCoR and by a faculty research grant from University of Alabama at Birmingham

where $n \geq 3$ is the dimension of the space. When $f = t^p$ and $g = t^q$, the system (1.1) becomes

$$\begin{aligned} \Delta v + u^p = 0, \\ \Delta u + v^q = 0, \end{aligned} \qquad x \in \mathbf{R}^n, \tag{1.2}$$

which is a natural extension of the well-known Lane-Emden equation (here $p > 0$, $q > 0$).

For the Lane-Emden equation, it is well known that the Sobolev exponent $l = (n+2)/(n-2)$ is the dividing number for existence and non-existence of positive solutions on the entire space $\mathbf{R}^n$. This result was generalized in [6, 8, 9] to show the *non-existence* of positive radial solutions for the system (1.2) when the exponents p and q are *subcritical* in the sense that

$$\frac{1}{p+1} + \frac{1}{q+1} > \frac{n-2}{n}. \tag{1.3}$$

Moreover, in [10] the following complementary *existence* result was proved for the system (1.2).

Theorem. *Suppose that $p, q > 0$ and that*

$$\frac{1}{p+1} + \frac{1}{q+1} \leq \frac{n-2}{n}. \tag{1.4}$$

Then there exist infinitely many pairs $(\xi, \eta) \in \mathbf{R}^+ \times \mathbf{R}^+$ such that (1.2) admits a positive radial solution (u, v) on $\mathbf{R}^n$ with central values

$$u(0) = \xi, \quad v(0) = \eta.$$

Moreover $u, v \to 0$ as $|x| \to \infty$, so that the solution is in fact a ground state for (1.2).

A further result in [10] shows that the set of points (ξ, η) in $\mathbf{R}^+ \times \mathbf{R}^+$ such that (1.2) admits a positive radial ground state with central values $u(0) = \xi$, $v(0) = \eta$ is in fact a simple smooth curve, having the form

$$\eta = c\xi^{(q+1)/(p+1)},$$

where $c = c(n, p, q)$. This can be considered a type of uniqueness theorem as well; that is, for each central value $u(0)$ there is exactly one corresponding value $v(0)$ for which there is a radial ground state, and similarly with $u(0)$, $v(0)$ interchanged. This is exactly analogous to the situation for the Lane-Emden equation, where for each given central value $u(0)$ there is exactly one radial ground state (up to translation).

One can ask for similar existence and uniqueness results for the system (1.1), and even for the general Hamiltonian system (see [2, 6])

$$\begin{aligned} \Delta v + H_u(u,v) &= 0, \\ \Delta u + H_v(u,v) &= 0, \end{aligned} \qquad x \in \mathbf{R}^n. \tag{1.5}$$

We suppose throughout that $H(0,0) = 0$ (normalization) and

$$H(u,v) \in C^1 \quad \text{for} \quad u \geq 0,\ v \geq 0,$$

$$H(u,v) \in C^2 \quad \text{for} \quad u > 0,\ v > 0;$$

that[1]

$$\begin{aligned} H_u(u,v) > 0 \quad &\text{for} \quad u > 0,\ v \geq 0, \\ H_v(u,v) > 0 \quad &\text{for} \quad u \geq 0,\ v > 0; \end{aligned} \tag{1.6}$$

and that there exist constants a and b, with $a + b = (n-2)/n$, such that

$$auH_u(u,v) + bvH_v(u,v) \geq H(u,v) \tag{1.7}$$

when $u, v \geq 0$. Note that there is no need to define H for values of (u,v) outside the quadrant $\mathbf{R}_0^+ \times \mathbf{R}_0^+$, since we consider only positive solutions of (1.5).

Then we have the following main result.

Theorem 1.1 *For every sufficiently small positive value ξ there exists at least one positive radial solution $u = u(r),\ v = v(r)$ of (1.5) with $u(0) = \xi$.*

Since the system (1.5) and the conditions (1.6), (1.7) are invariant under interchange of u and v, it is evident that Theorem 1.1 also shows that for every sufficiently small positive value η there exists at least one positive radial solution of (1.5) with $v(0) = \eta$.

Theorem 1.1 can be extended further to state that, for any continuous curve Σ in $\mathbf{R}^+ \times \mathbf{R}^+$ with endpoints respectively on the positive ξ-axis and the positive η-axis, there is at least one positive radial solution u, v of (1.5) with $u(0) = \xi$, $v(0) = \eta$ and $(\xi, \eta) \in \Sigma$. See the remark after the proof of Theorem 1.1 in Section 3.

The question whether solutions exist whatever the value of ξ (or η) requires a further condition. The basic conclusion is as follows.

[1]In a recent paper [1], Chen and Lu study a related (special) Hamiltonian system for which (1.6) fails.

Theorem 1.2 *Suppose, in addition to the previous hypotheses* (1.6) *and* (1.7), *that for each fixed* $\xi > 0$ *we have*

$$\xi \sup_I H_u(u,v) \le \eta \inf_I H_v(u,v) \tag{1.8}$$

whenever η *is suitably large, where*

$$I = \{\, 0 < u < \xi,\ \eta < v < 2\eta \}. \tag{1.8$'$}$$

Then for every positive value ξ *there exists at least one positive radial solution* $u = u(r), v = v(r)$ *of* (1.5) *with* $u(0) = \xi$.

Remark. Note by (1.6) that $\inf_I H_v(u,v) > 0$ for all $\xi, \eta > 0$. Thus (1.8) is equivalent to

$$\frac{\xi \sup_I H_u(u,v)}{\eta \inf_I H_v(u,v)} \le 1.$$

Theorem 1.2 will be the main tool for the proof of our next main existence result. We first need the following elementary proposition.

Proposition 1.1 *There exist* α, $\beta > 0$ *such that*

$$\begin{aligned} H(u,0) &\ge \text{Pos. Constant } u^{\alpha+1} \qquad \text{as } u \to \infty \\ H(0,v) &\ge \text{Pos. Constant } v^{\beta+1} \qquad \text{as } v \to \infty. \end{aligned} \tag{1.9}$$

We can now state our second main existence result.

Theorem 1.3 *Let* α, $\beta > 0$ *be such that* (1.9) *holds. Suppose*

$$H_{uv}(u,v) \le o(u^\alpha v^\beta) \tag{1.10}$$

as $u + v \to \infty$. *Then for every positive value* ξ *(respectively* η*) there exists at least one positive radial solution* $u(r)$, $v(r)$ *such that* $u(0) = \xi$ *(respectively* $v(0) = \eta$*).*

It is remarkable that the above results hold under such minimal assumptions on the Hamiltonian, especially in view of the recent work of de Figueredo and Magalhaes [2] on the Dirichlet problem for (1.5) in a bounded domain - see also [3].

It is also worth noting explicitly that *any positive radial solution of* (1.5), *e.g., as obtained in Theorems* 1.1-1.3, *necessarily has the property*

$$u(r),\ v(r) \to 0 \qquad \text{as } r \to \infty.$$

See Proposition 2.1 in Section 2.

The structure of the set of central values (ξ, η) for which a positive solution can exist is also of interest.

Theorem 1.4 *Assume in addition to* (1.6), (1.7) *that*

$$H_{vv}(u,v) \geq 0, \quad H_{uu}(u,v) \geq 0, \quad H_{uv}(u,v) \leq 0 \qquad \text{for } u,v > 0. \tag{1.11}$$

Then for every positive value ξ there exists exactly one corresponding value $\eta = \eta(\xi)$ such that (1.5) *has a positive radial solution with central value (ξ, η). Moreover, the mapping $\xi \mapsto \eta(\xi)$ is continuous, strictly monotone increasing and 1-1 (onto) from $\mathbf{R}^+$ to $\mathbf{R}^+$.*

An important special case is equation (1.1), which arises from the separable Hamiltonian

$$H(u,v) = F(u) + G(v),$$

where

$$F(u) = \int_0^u f(s)ds, \qquad G(v) = \int_0^v g(s)ds.$$

In order that H_u and H_v should be positive one requires that

$$f(s),\ g(s) > 0 \qquad \text{for } s > 0. \tag{1.12}$$

In this case

$$H_{uu} = f'(u), \quad H_{vv} = g'(v), \qquad H_{uv} \equiv 0.$$

Theorem 1.5 *Let f, g be continuous on $[0, \infty)$ and satisfy* (1.12). *Assume that there exist constants a, b such that $a + b = (n-2)/n$ and*

$$auf(u) \geq F(u), \qquad bvg(v) \geq G(v) \tag{1.13}$$

for $u, v > 0$. Then

(i) *For each positive value ξ (respectively η) there exists at least one positive radial solution of* (1.1) *with $u(0) = \xi$ (respectively $v(0) = \eta$).*

(ii) *If additionally $f'(s) \geq 0$, $g'(s) \geq 0$ for $s > 0$ then there exists exactly one value $\eta = \eta(\xi)$ such that* (1.1) *has a positive radial solution with central value (ξ, η). Moreover*

(iii) *The function $\eta(\xi)$ is continuous, strictly monotone and 1-1 from $\mathbf{R}^+$ to $\mathbf{R}^+$.*

Theorem 1.5 is an immediate corollary of Theorems 1.3 and 1..4, since for $H = F + G$ it is evident that (1.7) is equivalent to (1.13).

The case $f(s) = s^p$, $g(s) = s^q$ was considered in [6, 8, 9], as noted above. Obviously (1.13) then requires

$$a \geq \frac{1}{p+1}, \qquad b \geq \frac{1}{q+1},$$

that is, since $a + b = (n-2)/n$,

$$\frac{1}{p+1} + \frac{1}{q+1} \le \frac{n-2}{n},$$

which is exactly the condition (1.4) given earlier. Moreover, for this case (see the remark at the beginning of the introduction) the function $\eta(\xi)$ takes the particular form

$$\eta(\xi) = c\xi^{(q+1)/(p+1)},$$

which of course is strictly monotone and 1-1 from $\mathbf{R}^+$ to $\mathbf{R}^+$.

In Section 3 we consider the non-separated Hamiltonian

$$H(u,v) = \frac{1}{p+1}u^{p+1} + \lambda u^s v^t + \frac{1}{q+1}v^{q+1},$$

where $\lambda > 0$, for which not only the exponents p, q but also s, t must be suitably restricted.

A final question of interest concerns the situation when the Hamiltonian is symmetric, i.e., $H(u,v) = H(v,u)$. One expects that in this case any positive radial solution would have the property $u(r) \equiv v(r)$, with

$$\Delta u + H_u(u,u) = 0, \qquad x \in \mathbf{R}^n.$$

This is in fact the case whenever one also has $H_{uu} \ge 0$, $H_{uv} \le 0$, see Theorem 4.1.

We remark that our approach also applies when the Hamiltonian depends on the radius $|x|$. Similar conclusions can then be obtained, though with some not entirely obvious modification of the proofs.

Theorems 1.1-1.3 will be proved in Section 3. A series of preliminary lemmas is given in Section 2. Theorem 1.4 is proved in Section 4, while Section 5 contains the proof of a technical lemma, required in Section 2 when H_u and H_v are not Lipschitz continuous on all $\mathbf{R}_0^+ \times \mathbf{R}_0^+$.

2 Preliminary Lemmas

With the aid of a topological argument going back to [4], we use the shooting method and the Pohozaev-Pucci-Serrin identity to show the existence of positive radial solutions of (1.5) when (1.6), (1.7) hold. Thus one is led to consider the ordinary differential system

$$\begin{aligned} v'' + \frac{n-1}{r}v' + H_u(|u|,|v|) = 0 \\ u'' + \frac{n-1}{r}u' + H_v(|u|,|v|) = 0 \end{aligned} \qquad (2.1)$$

with initial values

$$u(0) = \xi > 0, \quad v(0) = \eta > 0; \qquad u'(0) = v'(0) = 0. \tag{2.2}$$

In particular, any *positive* solution of (2.1)-(2.2) defines a positive radial solution of (1.5) with central value $(u(0), v(0))$.

We shall require several preliminary lemmas. The first is a standard local existence and uniqueness result.

Lemma 2.1 *(Local existence and uniqueness). For each* $(\xi, \eta) \in \mathbf{R}^+ \times \mathbf{R}^+$, *there exists a positive number* $\rho = \rho(\xi, \eta)$ *such that* (2.1)-(2.2) *admits a unique positive solution for* $0 \le r \le \rho$.

Proof. For $\rho > 0$ and fixed $(\xi, \eta) \in \mathbf{R}^+ \times \mathbf{R}^+$, we put

$$X = \left\{ \mathbf{u} \in C([0,\rho]) \times C([0,\rho]) \;\middle|\; \|\mathbf{u} - (\xi,\eta)\|_{[0,\rho]} < \min(\xi,\eta) \right\},$$

where

$$\|\mathbf{u}\|_{[0,\rho]} = \sup_{0 \le r \le \rho} \sqrt{u^2 + v^2}$$

is the usual norm. Consider the mapping $F(u,v) = (F_1, F_2)(u,v)$ on X given by

$$\begin{aligned} (F_1(u,v))(r) &= \xi - \int_0^r \frac{dt}{t^{n-1}} \int_0^t H_u(u,v) s^{n-1} ds, \\ (F_2(u,v))(r) &= \eta - \int_0^r \frac{dt}{t^{n-1}} \int_0^t H_v(u,v) s^{n-1} ds. \end{aligned} \tag{2.3}$$

It is clear that $F : X \to X$ for suitably small ρ; indeed, in view of the estimation

$$0 < \int_0^r \frac{dt}{t^{n-1}} \int_0^t H_u(u,v) s^{n-1} ds \le \frac{1}{2n} \cdot J \cdot r^2,$$

where

$$J = \sup_{[0,2\xi]\times[0,2\eta]} H_v(u,v) > 0,$$

and a similar one for the second integral in (2.3) (with $K > 0$ corresponding to J), the condition $F : X \to X$ becomes exactly

$$\rho^2 < 2n \min\left(\frac{1}{J}, \frac{1}{K}\right) \cdot \min(\xi, \eta).$$

Thus we can take

$$\rho^2 = n \min\left(\frac{1}{J}, \frac{1}{K}\right) \cdot \min(\xi, \eta). \tag{2.4}$$

Existence is then standard by the Banach fixed point theorem.

Clearly, the solution initiated by Lemma 2.1 (but no longer restricted to be positive) can be extended to a maximal interval $[0, T)$ with $T = T_{\xi,\eta} > \rho$ and possibly infinite. (If either H_u or H_v is not Lipschitz on the closed first quadrant $\mathbf{R}_0^+ \times \mathbf{R}_0^+$, the uniqueness of the solution could fail at any point $r = \bar{r}$ where either $u(\bar{r})$ or $v(\bar{r}) = 0$. Consequently the value T could in this case possibly depend on the particular solution itself, as well as on ξ and η. By abuse of notation, we do not explicitly indicate this when we write $T = T_{\xi,\eta}$.) Define

$$R = R_{\xi,\eta} = \sup\{r \in [0, T) \mid u(t) > 0,\ v(t) > 0 \text{ for } t \in (0, r)\},$$

again with R possibly infinite. Notice that the solution *is* unique on the range $[0, R]$, so in particular R is dependent *only* on the values ξ, η.

If R is finite, then $T > R$ by standard continuation theorems. Moreover in this case necessarily either $u(R) = 0$ or $v(R) = 0$, since otherwise the continuation past R is certainly positive at least for some distance, contradicting the definition of R. Moreover, it is easy to see that $R > \rho$. Otherwise $R \le \rho < T$, and in turn $u(R) = 0$ or $v(R) = 0$, contradicting the fact that the solution is positive on $[0, \rho]$.

Lemma 2.2 *We have*

$$u'(r) < 0, \quad v'(r) < 0 \quad \textit{for } r \in (0, T).$$

This can be proved by direct integration of the system (2.1)-(2.2), since $|u|, |v| \ge 0$ for all $r \in (0, T)$ and $u, v > 0$ on the interval $(0, \rho)$.

The following result shows further that if a solution stays positive, then it must go to zero as $r \to \infty$.

Proposition 2.1 *Let $u(r), v(r)$ be an entire positive solution of* (2.1)*. Then*

$$u, v \to 0 \quad \textit{as } r \to \infty.$$

Proof. Suppose for contradiction that $u, v \not\to 0$ as $r \to \infty$. Without loss of generality, we may assume $u \to l > 0$ and $v \to m \ge 0$ (by Lemma 2.2). Then along the solution (u, v)

$$\lim_{r\to\infty} H_u(u(r), v(r)) = H_u(l, m) > 0$$

by (1.6). Integrate $(2.1)_1$ from 0 to $r > 0$ to obtain

$$v'(r) = -\frac{1}{r^{n-1}} \int_0^r H_u(u(s), v(s)) s^{n-1} ds,$$

whence

$$\lim_{r\to\infty} v'(r) = -\infty.$$

This is impossible since $v(r)$ is bounded, completing the proof.

Now define the sets

$$A = \{(\xi,\eta) \in \mathbf{R}^+ \times \mathbf{R}^+ \mid R_{\xi,\eta} < \infty \text{ and } v(R_{\xi,\eta}) > u(R_{\xi,\eta}) = 0\},$$

and

$$B = \{(\xi,\eta) \in \mathbf{R}^+ \times \mathbf{R}^+ \mid R_{\xi,\eta} < \infty \text{ and } u(R_{\xi,\eta}) > v(R_{\xi,\eta}) = 0\}.$$

The next two lemmas show that both A and B are *non-empty* and *open* in $\mathbf{R}^+ \times \mathbf{R}^+$.

Lemma 2.3 *Let $\xi, \eta > 0$ be fixed such that* (1.8) *holds. Then $(\xi, 2\eta) \in A$.*

Proof. Let $u(r), v(r)$ be a solution of (2.1) with $u(0) = \xi$ and $v(0) = 2\eta$. By (2.1)-(2.2) together with Lemma 2.2, we get

$$u(r) \le \xi - \frac{r^2}{2n} \inf_I H_v(u,v) \tag{2.5}$$

$$v(r) \ge 2\eta - \frac{r^2}{2n} \sup_I H_u(u,v) \tag{2.6}$$

for all r such that the pair $u(r), v(r)$ is in I, say $0 < r < R_I$. (The set I is defined in (1.8)′).

Case 1. $R_I < \infty$. By Lemma 2.2 it is apparent that *either* $u(R_I) = 0$ *or* $v(R_I) = \eta$, for otherwise the solution pair $u(r), v(r)$ would be in I for $0 < r \le R_I + \epsilon$ for some small $\epsilon > 0$. In fact the case $u(R_I) = 0$ entails $v(R_I) \ge \eta$, so $(\xi, 2\eta) \in A$, and we are done.

The remaining possibility is

$$v(R_I) = \eta, \qquad u(R_I) > 0. \tag{2.7}$$

We assert that

$$R_I \ge \left(\frac{2n\eta}{\sup_I H_u(u,v)}\right)^{1/2} \equiv r_0.$$

Otherwise $R_I < r_0$, so from (2.6)

$$\begin{aligned} v(R_I) &\ge 2\eta - \frac{R_I^2}{2n} \sup_I H_u(u,v) \\ &> 2\eta - \frac{r_0^2}{2n} \sup_I H_u(u,v) = \eta, \end{aligned}$$

a contradiction. Thus $r_0 \le R_I$. Next by (2.5), since $r_0 \le R_I$,

$$u(r_0) \le \xi - \frac{1}{2n}\inf_I H_v(u,v)\frac{2n\eta}{\sup_I H_u(u,v)}.$$

Now let (ξ, η) satisfy (1.8). This gives

$$u(r_0) \le 0,$$

which is again a contradiction, since $0 < u(R_I) \le u(r_0)$. Thus (2.7) does not occur, and the proof for Case 1 is complete.

Case 2. $R_I = \infty$. Then $r_0 < R_I$ automatically, and

$$u(r_0) \le \xi - \frac{1}{2n}\inf_I H_v(u,v)\frac{2n\eta}{\sup_I H_u(u,v)} \le 0$$

by (1.8). On the other hand, since $R_I = \infty$ we have $u(r) > 0$ for all r. This is a contradiction, so Case 2 also cannot occur.

Lemma 2.3′. *For each $d > 0$ there exist positive increasing functions $\phi = \phi_d : (0,d] \to \mathbf{R}^+$, $\psi = \psi_d : (0,d] \to \mathbf{R}^+$ such that*

$$\{(\xi,\eta) \in (0,d] \times (0,d] \mid \xi \le \phi(\eta)\} \subset A,$$

$$\{(\xi,\eta) \in (0,d] \times (0,d] \mid \eta \le \psi(\xi)\} \subset B.$$

Hence A and B are non-empty.

Proof. It is enough to prove the result for the set A, since the conclusion for B can be obtained by interchanging u and v.

Define

$$C = \sup_{(0,d]\times(0,d]} H_u(u,v)$$

and

$$D(\eta) = \inf_{(0,d]\times[\eta,d]} H_v(u,v) > 0$$

by (1.6). Then for $0 < \xi \le d$, $0 < \eta \le d/2$ we have

$$\frac{\xi \sup_I H_u(u,v)}{\eta \inf_I H_v(u,v)} \le \frac{\xi C}{\eta D(\eta)}$$

and the right side does not exceed 1 if

$$\xi \le \frac{\eta D(\eta)}{C}, \qquad 0 < \eta \le d/2.$$

In view of Lemma 2.3, it follows that $(\xi,\eta)\in A$ whenever

$$\xi\le\min[d,\phi(\eta)],\qquad 0<\eta\le d,$$

where

$$\phi(\eta)=\frac{\eta}{2C}D(\eta/2). \tag{2.8}$$

This completes the proof, once we observe that $D(\cdot)$ is an increasing function.

Lemma 2.4 *Both sets A and B are open in $\mathbf{R}^+\times\mathbf{R}^+$.*

Proof. It suffices to show that A is open. For any fixed $(\xi_0,\eta_0)\in A$, let (u_0,v_0) be the solution of (2.1)-(2.2) with initial value (ξ_0,η_0). Then

$$u_0(R_0)>v_0(R_0)=0,\qquad R_0=R_{\xi_0,\eta_0}<\infty.$$

For $(\xi,\eta)\in\mathbf{R}^+\times\mathbf{R}^+$, denote by (u,v) a solution of (2.1) with initial value (ξ,η), extended of course to its maximal interval.

First consider the case when H_u,H_v are Lipschitz continuous on $\mathbf{R}_0^+\times\mathbf{R}_0^+$. Since $T>R_0$ and $v_0'(R_0)<0$ by Lemma 2.2 it is clear that we can find $\epsilon>0$ such that

$$u_0(R_0+\epsilon)>0,\quad v_0(R_0+\epsilon)<0.$$

But then by continuous dependence on initial data (recall that the right hand sides of (2.1) are Lipschitz continuous for all u,v when H_u,H_v are Lipschitz continuous on $\mathbf{R}_0^+\times\mathbf{R}_0^+$) we see that

$$u(R_0+\epsilon)>0,\quad v(R_0+\epsilon)<0 \tag{2.9}$$

for all (ξ,η) sufficiently near (ξ_0,η_0), say $|(\xi,\eta)-(\xi_0,\eta_0)|<\delta$. From $(2.9)_2$ there exists $\overline{R}$ in $(0,R_0+\epsilon)$ such that $v(\overline{R})=0$. At the same time, since $u'(r)<0$ for $r>0$, we get $u(\overline{R})>u(R_0+\epsilon)>0$. Hence the disk $|(\xi,\eta)-(\xi_0,\eta_0)|<\delta$ is in A (with $R_{\xi,\eta}=\overline{R}$). Thus A is open, proving the lemma when H_u,H_v are Lipschitz continuous on $\mathbf{R}_0^+\times\mathbf{R}_0^+$.

The case when either H_u or H_v is not Lipschitz continuous on the closed first quadrant $\mathbf{R}_0^+\times\mathbf{R}_0^+$ needs more attention and is delayed until Section 5.

The following identity will be crucial in completing the proof of Theorem 1.1.

Lemma 2.5 *Let $u=u(r)$, $v=v(r)$ be a solution of (2.1) such that*

$$u(r),v(r)>0\quad \textit{for } 0\le r<R,\quad u(R)=v(R)=0$$

for some $R\in(0,\infty)$. Then

$$R^n u'(R)v'(R)=n\int_0^R\big\{H-auH_u-bvH_v\big\}r^{n-1}dr \tag{2.10}$$

where a and b are constants satisfying

$$a+b=(n-2)/n.$$

Proof. This is a special case of the Pohozaev-Pucci-Serrin identity [7]. In particular, the system (1.5) has variational structure with integrand

$$\mathcal{F}(u,v) = \nabla u \cdot \nabla v - H(u,v).$$

Hence by taking $\mathbf{h} = x$, $\mathbf{a} = (na, nb)$ in Proposition 3 of [6], we get after a short calculation

$$\{r^n[(u'v' - H) - 2u'v'] - nr^{n-1}(auv' + bu'v)\}\Big|_{r=R}$$
$$= \int_0^R \Big\{n(u'v' - H) - 2u'v' - n(a+b)u'v' + nauH_u + nbvH_v\Big\} r^{n-1} dr.$$

Using the fact $u = v = 0$ when $r = R$, the result follows since $n(a+b) = n-2$. (This calculation is essentially one given earlier in [8] - see also [4, 11]).

The next two lemmas are not needed until Section 5, but it is convenient to place them here.

Lemma 2.6 *Suppose $R = R_{\xi,\eta} < \infty$. Then for any $\sigma \in (\rho, R)$ the maximal extension of $\mathbf{u}$ includes the interval $[0, \sigma+\beta]$, where $\beta = \rho^2/2\sigma > 0$. Moreover*

$$\|\mathbf{u}\|_{[0,\sigma+b]} < \min\{\xi, \eta\}.$$

Lemma 2.6 is a local existence result, with initial data $u(\sigma)$, $v(\sigma)$, $u'(\sigma)$ and $v'(\sigma)$ at $r = \sigma$. The proof is essentially the same as that of Lemma 2.1, by using the Banach theorem again; indeed we can take

$$X = \Big\{\mathbf{u} \in C([\sigma, \sigma+\beta]) \times C([\sigma, \sigma+\beta]) \;\Big|\; \|\mathbf{u} - \mathbf{u}(\sigma)\|_{[\sigma,\sigma+b]} < \min(\xi, \eta)\Big\}$$

and F_1 and F_2 as in (2.3) - with the exception that ξ, η are replaced by $u(\sigma), v(\sigma)$ and that the t integrations are carried out on the subinterval from σ to r. Now $\sigma < R$ so that $u(\sigma)$, $v(\sigma) > 0$. But also $(u(\sigma),\ v(\sigma)) < (\xi, \eta)$ by Lemma 2.2; consequently

$$|u(r)| < 2\xi, \qquad |v(r)| < 2\eta$$

when $(u, v) \in X$. Then the required estimations are

$$\int_\sigma^r \frac{dt}{t^{n-1}} \int_0^t H_u(|u|, |v|) s^{n-1} ds \le \frac{1}{2n} \cdot J \cdot (r^2 - \sigma^2), \qquad \sigma < r \le \sigma + b,$$

and a similar one for the remaining integral. The condition $F : X \to X$ then becomes $\beta^2 + 2\sigma\beta < 2\rho^2$, where ρ is given in (2.4). Then since $\sigma > \rho$, we can take

$$\beta = \rho^2/2\sigma. \tag{2.11}$$

Note that if either H_u or H_v is not Lipschitz then one may lose the uniqueness of the extension beyond R due to the lack of Lipschitz continuity at $u = 0$ or $v = 0$.

Lemma 2.7 *If $R < \infty$ then $T \geq R + \rho^2/2R$.*

Proof. If $R < \infty$ then for all $\sigma \in (\rho, R)$ we have $T > \sigma + \rho^2/2\sigma$; thus $T \geq R + \rho^2/2R$ by letting $\sigma \to R$.

3 Proof of the main Theorems 1.1, 1.2, 1.3

Proof of Theorem 1.1. Let $d > 0$ be fixed and let $\phi = \phi_d$, $\psi = \psi_d$ be the functions in Lemma 2.3′. Then by Lemma 2.3′ we have

$$(\xi, \eta) \in A \qquad \text{if } \phi^{-1}(\xi) \leq \eta \leq d, \ \ 0 < \xi \leq \min\{d, \phi(d)\} = \xi_0$$

and

$$(\xi, \eta) \in B \qquad \text{if } \psi^{-1}(\eta) \leq \xi \leq d, \ \ 0 < \xi \leq \min\{d, \psi(d)\} = \eta_0.$$

On the other hand, $A \cap B = \emptyset$ and A and B are open. Consequently for any fixed $\bar{\xi} \in (0, \xi_0)$ there must exist a value $\bar{\eta}$ satisfying

$$\psi(\bar{\xi}) < \bar{\eta} < \phi^{-1}(\bar{\xi})$$

such that $(\bar{\xi}, \bar{\eta})$ is in neither A nor B.

This being shown, we assert that the solution of (2.1) with $u(0) = \bar{\xi}$, $v(0) = \bar{\eta}$ and $u'(0) = v'(0) = 0$ is an everywhere positive radial solution of (1.5). Since $(\bar{\xi}, \bar{\eta})$ is in neither A nor B, it follows from the definition of these sets that *either*

$$R = \infty,$$

or else

$$R < \infty$$

and the two functions u and v simultaneously cross the r axis at the point R (here we write R for $R_{\bar{\xi}, \bar{\eta}}$). In the latter case, by Lemma 2.2

$$\begin{aligned} u(r),\ v(r) > 0 \qquad &\text{for} \quad 0 \leq r < R, \\ u'(r),\ v'(r) < 0 \qquad &\text{for} \quad 0 < r \leq R \end{aligned}$$

and

$$u(R) = v(R) = 0.$$

But from Lemma 2.5 we then get

$$0 < R^n u'(R) v'(R) = n \int_0^R [H - auH_u - bvH_v] r^{n-1} dr.$$

The left side is positive, while the right side is non-positive by (1.7). This being impossible, it follows that $R = \infty$, and consequently (u, v) is a positive radial solution of (1.5) on $\mathbf{R}^n$ with central value $(\bar{\xi}, \bar{\eta})$.

By the same argument, for any fixed $\bar{\eta} \in (0, \eta_0)$ there must exist a corresponding value $\bar{\xi}$ satisfying

$$\phi(\bar{\eta}) < \bar{\xi} < \psi^{-1}(\bar{\eta})$$

and for which $(\bar{\xi}, \bar{\eta})$ is in neither A nor B. The corresponding solution (u, v) is then a positive entire solution of (1.5) with central values $\bar{\xi}$, $\bar{\eta}$.

Remark. Let Σ be a continuous curve in $\mathbf{R}^+ \times \mathbf{R}^+$ with endpoints respectively on the positive ξ-axis and the positive η-axis. We show that there must be at least one point $(\bar{\xi}, \bar{\eta})$ on Σ which is in neither A nor B. In fact, following the proof of Theorem 1.1, let d be chosen larger than the distance from either endpoint to the origin. It is then clear by (3.1) and (3.2) that the sets A and B intersect Σ near the endpoints of Σ, and the assertion then follows since A and B are open and disjoint.

In turn the solution of (2.1) with $u(0) = \bar{\xi}$, $v(0) = \bar{\eta}$ is a positive radial entire solution of (1.5), as required.

Proof of Theorem 1.2. Let $\xi > 0$ be fixed and choose $d = \xi$. Then by Lemma 2.3′ the point $(\xi, \eta) \in B$ if $\eta \leq \min\{d, \psi(\xi)\}$. On the other hand, by the principal hypothesis of Theorem 1.2, condition (1.8) holds when η is suitably large, say $\eta \geq \omega(\xi)$. In turn $(\xi, 2\eta) \in A$ by Lemma 2.3. Then since $A \cap B = \emptyset$ it is clear, as in the previous proof, that the interval

$$\min\{d, \psi(\xi)\} < \eta < 2\omega(\xi)$$

must contain at least one value $\bar{\eta}$ such that $(\xi, \bar{\eta})$ is in neither A nor B. As in the previous proof, the pair $(\xi, \bar{\eta})$ then defines a ground state of (1.5) as required.

Before proving Theorem 1.3, we first prove Proposition 1.1.

Proof of Proposition 1.1. Evidently

$$\begin{aligned}\frac{d}{d\lambda} H(\lambda^a u, \lambda^b v) &= a\lambda^{a-1} u H_u(\lambda^a u, \lambda^b v) + b\lambda^{b-1} v H_v(\lambda^a u, \lambda^b v) \\ &\geq \lambda^{-1} H(\lambda^a u, \lambda^b v)\end{aligned}$$

by (1.7). By integration from 1 to λ, one immediately obtains

$$H(\lambda^a u, \lambda^b v) \geq \lambda H(u, v).$$

It follows that, by putting $u = 0$ and $v = 1$ in the above inequality,

$$H(0, \lambda^b) \geq \lambda H(0,1).$$

Consequently

$$H(0,v) \geq v^{1/b} H(0,1)$$

provided $v \geq 1$. Similarly, one has

$$H(u,0) \geq u^{1/a} H(1,0)$$

for $u \geq 1$.

Clearly, $a > 0$ in view of (1.7) and the fact (see (1.6)) that $H, H_u > 0$ when $u > 0, v = 0$. Similarly $b > 0$, and therefore necessarily a, $b < 1$ because $a + b = (n-2)/n < 1$. The proof is now complete by taking

$$\alpha = \frac{1}{a} - 1 > 0, \qquad \beta = \frac{1}{b} - 1 > 0,$$

since $H(0,1)$, $H(1,0) > 0$.

Remark. The particular values of α and β given here may not be the largest ones possible. The proof simply shows (by construction) that there *exist* positive values α, β satisfying (1.9).

Proof of Theorem 1.3. It is enough to show that, for each fixed value $\xi > 0$,

$$J = \frac{\xi \sup_I H_u(u,v)}{\eta \inf_I H_v(u,v)} \leq 1$$

when η is sufficiently large, where I is given by (1.8)$'$. Using (1.10) and the mean value theorem, there exists some $\zeta \in (0, v)$ such that

$$H_u(u,v) = H_u(u,0) + vH_{uv}(u,\zeta) \leq H_u(u,0) + o(u^\alpha v^{\beta+1}).$$

In turn, for any fixed $\xi > 0$,

$$\sup_I H_u(u,v) \leq \sup_{0<u<\xi} H_u(u,0) + o(\xi^\alpha \eta^{\beta+1}) = o(\eta^{\beta+1})$$

as $\eta \to \infty$. Thus it is enough to show that, for fixed ξ, there exists $C = C(\xi) > 0$ such that

$$\inf_I H_v(u,v) \geq C\eta^\beta \tag{3.1}$$

as $\eta \to \infty$.

On the other hand, for η sufficiently large, using (1.7) and (1.10) and the mean value theorem again, there exists some $\zeta \in (0, v)$ such that

$$\begin{aligned} bH_v(u,v) &\geq v^{-1}[H(u,v) - auH_u(u,v) \\ &= v^{-1}[H(u,v) - au(H_u(u,0) + vH_{uv}(u,\zeta))] \\ &\geq v^{-1}H(u,v) - auv^{-1}H_u(u,0) + o(u^{\alpha+1}v^{\beta}). \end{aligned}$$

But by (1.6) and Proposition 1.1,

$$v^{-1}H(u,v) \geq v^{-1}H(0,v) \geq \text{Pos. Constant } v^{\beta}$$

for v large. Therefore

$$\begin{aligned} bH_v(u,v) &\geq \text{Pos. Const. } v^{\beta} - auv^{-1}H_u(u,0) - o(v^{\beta}) \\ &\geq \text{Pos. Const. } v^{\beta} \end{aligned}$$

for $0 < u < \xi$ (fixed) and for v sufficiently large. It follows that

$$\inf_I H_v(u,v) \geq \text{Pos. Constant } \eta^{\beta}$$

for η sufficiently large, which is (3.1). This completes the proof.

Example. Consider the Hamiltonian

$$H(u,v) = \frac{1}{p+1}u^{p+1} + \lambda u^s v^t + \frac{1}{q+1}v^{q+1}, \qquad \lambda,\ s,\ t > 0. \tag{3.2}$$

Clearly $H_u > 0$ for $u > 0$ and $H_v > 0$ for $v > 0$. It remains to consider (1.7) and (1.9). For (1.10) we can clearly take

$$\alpha = p, \qquad \beta = q$$

so that one must require

$$s < p+1, \qquad t < q+1.$$

For (1.7) we find

$$a \geq \frac{1}{p+1}, \qquad b \geq \frac{1}{q+1} \tag{3.3}$$

and

$$as + bt \geq 1.$$

Hence as before

$$\frac{1}{p+1} + \frac{1}{q+1} \leq \frac{n-2}{n},$$

while moreover the point (s,t) must lie above the line $ax + by = 1$ for some pair (a,b) satisfying (3.3) (see Figure 1 below). This line, whatever choice one makes for a and b, passes through the point $(n/(n-2), n/(n-2))$ and has x-intercept $1/a$ and y-intercept $1/b$. Combining this with (3.3) one sees easily that any point (s,t) in the shaded region can be allowed (closed on the diagonal boundary, open on the horizontal and vertical ones).

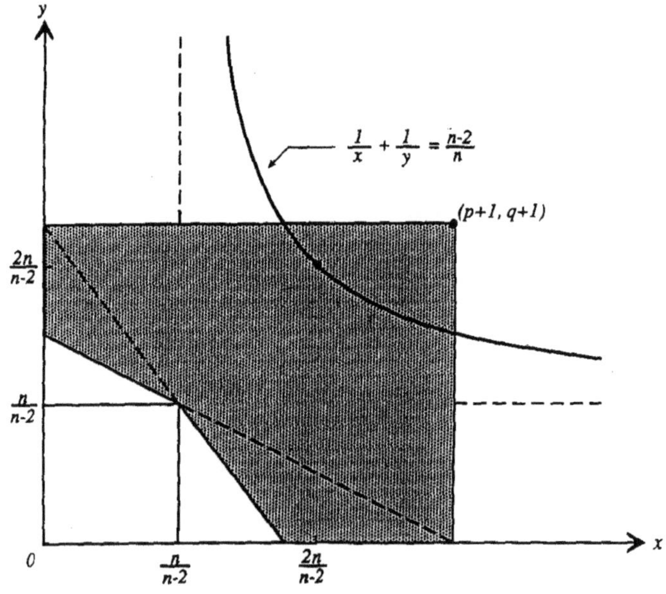

Figure 1. Existence region for the Hamiltonian (3.2). The point $(p+1, q+1)$ should lie above the hyperbola, as shown. The point (s, t) should lie in the shaded region.

4 The structure of the set Γ: Uniqueness and Symmetry

Let Γ denote the set of points in the first quadrant of the (ξ, η)-plane which are the central values for some ground state of (1.5). In this section we shall prove Theorem 1.4 together with a symmetry result concerning the set Γ.

Proof of Theorem 1.4. The existence part follows immediately from Theorem 1.3 since (1.11) implies (1.10). Next, if the principal conclusion were false, there would exist $\xi_1, \xi_2 \in \mathbf{R}^+$ and $\eta_1, \eta_2 \in \mathbf{R}^+$ such that (1.5) has two different positive radial solutions (u_1, v_1) and (u_2, v_2), with

$$u_i(0) = \xi_i, \quad v_i(0) = \eta_i, \qquad i = 1, 2$$

and

$$0 < \xi_1 \leq \xi_2, \qquad \eta_1 \geq \eta_2 > 0, \tag{4.1}$$

with at least one of the inequalities $\leq$ and $\geq$ being strict.

Case 1. $\xi_1 < \xi_2$ and $\eta_1 \geq \eta_2$. We claim that

$$u_1(r) < u_2(r), \qquad r \geq 0. \tag{4.2}$$

Otherwise, put

$$S = \sup\{r > 0 \mid u_1(s) < u_2(s),\ s \in [0, r)\}.$$

Then $0 < S < \infty$ since $\xi_1 < \xi_2$. By direct integration of the system (2.1)-(2.2), we have

$$v_i(r) = \eta_i - \int_0^r \frac{dt}{t^{n-1}} \int_0^t H_u(u_i, v_i) s^{n-1} ds, \quad i = 1, 2.$$

It follows that for $r \leq S$

$$\begin{aligned} v_2(r) - v_1(r) &= \eta_2 - \eta_1 + \int_0^r \frac{dt}{t^{n-1}} \int_0^t [H_u(u_1, v_1) - H_u(u_2, v_2)] s^{n-1} ds \\ &\leq \int_0^r \frac{dt}{t^{n-1}} \int_0^t [H_u(u_1, v_1) - H_u(u_1, v_2) \\ &\qquad + H_u(u_1, v_2) - H_u(u_2, v_2)] s^{n-1} ds \\ &\leq \int_0^r \frac{dt}{t^{n-1}} \int_0^t [H_u(u_1, v_1) - H_u(u_1, v_2)] s^{n-1} ds, \end{aligned}$$

since $H_u(u_1, v_2) - H_u(u_2, v_2) \leq 0$ by (1.11). Using the mean value theorem, there exists $\zeta = \zeta(s)$ between v_1 and v_2 such that

$$\begin{aligned} v_2(r) - v_1(r) &\leq \int_0^r \frac{dt}{t^{n-1}} \int_0^t [-H_{uv}(u_1, \zeta)][v_2(s) - v_1(s)] s^{n-1} ds \\ &\leq \int_0^r \frac{dt}{t^{n-1}} \int_0^t [-H_{uv}(u_1, \zeta)][v_2(s) - v_1(s)]_+ s^{n-1} ds \end{aligned}$$

since $-H_{uv} \geq 0$ by (1.11). In turn, there exists $C = C(S) > 0$ such that

$$[v_2(r) - v_1(r)]_+ \leq C \int_0^r [v_2(s) - v_1(s)]_+ ds.$$

It is standard by the Gronwall inequality that

$$[v_2(r) - v_1(r)]_+ \equiv 0, \quad \text{i.e.,} \quad v_1(r) \geq v_2(r), \qquad r \in [0, S]. \tag{4.3}$$

Therefore, for $|x| \leq S$,

$$\begin{aligned} \Delta(u_2 - u_1) &= H_v(u_1, v_1) - H_v(u_2, v_2) \\ &= [H_v(u_1, v_1) - H_v(u_1, v_2)] \\ &\qquad + [H_v(u_1, v_2) - H_v(u_2, v_2)] \geq 0, \end{aligned}$$

by (1.11) and (4.2), (4.3). Moreover, $u_1 - u_2 = 0$ at $|x| = S$. By the maximum principle this gives $u_2 - u_1 \leq 0$ for $|x| \leq S$, a contradiction.

Therefore (4.2) holds and it follows as in the previous argument (see (4.3)) that

$$v_1(r) \geq v_2(r), \qquad 0 \leq r < \infty.$$

In turn

$$\Delta(u_2 - u_1) = H_v(u_1, v_1) - H_v(u_2, v_2) \geq 0, \qquad x \in \mathbf{R}^n.$$

On the other hand, $u_1 \to 0$ and $u_2 \to 0$ as $r \to \infty$, by Proposition 2.1. Therefore, again by the maximum principle, $u_2 \leq u_1$ for $x \in \mathbf{R}^n$, a contradiction again. This completes the proof of Case 1.

Case 2. $\xi_1 = \xi_2$ and $\eta_1 > \eta_2$. The proof is essentially the same by interchanging the roles of u and v.

The theorem now follows at once, since any strictly monotone, 1-1 (onto) mapping $\eta = \eta(\xi) : \mathbf{R}^+ \to \mathbf{R}^+$ must be continuous.

The next result shows the symmetry of the set Γ, mentioned in the introduction.

Theorem 4.1 *Suppose that H is symmetric, $H(u,v) = H(v,u)$, and that $H_{uu}, H_{vv} \geq 0$, $H_{uv} \leq 0$. Then the set Γ is the diagonal line $\xi = \eta$ in the first quadrant of the (ξ,η) plane. Moreover, all positive radial solutions of* (1.5) *satisfy $u(r) \equiv v(r)$ for $r \geq 0$.*

Proof. Since $H(u,v) = H(v,u)$ we get

$$H_u(u,v) = H_u(v,u), \qquad H_v(u,v) = H_v(v,u).$$

It follows that the set Γ is symmetric with respect to the diagonal line $\xi = \eta$ in the first quadrant of the (ξ,η) plane (interchanging u and v). Therefore Γ must be contained in the diagonal line since it is monotonically increasing. Consequently, Γ must be the diagonal line itself since it is continuous.

Finally, the statement that $u(r) \equiv v(r)$ for $r \geq 0$ follows from the uniqueness Theorem 1.4, and the proof is complete.

5 The non-Lipschitz case

We complete the proof of Lemma 2.4 by considering the case not treated earlier, namely, when either H_u or H_v is non-Lipschitz on $\mathbf{R}_0^+ \times \mathbf{R}_0^+$ (closed). (The difficulty here is that the right sides of (2.1) are no longer Lipschitz continuous at the values $u = 0$ or $v = 0$; thus one loses continuous dependence on the data exactly at the critical point $r = R$, a fact which was used crucially in the earlier proof.)

For the proof of Lemma 2.4 in the present case, we begin as before, with $(\xi_0, \eta_0) \in A$ and with $\mathbf{u}_0$, $\mathbf{u}$ respectively denoting solutions of the initial value problem (2.1)-(2.2) with initial values (ξ_0, η_0) and (ξ, η). We assume that $\mathbf{u}_0$ and $\mathbf{u}$ are extended to their maximal intervals of existence, and let $R_0 = R_{\xi_0, \eta_0}$, $R = R_{\xi,\eta}$. The argument then divides into three steps.

Step 1. There exist $c > 0$ and $\delta' \in (0, 1)$ such that when $|(\xi, \eta) - (\xi_0, \eta_0)| < \delta'$ the solutions $\mathbf{u}_0$ and $\mathbf{u}$ are defined on $[0, R_0 + c]$.

By Lemma 2.7, one can extend $\mathbf{u}_0$ to the interval $[0, R_0 + \beta_0)$ where

$$\beta_0 = \rho_0^2/2R_0$$

and ρ_0 is given by (2.4) with (ξ, η) replaced by (ξ_0, η_0). Using Lemma 2.2, we have moreover

$$u_0'(r) < 0, \quad v_0'(r) < 0, \qquad r \in (0, R_0].$$

Fix $\alpha \in (0, R_0 - \rho_0)$ and $\sigma = R_0 - \alpha$. Then $\rho_0 < \sigma < R_0$ and

$$u_0(r) \geq u_0(\sigma) > 0, \quad v_0(r) \geq v_0(\sigma) > 0 \quad \text{for } r \in [0, \sigma].$$

Obviously the system (2.1) is Lipschitz continuous in u and v in a neighborhood of the rectangle $[u_0(\sigma), \xi_0] \times [v_0(\sigma), \eta_0]$ in $\mathbf{R}^+ \times \mathbf{R}^+$. Using continuous dependence on initial data, it follows that there exists $\delta' > 0$ such that when $|(\xi, \eta) - (\xi_0, \eta_0)| < \delta'$ the solution $\mathbf{u}$ is defined and positive on $[0, \sigma]$, that is $R = R_{\xi,\eta} > \sigma$.

We now wish to apply Lemma 2.3 to the solution $\mathbf{u}$. By making δ' smaller if necessary we can assume that ρ is sufficiently near ρ_0 so that

$$\rho_0^2/2 < \rho^2 < \sigma^2$$

where ρ is given by (2.4). In particular $\rho < \sigma < R$, so by Lemma 2.6 we can extend $\mathbf{u}$ to $[0, \sigma + \beta]$, where $\beta = \rho^2/2\sigma$. In particular

$$\beta = \frac{\rho^2}{2\sigma} > \frac{\rho_0^2}{4\sigma} > \frac{\rho_0^2}{4R_0} = 2c \quad \text{for} \quad c = \rho_0^2/8R_0.$$

Thus if we take α to satisfy also $\alpha \leq c$ (which can certainly be done), then

$$\sigma + \beta = R_0 - \alpha + \beta > R_0 + c.$$

Hence $\mathbf{u}$ extends to the interval $[0, R_0 + c]$. But $c < \beta_0$, so that $\mathbf{u}_0$ also exists on $[0, R_0 + c]$, completing the proof of Step 1. (Note also $c < R_0/8$, since $\rho_0 < R_0$.)

Step 2. We claim that there exist $\epsilon_0 > 0$ and $\delta_0 \in (0, 1)$ such that

$$|u'(r) - u_0'(R_0)| \leq \frac{1}{2}|u_0'(R_0)|, \qquad |v'(r) - v_0'(R_0)| \leq \frac{1}{2}|v_0'(R_0)| \qquad (5.1)$$

when $|(\xi,\eta)-(\xi_0,\eta_0)|<\delta_0$ and $|r-R_0|\le\epsilon_0$. We only prove $(5.1)_1$, with $(5.2)_2$ being obtained in the same way. For $\epsilon<c$ and for $|r-R_0|\le\epsilon$ we have by integration of (2.1)

$$\begin{aligned} u'(r)-u_0'(R_0) &= (u'(r)-u_0'(r))+(u_0'(r)-u_0'(R_0)) \\ &= [u'(R_0-\epsilon)-u_0'(R_0-\epsilon)]\frac{(R_0-\epsilon)^{n-1}}{r^{n-1}} \\ &\quad + \int_{R_0-\epsilon}^{r}\frac{s^{n-1}}{r^{n-1}}[H_v(|u|,|v|)-H_v(|u_0|,|v_0|)]ds \\ &\quad + u_0'(R_0)\Big(\frac{R_0^{n-1}}{r^{n-1}}-1\Big)-\int_{R_0}^{r}\frac{s^{n-1}}{r^{n-1}}H_v(|u_0|,|v_0|)ds. \end{aligned}$$

It follows that

$$|u'(r)-u_0'(R_0)|\le|u'(R_0-\epsilon)-u_0'(R_0-\epsilon)|+|u_0'(R_0)|\Big|\frac{R_0^{n-1}}{r^{n-1}}-1\Big|+I+II, \quad (5.2)$$

where

$$I=\int_{R_0-\epsilon}^{r}\frac{s^{n-1}}{r^{n-1}}H_v(|u|,|v|)ds, \qquad II=\int_{R_0-\epsilon}^{R_0}\frac{s^{n-1}}{r^{n-1}}H_v(|u_0|,|v_0|)ds.$$

We now estimate (5.2) term by term. Since $|r-R_0|\le\epsilon$ we have

$$I=\int_{R_0-\epsilon}^{r}\frac{s^{n-1}}{r^{n-1}}H_v(|u|,|v|)ds\le 2\sup_{I_1}H_v(u,v)\epsilon\le\frac{1}{8}|u_0'(R_0)|,$$

where $I_1=\{0<u<2(\xi_0+1), 0<v<2(\eta_0+1)\}$,

$$II=\int_{R_0-\epsilon}^{R_0}\frac{s^{n-1}}{r^{n-1}}H_v(|u_0|,|v_0|)ds\le(8/7)^{n-1}\sup_{I_2}H_v(u,v)\epsilon\le\frac{1}{8}|u_0'(R_0)|,$$

where $I_2=\{0<u<2\xi_0, 0<v<2\eta_0\}$; and where finally we have used the fact that

$$\Big|\frac{R_0^{n-1}}{r^{n-1}}-1\Big|\le(n-1)(8/7)^{n-1}\frac{\epsilon}{R_0}\le\frac{1}{8}$$

if ϵ is sufficiently small, say $\epsilon\le\epsilon_0<c$. Then from (5.2) we get

$$|u'(r)-u_0'(R_0)|\le|u'(R_0-\epsilon)-u_0'(R_0-\epsilon)|+\frac{3}{8}|u_0'(R_0)|, \qquad |r-R_0|\le\epsilon\le\epsilon_0.$$

Let ϵ be fixed, say $\epsilon=\epsilon_0$. Using continuous dependence on initial data and the fact $u_0(r)\ge u_0(R_0-\epsilon)>0$ for $r\in[0,R_0-\epsilon]$, we infer that there exists $\delta''>0$ such that also

$$|u'(R_0-\epsilon)-u_0'(R_0-\epsilon)|\le\frac{1}{8}|u_0'(R_0)|$$

for $|\xi - \xi_0| < \delta''$. Our claim now follows with the choice $\delta_0 = \min(\delta', \delta'')$.

In what follows we assume (without loss of generality) that

$$\epsilon_0 < \frac{u_0(R_0)}{6|u_0'(R_0)|}.$$

Step 3. Since $v_0(R_0) = 0$ it is clear from continuity that there exists a value $\alpha' \leq \epsilon_0$ satisfying

$$0 < v_0(R_0 - \alpha') < \frac{1}{4}|v_0'(R_0)|\epsilon, \tag{5.3}$$

where as before we put $\epsilon = \epsilon_0$. Then, again by continuous dependence on initial data, there exists $\delta_1 > 0$ such that

$$u(R_0 - \alpha') > \frac{1}{2}u_0(R_0 - \alpha'), \qquad v(R_0 - \alpha') < 2v_0(R_0 - \alpha') \tag{5.4}$$

when $|(\xi, \eta) - (\xi_0, \eta_0)| < \delta_1$.

Next, from (5.1), recalling that $v' < 0$, we get for $|(\xi, \eta) - (\xi_0, \eta_0)| < \delta_0$ and $|r - R_0| < \epsilon$

$$|v'(r)| = |v'(R_0)| + v'(R_0) - v'(r) \geq \frac{1}{2}|v'(R_0)|. \tag{5.5}$$

But from the mean value theorem, (5.1), (5.3)-(5.5),

$$\begin{aligned} v(R_0 + \epsilon) &\leq v(R_0 - \alpha') - \min_{|r - R_0| \leq \epsilon} |v'(r)|(\epsilon + \alpha') \\ &< 2v_0(R_0 - \alpha') - \frac{1}{2}|v_0'(R_0)|\epsilon < 0 \end{aligned}$$

when $|(\xi, \eta) - (\xi_0, \eta_0)| < \min(\delta_0, \delta_1)$. Similarly,

$$\begin{aligned} u(R_0 + \epsilon) &\geq u(R_0 - \alpha') - \max_{|r - R_0| \leq \epsilon} |u'(r)|(\epsilon + \alpha') \\ &> \frac{1}{2}u_0(R_0 - a\alpha') - 3|u_0'(R_0)|\epsilon \\ &> \frac{1}{2}u_0(R_0) - 3|u_0'(R_0)|\epsilon > 0 \end{aligned}$$

when $|(\xi, \eta) - (\xi_0, \eta_0)| < \min(\delta_0, \delta_1)$. Therefore (as earlier) we find that A is open; namely, if $\xi_0 \in A$, then

$$\{|(\xi, \eta) - (\xi_0, \eta_0)| < \min(\delta_0, \delta_1)\} \subset A,$$

and the proof is complete.

Acknowledgement. The authors wish to thank the referee for his careful reading of the manuscript and in particular for suggesting a new proof for Theorem 4.1.

References

[1] Chen, S. and G. Lu, *Existence and non-existence of positive radial solutions for a class of semilinear elliptic systems*, preprint.

[2] de Figueredo, D., and C.A. Magalhaes, *On nonquadratic Hamiltonian elliptic systems*, Advances Differential Eqs., **1** (1996), 881-898.

[3] Hulshof, J., E. Mitidieri and R.C.A.M. van der Vorst, *Strongly indefinite systems with critical Sobolev exponents*, preprint.

[4] McLeod, J.B. and J. Serrin, *The existence of similar solutions of some laminar boundary layer problems*, Arch. Rational Mech. Anal., **31** (1968), 288-303.

[5] Mitidieri, E., *A Rellich type identity and applications*, Comm. PDE, **18** (1993), 125-151.

[6] Mitidieri, E., *Non-existence of positive solutions of semilinear elliptic systems in* $\mathbf{R}^n$, Quaderno Matematico, **285** (1992).

[7] Pucci, P. and J. Serrin, *A general variational identity*, Indiana J. Math., **35** (1986), 681-703.

[8] Serrin, J. and H. Zou, *Non-existence of positive solutions of semilinear elliptic systems*, Discourses in Mathematics and Its Applications, No. 3, Department of Mathematics, Texas A&M University, College Station, Texas, 1994, pp. 55-68.

[9] Serrin, J., and H. Zou, *Non-existence of positive solutions of the Lane-Emden system*, Differential Integral Eqs., **9** (1996), 635-653.

[10] Serrin, J. and H. Zou, *Existence of positive solutions of the Lane-Emden system*, to appear, Atti del Sem. Mat. Univ. Modena, 1997.

[11] van der Vorst, R.C.A.M., *Variational identities and applications to differential systems*, Arch. Rational Mech. Anal., **116** (1991), 375-398.

Received: May 1997
Revised: January 1998

James Serrin's Publications

[**1**] *Free boundaries and jets in the theory of cavitation* (with D. Gilbarg), J. Math. and Physics, **29** (1950), pp. 1–12.

[**2**] *Commentary on Eberhard Hopf's hydrodynamic work*, Math. Nachr., **4** (1950), pp. 213–231.

[3] *Uniqueness theorems for two free boundary problems*, Amer. J. Math., **74** (1952), pp. 492–506.

[4] *Existence theorems for some hydrodynamical free boundary problems*, J. Rational Mech. Anal., **1** (1952), pp. 1–48.

[5] *Two hydrodynamical comparison theorems*, J. Rational Mech. Anal., **1** (1952), pp. 563–572.

[6] *Notes on hydrodynamics*, Photoprinted by Princeton University, 1952, 218 pages.

[7] *On plane and axially symmetric free boundary problems*, J. Rational Mech. Anal., **2** (1953), pp. 563–575.

[**8**] *A note on the wave equation*, Proc. Amer. Math. Soc., **5** (1954), pp. 307–308.

[**9**] *Comparison theorem for subsonic flows*, J. Math. and Physics, **23** (1954), pp. 27–45.

[10] *On the Phragmen–Lindelőf theorem for elliptic partial differential equations*, J. Rational Mech. Anal., **3** (1954), pp. 395–413.

[11] *A uniqueness theorem for the parabolic equation* $u_t = a(x)u_{xx} + b(x)u_x + c(x)u$, Bull. Amer. Math. Soc., **60** (1954), p. 344.

[12] *Uniqueness of axially symmetric subsonic flow past a finite body* (with D. Gilbarg), J. Rational Mech. Anal., **4** (1955), pp. 169–175.

[13] *A characterization of regular boundary points for second order linear differential equations*, Bull. Amer. Math. Soc., **61** (1955), p. 224.

[**14**] *On the Harnack inequality for linear elliptic equations*, J. d'Analyse Math., **4** (1956), pp. 292–308.

[**15**] *On isolated singularities of solutions of second order linear elliptic equations* (with D. Gilbarg), J. d'Analyse Math., **4** (1956), pp. 309–340.

[16] *A note on harmonic functions defined in a half-plane*, Duke Math. J., **24** (1956), pp. 523–526.

[17] *On the Hölder continuity of quasi-conformal and elliptic mappings* (with R. Finn), Trans. Amer. Math. Soc., **89** (1958), pp. 1–15.

[18] *Mathematical principles of classical fluid mechanics* (Monograph), Handbuch der Physik, **VIII/1** (1959), pp. 125–263; Russian translation: Foreign Literature Publ. House, Moscow 1963, 265 pages.

[**19**] *On the stability of viscous fluid motion*, Archive Rational Mech. Anal., **3** (1959), pp. 1–13.

DOI 10.1007/978-3-0348-0845-3, © Springer Basel 2014

[**20**] *A note on the existence of periodic solutions of the Navier–Stokes equations*, Archive Rational Mech. Anal., **3** (1959), pp. 120–122.

[**21**] *On the uniqueness of compressible fluid motions*, Archive Rational Mech. Anal., **3** (1959), pp. 271–288.

[22] *On the derivation of stress-deformation relations for a Stokesian fluid*, J. Math. and Mech., **8** (1959), pp. 459–470.

[**23**] *Poiseuille and Couette flow of non-Newtonian fluids*, Zeit. Ang. Math. Mech., **39** (1959), pp. 295–299.

[**24**] *On a fundamental theorem of the calculus of variations*, Acta Math., **102** (1959), pp. 1–22.

[25] *A new definition of the integral for non-parametric problems in the calculus of variations*, Acta Math., **102** (1959), pp. 23–32.

[**26**] *The exterior Dirichlet problem for second order elliptic partial differential equations* (with N. Meyers), J. Math. Mech., **9** (1960), pp. 513–538.

[27] *On the area of curved surfaces*, Amer. Math. Monthly, **68** (1961), pp. 435–440.

[**28**] *On the differentiability of functions of several variables*, Archive Rational Mech. Anal., **7** (1961), pp. 359–372.

[**29**] *On the definition and properties of certain variational integrals*, Trans. Amer. Math. Soc., **101** (1961), pp. 139–167.

[**30**] *On the entropy change through a shock layer* (with Y.C. Whang), J. Aerospace Sciences, **28** (1961), pp. 990–991.

[31] *Dirichlet's principle in the calculus of variations*, Proc. Symposia in Pure Math., vol. 4, Providence, 1961, pp. 17–22.

[32] *Interior estimates for solutions of the Navier–Stokes equations*, in Partial Differential Equations and Continuum Mechanics, edited by R. Langer, University of Wisconsin Press, 1961, pp. 376–378.

[**33**] *On the interior regularity of weak solutions of the Navier–Stokes equations*, Archive Rational Mech. Anal., **9** (1962), pp. 187–195.

[**34**] *Strong convergence in a product space*, Proc. Amer. Math. Soc., **13** (1962), pp. 651–655.

[**35**] *The initial value problem for the Navier–Stokes equations*, in Nonlinear problems, edited by R.E. Langer, University of Wisconsin Press, 1963, pp. 69–98.

[36] *Variational problems of minimal surface type, I* (with H. Jenkins), Archive Rational Mech. Anal., **12** (1963), pp. 185–212. See also entries 51, 60.

[37] *A Harnack inequality for non-linear equations*, Bull. Amer. Math. Soc., **69** (1963), pp. 481–486.

[38] *Comparison and averaging methods in mathematical physics*, in *Proprietà di Media e Teoremi di Confronto in Fisica Matematica*, Centro Internazionale Matematico Estivo, Rome, Edizioni Cremonese, 1965, pp. 1–87.

[39] *A priori estimates for solutions of the minimal surface equation*, Archive Rational Mech. Anal., **14** (1963), pp. 376–383. See also entry 58.

[40] *Mathematical Aspects of Boundary Layer Theory*, notes taken by H.K. Wilson, Department of Mathematics, University of Minnesota, 1963 (131 pages).

[**41**] *Sublinear functions of measures and variational integrals* (with C. Goffman), Duke Math. J., **31** (1964), pp. 159–178.

[**42**] *Local behavior of solutions of quasi-linear equations*, Acta Math., **111** (1964), pp. 247–302.

[**43**] $H = W$ (with Norman Meyers), Proc. Nat. Acad. Sciences, **51** (1964), pp. 1055–1056.

[44] *Removable singularities of solutions of elliptic equations*, Archive Rational Mech. Anal., **17** (1964), pp. 67–78. See also entry 49.

[**45**] *Pathological solutions of elliptic differential equations*, Ann. Sc. Normale Sup. Pisa, Sci. Fis. Math., (3) **18** (1964), pp. 385–387.

[46] *Singularities of solutions of nonlinear equations*, Proc. Symposia in Pure Math., **17**, pp. 68–88, Providence, 1965.

[**47**] *Isolated singularities of solutions of quasi-linear equations*, Acta Math., **113** (1965), pp. 219–240.

[48] *Theory of differentiation*, notes taken by T. Hatcher, Department of Mathematics, University of Minnesota, 1965 (135 pages).

[**49**] *Removable singularities of solutions of elliptic equations, II*, Archive Rational Mech. Anal., **20** (1965), pp. 163–169.

[50] *The Dirichlet problem for the minimal surface equation with infinite data* (with H. Jenkins), Bull. Amer. Math. Soc., **72** (1966), pp. 102–106. See also entry 60.

[51] *Variational problems of minimal surface type, II: Boundary value problems for the minimal surface equation* (with H. Jenkins), Archive Rational Mech. Anal., **21** (1966), pp. 321–342.

[**52**] *Isolated singularities of solutions of linear elliptic equations* (with H. Weinberger), Amer. J. Math., **88** (1966), pp. 258–272.

[**53**] *Local behavior of solutions of quasi-linear parabolic equations* (with D.G. Aronson), Archive Rational Mech. Anal., **25** (1967), pp. 81–122.

[54] *A maximum principle for nonlinear parabolic equations* (with D.G. Aronson), Ann. Sc. Normale Sup. Pisa, Sci. Fis. Mat., **21** (1967), pp. 291–305.

[**55**] *The Dirichlet problem for quasi-linear elliptic equations with many independent variables*, Proc. Nat. Acad. Science, **58** (1967), pp. 1829–1835. See also entry 65.

[**56**] *On the asymptotic behavior of velocity profiles in the Prandtl boundary layer theory*, Proc. Royal Soc., Ser. A, **299** (1967), pp. 491–507.

[**57**] *The Dirichlet problem for the minimal surface equation in higher dimensions* (with H. Jenkins), J. Reine Ang. Math., **223** (1968), pp. 170–187.

[58] *Addendum to A priori estimates for solutions of the minimal surface equation*, Archive Rational Mech. Anal., **28** (1968), pp. 149–154.

[**59**] *On the mathematical basis of Prandtl's boundary layer theory: an example*, Archive Rational Mech. Anal., **28** (1968), pp. 217–225.

[**60**] *Variational problems of minimal surface type, III: The Dirichlet problem with infinite data* (with H. Jenkins), Archive Rational Mech. Anal., **29** (1968), pp. 304–322.

[**61**] *The behavior of similar solutions in a compressible boundary layer* (with J.B. McLeod), J. Fluid Mech., **34** (1968), pp. 337–342.

[62] *A new proof in differentiation theory*, Notices Amer. Math. Soc., **15** (1968), p. 1036, Abstract; see also W. Rudin, *Real and Complex Analysis*, Second Edition, McGraw–Hill, 1974, pp. xii, 435, and 162–167.

[63] *A general chain rule for derivatives and the change of variable formula for the Lebesgue integral* (with D.E. Varberg), Amer. Math. Monthly, **76** (1969), pp. 514–520.

[64] *The existence of similar solutions for some laminar boundary layer problems* (with J.B. McLeod), Archive Rational Mech. Anal., **31** (1969), pp. 288–303.

[**65**] *The problem of Dirichlet for quasilinear elliptic differential equations with many independent variables*, Philosophical Trans. Royal Soc. London, **A 264** (1969), pp. 413–496.

[**66**] *On surfaces of constant mean curvature which span a given space curve*, Math. Zeit., **112** (1969), pp. 77–88.

[**67**] *Existence theorems for some compressible boundary layer problems*, in *Qualitative Theory of Nonlinear Differential and Integral Equations*, SIAM Studies in Appl. Math., **5** (1969), pp. 35–42.

[68] *The Dirichlet problem for surfaces of constant mean curvature*, Proc. London Math. Soc., **21** (1970), pp. 361–384.

[69] *On the strong maximum principle for nonlinear second order differential inequalities*, J. Funct. Anal., **5** (1970), pp. 184–193.

[70] *Curvature inequalities for surfaces over a disk* (with H.F. Weinberger), in *Some problems of Mathematics and Mechanics – M.A. Lavrentieff Anniversary Volume*, "Nauka" Leningrad, 1970, pp. 242–250 (in Russian).

[**71**] *Boundary curvatures and the solvability of Dirichlet's Problem*, Actes du Congrès International des Mathématiciens (Nice, 1970), Tome **2**, pp. 867–875, Gauthier–Villars, Paris, 1971.

[**72**] *Recent developments in the mathematical aspects of boundary layer theory*, International J. Engineering Sciences, **9** (1971), pp. 233–240.

[73] *Uniqueness and comparison theorems for nonlinear elliptic equations in divergence form* (with J. Douglas, Jr., and T. Dupont), Archive Rational Mech. Anal., **42** (1971), pp. 157–168.

[**74**] *A symmetry problem in potential theory*, Archive Rational Mech. Anal., **43** (1971), pp. 304–318.

[75] *Gradient estimates for solutions of nonlinear elliptic and parabolic equations*, in *Contributions to Nonlinear Functional Analysis*, edited by E. Zarantonello, Univ. of Wisconsin Press, 1971, pp. 565–601.

[76] *Nonlinear elliptic equations of second order*, Lectures at *Symposium on Partial Differential Equations*, Berkeley, 1971, Mimeographed notes (57 pages).

[**77**] *The swirling vortex*, Philosophical Trans. Royal Soc. London, **A271** (1972), pp. 325–360.

[**78**] *Entire solutions of nonlinear Poisson equations*, Proc. London Math. Soc., **24** (1972), pp. 348–366.

[79] *A note on the preceding paper of Amann*, Archive Rational Mech. Anal., **44** (1972), pp. 182–186.

[**80**] *Rectilinear steady flow of simple fluids* (with R.L. Fosdick), Proc. Royal Soc. London, **332** (1973), pp. 311–333.

[81] *Global properties of continuum thermodynamic processes* (with R.L. Fosdick), Archive Rational Mech. Anal., **59** (1975), pp. 97–109.

[82] *On the axioms of classical mechanics*, Department of Mathematics, University of Minnesota, 1974 (74 pages).

[83] *Liouville theorems for nonlinear Poisson equations*, in *Convegno Internazionale sui Metodi Valutativi nella Fisica-Matematica*, Accad. Naz. Lincei, Problemi Attuali di Scienza e di Cultura, **217** (1975), pp. 207–215.

[84] *Doomsday: on target?*, Science Magazine, **189**, July 11, 1975, pp. 86–87.

[**85**] *The solvability of boundary value problems*, in *Mathematical Developments arising from Hilbert Problems* (Proc. Sympos. Pure Math., Northern Illinois Univ., De Kalb, Ill., 1974), Proc. Sympos. Pure Math., **28**, Amer. Math. Soc., Providence, R.I., 1976, pp. 507–524.

[**86**] *Curvature inequalities for surfaces over a disk* (with H.F. Weinberger), Amer. Math. Soc. Translation, (2) **104** (1976), pp. 223–231 (English translation of item [70]).

[87] *Liouville theorems and gradient bounds for quasilinear elliptic systems*, Archive Rational Mech. Anal., **66** (1977), pp. 295–310.

[88] *Optimal shapes for brittle beams under torsion* (with H.F. Weinberger), in *Complex Analysis and its Applications – jubilee volume for I.Y. Vekua*, "Nauka" Moscow, 1978, pp. 88–91.

[89] *Gradient estimates and Liouville theorems for quasilinear elliptic equations* (with L.A. Peletier), Ann. Sc. Normale Sup. Pisa, **5**, Ser. IV (1978), pp. 65–104.

[90] *The concepts of thermodynamics*, in *Continuum Mechanics and Partial Differential Equations*, edited by G.M. de la Penha et al., North–Holland Publ., Amsterdam, 1978, pp. 411–451.

[91] *On the impossibility of linear Cauchy and Piola–Kirchhoff constitutive theories for stress in solids* (with R.L. Fosdick), J. Elasticity, **9** (1979), pp. 83–89.

[92] *Conceptual analysis of the classical second laws of thermodynamics*, Archive Rational Mech. Anal., **70** (1979), pp. 254–272.

[93] *Foundations of thermodynamics*, Lecture Notes, University of Naples, 1979 (150 pages).

[94] *Phase transitions and interfacial layers for van der Waals fluids*, in *Recent Methods in Nonlinear Analysis and Applications* (Proc. Fourth International Meeting of SAFA), edited by A. Canfora et al., Liguori Editore, Naples, 1981, pp. 169–175.

[95] *The second law of thermodynamics for systems with approximate cycles* (with B.D. Coleman and D.R. Owen), Archive Rational Mech. Anal., **77** (1981), pp. 103–142.

[96] *Uniqueness of solutions of semilinear Poisson equations* (with K. McLeod), Proc. Nat. Acad. Sciences USA, **78** (1981), pp. 6592–6598. See also entry 116.

[97] *Uniqueness of positive solutions of semilinear equations in* $\mathbb{R}^n$ (with L.A. Peletier), Archive Rational Mech. Anal., **81** (1983), pp. 181–197. See also entry 112.

[98] *The form of interfacial surfaces in Korteweg's theory of phase equilibria*, Quart. Appl. Math., **41** (1983), pp. 357–364.

[99] *The mechanical theory of fluid interfaces and Maxwell's rule* (with E. Aifantis), J. Coll. and Interface Sci., **96** (1983), pp. 519–547.

[100] *The structure and laws of thermodynamics*, Proc. International Congress of Mathematicians (Warsaw 1983), pp. 1717–1728.

[**101**] *Applied mathematics and scientific thought*, in *Nonlinear analysis and optimization* (Bologna, 1982), Lecture Notes in Math. **1107**, Springer, 1984, pp. 19–27.

[102] *One-dimensional shock layers in Korteweg fluids* (with R. Hagan), in *Phase Transformations and Material Instabilities in Solids*, Academic Press, 1984, pp. 113–128.

[**103**] *Extensions of the mountain pass theorem* (with P. Pucci), J. Funct. Anal., **59** (1984), pp. 185–210.

[**104**] *A mountain pass theorem* (with P. Pucci), J. Diff. Equations, **60** (1985), pp. 142–149.

[105] *Esistenza ed unicità degli stati fondamentali per equazioni ellittiche quasilineari* (with B. Franchi and E. Lanconelli), Rend. Accad. Naz. Lincei, Serie 8, **79** (1985), pp 121–126. See also entries 121, 151.

[**106**] *A continuum model for chemical mixture dynamics*, (Proc. Nineteenth Midwestern Mechanics Conference, Columbus, Ohio) *Developments in Mechanics*, **13** (1985), pp. 59–64.

[107] *Nonexistence theorems for quasilinear partial differential equations* (with W.-M. Ni), Suppl. Rend. Circolo Mat. Palermo, **8** (1985), pp. 171–185.

[108] *On the thermomechanics of interstitial working* (with J.E. Dunn), Archive Rational Mech. Anal., **88** (1985), pp. 95–133.

[109] *An outline of thermodynamical structure*, in *New Perspectives in Thermodynamics*, edited by J. Serrin, Springer-Verlag, 1986, pp. 3–32.

[110] *Dynamic changes of phase in a van der Waals fluids* (with R. Hagan), in *New Perspectives in Thermodynamics*, edited by J. Serrin, Springer-Verlag, 1986, pp. 241–260.

[**111**] *Existence and non-existence theorems for ground states of quasilinear partial differential equations. The anomalous case* (with W.-M. Ni), Accad. Naz. Lincei, Atti dei Convegni, **77** (1986), pp. 231–257.

[**112**] *Uniqueness of nonnegative solutions of semilinear equations in* $\mathbb{R}^n$ (with L.A. Peletier), J. Diff. Equations, **61** (1986), pp. 380–397.

[**113**] *Nonexistence theorems of singular solutions for quasilinear partial differential equations* (with W.-M. Ni), Comm. Pure Appl. Math., **39** (1986), pp. 379–399.

[**114**] *A general variational identity* (with P. Pucci), Indiana Univ. Math. J., **35** (1986), pp. 681–703.

[**115**] *The laws of thermodynamics*, Res Mechanica, **21** (1987), pp. 361–379.

[116] *Uniqueness of positive radial solutions of* $\Delta u + f(u) = 0$ *in* $\mathbb{R}^n$ (with K. McLeod), Archive Rational Mech. Anal., **99** (1987), pp. 115–145.

[**117**] *The structure of the critical set in the mountain pass theory* (with P. Pucci), Trans. Amer. Math. Soc., **229** (1987), pp. 115–132.

[118] *Ground states for a prescribed mean curvature equation* (with L.A. Peletier), Proc. Amer. Math. Soc., **100** (1987), pp. 694–700.

[119] *Positive solutions of a prescribed mean curvature equation*, in *Calculus of Variations and Partial Differential Equations*, edited by S. Hildebrandt et al., Springer Lecture Notes in Math. **1340** (1988), pp. 248–255.

[120] *Ground states for the prescribed mean curvature equation: the supercritical case* (with F.V. Atkinson and L.A. Peletier), in *Nonlinear Diffusion Equations and their*

Equilibrium States (I), edited by W.-M. Ni et al., MSRI Publ., **12**, Springer-Verlag, 1988, pp. 51–74.

[121] *Existence and uniqueness of ground states of quasilinear elliptic equations* (with B. Franchi and E. Lanconelli), in *Nonlinear Diffusion Equations and their Equilibrium States (I)*, MSRI Publ., **12**, Springer-Verlag, 1988, pp. 293–300. See also entry 151.

[122] *Remarks on the first eigenspace for polyharmonic operators* (with P. Pucci), Atti Sem. Mat. Fis. Univ. Modena, **36** (1988), pp. 107–118.

[123] *Asymptotic properties for solutions of strongly nonlinear ordinary differential equations* (with P. Pucci), in *Conference on Differential Equations and Geometry*, Rend. Sem. Mat. Univ. Politec. Torino, Fascicolo Speciale (1989), pp. 121–129.

[**124**] *Critical exponents and critical dimensions for polyharmonic operators* (with P. Pucci), J. Math. Pures Appl., **69** (1990), pp. 55–83.

[**125**] *Continuation and limit properties for solutions of strongly nonlinear second order differential equations* (with P. Pucci), Asymptotic Anal., **4** (1991), pp. 97–160.

[**126**] *Estimates for vertical points of solutions of prescribed mean curvature type equations, II* (with F.V. Atkinson and L.A. Peletier), Asymptotic Anal., **5** (1992), pp. 283–310.

[127] *The nature of thermodynamics*, Atti Sem. Mat. Fis. Univ. Modena, **39** (1991), pp. 445–472; Italian translation: *Sulla natura della termodinamica*, Atti e Memorie Accad. Naz. Scienze Lettere e Arti di Modena, Ser. VII, **7** (1992), pp. 185–202.

[128] *Global asymptotic stability for strongly nonlinear second order systems* (with P. Pucci), in *Nonlinear Diffusion Equations and their Equilibrium States (III)*, edited by N.G. Lloyd et al., Birkhäuser, 1992, pp. 437–449.

[129] *Existence and non-existence for ground states of quasilinear elliptic equations* (with H. Zou), Archive Rational Mech. Anal., **121** (1992), pp. 101–130.

[**130**] *Precise damping conditions for global asymptotic stability for nonlinear second order systems* (with P. Pucci), Acta Math., **170** (1993), pp. 275–307. See also entry 135.

[131] *Continuation and limit behavior for damped quasi-variational systems* (with P. Pucci), in *Degenerate Diffusions*, edited by W.-M. Ni et al., IMA Vols. in Math. Appl., **47**, Springer-Verlag, 1993, pp. 157–174.

[**132**] *The equations of continuum mechanics as a consequence of group invariance*, in *Advances in Continuum Mechanics*, ed. by G. Ferrarese, Pitagora Editrice, Bologna, 1993, pp. 217–225.

[133] *Asymptotic behavior of solutions of a nonstandard second order differential equation* (with W.A. Harris and P. Pucci), Diff. Integral Equations, **6** (1993), pp. 1201–1215.

[134] *Existence and non-existence results for ground states of degenerate Laplace equations* (with H. Zou), in *Differential Equations with Applications to Mathematical Physics*, **192**, Academic Press, 1993, pp. 287–305.

[135] *Precise damping conditions for global asymptotic stability of nonlinear second order systems, II* (with P. Pucci), J. Diff. Equations, **113** (1994), pp. 505–534.

[**136**] *Asymptotic stability for intermittently controlled nonlinear oscillators* (with P. Pucci), SIAM J. Math. Anal., **25** (1994), pp. 815–835.

[137] *On the derivation of Hamilton's equations* (with P. Pucci), Archive Rational Mech. Anal., **125** (1994), pp. 297–310.

[**138**] *Ground states of a quasilinear equation* (with L.A. Peletier and H. Zou), Diff. Integral Equations, **7** (1994), pp. 1063–1082.

[139] *Asymptotic estimates for a nonstandard second order differential equation* (with W.A. Harris and P. Pucci), in *Differential Equations, Dynamical Systems, and Control Science: A Festschrift in Honor of Lawrence Markus*, edited by K.D. Elworthy et al., Lecture Notes in Pure and Appl. Math., M. Dekker, New York, 1994, pp. 75–85.

[140] *Classification of positive solutions of quasilinear elliptic equations* (with H. Zou), Topological Methods in Nonlinear Analysis, **3** (1994), pp. 1–26.

[**141**] *Asymptotic integration of second order systems* (with H. Zou), Amer. J. Math., **116** (1994), no. 5, pp. 1241–1264.

[**142**] *A bifurcation problem at a singular limit* (with F. Merle and L.A. Peletier), Indiana Univ. Math. J., **43** (1994), pp. 585–609.

[143] *Non-existence of positive solutions of semilinear elliptic systems* (with H. Zou), in *A Tribute to Ilya Bakelman*, edited by S.A. Fulling et al., Discourses in Math. and its Applications, **3**, Department of Mathematics, Texas A&M University, 1994, pp. 56–69.

[144] *A numerical study of the existence and non-existence of ground states and their bifurcation for the equations of Chipot and Weissler* (with Y. Yan and H. Zou), unpublished manuscript, 1994.

[**145**] *Reflections on mathematics*, Lectio Doctoralis at the University of Padova. In "*L'Anno Galileiano*", Vol. I, Ed. Lint, Trieste, 1995, pp. 120–124.

[146] *Remarks on Lyapunov stability* (with P. Pucci), Differential and Integral Equations, **8** (1995), pp. 1265–1278.

[147] *On the elementary thermodynamics of quasistatic systems and other remarks*, in *Thermoelastic Problems and the Thermodynamics of Continua*, AMD – Vol. **198**, pp. 53–62, Amer. Soc. Mech. Eng., Applied Mech. Div., 1995.

[**148**] *Asymptotic stability for ordinary differential systems with time dependent restoring potentials* (with P. Pucci), Arch. Rational Mech. Anal., **132** (1995), pp. 207–232.

[**149**] *Asymptotic stability for nonautonomous dissipative wave systems* (with P. Pucci), Comm. Pure Applied Math., **49** (1996), pp. 177–216.

[150] *Quasi-variational systems*, in *First World Congress of Nonlinear Analysts*, ed. V Lakshmikantham, De Gruyter, Berlin, 1996, pp. 2055–2068.

[**151**] *Existence and uniqueness of ground states of quasilinear elliptic equations* (with B. Franchi and E. Lanconelli), Advances in Math., **118** (1996), pp. 177–243.

[**152**] *The equations of continuum mechanics and the laws of thermodynamics*, Meccanica, **31** (1996), pp. 547–563.

[153] *Stability for abstract evolution equations* (with P. Pucci), in *Partial Differential Equations and Applications – Collected papers in Honor of Carlo Pucci*, P. Marcellini, G. Talenti and E. Vesentini eds., Lecture Notes in Pure and Appl. Math. **177**, 1996, M. Dekker, Inc. New York, pp. 279–288.

[**154**] *Non-existence of positive solutions of Lane–Emden systems* (with H. Zou), Differential and Integral Equations, **9** (1996), pp. 635–653.

[155] *Asymptotic stability for nonlinear parabolic systems* (with P. Pucci), in *Energy Methods in Continuum Mechanics*, S.N. Antonsev, J.L. Diaz and S.I. Shmarev, eds., Kluwer Acad. Publ., Dordrecht, 1996, pp. 66–74.

[156] *Some remarks on global nonexistence for nonautonomous abstract evolution equations*, in *Harmonic analysis and nonlinear differential equations*, (Riverside, CA, 1995), pp. 253–263, *Contemp. Math.*, **208**, Amer. Math. Soc., Providence, RI, 1997.

[157] *Existence, stability and blow-up for dissipative evolution equations* (with P. Pucci), *Partial Differential Equations and Applications*, ed. G. Caristi and E. Mitidieri, Lecture Notes in Pure and Applied Math., **194**, Marcel Dekker, New York 1997, pp. 299–317.

[**158**] *Some remarks on global nonexistence for nonautonomous abstract evolution equations* (with H. Levine and P. Pucci), in *Harmonic analysis and nonlinear differential equations* (Riverside, CA, 1995), pp. 253–263, *Contemporary Math.*, **208**, Amer. Math. Soc., Providence, RI, 1997.

[159] *Global non-existence theorems for quasilinear evolution equations with dissipation* (with H. Levine), Arch. Rational Mech. Anal., **137** (1997), pp. 341–361.

[**160**] *Space, time and energy*, Lectio Doctoralis at the University of Ferrara, in *Nonlinear Analysis and Continuum Mechanics, Papers for the 65th Birthday of James Serrin*, G. Buttazzo, G.P. Galdi, E. Lanconelli and P. Pucci, eds., Springer Verlag, 1998, pp. 195–208.

[161] *Global existence and global nonexistence of solutions of the Cauchy problem for a nonlinearly damped wave equation* (with H.A. Levine and S.R. Park), J. Math. Anal. Appl., **228** (1998), pp. 181–205.

[162] *Global existence and nonexistence theorems for quasilinear evolution equations of formally parabolic type* (with H.A. Levine and S.R. Park), J. Differential Equations, **142** (1998), pp. 212–229.

[163] *Existence of positive solutions of the Lane–Emden system* (with H. Zou), Atti Sem. Mat. Fis. Univ. Modena, supplemento, **46** (1998), pp. 368–380.

[164] *Local asymptotic stability for nonlinear evolution equations with amplification forces* (with P. Pucci), Israel J. Math., **104** (1998), pp. 29–50.

[165] *Uniqueness of ground states for quasilinear elliptic operators* (with P. Pucci), Indiana Univ. Math. J., **47** (1998), pp. 501–525.

[166] *Uniqueness of ground states for quasilinear elliptic equations in the exponential case* (with P. Pucci), Indiana Univ. Math. J., **47** (1998), pp. 523–539.

[**167**] *Existence of positive entire solutions of elliptic Hamiltonian systems* (with H. Zou), Comm. Partial Differential Equations, **23** (1998), pp. 577–599.

[**168**] *Global nonexistence for abstract evolution equations with positive initial energy* (with P. Pucci), J. Differ. Equations, **150** (1998), pp. 203–214.

[**169**] *Symmetry of ground states of quasilinear elliptic equations* (with H. Zou), Archive Rational Mech. Anal., **148** (1999), pp. 265–290.

[170] *A strong maximum principle and a compact support principle for singular elliptic inequalities* (with P. Pucci and H. Zou), *Papers in memory of Ennio De Giorgi*, Ricerche Mat., **48** (1999), supplemento, pp. 373–398.

[**171**] *A strong maximum principle and a compact support principle for singular elliptic inequalities* (with P. Pucci and H. Zou), J. Math. Pures Appl., **78** (1999), pp. 769–789.

[172] *Existence of ground states and free boundary problems for quasilinear elliptic operators*(with F. Gazzola and M. Tang), Adv. Differential Equations, **5** (2000), pp. 1–30.

[173] *A note on the strong maximum principle for elliptic differential inequalities* (with P. Pucci), J. Math. Pure Appl., **79** (2000), pp. 57–71.

[174] *Ground states and free boundary problems for the n-Laplacian in n dimensional space* (with M. García–Huidobro, R. Manásevich, M. Tang and C. Yarur), J. Funct. Anal., **172** (2000), pp. 177–201.

[**175**] *Uniqueness of ground states for quasilinear elliptic equations* (with M. Tang), Indiana Univ. Math. J., **49** (2000), pp. 897–923.

[**176**] *Jean Leray's contributions to nonlinear differential equations, fixed point theory and fluid mechanics*, Gaz. des Mathématiciens, Supplement to volume **84** (2000), pp. 61–70.

[**177**] *The Harnack inequality for quasilinear elliptic equations in* $\mathbb{R}^2$ (with P. Pucci), J. d'Anal. Math., **85** (2001), pp. 307–321.

[**178**] *Cauchy–Liouville and universal boundedness theorems for quasilinear elliptic equations and inequalities* (with H. Zou), Acta Math., **189**, (2002), pp. 79–142.

[179] *Asymptotic behavior of ground states of quasilinear elliptic equations with two vanishing parameters* (with F. Gazzola), Ann. Inst. H. Poincaré Anal. Non Linéaire, **19** (2002), pp. 477–504.

[180] *Commentary on the Hopf strong maximum principle*, in *Selected Works of Eberhard Hopf with commentaries*, ed. by C.S. Morawetz, J.B. Serrin and Y.G. Sinai, *Amer. Math. Soc.*, Providence, 2002.

[**181**] *Existence for a nonlinear wave equation with damping and source terms* (with G. Todorova and E. Vitillaro), Differential and Integral Equations, **16** (2003), pp. 13–50.

[182] *Asymptotic behavior of ground states of quasilinear elliptic equations with two vanishing parameters, II* (with F. Gazzola, L.A. Peletier and P. Pucci), Ann. Inst. H. Poincaré Anal. Non Linéaire, **20** (2003), pp. 947–974.

[**183**] *An appreciation of Clifford Truesdell*, in *Essays and papers dedicated to the memory of Clifford Ambrose Truesdell III*, Vol. I, J. Elasticity, **70** (2003), pp. 31–38.

[184] *The strong maximum principle revisited* (with P. Pucci), J. Differ. Equations, **196** (2004), pp. 1–66; *Erratum*, J. Differ. Equations **207** (2004), pp. 226–227.

[**185**] *Elliptic equations and products of positive definite matrices* (with C. H. Conley and P. Pucci), Math. Nachrichten, **278** (2005), pp. 1490–1508.

[186] *Qualitative properties of ground states for singular elliptic equations with weights* (with P. Pucci, M. García–Huidobro and R. Manásevich), Annali Mat. Pura Appl., **185** (2006), pp. 205–243.

[**187**] *Dead cores and bursts for quasilinear singular elliptic equations* (with P. Pucci), SIAM J. Math. Anal., **38** (2006), pp. 259–278.

[**188**] *A maximum principle for mean-curvature type elliptic inequalities*, J. Eur. Math. Soc. (JEMS), **8** (2006), pp. 389–398.

[189] *Qualitative properties for solutions of singular elliptic inequalities on complete manifolds* (with P. Pucci and M. Rigoli), J. Diff. Equations, **234** (2007), pp. 507–543.

[190] *A remark on the Morrey potential*, Contemp. Math., Special Volume "Control methods in PDE-dynamical systems", **426** (2007), pp. 307–315.

[191] *Maximum Principles for Elliptic Partial Differential Equations* (with P. Pucci), in *Handbook of Differential Equations – Stationary Partial Differential Equations*, Ed. M. Chipot, Elsevier BV, **4** (2007), pp. 355–483.

[192] *Partial and full symmetry of solutions of quasilinear elliptic equations, via the comparison principle* (with P. Pucci and B. Sciunzi), Contemp. Math., special volume dedicated to H. Brezis "*Perspectives in nonlinear partial differential equations*", **446** (2007), 437–444.

[193] *The Maximum Principle* (with P. Pucci), in "*Progress in Nonlinear Differential Equations and their Applications*", Vol. **73**, Birkhäuser Publ., Switzerland, 2007, X, 234 pages.

[**194**] *The area rule for simple fluid phase transitions*, J. Elasticity, **90** (2008), pp. 129–159.

[195] *Entire solutions of quasilinear elliptic equations*, J. Math. Anal. Appl., **352** (2009), pp. 3–14.

[196] *Entire solutions of completely coercive quasilinear elliptic equations* (with A. Farina), J. Differential Equations, **250** (2011), pp. 4367–4408.

[197] *Entire solutions of completely coercive quasilinear elliptic equations, II* (with A. Farina), J. Differential Equations, **250** (2011), pp. 4409–4436.

[198] *A remark on entire solutions of completely coercive quasilinear elliptic equations* (with P. Pucci), J. Differential Equations, **250** (2011), pp. 675–689.

[199] *The Liouville theorem for homogenous elliptic differential inequalities*, J. Math. Sciences (St. Petersberg, Russia), **179** (2011), pp. 176–185.

[200] *Weakly subharmonic functions*, Boll. Unione Mat. Ital., (9) **4** (2011), pp. 347–361.

[201] *Comparison principles, uniqueness and symmetry results of solutions of quasilinear elliptic equations and inequalities* (with L. D'Ambrosio, A. Farina and E. Mitidieri), Nonlinear Analysis, **90** (2013), pp. 135–158.

[202] *The splitting of intrinsic energy and the origin of mass density in continuum mechanics* (with R. Fosdick), Contin. Mech. Thermodyn., DOI 10.1007/s00161-013-0301-1.

[203] *Foundations of Thermodynamics* (with C.-S. Man), monograph in preparation.

GPSR Compliance

The European Union's (EU) General Product Safety Regulation (GPSR) is a set of rules that requires consumer products to be safe and our obligations to ensure this.

If you have any concerns about our products, you can contact us on ProductSafety@springernature.com

In case Publisher is established outside the EU, the EU authorized representative is:

Springer Nature Customer Service Center GmbH
Europaplatz 3
69115 Heidelberg, Germany

Batch number: 10747675

Printed by Printforce, the Netherlands